THE SELECTED WORKS OF

SIGURÐUR HELGASON

(Photo courtesy of Anna Loa Helgason.)

Sigurður Helgason

THE SELECTED WORKS OF SIGURÐUR HELGASON

Gestur Ólafsson
Henrik Schlichtkrull
Editors

American Mathematical Society
Providence, Rhode Island

2000 *Mathematics Subject Classification.* Primary 22E30, 33C67, 43A85, 46E20, 53C35, 58J35.

Library of Congress Cataloging-in-Publication Data

Helgason, Sigurður, 1927–
[Selections. 2009]
The selected works of Sigurður Helgason / Gestur Ólafsson, Henrik Schlichtkrull, editors.
p. cm.
Includes bibliographical references.
ISBN 978-0-8218-4753-4 (alk. paper)
1. Riemannian manifolds. 2. Geometry, Riemannian. I. Ólafsson, Gestur. II. Schlichtkrull, Henrik, 1954– III. Title.

QA649.H427A25 2009
516.3′73—dc22 2009010403

∞ The paper used in this book is acid-free and falls within the guidelines
established to ensure permanence and durability.
Visit the AMS home page at `http://www.ams.org/`

10 9 8 7 6 5 4 3 2 1 14 13 12 11 10 09

Contents

Preface

The present volume contains a selection of 32 articles by Sigurður Helgason, covering various aspects of geometric analysis on Riemannian symmetric spaces. The papers are arranged chronologically and cover a time span of more than 50 years, from 1956 to 2007.

Sigurður Helgason has had a profound impact on geometric analysis, more particularly in the geometry of homogeneous spaces, harmonic analysis on symmetric spaces, the theory of the Radon transform, and the study of invariant differential operators and their eigenfunctions. The collection is preceded by an introduction by Helgason. Here some background and perspective is given on the individual papers, and they are brought into a historical perspective as seen by the author himself today. The introduction organizes the articles according to the following list of topics, which describes the evolution of Helgason's work over the period:

1. Invariant differential operators,
2. Geometric properties of solutions,
3. Double fibrations in integral geometry, Radon transforms,
4. Spherical functions and spherical transforms,
5. Duality for symmetric spaces,
6. Representation theory,
7. Fourier transform on Riemannian symmetric spaces,
8. Multipliers.

Many people know the work of Helgason primarily from his outstanding books. The writing of Sigurður Helgason has been a model of exposition since the publication in 1962 of *Differential Geometry and Symmetric Spaces*, [B1], which quickly became a classic. It was not only used by students and researchers in Lie group theory, the first chapter was-and still is-an excellent introduction to Riemannian geometry. This book, and its successor from 1984, *Groups and Geometric Analysis*, [B6], was awarded the Steele price for expository writing in 1988.

In addition to these two classical and prize-winning books, Helgason has published several other books, notably *The Radon Transform* [B4] and *Geometric Analysis on Symmetric Spaces* [B7]. A complete list of books and articles up to 2008 is enclosed in this volume. The above citations refer to that list.

The exposition in the original papers is of the same outstanding clarity as that in the books. It is our hope that by collecting these papers, we can create renewed attention on all those aspects of his work which are not covered by the books.

We would like to thank Ed Dunne and Cristin Zannella at AMS for their invaluable help to get this project started and going. We also thank the publishers of the original articles for their cooperation and permission to include the articles in this volume.

Gestur Ólafsson and Henrik Schlichtkrull

Biographical note

Sigurður Helgason was born on September 30, 1927, in Akureyri, Iceland. He went to the Gymnasium in Akureyri 1939-1945. After a short period at the University of Iceland in Reykjavík, he began studies at the University of Copenhagen in 1946. Here he received the Gold Medal in 1951 for a paper summarized in [60] and the M.Sc. degree in 1952. He went to Princeton as a graduate student in 1952 and received the ph.D. from there in 1954. He was a Moore instructor at M.I.T. 1954-56, after which he returned to Princeton 1956-57. He was at University of Chicago 1957-59, and at Columbia University 1959-60. Since then he has been at M.I.T., where he was appointed full professor in 1965. The periods 1964-66, 1974-75, 1983 (fall) and 1998 (spring) he spent at the Institute for Advanced Study, Princeton, and the periods 1970-71 and 1995 (fall) at the Mittag-Leffler Institute, Stockholm. He has been awarded honorary doctor by University of Iceland, University of Copenhagen and University of Uppsala. In 1988 he received the Steele prize of the A.M.S. for expository writing. He carries the Major Knights Cross of the Icelandic Falcon since 1991.

He and his wife live in Belmont, Massachusetts. They have a son and a daughter, and 3 grandchildren.

Introduction

I am deeply grateful to the American Mathematical Society and to Gestur Ólafsson and Henrik Schlichtkrull for their interest and their proposal to produce this volume with selections of my mathematical papers. The editors and publisher kindly asked me to write some explanatory comments on these papers.

My interest in mathematics I trace to my mathematics teacher in the Gymnasium in Akureyri in North Iceland, Trausti Einarsson. Although an astronomer, his deep respect for mathematics was highly infectious. The geometry textbooks by the remarkable mathematician Ólafur Danielsson, the pioneering founder of mathematics education in Iceland, were written by a man with a real mission. The program was on a respectable level — for example the final exam, for students at age 17, included finding the radius of a sphere inscribed in a regular dodecahedron of edge length 1. Danielsson, who wrote several beautiful papers in geometry along with his demanding Gymnasium teaching, was profoundly influenced by the rich geometry tradition in Denmark, which started around 1870 with J. Petersen and H. Zeuthen.

When I entered the Gymnasium in Akureyri around 1940, the Icelandic population totaled about 125,000. Thus the Gymnasium program put considerable emphasis on the teaching of several foreign languages. Our French teacher, Thórarinn Björnsson, was particularly memorable because of his outspoken admiration of French authors like Maupassant, Daudet and Rolland, in addition to his fondness for the French language in which he left us with a decent reading ability. Thus my later difficulties in understanding Élie Cartan's papers were not of linguistic nature.

Later, at the University of Copenhagen, I benefited from splendid lectures by Harald Bohr, Børge Jessen and Werner Fenchel. The first two gave substantial courses in complex function theory and real analysis, respectively. With Fenchel I had also a kind of reading course in projective geometry. At that time I obtained a copy of Julius Petersen's remarkable book, "Methods and Theories for Solutions of Geometric Construction Problems" (1866). I worked on many of the 300 problems there. Example of such problems: Construct a triangle ABC from h_a, m_a and r. This was consistent with my expectation later to teach at a Gymnasium in Iceland.

One day I happened to notice a posting of the collection of prize questions which were posed by the university each year. The mathematics problem on the list called for a generalization to analytic almost periodic functions of the value-distribution theory of Ahlfors and Nevanlinna. Everyone under 30 was free to try for a year. A solution was to be submitted and judged anonymously so I could not consult anybody. Thus four months went by before I understood what the problem was about. My paper solving the problem became my masters thesis. A summary is given in paper [60] in this volume.

After completing my thesis in Princeton enjoying supportive and stimulating supervision by Bochner, my interest shifted to Lie group theory with the intention of studying Harish–Chandra's work. A related interest was geometric analysis, having received from my friend Leifur Ásgeirsson the page proofs of Fritz John's famous book *Plane Waves and Spherical Means.* Although the term 'group' does not appear in this book I sensed quickly that some of its themes were ripe for group-theoretic variations.

At an early stage it was a source of inspiration to me how at the hands of Sophus Lie, the study of differential equations led to the concept of Lie groups which in turn went on to penetrate numerous other fields of mathematics. Since Lie group theory had in the mid-fifties become so well developed I became interested in a kind of converse to Lie's program, namely to investigate differential operators and differential equations invariant under known groups. Examples were readily at hand: the Laplace–Beltrami operator on Riemannian or pseudo-Riemannian manifolds, the Dirac operator, the wave equation and its variations.

In the account that follows I describe some of my own work connected to analysis on homogeneous spaces. This is accompanied by brief mention of some prior and subsequent work which is *closely related.* These personal viewpoints do *not* represent any attempt at a historical account, the feature of relevance and of value being highly subjective. As Peter Lax aptly observed: "For a mathematician the central problem is the one he happens to be working on".

Inspired by Harish–Chandra's brilliant work on the representation theory of semisimple Lie groups G, my taste for geometry led me to a related topic, namely a study of the symmetric spaces of É. Cartan. Here enter several differential geometric features: geodesics, curvature, K-orbits, totally geodesic submanifolds and horocycles; for the compact dual spaces U/K we have also closed geodesics, especially those of minimal length, equators, antipodal manifolds, and totally geodesic spheres. Geometric analysis means, in this context, analysis formulated in terms of these geometric objects.

I have had the fortune of teaching for 50 years at M.I.T. but have also enjoyed the privilege of extended, stimulating visits at the Institute for Advanced Study in Princeton and at the Mittag–Leffler Institute in Djursholm, Sweden.

In conclusion I want to emphasize my wife's role in my life and work. Her support during these 50 years has been precious, and her tolerance, at times when nothing but mathematics mattered, has been admirable. She even typed my first two books in pre-TEX times, giving stylistic suggestions along the way.

1. Invariant Differential Operators.

Articles:[1] [6], *[9], [13], *[16], *[23], *[32], *[35], *[37], *[43], [44], [46], [49], [55], *[59], *[63], [64], *[71], *[81], *[82].
Books: [B2], [B6], [B7]

The general project outlined above suggested the task of describing the algebra $\mathbf{D}(G/H)$ of G-invariant differential operators on a homogeneous space G/H. For G/H reductive some preliminaries are done in [9]; for G/H of maximum mobility, namely two-point homogeneous, the algebra is easily shown to be generated by the Laplace–Beltrami operator. For G/K symmetric, Harish–Chandra's results for G implied that $\mathbf{D}(G/K)$ is commutative and could in fact be described rather explicitly. However, contrary to a statement appearing frequently in the literature, the operators in $\mathbf{D}(G/K)$ are not all induced from operators in the algebra $\mathbf{Z}(G)$ of bi-invariant differential operators on G. While this was verified in [16] to be true for all classical irreducible G/K, it fails for exactly four of the exceptional G/K [63]. However, even here, each D is induced by a fraction Z_1/Z_2 ($Z_i \in \mathbf{Z}(G)$), that is $\widetilde{Z}_1 D = \widetilde{Z}_2$ where $\widetilde{Z}_i \in \mathbf{D}(G/K)$ is induced by Z_i via the action of G on G/K. The algebra $\mathbf{D}(G/K)$ is the subject of Chapter II in [B6] and is related to the Iwasawa decomposition

$$G = NAK\,, \quad \mathfrak{g} = \mathfrak{n} + \mathfrak{a} + \mathfrak{k}\,, \quad g = n \exp A(g) k\,, \quad A(g) \in \mathfrak{a}\,.$$

Given $D \in \mathbf{D}(G/K)$ and its radial part under the action of N on $X = G/K$, it is proved in [16] that the Abel transform (the Radon transform on K-invariant functions on X) is a simultaneous transmutation operator for all $D \in \mathbf{D}(G/K)$, that is, "converts" them into constant coefficient operators on A. Combining this with Harish–Chandra's work on the spherical transform and Hörmander's divisibility theorem for tempered distribution on $\mathbf{R}^n$, I showed in [16] that each $D \in \mathbf{D}(G/K)$ has a fundamental solution. The later Paley–Wiener theorem furnishes a shorter proof but requires the surjectivity of a constant coefficient differential operator on $\mathcal{D}'(\mathbf{R}^n)$.

The next step was to prove the surjectivity of each $D \in \mathbf{D}(G/K)$ on the space $\mathcal{C}^\infty(G/K)$, i.e., the existence of a global solution to the equation $Du = f$ for arbitrary smooth f. Already in some notes in 1964 I had reduced this to a support theorem for the horocycle Radon transform on X. I used Harish–Chandra's expansion for Eisenstein integrals but was stymied by singularities on $\mathfrak{a}_c^*$ whose location seemed unpredictable. But by looking

[1]Papers in this volume are indicated by an asterisk.

at the simplest possible examples I discovered, but not until 1972, that these singularities could be canceled out by a certain intertwining procedure. This yielded the support theorem for the horocycle Radon transform and thereby the surjectivity

$$D\mathcal{C}^\infty(X) = \mathcal{C}^\infty(X)$$

follows in [37]. The local solvability of each $Z \in \mathbf{Z}(G)$ is also proved there invoking methods of Raïs and Harish-Chandra. General local solvability was later proved by Duflo. The global solvability however fails (Cerèzo–Rouvière, *Sem. École Polytech.* (1973)) although it holds for the Casimir operator (Cerèzo–Rouvière loc. cit., Rauch and Wigner, *Ann. of Math.* (1976)). Other surjectivity results are proved in [35], [37], [43]. The non-Riemannian symmetric case has been investigated by Duflo, van den Ban and Schlichtkrull and others. The support theorem above was given a new proof by Gonzalez and Quinto. This approach has led Quinto and collaborators to various interesting support theorems in much more general context.

2. Geometric Properties of Solutions.

Articles: *[9], *[15], *[16], *[23], [25], *[32], *[40], [55], *[71], [78], *[81], *[82].

In his paper (*C.R. Acad. Sci.* Paris (1952)) Godement defined a harmonic function on G/K as a solution to $Du = 0$ for all $D \in \mathbf{D}(G/K)$ annihilating the constants. He proved that they had a characterization in terms of mean values over orbits of K and their translates. In [9] I proved a similar extension of Ásgeirsson's mean-value theorem for solutions of $L_x(u(x,y)) = L_y(u(x,y))$ on $\mathbf{R}^n \times \mathbf{R}^n$ (L-Laplacian) to the system $D_x u = D_y u$ on $G/K \times G/K$ for $D \in \mathbf{D}(G/K)$. Inspired by work of Ásgeirsson and Günther on Huygens' principle I became interested in studying the wave equation on symmetric spaces G/K and on Lorentzian manifolds of constant curvature. In both cases the natural equation is $L_x + c = \frac{\partial^2}{\partial t^2}$, where c is a constant related to the scalar curvature. Papers [16], [45], [55], [58], [71], [78] and a chapter in [B7] deal with explicit solution formulas for the natural wave equation on G/K and the occurrence of Huygens' principle. My work on these wave equations is related to extensive contributions by Branson, Ólafsson, Lax, Phillips, Schlichtkrull, Shahshahani, Solomatina and Ørsted. Papers [81] and [82] (the latter with Schlichtkrull) deal with a hyperbolic system on G/K introduced by Semenov–Tian–Shansky, *Izvestija* (1976). The time variable has dimension rank(X) and for rank$(X) = 1$ it reduces to the above wave equation. Developing further the work of Semenov–Tian–Shansky, Shahshahani and Phillips, paper [81], gives an explicit solution of the Cauchy problem in terms of the Radon transform and in terms of the Fourier transform. Paper [82] with Schlichtkrull deals with an associated range problem.

For a Lorentzian manifold X of constant curvature, paper [9] sets up an analog of Riesz potentials and used properties of these to show that a function u on X is determined by a formula

$$u = c \lim_{r \to 0} r^{n-2} Q(\square) M^r u \,,$$

where $(M^r u)(x)$ is the "mean value" of u over a "Lorentzian sphere" of radius r and center x, $Q(\square)$ being a polynomial in the Laplacian $\square$ on X. This can be used to deduce a result of Schimming and Schlichtkrull (*Acta Math.* (1994)) that each factor $\square + c$ in $Q(\square)$ satisfies Huygens' principle.

It was proved by Furstenberg and Karpelevich that a bounded solution of the Laplace equation $Lu = 0$ on the symmetric space G/K is necessarily harmonic (in Godement's sense above). This led them to a Poisson type integral formula for such functions. While the analog of the Schwarz theorem for continuous boundary values is easy, in [23] Korányi and I proved the corresponding Fatou theorem for such functions $u(x)$; namely

$$\lim_{t \to \infty} u(\gamma(t)) \quad \text{exists}$$

for almost all geodesics, starting at the origin $o = eK$. The result has many refinements, variations and generalizations due to Knapp, Korányi, Lindahl, Michelson, Schlichtkrull, Sjögren, Stein, Williamson and others.

3. Double Fibrations in Integral Geometry. Radon Transforms.

Articles: *[9], *[15], [17], *[18], *[19], *[20], *[28], [52], [53], *[62], *[69], *[74], [78], *[91], *[92].

Books: [B4], [B6], [B7].

In Radon's pioneering paper (1917) a function f on $\mathbf{R}^2$ is explicitly determined by its line integrals. In projective geometry points and lines are put on equal footing. Observing that $\mathbf{R}^2$ and the set $\mathbf{P}^2$ of lines are both permuted transitively by the same group $\mathbf{M}(2)$, the isometry group of $\mathbf{R}^2$, the following generalization was proposed in [20]. Given a locally compact group and K and H two closed subgroups satisfying some natural conditions we consider the coset spaces $X = G/K$, $\Xi = G/H$. Elements $x = gK$, $\xi = \gamma H$ are (by Chern (1942)) said to be *incident* if the sets gK and γH intersect as subsets of G. Given $x \in X$, $\xi \in \Xi$ let

$$\check{x} = \{\xi \in \Xi : x, \xi \ \text{ incident}\} \,,$$
$$\widehat{\xi} = \{x \in X : x, \xi \ \text{ incident}\} \,.$$

Then we have two integral transforms

$$f \in C_c(X) \to \widehat{f} \in C(\Xi) \,, \; \varphi \in C_c(\Xi) \to \check{\varphi} \in C(X)$$

given by

$$\widehat{f}(\xi) = \int\limits_{\widehat{\xi}} f(x)\, dm(x)\,, \quad \check{\varphi}(x) = \int\limits_{\check{x}} \varphi(\xi)\, d\mu(\xi)$$

with canonically defined measures dm and $d\mu$. In group-theoretic terms with $L = K \cap H$, dk_L, dh_L invariant measures on K/L and H/L we have

$$\widehat{f}(\gamma H) = \int\limits_{H/L} f(\gamma h K)\, dh_L\,, \quad \check{\varphi}(gK) = \int\limits_{K/L} \varphi(gkH)\, dk_L\,.$$

These geometrically dual transforms are also adjoint:

$$\int\limits_X f(x)\check{\varphi}(x)\, dx = \int\limits_\Xi \widehat{f}(\xi)\varphi(\xi)\, d\xi\,,$$

where dx and $d\xi$ are invariant measures, $dx = dg_K$, $d\xi = dg_H$.

The principal problems for these transforms $f \to \widehat{f}$, $\varphi \to \check{\varphi}$ would be

(i) Injectivity
(ii) Inversion Formulas
(iii) Range Questions
(iv) Support Theorems. These are results of the following type:

$C_c(X) = \{f \in C(X) : \widehat{f} \in C_c(\Xi)\}$, but there are many variations.

In addition to considering the classical case $X = \mathbf{R}^n$, $\Xi = \mathbf{P}^n$, the space of hyperplanes, where I determined the image of $\mathcal{D}(\mathbf{R}^n)$ under the Radon transform, I obtained in [9] and [62] inversion formulas for the case when X is a Riemannian manifold of constant curvature and Ξ the family of totally geodesic submanifolds of a given dimension. This was later extensively studied by Semyanisty, Rubin, Palamodov, Rouvière, Kurusa, Berenstein and Casadio-Tarabusi. Substantial generalization to totally geodesic submanifolds in a noncompact symmetric space $X = G/K$ was done by Ishikawa (2003). Paper [18] contains an inversion formula for the case when X is a compact symmetric space of rank 1 and Ξ the set of antipodal submanifolds. The extensive calculation was later simplified (see *Integral Geometry and Radon Transforms*, book to appear) using Rouvière's method for the noncompact analog (*Enseign. Math.* (2001)). This implies the injectivity of the X-ray transform on any compact symmetric space (the case rank$(X) > 1$ being settled by Strichartz and Gindikin using Fourier series on a torus). For $X = G/K$ noncompact there is, in addition to the injectivity, a support theorem for the X-ray transform [51].

In his paper (*C.R. Acad. Sci.* Paris **342** (2006), 1–6) Rouvière proved an inversion formula for the X-ray transform on the noncompact symmetric space $X = G/K$. His method is based on a skillful reduction to the hyperbolic plane $\mathbf{H}^2$. The method does not work for the compact space U/K because of the antipodal manifolds. While the X-ray transform on the compact space U/K is injective (as mentioned) one can still ask:

(a) Is there a general inversion formula?

(b) If one only integrates over minimal geodesics (this transform is called the *Funk transform*), is this transform injective?

(a) A global formula from [92] implies an inversion formula for the Funk transform of f provided f is supported in a ball of radius $\pi/4$ (with U/K of maximum sectional curvature 1; here U/K is assumed irreducible and simply connected). The proof is based on results from [21] according to which all the minimal geodesics are conjugate under U. In the absence of the support condition the global formula mentioned also involves integration over the antipodal manifolds.
(b) In a recent preprint by Klein, Thorbergsson and Verhóczki, the Funk transform is shown to be injective without the support condition.

The case of $X = G/K$ and $\Xi = G/MN$, the space of horocycles, is discussed in a later section.

One particularly simple instance of the dual setup is when Ξ is the unit disk, X its boundary, both having the group $G = \mathbf{SU}(1,1)$ acting transitively by

$$g : z \to \frac{az+b}{\bar{b}z+\bar{a}} \qquad |z| \leq 1 \qquad |a|^2 - |b|^2 = 1\,.$$

Here the transform $f \to \widehat{f}$ from $C(X)$ to $C(\Xi)$ turns out to be just the usual Poisson integral and the inversion formula is just the classical Schwarz theorem for the boundary values of a harmonic function. Other classical theorems in Potential Theory furnish answers to the range question (iii). While the range here consists of the harmonic functions there is a striking analogy with the case $X = \mathbf{R}^3$, $\Xi =$ space of lines in X, where John (*Duke Math. J.* (1938)) gave a precise description of the range as solutions of Ásgeirsson's ultrahyperbolic equation in four variables.

4. Spherical Functions. Spherical Transforms.

Articles: *[22], *[26], *[28], *[37], [38], *[43], *[48], *[91]
Books: [B6], [B7]

Let G be a connected noncompact semisimple Lie group with finite center, K a maximal compact subgroup. As proved by Gelfand, the convolution algebra $C_c(K\backslash G/K)$ of K-bi-invariant functions on G is commutative under convolution. Given its standard topology, the continuous homomorphisms of the algebra $C_c(K\backslash G/K)$ onto $\mathbf{C}$ are functionals

$$f \to \int_G f(x)\varphi(x)\,dx$$

where the functions φ are the so-called *zonal spherical functions* on G. Harish–Chandra gave a formula for these φ in terms of the Iwasawa decomposition of G whereby the set of these $\varphi = \varphi_\lambda$ is parametrized by λ in $\mathfrak{a}_c^*/W$. On the other hand the space $L^1(K\backslash G/K)$ is a commutative Banach algebra under convolution and its maximal ideal space corresponds to those spherical functions φ_λ which are bounded. Johnson and I found in [26] that the bounded spherical functions are those φ_λ (in the parametrization) for which λ lies in a certain explicit tube in $\mathfrak{a}_c^*$ over a certain convex set in $\mathfrak{a}^*$. Flensted–Jensen has discovered an interesting relationship between the spherical functions on G and its complexification G^C (*J. Funct. Anal.* (1978)). This leads, for example, to a somewhat simplified proof of the boundedness criterion above.

The paper [22] is devoted to the problem of finding the range of the space $\mathcal{D}(K\backslash G/K)$ under the spherical transform. For this I used Harish–Chandra's expansion of the spherical function on the positive Weyl chamber $\mathfrak{a}^+$. The coefficients are certain meromorphic functions on $\mathfrak{a}^*$ and the problem is to prove uniform compact support of the integral

$$\int_{\mathfrak{a}^*} \varphi_\lambda(\exp H)F(\lambda)\,|\mathbf{c}(\lambda)|^{-2}\,d\lambda\,.$$

This is proved in [22] for each term $\psi_\mu(\lambda, H)$ in the series for φ_λ by shifting the contour $\mathfrak{a}^*$ into the complex space $\mathfrak{a}_c^*$ yet avoiding the singularities in the coefficients $\psi_\mu(\lambda, H)$ as well as $c(\lambda)$. Fortunately, there was a half-space in $\mathfrak{a}_c^*$ free of singularities. It remained to justify the term-by-term integration

$$\int_{\mathfrak{a}^*} \sum_{\mu} \psi_\mu(\lambda, H)F(\lambda)\,|\mathbf{c}(\lambda)|^{-2}\,d\lambda = \sum_{\mu} \int_{\mathfrak{a}^*} \psi_\mu(\lambda, H)F(\lambda)\,|\mathbf{c}(\lambda)|^{-2}\,d\lambda$$

and this was done by Gangolli (*Ann. of Math.* (1971) 150–165) by the device of multiplying the expansion of φ_λ by a suitable factor. A simplified justification is done in my paper [28], p. 38 using the same device.

An exposition of Harish–Chandra's theory with some simplifications is given in [B6], incorporating also Rosenberg's method for the inversion formula (*Proc. Amer. Math. Soc.* (1977)). Another major simplification is Anker's proof (*J. Funct. Anal.* (1991)) of Harish-Chandra's determination of the image of the Schwartz space under the spherical transform.

The Abel transform mentioned earlier is considered again in [91] and some new operational properties established, for example that its adjoint $\mathcal{A}^*$ is a bijection of $\mathcal{C}_W^\infty(A)$ onto $\mathcal{C}^\infty(K\backslash G/K)$, the subscript denoting Weyl group invariance.

Harish–Chandra's Eisenstein integrals (generalized spherical functions) defined for each $\delta \in \widehat{K}$ are studied in [28] and [43]. Here I proved that they satisfy functional equations, one for each $s \in W$, and that they can be expressed in terms of the zonal spherical function φ_λ. This plays a role in the proof of the Paley–Wiener theorem for the Fourier transform.

5. Duality for Symmetric Spaces.

Articles: *[15], [17], *[18], *[19], *[20], *[27], *[28], *[43], *[48], *[74],
Book: [B7]

The duality referred to here is between the symmetric space $X = G/K$ and the space $\Xi = G/MN$ of horocycles in X. Here G has the Iwasawa decomposition $G = NAK$ and M is the centralizer of A in K. In the framework of an earlier section, we have the double fibration

$$\begin{array}{ccc} & G/M & \\ \swarrow & & \searrow \\ X = G/K & & \Xi = G/MN \end{array}$$

and the transforms $f \to \widehat{f}$, $\varphi \to \check{\varphi}$ are

$$\widehat{f}(\xi) = \int_{\xi} f(x)\, dm(x) \quad \text{(the horocycle transform)}$$

$$\check{\varphi}(x) = \int_{\xi \ni x} \varphi(\xi)\, d\mu(\xi) \quad \text{(the dual transform).}$$

The injectivity of $f \to \widehat{f}$ on $L^1(X)$ can be deduced from the boundedness criterion of φ_λ mentioned earlier. For G a complex classical group Gelfand, Naimark and Graev had shown how a function $F \in \mathcal{C}_c^\infty(G)$ can be explicitly given in terms of its integrals over the conjugacy classes in G. Combined with the decomposition $G = NAK$ this proves an inversion formula for $f \to \widehat{f}$ for complex classical G. For general G the inversion and Plancherel formula for $f \to \widehat{f}$ were given in [15], [17], [19]. The kernel of the map $\varphi \to \check{\varphi}$ and the surjectivity $\mathcal{C}^\infty(\Xi)^\vee = \mathcal{C}^\infty(X)$ are established in [B7].

The main theme in [20], [28] is the pattern of analogies between G/K and G/MN. The algebra $\mathbf{D}(G/MN)$ is commutative as is the algebra $\mathbf{D}(G/K)$ and the transforms $f \to \widehat{f}$, $\varphi \to \check{\varphi}$ intertwine the operators in $\mathbf{D}(G/K)$ and $\mathbf{D}(G/MN)$. The analogs for G/MN of spherical functions on G/K are the *conical distributions* on G/MN which by definition are

(i) MN-invariant.
(ii) Eigendistributions of each $D \in \mathbf{D}(G/MN)$.

While the set of spherical functions is parametrized by $\mathfrak{a}_c^*/W$ (corresponding to $K\backslash G/K \sim A/W$) I proved in [28] and [43] that the set of conical distributions is parametrized by $\mathfrak{a}_c^* \times W$, corresponding to the decomposition $\Xi = MN \cdot (A \times W) \cdot \xi_0$ via the Bruhat decomposition of G. The injectivity criterion for the Poisson transform $\mathcal{P}_\lambda$ mentioned below is of central importance for this problem. The parametrization required the exclusion of certain singular eigenvalues. For X of rank 1 this restriction was removed by Men Cheng Hu (MIT thesis of 1973; *Bull. Am. Math.*

Soc. (1975)), thus finishing the determination of all conical distributions for this case.

A representation of G is *spherical* if there is a fixed vector under K and *conical* if there is a fixed vector under MN. It is proved in [17] and [28] that an irreducible finite-dimensional representation is *spherical* if and only if it is *conical*. This leads to an explicit characterization of their highest weights. It also leads to simultaneous imbeddings of X and Ξ into the same finite-dimensional vector space. The horocycles then appear as certain plane sections with X ([B7], Ch. II, §4).

With π an irreducible finite-dimensional spherical representation of G on V let v_K be a $\pi(K)$ fixed vector and v'_K a K fixed vector under the contragredient representation $\check{\pi}$. Then π is equivalent to the natural representation of G on the space of functions

$$g \to \langle \pi(g^{-1})v, v'_K \rangle \qquad v \in V$$

and in this model the spherical and conical vectors are respectively given by

$$\varphi_K(gK) = \langle \pi(g^{-1})v_K, v'_K \rangle \,,$$
$$\psi_{MN}(gK) = e^{-\Lambda(A(g))} \,,$$

Λ being the highest restricted weight of π and $A(g)$ as in this Introduction §1. In a lecture in Iceland, 2007, David Vogan interpreted the identity above of the spherical and conical representation in terms of a "deformation" of K into MN. Also Adam Korányi has shown that the MN-fixed vector is a suitable limit of an orbit of the K-fixed vector.

The continuous decompositions of $L^2(X)$ and $L^2(\Xi)$ into irreducibles correspond under the maps $f \to \widehat{f}$ and $\varphi \to \check{\varphi}$. The map $\varphi \to \check{\varphi}$ induces for each $\lambda \in \mathfrak{a}_c^*$ a "Poisson transform" $\mathcal{P}_\lambda$ which maps functions (and functionals) on $B = K/M$ into a joint eigenspace $\mathcal{E}_\lambda(X)$ of the operators in $\mathbf{D}(G/K)$. In [27], [28] and [43] it is stated and proved that $\mathcal{P}_\lambda$ is injective if and only if $\Gamma_X^+(\lambda)^{-1} \neq 0$ where Γ_X^+, the Gamma function of X, is the denominator in the Gindikin–Karpelevich formula for Harish–Chandra's $\mathbf{c}(\lambda)$ function. For X of rank one this was done in [28] by analyzing the Poisson transform at ∞. For X of higher rank this was done in [43] by studying its behavior at the origin, using methods from Kostant's work on the spherical principal series. The surjectivity of $\mathcal{P}_\lambda$ for hyperbolic spaces X (with analytic functionals on B) and the surjectivity of $\mathcal{P}_\lambda$ for general X on the K-finite level was proved in [40] and [43]. The surjectivity without the K-finiteness assumption was proved by Kashiwara, Kowata, Minemura, Okamoto, Oshima and Tanaka (*Ann. of Math.* (1978)). Professor Okamoto has kindly explained how this remarkable collaboration came about.

"I still remember vividly how we reached our proof of your great conjecture. We tried very hard in vain to prove your conjecture in the higher rank case, using the concept of analytical functionals. To understand your conjecture in a different way, we decided to apply directly the original idea of Prof. Sato about hyperfunctions to the new approach.

I took advantage of the fact that Prof. Sato is an uncle of Minemura so that Minemura asked his uncle (Prof. Sato) to spare some time on Saturdays or Sundays for discussions about the possibility of applying directly the original idea of Sato's hyperfunctions to prove your conjecture. We traveled to Kyoto University every weekend. During this time Kashiwara (Kyoto University) and Oshima (Tokyo University) came to join in our discussions.

To carry out our idea in the higher rank case, we realized that it is important to extend the notion of the ordinary differential equation with regular singularity to the notion of "the system of partial differential equations with regular singularity".

The splendid aspect of your conjecture is to have given birth to a new branch of Mathematics: Kashiwara–Oshima: Systems of differential equations with regular singularities and their boundary value problems, *Ann. of Math.* (1977)."

An engaging account of this work is contained in a book by Schlichtkrull (Birkhäuser (1984)).

Schmid has indicated how the rank 1 method (which relied on hypergeometric function estimates) might extend to the general rank.

The Poisson transform of distributions on B was investigated by Lewis and by Oshima–Sekiguchi. The latter authors also treat the case of affine symmetric G/H. Generalization to vector bundles was done by An Yang, (Thesis MIT, (1994)) and by Martin Olbrich (Dissertation, Berlin (1994)).

The flat analog of the double fibration above, replacing horocycles by their tangent planes, is investigated in [48] and [69]. The results are analogous but the expected tools like spherical function theory become degenerate so the results are obtained by looking at the flat case as a limiting case of the curved case.

The complexification of the double fibration above, replacing each group L by L^c has been actively studied in recent years (Gindikin, Krötz, Ólafsson, Stanton), and the extension of the joint eigenfunctions of $\mathbf{D}(G/K)$ to a tube around G/K in G^c/K^c plays a central role.

6. Representation Theory.

Articles: [17], *[19], *[20], *[27], *[28], [29], *[30], [31], *[43], [47], *[48], [49]

Book: [B7]

Given a coset space L/H with $\mathbf{D}(L/H)$ commutative the group L acts naturally on each joint eigenspace of the operators in $\mathbf{D}(L/H)$. I have called these representations *eigenspace representations.* If U/K is a symmetric space of the compact type the eigenspace representations are precisely the irreducible spherical representations π of U. *Spherical* means that there is a fixed vector under $\pi(K)$. If U is simply connected the highest weights of these π are given by the characterization mentioned before. This was further generalized by Schlichtkrull (1984) and by Johnson (1987).

In the papers listed above the irreducibility question for the eigenspace representations is considered for the cases $\mathbf{R}^n = \mathbf{M}(n)/\mathbf{O}(n)$, the noncompact symmetric space G/K, the dual space G/MN and G/N for G complex. For G/K the joint eigenspaces are parametrized by $\mathfrak{a}_c^*/W$ (since each joint eigenspace contains a unique spherical function); the irreducibility turns out to be equivalent to $1/\Gamma_X(\lambda) \neq 0$ where $\Gamma_X(\lambda)$ is the denominator of $\mathbf{c}(\lambda)\mathbf{c}(-\lambda)$.

Using the structure of $\mathbf{D}(G/MN)$ the eigenspace representations for G/MN turn out to be just the members of the spherical principal series of G; the intertwining operators, realizing the equivalence between two equivalent representations from this series, consist of convolutions with conical distributions ([27], Theorem 4.6). Another parallel determination of these operators was done by Schiffmann and by Kunze, Knapp and Stein. The irreducibility of these representations is equivalent to $1/\Gamma_X(\lambda) \neq 0$; this gives another proof and interpretation of the algebraic irreducibility criterion of Wallach, Parthasarathy–Rao–Varadarajan and Kostant though stated in quite different terms. For G complex, $\mathbf{D}(G/N)$ is still commutative, and the eigenspace representations form the full principal series.

For a non-Riemannian symmetric space G/H the eigenspace representations have been studied by Huang (*Ann. of Math.* (2001)). For the hyperbolic spaces over various fields this had been done very completely by Schlichtkrull (1987). For G nilpotent or solvable they have been effectively studied by Hole, Jacobsen and Stetkær (*Math. Scand.* (1975, 1983, 1981)).

7. Fourier Transform on G/K.

Articles: *[19], *[20], [24], *[28], *[35], [36], *[37], *[43], [54], [57], [79], *[81], *[82] *[91]

The spherical function φ_λ on G is K-bi-invariant, that is K-invariant on $X = G/K$ and the theory of the spherical transform $\widetilde{f}(\lambda) = \int_X f(x)\varphi_\lambda(x)\,dx$ only concerns functions f on X which are K-invariant. The analog of φ_λ for $\mathbf{R}^2$ is the Bessel function $J_0(\lambda|x|)$ which is a radial eigenfunction of the Laplacian L on $\mathbf{R}^2$ and the Bessel transform on $\mathbf{R}^+$ becomes the analog of the spherical transform. In [19] I proposed a definition of a *Fourier transform* on $X = G/K$, applying to all functions on X but reducing to the spherical transform for f K-invariant in the same way as the Fourier transform on $\mathbf{R}^2$ when used on radial functions reduces to the Bessel transform. Given the Iwasawa decomposition $G = NAK$ $g = n \exp A(g) k$, $A(g) \in \mathfrak{a}$; we define the vector-valued inner product $A(x, b)$ on $X \times B$ by

$$A(gK, kM) = A(k^{-1}g)\,.$$

Putting $e_{\lambda,b}(x) = e^{(i\lambda+\rho)(A(x,b))}$ we define the Fourier transform of a function f on X by

$$\widetilde{f}(\lambda, b) = \int_X f(x)e_{-\lambda,b}(x)\,dx$$

for all $(\lambda, b) \in \mathfrak{a}_c^* \times B$ for which the integral converges. The definition is suggested by the geometric analogy between horocycles and hyperplanes; $A(x, b)$ is a "vector-valued distance" from o to the horocycle through x with normal b. A representation-theoretic interpretation was pointed out by Sitaram (*Pacific J. Math.* (1988)) and Ólafsson–Schlichtkrull (Tribute to Mackey Volume, AMS (2008)). All the principal theorems from classical Fourier transform theory have their analogs for this Fourier transform. Thus [19] and [28] contain the Plancherel theorem (with range) and the inversion formula which can be written

$$\delta_o = \int_{\mathfrak{a}^* \times B} e_{\lambda,b}\, d\mu(\lambda, b)\,,$$

in analogy with the Euclidean version

$$\delta_o = \int_{\mathbf{R}^+ \times \mathbf{S}^{n-1}} e_{\lambda,\omega}\lambda^{n-1}\, d\lambda\, d\omega\,, \qquad e_{\lambda,\omega}(x) = e^{i\lambda(x,\omega)}\,.$$

The Paley–Wiener theorem (description of $\mathcal{D}(X)^{\sim}$) is proved in [19] and [28]. This theorem was used in [43] to prove the necessary and sufficient condition for the bijectivity of the Poisson transform for K-finite functions on B. In turn this was used to prove the irreducibility criterion for the eigenspace representations mentioned earlier. Another proof of the Paley–Wiener theorem was given by Torasso (1977). The paper [91] contains a strong form of the Riemann–Lebesgue lemma, proved jointly with Sengupta and Sitaram. These two and several others, Mohanty, Ray, Cowling, Bagci, Sarkar, Sundari, Rawat, have made great progress in topics such as the analog of the Wiener–Tauberian theorem and Hardy's theorem concerning simultaneous exponential decay of a function and its Fourier transform. Okamoto and especially Eguchi have proved deep theorems for the Fourier transform on the Schwartz spaces $\mathcal{S}(X)$ and their L^p analogs (Eguchi, *J. Funct. Anal.* (1979)).

The analog of the above Fourier transform for the dual compact symmetric space U/K was developed by Sherman (*Acta Math.* (1990)), and for vector bundles over X by Camporesi (*Pacific J. Math* (1997)). A Paley–Wiener type theorem for the compact case has been obtained (2000) by Gonzalez (on U) and more generally (2008) by Ólafsson and Schlichtkrull (on U/K).

In more recent years Fourier analysis theory for the non-Riemannian symmetric spaces G/H has been vigorously developed by van den Ban, Carmona, Delorme, van Dijk, Faraut, Flensted–Jensen, Matsuki, Molchanov, Ólafsson, Oshima, Rossmann, Schlichtkrull and many others.

8. Multipliers.

[1], [2], *[3], [4], [5]

Multipliers appear naturally in classical Fourier analysis on groups, where they correspond to operators on function spaces commuting with translations. In the papers listed the setup is a semisimple Banach algebra A; a subalgebra A_0, *the derived algebra*, is introduced as the algebra of elements $x \in A$ for which the multiplication $y \to xy$ is continuous from the topology of the *spectral norm* $\|y\|_{sp} = \|\widehat{y}\|_\infty$ to the original topology. Several properties of A_0 are proved in [3], for example the determination of the representative algebra $\widehat{A}_0$ as the intersection of certain L^1 spaces.

For various function algebras on a compact group G the derived algebra A_0 is explicitly determined. For G the circle group the results specialize to classical results of Littlewood, Sidon and Zygmund in Fourier series theory; for G the Bohr compactification of $\mathbf{R}$ they solve the multiplier problem for almost periodic functions on $\mathbf{R}$. My Ph.D. thesis consisted of [2] and [3].

For $A = L^1(G)$, where G is a compact group (abelian or not) A_0 turns out to be $L^2(G)$ and in fact,

$$\|f\|_2 \leq \sqrt{2} \sup_g \{\|f * g\|_1 : \|g\|_{sp} \leq 1\}.$$

For G abelian it follows from a combination of the papers Edwards and Ross (*Helgason's number and lacunarity constants*, *Bull. Austr. Math. Soc.* (1973)) and Sawa (*Studia Math.* (1985)) that $\sqrt{2}$ can be replaced by $2/\sqrt{\pi}$ and that this is the best possible constant. For G nonabelian or G/K symmetric it is not known whether the best constant is independent of G.

Photos

Akureyri Gymnasium. (Photo courtesy of F. Rouvière.)

With family, Artie (wife), children Thor and Annie.

With student J. Dadok and Adviser S. Bochner.

Conference Roskilde Denmark (for Helgason's 65th birthday).

At the Lie Memorial Conference, Oslo, 1992. (Photo courtesy of Artie Helgason.)

With Birgitte Burkovics, J. Radon's daughter.

With H. Schlichtkrull and M. Flensted-Jensen.

Receiving Major Knights Cross from President
Vigdís Finnbógadottir, 1991.

From R. Sulanke's retirement. With Th. Friederich, U. Simon, R. Sulanke, A. Onishchik.

At Dr.h.c. in Uppsala, 1996, with D. Hejhal and C. Kieselman. (Photo courtesy of Artie Helgason.)

With Gestur Ólafsson, Tokyo, 2006. (Photo courtesy of B. Rubin.)

With A. Korányi, Conference in Icleand, 2007. (Photo courtesy of F. Rouvière.)

Workshop in Tambara, Japan, 2006.

Bibliography, up to 2008, of S. Helgason

Books

[B1] *Differential Geometry and Symmetric Spaces*, Academic Press, New York, 1962. Russian translation, MIR, Moscow, 1964. Reprint by American Mathematical Society, 2001.

[B2] *Analysis on Lie Groups and Homogeneous Spaces*, Conference Board of Math. Science Lectures, Dartmouth 1971. Amer. Math. Soc. CBMS Monographs 1972. Corrected reprint 1977.

[B3] *Differential Geometry, Lie Groups and Symmetric Spaces*, Academic Press, New York, 1978. Russian translation, Factorial Press, 2005. Corrected reprint by American Mathematical Society in 2001.

[B4] *The Radon Transform*, Birkhäuser, Boston 1980, Russian translation, MIR, Moscow, 1983. 2nd edition, 1999.

[B5] *Topics in Harmonic Analysis*, Birkhäuser, Boston, 1981.

[B6] *Groups and Geometric Analysis*, Academic Press, New York and Orlando, 1984. Russian translation, MIR, Moscow, 1987. Corrected reprint by American Mathematical Society in 2000.

[B7] *Geometric Analysis on Symmetric Spaces*, Mathematical Surveys and Monographs. Amer. Math. Soc. 1994. 2nd edition, 2008.

[B8] (with Guðmundur Arnlaugsson) *Stærðfræðingurinn Ólafur Daníelsson. Saga Brautryðjanda.* Háskólaútgáfan, 1996.

Papers

Asterisks indicate the paper is included in this volume

[1] *The derived algebra of a Banach algebra*, Proc. Nat. Acad. Sci. USA **40** (1954), 994-995.

[2] *Some problems in the theory of almost periodic functions*, Math. Scand. **3** (1955), 49-67.

*[3] *Multipliers of Banach algebras*, Ann. of Math. **64** (1956), 240-254.

[4] *A characterization of the intersection of L^1 spaces*, Math. Scand. **4** (1956), 5-8.

[5] *Topologies of group algebras and a theorem of Littlewood*, Trans. Amer. Math. Soc. **86** (1957), 269-283.

[6] *Partial differential equations on Lie groups*, Thirteenth Scand. Math. Congr. Helsinki, 1957, 110-115.

[7] *Lacunary Fourier series on noncommutative groups*, Proc. Amer. Math. Soc. **9** (1958), 782-790.

[8] *On Riemannian curvature of homogeneous spaces*, Proc. Amer. Math. Soc. **9** (1958), 831-838.

*[9] *Differential operators on homogeneous spaces*, Acta Math. **102** (1959), 239-299.

[10] *Some remarks on the exponential mapping of an affine connection*, Math. Scand. **9** (1961), 129-146.

[11] *Some results on invariant theory*, Bull. Amer. Math. Soc. **68** (1962), 367-371.

[12] *Invariants and fundamental functions*, Acta Math. **109** (1963), 241-258.

*[13] *Fundamental solutions of invariant differential operators on symmetric spaces*, Bull. Amer. Math. Soc. **68** (1963), 778-781.

[14] *Duality and Radon transform for symmetric spaces*, Bull. Amer. Math. Soc. **69** (1963), 782-788.

*[15] *Duality and Radon transform for symmetric spaces*, Amer. J. Math. **85** (1963), 667-692.

*[16] *Fundamental solutions of invariant differential operators on symmetric spaces*, Amer. J. Math. **86** (1964), 565-601.

[17] *A duality in integral geometry; some generalizations of the Radon transform*, Bull. Amer. Math. Soc. **70** (1964), 435-446.

*[18] *The Radon transform on Euclidean spaces, compact two-point homogeneous spaces and Grassmann manifolds*, Acta Math. **113** (1965), 153-180.

*[19] *Radon-Fourier transforms on symmetric spaces and related group representations*, Bull. Amer. Math. Soc. **71** (1965), 757-763.

*[20] *A duality in integral geometry on symmetric spaces*, Proc. U.S.-Japan Seminar in Differential Geometry, Kyoto, Japan, 1965, pp. 37-56. Nippon Hyronsha, Tokyo 1966.

*[21] *Totally geodesic spheres in compact symmetric spaces*, Math. Ann. **165** (1966), 309-317.

*[22] *An analog of the Paley-Wiener theorem for the Fourier transform on certain symmetric spaces*, Math. Ann. **165** (1966), 297-308.

*[23] (with A. Korányi) *A Fatou- type theorem for harmonic functions on symmetric spaces*, Bull. Amer. Math. Soc. **74** (1968), 258-263.

[24] *Lie groups and symmetric spaces*, Battelle Rencontres 1967, 1-71. W. A. Benjamin, Inc., New York 1968.

[25] *Some geometric properties of differential operators*, Scientia Islandica, Anniversary Volume 1968, 41-43.

*[26] (with K. Johnson) *The bounded spherical functions on symmetric spaces*, Adv. in Math. **3** (1969), 586-593.

*[27] *Applications of the Radon transform to representations of semisimple Lie groups*, Proc. Nat. Acad. Sci. USA **63** (1969), 643-647.

*[28] *A duality for symmetric spaces with applications to group representations*, Adv. in Math. **5** (1970), 1-154.

[29] *Representations of semisimple Lie groups*, pp. 65-118, CIME lectures, Montecatini 1970, Edition Cremonese, Rome, 1971.

*[30] *Group representations and symmetric spaces*, Proc. Internat. Congress Math. Nice 1970, Vol. 2. Gauthier-Villars, Paris 1971, 313-320.

[31] *Conical distributions and group representations*, Symmetric Spaces, Short courses presented at Washington Univerity, Marcel Dekker, Inc. New York, 1972, 141-156.

*[32] *A formula for the radial part of the Laplace-Beltrami operator*, J. Diff. Geometry **6** (1972), 411-419.

[33] *Harmonic analysis in the non-Euclidean disk*, Proc. Internat. Conf. on Harmonic Analysis, Univ. of Maryland 1971. Lecture Notes in Mathematics **266**, 151-156.

[34] *Geometric reduction of differential operators*, Symposia Mathematica **10** (1972), 107-115.

*[35] *Paley-Wiener theorems and surjectivity of invariant differential operators on symmetric spaces and Lie groups*, Bull. Amer. Math. Soc. **79** (1973), 129-132.

[36] *Functions on symmetric spaces*, Proc. Sympos. Pure Math. **26**, Amer. Math. Soc., Providence, R. I., 1973, 101-146.

*[37] *The surjectivity of invariant differential operators on symmetric spaces*, Ann. of Math. 98 (1973), 451-480.

[38] *Spherical functions, spherical representations and correspondence theorems for the spherical Fourier transform*, Colloque sur les fonctions sphériques et la théorie des groupes, Nancy, 5-9, 1971. Russian translation, Mathematika **17** (1973).

[39] *Elie Cartan's symmetriske rum*, Danish Math. Soc. Centennial Publication, 1973, pp. 93-101.

*[40] *Eigenspaces of the Laplacian; integral representations and irreducibility*, J. Funct. Anal. **17** (1974), 328-353.

[41] *The eigenfunctions of the Laplacian on a two-point homogeneous space; integral representations and irreducibility*, Proc. Sympos. Pure Math. **27**, 357-360. Amer. Math. Soc., Providence 1975.

[42] *Remarque sur R.G. McLenaghan: On the validity of Huygens principle for second order partial differential equations with four independent variables. I.* Ann. Inst. Poincaré Sect. A **21** (1974), 347.

*[43] *A duality for symmetric spaces with applications to group representations II. Differential equations and eigenspace representations*, Adv. in Math. **22** (1976), 187-219.

[44] *Solvability of invariant differential operators on homogeneous manifolds*, C.I.M.E. seminar lecture (Varenna 1975), 281-310. Cremonese, Roma 1975.

[45] *Solvability questions for invariant differential operators*, Proc. Fifth Internat. Colloq. on Group Theoretical Methods in Physics (Montreal 1976), pp. 517-527. Academic Press 1977.

[46] *Invariant differential equations on homogeneous manifolds*, Bull. Amer. Math. Soc. **83** (1977), 751-774.

[47] *Some results on eigenfunctions on symmetric spaces and eigenspace representations*, Math. Scand. **41** (1977), 79-89.

*[48] *A duality for symmetric spaces with applications to group representations III. Tangent space analysis*, Adv. in Math. **36** (1980), 297-323.

[49] *Invariant differential operators and eigenspace representations*, In *Representations of Lie groups* (G. Luke, ed.), London Math. Soc. Lecture Note Series **34**, Cambridge Univ. Press 1979, 236-286.

[50] *Support of Radon transforms*, Adv. in Math. **38** (1980), 91-100.

[51] *The X-ray transform on a symmetric space*, Proc. Conf. Differential Geometry and Global Analysis (Berlin 1979). Lecture Notes in Math. **838** (1981), 145-148.

[52] *Ranges of radon transforms*, In Computed tomography (Cincinnati, 1982), Proc. Symp. Appl. Math. **27**, 63-70, Amer. Math. Soc. 1982.

[53] *The range of the radon transform on a symmetric space*, Proc. Conf. *Representations of Reductive Lie Groups* (Park City, 1982). Birkhäuser, 1983, 145-151.

[54] *Topics in geometric Fourier analysis.* In *Topics in Modern Harmonic Analysis* (Turin/Milan, 1982). Ist. Naz. Alta Mat. Francesco Severi, Rome 1983, 301-351.

[55] *Wave equations on homogeneous spaces*, Lie Group Representations III (College Park, 1983). Lecture Notes in Math. **1077** (1984), 254-287.

[56] *Operational properties of the Radon transform with applications*, Proc. Conf. Differential Geometry and its Applications, (Nove Mesto Morave Czechoslovakia 1983). Publ. Univ. Karlova Praha 1984, 59-75.

[57] *The Fourier transform on symmetric spaces*, The mathematical heritage of Élie Cartan, Astérisque 1985, Numero Hors Série, 151-164.

[58] *Some results on Radon transforms, Huygens Principle and X-ray transforms*, Integral geometry (Brunswick, 1984). Contemp. Math. **63** (1987), 151-177.

*[59] (with F. Gonzalez) *Invariant differential operators on Grassmann manifolds*, Adv. in Math. **60** (1986), 81-91.

*[60] *Value-distribution theory for holomorphic almost periodic functions*, The Harald Bohr Centenary (Copenhagen 1987). Mat.-Fys. Medd. Danske Vid. Selsk. **42** (1989), 93-102.

[61] *Response to 1988 Steele Prize Award*, Notices Amer. Math. Soc. **35** (1988), 965-967.

*[62] *The totally geodesic Radon transform on constant curvature spaces*, Integral geometry and tomography (Arcata, 1989). Contemp. Math. **113** (1990), 141-149.

*[63] *Some results on invariant differential operators on symmetric spaces*, Amer. J. Math. **114** (1992), 789-811.

[64] *Invariant differential operators and Weyl group invariants.* In (Ed. W. Barker and P. Sally) *Harmonic Analysis on Reductive Groups*, 193-200, Birkhäuser, Boston 1991.

[65] *A centennial: Wilhelm Killing and the exceptional groups*, Math. Intelligencer, **12** (1990) 54-57.

[66] *Geometric and analytic features of the radon transform*, Mitt. Math. Ges. Hamburg, **12** (1991), 977-988 (1992).

[67] *Support theorems in integral geometry and their applications.* Proc. AMS Summer Inst. on Differential Geometry (Los Angeles 1990), Proc. Sympos. Pure Math. **54** (1993), 317-323.

[68] *Tveir töframenn rúmfræðinnar, Poncelet og Jacobi.* Icelandic Math. Soc. Bulletin 1991, 6-17.

*[69] *The flat horocycle transform for a symmetric space*, Adv. in Math. **91** (1992), 232-251.

[70] *Radon vörpunin, afbrigði og notkun*, Leifur Asgeirsson Memorial Volume. Háskólaútgáfan, Reykjavik 1998.

*[71] *Huygens' principle for wave equations on symmetric spaces*, J. Funct. Anal. **107** (1992) 279-288.

[72] *The wave equation on symmetric spaces.* In *Classical and Quantum Systems*, Int. Wigner Symp. (Goslar 1991). World Sci. Publ. 1993, 149-157.

[73] *Sophus Lie and the role of Lie groups in mathematics*, Report of the 14. Nordic LMFK-congress Reykjavík, 1990.

*[74] *Radon transforms for double fibrations. Examples and viewpoints*, Proc. of Symposium *75 years of Radon Transform*, Vienna 1992. Int. Press, Boston 1994, 175-188.

[75] *The Fourier transform on symmetric spaces and applications*, Proc. Conf. Differential Geometry and its Applications (Opava 1992), Silesian Univ. Opava 1993, 23-28.

[76] *Sophus Lie, the mathematician*, Proc. of The Sophus Lie Memorial Conference (Oslo 1992), Scand. Univ. Press, Oslo 1994, 3-21.

[77] *Harish-Chandra's c-function; a mathematical jewel.* In *Noncompact Lie Groups and Some of Their Applications.* NATO Adv. Sci. Inst. Ser. C Math. Phys. Sci., 429 (San Antonio 1993). Kluwer Acad. Publ. 1994, 55-67.

[78] *Radon transforms and wave equations*, C.I.M.E. 1996. Lecture Notes in Math. **1684** (1998), 99-121

[79] *Orbital integrals, symmetric Fourier analysis and eigenspace representations*, Proc. Sympos. Pure Math. **61** (1997), 167-189.

[80] *Paul Günther's work on Hadamard's conjecture*, Zeitschr. Anal. Anw. **16** (1997), 5-6.

*[81] *Integral geometry and multitemporal wave equations*, Comm. Pure and Appl. Math. **51** (1998) 1035-1071.

*[82] (with H. Schlichtkrull) *The Paley-Wiener space for the multitemporal wave equation*, Comm. Pure and Appl. Math. **52** (1999) 49-52.

[83] *Harish-Chandra and his mathematical legacy. Some personal recollections*, Current Sci. **74** (1998), 921-924.

[84] *Harish-Chandra's c-function; a mathematical jewel.* Reprint of [77], The mathematical legacy of Harish-Chandra, Proc. Sympos. Pure Math. **68** (2000), 273-283.

[85] *Harish-Chandra memorial talk*, ibid. 43-45.

[86] *Hinn óevklíðski heimur, saga, rúmfræði og greining*, Verpill, 2000.

[87] *The Fourier transform on symmetric spaces. Applications to classical and multitemporal wave equations*, Amer. Math. Soc. Proc. Symp. Pure Math. (to appear).

[88] *Rúmfræði og raunveruleikinn*, Icelandic Math Society Bulletin, 2000, 1-12.

[89] *Non-Euclidean analysis.* In (Eds. Prékopa and Molnár) *Non-Euclidean Geometries*, Proceedings of the Janos Bolyai Mem. Conf., Budapest, July, 2002. Springer, 2006. p. 367-384.

[90] *Remarks on B. Rubin, radon, cosine and sine transforms on real hyperbolic space*, Adv. in Math. **192** (2005), 225.

*[91] *The Abel, Fourier and radon transforms on symmetric spaces*, Indag. Math. **16** (2005), 531-551.

*[92] *The Inversion of the X-ray transform on a compact symmetric space*, J. Lie Theory **17** (2007), 307-315.

Annals of Mathematics
Vol. 64, No. 2, September, 1956
Printed in U.S.A.

MULTIPLIERS OF BANACH ALGEBRAS

By S. Helgason

(Received June 8, 1955)

Introduction

In Gelfand's theory of commutative Banach algebras the abstract algebra A is represented homomorphically by an algebra $\hat{A}$ of continuous functions vanishing at infinity on a locally compact Hausdorff space $\mathfrak{M}$, the maximal ideal space of A. Although some general features of $\hat{A}$ are known, a closer description of $\hat{A}$ remains a central problem in the theory.

One approach to this problem is to study the *multipliers* of A, that is functions f on $\mathfrak{M}$ such that $f\hat{x} \in \hat{A}$ for every $\hat{x} \in \hat{A}$. Examples from harmonic analysis show that the class of multipliers can sometimes be determined more or less explicitly, even when little is known about $\hat{A}$ itself.

In this paper we consider a certain subalgebra A_0, *the derived algebra,* of a semi-simple commutative Banach algebra A. A_0 is the set of elements in A with which the multiplication is a continuous mapping from the spectral norm topology of A to the original topology of A. A_0 is again a Banach algebra which inherits many general properties from A, notably the topology of the maximal ideal space. But A_0 has some special properties which become apparent when the multipliers of A_0 are considered. All continuous functions vanishing at infinity on $\mathfrak{M}_0$ (the maximal ideal space of A_0) are multipliers of A_0. In the case when A is weakly complete we obtain a concrete representation of $\hat{A}_0$ as the algebra of all continuous functions vanishing at infinity on $\mathfrak{M}_0$ that are integrable with respect to certain measures on $\mathfrak{M}_0$. In particular, all bounded continuous functions on $\mathfrak{M}_0$ are multipliers, that is, $\hat{A}_0$ is an ideal in $\mathfrak{C}^\infty(\mathfrak{M}_0)$. This last statement also holds under other restrictions on A_0. The main tool in the proofs is a representation of the dual space of A_0, namely a decomposition of every bounded linear functional on A_0 into homomorphisms.

Examples are given in Section 3. If $A = L^1(G)$, where G is a locally compact abelian group, $A_0 = L^2(G)$ if G is compact, and $A_0 = 0$ if G is non-compact and connected. The convolution algebra $L^\infty(G)$ where G is a compact abelian group has for derived algebra the set of functions on G that have absolutely convergent Fourier series.

The author is grateful to Professor S. Bochner for many stimulating discussions.

1. Notations and preliminaries

If S is a locally compact space, $C^\infty(S)$ shall denote the space of bounded complex valued functions on S with the topology of the uniform norm $\|f\|_\infty = \sup_{t \in S} |f(t)|$. $\mathcal{K}(S)$ shall denote the set of continuous functions on S that have compact support. The closure of $\mathcal{K}(S)$ in $\mathfrak{C}^\infty(S)$ is the set of continuous functions vanishing at infinity on S and is denoted by $\mathcal{C}(S)$. We adopt the measure-theoretic terminology of Bourbaki [1]. In particular, a positive Radon measure refers to a positive linear functional on $\mathcal{K}(S)$ as well as to the corresponding set function on S. If μ is a positive Radon measure on S and $p \geqq 1$, we write $L^p(\mu)$ (or $L^p(S)$ when no misunderstanding can arise) for the space of complex valued functions f on S for which $|f|^p$ is μ-integrable with the norm

$$\|f\|_p = \left(\int_S |f|^p \, d\mu\right)^{1/p}.$$

Let A be a commutative Banach algebra over the complex numbers C. We do not assume that A has an identity element. The regular maximal ideals are precisely the kernels of the homomorphisms of A onto C. If $M = h^{-1}(0)$ we denote the image of an element $x \in A$ under h by $h(x)$ or $\hat{x}(M)$. The set of all homomorphisms of A onto C is denoted by Δ and will be identified with $\mathfrak{M}$, the set of all regular maximal ideals in A. $\mathfrak{M}$ is usually given the *weak topology*, the weakest topology for which all the functions $\hat{x}(M)$ are continuous. $\mathfrak{M}$ is locally compact in this topology and all the functions $\hat{x}(M)$ vanish at infinity on $\mathfrak{M}$. A is thus homomorphically represented by a subalgebra $\hat{A}$ of $\mathcal{C}(\mathfrak{M})$. The representation algebra $\hat{A}$ is a normed algebra under the uniform norm $\| \hat{x} \|_\infty = \sup_M | \hat{x}(M) |$, called the *spectral norm* of x. Between the norms in $\hat{A}$ and A we have the relation $\| \hat{x} \|_\infty \leqq \| x \|$ and the Beurling-Gelfand formula

$$\| \hat{x} \|_\infty = \lim_{n\to\infty} \| x^n \|^{1/n}. \tag{1.1}$$

If the representation $x \to \hat{x}$ is one-to-one, the algebra is called *semi-simple*, and A is called *self adjoint*, if $\hat{A}$ is closed under complex conjugation. If A is self adjoint, then $\hat{A}$ is dense in $\mathcal{C}(\mathfrak{M})$. The maximal ideal space is sometimes given another topology, the *hull-kernel topology*, which is described in the following way. If I is an ideal, the hull of I is the set of regular maximal ideals in A that contain I. If $\mathfrak{N} \subset \mathfrak{M}$, the kernel of $\mathfrak{N}$ is the intersection of the members of the subset $\mathfrak{N}$. The topology is then defined by the closure operation $\overline{\mathfrak{N}} = \text{hull}(\text{kernel}(\mathfrak{N}))$. $\mathfrak{M}$ will always be taken with the weak topology unless the converse is explicitly stated. If the two mentioned topologies on $\mathfrak{M}$ coincide, then A is called *regular*.

2. The derived algebra

Throughout this paper A will denote a commutative Banach algebra over C satisfying the following conditions:

(i) A is semi-simple;

(ii) A is self adjoint;

(iii) For every $x \in A$ there exists[1] a sequence $y_n \in A$ such that $\| \hat{y}_n \|_\infty \to 1$ and $xy_n \to x$.

[1] Condition (iii) could easily be dispensed with by adjoining a unit to the algebra if necessary. In case the condition (iii) is fulfilled, it makes no essential difference whether we adjoin a unit or not, due to the relation $(A_e)_0 \cap A = A_0$, which is easily proved.

The norm $\| \hat{x} \|_\infty$ is a proper norm on A due to (i) and induces a topology on A, say A_w, which is weaker than the original one.

DEFINITION. Let A_0 be the set of elements $x \in A$ for which the mapping $y \to xy$ is continuous from A_w to A, that is A_0 is the set of elements in A for which the quantity

$$\| x \|_0 = \sup_y \frac{\| xy \|}{\| \hat{y} \|_\infty}$$

is finite. A_0 is called *the derived algebra of A*.

THEOREM 1. *A_0 is a subalgebra of A and a Banach algebra under the norm $\| \|_0$.*

The proof follows easily from the definition and the relation $\| x \|_0 \geqq \| x \|$ which is a simple consequence of (iii).

We show next that the derived algebra can be characterized in a somewhat different way.

THEOREM 2. *A_0 is the set of elements $x \in A$ such that $\hat{x}f \in \hat{A}$ for all $f \in \mathfrak{C}(\mathfrak{M})$.*

PROOF. We denote by x^f the element in A for which $(x^f)^\wedge = \hat{x}f$. Then if x is an element in A such that $x^f \in A$ for all $f \in \mathfrak{C}(\mathfrak{M})$, the mapping $f \to x^f$ is a linear mapping from the complete space $\mathfrak{C}(\mathfrak{M})$ to the complete space A, and is easily seen to be continuous by the closed graph theorem. Clearly $\| x \|_0 < \infty$. On the other hand, if $x \in A_0$, then the mapping $\hat{y} \to xy$ is continuous from $\hat{A}$ to A and extends uniquely to a continuous mapping of the completion of $\hat{A}$, namely $\mathfrak{C}(\mathfrak{M})$, into A.

We will denote the maximal ideal spaces of A and A_0 by $\mathfrak{M}$ and $\mathfrak{M}_0$ respectively and the corresponding homomorphism spaces by Δ and Δ_0. If $h_0 \in \Delta_0$, $h_0^{-1}(0) = M_0$, we put $\mathring{x}(M_0) = h_0(x)$.

THEOREM 3. *There exists a mapping ϕ of $\mathfrak{M}_0$ into $\mathfrak{M}$ such that:*

(i) *If $\mathfrak{M}_0$ and $\mathfrak{M}$ have the weak topologies, ϕ is a homeomorphism, and $\phi(\mathfrak{M}_0)$ is open in $\mathfrak{M}$.*

(ii) *Statement* (i) *holds also if $\mathfrak{M}_0$ and $\mathfrak{M}$ have the hull-kernel topologies.*

PROOF. We first remark that A_0 is an ideal in A. (For later use we remark that this is the only relation between A_0 and A that Theorem 3 depends on.) Let h_0 be an arbitrary element of Δ_0 and $a \in A_0$ such that $h_0(a) \neq 0$. Then

$$h(z) = h_0(az)/h_0(a) \qquad \text{for all } z \in A$$

defines a unique extension of h_0 to a homomorphism of A onto C. We put $\phi(h_0) = h$. Then the image $\phi(\Delta_0) = \Delta - \bigcap_{x \in A_0}\{h \mid h(x) = 0\}$ is open in Δ in the weak topology and the hull-kernel topology since $\bigcap_{x \in A_0}\{M \mid \hat{x}(M) = 0\}$ is closed in both cases. Now, if $M = \phi(M_0)$, then $\mathring{x}(M_0) = \hat{x}(M)$ for $x \in A_0$, and it follows in a straightforward manner that ϕ is a homeomorphism in the weak topologies of $\mathfrak{M}_0$ and $\mathfrak{M}$. Now suppose $\mathfrak{M}_0$ and $\mathfrak{M}$ have the hull-kernel topologies. Let $\mathfrak{N}_0 \subset \mathfrak{M}_0$ and $\phi(\mathfrak{N}_0) = \mathfrak{N}$. Then kernel$(\mathfrak{N}_0) = A_0 \cap$ kernel$(\mathfrak{N})$ and since h_0

vanishes on the kernel of $\mathfrak{N}_0$ if and only if the extension $\phi(h_0)$ vanishes on kernel $(\mathfrak{N})$, we have

$$\phi[\text{hull}(\text{kernel}(\mathfrak{N}_0))] = \phi(\mathfrak{N}_0) \cap \text{hull}(\text{kernel}(\mathfrak{N}))$$

so ϕ is a homeomorphism.

We now prove some further elementary properties of A_0.

THEOREM 4. *Let A_0 be the derived algebra of A. Then*

(i) *A_0 is semi-simple*

(ii) *A_0 is self adjoint*

(iii) *A_0 is regular*

(iv) $\mathcal{K}(\mathfrak{M}_0) \subset \hat{A}_0$

(v) *$\hat{A}_0$ is an ideal in $\mathcal{C}(\mathfrak{M}_0)$*

(vi) *The canonical mapping $\mathring{x} \to \hat{x}$ imbeds $\hat{A}_0$ isometrically in $\hat{A}$ (spectral norms).*

PROOF. (i): Immediate from Theorem 3. (v): Let $x \in A_0$ and $f \in \mathcal{C}(\mathfrak{M}_0)$. We extend f to $\tilde{f} \in \mathcal{C}(\mathfrak{M})$ by putting $\tilde{f} = 0$ on $\mathfrak{M} - \mathfrak{M}_0$. Then if $g \in \mathcal{C}(\mathfrak{M})$ we get from Theorem 2 $\hat{x}(\tilde{f}g) \in \hat{A}$; applying the theorem again we deduce that $\mathring{x}f \in \hat{A}_0$. (ii): Let $x \to x^*$ be the involution of A that corresponds to complex conjugation of $\hat{A}$. From the closed graph theorem it follows (see e.g. Loomis [5], p. 90) that this involution is continuous, and therefore $x \in A_0$ implies $x^* \in A_0$, and by Theorem 3 A_0 is self adjoint. (iii): In view of (i) and (v), we can regard the Banach algebra A_0 as an ideal in the Banach algebra $\mathcal{C}(\mathfrak{M}_0)$, and the conclusion of Theorem 3 holds for the maximal ideal spaces of A_0 and $\mathcal{C}(\mathfrak{M}_0)$. Since $\mathcal{C}(\mathfrak{M}_0)$ is regular and has the maximal ideal space homeomorphic to $\mathfrak{M}_0$, (iii) follows. (iv): If the function $f \in \mathcal{C}(\mathfrak{M}_0)$ has compact support, say S, we can, using the fact that A_0 is self adjoint, find $x \in A_0$ such that $\mathring{x} > 0$ on S and then a function $g \in \mathcal{C}(\mathfrak{M}_0)$ such that $g = 1/\mathring{x}$ on S. Then by (v) $f = f\mathring{x}g \in A_0$. (vi): The relation $\| \mathring{x} \|_\infty \leqq \| \hat{x} \|_\infty$ for $x \in A_0$ is contained in the proof of Theorem 3 and the converse inequality follows from $\| x \|_0 \geqq \| x \|$ in connection with (1.1). Theorem 4 is proved.

3. Examples

Let G be an abelian locally compact group with the group operation denoted multiplicatively. The character group of G is denoted by $\hat{G}$ and (t, χ) will denote the function on $G \times \hat{G}$ that sets up the duality between G and $\hat{G}$. By dt we denote a fixed Haar measure in G which we assume normalized in case G is compact. The identity element in G will be denoted by e.

EXAMPLE 1. Let A be the algebra $L^1(G)$ with usual addition and convolution multiplication $x * y(t) = \int_G x(ts^{-1})y(s)\, ds$. Here $\mathfrak{M}$ and $\hat{G}$ can be put in a one-to-one correspondence $M \leftrightarrow \chi$ such that $\hat{x}(M) = \int_G x(s)\overline{(s, \chi)}\, ds$. A is clearly self adjoint. If $x \in A$ and $\varepsilon > 0$ there exists a neighbourhood N of e in G such that if $u(t)$ is a non-negative continuous function with support contained in N and $\| u \|_1 = 1$, then

$$(3.1) \qquad \| x * u - x \|_1 < \varepsilon.$$

Conditions (ii) and (iii) Section 2 are then satisfied. The condition (i), the semi-simplicity of $L^1(G)$, is a well known fact of which we give here a direct proof, using standard techniques. Suppose there existed a $z \in L^1(G)$ such that $z \neq 0$ but $\hat{z}(M) \equiv 0$. The function $z(t)$ can be assumed bounded, because otherwise we take a suitable convolution $z * u(t)$ instead of $z(t)$. Consider the functional $F(x) = z * z^* * x(e)$ for all $x \in L^1(G)$ which is well defined since $z(t)$ is bounded and $z * z^* * x(t)$ exists for *every* t. Furthermore $F(x * x^*) \geqq 0$ for every $x \in L^1(G)$ so that we have the Schwarz inequality

$$(3.2) \qquad | F(x * u) |^2 \leqq F(x * x^*) F(u * u^*)$$

and since F is continuous on $L^1(G)$ we get from (3.1): $| F(x) | \leqq F(x * x^*)^{\frac{1}{2}} K$ for all $x \in L^1(G)$, where K is independent of x. Applying this successively we obtain

$$\begin{aligned} | F(x) | \leqq KF(x * x^*)^{\frac{1}{2}} &\leqq K^{1+\frac{1}{2}} F(x * x^* * x * x^*)^{\frac{1}{4}} \\ &\leqq K^{1+\frac{1}{2}+\cdots 2^{-n}} F(x * x^* * \cdots * x^*)^{2^{-n-1}} \\ &\leqq K^2 \| F \|^{2^{-n-1}} \| x * \cdots * x^* \|_1^{2^{-n-1}}. \end{aligned}$$

Here we let n tend to infinity, and we get from (1.1): $| F(x) | \leqq K^2 \| \hat{x} \|_\infty$. This last inequality gives, when applied to the function $x = z * z^*$, the relation $z * z^* * z * z^*(e) = 0$; hence $z = 0$, contrary to assumption.

Since the conditions (i) (ii) and (iii) Section 2 are satisfied, we can consider the derived algebra A_0.

THEOREM 5. *Let* $A = L^1(G)$. *For the derived algebra we have*

(i) *If* G *is compact, then* $A_0 = L^2(G)$ *(algebraically and topologically), and*

$$2^{-\frac{1}{2}} \| x \|_2 \leqq \| x \|_0 \leqq \| x \|_2 \textit{ for every } x \in A_0.$$

(ii) *If* G *is non-compact and connected* $A_0 = 0$.

PROOF. Let G be compact and let x be an element of $L^1(G)$. $x(t)$ has a Fourier series $x(t) \sim \sum a_n(t, \chi_n)$ (countable sum). For each $y \in L^1(G)$, $y(t) \sim \sum b_m(t, \chi_m)$, we have

$$\| x * y \|_1 \leqq \| x * y \|_2 = (\sum | a_n b_n |^2)^{\frac{1}{2}} \leqq (\sum | a_n |^2)^{\frac{1}{2}} \max_m | b_m | = \| x \|_2 \| \hat{y} \|_\infty$$

by the generalized Riesz-Fischer theorem. Hence $\| x \|_0 \leqq \| x \|_2$ and $L^2(G) \subset A_0$. Now let $x(t) = \sum_1^n a_k(t, \chi_k)$ be an arbitrary trigonometric polynomial. Then, if T^n denotes the n-fold torus $\{u = (u_1, u_2, \cdots u_n) \mid 0 \leqq u_j < 1, j = 1, \cdots n\}$ with normalized Haar measure du, we have

$$\sup_y \frac{\| x * y \|}{\| \hat{y} \|_\infty} \geqq \sup_{u \in T^n} \int_G | \textstyle\sum_1^n a_k \exp(2\pi i u_k)(t, \chi_k) | \, dt$$

$$\geqq \int_{T^n} du \int_G | \sum_1^n a_k \exp (2\pi i u_k)(t, \chi_k) | \, dt$$

$$= \int_G dt \int_{T^n} | \sum_1^n a_k \exp (2\pi i u_k)(t, \chi_k) | \, du$$

$$= \int_{T^n} | \sum_1^n a_k \exp (2\pi i u_k) | \, du$$

since du is translation-invariant. The last integral can be estimated in several ways, giving a lower bound $\geqq M(\sum_1^n | a_k |^2)^{\frac{1}{2}}$ where M is an absolute constant. The following method, suggested to me by Professor R. Salem, gives a better estimate on M than my original proof did.

We have by Hölders inequality

$$\int_{T^n} |f|^2 \, du \leqq \left(\int_{T^n} |f| \, du\right)^{\frac{2}{3}} \left(\int_{T^n} |f|^4 \, du\right)^{\frac{1}{3}}$$

and it is easily verified, that

$$\int_{T^n} | \sum_1^n a_k \exp (2\pi i u_k) |^4 \, du = 2(\sum_1^n | a_k |^2)^2 - \sum_1^n | a_k |^4$$

hence

$$\int_{T^n} | \sum_1^n a_k \exp (2\pi i u_k) | \, du \geqq 2^{-\frac{1}{2}}(\sum_1^n | a_k |^2)^{\frac{3}{2}}(\sum_1^n | a_k |^2)^{-1} = 2^{-\frac{1}{2}}(\sum_1^n | a_k |^2)^{\frac{1}{2}}.$$

We have therefore proved the inequalities

$$2^{-\frac{1}{2}}(\sum_1^n |a_k |^2)^{\frac{1}{2}} \leqq \sup_y \frac{\| x * y \|_1}{\| \hat{y} \|_\infty} \leqq (\sum | a_k |^2)^{\frac{1}{2}} \tag{3.3}$$

for every trigonometric polynomial $x(t) = \sum a_k(t, \chi_k)$. Since these are dense in $L^2(G)$ and A_0, we conclude that $L^2(G) = A_0$ and (3.3) holds for all $x \in A_0$.

The question arises whether the equality sign holds on the right side in (3.3). (Compare Theorem 6.) The answer is negative. Actually one finds by computations (involving elliptic integrals) that for the algebra $A = L^1(0, 2\pi)$, $\| 1 + \exp (it) \|_0 < 2^{\frac{1}{2}}$. We do not know what the best constant in (3.3) is, although it seems likely to be $2^{-\frac{1}{2}}$.

Next we consider the case (ii) where G is non-compact and connected. By a theorem of Pontrjagin, (see Weil [7], p. 110), G is a direct product $G = R^p \times H$ where R^p is the additive group of the p-dimensional euclidean space and H is compact. Clearly $p > 0$. We want to prove that $A_0 = 0$. Suppose there were an element $x \in A_0$ with $\| x \|_0 \neq 0$. We consider first the case that $x(t)$ vanishes outside a compact subset C in G. Given $\varepsilon > 0$ we first shall find a large family $\mathfrak{F}(\varepsilon)$ of functions $y(t) \in L^1(G)$ such that

$$| \| x * y \|_1 - \| x \|_1 \| y \|_1 | < \varepsilon \| y \|_1 \qquad \text{for } y \in \mathfrak{F}(\varepsilon). \tag{3.4}$$

By the above representation of G there exists an element $d \in G$ such that the translates $d^n C$ $(n = 0, \pm 1, \pm 2, \cdots)$ are all disjoint. If $u(t) \geqq 0$ and $u(t)$ vanishes outside a sufficiently small neighbourhood $N(\varepsilon)$ of e and $\| u \|_1 = 1$, it is not hard to see that every function of the form[2]

$$(3.5) \qquad y(t) = \sum{}^* b_n u(d^n t),$$

where b_n are complex numbers, has norm $\| y \|_1 = \sum{}^* | b_n |$ and satisfies (3.4). The functions (3.5) form the desired family $\mathfrak{F}(\varepsilon)$. Since $\| x * y \|_1 \leqq \| x \|_0 \| \hat{y} \|_\infty$, it follows from (3.4) for $\varepsilon < \| x \|_1$ that

$$(3.6) \qquad \| y \|_1 \leqq K \| \hat{y} \|_\infty \qquad \text{for all } y \in \mathfrak{F}(\varepsilon),$$

where K is a constant. Now $\hat{y}(\chi) = [\sum{}^* b_n (d^n, \chi)] \hat{u}(\chi)$ so (3.6) implies

$$(3.7) \qquad \sum{}^* | b_n | \leqq K \sup_{\chi \in \hat{G}} | \sum{}^* b_n (d^n, \chi) | \, \| \hat{u} \|_\infty .$$

If D denotes the cyclic group $\{d^n\}$ with the discrete topology, then, since the restriction of every $\chi \in \hat{G}$ is a character on D, (3.7) implies $\| \phi \|_1 \leqq K_1 \| \phi \|_\infty$ for every $\phi \in L^1(D)$ with Fourier transform $\phi \in \mathfrak{C}(\hat{D})$ where K_1 is another constant. From that follows that every function in $\mathfrak{C}(\hat{D})$ is a Fourier transform of a function in $L^1(D)$. But by a theorem of Segal [6] this can not happen in an infinite group, and since D is infinite we have a contradiction.

Now we drop the assumption that $x(t)$ has compact support. Suppose $x \in A_0$ and $\| x \|_0 \neq 0$. Choose $\varepsilon > 0$, $\varepsilon < \frac{1}{4} \| x \|_1$ then choose $z(t)$ continuous with compact support C such that $\| x - z \|_1 < \varepsilon$, then $d \in G$ such that the sets $d^n C$ are disjoint and finally a neighbourhood $V = V^{-1}$ of e such that[3] $\| z_s - z \| < \varepsilon$ if $s \in V$. Let $u(t)$ have support contained in V, $u(t) \geqq 0$ and $\| u \|_1 = 1$. If $y(t)$ is of the form (3.5) we have

$$\| z * y - \sum{}^* b_n z_{dn} \|_1 \leqq \sum{}^* | b_n | \, \| z * u - z \|_1 \leqq \varepsilon \| y \|_1$$

and therefore

$$\| x \|_0 \| \hat{y} \|_\infty \geqq \| x * y \|_1 \geqq \| z * y \|_1 - \varepsilon \| y \|_1 \geqq \| z \|_1 \| y \|_1 - 2\varepsilon \| y \|_1 \geqq \varepsilon \| y \|_1$$

which gives a relation of the form (3.6), and we get a contradiction as before. Theorem 5 is proved.

For more information on this example see Corollary 2 of Theorem 8 and subsequent Remark 1.

Remark. In view of Theorem 2, the second part of Theorem 5 amounts to the fact that a Fourier transform $\hat{x}(\chi) = \int_G x(s) \overline{(s, \chi)} \, ds$ which remains a Fourier transform of a function in $L^1(G)$ after multiplication by an arbitrary function in $\mathfrak{C}(\hat{G})$ must vanish identically.

Example 2. Let G be an abelian compact group and let A be the algebra $\mathfrak{C}(G)$ under usual addition, uniform norm and convolution multiplication.

[2] Here and in the sequel, $\sum{}^*$ denotes a finite sum.

[3] z_s denotes the function $t \to z(ts)$.

$\mathcal{C}(G)$ is an ideal in $L^1(G)$ and from Example 1 and Theorem 3 it follows that the set $\mathfrak{M}$ of regular maximal ideals in $\mathcal{C}(G)$ can be identified with $\hat{G}$, such that $\hat{x}(M) = \int_G x(t)\overline{(t, \chi)}\, dt$, and that $\mathcal{C}(G)$ is semi-simple. Also $\mathcal{C}(G)$ is self adjoint. It can be verified in the same way as in Example 1 that the condition (iii) in Section 2 is fulfilled. For the derived algebra A_0 we have

THEOREM 6. *If G is compact and $A = \mathcal{C}(G)$ (convolution product) then A_0 consists of those functions on G that have absolutely convergent Fourier series. Furthermore, if $x(t) \sim \sum a(\chi)(t, \chi)$ then $\| x \|_0 = \sum | a(\chi) |$.*

PROOF. We have, if $y \varepsilon \mathcal{C}(G)$ and $y(t) \sim \sum b(\chi)(t, \chi)$,

$$\| x \|_0 = \sup_y \frac{\| x * y \|_\infty}{\| \hat{y} \|_\infty} = \sup_y \frac{\| \sum a(\chi) b(\chi)(t, \chi) \|_\infty}{\max | b(\chi) |} = \sum | a(\chi) |.$$

REMARK. The same derived algebra is obtained for the convolution algebra $L^\infty(G)$.

EXAMPLE 3. Let G be abelian and compact and $1 < p < \infty$. The space $L^p(G)$ with norm $\| x \|_p$ is a Banach algebra under usual addition and convolution multiplication. $L^p(G)$ is an ideal in $L^1(G)$ so we verify as in Example 2 that $\mathfrak{M} = \hat{G}$, $\hat{x}(M) = \int_G x(s)\overline{(s, \chi)}\, ds$, $L^p(G)$ is semi-simple and self adjoint and the condition (iii) in Section 2 is satisfied.

THEOREM 7. *If G is compact and $A = L^p(G)$ $(1 < p < \infty)$, then the derived algebra A_0 has the following properties:*

(i) *If $a(\chi)$ are the Fourier coefficients for a function in A_0 and $| b(\chi) | \leqq | a(\chi) |$ for all χ, then there exists a function in A_0 whose Fourier coefficients are $b(\chi)$.*

(ii) *If $1 < p \leqq 2$, then $A_0 = L^2(G)$ (algebraically and topologically).*

PROOF. Let $x \in A_0$ and $z \in L^q(G)$ where $1/p + 1/q = 1$. Then the linear form $F(y) = \int_G \{x * y(t)\} z(t^{-1})\, dt$ on $L^p(G)$ satisfies the inequalities $| F(y) | \leqq \| x * y \|_p \| z \|_q \leqq \| x \|_0 \| z \|_q \| \hat{y} \|_\infty$ and there exists therefore a bounded measure μ on the discrete space $\hat{G}$ such that

$$\int_G \{x * y(t)\} z(t^{-1})\, dt = \sum_{\chi \in G} \mu(\chi) \hat{y}(\chi) \qquad \text{for all } y \in L^p(G).$$

On interchanging intergrations and putting $y(t) = (t, \chi)$ we see that the function $x * z(t)$ has the Fourier series $\sum \mu(\chi)(t, \chi)$ which is absolutely convergent since μ is a bounded measure. Denoting by AC the set of functions in $L^1(G)$ with absolutely convergent Fourier series we can express the result by

$$A_0 * L^q \subset AC. \tag{3.8}$$

Now suppose $f \in A_0$ has the Fourier series $f(t) \sim \sum a(\chi)(t, \chi)$ and that the function $| b(\chi) |$ is majorized by $| a(\chi) |$. For each $h \in L^q(G)$ with Fourier series $h(t) \sim \sum c(\chi)(t, \chi)$ we have by (3.8) $\sum | a(\chi) c(\chi) | < \infty$, hence

$$\sum | b(\chi) c(\chi) | < \infty. \tag{3.9}$$

Clearly $b(\chi)$ vanishes except for a countable number of χ, say $\chi_1, \chi_2, \cdots$. Putting $g_N(t) = \sum_1^N b(\chi_\nu)(t, \chi_\nu)$ we see from the equation

$$\int_G g_N(t)h(t^{-1})\, dt = \sum_1^N b(\chi_\nu)c(\chi_\nu)$$

in connection with (3.9), that $y_N(t)$ is a weakly convergent sequence in $L^p(G)$. Since $L^p(G)$ is reflexive, hence weakly complete, there exists a function $g(t)$ in $L^p(G)$ such that

$$\lim_{N\to\infty} \int_G g_N(t)h(t^{-1})\, dt = \int_G g(t)h(t^{-1})\, dt \quad \text{for each } h \in L^q(G).$$

The function $g(t)$ has the Fourier series $\sum b(\chi)(t, \chi)$. Since $f(t) \sim \sum a(\chi)(t, \chi)$ belongs to A_0 we have $\| f * x \|_p \leqq \| f \|_0 \| \hat{x} \|_\infty$ for every $x \in A$. For an arbitrary trigonometric polynomial $q(t) \sim \sum^* \hat{q}(\chi)(t, \chi)$ we then obtain

$$\| g * q \|_p = \| f * \textstyle\sum^* b(\chi)[a(\chi)]^{-1}\hat{q}(\chi)(t, \chi) \|_p \leqq \| f \|_0 \| ba^{-1}\hat{q} \|_\infty \leqq \| f \|_0 \| \hat{q} \|_\infty .$$

But since the trigonometric polynomials are dense in $L^p(G)$ we get $\| g * y \|_p \leqq \| f \|_0 \| \hat{y} \|_\infty$ for every $y \in L^p(G)$. Hence $g \in A_0$ and the first part of Theorem 7 is proved.

For the second part observe first that if $1 < p \leqq 2$,

$$\| x * y \|_p \leqq \| x * y \|_2 \leqq \| x \|_2 \| \hat{y} \|_\infty \qquad \text{for all } x, y \in L^p(G),$$

since $\| \hat{y} \|_\infty$ is independent of whether y is considered as an element in $L^2(G)$ or $L^p(G)$. Hence $\| x \|_0 \leqq \| x \|_2$ and $L^2(G) \subset A_0$. On the other hand, by (3.3)

$$\sup_{y\in L^p} \frac{\| x * y \|_p}{\| y \|_\infty} \geqq \sup_{y \in L^p} \frac{\| x * y \|_1}{\| y \|_\infty} = \sup_{y\in L^1} \frac{\| x * y \|_1}{\| \hat{y} \|_\infty} \geqq 2^{-\frac{1}{2}} \| x \|_2$$

so $A_0 \subset L^2(G)$. Hence the identity mapping is an algebraic isomorphism between A_0 and $L^2(G)$ and this mapping is a homeomorphism due to a theorem of Gelfand [2], stating that the norm in a semi-simple Banach algebra is uniquely determined up to equivalence.

REMARK. A more explicit description of A_0 will be given in Section 5.

4. Representation theorems

DEFINITION. If A is a Banach algebra with maximal ideal space $\mathfrak{M}$ we call a complex valued function f on $\mathfrak{M}$ a *multiplier* of A if $f\hat{x} \in \hat{A}$ for each $\hat{x} \in \hat{A}$.

The algebra of multipliers is the largest algebra of continuous functions on $\mathfrak{M}$ in which $\hat{A}$ is an ideal. If A is semi-simple, every multiplier f of A induces in an obvious way a linear transformation T_f of A into itself and by the closed graph theorem it is easy to see that T_f is continuous. The operators T_f form a Banach algebra $\mathbf{A}$ (under the natural norm) in which A is imbedded as an ideal and accordingly $\mathfrak{M}$ is topologically imbedded in $\mathbf{M}$.

In most non-trivial cases the multipliers will be bounded functions,[4] i.e. form a

[4] This is always the case if A is semi-simple as pointed out to the author by H. Mirkil.

subalgebra of $\mathcal{C}^{\infty}(\mathfrak{M})$. In the examples for the derived algebra A_0 in the preceding section we have the extreme case, namely the multipliers fill up the entire $\mathcal{C}^{\infty}(\mathfrak{M}_0)$. Now we shall show that more general derived algebras have this property.

DEFINITION. We shall call a Banach space B *weakly complete* if for every sequence $x_n \in B$ for which $F(x_n)$ converges for all $F \in B'$ (the dual of B) there exists an element $x \in B$ such that $F(x_n)$ converges to $F(x)$ for all $F \in B'$.

THEOREM 8. *Suppose the algebra A is weakly complete and let A_0 be the derived algebra of A. Then there exists a family of positive Radon measures μ_k on $\mathfrak{M}_0$ such that*[5]

$$\hat{A}_0 = \bigcap_k L^1(\mu_k) \cap \mathcal{C}(\mathfrak{M}_0). \tag{4.1}$$

In particular, $\hat{A}_0$ is an ideal in $\mathcal{C}^{\infty}(\mathfrak{M}_0)$.

The proof depends mainly on the following lemma which is true without the weak completeness assumption on A.

LEMMA 1. *Let A_0 be the derived algebra of A. To each continuous linear functional F on A_0 there corresponds a unique complex Radon measure μ_F on $\Delta_0 = \mathfrak{M}_0$ such that for every x, $y \in A_0$*

$$F(xy) = \int_{\Delta_0} h_0(xy)\, d\mu_F(h_0). \tag{4.2}$$

Furthermore, the function $\mathring{x}$ is integrable[6] *with respect to each μ_F, and if $\mathring{x}$ has compact support then*

$$F(x) = \int_{\Delta_0} h_0(x)\, d\mu_F(h_0). \tag{4.3}$$

PROOF. We have for x, $y \in A_0$

$$\| xy \|_0 = \sup_z \frac{\| xyz \|}{\| \hat{z} \|_\infty} \leqq \sup_z \frac{\| xyz \|}{\| \hat{y}\hat{z} \|_\infty} \| \hat{y} \|_\infty \leqq \| x \|_0 \| \hat{y} \|_\infty .$$

Hence, by theorem 4 (vi)

$$\| xy \|_0 \leqq \| x \|_0 \| \mathring{y} \|_\infty . \tag{4.4}$$

If F is a continuous linear functional on A_0 we have $| F(xy) | \leqq \| F \| \| xy \|_0 \leqq \| F \| \| x \|_0 \| \mathring{y} \|_\infty$. Therefore, for fixed F and x, $F(xy)$ is a continuous functional on $\hat{A}_0$ and since A_0 is self adjoint and $\hat{A}_0$ therefore dense in $\mathcal{C}(\mathfrak{M}_0)$ there exists a unique bounded Radon measure μ_x on $\mathfrak{M}_0$ such that

$$F(xy) = \int_{\mathfrak{M}_0} \mathring{y}(M_0)\, d\mu_x(M_0) \qquad \text{for all } y \in A_0 \tag{4.5}$$

[5] Here $L^1(\mu)$ denotes a space of functions but not of equivalence-classes of functions.

[6] "Integrable with respect to μ_F" means integrable with respect to each component in the Jordan decomposition of μ_F.

and

(4.6) $$\| \mu_x \| \leqq \| F \| \, \| x \|_0 .$$

Then obviously

$$F(xyz) = \int_{\mathfrak{M}_0} \mathring{y}\mathring{x} \, d\mu_z = \int_{\mathfrak{M}_0} \mathring{y}\mathring{z} \, d\mu_x$$

and by the uniqueness

(4.7) $$\mathring{x}\mu_z = \mathring{z}\mu_x \qquad \text{for } x,\ z \in A_0 .$$

Now we consider the set $S_x = \{M_0 \mid \mathring{x}(M_0) \neq 0\}$. S_x is open, hence a locally compact subspace of $\mathfrak{M}_0$. For every $\phi \in \mathcal{K}(S_x)$ the integral

(4.8) $$\int_{S_x} \phi(M_0)[\mathring{x}(M_0)]^{-1} \, d\mu_x(M_0)$$

exists. Let $\mathcal{K}_K(S_x)$ denote the subspace of $\mathcal{K}(S_x)$ consisting of those functions that vanish outside the subset $K \subset S_x$. For any such compact set K, the integral (4.8) is continuous on $\mathcal{K}_K(S_x)$ in the topology of uniform convergence and therefore defines a complex Radon measure μ_F^x on S_x such that

(4.9) $$\int_{S_x} \phi(M_0)[\mathring{x}(M_0)]^{-1} \, d\mu_x(M_0) = \int_{S_x} \phi(M_0) \, d\mu_F^x(M_0) \qquad \text{for all } \phi \in \mathcal{K}(S_x).$$

When x runs through A_0, the sets S_x constitute an open covering of $\mathfrak{M}_0$. For every pair x, $z \in A_0$, the measures μ_F^x and μ_F^z have identical restrictions to $S_x \cap S_z$ as follows easily from (4.7). By the principle of localization (Bourbaki [1], p. 70) there exists a unique Radon measure μ_F on $\mathfrak{M}_0$ whose restriction to S_x is μ_F^x for every $x \in A_0$. Now

$$F(xx) = \int_{\mathfrak{M}_0} \mathring{x} \, d\mu_x = \int_{S_x} \mathring{x} \, d\mu_x$$

and from (4.9) we obtain

$$\int_{S_x} \phi \, d\mu_x = \int_{S_x} \phi\mathring{x} \, d\mu_F \qquad \text{for all } \phi \in \mathcal{K}(S_x)$$

and since $\mathring{x}$ is a uniform limit on S_x of such ϕ we conclude that

$$F(xx) = \int_{S_x} \mathring{x} \, d\mu_x = \int_{S_x} \mathring{x}\mathring{x} \, d\mu_F = \int_{\mathfrak{M}_0} \mathring{x}\mathring{x} \, d\mu_F .$$

Using the identity $4xy = (x + y)^2 - (x - y)^2$, (4.2) follows immediately.

If $\mathring{x}$ has compact support, we can, using (iv) Theorem 4, write $\mathring{x} = \mathring{x}\mathring{y}$ for a suitable $y \in A_0$. Clearly $x = xy$ and (4.3) follows. This argument also proves the uniqueness of μ_F since $\mathcal{K}(\mathfrak{M}_0) \subset A_0$.

Let $x \to x^*$ be the involution on A_0 that corresponds to complex conjugation of $\hat{A}_0$. As remarked earlier this involution is continuous and the linear functional

$F^*:x \to \overline{F(x^*)}$ is continuous on A_0. From the uniqueness of μ_F just proved it follows that $\mu_{F^*} = \bar{\mu}_F$ and $\mu_{F+G} = \mu_F + \mu_G$. To prove that $\mathring{x}$ is integrable with respect to μ_F we may assume that $\mathring{x}$ is real valued and μ_F a real measure due to the decompositions $2x = x + x^* - i(i(x - x^*))$ and $2\mu_F = \mu_{F+F^*} - i\mu_{i(F-F^*)}$. The Jordan decomposition of μ_F is $\mu_F = \mu^+ - \mu^-$ where μ^+ and μ^- are positive measures defined for every Borel set M by $\mu^+(M) = \sup \mu_F(N)$ and $-\mu^-(M) = \inf \mu_F(N)$, N ranging over all Borel subsets of M. It is easy to prove that $\sup \mu^{\pm}(C) = \mu^{\pm}(M)$ where C varies over the compact subsets of M. The set S_x can be written as a countable union of increasing compact sets $C_n (n = 0, 1, 2, \cdots)$, C_0 denoting the empty set. Each set $C_n - C_{n-1}$ has finite μ^+ (and μ^-) measure and can by the Hahn decomposition theorem be written $C_n - C_{n-1} = E_n^+ \cup E_n^-$ where E_n^+ and E_n^- are disjoint Borel sets for which $\mu^+(E_n^-) = 0$ and $\mu^-(E_n^+) = 0$. We put $E^+ = \bigcup E_n^+$ and $E^- = \bigcup E_n^-$. We shall prove that $\mathring{x}$ is integrable with respect to μ^+ and μ^-. If for example $\int |\mathring{x}| \, d\mu^+$ were infinite we could find a compact set $C \subset E^+$ such that

$$\int_C |\mathring{x}| \, d\mu^+ = \int_C |\mathring{x}| \, d\mu_F \geqq 2 \| x \|_0 \| F \|$$

and then a continuous function ϕ with $\| \phi \|_\infty < 2$ and vanishing outside C such that

$$\left| \int_{\mathfrak{M}_0} \mathring{x}\phi \, d\mu_F \right| \geqq \int_C |\mathring{x}| \, d\mu_F$$

contradicting the estimate $\left| \int \phi\mathring{x} \, d\mu_F \right| \leqq \| \phi \|_\infty \| x \|_0 \| F \|$. The lemma is hereby proved.

We now proceed to prove Theorem 8. The measures μ_F from the lemma have Jordan decompositions, say $\mu_F = \overset{1}{\mu}_F - \overset{2}{\mu}_F + i\overset{3}{\mu}_F - i\overset{4}{\mu}_F$, into positive measures. The measures $\overset{1}{\mu}_F + \overset{2}{\mu}_F + \overset{3}{\mu}_F + \overset{4}{\mu}_F$ will serve as our measures μ_k in (4.1). In fact, the relation $\hat{A}_0 \subset \bigcap_k L^1(\mu_k) \cap \mathfrak{C}(\mathfrak{M}_0)$ is contained in the lemma and to prove the converse inclusion suppose that $g \in \mathfrak{C}(\mathfrak{M}_0)$ is integrable with respect to all $\overset{1}{\mu}_F + \overset{2}{\mu}_F + \overset{3}{\mu}_F + \overset{4}{\mu}_F$. It is clear from Theorem 4 (iv) that there exists a sequence $x_n \in A_0$ such that each function $\mathring{x}_n$ has compact support and $\mathring{x}_n$ converges to g uniformly on $\mathfrak{M}_0$ and $|\mathring{x}_n(M_0)| \leqq |g(M_0)|$ for all M_0 and all n. By the Lebesgue theorem on dominated convergence

$$\int_{\mathfrak{M}_0} g(M_0) \, d\mu_F = \lim_{n\to\infty} \int_{\mathfrak{M}_0} \mathring{x}_n(M_0) \, d\mu_F = \lim_{n\to\infty} F(x_n)$$

for every continuous linear functional F on A_0. Since every continuous linear functional on A is continuous on A_0 we see from the above, that (x_n) is a weakly convergent sequence in A, and by the assumption on A there exists an element $y \in A$ such that $F(x_n) \to F(y)$ for all $F \in A'$. (A' denotes the dual space of A). Clearly $\hat{y}$ restricted to $\mathfrak{M}_0$ is identical to g. Now let $f \in \mathfrak{C}(\mathfrak{M})$ and $F \in A'$, and as before we denote by x^f the element in A (if it exists) for which $(x^f)^\wedge = \hat{x}f$. Then $x \to F(x^f)$ is by Theorem 2 a continuous linear functional on A_0 since $\| x^f \| \leqq \| x \|_0 \| f \|_\infty$. As (x_n) converges weakly in A_0, we see that $F(x_n^f)$ converges for each $F \in A'$. The weak limit of x_n^f exists and is readily seen to be equal to y^f. By Theorem 2 y belongs to A_0 and Theorem 8 is proved.

Corollary 1. *If A is separable and weakly complete then $\mathfrak{M}_0$ is discrete.*

Proof. Clearly $\mathfrak{M}$ and $\mathfrak{M}_0$ are separable. Let M_0 be an arbitrary point in $\mathfrak{M}_0$ and $K_1, K_2, \cdots$ a sequence of compact sets in $\mathfrak{M}_0$ such that for $n = 2, 3, \cdots$, K_n is contained in the interior of K_{n-1} and such that $M_0 = \bigcap K_n$. Choose for each n an element x_n in A_0 such that $\mathring{x}_n \equiv 1$ on K_{n+1} and $\mathring{x}_n \equiv 0$ on K_n. The sequence x_n converges weakly in A_0, and a fortiori converges weakly in A, and since A is weakly complete we see that the point M_0 is open in $\mathfrak{M}$, hence open in $\mathfrak{M}_0$.

Remark. The conclusions of Corollary 1 does not hold if the weak completeness assumption is dropped. Example: Let μ be a positive Radon measure on a locally compact space S and let $A = L^1(\mu) \cap C(S)$ with norm $\| x \| = \| x \|_1 + \| x \|_\infty$. Here $\mathfrak{M}_0 = S$.

Corollary 2. *If G is an abelian separable locally compact non-compact group then the derived algebra of $L^1(G)$ is 0.*

Proof. If the derived algebra A_0 of $L^1(G)$ were different from 0 it is easy to see that $\mathfrak{M}_0$ would be homeomorphic to $\hat{G}$ and Corollary 1 would contradict Pontrjagin's duality theorem.

Remark 1. It can be proved similarly that the statement of the above corollary remains valid if the hypothesis that G is separable is replaced by the assumption that the character group $\hat{G}$ is connected.

Remark 2. We have seen in Theorem 5 (i) that if A is the group algebra $L^1(G)$ of a compact abelian group G, then $A_0 = L^2(G)$ and $\hat{A}_0 = L^2(\mathfrak{M}_0)$. From Theorem 8 we see that $L^2(\mathfrak{M}_0)$ in this case occurs as an intersection of L^1-spaces but not as a Hilbert space, which is reflected in the previously mentioned fact, that the norm of A_0 is in general not identical to the Hilbert space norm of $L^2(G)$.

Theorem 9. *Let A_0 be the derived algebra of A and suppose that the linear span of A_0^2 is dense in A_0. Then $\hat{A}_0$ is an ideal in $\mathfrak{C}^\infty(\mathfrak{M}_0)$.*

Proof. We first prove that the set of elements z in A_0 for which $\mathring{z}$ has compact support is dense in A_0. In fact, if $x \in A_0$ we can to each $\varepsilon > 0$ find a $y \in A_0$ such that $\mathring{y}$ has compact support and $\| \mathring{x} - \mathring{y} \|_\infty < \varepsilon / \| x \|_0$. Then $\| xy - x^2 \|_0 \leqq \| x \|_0 \| \mathring{y} - \mathring{x} \|_\infty < \varepsilon$ and the statement follows.

Now to prove Theorem 9 suppose $x \in A_0$ and $f \in \mathfrak{C}^\infty(\mathfrak{M}_0)$. Let x_n be a sequence in A_0 converging to x such that each $\mathring{x}_n$ has compact support. For each n there exists an element $z_n \in A_0$ such that $\mathring{z}_n = f\mathring{x}_n$. For every $F \in A_0'$ (the dual space of A_0) we use Lemma 1 and consider the linear functional

$$G(y) = \int_{\mathfrak{M}_0} \mathring{y} f \, d\mu_F$$

which is continuous on A_0 in virtue of (4.6). The mapping T_f that sends F into G is a linear mapping of A_0' into itself. Also T_f is continuous as follows easily from the closed graph theorem. In fact, let $F_n \to F$ and $T_f(F_n) \to H$. Using (4.6) again we get the relation

$$\left| \int f\mathring{y}\, d\mu_{F_n} - \int f\mathring{y}\, d\mu_F \right| = \left| \int f\mathring{y}\, d\mu_{F_n - F} \right| \leqq \| f \|_\infty \| y \|_0 \| F_n - F \|$$

which together with $\int f\mathring{y}\, d\mu_{F_n} \to H(y)$ implies $T_f(F) = H$, and hence the continuity of T_f. The adjoint T_f^* is an endomorphism of the second dual space A_0''. We imbed A_0 isometrically into A_0'' and consider the effect of T_f^* on A_0. If $x'' \in A_0''$ and $F \in A_0'$ we denote by $\langle x'', F \rangle$ the value of the linear form x'' at F. We have then

$$\langle T_f^*(x_n), F \rangle = \langle x_n, T_f(F) \rangle = \int f\mathring{x}_n\, d\mu_F = \int \mathring{z}_n\, d\mu_F = \langle z_n, F \rangle$$

for each $F \in A_0'$. It follows that $T_f^*(x_n) = z_n$. The sequence z_n converges to the element $z = T_f^*(x)$ which belongs to A_0, and $z = f\mathring{x}$.

5. Concluding remarks

We formulate some classical facts about Fourier series in $L^p(0, 2\pi)$ which (together with the generalizations to compact abelian groups) are contained in the preceding theory.

THEOREM 10. *Let* $1 \leqq p \leqq \infty$. *We denote by* $(L^p)^*$ *the set of functions* $x(t) \sim \sum a_n e^{itn}$ in $L^p(0, 2\pi)$ *that remain in* $L^p(0, 2\pi)$ *after every change in the arguments of the Fourier coefficients* (a_n). *Then*

(i) $(L^p)^* = L^2(0, 2\pi)$ *if*[7] $1 \leqq p \leqq 2$

(ii) $(L^\infty)^* = AC$

(iii) *For each fixed* $q (2 < q < \infty)$ *there exists a system of sequences* μ_n^k *such that* $\sum | a_n \mu_n^k | < \infty$ *for all* k *is a necessary and sufficient condition*[8] *for* $x(t) \in (L^q)^*$.

PROOF. (iii): $(L^q)^* \subset (L^q)_0$ follows easily from Theorem 2 and the converse inclusion is contained in Theorem 7. Since L^q is weakly complete Theorem 8 applies. (i) and (ii) are obtained similarly.

It is interesting to note that for $1 \leqq p \leqq 2$ and $p = \infty$ the possibility of arbitrary change of arguments of (a_n) without $x(t) \sim \sum a_n e^{itn}$ leaving $L^p(0, 2\pi)$ implies that arbitrary rearrangement of the coefficients (a_n) is also permitted. Whether the same is true for $2 < q < \infty$ is apparently unknown.

MASSACHUSETTS INSTITUTE OF TECHNOLOGY

[7] *Added in proof:* Theorem (i) is due to Littlewood. Another proof has been shown to me by Dr. A. Grothendieck, see his notes, C. R. Acad. Sci. Paris, 1954, pp. 577–579 and pp. 607–609. Stronger versions of (i) are known, with probabilistic interpretations, see e.g. Zygmund [8] p. 125 and for the case of almost periodic functions Jessen [3]. Theorem 5 in the present paper gives a sharpening of (i) in another direction.

[8] See Kaczmarz and Steinhaus [4].

References

1. N. Bourbaki, Intégration, Chap. I-IV, Paris, 1952.
2. I. Gelfand, *Normierte Ringe*, Rec. Math. (Mat. Sbornik) N.S., 9 (1941), pp. 3–24.
3. B. Jessen, *The theory of integration in infinite number of dimensions*, Acta Math., 63 (1934), pp. 249–232.
4. S. Kaczmarz, and H. Steinhaus, Theorie der Orthogonalreihen, Warsaw, 1935.
5. L. Loomis, Abstract harmonic analysis, New York, 1953.
6. I. Segal, *The class of functions which are absolutely convergent Fourier transforms*, Acta Sci. Math. Szeged., 12 (1950), pp. 157–161.
7. A. Weil, L'intégration dans les groupes topologiques et ses applications, Paris, 1940.
8. A. Zygmund, Trigonometrical series, Warsaw, 1935.

DIFFERENTIAL OPERATORS ON HOMOGENEOUS SPACES

BY

SIGURÐUR HELGASON

Chicago

Tileinkað foreldrum mínum

Introduction. Among all linear differential operators in Euclidean space $\mathbf{R}^n$, those that have constant coefficients are characterized by their invariance under the transitive group of all translations. The special role played by Laplace's equation is partly due to its invariance under all rigid motions. Another example of physical importance is the wave equation which can essentially be characterized by its invariance under the Lorentz group. This implicit physical significance of the Lorentz group so far as electromagnetic phenomena is concerned is made explicit in Einstein's special theory of relativity. Here the Lorentz group is given an interpretation in terms of pure mechanics.

In the present paper a study is made of differential operators on a manifold under the assumption that these operators are invariant under a transitive group G of "automorphisms" of this manifold M. Let p be a point of M, H the subgroup of G leaving p fixed and M_p the tangent space to M at p. It is easy to set up a linear correspondence between the set of invariant differential operators on M and the set of all polynomials on M_p that are invariant under the action of the isotropy group H at p. However, the multiplicative properties of this correspondence are complicated and are better understood (at least in case G/H is reductive) by describing the differential operators by means of the Lie algebras of G and H (Theorem 10).

Our purpose is to study various geometrical properties of solutions of differential equations involving these invariant operators. We give now a summary of the different chapters.

Chapter I contains a general discussion of linear differential operators on manifolds. On pseudo-Riemannian manifolds there is always one differential operator, the Laplace-

16 – 593805. *Acta mathematica.* 102. Imprimé le 16 décembre 1959

Beltrami operator, which is invariant under all isometries but under no other diffeomorphisms.

Chapter II. In § 1 we recall some essentially known results on transitive transformation groups and homogeneous spaces. Two-point homogeneous spaces admit essentially only one invariant differential operator, the Laplace-Beltrami operator. Potential theory has a particularly explicit character. In § 3 we prove some properties of these spaces which are used later, e.g., the symmetry of non-compact two-point homogeneous spaces. A fairly direct proof of this fact is possible, but for the compact spaces such a proof seems to be unknown although the symmetry can be verified by means of Wang's classification. In § 4 we investigate in some detail Lorentzian spaces of constant curvature and the behavior of the geodesics on these spaces. For the spaces of negative curvature (simply connected) the timelike geodesics through a given point are infinite and do not intersect each other. The spaces of positive curvature that we consider have infinite cyclic fundamental group. Their timelike geodesics through a given point are all closed and do not intersect each other.

Chapter III. In § 1 we represent the algebra $\mathbf{D}(G/H)$ of invariant differential operators by means of the symmetric invariants of the group $Ad_G(H)$. Thus if H is semi-simple, $\mathbf{D}(G/H)$ has a finite system of generators. If G/K is a Riemannian symmetric space, $\mathbf{D}(G/K)$ is finitely generated and commutative (Gelfand [11], Selberg [36]). For Lorentz spaces of constant curvature (or two-point homogeneous spaces) $\mathbf{D}(G/H)$ is generated by the Laplace-Beltrami operator.

Chapter IV. We consider in § 1 the mean value operators M^x which in a natural way generalize the operation M^r of averaging over spheres in $\mathbf{R}^n$ of fixed radius r. It is well known that M^r is formally a function (Bessel function) of the Laplacian Δ. The analogue holds for the space G/K if K is compact. In fact M^x is formally a function of the generators $D^1, \ldots, D^l$ of $\mathbf{D}(G/K)$. This has some applications, for example a generalization of the mean value theorem of Ásgeirsson. For two-point homogeneous spaces we obtain more explicit results, for example a simple geometric solution of Poisson's equation. In § 4 is given for a Riemannian space of constant curvature a decomposition of a function into integrals over totally geodesic submanifolds. A somewhat analogous problem is treated in § 7 for a Lorentzian space of constant curvature. Here a function is represented by means of its integrals over Lorentzian spheres. We use here methods of analytic continuation introduced by M. Riesz in his treatment of the wave equation. In § 8 we verify that Huygens' principle in Hadamard's formulation is absent for non-flat harmonic Lorentz spaces.(1)

(1) This confirms, in a very special case, a well-known conjecture attributed to Hadamard.

An outline of the results of this paper (with the exception of Ch. II, § 4 and Ch. IV, §§ 5–8) was given [25] at the Scandinavian Mathematical Congress in Helsinki, August 1957. An exposition was given in a course at the University of Chicago, Spring 1958. I am grateful to Professor Ásgeirsson for advice concerning some problems dealt with in Chapter IV, § 3. I am also grateful to Professor Harish-Chandra for interesting conversations about the topic of his paper [21].

Contents

CHAPTER I

Preliminary remarks on differential operators

Let M be a locally connected topological space with the property that each connected component of M is a differentiable manifold of class C^∞ and dimension n. We shall then say that M is a C^∞-manifold of dimension n. We shall only be dealing with separable C^∞-manifolds and will simply refer to them as manifolds. If p is a point on the manifold M, the tangent space to M at p will be denoted by M_p. The set of real valued indefinitely differentiable functions on M constitutes an algebra $C^\infty(M)$ over the real numbers $\mathbf{R}$, the multiplication in $C^\infty(M)$ being given by pointwise multiplication of functions. The functions in $C^\infty(M)$ that have compact support form a subalgebra $C_c^\infty(M)$. We use the topology on $C_c^\infty(M)$ which is familiar from the theory of distributions (L. Schwartz [33] I, p. 67), and is based on uniform convergence of sequences of functions and their derivatives. The derivations $\mathfrak{D}$ of the algebra $C^\infty(M)$ are the C^∞-vector fields on M; each $X \in \mathfrak{D}$ leaves $C_c^\infty(M)$ invariant. An endomorphism of a vector space V is a linear mapping of V into itself. If D is an endomorphism of $C^\infty(M)$ and $f \in C^\infty(M)$ then $[Df](p)$ will always denote the value of Df at $p \in M$. If $X \in \mathfrak{D}$, then the linear functional X_p on $C^\infty(M)$ defined by $X_p(f) = [Xf](p)$ for $f \in C^\infty(M)$ is a tangent vector([1]) to M at p, that is $X_p \in M_p$. Let $\mathbf{R}^n$ denote the Euclidean n-space with a fixed coordinate system. If the mapping $\Psi : x \to (x_1, \ldots, x_n) \in \mathbf{R}^n$ is a local coordinate system valid in an open subset $U \subset M$, we shall often write f^* for the composite function $f \circ \Psi^{-1}$ defined on $\Psi(U)$. We also write D_i for the partial differentiation $\partial/\partial x_i$ and if $\alpha = (\alpha_1, \ldots, \alpha_n)$ is an n-tuple of indices $\alpha_i \geqslant 0$ we put $D^\alpha = D_n^{\alpha_1} \ldots D_n^{\alpha_n}$ and $|\alpha| = \alpha_1 + \cdots + \alpha_n$.

DEFINITION. A continuous endomorphism D of $C_c^\infty(M)$ is called a *differential operator* on M if it is of local character. This means that whenever U is an open set in M and $f \in C_c^\infty(M)$ vanishes on U, then Df vanishes on U.

PROPOSITION 1.([2]) *Let D be an endomorphism of $C_c^\infty(M)$ which has the following property. For each $p \in M$ and each open connected neighborhood U of p on which the local coordinate system $\Psi : x \to (x_1, \ldots, x_n)$ is valid there exists a finite set of functions a_α of class C^∞ such that for each $f \in C_c^\infty(M)$ with support contained in U*

$$[Df](x) = \sum_\alpha a_\alpha(x) [D^\alpha f^*](x_1, \ldots, x_n) \qquad \textit{for } x \in U$$

$$[Df](x) = 0 \qquad \textit{for } x \notin U.$$

Then D is a differential operator on M and each differential operator on M has the property above.

Proof. Let E be a differential operator, p, U and Ψ as above. Let V be an open subset of U whose closure $\bar{V}$ is compact and contained in U. Let $C_{\bar{V}}(M)$ and $C_V(M)$ denote the the set of functions $f \in C_c^\infty(M)$ with compact support contained in $\bar{V}$ and V respectively. The operator E induces a continuous endomorphism of $C_{\bar{V}}(M)$. This implies that for each $\varepsilon > 0$ there exists an integer m and a real number $\delta > 0$ such that

([1]) We use here and often in the sequel the terminology of Chevalley [7].

([2]) This proposition is attributed to L. Schwartz in A. Grothendieck, *Sur les espaces de solutions d'une classe générale d'équations aux dérivées partielles*. *J. Analyse Math.* 2 (1953) 243–280.

$$|Ef(x)| < \varepsilon \qquad \text{for all } x \in \overline{V}$$

whenever $[D^\alpha f^*](x_1, \ldots, x_n)$ is in absolute value less than δ for all $(x_1, \ldots, x_n) \in \Psi(\overline{V})$ and all α satisfying $|\alpha| \leqslant m$. For a fixed point $x \in V$ we put $T(f^*) = [Ef](x)$ for all $f \in C_V(M)$. The linear functional $f^* \to T(f^*)$ is then a distribution on $\Psi(V)$ of order $\leqslant m$ in the sense of [33] I, p. 25. From the local character of E it follows that this distribution has support at the point $\Psi(x)$. Due to Schwartz' theorem on distributions with point supports (*loc. cit.* p. 99), $T(f^*) = [Ef](x)$ can be written as a finite sum

$$[Ef](x) = \sum_{|\alpha| \leqslant m} a_\alpha(x) [D^\alpha f^*](x_1, \ldots, x_n) \tag{1.1}$$

where the coefficients $a_\alpha(x)$ are certain constants. Each constant $a_\alpha(x)$ varies differentiably with x as is easily seen by choosing f such that $D^\alpha f^*$ is constant in a neighborhood of $\Psi(x)$. The representation (1.1) holds for all $x \in V$ and all $f \in C_V(M)$. However, since U can be covered by a chain of open sets each of which has compact closure it is easily seen that (1.1) is valid for all $x \in U$ and all $f \in C_c^\infty(M)$ with support contained in U.

On the other hand, let D be an endomorphism of $C_c^\infty(M)$ with the properties described in the proposition. D is obviously of local character. Also D is continuous on the subspace of functions that have support inside a fixed coordinate neighborhood. Using the well-known technique of partition of unity (see for example [7], p. 163), D is seen to be continuous on the entire $C_c^\infty(M)$.

A differential operator on M can be extended to an endomorphism of $C^\infty(M)$ such that the condition of local character holds for all $f \in C^\infty(M)$. This extension is unique.

Let Φ be a homeomorphism of M onto itself such that Φ and Φ^{-1} are differentiable mappings. The mapping Φ is then called a diffeomorphism of M. If $p \in M$, the differential $d\Phi_p$ maps M_p onto $M_{\Phi(p)}$ in such a way that $d\Phi_p(X_p) f = X_p(f \circ \Phi)$. For each C^∞-vector field X on M we obtain a new vector field X^Φ by putting $X^\Phi f = (X(f \circ \Phi)) \circ \Phi^{-1}$ for $f \in C^\infty(M)$. It follows then that $(X^\Phi)_{\Phi(p)} = d\Phi_p(X_p)$ and we often write $d\Phi \cdot X$ instead of X^Φ. If A is an endomorphism of $C^\infty(M)$ we define the operator A^Φ in accordance with the notation above by $A^\Phi(f) = (A(f \circ \Phi)) \circ \Phi^{-1}$. If D is a differential operator on M, then so is the operator D^Φ. The transformation Φ is said to leave D *invariant* if $D^\Phi = D$. We sometimes write f^Φ for the composite function $f \circ \Phi^{-1}$. We have then the convenient rule $A^\Phi f = (A f \Phi^{-1})^\Phi$.

Let M be a connected manifold. A *linear connection* on M is a rule which assigns to each $X \in \mathfrak{D}$ a linear operator ∇_X on $\mathfrak{D}$ satisfying the following two conditions

(i) $$\nabla_{fX+gY} = f\nabla_X + g\nabla_Y$$

(ii) $$\nabla_X(fY) = f\nabla_X(Y) + (Xf)Y$$

for $f, g \in C^\infty(M)$, $X, Y \in \mathfrak{D}$. The operator ∇_X is called *covariant differentiation* with respect to X. This definition of a linear connection is adopted in K. Nomizu [33], and we refer to this paper for a treatment of concepts in the theory of linear connections such as parallelism, curvature and torsion tensor. A curve in M is called a *path* if it has a parameter representation such that all its tangent vectors are parallel. Let p be a point in M and $X \neq 0$ a vector in M_p. There exists a unique parametrized path $t \to \gamma_X(t)$ such that $\gamma_X(0) = p$ and $\gamma_X'(t) = X$. The parameter t is called the *canonical parameter* with respect to X. We put $\gamma_0(t) = p$. The mapping $X \to \gamma_X(1)$ is a one-to-one C^∞-mapping of a neighborhood of 0 in M_p onto a neighborhood of p in M. This mapping is called Exp (the Exponential mapping at p) and will often be used in the sequel.

A pseudo-Riemannian metric Q on a connected manifold M is a rule which in a differentiable way assigns to each $p \in M$ a non-singular symmetric real bilinear form Q_p on the tangent space M_p. Since M is connected the signature of Q_p is the same for all p. If the signature is $+ + \cdots +$ we call M a Riemannian space; if the signature is $+ - - \cdots -$ we speak of a Lorentzian space, otherwise of a pseudo-Riemannian manifold. On a pseudo-Riemannian manifold there exists one and only one linear connection (the pseudo-Riemannian connection) satisfying the conditions: 1°. The torsion is 0. 2.° The parallel displacement preserves the inner product Q_p on the tangent spaces. In the case of a Riemannian space, arc length can be defined for all differentiable curves. The space can then be metrized by defining the distance between two points as the greatest lower bound of length of curves joining the two points. For a Lorentzian space where this procedure fails we adopt the following terminology from the theory of relativity. The cone C_p in the tangent space M_p given by $Q_p(X, X) = 0$ is called the *null cone* or the *light cone* in M_p with vertex p. A vector $X \in M_p$ is called *timelike, isotropic,* or *spacelike* if $Q_p(X, X)$ is positive, 0, or negative respectively. Similarly we use the terms timelike, isotropic, and spacelike for rays (oriented half lines) or unoriented straight lines in M_p. A timelike curve is a curve each of whose tangent vectors is timelike. Such curves have well-defined arc length. If a path has timelike tangent vector at a point, then all of its tangent vectors are timelike and the path is called a timelike path. A curve is said to have length 0 if all its tangent vectors are isotropic.

A diffeomorphism Φ of a pseudo-Riemannian manifold M is called an *isometry* if

$$Q_p(X, Y) = Q_{\Phi(p)}(d\Phi_p X, d\Phi_p Y)$$

for each $p \in M$ and each $X, Y \in M_p$. The group of all isometries of M will be denoted by $\mathbf{I}(M)$. Let U be an open neighborhood of p on which local coordinates $x \to (x_1, \ldots, x_n)$ are valid. We put

$$q_{ij}(x) = Q_x\left(\frac{\partial}{\partial x_i}, \frac{\partial}{\partial x_j}\right)$$

and define the functions $q^{jk}(x)$, $q(x)$ on U by

$$\sum_k q^{jk}(x)\, q_{jk}(x) = \delta_i^j, \quad q(x) = |\det(q_{ij}(x))|,$$

δ_i^j being Kronecker deltas. For each function $f \in C_c^\infty$ (M) we set

$$\Delta f = \frac{1}{\sqrt{q(x)}} \sum_k \frac{\partial}{\partial x_k}\left(\sum_i q^{ik}(x) \sqrt{q(x)} \frac{\partial f^*}{\partial x_i}\right).$$

The expression on the right is invariant under coordinate changes due to the classical transformation formulas for the functions q_{ij}. It is easily seen that Δ is a differential operator on M. It is called the Laplace-Beltrami operator.

PROPOSITION 2. *Let Φ be a diffeomorphism of M. Then Φ leaves the Laplace-Beltrami operator invariant if and only if it is an isometry.*

Proof. Let $p \in M$ and let U be a neighborhood of p on which local coordinates $x \to (x_1, \ldots, x_n)$ are given. Then $\Phi(U)$ is a neighborhood of the point $q = \Phi(p)$ and $y \to (y_1, \ldots, y_n)$ where $y = \Phi(x)$, $y_i = x_i$ $(i = 1, 2, \ldots, n)$ is a local coordinate system on $\Phi(U)$. We also have

$$d\,\Phi_x\left(\frac{\partial}{\partial x_i}\right) = \frac{\partial}{\partial y_i} \qquad (i = 1, 2, \ldots, n).$$

For each function $f \in C_c^\infty(M)$ we have

$$[(\Delta f)\Phi^{-1}](x) = [\Delta f](y) = \frac{1}{\sqrt{q(y)}} \sum_k \frac{\partial}{\partial y_k}\left(\sum_i q^{ik}(y) \sqrt{q(y)} \frac{\partial f^*}{\partial y_i}\right) \tag{1.2}$$

$$[\Delta f \Phi^{-1}](x) = \frac{1}{\sqrt{q(x)}} \sum_k \frac{\partial}{\partial x_k}\left(\sum_i q^{ik}(x) \sqrt{q(x)} \frac{\partial (f \circ \Phi)^*}{\partial x_i}\right). \tag{1.3}$$

Due to the choice of coordinates we have

$$\frac{\partial f^*}{\partial y_i} = \frac{\partial (f \circ \Phi)^*}{\partial x_i}, \quad \frac{\partial^2 f^*}{\partial y_i \partial y_k} = \frac{\partial^2 (f \circ \Phi)^*}{\partial x_i \partial x_k} \qquad (i, k = 1, 2, \ldots, n).$$

Now, if Φ is an isometry, then $q_{ik}(x) = q_{ik}(y)$ for all i, k so the expressions (1.2) and (1.3) are the same and $\Delta^\Phi = \Delta$. On the other hand, if (1.2) and (1.3) agree we obtain by equating coefficients, $q_{ik}(x) = q_{ik}(y)$ for all i, k, which shows that Φ is an isometry.

For Lorentzian spaces the Laplace-Beltrami operator will always be denoted by $\square$.

Chapter II

Homogeneous spaces

1. The analytic structure of a coset space

Let G be a separable Lie group and H a closed subgroup. The identity element of a group will always be denoted by e. Let $L(g)$ and $R(g)$ denote the left and right translations of G onto itself given by $L(g)\cdot x = gx$, $R(g)\cdot x = xg^{-1}$. The system G/H of left cosets gH has a unique topology with the property that the natural projection π of G onto G/H is a continuous and open mapping. This is called the *natural topology* of the coset space G/H. In this topology G/H is a locally connected Hausdorff space. For each $x \in G$, the mapping $\tau(x) : gH \to xgH$ is a homeomorphism of G/H onto itself. The connected components of G/H are all homeomorphic to $G_0/G_0 \cap H$ where G_0 is the identity component of G. The point $\pi(e)$ will usually be denoted by p_0. For later purposes we need Lemma 1 below which gives a special local cross-section in G, considered as a fiber bundle over G/H. The group H is a Lie group, regularly imbedded in G and thus the Lie algebra $\mathfrak{h}$ $(= H_e)$ of H can be regarded as a subalgebra of the Lie algebra $\mathfrak{g}$ $(= G_e)$ of G. We choose a fixed complementary subspace to $\mathfrak{h}$ in $\mathfrak{g}$ and denote it by $\mathfrak{m}$. Let exp denote the usual exponential mapping of $\mathfrak{g}$ into G and Ψ its restriction to $\mathfrak{m}$.

Lemma 1. *There exists a neighborhood U of 0 in $\mathfrak{m}$ which is mapped homeomorphically under Ψ and such that π maps $\Psi(U)$ homeomorphically onto a neighborhood of p_0 in G/H.*

Proof. Consider the mapping $\Phi : (X, Y) \to \exp X \cdot \exp Y$ of the product space $\mathfrak{m} \times \mathfrak{h}$ into G. We choose a basis $X_1, \ldots, X_r$ of $\mathfrak{g}$ such that the first n elements form a basis of $\mathfrak{m}$ and the $r-n$ last elements form a basis of $\mathfrak{h}$. Let $x_1, \ldots, x_r$ be a system of canonical coordinates with respect to this basis, valid in a neighborhood V' of e in G. For sufficiently small t_i the element

$$\exp(t_1 X_1 + \cdots + t_n X_n) \exp(t_{n+1} X_{n+1} + \cdots + t_r X_r)$$

belongs to V' and its canonical coordinates are given by $x_j = \xi_i(t_1, \ldots, t_r)$ where ξ_i are analytic functions in a neighborhood of 0 in $\mathfrak{g}$. The Jacobian determinant of the transformation $(t_1, \ldots, t_r) \to (x_1, \ldots, x_r)$ is $\neq 0$ in a neighborhood of 0 in $\mathfrak{g}$. There exists therefore a neighborhood of 0 in $\mathfrak{g}$ of the form $N_1 \times N_2$ where $N_1 \subset \mathfrak{m}$, $N_2 \subset \mathfrak{h}$ which Φ maps homeomorphically onto an open subset V'' of V'. Choose a neighborhood V of e such that $V^{-1} \cdot V \subset V''$. Let U be a compact neighborhood of 0 in $\mathfrak{m}$ contained in $\Phi^{-1}(V) \cap N_1$. Then Ψ maps U homeomorphically onto $\Psi(U)$. Also π maps $\Psi(U)$ in a one-to-one fashion because otherwise there exist $X_1, X_2 \in U$ and $h \in H$, $h \neq e$ such that $\exp X_1 = \exp X_2 h$. It follows that $h \in V''$ and there exists a $Y \in N_2$ such that $h = \exp Y$, $Y \neq 0$. The elements $(X_1, 0)$ and (X_2, Y) belong to $N_1 \times N_2$ and have the same image under Φ which is a contradiction. The set $\Psi(U)$, being compact, is mapped homeomorphically by π and the image $\pi(\Psi(U))$ is a neighborhood of p_0 in G/H because $\pi(\Psi(U)) = \pi(\Psi(U) H)$, $\Psi(U) H$ is a neighborhood of e in G and π is an open mapping.

THEOREM 1. *The coset space G/H has a unique analytic structure with the property that the mapping $(x, gH) \to xgH$ is a differentiable mapping of the product manifold $G \times G/H$ onto G/H. A coset space G/H will always be given this analytic structure.*

Proof. We use the terminology of Lemma 1 and introduce coordinates in G/H as follows. For each $p \in G/H$ we can find a $g \in G$ such that $\pi(g) = p$; let N_p denote the interior of the set $\pi(g\Psi(U))$. Then the mapping

$$\pi(g \exp(t_1 X_1 + \cdots + t_n X_n)) \to (t_1, \ldots, t_n)$$

is a system of coordinates valid on N_p. It is not difficult to show that this procedure defines an analytic structure on G/H with the property that the mapping $(x, gH) \to xgH$ is an analytic mapping of $G \times G/H$ onto G/H (Chevalley [7], p. 111). The uniqueness statement is contained in the following theorem.

THEOREM 2. *Let G be a separable, transitive Lie group of diffeomorphisms of a manifold M. Assume that the mapping $(g, q) \to g \cdot q$ of $G \times M$ onto M is continuous. Let p be a point on M and G_p the subgroup of G that leaves p fixed. Then G_p is closed and the mapping $g \cdot p \to gG_p$ is a diffeomorphism of M onto G/G_p in the analytic structure defined above.*

If M is connected, then G_0, the identity component of G, is transitive on M.

Proof. We first prove (following R. Arens, "Topologies for homeomorphism groups", *Amer. J. Math.* 68 (1946), 593–610) that the coset space G/G_p, in its natural topology, is homeomorphic to M. For this it suffices to prove that the mapping $\Phi: g \to g \cdot p$ of G onto M is an open mapping. Let V be a compact symmetric neighborhood of e in G; then there exists a sequence $(g_n) \in G$ such that $G = \bigcup_n g_n V$. Thus $M = \bigcup_n g_n V \cdot p$ and it follows by a category argument that at least one of the summands has an inner point. Hence $V \cdot p$ has an inner point, say $h \cdot p$ where $h \in V$. Then p is an inner point of $h^{-1} V \cdot p \subset V^2 \cdot p$ so Φ is an open mapping. In particular $\dim G/G_p = \dim M$.

Consider now the interior B of the subset $\Psi(U)$ from Lemma 1. B is a submanifold of G because $(t_1, \ldots, t_n)$ and $(t_1, \ldots, t_r)$ are local coordinates of the points

$$\exp\,(t_1 X_1 + \cdots + t_n X_n)$$

and
$$\exp\,(t_1 X_1 + \cdots + t_n X_n)\ \exp\ (t_{n+1} X_{n+1} + \cdots + t_r X_r)$$

in B and G respectively and thus the injection i of B into G is regular. By the definition of the analytic structure of G/G_p, π is a differentiable transformation of B onto an open subset N of G/G_p. Due to a theorem of S. Bochner and D. Montgomery ("Groups of differentiable and real or complex analytic transformations", *Ann. of Math.* 46 (1945), 685–694), the continuous mapping $(g, q) \to g \cdot q$ is automatically differentiable. The mapping $gG_p \to g \cdot p$ is a homeomorphism of N onto an open set in M and is differentiable since it is of the form $\Phi \circ i \circ \pi^{-1}$. To show that the inverse is differentiable we just have to show that the Jacobian of Φ at $g = e$ has rank equal to dim M. Let $\mathfrak{g}$ and $\mathfrak{h}$ denote the Lie algebras of G and G_p respectively. We shall prove that if $X \in \mathfrak{g}$ and $X \notin \mathfrak{h}$ then $(d\Phi)_e X \neq 0$, in other words the Jacobian of Φ at e has rank equal to $\dim \mathfrak{g} - \dim \mathfrak{h} = \dim M$. Suppose to the contrary that $(d\Phi)_e X = 0$; then if $f \in C^\infty(M)$ we have

$$0 = (d\,\Phi)_e X \cdot f = X_e\,(f \circ \Phi) = \frac{d}{d\,t} f\ (\exp\ tX \cdot p)\Big\}_{t=0}.$$

If we use this relation on the function $f^*\,(q) = f\ (\exp\ sX \cdot q)$ we obtain

$$0 = \frac{d}{dt} f^*(\exp tX \cdot p)\Big\}_{t=0} = \frac{d}{d\,s} f\ (\exp\ sX \cdot p)$$

which shows that $f(\exp sX \cdot p)$ is constant in s. Hence $\exp sX \cdot p = p$ and $X \in \mathfrak{h}$. This shows that M is diffeomorphic to G/G_p. For the last statement of the theorem consider a sequence $(x_n) \in G$ such that $G = \bigcup_n G_0 x_n$. Each orbit $G_0 x_n \cdot p$ is an open subset of M; since M is connected we conclude that G_0 is transitive on M.

In general, if G is a group of diffeomorphisms of a manifold M, the *isotropy* group at $p \in M$, G_p, is the subgroup of G which leaves p fixed. The *linear isotropy group* at p is the group of linear transformations of M_p induced by G_p.

Suppose now G is a connected Lie group with Lie algebra $\mathfrak{g}$. Let $g \to \mathrm{Ad}(g)$ denote the adjoint representation of G on $\mathfrak{g}$ and $X \to \mathrm{ad}\,X$ the adjoint representation of $\mathfrak{g}$ on $\mathfrak{g}$. Then $\mathrm{ad}\,X(Y) = [X, Y]$ and $\mathrm{Ad}\,(\exp X) = e^{\mathrm{ad}X}$ for $X, Y \in \mathfrak{g}$. Let H be a closed subgroup of G with Lie algebra $\mathfrak{h}$. The coset space G/H is called *reductive* (Nomizu [31]) if there exists a subspace $\mathfrak{m}$ of $\mathfrak{g}$ complementary to $\mathfrak{h}$ such that $\mathrm{Ad}(h)\mathfrak{m} \subset \mathfrak{m}$ for all $h \in H$. We shall only be dealing with reductive coset spaces. All spaces G/H where H is compact or connected and semi-simple are reductive. For reductive coset spaces G/H, the mapping $(d\pi)_e$ maps $\mathfrak{m}$ isomorphically onto the tangent space to G/H at p_0 such that the action of $\mathrm{Ad}(h)$ on $\mathfrak{m}$ corresponds to the action of $d\tau(h)$ on the tangent space. It is customary to identify these spaces. If in a reductive coset space the subspace $\mathfrak{m}$ satisfies $[\mathfrak{m}, \mathfrak{m}] \subset \mathfrak{h}$ we say that G/H is *infinitesimally symmetric*. Suppose the group G has an involutive automorphism σ such that H lies between the group H_σ of fixed points of σ and the identity component of H_σ. The space G/H is then called a *symmetric* coset space. Such a space is infinitesimally symmetric as is easily seen by taking $\mathfrak{m}$ as the eigenspace for the eigenvalue -1 of the automorphism $d\sigma$ of $\mathfrak{g}$.

Let G/H be an infinitesimally symmetric coset space. Here one has the relations

$$\mathfrak{g} = \mathfrak{h} + \mathfrak{m}, \quad \mathrm{Ad}(h)\mathfrak{m} \subset \mathfrak{m} \text{ for all } h \in H, \quad [\mathfrak{m}, \mathfrak{m}] \subset \mathfrak{h}. \tag{2.1}$$

On G/H we consider the *canonical linear connection* which is defined in Nomizu [31] and has the following properties. It is torsion free, invariant under G and the paths (that is the autoparallel curves) through p_0 have the form $t \to \exp tX \cdot p_0$ where $X \in \mathfrak{m}$. This last property is usually expressed: paths in G/H are orbits of one-parameter groups of transvections. In terms of the Exponential mapping at p_0 we can express this property by the relation

$$\mathrm{Exp}\, X = \pi \circ \exp X \qquad \text{for } X \in \mathfrak{m}. \tag{2.2}$$

In particular G/H is *complete* in the sense that each path can be extended in both directions to arbitrary large values of the canonical parameter. Now it is known that the differential of the exponential mapping of the manifold $\mathfrak{g}$ into G is given by

$$d\, \exp_X = d\, L\, (\exp X) \circ \frac{1 - e^{-\mathrm{ad}\, X}}{\mathrm{ad}\, X} \qquad X \in \mathfrak{g}. \tag{2.3}$$

This is essentially equivalent to the formula of Cartan (proved in Chevalley [7]), which expresses the Maurer–Cartan forms in canonical coordinates. A different proof without the use of differential forms is given in Helgason [24]. To derive a similar formula for $d\, \mathrm{Exp}_X$ ($X \in \mathfrak{m}$) we observe, as a consequence of (2.1), that the linear mapping $(\mathrm{ad}\, X)^2$ maps $\mathfrak{m}$ into itself. Let T_X denote the restriction of $(\mathrm{ad}\, X)^2$ to $\mathfrak{m}$. From the relation $\pi \circ L(g) = \tau(g) \circ \pi$ and (2.2) we obtain for $Y \in \mathfrak{m}$

$$d\, \mathrm{Exp}_X(Y) = d\pi \circ d\, \exp_X(Y) = d\pi \circ d\, L\, (\exp X) \circ \frac{1 - e^{-\mathrm{ad}\, X}}{\mathrm{ad}\, X}\, (Y)$$

$$= d\,\tau\, (\exp X) \circ d\pi \sum_0^\infty (-1)^m \frac{(\mathrm{ad}\, X)^m}{(m+1)!} (Y).$$

From the relations (2.1) it follows that

$$d\pi \circ (\mathrm{ad}\, X)^m (Y) = \begin{cases} (T_X)^n (Y) & \text{if } m = 2n \\ 0 & \text{if } m \text{ is odd.} \end{cases}$$

We have then proved the desired formula

$$d\ \mathrm{Exp}_X = d\,\tau\ (\exp X) \circ \sum_0^\infty \frac{T_X^n}{(2n+1)!} \qquad \text{for } X \in \mathfrak{m} \tag{2.4}$$

which will be used presently.

2. Spherical areas

Let M be a Riemannian manifold such that the group $\mathbf{I}(M)$ of all isometries of M is transitive on M. M is then called a *Riemannian homogeneous space*. The group $\mathbf{I}(M)$, endowed with the compact-open topology, is a Lie group (S. Myers and N. Steenrod [30]). Let p_0 be a point in M and $\tilde{K}$ the subgroup of $\mathbf{I}(M)$ that leaves p_0 fixed. It is well known that $\tilde{K}$ is compact. Now M, and consequently the group $\mathbf{I}(M)$, are separable. By the definition of the topology of $\mathbf{I}(M)$, the mapping $\Phi : (g, p) \to g \cdot p$ of $\mathbf{I}(M) \times M$ onto M is continuous. Theorem 2 then implies that $\mathbf{I}(M)/\tilde{K}$ is homeomorphic to M, in particular connected. The group $\tilde{K}$, being compact, has finitely many components and it follows easily that the same is true of $\mathbf{I}(M)$. Let G denote the identity component of $\mathbf{I}(M)$ and let $K = G \cap \tilde{K}$. Then K is compact and due to Theorem 2 we can state

Lemma 2. *A Riemannian homogeneous space M can (with respect to the differentiable structure) be identified with the coset space G/K where G is the identity component of $\mathbf{I}'(M)$ and K is compact. Here $\mathbf{I}'(M)$ is any closed subgroup of $\mathbf{I}(M)$, transitive on M.*

On the other hand let G be a connected Lie group and H a closed subgroup. We assume that the group $\mathrm{Ad}_G(H)$ consisting of all the linear transformations $\mathrm{Ad}(h)$, $h \in H$, is compact. Then G/H is reductive and there exists a positive definite quadratic form on $\mathfrak{m}$ invariant under $\mathrm{Ad}_G(H)$. This form gives by translation a positive definite Riemannian metric on G/H which is invariant under the action of G. Such a space we shall call a *Riemannian coset space*.

Lemma 3. *Let G/H be a symmetric Riemannian coset space which is non-compact, simply connected and irreducible (that is, $\mathrm{Ad}_G(H)$ acts irreducibly on $\mathfrak{m}$). Let $A(r)$ denote the area of a geodesic sphere in G/H of radius r. Then $A(r)$ is an increasing function of r.*

Proof. We can assume that G acts effectively on G/H because if N is a closed normal subgroup of G contained in H then the coset space G^*/H^*, where $G^* = G/N$, $H^* = H/N$ satisfies all the conditions of the lemma. The G-invariant metric on G/H induces the canonical linear connection on G/H (K. Nomizu [31]), and the paths are now geodesics. Since G/H is irreducible and non-compact it has sectional curvature everywhere $\leqslant 0$ due to a theorem of E. Cartan [4]. (Another proof is given in [24]). Furthermore, since G/H is simply connected and has negative curvature, a well-known result of J. Hadamard and

E. Cartan ([5] and [17]) states that the mapping Exp is a one-to-one mapping of $\mathfrak{m}$ onto [1] G/H. Now, each $\tau(x)$, $x \in G$ is an isometry of G/H. From (2.4) it follows therefore that the ratio of the volume elements in G/H and $\mathfrak{m}$ is given by the determinant of the endomorphism

$$A_X = \sum_0^\infty \frac{T_X{}^n}{(2n+1)!}. \tag{2.5}$$

For the volume of a geodesic sphere in G/H with radius r we obtain the expression

$$V(r) = \int_{\|X\| \leqslant r} \det(A_X)\, dX.$$

Here dX and $\| \ \|$ denote the volume element and norm respectively in the space $\mathfrak{m}$. On differentiation with respect to r we obtain

$$A(r) = \int_{\|X\| = r} \det(A_X)\, d\omega_r(X) \tag{2.6}$$

where $d\omega_r$ is the Euclidean surface element of the sphere $\|X\| = r$ in $\mathfrak{m}$. Now it is known that the irreducibility of G/H implies that either $\mathfrak{g}$ is semi-simple or $[\mathfrak{m}, \mathfrak{m}] = 0$. (A proof can be found in K. Nomizu [31] p. 56; observe the slight difference in the definition of irreducibility). In the case $[\mathfrak{m}, \mathfrak{m}] = 0$, Lemma 3 is obvious so we shall now assume $\mathfrak{g}$ semi-simple. In the proof of Theorem 2 in [24] it is shown that the Killing form B is not only non-degenerate on $\mathfrak{g}$ but

$$B(X, X) > 0 \qquad \text{for } X \neq 0 \text{ in } \mathfrak{m} \tag{2.7}$$

$$B(Y, Y) < 0 \qquad \text{for } Y \neq 0 \text{ in } \mathfrak{h}. \tag{2.8}$$

Using the invariance of the Killing form we obtain also

$$B(\{\operatorname{ad} X\}^2 Z_1, Z_2) = -B([X, Z_1], [X, Z_2]) = B(Z_2, \{\operatorname{ad} X\}^2 (Z_1)) \tag{2.9}$$

which shows that for $X \in \mathfrak{m}$, T_X is symmetric with respect to B. Using (2.7), (2.8) and (2.9) for $Z_1 = Z_2$ we see also that the eigenvalues of T_X are all $\geqslant 0$. If we call these $\xi_1(X)$, $\ldots$, $\xi_n(X)$ and throw T_X into diagonal form we obtain the formula

$$\det(A_X) = \prod_{i=1}^n \frac{\sinh(\xi_i(X))^{\frac{1}{2}}}{(\xi_i(X))^{\frac{1}{2}}}. \tag{2.10}$$

The function $\sinh t/t$ is increasing; it follows then from (2.6) that the function $A(r)$ increases with r, in fact faster than r^{n-1}.

[1] Using the theory of symmetric spaces, the assumption in Lemma 3 could be reduced somewhat. In fact, either G/H is a Euclidean space or G is semi-simple. In the latter case it can be proved directly, without using the simple connectedness (Cartan [3], Mostow [29]) that Exp is a homeomorphism of $\mathfrak{m}$ onto G/H.

The example of a sphere shows that the hypothesis in Lemma 3 that G/H is non-compact cannot be dropped. However, it seems very likely that the conclusion of Lemma 3 holds for every simply connected Riemannian manifold of negative curvature. The proof above shows (after decomposition) that this is the case if the space is symmetric.

3. Two-point homogeneous spaces

DEFINITION. A connected differentiable manifold M with a positive definite Riemannian metric of class C^∞ is called a *two-point homogeneous space* if the group $\mathbf{I}(M)$ is transitive on the set of all equidistant point pairs of M.

We shall now outline a proof of a theorem which will be of use later. This theorem is known through the classification of the two-point homogeneous spaces. We aim at proving the theorem more directly.

THEOREM 3.

(i) *A two-point homogeneous space M is isometric to a symmetric Riemannian coset space G/K where G is the identity component of $\mathbf{I}(M)$ and K is compact.*

(ii) *If M has odd dimension it has constant sectional curvature.*

(iii) *The non-compact spaces M are all simply connected, in fact homeomorphic to a Euclidean space.*

REMARK. Considerably more is known about two-point homogeneous spaces even under less restrictive definition. A complete classification of the compact two-point homogeneous spaces was given by H. Wang [36]. He found that these are the spherical spaces, real elliptic spaces, complex elliptic spaces, quaternian elliptic spaces and the Cayley elliptic plane. The dimensions of these spaces are respectively d, $d+1$, $2d$, $4d$ and 16 ($d = 1, 2, \ldots$). These are known to be symmetric spaces, that is the geodesic symmetry with respect to each point extends to a global isometry of the whole space. We indicate briefly how (i) follows in the compact case.

Choose a fixed point $p_0 \in M$ and let s_0 denote the geodesic symmetry around p_0. In view of Lemma 2 we can identify M and G/K. (Here K is the subgroup of G that leaves p_0 fixed). The mapping $\sigma : g \to s_0 g s_0$ is an automorphism of $\mathbf{I}(M)$ which maps the identity component G into itself. Also $s_0 \cdot k \cdot s_0 = k$ since both sides are isometries which induce the same mapping on M_{p_0}. It follows that the involutive automorphism $(d\sigma)_e$ of $\mathfrak{g}$ is identity on $\mathfrak{k}$, the Lie algebra of K. On the other hand if $(d\sigma)_e X = X$ for some X in $\mathfrak{g}$ then $\sigma \cdot \exp X = \exp X$ and $\exp X \cdot p_0$ is a fixed point under s_0. Hence $X \in \mathfrak{k}$. Thus $\mathfrak{k}$ is the set of fixed points of $(d\sigma)_e$ and it follows immediately that G/K is a symmetric coset space.

The non-compact two-point homogeneous spaces were classified by J. Tits [35]. In the following we shall establish (i) more directly. When this is done Tits' classification could be obtained from Cartan's classification of non-compact symmetric spaces of rank 1 ([4], p. 385). Using Cartan's terminology, the spaces that occur are: A IV (the hermitian hyperbolic spaces), BD II (the real hyperbolic spaces), C II (for $q = 1$) (the quaternian hyperbolic spaces) and F II (the hyperbolic analogue of the Cayley plane).

Suppose now M is a two-point homogeneous space, p_0 a fixed point in M and $\tilde{K}$ the subgroup of $\mathbf{I}(M)$ that leaves p_0 fixed.

LEMMA 4. *Let G be a closed, connected subgroup of $\mathbf{I}(M)$, and assume that G is transitive on equidistant point pairs of M. If G' is a closed connected normal subgroup of G ($G' \neq e$) then G' is transitive on M.*

This lemma is essentially due to Wang and Tits. We give a proof for the reader's convenience. Let $p \in M$ and let H be the subgroup of G leaving p fixed. H is compact. The Lie algebra $\mathfrak{g}$ of G can be written $\mathfrak{g} = \mathfrak{h} + \mathfrak{m}$ where $\mathfrak{h}$ is the Lie algebra of H and $\mathfrak{m}$ is invariant under $\mathrm{Ad}_G(H)$. From Lemma 2 it is clear that M is isometric with the Riemannian coset space G/H and $\mathfrak{m}$ can be identified with the tangent space M_p. Now the group G, being a group of motions, acts effectively on M, so M', the orbit of p under G', does not consist of p alone. Due to S. Myers and N. Steenrod [30], we know that this orbit is a regularly imbedded submanifold of M. We can choose a one-parameter subgroup g_t of G' which does not keep p fixed. Let X be the tangent vector to the curve $g_t \cdot p$ at $t = 0$. Then $X \neq 0$. In fact, assume to the contrary that $X = 0$. Then we have for each $f \in C^\infty(M)$, $Xf = \frac{d}{dt} f(g_t \cdot p)\Big\}_{t=0} = 0$. Using this on the function f^* given by $f^*(q) = f(g_u \cdot q)$ we find $\frac{d}{du} f(g_u \cdot p) = 0$ so $g_u \cdot p = p$ which is a contradiction. If $h \in H$ the curve $hg_t h^{-1} \cdot p$ lies in M' and has tangent vector $\mathrm{Ad}(h)X$. But the group $\mathrm{Ad}_G(H)$ acts transitively on the directions in $\mathfrak{m}$. Therefore, if I denotes the imbedding of M' into M, dI_p is an isomorphism of M'_p onto M_p. Consequently some neighborhood of p in M lies in M'. By homogeneity this holds for each $p \in M'$ and M' is open in M. This proves that each orbit in M under G' is open. By the connectedness of M this is impossible unless $M' = M$ and the lemma is proved.

LEMMA 5. *Let G/H be a reductive coset space ($H \neq G$) and let H_0 denote the identity component of H. Let $\mathfrak{m}$ be a subspace of $\mathfrak{g}$ (the Lie algebra of G) such that $\mathfrak{g} = \mathfrak{m} + \mathfrak{h}$ and $\mathrm{Ad}(h)\mathfrak{m} \subset \mathfrak{m}$ for $h \in H$. Here $\mathfrak{h}$ is the Lie algebra of H.*

(i) *If $\mathrm{Ad}_G(H_0)$ acts irreducibly on $\mathfrak{m}$, then $\mathfrak{h}$ is a maximal proper subalgebra of $\mathfrak{g}$.*

(ii) *If $[\mathfrak{m}, \mathfrak{m}] \subset \mathfrak{h}$ (that is G/H is infinitesimally symmetric), the converse of* (i) *is true.*

This lemma which is undoubtedly known can be proved as follows. Suppose $\mathrm{Ad}_G(H_0)$

acts irreducibly on $\mathfrak{m}$ and that $\mathfrak{h}$ is not maximal. Then there exists a subalgebra $\mathfrak{h}^*$ of $\mathfrak{g}$ such that we have the proper inclusions $\mathfrak{h} \subset \mathfrak{h}^* \subset \mathfrak{g}$. The subspace $\mathfrak{m}^* = \mathfrak{h}^* \cap \mathfrak{m}$ satisfies $[\mathfrak{h}, \mathfrak{m}^*] \subset \mathfrak{h}^* \cap \mathfrak{m} = \mathfrak{m}^*$ so $\mathfrak{m}^*$ is invariant under $\mathrm{Ad}_G(H_0)$. Hence $\mathfrak{m}^* = 0$ or $\mathfrak{m}^* = \mathfrak{m}$. This last possibility is impossible because it implies $\mathfrak{h}^* = \mathfrak{g}$. But the relation $\mathfrak{m}^* = 0$ is also impossible because if X belongs to the complement of $\mathfrak{h}$ in $\mathfrak{h}^*$ we have $X = Y + Z$, where $Y \in \mathfrak{h}$, $Z \in \mathfrak{m}$ and $Z \neq 0$. But $Z = X - Y \in \mathfrak{h}^*$ so $Z \in \mathfrak{m} \cap \mathfrak{h}^* = 0$. This proves (i). In order to prove (ii) assume $\mathfrak{n}$ is a proper subspace of $\mathfrak{m}$, invariant under $\mathrm{Ad}_G(H_0)$. The relation $[\mathfrak{m}, \mathfrak{m}] \subset \mathfrak{h}$ shows that $\mathfrak{h} + \mathfrak{n}$ is a proper subalgebra of $\mathfrak{g}$, which properly contains $\mathfrak{h}$.

We shall now indicate a proof of Theorem 3 in case M is non-compact. Let G be the identity component of $\mathbf{I}(M)$. We know that M is isometric to G/K where $K = G \cap \tilde{K}$. We can assume $\dim M > 1$. Then a small geodesic sphere $\mathbf{S}_r$ around p_0 is connected and $\tilde{K}$ acts transitively on $\mathbf{S}_r$. From Theorem 2 we see that K, having the same dimension as $\tilde{K}$, acts transitively on $\mathbf{S}_r$ and thus G acts transitively on equidistant point pairs of M. If G is not semi-simple, G contains an abelian connected normal subgroup $\neq e$ which by Lemma 4 acts transitively on M. M is then a vector space for which Theorem 3 is obvious. If on the other hand G is semi-simple, we see from Lemma 5 that $\mathfrak{k}$, the Lie algebra of K, is a maximal proper subalgebra of $\mathfrak{g}$. Since maximal compact subgroups of connected semi-simple groups are connected, we conclude that K is a maximal compact subgroup of G and G/K is a symmetric coset space. Due to a well-known theorem of Cartan on semi-simple groups, G/K is homeomorphic to a Euclidean space. In our special case, this can be established as follows. Clearly G/K has an infinite geodesic and therefore all its geodesics are infinite. The mapping Exp of $\mathfrak{m}$ into G/K has Jacobian determinant at X given by (2.10) (the derivation of (2.10) did not use the simple connectedness of G/H). The expression (2.10) is always $\neq 0$ so Exp is everywhere regular. Since geodesics issuing from p_0 intersect the geodesic spheres around p_0 orthogonally we see that geodesics issuing from p_0 do not intersect again. Thus Exp is one-to-one and hence a homeomorphism.

Part (ii) of Theorem 3 which is due to Wang [36] depends on the fact that if a linear group of motions acts transitively on an even-dimensional sphere then the action is transitive on equidistant point pairs.

4. Harmonic Lorentz spaces

Let M be a Lorentz space with metric tensor Q. Let p_0 be an arbitrary but fixed point of M and let Exp be the Exponential mapping at p_0 which maps a neighborhood U_0 of 0 in M_{p_0} in a one-to-one manner onto a neighborhood U of p_0 in M. Let $X_1, \ldots, X_n$ be any basis of M_{p_0}. If $X = \sum x_i X_i$ and $x = \mathrm{Exp}\, X$ the mapping $x \to (x_1, \ldots, x_n)$ is a system of coordinates valid on U. Following Hadamard we consider now the distance function $\Gamma(x) = Q_{p_0}(X, X)$ defined for $x = \operatorname{Exp} X$ in U.

DEFINITION. Suppose M and Q are analytic. M is called *harmonic* if for each $p_0 \in M$ there exists a neighborhood of p_0 in which $\Box \Gamma$ is a function of Γ only, $\Box \Gamma = f(\Gamma)$.

We shall now study in some detail three types of harmonic homogeneous spaces. These are denoted G^0/H, G^-/H and G^+/H below. For each integer $n \geqslant 1$ there is one space of each class with dimension n. If $n = 1$, $G^0/H = G^-/H$. If $n = 2$, G^-/H and G^+/H are diffeomorphic but not isometric. Otherwise the spaces are all different (even topologically). Due to Theorem 9 these spaces exhaust the class of harmonic Lorentz spaces up to local isometry.

G^0/H. *Flat Lorentz spaces.* We consider the Euclidean space $\mathbf{R}^n$ as a manifold in the usual way that the tangent space at each point is identified with $\mathbf{R}^n$ under the usual identification of parallel vectors. We define a Lorentzian metric Q^0 on $\mathbf{R}^n$ by

$$Q_p^0(Y, Y) = y_1^2 - y_2^2 - \cdots - y_n^2$$

if $Y = (y_1, \ldots, y_n)$ is a vector at $p \in \mathbf{R}^n$. We have then obtained a Lorentz space M. Let L_n denote the general Lorentz group, that is the group of all linear homogeneous transformations h of $\mathbf{R}^n$ such that $Q_0^0(h \cdot X, h \cdot X) = Q_0^0(X, X)$ for all $X \in \mathbf{R}^n$. Each isometry $g \in \mathbf{I}(M)$ can be uniquely decomposed $g = th$ where t is a translation and $h \in L_n$. Hence $\mathbf{I}(M) = \mathbf{R}^n \cdot L_n$. $\mathbf{R}^n$ is a normal subgroup of $\mathbf{I}(M)$. If G^0 is the identity component of $\mathbf{I}(M)$ and H is the subgroup of G^0 that leaves 0 fixed, then M is diffeomorphic to G^0/H and H is connected. G^0/H is a symmetric coset space under the mapping $th \to t^{-1}h$, $t \in \mathbf{R}^n$, $h \in H$.

The group L_n acts transitively on the set of timelike rays from 0; L_n also acts transitively on the set of spacelike rays from 0. Furthermore L_n acts transitively on the punctured cone $\mathbf{C}_0 - 0$. Since $\Box$ is invariant under L_n, it follows in particular that $\Box\Gamma$ is a function of Γ only; G^0/H is harmonic.

G^-/H. *Negatively curved harmonic Lorentz spaces.* We consider now the quadratic form

$$T(Y, Y,) = -y_1^2 + y_2^2 + \cdots + y_{n+1}^2 \qquad Y = (y_1, \ldots, y_{n+1})$$

and let G^- denote the identity component of the group L_{n+1} which leaves the form $T(Y, Y)$ invariant. Let H be the subgroup of G^- that leaves the point $(0, 0, \ldots, 1)$ fixed. If the transformations $g \in L_{n+1}$ are represented in matrix form $g = (g_{ij})$ then $g \in G^-$ if and only if $g_{11} > 0$ and $\det g = 1$. From this well-known fact follows immediately that H is connected and actually the same as the group H above. The coset space G^-/H can be identified with the orbit of the point $(0, 0, \ldots, 1)$ under G^-. This is the hyperboloid $-y_1^2 + y_2^2 + \ldots + y_{n+1}^2 = 1$ which is homeomorphic to $S^{n-1} \times \mathbf{R}$ (S^m denotes the m-dimensional sphere). It is clear that

17 – 593805. *Acta mathematica.* 102. Imprimé le 16 décembre 1959

G^- acts effectively on G^-/H. Let $\mathfrak{g}^-$ and $\mathfrak{h}$ denote the Lie algebras of G^- and H respectively. If J denotes the matrix of the quadratic form T then a matrix A belongs to G^- if and only if ${}^tAJA = J$ (${}^tA$ is the transpose of A). Using this on matrices of the form $A = \exp X$, $X \in \mathfrak{g}^-$ we find that a basis of $\mathfrak{g}^-$ is given by

$$X_i = E_{1i} + E_{i1} \quad (2 \leqslant i \leqslant n+1), \; X_{ij} = E_{ij} - E_{ji} \quad (2 \leqslant i < j \leqslant n+1). \tag{2.11}$$

Here E_{ij} denotes as usual the matrix (a_{km}) where all $a_{km} = 0$ except $a_{ij} = 1$. A basis of $\mathfrak{h}$ is given by

$$Y_i = E_{1i} + E_{i1} \, (2 \leqslant i \leqslant n), \quad Y_{ij} = E_{ij} - E_{ji} \qquad (2 \leqslant i < j \leqslant n).$$

Let $B^-(X, X)$ denote the Killing form $\mathrm{Tr}\,(\mathrm{ad}\, X \;\mathrm{ad}\, X)$ on $\mathfrak{g}^-$.

LEMMA 6. *The Killing form $\mathfrak{g}^-$ is given by*

$$B^-(X, X) = (n-1)\,\mathrm{Tr}\,(XX) = 2\,(n-1)\left\{\sum_{2 \leqslant i \leqslant n+1} x_1^2 - \sum_{2 \leqslant i < j \leqslant n+1} x_{ij}^2\right\}$$

if

$$X = \sum_{2 \leqslant i \leqslant n+1} x_i X_i + \sum_{2 \leqslant i < j \leqslant n+1} x_{ij} X_{ij}.$$

Proof. The complexification $\mathfrak{g}^c$ of $\mathfrak{g}^-$ is the Lie algebra of complex linear transformations which leave invariant the form $-z_1^2 + z_2^2 + \ldots + z_{n+1}^2$. However within the complex number field the signature $-++\ldots+$ is equivalent to the signature $++\ldots+$ and thus $\mathfrak{g}^c$ is isomorphic to the Lie algebra $\mathfrak{o}(n+1, C)$ which consists of all skew symmetric complex matrices. The isomorphism $X \to X'$ in question is given by the mapping $X_i \to i\,(E_{i1} - E_{1i})$ and $X_{ij} \to X_{ij}$. Now the Killing form B' on $\mathfrak{o}(n+1, C)$ is well known to be $B'(X', X') = (n-1)\mathrm{Tr}(X'X')$. Since $\mathrm{Tr}(XX) = \mathrm{Tr}(X'X')$ and since Killing forms are preserved by isomorphisms we see that the Killing form B^c on $\mathfrak{g}^c$ i given by $B^c(X, X) = (n-1)\;\mathrm{Tr}(X, X)$. Now the restriction of B^c to $\mathfrak{g}^-$ coincides with B^- and Lemma 6 follows.

Let s_0 be the linear transformation

$$s_0 : (y_1, \ldots, y_{n+1}) \to (-y_1, -y_2, \ldots, -y_n, y_{n+1}).$$

s_0 leaves the form T invariant and the mapping $\sigma : g \to s_0 g s_0$ is an involutive automorphism of G^-. The corresponding automorphism of $\mathfrak{g}$ is $d\sigma : X \to s_0 X s_0$ and it is easy to see that $\mathfrak{h}$ is the set of all fixed points of $d\sigma$. Thus G^-/H is a symmetric coset space. Let $\mathfrak{p}$ be the eigenspace for the eigenvalue -1 of $d\sigma$. $\mathfrak{p}$ is the subspace of $\mathfrak{g}^-$ spanned by the basis vectors X_{n+1} and $X_{i,n+1}$ $(2 \leqslant i \leqslant n)$, and we have the relations

$$\mathfrak{g}^- = \mathfrak{h} + \mathfrak{p}, \; [\mathfrak{h}, \mathfrak{p}] \subset \mathfrak{p}, \; [\mathfrak{p}, \mathfrak{p}] \subset \mathfrak{h} \tag{2.12}$$

and since H is connected, $\mathrm{Ad}(h)\,\mathfrak{p} \subset \mathfrak{p}$. As usual we identify $\mathfrak{p}$ with the tangent space to G^-/H at p_0.

Since the Killing form B^- is invariant under all $\mathrm{Ad}(g)$, $g \in G^-$ we see that the quadratic form Q^- on $\mathfrak{p}$ given by

$$Q^-(X, X) = x_{n+1}^2 - \sum_2^n x_{i,n+1}^2, \quad X = x_{n+1} X_{n+1} + \sum_2^n x_{i,n+1} X_{i,n+1} \tag{2.13}$$

is invariant under the action of $\mathrm{Ad}(H)$ on $\mathfrak{p}$. The form Q^- "extends" uniquely to a G^--invariant Lorentzian metric on G^-/H which induces the canonical linear connection on G^-/H (Nomizu [31]). We denote the metric tensor also by Q^-. Consider now the action of the group $\mathbf{I}(G^-/H)$. Let H^* denote the corresponding linear isotropy group at p_0 which consists of certain linear transformations leaving the form Q^- invariant.

LEMMA 7. *H^* acts transitively on* 1°. *The punctured cone* $C_{p_0} - 0$; 2°. *The set of all timelike rays from* 0; 3°. *The set of all spacelike rays from* 0.

Proof. H^* contains the restriction of the group $\mathrm{Ad}_{G^-}(H)$ to $\mathfrak{p}$ which is isomorphic to H. H^* contains also the symmetry $X \to -X$. As remarked earlier L_n acts transitively on the set

$$M_c = \{X \in \mathfrak{p} \mid X \neq 0, \; Q^-(X, X) = c\}.$$

Here c is any real number. Due to Theorem 2, H acts transitively on each component of M_c. If $(n, c) \neq (2, 0)$, M_c consists of one or two components, symmetric with respect to 0. If we exclude for a moment the case $(n, c) = (2, 0)$, H^* acts transitively on M_c, as stated in the lemma. If $n = 2$, M_0 consists of four components which are the rays

$$t(X_3 + X_{23}), \; t(X_3 - X_{23}), \; t(-X_3 + X_{23}), \; t(-X_3 - X_{23})$$

where $0 < t < \infty$. H^* will clearly be transitive on M_0 if we can prove that the mapping

$$A: \; x_3 X_3 + x_{23} X_{23} \to -x_3 X_3 + x_{23} X_{23}$$

belongs to H^*. The Killing form on $\mathfrak{g}^-$ is

$$B^-(X, X) = 2(x_2^2 + x_3^2 - x_{23}^2), \quad X = x_2 X_2 + x_3 X_3 + x_{23} X_{23}$$

G^- is the group leaving B^- invariant and H is the subgroup of G^- which leaves the point $(1, 0, 0)$ fixed. G^-/H can thus be identified with the hyperboloid $B^-(X, X) = 2$. Hence G^-/H is isometrically imbedded in the flat Lorentz space $\mathfrak{g}^-$ with metric B^-. Now the transformation $(x_2, x_3, x_{23}) \to (x_2, -x_3, x_{23})$ is an isometry of $\mathfrak{g}^-$ which maps the hyperboloid onto itself and leaves the point $(1, 0, 0)$ fixed. Hence A belongs to H^*, as we wanted to prove.

COROLLARY. *G^-/H is harmonic.*

In fact $\square\Gamma$ is invariant under the isotropy subgroup of $\mathbf{I}(G^-/H)$. Due to Lemma 7 $\square\Gamma$ is a function of Γ only.

LEMMA 8. *The timelike paths in G^-/H are infinite and have no double points.*

Proof. Consider the vector $X_{n+1} \in \mathfrak{p}$ which lies inside the cone $Q^-(X, X) = 0$. The path with tangent vector X_{n+1} has the form $\pi \circ \exp t X_{n+1}$, $(t \in \mathbf{R})$. If we use the matrix representation (2.11) we get

$$\exp t X_{n+1} = I + (\cosh t - 1)(E_{11} + E_{n+1, n+1}) + (\sinh t) X_{n+1}$$

and this one-parameter subgroup intersects H only for $t = 0$. It follows easily that the path in question has no double points and since $\mathbf{I}(G^-/H)$ is transitive on the timelike paths the lemma follows.

As before, let Exp denote the Exponential mapping of $\mathfrak{p}$ into G^-/H and A_X the linear transformation (2.5).

LEMMA 9.

$$\det A_X = \left\{\frac{\sinh (Q^- (X, X))^{\frac{1}{2}}}{(Q^- X, X))^{\frac{1}{2}}}\right\}^{n-1}$$

if $Q^-(X, X) > 0$. *In particular,* Exp *is regular in the cone* $Q^-(X, X) > 0$.

Proof. Let as before T_X be the restriction of $(\operatorname{ad} X)^2$ to $\mathfrak{p}$. If $n = 1$, $T_X = 0$ and $G^-/H = \mathbf{R}$; hence we assume $n > 1$. Suppose now $Q^-(X, X) > 0$ and that $Y \neq 0$ is an eigenvector of T_X with eigenvalue ξ. There exists an element $h \in H$ such that $\operatorname{Ad}(h) X = c X_{n+1}$ where $c^2 = Q^-(X, X)$. The relation $T_X \cdot Y = \xi Y$ implies

$$T_{X_{n+1}} Y^* = \frac{\xi}{c^2} Y^* \qquad \text{where } Y^* = \operatorname{Ad}(h)\, Y. \tag{2.14}$$

Writing $Y^* = y_{n+1} X_{n+1} + \sum_2^n y_{i, n+1} X_{i, n+1}$ we find easily $[X_{n+1}, Y^*] = -\sum_2^n y_{i, n+1} X_i$ and

$$T_{X_{n+1}} \cdot Y^* = \sum_2^n y_{i, n+1} X_i. \tag{2.15}$$

From (2.14) and (2.15) we obtain

$$\xi \left(y_{n+1} X_{n+1} + \sum_2^n y_{i, n+1} X_{i, n+1}\right) = c^2 \left(\sum_2^n y_{i, n+1} X_{i, n+1}\right).$$

This shows that either $\xi = 0$ (in which case Y is a non-zero multiple of X) or $\xi = c^2$ (in which case $y_{n+1} = 0$, $y_{i, n+1}$ arbitrary). This shows that the eigenvalues of T_X are 0 and $Q^-(X, X)$; the latter is an $(n-1)$-tuple eigenvalue. The lemma now follows from the relation (2.10).

Suppose now M is an arbitrary complete Lorentz space with metric tensor Q. For a given point $p \in M$ let $\mathbf{S}_r(p)$ be a "sphere" in M_p of radius r and center p; that is $\mathbf{S}_r(p)$ is one of the two components of the set of vectors $\{X \mid X \in M_p, Q_p(X, X) = r^2\}$. If Exp is the Exponential mapping at p we put $\mathfrak{S}_r(p) = \text{Exp } S_r(p)$. For the present considerations it is convenient not to specify which of the two components is chosen. In Chapter IV, § 5 we shall (for the special cases treated there) make such a choice in a continuous manner over the entire manifold.

LEMMA 10. *The timelike paths in G^-/H issuing from p_0 intersect the manifold $\mathfrak{S}_r(p_0)$ at a right angle (in the Lorentzian sense).*

Proof. $\mathfrak{S}_r(p_0)$ is a manifold since Exp is regular in an open set containing $S_r(p_0)$. Let p be a point on $S_r(p_0)$, X the vector $\overrightarrow{p_0 p}$ and Y a tangent vector to $S_r(p_0)$ at p. Clearly $Q_p^-(X, Y) = 0$. To prove the lemma we have to prove

$$Q_q^-(d\,\text{Exp}_X(X),\ d\,\text{Exp}_X(Y)) = 0 \qquad (q = \text{Exp } X). \tag{2.16}$$

(Here we have considered X as a tangent vector to $\mathfrak{p}$ at p, parallel to $\overrightarrow{p_0 p}$.) Using (2.4) and the fact that $\tau(g)$, $g \in G$ is an isometry of G^-/H we see that (2.16) amounts to

$$B^-(A_X(X),\ A_X(Y)) = 0.$$

This relation, however, is immediate from the invariance of B^-.

It is possible to extend Lemma 9 to arbitrary Lorentz spaces by using the structural equations for pseudo-Riemannian connections. We do not do this here since the proof in the special case above is much simpler.

LEMMA 11. *Let Z be a non-vanishing tangent vector to $\mathfrak{S}_r(p_0)$ at q. Then $Q_q^-(Z, Z) < 0$.*

Proof. It suffices to prove this when $q = \text{Exp } X_{n+1}$ in which case

$$Z = d\,\text{Exp}_{X_{n+1}}(Y) \text{ with } Y = \sum_2^n y_{i,n+1} X_{i,n+1}$$

To prove $Q_q^-(Z, Z) < 0$ we just have to prove

$$Q^-(A_{X_{n+1}}(Y),\ A_{X_{n+1}}(Y)) < 0. \tag{2.17}$$

This however is obvious since $T_{X_{n+1}} \cdot Y = Y$ and $Q^-(Y, Y) < 0$.

From Lemma 11 it follows that $\mathfrak{S}_r(p_0)$ has at each point a unique Lorentzian normal direction. Combining this with Lemmas 8, 9, and 10 we obtain

THEOREM 4. *The Exponential mapping at p_0 which maps $\mathfrak{p}$ into G^-/H is a diffeomorphism of the interior of C_{p_0} into G^-/H. (By the interior of C_{p_0} we mean the set of points $p \in M_{p_0}$ such that $\overrightarrow{p_0 p}$ is timelike).*

On the manifold $\mathbf{S}_r(p_0)$ the tensor $-Q^-$ induces a positive definite Riemannian metric. The same applies clearly to $\mathbf{S}_r(0)$ in the flat Lorentz space $\mathbf{R}^n$.

THEOREM 5. *Suppose the space G^-/H has dimension $n > 2$. With the metric induced by $-Q^-$, $\mathbf{S}_r(p_0)$ is a Riemannian manifold of constant negative curvature. The same statement holds for $\mathbf{S}_r(0)$ in the flat Lorentz space $\mathbf{R}^n (n > 2)$.*

Proof. Let $q_0 = \operatorname{Exp}(rX_{n+1})$. The group H acts transitively on $\mathbf{S}_r(p_0)$ and leaves invariant the positive definite metric on $\mathbf{S}_r(p_0)$. Let H_1 be the subgroup of H leaving q_0 fixed. H_1 is connected since $\mathbf{S}_r(p_0)$ is simply connected. The group $\operatorname{Ad}_H(H_1)$ is the group of all proper rotations in the tangent space to $\mathbf{S}_r(p_0)$ at q_0. In particular, $\operatorname{Ad}_H(H_1)$ acts transitively on the set of two-dimensional subspaces through q_0. Thus $\mathbf{S}_r(p_0)$ has constant sectional curvature at q_0 and, due to the homogeneity, at all points. Since $\mathbf{S}_r(p_0)$ is non-compact the curvature is non-positive. If $n = 2$, $\mathbf{S}_r(p_0)$ is flat, but for $n > 2$ we see from Lemma 6 that H is semi-simple (actually simple), and $\mathbf{S}_r(p_0)$ cannot be flat.

Let M be a connected manifold with a linear connection $X \to \nabla_X$. The curvature tensor R of this connection is a mapping of $\mathfrak{D} \times \mathfrak{D}$ into the space of linear mappings of $\mathfrak{D}$ into itself given by $(X, Y) \to R(X, Y)$ where

$$R(X, Y) = \nabla_X \nabla_Y - \nabla_Y \nabla_X - \nabla_{[X, Y]}.$$

Here $[X, Y]$ is the usual Poisson bracket of vector fields. If $x \to (x_1, \ldots, x_n)$ is a system of coordinates valid in an open subset of M the coefficients $R_l{}^i{}_{jk}$ of R are defined by

$$R\left(\frac{\partial}{\partial x_j}, \frac{\partial}{\partial x_k}\right) \cdot \frac{\partial}{\partial x_l} = \sum_i R_l{}^i{}_{jk} \frac{\partial}{\partial x_i}.$$

Suppose the connection $X \to \nabla_X$ is the connection induced by a pseudo-Riemannian metric Q on M. If q_{ij} is defined as in Chapter I, the coefficients $R_l{}^i{}_{jk}$ are given by

$$R_l{}^i{}_{jk} = \frac{\partial \Gamma_k{}^i{}_l}{\partial x_j} - \frac{\partial \Gamma_j{}^i{}_l}{\partial x_k} + \sum_m (\Gamma_j{}^i{}_m \Gamma_k{}^m{}_l - \Gamma_k{}^i{}_m \Gamma_j{}^m{}_l)$$

where Γ^j_{ik} are the Christoffel symbols

$$\Gamma_i{}^j{}_k = \tfrac{1}{2} \sum_l q^{jl} \left(\frac{\partial q_{il}}{\partial x_k} + \frac{\partial q_{lk}}{\partial x_i} - \frac{\partial q_{ik}}{\partial x_l}\right).$$

As usual, we put

$$R_{ijkl} = \sum_m q_{jm} R_j{}^m{}_{kl}.$$

The pseudo-Riemannian manifold is said to have *constant curvature* $\varkappa$ if the relation

$$R_{ijkl} = \varkappa (q_{jk} q_{il} - q_{ik} q_{jl}) \tag{2.18}$$

holds on M. For a Riemannian manifold (with positive definite metric) the relation (2.18) is a necessary and sufficient condition for the manifold to have constant sectional curvature in the ordinary sense.

THEOREM 6. *The space G^-/H $(n>1)$ has constant curvature $\varkappa=-1$.*

Proof. The G^--invariant Lorentzian metric induces the canonical linear connection on the symmetric space G^-/H. The curvature tensor at p_0 is given by $R(X, Y)\cdot Z = -[[X, Y], Z]$ for $X, Y, Z \in \mathfrak{p}$; see e.g. K. Nomizu [31]. We choose coordinates $x_1, \ldots, x_n$ in a neighborhood of p_0 such that

$$\left(\frac{\partial}{\partial x_1}\right)_{p_0} = X_{n+1}, \quad \left(\frac{\partial}{\partial x_i}\right)_{p_0} = X_{i,\,n+1} \qquad (2 \leqslant i \leqslant n).$$

At p_0 we have $q_{11}=1$, $q_{22}=\cdots=q_{nn}=-1$. The coefficients of the curvature tensor at p_0 can be found by routine computation. The result is

$$\left.\begin{aligned} &R_{ijkl} = \delta_i^k\,\delta_l^j = \delta_k^j\,\delta_l^i \\ &R_{1i1j} = -R_{i11j} = R_{i1j1} = -R_{1ij1} = -\delta_j^i \end{aligned}\right\} \; (2 \leqslant i, j, k, l \leqslant n).$$

All other coefficients vanish. It is immediate to verify that (2.18) holds with $\varkappa=-1$. Since the validity of (2.18) is independent of the choice of coordinates and since G^-/H is homogeneous, the theorem follows.

G^+/H. *Positively curved harmonic Lorentz space.* Still maintaining the notation from above, we consider the complexification $\mathfrak{g}^c$ of the Lie algebra $\mathfrak{g}^-$. If we consider $\mathfrak{g}^c$ as a real Lie algebra, it is clear that $\mathfrak{g}^+ = \mathfrak{h} + i\mathfrak{p}$ is a real subspace, and in fact a real subalgebra due to the relations (2.12). Let G^+ denote the corresponding real analytic subgroup of the general linear group $\mathbf{GL}(n+1, C)$, considered as a real group. H is then a closed subgroup of G^+ and we shall now investigate the space G^+/H of left cosets gH. A basis for $\mathfrak{g}^+$ is given by

$$X_i \;(2 \leqslant i \leqslant n),\;\; X_{ij} \;(2 \leqslant i < j \leqslant n),\;\; i\,X_{n+1},\;\; i\,X_{j,\,n+1} \quad (2 \leqslant j \leqslant n).$$

and the bracket operation in $\mathfrak{g}^+$ is the ordinary matrix bracket $[A, B] = AB - BA$. The relations

$$\mathfrak{g}^+ = \mathfrak{h} + i\,\mathfrak{p}, \;\; [\mathfrak{h}, i\,\mathfrak{p}] \subset i\,\mathfrak{p}, \;\; [i\,\mathfrak{p}, i\,\mathfrak{p}] \subset \mathfrak{h} \tag{2.19}$$

are obvious from (2.12) and, since H is connected, $\mathrm{Ad}_{G^+}(h)\, i\mathfrak{p} \subset i\mathfrak{p}$ for each $h \in H$. Thus G^+/H is an infinitesimally symmetric coset space. To see that G^+/H is a symmetric coset space, let s_0 denote the linear transformation

$$s_0 : (y_1, \ldots, y_n, y_{n+1}) \to (-y_1, -y_2, \ldots, -y_n, y_{n+1}).$$

It is easy to see that the mapping $\sigma: g \to s_0 g s_0$ is an involutive automorphism of G^+ and $\mathfrak{h}$ is the set of fixed points of $d\sigma$.

LEMMA 12. *The Killing form on $\mathfrak{g}^+$ is given by*

$$B^+(X, X) = (n-1)\operatorname{Tr}(XX) = 2(n-1)\left\{\sum_{2\leqslant i\leqslant n} x_i^2 - \sum_{2\leqslant i<j\leqslant n} x_{ij}^2 - x_{n+1}^2 + \sum_{2\leqslant j\leqslant n} x_{j,n+1}^2\right\}$$

in terms of the basis above.

Proof. Let B^c denote the Killing form on the complex Lie algebra $\mathfrak{g}^c$. The forms B^+ and B^- are the restrictions of B^c to $\mathfrak{g}^+$ and $\mathfrak{g}^-$ respectively. If we write $X = Y + Z$, $Y \in \mathfrak{h}$, $Z \in \mathfrak{p}$ we have

$$B^+(X, X) = B^c(X, X) = 2i\,B^-(Y, Z) + B^-(Y, Y) - B^-(Z, Z)$$
$$= (n-1)\operatorname{Tr}(Y + iZ)(Y + iZ)$$

Due to the invariance of the Killing form the quadratic form on $i\mathfrak{p}$ given by

$$Q^+(X, X) = x_{n+1}^2 - \sum_2^n x_{j,n+1}^2, \quad X = x_{n+1}(i X_{n+1}) + \sum_2^n x_{j,n+1}(i X_{j,n+1})$$

is invariant under the action of $\operatorname{Ad}_{G^+}(H)$ on $i\mathfrak{p}$. The tangent space to G^+/H at p_0 can be identified with the subspace $i\mathfrak{p}$ of $\mathfrak{g}^+$. As before Q^+ extends to a G^+-invariant Lorentzian metric on G^+/H. If $n = 1$, G^+/H can be identified with $\mathbf{S}^1$. If $n > 1$, G^+ is semi-simple and from the signature of B^+, ($\frac{1}{2}(n^2 - 3n + 4)$ minus signs), one knows that G^+ has a maximal compact subgroup of dimension $\frac{1}{2}(n^2 - 3n + 4)$. This group is generated by X_{ij} $(2 \leqslant i < j \leqslant n)$ and iX_{n+1}. The vectors X_{ij} $(2 \leqslant i < j \leqslant n)$ generate a maximal compact subgroup of H. From this it can be concluded that G^+/H is homeomorphic to $\mathbf{S}^1 \times \mathbf{R}^{n-1}$ (also for $n = 1$) but we shall not need this fact. Lemma 7 extends easily to the space G^+/H, and G^+/H is a harmonic Lorentz space. Note that for $n = 2$, G^+/H and G^-/H are diffeomorphic to a hyperboloid F: $-y_1^2 + y_2^2 + y_3^2 = 1$ such that $Q^-_{p_1} = -Q^-_{p_2}$ if p_1 and p_2 correspond to the same $p \in F$.

LEMMA 13. *All the timelike paths issuing from p_0 are closed and have length 2π.*

Proof. We consider the one parameter subgroup of G^+ generated by the timelike vector iX_{n+1}. We find

$$\exp t i X_{n+1} = I + (\cos t - 1)(E_{11} + E_{n+1,n+1}) + (\sin t)(i X_{n+1}).$$

The path in G^+/H with tangent vector iX_{n+1} at p_0 has the form $\pi \circ \exp tiX_{n+1}$ and this is clearly a closed path of length 2π. (The matrix $I - 2(E_{11} + E_{n+1,n+1})$ does not belong to H). Since $\operatorname{Ad}_{G^+}(H)$ acts transitively on the set of timelike lines through p_0, the lemma follows.

LEMMA 14.

$$\det(A_Z) = \left\{ \frac{\sin (Q^+(Z,Z))^{\frac{1}{2}}}{(Q^+(Z,Z))^{\frac{1}{2}}} \right\}^{n-1}$$

for all $Z \in i\mathfrak{p}$ *that lie in the cone* $Q^+(Z,Z) > 0$. *From* (6) *it follows that* Exp *is regular in the set* $0 < Q^+(Z,Z) < \pi^2$.

The proof is entirely analogous to that of Lemma 8 and will be omitted. Just as before it can also be proved that $Q_q^+(Z,Z) < 0$ if Z is a non-vanishing tangent vector at q to $S_r(p_0)$, $(r < \pi)$, and Lemma 10 remains valid here if $r < \pi$. Combining these results we have

THEOREM 7. *The Exponential mapping at* p_0 *which maps* $i\mathfrak{p}$ *into* G^+/H *is a diffeomorphism of the open set* $0 < Q^+(Z,Z) < \pi^2$ *into* G^+/H.

The situation is thus somewhat analogous to the sphere in Euclidean space. The following question arises. Do the timelike paths issuing from p_0 all meet at the point

$$p^* = \pi(I - 2(E_{11} + E_{n+1,\,n+1}))$$

in G^+/H which corresponds to the antipodal point on the sphere? The answer is no and the timelike paths behave more like geodesics in a real elliptic space.

LEMMA 15. *Two different timelike paths issuing from* p_0 *have no other point in common.*

Proof. We can assume that one of the paths is $t \to \pi(\exp t i X_{n+1})$. The other then has the form $t \to \pi(\exp t \operatorname{Ad}(h)\, i X_{n+1})$ with $h \in H$. By Theorem 7 it is clear that the only possible point of intersection other than p_0 would be the point p^* above, occurring for $t = \pi$. Then there exists $h_1 \in H$ such that

$$(E_{11} + E_{n+1,\,n+1})\, h_1 = h\, (E_{11} + E_{n+1,\,n+1})\, h^{-1}.$$

We can represent h_1, h and h^{-1} in the form

$$h_1 = E_{n+1,\,n+1} + \sum_{i,j=1}^{n} a_{ij} E_{ij}$$

$$h = E_{n+1,\,n+1} + \sum_{i,j=1}^{n} b_{ij} E_{ij}$$

$$h^{-1} = E_{n+1,\,n+1} + \sum_{i,j=1}^{n} c_{ij} E_{ij}.$$

Then the relation above implies

$$b_{11} c_{1j} = a_{1j} \qquad (1 \leqslant j \leqslant n)$$
$$b_{i1} c_{1j} = 0 \qquad (1 < i \leqslant n,\ 1 \leqslant j \leqslant n).$$

Also $c_{11}^2 - \sum_1^n c_{i1}^2 = 1$ so $c_{11} \neq 0$ and therefore $b_{i1} = 0$ for $1 < i \leqslant n$. On the other hand,

$$b_{11} b_{1j} - \sum_{i=2}^{n} b_{i1} b_{ij} = 0 \qquad (1 < j)$$

so $b_{1j} = 0$. Hence

$$h = E_{11} + E_{n+1,\, n+1} + \sum_{i,j=2} b_{ij} E_{ij}$$

which obviously commutes with $i X_{n+1}$; this implies that the paths coincide, contrary to assumption.

THEOREM 8. *The space G^+/H has constant curvature $\varkappa = +1$.*

The proof is entirely analogous to that of Theorem 6 and will be omitted.

Now let M be an arbitrary harmonic Lorentz space. An important theorem of A. Lichnerowicz and A. G. Walker [28] states that such a space has constant curvature in the sense of the relation (2.18). Using a similarity transformation (i.e. a multiplication of Q by a positive constant) we can assume that the curvature $\varkappa$ is 0, 1 or -1. In particular, the covariant derivatives of the curvature tensor all vanish, $\nabla_X R = 0$ for all $X \in \mathfrak{D}$. A torsion-free linear connection with this last property is uniquely determined in a suitable neighborhood U_p of a given point p, by the value R_p (see e.g. [31]). Furthermore, a diffeomorphism Φ leaving invariant a pseudo-Riemannian connection is an isometry if $(d\Phi)_p$ is an isometry for some point p. From the quoted result of Lichnerowicz and Walker follows

THEOREM 9. *The spaces G^0/H, G^-/H* and *G^+/H exhaust the class of harmonic Lorentz spaces up to local isometry.*

It is customary to denote by $\mathbf{S\,O}^h(n)$ the identity component of the group of non-singular real $n \times n$ matrices that leave invariant the quadratic form $-\sum_1^h x_i^2 + \sum_{h+1}^n x_i^2$. $\mathbf{S\,O}^0(n)$ is the usual rotation group $\mathbf{S\,O}(n)$. In this terminology we have

$$G^0/H = R^n \cdot \mathbf{S\,O}^1(n)/\mathbf{S\,O}^1(n), \quad G^-/H = \mathbf{S\,O}^1(n+1)/\mathbf{S\,O}^1(n).$$

CHAPTER III

Invariant differential operators

1. A general representation theorem

To begin with we introduce some notation which will be used in the rest of the paper. Let G/H be a reductive coset space with a fixed decomposition $\mathfrak{g} = \mathfrak{h} + \mathfrak{m}$, where $\mathrm{Ad}(h)\,\mathfrak{m} \subset \mathfrak{m}$ for all $h \in H$. We shall in this chapter study the set $\mathbf{D}(G/H)$ of differential operators on

G/H that are invariant under the action of G; a differential operator D on G/H belongs to $\mathbf{D}(G/H)$ if and only if $D^{\tau(g)} = D$ for all $g \in G$. We shall write $\mathbf{D}(G)$ instead of $\mathbf{D}(G/e)$. Let $L(g)$ and $R(g)$ denote the left and right translations of G onto itself given by $L(g)\cdot x = gx$, $R(g)\cdot x = xg^{-1}$. For each $f \in C^\infty(G/H)$ we put $\tilde{f} = f \circ \pi$. Then $\tilde{f} \in C^\infty(G)$ and $\tilde{f}$ is constant on each coset gH. The set of all such functions will be denoted by $C_0^\infty(G)$. Finally let $\mathbf{D}_0(G)$ denote the subset of $\mathbf{D}(G)$ consisting of operators that are invariant under right translations by H, that is $D \in \mathbf{D}_0(G)$ if and only if $D^{L(g)} = D$ and $D^{R(h)} = D$ for all $g \in G$ and all $h \in H$. Each $D \in \mathbf{D}_0(G)$ leaves the space $C_0^\infty(G)$ invariant.

LEMMA 16. *The algebra $\mathbf{D}(G/H)$ is isomorphic with the algebra of restrictions of $\mathbf{D}_0(G)$ to $C_0^\infty(G)$.*

Proof. The mapping $f \to f \circ \pi$ is an isomorphism of $C^\infty(G/H)$ onto $C_0^\infty(G)$. Let $D_0 \in \mathbf{D}_0(G)$; we define $D \in \mathbf{D}(G/H)$ by the requirement $(Df)^\sim = D_0\tilde{f}$ for all $f \in C^\infty\, G/H)$. This gives a mapping Ψ: $D_0 \to D$ of the algebra of restriction of $\mathbf{D}_0(G)$ to $C_0^\infty(G)$ into $\mathbf{D}(G/H)$. It is easy to see that Ψ is one-to-one, linear and preserves multiplication. To see that the image of Ψ is all of $\mathbf{D}(G/H)$, let $D' \in \mathbf{D}(G/H)$. We choose a basis $X_1, \ldots, X_n$ of $\mathfrak{m}$; Lemma 1 shows that for small t, $\exp(t_1 X_1 + \cdots + t_n X_n)$ is a local cross section in G over a neighborhood N of p_0 in G/H and the mapping

$$\pi(\exp(t_1 X_1 + \cdots + t_n X_n)) \to (t_1, \ldots, t_n)$$

defines a local coordinate system on G/H valid in N. There exists by Proposition 1 a polynomial P in n variables such that

$$[D'f](p_0) = \left[P\left(\frac{\partial}{\partial t_1}, \ldots, \frac{\partial}{\partial t_n}\right) \tilde{f}(\exp(t_1 X_1 + \cdots + t_n X_n))\right]_{t=0} \tag{3.1}$$

for $f \in C^\infty(G/H)$. Using $(D')^{\tau(g)} = D'$ we find easily that if $g \cdot p_0 = p$

$$[D'f](p) = \left[P\left(\frac{\partial}{\partial t_1}, \ldots, \frac{\partial}{\partial t_n}\right) \tilde{f}(g \exp(t_1 X_1 + \cdots + t_n X_n))\right]_{t=0}. \tag{3.2}$$

If $X_{n+1}, \ldots, X_r$ is a basis of $\mathfrak{h}$, the mapping

$$g \exp(t_1 X_1 + \cdots + t_r X_r) \to (t_1, \ldots, t_r)$$

is a coordinate system valid in a neighborhood of $g \in G$ and the operator D_0' defined by

$$[D_0' F](g) = \left[P\left(\frac{\partial}{\partial t_1}, \ldots, \frac{\partial}{\partial t_n}\right) F(g \exp(t_1 X_1 + \cdots + t_r X_r))\right]_{t=0} \tag{3.3}$$

for $F \in C^\infty(G)$, is a differential operator on G. Now if $h \in H$ we know $(D')^{\tau(h)} = D'$ so for $f \in C^\infty(G/H)$

$$[D'f](p_0) = [D' f^{\tau(h^{-1})}](p_0)$$
$$= \left[P\left(\frac{\partial}{\partial t_1}, \ldots, \frac{\partial}{\partial t_n}\right) \tilde{f}(h \exp(t_1 X_1 + \cdots + t_n X_n) h^{-1})\right]_{t=0}$$
$$= \left[P\left(\frac{\partial}{\partial t_1}, \ldots, \frac{\partial}{\partial t_n}\right) \tilde{f}(\exp \operatorname{Ad}(h)(t_1 X_1 + \cdots + t_n X_n))\right]_{t=0}$$

which in view of (3.1) implies

$$P(X_1, \ldots, X_n) = P(\operatorname{Ad}(h) \cdot X_1, \ldots, \operatorname{Ad}(h) \cdot X_n) \text{ for all } h \in H,$$

that is, P is invariant under H. It follows quickly that D_0' is invariant under all $R(h)$, $h \in H$. Similarly, if $x \in G$

$$[(D_0')^{L(x)} F](g) = [D_0' F^{L(x^{-1})}](x^{-1} g)$$
$$= \left[P\left(\frac{\partial}{\partial t_1}, \ldots, \frac{\partial}{\partial t_n}\right) F^{L(x^{-1})}(x^{-1} g \exp(t_1 X_1 + \ldots + t_r X_r))\right]_{t=0} = [D_0' F](g)$$

so $D_0' \in \mathbf{D}_0(G)$. The relations (3.2) and (3.3) imply that

$$(D'f)^{\sim} = D_0' \tilde{f} \qquad \text{for } f \in C^\infty(G/H)$$

so the image of Ψ' is all of $\mathbf{D}(G/H)$.

Now each $X \in \mathfrak{g}$ defines uniquely a left invariant vector field on G. This vector field is a differential operator on G (again denoted X) satisfying $X^{L(g)} = X$ for all $g \in G$. It follows easily that

$$[Xf](g) = \left[\frac{d}{dt} f(g \exp tX)\right]_{t=0} \qquad \text{for } f \in C^\infty(G). \tag{3.4}$$

This mapping of $\mathfrak{g}$ into $\mathbf{D}(G)$ sends the Lie algebra element $[X, Y]$ in the operator $X \cdot Y - Y \cdot X$ and extends uniquely to a homomorphism ξ of $U(\mathfrak{g})$, the universal enveloping algebra of $\mathfrak{g}$, into $\mathbf{D}(G)$. More crucially, ξ is an isomorphism of $U(\mathfrak{g})$ onto $\mathbf{D}(G)$. (See Harish-Chandra [22]). On the other hand, let $X_1, \ldots, X_r$ be a basis of $\mathfrak{g}$ and $S(\mathfrak{g})$ the symmetric algebra over $\mathfrak{g}$, that is the set of polynomials over $\mathbf{R}$ in the letters $X_1, \ldots, X_r$. Harish-Chandra's version [19] of the Poincaré-Birkhoff-Witt theorem gives a one-to-one linear mapping λ of $S(\mathfrak{g})$ onto $\mathbf{D}(G)$ with the property that for arbitrary elements $Y_1, \ldots Y_p$

$$\lambda(Y_1 Y_2 \ldots Y_p) = \frac{1}{p!} \sum_{\sigma} Y_{\sigma(1)} \cdot Y_{\sigma(2)} \ldots Y_{\sigma(p)} \tag{3.5}$$

where σ runs over the symmetric group on p letters. (Note the difference in the notation for multiplication in $S(\mathfrak{g})$ and $\mathbf{D}(G)$.) We shall refer to a mapping with the property (3.5) as "symmetrization".

For each $g \in G$, $\mathrm{Ad}(g)$ is an automorphism of $\mathfrak{g}$ and extends uniquely to an automorphism of $U(\mathfrak{g})$ which combined with ξ gives an automorphism of $\mathbf{D}(G)$. Denoting this automorphism again by $\mathrm{Ad}(g)$ we have

$$\mathrm{Ad}(g) \cdot D = D^{R(g)} \qquad \text{for } D \in \mathbf{D}(G). \tag{3.6}$$

In fact, since $D \to D^{R(g)}$ is an automorphism of $\mathbf{D}(G)$ it suffices, due to the uniqueness mentioned, to prove (3.6) when D is a vector field X. But $\mathrm{Ad}(g) \cdot X = X(X^{L(g)})^{R(g)} = X^{R(g)}$. Now if f is analytic in a neighborhood of $g \in G$, (3.4) implies that

$$f(g \exp tX) = \sum_0^\infty \frac{t^n}{n!} [X^n f](g) \tag{3.7}$$

for sufficiently small t. Using the fact that $D \in \mathbf{D}(G)$ has analytic coefficients we obtain from (3.4) and (3.7)

$$D \cdot X = X \cdot D \text{ if and only if } D^{R(\exp tX)} = D \text{ for all } t. \tag{3.8}$$

Let $\mathbf{Z}(G)$ denote the center of $\mathbf{D}(G)$; from (3.6) and (3.8) we see (¹) that $D \in \mathbf{Z}(G)$ if and only if $\mathrm{Ad}(g) \cdot D = D$ for all $g \in G$.

If V is a finite dimensional vector space over $\mathbf{R}$, $X_1, \ldots, X_l$ a basis of V, $S(V)$ shall denote the symmetric algebra over V, that is the algebra of polynomials over $\mathbf{R}$ in the letters $X_1, \ldots, X_l$. Let A be an endomorphism of V. A induces a homomorphism of $S(V)$, say $P \to A \cdot P$ where $(A \cdot P)(X_1, X_2, \ldots, X_l) = P(AX_1, AX_2, \ldots, AX_l)$. Using (3.5) it follows that $\lambda^{-1}(\mathbf{Z}(G))$ is the subset $I(\mathfrak{g})$ of $S(\mathfrak{g})$ consisting of all polynomials that are invariant under $\mathrm{Ad}(G)$. In the same manner we obtain

LEMMA 17. *$\lambda^{-1}(\mathbf{D}_0(G))$ is the set of polynomials $P \in S(\mathfrak{g})$ such that* $\mathrm{Ad}(h)P = P$ *for all $h \in H$.*

LEMMA 18. *$\mathbf{D}(G) = \mathbf{D}(G)\mathfrak{h} + \lambda(S(\mathfrak{m}))$ where the sum is a direct sum of vector spaces. (Here $\mathbf{D}(G)\mathfrak{h}$ denotes the left ideal in $\mathbf{D}(G)$ generated by $\mathfrak{h}$).*

Proof. To begin with we shall prove by induction that for each $P \in S(\mathfrak{g})$ there exists $Q \in S(\mathfrak{m})$ such that $\lambda(P-Q) \in \mathbf{D}(G)\mathfrak{h}$. This is obvious if P has degree 1 and we assume it true for all $P \in S(\mathfrak{g})$ of degree $< d$. To prove it for P of degree d we can assume P has the form $X_1^{e_1} \ldots X_r^{e_r}$, $e_1 + \cdots + e_r = d$ where $X_1, \ldots, X_r$ is a basis of $\mathfrak{g}$ such that $X_i \in \mathfrak{m}$ for $1 \leqslant i \leqslant n$ and $X_j \in \mathfrak{h}$ for $n+1 \leqslant j \leqslant r$. If $e_{n+1} = \cdots = e_r = 0$ there is nothing to prove; otherwise $\lambda(P)$ is a linear combination of terms of the form

(¹) Equivalent result is given in Harish-Chandra [20] Cor. of Lemma 11 and in I. Gelfand [12].

$X_{\alpha_1} \cdot X_{\alpha_2} \dots X_{\alpha_d}$ where for some i, $X_{\alpha_i} \in \mathfrak{h}$. Let $S_e(\mathfrak{g})$ denote the set of homogeneous polynomials in $S(\mathfrak{g})$ of degree e and put $\mathbf{D}_d(G) = \lambda(\sum_0^d S_e(\mathfrak{g}))$. Then

$$(X_{\alpha_1} \cdot X_{\alpha_2} \dots \cdot X_{\alpha_d}) - (X_{\alpha_1} \dots \cdot X_{\alpha_{i-1}} \cdot X_{\alpha_{i+1}} \dots X_{\alpha_d} X_{\alpha_i}) \in \mathbf{D}_{d-1}(G).$$

Therefore, there is an element $D \in \mathbf{D}_{d-1}(G)$ such that

$$\lambda(P) \equiv D \quad \bmod (\mathbf{D}(G)\mathfrak{h}).$$

Using the inductive assumption we obtain a $Q \in S(\mathfrak{m})$ such that $\lambda(P-Q) \in \mathbf{D}(G)\mathfrak{h}$ as desired. To prove the uniqueness we note first that if $P \in S(\mathfrak{g})$, $f \in C^\infty(G)$

$$[\lambda(P) f](e) = \left[P\left(\frac{\partial}{\partial t_1}, \dots, \frac{\partial}{\partial t_r}\right) f(\exp(t_1 X_1 + \dots + t_r X_r))\right]_{t=0}. \tag{3.9}$$

In fact, if f is analytic on a neighborhood of e in G, (3.7) shows that for sufficiently small t_i

$$f(\exp(t_1 X_1 + \dots + t_r X_r)) = \sum_0^\infty \frac{1}{m!} [(t_1 X_1 + \dots + t_r X_r)^m f](e)$$

$$= \sum_{M=0}^\infty \frac{1}{M!} \sum_{m_1 + \dots + m_r = m} t_1^{m_1} \dots t_r^{m_r} \frac{M!}{m_1! \dots m_r!} [\lambda(X_1^{m_1} \dots X_r^{m_r}) f](e).$$

Comparison with the usual Taylor formula yields (3.9). Now, by Lemma 1, exp $(t_1 X_1 + \dots + t_n X_n)$ defines for small t_i a local cross section in G over a neighborhood N of p_0 and $(t_1, \dots, t_n)$ are local coordinates on N. If $P \in S(\mathfrak{m})$, $P \neq 0$ we can choose $f^* = f^*(t_1, \dots, t_n)$ of class C^∞ such that

$$\left[P\left(\frac{\partial}{\partial t_1}, \dots, \frac{\partial}{\partial t_n}\right) f^* \right]_{t=0} \neq 0$$

and there exists a function $f \in C_0^\infty(G)$ such that

$$f(\exp(t_1 X_1 + \dots + t_n X_n)) = f^*(t_1, \dots, t_n)$$

for sufficiently small t_i. From (3.9) we have

$$[\lambda(P) f](e) = \left[P\left(\frac{\partial}{\partial t_1}, \dots, \frac{\partial}{\partial t_n}\right) f(\exp(t_1 X_1 + \dots + t_r X_r))\right]_{t=0}$$

$$= \left[P\left(\frac{\partial}{\partial t_1}, \dots, \frac{\partial}{\partial t_n}\right) f(\exp(t_1 X_1 + \dots + t_n X_n))\right]_{t=0} \neq 0.$$

Since each operator in $\mathbf{D}(G)\mathfrak{h}$ annihilates all of $C_0^\infty(G)$ we see that $\lambda(S(\mathfrak{m})) \cap \mathbf{D}(G)\mathfrak{h} = 0$. This proves Lemma 18.

LEMMA 19. *Each* $P \in S(\mathfrak{g})$ *which is invariant under* $\mathrm{Ad}_G(H)$ *is congruent* mod $\lambda^{-1}(\mathbf{D}(G)\mathfrak{h})$ *to a polynomial in* $S(\mathfrak{m})$ *invariant under* $\mathrm{Ad}_G(H)$.

Proof. By Lemma 18, $P = Q + Q_0$ where $Q \in S(\mathfrak{m})$ and $\lambda(Q_0) \in \mathbf{D}(G)\mathfrak{h}$. For each $h \in H$ we obtain $P = \mathrm{Ad}(h) \cdot Q + \mathrm{Ad}(h) \cdot Q_0$ and by (3.5) and (3.6), $\lambda(\mathrm{Ad}(h) \cdot Q_0) = \lambda(Q_0)^{R(h)}$. Now the mapping $D \to D^{R(h)}$ is an automorphism of $\mathbf{D}(G)$ leaving $\mathfrak{h}$ invariant. Hence it leaves $\mathbf{D}(G)\mathfrak{h}$ invariant and $\lambda(\mathrm{Ad}(h) \cdot Q_0) \in \mathbf{D}(G)\mathfrak{h}$. On the other hand $\mathrm{Ad}(h)Q_0 \in S(\mathfrak{m})$ and Lemma 19 follows from the uniqueness statement in Lemma 18.

Let $I(\mathfrak{g}/\mathfrak{h})$ denote the set of polynomials in $S(\mathfrak{m})$ that are invariant under $\mathrm{Ad}_G(H)$. We define a mapping of $I(\mathfrak{g}/\mathfrak{h})$ into $\mathbf{D}(G/H)$ as follows. If $P \in I(\mathfrak{g}/\mathfrak{h})$, then $\lambda(P) \in \mathbf{D}_0(G)$ and the restriction of $\lambda(P)$ to $C_0^\infty(G)$ gives by Lemma 16 rise to a well-defined operator $D_P \in \mathbf{D}(G/H)$. This mapping $P \to D_P$ is linear. It maps $I(\mathfrak{g}/\mathfrak{h})$ onto $\mathbf{D}(G/H)$ because Lemma 19 shows that if $P \in S(\mathfrak{g})$ is invariant under $\mathrm{Ad}_G(H)$ there exists a $Q \in I(\mathfrak{g}/\mathfrak{h})$ such that $\lambda(P)$ and $\lambda(Q)$ have the same restrictions to $C_0^\infty(G)$. Finally the mapping $P \to D_P$ is one-to-one. In fact, let $P \in I(\mathfrak{g}/\mathfrak{h})$, $P \neq 0$. As shown in the proof of Lemma 18 there exists a function $f \in C_0^\infty(G)$ such that $[\lambda(P)f](e) \neq 0$. The following theorem gives the desired representation of $\mathbf{D}(G/H)$.

THEOREM 10. *Let* G/H *be a reductive coset space,* $\mathfrak{g} = \mathfrak{h} + \mathfrak{m}$, $\mathrm{Ad}(h)(\mathfrak{m} \subset \mathfrak{m}$ *for* $h \in H$. *Let* $X_1, \ldots, X_n$ *be a basis of* $\mathfrak{m}$, *and let* $\tilde{f} = f \circ \pi$ *for* $f \in C^\infty(G/H)$. *There is a one-to-one linear correspondence* $Q \to D_Q$ *between* $I(\mathfrak{g}/\mathfrak{h})$ *and* $\mathbf{D}(G/H)$ *such that*

$$[D_Q f](p) = \left[Q\left(\frac{\partial}{\partial t_1}, \ldots, \frac{\partial}{\partial t_n}\right) \tilde{f}(g \exp(t_1 X_1 + \cdots + t_n X_n)) \right]_{t=0}$$

where $p = \pi(g)$. D_Q *is obtained from* $Q(X_1, \ldots, X_n)$ *by symmetrization (followed by the mapping* $\Psi^{\cdot}$ *from Lemma* 16).

REMARK. If $P = X_1^{e_1} \ldots \cdot X_r^{e_r}$ then (3.5) shows easily that

$$\lambda(P) = X_1^{e_1} X_2^{e_2} \ldots \cdot X_r^{e_r} + \lambda(Q)$$

where Q is of lower degree than P. It follows that if $P_1, P_2 \in I(\mathfrak{g}/\mathfrak{h})$ then $D_{P_1 P_2} = D_{P_1} D_{P_2} + D$ where the order of D is less than the sum of the degrees of P_1 and P_2.

COROLLARY *If* $I(\mathfrak{g}/\mathfrak{h})$ *has a finite system of generators, say* $P_1, \ldots, P_l$, *and we put* $D_i = D_{P_i}$, *then each* D *can be written*

$$D = \sum \alpha_{n_1 \ldots n_l} D_1^{n_1} \ldots D_l^{n_l} \qquad \text{where } \alpha_{n_1 \ldots n_l} \in \mathbf{R}.$$

In fact, suppose $D = D_P$ where $P \in I(\mathfrak{g}/\mathfrak{h})$. Then P can be written

$$P = \sum \beta_{n_1 \ldots n_l} P_1^{n_1} \ldots P_l^{n_l}, \quad \beta_{n_1 \ldots n_l} \in \mathbf{R}.$$

If $\beta P_1^{N_1} \dots P_l^{N_l}$ is the term of highest degree, the preceding remark shows that

$$D - \beta_1^{N_1} \dots \cdot D_l^{N_l}$$

is of lower order than D and the corollary follows by induction.

2. Invariant differential operators on two-point homogeneous spaces and on harmonic Lorentz spaces

THEOREM 11. *Let M be a two-point homogeneous space. The only differential operators on M that are invariant under all isometries of M are the polynomials in the Laplace-Beltrami operator Δ.*

Proof. If $\dim M = 1$, M is isometric to the real line or to a circle and in both cases Theorem 11 is obvious. We can therefore assume that $\dim M > 1$. From Chapter II, § 3 we know that M is isometric to a homogeneous space G/K where K is compact, G is a connected Lie group of isometries which is pairwise transitive on G/K. The Lie algebra $\mathfrak{g}$ of G can be written $\mathfrak{g} = \mathfrak{k} + \mathfrak{m}$ where $\mathfrak{k}$ is the Lie algebra of K, the group $\mathrm{Ad}_G(K)$ leaves $\mathfrak{m}$ invariant and acts transitively on the directions in $\mathfrak{m}$. $\mathrm{Ad}_G(K)$ leaves invariant a positive definite inner product on $\mathfrak{m}$; let $X_1, \dots, X_n$ be an orthonormal basis with respect to this inner product. Each $D \in \mathbf{D}(G/K)$ has by Theorem 10 the form D_P where $P \in I(\mathfrak{g}/\mathfrak{k})$. Explicitly, we write $P = \sum a_{r_1 \dots r_n} X_1^{r_1} \dots . X_n^{r_n}$ and consider the corresponding polynomial function P^* on $\mathfrak{m}$ given by $P^*(X) = \sum a_{r_1 \dots r_n} \cdot x_1^{r_1} \dots . x_n^{r_n}$ if $X = \sum x_i X_i$. Since $P \in I(\mathfrak{g}/\mathfrak{k})$ we have $P^*(\mathrm{Ad}(k)X) = P^*(X)$ for all $k \in K$, and it follows that P^* is constant on each sphere around the origin in $\mathfrak{m}$. Thus P^* can be written

$$P^*(X) = \sum_1^N a_k (x_1^2 + \dots + x_n^2)^k \qquad \text{where } a_k \in \mathbf{R}$$

and

$$P = \sum_1^N a_k (X_1^2 + \dots + X_n^2)^k.$$

Let Δ denote the member of $\mathbf{D}(G/K)$ that corresponds to the invariant polynomial $X_1^2 + \dots + X_n^2$. From the remark following Theorem 10 we know that $D_P - \sum_1^N a_k \Delta^k = D_Q$ where Q belongs to $I(\mathfrak{g}/k)$ and has degree lower than P. Theorem 11 now follows by a simple induction.

It is to be expected in view of Theorem 11 that potential theory on two-point homogeneous spaces parallels potential theory in Euclidean spaces very closely. This agrees also with the fact that two-point homogeneous spaces are harmonic spaces and as Willmore

[37] has shown, harmonic spaces can be characterized by the fact that the usual mean value theorem for solutions of Laplace's equation $\Delta u = 0$ remains valid.

We shall next consider the case of a harmonic Lorentz space M with metric tensor Q.

THEOREM 12. *The algebras* $\mathbf{D}(G^0/H)$, $\mathbf{D}(G^-/H$ *and* $\mathbf{D}(G^+/H)$ *consist of all polynomials in the Laplace-Beltrami operator* $\square$.

It is easy to adapt the proof of Theorem 11 to the present case. The essential point is that $\mathrm{Ad}_G(H)$, $(G = G^0,\ G^-$ or $G^+)$, acts transitively on each component of the set $\{X \in \mathfrak{m} \mid X \neq 0,\ Q(X, X) = c\}$. Here Q is the quadratic form on $\mathfrak{m}$ invariant under $\mathrm{Ad}_G(H)$.

3. The case of a symmetric coset space

The assumption that M is symmetric also has important consequences as Theorem 13 shows. This theorem is essentially known from Gelfand's paper [11], and in [34] A. Selberg gave a very direct and transparent proof.

THEOREM 13. *Let* G/K *be a symmetric coset space,* K *compact. Then* $\mathbf{D}(G/K)$ *is commutative.*

In the special case when G is a complex semi-simple Lie group and K is a maximal compact subgroup, the algebra $\mathbf{D}(G/K)$ can be described more explicitly. It is known that K is connected and the Lie algebra $\mathfrak{g}$ of G is the complexification of $\mathfrak{k}$, the Lie algebra of K. We express this by the relation $\mathfrak{g} = \mathfrak{k} + i\,\mathfrak{k}$ where $\mathfrak{g}$ and $\mathfrak{k}$ are considered as Lie algebras over $\mathbf{R}$. As is well known G/K is a symmetric coset space and thus $\mathbf{D}(G/K)$ is commutative. Let $I(\mathfrak{k})$ denote the set of polynomials in $S(\mathfrak{k})$ that are invariant under the adjoint group of K. Then it is easy to see that the mapping $iX \to X$ of $i\,\mathfrak{k}$ onto $\mathfrak{k}$ induces an isomorphism of $I(\mathfrak{g}\,\mathfrak{k}/)$ onto $I(\mathfrak{k})$. The algebra $I(\mathfrak{k})$ has significance in topological study of the group K (see e.g. C. Chevalley [8]) during which the following results have been proved. Let 1 be the rank of K (dimension of the maximal tori) and let p_i be the indices occurring in the Hopf-splitting of the Poincaré polynomial of K

$$\sum_p B_p\, t^p = \prod_{i=1} (1 + t^{p_i}).$$

Then $I(\mathfrak{k})$ is generated by l algebraically independent polynomials $P_1, \ldots, P_l$ of degrees $\frac{1}{2}(p_i + 1)$, $i = 1, \ldots, l$. The corresponding operators $D_{P_1}, \ldots, D_{P_l}$ form a system of generators of $\mathbf{D}(G/K)$.

18 – 593805. *Acta mathematica.* 102. Imprimé le 16 décembre 1959

Chapter IV

Mean value theorems

1. The mean value operator

Suppose now that G is a connected Lie group and K a compact subgroup. We fix a G-invariant Riemannian metric tensor Q on G/K, and denote the distance function by d. There exists in this case a finite system $D^1, \ldots, D^l$ of generators(¹) for $\mathbf{D}(G/K)$. Let dk denote a normalized invariant measure on K. If π is the natural projection of G onto G/K we put as before $\tilde{f} = f \circ \pi$ for each $f \in C^\infty(G/K)$. Let x be a fixed element of G. The function

$$g \to \int_K \tilde{f}(g k x)\, dk$$

is constant on each coset gK and determines a C^∞-function on G/K which we call $M^x f$. M^x is therefore the linear operator on $C^\infty(G/K)$ given by

$$[M^x f](p) = \int_K \tilde{f}(g k x)\, dk \qquad \text{if } \pi(g) = p.$$

The set $\{\pi(gkx) \mid k \in K\}$ is the orbit of the point $\pi(gx)$ under the group gKg^{-1} and lies on a sphere in G/K with center $\pi(g)$. $[M^x f](p)$ is the average of the values of f on this orbit. In the case that G is pairwise transitive on G/K, M^x is the operation of averaging over a sphere of fixed radius equal to $d(\pi(e), \pi(x))$. Next theorem shows that M^x can be represented as a function of the operators $D^1, \ldots, D^l$. This was proved by Berezin and Gelfand in [2] for the case when G/K is symmetric. Their proof, which does not seem to generalize to the non-symmetric case, is different from ours, which was found independently.

Theorem 14. *Let $p \in G/K$ and let U be a neighborhood of p. Suppose $X \in \mathfrak{g}$ is so small that U contains the sphere with center p and radius $d(\pi(e), \pi(\exp X))$. Then there exists a neighborhood V of p, $V \subset U$, and certain polynomials without constant term, say p_n, such that*

$$[M^{\exp X} f](q) = f(q) + \sum_n [p_n(D^1, \ldots, D^l) f](q)$$

for each f analytic on U and each $q \in V$.

Proof. Choose $g_0 \in G$ such that $\pi(g_0) = p$, and let $x = \exp X$. Then $\pi(g_0 k x) \in U$ for all $k \in K$, and there exists a neighborhood U^* of g_0 in G such that $\pi(gkx) \in U$ for all $g \in U^*$ and all $k \in K$. Put $V = \pi(U^*)$. Now suppose f is analytic in U and $q \in V$. Select $g \in G$ such that $\pi(g) = q$. Then

(¹) This is also proved in [34].

$$[M^x f](q) = \int_K \tilde{f}(g k x k^{-1})\, dk = \int_K \tilde{f}(g \exp Ad(k) X)\, dk$$

which by (3.4) is equal to

$$\int_K \sum_0^\infty \frac{1}{m!} [(Ad(k) X)^m \tilde{f}](g)\, dk \tag{4.1}$$

Let $X_1, \ldots, X_n$ be an orthonormal basis of $\mathfrak{M}$ and write $Ad(k) = r \sum_1^n k_i X_i$ where $r^2 = Q_{p_0}(X, X)$ and $\sum k_i^2 = 1$. Put

$$f^*(t_1, \ldots t_n) = \tilde{f}(g \exp (t_1 X_1 + \cdots + t_n X_n).$$

Then by (3.9)

$$[(Ad(k) X)^m \tilde{f}](g) = \left[\left(r k_1 \frac{\partial}{\partial t_1} + \cdots + r k_n \frac{\partial}{\partial t_n}\right)^m f^*(t_1, \ldots, t_n)\right]_{t=0}$$

and (4.1) is just the ordinary Taylor series for $f^*(r k_1 \ldots r k_n)$. Thus the series (4.1) converges uniformly in k so the summation and integration can be interchanged; also $(\mathrm{Ad}(k) X)^m = \mathrm{Ad}(k) \cdot X^m$ and the operator $\int \mathrm{Ad}(k) \cdot X^m\, dk$ belongs to $\mathbf{D}_0(G)$. By Lemma 16 this corresponds to an operator $D^m \in \mathbf{D}(G/K)$ which can be written $p^m(D^1, \ldots, D^l)$ as we have seen, and the theorem follows.

We shall now generalize the well-known mean value theorem of Ásgeirsson [1] for solutions of the ultrahyperbolic equation

$$\frac{\partial^2 u}{\partial x_1^2} + \cdots + \frac{\partial^2 u}{\partial x_n^2} = \frac{\partial^2 u}{\partial y_1^2} + \cdots + \frac{\partial^2 u}{\partial y_n^2}$$

which states that each solution $u(x_1, \ldots, x_n; y_1, \ldots, y_n) = u(X, Y)$ satisfies the relation

$$\int_{\mathrm{S}_r(X_0)} u(X, Y_0)\, d\boldsymbol{\omega}_r(X) = \int_{\mathrm{S}_r(Y_0)} u(X_0, Y)\, d\boldsymbol{\omega}_r(Y)$$

for every $X_0, Y_0 \in \mathbf{R}^n$. Here $d\boldsymbol{\omega}_r$ stands for the Euclidean area element of the sphere S_r.

Definition. Let u be a function in $C^\infty(G/K \times G/K)$. We say u is of *slow growth* if $D_1 u$ and $D_2 u$ are bounded for each $D \in \mathbf{D}(G/K)$.

Theorem 15. *Let u be a function on $G/K \times G/K$ which is either of slow growth or analytic. Suppose u satisfies the differential equations*

$$D_1 u = D_2 u \quad \text{for all} \quad D \in \mathbf{D}(G/K) \tag{4.2}$$

Then

$$M_1^x u = M_2^x u \quad \text{for all} \quad x \in G. \tag{4.3}$$

(*Here the subscripts* 1, 2 *on an operator indicate that it operates on the first and second variable respectively.*) *Conversely, if* (4.3) *holds for a function of class* C^∞, *then* (4.2) *follows.*

18* – 593805

Proof. We first prove the theorem under the assumption that u is analytic (but not necessarily bounded). If (4.2) holds, it follows from Theorem 14 that (4.3) is valid at least if x lies in a suitable neighborhood of e in G. Since however both sides of (4.3) are analytic in x, (4.3) holds for all x. On the other hand, if u belongs to $C^\infty(G/K \times G/K)$ and $\tilde{u}$ is defined by $\tilde{u}(g, g') = u(\pi(g), \pi(g'))$ then $\tilde{u} \in C^\infty(G \times G)$ and the relation (4.3) can be written

$$\int_K \tilde{u}(g k x, g')\, dk = \int_K \tilde{u}(g, g' k x)\, dk. \tag{4.4}$$

We take now an operator $T \in \mathbf{D}_0(g)$ and apply to both sides of (4.4) considered as functions of x, and put $x = e$. It follows that

$$[T_1 \tilde{u}](g, g') = [T_2 \tilde{u}](g, g')$$

which is equivalent to (4.2).

Let us consider the case when u is constant in the second argument, i.e., $\tilde{u}(g, g') = \tilde{u}(g, e)$ and put $\tilde{v}(g) = \tilde{u}(g, e)$. The algebra $\mathbf{D}(G/K)$ always contains an elliptic operator, e.g., the Laplace-Beltrami operator with respect to the G-invariant metric. By S. Bernstein's theorem, a function v that satisfies the equation

$$Dv = 0 \tag{4.5}$$

for all D that annihilate constants, is automatically analytic. Using (4.3) we see that each solution of (4.5) is characterized by the mean value relation

$$M^x v = v \text{ for all } x \in G.$$

This result was proved somewhat differently by Godement [15]. It generalizes the mean value theorem for harmonic functions in $\mathbf{R}^n$. Earlier Feller [10] had extended this theorem to certain non-Euclidean spaces in connection with mean value theorems for more general elliptic equations. Whereas the assumption of analyticity is no restriction in Godement's theorem, this is not so in Theorem 15 where the most interesting solutions are the non-analytic ones.

Let dg denote a left invariant Haar measure on G. The convolution $f_1 * f_2$ of two functions f_1 and f_2 on G is defined by

$$f_1 * f_2(x) = \int_G f_1(y) f_2(y^{-1} x)\, dy$$

whenever this integral exists. We shall use the following lemma to prove Theorem 15 in full generality.

LEMMA 20. *Let f be a bounded continuous function on G, ε a number > 0 and C a compact subset of G. Then there exists a function φ on G such that*

$$\varphi * f \quad \textit{is analytic} \tag{4.6}$$

$$|(\varphi * f)(x) - f(x)| < \varepsilon \quad \textit{for all } x \in C. \tag{4.7}$$

Proof. If G is compact the lemma is an easy consequence of the Peter-Weyl theory and it can also be proved directly for a commutative Lie group. The general case is handled by using the fact that as a manifold G is analytically isomorphic to a product manifold $K \times N$ where K is a compact subgroup of G and N is a submanifold of G analytically isomorphic to a Euclidean space.

An analogous procedure is followed in Harish-Chandra's theory of well-behaved vectors (see Harish-Chandra [20] and the generalization given by Cartier-Dixmier [6]). As we shall indicate, Lemma 20 is essentially contained in the theorem which states that the well-behaved vectors are dense in the representation space.

Let π denote the left regular representation of G on the Banach space $L^1(G)$ of functions on G that are integrable with respect to left invariant Haar measure, that is $[\pi(x)h](y) = h(x^{-1}y)$ for $h \in L^1(G)$. If h is a well-behaved vector in $L^1(G)$ then so is $\pi(x)h$ and from Lemma 18 in [20] it follows that if f is bounded and continuous on G, the function

$$x \to \int_G f(y)\,[\pi(x)h](y)\,dy$$

is analytic on G and the function $h * f$ likewise. Now to prove Lemma 20 we select a continuous function γ on G of compact support such that

$$|\gamma * f(x) - f(x)| < \frac{\varepsilon}{2} \quad \text{for } x \in C;$$

next we select a sequence (φ_n) of well-behaved vectors converging to γ. Then the sequence $(\varphi_n * f)(g)$ converges to $(\gamma * f)(g)$ uniformly on G and a suitable φ_N satisfies (4.6) and (4.7).

Now we can finish the proof of Theorem 15. Let u be a solution of (4.2) of slow growth. The function $\tilde{u}$ on $G \times G$ introduced earlier satisfies

$$T_1 \tilde{u} = T_2 \tilde{u} \quad \text{for each } T \in \mathbf{D}_0(G).$$

If φ belongs to $C^\infty(G \times G)$ and $L^1(G \times G)$ the convolution

$$(\varphi * u)(x_1, x_2) = \int_{G \times G} \varphi(y_1, y_2)\,\tilde{u}(y_1^{-1}x_1, y_2^{-1}x_2)\,dy_1\,dy_2$$

exists, and since u is of slow growth

$$T_1(\varphi * \tilde{u}) = \varphi * T_1 \tilde{u}$$
$$T_2(\varphi * \tilde{u}) = \varphi * T_2 \tilde{u}$$

and therefore

$$T_1(\varphi * \tilde{u}) = T_2(\varphi * \tilde{u}) \quad \text{for all } T \in \mathbf{D}_0(G). \tag{4.8}$$

The function $\varphi * \tilde{u}$, being constant on left cosets mod $(K \times K)$ determines a function $v \in C^\infty(G/K \times G/K)$ such that

$$v(\pi(g), \pi(g')) = (\varphi * \tilde{u})(g, g')$$

and

$$D_1 v = D_2 v \quad \text{for all } D \in \mathbf{D}(G/K).$$

If we choose φ in accordance with Lemma 20, v is an analytic solution of (4.2), and a suitable sequence of such solutions approximates u uniformly on compact subsets of G/K and (4.3) follows.

Remark. The relation $M^x v = v$ which characterizes the solutions of (4.5) can be written differently. Let h have compact support on G and satisfy the conditions:

$$\text{(i)} \quad h(xk) = h(x) \quad \text{for all } x \in G \text{ and all } k \in K$$

$$\text{(ii)} \quad \int_G h(x^{-1})\, dx = 1.$$

The relation $M^x v = v$ for all x is then equivalent to

$$\tilde{v} * h = \tilde{v}$$

for every h with the properties (i) and (ii). This is easily proved by using the integration theory on homogeneous spaces and shows how the operators in $\mathbf{D}(G/K)$ appear as infinitesimal generators for the convolution operators $\tilde{f} \to \tilde{f} * h$ considered as operators on $C^\infty(G/K)$.

2. The Darboux equation in a symmetric space

We shall now suppose G/K is a symmetric coset space and K compact. Here the algebra $\mathbf{D}(G/K)$ is commutative; we shall give certain consequences of this fact.

Theorem 16. *For each $x \in G$, M^x commutes with all the operators in $\mathbf{D}(G/K)$.*

Proof. It is clear from Theorems 13 and 14 that if f is analytic on G/K and $D \in \mathbf{D}(G/K)$, then

$$D M^x f = M^x D f \tag{4.9}$$

if x is sufficiently close to e in G. However f and Df are analytic so (4.9) holds for all $x \in G$. Let T be the operator in $\mathbf{D}_0(G)$ that corresponds to D according to Lemma 16, and N^x the operator on $C^\infty(G)$ given by

$$[N^x F](g) = \int_K F(gkx)\, dk.$$

We shall now prove (4.9) for $f \in C_c^\infty(G/K)$ and fixed $x \in G$. The arguments used in the proof of Lemma 20 show that there exists a sequence φ_n of functions on G such that $\varphi_n * \tilde{f}$, $\varphi_n * TN^x \tilde{f}$ and $\varphi_n * T\tilde{f}$ are all analytic functions on G and the sequences $(\varphi_n * \tilde{f})$, $(\varphi_n * TN^x \tilde{f})$, $(\varphi_n * T\tilde{f})$ converge to the functions $\tilde{f}$, $TN^x \tilde{f}$ and $T\tilde{f}$ uniformly on G. Using the obvious relations

$$\varphi_n * N^x \tilde{f} = N^x (\varphi_n * \tilde{f})$$

$$\varphi_n * T \tilde{f} = T (\varphi_n * \tilde{f}),$$

(4.9) follows easily for each $f \in C_c^\infty(G/K)$. Finally, to prove (4.9) for each $f \in C^\infty(G/K)$, one just has to observe that for each compact subset M of G/K there exists a function $f_M \in C_c^\infty(G/K)$ which agrees with f on an open set containing M.

The following corollary is proved in [2] in a different way.

COROLLARY. *Let $f \in C^\infty(G/K)$ and put*

$$V(x, g) = \int_K \tilde{f}(gkx)\, dk.$$

Then V satisfies the "Darboux Equation"

$$T_1 V = T_2 V \quad \text{for each } T \in \mathbf{D}_0(G).$$

In fact, write $T\tilde{f} = F$. Then

$$[T_1 V](x, g) = \int_K F(gkx)\, dk = [N^x T \tilde{f}](g) = [T N^x \tilde{f}](g) = [T_2 V](x, g).$$

The zonal spherical functions φ on G/K introduced by E. Cartan and I. Gelfand are by definition the (analytic) eigenfunctions of all $D \in \mathbf{D}(G/K)$ which are invariant under K, that is $\varphi^{\tau(k)} = \varphi$ for all $k \in K$. Since M^x (for x near e in G) is a power series in the generators $D^1, \ldots, D^l$, $M^x = P(D^1, \ldots, D^l)$, it is clear that φ is an eigenfunction of M^x, $M^x \varphi = \lambda \varphi$. It follows that if φ is not identically 0, then $\varphi(\pi(e)) \neq 0$ so we assume the zonal spherical functions normalized by $\varphi(\pi(e)) = 1$. These functions then satisfy the functional equation

$$M^x \varphi = \varphi(\pi(x))\, \varphi.$$

On the other hand, there exist constants $\lambda^1, \ldots, \lambda^l$ such that $D^i \varphi = \lambda^i \varphi$. Hence $M^x \varphi = P(\lambda^1, \ldots, \lambda^l)\varphi$ so

$$\varphi(\pi(x)) = P(\lambda^1, \ldots, \lambda^l).$$

This shows that φ is determined by the ordered system $(\lambda^1, \ldots, \lambda^l)$ of eigenvalues. Formally M^x is a zonal spherical function of the operators $D^1, \ldots, D^l$.

3. Invariant differential equations on two-point homogeneous spaces

We shall now combine the previous group theoretic methods with special geometric properties of two-point homogeneous spaces. This leads naturally to more explicit results.

We shall now assume that M is a two-point homogeneous space, and we exclude in advance the trivial case when M has dimension 1. Let G be the connected component of e in the group of all isometries of M. Then M can be represented G/K where K is compact and G is pairwise transitive on M. $\mathbf{D}(G/K)$ consists of all polynomials in the Laplace-Beltrami operator Δ. We see also that the mean value operators M^x and M^y are the same if $d(\pi(e), \pi(x)) = d(\pi(e), \pi(y))$ and consequently we write M^r instead of M^x if $r = d(\pi(e), \pi(x))$. Let p be a point in M, $\mathbf{S}_r(p)$ the geodesic sphere around p with radius r, $d\omega_r$ the volume element on $\mathbf{S}_r(p)$ and $A(r)$ the area of $\mathbf{S}_r(p)$.

LEMMA 21. *In geodesic polar coordinates around p, Δ has the form*

$$\Delta = \frac{\partial^2}{\partial r^2} + \frac{1}{A(r)} \frac{dA}{dr} \frac{\partial}{\partial r} + \Delta'$$

where Δ' is the Laplace-Beltrami operator on $\mathbf{S}_r(p)$.

Proof. Let the geodesic polar coordinates be denoted by $r, \theta_1, \ldots, \theta_{n-1}$. Due to the fact that the geodesics emanating from p are perpendicular to $\mathbf{S}_r(p)$ the metric tensor must have the form

$$ds^2 = dr^2 + \sum_{i,j=1}^{n-1} g_{ij}\, d\theta_i\, d\theta_j$$

and the Laplace-Beltrami operator is given by

$$\Delta = \frac{\partial^2}{\partial r^2} + \frac{1}{\sqrt{g}} \frac{\partial \sqrt{g}}{\partial r} \frac{\partial}{\partial r} + \frac{1}{\sqrt{g}} \sum_k \frac{\partial}{\partial \theta_k} \left(\sum_i g^{ik} \sqrt{g} \frac{\partial}{\partial \theta_i} \right).$$

Since r and Δ are invariant under the subgroup of G that leaves p fixed, Δr is also invariant under this subgroup which acts transitively on the geodesics emanating from p. Hence

$$\frac{1}{\sqrt{g}} \frac{\partial \sqrt{g}}{\partial r}$$

is a function of r alone so

$$\log \sqrt{g} = \alpha(r) + \beta(\theta_1, \ldots, \theta_{n-1})$$

and

$$\sqrt{g} = e^{\alpha(r)} e^{\beta(\theta_1, \ldots, \theta_{n-1})}.$$

On the other hand, the volume of $\mathbf{S}_r(p)$ is given by

$$V(r) = \int \sqrt{g}\, dr\, d\theta_1, \ldots, d\theta_{n-1}$$

and thus we find for $A(r) = dV/dr$ the formula

$$A(r) = \int \sqrt{g}\, d\theta_1, \ldots, d\theta_{n-1} = C e^{\alpha(r)} \quad (C = \text{constant})$$

and
$$\frac{1}{\sqrt{g}} \frac{\partial \sqrt{g}}{\partial r} = \frac{1}{A(r)} \frac{dA}{dr}.$$

The lemma now follows by observing that the induced metric on $S_r(p)$ is given by

$$ds_1^2 = \sum_{i,j=1}^{n-1} g_{ij}\, d\theta_i\, d\theta_j.$$

The next lemma, which also is proved by Günther [16], is just a special case of the corollary of Theorem 16.

LEMMA 22. *Let $f \in C^\infty(M)$ and put $F(p,q) = [M^r f](p)$ if $p, q \in M$, $d(p,q) = r$. Then $\Delta_1 F = \Delta_2 F$.*

We shall now state and give a different proof for the extension of Ásgeirsson's theorem to two-point homogeneous spaces. The proof is based on an ingenious method used in Ásgeirsson's original proof ([1], p. 334).

THEOREM 17. *Let M be a two-point homogeneous space and let u be a twice continuously differentiable function on $M \times M$ which satisfies the equation*

$$\Delta_1 u = \Delta_2 u \tag{4.10}$$

Then for each $(x_0, y_0) \in M \times M$

$$\int_{S_r(x_0)} u(x, y_0)\, d\omega_r(x) = \int_{S_r(y_0)} u(x_0, y)\, d\omega_r(y). \tag{4.11}$$

Proof. We assume first M is non-compact. From Theorem 3 we know that M is isometric to a symmetric Riemannian space G/K. $\mathrm{Ad}_G(K)$ is transitive on the directions in the tangent space to G/K at $\pi(e)$, in particular G/K is irreducible. As we saw at the end of the proof of Theorem 3, geodesic polar coordinates with origin at a point $p \in M$ are valid on the entire M.

Now, suppose the function u satisfies (4.10) and let (x_0, y_0) be an arbitrary point in $M \times M$. Consider the function U defined by

$$U(r, s) = [M_1^r M_2^s u](x_0, y_0) \quad \text{for } r, s \geqslant 0$$

We view U as a function on $M \times M$ by giving it the value $U(r,s)$ on the set $S_r(x_0) \times S_s(y_0)$. Since Δ commutes [1] with M^r we obtain from (4.10) and Lemma 21

---

(1) Theorem 16 shows that Δ and M^r commute when applied to C^∞-functions. In the same way it can be shown that they commute when applied to C^2-functions.

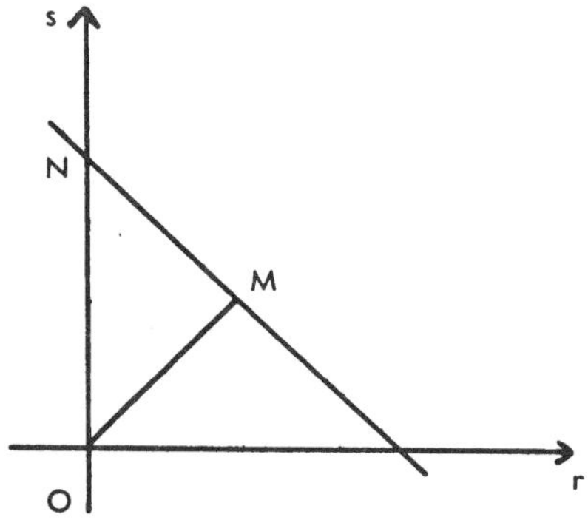

$$\frac{\partial^2 U}{\partial r^2} + \frac{1}{A(r)} \frac{dA}{dr} \frac{\partial U}{\partial r} = \frac{\partial^2 U}{\partial s^2} + \frac{1}{A(s)} \frac{dA}{ds} \frac{\partial U}{\partial s} .$$

If we put $F(r, s) = U(r, s) - U(s, r)$ we obtain the relations

$$\frac{\partial^2 F}{\partial r^2} + \frac{1}{A(r)} \frac{dA}{dr} \frac{\partial F}{\partial r} - \frac{\partial^2 F}{\partial s^2} - \frac{1}{A(s)} \frac{dA}{ds} \frac{\partial F}{\partial s} = 0 \tag{4.12}$$

$$F(r, s) = -F(s, r).$$

After multiplication of (4.12) by $2\, A(r) \partial F/\partial s$ and some manipulation we obtain

$$-A(r) \frac{\partial}{\partial s}\left[\left(\frac{\partial F}{\partial r}\right)^2 + \left(\frac{\partial F}{\partial s}\right)^2\right] + 2 \frac{\partial}{\partial r}\left(A(r) \frac{\partial F}{\partial r} \frac{\partial F}{\partial s}\right) - \frac{2A(r)}{A(s)} \frac{dA}{ds} \left(\frac{\partial F}{\partial s}\right)^2 = 0. \tag{4.13}$$

Now consider the line MN with equation $r + s =$ constant in the (r, s)-plane and form the plane integral of (4.12) over the triangle OMN, (see figure), and use Green's formula. If $\partial/\partial n$ denotes derivation in the direction of the outgoing normal and dl is the element of arc length, we obtain

$$\oint_{OMN} \left\{-A(r) \left[\left(\frac{\partial F}{\partial r}\right)^2 + \left(\frac{\partial F}{\partial s}\right)^2\right] \frac{\partial s}{\partial n} + 2A(r) \frac{\partial F}{\partial r} \frac{\partial F}{\partial s} \frac{\partial r}{\partial n}\right\} dl - \iint_{OMN} \frac{2A(r)}{A(s)} \frac{dA}{ds} \left(\frac{\partial F}{\partial r}\right)^2 dr\, ds = 0. \tag{4.14}$$

On OM: $\left(\frac{\partial r}{\partial n}, \frac{\partial s}{\partial n}\right) = \left(\frac{1}{\sqrt{2}}, -\frac{1}{\sqrt{2}}\right)$, $F(r, r) = 0$ so $\frac{\partial F}{\partial r} + \frac{\partial F}{\partial s} = 0.$

On MN: $\left(\frac{\partial r}{\partial n}, \frac{\partial s}{\partial n}\right) = \left(\frac{1}{\sqrt{2}}, \frac{1}{\sqrt{2}}\right).$

On ON: $A(r) = 0.$

From (4.14) follows the relation

$$\frac{1}{\sqrt{2}} \int_{MN} A(r) \left(\frac{\partial F}{\partial r} - \frac{\partial F}{\partial s}\right)^2 dl + \iint_{OMN} \frac{2A(r)}{A(s)} \frac{dA}{ds} \left(\frac{\partial F}{\partial r}\right)^2 dr\, ds = 0. \tag{4.15}$$

Now Lemma 3 shows that $dA/ds \geqq 0$ for all s and (4.15) shows therefore that

$$\frac{\partial F}{\partial r} - \frac{\partial F}{\partial s},$$

which is the directional derivative of F in the direction MN, vanishes on MN. Consequently $F \equiv 0$ so U is symmetric. In particular $U(r, 0) = U(0, r)$ and this is (4.11). If M is compact the proof above fails since $A(r)$ is no longer an increasing function of r. To show that (4.11) is valid even if the solution u is not analytic we resort again to Lemma 20 to approximate u by analytic solutions. Since M is compact this requires only the Peter-Weyl theory and not the theory of well-behaved vectors.

We recall now some facts from [4] about the behaviour of the geodesics on $M = G/K$. Let $p_0 = \pi(e)$ and dim $M = n$. If M is non-compact Exp maps $M_{p_0} (= \mathfrak{m})$ homeomorphically onto M. If M is compact all geodesics are closed and have the same length 2σ. The mapping Exp maps the ball $0 \leqslant Q_{p_0}(X, X) \leqslant \sigma^2$ onto M and is one-to-one on the open ball $0 \leqslant Q_{p_0}(X, X) < \sigma^2$. Except for the real elliptic spaces, Exp becomes singular on the sphere $Q_{p_0}(X, X) = \sigma^2$, and thus the set $S_\sigma(p_0)$, which Cartan calls the *antipodal variety associated to* p_0, will in general have dimension inferior to $n-1$. For the various n-dimensional two-point homogeneous spaces the dimension of $S_\sigma(p_0)$ is given in [4] as 0, $n-1$, $n-2$, $n-4$ for the spheres, real elliptic spaces, hermitian elliptic spaces and quaternian elliptic spaces respectively. For the Cayley elliptic plane $S_\sigma(p_0)$ has dimension 8.

The following theorem gives a generalization of the Poisson equation to two-point homogeneous spaces. Consider the function

$$\varphi(r) = \int_a^r \frac{1}{A(t)} dt \text{ where } \begin{cases} a > 0 \text{ if } M \text{ is non-compact} \\ a = \sigma \text{ if } M \text{ is real elliptic} \\ 0 < a < \sigma \text{ otherwise} \end{cases}$$

We define the function Ψ by

$$\Psi(p, q) = \varphi(r) \text{ if } d(p, q) = r.$$

In view of Lemma 21, Ψ satisfies the equations $\Delta_1 \Psi = \Delta_2 \Psi = 0$ and as the following theorem shows Ψ can be regarded as a fundamental solution.

THEOREM 18. *Let f be a twice continuously differentiable function on M with compact support. Then the function u given by (dq is the volume element on M)*

$$u(p) = \int_M f(q) \Psi(p, q) dq$$

satisfies the "Poisson equation"

$$\Delta u = f \quad \textit{if } M \textit{ is non-compact} \tag{4.16}$$

$$\Delta u = f - M^\sigma f \quad \textit{if } M \textit{ is compact.} \tag{4.17}$$

In the compact case the compensating term $[M^\sigma f](p)$ is the average of f on the antipodal variety associated to p. In the case when M is a sphere, $M^\sigma f = f \circ A$ where A is the antipodal mapping.

We first prove (4.16). Since $\varphi(r) = O(r^{2-n})$ as $r \to 0$ the integral $\int f(q)\psi(p, q)\,dq$ is convergent and

$$u(p) = \int_M f(q)\Psi(p, q)\,dq = \int_0^\infty dr \int_{S_r(p)} f(q)\Psi(p, q)\,d\omega_r(q)$$

$$= \int_0^\infty A(r)\varphi(r)[M^r f](p)\,dr.$$

We apply Δ to this relation and make use of Lemma 22. Then we obtain

$$[\Delta u](p) = \int_0^\infty \varphi(r) A(r)[\Delta M^r f](p)\,dr = \int_0^\infty \varphi(r) A(r)\Delta_r([M^r f](p))\,dr.$$

Now we keep p fixed (and omit writing it in the formulas below) and use Lemma 21. Then

$$\Delta u = \int_0^\infty \varphi(r) A(r) \left\{\frac{\partial^2 M^r f}{\partial r^2} + \frac{1}{A(r)}\frac{dA}{dr}\frac{\partial M^r f}{\partial r}\right\} dr$$

$$= \lim_{\varepsilon \to 0}\left[\varphi(r) A(r) \frac{\partial}{\partial r} M^r f\right]_\varepsilon^\infty - \lim_{\varepsilon \to 0} \int_\varepsilon^\infty \varphi'(r) A(r) \frac{\partial}{\partial r} M^r f\,dr.$$

Since $\lim_{\varepsilon \to 0} \varphi(\varepsilon) A(\varepsilon) = 0$ and $\varphi'(r) A(r) = 1$, the relation (4.16) follows.

We next consider the case when M is compact. Here $\varphi(r) \to \infty$ as $r \to \sigma$ (except for the real elliptic space). Nevertheless $A(r)\varphi(r)$ is bounded as $r \to \sigma$ and the integral $\int f(q)\Psi(p, q)\,dq$ exists. As before we obtain

$$u(p) = \int_0^\sigma A(r)\varphi(r)[M^r f](p)\,dr.$$

Using Lemma 21 and 22 it follows that

$$\Delta u = \int_0^\sigma \varphi(r) A(r) \left\{\frac{\partial^2 M^r f}{\partial r^2} + \frac{1}{A(r)}\frac{dA}{dr}\frac{\partial M^r f}{\partial r}\right\} dr$$

$$= \lim_{\varepsilon_1, \varepsilon_2 \to 0}\left[\varphi(r) A(r) \frac{\partial M^r f}{\partial r}\right]_{\varepsilon_1}^{\sigma-\varepsilon_2} - \lim_{\varepsilon_1, \varepsilon_2 \to 0} \int_{\varepsilon_1}^{\sigma-\varepsilon_2} \varphi'(r) A(r) \frac{\partial M^r f}{\partial r}\,dr.$$

If M is real elliptic,

$$\lim_{r \to \sigma} A(r) \neq 0 \text{ and } \lim_{r \to \sigma} \varphi(r) = 0$$

due to the choice of a. If, on the other hand, M is not a real elliptic space

$$\lim_{r \to \sigma} A(r) = 0 \text{ and } \lim_{r \to \sigma} \varphi(r) = \infty .$$

Now $\varphi'(r) A(r) = 1$ and $\lim \varphi(\varepsilon_1) A(\varepsilon_1) = 0$ as ε_1 tends to 0. To determine $\lim_{r \to \sigma} \varphi(r) A(r)$ as r converges to σ we observe that $A(r)$ is given by the formula

$$A(r) = \int_{\|X\|=r} \det(A_X)\, d\omega_r(X) \quad 0 \leqq r < \sigma, \tag{4.18}$$

where A_X is the linear transformation (2.5). Since $\det(A_X)$ is invariant under the group $Ad_G(K)$ it is a function of r only and

$$A(r) = \det(A_X)\, r^{n-1} \Omega_n$$

where Ω_n is the surface area of the unit sphere in $\mathbf{R}^n$. We can use this last formula to continue $A(r)$ to an analytic function in an open interval containing $r = \sigma$. Consequently $A(r)$ has the form $A(r) = (r - \sigma)^m h(r)$ in such an interval. Here m is an integer and $h(r)$ is an analytic function, $h(\sigma) \neq 0$. This being established, the relation

$$\lim_{r \to \sigma} \varphi(r) A(r) = 0$$

follows easily. We find therefore, whether M is real elliptic or not,

$$\Delta u = -M^\sigma f + M^0 f = f - M^\sigma f.$$

4. Decomposition of a function into integrals over totally geodesic submanifolds

The formula of J. Radon determining a function on $\mathbf{R}^n$ by means of its integrals over hyperplanes has had considerable importance for partial differential equations, particularly in G. Herglotz' treatment of hyperbolic equations with constant coefficients (G. Herglotz [26], F. John [27]). We give below an extension of Radon's formula to spherical and hyperbolic spaces. The proof seems to be new in the Euclidean case.

DEFINITION. Let S be a connected submanifold of a Riemannian manifold M. S is called *totally geodesic* if each geodesic in M which touches S lies entirely in S.

Let M be a simply connected Riemannian manifold of constant curvature $\varkappa$ and dimension $n > 1$. Such a space is either a hyperbolic, Euclidean or a spherical space. It is well known that for each integer d, $0 < d < n$ there exist totally geodesic submanifolds of M of dimension d. Using the notation from the end of Chapter II, M can be written

$$\mathbf{SO}(n+1)/\mathbf{SO}(n), \quad \mathbf{R}^n \cdot \mathbf{SO}(n)/\mathbf{SO}(n), \quad \mathbf{SO}^1(n+1)/\mathbf{SO}(n) \tag{4.19}$$

according as $\varkappa$ is positive, 0 or negative. Let $\mathbf{M}_{n,d}(p)$ denote the set of d-dimensional totally geodesic submanifolds of M passing through some fixed point p. Since $\mathbf{O}(n)$ acts transitively on the set of d-dimensional subspaces of $\mathbf{R}^n$ we see that $\mathbf{M}_{n,d}(p)$ can be identified with the coset space $\mathbf{O}(n)/\mathbf{O}(d) \times \mathbf{O}(n-d)$. In particular $\mathbf{M}_{n,d}(p)$ has a unique normalized measure invariant under the action of $\mathbf{O}(n)$.

THEOREM 19. *Let M be a simply connected Riemannian manifold of constant curvature $\varkappa$ and dimension $n > 1$. For d even, $0 < d < n$, let $Q_d(x)$ denote the polynomial*

$$Q_d(x) = [x - \varkappa(d-1)(n-d)][x - \varkappa(d-3)(n-d+2)] \ldots .[x - \varkappa \cdot 1(n-2)]$$

of degree $d/2$. For each function $f \in C_c^\infty(M)$, let $[I_d f](p)$ denote the average of the values of the integrals of f over all d-dimensional totally geodesic submanifolds through p. Then

$$Q_d(\Delta) I_d f = \gamma f \qquad \text{if } M \text{ is non-compact}$$

$$Q_d(\Delta) I_d f = \gamma(f + f \circ A) \qquad \text{if } M \text{ is compact.}$$

In the latter case $M = \mathbf{S}^n$ and A denotes the antipodal mapping. The constant γ equals

$$\frac{\Gamma\left(\frac{n}{2}\right)}{\Gamma\left(\frac{n-d}{2}\right)} (-4\pi)^{\frac{1}{2}d}.$$

Proof. We consider first the non-compact case, $M = G/K$, $K = \mathbf{SO}(n)$. In geodesic polar coordinates which are valid on the entire M the metric is given by

$$ds^2 = dr^2 + \frac{\sinh^2(r\sqrt{-\varkappa})}{(r\sqrt{-\varkappa})^2} r^2 d\sigma^2$$

where $d\sigma^2$ is the fundamental metric form on the unit sphere in $\mathbf{R}^n$. Let $p_0 = \pi(e)$ and choose g such that $g \cdot p_0 = p$. If E is a fixed element in $\mathbf{M}_{n,d}(p_0)$ we consider the integral

$$F(k) = \int_E f(gk \cdot q)\, dq \qquad k \in K$$

where dq denotes the volume element in E. If K_0 is the subgroup of K that transforms E into itself then $F(kk_0) = F(k)$ for $k_0 \in K_0$; consequently the average $[I_d f](p) = \int_K F(k)\, dk$ where dk is the normalized Haar measure on K.

$$[I_d f](p) = \int_K dk \int_E f(gk \cdot q)\, dq$$

$$= \int_E dq \int_K f(gk \cdot q)\, dk = \int_E [M^r f](p)\, dq$$

where $r = d(p_0, q)$. Now we make use of the fact that E is totally geodesic. Let γ be an E-geodesic in E; let Γ be an M-geodesic touching γ at p. Then $\Gamma \subset E$ and due to the local minimizing property of geodesics, $\Gamma = \gamma$. It follows immediately that E is complete and thus two arbitrary points $q_1, q_2 \in E$ can be joined by a minimizing E-geodesic arc $\gamma_{q_1 q_2}$. Let Γ_{q_1} be an M-geodesic touching $\gamma_{q_1 q_2}$ at q_1. Then by the previous remark $\gamma_{q_1 q_2} \subset \Gamma_{q_1}$. Since two arbitrary points in M can be joined by exactly one geodesic the same is true of E and the distance between q_1 and q_2 is the same whether it is measured in the E-metric or the M-metric. In particular E and M have the same constant sectional curvature $\varkappa$. Let S_r^{d-1} and S_r^{n-1} be geodesic spheres in E and M respectively with radius r. Their areas are

$$A_d(r) = \left[\frac{\sinh (r\sqrt{-\varkappa})}{\sqrt{-\varkappa}}\right]^{d-1} \Omega_d$$

$$A(r) = \left[\frac{\sinh (r\sqrt{-\varkappa})}{\sqrt{-\varkappa}}\right]^{n-1} \Omega_n.$$

From this we find
$$[I_d f](p) = \int_0^\infty A_d(r)\,[M^r f](p)\,dr \tag{4.20}$$

Now we apply Δ to both sides of (4.20) and make use of Lemma 22;

$$[\Delta I_d f](p) = \int_0^\infty A_d(r)\,[\Delta M^r f](p)\,dr = \int_0^\infty A_d(r)\,\Delta_r([M^r f](p))\,dr.$$

We shall now keep p fixed and write $F(r) = [M^r f](p)$.

LEMMA 23. *Let m be an integer, $0 < m < n = \dim M$. Put $\lambda = \sqrt{-\varkappa}$. Then*

$$\int_0^\infty \sinh^m \lambda r\,\Delta_r F\,dr = (-\lambda^2)(n-m-1)\left[m\int_0^\infty \sinh^m \lambda r\,F(r)\,dr + (m-1)\int_0^\infty \sinh^{m-2}\lambda r\,F(r)\,dr\right].$$

If $m = 1$ the term $(m-1)\int_0^\infty \sinh^{m-2}\lambda r\,F(r)\,dr$ should be replaced by $\frac{1}{\lambda}F(0)$.

Proof. Using Lemma 21 we have

$$\int_0^\infty \sinh^m \lambda r\,\Delta_r F\,dr = \int_0^\infty \sinh^m \lambda r\left(\frac{d^2 F}{dr^2} + \frac{1}{A(r)}\frac{dA}{dr}\frac{dF}{dr}\right)dr$$

and the result follows after repeated integration by parts. From Lemma 23 we see that

19 – 593805. *Acta mathematica*. 102. Imprimé le 16 décembre 1959

$$[\Delta_p + \lambda^2 m (n-m-1)] \int_0^\infty \sinh^m \lambda r [M^r f](p)\, dr$$

$$= (-\lambda^2)(n-m-1)(m-1) \int_0^\infty \sinh^{m-2} \lambda r\, [M^r f](p)\, dr.$$

Applying this repeatedly to (4.20) the first relation of Theorem 19 follows.

If M is compact it is a sphere and we can proceed in a similar way as in the non-compact case, but here we have to observe that the geodesics emanating from p all intersect at the antipodal point $A(p)$. In geodesic polar coordinates the metric on M is given by

$$d s^2 = d r^2 + \frac{\sin^2 (r\sqrt{\varkappa})}{(r\sqrt{\varkappa})^2}\, r^2\, d\sigma^2$$

where $d\sigma^2$ is the fundamental metric form on the unit sphere in $\mathbf{R}^n$. As in the non-compact case we prove the formula

$$[I_d f](p) = \int_0^{\frac{\pi}{\sqrt{\varkappa}}} A_d(r)\, [M^r f](p)\, dr \tag{4.21}$$

where
$$A_d(r) = \left[\frac{\sin (r\sqrt{\varkappa})}{\sqrt{\varkappa}}\right]^{d-1} \Omega_d.$$

For a fixed p, we put $F(r) = [M^r f](p)$. The analogue of Lemma 23 is here

LEMMA 24. *Let m be an integer satisfying $0 < m < n = \dim M$. We put $\lambda = \sqrt{\varkappa}$. Then*

$$\int_0^{\frac{\pi}{\lambda}} \sin^m \lambda r\, \Delta_r F\, dr = \lambda^2 (n-m-1) \left[m \int_0^{\frac{\pi}{\lambda}} \sin^m \lambda r\, F(r)\, dr - (m-1) \int_0^{\frac{\pi}{\lambda}} \sin^{m-2} \lambda r\, F(r)\, dr \right].$$

If $m = 1$, the term $(m-1) \int_0^{\frac{\pi}{\varkappa}} \sin^{m-2} \lambda r\, F(r)\, dr$ should be replaced by $\frac{1}{\lambda}\left\{f(p) + f(A(p))\right\}$.

This is easily verified by using the formula

$$A(r) = \left[\frac{\sin (r\sqrt{\varkappa})}{\sqrt{\varkappa}}\right]^{n-1} \Omega_n.$$

Lemma 24 can be rewritten by using Lemma 22 and we obtain

$$[\Delta_p - m\lambda^2 (n-m-1)] \int_0^{\frac{\pi}{\lambda}} \sin^m \lambda r [M^r f](p)\, dr$$

$$= (-\lambda^2)(n-m-1)(m-1) \int_0^{\frac{\pi}{\lambda}} \sin^{m-2} \lambda r\, [M^r f](p)\, dr.$$

If we apply this repeatedly to (4.21) the latter part of Theorem 19 follows.

5. Wave equations on harmonic Lorentz spaces

In the following sections we shall show that certain mean value theorems connected with the Laplace operator are not restricted to a positive definite metric as given in ordinary potential theory. We extend the definition of the mean value operator M^r to harmonic Lorentz spaces and establish various relations between $\square$ and M^r. The situation changes considerably as we pass to Lorentzian metric. "Spheres" are no longer compact and a family of concentric spheres does not shrink to a point as the radius converges to 0. Also the analyticity of the solution of Laplace's equation is lost.

We consider the Lorentz spaces of constant curvature studied in Chapter II, § 4, where the wave operator has a simple characterization (Theorem 12). Let $M = G/H$ be such a Lorentz space of dimension $n > 1$, carrying the metric tensor Q. Here $H = \mathbf{SO}^1(n)$ and G is either $G^0 = \mathbf{R}^n \cdot \mathbf{SO}^1(n)$, $G^- = \mathbf{SO}^1(n+1)$ or G^+ as defined in Chapter II, § 4. Let s_0 be the geodesic symmetry of G/H with respect to the point p_0. Then s_0 extends to an isometry of G/H as we have seen in Chapter II. The mapping $\sigma : g \to s_0 g s_0$ is an involutive automorphism of G which is identity on H. Let $\mathfrak{m}$ be the eigenspace for the eigenvalue -1 of the automorphism $d\sigma$ of the Lie algebra $\mathfrak{g}$. If $\mathfrak{h}$ denotes as before the Lie algebra of H we have

$$\mathfrak{g} = \mathfrak{h} + \mathfrak{m}, \quad [\mathfrak{m}, \mathfrak{m}] \subset \mathfrak{h}, \quad [\mathfrak{h}, \mathfrak{m}] \subset \mathfrak{m} \tag{4.22}$$

As before we identify $\mathfrak{m}$ with M_{p_0} and denote by C_{p_0} the light cone in M_{p_0} at p_0. The interior of the cone C_{p_0} has two components; the component that contains the timelike vectors $(-1, 0, \ldots, 0)$, $-X_{n+1}$, $-iX_{n+1}$ in the cases G^0/H, G^-/H, G^+/H respectively we call the *retrograde cone* in $\mathfrak{m}$ at p_0. It will be denoted by D_{p_0}. The component of the hyperboloid $Q_{p_0}(X, X) = r^2$ that lies in D_{p_0} will be denoted by $S_r(p_0)$ in agreement with previous terminology. If p is any other point of M, we define the light cone C_p in M_p at p, and the retrograde cone D_p in M_p at p as follows. We choose $g \in G$ such that $\tau(g) \cdot p_0 = p$ and put $C_p = d\tau(g) \cdot C_{p_0}$, $D_p = d\tau(g) \cdot D_{p_0}$. Due to the connectedness of H this is a valid definition. Similarly the "sphere" $S_r(p)$ (the ball $B_r(p)$) is the component of the hyperboloid $Q_p(X, X) = r^2$ $(0 < Q_p(X, X) < r^2)$ which lies in D_p. Finally, if Exp is the Exponential mapping of M_p into M we put

$$\mathbf{D}_p = \operatorname{Exp} D_p \qquad \mathbf{C}_p = \operatorname{Exp} C_p$$

$$\mathbf{S}_r(p) = \operatorname{Exp} S_r(p) \qquad \mathbf{B}_r(p) = \operatorname{Exp} B_r(p)$$

$\mathbf{C}_p$ and $\mathbf{D}_p$ are called the light cone in M with vertex p and the retrograde cone in M with vertex p. For the spaces G^+/H we tacitly assume $r < \pi$ in order that Exp will be one-to-one.

We wish now to study solutions of various equations involving $\square$ inside the retrograde cone $\mathbf{D}_p$ for $p \in M$. This emphasis on $\mathbf{D}_p$ is in agreement with the physical and geometric situation occurring in relativity theory and in Hadamard's theory of hyperbolic equations.

Let dh denote a two-sided invariant measure on the unimodular group H. Let p be a point in M, and u a function defined in the retrograde cone $\mathbf{D}_p$. Let $q \in \mathbf{S}_r(p)$ $(r > 0)$ and consider the integral

$$\int_H u(g h g^{-1} \cdot q)\, dh$$

where g is an arbitrary element in G such that $\pi(g) = p$. The choice of g in the coset gH and of $q \in \mathbf{S}_r(p)$ is immaterial due to the invariance of dh. The integral is thus an invariant integral of u over $\mathbf{S}_r(p)$ and in analogy with the previous mean value we write

$$[M^r u](p) = \int_H u(g h g^{-1} \cdot q)\, dh$$

Now $\mathbf{S}_r(p)$ has a positive definite Riemannian metric induced by the Lorentzian metric on M. Let $d\boldsymbol{\omega}_r$ denote the volume element on $\mathbf{S}_r(p)$. Then if K denotes the (compact) subgroup of gHg^{-1} which leaves the point q fixed, $\mathbf{S}_r(p)$ can be identified with coset space gHg^{-1}/K and

$$\int_H u(g h g^{-1} \cdot q)\, dh = \frac{1}{A(r)} \int_{\mathbf{S}_r(p)} u(q)\, d\boldsymbol{\omega}_r(q)$$

where $A(r)$ is a positive scalar depending on r only. We have thus $dh = d\boldsymbol{\omega}_r\, dk / A(r)$ where dk is the normalized Haar measure on K. Now the Exponential mapping at p which maps D_p onto $\mathbf{D}_p$ is length preserving on the geodesics through p and maps $S_r(p)$ onto $\mathbf{S}_r(p)$. Consequently, if $s \in S_r(p)$ and X denotes the vector $\overrightarrow{ps}$ in M_p, the ratio of the volume elements of $\mathbf{S}_r(p)$ and $S_r(p)$ at s is given by $\det(d\,\mathrm{Exp}_X)$. By Lemma 8 and 13 this equals 1, $(\sinh r/r)^{n-1}$, $(\sin r/r)^{n-1}$ in the flat, negatively curved and positively curved case respectively. It follows that $A(r) = c r^{n-1}$, $c(\sinh r)^{n-1}$, $c(\sin r)^{n-1}$ in the three cases. Here c is a constant which depends on the choice of dh. We normalize dh in such a way that $c = 1$ and have then the relation

$$[M^r u](p) = \int_H u(g h g^{-1} \cdot q)\, dh = \int_{\mathbf{S}_r(p)} u(q)\, d\sigma(q) \tag{4.23}$$

where $d\sigma = 1/A(r)\, d\boldsymbol{\omega}_r$. Suppose now $x_1, \ldots, x_n$ are coordinates in M_p such that the cone C_p has equation $x_1^2 - x_2^2 - \cdots - x_n^2 = 0$ and the axis in the retrograde cone D_p is the negative x_1-axis. If $\theta_1, \ldots, \theta_{n-2}$ are geodesic polars on the unit sphere in $\mathbf{R}^{n-1}$ we obtain coordinates in D_p by

$$x_1 = -r \cosh \zeta \qquad 0 \leqq \zeta < \infty, \ 0 < r < \infty$$
$$x_2 = r \sinh \zeta \cos \theta_1$$
$$\cdots\cdots\cdots\cdots$$
$$x_n = r \sinh \zeta \sin \theta_1 \sin \theta_2, \ldots, \sin \theta_{n-2}.$$

The volume element on $S_r(p)$ is then given by

$$d\omega_r = r^{n-1} \sinh^{n-2} \zeta \, d\zeta \, d\omega^{n-2}$$

where $d\omega^{n-2}$ is the volume element on the unit sphere in $\mathbf{R}^{n-1}$. Using the Exponential mapping at p we can consider $(r, \zeta, \theta_1, \ldots, \theta_{n-2})$ as coordinates(1) on $\mathbf{D}_p$. Let u be a function defined in $\mathbf{B}_{r_0}(p_0)$. We shall say u has *order* a if there exists a continuous (not necessarily bounded) function $C(r)$, $(0 < r < r_0)$ such that

$$|(u \circ \operatorname{Exp})(q)| \leqslant C(r) e^{-a\zeta} \qquad \text{for } q \in B_{r_0}(p_0) \tag{4.24}$$

in terms of the coordinates above.

For R^2 the following result has also been noted by Ásgeirsson (letter to the author).

THEOREM 20. *Suppose u satisfies the equation $\square u = 0$ in $\mathbf{B}_{r_0}(p_0)$. We assume that u and its first and second order partial derivatives have order $a > n-2$. Then*

$$[M^r u](p_0) = \alpha \int_r^\beta \frac{1}{A(r)} \, dr$$

where α and β are constants.

REMARK. If u converges to 0 fast enough in an immediate neighborhood of the cone $\mathbf{C}_{r_0}$ so that

$$[M^r u](p_0) = O\left(\left(\frac{|\log r|}{r}\right)^{n-2}\right)$$

then $[M^r u](p_0)$ is constant. We get thus an analogue of the mean value theorem for harmonic functions.

To prove the relation above we consider the integral

$$F(q) = \int_H u(h \cdot q) \, dh.$$

The measure dh has been normalized such that

$$dh = \sinh^{n-2} \zeta \, d\zeta \, d\omega^{n-2} \, dk.$$

(1) We call these the geodesic polar coordinates on $\mathbf{D}_p$.

Due to the growth condition on u it is clear that the integral is convergent and the operator $\square_q$ can be applied to the integral by differentiating under the integral sign. Since $\square$ is invariant under H we obtain $\square F = 0$. We now need a lemma whose statement and proof are entirely analogous to that of Lemma 21.

LEMMA 25. *In geodesic polarcoordinates on* $\mathbf{D}_p$, $\square$ *can be expressed*

$$\square = \frac{\partial^2}{\partial r^2} + \frac{1}{A(r)} \frac{dA}{dr} \frac{\partial}{\partial r} - \Delta'$$

where Δ' *is the Laplace-Beltrami operator on* $\mathbf{S}_r(p)$.

The minus sign is due to the circumstance that Q induces a negative definite metric on $\mathbf{S}_r(p)$ whereas Δ' is taken with respect to the positive definite metric.

The function $F(q)$ is constant on each sphere $\mathbf{S}_r(p_0)$. Due to Lemma 25, $F(q) = [M^r u](p_0)$ is a solution of the differential equation

$$\frac{d^2 v}{d r^2} + \frac{1}{A(r)} \frac{dA}{dr} \frac{dv}{dr} = 0$$

and can therefore be written

$$[M^r u](p_0) = \alpha \int_r^{\beta} \frac{1}{A(r)} dr$$

where α and β are constants.

6. Generalized Riesz potentials

For two-point homogeneous spaces M^r can be expressed as a power series in Δ when applied to analytic functions. This does not hold for the operators M^r and $\square$ in a harmonic Lorentz space; nevertheless we shall now establish various relations between M^r and $\square$. For this purpose it is convenient to generalize certain facts concerning Riesz potentials (M. Riesz [32]) to harmonic Lorentz spaces. These potentials, defined below, do not however coincide with the generalization to arbitrary Lorentzian spaces given by Riesz himself in [32].

We consider first the case $M = G^-/H$. Let $f \in C_c^\infty(M)$. The integral

$$\int_{\mathbf{D}_p} f(q) \sinh^{\lambda-n} r_{pq}\, dq \qquad dq = dr\, d\omega_r$$

converges absolutely if the complex number λ has real part $\geqslant n$.

We define
$$[I_-^\lambda f](p) = \frac{1}{H_n(\lambda)} \int_{\mathbf{D}_p} f(q) \sinh^{\lambda-n} r_{pq}\, dq. \tag{4.25}$$

Here $$H_n(\lambda) = \pi^{\frac{1}{2}(n-2)} 2^{\lambda-1} \Gamma\left(\frac{\lambda}{2}\right) \Gamma\left(\frac{\lambda+2-n}{2}\right)$$

just as for the ordinary Riesz potentials. The right-hand side of (4.25) can be written

$$\frac{1}{H_n(\lambda)} \cdot \int_{D_p} (f \circ \operatorname{Exp})(Q) \frac{\sinh^{\lambda-1} r_{0Q}}{r_{0Q}^{\lambda-1}} r_{0Q}^{\lambda-n} \, dQ$$

which is of the form $$\frac{1}{H_n(\lambda)} \cdot \int_{D_p} h(Q, \lambda) \, r_{0Q}^{\lambda-n} \, dQ \tag{4.26}$$

where $h(Q, \lambda)$ as well as all its partial derivatives with respect to the first argument are holomorphic in λ and $h(Q, \lambda) \in C_c^\infty(M_p)$ for each λ. The methods of M. Riesz ([32], Ch. III, III) can be applied to such integrals. We find in particular that (4.26), which by its definition is holomorphic in the half plane $\Re\lambda > n$, admits an analytic continuation in the entire plane and the value for $\lambda = 0$ of this entire function is $h(0, 0) = f(p)$. We denote the analytic continuation of (4.25) by $[I^\lambda f](p)$ and have then

$$I_-^0 f = f \tag{4.27}$$

We can differentiate (4.25) with respect to p and carry out the differentiation under the integral sign (for large λ), treating $\mathbf{D}_p$ as a region independent of p. This can be seen ([32] p. 68) by writing the integral (4.25) as $\int_F f(q) \, K(p, q) dq$ over a region F which properly contains the intersection of the support of f and the closure of $\mathbf{D}_p$. $K(p, q)$ is defined as $\sinh^{\lambda-n} r_{pq}$ if $q \in \mathbf{D}_p$, otherwise 0. We obtain thus

$$[\Box I_-^\lambda f](p) = \frac{1}{H_n(\lambda)} \int_{\mathbf{D}_p} f(q) \, \Box_p \sinh^{\lambda-n} r_{pq} \, dq.$$

Using Lemma 25 and the relation

$$\frac{1}{A(r)} \frac{dA}{dr} = \frac{(n-1)\cosh r}{\sinh r}$$

we find that

$$\Box_p \sinh^{\lambda-n} r_{pq} = \Box_q \sinh^{\lambda-n} r_{pq} = (\lambda-n)(\lambda-1) \sinh^{\lambda-n} r_{pq} + (\lambda-n)(\lambda-2) \sinh^{\lambda-n-2} r_{pq}.$$

We also have $H_n(\lambda) = (\lambda-2)(\lambda-n) H_n(\lambda-2)$ and therefore

$$\Box I_-^\lambda f = (\lambda-n)(\lambda-1) I_-^\lambda f + I_-^{\lambda-2} f.$$

On the other hand, we can use Green's formula to express,

$$\int_{\mathbf{D}_p} (f(q) \, \Box_q (\sinh^{\lambda-n} r_{pq}) - \sinh^{\lambda-n} r_{pq} [\Box f](q)) \, dq$$

as a surface integral stretching over a part of $\mathbf{C}_p$ and a surface inside $\mathbf{D}_p$ on which f and its derivatives vanish. It is obvious that these surface integrals vanish (for large λ). This proves the relations

$$\square I_-^\lambda f = I_-^\lambda \square f = (\lambda - n)(\lambda - 1) I_-^\lambda f + I_-^{\lambda-2} f \tag{4.28}$$

for all complex λ with sufficiently large real part; due to the uniqueness of the analytic continuation, (4.28) holds for all λ. In particular we have $I_-^{-2} f = \square f - nf$. Thus our definition (4.25) differs from Riesz' own generalized potential ([32], p. 190) which is suited to obey the law $I^{-2} f = \square f$.

We consider next the case $M = G^+/H$ and define for $f \in C_c^\infty(M)$

$$[I_+^\lambda f](p) = \frac{1}{H_n(\lambda)} \int_{\mathbf{D}_p} f(q) \sin^{\lambda-n} r_{pq}\, dq,$$

where $dq = dr\, d\boldsymbol{\omega}_r$. In order to bypass the difficulties caused by the fact that the function $q \to \sin r_{pq}$ vanishes on the antipodal variety $\mathbf{S}_\pi(p)$, we assume that the support of f is disjoint from the antipodal variety $\mathbf{S}_\pi(p)$; this suffices for the present applications. We can then prove just as before

$$[I_+^0 f](p) = f(p)$$

$$[\square I_+^\lambda f](p) = [I_+^\lambda \square f](p) = -(\lambda - n)(\lambda - 1)[I_+^\lambda f](p) + [I_+^{\lambda-2}](p). \tag{4.29}$$

In the flat case $M = G^0/H$ we define

$$[I_0^\lambda f](p) = \frac{1}{H_n(\lambda)} \int_{\mathbf{D}_p} f(q) r_{pq}^{\lambda-n}\, dq, \qquad f \in C_c^\infty(M).$$

Then, as proved by M. Riesz,

$$\square I_0^\lambda f = I_0^\lambda \square f = I_0^{\lambda-2} f, \qquad I_0^0 f = f. \tag{4.30}$$

THEOREM 21. *For each of the spaces G^0/H, G^-/H and G^+/H $\square$ and M^r commute*, i.e.

$$\square M^r u = M^r \square u \qquad \textit{for } u \in C_c^\infty(M)$$

(for G^+/H we assume $r < \pi$).

Proof. We restrict ourselves to the case G^+/H. When proving the relation $[\square M^r u](p^*) = [M^r \square u](p^*)$ for $r < \pi$ we can assume without loss of generality that the support of u is disjoint from the antipodal variety $\mathbf{S}_\pi(p^*)$. Now we have for $\Re \lambda > n$

$$\int_{\mathbf{D}_p} u(q) \sin^{\lambda-n} r_{pq}\, dq = \int_0^a [M^r u](p) \sin^{\lambda-1} r\, dr$$

where a is a constant as p varies in some neighborhood of p^*. We now apply $\square$ and make use of (4.29). Then we obtain

$$\int_0^a [M^r \square u](p) \sin^{\lambda-1} r\, dr = \int_0^a [\square M^r u](p) \sin^{\lambda-1} r\, dr.$$

In the same way one can prove

$$\int_0^a [M^r \square u](p) \sin^{\lambda-1} r\, \varphi(r)\, dr = \int_0^a [\square M^r u](p) \sin^{\lambda-1} r\, \varphi(r)\, dr$$

where $\varphi(r)$ is an arbitrary continuous function. It follows that $[\square M^r u](p) = [M^r \square u](p)$.

The following Corollary is obtained just as the Corollary of Theorem 16.

COROLLARY (*The Darboux equation*). *Let* $f \in C_c^\infty(M)$ *and put* $F(p, q) = [M^r f](p)$ *if* $q \in S_r(p)$. *Then*

$$\square_1 F = \square_2 F.$$

7. Determination of a function in terms of its integrals over Lorentzian spheres

In a Riemannian manifold a function is determined in terms of its spherical mean values by the simple relation $u = \lim_{r \to 0} M^r u$. We shall now consider the problem of expressing a function u in a harmonic Lorentz space by means of its mean values $M^r u$ over Lorentzian spheres. Here the situation is naturally quite different because the "spheres" S_r do not shrink to a point as $r \to 0$. For this purpose we use the potentials I_-, I_+ and I_0 defined above; a similar method was used by I. Gelfand and M. Graev [13] in determining a function on a complex classical group by means of the family of integrals I_δ over the conjugacy class given by the diagonal matrix δ. Here I_δ is *bounded* as $\delta \to e$ whereas $M^r u$ is in general unbounded as $r \to 0$. For another related problem see Harish-Chandra's paper [23].

We consider first the negatively curved space $M = G^-/H$ and assume that $n = \dim M$ is even. Let $f \in C_c^\infty(M)$. The potential $I_-^\lambda f(p)$ can be expressed

$$[I_-^\lambda f](p) = \frac{1}{H_n(\lambda)} \int_{D_p} \sinh^{\lambda-1} r\, F(r)\, dr \tag{4.31}$$

where $F(r) = [M^r f](p)$. We use now the coordinates $x_1, \ldots, x_n$ from Chapter IV, § 5. Let R be such that $f \circ \mathrm{Exp}$ vanishes outside the surface $x_1^2 + x_2^2 + \ldots + x_n^2 = R^2$ in M_p. It is easy to see that in the integral

$$F(r) = \iint (f \circ \mathrm{Exp})(-r \cosh \zeta, r \sinh \zeta \cos \theta_1, \ldots, r \sinh \zeta \sin \theta_1 \ldots \sin \theta_{n-2}) \sinh^{n-2} \zeta\, d\zeta\, d\omega^{n-2}$$

the range of ζ is contained in the interval $(0, \zeta_0)$ where $r^2 \cosh^2 \zeta_0 + r^2 \sinh^2 \zeta_0 = R^2$. If $n \neq 2$ we see by the substitution $y = r \sinh \zeta$ that the integral expression for $F(r)$ behaves for small r like

$$\int_0^K \varphi(y) \left(\frac{y}{r}\right)^{n-2} \frac{1}{(r^2+y^2)^{\frac{1}{2}}} dy$$

where φ is bounded. If $n=2$ we see in the same way that $F'(r)$ behaves for small r like

$$\int_0^K \varphi(y) \cdot \frac{y}{r} \frac{1}{(r^2+y^2)^{\frac{1}{2}}} dy.$$

Therefore, the limits

$$a = \lim_{r \to 0} (\sinh^{n-2} r) F(r) \qquad (n \neq 2) \tag{4.32}$$

$$b = \lim_{r \to 0} (\sinh r) F'(r) \qquad (n = 2) \tag{4.33}$$

do exist. Consider now the first case $n \neq 2$. We can rewrite (4.31) as

$$[I_-^\lambda f](p) = \frac{1}{H_n(\lambda)} \int_0^R \sinh^{n-2} r\, F(r) \sinh^{\lambda-n+1} r\, dr$$

where $F(R) = 0$. We now evaluate both sides for $\lambda = n-2$. Since $H_n(\lambda)$ has a simple pole for $\lambda = n-2$ the same is true of the integral and the residue is

$$\lim_{\lambda \to n-2} \int_0^R \sinh^{n-2} r\, F(r) (\lambda - n + 2) \sinh^{\lambda-n+1} r\, dr.$$

Here λ can be restricted to be real and $> n-2$ which is convenient since the integral above is then absolutely convergent and we do not have to think of it as an implicitly given holomorphic extension. We split the integral into two parts

$$\int_0^R (\sinh^{n-2} r\, F(r) - a)(\lambda - n + 2) \sinh^{\lambda-n+1} r\, dr + a \int_0^R (\lambda - n + 2) \sinh^{\lambda-n+1} r\, dr.$$

Concerning the last term we note that

$$\lim_{\mu \to 0+} \mu \int_0^R \sinh^{\mu-1} r\, dr = \lim_{\mu \to 0+} \mu \int_0^R r^{\mu-1} dr = 1.$$

As for the first term, we can for each $\varepsilon > 0$ find a $\delta > 0$ such that

$$|(\sinh^{n-2} r) F(r) - a| < \varepsilon \text{ for } 0 < r < \delta$$

If $N = \max |(\sinh^{n-2} r) F(r)|$ we have for $n-2 < \lambda < n-1$ the estimates

$$\left|\int_0^R (\sinh^{n-2} r\, F(r) - a)(\lambda - n + 2) \sinh^{\lambda-n+1} r\, dr\right| \leqq 2(\lambda - n + 2) N (R - \delta) \sinh^{-1} \delta$$

$$\left|\int_0^\delta (\sinh^{n-2} r\, F(r) - a)(\lambda - n + 2) \sinh^{\lambda-n+1} r\, dr\right| \leqq \varepsilon (\lambda - n + 2) \int_0^\delta r^{\lambda-n+1} dr.$$

We conclude easily that

$$\lim_{\lambda \to (n-2)} \int_0^\infty \sinh^{\lambda-1} r\, F(r)(\lambda - n + 2)\, dr = \lim_{x \to 0} \sinh^{n-2} r\, F(r).$$

Taking into account the formula for $H_n(\lambda)$ we obtain

$$I_-^{n-2} f = (4\pi)^{\frac{1}{2}(2-n)} \frac{1}{\Gamma(\frac{1}{2}(n-2))} \lim_{r \to 0} \sinh^{n-2} r\, M^r f. \tag{4.34}$$

On the other hand, if we use the formula (4.28) recursively, we obtain for arbitrary $u \in C_c^\infty(M)$

$$I^{n-2}(Q(\square)u) = u$$

where

$$Q(\square) = (\square + (n-3)2)(\square + (n-5)4) \ldots . (\square + 1(n-2)).$$

We combine this with (4.34) and use on the right-hand side the commutativity of $\square$ and M^r. This yields the desired formula

$$u = (4\pi)^{\frac{1}{2}(2-n)} \frac{1}{\Gamma(\frac{1}{2}(n-2))} \lim_{r \to 0} \sinh^{n-2} r\, Q(\square_r)(M^r u)$$

where

$$\square_r = \frac{d^2}{dr^2} + (n-1) \frac{\cosh r}{\sinh r} \frac{d}{dr}.$$

It remains to consider the case $n = 2$. Here we have by (4.31)

$$I_-^2 f = \frac{1}{H_2(2)} \int_0^\infty \sinh r\, F(r)\, dr \quad f \in C_c^\infty(M)$$

where the integral converges absolutely. In fact $F(r) \leqq C |\log r|$ for small r. We apply this relation to the function $f = \square u$ where u is an arbitrary function in $C_c^\infty(M)$. We also make use of (4.28) and Theorem 21. It follows that

$$I_-^2 \square u = u = \tfrac{1}{2} \int_0^\infty \sinh r\, M^r \square u\, dr = \tfrac{1}{2} \int_0^\infty \sinh r \left(\frac{d^2}{dr^2} + \frac{\cosh r}{\sinh r} \frac{d}{dr} M^r u\right) dr$$

$$= \tfrac{1}{2} \int_0^\infty \frac{\partial}{\partial r}\left(\sinh r \frac{d M^r u}{dr}\right) dr = -\tfrac{1}{2} \lim_{r \to 0} \left(\sinh r \frac{d M^r u}{dr}\right)$$

The spaces G^+/H and G^0/H can be treated in the same manner. The combined result is as follows.

THEOREM 22. *Let M be one of the spaces $G^0/H, G^-/H, G^+/H$. Let $\varkappa$ denote the curvature of M ($\varkappa = 0, -1, +1$) and assume $n = \dim \mathrm{M}$ is even. We also put*

$$Q(x) = (x - \varkappa (n-3)\, 2)\,(x - \varkappa (n-5)\, 4) \,\dots\, (x - \varkappa\, 1\,(n-2)).$$

Then if $u \in C_c^\infty(M)$

$$u = (4\pi)^{\frac{1}{2}(2-n)} \frac{1}{\Gamma(\frac{1}{2}(n-2))} \lim_{r\to 0} r^{n-2}\, Q(\square_r)\,(M^r u) \qquad (n \neq 2)$$

and

$$u = -\tfrac{1}{2} \lim_{r\to 0} r \frac{d}{dr} M^r u. \qquad (n = 2)$$

8. Huygens' principle

We consider now an arbitrary Lorentzian space M with metric tensor Q and dimension n. Let U be an open subset of M with the property that arbitrary two points $p, q \in U$ can be joined by exactly one path segment contained in U. All considerations will now take place inside U. The paths of zero length through a point $p \in U$ generate the light cone $\mathbf{C}_p$ in U with vertex p. A submanifold S of U is called spacelike if each tangent vector to S is spacelike. Suppose now that a Cauchy problem is posed for the wave equation $\square u = 0$ with initial data on a spacelike hypersurface $S \subset U$. From Hadamard's theory it is known that the value $u(p)$ of the solution at $p \in U$ only depends on the initial data on the piece $S^* \subset S$ that lies inside the light cone $\mathbf{C}_p$. Huygens' principle is said to hold for $\square u = 0$ if the value $u(p)$ only depends on the initial data in an arbitrary small neighborhood of the edge s of S^*, $s = \mathbf{C}_p \cap S$. Hadamard has shown that Huygens' principle can never hold if n is odd. On the other hand the wave equation $\square u = 0$ in $\mathbf{R}^n$ (n even > 2) is of Huygens' type. A long-standing conjecture, attributed(1) to Hadamard, states that these are essentially the only hyperbolic equations of Huygens' type. A counterexample of the form $\square u + cu = 0$ was given by K. Stellmacher (Ein Beispiel einer Huygenschen Differentialgleichung, *Nachr. Akad. Wiss. Göttingen* 1953) but for the pure equation $\square u = 0$ the problem is, to my knowledge, unsettled. For harmonic Lorentz spaces the problem is easily answered by using properties of these spaces obtained in Chapter II.

(1) Courant-Hilbert, *Methoden der mathematischen Physik*, Vol. II, p. 438. An interesting discussion and results concerning this problem are given in L. Ásgeirsson, Some hints on Huygens' principle and Hadamard's conjecture. *Comm. Pure Appl. Math. IX* (1956), 307–326.

THEOREM 23. *The wave equation* $\square u = 0$ *in a harmonic Lorentz space* M *satisfies Huygens' principle if and only if* M *is flat and has even dimension* > 2.

Proof. Since Huygens' principle is a local property, we can, due to Theorem 9, assume $M = G^-/H$ or $M = G^+/H$. In either case we can find a solution of $\square u = 0$, valid in $\mathbf{D}_{p_0}$, by solving the equation

$$\frac{d^2 v}{d r^2} + \frac{1}{A(r)} \frac{dA}{dr} \frac{dv}{dr} = 0$$

and putting $u(p) = v(r_{p_0 p})$. We find immediately a solution of the form

$$v(r) = \int_a^r \frac{1}{\sinh^{n-1} r} dr \qquad \text{if } M = G^-/H$$

$$v(r) = \int_a^r \frac{1}{\sin^{n-1} r} dr, \qquad \text{if } M = G^+/H.$$

Due to Hadamard's result already quoted we can assume n to be even. Under this assumption it follows by easy computation that v can be written

$$v(r) = \frac{P(r)}{r^{n-2}} + Q(r) \log r, \qquad Q(0) \neq 0$$

where P and Q are regular functions. u is thus an elementary solution and since it contains a non-vanishing logarithmic term, Huygens' principle is absent (Hadamard [18] p. 236, Courant-Hilbert, *loc. cit.*, p. 438).

References

[1]. ÁSGEIRSSON, L., Über eine Mittelwerteigenschaft von Lösungen homogener linearer partieller Differentialgleichungen 2. Ordnung mit konstanten Koefficienten. *Math. Ann.* 113 (1936), 321–346.

[2]. BEREZIN, F. A. & GELFAND, I. M., Some remarks on the theory of spherical functions on symmetric Riemannian manifolds. *Trudy Moskov. Mat. Obšč.* 5 (1956), 311–351.

[3]. CARTAN, E., Groupes simples clos et ouverts et géométrie riemannienne. *J. Math. Pures Appl.* 9, vol. 8 (1929), 1–33.

[4]. ——, Sur certaines formes riemanniennes remarquables des géométries à groupe fondamental simple. *Ann. Ec. Normale* 44 (1927), 345–467.

[5]. ——, Leçons sur la géométrie des espaces de Riemann. Deuxième édition, Paris 1951.

[6]. CARTIER, P. & DIXMIER, J., Vecteurs analytiques dans les représentations de groupes de Lie. *Amer. J. Math.* 80 (1958), 131–145.

[7]. CHEVALLEY, C., *Theory of Lie Groups I*, Princeton, 1946.

[8]. ——, The Betti numbers of the exceptional simple Lie Groups. *Proc. Int. Congress of Math.* Cambridge, Mass. 1950, Vol. 2, 21–24.

[9]. ——, Invariants of finite groups generated by reflections. *Amer. J. Math.* 77 (1955), 778–782.

[10]. FELLER, W., Über die Lösungen der linearen partiellen Differentialgleichungen zweiter Ordnung vom elliptischen Typus. *Math. Ann.* 102 (1930), 633–649.

[11]. GELFAND, I. M., Spherical functions in symmetric Riemannian spaces. *Doklady Akad. Nauk. SSSR (N.S.)* 70 (1950), 5–8.

[12]. ——, The center of an infinitesimal group ring. *Mat. Sbornik N.S.* 26 (68) (1950), 103–112.

[13]. GELFAND, I. M. & GRAEV, M. I., An analogue of the Plancherel formula for the classical groups, *Trudy Moskov. Mat. Obšč.* 4 (1955), 375–404.

[14]. GODEMENT, R., A theory of spherical functions I, *Trans. Amer. Math. Soc.* 73 (1952), 496–556.

[15]. ——, Une généralisation du théorème de la moyenne pour les fonctions harmoniques. *C. R. Acad. Sci. Paris* 234 (1952), 2137–2139.

[16]. GÜNTHER, P., Über einige spezielle probleme aus der Theorie der linearen partiellen Differentialgleichungen 2. Ordnung. *Ber. Verh. Sächs. Akad. Wiss. Leipzig* 102 (1957), 1–50.

[17]. HADAMARD, J., Les surfaces à courbures opposées et leur lignes géodésiques, *J. Math. Pures Appl.* 4 (1898), 27–73.

[18]. ——, *Lectures on Cauchy's problem in linear partial differential equations.* Yale Univ. Press 1923. Reprinted by Dover, New York 1952.

[19]. HARISH-CHANDRA, On representations of Lie algebras, *Ann. of Math.* 50 (1949), 900–915.

[20]. ——, Representations of Lie groups on a Banach space I. *Trans. Amer. Math. Soc.* 75 (1953), 185–243.

[21]. ——, On the Plancherel formula for the right-invariant functions on a semi-simple Lie group. *Proc. Nat. Acad. Sci.* 40 (1954), 200–204.

[22]. ——, The characters of semi-simple Lie groups. *Trans. Amer. Math. Soc.* 83 (1956), 98–163.

[23]. ——, Fourier transforms on a semi-simple Lie algebra I. *Amer. J. Math.* 79 (1957), 193–257.

[24]. HELGASON, S., On Riemannian curvature of homogeneous spaces. *Proc. Amer. Math. Soc.* 9 (1958), 831–838.

[25]. ——, Partial differential equations on Lie Groups. Scand. Math. Congress XIII, Helsinki 1957, 110–115.

[26]. HERGLOTZ, G., Über die Integration linearer partieller Differentialgleichungen mit konstanten Koefficienten. *Ber. Math. Phys. Kl. Sächs. Akad. Wiss. Leipzig* Part I 78 (1926), 41–74, Part II 78 (1926), 287–318, Part III 80 (1928), 69–114.

[27]. JOHN, F., *Plane waves and spherical means applied to partial differential equations.* New York 1955.

[28]. LICHNEROWICZ, A. & WALKER, A. G., Sur les espaces Riemanniens harmoniques de type hyperbolique normal. *C. R. Acad. Sci. Paris* 221 (1945), 394–396.

[29]. MOSTOW, G. D., A new proof of E. CARTAN's theorem on the topology of semi-simple groups. *Bull. Amer. Math. Soc.* 55 (1949), 969–980.

[30]. MYERS, S. B. & STEENROD, N. E., The group of isometries of a Riemannian manifold. *Ann. of Math.* 40 (1939), 400–416.

[31]. NOMIZU, K., Invariant affine connections on homogeneous spaces. *Amer. J. Math.* 76 (1954), 33–65.

[32]. RIESZ, M., L'intégrale de Riemann-Liouville et le problème de Cauchy. *Acta Math.* 81 (1949), 1–223.

[33]. SCHWARTZ, L., *Théorie des distributions* Vol. I, II. Paris 1950–1951.

[34]. SELBERG, A., Harmonic analysis and discontinuous groups in weakly symmetric Riemannian spaces with applications to Dirichlet series. *J. Ind. Math. Soc.* 20 (1956), 47–87.

[35]. TITS, J., Sur certaines classes d'espaces homogènes de groupes de Lie. *Acad. Roy. Belg. Cl. Sci. Mém. Coll. in 8°* 29 (1955), no. 3.

[36]. WANG, H. C., Two-point homogeneous spaces. *Ann. of Math.* 55 (1952), 177–191.

[37]. WILLMORE, T. J., Mean value theorems in harmonic Riemannian spaces. *J. London Math. Soc.* 25 (1950), 54–57.

FUNDAMENTAL SOLUTIONS OF INVARIANT DIFFERENTIAL OPERATORS ON SYMMETRIC SPACES

BY S. HELGASON[1]

Communicated by I. M. Singer, June 27, 1963

1. **Introduction and notation.** Let S be a Riemannian globally symmetric space, G the largest connected group of isometries of S in the compact open topology. We assume that S is of the noncompact type, that is, G is semisimple and has no compact normal subgroup $\neq \{e\}$. Let o be any point in S, K the isotropy subgroup of G at o, $\mathfrak{k}$ and $\mathfrak{g}$ their respective Lie algebras, and $\mathfrak{p}$ the orthogonal complement of $\mathfrak{k}$ in $\mathfrak{g}$ with respect to the Killing form B of $\mathfrak{g}$. Let $\mathfrak{a}$ be any maximal abelian subspace of $\mathfrak{p}$ and let $A = \exp(\mathfrak{a})$. For each λ in the dual space of $\mathfrak{a}$ (which we identify with $\mathfrak{a}$, via B) let $\mathfrak{g}_\lambda = \{X \in \mathfrak{g} \mid [H, X] = \lambda(H)X \text{ for all } H \in \mathfrak{a}\}$. Let $d_\lambda = \dim(\mathfrak{g}_\lambda)$. Choose some order on $\mathfrak{a}$ and let

$$\mathfrak{n} = \sum_{\lambda>0} \mathfrak{g}_\lambda, \qquad \rho = \sum_{\lambda>0} d_\lambda \lambda, \qquad \pi' = \prod_{\lambda>0} \lambda^{d_\lambda}$$

and let π denote the product of the distinct prime factors in π'. Then we have the Iwasawa decompositions $\mathfrak{g} = \mathfrak{k} + \mathfrak{a} + \mathfrak{n}$, $G = KAN$ where N is the nilpotent group $\exp(\mathfrak{n})$. Given $g \in G$, let $H(g)$ denote the unique element in $\mathfrak{a}$ for which $g \in K \exp H(g) N$. Let W denote the Weyl group M'/M where M and M', respectively, denote the centralizer and normalizer of $\mathfrak{a}$ in K.

For each $\lambda \in \mathfrak{a}$ consider the spherical function

$$\phi_\lambda(x) = \int_K e^{(i\lambda - \rho)(H(xk))} dk \qquad (x \in G)$$

dk being the normalized Haar measure on K. Let $c(\lambda)$ denote Harish-Chandra's function on $\mathfrak{a}$ which occurs in the leading term of the asymptotic expansion of ϕ_λ [2, p. 283], i.e.,

$$\phi_\lambda(\exp H) \sim \sum_{s \in W} c(s\lambda) e^{(is\lambda - \rho)(H)}$$

where λ and H are suitably restricted in $\mathfrak{a}$.

Each $x \in G$ can be written uniquely in the form $x = k \exp X (k \in K, X \in \mathfrak{p})$. We put $|X| = (B(X, X))^{1/2}$ and $\omega(X) = \{\det (\sinh ad X / ad X)_{\mathfrak{p}}\}^{1/2}$ where the subscript $\mathfrak{p}$ indicates restriction to $\mathfrak{p}$ of the linear

[1] Work supported in part by the National Science Foundation NSF GP-149.

transformation of $\mathfrak{g}$ given by

$$\sinh ad\, X/ad\, X = \sum_{n \geq 0} (ad\, X)^{2n}/(2n+1)!.$$

Let $D(G)$ and $\boldsymbol{D}(S)$ denote the set of left invariant (resp. G-invariant) differential operators on G (resp. S). Let $S(\mathfrak{a})$ denote the symmetric algebra over $\mathfrak{a}$, $\mathcal{S}(\mathfrak{a})$ the space of C^∞ functions on $\mathfrak{a}$ which are rapidly decreasing together with all their derivatives. Let $I(\mathfrak{a})$ and $\mathcal{I}(\mathfrak{a})$ denote the set of W-invariants in $S(\mathfrak{a})$ and $\mathcal{S}(\mathfrak{a})$, respectively; $\mathcal{S}(\mathfrak{a})$ is taken with its usual locally convex topology [7, p. 90] and then $\mathcal{I}(\mathfrak{a})$ is a closed subspace. According to a theorem of Harish-Chandra (cf. [2, Theorem 1, p. 260], also [3, p. 432]) there exists an isomorphism Γ of the algebra $\boldsymbol{D}(S)$ onto $I(\mathfrak{a})$. Let $I(G)$ denote the set of C^∞ functions f on G which are bi-invariant under K and for each integer $q \geqq 0$ and each $D \in \boldsymbol{D}(G)$ satisfy

$$\tau_{D,q}(f) = \sup_{H \in \mathfrak{a}} (1 + |H|)^q \omega(H) |(Df)(\exp H)| < \infty.$$

Let $I_0(G)$ denote the space of functions of the form

$$\phi_a(x) = \int_{\mathfrak{a}} \pi(\lambda) a(\lambda) \phi_\lambda(x) d\lambda \qquad (x \in G)$$

where $a \in \mathcal{S}(\mathfrak{a})$ and $d\lambda$ is the Euclidean measure on $\mathfrak{a}$. Then, by [2, p. 586], $I_0(G) \subset I(G)$ (it can be shown using [1] that $\pi = \pi_0$). The seminorms $\tau_{D,q}$ turn $I_0(G)$ and $I(G)$ into locally convex spaces.

LEMMA 1. *$I_0(G)$ is an algebra under convolution on G.*

Under the restriction from G to A, $I_0(G)$ and $I(G)$ are mapped isomorphically onto spaces $I_0(A)$ and $I(A)$ of W-invariant C^∞ functions on A. We carry the algebraic and topological structure of $I_0(G)$ over on $I_0(A)$ by means of this mapping. The space $\mathcal{I}(\mathfrak{a})$ is an algebra under convolution on $\mathfrak{a}$.

2. **Transmutation operators.** A function f on the space $S = G/K$ is called a *radial function* if $f(k \cdot p) = f(p)$ for all $k \in K$, $p \in S$. The set of continuous (resp. C^∞) radial functions on G/K is in one-to-one correspondence $f \to \bar{f}$ with the set of all continuous (resp. C^∞) W-invariant functions on A. Here $f(aK) = \bar{f}(a)$ for $a \in A$. Let $D \in \boldsymbol{D}(S)$; then by [2, p. 265] there exists a unique differential operator $\delta'(D)$ on A' (the set of regular elements in A) such that $(Df)^- = \delta'(D)\bar{f}$ for all C^∞ radial functions f. The operator $\delta'(D)$ is called the radial part of D in analogy with the radial part $D_r^2 + (n-1)/r\, D_r$ of the Laplacian on $\boldsymbol{R}^n$, $(D_r = d/dr)$. It is known [5] that there exists an isomorphism X

("transmutation operator") of the vector space of even C^∞ functions on $\boldsymbol{R}$, onto itself, under which the singular operator $D_r^2+(n-1)/r\, D_r$ corresponds to D_r^2. The operators $\delta'(D)$ $(D\in \boldsymbol{D}(S))$ are singular when considered as differential operators on A but Theorem 1 shows that they have a simultaneous transmutation operator X under which they correspond to differential operators on the Euclidean space $\mathfrak{a}$ with constant coefficients.

Given a W-invariant function ϕ on A, let $\tilde{\phi}$ denote the corresponding radial function on S. For $\phi\in I(A)$, put

$$(X\phi)(H) = e^{\rho(H)}\int_N \tilde{\phi}((\exp H)n\cdot o)dn \qquad (H \in \mathfrak{a})$$

where dn is a suitably normalized invariant measure on N. As proved by Harish-Chandra [**11**, p. 595], X is a continuous mapping of $I(A)$ into $\mathfrak{g}(\mathfrak{a})$.

THEOREM 1. *The mapping X is a topological isomorphism of the algebra $I_0(A)$ onto the algebra $\mathfrak{g}(\mathfrak{a})$. Moreover, if $D\in \boldsymbol{D}(S)$ then*

$$X\delta'(D)\phi = \Gamma(D)X\phi, \qquad \phi \in I_0(A).$$

Here $\Gamma(D)$ is considered as a differential operator on $\mathfrak{a}$.

The proof is based on the Plancherel formula for functions in $I_0(G)$, proved by Harish-Chandra [**2**]. It also uses the recent result of Gindikin and Karpelevič [**1**] according to which the function $\mathbf{c}(\lambda)$ above can be expressed in terms of Gamma functions.

REMARKS. At the end of [**2**], Harish-Chandra states the following two conjectures which would imply that $I_0(A)$ contains all the W-invariant C^∞ functions on A with compact support.

(I) There exists a polynomial $p\in S(\mathfrak{a})$ such that $|\mathbf{c}(\lambda)\pi(\lambda)p(\lambda)|\geqq 1$ for all $\lambda\in\mathfrak{a}$. (Here we have used the fact that $\pi=\pi_0$.)

(II) The mapping X is one-to-one on $I(A)$.

Now (I) can be verified on the basis of the mentioned result of Gindikin and Karpelevič. Theorem 1 shows that (II) is equivalent to $I_0(G)=I(G)$. On the other hand, (II) is easily implied by the Plancherel formula for the functions in $I(G)$. This formula is not proved in [**2**] but I understand that Harish-Chandra has proved it in recent, as yet unpublished, work. In the next section we shall therefore assume that $I_0(G)=I(G)$.

3. **Fundamental solutions.** Let $C_c^\infty(S)$ denote the space of C^∞ functions on S with compact support. Let δ denote the distribution on S

given by $\delta(f)=f(o)$ where $f\in C_c^\infty(S)$. Let $D\in \boldsymbol{D}(S)$. A distribution T on S is called a *fundamental solution* of D if $DT=\delta$. If $f\in C_c^\infty(S)$, then a fundamental solution T of D gives a solution of the equation $Du=f$ by putting $u=f*T$ where $*$ is the operation on distributions on G/K induced by the convolution product of distributions on G.

THEOREM 2. *Each invariant differential operator $D\in \boldsymbol{D}(S)$ $(D\neq 0)$ on the symmetric space S has a fundamental solution.*

This is a consequence of Theorem 1 and the fact that a nonzero differential operator on $\boldsymbol{R}^n$ with constant coefficients always has a tempered fundamental solution [**4**; **6**].

Added in proof. In the case when G is complex the following formula (which is a simple consequence of Lemma 55 in [**2**]) gives a simpler proof of Theorem 2.

$$(DF)\circ \operatorname{Exp} = \frac{1}{\omega}\lambda(D)(\omega(F\circ \operatorname{Exp})) \qquad (D\in \boldsymbol{D}(S)).$$

Here Exp is the usual Exponential mapping of $\mathfrak{p}$ onto S, λ is a certain isomorphism of $\boldsymbol{D}(S)$ onto the algebra of $\mathrm{Ad}(K)$-invariant polynomials on $\mathfrak{p}$ and F is any radial function on S.

REFERENCES

1. S. G. Gindikin and F. I. Karpelevič, *Plancherel measure of Riemannian symmetric spaces of nonpositive curvature*, Dokl. Akad. Nauk. SSSR **145** (1962), 252–255= Soviet Math. Dokl. **3** (1962), 962–965.

2. Harish-Chandra, *Spherical functions on a semisimple Lie group*. I, II, Amer. J. Math. **80** (1958), 241–310, 553–613.

3. S. Helgason, *Differential geometry and symmetric spaces*, Academic Press, New York, 1962.

4. L. Hörmander, *On the division of distributions by polynomials*, Ark. Mat. **3** (1958), 555–568.

5. J. L. Lions, *Équations différentielles opérationelles*, Springer, Berlin, 1961.

6. S. Lojasiewicz, *Division d'une distribution par une fonction analytique de variables réelles*. C. R. Acad. Sci. Paris **246** (1958), 683–686.

7. L. Schwartz, *Théorie des distributions*. II, Hermann, Paris, 1951.

MASSACHUSETTS INSTITUTE OF TECHNOLOGY

DUALITY AND RADON TRANSFORM FOR SYMMETRIC SPACES.*

By S. Helgason.[1]

1. Introduction. The classical Radon transform for the Euclidean space $\boldsymbol{R}^n$ associates to each continuous function f of compact support the integrals of f over the hyperplanes in $\boldsymbol{R}^n$. If f is of class C^∞ then it can be recovered from these integrals by means of a simple formula. This inversion formula amounts to a decomposition of f into plane waves, that is into functions which have parallel planes as level surfaces. A plane wave is just a function of one variable so the decomposition mentioned can sometimes reduce a problem for functions in n variables to a similar problem for functions in one variable.

Now let S be a Riemannian globally symmetric space of the noncompact type and $G = I_0(S)$ the largest connected group of isometries of S in the compact open topology. The group G contains maximal unipotent subgroups (all mutually conjugate) and their orbits in S are called horocycles. Each horocycle has dimensional equal to $\dim S - \operatorname{rank} S$. The space $\hat{S}$ of all horocycles in S is another homogeneous space acted on transitively by G and will be called the dual of S in view of the analogy with the projective duality between points and hyperplanes in $\boldsymbol{R}^n$. Although totally geodesic subspaces in S are the closest analogy to planes in $\boldsymbol{R}^n$, the horocycles serve better in generalizing the Radon transform because S is a disjoint union of parallel horocycles. The "polar coordinate" decomposition of S (Prop. 3.7) has an analogue for $\hat{S}$ (Prop. 3.6). Moreover, if we consider the isotropy groups for the action of G on S and $\hat{S}$, respectively, the orbit spaces of these isotropy groups have a dual structure (Theorem 3.8). The algebras $\boldsymbol{D}(S)$ and $\boldsymbol{D}(\hat{S})$ of G-invariant differential operators on S and $\hat{S}$, respectively, also have similar properties.

By analogy with the classical Radon transform, the Radon transform of a continuous function f of compact support on S is defined to be the function $\hat{f}$ on $\hat{S}$ such that for $\xi \in \hat{S}$, $\hat{f}(\xi)$ is the integral of f over ξ. This mapping $f \to \hat{f}$ is one-to-one and induces an isomorphism $D \to \hat{D}$ of $\boldsymbol{D}(S)$ into $\boldsymbol{D}(\hat{S})$ described in Theorem 5.3. On the other hand there is a natural dual to the

* Received August 1, 1963.

[1] This work was supported in part by the National Science Foundation, NSF GP-149.

transform $f \to \hat{f}$. This dual transform associates to a continuous function ψ on $\hat{S}$ the function $\check{\psi}$ on S such that for $p \in S$, $\check{\psi}(p)$ is the average of ψ over the (compact) set of horocycles passing through p. A natural problem is now to relate f and $(\hat{f})^{\vee}$. This is solved by Theorem 6.1 for the case when G is a complex Lie group, the principal tool in the proof being the Plancherel formula for the group G. This theorem can be interpreted as a decomposition of a function on S into plane waves.

In § 7 the preceding results (G assumed complex) are applied to invariant differential equations on S. It is shown that a differential equation $Du = f$ where $D \in \boldsymbol{D}(S)$ and f is a given function can be reduced to a differential equation with constant coefficients on a Euclidean space (and is thus, in principle, solvable). The Euclidean space in question is a maximal flat totally geodesic submanifold of S. In [13b] a different method is given, applying also to real G.

A short summary of this paper is published in [13a].

2. Notation. As usual, $\boldsymbol{R}$ and $\boldsymbol{C}$ denote the fields of real and complex numbers respectively. The identity element of a group will be denoted by e. If A is a group and B a subgroup the system of left cosets aB will be denoted by A/B. The subgroup B will, when viewed as an element in A/B, be denoted $\{B\}$. For each $x \in A$, the mapping $aB \to xaB$ of A/B onto itself will be denoted $\tau(x)$; the left translation $a \to xa$ and the right translation $a \to ax$ will be denoted L_x (or $L(x)$) and R_x (or $R(x)$) respectively. If M is a topological space, $C(M)$ (respectively, $C_c(M)$) denotes the space of continuous functions (respectively, continuous functions of compact support) on M. If $L(M)$ is any linear space of functions f on M and D is an endomorphism of $L(M)$ then $[Df](p)$ and sometimes $D_p(f(p))$ denotes the value of the function Df at p. If M_1 and M_2 are two topological spaces and τ a homeomorphism of M_1 onto M_2 we put f^τ for the composite function $f \circ \tau^{-1}$. If $M_1 = M_2 = M$ and the maps $f \to f^\tau$ and $f \to f^{\tau^{-1}}$ leave $L(M)$ invariant we define D^τ by $D^\tau f = (Df^{\tau^{-1}})^\tau$ for $f \in L(M)$. If $D^\tau = D$ then D is called invariant under τ. If M is a manifold, $C^\infty(M)$ (respectively $C^\infty{}_c(M)$) denotes the space of differentiable functions (respectively, differentiable functions with compact support) on M. If D is a differential operator on M and τ a diffeomorphism of M then D^τ is also a differential operator on M.

Lie group will be denoted by capital Roman letters and their Lie algebras by corresponding lower case German letters. The adjoint representation of a Lie group G (respectively, Lie algebra $\mathfrak{g}$) will be denoted Ad_G (respectively, $\mathrm{ad}_{\mathfrak{g}}$). The subscripts are ommited when no confusion is likely.

3. The dual space of a symmetric space. Let S be a symmetric space (that is a Riemannian globally symmetric space), and let $I_0(S)$ denote the largest connected group of isometries of S in the compact open topology. It will always be assumed that S is of *the noncompact type,* that is $I_0(S)$ is semisimple and has no compact normal subgroup $\neq \{e\}$. The group $I_0(S)$ is generated by all products $s_p \circ s_q$ where s_p and s_q denote the geodesic symmetries with respect to the points $p, q \in S$. If we change the Riemannian structure on S to another one which is also invariant under $I_0(S)$ the geodesics remain the same, the geodesic symmetries s_p are still isometries so $I_0(S)$ remains the same. In view of this fact it is natural to choose the Riemannian structure as follows: For each $p \in S$ let K_p denote the isotropy subgroup of $G = I_0(S)$ at p, let $\mathfrak{k}_p$ and $\mathfrak{g}_0$ denote the corresponding Lie algebras and let $\mathfrak{p}_p$ denote the orthogonal complement of $\mathfrak{k}_p$ in $\mathfrak{g}_0$ with respect to the Killing form B of $\mathfrak{g}_0$. Then B is strictly positive definite on $\mathfrak{p}_p$ and the differential of the mapping $g \rightarrow g \cdot p$ of G onto S induces a linear one-to-one mapping of $\mathfrak{p}_p$ onto the tangent space S_p. The Riemannian structure on S is chosen such that this mapping of $\mathfrak{p}_p$ onto S_p is an isometry for each $p \in S$.

Let l denote the rank of S. Then S contains flat, totally geodesic submanifolds of dimension l. Let us simply call these manifolds *planes* in S. Now fix a point $o \in S$, let π denote the mapping $g \rightarrow g \cdot o$ of G onto S and let θ denote the automorphism $g \rightarrow s_o g s_o$ of G as well as the corresponding automorphism of $\mathfrak{g}_0$. The Exponential mapping Exp at o is a diffeomorphism of S_o onto S. Let E be any plane containing o and let $\mathfrak{h}_{\mathfrak{p}_0}$ denote the (maximal abelian) subspace of vectors $X \in \mathfrak{p}_0$ for which $\operatorname{Exp}(d\pi X) \in E$. The subalgebra $\mathfrak{h}_{\mathfrak{p}_0} \subset \mathfrak{g}_0$ is often called a *Cartan subalgebra* for the space S. Let K stand for the isotropy group K_o of o and let $A_\mathfrak{p}$ denote the analytic subgroup of G corresponding to $\mathfrak{h}_{\mathfrak{p}_0}$. An element $H \in \mathfrak{h}_{\mathfrak{p}_0}$ is called regular if its centralizer in $\mathfrak{p}_0$ equals $\mathfrak{h}_{\mathfrak{p}_0}$. Let $\mathfrak{h}'_{\mathfrak{p}_0}$ denote the set of regular elements in $\mathfrak{h}_{\mathfrak{p}_0}$. The connected components of $\mathfrak{h}'_{\mathfrak{p}_0}$ are called the Weyl chambers in $\mathfrak{h}_{\mathfrak{p}_0}$. Select any Weyl chamber C in $\mathfrak{h}_{\mathfrak{p}_0}$. Then the dual space of $\mathfrak{h}_{\mathfrak{p}_0}$ can be ordered by calling a linear function λ on $\mathfrak{h}_{\mathfrak{p}_0}$ *positive* if $\lambda(H) > 0$ for all $H \in C$. This ordering gives rise to an Iwasawa decomposition of G, $G = KA_\mathfrak{p}N$ where N is a connected nilpotent subgroup of G (see e.g. [13], Ch. VI). The Lie algebra $\mathfrak{n}_0$ of N is given by $\mathfrak{n}_0 = \sum_{\lambda > 0} \mathfrak{g}_{0\lambda}$, where for each linear function λ on $\mathfrak{h}_{\mathfrak{p}_0}$ we put $\mathfrak{g}_{0\lambda} = \{X \in \mathfrak{g}_0 \mid [H, X] = \lambda(H)X,\ H \in \mathfrak{h}_{\mathfrak{p}_0}\}$.

The group N depends on the choices of o, E and C. However, since all pairs (o, E) where $o \in E$ are conjugate under G and all Weyl chambers C in $\mathfrak{h}_{\mathfrak{p}_0}$ are conjugate under K it follows that the nilpotent subgroups of the various

9

Iwasawa decompositions of G are all conjugate in G, i. e. all of the form gNg^{-1} $(g \in G)$.

Definition. A *horocycle* in S is an orbit of a subgroup of the form gNg^{-1}, g being any element in G.

Remark. Identifying G with its adjoint group, the groups gNg^{-1} are the maximal unipotent subgroups of G ([1], [7], [17], [21]). Thus the horocycles in S can be defined quite directly, that is without invoking the Iwasawa decomposition.

We shall now turn the set of horocycles into a manifold and study this manifold in some detail. First we introduce some notation.

Let M and M', respectively, denote the centralizer and normalizer of $\mathfrak{h}_{\mathfrak{p}_0}$ in K. Let W denote the (finite) Weyl group M'/M. Let $\mathfrak{h}_0$ be any maximal abelian subalgebra of $\mathfrak{g}_0$ containing $\mathfrak{h}_{\mathfrak{p}_0}$, let $\mathfrak{g}$ denote the complexification of $\mathfrak{g}_0$ and $\mathfrak{h}$ the subspace of $\mathfrak{g}$ spanned by $\mathfrak{h}_0$. Then $\mathfrak{h}$ is a Cartan subalgebra of $\mathfrak{g}$. Let Δ denote the set of nonzero roots of $\mathfrak{g}$ with respect to $\mathfrak{h}$ and let $\mathfrak{h}_{\mathfrak{k}_0} = \mathfrak{h} \cap \mathfrak{k}_0$, $\mathfrak{h}^* = \mathfrak{h}_{\mathfrak{p}_0} + i\mathfrak{h}_{\mathfrak{k}_0}$. All the roots in Δ are real on $\mathfrak{h}^*$. Let C^* be any Weyl chamber in $\mathfrak{h}^*$ whose closure contains the Weyl chamber C in $\mathfrak{h}_{\mathfrak{p}_0}$. We order the dual space of $\mathfrak{h}^*$ by means of the Weyl chamber C^*. Then if $\bar{\alpha}$ denotes the restriction to $\mathfrak{h}_{\mathfrak{p}_0}$ of a root $\alpha \in \Delta$ then $\alpha > 0 \Rightarrow \bar{\alpha} \geqq 0$ and $\bar{\alpha} > 0 \Rightarrow \alpha > 0$. The set Δ^+ of positive roots in Δ is a disjoint union, $\Delta^+ = P_+ \cup P_-$, where $\alpha \in \Delta^+$ belongs to P_+ or P_- respectively according to whether $\bar{\alpha} > 0$ or $\bar{\alpha} = 0$. Let $\rho = \frac{1}{2} \sum_{\alpha \in P_+} \alpha$. Now $H \in \mathfrak{h}'_{\mathfrak{p}_0}$ if and only if $\alpha(H) \neq 0$ for all $\alpha \in P_+$. The set $\{H \in \mathfrak{h}_{\mathfrak{p}_0} \mid \alpha(H) > 0 \text{ for } \alpha \in P_+\}$ is therefore a subset of $\mathfrak{h}'_{\mathfrak{p}_0}$ which is convex, hence connected, and contains C. Consequently

$$C = \{H \in \mathfrak{h}_{\mathfrak{p}_0} \mid \alpha(H) > 0 \text{ for } \alpha \in P_+\}.$$

LEMMA 3.1. *Let H be any element in C. Then*

$$N = \{z \in G \mid \lim_{t \to \infty} \exp(-tH) z \exp(tH) = e\}. \tag{1}$$

This lemma is presumably known (cf. [16]) but no proof seems available. Denoting the right hand side of (1) by N_H the previous expression for $\mathfrak{n}_0$ shows that $N \subset N_H$. The converse requires some matrix computation. As is well known (see e.g. [13], p. 223) there exists a basis of $\mathfrak{g}$ such that the matrices representing $\mathrm{ad}_{\mathfrak{g}}$ have the following properties:

$\mathrm{ad}(\mathfrak{k}_0)$: skew Hermitian;
$\mathrm{ad}(\mathfrak{n}_0)$: lower triangular with zeros in the diagonal;

$\mathrm{ad}(\theta\mathfrak{n}_0)$: upper triangular with zeros in the diagonal;

$\mathrm{ad}_{\mathfrak{g}}(H)$ $(H \in \mathfrak{h})$ is a diagonal matrix with diagonal $\lambda_1(H), \lambda_2(H), \cdots$, where $\lambda_1, \lambda_2, \cdots$ are all the roots of $\mathfrak{g}$ with respect to $\mathfrak{h}$ in increasing order (here the root 0 is counted with multiplicity $\dim \mathfrak{h}$).

Now suppose $ka \in N_H$ for some $k \in K$, $a \in A_{\mathfrak{p}}$. Then

$$(2) \qquad \mathrm{Ad}(\exp(-tH)k\exp(tH)) \rightarrow \mathrm{Ad}(a^{-1})$$

for $t \rightarrow \infty$. Consider the matrices in (2),

$$(3) \qquad (\mathrm{Ad}(\exp(-tH)k\exp(tH)))_{ij} = \exp((\lambda_j - \lambda_i)(tH))k_{ij}$$

where $(\mathrm{Ad}(k))_{ij} = k_{ij}$. Then it follows from (2) that $k_{ij} = 0$ for $i < j$ and $k_{ii} > 0$ for all i. But being also unitary, $\mathrm{Ad}(k) = e$, so we find $ka = e$; q. e. d.

LEMMA 3.2. $(N\theta(N)) \cap K = \{e\}$.

Proof. Suppose $n_1\theta(n_2) \in K$ for some $n_1, n_2 \in N$. Let $\mathrm{Ad}_G(n_1\theta(n_2))_{ij} = u_{ij}$ in terms of the basis of $\mathfrak{g}$ used above. Clearly $u_{11} = 1$ and therefore $u_{1j} = 0$ for $j \neq 1$, the matrix (u_{ij}) being unitary. This implies $\mathrm{Ad}_G(\theta(n_2))_{1j} = 0$ for $j > 1$. Hence $u_{22} = 1$ and $u_{2j} = 0$ for $j \neq 2$. By induction we find that $u_{ij} = \delta_{ij}$ so $n_1\theta(n_2)$ lies in the center of G. But this center lies in K and G acts effectively so the lemma is reduced to proving $N \cap \theta(N) = \{e\}$. This however is clear from the matrix representation already used.

LEMMA 3.3. *The plane E is orthogonal to the horocycle $N \cdot o$ at o.*

Let $\mathfrak{q}_0$ denote the subspace of $\mathfrak{p}_0$ which $d\pi$ maps onto the tangent space to the orbit $N \cdot o$ at o. This tangent space consists of the vectors $d\pi(X)$ as X runs through $\mathfrak{n}_0$. Let $X_1 \in \mathfrak{q}_0$, $H \in \mathfrak{h}_{\mathfrak{p}_0}$; then $d\pi(X_1) = d\pi(X)$ for some $X \in \mathfrak{n}_0$. Hence $X_1 - X \in \mathfrak{k}_0$ so $B(X_1, H) = B(X, H) = 0$, proving the lemma.

THEOREM 3.4. *The group G acts transitively on the set of horocycles in S. The subgroup of G which maps the horocycle $N \cdot o$ into itself equals MN.*

Proof. Let $g, h \in G$ and consider the orbits $gNg^{-1}h \cdot o$ and $N \cdot o$. Writing $h^{-1}g = kan$ and using $aNa^{-1} = N$ we find that $hkN \cdot o = gNg^{-1}h \cdot o$ so G is transitive.

Now $mNm^{-1} \subset N$ for each $m \in M$ so MN is a group which maps $N \cdot o$ into itself. On the other hand let $g \in G$ be such that $gN \cdot o = N \cdot o$. Writing $g = kan$ we conclude that $kann_1 = k_1$ for some $n_1 \in N$, $k_1 \in K$. Hence $a = e$ so $kN \cdot o = N \cdot o$. Lemma 3.3 now implies that $\mathrm{Ad}_G(k)\mathfrak{h}_{\mathfrak{p}_0} \subset \mathfrak{h}_{\mathfrak{p}_0}$, that is, $k \in M'$. Suppose $k \notin M$. Then the Weyl group element $s = kM$ does not leave C invariant so given $H_s \in s^{-1}C$ there exists a root $\alpha \in P_+$ such that $\alpha(H_s) < 0$.

Then the hyperplane $\alpha(H)=0$ in $\mathfrak{h}_{\mathfrak{p}_0}$ separates the Weyl chambers C and $s^{-1}C$. Select $X\neq 0$ in $\mathfrak{n}_0$ such that $[H,X]=\alpha(H)X$ for $H\in\mathfrak{h}_{\mathfrak{p}_0}$. Since $\alpha(s^{-1}H)<0$ for $H\in C$ we have $\mathrm{Ad}(k)X\in\theta(\mathfrak{n}_0)$ so $kNk^{-1}\cap\theta(N)\neq\{e\}$. But $kN\cdot o=N\cdot o$ implies $kNk^{-1}\subset NK$ so Lemma 3.2 gives a contradiction. Thus $k\in M$ and Theorem 3.4 is proved.

Since M is compact and N is closed in G, the subgroup MN is closed in G.

Definition. The set of horocycles in S, with the differentiable structure of G/MN, will be called the *dual space of* S and denoted by $\hat{S}$.

Although the differentiable structure of $\hat{S}$ involves a choice of o, E and C it is clear that a different choice only changes the subgroup MN into a conjugate subgroup $g(MN)g^{-1}$ and this gives the same differentiable structure on $\hat{S}$.

Let $t\to\gamma(t)$ $(t\in\mathbf{R})$ be any geodesic in S. Let s_τ denote the geodesic symmetry of S with respect to the point $\gamma(\tau)$ and put $T_t=s_{t/2}s_0$. Then $T_t(t\in\mathbf{R})$ is a one-parameter subgroup of G. Its element are called *transvections* along the geodesic γ. Two horocycles ξ_1 and ξ_2 are called parallel if there exists a geodesic γ intersecting ξ_1 and ξ_2 under a right angle and such that $T(\xi_1)=\xi_2$ for a suitable transvection T along γ. If $\xi\in\hat{S}$, $\{\xi\}$ shall denote the family of horocycles parallel to ξ.

LEMMA 3.5. *For a fixed $g\in G$ the orbits of the group gNg^{-1} form a parallel family of horocycles. Each parallel family of horocycles arises in this fashion.*

Proof. Since $G=NA_{\mathfrak{p}}K$ each orbit of N intersects E. If $a\in A_{\mathfrak{p}}$, the orbit $N\cdot(a\cdot o)$ equals $aN\cdot o$ which is parallel to $N\cdot o$ (Lemma 3.3). Since G acts transitively on the set $\{\{\xi\}\mid\xi\in\hat{S}\}$ the lemma follows.

Consider now the set of horocycles which pass through o. In view of Theorem 3.4 this set is in a natural correspondence with the coset space K/M, which again is in an obvious natural correspondence with the set of all Weyl chambers in all maximal abelian subspaces of $\mathfrak{p}_0$.

PROPOSITION 3.6. *The mapping $\psi\colon(kM,a)\to ka(N\cdot o)$ is a diffeomorphism of $K/M\times A_{\mathfrak{p}}$ onto $\hat{S}$.*

Proof. Let $k\in K$ and put $\xi=k(N\cdot o)$; since $(kak^{-1})kNk^{-1}\cdot o$ is parallel to ξ it is clear that the parallel family $\{\xi\}$ can be written $\{kaN\cdot o\mid a\in A_{\mathfrak{p}}\}$. Thus ψ is onto. Next suppose $k_1,k_2\in K$, $a_1,a_2\in A_{\mathfrak{p}}$ such that $k_1a_1N\cdot o=k_2a_2N\cdot o$. By Theorem 3.4, $(k_2a_2)^{-1}k_1a_1\in MN$ and since $aNa^{-1}=N$, $aMa^{-1}=M$ $(a\in A_{\mathfrak{p}})$ we obtain $k_2^{-1}k_1a_1a_2^{-1}\in MN$ so $k_1M=k_2M$ and $a_1=a_2$.

Thus ψ is one-to-one. Now let $\mathfrak{m}_0$ denote the Lie algebra of M and let $\mathfrak{l}_0$ denote the orthogonal complement of $\mathfrak{m}_0$ in $\mathfrak{k}_0$ with respect to B. Consider the natural projections

$$\pi_0: K \to K/M, \quad \pi: G \to G/K, \quad \hat{\pi}: G \to G/MN$$

and let $\tau(x)$ be as described in Section 2. The differentials $d\pi_0$ and $d\hat{\pi}$ map $\mathfrak{l}_0$ and $\mathfrak{l}_0 + \mathfrak{h}_{\mathfrak{p}_0}$, respectively, isomorphically onto the tangent spaces $(K/M)_{\pi_0(e)}$ and $(G/MN)_{\pi^\wedge(e)}$. Let $a \in A_{\mathfrak{p}}$, $k \in K$, $H \in \mathfrak{h}_{\mathfrak{p}_0}$, $L \in \mathfrak{l}_0$. Then for $t \in \mathbf{R}$,

$$\psi(k(\exp tL)M, a) = ka \exp(\mathrm{Ad}(a^{-1})tL)MN,$$
$$\psi(kM, a \exp tH) = ka(\exp tH)MN,$$

so

$$d\psi(d\tau(k)\, d\pi_0(L), d\tau(a)H) = d\tau(ka)\,(d\hat{\pi}(\mathrm{Ad}(a^{-1})L + H)).$$

The right hand side is $\neq 0$ unless $\mathrm{Ad}(a^{-1})L + H \in \mathfrak{m}_0 + \mathfrak{n}_0$. But this happens only if $L = H = 0$ so ψ is regular and the proposition is proved.

In comparison we recall the following well known fact.

PROPOSITION 3.7. *The mapping $\phi: (kM, a) \to kaK$ is a differentiable mapping of $(K/M) \times A_{\mathfrak{p}}$ onto S and a regular w-to-one mapping of $K/M \times A'_{\mathfrak{p}}$ onto S'.*

Here w denotes the order of the Weyl group M'/M, $A'_{\mathfrak{p}} = \exp \mathfrak{h}'_{\mathfrak{p}_0}$ and S' is the set of points in S which lie on only one plane through o.

We shall now describe the double coset spaces $K\backslash G/K$ and $MN\backslash G/MN$ which of course can be regarded as the orbit spaces for the action of K on G/K and of MN on G/MN. Let $A_{\mathfrak{p}}/W$ be the space obtained from $A_{\mathfrak{p}}$ by identifying points which are conjugate under W.

THEOREM 3.8. *The following relations are natural identifications:*

(i) $K\backslash G/K = A_{\mathfrak{p}}/W$;

(ii) $MN\backslash G/MN = A_{\mathfrak{p}} \times W$.

Proof. The first relation is well known and is proved as follows: Each $g \in G$ can be written $g = k_1 a k_2$ $(k_1, k_2 \in K, a \in A_{\mathfrak{p}})$. If $g = k'_1 a' k'_2$ is another such representation one finds by applying the automorphism θ, $ka^2k^{-1} = (a')^2$, where $k \in K$. As is known (see e.g. [13], p. 245) this implies that a^2 and $(a')^2$ (and therefore a and a') are conjugate under the Weyl group. This proves (i).

In order to prove (ii) we recall a result of Bruhat-Harish-Chandra [9] which identifies the double coset space $MNA_{\mathfrak{p}}\backslash G/MNA_{\mathfrak{p}}$ with the Weyl group.

More precisely, each $g \in G$ can be written

$$g = m_1 n_1 a_1 m_w a_2 n_2 m_2 \quad (m_1, m_2 \in M, n_1, n_2 \in N, a_1, a_2 \in A_{\mathfrak{p}}) \tag{4}$$

where $m_w \in M'$ is uniquely determined $(\operatorname{mod} M)$. Since $M' A_{\mathfrak{p}} (M')^{-1} \subset A_{\mathfrak{p}}$, (4) can be written

$$g = m_1 n_1 m_w a n_2 m_2 \tag{5}$$

and (ii) therefore amounts to, that not only is $m_w M$ uniquely determined by g, but that the same is true of a. For this we need a lemma.

LEMMA 3.9. *Let C and C' be two Weyl chambers in $\mathfrak{h}_{\mathfrak{p}_0}$ and $G = KA_{\mathfrak{p}}N$, $G = KA_{\mathfrak{p}}N'$ the corresponding Iwasawa decompositions. Then*

$$(NN') \cap (MA_{\mathfrak{p}}) = \{e\}.$$

Proof. We first consider the case $C' = -C$; here $N' = \theta(N)$. Let n_1, $n_2 \in N$, $m \in M$, $a \in A_{\mathfrak{p}}$ such that

$$n_1 \theta(n_2) = ma. \tag{6}$$

Now we apply θ and use $ma = am$. Then

$$\theta(n_1) n_2 n_1 \theta(n_2) = m^2.$$

But $m\theta(N)m^{-1} = \theta(N)$ so we get from Lemma 3.2, $m^2 = e$ and $n_2 n_1 = e$. Let b denote the unique square root of a in $A_{\mathfrak{p}}$. Since m commutes with b we obtain from (6), $b^{-1} n_1 \theta(n_1^{-1}) b^{-1} = m$, or $s\theta(s^{-1}) = m$ where $s = b^{-1} n_1$. Writing $s = (\exp X)k$ where $X \in \mathfrak{p}_0$ and $k \in K$ it follows that $m = \exp 2X$ so $m^2 = e$ implies $m = e$. Now, in terms of the basis of $\mathfrak{g}$ used earlier, $\mathrm{Ad}_G(n_1)$ is lower triangular and $\mathrm{Ad}_G(\theta(n_1^{-1}))$ is upper triangular. Their diagonals are 1 and their product is (by (6)) a diagonal matrix. It follows that $n_1 = e$, so the relation $(N\theta(N)) \cap (MA_{\mathfrak{p}}) = \{e\}$ is proved.

The proof of the lemma will be finished by proving $N' \subset N\theta(N)$. For each $\alpha \in \Delta$, let $\mathfrak{g}^\alpha$ denote the root subspace

$$\mathfrak{g}^\alpha = \{X \in \mathfrak{g} \mid [H, X] = \alpha(H)X \text{ for all } H \in \mathfrak{h}\}.$$

Let P' denote the set of roots in Δ which are > 0 on C'. Then if we put $\mathfrak{n}' = \sum_{\alpha \in P'} \mathfrak{g}^\alpha$, the Lie algebra of N' is $\mathfrak{n}'_0 = \mathfrak{n}' \cap \mathfrak{g}_0$. If $-P_+$ denotes the set of negatives of the roots in P_+ it is clear that $P' \subset P_+ \cup (-P_+)$ so

$$\mathfrak{n}' = \sum_{\alpha \in P' \cap P_+} \mathfrak{g}^\alpha + \sum_{\alpha \in P' \cap (-P_+)} \mathfrak{g}^\alpha.$$

Taking intersections with $\mathfrak{g}_0$ we obtain the direct decomposition

$$\mathfrak{n}'_0 = \mathfrak{n}_1 + \mathfrak{n}_2.$$

Here $\mathfrak{n}_1 = \mathfrak{n}'_0 \cap \mathfrak{n}_0$ so it is a subalgebra of $\mathfrak{n}'_0$. Now if $\beta_1 < \beta_2 < \cdots < \beta_p$ are the roots in P' in increasing order, consider the space

$$\mathfrak{n}^{(k)} = \Big(\sum_{k \leqq i \leqq p} \mathfrak{g}^{\beta_i} \Big) \cap \mathfrak{g}_0 \qquad (1 \leqq k \leqq p).$$

Then $\mathfrak{n}^{(k)}$ is an ideal in $\mathfrak{n}'_0$, $[\mathfrak{n}'_0, \mathfrak{n}^{(k)}] \subset \mathfrak{n}^{(k+1)}$, $\mathfrak{n}^{(1)} = \mathfrak{n}'_0$, and $\mathfrak{n}^{(k)}$ is the direct sum of its intersections with $\mathfrak{n}_1$ and $\mathfrak{n}_2$. Using a result of [10], p. 736 we conclude that each $n \in N'$ can be written $n = \exp X \exp Y$ where $X \in \mathfrak{n}_1$, $Y \in \mathfrak{n}_2$. Hence $n \in N\theta(N)$ and Lemma 3.9 is proved.

Passing now to the proof of the theorem consider a second decomposition $g = m_3 n_3 m_w a' n_4 m_4$ of the element g in (5). Now $m_w^{-1} N m_w = N'$ for a suitable Weyl chamber C'. The lemma shows that $a = a'$ and the theorem is proved.

4. Invariant differential operators on the space of horocycles. Let A be a Lie group, B a closed subgroup. Then $\boldsymbol{D}(A/B)$ shall denote the algebra of differential operators on A/B which are invariant under the action of A. We write $\boldsymbol{D}(A)$ instead of $\boldsymbol{D}(A/\{e\})$ and sometimes we shall also write $\boldsymbol{D}(S)$ for $\boldsymbol{D}(G/K)$ and $\boldsymbol{D}(\hat{S})$ for $\boldsymbol{D}(G/MN)$. As will be explained in more detail in the next section, the algebra $\boldsymbol{D}(S)$ is isomorphic to the algebra $I(\mathfrak{h}_{\mathfrak{p}_0})$ of all polynomials on $\mathfrak{h}_{\mathfrak{p}_0}$ which are invariant under the Weyl group. At present our problem is to analyze $\boldsymbol{D}(\hat{S})$. For this purpose it is convenient to consider $\hat{S}$ as a fibre bundle with base K/M, fibre $A_{\mathfrak{p}}$, the projection $p: \hat{S} \to K/M$ being the mapping which to each horocycle associates the parallel horocycle through o. The fibre F_k over the base point kM is the set $\{kaN \cdot o \mid a \in A_{\mathfrak{p}}\}$.

Now let U be a left invariant differential operator on the Lie group $A_{\mathfrak{p}}$. We carry this operator over on F_k by means of the diffeomorphism $a \to ka(N \cdot o)$ $(a \in A_{\mathfrak{p}})$ and obtain a differential operator U_{F_k} on F_k. Now if $f \in C(\hat{S})$ let $f \mid F_k$ denote the restriction of f to the fibre F_k. Consider the endomorphism D_U of $C^\infty(\hat{S})$ given by

$$(1) \qquad (D_U f) \mid F = U_F(f \mid F),$$

where F stands for an arbitrary fibre F_k. Now for each $g \in G$ the diffeomorphism $\tau(g)$ of $\hat{S}$ is fibre-preserving. In fact it is geometrically clear that each isometry g of S preserves parallelism of horocycles; analytically,

$$\tau(g)(kaN \cdot o) = k_1 a_1 a N \cdot o$$

if $gk = k_1 n_1 a_1$ so $\tau(g)$ maps the fibre $\{kaN \cdot o \mid a \in A_{\mathfrak{p}}\}$ onto the fibre $\{k_1 aN \cdot o \mid a \in A_{\mathfrak{p}}\}$. Let us now prove

(2) $$(U_F(\phi\circ\tau(g)))\circ\tau(g^{-1})=U_{\tau(g)\cdot F}(\phi) \qquad (g\in G)$$

for each fibre F and each $\phi\in C^\infty(\tau(g)\cdot F)$. In view of the definition of the U_{F_k} is suffices to prove (2) in the case when $F=\{aN\cdot o \mid a\in A_{\mathfrak{p}}\}$ and g maps F into itself. But such a g is necessarily of the form $g=a_1mn$ where $a_1\in A_{\mathfrak{p}}$ $m\in M$, $n\in N$. In this case, however, (2) reduces to the left invariance of U.

The operator D_U defined by (1) is invariant under each $g\in G$; in fact, using (2) we have

$$\begin{aligned}(D_U{}^{\tau(g)}f)\mid F&=(D_Uf^{\tau(g^{-1})})^{\tau(g)}\mid F=((D_Uf^{\tau(g^{-1})})\mid\tau(g^{-1})\cdot F)^{\tau(g)}\\&=(U_{\tau(g^{-1})F}(f^{\tau(g^{-1})}\mid\tau(g^{-1})F))^{\tau(g)}=U_F(f\mid F)=(D_Uf)\mid F.\end{aligned}$$

Similarly, one proves that the mapping $U\to D_U$ $(U\in \boldsymbol{D}(A_{\mathfrak{p}}))$ is a homomorphism. It is clear that D_U is a continuous linear mapping of $C^\infty{}_c(\hat{S})$ into itself (taken with the usual Schwartz topology) and that for each $f\in C^\infty{}_c(\hat{S})$ the support of D_Uf is contained in the support of f. Hence, by well-known facts (see [2] e.g. [12], p. 242) D_U is a differential operator. It is clear since $A_{\mathfrak{p}}$ is abelian that $\boldsymbol{D}(A_{\mathfrak{p}})$ is canonically isomorphic to the symmetric algebra $S(\mathfrak{h}_{\mathfrak{p}_0})$ over $\mathfrak{h}_{\mathfrak{p}_0}$.

THEOREM 4.1. *The mapping $U\to D_U$ defined by* (1) *is an isomorphism of $S(\mathfrak{h}_{\mathfrak{p}_0})$ onto $\boldsymbol{D}(\hat{S})$. In particular $\boldsymbol{D}(\hat{S})$ is commutative.*

For simplicity, put $H=MN$ and let o denote the point $\{H\}$ in G/H. If G/H were a reductive coset space we know from [12], p. 269 that $\boldsymbol{D}(G/H)$ is determined by the invariants of the linear group $d\tau(h)$ $(h\in H)$ acting on the symmetric algebra over the tangent space $(G/H)_o$. Although G/H is not in general reductive we shall nevertheless determine $\boldsymbol{D}(G/H)$ by means of the mentioned invariants.

Consider the subspaces $\mathfrak{k}_0$, $\mathfrak{p}_0$, $\mathfrak{n}_0$, $\mathfrak{h}_{\mathfrak{p}_0}$, $\mathfrak{h}_{\mathfrak{k}_0}$, $\mathfrak{m}_0$, $\mathfrak{l}_0$ introduced already and their respective complexifications, $\mathfrak{k}$, $\mathfrak{p}$, $\mathfrak{n}$, $\mathfrak{h}_{\mathfrak{p}}$, $\mathfrak{h}_{\mathfrak{k}}$, $\mathfrak{m}$, $\mathfrak{l}$ in $\mathfrak{g}$. For each root subspace $\mathfrak{g}^\alpha$ $(\alpha\in\Delta)$ select a nonzero vector $X_\alpha\in\mathfrak{g}^\alpha$. Then

(3) $$\mathfrak{l}=\sum_{\alpha\in P_+}\boldsymbol{C}(X_\alpha+\theta X_\alpha),\quad \mathfrak{m}=\mathfrak{h}_{\mathfrak{k}}+\sum_{\alpha\in P_-}(\mathfrak{g}^\alpha+\mathfrak{g}^{-\alpha}),\quad \mathfrak{n}=\sum_{\alpha\in P_+}\mathfrak{g}^\alpha.$$

Let σ_0 denote the projection of $\mathfrak{g}_0$ onto $\mathfrak{l}_0+\mathfrak{h}_{\mathfrak{p}_0}$ given by the direct decomposition $\mathfrak{g}_0=(\mathfrak{l}_0+\mathfrak{h}_{\mathfrak{p}_0})+(\mathfrak{m}_0+\mathfrak{n}_0)$. Let σ denote the analogous projection of $\mathfrak{g}$ onto $\mathfrak{l}+\mathfrak{h}_{\mathfrak{p}}$. Under the isomorphism

$$d\hat{\pi}\colon \mathfrak{l}_0+\mathfrak{h}_{\mathfrak{p}_0}\to(G/H)_o$$

[2] Note that in Prop. 1 in [12] p. 242, the word "finite" should be replaced by "locally finite."

the endomorphism $d\tau(h)$ $(h \in H)$ of $(G/H)_0$ corresponds to the endomorphism $\sigma_0 \circ \mathrm{Ad}_G(h)$ of $\mathfrak{l}_0 + \mathfrak{h}_{\mathfrak{p}_0}$ (cf. [13], p. 367-368). In order to determine the invariants of the group $d\tau(h)$ $(h \in H)$ acting on $(G/H)_0$ it suffices to determine the invariants of $\sigma_0 \circ \mathrm{Ad}_G(H)$ on $\mathfrak{l}_0 + \mathfrak{h}_{\mathfrak{p}_0}$ or equivalently, the invariants of $\sigma \circ \mathrm{Ad}_G(H)$ on $\mathfrak{l} + \mathfrak{h}_{\mathfrak{p}}$.

Let $\alpha_1 > \alpha_2 > \cdots > \alpha_p$ be the roots in P_+ in decreasing order and put $E_\alpha = X_\alpha + \theta X_\alpha$ $(\alpha \in P_+)$. If $H_1, \cdots, H_l$ is a basis of $\mathfrak{h}_{\mathfrak{p}_0}$ the elements in the symmetric algebra $S(\mathfrak{l} + \mathfrak{h}_{\mathfrak{p}})$ over $\mathfrak{l} + \mathfrak{h}_{\mathfrak{p}}$ can be regarded as polynomials in $H_1, \cdots, H_l, E_{\alpha_1}, \cdots, E_{\alpha_p}$. For each $X \in \mathfrak{m} + \mathfrak{n}$ let $d(X)$ denote the derivation of $S(\mathfrak{l} + \mathfrak{h}_{\mathfrak{p}})$ which extends the endomorphism $\sigma \circ \mathrm{ad}_{\mathfrak{g}}(X)$ of $\mathfrak{l} + \mathfrak{h}_{\mathfrak{p}}$. If $\alpha \in \Delta$, let $H_\alpha \in \mathfrak{h}$ be determined by $B(H, H_\alpha) = \alpha(H)$ $(H \in \mathfrak{h})$ and let α^θ denote the root $H \to \alpha(\theta H)$. Also put $H'_\alpha = \sigma(H_\alpha)$; if $\alpha \in P_+$, then $H'_\alpha \neq 0$ because $\mathfrak{h} = \mathfrak{h}_{\mathfrak{p}} + \mathfrak{h}_{\mathfrak{k}}$. The mapping $\alpha \to -\alpha^\theta$ is a permutation of P_+. Now, let $\alpha, \beta \in P_+$. Then

$$d(X_\beta)(E_\alpha) = \sigma([X_\beta, X_\alpha + \theta X_\alpha]) = \sigma([X_\beta, \theta X_\alpha]) = C'_\alpha\, \sigma([X_\beta, X_{\alpha^\theta}])$$

where $C'_\alpha \neq 0$. Since $[\mathfrak{g}^\alpha, \mathfrak{g}^{-\alpha}] = \boldsymbol{C} H_\alpha$ we have

$$d(X_\beta) E_\alpha = \begin{cases} 0 & \text{if } \beta > -\alpha^\theta \\ C_\beta H'_\beta & \text{if } \beta = -\alpha^\theta;\ \text{here } C_\beta \neq 0. \end{cases}$$

Now suppose $q \in S(\mathfrak{l} + \mathfrak{h}_{\mathfrak{p}})$ is invariant under $\sigma \circ \mathrm{Ad}_G(H)$. Then $d(X)q = 0$ for each $X \in \mathfrak{m} + \mathfrak{n}$. We write q as a polynomial in $E_{\alpha_1}, \cdots, E_{\alpha_p}$ whose coefficients are polynomials in $H_1, \cdots, H_l$:

$$(4) \qquad q(H_1, \cdots, H_l, E_{\alpha_1}, \cdots, E_{\alpha_p}) = \sum r_{(n)}(H_1, \cdots, H_l) E_{\alpha_1}^{n_1} \cdots E_{\alpha_p}^{n_p}.$$

The derivation $d(X_{\alpha_1})$ annihilates each H_i and all $E_{\alpha_1}, \cdots, E_{\alpha_p}$ except one, say E_{α_j}, for which $-\alpha_j{}^\theta = \alpha_1$. Hence

$$0 = d(X_{\alpha_1}) q = (C_{\alpha_1} n_j H'_{\alpha_1}) \textstyle\sum' r_{(n)}(H_1, \cdots, H_l) E_{\alpha_1}^{n_1} \cdots E_{\alpha_j}^{n_j - 1} \cdots E_{\alpha_p}^{n_p},$$

where $\sum'$ ranges over those terms in (4) for which $n_j > 0$. Since $C_{\alpha_1} H'_{\alpha_1} \neq 0$ the last equation shows that $\sum'$ is an empty sum so E_{α_j} does not occur in (4). Applying $d(X_{\alpha_2})$ to (4) we see that E_{α_k} $(-\alpha_k{}^\theta = \alpha_2)$ does not occur in (4), etc. Since $\alpha \to -\alpha^\theta$ is a permutation of P_+ this proves that the right hand side of (4) lies in $S(\mathfrak{h}_{\mathfrak{p}})$.

LEMMA 4.2. *Let $I(\mathfrak{l}_0 + \mathfrak{h}_{\mathfrak{p}_0})$ denote the set of polynomials in $S(\mathfrak{l}_0 + \mathfrak{h}_{\mathfrak{p}_0})$ which are invariant under $\sigma \circ \mathrm{Ad}_G(MN)$. Then*

$$I(\mathfrak{l}_0 + \mathfrak{h}_{\mathfrak{p}_0}) = S(\mathfrak{h}_{\mathfrak{p}_0}).$$

It has been proved above that $I(\mathfrak{l}_0 + \mathfrak{h}_{\mathfrak{p}_0}) \subset S(\mathfrak{h}_{\mathfrak{p}_0})$. The converse inclusion follows from the fact that each $\mathrm{Ad}(m)$ $(m \in M)$ is the identity on $\mathfrak{h}_{\mathfrak{p}_0}$ and that if $n \in N$, then $\mathrm{Ad}(n)H \equiv H \pmod{\mathfrak{n}_0}$ for each $H \in \mathfrak{h}_{\mathfrak{p}_0}$.

Turning now to the proof of Theorem 4.1, let $L_1, \cdots, L_p$ be a basis of $\mathfrak{l}_0$. If $a > 0$ is sufficiently small, the mapping

$$(t_1, \cdots, t_{p+l}) \rightarrow \exp(t_1 H_1 + \cdots + t_l H_l + t_{l+1} L_1 + \cdots + t_{l+p} L_p) H$$

is a diffeomorphism of the cube $|t_i| < a$ onto an open neighborhood V of o in G/H. The inverse mapping is a local coordinate system on V and the coordinate vector fields $\partial/\partial t_i$ satisfy

$$(\partial/\partial t_i)_o = d\hat{\pi}(H_i), \qquad (\partial/\partial t_{j+l})_o = d\hat{\pi}(L_j), \qquad (1 \leqq i \leqq l, 1 \leqq j \leqq p).$$

Let $D \in \boldsymbol{D}(G/H)$. Then there exists a unique polynomial P in $p + l$ variables such that

$$(5) \qquad [Df](o) = [P\Big(\frac{\partial}{\partial t_1}, \cdots, \frac{\partial}{\partial t_{p+l}}\Big) f(\exp(t_1 H_1 + \cdots + t_{p+l} L_p) H)](0)$$

for all $f \in C^\infty(G/H)$. In the decomposition of P into homogeneous components let P_D denote the component of highest degree. The invariance of D of course implies

$$(6) \qquad [Df^{\tau(h)}](o) = [Df](o) \qquad (h \in H).$$

Considering the highest order derivatives in (5) one deduces fom (6) that

$$P_D(d\tau(h)\, d\hat{\pi}(H_1), \cdots, d\tau(h)\, d\hat{\pi}(L_p)) = P_D(d\hat{\pi}(H_1), \cdots, d\hat{\pi}(L_p)) \quad (h \in \boldsymbol{H}),$$

so P_D is invariant under $d\tau(H)$. In view of Lemma 4.2, this implies that P_D is a polynomial in the basis vectors $d\hat{\pi}(H_1), \cdots, d\hat{\pi}(H_l)$ alone. Let $P(D)$ denote the corresponding polynomial in $H_1, \cdots, H_l$ and consider the differential operator $D_{P(D)}$ defined by (1). Then, if $f \in C^\infty(G/H)$,

$$(7) \quad [D_{P(D)} f](o) = [P(D)\Big(\frac{\partial}{\partial t_1}, \cdots, \frac{\partial}{\partial t_l}\Big) f(\exp(t_1 H_1 + \cdots + t_l H_l) H)](0).$$

On comparison of (5) and (7) it is clear that $D - D_{P(D)}$ is of lower order than D at the point o, hence of lower order everywhere, both operators being invariant. This proves, by induction, that every $D \in \boldsymbol{D}(G/H)$ has the form $D = D_U$ for some $U \in S(\mathfrak{h}_{\mathfrak{p}_0})$. Now by (1),

$$(8) \quad \begin{aligned} [D_U f](o) &= [U\Big(\frac{\partial}{\partial t_1}, \cdots, \frac{\partial}{\partial t_l}\Big) f(\exp(t_1 H_1 + \cdots + t_l H_l) H)](0) \\ &= [U\Big(\frac{\partial}{\partial t_1}, \cdots, \frac{\partial}{\partial t_l}\Big) (f \circ \hat{\pi})(\exp(t_1 H_1 + \cdots + t_l H_l))](0) \end{aligned}$$

for all $U \in S(\mathfrak{h}_{\mathfrak{p}_0})$. This shows that the mapping $U \to D_U$ is one-to-one and Theorem 4.1 is proved.

5. The Radon transform. Let ξ be any horocycle in S, ds_ξ the volume element on ξ in the Riemannian structure on ξ induced by S. For $f \in C_c(S)$ we put

$$(1) \qquad \hat{f}(\xi) = \int_\xi f(s)\, ds_\xi, \qquad \xi \in \hat{S}.$$

Definition. The function $\hat{f}$ on $\hat{S}$ given by (1) will be called the *Radon transform* of f.

It will be more convenient in what follows to express (1) in terms of functions on G. Let

$$\pi\colon G \to G/K, \qquad \hat{\pi}\colon G \to G/MN$$

denote the natural projections, and let

$$(2) \qquad F = f \circ \pi, \qquad \hat{F} = \hat{f} \circ \hat{\pi} \qquad (f \in C_c(S)).$$

The group N preserves the measure ds_{ξ_0} on the horocycle $\xi_0 = N \cdot o$; hence we can select a left invariant measure dn on N such that the mapping $n \to n \cdot o$ of N onto ξ_0 is measure-preserving. Then

$$\hat{f}(g \cdot \xi_0) = \int_{g \cdot \xi_0} f(s)\, ds_{g \cdot \xi_0} = \int_{\xi_0} f(g \cdot s_o)\, ds_o = \int_N f(gn \cdot o)\, dn$$

so

$$(3) \qquad \hat{F}(g) = \int_N F(gn)\, dn.$$

PROPOSITION 5.1. *The mapping $f \to \hat{f}$ is a one-to-one linear mapping of $C^\infty{}_c(S)$ into $C^\infty{}_c(\hat{S})$.*

Proof. Consider the mapping ψ of $(K/M) \times A_\mathfrak{p}$ onto G/MN given by $\psi(kM, a) = kaMN$. Then

$$(\hat{f} \circ \psi)(kM, a) = \int_N F(kan)\, dn.$$

Now $F \in C^\infty{}_c(G)$ and the mapping $(k, a, n) \to kan$ is a diffeomorphism of $K \times A_\mathfrak{p} \times N$ onto G. Hence $\hat{f} \circ \psi \in C^\infty{}_c((K/M) \times A_\mathfrak{p})$ and since ψ is a diffeomorphism, $\hat{f} \in C^\infty{}_c(\hat{S})$. Next, suppose $\hat{f} \equiv 0$ and consider the function

$$f_1(g) = \int_K F(kg)\, dk,$$

where dk is the normalized Haar measure on K. Then f_1 is bi-invariant under K. Now let u be any function in $C_c(G)$, bi-invariant under K, and put

$$F_u(a) = e^{\rho(\log a)} \int_N u(an)\,dn, \qquad a \in A_{\mathfrak{p}},$$

where $\log a$ is the unique element H in $\mathfrak{h}_{\mathfrak{p}_0}$ such that $\exp H = a$. Let ν be any linear mapping of $\mathfrak{h}_{\mathfrak{p}_0}$ into $\boldsymbol{C}$ and consider the corresponding spherical function

$$\phi_\nu(x) = \int_K e^{(i\nu-\rho)(H(xk))}\,dk,$$

where for each $g \in G$, $H(g)$ is the unique element in $\mathfrak{h}_{\mathfrak{p}_0}$ such that $g \in K \exp H(g) N$. Then, for a suitable normalization of the invariant measure dx on G,

$$\int_G u(x)\phi_\nu(x)\,dx = \int_{A\mathfrak{p}} F_u(a)\,e^{i\nu(\log a)}da$$

(see [11], p. 262). Now since $\hat{f} \equiv 0$ we have $F_{f_1} = 0$. Then by [13], p. 410 and p. 453 we conclude that $f_1 \equiv 0$. In particular, $F(e) = f(o) = 0$. Since $(f^{\tau(x)})^\wedge = (\hat{f})^{\tau(x)}$ for each $x \in G$, it follows that $f \equiv 0$ and the proposition is proved.

Remark. The image of the space $C^\infty{}_c(S)$ under the Radon transform is properly contained in $C^\infty{}_c(\hat{S})$. In fact, let ϕ be an arbitrary function in $C^\infty{}_c(A_{\mathfrak{p}})$ and consider the function $\Phi \in C^\infty{}_c(\hat{S})$ defined by $\Phi(\psi(kM, a)) = \phi(a)$ for $k \in K$, $a \in A_{\mathfrak{p}}$, where ψ is the diffeomorphism from Prop. 3. 6. Suppose $f \in C^\infty{}_c(S)$ were a function such that $\hat{f} = \Phi$. Since $\Phi^{\tau(k)} = \Phi$ and $(f^{\tau(k)})^\wedge = \hat{f}^{\tau(k)}$ for all $k \in K$ we may, replacing f by the function $p \to \int f(k \cdot p)\,dk$, assume that f is invariant under the action of K on S. Since the function F_u above is invariant under W (Lemma 17, [11], p. 261) we deduce that the function $a \to \phi(a) e^{-\rho(\log a)}$ is invariant under W. This is impossible, ϕ being arbitrary.

We shall now require a theorem which is essentially a restatement of some results of Harish-Chandra [11], (Lemma 3, p. 247 and Theorem 1, p. 260). Subsequently a different proof was indicated by Karpelevič [16].

If ν is a linear function on $\mathfrak{h}_{\mathfrak{p}_0}$ then $e^\nu \in C^\infty(\mathfrak{h}_{\mathfrak{p}_0})$. For simplicity, the function $a \to e^{\nu(\log a)}$ on $A_{\mathfrak{p}}$ shall also be denoted e^ν.

THEOREM 5. 2. *The notation being as in the preceding sections, let* $\boldsymbol{D}_0(G)$ *denote the set of differential operators in* $\boldsymbol{D}(G)$ *which are invariant under all right translations from* K. *Then*

(i) *For each* $D \in \boldsymbol{D}(G)$ *there exists a unique element* $D_{\mathfrak{a}} \in \boldsymbol{D}(A_{\mathfrak{p}})$ *such that*

$$D - D_{\mathfrak{a}} \in \mathfrak{n}_0 \boldsymbol{D}(G) + \boldsymbol{D}(G)\mathfrak{k}_0.$$

(ii) *If* $\phi \in C^{\infty}(G)$ *such that* $\phi(ngk) = \phi(g)$ *for all* $n \in N$, $g \in G$, $k \in K$ *then*

$$(D\phi)^{-} = D_{\mathfrak{a}}\bar{\phi}, \qquad D \in \boldsymbol{D}(G),$$

where the bar denotes restriction to $A_{\mathfrak{p}}$.

(iii) *The mapping* $D \to e^{-\rho} \circ D_{\mathfrak{a}} \circ e^{\rho}$ *is a homomorphism of* $\boldsymbol{D}_0(G)$ *onto* $I(\mathfrak{h}_{\mathfrak{p}_0})$ *and the kernel is* $\boldsymbol{D}_0(G) \cap \boldsymbol{D}(G)\mathfrak{k}_0$.

In this connection we recall that a C^{∞} function f on a manifold can be regarded as a differential operator $F \to fF$ and that $\circ$ denotes composition of differential operators. Also $\boldsymbol{D}(A_{\mathfrak{p}})$ is canonically isomorphic to $S(\mathfrak{h}_{\mathfrak{p}_0})$. Under this isomorphism, the unique automorphism $p \to 'p$ of $S(\mathfrak{h}_{\mathfrak{p}_0})$ given by $'H = H - \rho(H)$ $(H \in \mathfrak{h}_{\mathfrak{p}_0})$ corresponds to the automorphism $D \to e^{\rho} \circ De^{-\rho}$ of $\boldsymbol{D}(A_{\mathfrak{p}})$. Statements (i) and (iii) follow immediately from the cited results of Harish-Chandra if we apply the anti-automorphism $D \to D^*$ of $\boldsymbol{D}(G)$, D^* denoting the adjoint of D. For (ii) one has to prove

$$[DT\phi](a) = [XD\phi](a) = 0$$

for $D \in \boldsymbol{D}(G)$, $T \in \mathfrak{k}_0$, $X \in \mathfrak{n}_0$, $a \in A_{\mathfrak{p}}$. But $T\phi \equiv 0$ so $DT\phi \equiv 0$; secondly

$$[XD\phi](a) = \lim_{t \to 0} 1/t([D\phi](a \exp tX) - [D\phi](a)).$$

This expression vanishes since $a \exp tXa^{-1} \in N$, whence

$$[D\phi](a \exp tX) = [(D\phi)^{L(a \exp(-tX)a^{-1})}](a) = [D\phi](a).$$

The factor algebra $\boldsymbol{D}_0(G)/\boldsymbol{D}_0(G) \cap \boldsymbol{D}(G)\mathfrak{k}_0$ is canonically isomorphic to $\boldsymbol{D}(G/K)$, (see [13], p. 432). In view of (iii) and Theorem 4.1 we have two isomorphisms onto,

$$\Gamma: \boldsymbol{D}(S) \to I(\mathfrak{h}_{\mathfrak{p}_0}), \qquad \hat{\Gamma}: \boldsymbol{D}(\hat{S}) \to S(\mathfrak{h}_{\mathfrak{p}_0}).$$

THEOREM 5.3. *Let* $'\boldsymbol{D}(\hat{S}) = '\boldsymbol{D}(G/MN)$ *denote the set of* $E \in \boldsymbol{D}(\hat{S})$ *such that* $'\hat{\Gamma}(E)) \in I(\mathfrak{h}_{\mathfrak{p}_0})$. *Let* $D \to \hat{D}$ *denote the isomorphism of* $\boldsymbol{D}(S)$ *onto* $'\boldsymbol{D}(\hat{S})$ *given by*

$$'(\hat{\Gamma}(\hat{D})) = \Gamma(D), \qquad D \in \boldsymbol{D}(S).$$

Then

$$(Df)^{\wedge} = \hat{D}\hat{f} \qquad f \in C^{\infty}(S).$$

Proof. If $a \in A_{\mathfrak{p}}$, the automorphism $n \to ana^{-1}$ of N has Radon-Nikodym derivative

$$\det(\operatorname{Ad}(a) \mid \mathfrak{n}) = e^{2\rho(\log a)}$$

so

$$\int_N \phi(na)\,dn = e^{2\rho(\log a)} \int_N \phi(an)\,dn, \qquad \phi \in C^\infty{}_c(A_{\mathfrak{p}}N). \tag{4}$$

The function $F = f \circ \pi$ and $\hat{F} = f \circ \hat{\pi}$ are related by

$$\hat{F}(g) = \int_N F(gn)\,dn$$

so

$$\hat{F}(a) = e^{-2\rho(\log a)} \int_N F(na)\,dn.$$

For $g \in G$, let $A(g)$ denote the unique element in $\mathfrak{h}_{\mathfrak{p}_0}$ such that $g = n \exp A(g) k$ for some $n \in N$ and $k \in K$. Then the function F^* given by

$$F^*(g) = e^{-2\rho(A(g))} \int_N F(ng)\,dn$$

satisfies $F^*(ngk) = F^*(g)$ for all $g \in G$, $n \in N$, $k \in K$. Let ξ denote the function $g \to e^{2\rho(A(g))}$ and as before let $F \to \bar{F}$ denote the restriction of a function on G to $A_{\mathfrak{p}}$. Then if $E \in \boldsymbol{D}(G)$ we have by Theorem 5.2 and (4)

$$\int_N [EF](na)\,dn = [E(\xi F^*)](a) = [E_{\mathfrak{a}}(\bar{\xi}(\hat{F})^-)](a). \tag{5}$$

If $E \in \boldsymbol{D}_0(G)$ then the transform $(EF)^\wedge$ is defined and by (5)

$$(EF)^\wedge(a) = \xi^{-1}(a)[E_{\mathfrak{a}}(\bar{\xi}(F^\wedge)^-)](a). \tag{6}$$

Let $D \in \boldsymbol{D}(S)$ be determined by $E(f \circ \pi) = (Df) \circ \pi$ for all $f \in C^\infty{}_c(S)$. Then $\Gamma(D) = e^{-\rho} \circ E_{\mathfrak{a}} \circ e^{\rho}$, by the definition of Γ. Hence, by (6),

$$\begin{aligned}(Df)^\wedge(\xi_0) = (EF)^\wedge(e) &= [(e^{-\rho}\Gamma(D)e^{\rho})(\hat{F})^-](e) \\ &= [\hat{\Gamma}(\hat{D})((\hat{F})^-)](e) = [\hat{D}\hat{f}](\xi_0),\end{aligned}$$

the last equality following from (8) § 4. Since $(f^{\tau(g)})^\wedge = (f^\wedge)^{\tau(g)}$ for all $g \in G$ we conclude that $(Df)^\wedge(g \cdot \xi_0) = [\hat{D}\hat{f}](g \cdot \xi_0)$ and the theorem is proved.

Now let Σ denote the set of all linear functions on $\mathfrak{h}_{\mathfrak{p}_0}$ which are restrictions of some member of P_+. Let

$$\Sigma_0 = \{\lambda \in \Sigma \mid \lambda/n \notin \Sigma \text{ for all integers } n \neq 1\}.$$

Let π denote the product of all elements in Σ_0. Each element s in the Weyl group W induces a permutation of $\Sigma \cup (-\Sigma)$ (see [13], Cor. 2.11, Ch. VII). It follows that $\pi^s = \epsilon(s)\pi$ where $\epsilon(s)$ is a real number. But $s \to \epsilon(s)$ is a homomorphism of W into $\boldsymbol{R}$ so $\epsilon(s) = \pm 1$. Consequently π^2 is a Weyl group invariant. Since B is nondegenerate on $\mathfrak{h}_{\mathfrak{p}_0}$ we identify $\mathfrak{h}_{\mathfrak{p}_0}$ with its dual and therefore $S(\mathfrak{h}_{\mathfrak{p}_0})$ with the algebra of polynomial functions on $\mathfrak{h}_{\mathfrak{p}_0}$. Since B is invariant under W, $I(\mathfrak{h}_{\mathfrak{p}_0})$ is hereby identified with the algebra of W-invariant polynomial functions on $\mathfrak{h}_{\mathfrak{p}_0}$. Let $\square$ denote the operator in $\boldsymbol{D}(S)$ which satisfies $\Gamma(\square) = \pi^2$ and let $\tilde{\square}$ be any operator in $\boldsymbol{D}_0(G)$ which goes into $\square$ by the natural homomorphism μ of $\boldsymbol{D}_0(G)$ onto $\boldsymbol{D}(G/K)$ ([13], p. 390).

6. The inversion formula for complex G. Since the Radon transform $f \to \hat{f}$ is one-to-one it is a natural problem to try to find an inversion formula for it. In this section such a formula will be proved in the case when the group $I_0(S)$ is a complex semisimple Lie group.

In view of the duality between points and horocycles there is a natural dual to the transform $f \to \hat{f}$. This dual transform associates to each function $\psi \in C(\hat{S})$ a function $\check{\psi}$ on S given by

$$\check{\psi}(p) = \int_{\xi \cap p = p} \psi(\xi)\, dm(\xi), \qquad p \in S,$$

where the integral on the right is the average of ψ over the (compact) set of horocycles passing through p. Select $g \in G$ such that $g \cdot o = p$ and as before let $\xi_0 = N \cdot o$. Then

$$\check{\psi}(g \cdot o) = \int_N \psi(gk \cdot \xi_0)\, dk.$$

Let $f \in C_c(S)$ and put $I_f = (\hat{f})^{\vee}$. Our problem is to relate f and I_f. We have

$$I_f(p) = \int_K \Big(\int_{\xi_0} f(gk \cdot s_0)\, ds_0 \Big)\, dk = \int_K \int_N f(gkn \cdot o)\, dn\, dk.$$

For the function $F = f \circ \pi$ we put

$$I_F(g) = \int_K \int_N F(gkn)\, dn\, dk$$

and of course we have $I_F(g) = I_f(g \cdot o)$.

THEOREM 6.1. *Let S be a Riemannian globally symmetric space of the noncompact type. Assume that the group $I_0(S)$ has a complex structure. Then*

$$\square I_f = cf, \qquad f \in C^{\infty}{}_c(S),$$

where c is a constant $\neq 0$, independent of f.

By assumption, the Lie algebra $\mathfrak{g}_0$ has a complex structure, say J. In this case the decomposition $\mathfrak{g}_0 = \mathfrak{k}_0 + J\mathfrak{k}_0$ is a Cartan decomposition of $\mathfrak{g}_0$ so $\mathfrak{p}_0 = J\mathfrak{k}_0$. For $\mathfrak{h}_0$ we can now take the subalgebra $\mathfrak{h}_{\mathfrak{p}_0} + J\mathfrak{h}_{\mathfrak{p}_0}$ of $\mathfrak{g}_0$ so in this case $\mathfrak{h}_{\mathfrak{p}_0} = J\mathfrak{h}_{\mathfrak{k}_0}$. Now, under the scalar multiplication $(a+ib)X = aX + bJX$, $\mathfrak{g}_0$ is a Lie algebra over $\boldsymbol{C}$ and $\mathfrak{h}_0$ is a Cartan subalgebra. Let Δ' denote the set of nonzero roots of $\mathfrak{g}_0$ with respect to $\mathfrak{h}_0$. For each $\alpha \in \Delta'$, let H'_α denote the vector in $\mathfrak{h}_0$ such that $B'(H'_\alpha, H) = \alpha(H)$ $(H \in \mathfrak{h}_0)$, B' denoting the Killing form of $\mathfrak{g}_0$, considered as a Lie algebra over $\boldsymbol{C}$. Then $H'_\alpha \in \mathfrak{h}_{\mathfrak{p}_0}$ and α is real valued on $\mathfrak{h}_{\mathfrak{p}_0}$ ([13], pp. 220, 238). Fix a basis $H_1, \cdots, H_l$ of $\mathfrak{h}_{\mathfrak{p}_0}$ over $\boldsymbol{R}$ and for each root $\alpha \in \Delta'$ write $H'_\alpha = \sum_{i=1}^{l} \alpha^i H_i$. Here we define $\alpha \gg 0$ if $\alpha \neq 0$ and $\alpha^j > 0$ if j is the least index such that $\alpha^j \neq 0$. Let Q denote the set of roots $\alpha \in \Delta'$ such that $\alpha \gg 0$ and put $\rho' = \frac{1}{2}\sum_{\alpha \in Q} \alpha$. For any complex numbers $a_j = b_j + ic_j$ $(1 \leq j \leq l)$ we put $H_a = \sum_{j=1}^{l} b_j H_j + c_j(JH_j)$ and

$$\left(\frac{\partial}{\partial a_j} = \tfrac{1}{2}\,\frac{\partial}{\partial b_j} - i\frac{\partial}{\partial c_j}\right), \quad \frac{\partial}{\partial \bar{a}_j} = \tfrac{1}{2}\left(\frac{\partial}{\partial b_j} + i\frac{\partial}{\partial c_j}\right);$$

$$D_\alpha = \sum_{j=1}^{l} \alpha^j \frac{\partial}{\partial a_j}, \qquad \bar{D}_\alpha = \sum_{j=1}^{l} \alpha^j \frac{\partial}{\partial \bar{a}_j}.$$

The following formula is the principal step in the proof of the Plancherel formula for G: (see Gelfand-Neumark [6], Harish-Chandra [8], Gelfand-Graev [3])

(1) $$F(e) = c \lim_{H_a \to 0} \prod_{\alpha \in Q} D_\alpha \bar{D}_\alpha \exp(\rho'(H_a) + (\rho'(H_a))^-) \int_{K \times N} F(k \exp H_a n k^{-1})\, dn dk.$$

Here F is any function in $C^\infty{}_c(G)$ and c is a constant $\neq 0$ independent of F.

Now regarding $\mathfrak{g}_0$ as a Lie algebra over $\boldsymbol{R}$ we consider $\mathfrak{g}$, $\mathfrak{h}$, Δ as in § 3. Then it is known (see [8], p. 513) that the relation between Δ and Δ' is as follows: For each $\alpha \in \Delta'$ let α_+ and α_- denote the $\boldsymbol{C}$-linear functions on $\mathfrak{h}$ such that

(2) $$\alpha_+(H) = \alpha(H), \qquad \alpha_-(H) = (\alpha(H))^- \qquad H \in \mathfrak{h}_0$$

the bar denoting complex conjugation. Then Δ consists of all α_+, α_- as α runs through Δ'. Since each $\alpha \in \Delta'$ is real on $\mathfrak{h}_{\mathfrak{p}_0}$ the set of restrictions of elements in Δ' to $\mathfrak{h}_{\mathfrak{p}_0}$ is the same as the set of restrictions of elements in Δ to $\mathfrak{h}_{\mathfrak{p}_0}$. Consequently there is a unique Weyl chamber in $\mathfrak{h}_{\mathfrak{p}_0}$, say C, where all $\alpha \in Q$ are > 0. A root $\alpha \in \Delta'$ satisfies $\alpha \gg 0$ if and only if $\alpha > 0$ with respect to the ordering of Δ' induced by C. Now we consider C^*, Δ, P_+, P_- and ρ, defined in § 3. Let $\alpha \in \Delta'$. Since $\mathfrak{h}_{\mathfrak{k}_0} = J\mathfrak{h}_{\mathfrak{p}_0}$ and since α is complex-linear it cannot vanish identically on $\mathfrak{h}_{\mathfrak{p}_0}$. From (2) we deduce then that $P_- = \emptyset$ and $P_+ = \{\alpha_+, \alpha_- \mid \alpha \in Q\}$. Now $H_a = H_b + H_c$ where $H_b \in \mathfrak{h}_{\mathfrak{p}_0}$, $H_c \in \mathfrak{h}_{\mathfrak{k}_0}$. Each $\alpha \in \Delta'$ is real on $\mathfrak{h}_{\mathfrak{p}_0}$ and purely imaginary on $\mathfrak{h}_{\mathfrak{k}_0}$. Hence by (2)

$$(3) \qquad \rho'(H_a) + (\rho'(H_a))^- = 2\rho'(H_b) = \rho(H_b).$$

Now suppose the function F in (1) is invariant under all right translations from K. Using (3) we see that the expression inside the braces in (1) is equal to

$$(4) \qquad \exp(\rho(H_b)) \int_{K \times N} F(k \exp H_b n)\, dkdn = e^{\rho(\log b)} \int_{K \times N} F(kbn)\, dkdn,$$

where $b = \exp H_b = \exp(\sum_{j=1}^{l} b_j H_j)$. Now, if $\phi \in C^\infty(A_\mathfrak{p})$ and $\Phi \in C^\infty(\boldsymbol{R}^l)$ such that

$$\phi(b) = \Phi(b_1, \cdots, b_l), \qquad b = \exp(\sum_{j=1}^{l} b_j H_j)$$

then

$$[H'_\alpha \phi](b) = \sum_{j=1}^{l} \alpha^j \frac{\partial \Phi}{\partial b_j}.$$

Hence we obtain from (1), (4), and (4) § 5

$$(5) \qquad F(e) = c_1 \lim_{b \to e} \prod_{\alpha \in Q} (H'_\alpha)^2 \{e^{-\rho(\log b)} \int_{K \times N} F(knb)\, dk\, dn\},$$

where $b \to e$ in $A_\mathfrak{p}$ and $c_1 = 2^{-p}c$ where p, as before, denotes the number of elements in P_+. As earlier, let B denote the Killing form of $\mathfrak{g}_0$ (as a Lie algebra over $\boldsymbol{R}$). Since B is nondegenerate on $\mathfrak{h}_{\mathfrak{p}_0} \times \mathfrak{h}_{\mathfrak{p}_0}$ we identify $\mathfrak{h}_{\mathfrak{p}_0}$ with its dual via B. Now $B(X, Y) = 2\mathrm{Re}(B'(X, Y))$ for $X, Y \in \mathfrak{g}_0$ so under the identification of $\mathfrak{h}_{\mathfrak{p}_0}$ with its dual, H'_α corresponds to $2\bar{\alpha}$ ($\alpha \in \Delta'$) where $\bar{\alpha}$ denotes the restriction of α to $\mathfrak{h}_{\mathfrak{p}_0}$. On the other hand, since α_+ and α_- have the same restriction to $\mathfrak{h}_{\mathfrak{p}_0}$ the function π^2 from § 5 coincides with $\prod_{\alpha \in Q} \bar{\alpha}^2$. This last product is therefore invariant under the Weyl group W and since B is also invariant, the same is the case for the operator $\prod_{\alpha \in Q} (H'_\alpha)^2 \in S(\mathfrak{h}_{\mathfrak{p}_0})$.

The function F_1 defined by

$$F_1(g) = \int_{K \times N} F(kng)\, dk\, dn, \qquad g \in G,$$

satisfies $F_1(ngk) = F_1(g)$ so $[DF_1](b) = [D_a \bar{F}_1](b)$ for $D \in \boldsymbol{D}(G)$ by Theorem 5.2. In particular,

$$[\tilde{\square} F_1](b) = (\tfrac{1}{2})^p [(e^\rho \circ \prod_{\alpha \in Q} (H'_\alpha)^2 \circ e^{-\rho}) \bar{F}_1](b).$$

10

Hence by (5)

$$F(e)=c\lim_{b\to e}[\tilde{\Box}F_1](b). \tag{6}$$

Now let x be a fixed element in G and consider the functions $F^x=f^{\tau(x^{-1})}\circ\pi$ and $F_1{}^x$ given by

$$F_1{}^x(g)=\int_{K\times N}F^x(kng)\,dk\,dn.$$

Then (6) implies

$$F(x)=c\lim_{b\to e}[\tilde{\Box}F_1{}^x](b). \tag{7}$$

Also, since

$$F_1{}^x(g)=\int_N\Big(\int_K F(xkng)\,dk\Big)\,dn$$

we have

$$[\tilde{\Box}F_1{}^x](g)=\int_N\Big(\int_K[\tilde{\Box}F](xkng)\,dk\Big)\,dn,$$

because as g runs through a compact set in G, the integration can be taken over a compact set, independent of g. Now let $g=b$ and let $b\to e$ in $A_{\mathfrak{p}}$. Then, by (7)

$$f(x\cdot o)=F(x)=c\int_N\Big(\int_K[\tilde{\Box}F](xkn)\,dk\Big)\,dn$$

which, by [13], pp. 442-443, equals

$$c\int_N\Box_x\Big(\int_K F(xkn)\,dk\Big)\,dn=c\tilde{\Box}_x\Big(\int_{K\times N}F(xkn)\,dk\,dn\Big),$$

where the subscript x denotes the argument on which $\tilde{\Box}$ acts. Since the last integral is $I_F(x)$ it follows that

$$f(x\cdot o)=c[\tilde{\Box}I_F](x)=c[\Box I_F](x\cdot o)$$

and the theorem is proved.

Remarks. The classical Radon transform in $\boldsymbol{R}^n$ associates to each function $f\in C_c(\boldsymbol{R}^n)$ the integrals of f over hyperplanes in $\boldsymbol{R}^n$ and the problem of representing f by means of these integrals was solved by Radon [18] and John [14]. The definition of Radon transform on S is suggested by the analogy between horocycles and hyperplanes and the inversion formula in Theorem 6.1 is analogous to the inversion formula of Radon-John for $\boldsymbol{R}^n$ where n is odd. For n even the inversion formula is more complicated and involves an integral operator in place of a differential operator. Such complications are to be expected for S in the case when the group $I_0(S)$ is real.

The Radon transform for $\boldsymbol{R}^n$ was generalized in different directions to Riemannian manifolds of constant curvature by Helgason [12], Semyanistyi [21] and Gelfand-Graev-Vilenkin [5]. In the case when S is the space of positive definite Hermitian $n \times n$ matrices a formula closely related to that of Theorem 6.1 was given by Gelfand-Graev [4] and Gelfand [2].

7. Applications to invariant differential equations. In this section, the group $I_0(S)$ will be assumed complex. We shall now indicate how Theorem 6.1 can be used to reduce a differential equation

$$(1) \qquad Du = f$$

where $D \in \boldsymbol{D}(S)$, $u \in C^\infty(S)$, $f \in C^\infty{}_c(S)$, to a differential equation with constant coefficients on a Euclidean space. The procedure is reminiscent of the method of plane waves for solving homogeneous hyperbolic equations with constant coefficients (see e.g. John [15]). The heuristic discussion will be followed by a concrete example, carried out rigorously.

For $x \in G$, $k \in K$, $a \in A_{\mathfrak{p}}$ we put

$$U_{k,x}(a) = \int_N u(xkan \cdot o)\, dn, \qquad F_{k,x}(a) = \int_N f(xkan \cdot o)\, dn,$$

ignoring convergence questions for the time being. Then ($A(g)$ being as in § 5) the functions

$$U_{k,x}{}^*(g) = e^{-2\rho(A(g))} \int_N u(xkng \cdot o)\,dn$$

$$F_{k,x}{}^*(g) = e^{-2\rho(A(g))} \int_N f(xkng \cdot o)\, dn$$

are invariant under each substitution $g \to n_1 g k_1$ and coincide with $U_{k,x}$ and $F_{k,x}$ on $A_{\mathfrak{p}}$. Using Theorem 5.2 we deduce that

$$(2) \qquad (e^{-\rho} \circ \Gamma(D) \circ e^{\rho}) U_{k,x} = F_{k,x}.$$

Now by Theorem 6.1

$$(3) \qquad f(x \cdot o) = c \tilde{\square}_x \Big(\int_K F_{k,x}(e)\, dk \Big).$$

This suggests the following way to write down a solution of (1) : The function f being known, equation (2) is a differential equation with constant coefficients on the Euclidean space $A_{\mathfrak{p}}$. Assume now we have some "boundary conditions" such that (2) has a unique solution $U_{k,x}$ depending continuously on k and differentiably on x. Then, as suggested by (3), a solution to (1) is given by

$$u(x\cdot o)=c\tilde{\square}_x\left(\int_K U_{k,x}(e)\,dk\right). \tag{4}$$

This will be verified below in a special case, where the stated assumptions are indeed satisfied.

Example: The wave equation on S. Let Δ denote the Laplace-Beltrami operator on S and let $f\in C^\infty{}_c(S)$. We shall write down a global solution of the wave equation

$$\Delta u=\frac{\partial^2 u}{\partial t^2}, \qquad (t\geqq 0) \tag{5}$$

with initial data

$$u(p,0)=0;\quad \left\{\frac{\partial}{\partial t}u(p,t)\right\}_{t=0}=f(p) \qquad (p\in S). \tag{6}$$

Let $|\rho|^2=B(\rho,\rho)$ (under the identification of $\mathfrak{h}_{\mathfrak{p}_0}$ with its dual). Now the Laplacian Δ on S is induced by the Casimir operator in $\boldsymbol{D}(G)$ ([13], p. 451) and we see therefore from [11], Cor. 2, p. 271 that $\Gamma(\Delta)=\Delta_A-|\rho|^2$, where Δ_A denotes the Laplacian on $A_{\mathfrak{p}}$. We are therefore led to consider the differential equation on $A_{\mathfrak{p}}\times\boldsymbol{R}^+$,

$$(\Delta_A-|\rho|^2)V^t{}_{k,x}=\frac{\partial^2}{\partial t^2}V^t{}_{k,x} \tag{7}$$

with initial data

$$V^0{}_{k,x}=0;\quad \left\{\frac{\partial}{\partial t}V^t{}_{k,x}\right\}_{t=0}=e^\rho F_{k,x}, \tag{8}$$

where $F_{k,x}$ is the same as above. The initial value problem (7), (8) is the Cauchy problem for damped waves on the Euclidean space $A_{\mathfrak{p}}$. The solution can be found in [19], p. 88 and we obtain

$$V^t{}_{k,x}(a)=\int_{|\log b-\log a|\leqq t}(e^\rho F_{k,x})(b)\,W(t,a,b)\,db, \tag{9}$$

where

$$W(t,a,b)=2^{-\frac{1}{2}l}(\pi\,|\rho|\,d)^{1-\frac{1}{2}l}J_{1-\frac{1}{2}l}(|\rho|\,d)$$

and $d=(t^2-|\log a-\log b|^2)^{\frac{1}{2}}$. Here J_n refers to the usual Bessel function and as before l denotes the rank of S. The integral (9) is to be understood as its regularized value by means of analytic continuation as in [19], p. 88. As shown in [20], p. 35, $V^t{}_{k,x}$ is then a convolution (in the group $\boldsymbol{R}\times A_{\mathfrak{p}}$)

of a distribution (independent of k and x) and the function $e^{\rho}F_{k,x}$. It follows that $V^t{}_{k,x}(a)$ depends differentiably on k and x and, as suggested by (4) and the fact that $e^{\rho}(e)=1$, we consider the function

$$u(x\cdot o,t)=c\tilde{\Box}_x\Big(\int_K V^t{}_{k,x}(e)\,dk\Big).$$

We shall now verify that this function is a solution of the initial value problem (5), (6). For this we need a lemma. Let $\tilde{\Delta}$ be an operator in $\boldsymbol{D}_0(G)$ such that $\mu(\tilde{\Delta})=\Delta$, μ being as in § 5.

LEMMA 7.1. *Let $a\in A_{\mathfrak{p}}$. Then*

$$\tilde{\Delta}_x\Big(\int_K V^t{}_{k,x}(a)\,dk\Big)=(e^{-\rho}\circ\tilde{\Delta}_{\mathfrak{a}}\circ e^{\rho})_a\Big(\int_K V^t{}_{k,x}(a)\,dk\Big), \tag{10}$$

where $\tilde{\Delta}_{\mathfrak{a}}$ is given by Theorem 5.2.

It will first be shown that both sides of the equation (10) are solutions to the initial value problem

$$(\Delta_A-|\rho|^2)W^t{}_x=\frac{\partial^2}{\partial t^2}W^t{}_x \qquad (t\geqq 0) \tag{11}$$

$$W^0{}_x\equiv 0,\quad \Big[\frac{\partial}{\partial t}W^t{}_x\Big](a,0)=\tilde{\Delta}_x\Big(\int_K (e^{\rho}F_{k,x})(a)\,dk\Big), \tag{12}$$

so the lemma will follow from the uniqueness for the Cauchy problem (11)-(12). There is no difficulty in verifying that the function

$$(t,a)\to\tilde{\Delta}_x\Big(\int_K V^t{}_{k,x}(a)\,dk\Big)$$

satisfies (11) and (12) so now let us put

$$W^t{}_x(a)=(e^{-\rho}\circ\tilde{\Delta}_{\mathfrak{a}}\circ e^{\rho})_a\Big(\int_K V^t{}_{k,x}(a)\,dk\Big).$$

Then $W^0{}_x\equiv 0$ and

$$\Big[\frac{\partial}{\partial t}W^t{}_x\Big](a,0)=(e^{-\rho}\circ\tilde{\Delta}_{\mathfrak{a}}\circ e^{\rho})_a\Big(\int_K (e^{\rho}F_{k,x})(a)\,dk\Big).$$

In order to show that this equals the right hand side of (12) it suffices to prove that for each $D\in\boldsymbol{D}_0(G)$,

$$D_x\Big(\int_K F_{k,x}(a)\,dk\Big)=(e^{-2\rho}\circ D_{\mathfrak{a}}\circ e^{2\rho})_a\Big(\int_K F_{k,x}(a)\,dk\Big). \tag{13}$$

Since

$$\int_K F_{k,x}{}^*(g)\,e^{2\rho(A(g))}dk = \int_N \Big(\int_K f(xkng\cdot o)\,dk\Big)\,dn$$

we have

$$e^{2\rho(A(g))}D_x\Big(\int_K F_{k,x}{}^*(g)\,dk\Big) = \int_N \Big(D_x\Big(\int_K f(xkng\cdot o)\,dk\Big)\Big)\,dn,$$

which by [13], p. 443 equals

$$\int_N\Big(\int_K [\mu(D)f](xkng\cdot o)\,dk\Big)\,dn = D_g\Big(\int_N \int_K f(xkng\cdot o)\,dk\,dn\Big).$$

Putting $g = a \in A_{\mathfrak{p}}$ we have by Theorem 5.2,

$$e^{2\rho(\log a)}D_x\Big(\int_K F_{k,x}(a)\,dk\Big) = (D_{\mathfrak{a}})_a\Big(\int_N \int_K f(xkna\cdot o)\,dk\,dn\Big)$$

and now (13) follows from (4) § 5. Finally (7) shows after a straightforward computation that $W^t{}_x$ satisfies (11) so the lemma follows.

Since (6) is an immediate consequence of (3) and (8) it remains just to verify (5). Using $\square\Delta = \Delta\square$ and Lemma 7.1 we have

$$\begin{aligned}\tilde{\Delta}_x(u(x\cdot o,t)) &= c\tilde{\square}_x\Big(\tilde{\Delta}_x\Big(\int_K V^t{}_{k,x}(e)\,dk\Big)\Big)\\ &= c\tilde{\square}_x\Big(\Big[(e^{-\rho}\circ\tilde{\Delta}_{\mathfrak{a}}\circ e^{\rho})\int_K V^t{}_{k,x}dk\Big](e)\Big),\end{aligned}$$

which equals

$$c\tilde{\square}_x\Big(\int_K \frac{\partial^2}{\partial t^2}(V^t{}_{k,x}(e))\,dk\Big) \tag{14}$$

as a consequence of (7) and the relation $e^{-\rho}\circ\Delta_{\mathfrak{a}}\circ e^{\rho} = \Delta_A - |\rho|^2$ already used. In (14) we can take $\partial^2/\partial t^2$ outside the integral sign and interchange the two differential operators since they act on different arguments. Hence the expression (14) reduces to

$$\frac{\partial^2}{\partial t^2}(u(x\cdot o,t)).$$

This completes the verification of the solution for (5) and (6).

REFERENCES.

[1] A. Borel, "Ensembles fondamentaux pour les groupes algébriques," *Colloque sur la théorie des groupes algébriques*, Bruxelles, 1962.

[2] I. M. Gelfand, "Integral geometry and its relation to the theory of grouh representations," *Uspekhi Matematičeskih Nauk*, vol. 15 (1960), pp. 155-164; Russian Mathematical Surveys 1960, pp. 143-151.

[3] I. M. Gelfand and M. I, Graev, "Ein Analogon der Plancherelschen Formel für die klassischen Gruppen," *Trudy Moscovskogo Matematičeskogo Obščestva*, vol. 4 (1955), pp. 375-404. (German transl. in [6]).

[4] ———, "The geometry of homogeneous spaces, group representations in homogeneous spaces and questions in integral geometry related to them, I," *Trudy Moscovskogo Matematičeskogo Obščestva*, vol. 8 (1959), pp. 321-390.

[5] I. M. Gelfand, M. I. Graev, and N. Vilenkin, *Integral geometry and its relation to problems in the theory of group representations* (Generalized Functions Vol. 5) Moscov, 1962.

[6] I. M. Gelfand and M. A. Neumark, "Unitäre Darstellungen der klassischen Gruppen," *Trudy Matematičeskogo Institute im. V. A. Steklova*, vol. 36, Moscov-Leningrad, 1950. German transl. Akademie-Verlag, Berlin, 1957.

[7] R. Godement, "Groupes linéaires algébriques sur un corps parfait," *Séminaire Bourbaki*, 1960.

[8] Harish-Chandra, "The Plancherel formula for complex semisimple Lie groups," *Transactions of the American Mathematical Society*, vol. 76 (1954), pp. 485-528.

[9] ———, "On a lemma of Bruhat," *Journal de Mathématiques Pures et Appliques*, vol. 35 (1956), pp. 203-210.

[10] ———, "A formula for semisimple Lie groups," *American Journal of Mathematics*, vol. 79 (1957), pp. 733-760.

[11] ———, "Spherical functions on a semisimple Lie group I, II," *American Journal of Mathematics*, vol. 80 (1958), pp. 241-310, 553-613.

[12] S. Helgason, "Differential operators on homogeneous spaces," *Acta Mathematica*, vol. 102 (1959), pp. 239-299.

[13] ———, *Differential Geometry and Symmetric Spaces*, Academic Press, New York, 1962.

[13a] ———, "Duality and Radon transform for symmetric spaces," *Bulletin of the American Mathematical Society* (to appear).

[13b] ———, "Fundamental solutions of invariant differential operators on symmetric spaces," *Bulletin of the American Mathematical Society* (to appear).

[14] F. John, "Bestimmung einer Funktion aus ihren Integralen über gewisse Mannigfaltigkeiten," *Mathematische Annalen*, vol. 100 (1934), pp. 488-520.

[15] ———, *Plane Waves and Spherical Means Applied to Partial Differential Equations*, Interscience, New York, 1955.

[16] F. I. Karpelevič, "Orispherical radial parts of Laplace operators on symmetric spaces," *Doklady Akademii Nauk*, vol. 143 (1962), pp. 1034-1037; *Soviet Mathematics*, vol. 3 (1962), pp. 528-531.

[17] G. D. Mostow, "On maximal subgroups of real Lie groups," *Annals of Mathematics*, vol. 74 (1961), pp. 503-517.

[18] J. Radon, "Über die Bestimmung von Funktionen durch ihre Integralwerte längs gewisser Mannigfaltigkeiten, *Berichte über die Verhandlungen der Sächsischen Akademie*, vol. 69 (1917), pp. 262-277.
[19] M. Riesz, "L'integrale de Riemann-Liouville et le problème de Cauchy," *Acta Mathematica*, vol. 81 (1949), pp. 1-223.
[20] L. Schwartz, *Théorie des Distributions II*, Hermann et Cie, Paris, 1951.
[21] E. B. Vinberg, "The Morosov-Borel theorem for real Lie groups," *Soviet Mathematics*, vol. 2 (1961), pp. 1416-1419.

FUNDAMENTAL SOLUTIONS OF INVARIANT DIFFERENTIAL OPERATORS ON SYMMETRIC SPACES.

By S. Helgason.*

1. Introduction and notation. Let D be a differential operator on $\boldsymbol{R}^n$ with constant coefficients. A fundamental solution for D is a distribution T on $\boldsymbol{R}^n$ such that $DT = \delta$ where δ is the delta function on $\boldsymbol{R}^n$. Such a distribution, whose existence is well known, gives by convolution a solution of any equation $Du = f$ on $\boldsymbol{R}^n$, f being any C^∞ function of compact support.

Let D be an arbitrary differential operator on an arbitrary manifold M. In general the differential equation $Du = 0$ has no solution, not even locally. This remains true when M is specialized, unless D is appropriately specialized, too.

In this paper we consider the case when M is a symmetric space of the noncompact type, that is a manifold G/K, where G is a connected semisimple Lie group with center $\{e\}$ and K a maximal compact subgroup. Since a differential operator on $\boldsymbol{R}^n$ with constant coefficients is characterized by its invariance under translations, a natural restriction on the differential operators on G/K is invariance under the action of G. A fundamental solution for such a differential operator D is a distribution T on G/K such that $DT = \delta$, where δ now denotes the unit mass at the point left fixed by K. In this paper the existence of such a fundamental solution is proved (Theorem 4.2; see, however, the introduction to § 4). A fundamental solution for D immediately gives a global solution of the differential equation $Du = f$, where f is a C^∞ function with compact support.

The definition above of a fundamental solution can of course be extended to more general coset spaces. However, for compact spaces, a modification is necessary in order to take into account the nullspace of the operator (cf. § 6).

The proof of Theorem 4.2 leans heavily on Harish-Chandra's theory of spherical functions on the group G. Section 2 summarizes the required results from this theory, combines these with a recent important contribution of Gindikin and Karpelevič, and states some simple consequences. Theorem 3.3 shows that the action of the G-invariant differential operators on G/K on

Received December 26, 1963.

* This work was supported by the National Science Foundation, NSF GP-149.

the radial functions on G/K is isomorphic with the action of certain differential operators with constant coefficients. The isomorphism in question is used in Harish-Chandra's work on the Fourier analysis on G ([13], p. 595) and is related to the Radon transform on G/K. For the case when G is complex (§ 5) a more direct isomorphism of this type is given in Theorem 5.2, again as a consequence of results of Harish-Chandra. As an application we verify the absence of Huygens' principle for the wave equation on G/K, thus supporting the classical conjecture attributed to Hadamard ([6], p. 438), that Laplace's equation on a non-flat Lorentzian manifold does not satisfy Huygens' principle. In § 6 we give a formula for the Green's function (i.e. the fundamental solution of the Laplacian) of a compact, simply connected, semisimple Lie group. In § 7 we investigate whether all the G-invariant differential operators on G/K are induced by the two-sided invariant differential operators on G. It is shown that this is the case when G is a classical group (real or complex) but fails in general for the exceptional groups. However, a weaker result (Theorem 7.6) is generally true.

The main results of this paper were announced in [16] and [18].

The following notation will be used throughout. As usual $\boldsymbol{R}$ and $\boldsymbol{C}$ denote the fields of real and complex numbers, respectively. The identity of a group will be denoted by e. If A is a group and B a subgroup the set of left cosets aB will be denoted by A/B. For each $x \in A$, the mapping $aB \rightarrow xaB$ of A/B onto itself will be denoted $\tau(x)$; the left translation $a \rightarrow xa$ and the right translation $a \rightarrow ax$ are denoted by $L(x)$ and $R(x)$, respectively. If M is a manifold, $C^\infty(M)$ (respectively $C_c^\infty(M)$) denote the set of C^∞ functions (respectively C^∞ functions of compact support) on M. If $p \in M$, M_p denotes the tangent space to M at p. If τ is a differentiable mapping of M into another manifold N and $p \in M$ then $d\tau_p$ (or simply $d\tau$) denotes the differential of τ at p, mapping M_p into $N_{\tau(p)}$. If τ is a diffeomorphism of M onto N and $f \in C^\infty(M)$ then f^τ denotes the composite function $f \circ \tau^{-1}$. If D is a differential operator on M, $[Df](p)$ (or $D_p(f(p))$) denotes the value of Df at p. If τ is a diffeomorphism of M onto itself then D^τ denotes the differential operator $f \rightarrow (Df^{\tau^{-1}})^\tau$ on M.

If A is a Lie group and B a closed subgroup then $\boldsymbol{D}(A/B)$ denotes the algebra of real linear differential operators on the manifold A/B which are invariant under the action of A. We write $\boldsymbol{D}(A)$ for $\boldsymbol{D}(A/\{e\})$. Lie groups will be denoted by capital Roman letters and their Lie algebras by the corresponding lower case German letters. The adjoint representation of a Lie group A is denoted by Ad_A (or simply Ad). The dual space of a vector space E will usually be denoted $\hat{E}$ or E'.

2. Fourier transforms on symmetric spaces. Let S be a Riemannian globally symmetric space, G the largest connected group of isometries of S in the compact open topology. We assume S is of the noncompact type, that is, G is semisimple and has no compact normal subgroup $\neq \{e\}$. Let o be a fixed point in S, K the (maximal compact) isotropy subgroup of G at o so S can be identified with the left coset space G/K. Let $\mathfrak{k}$ and $\mathfrak{g}$ denote the corresponding Lie algebras, $\mathfrak{p}$ the orthogonal complement of $\mathfrak{k}$ in $\mathfrak{g}$ with respect to the Killing form $\langle\ ,\ \rangle$ of $\mathfrak{g}$. Let $\exp: \mathfrak{g} \to G$ denote the exponential mapping for G and $\mathrm{Exp}: S_o \to S$ the Exponential mapping for S; the latter is a diffeomorphism. If π denotes the natural mapping of G onto G/K then $d\pi$ maps $\mathfrak{p}$ isomorphically onto S_o and $\pi(\exp X) = \mathrm{Exp}\, d\pi(X)$ for $X \in \mathfrak{p}$. As is customary we identify these two spaces and write $\mathfrak{p}$ for S_o. Let $|X| = \langle X, X\rangle^{\frac{1}{2}}$ for $X \in \mathfrak{p}$. Choose a maximal abelian subspace $\mathfrak{a}$ of $\mathfrak{p}$ (a Cartan subalgebra for S) and put $A = \exp(\mathfrak{a})$. Let $\mathfrak{a}'$ denote the set of regular elements in $\mathfrak{a}$ and put $A' = \exp(\mathfrak{a}')$, $S' = K \cdot \mathrm{Exp}(\mathfrak{a}')$. We fix a Weyl chamber C in $\mathfrak{a}$, i.e. a connected component of $\mathfrak{a}'$. Let $\mathfrak{F}$ denote the set of linear maps of $\mathfrak{a}$ into $\boldsymbol{C}$, so $\mathfrak{F}_{\boldsymbol{R}} = \{\lambda \in \mathfrak{F}: \lambda(\mathfrak{a}) \subset \boldsymbol{R}\}$ is the dual space of $\mathfrak{a}$. An element $\lambda \in \mathfrak{F}_{\boldsymbol{R}}$ is called *positive* if $\lambda(H) > 0$ for all $H \in C$. For each $\lambda \in \mathfrak{F}_{\boldsymbol{R}}$ let $\mathfrak{g}_\lambda = \{X \in \mathfrak{g}: [H, X] = \lambda(H)X \text{ for } H \in \mathfrak{a}\}$ and let $d_\lambda = \dim(\mathfrak{g}_\lambda)$; λ is called a *root* of S with respect to $\mathfrak{a}$ if $d_\lambda \neq 0$. Let $\Sigma = \{\lambda \in \mathfrak{F}_{\boldsymbol{R}}: \lambda > 0, d_\lambda \neq 0\}$, let $2\rho = \Sigma_{\lambda > 0} d_\lambda \lambda$ and let Σ_0 denote the set of elements in Σ which are not integral multiples of other members of Σ. The algebra $\mathfrak{n} = \Sigma_{\lambda>0} \mathfrak{g}_\lambda$ is a nilpotent subalgebra of $\mathfrak{g}$ and we have Iwasawa decompositions $\mathfrak{g} = \mathfrak{k} + \mathfrak{a} + \mathfrak{n}$, $G = KAN$ where $N = \exp(\mathfrak{n})$. For $g \in G$ let $H(g)$ denote the unique element in $\mathfrak{a}$ satisfying $g \in K \exp H(g) N$. If $h = \exp H$ $(H \in \mathfrak{a})$ we write $H = \log h$.

The Weyl group W of S is the group of linear transformations of $\mathfrak{a}$ induced by those elements of $\mathrm{Ad}_G(K)$ which leave $\mathfrak{a}$ invariant; W also acts on A via the mapping exp and on $\mathfrak{F}_{\boldsymbol{R}}$ by duality.

Let $\boldsymbol{D}_0(G)$ denote the set of elements in $\boldsymbol{D}(G)$ which are invariant under right translations by the elements of K. Then we have a linear mapping μ of $\boldsymbol{D}_0(G)$ onto $\boldsymbol{D}(G/K)$ given by $(\mu(D)f) \circ \pi = D(f \circ \pi)$ for $f \in C^\infty(G/K)$ ([15], p. 390). We have also a homomorphism γ of the algebra $\boldsymbol{D}_0(G)$ onto the algebra $I(\mathfrak{a})$ of W-invariants in the symmetric algebra $S(\mathfrak{a})$ over $\mathfrak{a}$ (Theorem 1, [13], p. 260). The elements in $S(\mathfrak{a})$ will be considered as differential operators on A in the usual manner.

We shall now state the Plancherel formula in the form proved by Harish-Chandra [13] and draw some consequences of a more recent paper of Gindikin-Karpelevič [8].

For $\lambda \in \mathfrak{F}$ let ϕ_λ denote the corresponding spherical function ([13], p. 261) and let $\lambda \rightarrow \boldsymbol{c}(\lambda)$ denote Harish-Chandra's function for S which occurs in the leading term in the asymptotic expansion of ϕ_λ, that is, (cf. [13], p. 283), if dk is the normalized Haar measure on K,

$$\phi_\lambda(\exp H) = \int_K e^{(i\lambda-\rho)(H(\exp H)k)}dk \sim e^{-\rho(H)} \sum_{s \in W} \boldsymbol{c}(\lambda(sH))e^{i\lambda(sH)}$$

for H varying in a subset of the Weyl chamber C and λ varying in a certain subset of $\mathfrak{F}$. The function $\boldsymbol{c}^{-1}$ is holomorphic in an open subset of $\mathfrak{F}$ containing $\mathfrak{F}_R$ ([13], p. 580).

For a certain space $I_0(G)$, defined below, of complex-valued C^∞ functions on G, bi-invariant under K, the following Plancherel formula holds. For $f \in I_0(G)$, put

$$(1) \qquad \tilde{f}(\lambda) = \int_G f(x)\phi_\lambda(x^{-1})\,dx \qquad (\lambda \in \mathfrak{F}_R).$$

Then ([13], p. 611),

$$(2) \qquad f(x) = w^{-1}\int_{\mathfrak{F}_R} \tilde{f}(\lambda)\phi_\lambda(x)\,|\,\boldsymbol{c}(\lambda)\,|^{-2}\,d\lambda,$$

$$(3) \qquad \int_G |\,f(x)\,|^2\,dx = w^{-1}\int_{\mathfrak{F}_R} |\,\tilde{f}(\lambda)\,|^2\,|\,\boldsymbol{c}(\lambda)\,|^{-2}\,d\lambda,$$

where dx and $d\lambda$ are suitably normalized invariant measures on G and $\mathfrak{F}_R$, respectively, and w is the order of W.

Now let $\alpha \in \Sigma_0$ and consider the subalgebra $\mathfrak{g}^\alpha$ of $\mathfrak{g}$ generated by the subspaces $\mathfrak{g}_\alpha$ and $\mathfrak{g}_{-\alpha}$. Let $\mathfrak{k}^\alpha = \mathfrak{g}^\alpha \cap \mathfrak{k}$, $\mathfrak{p}^\alpha = \mathfrak{g}^\alpha \cap \mathfrak{p}$. Then $\mathfrak{g}^\alpha$ is semisimple, $\mathfrak{g}^\alpha = \mathfrak{k}^\alpha + \mathfrak{p}^\alpha$ is a Cartan decomposition and the corresponding symmetric space, say S^α, has rank one. For a Cartan subalgebra of this symmetric space one can choose $\mathfrak{a}^\alpha = \mathfrak{p}^\alpha \cap \mathfrak{a}$. For $\lambda \in \mathfrak{F}$, let λ_α denote its restriction to $\mathfrak{a}^\alpha$. If $\boldsymbol{c}_\alpha$ denotes the Harish-Chandra function for S^α then, according to Gindikin-Karpelevič [8],

$$(4) \qquad \boldsymbol{c}(\lambda) = \prod_{\alpha \in \Sigma_0} \boldsymbol{c}_\alpha(\lambda_\alpha).$$

For each $\alpha \in \Sigma_0$, select $H_\alpha \in \mathfrak{a}$ such that $\langle H_\alpha, H\rangle = \alpha(H)$ for all $H \in \mathfrak{a}$. Then $H_\alpha \in \mathfrak{a}^\alpha$. As shown in [8], $\boldsymbol{c}(\lambda) = I(i\lambda)/I(\rho)$ where

$$I(\nu) = \prod_{\alpha \in \Sigma} B\left(\frac{p_\alpha}{2}, \frac{p_{\alpha/2}}{4} + \frac{(\nu,\alpha)}{(\alpha,\alpha)}\right),$$

B is the Beta function $(x,y) \rightarrow \Gamma(x)\Gamma(y)/\Gamma(x+y)$, and p_α is the multiplicity of α. For the space S^α, Σ_0 consists of just one element, say γ. Then 2γ is the only other possible element in Σ. Denoting their respective multiplicities by p_1 and p_2, the Killing form of $\mathfrak{g}^\alpha$ by $(\ , \ )_\alpha$, the function $\boldsymbol{c}_\alpha(\lambda_\alpha)$ becomes

$$\boldsymbol{c}_\alpha(\lambda_\alpha) = I^{-1}(\rho^\alpha) B\left(\frac{p_1}{2}, i\,\frac{(\lambda_\alpha, \gamma)_\alpha}{(\gamma,\gamma)_\alpha}\right) B\left(\frac{p_2}{2}, \frac{p_1}{4} + i\,\frac{(\lambda_\alpha, 2\gamma)_\alpha}{(2\gamma, 2\gamma)_\alpha}\right),$$

where ρ^α is the function ρ for S^α. If $p_2 = 0$, the last factor does not occur. In any case we have

$$(5) \qquad \boldsymbol{c}_\alpha(\lambda_\alpha) = a_1 \frac{\Gamma(2ia_2\lambda_\alpha(H_\alpha))}{\Gamma(a_3 + 2ia_2\lambda_\alpha(H_\alpha))} \frac{\Gamma(a_4 + ia_2\lambda_\alpha(H_\alpha))}{\Gamma(a_5 + ia_2\lambda_\alpha(H_\alpha))},$$

where the a_i are constants > 0.

Consider now the polynomial function $\pi = \prod_{\alpha \in \Sigma_0} H_\alpha$ on $\mathfrak{F}$ (cf. [13], p. 252). Then

$$(6) \qquad \boldsymbol{c}(\lambda)\pi(\lambda) = \prod_{\alpha \in \Sigma_0} \boldsymbol{c}_\alpha(\lambda_\alpha) H_\alpha(\lambda) = \prod_{\alpha \in \Sigma_0} \boldsymbol{c}_\alpha(\lambda_\alpha)\lambda_\alpha(H_\alpha).$$

Because of (5), $\boldsymbol{c}_\alpha$ has a simple pole at the origin. Thus $\boldsymbol{c}(\lambda)\pi(\lambda)$ is holomorphic and $\neq 0$ for $\lambda = 0$. This means that π_0 in Lemma 29 in [13] p. 580 actually is a constant multiple of π. This shows that the space $I_0(G)$ in [13], pp. 611, 586, consists of all functions ϕ_a of the form

$$(7) \qquad \phi_a(x) = \int_{\mathfrak{F}_R} a(\lambda)\pi(\lambda)\phi_\lambda(x)\,d\lambda,$$

where a belongs to the space $\mathcal{S}(\mathfrak{F}_R)$ of complex-valued rapidly decreasing C^∞ functions on $\mathfrak{F}_R$ ([27], p. 89). Let $\mathfrak{I}(\mathfrak{F}_R)$ denote the set of W-invariants in $\mathcal{S}(\mathfrak{F}_R)$. Let $D \in \boldsymbol{D}_0(G)$; then $D\phi_\lambda = \gamma(D)(i\lambda)\phi_\lambda$ ([13], p. 262) so D leaves the space $I_0(G)$ invariant (cf. Theorem 3, [13], p. 586). Let the positive functions ω on $\mathfrak{p}$, Λ and Λ_0 on $\mathfrak{a}$ be determined by

$$(8) \qquad \int_{G/K} f(p)\,dp = \int_{\mathfrak{p}} f(\operatorname{Exp} X)\omega^2(X)\,dX = \int_{\mathfrak{a}} \Lambda(H)\,dH \int_K f(k \exp H \cdot o)\,dk$$

$$(9) \qquad \int_{\mathfrak{p}} F(X)\,dX = \int_{\mathfrak{a}} \Lambda_0(H)\,dH \int_K F(\operatorname{Ad}(k)H)\,dk,$$

f and F being continuous functions of compact support and dp, dX, dH being the volume elements induced by the Killing form. For each integer $q \geqq 0$ let $\omega_q(x) = \omega(X)(1 + |X|)^q$ if $x = k \exp X$ $(k \in K, X \in \mathfrak{p})$. As in [13], p. 585, let $I(G)$ denote the set of all complex-valued C^∞ functions f on G which satisfy the following two conditions: a) f is bi-invariant under K; b) for each $D \in \boldsymbol{D}(G)$ and each integer $q \geqq 0$

$$\tau_{D,q}(f) = \sup_{x \in G} |\omega_q(x)[Df](x)| < \infty.$$

The collection of all seminorms $\tau_{D,q}$ turns $I(G)$ into a locally convex space. Also $I_0(G) \subset I(G)$ and the mapping $a \to \phi_a$ is a continuous mapping of $\mathcal{S}(\mathfrak{F}_R)$ into $I(G)$ (cf. [13], p. 586).

Next we consider the integral transform

$$F_f(h) = e^{\rho(\log h)} \int_N f(hn)\, dn, \qquad f \in I(G), h \in A,$$

dn being a suitably normalized invariant measure on N. This integral transform is used in [13], pp. 595-596 where it is shown that

(i) The mapping $f \to F_f$ is a continuous mapping of $I(G)$ into the space $I(A)$ of W-invariant rapidly decreasing C^∞ functions on A with the usual topology.

(ii)

$$\tilde{f}(\lambda) = \int_A F_f(h) e^{-i\lambda(\log h)} dh, \qquad f \in I(G), h \in A.$$

with an appropriate normalization of the invariant measure on A.

The space $I(A)$ is of course an algebra under the convolution product on A.

LEMMA 2.1. *Let $D \in \mathbf{D}(G)$ and if $f \in C^\infty(G)$ let*

$$f^\natural(x) = \int_{K \times K} f(kxk')\, dk\, dk' \qquad (x \in G).$$

Then there exists a constant M_D and finitely many operators $D_1, \cdots, D_{p(D)}$ in $\mathbf{D}(G)$ such that the uniform norm $\| \ \|_\infty$ on G satisfies

$$(10) \qquad \| Df^\natural \|_\infty \leqq M_D \sum_{i=1}^{p(D)} \| D_i f \|_\infty \qquad \textit{for all } f \in C_c^\infty(G).$$

Proof. Let $D_1, \cdots, D_p$ be a basis of the subspace of operators in $\mathbf{D}(G)$ of order $\leqq$ order (D). If $x \in G$, then $D^{R(x^{-1})} = \sum_{i=1}^{p} a_i(x) D_i$ where $a_i \in C^\infty(G)$. Hence

$$[Df^\natural](x) = \sum_{i=1}^{p} \int_{K \times K} a_i(k') [D_i f](kxk')\, dk\, dk'$$

and (10) follows.

LEMMA 2.2. *Let $I_c(G) = I(G) \cap C_c^\infty(G)$. Then $I_c(G)$ is a dense subspace of $I(G)$.*

Proof. There exists a sequence $(\phi_i) \subset C_c^\infty(G)$ and an increasing sequence (U_i) of open subsets of G with union G such that

(i) $\phi_i(x) = 1$ for $x \in U_i$;

(ii) For each $D \in \mathbf{D}(G)$, $\| D\phi_i \|_\infty \leqq c(D)$ where $c(D)$ is constant which only depends on D.

A proof can be found in Langlands [22], p. 119. Replacing ϕ_i by $\phi_i^\natural$ and U_i by the open set $\{x \in G\colon KxK \subset U_i\}$ we may, by (10), assume that each ϕ_i and each U_i is bi-invariant under K.

Now let $f \in I(G)$, $D \in \mathbf{D}(G)$ and q an integer $\geqq 0$. We shall show that $\tau_{D,q}(f - f\phi_i) \to 0$ as $i \to \infty$. It is clear that

$$D(f - f\phi_i) = \sum_{\alpha,\beta} (D_\alpha f) E_\beta (1 - \phi_i) \qquad \text{(finite sum)} \tag{11}$$

where D_α, $E_\beta \in \mathbf{D}(G)$. Let $\epsilon > 0$ be given and let N (independent of i) denote the number of terms in (11). Then by (ii) there exists an index i_0 such that for all α, β, i

$$\sup_{x \in G - U_{i_0}} | \omega_q(x) [D_\alpha f](x) [E_\beta(1 - \phi_i)](x) | < \frac{\epsilon}{N}. \tag{12}$$

Since $E_\beta(1 - \phi_i)$ vanishes identically on U_{i_0} if $i \geqq i_0$ it follows from (11) and (12) that

$$\tau_{D,q}(f - f\phi_i) < \epsilon \qquad \text{if } i \geqq i_0.$$

3. The transmutation operator. Since the Weyl group W permutes the elements in $\Sigma \cup (-\Sigma)$ it will also permute the elements in $\Sigma_0 \cup (-\Sigma_0)$. It follows that for each $s \in W$, $\pi^s = \epsilon(s)\pi$ where $\epsilon(s) = \pm 1$. A function a on $\mathfrak{F}_R$ is called *skew* if $a^s = \epsilon(s)a$ for all $s \in W$.

LEMMA 3.1. *Suppose $a \in \mathcal{S}(\mathfrak{F}_R)$ is skew. Then $a = \pi a_1$ where $a_1 \in \mathcal{S}(\mathfrak{F}_R)$.*

Since a is skew it vanishes identically on each hyperplane $H_\alpha(\lambda) = 0$ $(\alpha \in \Sigma)$ in $\mathfrak{F}_R$; (cf. [13], Lemma 10, p. 252). Choose a basis $(x_1, \cdots, x_l)$ of $\mathfrak{a}$ $(=$ dual of $\mathfrak{F}_R)$ such that $x_1 = H_\alpha$. Then $a(0, x_2, \cdots, x_l) = 0$ so $a(x_1, \cdots, x_l) = x_1 b(x_1, \cdots, x_l)$ for

$$b(x_1, \cdots, x_l) = \int_0^1 [D_1 a](x_1 s, x_2, \cdots, x_l)\, ds,$$

where D_1 denotes differentiation with respect to the first argument. Thus $a = H_\alpha b$ where $b \in \mathcal{S}(\mathfrak{F}_R)$ and the lemma follows by iteration, since no two elements in Σ_0 are proportional.

PROPOSITION 3.2. *The space $I_0(G)$ is an algebra with respect to the convolution on G.*

Let ϕ_a, ϕ_b be arbitrary elements in $I_0(G)$, where $a, b \in \mathcal{S}(\mathfrak{F}_R)$. Let $A^+ = \exp C$. Then for a suitable normalization of the invariant measure dl on A we have for $z \in G$, ((8), (9) § 2),

$$\phi_a * \phi_b(z) = \int_G \phi_a(zx^{-1})\phi_b(x)\,dx = \int_{A^+} \phi_b(l)\Lambda(l) \int_K \phi_a(zkl^{-1})\,dk\,dl.$$

Now

$$\int_K \phi_a(zkl^{-1})\,dk = \int_K dk \int_{\mathfrak{F}_R} a(\lambda)\pi(\lambda)\phi_\lambda(zkl^{-1})\,d\lambda \tag{1}$$

and since $|\phi_\lambda(x)| \leqq 1$, and $a(\lambda)$ rapidly decreasing we may interchange the order of integration in (1). Using the functional equation for the spherical function it follows that

$$\phi_a * \phi_b(z) = \int_{A^+} \phi_b(l)\Lambda(l) \int_{\mathfrak{F}_R} a(\lambda)\pi(\lambda)\phi_\lambda(z)\phi_\lambda(l^{-1})\,d\lambda\,dl. \tag{2}$$

Let the norm of the linear form $\lambda \in \mathfrak{F}$ be denoted by $|\lambda|$. Now as shown in [13] p. 611 there exists an integer $k \geqq 0$ and, for a fixed b, a positive constant L such that for $\lambda \in \mathfrak{F}_R$

$$|\pi(\lambda)| \int_{A^+} |\phi_b(l)\phi_\lambda(l^{-1})|\Lambda(l)\,dl \leqq L(1 + |\lambda|)^k.$$

Consequently

$$\int_{A^+}\int_{\mathfrak{F}_R} |\phi_b(l)\Lambda(l)a(\lambda)\pi(\lambda)\phi_\lambda(z)\phi_\lambda(l^{-1})|\,d\lambda\,dl$$
$$\leqq \int_{\mathfrak{F}_R} L(1 + |\lambda|)^k |a(\lambda)\pi(\lambda)|\,d\lambda$$

which is finite; thus the integrals on the right hand side of (2) can be interchanged so we obtain

$$\phi_a * \phi_b(z) = \int_{\mathfrak{F}_R} a(\lambda)\pi(\lambda)\phi_\lambda(z)\tilde{\phi}_b(\lambda)\,d\lambda.$$

But (i) and (ii) in § 2 show that $\tilde{\phi}_b \in \mathcal{S}(\mathfrak{F}_R)$ so

$$\phi_a * \phi_b = \phi_c \qquad \text{where } c = a\tilde{\phi}_b \tag{3}$$

and the proposition follows.

THEOREM 3.3. *The mapping $f \to F_f$ is a topological isomorphism of the algebra $I_0(G)$ onto the algebra $I(A)$. Furthermore,*

$$F_{Df} = \gamma(D) F_f \tag{4}$$

for $D \in \mathbf{D}_0(G)$, $f \in I_0(G)$.

LEMMA 3.4. *Let* $f \in I_0(G)$ *and* $D \in \mathbf{D}(G)$. *For each compact set* $\Omega \subset G$ *the integral*

$$\int_N [Df](ng)\, dn \tag{5}$$

converges uniformly for $g \in \Omega$.

Writing $g = n_0 a k$, where $n_0 \in N$, $a \in A$, $k \in K$, we have with the notation from the proof of Lemma 2.1,

$$\int_N [Df](ng)\, dn = \int_N [(Df)^{R(k^{-1})}](nn_0 a)\, dn$$
$$= \sum_{i=1}^{p} a_i(k) \int_N [D_i f](nn_0 a)\, dn.$$

Therefore it suffices to prove that (5) converges uniformly as g varies in a compact subset of NA. But this is guaranteed by Cor. 2, p. 593 in [13].

In order to prove (4) let $g \in G$ and denote by $A(g)$ the unique element in $\mathfrak{a}$ such that $g \in N \exp A(g) K$. If $h \in A$, the automorphism $n \rightarrow hnh^{-1}$ of N has Jacobian at e equal to the determinant of $\mathrm{Ad}(h)$ restricted to $\mathfrak{n}$, i.e., $e^{2\rho(\log h)}$. Hence

$$\int_N f(nh)\, dn = e^{2\rho(\log h)} \int_N f(hn)\, dn \qquad (h \in A) \tag{6}$$

so the function

$$F^*(g) = e^{-\rho(\log A(g))} \int_N f(ng)\, dn$$

is an extension of F_f to a C^∞ function on G. Using Lemma 3.4 we have

$$D_g(e^{\rho(\log A(g))} F^*(g)) = \int_N [Df](ng)\, dn.$$

The function $g \rightarrow e^{\rho(\log A(g))} F^*(g)$ is invariant under each substitution $g \rightarrow ngk$ $(n \in N, k \in K)$. From Theorem 5.2 in [17] we therefore obtain

$$(D_{\mathfrak{a}})_h (e^{\rho(\log h)} F^*(h)) = \int_N [Df](nh)\, dn \qquad (h \in A)$$

where $D_{\mathfrak{a}} = e^{\rho} \circ \gamma(D) \circ e^{-\rho}$. Since $F_f(h) = F^*(h)$ we obtain (4) by a second application of (6). Next (2) and (ii) in § 2 show immediately that the mapping $f \rightarrow F_f$ is one-to-one on $I_0(G)$.

Next we prove that $f \to F_f$ maps $I_0(G)$ onto $I(A)$. Let $\psi \in I(A)$. Then the Fourier transform

$$\hat{\psi}(\lambda) = \int_A \psi(a) e^{-i\lambda(\log a)} da \qquad (\lambda \in \mathfrak{F}_R)$$

is a W-invariant function in $\mathcal{S}(\mathfrak{F}_R)$. Consider now the integral

$$f(x) = w^{-1} \int_{\mathfrak{F}_R} \hat{\psi}(\lambda) \phi_\lambda(x) |\mathbf{c}(\lambda)|^{-2} d\lambda \qquad (x \in G)$$

whose existence is assured by the following lemma.

LEMMA 3.5. *There exist positive integers a and b such that*

$$\frac{1}{|\mathbf{c}(\lambda)|} \leqq a(1 + |\lambda|)^b, \qquad \lambda \in \mathfrak{F}_R.$$

In view of (5) and (6) § 2 this is contained in the following lemma.

LEMMA 3.6. *Let a and b be any positive real numbers. The function s on $\boldsymbol{R}$ given by*

$$s(x) = \Gamma(a + ix)(\Gamma(b + ix))^{-1}$$

has each of its derivatives majorized uniformly by a polynomial.

Proof. Using the formula

$$\frac{\Gamma'(z)}{\Gamma(z)} = \sum_1^\infty \frac{1}{n(n+z)} - \gamma - \tfrac{1}{2},$$

where γ is Euler's constant, we find that

$$-i\frac{s'(x)}{s(x)} = -\frac{1}{a+ix} + \frac{1}{b+ix} + (a+ix)\sum_1^\infty \frac{1}{n(n+a+ix)} - (b+ix)\sum_1^\infty \frac{1}{n(n+b+ix)}$$

so

(7) $$s'(x) = s(x)t(x)$$

where $t(x)$ is a function each of whose derivatives is bounded by a polynomial. On the other hand, for each fixed real number c we have for x real,

(8) $$\lim_{|x| \to \infty} |\Gamma(c + ix)| \exp(\tfrac{1}{2}\pi |x|) |x|^{\frac{1}{2}-c} = (2\pi)^{\frac{1}{2}}$$

(see e. g. [30], p. 151). This equation shows that $s(x)$ is uniformly bounded by a polynomial. Now the lemma follows from (7) by induction.

The formula for $f(x)$ can be written

$$f(x) = w^{-1} \int_{\mathfrak{F}_R} \hat{\psi}(\lambda) \phi_\lambda(x) \,|\, \boldsymbol{c}(\lambda) \,|^{-2} d\lambda = \int_{\mathfrak{F}_R} a(\lambda) \pi(\lambda) \phi_\lambda(x) \, d\lambda,$$

where

$$a(\lambda) = w^{-1} \hat{\psi}(\lambda) \pi(\lambda) \{ \boldsymbol{c}(\lambda) \pi(\lambda) \operatorname{conj}(\boldsymbol{c}(\lambda) \pi(\lambda)) \}^{-1},$$

conj denoting complex conjugation. In view of (5), (6) § 2 and Lemma 3.6 the function a belongs to $\mathcal{S}(\mathfrak{F}_R)$ so

$$f = \phi_a. \tag{9}$$

Now the following lemma implies $\tilde{f} = \hat{\psi}$ and therefore $F_f = \psi$.

LEMMA 3.7. *Suppose $a \in \mathcal{S}(\mathfrak{F}_R)$ is skew. Then*

$$\pi(\lambda) \hat{\phi}_a(\lambda) = w \,|\, \boldsymbol{c}(\lambda) \,|^2 \,|\, \pi(\lambda) \,|^2 a(\lambda).$$

This lemma is an immediate consequence of Theorem 5 in [13] p. 610 (since $\pi_0 = \pi$).

Since $(f_1 * f_2)^\sim = \tilde{f}_1 \tilde{f}_2$ $(f_1, f_2 \in I_0(G))$ formula (ii) § 2 shows that the mapping $f \to F_f$ is a homomorphism of $I_0(G)$ into $I(A)$.

As remarked earlier the mapping $f \to F_f$ is continuous. Thus it just remains to prove that its inverse is continuous. Since $\phi_{s\lambda} = \phi_\lambda$ for $\lambda \in \mathfrak{F}_R$ and $s \in W$ we may, replacing $a(\lambda)$ by $w^{-1} \sum_{s \in W} \epsilon(s) a(\lambda)$ in (7) § 2, always take $a(\lambda)$ skew in the definition of ϕ_a. Since the Euclidean Fourier transform is a homeomorphism between the spaces of rapidly decreasing functions we just have to prove that the mapping $f \to \tilde{f}$ of $I_0(G)$ onto $\mathfrak{d}(\mathfrak{F}_R)$ has a continuous inverse. Now $f = \phi_a$ where a is a skew function in $\mathcal{S}(\mathfrak{F}_R)$ which is unique because $\tilde{f}(\lambda) = w \,|\, \boldsymbol{c}(\lambda) \pi(\lambda) \,|^2 a(\lambda) \pi(\lambda)^{-1}$. The mapping $h \to |\, \boldsymbol{c}\pi \,|^2 h$ is a homeomorphism of $\mathfrak{d}(\mathfrak{F}_R)$ onto itself so the question is reduced to the continuity of the mappings

$$d \to \pi d \qquad \text{of } \mathfrak{d}(\mathfrak{F}_R) \text{ into } \mathcal{S}(\mathfrak{F}_R).$$

$$a \to \phi_a \qquad \text{of } \mathcal{S}(\mathfrak{F}_R) \text{ into } I_0(G).$$

The first is obviously continuous and the continuity of the second was mentioned in § 2. All the statements in Theorem 3.3 have now been verified.

The operator $X : f \to F_f$ is a simultaneous "transmutation operator" for the differential operators in $\boldsymbol{D}_0(G)$ in the sense that it transforms a function space on which the $D \in \boldsymbol{D}_0(G)$ act isomorphically onto another function space, where these differential operators act with constant coefficients.

4. Applications to invariant differential operators. At the end of his paper [13], Harish-Chandra states the following two conjectures which would imply the Plancherel formula (3) § 2 for all $f \in I(G)$.

(I) *There exists a polynomial* p *on* $\mathfrak{F}_R$ *such that* $|\mathbf{c}(\lambda)\pi(\lambda)p(\lambda)| \geqq 1$ *for all* $\lambda \in \mathfrak{F}_R$. (Here we have used $\pi = \pi_0$, proved in § 2).

(II) *The mapping* $f \to F_f$ *is one-to-one on* $I(G)$.

Now (I) is a consequence of formulas (5) and (6) § 2 together with Lemma 3. 6. In view of Theorem 3. 3, (II) is equivalent to $I_0(G) = I(G)$. On the other hand, the Plancherel formula (3) § 2 for all $f \in I(G)$ together with (ii) § 2 clearly implies (II).

Now I understand that Harish-Chandra has proved the Plancherel formula (3) § 2 for all $f \in I(G)$ in a recent, as yet unpublished, work. In this section we shall therefore assume that $I_0(G) = I(G)$.

If E is any differential operator on G, the adjoint E^* of E (with respect to the measure dx on G) is given by

$$\int_G (Ef)(x)g(x)\,dx = \int_G f(x)(E^*g)(x)\,dx \qquad f, g \in C_c^\infty(G).$$

If T is a distribution on G then ET is by definition the distribution $f \to T(E^*f)$. If T_1, T_2 are two distributions on G, their convolution $T_1 * T_2$ is the functional on $C_c^\infty(G)$ formally given by (provided the right hand side makes sense)

$$(T_1 * T_2)(f) = \int_{G\times G} f(xy)\,dT_1(x)\,dT_2(y).$$

If at least one T_i has compact support then $T_1 * T_2$ is another distribution on G. For this and several other formal properties of distributions on G, see [9], § 4. In particular if δ_K denotes the distribution $f \to \int_K f(k)\,dk$ on G we have for $f \in C_c^\infty(G)$

$$f * \delta_K(x) = \int_K f(xk)\,dk, \qquad x \in G.$$

On the space G/K the adjoint operation $D \to D^*$ on differential operators is defined by means of the G-invariant measure on G/K induced by the invariant measure dx on G and the differentiation of distributions is defined just as above. If T is a distribution on G/K, $\tilde{T}$ denotes the distribution on G given by $\tilde{T}(f) = T(f_\pi)$ where $f \in C_c^\infty(G)$ and f_π is the function on G/K given by $f_\pi \circ \pi = f * \delta_K$. In particular, if $F \in C_c^\infty(G/K)$ then $\tilde{F} = F \circ \pi$ and if δ denotes the distribution $F \to F(o)$ on G/K then $\tilde{\delta} = \delta_K$. Moreover, the convolution of distributions on G induces an analogous operation $\times$ for distributions on G/K. In fact, if T_1, T_2 are two distributions on G/K, at least one of the compact support, then $T_1 \times T_2$ is the distribution $F \to (\tilde{T}_1 * \tilde{T}_2)(\tilde{F})$ for $F \in C_c^\infty(G/K)$.

Definition. Let $D \in \mathbf{D}(G/K)$. A distribution T on G/K is called a *fundamental solution* for D if $DT = \delta$.

PROPOSITION 4.1. *Let $D \in \mathbf{D}(G/K)$ and $F \in C_c^\infty(G/K)$. If T is a fundamental solution for D then the C^∞ function $u = F \times T$ is a solution of the equation*

$$Du = F.$$

Proof. Let $f \in C^\infty(G)$ and let S be a distribution on G. Let f' denote the function $x \to f(x^{-1})$ and S' the distribution $f \to S(f')$ $(f \in C_c^\infty(G))$. Let $E \in \mathbf{D}(G)$ and let ϵ_E the distribution $f \to [Ef'](e)$. Then

(1) $\quad Ef = f * \epsilon_E,$

(2) $\quad ES = S * \epsilon_E.$

In fact

$$f * \epsilon_E(x) = \int_G f(xy^{-1})\, d\epsilon_E(y) = \epsilon_E(f^{L(x^{-1})\prime}) = [Ef^{L(x^{-1})}](e) = [Ef](x).$$

Secondly,

$$\begin{aligned}(ES)(f) &= S(E^*f) = (S * (E^*f)')(e) = (S * (f * \epsilon_{E^*})')(e) \\ &= ((S * (\epsilon_{E^*})') * f')(e) = (S * (\epsilon_{E^*})')(f).\end{aligned}$$

If Y is a left invariant vector field on G then $Y^* = -Y$ so $(\epsilon_{Y^*})' = \epsilon_Y$. If $E_1, E_2 \in \mathbf{D}(G)$ then (1) shows that $\epsilon_{E_1E_2} = \epsilon_{E_2} * \epsilon_{E_1}$. Also $(E_1E_2)^* = E_2^*E_1^*$ and $(S_1 * S_2)' = S_2' * S_1'$ if S_1, S_2 are distributions on G with compact support. These formulas combined show that $(\epsilon_{E^*})' = \epsilon_E$ so (2) follows.

Since

$$f * S(x) = \int_G f(xy^{-1})\, dS(y)$$

it follows easily that $(F \times T) \circ \pi = \tilde{F} * \tilde{T}$. Select $E \in \mathbf{D}_0(G)$ such that $\mu(E) = D$. Then

$$(D(F \times T)) \circ \pi = E((F \times T) \circ \pi) = (\tilde{F} * \tilde{T}) * \epsilon_E = \tilde{F} * E\tilde{T}$$

Now $\mu(E^*) = (\mu(E))^*$ and $E^* \in \mathbf{D}_0(G)$. It follows that if $f \in C_c^\infty(G)$,

$$(E\tilde{T})(f) = \tilde{T}(E^*f) = T((E^*f)_\pi) = T(D^*f_\pi) = DT(f_\pi) = \delta_K(f).$$

Thus

$$(D(F \times T)) \circ \pi = \tilde{F} * \delta_K = F \circ \pi,$$

which proves the proposition.

THEOREM 4.2. *Let D be a nonzero differential operator in $\mathbf{D}(G/K)$. Then there exists a fundamental solution for D.*

Consider the transmutation operator $X: f \to F_f$ which is a linear homeomorphism of $I(G)$ onto $I(A)$. Its transpose, ${}^tX$, is a linear homeomorphism of the dual space $I(A)'$ onto the dual space $I(G)'$. Now $I(A)'$ can be regarded as the space of W-invariant tempered distributions on the Euclidean space A; in fact if $S \in I(A)'$, the linear mapping $f \to S(w^{-1} \sum_{s \in W} f^s)$ $(f \in C_c^\infty(A))$ is a W-invariant distribution on A, which determines S completely since the space $I_c(A) = I(A) \cap C_c^\infty(A)$ is dense in $I(A)$. This distribution can therefore be denoted by S. Each $D \in S(\mathfrak{a})$, considered as a differential operator on A, with adjoint D^*, acts on $I(A)'$ accordingly: $(DS)(f) = S(D^*f)$. Similarly, the elements of $I(G)'$ can be regarded as distributions on G, bi-invariant under K. In precise terms, let $f \in C_c^\infty(G)$ and let $f^\natural$ be as in Lemma 2.1. If $T \in I(G)'$ we see easily from (10) § 2 that the linear mapping $f \to T(f^\natural)$ is a distribution on G. Since $I_c(G)$ is dense in $I(G)$, (Lemma 2.2), this distribution determines T uniquely and will therefore be denoted by T. Now select $E \in \mathbf{D}_0(G)$ such that $\mu(E) = D$. The distribution δ_K clearly belongs to $I(G)'$ so there exists a unique element $\epsilon \in I(A)'$ such that ${}^tX\epsilon = \delta_K$. By theorem of Hörmander [19] and Łojasiewicz [23] each nonzero differential operator with constant coefficients on $\mathbf{R}^n$ maps the space of tempered distributions onto itself. In particular there exists a tempered distribution S on A such that

$$\gamma(E^*)^*S = \epsilon.$$

Both $\gamma(E^*)^*$ and ϵ are W-invariant so we can also take S invariant under W, that is, $S \in I(A)'$. Put $T_1 = {}^tX(S)$; then $T_1 \in I(G)'$ and if $f \in C_c^\infty(G)$ we have

$$\begin{aligned} T_1(E^*f) &= T_1((E^*f)^\natural) = {}^tXS(E^*f^\natural) = S(XE^*f^\natural) = S(\gamma(E^*)Xf^\natural) \\ &= (\gamma(E^*)^*S)(Xf^\natural) = ({}^tX\epsilon)(f^\natural) = \delta_K(f^\natural) = \delta_K(f). \end{aligned}$$

This proves that

$$ET_1 = \delta_K. \tag{3}$$

The proof of Theorem 4.2 is now easily finished. Let T denote the linear functional

$$F \to T_1(F \circ \pi) \qquad\qquad F \in C^\infty(G/K).$$

Then T is a distribution on G/K and $T_1 = \tilde{T}$. Also

$$\begin{aligned}(DT)(F) &= T(D^*F) = T_1((D^*F)\circ\pi)\\ &= T_1(E^*(F\circ\pi)) = (ET_1)(F\circ\pi) = \delta_K(F\circ\pi) = F(o).\end{aligned}$$

Thus

$$DT = \delta$$

and the theorem is proved. The proof of (3) has the following corollary.

Corollary 4.3. *Let E be any nonzero element in $\mathbf{D}_0(G)$. Then*

(i) *E maps $I(G)'$ onto itself.*

(ii) *E is a one-to-one mapping of $I(G)$ into itself.*

The first part is proved exactly as formula (3). For the second part, suppose $f \in I(G)$ such that $Ef = 0$. Then $X(Ef) = \gamma(E)Xf = 0$. Since Xf and $\gamma(E)Xf$ lie in $I(A)$ we obtain by taking Fourier transform, $Xf = 0$ so $f = 0$.

Proposition 4.4. *Each $D \neq 0$ in $\mathbf{D}(G/K)$ is a one-to-one mapping of $C_c^\infty(G/K)$ into itself.*

Proof. Suppose $F \in C_c^\infty(G/K)$ satisfies $DF = 0$. Let

$$f^*(g,x) = \int_K F(gkx\cdot o)\,dk, \qquad g, x \in G,$$

and select $E \in \mathbf{D}_0(G)$ such that $\mu(E) = D$. Then by the Darboux equation [15], p. 443

$$E_x(f^*(g,x)) = \int_K [DF](gkx\cdot o)\,dk = 0. \tag{4}$$

For each fixed g, the function $x \to f^*(g,x)$ is an element of $I(G)$. By Cor. 4.3 (ii) and (4), $f^*(g,x) = 0$ for all $x \in G$. Putting $x = e$, it follows that $F = 0$.

5. The case of a complex group G. In this section we shall assume that S is a Riemannian globally symmetric space of the noncompact type for which the group G is a complex Lie group. In this case we give an inversion formula for the transmutation operator $X\colon f \to F_f$ and a representation formula for the differential operators in $\mathbf{D}(G/K)$, which leads to a sharper version of Theorem 4.2 and certain applications.

Theorem 5.1. *Let $\tilde{\square}$ be any element in $\mathbf{D}_0(G)$ such that $\gamma(\tilde{\square}) = \pi^2$. Then the inverse of the operator $f \to F_f$ of $I(G)$ onto $I(A)$ is given by*

$$(1) \qquad \bar{\square}_g\left(\int_K F_f(\exp(H(gk)))\, e^{-\rho(H(gk))}\, dk\right) = cf(g)$$

where c is a constant $\neq 0$, *independent of f.*

Proof. Suppose first that f has compact support. Then

$$\int_K F_f(\exp(H(gk)))\, e^{-\rho(H(gk))}\, dk = \int_K dk \int_N f(\exp(H(gk))n)\, dn$$
$$= \int_K \int_N f(gkn)\, dn dk,$$

since dn is invariant and f is invariant under left translations from K. The last integral equals $I_f(g)$ in the notation of [17], § 6 and (1) follows from Theorem 6.1 in [17]. Now if $f \in I(G)$ is arbitrary there exists a sequence (f_n) of elements in $I_c(G)$ which converges to f (in $I(G)$). Then the inequality in Cor. 2 of [13], p. 593, shows that for each $D \in \mathbf{D}(A)$, DF_{f_n} converges to DF_f uniformly on each compact subset of A. For $a \in A$ the left hand side of (1) can be written

$$\delta'(\bar{\square})_a \int_K F_f(\exp(H(ak)))\, e^{-\rho(H(ak))}\, dk$$

where $\delta'(\bar{\square})$ is a certain differential operator on A' (cf. [13], Lemma 23, p. 265). From these statements it follows that (1) is valid on A' for each $f \in I(G)$. But $KA'K$ is dense in G so (1) holds on the entire G by continuity.

A function F on $S = G/K$ (or $\mathfrak{p}$) will be called a *radial function* if $F^{\tau(k)} = F$ (or $F^{\mathrm{Ad}(k)} = F$) for all $k \in K$. Similarly radial distributions are defined. Let $I(\mathfrak{p})$ denote the algebra of radial polynomial functions on $\mathfrak{p}$. If we combine the isomorphism of $\mathbf{D}(G/K)$ onto $I(\mathfrak{a})$ ([13], Theorem 1, [15], p. 432) with Chevalley's restriction isomorphism ([15], p. 430) of $I(\mathfrak{p})$ onto $I(\mathfrak{a})$ we obtain an isomorphism λ of $\mathbf{D}(G/K)$ onto $I(\mathfrak{p})$.

THEOREM 5.2. *Let F be any radial* C^∞ *function on G/K and* $D \in \mathbf{D}(G/K)$. *Then*

$$(2) \qquad (Df) \circ \mathrm{Exp} = \frac{1}{\omega} \lambda(D)(\omega(F \circ \mathrm{Exp})).$$

Proof. Let $\bar{F}$ denote the function on $\mathfrak{a}$ given by $\bar{F}(H) = F(\mathrm{Exp}\, H)$. Then Lemma 55, p. 305 in [13] (cf. also [3], Theorem 28), can be stated

$$(3) \qquad (DF)^- = \Lambda^{-\frac{1}{2}}\Gamma(D)(\Lambda^{\frac{1}{2}}\bar{F}) \qquad \text{(on } \mathfrak{a}'\text{)}.$$

and Theorem 1 in [12] implies

(4) $$(\lambda(D)(F\circ\mathrm{Exp}))^{-}=\Lambda_0^{-\frac{1}{2}}\Gamma(D)(\Lambda_0^{\frac{1}{2}}\bar{F}) \qquad (\text{on } \mathfrak{a}').$$

Now $\bar{\omega}^2=\Lambda/\Lambda_0$ so by (3) and (4)

$$(DF)^{-}=\Lambda^{-\frac{1}{2}}\Gamma(D)(\Lambda^{\frac{1}{2}}\bar{F})=\Lambda^{-\frac{1}{2}}\Gamma(D)(\Lambda_0^{\frac{1}{2}}\bar{\omega}\bar{F})=\bar{\omega}^{-1}(\lambda(D)(\omega(F\circ\mathrm{Exp}))^{-}.$$

This proves formula (2) on the set $\mathfrak{a}'$. By continuity, it holds on $\mathfrak{a}$. But both sides of (2) are radial functions so (2) holds on the entire space $\mathfrak{p}$.

Remark. I am indebted to B. Kostant for conversations which raised the question of a formula of the type of (2). For the bi-invariant differential operators on any real semisimple Lie group a local formula similar to (2) has also been proved by Harish-Chandra.

As a consequence of Theorem 5.2 we can now prove a sharper form of Theorem 4.2.

THEOREM 5.3. *Let $D\in\boldsymbol{D}(G/K)$. The radial fundamental solutions Q for D are the distributions*

$$Q: F\to Q_0(\omega(F\circ\mathrm{Exp})), \qquad F\in C_c^{\infty}(G/K),$$

where Q_0 is an arbitrary radial fundamental solution for $\lambda(D^*)^*$.

Proof. Let $I_c^{\infty}(G/K)$ and $I_c^{\infty}(\mathfrak{p})$ denote the spaces of radial C^{∞} functions on G/K and $\mathfrak{p}$, respectively, equipped with the topologies induced by $C_c^{\infty}(G/K)$ and $C_c^{\infty}(\mathfrak{p})$. Their duals can be indentified (compare § 4) with the spaces of radial distributions on G/K and $\mathfrak{p}$, respectively. The mapping

$$Y: F\to\omega(F\circ\mathrm{Exp})$$

is a linear homeomorphism of $I_c^{\infty}(G/K)$ onto $I_c^{\infty}(\mathfrak{p})$ and according to (2),

(5) $$YDF=\lambda(D)YF$$

so Y is a simultaneous transmutation operator for the differential operators in $\boldsymbol{D}(G/K)$. The transpose ${}^tY$ is a linear homeomorphism between the mentioned distribution spaces. In particular, the distribution $({}^tY)^{-1}(\delta)$ is the delta distribution $\delta_0: f\to f(0)$ on $\mathfrak{p}$. Now let Q_0 be any radial distribution on $\mathfrak{p}$. Then the equation

$$\lambda(D^*)^*Q_0=\delta_0$$

is easily seen (by (5)) to be equivalent to the equation

$$D({}^tYQ_0)=\delta;$$

this proves the theorem.

Examples.

I. The Laplace-Beltrami operator on S. Let Δ and Δ_0 denote the Laplace-Beltrami operators on S and $\mathfrak{p}$ respectively. As pointed out in [15], p. 451, Δ is the image under the mapping μ of the Casimir operator in $\boldsymbol{D}(G)$. Therefore, if we use Cor. 2, p. 271 in [13], we see that $\lambda(\Delta)=\Delta_0-|\rho|^2$. Now $\Delta^*=\Delta$, $\Delta_0{}^*=\Delta_0$, and a well known fundamental solution of $\Delta_0-|\rho|^2$ is given by

$$Q_0(X)=2^{-\frac{1}{2}(n+2)}\pi^{\frac{1}{2}(1-n)}(i|\rho||X|^{-1})^{\frac{1}{2}(n-2)}N_{\frac{1}{2}(n-2)}(i|\rho||X|), \tag{6}$$

where $n=\dim(\mathfrak{p})$ and N_p is the Neumann function (see e.g. Schwartz [28], Exposé 8). Thus $|X|^{n-2}Q_0(X)$ is bounded near the origin $X=0$ so Q_0 is a locally integrable function on $\mathfrak{p}$.

THEOREM 5.4. *The Laplace-Beltrami operator Δ on S has a locally integrable fundamental solution Q given by*

$$Q\circ\mathrm{Exp}=\omega^{-1}Q_0, \tag{7}$$

where Q_0 is given by (6).

In fact, the fundamental solution Q for Δ given by Theorem 5.2 is locally integrable on S and satisfies for all $F\in C_c{}^\infty(S)$

$$Q(F)=\int_S Q(p)F(p)\,dp=\int_{\mathfrak{p}} Q(\mathrm{Exp}\,X)F(\mathrm{Exp}\,X)\omega^2(X)\,dX,$$

$$Q(F)=Q_0(\omega(F\circ\mathrm{Exp}))=\int_{\mathfrak{p}} Q_0(X)\omega(X)(F\circ\mathrm{Exp})(X)\,dX,$$

so $Q\circ\mathrm{Exp}=\omega^{-1}Q_0$ as stated.

Remark. For the case when S is the space of positive definite Hermitian $m\times m$ matrices a formula of the type (7) was given without proof by Dynkin [7].

II. The wave equation on S. Absence of Huygens' Principle. In [17] the solution of the Cauchy problem for the wave equation $(\Delta-\partial^2/\partial t^2)u=0$ on S is written down. However, the solution formula, although explicit, seems too complicated to settle immediately the question whether Huygens' principle[1] holds for the wave equation on S. This can be answered in the negative by using the representation of Δ in Example I.

---

[1] For a discussion and results concerning Hadamard's problem about the occurrence of Huygen's principle, see e.g. Hadamard [10], p. 236, Stellmacher [26], Asgeirsson [2], Helgason [14], p. 296.

According to Hadamard, Huygens' principle cannot hold if $n = \dim S$ is even. Hence we may assume n odd. The equation

$$(\Delta_0 - |\rho|^2 - \partial^2/\partial t^2)u = 0$$

for damped waves in the Euclidean space $\mathfrak{p}$ has in the interior of the retrograde light cone

$$D_0 = \{(X,t) \in \mathfrak{p} \times \boldsymbol{R} : t^2 - |X|^2 > 0, t < 0\}$$

a solution of the form

$$(8) \qquad u(X,t) = U(X,t)(t^2 - |X|^2)^{\frac{1}{2}(1-n)} + V(X,t)\log(t^2 - |X|^2),$$

where the functions U and V are bounded for small X and t and $V(0,0) \neq 0$ (cf. Hadamard [10], p. 107). Replacing X in (8) by $\mathrm{Ad}(k)X$ and integrating over K we may assume that U and V are radial functions in the variable X. Now consider the function F given by $F(\operatorname{Exp} X, t) = \omega(X)^{-1}u(X,t)$ for $(X,t) \in D_0$. Since $\lambda(\Delta) = \Delta_0 - |\rho|^2$ we have from (2)

$$\begin{aligned}[(\Delta - \frac{\partial^2}{\partial t^2})F](\operatorname{Exp} X, t) &= [\Delta F](\operatorname{Exp} X, t) - \frac{\partial^2}{\partial t^2} F(\operatorname{Exp} X, t) \\ &= \omega(X)^{-1}[(\Delta_0 - |\rho|^2)u](X,t) - \omega(X)^{-1}\frac{\partial^2}{\partial t^2}u(X,t) = 0\end{aligned}$$

so the function F satisfies the wave equation in the interior of the retrograde light cone

$$\boldsymbol{D}_0 = \{(\operatorname{Exp} X, t) \in S \times \boldsymbol{R} : (X,t) \in D_0\}.$$

Let r denote the Lorentzian distance of $(\operatorname{Exp} X, t)$ from the origin in $S \times \boldsymbol{R}$, i.e., $r(\operatorname{Exp} X, t) = (t^2 - |X|^2)^{\frac{1}{2}}$. Then F is a solution of the wave equation in $\boldsymbol{D}_0$ of the form

$$F(p) = U_1(p)r(p)^{1-n} + V_1(p)\log r(p) \qquad (p \in \boldsymbol{D}_0)$$

where $V_1(0) \neq 0$. This means that Huygens' principle is absent (cf. Hadamard [10], p. 236, Courant-Hilbert [6], p. 438).

6. The Green's function of a compact semisimple Lie group. Let M be a compact Riemannian manifold of dimension m, $C_0^\infty(M)$ the subspace of $f \in C^\infty(M)$ which satisfy $\int_M f(x)\,dx = 0$, dx denoting the Riemannian measure on M. The Laplace-Beltrami operator Δ on M is an isomorphism of $C_0^\infty(M)$ onto itself. The inverse operator, say Γ, is an integral operator whose kernel, *the Green's kernel of M,* is a differentiable function outside the diagonal of $M \times M$ and is characterized by the following conditions (see de Rham [24], p. 157):

(i) $g(x,y)=g(y,x)$ for $x \neq y$ in M.

(ii) $d(x,y)^{m-2}g(x,y)$ is bounded, d denoting the distance on M. (If $m=2$ this condition should read "$(\log d(x,y))^{-1}g(x,y)$ is bounded").

(iii) $[\Gamma f](x)=\int_M g(x,y)f(y)\,dy$ for all $f \in C_0^\infty(M)$.

(iv) $\int_M g(x,y)\,dy \equiv 0$.

Now suppose M is globally symmetric and let U be the largest connected group of isometries of M in the compact open topology. Then U is compact. Fix a point $o \in M$ and let K be the isotropy subgroup of U at o, so $M=U/K$. If g is the Green's kernel of M it is easy to see that for each $u \in U$, the function $(x,y)\to g(u\cdot x, u\cdot y)$ also satisfies (i)-(iv) so $g(x,y)\equiv g(u\cdot x,u\cdot y)$. Now define the function γ on $M-\{o\}$ by $\gamma(x)=g(x,o)$. Then $\gamma(k\cdot x)=\gamma(x)$ for $k \in K$, $x \neq o$ in M. The function γ on $M-\{o\}$ will be called *the Green's function of* M. Because of (ii), γ is an integrable function on M. Let $|M|$ denote the total volume of M and δ the distribution $f\to f(o)$ on M.

THEOREM 6.1. *Let M be a compact Riemannian globally symmetric space. The Green's function γ, considered as a distribution on M, is characterized as the unique solution of the equations*

$$\Delta\gamma=\delta-\frac{1}{|M|}, \qquad \gamma(1)=0. \tag{1}$$

For each function f on $M=U/K$ let $\tilde{f}=f\circ\pi$, where $\pi\colon U\to U/K$ is the natural mapping. We normalize the invariant measure du on U such that $|U|=\int_U du$ equals $|M|$. For two distributions T_1, T_2 on M, $T_1\times T_2$ is defined as in § 4. Then if $f \in C^\infty(M)$ and $u \in U$,

$$\begin{aligned}(f\times\gamma)(u\cdot o)&=\int_U \tilde{f}(v)\tilde{\gamma}(v^{-1}u)\,dv=\int_U \tilde{f}(v)g(v^{-1}u\cdot o,o)\,dv\\&=\int_U f(v\cdot o)g(u\cdot o,v\cdot o)\,dv=\int_M f(x)g(u\cdot o,x)\,dx\\&=\int_M g(u\cdot o,x)f_0(x)\,dx=[\Gamma f_0](u\cdot o),\end{aligned}$$

where we have put $f_0=f-|M|^{-1}\int f(x)\,dx$. It follows that

$$(2) \qquad \Delta(f \times \gamma) = f - \frac{1}{|M|}\int f(x)\,dx \qquad (f \in C^\infty(M)).$$

Now there exists an operator $\tilde{\Delta} \in \boldsymbol{D}(U)$, invariant under all right translations by elements of K, such that $\tilde{\Delta}(f \circ \pi) = \Delta f \circ \pi$ for all $f \in C^\infty(M)$ (cf. [15], p. 390). Computations, analogous to those of the proof of Prop. 4.1 show that ($*$ denoting convolution on U),

$$(\Delta(f \times \gamma)) \circ \pi = \tilde{f} * \tilde{\Delta}\tilde{\gamma}, \qquad f \in C^\infty(M)$$
$$(\Delta\gamma)^{\sim} = \tilde{\Delta}\tilde{\gamma}$$

so

$$(3) \qquad \Delta(f \times \gamma) = f \times \Delta\gamma \qquad f \in C^\infty(M).$$

Next let s denote the geodesic symmetry of M with respect to o and let σ denote the automorphism $u \to sus$ of U. Let $f^\natural$ denote the function on M given by the average,

$$f^\natural(x) = \int_K f(k \cdot x)\,dk$$

and if F is a function on U let F' denote the function $u \to F(u^{-1})$ on U. If F is bi-invariant under K then $F' = F^\sigma$ since $\sigma(u) \in Ku^{-1}K$ for $u \in U$. Since $s \circ \pi = \pi \circ \sigma$ (cf. [15], p. 174) it is easy to verify $(((f^s)^\natural)^\sim)^\sigma = (f^\natural)^\sim$ for $f \in C^\infty(M)$. Consequently, by (2) and (3),

$$\begin{aligned}(f \times \Delta\gamma)(o) &= (f^s \times \Delta\gamma)(o) = ((f^s)^\natural \times \Delta\gamma)(o)\\ &= (((f^s)^\natural)^\sim * (\Delta\gamma)^\sim)(e) = (\Delta\gamma)^\sim((((f^s)^\natural)^\sim)') = (\Delta\gamma)^\sim((f^\natural)^\sim)\\ &= \Delta\gamma(f^\natural) = \Delta\gamma(f)\end{aligned}$$

so

$$(4) \qquad (f \times \Delta\gamma)(o) = \Delta\gamma(f) \qquad f \in C^\infty(M).$$

Now (1) results from formulas (2)-(4). On the other hand, suppose γ_1 is another distribution satisfying (1). Then $\gamma - \gamma_1$ satisfies the elliptic equation $\Delta(\gamma - \gamma_1) = 0$. By the regularity theorem for elliptic operators (see e.g. [19], p. 178) $\gamma - \gamma_1$ is a C^∞ function on M, hence a constant which must be 0, because $\gamma(1) = \gamma_1(1) = 0$.

In the case when M has rank one, γ is determined in more explicit terms in [15], p. 447.

Suppose now U is a compact, simply connected semisimple Lie group with Riemannian structure induced by the negative of the Killing form $\langle\ ,\ \rangle$ of the Lie algebra $\mathfrak{u}$ of U. We shall compute the Green's function of U. Let $\mathfrak{t}$ be any maximal abelian subalgebra of $\mathfrak{u}$ and let $T = \exp(\mathfrak{t})$; T is taken with the Riemannian structure induced by that of U. Let $\mathfrak{t}_e$ denote the unit lattice $\{H \in \mathfrak{t}: \exp H = e\}$ in $\mathfrak{t}$ and let Λ_e denote the lattice of complex-valued linear functions λ on $\mathfrak{t}$ satisfying $\lambda(\mathfrak{t}_e) \subset 2\pi i\mathbf{Z}$, where $\mathbf{Z}$ denotes the set of integers. For each $\lambda \in \Lambda_e$, the function e^λ on T determined by $e^\lambda(\exp H) = e^{\lambda(H)}$ $(H \in \mathfrak{t})$ is a character of T. On the other hand, each character χ of T arises in this manner from a unique $\lambda \in \Lambda_e$ so Λ_e can be identified with the character group of T. If f is a function on U, let $\bar{f}$ denote its restriction to T.

The superscript denoting complexification we consider the set of roots of the complex Lie algebra $\mathfrak{u}^c$ with respect to the Cartan subalgebra $\mathfrak{t}^c$. Suppose the real vector space spanned by the roots is ordered by means of some Weyl chamber C in $i\mathfrak{t}$ (cf. § 2) and let ρ denote one half of the sum of the positive roots. Let W denote the Weyl group of U; it acts on T and on $\mathfrak{t}$. If $s \in W$ let $\epsilon(s)$ denote the determinant of s acting on $\mathfrak{t}$. It is well known that $\rho \in \Lambda_e$ and $s(\mathfrak{t}_e) \subset \mathfrak{t}_e$ for each $s \in W$ so the function $j = \sum_{s \in W} \epsilon(s) e^{s\rho}$ is a well-defined function on T.

The extension of $\langle \; , \; \rangle$ to a bilinear form on $\mathfrak{u}^c \times \mathfrak{u}^c$ is the Killing form of $\mathfrak{u}^c$. If λ is a linear form on $\mathfrak{t}^c$, let $H_\lambda \in \mathfrak{t}^c$ be determined by $\langle H_\lambda, H\rangle = \lambda(H)$ for $H \in \mathfrak{t}^c$. If $H_\lambda \in i\mathfrak{t}$ we put $|\lambda| = \langle H_\lambda, H_\lambda\rangle^{\frac{1}{2}}$. Then $H_\rho \in C$ since $\rho > s\rho$ for $s \neq e$ in W.

LEMMA 6.2. *The orbit of ρ under W coincides with the set of elements in Λ_e of norm $|\rho|$*, i.e.,

$$\{s\rho : s \in W\} = \{\lambda \in \Lambda_e : |\lambda| = |\rho|\}.$$

Proof. Consider the subring A of $C^\infty(T)$ generated by the functions e^μ $(\mu \in \Lambda_e)$. A function $f \in C^\infty(T)$ is called *skew* if $f^s = \epsilon(s)f$. It is well known and easily proved (cf. Lemma 3.1) that if $f \in C^\infty(T)$ is skew then $f = jf_1$ where $f_1 \in C^\infty(T)$ and if $f \in A$ then $f_1 \in A$. Now suppose $\lambda \in \Lambda_e$ and $|\lambda| = |\rho|$. Then

$$\sum_{s \in W} \epsilon(s) e^{s\lambda} = \sum_{s \in W} \epsilon(s)^{s\rho} \sum_{\mu} a_\mu e^\mu.$$

Now suppose $a_\mu \neq 0$ for some $\mu > 0$ in Λ_e. Then if λ_0 is the highest among the linear forms $s\lambda$ $(s \in W)$ we have $\lambda_0 = \rho + \mu$. But $\mu > 0$ implies $\langle \mu, \rho\rangle = \mu(H_\rho) > 0$ so $|\lambda_0|^2 > |\rho|^2 + |\mu|^2$ which is a contradiction. Similarly, $a_\mu = 0$ if $\mu < 0$ so we find that $\lambda_0 = \rho$, proving the lemma.

To any distribution S on T we associate the Fourier series

$$S \sim \sum_{\lambda \in \Lambda_e} a(\lambda) e^\lambda, \qquad a(\lambda) = \frac{1}{|T|} S(e^{-\lambda})$$

in the usual manner ([27]). Since j divides each skew function in $C^\infty(T)$ it is easily seen that the linear form J on $C^\infty(T)$ given by

$$J(F)=\frac{1}{|U|}(-1)^{\frac{1}{2}(\dim U-\dim T)}\Big(\frac{1}{j}\sum_{s\in W}\epsilon(s)F^s\Big)(e)$$

for $F\in C^\infty(T)$ is a distribution on T. The distribution γ is an integrable function on U and being invariant under all automorphisms of U is completely determined by its restriction $\bar\gamma$ to T. Let T' denote the set of points $t\in T$ for which $j(t)\neq 0$.

THEOREM 6.3. *The Green's function* γ *on* U *is given by*

$$\bar\gamma=\frac{1}{j}(\gamma_\rho * J) \qquad\qquad \text{on } T',$$

where γ_ρ *is the distribution on* T *given by the Fourier series*

$$\gamma_\rho=\sum_{|\lambda|\neq|\rho|}\frac{1}{|\lambda|^2-|\rho|^2}e^\lambda .$$

Proof. Let $I(U)$ denote the space of all $f\in C^\infty(U)$ which are invariant under all automorphisms of U; the space $C^\infty(U)$ is given the usual Schwartz topology and $I(U)$ is a closed subspace. Let $S(T)$ denote the space of skew functions in $C^\infty(T)$, topologized similarly. The mapping $X\colon f\to j\bar f$ is a linear homeomorphism of $I(U)$ onto $S(T)$ and

$$X(\Delta f)=(\Delta_T+|\rho|^2)Xf \qquad\qquad f\in I(U),$$

where Δ is the Laplacian on U and Δ_T the Laplacian on T. This formula can be obtained from Weyl's classical character formula and also from Berezin [3], Theorem 2.4 or Harish-Chandra [13], p. 271. The transpose ${}^tX$ is a linear one-to-one mapping of the dual $S(T)'$ (which can be identified with the space of skew distributions on T) onto the dual $I(U)'$ (which can be regarded as the space of invariant distributions on U). If we put

$$Y_0=({}^tX)^{-1}\Big(\delta-\frac{1}{|U|}\Big)$$

and find a solution $S_0\in S(T)'$ to the equation

$$(\Delta_T+|\rho|^2)S_0=Y_0$$

then $\gamma={}^tX(S_0)$. Using Weyl's integral formula

$$(5)\qquad \frac{1}{|U|}\int_U f(u)\,du=\frac{1}{w}\frac{1}{|T|}\int_T |j(t)|^2\bar f(t)\,dt \qquad\qquad (w=\text{order of } W)$$

Y_0 is easily found whereupon S_0 can be written down as a Fourier series on T. Using (5) once more, and Lemma 6.2, the theorem follows. We omit the routine details of the computation.

7. Restriction of the central operators in $\boldsymbol{D}(G)$. Returning now to the notation of § 1 and § 2 let $\boldsymbol{Z}(G)$ denote the center of $\boldsymbol{D}(G)$. As is well known, $\boldsymbol{Z}(G)$ consists of all differential operators on G which are invariant under all right and all left translations on G. We shall now consider the question whether $\boldsymbol{Z}(G)$ is mapped onto $\boldsymbol{D}(G/K)$ by means of the natural homomorphism μ of $\boldsymbol{D}_0(G)$ onto $\boldsymbol{D}(G/K)$ given by $(\mu(D)f)\circ\pi = D(f\circ\pi)$ for all $D \in \boldsymbol{D}_0(G)$, $f \in C^\infty(G/K)$.

At the same time we shall consider another homomorphism of $\boldsymbol{Z}(G)$ into $\boldsymbol{D}(G/K)$. Since G is a Lie transformation group of G/K we have a representation ν of G on $C^\infty(G/K)$ by putting $\nu(g)f = f^{\tau(g)}$ for $g \in G$, $f \in C^\infty(G/K)$ and $\tau(g)$ as in § 1. This induces a representation of $\mathfrak{g}$ on $C^\infty(G/K)$ which extends to a homomorphism, also denoted ν, of $\boldsymbol{D}(G)$ into the algebra of all differential operators on G/K.

LEMMA 7.1. *Let $D \in \boldsymbol{D}(G)$, $g \in G$. Then*

$$\nu(D)\nu(g) = \nu(g)\nu(D^{R(g)}).$$

Proof. It suffices to prove this for $D = X \in \mathfrak{g}$. If $f \in C^\infty(G)$ we have for $p \in G/K$,

$$[\nu(X)\nu(g)f](p) = [\frac{d}{dt}f(g^{-1}\exp(-tX)\cdot p)](0) = [\nu(\mathrm{Ad}(g^{-1})X)f](g^{-1}\cdot p)$$
$$= [\nu(g)\nu(X^{R(g)})f](p).$$

LEMMA 7.2. *Let $D \in \boldsymbol{D}(G)$, $f \in C^\infty(G)$. Then*

$$[\nu(D)f](\pi(g)) = [(D^*)^{R(g)}(f\circ\pi)](g),$$

where D^* is the adjoint of D.

Proof. If $g \in G$ and $X \in \mathfrak{g}$ we have

$$[\nu(X)f](\pi(g)) = \{\frac{d}{dt}(f(\exp(-tX)\cdot\pi(g)))\}_{t=0}$$
$$= \{\frac{d}{dt}((f\circ\pi)(g\exp(\mathrm{Ad}(g^{-1})(-tX)))\}_{t=0}$$

so

(1) $$[\nu(X)f](\pi(g)) = [\mathrm{Ad}(g^{-1})(-X)(f\circ\pi)](g).$$

Let $X_1, \cdots, X_r \in \mathfrak{g}$. We shall then prove by induction on r that

$$(2) \qquad [\nu(X_1 \cdots X_r)f](\pi(g)) = (-1)^r [\mathrm{Ad}(g^{-1})[X_r \cdots X_1)(f \circ \pi)](g).$$

For $r = 1$ this is proved and assuming (2) for a fixed r we have for $X_0 \in \mathfrak{g}$

$$\begin{aligned}
[\nu(X_0 \cdots X_r)f](\pi(g)) &= [\nu(X_0)(\nu(X_1 \cdots X_r)f)](\pi(g)) \\
&= \{\frac{d}{dt}(\nu(X_1 \cdots X_r)f(\pi(\exp(-tX_0)g)))\}_{t=0} \\
&= (-1)^r \{\frac{d}{dt}([(\mathrm{Ad}(g^{-1})\mathrm{Ad}(\exp(tX_0)(X_r \cdots X_1))(f \circ \pi)] \\
&\qquad\qquad (g \exp \mathrm{Ad}(g^{-1})(-tX_0)))\}_{t=0} \\
&= (-1)^r [\mathrm{Ad}(g^{-1})(\{X_0, X_r \cdots X_1\})(f \circ \pi)](g) \\
&\quad + (-1)^r [\mathrm{Ad}(g^{-1})(-X_0)(\mathrm{Ad}(g^{-1})(X_r \cdots X_1)(f \circ \pi))](g),
\end{aligned}$$

where the unique derivation of $\boldsymbol{D}(G)$ extending the endomorphism $X \to [X_0, X]$ of $\mathfrak{g}$ has been denoted $D \to \{X_0, D\}$. But it is clear that $\{X_0, D\} = X_0 D - D X_0$ so the two last expressions add up to

$$(-1)^{r+1}[\mathrm{Ad}(g^{-1})(X_r \cdots X_1 X_0)(f \circ \pi)](g).$$

This proves (2). Since G is unimodular, the adjoint X^* equals $-X$ ($X \in \mathfrak{g}$) and the lemma is proved.

COROLLARY 7.3. *$\mu(D^*) = \nu(D)$ for each $D \in \boldsymbol{Z}(G)$.*

In fact $\boldsymbol{Z}(G)$ is invariant under the mapping $D \to D^*$, so $(D^*)^{R(g)} = D^*$.

Remark. The mapping μ of $\boldsymbol{D}_0(G)$ into $\boldsymbol{D}(G/K)$ can be defined for any connected Lie group G and a closed subgroup K, and if G is unimodular, Cor. 7.3 holds.

The image of $\boldsymbol{Z}(G)$ under the mapping μ (or ν) will be denoted $\boldsymbol{Z}(G/K)$.

The Killing form $\langle \; , \; \rangle$ of $\mathfrak{g}$ induces an isomorphism of the vector space $\mathfrak{g}$ onto its dual $\hat{\mathfrak{g}}$, which then extends to an isomorphism of the symmetric algebra $S(\mathfrak{g})$ onto $S(\hat{\mathfrak{g}})$, the algebra of all polynomial functions on $\mathfrak{g}$. Under this isomorphism the ideal $S(\mathfrak{g})\mathfrak{k}$ goes into the ideal of polynomial functions on $\mathfrak{g}$, which vanish identically on $\mathfrak{p}$. The subalgebra $S(\mathfrak{p}) \subset S(\mathfrak{g})$ goes into an algebra of functions on $\mathfrak{g}$ which under the restriction from $\mathfrak{g}$ to $\mathfrak{p}$, is isomorphic to $S(\mathfrak{p})$. Let $I(\mathfrak{g})$ and $I(\hat{\mathfrak{g}})$ denote the set of $\mathrm{Ad}(G)$-invariants in $S(\mathfrak{g})$ and $S(\hat{\mathfrak{g}})$, respectively, $I(\mathfrak{p})$ and $I(\hat{\mathfrak{p}})$ the set of $\mathrm{Ad}_G(K)$-invariants in $S(\mathfrak{p})$ and $S(\hat{\mathfrak{p}})$, respectively.

Let $\mathfrak{u}$ denote the compact real form $\mathfrak{u} = \mathfrak{k} + i\mathfrak{p}$ of the complexification $\mathfrak{g}^c$ of $\mathfrak{g}$. We put $\mathfrak{p}_* = i\mathfrak{p}$ and $U = \mathrm{Int}(\mathfrak{u})$, the adjoint group of $\mathfrak{u}$. Since G has center $\{e\}$ it can be identified with its adjoint group, whereby K is identified with the analytic subgroup of U with Lie algebra $\mathfrak{k}$. Let $I(\hat{\mathfrak{u}})$

denote the ring of $\mathrm{Ad}(U)$-invariant polynomial functions on $\mathfrak{u}$ and $I(\hat{\mathfrak{p}}_*)$ the ring of $\mathrm{Ad}_U(K)$-invariant polynomial functions on $\mathfrak{p}_*$.

The subspace $\mathfrak{a}_* = i\mathfrak{a}$ is a maximal abelian subspace of $\mathfrak{p}_*$. We extend $\mathfrak{a}_*$ to a maximal abelian subalgebra $\mathfrak{t}$ of $\mathfrak{u}$. Let θ denote the involutive automorphism $T + X \rightarrow T - X (T \in \mathfrak{k}, X \in \mathfrak{p}_*)$ of $\mathfrak{u}$. If $H \in \mathfrak{t}$, $\theta H - H \in \mathfrak{p}_*$ and commutes elementwise with $\mathfrak{a}_*$. Hence $\theta\mathfrak{t} \subset \mathfrak{t}$ and $\mathfrak{t} = \mathfrak{t} \cap \mathfrak{k} + \mathfrak{a}_*$. Let $\mathfrak{h} = \mathfrak{t}^c$ $\mathfrak{h}_\mathfrak{p} = \mathfrak{a}^c = \mathfrak{a}_*{}^c$, the superscript c denoting complexification in $\mathfrak{g}^c$. Then $\mathfrak{h}_\mathfrak{p} = \mathfrak{h} \cap \mathfrak{p}^c$. Extending the members of $\mathrm{Ad}(U)$ to $\mathfrak{g}^c$, the Weyl group $W(\mathfrak{h})$ is the group of linear transformations of $\mathfrak{h}$ induced by those members of $\mathrm{Ad}(U)$ which leave $\mathfrak{h}$ invariant, and the Weyl group $W(\mathfrak{h}_\mathfrak{p})$ is the group of linear transformations of $\mathfrak{h}_\mathfrak{p}$ induced by those members of $\mathrm{Ad}_U(K)$ which leave $\mathfrak{h}_\mathfrak{p}$ invariant. Let $I(\hat{\mathfrak{h}})$ and $I(\hat{\mathfrak{h}}_\mathfrak{p})$ denote the corresponding algebras of invariant polynomial functions.

PROPOSITION 7.4. *The following properties of the symmetric space G/K are equivalent:*

(i) $\boldsymbol{Z}(G/K) = \boldsymbol{D}(G/K)$.

(ii) *The restriction from $\mathfrak{g}$ to $\mathfrak{p}$ maps $I(\hat{\mathfrak{g}})$ onto $I(\hat{\mathfrak{p}})$.*

(iii) *The restriction from $\mathfrak{u}$ to $\mathfrak{p}_*$ maps $I(\hat{\mathfrak{u}})$ onto $I(\hat{\mathfrak{p}}_*)$.*

(iv) *The restriction from $\mathfrak{h}$ to $\mathfrak{h}_\mathfrak{p}$ maps $I(\hat{\mathfrak{h}})$ onto $I(\hat{\mathfrak{h}}_\mathfrak{p})$.*

Proof. To prove the equivalence of (i) and (ii) we consider the canonical mapping λ of $S(\mathfrak{g})$ onto $\boldsymbol{D}(G)$ which maps $I(\mathfrak{g})$ onto $\boldsymbol{Z}(G)$ (see [11], p. 192). Let X_i $(1 \leqq i \leqq n)$ be a basis of $\mathfrak{g}$ such that X_i $(1 \leqq i \leqq r)$ is a basis of $\mathfrak{p}$. Then, if $g \in G$, $f \in C^\infty(G)$, $P \in S(\mathfrak{g})$, $\partial_i = \partial/\partial x_i$ we have

$$[\lambda(P)f](g) = [P(\partial_1, \cdots, \partial_n)f(g \exp(x_1X_1 + \cdots + x_nX_n)](0).$$

Now suppose $P \in I(\mathfrak{g})$ and let $\bar{P} \in S(\mathfrak{p})$ be determined by $P - \bar{P} \in S(\mathfrak{g})\mathfrak{k}$. Then $\bar{P} \in I(\mathfrak{p})$. On the other hand there exists a unique $P' \in I(\mathfrak{p})$ such that (see [14], Theorem 10)

$$[\mu(\lambda(P))f](\pi(g)) = [P'(\partial_1, \cdots, \partial_r)(f \circ \pi)(g \exp(x_1X_1 + \cdots + x_rX_r))](0),$$

and $\mu(\lambda(P')) = \mu(\lambda(P))$. It follows that

$$\lambda(\bar{P} - P') + \lambda(P - \bar{P}) = \lambda(P - P') \in \boldsymbol{D}(G)\mathfrak{k}, \tag{3}$$

since μ has kernel $\boldsymbol{D}_0(G) \cap \boldsymbol{D}(G)\mathfrak{k}$. Now suppose P is homogeneous of degree $d > 0$. Then the same holds for $P - \bar{P}$ (0 has any degree) and P' has degree d. But there exists an operator $D_1 \in \boldsymbol{D}(G)$ of order $< d$ such that

$\lambda(P-\bar{P})-D_1\in \boldsymbol{D}(G)\mathfrak{k}$. Thus, by (3) and the fact that $\lambda(S(\mathfrak{p}))\cap \boldsymbol{D}(G)\mathfrak{k}=0$ we obtain

$$\text{(4)} \qquad \text{degree}(\bar{P}-P')<\text{degree}(P).$$

The mapping μ maps $\boldsymbol{Z}(G)$ onto $\boldsymbol{D}(G/K)$ if and only if the mapping $P\to P'$ maps $I(\mathfrak{g})$ onto $I(\mathfrak{p})$, which, by (4), happens only if $P\to\bar{P}$ maps $I(\mathfrak{g})$ onto $I(\mathfrak{p})$. This proves the equivalence of (i) and (ii).

If the elements of $\mathrm{Ad}(G)$ and $\mathrm{Ad}(U)$ are extended to $\mathfrak{g}^c$, a polynomial function on $\mathfrak{g}^c$ is invariant under $\mathrm{Ad}(G)$ if and only if it is invariant under $\mathrm{Ad}(U)$. Since the decompositions $\mathfrak{u}=\mathfrak{k}+\mathfrak{p}_*$ and $\mathfrak{g}=\mathfrak{k}+\mathfrak{p}$ have a common complexification, namely $\mathfrak{g}^c=\mathfrak{k}^c+\mathfrak{p}^c$, the equivalence of (ii) and (iii) is clear.

Let $I(\hat{\mathfrak{g}}^c)$ and $I(\hat{\mathfrak{p}}^c)$, respectively, denote the algebras of $\mathrm{Ad}(U)$-invariant and $\mathrm{Ad}_U(K)$-invariant polynomial functions on $\mathfrak{g}^c$ and $\mathfrak{p}^c$. By a theorem of Chevalley ([15], p. 430) the restrictions from $\mathfrak{g}^c$ to $\mathfrak{h}$ and from $\mathfrak{p}^c$ to $\mathfrak{h}_\mathfrak{p}$ map $I(\hat{\mathfrak{g}}^c)$ isomorphically onto $I(\hat{\mathfrak{h}})$ and $I(\hat{\mathfrak{p}}^c)$ isomorphically onto $I(\hat{\mathfrak{h}}_\mathfrak{p})$. The restriction from $\mathfrak{g}^c$ to $\mathfrak{p}^c$ maps $I(\hat{\mathfrak{g}}^c)$ into $I(\hat{\mathfrak{p}}^c)$ so the restriction from $\mathfrak{h}$ to $\mathfrak{h}_\mathfrak{p}$ maps $I(\hat{\mathfrak{h}})$ into $I(\hat{\mathfrak{h}}_\mathfrak{p})$ and these maps are both onto if and only if one of them is. This proves the equivalence of (iii) and (iv) so Prop. 7.4 is proved.

THEOREM 7.5. (i) *Suppose* $\mathfrak{g}$ *is a classical semisimple Lie algebra (real or complex). Then the restriction from* $\mathfrak{h}$ *to* $\mathfrak{h}_\mathfrak{p}$ *maps* $I(\hat{\mathfrak{h}})$ *onto* $I(\hat{\mathfrak{h}}_\mathfrak{p})$.

(ii) *Part* (i) *does not hold*[2] *in general for the real forms of the exceptional Lie algebras* $\mathfrak{e}_6$, $\mathfrak{e}_7$, $\mathfrak{e}_8$.

Let Δ denote the set of nonzero roots of $\mathfrak{g}^c$ with respect to $\mathfrak{h}$. Let $\mathfrak{h}_\mathfrak{p}{}^*$ denote the annihilator of $\mathfrak{h}_\mathfrak{p}$ in the dual space $\hat{\mathfrak{h}}$ of $\mathfrak{h}$, and if f is a function on $\mathfrak{h}$, let $\bar{f}$ denote its restriction to $\mathfrak{h}_\mathfrak{p}$.

In the proof below we shall use the well known fact ([21], Theorem 3, [25], p. 108) that $W(\mathfrak{h}_\mathfrak{p})$ consist of the restrictions to $\mathfrak{h}_\mathfrak{p}$ of those elements in $W(\mathfrak{h})$ which leave $\mathfrak{h}_\mathfrak{p}$ invariant.

Proof of (i). It suffices to verify (i) in the case when $\mathfrak{g}$ is a simple Lie algebra over $\boldsymbol{R}$ and does not have a complex structure. We shall use Cartan's classification of such algebras, listed in [15], Ch. IX. Let $D(a_1,\cdots,a_n)$ denote a diagonal matrix with diagonal elements $a_1,\cdots,a_n$.

If $\mathfrak{g}$ is a normal real form of its complexification, then (i) is obvious.

[2] The homomorphisms $\nu\colon \boldsymbol{Z}(G)\to \boldsymbol{D}(G/K)$ and $\mu\colon \boldsymbol{Z}(G)\to \boldsymbol{D}(G/K)$ are considered in F. A. Berezin [3]. On page 243 it is stated that ν is onto and on page 248 it is stated that μ is onto. Each of these statements is contradicted by Theorems 7.4 and 7.5.

These are the real forms **AI, BDI** ($p=q$ and $p+1=q$) and **CI.** We shall now verify (i) for the other real simple Lie algebras.

Type A II. Here

$$(4)\qquad \begin{aligned} \mathfrak{h}_{\mathfrak{p}} &= \{D(h_1,\cdots,h_n,h_1,\cdots,h_n) : h_i \in \boldsymbol{C},\ \sum_1^n h_i = 0\}; \\ \mathfrak{h} &= \{D(h_1,\cdots,h_n,h_{n+1},\cdots,h_{2n}) : h_i \in \boldsymbol{C},\ \sum_1^{2n} h_i = 0\}. \end{aligned}$$

Let e_i $(1 \leqq i \leqq 2n)$ be the linear form on $\mathfrak{h}$ determined by

$$(5)\qquad e_i(H) = h_i \qquad\qquad \text{if } H = D(h_1,\cdots,h_{2n}).$$

Then the root system Δ is given by

$$\Delta = \{\pm(e_i - e_j) : 1 \leqq i < j \leqq 2n\}.$$

The Weyl group $W(\mathfrak{h})$, acting on the dual of $\mathfrak{h}$, consists of all permutations of the e_i $(1 \leqq i \leqq 2n)$. Hence $I(\hat{\mathfrak{h}})$ consists of those polynomial functions on $\mathfrak{h}$ which can be written as a symmetric function of the e_i. Now if $s \in W(\mathfrak{h})$ leaves $\mathfrak{h}_{\mathfrak{p}}$ invariant its action on $\hat{\mathfrak{h}}$ leaves $\mathfrak{h}_{\mathfrak{p}}{}^*$ invariant so $s(e_i + e_{i+n}) = e_j + e_{j+n}$. Putting

$$f_i(H) = \tfrac{1}{2}(e_i(H) + e_{i+n}(H)) \qquad\qquad (H \in \mathfrak{h}_{\mathfrak{p}}, 1 \leqq i \leqq n)$$

so $f_i = \bar{e}_i = \bar{e}_{i+n}$, one finds that $W(\mathfrak{h}_{\mathfrak{p}})$, acting on $\hat{\mathfrak{h}}_{\mathfrak{p}}$, consists of all permutations of the f_i $(1 \leqq i \leqq n)$. Thus $I(\hat{\mathfrak{h}}_{\mathfrak{p}})$ consists of those polynomial functions on $\mathfrak{h}_{\mathfrak{p}}$ which can be written as a symmetric function of the f_i $(1 \leqq i \leqq n)$. It is now clear that $I(\hat{\mathfrak{h}})^- = I(\hat{\mathfrak{h}}_{\mathfrak{p}})$.

Type A III. Here we can take $p \leqq q$ and

$$\mathfrak{h}_{\mathfrak{p}} = \sum_1^p \boldsymbol{C}(E_{ip+i} + E_{p+ii});$$

$$\mathfrak{h} = \{H = \sum_1^p h_i(E_{ip+i} + E_{p+ii}) + \sum_1^p u_i(E_{ii} + E_{p+ip+i}) + \sum_{2p+1}^{p+q} u_i E_{ii} : h_i, u_i \in \boldsymbol{C}, \operatorname{Tr}(H) = 0\}.$$

In fact, this $\mathfrak{h}$ is the image of (4) ($p+q$ for $2n$) under the automorphism $X \to AXA^{-1}$ of $\mathfrak{sl}(p+q,\boldsymbol{C})$, where $A = I - \sum_1^p (E_{ip+i} - E_{p+ii})$. Hereby, the linear forms (5) correspond to the linear forms

$$e_i(H) = \begin{cases} \tfrac{1}{2}(u_i + h_i) & 1 \leqq i \leqq p, \\ \tfrac{1}{2}(u_i - h_i) & p+1 \leqq i \leqq 2p, \\ u_i & 2p+1 \leqq i \leqq p+q. \end{cases}$$

Thus

$$\Delta = \{\pm (e_i - e_j) : 1 \leqq i, j \leqq p+q\},$$

and $W(\mathfrak{h})$, acting on the dual, consists of all permutations of the e_i $(1 \leqq i \leqq p+q)$. Since

$$\mathfrak{h}_{\mathfrak{p}}^* = \sum_{2p+1}^{p+q} \boldsymbol{C} e_i$$

it is clear that $\sigma \in W(\mathfrak{h})$ leaves $\mathfrak{h}_{\mathfrak{p}}$ invariant if and only if it permutes the e_i $(2p+1 \leqq i \leqq p+q)$ and therefore also the e_i $(1 \leqq i \leqq 2p)$. The restrictions $f_i = \bar{e}_i$ $(1 \leqq i \leqq p+q)$ satisfy

$$\bar{e}_i = -\bar{e}_{i+p} \quad (1 \leqq i \leqq p), \qquad \bar{e}_i = 0 \quad (2p+1 \leqq i \leqq p+q).$$

Therefore $W(\mathfrak{h}_{\mathfrak{p}})$, acting on the dual of $\mathfrak{h}_{\mathfrak{p}}$, consists of all the maps $f_i \to \epsilon_i f_{\sigma(i)}$, where $\epsilon_i = \pm 1$, and σ is any permutation of $\{1, \cdots, p\}$. Thus $I(\hat{\mathfrak{h}}_{\mathfrak{p}})$ is generated by the polynomials $\sum_1^p f_i^{2k}$ $(1 \leqq k \leqq p)$ which are restrictions of the polynomials $\frac{1}{2} \sum_{i=1}^{p+q} e_i^{2k}$.

Type BDI. Here we use the description in [29], p. 401. The quadratic form corresponding to the symmetric matrix

$$J_{p,q} = \begin{pmatrix} 0 & I_p & 0 \\ I_p & 0 & 0 \\ 0 & 0 & -I_{q-p} \end{pmatrix} \qquad (q \geqq p)$$

has signature (p, q). The corresponding orthogonal group has Lie algebra

$$\mathfrak{g} = \{X \in \mathfrak{gl}(p+q, \boldsymbol{R}) : {}^tXJ_{p,q} + J_{p,q}X = 0\}$$

and a Cartan involution of this Lie algebra is given by $X \to J_{p,q}XJ_{p,q}$, inducing a Cartan decomposition $\mathfrak{g} = \mathfrak{k} + \mathfrak{p}$. Here we can take

$$\mathfrak{h}_{\mathfrak{p}} = \{D(h_1, \cdots, h_p, -h_1, \cdots, -h_p, 0, \cdots, 0) : h_i \in \boldsymbol{C}\}$$

and if $t = [\frac{1}{2}(q-p)]$,

$$\mathfrak{h} = \{H = \sum_1^p h_i(E_{ii} - E_{i+ip+p}) + \sum_1^p u_i(E_{2p+t+i\,2p+i} - E_{2p+i\,2p+t+i}) : h_i, u_i \in \boldsymbol{C}\}.$$

Let e_i denote the linear form on $\mathfrak{h}$ given by

$$e_j(H) = \begin{cases} h_j & 1 \leqq j \leqq p \\ iu_{j-p} & p+1 \leqq j \leqq [\frac{1}{2}(p+q)]. \end{cases}$$

Then

$$\Delta = \{\pm (e_i \pm e_j) : 1 \leqq i < j \leqq [\tfrac{1}{2}(p+q)]\}.$$

Now we treat separately the cases $p+q=$ even, $p+q=$ odd.

Type DI, $p+q=2l$. Here $W(\mathfrak{h})$, acting on $\hat{\mathfrak{h}}$, consists of all maps $e_i \to \epsilon_i e_{\sigma(i)}$, where σ is any permutation of $\{1,\cdots,l\}$, $\epsilon_i = \pm 1$, and $\prod_1^l \epsilon_i = 1$. Hence $I(\mathfrak{h})$ is generated by $e_1 \cdots e_l$ and the symmetric polynomials in $e_1^2, \cdots, e_l^2$.

Now if $p=q$, $\mathfrak{g}$ is the normal form of $\mathfrak{g}^c$ so we may assume $p<q$, so $p+1 \leqq l$. Then $s \in W(\mathfrak{h})$ leaves $\mathfrak{h}_\mathfrak{p}$ invariant if and only if it maps $\sum_{p+1}^{l} \boldsymbol{C} e_j$, and therefore also $\sum_1^p \boldsymbol{C} e_j$, into itself. Putting $f_i = \bar{e}_i$ $(1 \leqq i \leqq p)$ we see therefore that $W(\mathfrak{h}_\mathfrak{p})$ consists of all the maps $f_i \to \epsilon_i f_{\sigma(i)}$ where σ is any permutation of $\{1,\cdots,p\}$ and $\epsilon_i = \pm 1$. It is now clear that $(I(\hat{\mathfrak{h}}))^- = I(\hat{\mathfrak{h}}_\mathfrak{p})$.

Type BI, $p+q=2l+1$. Here $W(\mathfrak{h})$ consists of the maps $e_i \to \epsilon_i e_{\sigma(i)}$, ($\sigma$ any permutation of $\{1,\cdots,l\}$, $\epsilon_i = \pm 1$). That $(I(\hat{\mathfrak{h}}))^- = I(\hat{\mathfrak{h}}_\mathfrak{p})$ is seen in the same way as for **DI**.

Type DIII. Using again the description in [29], p. 406, $\mathfrak{g}$ can be regarded as the Lie algebra of the group of linear transformations in $2l$ complex variables leaving invariant

$$z_1 z_{k+1} + \cdots + z_{3k} z_{4k} \; (+ z_{2l-1} z_{2l}),$$
$$z_1 \bar{z}_{2k+1} + \cdots + z_{2k} \bar{z}_{4k} + \bar{z}_1 z_{2k+1} + \cdots + \bar{z}_{2k} z_{4k} \; (+ z_{2l-1} \bar{z}_{2l-1} - z_{2l} \bar{z}_{2l}),$$

where $k = [\tfrac{1}{2} l]$, and the terms in the parentheses occur only if l is odd. The mapping $X \to -{}^t\bar{X}$ is a Cartan involution of $\mathfrak{g}$, inducing a Cartan decomposition $\mathfrak{g} = \mathfrak{k} + \mathfrak{p}$ of $\mathfrak{g}$. Here we can take (cf. [29], p. 408)

$$\mathfrak{h}_\mathfrak{p} = \{D(h_1,\cdots,h_k,-h_1,\cdots,-h_k,-h_1,\cdots,$$
$$-h_k,h_1,\cdots,h_k,(0,0)) : h_i \in \boldsymbol{C}\},$$
$$\mathfrak{h} = \{H = D(h_1,\cdots,h_k,-h_1,\cdots,-h_k,h_{k+1},\cdots,$$
$$h_{2k},-h_{k+1},\cdots,-h_{2k},(h_{2k+1},-h_{k+1})) : h_i \in \boldsymbol{C}\},$$

the terms in the parentheses occurring only if n is odd. If e_i is the linear form on $\mathfrak{h}$ given by

$$e_i(H) = h_i \qquad (1 \leqq i \leqq l)$$

then

$$\Delta = \{\pm (e_i \pm e_j) : 1 \leqq i < j \leqq l\}.$$

Here $W(\mathfrak{h})$ consists of all the maps $e_i \to \epsilon_i e_{\sigma(i)}$, where σ is any permutation of $\{1, \cdots, l\}$, $\epsilon_i = \pm 1$ and $\prod_1^l \epsilon_i = 1$. Hence $I(\mathfrak{h})$ is the same as for Type **DI**. Now $\mathfrak{h}_\mathfrak{p}^* = \sum_{i=1}^k \boldsymbol{C}(e_i + e_{i+k}) + (\boldsymbol{C}e_{2k+1})$; if $s \in W(\mathfrak{h})$ leaves $\mathfrak{h}_\mathfrak{p}$ invariant, it leaves $\mathfrak{h}_\mathfrak{p}^*$ invariant and consequently $s(e_j + e_{j+k}) = \pm (e_i + e_{i+k})$. Hence $s(e_j - e_{j+k}) = \pm (e_i - e_{i+k})$. If we put $f_j = \bar{e}_j - \bar{e}_{j+k}$ $(1 \leqq j \leqq k)$ we find therefore that the restriction of s to $\mathfrak{h}_\mathfrak{p}$, acting on $\hat{\mathfrak{h}}_\mathfrak{p}$, is of the form $f_j \to \epsilon_j f_{\sigma(j)}$ where $\epsilon_j = \pm 1$ and σ is a permutation of $\{1, \cdots, k\}$. It is also clear that each such σ occurs in this manner; the ϵ_j are also unrestricted since a change of sign of some f_j is induced by a change of sign of an even number of e_i. The transformations $f_j \to \epsilon_j f_{\sigma(j)}$ $(\epsilon_j = \pm 1$, σ arbitrary permutation of $\{1, \cdots, k\})$ which are all different, the f_j being linearly independent, exhaust the group $W(\mathfrak{h}_\mathfrak{p})$; the algebra $I(\mathfrak{h}_\mathfrak{p})$ is generated by $\sum_{i=1}^k f_i^{2n} (n = 1, \cdots, k)$. Also, if $H \in \mathfrak{h}_\mathfrak{p}$

$$\bar{e}_i(H) = h_i \quad (1 \leqq i \leqq k), \qquad \bar{e}_i(H) = -h_{i-k} \quad (k+1 \leqq i \leqq 2k),$$

so

$$\sum_{i=1}^{l} (\bar{e}_i)^{2n} = 2 \sum_{i=1}^{k} f_i^{2n},$$

proving $(I(\hat{\mathfrak{h}}))^- = I(\hat{\mathfrak{h}}_\mathfrak{p})$.

Type CII. Here $\mathfrak{g} = \mathfrak{sp}(p, q)$ and we suppose $\mathfrak{g} = \mathfrak{k} + \mathfrak{p}$ is the Cartan decomposition given in [15], p. 351. We assume $p \leqq q$. The complexification $\mathfrak{g}^c = \mathfrak{sp}(p+q, \boldsymbol{C})$ has a Cartan subalgebra

$$\mathfrak{h}' = \{H' = D(h_1, \cdots, h_{p+q}, -h_1, \cdots, -h_{p+q}) : h_i \in \boldsymbol{C}\}.$$

Let J denote the matrix in $\boldsymbol{Sp}(n, \boldsymbol{C})$ given by

$$J = 2^{-\frac{1}{2}}\{I + \sum_1^p (E_{i\,p+i} - E_{p+i\,i}) + \sum_1^p (E_{p+q+i\ 2p+q+i} - E_{2p+q+i\ p+q+i})\}$$

and let $\mathfrak{h}$ denote the image of $\mathfrak{h}'$ under the automorphism $X \to JXJ^{-1}$ of $\mathfrak{sp}(p+q, \boldsymbol{C})$. We have

$$H = JH'J^{-1} = D(u_1, \cdots, u_{p+q}, -u_1, \cdots, -u_{p+q}) + \sum_1^p k_i(E_{i\,p+i} + E_{p+i\,i})$$
$$- \sum_1^p k_i (E_{p+q+i\ 2p+q+i} + \ + E_{2p+q+i\ p+q+i})$$

where

$$u_i = \tfrac{1}{2}(h_i + h_{i+p}), \quad (1 \leqq i \leqq p), \qquad u_j = \tfrac{1}{2}h_j, \quad (2p+1 \leqq j \leqq p+q)$$
$$k_i = \tfrac{1}{2}(-h_i + h_{i+p}), \quad (1 \leqq i \leqq p).$$

9

The intersection $\mathfrak{h} \cap \mathfrak{p}^c$ is

$$\sum_1^p \boldsymbol{C}(E_{i\,p+i} + E_{p+i\,i} - E_{p+q+i\;2p+q+i} - E_{2p+q+i\;p+q+i}),$$

which can serve as $\mathfrak{h}_\mathfrak{p}$. Now consider the linear forms e_i' on $\mathfrak{h}'$ defined by

$$e_i'(H) = h_i \quad (1 \leqq i \leqq p+q).$$

Then if Δ' denotes the set of nonzero roots of $\mathfrak{g}^c$ with respect to $\mathfrak{h}'$,

$$\Delta' = \{\pm(e_i' \pm e_j'), 1 \leqq i < j \leqq p+q; \pm 2e_i' \, (1 \leqq i \leqq p+q)\}.$$

The Weyl group $W(\mathfrak{h}')$ consists of all maps $e_i' \to \epsilon_i e'_{\sigma(i)}$, where $\epsilon_i = \pm 1$ and σ is an arbitrary permutation of $\{1, 2, \cdots, n\}$. Consider the linear form e_i on $\mathfrak{h}$ defined by $e_i(JH'J^{-1}) = e_i'(H')$, $H' \in \mathfrak{h}'$. Then

$$e_i(H) = \begin{cases} u_i - k_i & 1 \leqq i \leqq p \\ u_{i-p} + k_{i-p} & p+1 \leqq i \leqq 2p \\ 2u_i & 2p+1 \leqq i \leqq p+q \end{cases}$$

and

$$\Delta = \{\pm(e_i \pm e_j), 1 \leqq i < j \leqq p+q; \pm 2e_i \, (1 \leqq i \leqq p+q)\},$$

and $W(\mathfrak{h})$ consists of all maps $e_i \to \epsilon_i e_{\sigma(i)}$, where $\epsilon_i = \pm 1$ and σ arbitrary permutation of $\{1, 2, \cdots, n\}$. Also

$$\mathfrak{h}_\mathfrak{p}^* = \sum_1^p \boldsymbol{C}(e_i + e_{i+p}) + \sum_{2p+1}^{p+q} \boldsymbol{C} e_i. \tag{6}$$

If $s \in W(\mathfrak{h})$ leaves $\mathfrak{h}_\mathfrak{p}^*$ invariant it is easily seen that s, acting on $\hat{\mathfrak{h}}$, leaves each of the sums in (6) invariant. Consequently $s(e_i - e_{i+p}) = \pm(e_j - e_{j+p})$. If we put $f_j = \bar{e}_j - \bar{e}_{j+p}$ $(1 \leqq j \leqq p)$, the restriction of s to $\mathfrak{h}_\mathfrak{p}$, acting on $\hat{\mathfrak{h}}_\mathfrak{p}$, is of the form $f_j \to \epsilon_j f_{\sigma(j)}$ ($\epsilon_j = \pm 1$, σ any permutation of $\{1, \cdots, p\}$ and any such transformation of $\hat{\mathfrak{h}}_\mathfrak{p}$ arises from some $s \in W(\mathfrak{h})$. The f_j $(1 \leqq j \leqq p)$ being linearly independent, these transformations exhaust $W(\mathfrak{h}_\mathfrak{p})$; the algebra $I(\hat{\mathfrak{h}}_\mathfrak{p})$ is generated by $\sum_1^p f_j^{2n}$ $(n = 1, \cdots, p)$ and $I(\hat{\mathfrak{h}})$ is generated by $\sum_1^{p+q} e_i^{2n}$ $(n = 1, \cdots, p+q)$. Since

$$\bar{e}_i = \begin{cases} \tfrac{1}{2} f_i & (1 \leqq i \leqq p), \\ \tfrac{1}{2} f_{i-p} & (p+1 \leqq i \leqq 2p), \\ 0 & (2p+1 \leqq i \leqq p+q), \end{cases}$$

it is clear that $I(\hat{\mathfrak{h}})^- = I(\hat{\mathfrak{h}}_\mathfrak{p})$.

Proof of (ii). Consider first the noncompact symmetric space of type E IV, that is E_6/F_4. Here $I(\hat{\mathfrak{h}})$ is generated by algebraically independent generators of degrees 2, 5, 6, 8, 9, 12 ([31], [5]). For this symmetric space the set Σ_0 from § 2 is the set of positive roots for the Lie algebra of $\boldsymbol{SU}(3)$ (see [4], p. 422 or [1], p. 33). It follows that $W(\mathfrak{h}_{\mathfrak{p}})$ is the Weyl group of $\boldsymbol{SU}(3)$ so $I(\hat{\mathfrak{h}}_{\mathfrak{p}})$ has generators i_2, i_3 of degrees 2 and 3, respectively. However $I(\hat{\mathfrak{h}})$ cannot contain any polynomial of degree 3 so $I(\hat{\mathfrak{h}})^- \neq I(\hat{\mathfrak{h}}_{\mathfrak{p}})$. For the noncompact symmetric space of type E VII, that is $E_7/E_6 \cdot \boldsymbol{T}$ ($\boldsymbol{T} =$ 1-torus) we have $I(\hat{\mathfrak{h}})$ generated by algebraically independent generators of degrees 2, 6, 8, 10, 12, 14, 18. Here the set Σ_0 is the set of positive roots for the Lie algebra of $\boldsymbol{Sp}(3)$ (see [4], p. 423 or [1], p. 33). Thus $W(\mathfrak{h}_{\mathfrak{p}})$ is the Weyl group of $\boldsymbol{Sp}(3)$ and $I(\hat{\mathfrak{h}}_{\mathfrak{p}})$ is generated by algebraically independent generators i_2, i_4, i_6 of degrees 2, 4, 6 respectively. By the algebraic independence, $i_4 \notin I(\hat{\mathfrak{h}})^-$. For the noncompact symmetric space of type E IX, that is $E_8/E_7 \cdot \boldsymbol{SU}(2)$, we have $I(\hat{\mathfrak{h}})$ generated by algebraically independent generators of degrees 2, 8, 12, 14, 18, 20, 24, 30. The set Σ_0 is now the set of positive roots for the Lie algebra of F_4 so $I(\hat{\mathfrak{h}}_{\mathfrak{p}})$ has algebraically independent generators of degrees 2, 6, 8, 12. The one of degree 6 cannot belong to $I(\hat{\mathfrak{h}})^-$.

This concludes the proof of Theorem 7. 5.

We observe now that in each of the three examples above the generators of $I(\hat{\mathfrak{h}}_{\mathfrak{p}})$ which do not lie in $(I(\hat{\mathfrak{h}}))^-$ have degrees $d - e$ where d and e are degrees of certain generators of $I(\hat{\mathfrak{h}})$. This raises the question whether the elements in $I(\hat{\mathfrak{h}}_{\mathfrak{p}})$ are all obtained from rational invariants on $\mathfrak{h}$ by restriction. We shall now prove that this is indeed the case.

THEOREM 7. 6. *Let* $\boldsymbol{C}(I(\hat{\mathfrak{h}}))$ *and* $\boldsymbol{C}(I(\hat{\mathfrak{h}}_{\mathfrak{p}}))$ *denote the quotient fields of* $I(\hat{\mathfrak{h}})$ *and* $I(\hat{\mathfrak{h}}_{\mathfrak{p}})$, *respectively. Then the restriction* $f \to \bar{f}$ *from* $\mathfrak{h}$ *to* $\mathfrak{h}_{\mathfrak{p}}$ *induces a mapping of* $\boldsymbol{C}(I(\hat{\mathfrak{h}}))$ *onto* $\boldsymbol{C}(I(\hat{\mathfrak{h}}_{\mathfrak{p}}))$.

Let $U = \mathrm{Int}(\mathfrak{u})$ as before and let θ denote the automorphism $T + X \to T - X$ ($T \in \mathfrak{t}, X \in \mathfrak{p}_*$) of $\mathfrak{u}$. The corresponding automorphism of U will also be denoted by θ, and K_* shall denote the group of fixed points of θ. Let A_* denote the analytic subgroup $\exp(\mathfrak{a}_*)$ of U and put $P_* = \exp(\mathfrak{p}_*)$.

LEMMA 7. 7. *A_* equals its centralizer in P_*.*

Proof. Suppose $p \in P_*$ commutes elementwise with A_*. Let B' denote the subgroup of U generated by p and A_*, B the closure of B' in U. Clearly, B consists of all products ap^n ($a \in A_*$, $n =$ integer). Now $a = b^2$ ($b \in A_*$) and $p = q^2$ ($q \in P_*$) so $ap^n = b^2p^n = bq^nq^nb = bq^n\theta((bq^n)^{-1})$. Since $P* = \{u\theta(u^{-1}) : u \in U\}$ it follows that $B' \subset P_*$; but P_* being closed we have

$B \subset P_*$. The identity component B_0 of B contains some power of p, so B_0 is a torus and B/B_0 is cyclic. As is well known this implies that there exists an element $b \in B$ such that the sequence $e, b, b^2, \cdots$ is dense in B. Select $X \in \mathfrak{p}_*$ such that $b = \exp X$. Then the closure of the one-parameter subgroup $\{\exp tX : t \in \boldsymbol{R}\}$ is a torus, contained in P_*, but containing p and A_*. But $\mathfrak{a}_*$ is maximal abelian in $\mathfrak{p}_*$ so this torus equals A_*, whence $p \in A_*$.

LEMMA 7.8. *Let M' denote the normalizer of $\mathfrak{a}_*$ in K_*. Then $M'A_*$ is the normalizer of $\mathfrak{a}_*$ in U.*

Proof. Suppose $u \in U$ such that $\mathrm{Ad}(u)\mathfrak{a}_* \subset \mathfrak{a}_*$. Then $u = kp$, ($k \in K_*$, $p = \exp X, X \in \mathfrak{p}_*$) so we conclude $e^{\mathrm{ad}\, X}(\mathfrak{a}_*) \subset \mathfrak{p}_*$, and therefore

$$(\sinh(\mathrm{ad}\, X))(H) = 0 \qquad \text{for } H \in \mathfrak{h}_{\mathfrak{p}}.$$

Now $\mathrm{ad}\, X$, acting on the complexification $\mathfrak{u}^c$ of $\mathfrak{u}$, is completely reducible and has purely imaginary eigenvalues. Let $e_1, \cdots, e_n$ be eigenvectors of $\mathrm{ad}\, X$ spanning $\mathfrak{u}^c$ and $\lambda_1, \cdots, \lambda_n$ the corresponding eigenvalues. Let

$$H = \sum_1^n a_j(H) e_j \qquad (H \in \mathfrak{h}_{\mathfrak{p}}).$$

Suppose $e_1, \cdots, e_m$ are all the eigenvectors for which $a_j \not\equiv 0$. Since

$$0 = \sinh(\mathrm{ad}\, X)(H) = \sum_1^m (\sinh \lambda_j) a_j(H) e_j$$

it follows that each λ_j $(1 \leqq j \leqq m)$ is an integral multiple of πi. Consequently

$$e^{\mathrm{ad}(2X)}(H) = H \qquad H \in \mathfrak{h}_{\mathfrak{p}}$$

so p^2 commutes elementwise with A_*. By Lemma 7.7, $p^2 = \exp 2H_0$ for some $H_0 \in \mathfrak{a}_*$. Put $p_0 = \exp H_0$. Then $\theta(p_0 p^{-1}) = p_0^{-1} p = p_0 p^{-1}$ so $p = k_0 p_0$ where $k_0 \in K_*$. Thus $u = k' \exp H_0$ $(k' \in K_*)$ whence $k' \in M'$.

LEMMA 7.9. *Let J denote the set of restrictions of the polynomials in $I(\hat{\mathfrak{h}})$ to $\mathfrak{h}_{\mathfrak{p}}$. Let $S(\hat{\mathfrak{h}}_{\mathfrak{p}})$ denote the symmetric algebra over $\hat{\mathfrak{h}}_{\mathfrak{p}}$. Then $S(\hat{\mathfrak{h}}_{\mathfrak{p}})$ is integral over J.*

Proof. Let $p \in S(\hat{\mathfrak{h}})$, the symmetric algebra over $\hat{\mathfrak{h}}$. Then p satisfies the equation $\prod_{s \in W(\mathfrak{h})} (x - p^s) = 0$ with leading coefficients 1 and coefficients in $I(\hat{\mathfrak{h}})$. This means that the ring $S(\hat{\mathfrak{h}})$ is integral over $I(\hat{\mathfrak{h}})$. The restriction $p \to \bar{p}$ gives a homomorphism of $S(\hat{\mathfrak{h}})$ onto $S(\hat{\mathfrak{h}}_{\mathfrak{p}})$ mapping $I(\hat{\mathfrak{h}})$ onto J. Hence the lemma.

Passing now to the proof of Theorem 7.6, suppose the quotient field $\boldsymbol{C}(J)$ were a proper subset of $\boldsymbol{C}(I(\hat{\mathfrak{h}}_{\mathfrak{p}}))$. Then there exists $\alpha \in I(\hat{\mathfrak{h}}_{\mathfrak{p}})$ such that $\alpha \notin \boldsymbol{C}(J)$. By Lemma 7.9, α is algebraic over $\boldsymbol{C}(J)$. Since $I(\hat{\mathfrak{h}})$ is finitely generated, the homomorphic image J also has finitely many generators, say $\xi_1, \cdots, \xi_r$. Let

$$p_\alpha(x) = x^n + f_1(\xi)x^{n-1} + \cdots + f_n(\xi) \qquad (n \geqq 2)$$

be the polynomial with coefficients in the field $\boldsymbol{C}(J) = \boldsymbol{C}(\xi_1, \cdots, \xi_r)$ of lowest degree having α as a zero and leading coefficient 1. Let $f(\xi)$ denote the product of the denominators of the $f_i(\xi)$ and let $q(\xi)$ denote the product of $f(\xi)$ with the discriminant of the polynomial $f(\xi)p_\alpha(x)$. Then $q(\xi)$ is a polynomial function on $\mathfrak{h}_{\mathfrak{p}}$, not identically 0. Select $H_* \in \mathfrak{a}_*$ such that

$$(7) \qquad q(\xi)(H_*) \neq 0, \{\exp tH_* : t \in \boldsymbol{R} \text{ dense in } A_*\}$$

Consider the homomorphism $\lambda: p \to p(H_*)$ of J into $\boldsymbol{C}$. The image of $p_\alpha(x)$ under λ is a polynomial with discriminant $\neq 0$, which therefore has n distinct roots, say $\alpha_1, \cdots, \alpha_n \in \boldsymbol{C}$. For each fixed i $(1 \leqq i \leqq n)$ we wish to extend λ to a homomorphism $\lambda_i: \boldsymbol{C}[\xi_1, \cdots, \xi_r, \alpha] \to \boldsymbol{C}$ by putting $\lambda_i(\alpha) = \alpha_i$. To see that this is possible, suppose $p(\xi, x)$ is a polynomial with coefficients in $J = \boldsymbol{C}[\xi_1, \cdots, \xi_r]$ which has α as a zero. Then $p_\alpha(x)$ divides $p(\xi, x)$, i. e. $p(\xi, x) = p_\alpha(x)q(\xi, x)$ and the coefficient of $q(\xi, x)$ are rational expressions in $\xi_1, \cdots, \xi_r$ whose denominators divide $f(\xi)$. Since $\lambda(f(\xi)) \neq 0$ the image of $p(\xi, x)$ under λ has $x = \alpha_i$ as a zero. Thus λ_i is well defined. Since $S(\hat{\mathfrak{h}}_{\mathfrak{p}})$ is integral over J, and *a fortiori* over $\mathbf{C}[\xi_1, \cdots, \xi_r, \alpha]$, there exists a homomorphism $\Lambda_i: S(\hat{\mathfrak{h}}_{\mathfrak{p}}) \to \boldsymbol{C}$ extending λ_i. Then there exists $H_i \in \mathfrak{h}_{\mathfrak{p}}$ such that $\Lambda_i(p) = p(H_i)$ for all $p \in S(\hat{\mathfrak{h}}_{\mathfrak{p}})$. In particular

$$p(H_*) = p(H_1) = p(H_2) \qquad \text{for } p \in I(\mathfrak{h}),$$

from which it follows easily that H_*, H_1, H_2 are all conjugate under $W(\mathfrak{h})$. Since $H_* \in \mathfrak{a}_*$, and since $\mathfrak{t}$ is invariant under $W(\mathfrak{h})$, we obtain $H_1, H_2 \in \mathfrak{t} \cap \mathfrak{h}_{\mathfrak{p}} = \mathfrak{a}_*$. Now select $u_1, u_2 \in U$ such that $\mathrm{Ad}(u_i)H_* = H_i$ $(i = 1, 2)$. Then $u_i \exp tH_* u_i^{-1} = \exp tH_i$ $(i = 1, 2)$ so by (7) u_1 and u_2 belong to the normalizer of $\mathfrak{a}_*$ in U. Using Lemma 7.8 we have $\mathrm{Ad}(m)H_1 = H_2$ for some $m \in M'$. But the restrictions of $\mathrm{Ad}(M')$ to $\mathfrak{a}_*$ constitute the Weyl group $W(\mathfrak{a}_*)$ although K_* may be larger than K (see [15], p. 244 and p. 273). Hence $sH_1 = H_2$ for some $s \in W(\mathfrak{a}_*)$. But then

$$\alpha_1 = \Lambda_1(\alpha) = \alpha(H_1) = \alpha(H_2) = \Lambda_2(\alpha) = \alpha_2,$$

which is a contradiction.

Remarks. (i) Theorem 7.6 had been proved independently by Harish-Chandra (cf. also his proof in [15], p. 433 of the quoted result of Chevalley). (ii) (*Added in proof*). Since $W(\mathfrak{h}_\mathfrak{p})$ is the group of linear transformations of $\mathfrak{h}_\mathfrak{p}$ induced by those elements of $W(\mathfrak{h})$ which leave $\mathfrak{h}_\mathfrak{p}$ invariant, Theorem 7.6 raises the following more general question: Let V be a vector space over C and G a finite group of linear transformations of V. Let W be a vector subspace of V and H the group of linear transformations of W induced by those members of G which leave W invariant. Does the restriction from V to W map the set of rational G-invariants on V onto the set of rational H-invariants on V onto the set of rational H-invariants on W? According to a communication from D. Mumford, the answer is affirmative.

REFERENCES.

[1] S. Araki, "On root systems and an infinitesimal classification of irreducible symmetric spaces," *J. Math. Osaka City Univ.*, vol. 13 (1962), pp. 1-34.

[2] L. Ásgeirsson, "Some hints on Huygens' principle and Hadamard's conjecture," *Communications on Pure and Applied Mathematics*, vol. 9 (1956), pp. 307-326.

[3] F. A. Berezin, "Laplace operators on semisimple Lie groups," *Trudy Moscovskogo Matematičeskogo Obščestva*, vol. 6 (1957), pp. 371-463 or *American Mathematical Society Translations*, vol. 21, pp. 239-339.

[4] É. Cartan, "Sur certaines formes riemanniennes remarquables des géométries a groupe fondamental simple," *Annales Scientiques de l'École Normale Supérieure*, vol. 44 (1927), pp. 345-467.

[5] C. Chevalley, "The Betti numbers of the exceptional simple Lie groups," *Proceedings of the International Congress of Mathematicians*, Harvard II (1950), pp. 21-24.

[6] R. Courant and D. Hilbert, *Methoden der mathematischen Physik*, II, Springer, Berlin, 1937.

[7] E. B. Dynkin, "Non-negative eigenfunctions of the Laplace-Beltrami operator and Brownian motion in certain symmetric spaces," *Doklady Akademii Nauk SSSR*, vol. 141 (1961). pp. 288-291; *Soviet Mathematics*, vol. 2 (1961), pp. 1433-1436.

[8] S. G. Gindikin and F. I. Karpelevič, "Plancherel measure of Riemannian symmetric spaces of nonpositive curvature," *Doklady Akademii Nauk SSSR*, vol. 145 (1962), pp. 252-255; *Soviet Mathematics*, vol. 3 (1962), pp. 962-965.

[9] R. Godement, "A theory of spherical functions, I," *Transactions of the American Mathematical Society*, vol. 73 (1952), pp. 496-556.

[10] J. Hadamard, *Lectures on Cauchy's problem in linear partial differential equations*, Yale Univ. Press, 1923.

[11] Harish-Chandra, "Representations of semisimple Lie groups I," *Transactions of the American Mathematical Society*, vol. 75 (1953), pp. 185-243.
[12] ———, "Differential operators on a semisimple Lie algebra," *American Journal of Mathematics*, vol. 79 (1957), pp. 87-120.
[13] ———, "Spherical functions on a semisimple Lie group I, II," *American Journal of Mathematics*, vol. 80 (1958), pp. 241-310, 553-613.
[14] S. Helgason, "Differential operators on homogeneous spaces," *Acta Mathematica*, vol. 102 (1959). pp. 239-299.
[15] ———, *Differential geometry and symmetric spaces*, Academic Press, New York, 1962.
[16] ———, "Some results on invariant theory," *Bulletin of the American Mathematical Society*, vol. 68 (1962), pp. 367-371.
[17] ———, "Duality and Radon transform for symmetric spaces," *American Journal of Mathematics*, vol. 85 (1963), pp. 667-692.
[18] ———, "Fundamental solutions for invariant differential operators on symmetric spaces," *Bulletin of the American Mathematical Society*, vol. 69 (1963), pp. 778-781.
[19] L. Hörmander, "On the division of distributions by polynomials," *Arkiv för Matematik*, vol. 3 (1958), pp. 555-568.
[20] ———, *Linear partial differential operators*, Springer, Berlin, 1963.
[21] B. Kostant, "On the conjugacy of real Cartan subalgebras I," *Proceedings of the National Academy of Sciences, U.S.A.*, vol. 41 (1955), pp. 967-970.
[22] R. P. Langlands, "The dimension of spaces of automorphic forms," *American Journal of Mathematics*, vol. 85 (1963), pp. 99-125.
[23] S. Łojasiewicz, "Division d'une distribution par une fonction analytique de variables reelles," *Comptes Rendus des Séances de l'Académie des Sciences*, Paris, vol. 246 (1958), pp. 683-686.
[24] G. de Rham, *Variétés différentiables*, Hermann, Paris, 1955.
[25] I. Satake, "On representations and compactifications of symmetric Riemann spaces," *Annals of Mathematics*, vol. 71 (1960), pp. 77-110.
[26] K. Stellmacher, "Ein Beispiel einer Huygensschen Differentialgleichung," *Göttingen Nachrichten*, 1953, pp. 133-138.
[27] L. Schwartz, *Théorie des Distributions II*, Hermann, Paris, 1951.
[28] ———, *Séminaire* 1954-55 (Equations aux dérivées partielles), Paris, 1955.
[29] M. Suguira, "Conjugate classes of Cartan subalgebras in real semisimple Lie algebras," *Journal of the Mathematical Society of Japan*, vol. 11 (1959), pp. 374-434.
[30] E. C. Titchmarsh, *The theory of functions*, 2nd ed., Oxford Univ. Press, London, 1939.
[31] Chih-Ta Yen, "Sur les polynomes de Poincaré des groupes de Lie exceptionnels," *Comptes Rendus des Séances de l'Académie des Sciences*, Paris, vol. 228 (1949), pp. 628-630.

THE RADON TRANSFORM ON EUCLIDEAN SPACES, COMPACT TWO-POINT HOMOGENEOUS SPACES AND GRASSMANN MANIFOLDS

BY

SIGURÐUR HELGASON

The Institute for Advanced Study, Princeton, N. J., U.S.A.(1)

§ 1. Introduction

As proved by Radon [16] and John [13], a differentiable function f of compact support on a Euclidean space $\mathbf{R}^n$ can be determined explicitly by means of its integrals over the hyperplanes in the space. Let $J(\omega, p)$ denote the integral of f over the hyperplane $\langle x, \omega\rangle = p$ where ω is a unit vector and $\langle , \rangle$ the inner product in $\mathbf{R}^n$. If Δ denotes the Laplacian on $\mathbf{R}^n$, $d\omega$ the area element on the unit sphere $\mathbf{S}^{n-1}$ then (John [14], p. 13)

$$f(x) = \tfrac{1}{2}(2\pi i)^{1-n} (\Delta_x)^{\frac{1}{2}(n-1)} \int_{\mathbf{S}^{n-1}} J(\omega, \langle \omega, x\rangle)\, d\omega, \quad (n \text{ odd}); \tag{1}$$

$$f(x) = (2\pi i)^{-n} (\Delta_x)^{\frac{1}{2}(n-2)} \int_{\mathbf{S}^{n-1}} d\omega \int_{-\infty}^{\infty} \frac{dJ(\omega, p)}{p - \langle \omega, x\rangle}, \quad (n \text{ even}), \tag{2}$$

where, in the last formula, the Cauchy principal value is taken.

Considering now the simpler formula (1) we observe that it contains two dual integrations: the first over the set of points in a given hyperplane, the second over the set of hyperplanes passing through a given point. Generalizing this situation we consider the following setup:

(i) Let X be a manifold and G a transitive Lie transformation group of X. Let Ξ be a family of subsets of X permuted transitively by the action of G on X, whence Ξ acquires a G-invariant differentiable structure. Here Ξ will be called the *dual* space of X.

(ii) Given $x \in X$, let $\check{x}$ denote the set of $\xi \in \Xi$ passing through x. It is assumed that each ξ and each $\check{x}$ carry measures μ and ν, respectively, such that the action of G on X and Ξ permutes the measures μ and permutes the measures ν.

(1) Work supported in part by the National Science Foundation, NSF GP 2600, U.S.A.

11 – 652923. *Acta mathematica*. 113. Imprimé le 11 mai 1965.

(iii) If f and g are suitably restricted functions on X and Ξ, respectively, we can define functions $\hat{f}$ on Ξ, $\check{g}$ on X by

$$\hat{f}(\xi) = \int_{\xi} f(x)\, d\mu(x), \qquad \check{g}(x) = \int_{\check{x}} g(\xi)\, d\nu(\xi).$$

These three assumptions have not been made completely specific because they are not intended as axioms for a general theory but rather as framework for special examples. In this spirit we shall consider the following problems.

A. *Relate function spaces on X and Ξ by means of the transforms $f \to \hat{f}$ and $g \to \check{g}$.*

B. *Let $\mathbf{D}(X)$ and $\mathbf{D}(\Xi)$, respectively, denote the algebras of G-invariant differential operators on X and Ξ. Does there exist a map $D \to \hat{D}$ of $\mathbf{D}(X)$ into $\mathbf{D}(\Xi)$ and a map $E \to \check{E}$ of $\mathbf{D}(\Xi)$ into $\mathbf{D}(X)$ such that*

$$(Df)^{\widehat{}} = \hat{D}\hat{f}, \quad (Eg)^{\vee} = \check{E}\check{g}$$

for all f and g above?

C. *In case the transforms $f \to \hat{f}$ and $g \to \check{g}$ are one-to-one find explicit inversion formulas. In particular, find the relationships between f and $(\hat{f})^{\vee}$ and between g and $(\check{g})^{\widehat{}}$.*

In this article we consider three examples within this framework: (1) The already mentioned example of points and hyperplanes (§ 2–§ 4); (2) points and antipodal manifolds in compact two-point homogeneous spaces (§ 5–§ 6); p-planes and q-planes in $\mathbf{R}^{p+q+1}$ (§7–§ 8). Other examples are discussed in [11] which also contains a bibliography on the Radon transform and its generalizations. See also [5].

The following notation will be used throughout. The set of integers, real and complex numbers, respectively, is denoted by $\mathbf{Z}$, $\mathbf{R}$ and $\mathbf{C}$. If $x \in \mathbf{R}^n$, $|x|$ denotes the length of the vector x; Δ denotes the Laplacian on $\mathbf{R}^n$. If M is a manifold, $C^\infty(M)$ (respectively $\mathcal{D}(M)$) denotes the space of differentiable functions (respectively, differentiable functions with compact support) on M. If $L(M)$ is a space of functions on M, D and endomorphism of $L(M)$ and $p \in M$, $f \in L(M)$ then $[Df](p)$ (and sometimes $D_p(f(p))$) denotes the value of Df at p. The tangent space to M at p is denoted M_p. If τ is a diffeomorphism of a manifold M onto a manifold N and if $f \in C^\infty(M)$ then f^τ stands for the function $f \circ \tau^{-1}$ in $C^\infty(N)$. If D is a differential operator on M then the linear transformation of $C^\infty(N)$ given by D^τ: $f \to (Df^{\tau^{-1}})^\tau$ is a differential operator on N. For $M = N$, D is called *invariant* under τ if $D^\tau = D$.

The adjoint representation of a Lie group G (respectively, Lie algebra $\mathfrak{G}$) will be denoted Ad_G (respectively $\mathrm{ad}_{\mathfrak{G}}$). These subscripts are omitted when no confusion is likely.

§ 2. The Radon transform in Euclidean space

Let $\mathbf{R}^n$ be a Euclidean space of arbitrary dimension n and let Ξ denote the manifold of hyperplanes in $\mathbf{R}^n$.

If f is a function on $\mathbf{R}^n$, integrable on each hyperplane in $\mathbf{R}^n$, the *Radon transform* of f is the function $\hat{f}$ on Ξ given by

$$\hat{f}(\xi) = \int_{\xi} f(x)\, d\sigma(x), \quad \xi \in \Xi, \tag{1}$$

where $d\sigma$ is the Euclidean measure on the hyperplane ξ. In this section we shall prove the following result which shows, roughly speaking, that f has compact support if and only if $\hat{f}$ does.

THEOREM 2.1. *Let $f \in C^\infty(\mathbf{R}^n)$ satisfy the following conditions:*

(i) *For each integer $k > 0$ $|x|^k f(x)$ is bounded.*
(ii) *There exists a constant $A > 0$ such that $\hat{f}(\xi) = 0$ for $d(0,\xi) > A$, d denoting distance.*

Then $$f(x) = 0 \quad \text{for} \quad |x| > A.$$

Proof. Suppose first that f is a radial function. Then there exists an even function $F \in C^\infty(\mathbf{R})$ such that $f(x) = F(|x|)$ for $x \in \mathbf{R}^n$. Also there exists an even function $\hat{F} \in C^\infty(\mathbf{R})$ such that $\hat{F}(d(0,\xi)) = \hat{f}(\xi)$. Because of (1) we find easily

$$\hat{F}(p) = \int_{\mathbf{R}^{n-1}} F((p^2 + |y|^2)^{\frac{1}{2}})\, dy = \Omega_{n-1} \int_0^\infty F((p^2 + t^2)^{\frac{1}{2}}) t^{n-2}\, dt, \tag{2}$$

where Ω_{n-1} is the area of the unit sphere in $\mathbf{R}^{n-1}$. Here we substitute $s = (p^2 + t^2)^{-\frac{1}{2}}$ and then put $u = p^{-1}$. Formula (2) then becomes

$$u^{n-3} \hat{F}(u^{-1}) = \Omega_{n-1} \int_0^u (F(s^{-1})\, s^{-n})\, (u^2 - s^2)^{\frac{1}{2}(n-3)}\, ds. \tag{3}$$

This formula can be inverted (see e.g. John [14], p. 83) and we obtain

$$F(s^{-1})\, s^{-n} = c\, s \left(\frac{d}{d(s^2)}\right)^{n-1} \int_0^s (s^2 - u^2)^{\frac{1}{2}(n-3)} u^{n-2} \hat{F}(u^{-1})\, du, \tag{4}$$

where c is a constant. Now by (ii), $\hat{F}(u^{-1}) = 0$ for $0 < u \leqslant A^{-1}$ so by (4), $F(s^{-1}) = 0$ for $0 < s \leqslant A^{-1}$, proving the theorem for the case when f is radial.

Now suppose $f \in C^\infty(\mathbf{R}^n)$ arbitrary, satisfying (i) and (ii). Let K denote the orthogonal group $\mathbf{O}(n)$. For $x, y \in \mathbf{R}^n$ we consider the spherical average

$$f^*(x, y) = \int_K f(x + k \cdot y)\, dk,$$

where dk is the Haar measure on $\mathbf{O}(n)$, with total measure 1. Let $R_2 f^*$ be the Radon transform of f^* in the second variable. Since $(f^\tau)\hat{} = (\hat{f})^\tau$ for each rigid motion τ of $\mathbf{R}^n$ it is clear that

$$[R_2 f^*](x, \xi) = \int_K \hat{f}(x + k \cdot \xi)\, dk, \quad x \in \mathbf{R}^n,\ \xi \in \Xi, \tag{5}$$

where $x + k \cdot \xi$ is the translate of $k \cdot \xi$ by x. Now it is clear that the distance d satisfies the inequality

$$d(0, x + k \cdot \xi) \geqslant d(0, \xi) - |x|$$

for all $x \in \mathbf{R}^n$, $k \in K$. Hence we conclude from (5)

$$[R_2 f^*](x, \xi) = 0 \quad \text{if} \quad d(0, \xi) > A + |x|. \tag{6}$$

For a fixed x, the function $y \to f^*(x, y)$ is a radial function in $C^\infty(\mathbf{R}^n)$ satisfying (i). Since the theorem is proved for radial functions, (6) implies that

$$\int_K f(x + k \cdot y)\, dk = 0 \quad \text{if} \quad |y| > A + |x|.$$

The theorem is now a consequence of the following lemma.

LEMMA 2.2. *Let f be a function in $C^\infty(\mathbf{R}^n)$ such that $|x|^k f(x)$ is bounded on $\mathbf{R}^n$ for each integer $k > 0$. Suppose f has surface integral 0 over every sphere which encloses the unit sphere. Then $f(x) \equiv 0$ for $|x| > 1$.*

Proof. The assumption about f means that

$$\int_{S^{n-1}} f(x + L\omega)\, d\omega = 0 \quad \text{for} \quad L > |x| + 1. \tag{7}$$

This implies that
$$\int_{|y| \geqslant L} f(x + y)\, dy = 0 \quad \text{for} \quad L > |x| + 1. \tag{8}$$

Now fix $L > 1$. Then (8) shows that

$$\int_{|y| \leqslant L} f(x + y)\, dy$$

is constant for $0 \leqslant |x| < L - 1$. The identity

$$\int_{S^{n-1}} f(x+L\omega)\,(x_i+L\omega_i)\,d\omega = x_i \int_{S^{n-1}} f(x+L\omega)\,d\omega + L^{2-n}\frac{\partial}{\partial x_i}\int_{|y|<L} f(x+y)\,dy$$

then shows that the function $x_i f(x)$ has surface integral 0 over each sphere with radius L and center x $(0 \leqslant |x| < L-1)$. In other words, we can pass from $f(x)$ to $x_i f(x)$ in the identity (7). By iteration, we find that on the sphere $|y| = L$ $(L>1)$ $f(y)$ is orthogonal to all polynomials, hence $f(y) \equiv 0$ for $|y| = L$. This concludes the proof.

Remark. The proof of this lemma was suggested by John's solution of the problem of determining a function on $\mathbf{R}^n$ by means of its surface integrals over all spheres of radius 1 (John [14], p. 114).

§ 3. Rapidly decreasing functions on a complete Riemannian manifold

Let M be a connected, complete Riemannian manifold, $\tilde{M}$ its universal covering manifold with the Riemannian structure induced by that of M, $\tilde{M} = \tilde{M}_1 \times \ldots \times \tilde{M}_l$ the de Rham decomposition of $\tilde{M}$ into irreducible factors ([17]) and let $M_i = \pi(\tilde{M}_i)$ $(1 \leqslant i \leqslant l)$ where π is the covering mapping of $\tilde{M}$ onto M. Let Δ, $\tilde{\Delta}$, Δ_i, $\tilde{\Delta}_i$ denote the Laplace-Beltrami operators on M, $\tilde{M}$, M_i, $\tilde{M}_i$, respectively. It is clear that $\tilde{\Delta}_i$ can be regarded as a differential operator on $\tilde{M}$. In order to consider Δ_i as a differential operator on M, let $f \in C^\infty(M)$, $\tilde{f} = f \circ \pi$. Any covering transformation τ of $\tilde{M}$ is an isometry so $(\tilde{\Delta}_i(f \circ \pi))^\tau = \tilde{\Delta}_i(f \circ \pi)$; hence $\tilde{\Delta}_i(f \circ \pi) = F \circ \pi$, where $F \in C^\infty(M)$. We define $\Delta_i f = F$. Because of the decomposition of $\tilde{M}$ each $m \in M$ has a coordinate neighborhood which is a product of coordinate neighborhoods in the spaces M_i. In terms of these coordinates, $\Delta = \sum_i \Delta_i$; in particular Δ_i is a differential operator on M, and the operators Δ_i $(1 \leqslant i \leqslant l)$ commute.

Now fix a point $o \in M$ and let $r(p) = d(o, p)$. A function $f \in C^\infty(M)$ will be called *rapidly decreasing* if for each polynomial $P(\Delta_1, \ldots, \Delta_l)$ in the operators $\Delta_1, \ldots, \Delta_l$ and each integer $k \geqslant 0$

$$\sup_p |(1+r(p))^k [P(\Delta_1, \ldots, \Delta_l f](p)| < \infty. \tag{1}$$

It is clear that condition (1) is independent of the choice of o. Let $\mathcal{S}(M)$ denote the set of rapidly decreasing functions on M.

In the case of a Euclidean space a function $f \in C^\infty(\mathbf{R}^n)$ belongs to $\mathcal{S}(\mathbf{R}^n)$ if and only if for each polynomial P in n variables the function $P(D_1^2, \ldots, D_n^2) f$ (where $D_i = \partial/\partial x_i$) goes to zero for $|x| \to \infty$ faster than any power of $|x|$. Then the same holds for the function $P(D_1, \ldots, D_n) f$ (so $\mathcal{S}(\mathbf{R}^n)$ coincides with the space defined by Schwartz [18], II, p. 89) as a consequence of the following lemma which will be useful later.

LEMMA 3.1. *Let f be a function in $C^\infty(\mathbf{R}^n)$, which for each pair of integers $k, l \geq 0$ satisfies*

$$\sup_{x \in \mathbf{R}^n} |(1+|x|)^k [\Delta^l f](x)| < \infty. \tag{2}$$

Then the inequality is satisfied when Δ^l is replaced by an arbitrary differential operator with constant coefficients.

This lemma is easily proved by using Fourier transforms.

LEMMA 3.2. *A function $F \in C^\infty(\mathbf{R} \times \mathbf{S}^{n-1})$ lies in $\mathcal{S}(\mathbf{R} \times \mathbf{S}^{n-1})$ if and only if for arbitrary integers $k, l \geq 0$ and any differential operator D on $\mathbf{S}^{n-1}$,*

$$\sup_{\omega \in \mathbf{S}^{n-1},\, r \in \mathbf{R}} \left| (1+|r|)^k \frac{d^l}{dr^l} (DF)(\omega, r) \right| < \infty. \tag{3}$$

Proof. It is obvious that (3) implies that F is rapidly decreasing. For the converse we must prove ($\mathbf{S}^{n-1}$ being irreducible) that (3) holds provided it holds when $l \geq 0$ is even and D an arbitrary power $(\Delta_S)^m$ $(m \geq 0)$ of the Laplacian Δ_S on $\mathbf{S}^{n-1}$. Let $G(\omega, r) = d^l/dr^l(F(\omega, r))$. Of course it suffices to verify (3) as $\omega = (\omega_1, \ldots, \omega_n)$ varies in some coordinate neighborhood on $\mathbf{S}^{n-1}$. Let $x_i = |x|\omega_i$ $(1 \leq i \leq n)$ and suppose G extended to a C^∞ function $\tilde{G}$ in the product of an annulus A_ε: $\{x \in \mathbf{R}^n \mid |x_1^2 + \ldots + x_n^2 - 1| < \varepsilon < 1\}$ with $\mathbf{R}$. Regardless how this extension is made, (3) would follow (for even l) if we can prove an estimate of the form

$$\sup_{\omega \in \mathbf{S}^{n-1},\, r \in \mathbf{R}} |(1+|r|)^k [D^\gamma \tilde{G}](\omega, r)| < \infty \tag{4}$$

for an arbitrary derivative $D^\gamma = \partial^{|\gamma|}/\partial x_1^{\gamma_1} \ldots \partial x_n^{\gamma_n}$ $(|\gamma| = \gamma_1 + \ldots + \gamma_n)$. Now, by Sobolev's lemma (see e.g. [3], Theorem 6′, p. 243) $[D^\gamma \tilde{G}](\omega, r)$ can be estimated by means of L^2 norms over A_ε of finitely many derivatives $D_x^\alpha D_x^\gamma(\tilde{G}(x, r))$. But the L^2 norm over A_ε of $D_x^\alpha D_x^\gamma(\tilde{G}(x, r))$ is estimated by the L^2 norm over A_ε of $\Delta_x^m(\tilde{G}(x, r))$, m being a suitable integer (see [12], p. 178–188). Now suppose $\tilde{G}$ was chosen such that for each r, the function $x \to \tilde{G}(x, r)$ is constant on each radius from 0. Then

$$\Delta_x(\tilde{G}(x, r)) = |x|^{-2}[\Delta_S G](\omega, r) \quad (x = |x|\omega)$$

and

$$\Delta_x^m(\tilde{G}(x, r)) = \sum_i f_i(|x|)\, [(\Delta_S)^i G](\omega, r),$$

where the sum is finite and each f_i is bounded for $||x| - 1| < \varepsilon$. Hence the L^2 norm over A_ε of $(\Delta^m)_x(\tilde{G}(x, r))$ is estimated by a linear combination of the L^2 norms over $\mathbf{S}^{n-1}$ of $[(\Delta_S)^i G](\omega, r)$. But these last derivatives satisfy (3), by assumption, so we have proved (4).

This proves (3) for l even. Let $H(\omega, s)$ be the Fourier transform (with respect to r) of the function $(DF)(\omega, r)$. Then one proves by induction on k that

$$\sup_{\omega\in \mathbf{S}^{n-1},\, s\in\mathbf{R}} \left|(1+|s|)^l \frac{d^k}{ds^k} H(\omega, s)\right| < \infty$$

for all k, $l\geqslant 0$ and now (3) follows for all k, $l\geqslant 0$ by use of the inverse Fourier transform.

§ 4. The Radon transforms of $\mathcal{S}(\mathbf{R}^n)$ and $\mathcal{D}(\mathbf{R}^n)$

If $\omega\in\mathbf{S}^{n-1}$, $r\in\mathbf{R}$ let $\xi(\omega, r)$ denote the hyperplane $\langle x, \omega\rangle = r$ in $\mathbf{R}^n$. Then the mapping $(\omega, r)\rightarrow\xi(\omega, r)$ is a two-fold covering map of the manifold $\mathbf{S}^{n-1}\times\mathbf{R}$ onto the manifold Ξ of all hyperplanes in $\mathbf{R}^n$; the (differentiable) functions on Ξ will be identified with the (differentiable) functions F on $\mathbf{S}^{n-1}\times\mathbf{R}$ which satisfy $F(\omega, r) = F(-\omega, -r)$. Thus $\mathcal{S}(\Xi)$ is, by definition, a subspace of $\mathcal{S}(\mathbf{S}^{n-1}\times\mathbf{R})$. We also need the linear space $\mathcal{S}_H(\Xi)$ of functions $F\in\mathcal{S}(\Xi)$ which have the property that for each integer $k\geqslant 0$ the integral $\int F(\omega, r) r^k dr$ can be written as a homogeneous kth degree polynomial in the components $\omega_1, \ldots, \omega_n$ of ω. Such a polynomial can, since $\omega_1^2+\ldots+\omega_n^2=1$, also be written as a $(k+2l)$th degree polynomial in the ω_i.

We shall now consider the situation outlined in the introduction for $X=\mathbf{R}_n$, Ξ as above and G the group of rigid motions of X. If $x\in X$, $\xi\in\Xi$, the measure μ is the Euclidean measure $d\sigma$ on the hyperplane ξ, ν is the unique measure on $\check{x}$ invariant under all rotations around x, normalized by $\nu(\check{x})=1$. We shall now consider problems A, B, C from § 1. If f is a function on X, integrable along each hyperplane in X then according to the conventions above

$$\hat{f}(\omega, r) = \int_{\langle x, \omega\rangle = r} f(x)\, d\sigma(X), \quad \omega\in\mathbf{S}^{n-1},\ r\in\mathbf{R}. \tag{1}$$

THEOREM 4.1. *The Radon transform $f\rightarrow\hat{f}$ is a linear one-to-one mapping of $\mathcal{S}(X)$ onto $\mathcal{S}_H(\Xi)$.*

Proof. Let $f\in\mathcal{S}(X)$ and let $\tilde{f}$ denote the Fourier transform

$$\tilde{f}(u) = \int f(x)\, e^{-i\langle x, u\rangle}\, dx, \quad u\in\mathbf{R}^n.$$

If $u\neq 0$ put $u=s\omega$, where $s\in\mathbf{R}$ and $\omega\in\mathbf{S}^{n-1}$. Then

$$\tilde{f}(s\omega) = \int_{-\infty}^{\infty} dr \int_{\langle x, \omega\rangle = r} f(x)\, e^{-i\langle x, \omega\rangle s}\, d\sigma(x)$$

so we obtain
$$\hat{f}(s\omega) = \int_{-\infty}^{\infty} \hat{f}(\omega, r)\, e^{-isr} dr, \tag{2}$$

for $s \neq 0$ in $\mathbf{R}$, $\omega \in \mathbf{S}^{n-1}$. But (2) is obvious for $s=0$ so it holds for all $s \in \mathbf{R}$. Now according to Schwartz [18], II, p. 105, the Fourier transform $f \to \tilde{f}$ maps $\mathcal{S}(\mathbf{R}^n)$ onto itself. Since

$$\frac{d}{ds}(\tilde{f}(s\omega)) = \sum_{i=1}^{n} \omega_i \frac{\partial \tilde{f}}{\partial u_i} \qquad (u = (u_1, \ldots, u_n))$$

it follows from (2) that for each fixed ω, the function $r \to \hat{f}(\omega, r)$ lies in $\mathcal{S}(\mathbf{R})$. For each $\omega_0 \in \mathbf{S}^{n-1}$, a subset of $\{\omega_1, \ldots, \omega_n\}$ will serve as local coordinates on a neighborhood of ω_0. To see that $\hat{f} \in \mathcal{S}(\Xi)$, it therefore suffices to verify (3) § 3 for $F = \hat{f}$ on an open subset N of $\mathbf{S}^{n-1}$ where ω_n is bounded away from 0 and $\omega_1, \ldots, \omega_{n-1}$ serve as coordinates, in terms of which D is expressed. Putting $\mathbf{R}^+ = \{s \in \mathbf{R} \mid s > 0\}$ we have on $N \times \mathbf{R}^+$

$$u_1 = s\omega_1, \ldots, u_{n-1} = s\omega_{n-1}, \ u_n = s(1 - \omega_1^2 - \ldots - \omega_{n-1}^2)^{\frac{1}{2}}, \tag{3}$$

so
$$\frac{\partial}{\partial \omega_i}(\tilde{f}(s\omega)) = s \sum_{i=1}^{n-1} \frac{\partial \tilde{f}}{\partial u_i} - s\omega_i (1 - \omega_1^2 - \ldots - \omega_{n-1}^2)^{\frac{1}{2}} \frac{\partial \tilde{f}}{\partial u_n}.$$

It follows that if D is any differential operator on $\mathbf{S}^{n-1}$ and k, l integers $\geqslant 0$ then

$$\sup_{\omega \in N, s \in \mathbf{R}} \left| (1 + s^{2k}) \left[\frac{d^l}{ds^l} D\tilde{f}\right] (\omega, s) \right| < \infty. \tag{4}$$

We can therefore apply D under the integral sign in the inversion formula

$$\hat{f}(\omega, r) = \frac{1}{2\pi} \int_{-\infty}^{\infty} \tilde{f}(s\omega)\, e^{isr}\, ds \tag{5}$$

and obtain

$$(1 + r^{2k}) \frac{d^l}{dr^l} (D_\omega(\hat{f}(\omega, r))) = \frac{1}{2\pi} \int \left(1 + (-1)^k \frac{d^{2k}}{ds^{2k}}\right) ((is)^l D_\omega(\tilde{f}(s\omega)))\, e^{isr}\, ds.$$

Now (4) shows that $\hat{f} \in \mathcal{S}(\Xi)$. Finally, if k is an integer $\geqslant 0$ then

$$\int_{-\infty}^{\infty} \hat{f}(\omega, r)\, r^k dr = \int_{-\infty}^{\infty} r^k dr \int_{\langle x, \omega \rangle = r} f(x)\, d\sigma(x) = \int_{\mathbf{R}^n} f(x) \langle x, \omega \rangle^k dx \tag{6}$$

so $\hat{f} \in \mathcal{S}_H(\Xi)$. The Fourier transform being one-to-one it remains to prove that each $g \in \mathcal{S}_H(\Xi)$ has the form $g = \hat{f}$ for some $f \in \mathcal{S}(\mathbf{R}^n)$. We put

$$G(s, \omega) = \int_{-\infty}^{\infty} g(\omega, r)\, e^{-irs} dr.$$

Then $G(-s, -\omega)=G(s, \omega)$ and $G(0, \omega)$ is a homogeneous polynomial of degree 0 in ω, hence independent of ω. Hence there exists a function F on $\mathbf{R}^n$ such that

$$F(s\omega)=\int_{-\infty}^{\infty} g(\omega, r)\, e^{-irs}dr, \quad s \in \mathbf{R}, \omega \in \mathbf{S}^{n-1}. \tag{7}$$

It is clear that F is C^∞ in $\mathbf{R}^n-\{0\}$. To prove that F is C^∞ in a neighborhood of 0 we consider the coordinate neighborhood N on $\mathbf{S}^{n-1}$ as before. Let $h(u_1, ..., u_n)$ be any function of class C^∞ in $\mathbf{R}^n-\{0\}$ and let $h^*(\omega_1, ..., \omega_{n-1}, s)$ be the function on $N \times \mathbf{R}^+$ obtained by means of the substitution (3). Then

$$\frac{\partial h}{\partial u_i}=\sum_{j=1}^{n-1} \frac{\partial h^*}{\partial \omega_j} \cdot \frac{\partial \omega_j}{\partial u_i}+\frac{\partial h^*}{\partial s} \cdot \frac{\partial s}{\partial u_i} \quad (1 \leqslant i \leqslant n)$$

and
$$\frac{\partial \omega_j}{\partial u_i}=\frac{1}{s}\left(\delta_{ij}-\frac{u_i u_j}{s^2}\right) \quad (1 \leqslant i \leqslant n, \ 1 \leqslant j \leqslant n-1),$$

$$\frac{\partial s}{\partial u_i}=\omega_i \ (1 \leqslant i \leqslant n-1), \qquad \frac{\partial s}{\partial u_n}=(1-\omega_1^2-\ldots-\omega_{n-1}^2)^{\frac{1}{2}}.$$

Hence
$$\frac{\partial h}{\partial u_i}=\frac{1}{s} \frac{\partial h^*}{\partial \omega_i}+\omega_i\left(\frac{\partial h^*}{\partial s}-\frac{1}{s} \sum_{j=1}^{n-1} \omega_j \frac{\partial h^*}{\partial \omega_j}\right) \quad (1 \leqslant i \leqslant n-1),$$

$$\frac{\partial h}{\partial u_n}=(1-\omega_1^2-\ldots-\omega_{n-1}^2)^{\frac{1}{2}}\left(\frac{\partial h^*}{\partial s}-\frac{1}{s} \sum_{j=1}^{n-1} \omega_j \frac{d h^*}{\partial \omega_j}\right).$$

In order to use these formulas for $h=F$ we write

$$F(s\omega)=\int_{-\infty}^{\infty} g(\omega, r)\, dr+\int_{-\infty}^{\infty} g(r, \omega)\,(e^{-irs}-1)\, dr$$

and by assumption, the first integral is independent of ω. Thus, for a constant $K>0$,

$$\left|\frac{1}{s} \frac{\partial}{\partial \omega_i}(F(s\omega))\right| \leqslant K \int (1+r^4)^{-1} s^{-1}\left|e^{-isr}-1\right| dr \leqslant K \int \frac{|r|}{1+r^4}\, dr.$$

This shows that all the derivatives $\partial F / \partial u_i$ $(1 \leqslant i \leqslant n)$ are bounded in a punctured ball $0<|u|<\varepsilon$ so F is continuous in a neighborhood of $u=0$. More generally, let q be any integer >0. Then we have for an arbitrary qth order derivative,

$$\frac{\partial^q h}{\partial u_{i_1} \ldots \partial u_{i_q}}=\sum_{i+j \leqslant q} A_{i,j}(\omega, s) \frac{\partial^{i+j} h^*}{\partial \omega_{k_1} \ldots \partial \omega_{k_i} \partial s^j}, \tag{8}$$

where the coefficient $A_{i,j}(\omega, s)=O(s^{j-q})$ near $s=0$. Also

$$F(s\omega)=\int_{-\infty}^{\infty} g(\omega,r)\sum_{k=0}^{q-1}\frac{(-isr)^k}{k!}\,dr+\int_{-\infty}^{\infty} g(\omega,r)\,e_q(-irs)\,dr, \tag{9}$$

where
$$e_q(t)=\frac{t^q}{q!}+\frac{t^{q+1}}{(q+1)!}+\dots.$$

Then it is clear that the first integral in (9) is a polynomial in $u_1, \dots, u_n$ of degree $\leqslant q-1$ and is therefore annihilated by the differential operator (8). Now, if $0\leqslant j\leqslant q$,

$$\left|s^{j-q}\frac{\partial^j}{\partial s^j}(e_q(-irs))\right|=|(-ir)^q(-irs)^{j-q}\,e_{q-j}(-irs)|\leqslant K_j\,r^q, \tag{10}$$

where K_j is a constant, because the function $t\to(it)^{-p}e_p(it)$ is obviously bounded on $\mathbf{R}$ $(p\geqslant 0)$. Since $g\in\mathcal{S}(\Xi)$ it follows from (8), (9), (10) that each qth order derivative of F with respect to $u_1, \dots, u_n$ is bounded in a punctured ball $0<|u|<\varepsilon$. Hence $F\in C^\infty(\mathbf{R}^n)$. That F is rapidly decreasing is now clear from formula (7), Lemma 3.1 and the fact that ([8], p. 278)

$$\Delta h=\frac{\partial^2 h^*}{\partial s^2}+\frac{n-1}{s}\frac{\partial h^*}{\partial s}+\frac{1}{s^2}\Delta_S h^*,$$

where Δ_S is the Laplace–Beltrami operator on $\mathbf{S}^{n-1}$. If f is the function in $\mathcal{S}(X)$ whose Fourier transform is F then $\hat{f}=g$ and the theorem is proved.

Let $\mathcal{S}^*(X)$ denote the space of all functions $f\in\mathcal{S}(X)$ which satisfy $\int f(x)P(x)\,dx=0$ for all polynomials $P(x)$. Similarly, let $\mathcal{S}^*(\Xi)$ denote the space of all functions $g\in\mathcal{S}(\Xi)$ which satisfy $\int g(\omega, r)P(r)\,dr\equiv 0$ for all polynomials $P(r)$. Note that under the Fourier transform, $\mathcal{S}^*(X)$ corresponds to the space $\mathcal{S}_0(\mathbf{R}^n)$ of functions in $\mathcal{S}(\mathbf{R}^n)$ all of whose derivatives vanish at the origin.

COROLLARY 4.2. *The transforms $f\to\hat{f}$ and $g\to\check{g}$, respectively, are one-to-one linear maps of $\mathcal{S}^*(X)$ onto $\mathcal{S}^*(\Xi)$ and of $\mathcal{S}(\Xi)$ onto $\mathcal{S}^*(X)$.*

The first statement follows from (6) and the well-known fact that the polynomials $\langle x, \omega\rangle^k$ span the space of homogeneous polynomials of degree k. As for the second, we observe that for $f\in\mathcal{S}(X)$ and ξ_0 a fixed plane through 0

$$(\hat{f})^\vee(x)=\int_K \hat{f}(x+k\cdot\xi_0)\,dk=\int_K\left(\int_{\xi_0} f(x+k\cdot y)\,dy\right)dk$$
$$=\int_{\xi_0}dy\int_K f(x+k\cdot y)\,dk=\Omega_{n-1}\int_0^\infty r^{n-2}\left(\frac{1}{\Omega_n}\int_{\mathbf{S}^{n-1}} f(x+r\omega)\,d\omega\right)dr,$$

so
$$(\hat{f})^\vee(x)=\frac{\Omega_{n-1}}{\Omega_n}\int_X |x-y|^{-1}f(y)\,dy. \tag{11}$$

This formula is also proved in [4]. Now the right-hand side is a tempered distribution, being the convolution of a tempered distribution and a member of $\mathcal{S}(X)$. By [18], II, p. 124, the Fourier transform is given by the product of the Fourier transforms so if $f \in \mathcal{S}^*(X)$ we see that $(\hat{f})^\vee$ has Fourier transform belonging to $\mathcal{S}_0(X)$. Hence $(\hat{f})^\vee \in \mathcal{S}^*(X)$ and the second statement of Cor. 4.2 follows.

Remarks. A characterization of the Radon transform of $\mathcal{S}(X)$ similar to that of Theorem 3.1 is stated in Gelfand–Graev–Vilenkin [5], p. 35. Their proof, as outlined on p. 36–39, is based on the inversion formula (1) § 1 and therefore leaves out the even-dimensional case. Corollary 4.2 was stated by Semyanistyi [19].

Now let $\mathcal{D}(X)$ and $\mathcal{D}(\Xi)$ be as defined in § 1, and put $\mathcal{D}_H(\Xi) = \mathcal{S}_H(\Xi) \cap \mathcal{D}(\Xi)$. The following result is an immediate consequence of Theorem 2.1 and 4.1.

COROLLARY 4.3. *The Radon transform $f \to \hat{f}$ is a linear one-to-one mapping of $\mathcal{D}(X)$ onto $\mathcal{D}_H(\Xi)$.*

Concerning problem B in § 1 we have the following result which is a direct consequence of Lemmas 7.1 and 8.1, proved later.

PROPOSITION 4.4. *The algebra $\mathbf{D}(X)$ is generated by the Laplacian Δ, the algebra $\mathbf{D}(\Xi)$ is generated by the differential operator $\square$: $g(\omega, r) \to (d^2/dr^2)g(\omega, r)$ and*

$$(\Delta f)^\wedge = \square \hat{f}, \qquad (\square g)^\vee = \Delta \check{g}$$

for $f \in \mathcal{S}(X)$, $g \in C^\infty(\Xi)$.

The following reformulation of the inversion formulas (1), (2) § 1 gives an answer to problem C.

THEOREM 4.5. (i) *If n is odd,*

$$f = c\Delta^{\frac{1}{2}(n-1)}((\hat{f})^\vee), \quad f \in \mathcal{S}(X);$$
$$g = c\square^{\frac{1}{2}(n-1)}((\check{g})^\wedge), \quad g \in \mathcal{S}^*(\Xi),$$

where c is a constant, independent of f and g.

(ii) *If n is even,*

$$f = c_1 J_1((\hat{f})^\vee), \quad f \in \mathcal{S}(X);$$
$$g = c_2 J_2((\check{g})^\wedge), \quad g \in \mathcal{S}^*(\Xi),$$

where the operators J_1 and J_2 are given by analytic continuation

$$J_1 : f(x) \to \underset{\alpha = 1-2n}{\text{anal. cont.}} \int_{R^n} f(y) \, |x-y|^\alpha dy,$$

$$J_2 : g(\omega, p) \to \underset{\beta=-n}{\text{anal. cont.}} \int_R g(\omega, q) \, |p-q|^\beta \, dq,$$

and c_1, c_2 *are constants, independent of* f *and* g.

Proof. In (i) the first formula is just (1) § 1 and the second follows by Prop. 4.4. We shall now indicate how (ii) follows from (2) § 1. Since the Cauchy principal value is the derivative of the distribution $\log |p|$ on $\mathbf{R}$ whose successive derivatives are the distributions $Pf\cdot(p^{-k})$ (see [18], I, p. 43) we have by (2) § 1

$$f(x) = (2\pi i)^{-n}(n-1)! \int_{S^{n-1}} (Pf\cdot(p - \langle \omega, x \rangle^{-n})) \, (\hat{f}(\omega, p)) \, d\omega. \tag{12}$$

On the other hand, if $\varphi \in C^\infty(X)$ is bounded we have by Schwartz [18], I, p. 45

$$[J_1 \varphi](0) = \lim_{\varepsilon \to 0} \left[\int_{|x|\geqslant\varepsilon} |x|^{1-2n} \varphi(x) \, dx + \varepsilon(\varphi) \right], \tag{13}$$

where $$\varepsilon(\varphi) = \sum_k H_k[\Delta^k \varphi](0) \, \frac{\varepsilon^{1-n+2k}}{1-n+2k}, \quad H_k = \frac{\pi^{\frac{1}{2}n}}{2^{2k-1} k! \, \Gamma(\frac{1}{2}n + k)}.$$

In particular $$[J_1((\hat{f})^\vee](0) = \lim_{\varepsilon\to 0} \left[\Omega_n \int_\varepsilon^\infty r^{-n} F(r) \, dr + \varepsilon(\hat{f})^\vee) \right], \tag{14}$$

where $F(r)$ is the average of $(\hat{f})^\vee$ on the sphere $|x| = r$. In order to express (14) in terms of f we assume f is a radial function and write $\hat{f}(p)$ for $\hat{f}(\omega, p)$. Then

$$F(r) = C \int_0^{\frac{1}{2}\pi} \hat{f}(r \cos \theta) \sin^{n-2}\theta \, d\theta, \quad C^{-1} = \int_0^{\frac{1}{2}\pi} \sin^{n-2}\theta \, d\theta, \tag{15}$$

$$[\Delta^k (\hat{f})^\vee](0) = \left(\frac{d^{2k}}{dp^{2k}} \hat{f} \right)(0). \tag{16}$$

If $q_n(p)$ is the Taylor series of $\hat{f}(p)$ around 0 up to order $n-2$ we get upon substituting (15) and (16) into (14),

$$(J_1((\hat{f})^\vee))(0) = C\,\Omega_n \lim_{\varepsilon\to 0} \int_0^{\frac{1}{2}\pi} \sin^{n-2}\theta \; \cos^{n-1}\theta \, d\theta \int_{\varepsilon\cos\theta}^\infty p^{-n} (\hat{f}(p) - q_n(p)) \, dp,$$

which on comparison with (12) gives

$$f(0) = c_1 \, J_1((\hat{f})^\vee)(0), \quad c_1 = \text{const.} \tag{17}$$

Now put for $\varphi \in C^\infty(X)$, $x, y \in X$,

$$\varphi_x^*(y) = \int_K \varphi(x+ky)\,dk$$

and let us prove $\qquad [J_1\varphi_x^*](0) = ((J_1\varphi)_x^*)(0) \quad \text{if } \varphi \text{ is bounded.} \qquad (18)$

In view of (13) this is a consequence of the obvious formula

$$\int_{|y|\geq\varepsilon} |y|^{1-2n}\varphi_x^*(y)\,dy = \int_{|y|\geq\varepsilon} |y|^{1-2n}\varphi(x+y)\,dy$$

and the Darboux equation ([8], p. 279) $[\Delta^k\varphi_x^*](0) = [\Delta^k\varphi](x)$. Now, a direct computation shows that $((\hat{f})^{\vee})_x^* = ((f_x^*)^{\wedge})^{\vee}$ for $f \in \mathcal{S}(X)$ and since f_x^* is radial we get from (17), (18)

$$f(x) = f_x^*(0) = c_1[(J_1((\hat{f})^{\vee}))_x^*](0) = c_1[J_1(\hat{f})^{\vee}](x).$$

Finally the inversion formula for $g \in \mathcal{S}^*(\Xi)$ would follow from the first one if we prove

$$(J_1 f)^{\wedge} = c_0 J_2 \hat{f}, \quad f \in \mathcal{S}^*(X),\ c_0 \text{ constant.} \qquad (19)$$

To see this we take the one-dimensional Fourier transform on both sides. The function $J_1 f$ is the convolution of a tempered distribution with a rapidly decreasing function. Hence it is a tempered distribution (Schwartz [18], II, pp. 102, 124) whose Fourier transform is (since $f \in \mathcal{S}^*(X)$) a function in $\mathcal{S}(X)$. Hence $J_1 f \in \mathcal{S}(X)$. Similarly $(J_2\hat{f})(\omega, p)$ is a rapidly decreasing function of p. Using the relation between the 1-dimensional and the n-dimensional Fourier transform ((2) § 4) and the formula for the Fourier transform of $Pf \cdot r^{\lambda}$ (Schwartz [18], II, p. 113) we find that both sides of (19) have the same Fourier transform, hence coincide. This concludes the proof.

Remark (added in proof). Alternative proofs of most of the results of § 4 have been found subsequently by D. Ludwig.

§ 5. The geometry of compact symmetric spaces of rank one

In this section and the next one we shall study problems A, B and C for the duality between points and antipodal manifolds in compact two-point homogeneous spaces. In the present section we derive the necessary geometric facts for symmetric spaces of rank one, without use of classification.

Let X be a compact Riemannian globally symmetric space of rank one and dimension >1. Let $I(X)$ denote the group of isometries of X in the compact open topology, $I_0(X)$ the identity component of $I(X)$. Let o be a fixed point in X and s_o the geodesic symmetry of X with repsect to o. Let $\mathfrak{u}$ denote the Lie algebra of $I(X)$ and $\mathfrak{u} = \mathfrak{k} + \mathfrak{p}$ the decomposition of $\mathfrak{u}$ into eigenspaces of the involutive automorphism of $\mathfrak{a}$ which corresponds to the automorphism $u \to s_o u s_o$ of $I(X)$. Here $\mathfrak{k}$ is the Lie algebra of the subgroup K of $I(X)$ which leaves o fixed. Changing the distance function d on X by a constant factor we may, since $\mathfrak{u}$ is semisimple, assume that the differential of the mapping $u \to u \cdot o$ of $I(X)$ onto X gives an isometry of $\mathfrak{p}$ (with the metric of the negative of the Killing form of $\mathfrak{u}$) onto X_o, the tangent space to X at o. Let L denote the diameter of X and if $x \in X$ let A_x denote the corresponding *antipodal manifold*, that is the set of points $y \in X$ at distance L from x; A_x is indeed a manifold, being an orbit of K. The geodesics in X are all closed and have length $2L$ and the Exponential mapping Exp at o is a diffeomorphism of the open ball in X_o of center 0 and radius L onto the complement $X - A_o$ (see [10], Ch. X, § 5).

PROPOSITION 5.1. *For each $x \in X$, the antipodal manifold A_x, with the Riemannian structure induced by X, is a symmetric space of rank one, and a totally geodesic submanifold of X.*

Proof. Let $y \in A_x$. Considering a geodesic in X through y and x we see that x is fixed under the geodesic symmetry s_y; hence $s_y(A_x) = A_x$. If σ_y denotes the restriction of s_y to A_x, then σ_y is an involutive isometry of A_x with y as isolated fixed point. Thus A_x is globally symmetric and σ_y is the geodesic symmetry with respect to y. Let $t \to \gamma(t)$ $(t \in \mathbf{R})$ be a geodesic in the Riemannian manifold A_x. We shall prove that γ is a geodesic in X. Consider the isometry $s_{\gamma(t)} s_{\gamma(0)}$ and a vector T in the tangent space $X_{\gamma(0)}$. Let τ_r: $X_{\gamma(0)} \to X_{\gamma(r)}$ denote the parallel translation in X along the curve $\gamma(\varrho)$ $(0 \leq \varrho \leq r)$. Then the parallel field $\tau_r \cdot T$ $(0 \leq r \leq t)$ along the curve $r \to \gamma(r)$ $(0 \leq r \leq t)$ is mapped by $s_{\gamma(t)}$ onto a parallel field along the image curve $r \to s_{\gamma(t)} \gamma(r) = \sigma_{\gamma(t)} \gamma(r) = \gamma(2t - r)$ $(0 \leq r \leq t)$. Since $s_{\gamma(t)} \tau_t T = -\tau_t T$ we deduce that $s_{\gamma(t)} s_{\gamma(0)} T = -s_{\gamma(t)} T = \tau_{2t} T$. In particular, the parallel transport in X along γ maps tangent vectors to γ into tangent vectors to γ. Hence γ is a geodesic in X. Consequently, A_x is a totally geodesic submanifold of X, and by the definition of rank, A_x has rank one.

Let $Z \to \operatorname{ad}(Z)$ denote the adjoint representation of $\mathfrak{u}$. Select a vector $H \in \mathfrak{p}$ of length L. The space $\mathfrak{a} = \mathbf{R}H$ is a Cartan subalgebra of the symmetric space X and we can select a positive restricted root α of X such that $\frac{1}{2}\alpha$ is the only other possible positive restricted root (see [10], Exercise 8, p. 280 where Σ is by definition the set of positive restricted roots). This means that the eigenvalues of $\operatorname{ad}(H)^2$ are 0, $\alpha(H)^2$ and possibly $(\frac{1}{2}\alpha(H))^2$ ($\alpha(H)$ is purely imaginary). Let $\mathfrak{u} = \mathfrak{u}_0 + \mathfrak{u}_\alpha + \mathfrak{u}_{\frac{1}{2}\alpha}$ be the corresponding decomposition of $\mathfrak{u}$ into eigenspaces and put $\mathfrak{k}_\beta = \mathfrak{u}_\beta \cap \mathfrak{k}$, $\mathfrak{p}_\beta = \mathfrak{u}_\beta \cap \mathfrak{p}$ for $\beta = 0, \alpha, \frac{1}{2}\alpha$. Then $\mathfrak{p}_0 = \mathfrak{a}$ and $\mathfrak{k}_\beta = \operatorname{ad} H(\mathfrak{p}_\beta)$ for $\beta \neq 0$.

PROPOSITION 5.2. *Let S denote the subgroup of K leaving the point* Exp$H$ *fixed, and let $\mathfrak{s}$ denote the Lie algebra of S. Then*

(i) $\mathfrak{s}=\mathfrak{k}_0+\mathfrak{k}_\alpha$ *if* H *is conjugate to* 0;

(ii) $\mathfrak{s}=\mathfrak{k}_0$ *if* H *is not conjugate to* 0;

(iii) *If* $\frac{1}{2}\alpha$ *is a restricted root then* H *is conjugate to* 0.

Proof. If exp: $\mathfrak{u}\to I(X)$ is the usual exponential mapping then a vector T in $\mathfrak{k}$ belongs to $\mathfrak{s}$ if and only if $\exp(-H)\exp(tT)\exp(H)\in K$ for all $t\in\mathbf{R}$. This reduces to

$$T\in\mathfrak{s} \quad \text{if and only if ad } H(T)+\tfrac{1}{3!}(\text{ad}\quad H)^3(T)+\ldots=0.$$

In particular, $\mathfrak{s}$ is the sum of its intersections with $\mathfrak{k}_0$, $\mathfrak{k}_\alpha$ and $\mathfrak{k}_{\frac{1}{2}\alpha}$. If $T\neq 0$ in $\mathfrak{k}_\beta$ $(\beta=0, \alpha, \frac{1}{2}\alpha)$ the condition above is equivalent to $\sinh(\beta(H))=0$. Thus (ii) is immediate ([10], Ch. VII, Prop. 3.1). To prove (i) suppose H is conjugate to 0. Whether or not $\frac{1}{2}\alpha$ is a restricted root we have by the cited result, $\alpha(\mathrm{H})\in\pi i\mathbf{Z}$ so $\mathfrak{k}_\alpha\in\mathfrak{s}$. We have also $\mathfrak{s}\cap\mathfrak{k}_{\frac{1}{2}\alpha}=\{0\}$ because otherwise $\frac{1}{2}\alpha(H)\in\pi i\mathbf{Z}$ which would imply that $\frac{1}{2}H$ is conjugate to 0. This proves (i). For (iii) suppose H were not conjugate to 0. The sphere in X_0 with radius $2L$ and center 0 is mapped by Exp onto o. It follows that the differential $d\,\mathrm{Exp}_{2H}$ is 0 so using the formula for this differential ([10], page 251, formula (2)) it follows that $(\frac{1}{2}\alpha)(2H)\in\pi i\mathbf{Z}$ so $\alpha(H)\in\pi i\mathbf{Z}$ which is a contradiction.

PROPOSITION 5.3. *Suppose* H *is conjugate to* 0. *Then all the geodesics in* X *with tangent vectors in* $\mathfrak{a}+\mathfrak{p}_\alpha$ *at* o *pass through the point* $\mathrm{Exp}\,H$. *The manifold* $\mathrm{Exp}(\mathfrak{a}+\mathfrak{p}_\alpha)$, *with the Riemannian structure induced by that of* X, *is a sphere, totally geodesic in* X.

Proof. Let $\mathfrak{G}$ denote the complexification of $\mathfrak{u}$ and B the Killing form of $\mathfrak{G}$. Since the various root subspaces $\mathfrak{G}^\beta$, $\mathfrak{G}^\gamma$ $(\beta+\gamma\neq 0)$ are orthogonal with respect to B ([10], p. 141) it follows without difficulty (cf. [10], p. 224) that

$$B([\mathfrak{k}_0, \mathfrak{p}_\alpha], \mathfrak{p}_{\frac{1}{2}\alpha})=B([\mathfrak{k}_\alpha, \mathfrak{p}_\alpha], \mathfrak{p}_{\frac{1}{2}\alpha})=0.$$

Also, if $Z\in\mathfrak{u}_0$ then $B([H,Z],[H,Z])=-B(Z,(\mathrm{ad}\,H)^2Z)=0$ so $\mathfrak{u}_0$ equals the centralizer of H in $\mathfrak{u}$. Thus $[\mathfrak{k}_0, \mathfrak{a}]=0$. Also $[\mathfrak{k}_\alpha, \mathfrak{a}]=\mathfrak{p}_\alpha$. Combining these relations we get

$$[\mathfrak{s}, \mathfrak{a}+\mathfrak{p}_\alpha]\subset\mathfrak{a}+\mathfrak{p}_\alpha.$$

Let S_0 denote the identity component of S and Ad the adjoint representation of the group $I(X)$. Then the tangent space to the orbit $\mathrm{Ad}(S_0)H$ at the point H is $[\mathfrak{s}, \mathbf{R}H]$ which equals $\mathfrak{p}_\alpha$, and by the relation above this orbit lies in the subspace $\mathfrak{a}+\mathfrak{p}_\alpha$. It follows that $\mathrm{Ad}(S_0)H$ is the sphere in $\mathfrak{a}+\mathfrak{p}_\alpha$ of radius L and center 0. But if $s\in S$ the geodesic $t\to s\cdot\mathrm{Exp}\,tH=\mathrm{Exp}(\mathrm{Ad}(s)tH)$ passes through $\mathrm{Exp}\,H$ so the first statement of the proposition is proved.

By consideration of the root subspaces $\mathfrak{G}^{\beta}$ as above, it is easy to see that the subspace $\mathfrak{a}+\mathfrak{p}_{\alpha}$ of $\mathfrak{p}$ is a Lie triple system. Thus the Riemannian manifold $X_{\mathfrak{p}}=\text{Exp}(\mathfrak{a}+\mathfrak{p}_{\alpha})$ is a totally geodesic submanifold of X ([10], p. 189). It is homogeneous and is mapped into itself by the geodesic symmetry s_0 of X, hence it is globally symmetric, and being totally geodesic, has rank one. If Z is a unit vector in $\mathfrak{p}_{\alpha}$, the curvature of $X_{\mathfrak{p}}$ along the plane section spanned by H and Z, is (cf. [10], p. 206)

$$-L^{-2}B([H,Z],[H,Z]) = -L^{-2}\alpha(H)^2.$$

But since $X_{\mathfrak{p}}$ has rank one, every plane section is congruent to one containing H; hence $X_{\mathfrak{p}}$ has constant curvature. Finally, $X_{\mathfrak{p}}-\{\text{Exp}\, H\}$ is the diffeomorphic image of an open ball, hence simply connected. Since $\dim X_{\mathfrak{p}}>1$ it follows that $X_{\mathfrak{p}}$ is also simply connected, hence a sphere.

PROPOSITION 5.4. *The antipodal manifold $A_{\text{Exp}\,H}$ is given by*

$$A_{\text{Exp}\,H}=\text{Exp}\,(\mathfrak{p}_{\frac{1}{2}\alpha}) \textit{ if } H \textit{ is conjugate to } 0.$$

$$A_{\text{Exp}\,H}=\text{Exp}\,(\mathfrak{p}_{\alpha}) \textit{ if } H \textit{ is not conjugate to } 0.$$

Proof. The geodesics from $\text{Exp}\,H$ to o intersect $A_{\text{Exp}\,H}$ in o under a right angle (Gauss' lemma; see e.g. [1], p. 34 or [9], Theorem 3). By Propositions 5.2 and 5.3 we deduce that the tangent space $(A_{\text{Exp}\,H})_o$ equals $\mathfrak{p}_{\frac{1}{2}\alpha}$ if H is conjugate to 0 and equals $\mathfrak{p}_{\alpha}$ if H is not conjugate to 0. Now use Prop. 5.1.

The next result shows that there is a kind of projective duality between points and antipodal manifolds.

PROPOSITION 5.5. *Let $x, y \in X$. Then*

(i) *$x \neq y$ implies $A_x \neq A_y$;*
(ii) *$x \in A_y$ if and only if $y \in A_x$.*

Proof. If $z \in A_x$ then the geodesics which meet A_x in z under a right angle all pass through a point z^* at distance L from z (Prop. 5.3 and Prop. 5.4); among these are the geodesics joining x and z. Hence $z^*=x$ and the result follows.

PROPOSITION 5.6. *Let $A(r)$ denote the surface area of a sphere in X of radius r $(0<r<L)$. Then*

$$A(r) = \Omega_{p+q+1}\lambda^{-p}(2\lambda)^{-q}\sin^p(\lambda r)\sin^q(2\lambda r),$$

where $p=\dim \mathfrak{p}_{\frac{1}{2}\alpha}$, $q=\dim \mathfrak{p}_{\alpha}$, Ω_n is the area of the unit sphere in $\mathbf{R}^n$ and

$$\lambda=\frac{1}{2L}\,|\alpha(H)|.$$

Proof. As proved in [8], p. 251, the area is given by

$$A(r)=\int_{\|Z\|=r}\det(A_Z)\,d\omega_r(Z), \tag{1}$$

where $d\omega_r$ is the Euclidean surface element of the sphere $\|Z\|=r$ in $\mathfrak{p}$, and

$$A_Z=\sum_0^\infty \frac{T_Z^n}{(2n+1)!},$$

where T_Z is the restriction of $(\operatorname{ad} Z)^2$ to $\mathfrak{p}$. The integrand in (1) is a radial function so

$$A(r)=\Omega_{p+q+1}\cdot r^{p+q}\cdot\det(A_{H_r}),\qquad \left(H_r=\frac{r}{L}\,H\right).$$

Since the nonzero eigenvalues of T_{H_r} are $(\tfrac{1}{2}\alpha(H_r))^2$ with multiplicity p and $\alpha(H_r)^2$ with multiplicity q we obtain

$$A(r)=\Omega_{p+q+1}\,r^{p+q}\left(\frac{\sin\lambda r}{\lambda r}\right)^p\left(\frac{\sin 2\lambda r}{2\lambda r}\right)^q.$$

where $\lambda=\tfrac{1}{2}L^{-1}|\alpha(H)|$.

§ 6. Points and antipodal manifolds in two-point homogeneous spaces

Let X be a compact two-point homogeneous space, or, what is the same thing (Wang [21]) a compact Riemannian globally symmetric space of rank one. We preserve the notation of the last section and assume $\dim X>1$. Let $G=I(X)$ and let Ξ be the set of all antipodal manifolds in X, with the differentiable structure induced by the transitive action of G. On Ξ we choose a Riemannian structure such that the diffeomorphism $\varphi: x\to A_x$ of X onto Ξ (see Prop. 5.5) is an isometry. Let Δ and $\hat{\Delta}$ denote the Laplace–Beltrami operators on X and Ξ, respectively. The measures μ and ν on the manifolds ξ and $\check{x}$ (§ 1) are defined to be those induced by the Riemannian structures of X and Ξ. If $x\in X$, then by Prop. 5.5

$$\check{x}=\{\varphi(y)\,|\,y\in\varphi(x)\}.$$

Consequently, if g is a continuous function on Ξ,

$$\check{g}(x)=\int_{\check{x}} g(\xi)\,d\nu(\xi)=\int_{y\in\varphi(x)} g(\varphi(y))\,d\nu(\varphi(y))=\int_{\varphi(x)}(g\circ\varphi)(y)\,d\mu(y),$$

12 – 652923. *Acta mathematica*. 113. Imprimé le 10 mai 1965.

so

$$\check{g} = (g \circ \varphi)\hat{} \circ \varphi. \tag{1}$$

Because of this correspondence between the integral transforms $f \to \hat{f}$ and $g \to \check{g}$ it suffices to consider the first.

Problems A, B, and C now have the following answer.

THEOREM 6.1.

(i) *The algebras* $\mathbf{D}(X)$ *and* $\mathbf{D}(\Xi)$ *are generated by* Δ *and* $\hat{\Delta}$ *respectively.*

(ii) *The mapping* $f \to \hat{f}$ *is a linear one-to-one mapping of* $C^\infty(X)$ *onto* $C^\infty(\Xi)$ *and*

$$(\Delta f)\hat{} = \hat{\Delta}\hat{f}.$$

(iii) *Except for the case when* X *is an even-dimensional real projective space,*

$$f = P(\Delta)((\hat{f})\check{}), \quad f \in C^\infty(X),$$

where P *is a polynomial, independent of* f, *explicitly given below.*

Proof. Part (i) is proved in [8], p. 270. Let $[M^r f](x)$ be the average of f over a sphere in X of radius r and center x. Then

$$\hat{f}(\varphi(x)) = c[M^L f](x), \tag{2}$$

where c is a constant. Since Δ commutes with the operator M^r ([8], Theorem 16, p. 276) we have

$$(\hat{\Delta}\hat{f}) \circ \varphi = \Delta(\hat{f} \circ \varphi) = cM^L \Delta f = (\Delta f)\hat{} \circ \varphi,$$

proving the formula in (ii). For (iii) we have to use the following complete list of compact Riemannian globally symmetric spaces of rank 1: The spheres $\mathbf{S}^n$, $(n=1, 2, \ldots)$, the real projective spaces $\mathbf{P}^n(\mathbf{R})$, $(n=2, 3, \ldots)$, the complex projective spaces $\mathbf{P}^n(\mathbf{C})$, $(n=4, 6, \ldots)$, the quaternion projective spaces $\mathbf{P}^n(\mathbf{H})$, $(n=8, 12, \ldots)$ and the Cayley projective plane $\mathbf{P}^{16}(\mathbf{Cay})$. The superscripts denote the real dimension. The corresponding antipodal manifolds are also known ([2], pp. 437–467, [15], pp. 35 and 52) and are in the respective cases: A point, $\mathbf{P}^{n-1}(\mathbf{R})$, $\mathbf{P}^{n-2}(\mathbf{C})$, $\mathbf{P}^{n-4}(\mathbf{H})$, and $\mathbf{S}^8$. For the lowest dimensions, note that $\mathbf{P}^1(\mathbf{R}) = \mathbf{S}^1$, $\mathbf{P}^2(\mathbf{C}) = \mathbf{S}^2$, $\mathbf{P}^4(\mathbf{H}) = \mathbf{S}^4$. Let $A_1(r)$ denote the area of a sphere of radius r in an antipodal manifold in X. Then by Prop. 5.6,

$$A_1(r) = C_1 \sin^{p_1}(\lambda_1 r) \sin^{q_1}(2\lambda_1 r),$$

where C_1 is a constant and p_1, q_1, λ_1 are the numbers p, q, λ for the antipodal manifold.

The multiplicities p and q are determined in Cartan [2], and show that $\frac{1}{2}\alpha$ is a restricted root unless X is a sphere or a real projective space. Ignoring these exceptions we have by virtue of the results of § 5:

L = diameter X = diameter A_x
= distance of 0 to the nearest conjugate point in X_0
= smallest number $M>0$ such that $\lim_{r\to M} A(r)=0$.

We can now derive the following list:

$X=\mathbf{S}^n$: $p=0,\ q=n-1,\ \lambda=\pi/2L,\ A(r)=C\sin^{n-1}(2\lambda r),\ A_1(r)\equiv 0$.

$X=\mathbf{P}^n(\mathbf{R})$: $p=0,\ q=n-1,\ \lambda=\pi/4L,\ A(r)=C\sin^{n-1}(2\lambda r),\ A_1(r)=C_1\sin^{n-2}(2\lambda r)$.

$X=\mathbf{P}^n(\mathbf{C})$: $p=n-2,\ q=1,\ \lambda=\pi/2L,\ A(r)=C\sin^{n-2}(\lambda r)\sin(2\lambda r),\ A_1(r)=C_1\sin^{n-4}(\lambda r)\sin(2\lambda r)$.

$X=\mathbf{P}^n(\mathbf{H})$: $p=n-4,\quad q=3,\quad \lambda=\pi/2L,\quad A(r)=C\sin^{n-4}(\lambda r)\sin^3(2\lambda r)$,
$A_1(r)=C_1\sin^{n-8}(\lambda r)\sin^3(2\lambda r)$.

$X=\mathbf{P}^{16}(\mathbf{Cay})$: $p=8,\ q=7,\ \lambda=\pi/2L,\ A(r)=C\sin^8(\lambda r)\sin^7(2\lambda r),\ A_1(r)=C_1\sin^7(2\lambda r)$.

In each case, C and C_1 are constants, not necessarily the same for all cases. Now if $x\in X$ and $f\in C^\infty(X)$ let $[If](x)$ denote the average of the integrals of f over the antipodal manifolds which pass through x. Then $(\hat{f})^\vee$ is a constant multiple of If. Fix a point $o\in X$ and let K be the subgroup of G leaving o fixed. Let ξ_o be a fixed antipodal manifold through o and let $d\sigma$ be the volume element on ξ_o. Then

$$[If](g\cdot o)=\int_K\left(\int_{\xi_o} f(gk\cdot y)\,d\sigma(y)\right)dk=\int_{\xi_o}[M^r f](g\cdot o)\,d\sigma(y),$$

where r is the distance $d(o,y)$ in the space X between the points o and y. Now if $d(o,y)<L$ there is a unique geodesic in X of length $d(o,y)$ joining o to y and since ξ_o is totally geodesic, $d(o,y)$ is also the distance between o and y in ξ_o. Hence, using geodesic polar coordinates in the last integral we find

$$[If](x)=\int_0^L A_1(r)\,[M^r f](x)\,dr. \tag{3}$$

In geodesic polar coordinates on X, the Laplace–Beltrami operator Δ equals $\Delta_r+\Delta'$ where Δ' is the Laplace-Beltrami operator on the sphere in X of radius r and ([10], p. 445)

$$\Delta_r=\frac{d^2}{dr^2}+\frac{1}{A(r)}\frac{dA}{dr}\frac{r}{dr}\quad (0<r<L).$$

The function $(x,r)\to[M^r f](x)$ satisfies

$$\Delta M^r f=\Delta_r(M^r f) \tag{4}$$

([8], p, 279 or [6]). Using Prop. 5,6, we have

$$\Delta_r = \frac{\partial^2}{\partial r^2} + \lambda(p \cot(\lambda r) + 2q \cot(2\lambda r)) \frac{\partial}{\partial r} \quad (0 < r < L) \tag{5}$$

(compare also [7], p. 302). Now (iii) can be proved on the basis of (3) (4) (5) by the method in [8], p. 285, where the case $\mathbf{P}^n(\mathbf{R})$ (n odd) is settled. The case $X = \mathbf{S}^n$ being trivial we shall indicate the details for $X = \mathbf{P}^n(\mathbf{C})$, $\mathbf{P}^n(\mathbf{H})$ and $\mathbf{P}^{16}(\mathbf{Cay})$.

LEMMA 6.2. *Let $X = \mathbf{P}^n(\mathbf{C})$, $f \in C^\infty(X)$. If m is an even integer, $0 \leqslant m \leqslant n-4$ then*

$$(\Delta - \lambda^2(n-m-2)(m+2)) \int_0^L \sin^m(\lambda r) \sin(2\lambda r) [M^r f](x)\, dr$$

$$= -\lambda^2(n-m-2)\, m \int_0^L \sin^{m-2}(\lambda r) \sin(2\lambda r) [M^r f](x)\, dr.$$

For $m = 0$ the right-hand side should be replaced by

$$-2\lambda(n-2) f(x).$$

LEMMA 6.3. *Let $X = \mathbf{P}^n(\mathbf{H})$, $f \in C^\infty(X)$. Let m be an even integer, $0 < m \leqslant n-8$. Then*

$$(\Delta - \lambda^2(n-m-4)(m+6)) \int_0^L \sin^m(\lambda r) \sin^3(2\lambda r) [M^r f](x)\, dr$$

$$= -\lambda^2(n-m-4)(m+2) \int_0^L \sin^{m-2}(\lambda r) \sin^3(2\lambda r) [M^r f](x)\, dr.$$

Also

$$(\Delta - 4\lambda^2(n-4))(\Delta - 4\lambda^2(n-2)) \int_0^L \sin^3(2\lambda r) [M^r f](x)\, dr = 16\lambda^3(n-2)(n-4) f(x).$$

LEMMA 6.4. *Let $X = \mathbf{P}^{16}(\mathbf{Cay})$, $f \in C^\infty(X)$. Let $m > 1$ be an integer. Then*

$$(\Delta - 4\lambda^2 m(11-m)) \int_0^L \sin^m(2\lambda r) [M^r f](x)\, dr$$

$$= -32\lambda^2(m-1) \int_0^L \sin^{m-2}(2\lambda r) \cos^2(\lambda r) [M^r f](x)$$

$$+ 4\lambda^2(m-1)(m-7) \int_0^L \sin^{m-2}(2\lambda r) [M^r f](x)\, dr;$$

$$(\Delta - 4\lambda^2(m+1)(10-m)) \int_0^L \sin^m(2\lambda r) \cos^2(\lambda r) [M^r f](x)\, dr$$

$$= 4\lambda^2(3m-5) \int_0^L \sin^m(2\lambda r) [M^r f](x)\, dr$$

$$+ 4\lambda^2(m-1)(m-15) \int_0^L \sin^{m-2}(2\lambda r) \cos^2(\lambda r) [M^r f](x)\, dr.$$

Iteration of these lemmas gives part (iii) of Theorem 6.1 where the polynomial $P(\Delta)$ has degree equal to one half the dimension of the antipodal manifold and is a constant multiple of

$$
\begin{array}{ll}
1 \text{ (the identity)}, & X = \mathbf{S}^n \\
(\Delta - \varkappa(n-2)1)(\Delta - \varkappa(n-4)3)\ldots(\Delta - \varkappa 1(n-2)), & X = \mathbf{P}^n(\mathbf{R}) \\
(\Delta - \varkappa(n-2)2)(\Delta - \varkappa(n-4)3)\ldots(\Delta - \varkappa 2(n-2)), & X = \mathbf{P}^n(\mathbf{C}) \\
[(\Delta - \varkappa(n-2)4)(\Delta - \varkappa(n-4)6)\ldots.(\Delta - \varkappa 8(n-6))][(\Delta - \varkappa 4(n-4))(\Delta - \varkappa 4(n-2))], & X = \mathbf{P}^n(\mathbf{H}) \\
(\Delta - 112\varkappa)^2(\Delta - 120\varkappa)^2, & X = \mathbf{P}^{16}(\mathbf{Cay}).
\end{array}
$$

In each case $\varkappa = (\pi/2L)^2$.

Finally, we prove part (ii). From (1) and (2) we derive

$$M^L M^L f = c^{-2}(\hat{f})^{\vee}$$

so, if X is not an even-dimensional projective space, f is a constant multiple of $M^L P(\Delta) M^L f$ which shows that $f \to \hat{f}$ is one-to-one and onto. For the even-dimensional projective space a formula relating f and $(\hat{f})^{\vee}$ is given by Semyanistyi [20]. In particular, the mapping $f \to \hat{f}$ is one-to-one. To see that it is onto, let (φ_n) be the eigenfunctions of Δ. Then each φ_n is an eigenfunction of M^L ([10], Theorem 7.2, Ch. X). Since the eigenvalue is $\neq 0$ by the above it is clear that no measure on X can annihilate all of $M^L(C^\infty(X))$. This finishes the proof of Theorem 6.1.

Added in proof. Theorem 6.1 shows that $\hat{f} =$ constant implies $f =$ constant. For $\mathbf{P}^n(\mathbf{R})$ we thus obtain a (probably known) corollary.

Corollary. Let B be an open set in $\mathbf{R}^{n+1}$, symmetric and starshaped with respect to 0, bounded by a hypersurface. Assume area $(B \cap P) =$ constant for all hyperplanes P through 0. Then B is an open ball.

§ 7. Differential operators on the space of p-planes

Let p and n be two integers such that $0 \leqslant p < n$. A *p-plane* E_p in $\mathbf{R}^n$ is by definition a translate of a p-dimensional vector subspace of $\mathbf{R}^n$. The 0-planes are just the points of $\mathbf{R}^n$. The p-planes in $\mathbf{R}^n$ form a manifold $\mathbf{G}(p, n)$ on which the group $\mathbf{M}(n)$ of all isometries of $\mathbf{R}^n$ acts transitively. Let $\mathbf{O}(k)$ denote the orthogonal group in $\mathbf{R}^k$ and let $\mathbf{G}_{p,n}$ denote the manifold $\mathbf{O}(n)/\mathbf{O}(p) \times \mathbf{O}(n-p)$ of p-dimensional subspaces of $\mathbf{R}^n$. The manifold $\mathbf{G}(p, n)$ is a fibre bundle with base space $\mathbf{G}_{p,n}$, the projection π of $\mathbf{G}(p, n)$ onto $\mathbf{G}_{p,n}$ being the mapping which to any p-plane $E_p \in \mathbf{G}(p, n)$ associates the parallel p-plane through the origin. Thus the fibre of this bundle $(\mathbf{G}(p, n), \mathbf{G}_{p,n}, \pi)$ is $\mathbf{R}^{n-p}$. If F denotes an arbitrary fibre and $f \in C^\infty(\mathbf{G}(p, n))$ then the restriction of f to F will be denoted $f|F$. Consider now the linear transformation $\square_p$ of $C^\infty(\mathbf{G}(p, n))$ given by

$$(\square_p f)|F = \Delta_F(f|F), \quad f \in C^\infty(\mathbf{G}(p, n)),$$

for each fibre F, Δ_F denoting the Laplacian on F. It is clear that $\square_p$ is a differential operator on $\mathbf{G}(p, n)$. For simplicity we usually write $\square$ instead of $\square_p$.

LEMMA 7.1.

(i) *The operator $\square_p$ is invariant under the action of $\mathbf{M}(n)$ on $\mathbf{G}(p, n)$.*

(ii) *Each differential operator on $\mathbf{G}(p, n)$ which is invariant under $\mathbf{M}(n)$ is a polynomial in $\square_p$.**

Proof. We recall that if φ is an isometry of a Riemannian manifold M_1 onto a Riemannian manifold M_2 and if Δ_1, Δ_2 are the corresponding Laplace–Beltrami operators then (cf. [10], p. 387)

$$(\Delta_1 f^{\varphi^{-1}}) = \Delta_2 f, \quad f \in C^\infty(M_2). \tag{1}$$

Now each isometry $g \in \mathbf{M}(n)$ induces a fibre-preserving diffeomorphism of $\mathbf{G}(p, n)$, preserving the metric on the fibres. Let $f \in C^\infty(\mathbf{G}(p, n))$ and F any fibre. Writing for simplicity $\square$ instead of $\square_p$ we get from (1)

$$(\square^g f)|F = (\square f^{g^{-1}})^g|F = ((\square f^{g^{-1}})|g^{-1}\cdot F)^g = (\Delta_{g^{-1}F}(f^{g^{-1}}|g^{-1}F))^g = \Delta_F(f|F) = (\square f)|F,$$

so $\square^g = \square$, proving (i).

Let E_p^o be a fixed p-plane in $\mathbf{R}^n$, say the one spanned by the p first unit coordinate vectors, $Z_1, \ldots, Z_p$. The subgroup of $\mathbf{M}(n)$ which leaves E_p^o invariant can be identified with the product group $\mathbf{M}(p) \times \mathbf{O}(n-p)$. For simplicity we put $G = \mathbf{M}(n)$, $H = \mathbf{M}(p) \times \mathbf{O}(n-p)$ and let $\mathfrak{G}$ and $\mathfrak{h}$ denote the corresponding Lie algebras. If $\mathfrak{M}$ is any subspace of $\mathfrak{G}$ such that $\mathfrak{G} = \mathfrak{M} + \mathfrak{h}$ (direct sum) and $\mathrm{Ad}_G(h)\mathfrak{M} \subset \mathfrak{M}$ for each $h \in H$ then we know from [8] Theorem 10 that the G-invariant differential operators on the space $G/H = \mathbf{G}(p, n)$ are directly given by the polynomials on $\mathfrak{M}$ which are invariant under the group $\mathrm{Ad}_G(H)$. Let $\mathfrak{o}(k)$ denote the Lie algebra of $\mathbf{O}(k)$. Then $\mathfrak{G}$ is the vector space direct sum of $\mathfrak{o}(n)$ and the abelian Lie algebra $\mathbf{R}^n$. Also if $T \in \mathfrak{o}(n)$, $X \in \mathbf{R}^n$ then the bracket $[T, X]$ in $\mathfrak{G}$ is $[T, X] = T\cdot X$ (the image of X under the linear transformation T). The Lie algebra $\mathfrak{h}$ is the vector space direct sum of $\mathfrak{o}(p)$, $\mathfrak{o}(n-p)$ and $\mathbf{R}^p (= E_p^o)$; we write this in matrix-vector form

$$\mathfrak{h} = \left\{ \begin{pmatrix} A & 0 \\ 0 & B \end{pmatrix} + \begin{pmatrix} V \\ 0 \end{pmatrix} \middle| A \in \mathfrak{o}(p), B \in \mathfrak{o}(n-q), V \in E_p^o \right\}.$$

For $\mathfrak{M}$ we choose the subspace

*See *[59] (from "The Selected Works Bibliography", p. xxxix) for a correction.

$$\mathfrak{M}=\left\{\begin{pmatrix}0 & X\\ -{}^tX & 0\end{pmatrix}+\begin{pmatrix}0\\ Z\end{pmatrix}\,\middle|\,\begin{matrix}X \text{ any } p\times(n-p) \text{ matrix, } {}^tX\\ \text{its transpose, } Z\in\mathbf{R}^{n-p}\end{matrix}\right\}.$$

Then it is clear that $\mathfrak{G}=\mathfrak{h}+\mathfrak{M}$, Let $a\in\mathbf{O}(p)$, $b\in\mathbf{O}(n-p)$, $V\in E_p^o$. Then

$$\operatorname{Ad}_G\begin{pmatrix}a & 0\\ 0 & b\end{pmatrix}\cdot\left[\begin{pmatrix}0 & X\\ -{}^tX & 0\end{pmatrix}+\begin{pmatrix}0\\ Z\end{pmatrix}\right]=\begin{pmatrix}a & 0\\ 0 & b\end{pmatrix}\begin{pmatrix}0 & X\\ -{}^tX & 0\end{pmatrix}\begin{pmatrix}a^{-1} & 0\\ 0 & b^{-1}\end{pmatrix}+\begin{pmatrix}a & 0\\ 0 & b\end{pmatrix}\begin{pmatrix}0\\ Z\end{pmatrix}$$

$$=\begin{pmatrix}0 & aXb^{-1}\\ -b\,{}^tXa & 0\end{pmatrix}+\begin{pmatrix}0\\ bZ\end{pmatrix}. \tag{2}$$

$$\operatorname{Ad}_G\begin{pmatrix}V\\ 0\end{pmatrix}\cdot\left[\begin{pmatrix}0 & X\\ -{}^tX & 0\end{pmatrix}+\begin{pmatrix}0\\ Z\end{pmatrix}\right]=\begin{pmatrix}0 & X\\ -{}^tX & 0\end{pmatrix}+\begin{pmatrix}0\\ Z+{}^tXV\end{pmatrix}. \tag{3}$$

It follows immediately that $\operatorname{Ad}_G(h)\mathfrak{M}\subset\mathfrak{M}$ for all $h\in H$. Now let as usual E_{ij} denote the matrix $(\delta_{ai}\delta_{bj})_{1\leqslant a,\,b\leqslant n}$, put $X_{ij}=E_{ip+j}-E_{p+ji}$ $(1\leqslant i\leqslant p,\ 1\leqslant j\leqslant n-p)$ and let $Z_k\,(p+1\leqslant k\leqslant n)$ denote the kth coordinate vector in $\mathbf{R}^n$. Then $\{X_{ij}, Z_k\}$ is a basis of $\mathfrak{M}$. Any element q in the symmetric algebra $S(\mathfrak{M})$ over $\mathfrak{M}$ can be written as a finite sum

$$q(X_{11},\dots,X_{pn-p},Z_{p+1},\dots,Z_n)=\sum_i r_i(Z_{p+1},\dots,Z_n)\,s_i(X_{11},\dots,X_{pn-p}),$$

where the r_i and s_i are polynomials. Suppose q is homogeneous of degree m (say) and invariant under $\operatorname{Ad}_G(H)$. From (2) and (3) for $X=0$ we see that a polynomial in $Z_{p+1},\dots,Z_n$ is invariant under $\operatorname{Ad}_G(H)$ if and only if it is a polynomial in $|Z|^2=Z_{p+1}^2+\dots+Z_n^2$. Hence the invariant polynomial q can be written

$$q=\sum_{r=0}^{[\frac{1}{2}m]}|Z|^{2r}q_r(X_{11},\dots,X_{pn-p}), \tag{4}$$

where q_r is homogeneous of degree $m-2r$. Now, by (3), q is invariant under the substitution $T(v)$: $X_{ij}\to X_{ij}+v_iZ_{p+j}$ ($v_1,\dots,v_p$ being any real numbers, and $1\leqslant i\leqslant p$, $1\leqslant j\leqslant n-p$). We can write

$$q_r(X_{11}+v_1Z_{p+1},\dots,X_{pn-p}+v_pZ_n)=\sum_{(s)}a_{r,s_1,\dots,s_p}\otimes v_1^{s_1}\dots v_p^{s_p},$$

where $\otimes$ denotes the tensor product (over $\mathbf{R}$) of the polynomial rings $\mathbf{R}[X_{11},\dots,Z_n]$ and $\mathbf{R}[v_1,\dots,v_p]$. Using (4) and the invariance of q we obtain

$$\sum_{r,(s)}|Z|^{2r}a_{r,s_1,\dots,s_p}\otimes v_1^{s_1}\dots v_p^{s_p}=\sum_r a_{r,0,\dots,0}.$$

It follows that
$$\sum_r|Z|^{2r}a_{r,s_1,\dots,s_p}=0 \quad \text{if } s_1+\dots+s_p>0, \tag{5}$$

and since $a_{r,s_1,\dots,s_p}$ has degree $s_1+\dots+s_p$ in the Z_i (5) implies $a_{r,s_1,\dots,s_p}=0$ for $s_1+\dots+s_p>0$, whence each q_r is invariant under the substitution $T(v)$ above. This implies easily that each

q_r is a constant. Thus the elements $q \in S(\mathfrak{M})$ invariant under $\mathrm{Ad}_G(H)$ are the polynomials in $|Z|^2$. By [8], Theorem 10, the polynomial $|Z|^2$ induces a G-invariant differential operator D on G/H such that for each $f \in C^\infty(G/H)$,

$$[Df]\,(E_p^o) = \left\{\left(\frac{\partial^2}{\partial t_{p+1}^2} + \ldots + \frac{\partial^2}{\partial t_n^2}\right) f((t_{p+1} Z_{p+1} + \ldots + t_n Z_n) \cdot E_p^o)\right\}_{t=0}. \tag{6}$$

Thus $[Df](E_p^o) = [\Box f](E_p^o)$ and since D and $\Box$ are both G-invariant, $D = \Box$. Now (ii) follows from [8], Cor. p. 269.

§ 8. p-planes and q-planes in $\mathbf{R}^{p+q+1}$

The notation being as in the preceeding section put $q = n - p - 1$. Let $\mathbf{G}^*(p, n)$ and $\mathbf{G}^*(q, n)$, respectively, denote the sets of p-planes and q-planes in $\mathbf{R}^n$ not passing through the origin. The projective duality between points and hyperplanes in $\mathbf{R}^n$, realized by the polarity with respect to the unit sphere $\mathbf{S}^{n-1}$ generalizes to a duality between $\mathbf{G}^*(p, n)$ and $\mathbf{G}^*(q, n)$. In fact, if $a \neq 0$ in $\mathbf{R}^n$, let $E_{n-1}(a)$ denote the polar hyperplane. If a runs through a p-plane $E_p \in \mathbf{G}^*(p, n)$ then the hyperplanes $E_{n-1}(a)$ intersect in a unique q-plane $E_q \in \mathbf{G}^*(q, n)$ and the mapping $E_p \to E_q$ is the stated duality.

We have now an example of the framework in § 1. Let $X = \mathbf{G}(p, n)$, put $G = \mathbf{M}(n)$, acting on X. Given a q-plane E_q consider the family $\xi = \xi(E_q)$ of p-planes intersecting E_q. If $E_q' \neq E_q''$ then $\xi(E_q') \neq \xi(E_q'')$; thus the set of all families ξ—the dual space Ξ—can be identified with $\mathbf{G}(q, n)$. In accordance with this identification, if $E_p = x \in X$ then $\check{x} = \check{x}(E_p)$ is the set of q-planes intersecting x. Because of convergence difficulties we do not define the measures μ and ν (§ 1) directly but if f is any function on $\mathbf{G}(p, n)$ we put

$$\hat{f}(E_q) = \int_{E_q} \left(\int_{a \in E_p} f(E_p)\, d\sigma_p(E_p) \right) d\mu_q(a),$$

whenever these integrals exist. Here $d\sigma_p$ is the invariant measure on the Grassmann manifold of p-planes through a with total measure 1, $d\mu_q$ is the Euclidean measure on E_q. The transform $g \to \check{g}$ is defined by interchanging p and q in the definition of $\hat{f}$. It is convenient to consider the operators M_p and L_q defined by

$$[M_p f]\,(a) = \int_{a \in E_p} f(E_p)\, d\sigma_p(E_p), \quad f \in C^\infty(\mathbf{G}(p,n)) \tag{1}$$

$$[L_q F]\,(E_q) = \int_{E_q} F(a)\, d\mu_q(a), \qquad F \in \mathcal{S}(\mathbf{R}^n). \tag{2}$$

Then we have, formally, $\hat{f} = L_q M_p f$.

LEMMA 8.1.

(i) *M_p maps $C^\infty(\mathbf{G}(p,n))$ into $C^\infty(\mathbf{R}^n)$ and $M_p\Box_p = \Delta M_p$.*

(ii) *L_q maps $\mathcal{S}(\mathbf{R}^n)$ into $C^\infty(\mathbf{G}(q,n))$ and $L_q\Delta = \Box_q L_q$.*

Proof. (i) Put $K = \mathbf{O}(n) \subset \mathbf{M}(n) = G$. For $f \in C^\infty(\mathbf{G}(p,n))$ let $f^* \in C^\infty(G)$ be determined by $f^*(g) = f(g \cdot E_p^o)$, $(g \in G)$. Then for a suitably normalized Haar measure dk on K we have

$$\int_K f^*(gk)\,dk = [M_p f]\,(g\cdot 0),$$

which shows that $M_p f \in C^\infty(\mathbf{R}^n)$.

For each $X \in \mathfrak{G}$, let $\tilde{X}$ denote the left invariant vector field on G satisfying $\tilde{X}_e = X$. Since $\mathbf{R}^n \subset \mathfrak{G}$ we can consider the left invariant differential operator $\tilde{\Delta} = \sum_{i=1}^n \tilde{Z}_i^2$ on G. If $k \in K$, $\mathrm{Ad}_G(k)$ leaves the subspace $\mathbf{R}^n \subset \mathfrak{G}$ and the polynomial $\sum_{i=1}^n Z_i^2$ invariant. Hence, if $R(k)$ denotes the right translation $g \to gk$ on G,

$$(\tilde{\Delta})^{R(k)} = \sum_{i=1}^n ((\tilde{Z}_i)^{R(k)})^2 = \sum_{i=1}^n ((\mathrm{Ad}_G(k^{-1})\,Z_i)^\sim)^2 = \sum_{i=1}^n \tilde{Z}_i^2$$

so $\tilde{\Delta}$ is invariant under $R(k)$. If $F \in C^\infty(\mathbf{R}^n)$ let $\tilde{F} \in C^\infty(G)$ be determined by $\tilde{F}(g) = F(g\cdot 0)$ for $g \in G$. Then (cf. [10], p. 392, equation (16))

$$\begin{aligned}[\tilde{\Delta}\tilde{F}]\,(g) &= \left\{\frac{\partial^2}{\partial t_1^2} + \dots + \frac{\partial^2}{\partial t_n^2}\,\tilde{F}(g\exp(t_1 Z_1 + \dots + t_n Z_n))\right\}_{t=0}\\ &= \left\{\frac{\partial^2}{\partial t_1^2} + \dots + \frac{\partial^2}{\partial t_n^2}\,F(g\cdot(t_1 Z_1 + \dots + t_n Z_n))\right\}_{t=0}\\ &= [\Delta F^{g^{-1}}]\,(0) = [\Delta F]\,(g\cdot 0)\end{aligned}$$

by (1) § 7, that is

$$\tilde{\Delta}\tilde{F} = (\Delta F)^\sim, \quad F \in C^\infty(\mathbf{R}^n). \tag{3}$$

Since

$$(M_p f)^\sim = \int_K (f^*)^{R(k)} dk$$

and $(\tilde{\Delta})^{R(k)} = \tilde{\Delta}$ it follows from (3) that

$$(\Delta M_p f)^\sim = \int_K (\tilde{\Delta} f^*)^{R(k)} dk$$

so

$$\begin{aligned}[\Delta M_p f]\,(g\cdot 0) &= \int_K \left\{\left(\frac{\partial^2}{\partial t_1^2} + \dots + \frac{\partial^2}{\partial t_n^2}\right)(f^*(gk\exp(t_1 Z_1 + \dots + t_n Z_n)))\right\}_{t=0} dk\\ &= \int_K \left\{\left(\frac{\partial^2}{\partial t_{p+1}^2} + \dots + \frac{\partial^2}{\partial t_n^2}\right) f(gk\exp(t_{p+1} Z_{p+1} + \dots + t_n Z_n)\cdot E_p^o)\right\}_{t=0} dk.\end{aligned}$$

This shows that

$$[\Delta M_p f](g\cdot 0)=\int_K [\Box_p f](gk\cdot E_p^o)\,dk=\int_K (\Box_p f)^*(gk)\,dk=[M_p\Box_p f](g\cdot 0)$$

proving (i). For (ii) let V_q denote the q-plane through 0, parallel to E_q, and let $X_1, \ldots, X_q, \ldots, X_n$ be an orthogonal basis of $\mathbf{R}^n$ such that $X_i \in V_q$ $(1 \leqslant i \leqslant q)$. The orthogonal projection of 0 onto E_q has the form $s_{q+1}X_{q+1}+\ldots+s_nX_n$ and

$$[L_q F](E_q)=\int F(t_1X_1+\ldots+t_qX_q+\ldots+s_nX_n)\,dt_1\ldots dt_q$$

so

$$[\Box_q L_q F](E_q)=\left\{\frac{\partial^2}{\partial t_{q+1}^2}+\ldots+\frac{\partial^2}{\partial t_n^2}(L_q F((t_{q+1}X_{q+1}+\ldots+t_nX_n)\cdot E_q))\right\}_{t=0}$$

$$=\int\left(\frac{\partial^2}{\partial s_{q+1}^2}+\ldots+\frac{\partial^2}{\partial s_n^2}\right)(F(t_1X_1+\ldots+s_nX_n)\,dt_1\ldots dt_q=\int_{E_q}[\Delta F](x)\,d\mu_q(x)$$

since $\partial^2F/\partial t_i^2$ $(1\leqslant i\leqslant q)$ gives no contribution. This proves (ii).

Let $\mathcal{S}^*(\mathbf{R}^n)$ be as in § 4 and let $\mathcal{L}_{p,n}$ be the subspace $L_p(\mathcal{S}^*(\mathbf{R}^n))$ of $C^\infty(\mathbf{G}(p,n))$.

THEOREM 8.2. *Suppose n odd. The transform $f\to\hat{f}$ is a linear one-to-one mapping of $\mathcal{L}(\mathbf{G}(p,n))$ onto $\mathcal{L}(\mathbf{G}(q,n))$ such that*

$$(\Box_p f)^\wedge=\Box_q\hat{f}$$

$$(\Box_p)^{\frac{1}{2}(n-1)}(\hat{f})^\vee=cf,\quad f\in\mathcal{L}(\mathbf{G}(p,n)),$$

where c is a constant $\neq 0$, independent of f.

Proof. Let $r=(x_1^2+\ldots+x_n^2)^{1/2}$ and λ a complex number whose real part Re λ is $>-n$. Then the function r^λ is a tempered distribution on $\mathbf{R}^n$ and so is its Fourier transform, say R_λ. If $\varphi\in\mathcal{S}(\mathbf{R}^n)$ the convolution $R_\lambda*\varphi$ is a tempered distribution ([18], II, p. 102) whose Fourier transform is the product of the Fourier transforms of φ and R_λ. If $\varphi\in\mathcal{S}^*(\mathbf{R}^n)$ then this product lies in $\mathcal{S}_0(\mathbf{R}^n)$ so the operator Λ_λ: $\varphi\to R_\lambda*\varphi$ maps the space $\mathcal{S}^*(\mathbf{R}^n)$ into itself. Also if λ,μ are complex numbers such that Re λ, Re μ and Re$(\lambda+\mu)$ all are $>-n$ then $\Lambda_{\lambda+\mu}=\Lambda_\lambda\Lambda_\mu$. In particular, $(\Lambda_2\varphi)^\sim=(2\pi)^n r^2\tilde{\varphi}=-(2\pi)^n(\Delta\varphi)^\sim$ so

$$\Lambda_2=-(2\pi)^n\Delta,\quad \Lambda_0=I.$$

We shall now verify that

$$M_dL_dF=\gamma_dR_{-d}*F,\quad F\in\mathcal{S}(\mathbf{R}^n),\ 0\leqslant d<n, \tag{4}$$

where d is an integer and γ_d is a constant $\neq 0$. For this let $d\omega_k$ be the surface element of the unit sphere in $\mathbf{R}^k$ and put $\Omega_k=\int d\omega_k$. Let $g\in G$ and $x=g\cdot 0$. If $d=0$, (4) is obvious so assume $0<d<n$. Then for a fixed d-plane E_d through 0

$$[M_d L_d F](x) = \int_K dk \int_{E_d} F(gk\cdot z)\,dz = \int_{E_d} dz \int_K F(gk\cdot z)\,dk$$

$$= \int_0^\infty \Omega_d r^{d-1}dr \left\{\frac{1}{\Omega_n}\int_{|y|=1} F(x+ry)\,d\omega_{n-1}(y)\right\} = \frac{\Omega_d}{\Omega_n}\int F(y)\,|x-y|^{d-n}\,dy$$

and since R_{-d} is a constant multiple of r^{d-n} ([18], II, p. 113) (4) follows. As an immediate consequence of (4) we have

$$\Lambda_d M_d L_d \varphi = M_d L_d \Lambda_d \varphi = \gamma_d \varphi, \quad \varphi \in S^*(\mathbf{R}^n), \quad (0 \leqslant d < n). \tag{5}$$

Now let $f \in \mathcal{L}_{p,n}$. Then $f = L_p\varphi$ for $\varphi \in S^*(\mathbf{R}^n)$ and $\hat{f} = L_q M_p f = L_q M_p L_p \varphi \in \mathcal{L}_{q,n}$ since $M_p L_p \varphi \in S^*(\mathbf{R}^n)$. If $\hat{f} = 0$ then $0 = M_q \hat{f} = M_q L_q M_p L_p \varphi = \Lambda_{-q-p}\varphi$ so $f = 0$. Similarly, if $F \in \mathcal{L}_{q,n}$ then $F = L_q \Phi$ for $\Phi \in S^*(\mathbf{R}^n)$ and by (5), $F = L_q M_p L_p \varphi$ for $\varphi \in S^*(\mathbf{R}^n)$ so $F = (L_p\varphi)^\wedge$. This shows that $f \to \hat{f}$ is an isomorphism of $\mathcal{L}_{p,n}$ onto $\mathcal{L}_{q,n}$. Also, by Lemma 8.1,

$$(\square_p f)^\wedge = L_q M_p \square_p f = L_q \Delta M_p f = \square_q L_q M_p f = \square_q \hat{f}.$$

Since $p+q=n-1$ is even we have

$$\Lambda_p \Lambda_q = (\Lambda_2)^{\frac{1}{2}(n-1)} = ((-2\pi)^n)^{\frac{1}{2}(n-1)}\,\Delta^{\frac{1}{2}(n-1)} = c_n \Delta^{\frac{1}{2}(n-1)}$$

the last equation defining c_n. Let $f \in \mathcal{L}_{p,n}$, $f = L_p\varphi$, $\varphi \in S^*(\mathbf{R}^n)$. Then, using Lemma 8.1, and (5)

$$(\hat{f})^\vee = L_p M_q \hat{f} = L_p M_q L_q M_p L_p \varphi,$$

$$(\square_p)^{\frac{1}{2}(n-1)}(\hat{f})^\vee = L_p \Delta^{\frac{1}{2}(n-1)} M_q \hat{f} = c_n^{-1} L_p \Lambda_p \Lambda_q M_q L_q M_p L_p \varphi$$

$$= c_n^{-1}\gamma_q L_p \Lambda_p M_p L_p \varphi = c_n^{-1}\gamma_q \gamma_p L_p \varphi = c_n^{-1}\gamma_p\gamma_q f.$$

References

[1]. Ambrose, W., The Cartan structure equations in classical Riemannian geometry. *J. Indian Math. Soc.*, 24 (1960), 23–76.

[2]. Cartan, É., Sur certaines formes riemanniennes remarquables des géométries a groupe fondamental simple. *Ann. Sci. École Norm. Sup.*, 44 (1927), 345–467.

[3]. Friedman, A., *Generalized functions and partial differential equations.* Prentice Hall, N.J. 1963.

[4]. Fuglede, B., An integral formula. *Math. Scand.*, 6 (1958), 207–212.

[5]. Gelfand, I. M., Graev, M. I. & Vilenkin, N., *Integral Geometry and its relation to problems in the theory of Group Representations.* Generalized Functions, Vol. 5, Moscow 1962.

[6]. Günther, P., Über einige spezielle Probleme aus der Theorie der linearen partiellen Differentialgleichungen 2. Ordnung. *Ber. Verh. Sächs. Akad. Wiss. Leipzig, Math.-Nat. Kl.*, 102 (1957), 1–50.

[7]. Harish-Chandra, Spherical functions on a semisimple Lie group I. *Amer. J. Math.*, 80 (1958), 241–310.

[8]. HELGASON, S., Differential operators on homogeneous spaces. *Acta Math.*, 102 (1959), 239–299.
[9]. ——, Some remarks on the exponential mapping for an affine connection. *Math. Scand.*, 9 (1961), 129–146.
[10]. ——, *Differential Geometry and Symmetric Spaces*. Academic Press, New York, 1962.
[11]. ——, A duality in integral geometry; some generalizations of the Radon transform. *Bull. Amer. Math. Soc.*, 70 (1964), 435–446.
[12]. HÖRMANDER, L., On the theory of general partial differential operators. *Acta Math.*, 94 (1955), 161–248.
[13]. JOHN, F., Bestimmung einer Funktion aus ihren Integralen über gewisse Mannigfaltigkeiten. *Math. Ann.*, 100 (1934), 488–520.
[14]. ——, *Plane waves and spherical means, applied to partial differential equations*. Interscience, New York, 1955.
[15]. NAGANO, T., Homogeneous sphere bundles and the isotropic Riemannian manifolds. *Nagoya Math. J.*, 15 (1959), 29–55.
[16]. RADON, J., Über die Bestimmung von Funktionen durch ihre Integralwerte längs gewisser Mannigfaltigkeiten. *Ber. Verh. Sächs. Akad. Wiss. Leipzig, Math.-Nat. Kl.*, 69 (1917), 262–277.
[17]. DE RHAM, Sur la réductibilité d'un espace de Riemann. *Comment. Math. Helv.*, 26 (1952), 328–344.
[18]. SCHWARTZ, L., *Théorie des Distributions*, I, II. Hermann et Cie, 1950, 1951.
[19]. SEMYANISTYI, V. I., On some integral transformations in Euclidean space. *Dokl. Akad. Nauk SSSR*, 134 (1960), 536–539; *Soviet Math. Dokl.*, 1 (1960), 1114–1117.
[20]. ——, Homogeneous functions and some problems of integral geometry in spaces of constant curvature. *Dokl. Akad. Nauk SSSR*, 136 (1961), 288–291; *Soviet Math. Dokl.*, 2 (1961), 59–62.
[21]. WANG, H. C., Two-point homogeneous spaces. *Ann. of Math.*, 55 (1952), 177–191.

Received April 11, 1964

RADON-FOURIER TRANSFORMS ON SYMMETRIC SPACES AND RELATED GROUP REPRESENTATIONS[1]

BY S. HELGASON

Communicated by G. D. Mostow, May 7, 1965

In §2 we announce some results in continuation of [10], connected with the Radon transform. §1 deals with tools which also apply to more general questions and §§2–3 contain some applications to group representations. A more detailed exposition of §2 appears in Proceedings of the U. S.-Japan Seminar in Differential Geometry, Kyoto, June, 1965.

1. **Radial components of differential operators.** Let V be a manifold, v a point in V and V_v the tangent space to V at v. Let G be a Lie transformation group of V. A C^∞ function f on an open subset of V is called locally invariant if $Xf=0$ for each vector field X on V induced by the action of G.

Suppose now W is a submanifold of V satisfying the following transversality condition:

(T) For each $w \in W$, $V_w = W_w + (G\cdot w)_w$ (direct sum).

If f is a function on a subset of V its restriction to W will be denoted $\bar{f}$.

LEMMA 1.1. *Let D be a differential operator on V. Then there exists a unique differential operator $\Delta(D)$ on W such that*

$$(Df)^- = \Delta(D)\bar{f}$$

for each locally invariant f.

The operator $\Delta(D)$ is called the *radial component* of D. Many special cases have been considered (see e.g. [1, §2], [4, §5], [5, §3], [7, §7], [8, Chapter IV, §§3–5]).

Suppose now dv (resp. dw) is a positive measure on V (resp. W) which on any coordinate neighborhood is a nonzero multiple of the Lebesgue measure. Assume dg is a bi-invariant Haar measure on G. Given $u \in C_c^\infty(G \times W)$ there exists [7, Theorem 1] a unique $f_u \in C_c^\infty(G\cdot W)$ such that

$$\int_{G\times W} F(g\cdot w)u(g, w)\, dg dw = \int_V F(v) f_u(v)\, dv \qquad (F \in C_c^\infty(G\cdot W)).$$

Let $\phi_u \in C_c^\infty(W)$ denote the function $w \to \int u(g, w) dg$.

[1] Work supported in part by the National Science Foundation, NSF GP-2600.

THEOREM 1.2. *Suppose G leaves dv invariant. Let T be a G-invariant distribution on $G\cdot W$. Then there exists a unique distribution $\overline{T}$ on W such that*

$$\overline{T}(\phi_u) = T(f_u), \qquad u \in C_c^\infty(G \times W).$$

If D is a G-invariant differential operator on V then

$$(DT)^- = \Delta(D)\overline{T}.$$

The proof is partly suggested by the special case considered in [**7**, §9]. See also [**12**, §4].

2. **The Radon transform and conical distributions.** Let G be a connected semisimple Lie group, assumed imbedded in its simply connected complexification. Let K be a maximal compact subgroup of G and X the symmetric space G/K. Let $G=KAN$ be an Iwasawa decomposition of G (A abelian, N nilpotent) and let M and M', respectively, denote the centralizer and normalizer of A in K. The space Ξ of all horocycles ξ in X can be identified with G/MN [**10**, §3]. Let $\boldsymbol{D}(X)$ and $\boldsymbol{D}(\Xi)$ denote the algebras of G-invariant differential operators on X and Ξ, respectively; let $S(A)$ denote the symmetric algebra over the vector space A and $I(A)$ the set of elements in $S(A)$ which are invariant under the Weyl group $W=M'/M$. There are isomorphisms Γ of $\boldsymbol{D}(X)$ onto $I(A)$ [**6**, p. 260], [**9**, p. 432] and $\hat{\Gamma}$ of $\boldsymbol{D}(\Xi)$ onto $S(A)$ [**10**, p. 676].

The Radon transform $f\to\hat{f}$ ($f\in C_c^\infty(X)$) and its dual $\phi\to\check{\phi}$ ($\phi\in C^\infty(\Xi)$) are defined by

$$\hat{f}(\xi) = \int_\xi f(x)dm(x), \qquad \check{\phi}(x) = \int_{\check{x}} \phi(\xi)d\mu(\xi) \qquad (x \in X,\ \xi \in \Xi)$$

where dm is the measure on ξ induced by the canonical Riemannian structure of X, $\check{x}$ is the set of horocycles passing through x and $d\mu$ is the measure on $\check{x}$ invariant under the isotropy subgroup of G at x, satisfying $\mu(\check{x})=1$. The easily proved relation

$$(1) \qquad \int_X f(x)\check{\phi}(x)dx = \int_\Xi \hat{f}(\xi)\phi(\xi)d\xi \qquad (f \in C_c^\infty(X),\ \phi \in C_c^\infty(\Xi))$$

dx and $d\xi$ being G-invariant measures on X and Ξ, respectively, suggests immediately how to extend the integral transforms above to distributions.

Let $\mathfrak{G}$ and $\mathfrak{A}$ be the Lie algebras of G and A, respectively, and $\mathfrak{A}^*$

the dual space of $\mathfrak{A}$. Let $\lambda \to \mathbf{c}(\lambda)$ be the function on $\mathfrak{A}^*$ giving the Plancherel measure $|\mathbf{c}(\lambda)|^{-2}d\lambda$ for the K-invariant functions on X (Harish-Chandra [6, p. 612]). Let j be the operator on rapidly decreasing functions on A which under the Fourier transform on A corresponds to multiplication by $\mathbf{c}^{-1}$. Let 2ρ denote the sum (with multiplicity) of the restricted roots on $\mathfrak{A}$ which are positive in the ordering given by N. Let e^{ρ} denote the function on Ξ defined by $e^{\rho}(kaMN) = \exp[\rho(\log a)]$ $(k \in K,\ a \in A)$. Viewing Ξ as a fibre bundle with base K/M, fibre A [10, p. 675] we define the operator Λ on suitable functions ϕ on Ξ by $(e^{\rho}\Lambda\phi)|F = j((e^{\rho}\phi)|F)$, where $|F$ denotes restriction to any fibre F. Similarly, the complex conjugate of $\mathbf{c}^{-1}$ determines an operator $\bar{\Lambda}$. By means of the Plancherel formula mentioned one proves (cf. [11, §6]).

THEOREM 2.1. *There exist constants c, $c' > 0$ such that*

$$\int_X |f(x)|^2 dx = c' \int_{\Xi} |\Lambda\hat{f}(\xi)|^2 d\xi, \tag{2}$$

$$f = c(\Lambda\bar{\Lambda}\hat{f})^{\vee} \tag{3}$$

for all $f \in C_c^{\infty}(X)$.

If all Cartan subgroups of G are conjugate, the operators j and Λ are differential operators ($\mathbf{c}^{-1}$ is a polynomial). Considering $j\bar{j}$ is an element in $I(A)$ we put $\square = \Gamma^{-1}(j\bar{j}) \in \boldsymbol{D}(X)$. Then (3) can be written in the form

$$f = c\,\square\,((\hat{f})^{\vee}), \qquad f \in C_c^{\infty}(X),$$

which is more convenient for applications [10, §7]. For the case when G is complex a formula closely related to (3) was given by Gelfand-Graev [2, §5.5].

Let x_0 and ξ_0 denote the origins in X and Ξ, respectively. The space $B = K/M$ can be viewed as the set of Weyl chambers emanating from x_0 in X. If $\xi = ka \cdot \xi_0$ $(k \in K,\ a \in A)$ we say that the Weyl chamber kM is *normal* to ξ and that a is the *complex distance* from x_0 to ξ. If $x \in X$, $b \in B$ let $\xi(x, b)$ be the horocycle with normal b passing through x, and let $A(x, b)$ denote the complex distance from x_0 to $\xi(x, b)$.

THEOREM 2.2. *For $f \in C_c^{\infty}(X)$ define the Fourier transform $\tilde{f}$ by*

$$\tilde{f}(\lambda, b) = \int_X f(x) \exp[(-i\lambda + \rho)(A(x, b))]dx \qquad (\lambda \in \mathfrak{A}^*,\ b \in B).$$

Then

$$(4) \qquad f(x) = \int_{\mathfrak{A}^*\times B} \tilde{f}(\lambda, b) \exp (i\lambda + \rho)(A(x, b))] \mid c(\lambda) \mid^{-2} d\lambda db$$

$$\int_X \mid f(x) \mid^2 dx = \int_{\mathfrak{A}^*\times B} \mid \tilde{f}(\lambda, b) \mid^2 \mid c(\lambda) \mid^{-2} d\lambda db,$$

where db is a suitably normalized K-invariant measure on B.

REMARKS. (i) In view of the analogy between horocycles in X and hyperplanes in $\boldsymbol{R}^n$ formula (4) corresponds exactly to the Fourier inversion formula in $\boldsymbol{R}^n$ when written in polar coordinate form.

(ii) If f is a K-invariant function on X, Theorem 2.2 reduces to Harish-Chandra's Plancherel formula [**6**, p. 612]. Nevertheless, Theorem 2.2 can be derived from Harish-Chandra's formula.

(iii) A "plane wave" on X is by definition a function on X which is constant on each member of a family of parallel horocycles. Writing (4) in the form

$$(4') \qquad f(x) = \int_B f_b(x)\, db$$

we get a continuous decomposition of f into plane waves. On the other hand, if we write (4) in the form

$$(4'') \qquad f(x) = \int_{\mathfrak{A}^*} f_\lambda(x) \mid c(\lambda) \mid^{-2} d\lambda$$

we obtain a decomposition of f into simultaneous eigenfunctions of all $D \in \boldsymbol{D}(X)$.

We now define for Ξ the analogs of the spherical functions on X.

DEFINITION. A distribution (resp. C^∞ function) on $\Xi = G/MN$ is called *conical* if it is (1) MN-invariant; (2) eigendistribution (resp. eigenfunction) of each $D \in \boldsymbol{D}(\Xi)$.

Let $\xi_0 = MN$, $\xi^* = m^*MN$, where m^* is any element in M' such that the automorphism $a \rightarrow m^*am^{*-1}$ of A maps ρ into $-\rho$. By the Bruhat lemma, Ξ will consist of finitely many MNA-orbits; exactly one, namely $\Xi^* = MNA \cdot \xi^*$, has maximum dimension and given $\xi \in \Xi^*$ there exists a unique element $a(\xi) \in A$ such that $\xi \in MNa(\xi) \cdot \xi^*$ [**10**, p. 673]. Using Theorem 1.2 we find:

THEOREM 2.3. *Let T be a conical distribution on Ξ. Then there exists a $\psi \in C^\infty(\Xi^*)$ such that $T = \psi$ on Ξ^* and a linear function μ: $\mathfrak{A} \rightarrow \boldsymbol{C}$ such that*

$$(5) \qquad \psi(\xi) = \psi(\xi^*) \exp[\mu(\log a(\xi))] \qquad (\xi \in \Xi^*).$$

In general ψ is singular on the lower-dimensional MNA-orbits. However, we have:

THEOREM 2.4. *Let μ: $\mathfrak{A}\to \mathbf{C}$ be a linear function and let $\psi \in C^\infty(\Xi^*)$ be defined by* (5). *Then ψ is locally integrable on Ξ if and only if*

$$\text{Re}\,(\langle \alpha, \mu + \rho \rangle) > 0 \qquad (\text{Re} = \textit{real part}) \tag{6}$$

for each restricted root $\alpha > 0$; here $\langle\ ,\ \rangle$ denotes the inner product on $\mathfrak{A}^*$ induced by the Killing form of $\mathfrak{G}$. If (6) *is satisfied then ψ, as a distribution on Ξ, is a conical distribution.*

THEOREM 2.5. *The conical functions on Ξ are precisely the functions ψ given by* (5) *where for each restricted root $\alpha > 0$,*

$$\frac{\langle \mu, \alpha \rangle}{\langle \alpha, \alpha \rangle} \text{ is an integer } \geqq 0. \tag{7}$$

DEFINITION. A representation π of G on a vector space E will be called (1) *spherical* if there exists a nonzero vector in E fixed by $\pi(K)$; (2) *conical* if there exists a nonzero vector in E fixed by $\pi(MN)$.

The correspondence between spherical functions on X and spherical representations is well known. In order to describe the analogous situation for Ξ, for an arbitrary function ϕ on Ξ, let E_ϕ denote the vector space spanned by the G-translates of ϕ and let π_ϕ denote the natural representation of G on E_ϕ.

THEOREM 2.6. *The mapping $\psi \to \pi_\psi$ maps the set of conical functions on Ξ onto the set of finite-dimensional, irreducible conical representations of G. The mapping is one-to-one if we identify proportional conical functions and identify equivalent representations. Also*

$$\psi(g\cdot\xi_0) = (\pi_\psi(g^{-1})\mathbf{e}, \mathbf{e}'),$$

where $\mathbf{e}$ and $\mathbf{e}'$, respectively, are contained in the highest weight spaces of π_ψ and of its contragredient representation. Finally, μ in (5) *is the highest weight of π_ψ.*

COROLLARY 2.7. *Let π be a finite-dimensional irreducible representation of G. Then π is spherical if and only if it is conical.*

The highest weights of these representations are therefore characterized by (7). Compare Sugiura [**13**], where the highest weights of the spherical representations are determined.

3. The case of a complex G. If G is complex, M is a torus and some of the results of §2 can be improved. Let $\mathfrak{H}$ be a Cartan subalgebra of $\mathfrak{G}$ containing $\mathfrak{A}$ and H the corresponding analytic subgroup of G. Now we assume G simply connected.

Let $\boldsymbol{D}(G/N)$ denote the algebra of all G-invariant differential operators on G/N. Let $\nu_0, \nu^* \in G/N$ be constructed similarly as ξ_0 and ξ^* in §2. Then §1 applies to the submanifold $W = H \cdot \nu^*$ of $V = NH \cdot \nu^*$ and for each differential operator D on G/N, $\Delta(D)$ is defined and can be viewed as a differential operator on H.

THEOREM 3.1. *The mapping $D \to \Delta(D)$ is an isomorphism of $\boldsymbol{D}(G/N)$ onto the (real) symmetric algebra $S(\mathfrak{H})$. In particular, $\boldsymbol{D}(G/N)$ is commutative.*

As a consequence one finds that the N-invariant eigenfunctions $f \in C^\infty(G/N)$ of all $D \in \boldsymbol{D}(G/N)$ have a representation analogous to (5) in terms of the characters of H. Let E_f denote the vector space spanned by the G-translates of f and let π_f be the natural representation of G on E_f.

THEOREM 3.2. *The mapping $f \to \pi_f$ is a one-to-one mapping of the set of N-invariant holomorphic eigenfunctions of all $D \in \boldsymbol{D}(G/N)$ (proportional f identified) onto the set of all finite-dimensional[2] irreducible holomorphic representations of G (equivalent representations identified). Moreover*

$$f(g \cdot \nu_0) = (\pi_f(g^{-1})\mathbf{e}, \mathbf{e}'),$$

where $\mathbf{e}$ and $\mathbf{e}'$, respectively, are contained in the highest weight spaces of π_f and of its contragredient representation.

REFERENCES

1. F. A. Berezin, *Laplace operators on semisimple Lie groups*, Trudy Moskov. Mat. Obšč. **6** (1957), 371–463; Amer. Math. Soc. Transl. **21** (1962), 239–299.

2. I. M. Gelfand and M. I. Graev, *The geometry of homogeneous spaces, group representations in homogeneous spaces and questions in integral geometry related to them.* I, Trudy Moskov. Mat. Obšč. **8** (1959), 321–390; Amer. Math. Soc. Transl. **37** (1964), 351–429.

3. R. Godement, *A theory of spherical functions.* I, Trans. Amer. Math. Soc. **73** (1952), 496–556.

4. Harish-Chandra, *The characters of semisimple Lie groups*, Trans. Amer. Math. Soc. **83** (1956), 98–163.

5. ———, *Differential operators on a semisimple Lie algebra*, Amer. J. Math. **79** (1957), 87–120.

6. ———, *Spherical functions on a semisimple Lie group.* I, II, Amer. J. Math. **80** (1958), 241–310, 553–613.

[2] Compare the problem indicated in [3, p. 553].

7. ———, *Invariant distributions on Lie algebras*, Amer. J. Math. **86** (1964), 271–309.

8. S. Helgason, *Differential operators on homogeneous spaces*, Acta Math. **102** (1959), 239–299.

9. ———, *Differential geometry and symmetric spaces*, Academic Press, New York, 1962.

10. ———, *Duality and Radon transform for symmetric spaces*, Amer. J. Math. **85** (1963), 667–692.

11. ———, *A duality in integral geometry; some generalizations of the Radon transform*, Bull. Amer. Math. Soc. **70** (1964), 435–446.

12. P-D. Methée, *Sur les distributions invariantes dans le groupe des rotations de Lorentz*, Comment. Math. Helv. **28** (1954), 225–269.

13. M. Sugiura, *Representations of compact groups realized by spherical functions on symmetric spaces*, Proc. Japan Acad. **38** (**3**) (1962), 111–113.

INSTITUTE FOR ADVANCED STUDY

A Duality in Integral Geometry on Symmetric Spaces

Sigurdur HELGASON

§ 1. Homogeneous Spaces in Duality.

In order to motivate the subject of this lecture, I recall the formula of J. Radon [28] and F. John [24] which describes how a differentiable function f of compact support on the Euclidean space $\boldsymbol{R}^n$ is determined by its plane integrals. If ω is a unit vector in $\boldsymbol{R}^n$ and $p \in \boldsymbol{R}$, let $\hat{f}(\omega, p)$ denote the integral of f over the hyperplane $(x, \omega) = p$, where (,) denotes the inner product. Then if Δ denotes the Laplacian and $d\omega$ the surface element on the unit sphere S^{n-1}, the Radon-John theorem can be formulated (cf. [25], [21])

$$f(x) = c\Delta_x^{(n-1)/2} \int_{S^{n-1}} \hat{f}(\omega, (x, \omega)) d\omega \ , \tag{1}$$

where c is a constant. If n is even, the operator in front of the integral is the customary fractional power of the Laplacian. Besides some important applications (see e. g. [25], [10]) formula (1) has the following interesting feature. The right hand side contains two dual integrations: first one integrates over the set of points in a given hyperplane through x, then this integral is integrated over the set of hyperplanes through x. Thus the formula involves both functions on the space $\boldsymbol{R}^n$ and on the space $\boldsymbol{P}^n$ of all hyperplanes in $\boldsymbol{R}^n$; both of these spaces being homogeneous spaces of the group $\boldsymbol{M}(n)$ of rigid motions of $\boldsymbol{R}^n$, the following general framework[1)] suggests itself.

Let G be a locally compact topological group, X and Ξ two left coset spaces of G,

$$X = G/H_X \ , \quad \Xi = G/H_\Xi \ ,$$

H_X and H_Ξ being closed subgroups. For reasons which will appear

1) This is a modification of the setup in [20], § 2, using Chern's notion of incidence (see below) which was suggested by him in this context.

later we make the following assumptions on the groups:

(i) G, H_X, H_Ξ, $H_X \cap H_\Xi$ all unimodular.

(ii) If $h_X \in H_X$, $h_X H_\Xi \subset H_\Xi H_X$ then $h_X \in H_\Xi$; if $h_\Xi \in H_\Xi$, $h_\Xi H_X \subset H_X H_\Xi$ then $h_\Xi \in H_X$.

Two elements in X and Ξ, respectively, are called *incident* (Chern [4]) if as cosets in G they have a point in common. For $x \in X$, put

$$\check{x} = \{\xi \in \Xi \mid x, \xi \text{ incident}\} ,$$

and for $\xi \in \Xi$, put

$$\hat{\xi} = \{x \in X \mid x, \xi \text{ incident}\} .$$

These sets can be identified with coset spaces of certain subgroups of G. In fact, suppose $x = gH_X$ ($g \in G$) and if $A \subset G$ is a subgroup, put $A^g = gAg^{-1}$. Then

$$\begin{aligned} \check{x} &= \{\gamma H_\Xi \mid \gamma H_\Xi \cap gH_X \neq \phi\} \\ &= \{gh_X H_\Xi \mid h_X \in H_X\} \end{aligned}$$

so

$$\check{x} = H_X{}^g / H_X{}^g \cap H_\Xi{}^g \qquad \text{if } x = gH_X . \tag{2}$$

Similarly

$$\hat{\xi} = H_\Xi{}^\gamma / H_X{}^\gamma \cap H_\Xi{}^\gamma \qquad \text{if } \xi = \gamma H_\Xi . \tag{3}$$

In particular, if x_0 and ξ_0 are the origins in X and Ξ, respectively, then

$$\check{x} = g \cdot \check{x}_0, \qquad \hat{\xi} = \gamma \cdot \hat{\xi}_0 . \tag{4}$$

LEMMA. *The maps $x \longrightarrow \check{x}$ ($x \in X$) and $\xi \longrightarrow \hat{\xi}$ ($\xi \in \Xi$) are one-to-one.*

In fact, suppose for example $x = gH_X$, $x_1 = g_1 H_X$ such that $\check{x} = \check{x}_1$. By (4) we see that $g^* = g^{-1}g_1$ maps $\check{x}_0$ into itself, in particular $g^* \cdot \xi_0 \subset \check{x}_0$, whence $g^* = h_X h_\Xi$, where $h_X \in H_X$, $h_\Xi \in H_\Xi$. But then the relation $g^* \cdot \check{x}_0 = \check{x}_0$ together with (ii) implies $g^* \in H_X$ so $x = x_1$.

In view of this lemma the elements of Ξ can be regarded as subsets of X, the elements of X as subsets of Ξ, both systems of subsets being permuted transitively by G.

Because of (i), the spaces $\check{x}_0 = H_X/H_X \cap H_\Xi$ and $\hat{\xi}_0 = H_\Xi/H_X \cap H_\Xi$ have canonically determined invariant measures, say $d\mu$ and dm, respectively. By (4), these measures are carried over in measures on $\check{x}$ and $\hat{\xi}$; the argument proving the lemma shows that these measures on $\check{x}$ and $\hat{\xi}$ do not depend on the choices of g and γ in (4). For simplicity we denote these measures again by $d\mu$ and dm.

If f and ϕ are suitably restricted function on X and Ξ, respectively, we define functions $\hat{f}$ on Ξ, $\check{\phi}$ on X, by

$$\hat{f}(\xi) = \int_{\hat{\xi}} f(x)dm(x)\,, \quad \check{\phi}(x) = \int_{\check{x}} \phi(\xi)d\mu(\xi)$$

and the following problems now arise:

A. *Relate function spaces on X and Ξ by means of the integral transforms $f \longrightarrow \hat{f}$ and $\phi \longrightarrow \check{\phi}$.*

B. *Relate directly the functions f and $(\hat{f})^\vee$ on X, ϕ and $(\check{\phi})^\wedge$ on Ξ.*

This last problem gives the reason underlying assumption (ii) above. In fact, Problem B could not be solved (for a class of f and ϕ separating points) unless the conclusion of the lemma holds and this is substantially equivalent to (ii).

PROPOSITION. *For suitably normalized G-invariant measures dx and $d\xi$ on X and Ξ, respectively, the formula*

$$\int_X f(x)\check{\phi}(x)dx = \int_\Xi \hat{f}(\xi)\phi(\xi)d\xi \tag{5}$$

holds for all continuous f and ϕ of compact support.

For the proof put $H = H_X \cap H_\Xi$ and consider the natural projections $p_X : G/H \longrightarrow X$, $p_\Xi : G/H \longrightarrow \Xi$ and the function

$$P = (f \circ p_X)(\phi \circ p_\Xi) \qquad \text{on } G/H\,.$$

If the invariant measures on G/H, G/H_X, G/H_Ξ, H_X/H, and H_Ξ/H are suitably normalized, we have by [17], p. 449,

$$\int_{G/H} P(gH)dg_H = \int_{G/H_X} dg_{H_X} \int_{H_X/H} P(gh_X H)d(h_X)_H$$
$$= \int_{G/H_\Xi} dg_{H_\Xi} \int_{H_\Xi/H} P(gh_\Xi H)d(h_\Xi)_H\,.$$

Now relation (5) follows immediately.

Formula (5) shows how the integral transforms $f \longrightarrow \hat{f}$, $\phi \longrightarrow \check{\phi}$ should be extended to distributions; formally one puts

$$\hat{S}(\phi) = S(\check{\phi}) , \quad \check{\Sigma}(f) = \Sigma(\hat{f}) , \tag{6}$$

if S and Σ are distributions on X and Ξ, respectively, but of course the validity of these definitions depends on some kind of solution to Problem A.

Because of the great generality of the framework above one cannot expect a general unified solution to Problems A and B. This is also substantiated by the diversity of the solutions to these problems, which have been obtained in special cases.

EXAMPLE. Let $X = \boldsymbol{R}^n$ and $\Xi = \boldsymbol{P}^n$ (the set of hyperplanes in $\boldsymbol{R}^n$) both spaces represented as coset spaces of the group $G = \boldsymbol{M}(n)$ of all rigid motions of $\boldsymbol{R}^n$. Let H_X be the subgroup of G leaving the origin fixed and let H_Ξ be the subgroup of G which leaves the hyperplane $\boldsymbol{R}^{n-1} \times \boldsymbol{0}$ invariant. Then assumptions (i) and (ii) are satisfied and

$$\boldsymbol{R}^n = G/H_X , \quad \boldsymbol{P}^n = G/H_\Xi ,$$

and the incidence notion above reduces to the usual one. Formula (1) can therefore be written

$$f = c\Delta^{(n-1)/2}((\hat{f})^\vee) ,$$

which then solves the first part of Problem B. The second part is solved by a similar inversion formula ([24] or [21])

$$\phi = c\Box^{(n-1)/2}((\check{\phi})^\wedge) ,$$

where $\Box$ is the differential operator $\phi(\omega, p) \longrightarrow d^2/dp^2\ (\phi(\omega, p))$ and $\phi(\omega, p)$ is assumed rapidly decreasing in p (uniformly in ω) and orthogonal to all polynomials in p. For this case formula (5) was given by Ludwig [27].

§ 2. A Duality for Noncompact Symmetric Spaces.

In this section we describe our main example of the duality in § 1. In § 2–§ 4 we give an outline with some indication of proof of the results stated in [20] § 6 and [22] § 2.

It is convenient first to recall some standard notions concerning semisimple Lie groups. Let $\mathfrak{g}$ be a semisimple Lie algebra over $\boldsymbol{R}$, $\mathfrak{g}^c$ its complexification, G^c the simply connected Lie group with Lie algebra $\mathfrak{g}^c$ and G the analytic subgroup of G^c corresponding to $\mathfrak{g}$. Let K be any maximal compact subgroup of G, $\mathfrak{k}$ its Lie algebra and $\mathfrak{p}$ the orthogonal complement of $\mathfrak{k}$ in $\mathfrak{g}$ with respect to the Killing form $(\ ,\ )$ of $\mathfrak{g}$. Let $\mathfrak{a}$ be any maximal abelian subspace of $\mathfrak{p}$, $\mathfrak{a}^*$ its dual, and if $\lambda \in \mathfrak{a}^*$ let

$$\mathfrak{g}_\lambda = \{X \in \mathfrak{g} \mid [H, X] = \lambda(H)X \text{ for all } H \in \mathfrak{a}\} .$$

If $\lambda \neq 0$, and $\mathfrak{g}_\lambda \neq \{0\}$, λ is called a root of $\mathfrak{g}$ with respect to $\mathfrak{a}$. Let $\mathfrak{a}'$ be the part of $\mathfrak{a}$ where all roots are $\neq 0$. Then $\mathfrak{a}'$ consists of finitely many components called the Weyl chambers. Pick one chamber, say C. A root is called positive if its values on C are positive. Consider the subalgebra $\mathfrak{n} = \sum_{\alpha > 0} \mathfrak{g}_\alpha$ of $\mathfrak{g}$ and the corresponding analytic subgroup N of G. Then we have the Iwasawa decomposition $G = KAN$ where $A = \exp \mathfrak{a}$.

The space $X = G/K$ is the most general symmetric space of the noncompact type. As usual, the Riemannian structure taken on X is that induced by the restriction of the Killing from $(\ ,\ )$ to $\mathfrak{p}$. A horocycle in X is by definition an orbit in X of a group of the form gNg^{-1}, g being any element in G. It is not difficult to see that G permutes the horocycles transitively, and if x_0 is the origin $\{K\}$ in X, and ξ_0 is the horocycle $N \cdot x_0$ then the subgroup of G which maps ξ_0 into itself is MN, where M is the centralizer of A in K. If Ξ denotes the set of all horocycles, we have the identifications

$$X = G/K\,, \quad \Xi = G/MN$$

and the assumptions (i) and (ii) from § 1 are verified without difficulty. The abstract incidence notion in § 1 now reduces to the naive one: a point x and a horocycle ξ are incident if and only if $x \in \xi$. Consequently the integral transforms $f \longrightarrow \hat{f}$, $\phi \longrightarrow \check{\phi}$ of § 1 now become

$$\hat{f}(\xi) = \int_\xi f(x)\, dm(x)\,, \quad \check{\phi}(x) = \int_{\check{x}} \phi(\xi)\, d\mu(\xi)\ ,$$

where dm is the volume element on ξ (for the Riemannian structure on ξ induced by that of X), and $d\mu$ is the measure in $\check{x}$ invariant under the isotropy subgroup of G at x (which acts transitively on $\check{x}$) satisfying $\mu(\check{x})=1$.

Before considering these integral transforms in more detail let us explore a bit the analogy between the space X and its dual Ξ. First we note that

$$\dim X = \dim \Xi \ .$$

In addition to the group M above consider M', the normalizer of A in K and the factor group $W=M'/M$ which is a finite group, the so-called Weyl group. Let w denote the order of W. According to Cartan's well known "polar coordinate decomposition" of X, we have (see e. g. [17], p. 381),

(A_X) *The mapping*

$$(kM, a) \longrightarrow kaK$$

is a differentiable mapping of $(K/M)\times A$ *onto* X *and if we put* $A' = \exp(\mathfrak{a}')$ *the restriction to* $(K/M)\times A'$ *is a covering of order* w.

For Ξ one has a corresponding, but simpler result ([18], p. 672).

(A_Ξ) *The mapping*

$$(kM, a) \longrightarrow kaMN$$

is a diffeomorphism of $(K/M)\times A$ *onto* Ξ.

Consider for example the case when X is the open unit disk D with the non-Euclidean metric. In this case the horocycles are the intersections of D with the circles tangential to $|z|=1$, and statement (A_Ξ) is obvious.

As a sharpening of (A_X) we note the well known description of the orbit space of K on G/K,

(B_X) $$K\backslash G/K = A/W \ .$$

This means that each $g\in G$ can be written $g=k_1ak_2$ ($k_1, k_2\in K$) where $a\in A$ is uniquely determined (mod W).

Similarly, the orbit space of MN on G/MN is given by ([18], p. 673),

(B_Ξ) $$MN\backslash G/MN = A\times W \ .$$

More precisely, each $g\in G$ can be written (with obvious notation)

$g = m_1 n_1 a m_w m_2 n_2$ where $a \in A$ is unique and $m_w \in M'$ is unique $(\mathrm{mod}\, M)$. Except for the uniqueness of a, this is the so-called Bruhat lemma (see [14]).

Let $\boldsymbol{D}(X)$ and $\boldsymbol{D}(\Xi)$ denote the algebras of G-invariant differential operators on X and Ξ, respectively. Let $S(A)$ denote the algebra of translations-invariant differential operators on A and let $I(A)$ denote the set of W-invariant elements in $S(A)$. According to a result of Harish-Chandra ([15], p. 260, [17], p. 432),

(C_X) *The algebra* $\boldsymbol{D}(X)$ *is canonically isomorphic to* $I(A)$.

This isomorphism Γ: $\boldsymbol{D}(X) \longrightarrow I(A)$ can be roughly described as follows: The K-invariant functions on $X = G/K$ can by (B_X) above be regarded as the W-invariant functions on A; when an operator $D \in \boldsymbol{D}(X)$ is applied to such functions, it becomes a singular differential operator on A. (Compare with the radial part $\partial^2/\partial r^2 + (n-1)/r\partial/\partial r$ of the Laplacian on $\boldsymbol{R}^n$.) The principal part of this operator has constant coefficients, and this coincides with the highest order term in $\Gamma(D)$.

In analogy with (C_X) we have ([18]):

(C_Ξ) *The algebra* $\boldsymbol{D}(\Xi)$ *is canonically isomorphic to* $S(A)$.

We shall now describe this isomorphism. Let $w_1, w_2, \cdots$ be the elements in W. Select corresponding representatives $m_1, m_2, \cdots$ in M' and consider the elements

$$\xi_1 = m_1 \cdot \xi_0\,, \quad \xi_2 = m_2 \cdot \xi_0\,, \quad \cdots \qquad \text{in } \Xi\,.$$

Then by (B_Ξ),

(7) $$\Xi = \cup_i MNA \cdot \xi_i = \cup_i \Xi_i \qquad \text{(disjoint union)}.$$

Let w^* be the element in W which maps all positive roots into negative roots. The corresponding orbit, say $\Xi^* = MNA \cdot \xi^*$, has dimension equal to $\dim \Xi$; all the others have lower dimension. Consider now the orbit $A \cdot \xi^*$ which is a cross section for the action of MN on Ξ^*. If $\xi \in \Xi^*$ let $A(\xi)$ denote the unique element in $\mathfrak{a}$ such that

$$\xi \in MN \exp A(\xi) \cdot \xi^*\,.$$

Now, let D be a differential operator on Ξ; then it is easy to prove on general grounds (cf. [22], Lemma 1.1) that there exists a differential operator $\Delta(D)$ on $A\cdot\xi^*$ such that

$$(D\phi)\mid A\cdot\xi^* = \Delta(D)(\phi\mid A\cdot\xi^*) \tag{8}$$

for each MN-invariant $\phi\in C^\infty(\Xi^*)$, the vertical bar denoting restriction. If D is G-invariant, it can be shown ([18], p. 678) that $\Delta(D)$, considered as a differential operator on A, has constant coefficients and that $D\longrightarrow\Delta(D)$ gives an isomorphism of $\boldsymbol{D}(\Xi)$ onto $S(A)$ as stated.

We recall now that a spherical function f on X is by definition a K-invariant eigenfunction of all $D\in\boldsymbol{D}(X)$ satisfying $f(x_0)=1$ ([6]). Similarly we put:

DEFINITION. A distribution (resp. C^∞ function) on $\Xi = G/MN$ is called *conical* if it is: 1) MN-invariant; 2) eigendistribution (resp. eigenfunction) of each $D\in\boldsymbol{D}(\Xi)$.

Whereas the distinction between conical functions and conical distributions is significant, there is no need to consider "spherical distributions"; since $\boldsymbol{D}(X)$ contains elliptic elements, those distributions would automatically be functions.

If $g\in G$ let $H(g)$ be the unique element in $\mathfrak{a}$ such that $g\in K\exp H(g)N$. The spherical functions are then determined by the following formula of Harish-Chandra:

(D_X) *The spherical functions on X are precisely the functions given by*

$$\phi_\lambda(gK) = \int_K e^{(i\lambda-\rho)(H(gk))}dk\,,$$

where $\rho = \frac{1}{2}\sum_{\alpha>0}(\dim\mathfrak{g}_\alpha)\alpha$ *and λ is an arbitrary complex-valued linear function on $\mathfrak{a}$.*

For the conical distributions we have a preliminary result.

(D_Ξ') *Let Ψ be a conical distribution on Ξ. Then there exists a $\psi\in C^\infty(\Xi^*)$ such that*

$$\Psi = \psi \quad \text{on } \Xi^*.$$

This ψ is given by

$$\psi(\xi) = \psi(\xi^*)e^{-\mu(A(\xi))} \qquad (\xi\in\Xi^*)\,, \tag{9}$$

where μ is a complex-valued linear function on $\mathfrak{a}$.

This is proved by extending (8) to the situation where D is G-invariant and ϕ an MN-invariant distribution; of course the restriction of such a distribution to $A\cdot\xi^*$ has to be defined rather indirectly. Theorem (D_{Ξ}') is an analog of (D_X) except that it gives no information about which μ do occur. Secondly, the conical distribution Ψ is not completely determined by μ in (9); it is not difficult to construct conical distribution supported entirely on a lower-dimensional orbit Ξ_i in (7).

Although (D_{Ξ}') says nothing about the existence of conical distributions this is guaranteed by the next result where we use the following standard terminology. If μ is a complex-valued linear function on $\mathfrak{a}$, let $H_\mu \in \mathfrak{a} + i\mathfrak{a}$ be determined by $(H_\mu, H) = \mu(H)$ for all $H \in \mathfrak{a}$. The component of H_μ in $\mathfrak{a}$ is called $Re(H_\mu)$. Also ρ has the same meaning as in (D_X) and C is defined earlier.

(D_{Ξ}'') *Let* $\mu : \mathfrak{a} \longrightarrow \mathbf{C}$ *be a linear function and let* $\psi \in C^\infty(\Xi^*)$ *be determined by* (9). *Then* ψ *is a locally integrable function on* Ξ *if and only if*

$$Re(H_{\mu+\rho}) \in C. \tag{10}$$

If (10) *is satisfied, then* ψ, *as a distribution on* Ξ, *is a conical distribution.*

Here one arrives at condition (10) by first transferring the local integrability question on Ξ to the space $(K/M)\times A$ via (A_{Ξ}). Next, the integration over K/M is transferred to an integration over N ([15], I, p. 287) and then one can use the induction procedure of [11].

(D_{Ξ}''') *The conical functions on* Ξ *are precisely the functions* ψ *given by* (9) *where for each positive root* α *of* $\mathfrak{g}$ *with respect to* $\mathfrak{a}$,

$$\frac{(\mu, \alpha)}{(\alpha, \alpha)} \text{ is an integer } \geqq 0\,. \tag{11}$$

Here (,) *is the inner product on* $\mathfrak{a}^*$ *induced by the Killing form.*

As the integrality condition (11) suggests, the proof of (D_{Ξ}''') is tied up with representation theory. We shall now explain this connection.

DEFINITION. A representation π of G on a vector space E will be called: 1) *spherical* if there exists a nonzero vector in E

fixed by $\pi(K)$; 2) *conical* if there exists a nonzero vector in E fixed under $\pi(MN)$.

There is a well known connection between spherical functions on X and spherical representations of G. Roughly speaking the "irreducible" spherical representations are obtained by taking for each spherical function f the natural representation of G on the space "spanned" by the translates of f under G; then f is recovered as a suitable coefficient of the representation.

We now describe the analogous situation for Ξ. If ϕ is a function on Ξ let E_ϕ denote the smallest vector space containing all the G-translates of ϕ and let π_ϕ denote the natural representation of G on E_ϕ.

(E_Ξ) *The mapping $\psi \longrightarrow \pi_\psi$ is a bijection of the set of conical functions on Ξ (proportional ψ identified) onto the set of finite-dimensional irreducible conical representations of G (equivalent representations identified). Also*

$$\psi(g\cdot\xi_0) = (\pi_\psi(g^{-1})e, e') \qquad (g\in G) \tag{12}$$

where **e** *and* **e**$'$, *respectively, are contained in the (one-dimensional) highest weight spaces of π_ψ and of its contragradient representation. Finally, μ in* (9) *is the highest weight of π_ψ.*

In order to prove statements (D_Ξ''') and (E_Ξ), one first determines the conical functions in the case $G = \boldsymbol{SL}(2, \boldsymbol{R})$. For this case (9) takes the form

$$\psi\left(\begin{pmatrix} x_{11} & x_{12} \\ x_{21} & x_{22}\end{pmatrix}\cdot\xi_0\right) = |x_{21}|^r \qquad r\in \boldsymbol{C}$$

(for $G = \boldsymbol{SL}(n, \boldsymbol{R})$ all the minors with x_{n1} in the lower left hand corner appear). It is clear that ψ is C^∞ if and only if r is an even integer $\geqq 0$. For a general G and Ξ one constructs for each root α a subgroup $G_\alpha \subset G$ locally isomorphic to $\boldsymbol{SL}(2, \boldsymbol{R})$. When expressing that ψ has a C^∞ restriction to the corresponding space Ξ_α one arrives at condition (11). On the other hand if (11) is satisfied, one can construct a finite-dimensional representation π^μ with highest weight μ. Using the structure of the group of components of M one finds that the function $g \longrightarrow (\pi^\mu(g^{-1})e, e')$ defines a conical function on Ξ which then must be proportional to ψ.

COROLLARY. *Let π be a finite-dimensional irreducible representation of G. Then π is spherical if and only if it is conical.*

In particular, (11) gives a characterization of the highest weights of these representations. Thus we recover the characterization of the highest weights of the spherical representations, stated in Sugiura [29].

Since the correspondence between spherical functions and spherical representations extends to infinite-dimensional representations, one is led to look for an extension of (E_{Ξ}) to conical distributions Ψ and infinite-dimensional representations on distribution spaces with "highest weight" given by μ in (9). Such an extension of (E_{Ξ}) and of the Corollary is given in § 4, b) in a weakened from.

Concerning Problem B we now have the following result ([22], Theorem 2.1, [20], p. 444)

$(F_{X,\Xi})$ (i) *For a certain integro-differential operator Λ on Ξ, Λ^* its adjoint,*

$$f = (\Lambda\Lambda^*\hat{f})^{\vee}, \tag{13}$$

$$\int_X |f(x)|^2 dx = \int_\Xi |\Lambda\hat{f}(\xi)|^2 d\xi , \qquad f \in C_c^\infty(X) .$$

(ii) *In case all Cartan subgroups of G are conjugate we have*

$$f = \square((\hat{f})^{\vee}) , \qquad f \in C_c^\infty(X) , \tag{14}$$

where $\square$ is a certain explicit differential operator on X.

In order to indicate the proof and the definition of Λ and $\square$ we recall Harish-Chandra's inversion formula for the spherical Fourier transform

$$\tilde{F}(\lambda) = \int_X F(x)\phi_{-\lambda}(x)dx \tag{15}$$

defined for all K-invariant functions F in $C_c^\infty(X)$. For a certain analytic function $\mathbf{c}^{-1}$ on $\mathfrak{a}^*$ and suitable normalization of the measure on $\mathfrak{a}^*$,

$$F(x) = \int_{\mathfrak{a}^*} \tilde{F}(\lambda)\phi_\lambda(x) |\mathbf{c}(\lambda)|^{-2} d\lambda \tag{16}$$

and

$$\int_X |F(x)|^2 dx = \int_{\mathfrak{a}^*} |\tilde{F}(\lambda)|^2 |c(\lambda)|^{-2} d\lambda \tag{17}$$

(see [15], p. 612 and [16], § 23). The function $c(\lambda)$ is given by $c(\lambda) = I(i\lambda)/I(\rho)$ where $I(\nu)$ is given by the following formula (see Gindikin-Karpelevič [11]),

$$I(\nu) = \prod_{\alpha>0} B\left(\frac{1}{2}m_\alpha, \frac{1}{4}m_{\alpha/2} + \frac{(\nu, \alpha)}{(\alpha, \alpha)}\right), \tag{18}$$

where α runs over the positive roots of $\mathfrak{g}$ with respect to $\mathfrak{a}$, $m_\alpha = \dim(\mathfrak{g}_\alpha)$ and B is the Beta function.

Let j be the operator on rapidly decreasing functions on A which under the (Euclidean) Fourier transform on A corresponds to multiplication by c^{-1}. Let e^ρ denote the function on Ξ defined by $e^\rho(kaMN) = e^{\rho(\log a)}$ $(k \in K, a \in A)$. Considering Ξ as a fibre bundle with base K/M, fibre A (cf. (A_Ξ)) we define the operator Λ on suitable functions ϕ on Ξ by

$$(e^\rho \Lambda\phi) | E = j((e^\rho \phi) | E),$$

where $|E$ denotes restriction to any fibre E. Since the action of G on Ξ is fibre-preserving and j translation-invariant, the operator Λ is G-invariant. With this definition, (i) follows quickly from (16).

If all Cartan subgroups of G are conjugate all the multiplicities m_α are even integers and $m_{\alpha/2} = 0$. Because of the functional equation $\Gamma(1 + x) = x\Gamma(x)$ for the Gamma function, $c(\lambda)^{-1}$ reduces to a polynomial. Thus j and Λ are differential operators and so is the operator $\square = \Gamma^{-1}(jj^*)$, cf. (C_X). Now (14) follows just as in [18] § 6. The operator $\square$ can be defined without the assumption in (ii) and then it is an integro-differential operator. This is similar to the even-dimensional case in (1).

For the case when G is complex a formula of the type (13) was proved by Gelfand-Graev [8]; in this case it is a direct consequence of the representation of a function on G by means of its integrals over all conjugacy classes and their translates ([9], [13]). Of course formula (14) (of the type required for Problem B) is valid in this case. See also [12].

§ 3. The Fourier Transform on the Symmetric Space X.

The analog for $\boldsymbol{R}^n$ of the integral transform (15) of K-invariant functions F on X is an integral transform of radial functions on $\boldsymbol{R}^n$ given by the radial eigenfunctions of the Laplacian Δ on $\boldsymbol{R}^n$ (i. e., the spherical functions on the group $M(n)$, see *e. g.* [17], p. 422). When stated in terms of the radial variable this becomes the Hankel type transform

$$f(r) \longrightarrow \int_0^\infty f(s) J_\nu(rs)(rs)^{-\nu} s^{2\nu+1}\, ds \qquad (2\nu = n-2)\,, \tag{19}$$

where J_ν is the Bessel function of order ν.

Guided by the analogy between horocycles in X and hyperplanes in $\boldsymbol{R}^n$ we shall now define for X the analog of the Fourier transform on $\boldsymbol{R}^n$. To make the motivation clearer we write the Fourier transform in polar coordinate form

$$\tilde{f}(\lambda\omega) = \int_{\boldsymbol{R}^n} f(x) e^{-i\lambda(x,\,\omega)}\, dx \qquad (\lambda \in \boldsymbol{R}, |\,\omega\,| = 1)$$

and then

$$f(x) = \int_{\boldsymbol{R}^+} \int_{S^{n-1}} \tilde{f}(\lambda\omega) e^{i\lambda(x,\omega)} \,|\,\lambda\,|^{n-1}\, d\lambda d\omega \tag{20}$$

with a suitable normalization of the surface element $d\omega$ of S^{n-1}.

Here (x, ω) of course gives the distance from 0 to the hyperplane through x, with normal vector ω. What is the analog of (x, ω) for X? First note that the set of all Weyl chambers in all maximal abelian subspaces of $\mathfrak{p}$ is acted on transitively by K. The subgroup leaving C invariant is M. Viewing a Weyl chamber as an analog of a direction, K/M thus becomes the analog of S^{n-1}. Now, according to (A_3) an arbitrary horocycle ξ can be written

$$\xi = ka\cdot\xi_0 \qquad (\xi_0 = N\cdot x_0) \tag{21}$$

where $a \in A$ is unique and $k \in K$ is unique (mod M). We say the Weyl chamber kM is ***normal*** to the horocycle ξ and that a is the *complex distance* from x_0 to ξ. The term "normal" is reasonable because the manifold $kA\cdot x_0$ (to which the Weyl chamber kM is tangential) is the "orthogonal complement" of the horocycle ξ at the point $ka\cdot x_0$ ([18], Lemma 3.3). Through each point $x \in X$ pass finitely many horocycles (w in number) having the same tangent space at x. They are distinguished by means of their normals. (Here we have a contrast to the Euclidean case which explains why the geodesic symmetry induces a symmetric space structure on $\boldsymbol{P}^n$ (in § 1) but not on G/MN.) The term "complex distance" is reasonable because the point $ka \cdot x_0$ is the only one in ξ at minimum distance from x_0 ([23], Lemma 5.4).

Now if $x \in X$, $b \in B = K/M$ let $\xi(x, b)$ denote the horocycle with normal b passing through x, and let $A(x, b)$ denote the log of the complex distance from x_0 to $\xi(x, b)$. This element $A(x, b) \in \mathfrak{a}$ is the symmetric space analog of the inner product (x, ω).

THEOREM. *For $f \in C_c^\infty(X)$ define the Fourier transform $\tilde{f}$ by*

$$\tilde{f}(\lambda, b) = \int_X f(x) e^{(-i\lambda+\rho)(A(x,b))} \, dx \qquad (\lambda \in \mathfrak{a}^*, \ b \in B) .$$

Then

$$f(x) = \int_{\mathfrak{a}^* \times B} \tilde{f}(\lambda, b) e^{(i\lambda+\rho)(A(x,b))} \, | \boldsymbol{c}(\lambda) |^{-2} d\lambda db \tag{22}$$

and

$$\int_X | f(x) |^2 dx = \int_{\mathfrak{a}^* \times B} | \tilde{f}(\lambda, b) |^2 \, | \boldsymbol{c}(\lambda) |^{-2} \, d\lambda db$$

if db is a suitably normalized K-invariant measure on B.

It is clear that the inversion formula (22) is the exact analog of (20) except for the "Poisson kernel" $e^{2\rho(A(x,b))}$ on $X \times B$. (Here $B = K/M$ enters as Furstenberg's boundary [5].) The underlying reason can be found to be (by comparing with the inversion formula (13)) that whereas the group invariant measure μ on $\check{x}$ (for $x \in \boldsymbol{R}^n$) is the same as the group invariant measure on the set of corresponding normals the analogous situation does not hold for $X = G/K$. Here the ratio of the two measures $d\mu$ and db is $e^{2\rho(A(x,b))}$. This is familiar in the case of the non-Euclidean disk D ([3]); in fact if $x \in D \cap \xi_0$ and $\theta(b)$ is the angle between the horocycles $\xi(x, b)$ and ξ_0 then $d\theta/db$ equals the Poisson kernel $P(x, b)$.

In the case when f is a K-invariant function on X, $\tilde{f}(\lambda, b)$ in (22) is independent of b and by (D_X) formula (22) reduces to Harish-Chandra's formula (16), just as the Hankel type transform (19) is obtained by specializing the Fourier transform on $\boldsymbol{R}^n$ to radial functions. On the other hand one can for arbitrary $f \in C_c^\infty(X)$

apply Harish-Chandra's formula (16) to the function $x \longrightarrow \int_k f(k \cdot x)dk$. Using simple properties of $A(x, b)$ formula (22) is obtained by measure-theoretic computations.

EXAMPLE. $X =$ *The non-Euclidean Disk D.* In this case we shall write down the Fourier inversion formula in a more explicit form. It will simplify the formulas if we normalize the non-Euclidean distance by

$$d(0, z) = \frac{1}{2} \log \frac{1 + |z|}{1 - |z|} \qquad (z \in D) .$$

If $x \in D$ and ω is a point on the unit circle S, let $\xi(x, \omega)$ be the horocycle passing through x and ω and let (x, ω) denote the non-Euclidean distance from 0 to $\xi(x, \omega)$ (taken with sign).

In this case the theorem above becomes:

For $f \in C_c^\infty(X)$ *let*

$$\tilde{f}(\lambda, \omega) = \int_X f(x) e^{(-i\lambda + 1)(x, \omega)} dx \qquad \lambda \in \boldsymbol{R},\ \omega \in \boldsymbol{S}.$$

Then

$$f(x) = c \int_R \int_S \tilde{f}(\lambda, \omega) e^{(i\lambda + 1)(x, \omega)} \lambda \tanh\left(\frac{\pi}{2}\lambda\right) d\lambda d\omega ,$$

where c is a constant. The function $e^{2(x,\omega)}$ is the Poisson kernel. Hence Poisson's integral for harmonic functions can be written

$$u(x) = \int_S e^{2(x,\omega)} u(\omega) d\omega ,$$

and thus falls within the framework of Fourier Analysis on D.

§ 4. Applications.

a) *Plane Wave Decompositions.* We recall that two horocycles are called parallel ([18], p. 672) if they correspond under a transvection along a geodesic perpendicular to both. The following lemma is readily verified.

LEMMA. *Let ξ_1 and ξ_2 be two horocycles. Then the following properties are equivalent*:

(i) *ξ_1 and ξ_2 are parallel.*

(ii) *ξ_1 and ξ_2 have the same normal.*
(iii) *ξ_1 and ξ_2 are orbits of the same group gNg^{-1}.*

A plane *wave* on X is by definition a function which is constant on each member of a maximal family of parallel horocycles. If b is the normal to this family, we shall call b the normal to the plane wave. Writing (22) in the form

$$f(x)=\int_B f_b(x)db \tag{23}$$

we obtain a decomposition of f into plane waves, in fact f_b is constant on each horocycle with normal b.

To indicate the usefulness of (23) let D be a G-invariant differential operator on X and $f\in C_c^\infty(X)$. Consider the differential equation

$$Du=f\ . \tag{24}$$

Now fix $b\in B$ and consider the differential equation

$$Dv=f_b \tag{25}$$

and in fact we look for a solution $v=v^b$, which is a plane wave with normal b. If $b=kM$ then f_b and the sought function v^b are constant on each orbit of kNk^{-1}. But each such orbit intersects the Euclidean space $kAk^{-1}\cdot x_0$ in exactly one point. Thus the differential equation $Dv^b=f_b$ can be regarded as a differential equation on this Euclidean space; as a consequence of [26] or [15], p. 247 this last differential equation has constant coefficients (cf. (2) p. 687 in [18]). As is well known such a differential equation always has a solution v^b. Assuming $v^b(x)$ can be chosen so as to depend smoothly on b and x (for example as a result of some boundary conditions) we have a solution to (24) given by

$$u(x)=\int_B v^b(x)db\ .$$

The weakness of this method lies in the nonuniqueness of the solution v^b. However a modification of the method (see [19]) shows that a solution u of (24) always exists.

b) *Decomposition of the representations of G on $L^2(X)$ and on $L^2(\Xi)$.* Let $f\in C_c^\infty(X)$; if we apply Harish-Chandra's formula (16) to the function $x \longrightarrow \int_K f(k\cdot x)dk$ we immediately obtain the formula

$$f(x) = \int_{\mathfrak{a}^*} f \times \varphi_\lambda(x)\,|\,c(\lambda)\,|^{-2} d\lambda\,, \qquad f \in C_c^\infty(X)\,, \tag{26}$$

where $\times$ is the operation on functions on X induced by the convolution product on G. We indicate the ensuing decomposition of the natural representation T_X of G on $L^2(X)$ into irreducible parts. The function φ_λ is positive definite (when lifted to a function on G) and by the classical construction of Gelfand-Raikov and Godement gives rise to a positive definite inner product on the vector space V_λ spanned by the G-translates of φ_λ. The completion of V_λ is a Hilbert space $\mathscr{H}_\lambda$ and the natural action of G on $\mathscr{H}_\lambda$ is an irreducible, unitary, spherical representation T_λ of G (see *e.g.* [17], § 4, Ch. X). Now, it is easy to approximate each $f\times\varphi_\lambda$ ($f\in C_c^\infty(X)$) by elements in V_λ. But $\mathscr{H}_\lambda$ consists of equivalence classes of Cauchy sequences in V_λ so we obtain a canonical bijection of the space $C_c^\infty(X)\times\varphi_\lambda$ onto a dense subspace of $\mathscr{H}_\lambda$. The norm which $f\times\varphi_\lambda$ thereby acquires is found to be given by

$$\|f \times \varphi_\lambda\|_\lambda^2 = \int_B |\tilde{f}(\lambda, b)|^2\, db\,.$$

Thus (26) and the theorem in § 3 imply a direct integral decomposition,

$$L^2(X) = \int \mathscr{H}_\lambda\,|\,c(\lambda)\,|^{-2}\, d\lambda\,, \quad T_X = \int T_\lambda\,|\,c(\lambda)\,|^{-2}\, d\lambda\,, \tag{27}$$

λ running through $\mathfrak{a}^*$ (mod W).

It is easy to prove an analog of (26) for the space Ξ. If $\psi \in C_c^\infty(\Xi)$ and if η is an MN-invariant distribution on Ξ we define $\psi\times\eta$ in $C^\infty(\Xi)$ by

$$\psi \times \eta(g\cdot\xi_0) = \eta(\psi\circ\tau(g)) \qquad (g\in G)\,,$$

where $\tau(g)$ is the translation $xMN \longrightarrow gxMN$ in Ξ. The distribution

$$\eta_\lambda : \psi \longrightarrow \int_A \psi(a\cdot\xi_0)e^{-(i\lambda-\rho)(\log a)}da\,, \qquad \psi\in C_c^\infty(\Xi)\,,$$

is a conical distribution on Ξ and the decomposition formula

$$\psi(\xi) = \int_{\mathfrak{a}^*} \psi \times \eta_\lambda(\xi) d\lambda \tag{28}$$

follows by Fourier analysis on the group A.

Now if $\lambda \in \mathfrak{a}^*$ let $\mathscr{H}_\lambda$ denote the Hilbert space of functions ψ on Ξ satisfying the conditions

(i) $\psi(ga \cdot \xi_0) = \psi(g \cdot \xi_0) e^{(i\lambda - \rho)(\log a)}$, $\quad a \in A, g \in G$;

(ii) $\|\psi\|_\lambda^2 = \int_K |\psi(k \cdot \xi_0)|^2 dk < \infty$,

and let τ_λ denote the unitary representation of G on $\mathscr{H}_\lambda$ given by

$$(\tau_\lambda(g)\psi)(\xi) = \psi(g^{-1} \cdot \xi) \qquad g \in G, \xi \in \Xi .$$

This is the representation of G induced by the character $man \longrightarrow e^{i\lambda(\log a)}$ of MAN and is well known to be irreducible, at least if λ is regular. In this case it is easy to set up a unitary equivalence between T_λ and τ_λ. Formula (28) gives an expansion of each $\psi \in C_c^\infty(\Xi)$ into simultaneous eigenfunctions of the operators in $\boldsymbol{D}(\Xi)$. Moreover, the space $C_c^\infty(\Xi) \times \eta_\lambda$ is dense in $\mathscr{H}_\lambda$ and if the measure $d\xi$ is suitably normalized,

$$\int_\Xi |\psi(\xi)|^2 d\xi = \int_{\mathfrak{a}^*} \|\psi \times \eta_\lambda\|_\lambda^2 d\lambda .$$

Thus (28) gives a direct integral decomposition of the natural T_Ξ of G on $L^2(\Xi)$,

$$L^2(\Xi) = \int_{\mathfrak{a}^*} \mathscr{H}_\lambda d\lambda , \qquad T_\Xi = \int_{\mathfrak{a}^*} \tau_\lambda d\lambda , \tag{29}$$

(compare the decomposition of the representation of G on $L^2(G/N)$ for G complex, given in [8], § 3).

As one would expect, the decompositions (27) and (29) are related by means of the dual transforms $f \longrightarrow \hat{f}$, $\varphi \longrightarrow \check{\varphi}$. This is a consequence of the following two lemmas. The first is a simple computation, the second uses Harish-Chandra's formula (16). Let

$$W_\lambda = \{f \in C^\infty(X) \mid Df = \lambda(D)f \text{ for all } D \in \boldsymbol{D}(X)\} ,$$

where $\lambda(D)$ is given by $D\varphi_\lambda = \lambda(D)\varphi_\lambda$; thus $V_\lambda \subset W_\lambda$. Let

$$\mathscr{H}_\lambda^\infty = C^\infty(\Xi) \cap \mathscr{H}_\lambda .$$

LEMMA. *The mapping $\varphi \longrightarrow \check{\varphi}$ maps $\mathscr{H}_\lambda^\infty$ into W_λ; more precisely,*

$$\check{\varphi}(x) = \int_B \varphi(\xi(x_0, b)) e^{(i\lambda+\rho)(A(x,b))} db \qquad \varphi \in \mathscr{H}_\lambda^\infty .$$

LEMMA. *Let $f \in C_c^\infty(X)$; the function $|\mathbf{c}(\lambda)|^{-2} f \times \varphi_\lambda$ has Radon transform $\hat{f} \times \eta_\lambda$ in the following weak sense:*

$$\left[\int_{\mathfrak{a}^*} A(\lambda) |\mathbf{c}(\lambda)|^{-2} f \times \varphi_\lambda d\lambda\right]^\wedge = \int_{\mathfrak{a}^*} A(\lambda) \hat{f} \times \eta_\lambda d\lambda$$

for all W-invariant functions A in $C_c^\infty(\mathfrak{a}^*)$.

Bibliography

[1] Bhanu Murthy, T. S.: Plancherel's measure for the factor space $\boldsymbol{SL(n,R)/SO(n)}$, Dokl. Akad. Nauk. SSSR, **133** (1960), 503-506.

[2] ———: The asymptotic behavior of zonal spherical functions on the Siegel upper half-plane Dokl. Akad. Nauk. SSSR, **135** (1960), 1027-1029.

[3] Carathéodory, C.: Funktionentheorie, II. Basel, 1950.

[4] Chern, S. S.: On integral geometry in Klein spaces, Ann. of Math., **43** (1942), 178-189.

[5] Furstenberg, H.: A Poisson formula for semisimple Lie groups, Ann. of Math., **77** (1963), 335-386.

[6] Gelfand, I. M.: Spherical functions on symmetric spaces, Dokl. Akad. Nauk. SSSR, **70** (1950), 5-8—Amer. Math. Soc. Transl., **37** (1964), 39-43.

[7] ———: Integral geometry and its relation to the theory of group representations, Uspehi Mat. Nauk, **15** (1960), 155-164—Russian Math. Surveys, 1960.

[8] ——— and M. I. Graev: The geometry of homogeneous spaces, group representations in homogeneous spaces and questions in integral geometry related to them, I. Trudy Moscov. Mat. Obšč., **8** (1959), 321-390—Amer. Math. Soc. Transl., **37** (1964), 351-429.

[9] ——— and M. A. Naimark: Unitary representations of the classical groups, Trudy Mat. Inst. Steklov., **36** (1950).

[10] ——— and S. J. Shapiro: Homogeneous functions and their applications, Uspehi Mat. Nauk., **10** (1955), 3-70.

[11] Gindikin, S. G. and F. I. Karpelevič: Plancherel measure of Riemannian symmetric spaces of nonpositive curvature, Doklady Akad. Nauk. SSSR, **145** (1962), 252-255—Soviet Mathematics, **3** (1962), 962-965.

[12] ———: On a problem in integral geometry Chebotarev Memorial Volume, Kazan Univ., 1964.
[13] Harish-Chandra: The Plancherel formula for complex semisimple Lie groups, Trans. Amer. Math. Soc., **76** (1954), 485-528.
[14] ———: On a lemma of Bruhat, J. Math. Pures Appl., **35** (1956), 203-210.
[15] ———: Spherical functions on a semisimple Lie group I; II, Amer. J. Math., **80** (1958), 241-310; 553-613.
[16] ———: Two theorems on semisimple Lie groups (to appear).
[17] Helgason, S.: Differential geometry and symmetric spaces, Academic Press, New York, 1962.
[18] ———: Duality and Radon transform for symmetric spaces, Amer. J. Math., **85** (1963), 667-692.
[19] ———: Fundamental solutions of invariant differential operators on symmetric spaces, Amer. J. Math., **86** (1964), 565-601.
[20] ———: A duality in integral geometry; some generalizations of the Radon transform, Bull. Amer. Math. Soc., **70** (1964), 435-446.
[21] ———: The Radon transform on Euclidean spaces, compact two-point homogeneous spaces and Grassmann manifolds, Acta Math., **113** (1965), 153-180.
[22] ———: Radon-Fourier transforms on symmetric spaces and related group representations, Bull. Amer. Math. Soc., **71** (1965), 757-763.
[23] ———: An analog of the Paley-Wiener theorem for the Fourier transform on certain symmetric spaces, Math. Ann. (to appear).
[24] F. John: Bestimmung einer Funktion aus ihren Integralen über gewisse Mannigfaltigkeiten, Math. Ann., **100** (1934), 488-520.
[25] ———: Plane waves and spherical means, applied to partial differential equations, Interscience, New York 1955.
[26] Karpelevič, F. I.: Orispherical radial parts of Laplace operators on symmetric spaces, Dokl. Akad. Nauk, SSSR. **143** (1962), 1034-1037—Soviet Mathematics, **3** (1962), 528-531.
[27] Ludwig, D.: The Radon transform on Euclidean spaces (to appear).
[28] Radon, J.: Über die Bestimmung von Funktionen durch ihre Integralwerte längs gewisser Mannigfaltigkeiten, Ber. Verh. Sächs. Akad. Wiss. Leipzig Math.-Nat. Kl., **69** (1917), 262-277.
[29] Sugiura, M.: Representations of compact groups realized by spherical functions on symmetric spaces, Proc. Japan Acad., **38** (3) (1962), 111-113.
[30] Weil, A.: L'intégration dans les groupes topologiques et ses applications, Hermann, Paris, 1940.

Massachusetts Institute of Technology

Reprinted by TRUEXpress Oxford England

HELGASON, S.
Math. Annalen 165, 309—317 (1966)

Totally Geodesic Spheres in Compact Symmetric Spaces*

SIGURDUR HELGASON

§ 1. Introduction

Let M be a compact, irreducible Riemannian globally symmetric space, $I_o(M)$ the largest connected group of isometries of M. As customary, let the Riemannian structure of M be that induced by the negative of the Killing form of the Lie algebra of $I_o(M)$. Let κ be the maximum of the sectional curvatures of M whose values are then restricted to the interval $[0, \kappa]$. By a theorem of E. CARTAN, the maximum dimensional totally geodesic submanifolds of M of constant curvature 0 are all conjugate under $I_o(M)$. In this paper we shall prove an analogous statement for the maximal curvature κ.

Theorem 1.1. *The space M contains totally geodesic submanifolds of constant curvature κ. Any two such submanifolds of the same dimension are conjugate under $I_o(M)$. The maximal dimension of such submanifolds is $1+m(\bar{\delta})$ where $m(\bar{\delta})$ is the multiplicity of the highest restricted root $\bar{\delta}$* (see § 2). *Also, $\kappa = \|\bar{\delta}\|^2$, where $\| \ \|$ denotes length.*

Remark. Except for the case when M is a real projective space the submanifolds above of dimension $1+m(\bar{\delta})$ are actually spheres.

During the proof of Theorem 1.1 we shall obtain the following result:

Theorem 1.2. *Assume that the space M above is simply connected. Then the closed geodesics in M of minimal length are permuted transitively by $I_o(M)$. The minimum length is $2\pi/\|\bar{\delta}\|$.*

For Grassmann manifolds a result like Theorem 1.2 is proved and applied by ELÍASSON [3] in a study of closed geodesics on manifolds homeomorphic to Grassmann manifolds. Certain totally geodesic spheres in Grassmann manifolds have been studied by WOLF [6], [7]. See also RAUCH [8].

Notation. We shall use the customary notation $\boldsymbol{Z}$, $\boldsymbol{R}$, and $\boldsymbol{C}$ for the integers, real numbers and complex numbers, respectively. Lie groups are denoted by capital Roman letters and their Lie algebras by the corresponding lower case German letters. The adjoint representation of a Lie group (resp. Lie algebra) is denoted Ad (resp. ad). If N is a manifold, p a point in N then N_p denotes the tangent space to N at p.

* Work supported in part by the National Science Foundation, NSF GP-2600.

§ 2. Root space decompositions

Let $\mathfrak{u}$ be a compact semisimple Lie algebra over $\boldsymbol{R}$, θ an involutive automorphism of $\mathfrak{u}$. Let $\mathfrak{u}^c$ be the complexification of $\mathfrak{u}$, $\mathfrak{g}$ the real form of $\mathfrak{u}^c$ corresponding to $(\mathfrak{u}, \theta)$, that is $\mathfrak{g}=\mathfrak{k}+\mathfrak{p}$, where $\mathfrak{k}=\{T\in\mathfrak{u}|\theta T=T\}$ and $\mathfrak{p}=\{iX|X\in\mathfrak{u}, \theta X=-X\}$. Then $\mathfrak{u}=\mathfrak{k}+\mathfrak{p}_*$, where $\mathfrak{p}_*=i\mathfrak{p}$. Let $\mathfrak{a}\subset\mathfrak{p}$ be a maximal abelian subspace, put $\mathfrak{a}_*=i\mathfrak{a}$ and extend $\mathfrak{a}_*$ to a maximal abelian subalgebra $\mathfrak{t}$ of $\mathfrak{u}$. Then the complexification $\mathfrak{t}^c\subset\mathfrak{u}^c$ is a Cartan subalgebra of $\mathfrak{u}^c$. Let Δ denote the corresponding system of nonzero roots, $\Delta_{\mathfrak{p}}$ the set of $\alpha\in\Delta$ which do not vanish identically on $\mathfrak{a}^c$ (the complexification of $\mathfrak{a}$ in $\mathfrak{u}^c$). Let Σ denote the set of restrictions of elements of $\Delta_{\mathfrak{p}}$ to $\mathfrak{a}^c$; its elements are called restricted roots. Let $\mathfrak{t}_{\mathfrak{k}}=\mathfrak{k}\cap\mathfrak{t}$, $\mathfrak{t}_{\mathfrak{k}_*}=i\mathfrak{t}_{\mathfrak{k}}$. Then all $\alpha\in\Delta$ take real values on $\mathfrak{a}+\mathfrak{t}_{\mathfrak{k}_*}$. Select compatible orderings in the dual spaces of $\mathfrak{a}+\mathfrak{t}_{\mathfrak{k}_*}$ and $\mathfrak{a}$ and let Δ^+ and Σ^+ denote the set of positive elements in Δ and Σ, respectively. If f is a function on $\mathfrak{t}^c$ its restriction to $\mathfrak{a}^c$ is denoted $\bar{f}$. For each $\lambda\in\Sigma$ the number of $\alpha\in\Delta_{\mathfrak{p}}$ such that $\lambda=\bar{\alpha}$ is called the multiplicity of λ and is denoted by $m(\lambda)$.

For each linear form λ on $\mathfrak{a}^c$ put

$$\mathfrak{k}_\lambda=\{T\in\mathfrak{k}|(\operatorname{ad}H)^2T=\lambda(H)^2T \quad \text{for all} \quad H\in\mathfrak{a}_*\}$$
$$\mathfrak{p}_\lambda=\{X\in\mathfrak{p}_*|(\operatorname{ad}H)^2X=\lambda(H)^2X \quad \text{for all} \quad H\in\mathfrak{a}_*\}.$$

Then $\mathfrak{k}_\lambda=\mathfrak{k}_{-\lambda}$, $\mathfrak{p}_\lambda=\mathfrak{p}_{-\lambda}$ and $\mathfrak{k}_o$ and $\mathfrak{p}_o$ are the centralizers of $\mathfrak{a}_*$ in $\mathfrak{k}$ and $\mathfrak{p}_*$, respectively. The endomorphisms $(\operatorname{ad}H)^2$ $(H\in\mathfrak{a}_*)$ commute and are symmetric with respect to the Killing form B of $\mathfrak{u}^c$. But the eigenvalues of $(\operatorname{ad}H)^2$ are 0 and $\lambda(H)^2$ $(\lambda\in\Sigma)$ (cf. [4], p. 248). It follows that

$$\mathfrak{k}_\lambda=\mathfrak{p}_\lambda=\{0\} \quad \text{if} \quad \lambda\notin\Sigma\cup\{0\}$$

and that we have the direct decompositions

$$\mathfrak{k}=\mathfrak{k}_o+\sum_{\lambda\in\Sigma^+}\mathfrak{k}_\lambda, \tag{1}$$

$$\mathfrak{p}_*=\mathfrak{p}_o+\sum_{\lambda\in\Sigma^+}\mathfrak{p}_\lambda. \tag{2}$$

For each $\alpha\in\Delta$ select $X_\alpha\neq0$ in $\mathfrak{u}^c$ such that $[H, X_\alpha]=\alpha(H)X_\alpha$ for all $H\in\mathfrak{t}^c$. Extend θ to an automorphism of $\mathfrak{u}^c$, also denoted θ, and let

$$\mathfrak{u}^+(\alpha)=\boldsymbol{C}(X_\alpha+\theta X_\alpha), \quad \mathfrak{u}^-(\alpha)=\boldsymbol{C}(X_\alpha-\theta X_\alpha).$$

Then we have for $\lambda\in\Sigma$ (cf. Lemma 3.6, Ch. VI in [4])

$$\mathfrak{k}_\lambda+i\mathfrak{k}_\lambda=\sum_{\bar{\alpha}=\lambda}\mathfrak{u}^+(\alpha), \quad \mathfrak{p}_\lambda+i\mathfrak{p}_\lambda=\sum_{\bar{\alpha}=\lambda}\mathfrak{u}^-(\alpha). \tag{3}$$

Since λ is real on $\mathfrak{a}$, the two right hand sides in (3) are invariant under the conjugation of $\mathfrak{u}^c$ with respect to $\mathfrak{g}$, so

$$\mathfrak{k}_\lambda=\mathfrak{k}\cap\sum_{\bar{\alpha}=\lambda}\mathfrak{u}^+(\alpha), \quad \mathfrak{p}_\lambda=\mathfrak{p}_*\cap\sum_{\bar{\alpha}=\lambda}\mathfrak{u}^-(\alpha) \tag{4}$$

and

$$\mathfrak{k}_o=\mathfrak{t}_{\mathfrak{k}}+\mathfrak{k}\cap\sum_{\bar{\alpha}=0}\mathfrak{u}^+(\alpha), \quad \mathfrak{p}_o=\mathfrak{a}_*. \tag{5}$$

Using the well-known commutation relations between the X_α the following two lemmas are obtained from (3), (4), and (5) without difficulty.

Lemma 2.1. *Let* $\lambda, \mu \in \Sigma \cup \{0\}$. *Then*

$$[\mathfrak{k}_\lambda, \mathfrak{p}_\mu] \subset \mathfrak{p}_{\lambda+\mu} + \mathfrak{p}_{\lambda-\mu},$$
$$[\mathfrak{k}_\lambda, \mathfrak{k}_\mu] \subset \mathfrak{k}_{\lambda+\mu} + \mathfrak{k}_{\lambda-\mu},$$
$$[\mathfrak{p}_\lambda, \mathfrak{p}_\mu] \subset \mathfrak{k}_{\lambda+\mu} + \mathfrak{k}_{\lambda-\mu};$$

For each linear function ξ on $\mathfrak{t}^c$ let $H_\xi \in \mathfrak{t}^c$ be determined by $B(H_\xi, H) = \xi(H)$ $(H \in \mathfrak{t}^c)$ and if η is a linear function on $\mathfrak{a}^c$ let $A_\eta \in \mathfrak{a}^c$ be determined by $B(A_\eta, H) = \eta(H)$ for all $H \in \mathfrak{a}^c$. Put $\langle \xi_1, \xi_2 \rangle = B(H_{\xi_1}, H_{\xi_2})$, $\langle \eta_1, \eta_2 \rangle = B(A_{\eta_1}, A_{\eta_2})$ and $|\xi| = |H_\xi| = B(H_\xi, H_\xi)^{1/2}$ if $H_\xi \in i\mathfrak{t}$, $\|\eta\| = \|A_\eta\| = B(A_\eta, A_\eta)^{1/2}$ if $A_\eta \in \mathfrak{a}$. If $\lambda \in \Sigma$ let $\mathfrak{a}_\lambda$ denote the subspace $\boldsymbol{R} iA_\lambda$ of $\mathfrak{a}_*$. If $\alpha \in \Delta$ then $H_\alpha - A_{\bar{\alpha}} \in \mathfrak{t}_{\mathfrak{t}_*}$.

Lemma 2.2. *Let* $\lambda \in \Sigma$. *Then*

$$[\mathfrak{k}_\lambda, \mathfrak{p}_\lambda] \subset \mathfrak{p}_{2\lambda} + \mathfrak{a}_\lambda .$$

Let $W_\mathfrak{t}$ and $W_\mathfrak{a}$ be the Weyl groups of $(\mathfrak{u}, \mathfrak{t})$ and $(\mathfrak{u}, \mathfrak{a}_*)$, respectively. They act on $\mathfrak{t}^c$ and $\mathfrak{a}^c$ as well as on their duals ($sH_\xi = H_{s\xi}$ etc.) The sets Δ and Σ have been ordered. Let $C \subset \mathfrak{a} + \mathfrak{t}_{\mathfrak{t}_*}$ be the Weyl chamber where all $\alpha \in \Delta^+$ take positive values and let $C_\mathfrak{p} \subset \mathfrak{a}$ be the Weyl chamber where all $\lambda \in \Sigma^+$ take positive values. By the compatibility of the orderings, $C_\mathfrak{p} \subset \mathrm{Cl}(C)$, Cl denoting closure.

Lemma 2.3. *Let* $\delta \in \Delta^+$ *denote the highest root. Then* $H_\delta \in \mathrm{Cl}(C)$, $A_{\bar{\delta}} \in \mathrm{Cl}(C_\mathfrak{p})$ *and* $A_{\bar{\delta}} \neq 0$.

Proof. Since $W_\mathfrak{t}$ permutes Δ and $W_\mathfrak{a}$ permutes Σ, the two first statements are contained in the general statement that the closed Weyl chamber consists of those elements which dominate their transforms under the Weyl group. Finally $\delta \geqq \alpha$ for all $\alpha \in \Delta$ so, by the compatibility of the orderings, $\bar{\delta} \geqq \lambda$ for all $\lambda \in \Sigma$; hence $\bar{\delta} \neq 0$.

Lemma 2.4. *Let* $\alpha \in \Delta_\mathfrak{p}$. *Then*

$$\langle \alpha, \alpha \rangle = m \langle \bar{\alpha}, \bar{\alpha} \rangle, \quad \textit{where} \quad m = 1, 2 \text{ or } 4 .$$

This is well known (cf. ARAKI [1], p. 6—7) but we give a direct proof. The linear form $\alpha + \theta\alpha$ is 0 on $\mathfrak{a}$; it is not in Δ because then $[X_\alpha, \theta X_\alpha]$ would lie in $\mathfrak{k} + i\mathfrak{k}$, whereas clearly $[X_\alpha, \theta X_\alpha] \subset \mathfrak{p} + i\mathfrak{p}$. If $\alpha + \theta\alpha = 0$ then $m = 1$. If $\alpha + \theta\alpha \neq 0$ then $\alpha + \theta\alpha$ is not a root so the integer $2\langle \theta\alpha, \alpha \rangle / \langle \alpha, \alpha \rangle$ is $\geqq 0$, whence the angle between H_α and $H_{\theta\alpha}$ is 0, $\pi/3$ or $\pi/2$. But

(6) $$A_{\bar{\alpha}} = \frac{1}{2}(H_\alpha - H_{\theta\alpha})$$

so the lemma follows.

§ 3. Closed geodesics of minimal length. Proof of Theorem 1.2

Preserving the notation of § 2, we assume now that $\mathfrak{u}$ is simple. Let U be the simply connected Lie group with Lie algebra $\mathfrak{u}$, "extend" θ to an automorphism of U and let K be the group of fixed points of θ. As is well known, K is connected. The negative of the Killing form B of $\mathfrak{u}^c$ induces a Riemannian structure on U/K and on U. Let o denote the point $\{K\}$ in U/K and let Exp denote the Exponential mapping of $\mathfrak{p}_*$ onto U/K.

Let $\gamma(t)$ $(-\infty<t<\infty)$ be a geodesic in a Riemannian manifold. The geodesic is called *closed* if there exists a number $L>0$ such that $\gamma(t+L)=\gamma(t)$ for all t. The geodesic is said to be *simply closed* if in addition $\gamma(t_1)\neq\gamma(t_2)$ for $0<t_1<t_2\leqq L$. If $|t|$ is the arc parameter, L is then called the length of the simply closed geodesic.

If a geodesic in the symmetric space U/K intersects itself then since it is an orbit of a one-parameter group of U it is a simply closed geodesic (see e.g. [4], p. 355).

Lemma 3.1. *The shortest, periodic one-parameter subgroups in a simple, simply connected, compact Lie group U have length $4\pi/|\delta|$ and they are all conjugate in U.*

Let $C_*=iC\subset\mathfrak{t}$ and consider the polyhedron $P_*\subset C_*$ given by

$$P_*=\{H\subset\mathfrak{t}|(2\pi i)^{-1}\alpha(H)>0(\alpha\in\Delta^+),\ (2\pi i)^{-1}\delta(H)<1\}.$$

Since U is simply connected, the unit lattice $\mathfrak{t}_e=\{H\in\mathfrak{t}|\exp H=e\}$ satisfies the well-known relation

$$\mathfrak{t}_e\cap \mathrm{Cl}(P_*)=\{0\} \tag{1}$$

([4], p. 264). For $\alpha\in\Delta$ let $\mathfrak{t}_\alpha=\{H\in\mathfrak{t}|\alpha(H)\in 2\pi i\boldsymbol{Z}\}$ and let T_α be the centralizer of $\exp\mathfrak{t}_\alpha$ in U. In view of Lemma 4.5, Ch. VII in [4], T_α contains an element u such that the restriction of $\mathrm{Ad}(u)$ to $\mathfrak{t}$ is the reflection s_α in the hyperplane $\alpha=0$. It follows that

$$\exp(2H(\alpha))=e, \tag{2}$$

where $H(\alpha)=2\pi i H_\alpha/|\alpha|^2$ is the projection of 0 onto the hyperplane $\alpha=2\pi i$ in $\mathfrak{t}$. Each $H\in\mathfrak{t}_e$ satisfies $\beta(H)\in 2\pi i\boldsymbol{Z}$ for all $\beta\in\Delta$ so it is clear from (1), (2) and Lemma 2.3 that the one-parameter subgroup $t\to\exp(2tH(\delta))$ is periodic and has length $2|H(\delta)|$. Now let $t\to\exp tX$ be any periodic one-parameter subgroup of U of length $\leqq 2|H(\delta)|$, the parameter being fixed such that $\exp tX\neq e$ for $0<t<1$ and $\exp X=e$. Then for some $u\in U$, $H_1=\mathrm{Ad}(u)X$ lies in $\mathrm{Cl}(C_*)$ and since $\exp H_1=e$, $\delta(H_1)=n2\pi i$ where $n\geqq 0$ in $\boldsymbol{Z}$. Since $H_1\neq 0$ we have $n\neq 0$. Also $n\neq 1$ by (1). But the point $2H(\delta)$ is the only point which minimizes the distance from 0 to the union of the hyperplanes $\delta=n2\pi i (n\geqq 2)$ in $\mathfrak{t}$. Thus $H_1=2H(\delta)$ and the lemma is proved.

We can now prove Theorem 1.2 for the space U/K. First observe that if $H\in\mathfrak{a}_*$ then $\mathrm{Exp}H=o$ if and only if $\exp 2H=e$. Let $A(\bar\alpha)=\pi i A_{\bar\alpha}/\|\bar\alpha\|^2$, $(\alpha\in\Delta_{\mathfrak{p}})$, and let us verify the relation

$$\mathrm{Exp}(2A(\bar\alpha))=o. \tag{3}$$

Let $|\alpha|^2=m\|\bar\alpha\|^2$, so by Lemma 2.4, $m=1, 2, 4$. In the first case, $H_\alpha=A_{\bar\alpha}$ and (3) reduces to (2). In the cases $m=2, 4$ we use $iH_\alpha-iA_{\bar\alpha}\in\mathfrak{k}$, so by (2),

$$\exp(2\pi i A_{\bar\alpha}/\|\bar\alpha\|^2)\in\exp(2\pi i H_\alpha/\|\bar\alpha\|^2)K=K,$$

proving (3). In particular $4A(\bar{\delta}) \in \mathfrak{t}_e$, and since $\delta(4A(\bar{\delta})) = 4\pi i$ and $4A(\bar{\delta}) \in \mathrm{Cl}(C_*)$ we have by (1), $4tA(\bar{\delta}) \notin \mathfrak{t}_e$ for $0 < t < 1$. Consequently the geodesic

$$(4) \qquad t \to \mathrm{Exp}(2tA(\bar{\delta}))$$

is simply closed and has length $2\|A(\bar{\delta})\| = 2\pi/\|\bar{\delta}\|$.

Now let $t \to \mathrm{Exp}\, tX$ be any simply closed geodesic of length $\leqq 2\pi/\|\bar{\delta}\|$ such that $\mathrm{Exp} X = o$, $\mathrm{Exp}\, t_1 X \neq \mathrm{Exp}\, t_2 X$ for $0 < t_1 < t_2 \leqq 1$. Select $k \in K$ such that $H = \mathrm{Ad}_U(k) X \in \mathrm{Cl}(C_{\mathfrak{p}_*})$ where $C_{\mathfrak{p}_*} = iC_{\mathfrak{p}}$. Then $2H \in \mathfrak{t}_e$ so

$$\delta(H) = m\pi i,$$

where $m \geqq 0$ in $\mathbf{Z}$. But $H \neq 0$ so $m \neq 0$ and $m \neq 1$ because of (1). But the point $2A(\bar{\delta})$ is the only point which minimizes the distance from 0 to the union of the hyperplanes $\bar{\delta} = m\pi i (m \geqq 2)$ in $\mathfrak{a}_*$. Thus $H = 2A(\bar{\delta})$ and Theorem 1.2 is proved for the space U/K.

In order to complete the proof of Theorem 1.2 we must consider the group U as a symmetric space $U \times U/U^*$ (U^* = diagonal in $U \times U$) with Riemannian structure Q^* defined by the Killing form B^* of $\mathfrak{u} \times \mathfrak{u}$, the identification being made via the mapping $\tau: (u_1, u_2) U^* \to u_1 u_2^{-1}$ $(u_1, u_2 \in U)$. The tangent space to $U \times U/U^*$ at o is according to the usual identification given by the orthogonal complement of the Lie algebra of U^* in $\mathfrak{u} \times \mathfrak{u}$, hence equals the space $\{(X, -X) | X \in \mathfrak{u}\}$. Since

$$(5) \qquad B^*((X, -X), (Y, -Y)) = 2B(X, Y) \qquad X, Y \in \mathfrak{u}$$

and since $d\tau_o(X, -X) = 2X$ it is clear that for each tangent vector Z to $U \times U/U^*$,

$$(6) \qquad 2Q^*(Z, Z) = Q(d\tau(Z), d\tau(Z)).$$

Here Q is the Riemannian structure on U defined by B. The space $\{(H, -H) | H \in \mathfrak{t}\}$ is a maximal abelian subspace of the tangent space above and is contained in the Cartan subalgebra $\mathfrak{t} \times \mathfrak{t}$ of $\mathfrak{u} \times \mathfrak{u}$. The roots of $(\mathfrak{u} \times \mathfrak{u})^c$ with respect to $(\mathfrak{t} \times \mathfrak{t})^c$ are given by the linear forms

$$(H_1, H_2) \to \alpha(H_1), \quad (H_1, H_2) \to \alpha(H_2) \quad (H_1, H_2 \in \mathfrak{t}^c)$$

as α runs through the roots of $\mathfrak{u}^c$ with respect to $\mathfrak{t}^c$. The highest restricted root, say $\tilde{\delta}$, is given by $\tilde{\delta}(H, -H) = \delta(H)$ $(H \in \mathfrak{t})$ and by (5) we find $A_{\tilde{\delta}} = (\frac{1}{2} H_\delta, -\frac{1}{2} H_\delta)$ and $2\|\tilde{\delta}\|^2 = |\delta|^2$. In the Riemannian structure Q^* the one-parameter subgroups from Lemma 3.1 have length $2\pi\sqrt{2}/|\delta|$ which equals $2\pi/\|\tilde{\delta}\|$. This concludes the proof of Theorem 1.2.

We note some simple consequences of the proof. First we have from (2) and (3),

Corollary 3.2. *Let* $\alpha \in \Delta$. *Then* $|\alpha| \leqq |\delta|$ *and* $\|\bar{\alpha}\| \leqq \|\bar{\delta}\|$.

Corollary 3.3. *Suppose* $\alpha \in \Delta$ *and* $\bar{\alpha} \neq \pm \bar{\delta}$. *Then*

$$2 \frac{\langle \bar{\alpha}, \bar{\delta} \rangle}{\langle \bar{\delta}, \bar{\delta} \rangle} = 0 \quad or \quad \pm 1$$

so the value is 0 *if* $2\bar{\alpha} \in \Sigma$.

In fact, $\delta(A(\bar{\delta}))=\pi i$ and by (3), $\alpha(A(\bar{\delta}))=\frac{1}{2}\pi i n$ where $n\in \mathbf{Z}$. But $\delta-\alpha$ is a positive integral linear combination of the simple roots in Δ^+, so $n=0$, ± 1, ± 2. But if $n=\pm 2$ we deduce from Cor. 3.2 that $\bar{\alpha}=\pm\bar{\delta}$.

Let M be as in Theorem 1.2. For each $p\in M$ let A_p denote the set of midpoints of closed geodesics of minimal length passing through p. We call A_p the *midpoint locus* (associated with p).

Corollary 3.4. *For each $p\in M$ the midpoint locus*[1] *A_p is a totally geodesic submanifold of M and is an orbit of the isotropy subgroup of $I_o(M)$ at p.*

In fact let $q\in A_p$. Then the geodesic symmetry of M with respect to q leaves p fixed and consequently maps A_p into itself. In addition to Theorem 1.2 we now need the following more general lemma whose proof is contained in the proof of Prop. 5.1 in [5].

Lemma 3.5. *Let Q be a Riemannian globally symmetric space, N a submanifold of Q such that for each $n\in N$, N is invariant under the geodesic symmetry of Q with respect to n. Then N, with the Riemannian structure induced by Q, is totally geodesic in Q.*

We conclude this section with a further description of the midpoint locus and the space of closed geodesics of minimal length.

For $\varepsilon=0$, $1/2$, 1, put

$$\mathfrak{k}(\varepsilon)=\sum_{\langle\lambda,\bar{\delta}\rangle=\varepsilon\langle\bar{\delta},\bar{\delta}\rangle}\mathfrak{k}_\lambda;\quad \mathfrak{p}(\varepsilon)=\sum_{\langle\lambda,\bar{\delta}\rangle=\varepsilon\langle\bar{\delta},\bar{\delta}\rangle}\mathfrak{p}_\lambda.$$

Then by Cor. 3.3, $\mathfrak{k}(1)=\mathfrak{k}_{\bar{\delta}}$, $\mathfrak{p}(1)=\mathfrak{p}_{\bar{\delta}}$ and

$$\mathfrak{k}=\mathfrak{k}_o+\mathfrak{k}(0)+\mathfrak{k}\left(\frac{1}{2}\right)+\mathfrak{k}(1),\quad \mathfrak{p}_*=\mathfrak{p}_o+\mathfrak{p}(0)+\mathfrak{p}\left(\frac{1}{2}\right)+\mathfrak{p}(1).$$

Proposition 3.6. *Let S and S', respectively, denote the centralizers of $\exp 2A(\bar{\delta})$ and $A(\bar{\delta})$ in K. Their respective Lie algebras are given by*

$$\mathfrak{s}=\mathfrak{k}_o+\mathfrak{k}(0)+\mathfrak{k}_{\bar{\delta}},\quad \mathfrak{s}'=\mathfrak{k}_o+\mathfrak{k}(0). \tag{7}$$

In the space U/K the midpoint locus A_o can be identified with K/S and the space of closed geodesics of minimal length is identified with U/S'. Let $o'=\operatorname{Exp}A(\bar{\delta})$; then

$$A_{o'}=\operatorname{Exp}\mathfrak{p}\left(\frac{1}{2}\right).$$

Proof. The relations (7) are immediate. Using θ one finds that an element $k\in K$ commutes with $\exp 2A(\bar{\delta})$ if and only if k leaves the point $\operatorname{Exp}A(\bar{\delta})$ fixed. Hence $A_o=K/S$; the identification of U/S' is also immediate. Finally

$$A_{o'}=\exp(-A(\bar{\delta}))K\exp A(\bar{\delta})\cdot o.$$

If $T\in\mathfrak{k}$ it is clear that

$$\operatorname{Ad}_U(\exp(-A(\bar{\delta})))T\equiv-\sinh(\operatorname{ad}(A(\bar{\delta})))T \quad (\operatorname{mod}\mathfrak{k})$$

[1] This leads to a natural definition of the Radon transform for M, generalizing the one given in [5].

so the curve

$$t \to \exp(-A(\bar{\delta}))\exp tT\exp(A(\bar{\delta}))\cdot o$$

has tangent vector $-\sinh(\operatorname{ad}A(\bar{\delta}))T$ at $t=0$. But

$$\sinh(\operatorname{ad}A(\bar{\delta}))\mathfrak{k} = \mathfrak{p}\left(\frac{1}{2}\right).$$

which finishes the proof.

§ 4. Totally geodesic spheres. Proof of Theorem 1.1

In order to compute the maximal curvature κ for the space U/K from § 3 it suffices to consider plane sections $\subset \mathfrak{p}_*$ spanned by orthonormal vectors H, X where $H \in C_{\mathfrak{p}_*}$. The corresponding curvature is then given by a formula of Cartan (cf. [4], p. 205) as

(1) $$B((\operatorname{ad}H)^2 X, X).$$

Decomposing X by (2) § 2,

$$X = X_o + \sum_{\lambda\in\Sigma^+} X_\lambda$$

we have

$$B((\operatorname{ad}H)^2 X, X) = \sum_{\lambda\in\Sigma^+} \lambda(H)^2 B(X_\lambda, X_\lambda) \leqq \delta(H)^2$$

so $\kappa \leqq \|\bar{\delta}\|^2$.

Now since $2\bar{\delta}$ is not a restricted root it is easily seen from Lemmas 2.1 and 2.2 that the subspace $\mathfrak{a}_{\bar{\delta}} + \mathfrak{p}_{\bar{\delta}}$ of $\mathfrak{p}_*$ is a Lie triple system. Let S and $\mathfrak{s}$ be as in Prop. 3.6. If $\lambda \in \Sigma^+$ such that $\langle \lambda, \bar{\delta}\rangle = 0$ then by Cor. 3.3 neither $\bar{\delta}+\lambda$ nor $\bar{\delta}-\lambda$ is a restricted root. Hence by (7) § 3 and Lemmas 2.1 and 2.2,

(2) $$[\mathfrak{s}, \mathfrak{a}_{\bar{\delta}}] = \mathfrak{p}_{\bar{\delta}}, \quad [\mathfrak{s}, \mathfrak{p}_{\bar{\delta}}] = \mathfrak{a}_{\bar{\delta}} + \mathfrak{p}_{\bar{\delta}}.$$

Since $\mathfrak{a}_{\bar{\delta}}$ is a maximal abelian subspace of $\mathfrak{a}_{\bar{\delta}} + \mathfrak{p}_{\bar{\delta}}$, the globally symmetric space $M_\delta = \operatorname{Exp}(\mathfrak{a}_{\bar{\delta}} + \mathfrak{p}_{\bar{\delta}})$ has rank one. If H, X are orthonormal vectors, $H \in \mathfrak{a}_{\bar{\delta}}$, $X \in \mathfrak{p}_{\bar{\delta}}$, the corresponding sectional curvature of M_δ is by (1) equal to $\|\bar{\delta}\|^2$. It follows that $\kappa = \|\bar{\delta}\|^2$ and that M_δ has constant curvature. Let S_o denote the identity component of S. Since $[\mathfrak{s}, \mathfrak{a}_{\bar{\delta}}] = \mathfrak{p}_{\bar{\delta}}$ it is clear that the orbit $\operatorname{Ad}_U(S_o)\cdot A(\bar{\delta})$ is the sphere in $\mathfrak{a}_{\bar{\delta}} + \mathfrak{p}_{\bar{\delta}}$ with center 0, passing through $A(\bar{\delta})$. Since all $s \in S_o$ leave $\operatorname{Exp}A(\bar{\delta})$ fixed all geodesics in M_δ through o pass through $o' = \operatorname{Exp}A(\bar{\delta})$. Hence $M_\delta - o'$ is the diffeomorphic image of a ball ([4], p. 356) so M_δ is simply connected, thus a sphere.

Now since the geodesic symmetry $s_{o'}$ of U/K leaves o fixed it is clear that any geodesic segment of minimum length joining o and o' is a part of a closed geodesic of length $\leqq 2\pi/\|\bar{\delta}\|$. By Theorem 1.2 it is a simply closed geodesic of length $2\pi/\|\bar{\delta}\|$. The closed geodesics of this length, starting at o, and passing through o' are permuted transitively by S, hence form finitely many distinct spheres Σ_i of dimension $1+m(\bar{\delta})$, namely the images of M_δ under S. Let σ_i denote the unit sphere in the tangent space $(\Sigma_i)_o$. Each σ_i is an orbit of

$\mathrm{Ad}_U(S_o)$ so any two different σ_i are disjoint. Now let Σ be any totally geodesic sphere in U/K of curvature $\|\bar{\delta}\|^2$. Using a motion $u\in U$ we may assume that Σ passes through o and o'. Then the unit sphere σ in the tangent space $(\Sigma)_o$ is contained in the union of the σ_i. But σ is connected so it is contained in a single σ_i. Hence there exists an element $s\in S$ such that $s\cdot\Sigma\subset M_\delta$.

Next we observe that

$$\mathfrak{s}\supset[\mathfrak{a}_{\bar{\delta}}+\mathfrak{p}_{\bar{\delta}},\ \mathfrak{a}_{\bar{\delta}}+\mathfrak{p}_{\bar{\delta}}].$$

The right hand side of this relation is a Lie algebra $\mathfrak{k}'$ contained in $\mathfrak{k}$ and $\mathrm{ad}_{\mathfrak{u}}(\mathfrak{k}')$ restricted to $\mathfrak{a}_{\bar{\delta}}+\mathfrak{p}_{\bar{\delta}}$ is the Lie algebra of the special orthogonal group of the tangent space $(M_\delta)_o$. It follows that this group is contained in the restriction of $\mathrm{Ad}_U(S_o)$ to $(M_\delta)_o$. Thus any two totally geodesic spheres of the same dimension, contained in M_δ, are conjugate under a member of U; to finish the proof of Theorem 1.1 for the space U/K it remains to verify that any totally geodesic submanifold N of U/K of constant curvature $\|\bar{\delta}\|^2$ is a sphere. But if N were not a sphere it is clear, passing to the universal covering of N, that N would contain a simply closed geodesic of length $<2\pi/\|\bar{\delta}\|^2$ which is impossible.

If M is any compact Riemannian globally symmetric space such that $I_o(M)$ is simple then Theorem 1.1 can be applied to the universal covering space of M and the theorem follows for M.

Now let us establish the remark following Theorem 1.1 in the case when $I_o(M)$ is simple. Let Z denote the center of U and put

$$K_Z=\{u|u^{-1}\theta(u)\in Z\}.$$

Then U/K_Z is a Riemannian globally symmetric space and any such space, associated with $(\mathfrak{u},\theta)$, actually covers U/K_Z ([4], Theorem 8.1, Ch. VII). It is appropriate to call U/K_Z the *adjoint space* of the orthogonal symmetric Lie algebra $(\mathfrak{u},\theta)$ because if its construction (which does not require $\mathfrak{u}$ to be simple) is carried out for the group case $\mathfrak{u}=\mathfrak{v}\times\mathfrak{v}$, θ interchanging the two factors, then one obtains the adjoint group of $\mathfrak{v}$. Under the covering map $\varphi: uK\to uK_Z$ of U/K onto U/K_Z the sphere $\mathrm{Exp}(\mathfrak{a}_{\bar{\delta}}+\mathfrak{p}_{\bar{\delta}})$ is mapped onto a sphere if and only if $\varphi(\mathrm{Exp}A(\bar{\delta}))\neq o$. But $\varphi(\mathrm{Exp}A(\bar{\delta}))=o$ if and only if $\exp A(\bar{\delta})\in K_Z$, which is equivalent to $\exp 2A(\bar{\delta})\in Z$, which in turn implies $K=S$. But if $K=S$ then by Prop. 3.6, $\mathfrak{p}\left(\frac{1}{2}\right)=0$ so

$$\mathfrak{p}_*=\mathfrak{a}_*=\mathfrak{p}_{\bar{\delta}}+\sum_{\langle\lambda,\bar{\delta}\rangle=0}\mathfrak{p}_\lambda.$$

Thus Σ^+ has the property that all $\lambda\in\Sigma^+-\{\bar{\delta}\}$ are perpendicular to $\bar{\delta}$. This means $\Sigma^+=\{\bar{\delta}\}$ so U/K is a sphere and consequently, U/K_Z is a real projective space.

We have now established Theorem 1.1 and the subsequent remark for the case when $I_o(M)$ is simple. The remaining case when M is a group (cf. [4], Prop. 1.2, p. 327) is handled quite similarly. We only indicate the necessary changes.

Suppose U has Riemannian structure given by $-B$. Then if X, Y are orthonormal vectors in $\mathfrak{u}$ the corresponding sectional curvature of U is

$$\frac{1}{4}B((\operatorname{ad}X)^2 Y, Y),$$

and the maximum sectional curvature is readily found to be $\frac{1}{4}|\delta|^2$. However, with the conventions in Theorem 1.1 we must view U as $U \times U/U^*$, whereby the Riemannian structure is multiplied by $\frac{1}{2}$ (cf. (6) § 3). Accordingly all sectional curvatures are multiplied by 2 so $\kappa = 2|\delta|^2/4$ which, as we have seen, equals $\|\tilde{\delta}\|^2$. Since the multiplicity $m(\tilde{\delta})$ now equals 2 the totally geodesic spheres in U of curvature $\|\tilde{\delta}\|^2$ have dimension $\leqq 3$. In the adjoint group U/Z the totally geodesic submanifolds of constant curvature $\|\tilde{\delta}\|^2$ and dimension 3 are spheres unless the point $\exp H(\delta)$ lies in Z. As is well known, $\exp H(\delta) \in Z$ if and only if $\alpha(H(\delta)) \in 2\pi i \boldsymbol{Z}$ for all $\alpha \in \Delta$ (see e.g. [4], p. 268). But in view of Cor. 3.2, $\alpha(H(\delta)) \in 2\pi i \boldsymbol{Z}$ implies that all roots except $\pm\delta$ are orthogonal to δ, which implies $U = \boldsymbol{SU}(2)$ so U/Z is a real projective space.

References

[1] Araki, S.: On root systems and an infinitesimal classification of irreducible symmetric spaces. J. Math. Osaka City Univ. **13**, 1–34 (1962).

[2] Cartan, E.: Sur certaines formes riemanniennes remarquables des géométries a groupe fondamental simple. Ann. Sci. École Norm. Sup. **44**, 345–467 (1927).

[3] Elíasson, H.: Über die Anzahl geschlossener Geodätischer in gewissen Riemannschen Mannigfaltigkeiten. Math. Ann. (to appear).

[4] Helgason, S.: Differential geometry and symmetric spaces. New York: Academic Press, 1962.

[5] — The Radon transform on Euclidean spaces, compact two-point homogeneous spaces and Grassmann manifolds. Acta Math. **113**, 153—180 (1965).

[6] Wolf, J. A.: Geodesic spheres in Grassmann manifolds. Illinois J. Math. **7**, 425–446 (1963).

[7] — Elliptic spaces in Grassmann manifolds. Illinois J. Math. **7**, 447–462 (1963).

[8] Rauch, H. E.: The global study of geodesics in symmetric and nearly symmetric Riemannian manifolds. Comment. Math. Helv. **35**, 111–125 (1961).

Prof. S. Helgason
The Institute for Advanced Study
Princeton, N. J.

(Received January 29, 1965)

HELGASON, S.
Math. Annalen 165, 297—308 (1966)

An Analogue of the Paley-Wiener Theorem for the Fourier Transform on Certain Symmetric Spaces*

SIGURDUR HELGASON

§ 1. Introduction

The classical theorem of PALEY-WIENER [12] (see also SCHWARTZ [13]) characterizes the class of functions which are Fourier transforms of C^∞ functions of compact support on $\boldsymbol{R}^n$. In the formulation of HÖRMANDER [10], the theorem states that an entire analytic function U is the Fourier-Laplace transform of a C^∞ function on $\boldsymbol{R}^n$ with support in the ball $|x| \leqq A$ if and only if for each integer N there exists a constant C_N such that

$$|U(\xi + i\eta)| \leqq C_N(1 + |\xi + i\eta|)^{-N} e^{A|\eta|}, \qquad (\xi, \eta \in \boldsymbol{R}^n). \tag{1}$$

The present paper considers the analogous question for the Fourier transform on a symmetric space S of the noncompact type, as developed primarily by HARISH-CHANDRA [5]. A partial answer is given by Corollary 5.6. The problem is to prove that for each $U(\xi + i\eta)$ satisfying (1) the inverse Fourier transform ((7) § 4)

$$f(\exp H) = \int_{A} U(\lambda)\, \varphi_\lambda(\exp H)\, |c(\lambda)|^{-2}\, d\lambda \tag{2}$$

vanishes for $|H| > A$. In Prop. 5.1, such a statement is proved for each term in the expansion (Prop. 4.2)

$$\varphi_\lambda(\exp H) = \sum_{\mu \in L} \psi_\mu(\lambda, H)\,; \tag{3}$$

the proof is based on translating the integration in (2) into the complex domain in such a way that the singularities of $c(\lambda)^{-1}$ and of the coefficients in the asymptotic expansion (3) are simultaneously avoided. This is possible by using the formula of Gindikin-Karpelevič for $c(\lambda)$ and Harish-Chandra's recursion formula for the coefficients in (3). The Paley-Wiener theorem then results from passage to the function φ_λ itself; however this requires justification of the term-by-term integration in (2) and (3). In this paper we have carried out the detailed justification for the case rank$(G/K) = 1$ and the simpler case G complex.

* Work supported in part by the National Science Foundation, NSF GP-2600.

§ 2. Notation

As usual, $\boldsymbol{R}$ and $\boldsymbol{C}$ denote the fields of real and complex numbers, respectively. Lie groups are denoted by capital Roman letters and their Lie algebras by the corresponding lower case German letters. The identity element of a group is denoted e (and sometimes O). If M is a manifold, and p a point in M, the tangent space to M at p is denoted M_p; $\mathscr{E}(M)$ (resp. $\mathscr{D}(M)$) denotes the space of complex-valued C^∞ functions (resp. C^∞ functions of compact support) on M. If $f \in \mathscr{E}(M)$, $\operatorname{supp}(f)$ denotes the support of f.

§ 3. Preliminaries

Let S be a symmetric space of the noncompact type [7], G the largest connected group of isometries of S in the compact open topology, o a fixed point in S and K the isotropy subgroup of G at o. Let $\mathfrak{g}=\mathfrak{k}+\mathfrak{p}$ be the corresponding Cartan decomposition of the Lie algebra $\mathfrak{g}$ of G, $\mathfrak{k}$ denoting the Lie algebra of K, and $\mathfrak{p}$ being the orthogonal complement of $\mathfrak{k}$ with respect to Killing form B of $\mathfrak{g}$. We assume as usual that the Riemannian structure of S is induced by B. Let d denote the distance function on S, and $|\;|$ the norm in $\mathfrak{p}$. As usual we identify S_0 and $\mathfrak{p}$, S and G/K, so the Exponential mapping of S_0 into S can be regarded as a map $\operatorname{Exp}: \mathfrak{p} \to G/K$. Let the positive function ω on $\mathfrak{p}$ be determined by

$$\int_{G/K} f(p)\,dp = \int_{\mathfrak{p}} f(\operatorname{Exp} X)\,\omega(X)^2\,dX,$$

dp and dX being the volume elements of G/K and $\mathfrak{p}$. As in [5], p. 585 let $I(G)$ denote the set of all $f \in \mathscr{E}(G)$ satisfying the following two conditions: a) f is bi-invariant under K; b) for each left invariant differential operator D on G and each integer $q \geqq 0$

$$\tau_{D,q}(f) = \sup_{x \in G} |\omega_q(x)\,(Df)(x)| < \infty. \tag{1}$$

Here $\omega_q(x) = \omega(X)(1+|X|^q)$ if $x = k \exp X (k \in K, X \in \mathfrak{p})$. Let $I_c(G) = \mathscr{D}(G) \cap I(G)$.

Let $\mathfrak{a} \subset \mathfrak{p}$ be a maximal abelian subspace, $\mathfrak{a}'$ the set of regular elements in $\mathfrak{a}$, $\mathfrak{a}^+$ any connected component of $\mathfrak{a}'$ (a Weyl chamber in $\mathfrak{a}$). Let Λ denote the dual of $\mathfrak{a}$, Λ^c and $\mathfrak{a}^c$ their respective complexifications. Each $\lambda \in \Lambda^c$ can be viewed as a linear form on $\mathfrak{a}^c$ and $H_\lambda \in \mathfrak{a}^c$ is uniquely determined by the condition $B(H, H_\lambda) = \lambda(H)$ for $H \in \mathfrak{a}^c$. We put $\langle \lambda, \mu \rangle = B(H_\lambda, H_\mu)$, $|\nu| = \langle \nu, \nu \rangle^{1/2}$ for $\lambda, \mu \in \Lambda^c$, $\nu \in \Lambda$. An element $\lambda \in \Lambda$ is called a root if the space

$$\mathfrak{g}_\lambda = \{X \in \mathfrak{g} \mid [H, X] = \lambda(H) X \quad \text{for all} \quad H \in \mathfrak{a}\}$$

is $\neq 0$. A root is called positive if its values on $\mathfrak{a}^+$ are >0. Let Δ denote the set of roots $\neq 0$, Δ_+ the set of all roots >0, and Δ_0 the set of positive roots which are not integral multiples of other positive roots. If $\alpha \in \Delta$, the number $m_\alpha = \dim(\mathfrak{g}_\alpha)$ is called the multiplicity of α and we put as usual

$$\varrho = \frac{1}{2} \sum_{\alpha \in \Delta_+} m_\alpha \alpha, \quad \pi = \prod_{\alpha \in \Delta_0} \alpha.$$

Let $\mathfrak{n} = \sum_{\alpha>0} \mathfrak{g}_\alpha$, $N = \exp(\mathfrak{n})$, $A = \exp(\mathfrak{a})$. Then we have the Iwasawa decompositions $\mathfrak{g} = \mathfrak{k} + \mathfrak{a} + \mathfrak{n}$ and $G = KAN$. If $x \in G$, let $H(x) \in \mathfrak{a}$ be determined by $x \in K \exp H(x) N$. Let log denote the inverse of the mapping $\exp: \mathfrak{a} \to A$.

The Weyl group W of S is the group of linear transformations of $\mathfrak{a}$ induced by those elements of $\mathrm{Ad}_G(K)$ which leave $\mathfrak{a}$ invariant. Here Ad_G denotes the adjoint representation of G. The group W acts on A via the exponential mapping and on Λ^c by duality. Let $\mathscr{S}(\mathfrak{a})$ denote the set of complex-valued rapidly decreasing C^∞ functions on $\mathfrak{a}$ ([13], II, p. 89) and $\mathscr{I}(\mathfrak{a})$ (resp. $\mathscr{I}_c(\mathfrak{a})$) the set of W-invariant elements in $\mathscr{S}(\mathfrak{a})$ (resp. $\mathscr{D}(\mathfrak{a})$). The spaces $\mathscr{S}(A)$, $\mathscr{I}(A)$, $\mathscr{I}_c(A)$ and $\mathscr{S}(\Lambda)$, $\mathscr{I}(\Lambda)$, $\mathscr{I}_c(\Lambda)$ are defined similarly.

§ 4. The expansion for the spherical function

For $\lambda \in \Lambda^c$ we consider now the spherical function

(1) $$\varphi_\lambda(x) = \int_K e^{(i\lambda - \varrho)(H(xk))} dk \qquad (x \in G).$$

Since $K \exp(\mathfrak{a}^+) K$ is dense in G this function is determined by the values $\varphi_\lambda(\exp H)$ $(H \in \mathfrak{a}^+)$ and on this Weyl chamber $\mathfrak{a}^+$ one has a series expansion for φ_λ which we shall now describe. We put

$$\varphi(\lambda, H) = \varphi_\lambda(\exp H) \qquad (H \in \mathfrak{a}).$$

Let $\alpha_1, \ldots, \alpha_l$ be the simple roots in Δ_+ ([5], p. 252) and let $L \subset \Lambda$ consist of all $\mu = m_1\alpha_1 + \cdots + m_l\alpha_l$ where $m_1, \ldots, m_l$ are nonnegative integers. Put $m(\mu) = m_1 + \cdots + m_l$. The rational functions ${}'\Gamma_\mu (\mu \in L)$ on $\mathfrak{a}^c$ are defined recursively by ${}'\Gamma_0 = 1$ and

$$\{\langle \mu, \mu - 2\varrho\rangle - 2\mu\}\, {}'\Gamma_\mu = 2 \sum_{\alpha \in \Delta_+} m_\alpha \sum_{k \geq 1} {}'\Gamma_{\mu - 2k\alpha} \{\langle \mu - 2k\alpha, \alpha\rangle - \alpha\},$$

where k runs over all integers ≥ 1 for which $\mu - 2k\alpha \in L$. Let $L' = L - \{0\}$. For $\mu \in L'$ let σ_μ denote the set of $H \in \mathfrak{a}^c$ satisfying $\langle \mu, \mu\rangle = 2i\mu(H)$ and if $s, t \in W$ let $\tau_\mu(s, t)$ denote the set of all $H \in \mathfrak{a}^c$ satisfying $i(sH - tH) = H_\mu$. Let ${}^*\mathfrak{a}^c$ denote the complement in $\mathfrak{a}^c$ of the union of all the hyperplanes $r\sigma_\mu$, $\tau_\mu(s, t)$ $(r, s, t \in W, \mu \in L')$ and let ${}'\mathfrak{a}^c$ denote the set of $H \in {}^*\mathfrak{a}^c$, where $\pi(H) \neq 0$. If $M \geq 0$ let $\mathfrak{a}(M)$ denote the set of $H \in \mathfrak{a}$ satisfying $\alpha_i(H) > M$ for $1 \leq i \leq l$. We shall often identify Λ^c and $\mathfrak{a}^c$ via the mapping $\lambda \to H_\lambda$; thus a function on $\mathfrak{a}^c$ automatically becomes a function on Λ^c without change in notation. For a certain holomorphic function $\boldsymbol{c}$ on ${}'\mathfrak{a}^c$, described explicitly below, the following holds ([5], p. 282). For any compact subset U of ${}^*\mathfrak{a}^c$ there exists a number $M \geq 0$ such that

(2) $$\varphi(\lambda, H) = \sum_{\mu \in L} \psi_\mu(\lambda, H) \qquad (H_\lambda \in {}'\mathfrak{a}^c \cap U, H \in \mathfrak{a}(M))$$

where

$$\psi_\mu(\lambda, H) = \sum_{s \in W} \boldsymbol{c}(s\lambda)\, {}'\Gamma_\mu(is\lambda - \varrho) e^{(is\lambda - \varrho - \mu)(H)} \qquad (H_\lambda \in {}'\mathfrak{a}^c, H \in \mathfrak{a}^c).$$

We shall now strengthen this statement.

Lemma 4.1. *Suppose H_λ does not lie on any of the hyperplanes $\sigma_\nu (\nu \in L')$ and let $H \in \mathfrak{a}^+$. Then there exists a constant $K_{\lambda,H}$ such that*

$$|'\Gamma_\mu(i\lambda - \varrho)| \leqq K_{\lambda,H} e^{\mu(H)} \quad \textit{for all} \quad \mu \in L.$$

Proof. Put $b_\mu = '\Gamma_\mu(i\lambda - \varrho)$ and select constants c_1, c_2 such that for all $\mu \in L$, $\alpha \in \Delta_+$,

$$|\langle \mu + \varrho, \alpha \rangle - i\langle \alpha, \lambda \rangle| \leqq c_1(m(\mu) + 1), \quad |\langle \mu, \mu \rangle - 2i\langle \mu, \lambda \rangle| \geqq c_2 m(\mu)^2.$$

Then, by the definition of $'\Gamma_\mu(i\lambda - \varrho)$,

$$|b_\mu| \leqq c(m(\mu))^{-1} \sum_{\alpha \in \Delta_+} m_\alpha \sum_{k \geqq 1} |b_{\mu - 2k\alpha}|,$$

where $c = 4c_1 c_2^{-1}$. Let N_0 be an integer such that

$$c \sum_{\alpha \in \Delta_+} m_\alpha (1 - e^{-2\alpha(H)})^{-1} \leqq N_0$$

and select $K = K_{\lambda,H}$ such that

(4) $$|b_\nu| \leqq K e^{\nu(H)}$$

for all $\nu \in L$, $m(\nu) \leqq N_0$. We shall prove (4) by induction on $m(\nu)$. Let $N > N_0$ be an integer and suppose (4) holds for all $\nu \in L$ with $m(\nu) < N$. Let $\mu \in L$ such that $m(\mu) = N$. Then

$$|b_\mu| \leqq cN^{-1} \sum_{\alpha \in \Delta_+} m_\alpha \sum_{k \geqq 1} K e^{(\mu - 2k\alpha)(H)} \leqq K e^{\mu(H)},$$

and the lemma is proved.

Proposition 4.2. *The expansion*

(5) $$\varphi(\lambda, H) = \sum_{\mu \in L} \psi_\mu(\lambda, H)$$

is valid for all $H_\lambda \in {'\mathfrak{a}^c}$ and all $H \in \mathfrak{a}^+$.

Proof. Consider now fixed H_λ and H. Let $b = \min_{1 \leqq i \leqq l} \alpha_i(H)$ and determine $H_1 \in \mathfrak{a}^+$ such that $\alpha_i(H_1) = \frac{1}{2} b \, (1 \leqq i \leqq l)$. Select an open neighborhood U of H in $\mathfrak{a}^+$ such that

$$\alpha_i(A) > \alpha_i(H_1) \qquad (A \in U, 1 \leqq i \leqq l).$$

Let $B \in \mathfrak{a}$. Then by Lemma 4.1 there exists a constant $M_{\lambda,B}$ such that

$$|\sum_s c(s\lambda) \, '\Gamma_\mu(is\lambda - \varrho) \, e^{is\lambda(A + iB)}| \leqq M_{\lambda,B} \, e^{\mu(H_1)}$$

for all $\mu \in L$, $A \in U$. It follows that

$$|\psi_\mu(\lambda, A + iB)| \leqq M_{\lambda,B} e^{-\varrho(A)} e^{\mu(H_1 - A)}$$

and consequently, the series on the right hand side of (5) converges uniformly on a neighborhood of H in $\mathfrak{a}^c$. Thus both sides of (5) are analytic in $\mathfrak{a}^+$ and by (2) they coincide on an open subset of $\mathfrak{a}^+$. But $\mathfrak{a}^+$ is connected so (5) follows.

Remark. It had previously been pointed out to me by HARISH-CHANDRA that Prop. 4.2 can be derived from the results of [6], by use of the method of [5], p. 297–298.

The function $\boldsymbol{c}(\lambda)$ is given by $\boldsymbol{c}(\lambda)=I(i\lambda)/I(\varrho)$ where

$$(6)\qquad I(v)=\prod_{\alpha\in\Delta_+} B\left(\frac{m_\alpha}{2}, \frac{m_{\alpha/2}}{4}+\frac{\langle v,\alpha\rangle}{\langle\alpha,\alpha\rangle}\right), \qquad v\in\Lambda^c$$

B denoting the Beta function. In special cases such a formula was proved by HARISH-CHANDRA [5], p. 303, BHANU-MURTHY [1], [2] and in full generality by GINDIKIN-KARPELEVIČ [4]. As observed in [9], §§ 2–3, formula (6) implies that on Λ, $\boldsymbol{c}^{-1}$ is a tempered distribution, the function π_0 in [5], p. 579 is a constant multiple of π, and the mapping $A\to|\pi\boldsymbol{c}|^2A$ maps $\mathscr{I}(\Lambda)$ onto itself. The Fourier inversion formula of HARISH-CHANDRA ([5], pp. 612, 610, 586) can therefore be restated as follows. The Fourier transform of $f\in I(G)$ is defined by

$$\tilde{f}(\lambda)=\int_G f(x)\,\varphi_{-\lambda}(x)\,dx \qquad (\lambda\in\Lambda)$$

and $\tilde{f}\in\mathscr{I}(\Lambda)$. Each function of the form

$$(7)\qquad f(x)=w^{-1}\int_\Lambda A(\lambda)\,\varphi_\lambda(x)\,|\boldsymbol{c}(\lambda)|^{-2}\,d\lambda \qquad (A\in\mathscr{I}(\Lambda))$$

lies in $I(G)$ and

$$(8)\qquad \tilde{f}=A\,.$$

In (7), w is the order of the Weyl group W, the measures dx, $d\lambda$ and also da below, are regularly normalized ([5], p. 597). Finally, if $F\in\mathscr{S}(A)$ and F^* denotes its (Euclidean) Fourier transform

$$F^*(\lambda)=\int_A F(a)e^{-i\lambda(\log a)}\,da \qquad (\lambda\in\Lambda)$$

then by [5], p. 596,

$$(9)\qquad \tilde{f}=(F_f)^* \qquad (f\in I(G))$$

where the function $F_f\in\mathscr{I}(A)$ is given by

$$F_f(a)=e^{\varrho(\log a)}\int_N f(an)\,dn \qquad (a\in A)$$

dn being Haar measure on N, suitably normalized.

§ 5. The analog of the Paley-Wiener theorem

In any metric space, let $V^R(p)$ denote the closed ball of radius R and center p.

Consider now any $F\in\mathscr{I}_c(A)$. Select R such that $\operatorname{supp}(F)\subset V^R(O)$. The Fourier transform F^* on Λ can, by the classical Paley-Wiener theorem, be extended to a holomorphic function on Λ^c satisfying for each integer $N>0$ an inequality

$$(1)\qquad |F^*(\xi+i\eta)|\leqq C_N(1+|\xi+i\eta|)^{-N}\,e^{R|\eta|}$$

for all $\xi, \eta \in A$, where C_N is a constant, and $|\xi + i\eta| = (|\xi|^2 + |\eta|^2)^{1/2}$. As suggested by (7) § 4, we form the function

$$g(x) = w^{-1} \int_A F^*(\lambda)\, \varphi_\lambda(x)\, |c(\lambda)|^{-2}\, d\lambda \qquad (x \in G)$$

and as will be seen later, the analog of the Paley-Wiener theorem amounts to that $g \in I_c(G)$, more precisely, that the restriction $\bar{g}$ of g to A satisfies $\operatorname{supp}(\bar{g}) \subset V^R(O)$. We shall at any rate prove the analogous statement for each term in the expansion (5), § 4.

Proposition 5.1. *Let* $\mu \in L$, $F \in \mathscr{I}_c(A)$ *with* $\operatorname{supp}(F) \subset V^R(O)$. *Then*

$$\int_A F^*(\lambda)\, \psi_\mu(\lambda, H)\, |c(\lambda)|^{-2}\, d\lambda = 0$$

if $H \in \mathfrak{a}^+$ *and* $|H| > R$.

Proof. The integrand is certainly defined if $H_\lambda \in {}'\mathfrak{a}^c \cap \mathfrak{a} = \mathfrak{a}'$, $H \in \mathfrak{a}^+$. For such H_λ we have $|c(\lambda)|^2 = c(\lambda)\, c(-\lambda) = c(s\lambda)\, c(-s\lambda)$ for each $s \in W$ so the integrand is $e^{-(\varrho+\mu)(H)}$ times

$$\sum_{s \in W} F^*(s\lambda)\, c(-s\lambda)^{-1}\, e^{is\lambda(H)}\, {}'\Gamma_\mu(is\lambda - \varrho)\,.$$

This expression, however, is an analytic integrable function on the entire A and it suffices to prove

$$\int_A F^*(-\xi)\, c(\xi)^{-1}\, e^{-i\xi(H)}\, {}'\Gamma_\mu(-i\xi - \varrho)\, d\xi = 0 \tag{2}$$

if $|H| > R$ in $\mathfrak{a}^+$. To do this we shift the integration in (2) into the complex space A^c in such a way that the singularities of the functions $c(\lambda)^{-1}$ and ${}'\Gamma_\mu(-i\lambda - \varrho)$ are simultaneously avoided, Cauchy's theorem applicable, resulting in estimates for (2). By the recursive definition of the rational function $\lambda \to {}'\Gamma_\mu(-i\lambda - \varrho)$ on A^c it is clear that it is holomorphic outside the set where $\langle 2\nu, -i\lambda - \varrho \rangle = \langle \nu, \nu - 2\varrho \rangle$ for some $\nu \in L'$, that is outside the set

$$\left\{\lambda \in A^c \,|\, \langle \nu, \lambda \rangle = \frac{1}{2} i \langle \nu, \nu \rangle \quad \text{for some} \quad \nu \in L' \right\}. \tag{3}$$

On the other hand the formula for $c(\lambda)$ shows that the function $\lambda \to c(\lambda)^{-1}$ is holomorphic on the complement in A^c of the subset

$$\left\{\lambda \in A^c \,|\, \langle \alpha, \lambda \rangle = i\left(\frac{m_\alpha}{2} + \frac{m_{\alpha/2}}{4} + n\right) \langle \alpha, \alpha \rangle \quad \text{for some} \quad n \in \mathbf{Z}^+, \alpha \in \Delta_+ \right\}. \tag{4}$$

Here $\mathbf{Z}^+$ denotes the set of nonnegative integers. The fortunate circumstance is that both sets (3) and (4) are contained in the set

$$\{\lambda \in A^c \,|\, i\langle \nu, \lambda \rangle < 0 \quad \text{for some} \quad \nu \in L' \}. \tag{5}$$

The path of integration in (2) will now be shifted away from this set.

Consider now any $\xi, \eta \in A$ such that $-H_\eta \in \mathfrak{a}^+$. Let $H_1, \ldots, H_l$ be the basis of $\mathfrak{a}$, dual to the basis $\alpha_1, \ldots, \alpha_l$ of A and for each $\lambda \in A^c$ write $H_\lambda = \sum_j \lambda_j H_j (\lambda_j \in \mathbf{C})$. We shall identify a function φ on a subset of A^c with a function on a subset of C^l by the convention $\varphi(\lambda)=\varphi(\lambda_1, \ldots, \lambda_l)$. All ξ_j, η_k are real and $\eta_k<0$ $(1 \leqq k \leqq l)$. The function

$$G(\lambda)=G(\lambda_1, \ldots, \lambda_l)=F^*(-\lambda)\, c(\lambda)^{-1}\, e^{-i\lambda(H)}\, {}'\Gamma_\mu(-i\lambda-\varrho)$$

is holomorphic at each point $\lambda=(\xi_j+is_j\eta_j)$ $(0 \leqq s_j \leqq 1, 1 \leqq j \leqq l)$ because the condition in (5) is violated. In fact, if $v \in L'$, $v=n_1\alpha_1+\cdots+n_l\alpha_l$, then $\langle v, \lambda\rangle = \sum_j n_j(\xi_j+is_j\eta_j)$ has a nonpositive imaginary part. If the function $G(\lambda)$ behaves suitably at infinity we can therefore conclude from Cauchy's theorem that

$$(6) \qquad \int_{R^l} G(\xi_1, \ldots, \xi_l)\, d\xi_1 \ldots d\xi_l = \int_{R^l} G(\xi_1+i\eta_1, \ldots, \xi_l+i\eta_l)\, d\xi_1 \ldots d\xi_l .$$

In order to estimate G we first observe, by induction on $m(\mu)$, that

$${}'\Gamma_\mu(-i(\xi+i\eta)-\varrho)=p_\mu(\xi+i\eta)\,(q_\mu(\xi+i\eta))^{-1},$$

where p_μ and q_μ are polynomials and

$$q_\mu(\xi+i\eta)=\prod_v (\langle v, v\rangle+2i\langle v, \xi\rangle-2\langle v, \eta\rangle),$$

the product being taken over all $v \in L'$ such that $\mu - v \in L$. Consequently,

$$(7) \qquad |{}'\Gamma_\mu(-i(\xi+i\eta)-\varrho)| \leqq c_\mu |\, p_\mu(\xi+i\eta)| \qquad (\xi \in A, \; -H_\eta \in \mathfrak{a}^+)$$

where c_μ is a constant. In order to estimate $c(\lambda)^{-1}$ we use the formula (see e.g. [11], p. 6 or [15], p. 151)

$$(8) \qquad \lim_{|z|\to\infty} \frac{\Gamma(z+a)}{\Gamma(z)}\, e^{-a\log z}=1, \quad |\arg z| \leqq \pi-\delta,$$

where a is any constant, log being the principal value of the logarithm and $\delta>0$. Now formula (6) § 3 implies a product formula

$$c(\lambda)^{-1}=\prod_{\alpha\in\Delta_+} c_\alpha(\lambda)^{-1}$$

and the factor $c_\alpha(\lambda)^{-1}$ can be estimated by means of (8), taking $a=\frac{1}{2}m_\alpha$ and

$$z=\frac{1}{4}m_{\alpha/2}+i\frac{\langle\lambda, \alpha\rangle}{\langle\alpha, \alpha\rangle}.$$

In fact (8) implies

$$\left|\frac{\Gamma\left(\frac{1}{2}m_\alpha+z\right)}{\Gamma(z)}\right| \leqq K_1+K_2|z|^{\frac{1}{2}m_\alpha} \qquad (K_1, K_2 \text{ constants})$$

whenever the real part of z is $\geqq 0$ and this is the case if λ lies in the set

$$(9) \qquad \{\lambda=\xi+i\eta \mid \xi \in A, \; -H_\eta \in Cl(\mathfrak{a}^+)\},$$

(Cl denoting closure). It follows that

$$(10) \qquad |c(\xi+i\eta)|^{-1} \leqq C_1+C_2 \prod_{\alpha\in\Delta_+} |\xi+i\eta|^{\frac{1}{2}m_\alpha}$$

for all $\xi \in \Lambda$, $-H_\eta \in \mathfrak{a}^+$, C_1 and C_2 being constants. The inequalities (1), (7) and (10) show that the application (6) of Cauchy's theorem is legitimate, so for each $-H_\eta \in \mathfrak{a}^+$,

$$\begin{aligned}&\int_\Lambda F^*(-\xi)\, c(\xi)^{-1} e^{-i\xi(H)}\,{}^{\backprime}\Gamma_\mu(-i\xi-\varrho)\,d\xi \\ &= \int_\Lambda F^*(-\xi-i\eta)\, c(\xi+i\eta)^{-1} e^{-i\xi(H)+\eta(H)}\,{}^{\backprime}\Gamma_\mu(-i\xi+\eta-\varrho)\,d\xi\,.\end{aligned}$$

Let us denote this quantity by $Q(H)$. Choosing the integer N in (1) equal to

$$l+1+\operatorname{degree}(p_\mu)+\frac{1}{2}\sum_{\alpha\in\Delta_+} m_\alpha$$

we have

(11) $$|Q(H)| \leqq C e^{R|\eta|} e^{\eta(H)} \quad \text{for all} \quad H,\ -H_\eta \in \mathfrak{a}^+\,,$$

C being a constant. Now fix $H \in \mathfrak{a}^+$ with $|H|>R$. Put $H_\eta = -tH$ $(t>0)$. Then by (11),

$$|Q(H)| \leqq C e^{t|H|(R-|H|)}\,.$$

Letting $t\to\infty$ we obtain $Q(H)=0$, and the proposition is proved.

Definition. We shall say that the symmetric space G/K satisfies condition (Γ) if for each $F \in \mathscr{I}_c(A)$ and each $H \in \mathfrak{a}^+$, the term-by-term integration

$$\int_\Lambda \sum_{\mu\in L} \psi_\mu(\lambda, H)\, F^*(\lambda)\, |c(\lambda)|^{-2}\, d\lambda = \sum_{\mu\in L} \int_\Lambda \psi_\mu(\lambda, H)\, F^*(\lambda)\, |c(\lambda)|^{-2}\, d\lambda$$

is valid.

Lemma 5.2. *The space G/K satisfies condition (Γ) in the following cases:*

(i) *G/K has rank one* (i.e. $\dim(\mathfrak{a})=1$).

(ii) *G is complex.*

Proof. (i) Without any assumption on the rank l we shall prove that there exist constants c, d and e such that

(12) $$|{}^{\backprime}\Gamma_{m\alpha}(i\lambda-\varrho)| \leqq c(1+m^d)(1+|\lambda|^e) \qquad (\lambda\in\Lambda,\ m\in \mathbf{Z}^+)$$

where α is any of the roots $\alpha_1, \ldots, \alpha_l$. Let $p=m_\alpha$, $q=m_{2\alpha}$. It is readily seen from the recursion formula for ${}^{\backprime}\Gamma_\mu$ that ${}^{\backprime}\Gamma_{m\alpha}\equiv 0$ for odd m. Put $a_n(\lambda) = {}^{\backprime}\Gamma_{2n\alpha}(i\lambda-\varrho)$ for $n\in \mathbf{Z}^+$. Then

$$\begin{aligned}&(4n^2\langle\alpha,\alpha\rangle - 4in\langle\alpha,\lambda\rangle)\, a_n(\lambda) \\ &= 2p\sum_0^{n-1} a_r(\lambda)\,(2r\langle\alpha,\alpha\rangle + \langle\varrho,\alpha\rangle - i\langle\alpha,\lambda\rangle) + \\ &+ 2q \sum_{0\leqq r\equiv n (\mathrm{mod}\, 2)}^{n-2} a_r(\lambda)\,(4r\langle\alpha,\alpha\rangle + \langle\varrho,2\alpha\rangle - 2i\langle\alpha,\lambda\rangle)\,.\end{aligned}$$

Now define for all $r\in\mathbf{Z}^+$,

$$\begin{aligned}\gamma_r(\lambda) &= 2p|2r\langle\alpha,\alpha\rangle + \langle\varrho,\alpha\rangle - i\langle\alpha,\lambda\rangle| + \\ &\quad + 2q|4r\langle\alpha,\alpha\rangle + \langle\varrho,2\alpha\rangle - 2i\langle\alpha,\lambda\rangle|\,, \\ c_r(\lambda) &= |4r^2\langle\alpha,\alpha\rangle - 4ir\langle\alpha,\lambda\rangle|\end{aligned}$$

and define the functions $b_n(\lambda)$ inductively by

$$b_n(\lambda) = c_n(\lambda)^{-1} \sum_{r=0}^{n-1} b_r(\lambda)\,\gamma_r(\lambda) \qquad b_0(\lambda) \equiv 1 .$$

It is clear by induction on n that

$$|a_n(\lambda)| \leqq b_n(\lambda) \quad \text{for all} \quad n \in \mathbf{Z}^+, \lambda \in \Lambda .$$

Now

$$\begin{aligned} b_n(\lambda)\, c_n(\lambda) &= \sum_0^{n-2} b_r(\lambda)\,\gamma_r(\lambda) + b_{n-1}(\lambda)\,\gamma_{n-1}(\lambda) \\ &= b_{n-1}(\lambda)\, c_{n-1}(\lambda)\,(1 + \gamma_{n-1}(\lambda)\,(c_{n-1}(\lambda))^{-1}) \end{aligned}$$

so by iteration

$$b_n(\lambda)\, c_n(\lambda) = b_2(\lambda)\, c_2(\lambda) \prod_{r=2}^{n-1} \left(1 + \frac{\gamma_r(\lambda)}{c_r(\lambda)}\right).$$

But it is clear that

$$\gamma_r(\lambda) \leqq C r^{-1} c_r(\lambda) \quad \text{for} \quad r \geqq 1, \lambda \in \Lambda ,$$

where C is a constant. Since $\log(1+x) \leqq x$ for $x \geqq 0$ we conclude that

$$\prod_{r=2}^{n-1} \left(1 + \frac{\gamma_r(\lambda)}{c_r(\lambda)}\right) \leqq \exp\left(C \sum_2^{n-1} \frac{1}{r}\right) < n^C$$

and the inequality (12) follows. It is now clear from (1), (10) and (12) that if $\operatorname{rank} S = 1$ we have for $H \in \mathfrak{a}^+$,

$$\sum_{\mu \in L} \int_{\Lambda} |\psi_\mu(\lambda, H)\, F^*(\lambda)\, c(\lambda)^{-2}|\, d\lambda < \infty$$

so condition (Γ) is verified.

(ii) In the case when G is complex we shall prove that each ${}'\Gamma_\mu$ is constant. Then Lemma 4.1 implies that G/K satisfies condition (Γ).

Lemma 5.3. *Suppose G complex and let $\mu \in L$. Then*

(13) $\quad {}'\Gamma_\mu \equiv$ *the number of partitions of* $\frac{1}{2}\mu$ *into a sum of positive roots.*

Remark. This partition function is familiar from the representation theory of compact Lie groups (see B. KOSTANT: A formula for the multiplicity of a weight, Trans. Am. Math. Soc. **93**, 53–73 (1959) and P. CARTIER: On Weyl's character formula, Bull. Am. Math. Soc. **67**, 228–230 (1961)).

Proof. According to HARISH-CHANDRA [5], p. 304, if $H \in \mathfrak{a}^+$, $H_\lambda \in {}'\mathfrak{a}^c$,

$$\varphi_\lambda(\exp H) = \sum_{s \in W} c(s\lambda) e^{is\lambda(H)} \prod_{\alpha \in \Delta_+} (e^{\alpha(H)} - e^{-\alpha(H)})^{-1} .$$

In this case every $\alpha \in \Delta_+$ has multiplicity 2 so

$$\begin{aligned} e^{\varrho(H)} \varphi_\lambda(\exp H) &= \sum_{s \in W} c(s\lambda) e^{is\lambda(H)} \prod_{\alpha \in \Delta_+} (1 + e^{-2\alpha(H)} + e^{-4\alpha(H)} + \cdots) \\ &= \sum_{s \in W} c(s\lambda) e^{is\lambda(H)} \sum_{\mu \in L} M_\mu e^{-\mu(H)} , \end{aligned}$$

where M_μ is the number of partitions of $\frac{1}{2}\mu$ into a sum of positive roots (order immaterial). Thus by (5) § 4,

$$\sum_{s,\mu} c(s\lambda)\,{}'\Gamma_\mu(is\lambda-\varrho)\,e^{(is\lambda-\mu)(H)} = \sum_{s,\mu} c(s\lambda)\,M_\mu\,e^{(is\lambda-\mu)(H)}$$

for all $H\in\mathfrak{a}^+$, $H_\lambda\in{}'\mathfrak{a}^c$. Fix λ such that the linear forms $is\lambda-\mu$ ($s\in W$, $\mu\in L$) are all different and in addition $H_\lambda\in{}'\mathfrak{a}^c$. There exists a nonempty open subset $V\subset\mathfrak{a}^+$ such that $tV\subset V$ for $t\geqq 1$ and such that for each $\xi\in\Lambda^c$ the series

$$\sum_{s,\mu} c(s\lambda)\,{}'\Gamma_\mu(is\lambda-\varrho)\,e^{(is\lambda-\mu+\xi)(H)},$$

$$\sum_{s,\mu} c(s\lambda)\,M_\mu\,e^{(is\lambda-\mu+\xi)(H)}$$

converge uniformly for $H\in V$. In fact, the first series converges uniformly on a subset of $\mathfrak{a}^+$ of the form

$$\left\{H\in\mathfrak{a}^+\,|\,\alpha_i(H)>M,\ \alpha_i(H)>\frac{1}{2}\alpha_j(H),\ 1\leqq i,j\leqq l\right\}$$

(see [5], Lemma 31, p. 276) which is open and nonempty, and the equation introducing M_μ shows that for each $H_0\in\mathfrak{a}^+$, $M_\mu\leqq c e^{\mu(H_0)}$, where c is a constant. The existence of V is now clear so by [5], p. 307 we can conclude that $M_\mu={}'\Gamma_\mu(is\lambda-\varrho)$. This proves (13) on a dense subset of Λ^c, hence on Λ^c itself. Now Lemma 5.2 is also proved.

Lemma 5.4. *Let o, d, A and N be as in* § 3. *Then*

$$d(o,a\cdot o)<d(o,an\cdot o)\quad \textit{for}\quad a\in A,\ n\neq e\quad \textit{in}\quad N.$$

Proof. Let n^* be an element in N such that the point $an^*\cdot o$ minimizes the distance from o to the horocycle $aN\cdot o$. Then the geodesic in S joining o and $an^*\cdot o$ is perpendicular to this horocycle at $an^*\cdot o$ (see e.g. [7], Lemma 13.6, Ch. I). The manifold $A\cdot o$ contains all the geodesics intersecting the horocycle $N\cdot o$ orthogonally at o ([8], Lemma 3.3). Hence the manifold $an^*A\cdot o$ contains all the geodesics intersecting the horocycle $aN\cdot o$ orthogonally at $an^*\cdot o$. In particular, there exists an element $b\in A$ such that $an^*b\cdot o=o$. This implies $n^*=e$, proving the lemma.

Theorem 5.5. *Suppose G/K satisfies condition (Γ). Then*

(i) *The mapping $f\to F_f$ is an isomorphism of the convolution algebra $I(G)$ onto the convolution algebra $\mathscr{I}(A)$.*

(ii) *The mapping $f\to F_f$ maps $I_c(G)$ onto $\mathscr{I}_c(A)$; moreover, if $f\in I_c(G)$ and $\bar f$ denotes its restriction to A then for each $R>0$,*

$$\operatorname{supp}(F_f)\subset V^R(0)\Leftrightarrow\operatorname{supp}(\bar f)\subset V^R(0).$$

Proof. Let $F\in\mathscr{I}_c(A)$. Select R such that $\operatorname{supp}(F)\subset V^R(0)$. Then the Fourier transform F^* satisfies an inequality of the form (1) and we can define $g\in I(G)$ by

$$g(x)=w^{-1}\int_\Lambda F^*(\lambda)\,\varphi_\lambda(x)\,|c(\lambda)|^{-2}\,d\lambda \qquad (x\in G).$$

By Prop. 5.1 and condition (Γ) we have $g \in I_c(G)$ and $g(\exp H) = 0$ if $|H| > R$ in $\mathfrak{a}$. Also, by (8) and (9) § 4,

$$\tilde{g}(\lambda) = F_g^*(\lambda) = F^*(\lambda) \qquad (\lambda \in \Lambda^c). \tag{14}$$

In particular, we can use this on a function $F = F_f$ where f is an arbitrary function in $I_c(G)$. Then $F_f^* = F^* = F_g^*$ so, since the Euclidean Fourier transform is one-to-one, $F_f = F_g$. But $f, g \in I_c(G)$ and the mapping $f \to F_f$ is one-to-one on $I_c(G)$ ([7], p. 454). Hence $f = g$. This shows that $I_c(G)$ is contained in the space $I_0(G)$, defined in [5], p. 586 (cf. also the remarks at the end of § 4 above). As shown in [5], p. 612 this implies that the inversion formula

$$f(x) = w^{-1} \int_\Lambda \tilde{f}(\lambda)\, \varphi_\lambda(x)\, |c(\lambda)|^{-2}\, d\lambda$$

holds for all $f \in I(G)$. In view of Theorem 3.3 in [9], this gives (i).

Concerning (ii) we first observe that (14) implies $F_g = F$. This means that if $f \in I_c(G)$, $\operatorname{supp}(F_f) \subset V^R(0)$ then $\operatorname{supp}(\tilde{f}) \subset V^R(0)$. The converse is clear from Lemma 5.4 so the theorem is proved.

Remark. That $f \to F_f$ is one-to-one on $I(G)$ for every G/K was conjectured by Harish-Chandra in [5], p. 612 and has subsequently been proved by him (unpublished).

Corollary 5.6. *Suppose G/K satisfies condition (Γ). If $R > 0$ let $I_c^R(G)$ denote the set of functions $f \in I_c(G)$ such that* $\operatorname{supp}(\tilde{f}) \subset V^R(0)$. *A W-invariant function on Λ is the Fourier transform $\tilde{f}(\lambda)$ $(f \in I_c^R(G))$ if and only if it can be extended to a holomorphic function ψ on Λ^c satisfying for each integer $N > 0$,*

$$|\psi(\xi + i\eta)| \leqq C_N (1 + |\xi + i\eta|)^{-N} e^{R|\eta|} \qquad (\xi, \eta \in \Lambda)$$

where C_N is a constant.

This is the analog of the classical Paley-Wiener theorem (see e.g. [10], p. 21) and is an immediate consequence of this theorem combined with (9) § 4 and Theorem 5.5.

Remarks. For the spaces G/K of constant negative curvature, Theorem 5.5 (ii) was proved by Takahashi [14], p. 330. For the case $G = \boldsymbol{SL}(2, \boldsymbol{R})$, $K = \boldsymbol{SO}(2)$, the analogue of the Paley-Wiener theorem was proved by Ehrenpreis and Mautner [3]. If G is complex, Cor. 5.6 gives nothing really new because then the Fourier transform is, due to the formula for the spherical function, so closely related to the Euclidean Fourier transform.

References

[1] Bhanu Murthy, T. S.: Plancherel's measure for the factor space $\boldsymbol{SL}(n, \boldsymbol{R})/\boldsymbol{SO}(n, \boldsymbol{R})$. Doklady Akad. Nauk SSSR **133**, 503—506 (1960).

[2] — The assymptotic behavior of zonal spherical functions on the Siegel upper half-plane. Doklady Akad. Nauk SSSR **135**, 1027—1029 (1960).

[3] Ehrenpreis, L., and F. I. Mautner: Some properties of the Fourier transform on semi-simple Lie groups, I. Ann. Math. **61**, 406—439 (1955); II. Trans. Am. Math. Soc. **84**, 1—55 (1957).

[4] Gindikin, S. G., and F. I. Karpelevič: Plancherel measure of Riemannian symmetric spaces of nonpositive curvature. Doklady Akad. Nauk SSSR **145**, 252—255 (1962).

[5] HARISH-CHANDRA: Spherical functions on a semisimple Lie group, I, II. Am. J. Math. **80**, 241—310, 553—613 (1958).
[6] — Some results on differential equations and their applications. Proc. Nat. Acad. Sci. USA **45**, 1763—1764 (1959).
[7] HELGASON, S.: Differential geometry and symmetric spaces. New York: Academic Press 1962.
[8] — Duality and Radon transform for symmetric spaces. Am. J. Math. **85**, 667—692 (1963).
[9] — Fundamental solutions of invariant differential operators on symmetric spaces. Am. J. Math. **86**, 565—601 (1964).
[10] HÖRMANDER, L.: Linear partial differential operators. Berlin-Göttingen-Heidelberg: Springer 1963.
[11] MAGNUS, W., u. R. OBERHETTINGER: Formeln und Sätze für die speziellen Funktionen der mathematischen Physik. Berlin-Göttingen-Heidelberg: Springer 1948.
[12] PALEY, R., and N. WIENER: Fourier transforms in the complex domain. Am. Math. Soc. Colloquium, New York, 1934.
[13] SCHWARTZ, L.: Théorie des Distributions, I, II. Paris: Hermann 1950, 1951.
[14] TAKAHASHI, R.: Sur les répresentations unitaires des groupes de Lorentz généralisés. Bull. Soc. Math. France **91**, 289—433 (1963).
[15] TITCHMARSH, E. C.: The theory of functions. 2nd edition. Oxford University Press 1939.

Professor S. HELGASON
The Institute for Advanced Study
Princeton, N. J.

(Received January 29, 1965)

A FATOU-TYPE THEOREM FOR HARMONIC FUNCTIONS ON SYMMETRIC SPACES[1]

BY S. HELGASON AND A. KORÁNYI

Communicated by G. D. Mostow, August 31, 1967

1. **Introduction.** The result to be proved in this article is that if u is a bounded harmonic function on a symmetric space X and x_0 any point in X then u has a limit along almost every geodesic in X starting at x_0 (Theorem 2.3). In the case when X is the unit disk with the non-Euclidean metric this result reduces to the classical Fatou theorem (for radial limits). When specialized to this case our proof is quite different from the usual one; in fact it corresponds to transforming the Poisson integral of the unit disk to that of the upper half-plane and using only a homogeneity property of the Poisson kernel. The kernel itself never enters into the proof.

2. **Harmonic functions on symmetric spaces.** Let G be a semisimple connected Lie group with finite center, K a maximal compact subgroup of G and $\mathfrak{g}$ and $\mathfrak{k}$ their respective Lie algebras. Let B denote the Killing form of $\mathfrak{g}$ and $\mathfrak{p}$ the corresponding orthogonal complement of $\mathfrak{k}$ in $\mathfrak{g}$. Let Ad denote the adjoint representation of G. As usual we view $\mathfrak{p}$ as the tangent space to the symmetric space $X=G/K$ at the origin $o=\{K\}$ and accordingly give X the G-invariant Riemannian structure induced by the restriction of B to $\mathfrak{p}\times\mathfrak{p}$. Let Δ denote the corresponding Laplace-Beltrami operator.

Fix a maximal abelian subspace $\mathfrak{a}\subset\mathfrak{p}$ and let M denote the centralizer of $\mathfrak{a}$ in K. If λ is a linear function on $\mathfrak{a}$ and $\lambda\neq 0$ let $\mathfrak{g}_\lambda=\{X\in\mathfrak{g}\mid [H,X]=\lambda(H)X \text{ for all } H\in\mathfrak{a}\}$; λ is called a restricted root if $\mathfrak{g}_\lambda\neq 0$. Let $\mathfrak{a}'$ denote the open subset of $\mathfrak{a}$ where all restricted roots are $\neq 0$. Fix a Weyl chamber $\mathfrak{a}^+$ in $\mathfrak{a}$, i.e. a connected component of $\mathfrak{a}'$. A restricted root α is called positive (denoted $\alpha>0$) if its values on $\mathfrak{a}^+$ are positive. Let the linear function ρ on $\mathfrak{a}$ be determined by $2\rho=\sum_{\alpha>0}(\dim\mathfrak{g}_\alpha)\alpha$ and denote the subalgebras $\sum_{\alpha>0}\mathfrak{g}_\alpha$ and $\sum_{\alpha>0}\mathfrak{g}_{-\alpha}$ of $\mathfrak{g}$ by $\mathfrak{n}$ and $\bar{\mathfrak{n}}$ respectively. Let N and $\overline{N}$ denote the corresponding analytic subgroups of G.

By a Weyl chamber in $\mathfrak{p}$ we understand a Weyl chamber in some maximal abelian subspace of $\mathfrak{p}$. The *boundary* of X is defined as the set B of all Weyl chambers in the tangent space $\mathfrak{p}$ to X at o; since this boundary is via the map $kM\to\mathrm{Ad}(k)\mathfrak{a}^+$ identified with K/M, which by the Iwasawa decomposition $G=KAN$ equals G/MAN, this defi-

[1] This work was supported by the National Science Foundation, GP 7477 and GP 6155.

nition of boundary is equivalent to Furstenberg's [2] (see also [6] and [4]). In particular the group G acts transitively on B as well as on X. The two actions will be denoted $(g, b)\to g(b)$ and $(g, x)\to g\cdot x$ ($g\in G$, $b\in B$, $x\in X$). Let db denote the unique K-invariant measure on B normalized by $\int_B db=1$. Then according to Furstenberg [2], the mapping $f\to u$ where

$$(1) \qquad u(g\cdot o) = \int_B f(g(b))db \qquad (g \in G),$$

is a bijection of the set $L^\infty(B)$ of bounded measurable functions on B onto the set of bounded solutions of Laplace's equation $\Delta u=0$ on X. The function u in (1) is called the *Poisson integral* of f.

If $g\in G$ let $k(g)\in K$, $H(g)\in\mathfrak{a}$ be determined by $g=k(g)\exp H(g)n$ ($n\in N$). Observe that if g^h denotes hgh^{-1} for $h\in G$ then $k(\bar{n}^m)=k(\bar{n})^m$, $H(\bar{n}^m)=H(\bar{n})$ for $\bar{n}\in\overline{N}$, $m\in M$. According to Harish-Chandra [3, Lemma 44], the mapping $\bar{n}\to k(\bar{n})M$ is a bijection of $\overline{N}$ onto a subset of K/M whose complement is of lower dimension and if f is a continuous function on B, then

$$(2) \qquad \int_B f(b)db = \int_{\overline{N}} f(k(\bar{n})M)\exp(-2\rho(H(\bar{n})))d\bar{n}$$

for a suitably normalized Haar measure $d\bar{n}$ on $\overline{N}$. If $a\in A$ we have $ak(\bar{n})MAN=k(\bar{n}^a)MAN$ whence

$$(3) \qquad a(k(\bar{n})M) = k(\bar{n}^a)M$$

so the action of a on the boundary corresponds to the conjugation $\bar{n}\to\bar{n}^a$ on $\overline{N}$.

Let $E_1, \cdots, E_r$ be a basis of $\bar{\mathfrak{n}}$ such that each E_i lies in some $\mathfrak{g}_{-\alpha}$, say $\mathfrak{g}_{-\alpha_i}$. Since the map exp: $\bar{\mathfrak{n}}\to\overline{N}$ is a bijection we can, for each $H\in\mathfrak{a}^+$, consider the function $\bar{n}\to|\bar{n}|_H$ defined as follows: If $\bar{n}=\exp(\sum_1^r a_iE_i)$ ($a_i\in R$) we put

$$|\bar{n}|_H=\operatorname*{Max}_{1\leq i\leq r}(|a_i|^{1/\alpha_i(H)}).$$

Since

$$(4) \qquad \bar{n}^{\exp tH} = \exp\left(\sum_1^r a_i\exp(-\alpha_i(H)t)E_i\right)$$

we have

$$(5) \qquad |\bar{n}^{\exp tH}|_H = e^{-t}|\bar{n}|_H \qquad \text{for } \bar{n}\in\overline{N},\quad t\in R,\quad H\in\mathfrak{a}^+.$$

For $r>0$ let $B_{H,r}$ denote the set $\{\bar{n}\in\overline{N}\mid |\bar{n}|_H<r\}$ and let $V_{H,r}$ denote the volume of $B_{H,r}$ (with respect to the Haar measure on $\overline{N}$).

LEMMA 2.1. *Let $f \in L^\infty(B)$ and u the Poisson integral (1) of f. Put $F(\bar{n}) = f(k(\bar{n})M)$ for $\bar{n} \in \bar{N}$. Fix $\bar{n}_0 \in \bar{N}$ and $H \in \mathfrak{a}^+$ and assume*

$$\frac{1}{V_{H,r}} \int_{B_{H,r}} | F(\bar{n}_0\bar{n}) - F(\bar{n}_0) | \, d\bar{n} \to 0 \tag{6}$$

for $r \to 0$. Then

$$\lim_{t \to +\infty} u(k(\bar{n}_0) \exp tH(\cdot o)) = f(k(\bar{n}_0)M).$$

PROOF. By the Iwasawa decomposition we can write $\bar{n}_0 = k(\bar{n}_0) \cdot (a_1 n_1)^{-1}$ $(a_1 \in A,\ n_1 \in N)$ so

$$u(k(\bar{n}_0) \exp tH \cdot o) = u(\bar{n}_0 a_1 n_1 \exp tH \cdot o) = u(\bar{n}_0 \exp tH a_1 n_1^{\exp(-tH)} \cdot o).$$

But $G = A\bar{N}K$ so $n_1^{\exp(-tH)} = a(t)\bar{n}(t)k(t)$, each factor tending to e as $t \to +\infty$. If $H_t \in \mathfrak{a}$ is determined by

$$\exp tH_t = \exp tH a_1 a(t)$$

we have

$$u(k(\bar{n}_0) \exp tH \cdot o) = u(\bar{n}_0 \bar{n}(t)^{\exp tH_t} \exp tH_t \cdot o).$$

The function $f'(b) = f(\bar{n}_0 \bar{n}(t)^{\exp tH_t}(b))$ has Poisson integral $u'(x) = u(\bar{n}_0 \bar{n}(t)^{\exp tH_t} \cdot x)$; using (1) on u' and f' with $g = \exp tH_t$ we get from (2) and (3)

$$u(k(\bar{n}_0) \exp tH \cdot o) - f(k(\bar{n}_0)M)$$

$$= \int_{\bar{N}} (F(\bar{n}_0 \bar{n}(t)^{\exp tH_t} \bar{n}^{\exp tH_t}) - F(\bar{n}_0)) \exp(-2\rho(H(\bar{n}))) d\bar{n}$$

so

$$| u(k(\bar{n}_0) \exp tH \cdot o) - f(k(\bar{n}_0)M) |$$

$$\leqq \int_{\bar{N}} | F(\bar{n}_0 \bar{n}^{\exp tH_t}) - F(\bar{n}_0) | \exp(-2\rho(H(\bar{n}(t)^{-1}\bar{n}))) d\bar{n}. \tag{7}$$

Now if $c > 0$ let $\bar{N}_c$ denote the "square"

$$\bar{N}_c = \left\{ \exp\left(\sum_1^r a_i E_i \right) \middle| \ | a_i | \leqq c,\ 1 \leqq i \leqq r \right\}.$$

The integral on the right in (7) equals the sum

$$(8)\quad \begin{aligned}&\int_{\bar{N}_c} \left| F(\bar{n}_0\bar{n}^{\exp tH_t}) - F(\bar{n}_0)\right| \exp(-2\rho(H(\bar{n}(t)^{-1}\bar{n})))d\bar{n}\\ &\quad + \int_{\bar{N}-\bar{N}_c} \left| F(\bar{n}_0\bar{n}^{\exp tH_t}) - F(\bar{n}_0)\right| \exp(-2\rho(H(\bar{n}(t)^{-1}\bar{n})))d\bar{n}.\end{aligned}$$

Since $\rho(H(\bar{n}))\geqq 0$ for all $\bar{n}\in\bar{N}$ ([3, p. 287]) and since the mapping $\bar{n}\rightarrow\bar{n}^{\exp H}$ has Jacobian $\exp(-2\rho(H))$ (cf. (4)) we see that

$$(9)\quad \begin{aligned}&\int_{\bar{N}_c} \left| F(\bar{n}_0\bar{n}^{\exp tH_t}) - F(\bar{n}_0)\right| \exp(-2\rho(H(\bar{n}(t)^{-1}\bar{n})))d\bar{n}\\ &\qquad \leqq \exp(2\rho(tH_t)) \int_{\bar{N}_c^{\exp tH_t}} \left| F(\bar{n}_0\bar{n}) - F(\bar{n}_0)\right| d\bar{n}.\end{aligned}$$

Now $\bar{n}\in\bar{N}_c{}^{\exp tH_t}$ if and only if

$$\bar{n} = \exp\left(\sum a_i e^{-\alpha_i(tH_t)} E_i\right) \quad \text{where } |a_i| \leqq c$$

and $tH_t - tH$ is bounded (for fixed $\bar{n}_0$ and H). It follows that

$$\bar{N}_c^{\exp tH_t} \subset B_{H,de^{-t}} \qquad \text{for all } t \geqq 0,$$

$d=d(H, \bar{n}_0, c)$ being a constant. But since the map exp: $\bar{\mathfrak{n}}\rightarrow\bar{N}$ is measure-preserving it is clear that

$$V_{H,de^{-t}} = \exp(-2\rho(H)t)d_1 \qquad t\geqq 0$$

where $d_1=d_1(H, \bar{n}_0, c)$ is another constant. Also

$$\exp(2\rho(tH_t)) \leqq \exp(2\rho(tH))d_2$$

where $d_2(H, \bar{n}_0)$ is a constant. Thus the right hand side of (9) can be majorized for all $t\geqq 0$:

$$(10)\quad \begin{aligned}&\exp 2\rho(tH_t) \int_{\bar{N}_c^{\exp tH_t}} \left| F(\bar{n}_0\bar{n}) - F(\bar{n}_0)\right| d\bar{n}\\ &\qquad \leqq d_3 \frac{1}{V_{H,de^{-t}}} \int_{B_{H,de^{-t}}} \left| F(\bar{n}_0\bar{n}) - F(\bar{n}_0)\right| d\bar{n}\end{aligned}$$

where d and d_3 are constants depending on H, $\bar{n}_0$ and c.

On the other hand, if $\| \ \|_\infty$ denotes the uniform norm on $\bar{N}$ the second term in (8) is majorized by

$$(11)\quad \begin{aligned}&2\|F\|_\infty \int_{\bar{N}-\bar{N}_c} \exp(-2\rho(H(\bar{n}(t)^{-1}\bar{n})))d\bar{n}\\ &\qquad = 2\|F\|_\infty\left(1 - \int_{\bar{n}(t)\bar{N}_c} \exp(-2\rho(H(\bar{n})))d\bar{n}\right).\end{aligned}$$

Now given $\epsilon > 0$ we first choose c so large that

$$2\|F\|_\infty \left(1 - \int_{\overline{N}_{c/2}} \exp(-2\rho(H(\bar{n})))d\bar{n}\right) < \epsilon/2;$$

since $\bar{n}(t) \to e$ for $t \to +\infty$ we can choose t_1 such that $\bar{n}(t)\overline{N}_c \supset \overline{N}_{c/2}$ for $t \geqq t_1$. Then the expression in (11) is $< \epsilon/2$ for $t \geqq t_1$; by our assumption (6) we can choose t_2 such that the right hand side of (10) is $< \epsilon/2$ for $t > t_2$. In view of (7) and (8) this proves the lemma.

The next lemma shows that, for a fixed H, the assumption of Lemma 2.1 actually holds for almost all $\bar{n}_0 \in \overline{N}$.

LEMMA 2.2. *Let $F \in L^\infty(\overline{N})$ and fix $H \in \mathfrak{a}^+$. Then*

$$\lim_{r \to 0} \frac{1}{V_{H,r}} \int_{B_{H,r}} |F(\bar{n}_0\bar{n}) - F(\bar{n}_0)| \, d\bar{n} = 0 \tag{12}$$

for almost all $\bar{n}_0 \in \overline{N}$.

The proof of this result is essentially in the literature: In [1] Edwards and Hewitt give all the necessary arguments for the case of a discrete sequence tending to 0 and everything they do remains trivially valid in the case $r \to 0$. The result in the exact form required here was also proved by E. M. Stein independently of [1] (cf. his expository article [6]).

THEOREM 2.3. *Let u be a bounded solution of Laplace's equation $\Delta u = 0$ on the symmetric space X. Then for almost all geodesics $\gamma(t)$ starting at o*

$$\lim_{t \to \infty} u(\gamma(t)) \quad \textit{exists.} \tag{13}$$

PROOF. Let $S^+ = \{H \in \mathfrak{a}^+ \mid B(H, H) = 1\}$. Then the mapping $(kM, H) \to \mathrm{Ad}(k)H$ is a bijection of $(K/M) \times S^+$ onto a subset of the unit sphere S in $\mathfrak{p}$ whose complement has lower dimension. Since $\dim(K/M - k(\overline{N})/M) < \dim K/M$ the mapping $(\bar{n}, H) \to \mathrm{Ad}(k(\bar{n}))H$ is a bijection of $\overline{N} \times S^+$ onto a subset of S whose complement in S has lower dimension. If $\overline{N}_H$ denotes the set of $\bar{n}_0$ for which (12) holds (with $F(\bar{n}) = f(k(\bar{n})M)$) and if $S_0 = \bigcup_{H \in S^+} \mathrm{Ad}(k(\overline{N}_H))H$ it follows from the Fubini theorem that $S - S_0$ is a null set. This concludes the proof.

REMARKS. (i) If f is continuous the limit relation

$$\lim_{t \to +\infty} u(k \exp tH \cdot o) = f(kM) \qquad (H \in \mathfrak{a}^+, kM \in K/M)$$

follows immediately from (1), (2) and (3), by use of the dominated convergence theorem. (See also [**4**, Theorem 18.3.2.]) In particular, u has the same limit along all geodesics from o which lie in the same Weyl chamber in $\mathfrak{p}$.

(ii) In the case when X has rank one ($\dim \mathfrak{a} = 1$) A. W. Knapp [**5**] has proved (13), even under the weaker assumption that $f \in L^1(B)$.

References

1. R. E. Edwards and E. Hewitt, *Pointwise limits for sequences of convolution operators*, Acta Math. **113** (1965), 181–218.
2. H. Furstenberg, *A Poisson formula for semisimple Lie groups*, Ann. of Math. **77** (1963), 335–386.
3. Harish-Chandra, *Spherical functions on a semisimple Lie group*. I, Amer. J. Math. **80** (1958), 241–310.
4. F. I. Karpelevič, *Geometry of geodesics and eigenfunctions of the Laplace-Beltrami operator on symmetric spaces*, Trudy Moscov. Mat. Obšč. **14** (1965), 48–185.
5. A. W. Knapp, unpublished manuscript.
6. C. Moore, *Compactifications of symmetric spaces*, Amer. J. Math. **86** (1964), 201–218.
7. E. M. Stein, *Maximal functions and Fatou's theorem*, C.I.M.E. summer course on bounded homogeneous domains, Cremonese, 1967.

Massachusetts Institute of Technology and
Yeshiva University

Reprinted from ADVANCES IN MATHEMATICS
All Rights Reserved by Academic Press, New York and London

Vol. 3, No. 4, October 1969
Printed in Belgium

The Bounded Spherical Functions on Symmetric Spaces

SIGURDUR HELGASON* AND KENNETH JOHNSON**

Dedicated to W. T. MARTIN

1. Introduction

Let G be a connected semisimple Lie group with finite center, K a maximal compact subgroup of G. A spherical function on G is by definition a continuous function $\phi \not\equiv 0$ on G satisfying

$$\int_K \phi(xky)\, dk = \phi(x)\,\phi(y) \qquad x, y \in G,$$

dk denoting the normalized Haar measure on K.

Let $L^\natural(G)$ denote the space of functions on G which are integrable on G and bi-invariant under K. Under the L^1-norm and convolution product on G, $L^\natural$ is a commutative Banach algebra. Its maximal ideals are all of the form

$$M_\phi = \left\{ f \in L^\natural(G) \,\middle|\, \int_G f(g)\,\phi(g)\, dg = 0 \right\},$$

where ϕ is an arbitrary *bounded* spherical function on G [(*3*), p. 410]. The present paper gives an explicit parametrization of the set of bounded spherical functions on G (Theorem 2.1).

2. The bounded spherical functions

We begin by recalling Harish–Chandra's parametrization of the set of spherical functions on G. Let $\mathfrak{g}$ and $\mathfrak{k}$ denote the Lie algebras of G and K respectively, $\mathfrak{p}$ the orthogonal complement of $\mathfrak{k}$ in $\mathfrak{g}$ with respect to the Killing form $\langle , \rangle$ of $\mathfrak{g}$. Let $\mathfrak{a} \subset \mathfrak{p}$ be a maximal abelian subspace, Λ the dual space of $\mathfrak{a}$, Λ^c and $\mathfrak{a}^c$ their respective complexifications. If $\lambda \in \Lambda^c$ let

* Supported by the National Science Foundation, NSF GP-9003.
** Supported by AFOSR, Contract No. F 44620-67-0029.

ξ, $\eta \in \Lambda$ be determined by $\lambda = \xi + i\eta$. Fix a Weyl chamber $\mathfrak{a}^+$ in $\mathfrak{a}$, let Δ_+ denote the corresponding set of positive roots, and $\mathfrak{n}$ the nilpotent subalgebra $\sum_{\alpha\in\Delta_+} \mathfrak{g}_\alpha$, where

$$\mathfrak{g}_\alpha = \{X \in \mathfrak{g} \,|\, [H, X] = \alpha(H)X \qquad \text{for all} \quad H \in \mathfrak{a}\}.$$

Let A and N be the analytic subgroups of G which correspond to $\mathfrak{a}$ and $\mathfrak{n}$ and if $g \in G$ let $H(g) \in \mathfrak{a}$ be determined by $g \in K \exp H(g)N$. As usual we put

$$\rho = \frac{1}{2} \sum_{\alpha\in\Delta_+} m_\alpha \alpha,$$

where $m_\alpha = \dim(\mathfrak{g}_\alpha)$, the ***multiplicity*** of the root α. Let Ad_G denote the adjoint representation of G and W the group of linear transformations of $\mathfrak{a}$ induced by those elements of the group $Ad_G(K)$ which leave $\mathfrak{a}$ invariant. The Weyl group W acts on A via the exponential mapping and on Λ^c by duality: If $s \in W$, $\lambda \in \Lambda^c$, then $s\lambda$ is given by $s\lambda(H) = \lambda(s^{-1}H)$, $H \in \mathfrak{a}$; λ is called ***regular*** if $s\lambda \neq \lambda$ for all $s \neq e$ in W.

According to a theorem of Harish–Chandra, all the spherical functions on G are given by

$$\phi_\lambda(x) = \int_K e^{(i\lambda-\rho)(H(xk))}\, dk \qquad x \in G$$

and $\phi_\lambda = \phi_\mu$ if and only if $\lambda = s\mu$ for some $s \in W$.

Theorem 2.1. *Let C_ρ denote the convex hull of the points $s\rho (s \in W)$.*

Then ϕ_λ is bounded if and only if $\lambda = \xi + i\eta$ belongs to the tube $\Lambda + iC_\rho$.

Let ${}^+\mathfrak{a}$ denote the dual cone of $\mathfrak{a}^+$, i.e.

$${}^+\mathfrak{a} = \{H \in \mathfrak{a} \,|\, \langle H, H' \rangle > 0 \qquad \text{for} \quad H' \in \mathfrak{a}^+\}$$

and put ${}^-\mathfrak{a} = -({}^+\mathfrak{a})$, that is the set of negatives of the elements in ${}^+\mathfrak{a}$. For $\lambda \in \Lambda^c$ let $H_\lambda \in \mathfrak{a}^c$ be determined by $\langle H, H_\lambda \rangle = \lambda(H)$ for $H \in \mathfrak{a}$. We put also $\langle \lambda, \mu \rangle = \langle H_\lambda, H_\mu \rangle$ for $\lambda, \mu \in \Lambda^c$, and $|\lambda| = \langle \lambda, \lambda \rangle^{1/2}$ if $\lambda \in \Lambda$. Polynomials on $\mathfrak{a}^c$ can be viewed as polynomials on Λ^c by means of the mapping $\lambda \to H_\lambda$.

Lemma 2.2. *ϕ_λ is bounded if $-H_\eta \in Cl\{\mathfrak{a}^+ \cap (^-\mathfrak{a} + H_\rho)\}$, Cl denoting closure.*

Proof. Let da and $d\lambda$ denote Haar measures on A and Λ respectively such that the Fourier inversion formulas on A and on Λ hold without multiplicative constants. Let the Haar measure dn on N be normalized such that

$$\int_N e^{-2\rho(H(\theta(n)))}\, dn = 1,$$

θ denoting the Cartan involution $k \exp X \to k \exp(-X)$ $(k \in K,\ X \in \mathfrak{p})$ on G. Then we normalize the Haar measure dg on G such that

$$\int_G f(g)\, dg = \int_{KAN} f(kan)\, e^{2\rho(\log a)}\, dkdadn$$

for each integrable function on G. Here $\log a$ is the element in $\mathfrak{a}$ satisfying $\exp(\log a) = a$. In particular, let $f \in L^{\natural}(G)$ and put

$$F_f(a) = e^{\rho(\log a)} \int_N f(an)\, dn.$$

By the Fubini theorem F_f exists almost everywhere on A and

$$\int_A e^{\rho(\log a)} F_{|f|}(a)\, da < \infty. \tag{1}$$

Now for each $H \in \mathfrak{a}^+$,

$$-s\eta(H) \leqslant -\eta(H) \qquad s \in W$$

[see (2), p. 280] so if $A^+ = \exp \mathfrak{a}^+$, we have

$$\int_{A^+} e^{-s\eta(\log a)} F_{|f|}(a)\, da \leqslant \int_{A^+} e^{-\eta(\log a)} F_{|f|}(a)\, da$$
$$\leqslant \int_{A^+} e^{\rho(\log a)} F_{|f|}(a)\, da$$

since $\eta + \rho$ is nonnegative on $\mathfrak{a}^+$. Using (1) and the Weyl group invariance of $F_{|f|}$ [(2), p. 261] it follows that if w is the order of W, then

$$w^{-1} \int_A e^{-\eta(\log a)} F_{|f|}(a)\, da \leqslant \int_A e^{\rho(\log a)} F_{|f|}(a)\, da < \infty.$$

But calculating with the positive functions $\phi_{i\eta}$ and $|f|$ we have by the Fubini theorem,

$$\int_G \phi_{i\eta}(g)|f(g)|\,dg = \int_G \Big(\int_K e^{-(\eta+\rho)(H(gk))}\,dk\Big)|f(g)|\,dg$$
$$= \int_G e^{-(\eta+\rho)(H(g))}|f(g)|\,dg$$
$$\int_{AN} |f(an)|\, e^{-(\eta+\rho)(H(a))}e^{2\rho(\log a)}\,da\,dn = \int_A e^{-\eta(\log a)}F_{|f|}(a)\,da.$$

Thus

$$w^{-1}\int_G \phi_{i\eta}(g)|f(g)|\,dg \leqslant \int_A e^{\rho(\log a)}F_{|f|}(a)\,da < \infty \quad \text{for all } f\in L^\natural(G)$$

which obviously implies that $\phi_{i\eta}$, and therefore ϕ_λ, is bounded.

Lemma 2.3. *Suppose that ϕ_λ is bounded and that $-H_\eta \in Cl(\mathfrak{a}^+)$. Then $\eta+\rho$ is nonegative on $\mathfrak{a}^+$.*

Proof. Let $\bar N = \theta(N)$ and the Haar measure $d\bar n$ on $\bar N$ normalized by $d\bar n = \theta(dn)$. Then by (*2*), p. 288–291

$$\phi_\lambda(a) = e^{(i\lambda-\rho)(\log a)}\int_{\bar N} e^{(i\lambda-\rho)(H(a\bar n a^{-1}))}e^{(-i\lambda-\rho)(H(\bar n))}\,d\bar n.$$

Assume first $-H_\eta \in \mathfrak{a}^+$. If $H\in\mathfrak{a}^+$ and $a_t = \exp tH$ $(t\in\mathbf{R})$ then

$$\lim_{t\to\infty}\int_{\bar N} e^{(i\lambda-\rho)(H(a_t\bar n a_t^{-1}))}\,e^{-(i\lambda+\rho)(H(\bar n))}\,d\bar n = \int_{\bar N}\lim_{t\to\infty} e^{(i\lambda-\rho)(H(a_t\bar n a_t^{-1}))}\,e^{-(i\lambda+\rho)(H(\bar n))}\,d\bar n$$

This interchange is justified in (*2*), p. 287–293 by means of

$$H(\bar n)\in Cl(^+\mathfrak{a}),\qquad H(\bar n) - H(a_t\bar n a_t^{-1}) \in Cl(^+\mathfrak{a}). \tag{2}$$

For later purposes we use (2) to give a different justification of the interchange. Choose ϵ such that $0<\epsilon<1$ and $H_\rho + \epsilon H_\eta \in \mathfrak{a}^+$. The left integrand above is then majorized by

$$e^{-(\eta+\rho)(H(a_t\bar n a_t^{-1}))}\,e^{(\eta-\rho)(H(\bar n))} \leqslant e^{(-\eta+\epsilon\rho)(H(a_t\bar n a_t^{-1}))}\,e^{(\eta-\rho)(H(\bar n))}$$
$$\leqslant e^{(-\eta+\epsilon\eta)(H(\bar n))}\,e^{(\eta-\rho)(H(\bar n))} = e^{(\epsilon\eta-\rho)(H(\bar n))},$$

which is integrable ((*2*), p. 291 or (*1*)). Thus we may conclude

$$\lim_{t\to\infty} \int_{\bar N} e^{(i\lambda-\rho)(H(a_t\bar n a_t^{-1}))}\, e^{-(i\lambda-\rho)(H(\bar n))}\, d\bar n = \int_{\bar N} e^{-(i\lambda+\rho)(H(\bar n))}\, d\bar n.$$

The right hand side is Harish-Chandra's function $\mathbf{c}(\lambda)$ which by our assumption about λ is $\neq 0$ (cf. (*1*)). This shows that

$$\lim_{t\to\infty} \phi_\lambda(a_t)\, e^{(-i\lambda+\rho)(\log a_t)}$$

exists and is $\neq 0$. Taking absolute values we deduce $(\eta + \rho)(H) \geqslant 0$ as desired.

Finally, we consider the case when $-H_\eta$ lies on the boundary of $\mathfrak{a}^+$. We first prove a modified formula for the spherical function.

Let $H_1 \in Cl(\mathfrak{a}^+)$, let Δ_1 denote the set of $\alpha \in \Delta_+$ such that $\alpha(H_1) = 0$, let $\Delta_2 = \Delta_+ - \Delta_1$, and let $\bar N_1$ and $\bar N_2$, respectively, denote the analytic subgroups of $\bar N$ with Lie algebras

$$\sum_{\alpha\in\Delta_1} \mathfrak{g}_{-\alpha} \qquad \text{and} \qquad \sum_{\alpha\in\Delta_2} \mathfrak{g}_{-\alpha}\,.$$

Let $2\rho_i$ denote the sum of the roots in Δ_i with multiplicity, $(i = 1, 2)$. The groups $\bar N_1$ and $\bar N_2$ are closed nilpotent subgroups on which we can select Haar measures $d\bar n_1$, $d\bar n_2$ such that the product measure $d\bar n_1 \times d\bar n_2$ corresponds to $d\bar n$ under the diffeomorphism $(\bar n_1, \bar n_2) \to \bar n_1 \bar n_2$ of $\bar N_1 \times \bar N_2$ onto $\bar N$.

Proposition 2.4. *Let $a \in A$ satisfy $\alpha(\log a) = 0$ for $\alpha \in \Delta_1$. Then*

$$\phi_\lambda(a) = c e^{(i\lambda-\rho)(\log a)} \int_{\bar N_2} e^{(i\lambda-\rho)(H(a\bar n_2 a^{-1}))}\, e^{-(i\lambda+\rho)(H(\bar n_2))}\, d\bar n_2\,,$$

where the constant c equals

$$c = \int_{\bar N_1} e^{-2\rho_1(H(\bar n_1))}\, d\bar n_1\,.$$

Proof. Put $\mu = i\lambda - \rho$, $\nu = -i\lambda - \rho$. Then since a centralizes $\bar N_1$,

$$\int_{\bar N} e^{\mu(H(a\bar n a^{-1}))}\, e^{\nu(H(\bar n))}\, d\bar n = \int_{\bar N_1}\int_{\bar N_2} e^{\mu(H(\bar n_1 a \bar n_2 a^{-1}))}\, e^{\nu(H(\bar n_1 \bar n_2))}\, d\bar n_1\, d\bar n_2\,.$$

The subgroup of G generated by $\bar{N}_1$ and $\theta\bar{N}_1$ is semisimple, normalizes $\bar{N}_2$, and has an Iwasawa decomposition compatible with $G = KAN$. Decomposing $\bar{n}_1$ accordingly, $\bar{n}_1 = k_1 a_1 n_1$, the last expression is found to be

$$\int_{\bar{N}_1}\int_{\bar{N}_2} e^{\mu(H\bar{n}(a_1\bar{n}_2a^{-1}))}\, e^{\nu(H(a_1\bar{n}_2))}\, d\bar{n}_1\, d\bar{n}_2 .$$

Since the map $\bar{n}_2 \to a_1\bar{n}_2a_1^{-1}$ has Jacobian $\exp(-2\rho_2(H(\bar{n}_1)))$ the integral becomes

$$\int_{\bar{N}_1} e^{(\mu+\nu+2\rho_2)(H(\bar{n}_1))} \int_{\bar{N}_2} e^{\mu(H(a\bar{n}_2a^{-1}))}\, e^{\nu(H(\bar{n}_2))}\, d\bar{n}_2\, d\bar{n}_1$$

as was to be proved.

In order to finish the proof of Lemma 2.3 take $H_1 = -H_\eta$ in the discussion above. If $H \in Cl(\mathfrak{a}^+)$ such that $\alpha(H) = 0$ for $\alpha \in \Delta_1$, $\alpha(H) > 0$ for $\alpha \in \Delta_2$ we put $a = a_t = \exp tH$ $(t \in \mathbf{R})$ in Prop. 2.4 and let $t \to +\infty$. We can then take the limit under the integral sign because of the inequalities above combined with the integrability of $\exp(\epsilon\eta - \rho)(H(\bar{n}_2)))$ (cf. (1)). Consequently $(\eta + \rho)(H) \geqslant 0$ as before. Finally, let $H_0 \in \mathfrak{a}^+$ be arbitrary. Its projection H_0' on the plane

$$\pi_1 = \{H \in \mathfrak{a} \mid \alpha(H) = 0 \quad \text{for } \alpha \in \Delta_1\}$$

is given by $H_0' = n^{-1}\Sigma_{s\in W_1} sH_0$ where W_1 is the subgroup of W leaving π_1 pointwise fixed and n its order. Since $\rho(H_0 - sH_0) \geqslant 0$ ((*2*), p. 280) we deduce

$$\begin{aligned}(\eta + \rho)(H_0) &= (\eta + \rho)(H_0') + (\eta + \rho)(H_0 - H_0') \\ &= (\eta + \rho)(H_0') + \rho(H - H_0') \geqslant 0\end{aligned}$$

which proves Lemma 2.3.

Lemma 2.5. *The closure of the set*

$$\bigcup_{s\in W} s\{\mathfrak{a}^+ \cap (^-\mathfrak{a} + H_0)\}$$

is the convex hull of the set of points $H_{s\rho}$, $s \in W$.

Proof. Let $\alpha_1, \ldots, \alpha_l$ denote the simple roots such that

$$\mathfrak{a}^+ = \{H \mid \alpha_1(H) > 0, \ldots, \alpha_l(H) > 0\}.$$

Let $H_1, \ldots, H_l$ be a basis of $\mathfrak{a}$ dual to the basis $\alpha_1, \ldots, \alpha_l$ of Λ. Then the H_i form the edges of the closed Weyl chamber $Cl(\mathfrak{a}^+)$. Let P_i denote the open half space containing 0 and bounded by the hyperplane π_i through H_ρ perpendicular to H_i. For $s \in W$ let

$$Q_s = sP_1 \cap \cdots \cap sP_l \cap s\mathfrak{a}^+$$

and

$$Q = \bigcap_{s \in W} (sP_1 \cap \cdots \cap sP_l).$$

Then of course $Q_s = sQ_e$ and since

$$^-\mathfrak{a} = \left\{ \sum_1^l c_i H_{\alpha_i} \quad c_1 < 0, \ldots, c_l < 0 \right\}$$

and

$$Q_e = \{H \in \mathfrak{a}^+ \mid \langle H_\rho - H, H_i \rangle > 0 \qquad \text{for} \quad i = 1, \ldots, l\}$$

we have

(3) $$Q_e = \mathfrak{a}^+ \cap (^-\mathfrak{a} + H_\rho).$$

Next let us prove

(4) $$Q_e = Q \cap \mathfrak{a}^+.$$

The inclusion $Q \cap \mathfrak{a}^+ \subset Q_e$ being obvious suppose $H \in Q_e$. Then

$$\langle H_i, H \rangle < \langle H_i, H_\rho \rangle \qquad (1 \leqslant i \leqslant l)$$

and since H_i, $H \in Cl(\mathfrak{a}^+)$ we have by (*2*), p. 280

$$\langle sH_i, H \rangle \leqslant \langle H_i, H \rangle \qquad s \in W$$

so

$$\langle sH_i, H \rangle < \langle sH_i, H_{s\rho} \rangle$$

proving (4). Now since $\mathfrak{a} = \bigcup_s Cl(s\mathfrak{a}^+)$ we deduce from (4)

(5) $$Cl(Q) = \bigcup_s Cl(Q_s)$$

so in view of (3) it remains to prove

$$\text{(6)} \qquad Cl(Q) = \text{convex hull of } \bigcup_s H_{s\rho} .$$

The right hand side of (6) is clearly contained in the left; let us prove the converse. Let s_i denote the reflection in the hyperplane $\alpha_i = 0$. Then the hyperplane π_i passes through the points

$$H_\rho , H_{s_1\rho} ,..., H_{s_{i-1}\rho} , H_{s_{i+1}\rho} ,..., H_{s_l\rho} .$$

All these planes $\pi_1 ,..., \pi_l$ meet at H_ρ, the planes $\pi_1 ,..., \pi_{l-1}$ meet in the line joining H_ρ and $H_{s_l\rho}$ etc. It is therefore clear that the part of the boundary of Q which is inside $\mathfrak{a}^+$ belongs to the convex hull of the points $H_\rho , H_{s_1\rho} ,..., H_{s_l\rho}$. Now (6) follows from (4) and (5) so the lemma is proved.

Since $\phi_{s\lambda} \equiv \phi_\lambda (s \in W)$ Theorem 2.1 follows from Lemmas 2.2, 2.3 and 2.5.

Remark. It has been remarked by E. Stein that the boundedness of ϕ_λ for $\lambda \in \Lambda + iC_\rho$ is an immediate consequence of the maximum principle for the entire function $\lambda \to \phi_\lambda(g)$ and the fact that

$$| \phi_{\xi + is\rho}(g)| \leq 1 \qquad \xi \in \Lambda, s \in W, g \in G.$$

REFERENCES

1. S. G. Gindikin and F. I. Karpelevič, Plancherel measure for Riemannian symmetric spaces of nonpositive curvature, *Soviet Math.* **3**, 962–965 (1962).
2. Harish-Chandra, Spherical functions on a semisimple Lie group I, II, *Am. J. Math.* **80**, 241–310, 553–613 (1958).
3. S. Helgason, "Differential Geometry and Symmetric Spaces," Academic Press, New York, 1962.

PRINTED IN BRUGES, BELGIUM, BY THE ST. CATHERINE PRESS, LTD.

Reprinted from the PROCEEDINGS OF THE NATIONAL ACADEMY OF SCIENCES
Vol. 63, No. 3, pp. 643–647. July, 1969.

APPLICATIONS OF THE RADON TRANSFORM TO REPRESENTATIONS OF SEMISIMPLE LIE GROUPS*

BY SIGURDUR HELGASON

MASSACHUSETTS INSTITUTE OF TECHNOLOGY

Communicated by Lars V. Ahlfors, April 21, 1969

Abstract.—From the point of view of the duality between points and horocycles in a symmetric space, the counterparts to the spherical functions on the symmetric space are the conical distributions on the manifold of horocycles. While the conical functions are closely related to certain finite-dimensional representations of semisimple Lie groups, in the present work the conical distributions are found to play various roles in the principal series of infinite-dimensional representations of these groups.

---

1. *Introduction.*—The manifold of horocycles in a symmetric space carries a certain class of distributions introduced in an earlier paper[1] under the name conical distributions. In section 3 we indicate how analytic continuation techniques can be used to determine these distributions. The result is applied to give a complete solution (Theorem 4.6) of the problem of constructing the intertwining operators to irreducible equivalent representations from the spherical principal series of a semisimple Lie group[2] G.

The conical distributions obtained also provide an analytic approach to the irreducibility question for the spherical principal series whereby irreducibility is expressed directly by means of Harish-Chandra's $\mathbf{c}$-function. For complex G a criterion was obtained by Parthasarathy, Rao, and Varadarajan;[3] by an extension of their purely algebraic method, Kostant[4] obtained an irreducibility criterion for all real G, although in a form quite different from that of Corollary 4.3.

Finally, when the spherical principal series is realized on appropriate distribution spaces (over the space of horocycles), the conical distributions provide *the highest weight vectors* (cf. Corollary 4.4).

The results show that both the numerator and the denominator of the $\mathbf{c}$-function have their individual significance (cf. Theorem 3.2 and Corollary 4.3).

2. *Notation.*—If M is a manifold, following the book of Schwartz,[5] we denote by $\mathcal{E}(M)$ the set of C^∞ functions on M and by $\mathcal{D}(M)$ the set of C^∞ functions on M of compact support, both spaces taken with their usual topologies. The dual $\mathcal{D}'(M)$ (resp. $\mathcal{E}'(M)$) consists of all distributions on M (resp. distributions on M of compact support). Both spaces will be taken with the strong topology.

Let G be a connected semisimple Lie group with finite center, $G = KAN$ an Iwasawa decomposition of G where K is compact, A abelian, and N nilpotent. Let M be the centralizer of A in K, θ the Cartan involution of G which is identity on K, put $\bar{N} = \theta N$ and let $\mathfrak{g}$, $\mathfrak{k}$, $\mathfrak{a}$, $\mathfrak{n}$, $\mathfrak{m}$, and $\bar{\mathfrak{n}}$ denote the respective Lie algebras of the groups above. Let log denote the inverse of the map exp: $\mathfrak{a} \to A$. Let M' denote the normalizer of A in K, W the Weyl group M'/M, and B the coset space K/M. Let $\mathfrak{a}^*$ denote the dual space of $\mathfrak{a}$, $\Sigma \subset \mathfrak{a}^*$ the set of restricted roots; if $\alpha \in \Sigma$, let $\mathfrak{g}_\alpha$ denote the corresponding root space and put $m_\alpha = \dim \mathfrak{g}_\alpha$. Let Σ^+ denote the set of positive restricted roots such that

$$\mathfrak{n} = \Sigma_{\alpha \in \Sigma^+} \mathfrak{g}_\alpha.$$

Let $\Sigma_0^+ = \{\alpha \in \Sigma^+ | {}^1/_2\alpha \notin \Sigma\}$ and let Σ^- and Σ_0^- denote the set of negatives of the elements of Σ^+ and Σ_0^+, respectively. Let $\mathfrak{a}_c^*$ denote the complex dual of $\mathfrak{a}$ and $\langle , \rangle$ the bilinear form on $\mathfrak{g}$ and $\mathfrak{a}_c^*$ induced by the Killing form of $\mathfrak{g}$.

3. *Conical Distributions.*—Consider now the symmetric space $X = G/K$ and the space[6] $\Xi = G/MN$ of all horocycles in X. A distribution on Ξ is called *conical* if (1) it is MN-invariant and (2) it is an eigendistribution of each G-invariant differential operator on Ξ. Since $K\backslash G/K = A/W$, $MN\backslash G/MN = A \times W$, one can, because of the analogy with spherical functions, expect a natural parametrization of the set of conical distributions in terms of $\mathfrak{a}_c^* \times W$. Theorem 3.3 is a partial verification of this.

For each $s \in W$, fix $m_s \in M'$ in the coset s and put $\xi_s = m_sMN$. If Ξ_s denotes the orbit $MNA \cdot \xi_s$, the Bruhat lemma implies $\Xi = \cup_s \Xi_s$ (disjoint union). If $N_s' = N \cap m_s\bar{N}m_s^{-1}$, the mapping $(n_s', a) \to n_s'a \cdot \xi_s$ is a diffeomorphism of $N_s' \times A$ onto Ξ_s, giving Ξ_s a natural measure dv. Given $\xi \in \Xi_s$, there exists a unique $a(\xi) \in A$ such that $\xi \in MNa(\xi) \cdot \xi_s$; for $\lambda \in \mathfrak{a}_c^*$ we consider the functional

$$\Psi'_{\lambda,s}: \quad \phi \to \int_{\Xi_s} \phi(\xi) e^{(is\lambda + s\rho)(\log a(\xi))}\, dv(\xi), \tag{1}$$

where $\rho = \frac{1}{2} \sum_{\alpha > 0} m_\alpha \alpha$.

LEMMA 3.1. *The integral (1) converges absolutely for all $\varphi \in \mathfrak{D}(\Xi)$ if and only if*

$$Re(\langle \alpha, i\lambda \rangle) > 0 \qquad \textit{for } \alpha \in \Sigma^+ \cap s^{-1}\Sigma^-, \tag{2}$$

Re denoting real part. If (2) is satisfied, $\Psi'_{\lambda,s}$ is a conical distribution on Ξ.

The function $\lambda \to \Psi'_{\lambda,s}$ is holomorphic in the tube (2) with values in $\mathfrak{D}'(\Xi)$. One of our main results is that it can be extended to a meromorphic function on the entire $\mathfrak{a}_c^*$ and its values and "residues" are still conical distributions. These residues are certain transversal derivatives of other $\Psi'_{\lambda',s'}$ (corresponding to other MNA-orbits in the closure of Ξ_s). Consider the function

$$c_s(\lambda) = d_s(\lambda) e_s(\lambda),$$

where

$$d_s(\lambda) = \prod_{\alpha \in \Sigma_0^+ \cap s^{-1}\Sigma_0^-} 2^{-\langle i\lambda, \alpha_0 \rangle} \Gamma(\langle i\lambda, \alpha_0 \rangle),$$

$$e_s(\lambda)^{-1} = \prod_{\alpha \in \Sigma_0^+ \cap s^{-1}\Sigma_0^-} \Gamma\left[\frac{1}{2}\left(\frac{m_\alpha}{2} + 1 + \langle i\lambda, \alpha_0 \rangle\right)\right] \Gamma\left[\frac{1}{2}\left(\frac{m_\alpha}{2} + m_{2\alpha} + \langle i\lambda, \alpha_0 \rangle\right)\right].$$

Here $\alpha_0 = \langle \alpha, \alpha \rangle^{-1}\alpha$ and Γ is the usual Gamma function. For the Weyl group element $s = s^*$ which interchanges Σ^+ and Σ^-, $c_s(\lambda)$ is up to a constant factor equal to Harish-Chandra's c-function.[7]

THEOREM 3.2. *For each $s \in W$, the mapping*

$$\lambda \to d_s(\lambda)^{-1} \Psi'_{\lambda,s} = \Psi_{\lambda,s}$$

extends from (2) to a holomorphic function on the entire space $\mathfrak{a}_c^*$. *Its values are conical distributions on* Ξ.

Let $\boldsymbol{D}(X)$ and $\boldsymbol{D}(\Xi)$ denote the algebras of G-invariant differential operators on X and Ξ, respectively, and $\boldsymbol{D}(A)$ the set of left invariant differential operators on A. Let $\mu \in \mathfrak{a}_c^*$ and χ_μ: $\boldsymbol{D}(\Xi) \to \boldsymbol{C}$ the homomorphism obtained by extending μ to a homomorphism of $\boldsymbol{D}(A)$ into $\boldsymbol{C}$ and combining with the isomorphism[8] of $\boldsymbol{D}(A)$ onto $\boldsymbol{D}(\Xi)$. Let

$$\mathfrak{D}_\lambda'(\Xi) = \{\Psi \in \mathfrak{D}'(\Xi) \,|\, D\Psi = \chi_{i\lambda-\rho}(D)\Psi \quad \text{for} \quad D \in \boldsymbol{D}(\Xi)\},$$

and let $\mathcal{K}_\lambda$ denote the Hilbert space of locally square integrable functions in $\mathfrak{D}'_\lambda$ with norm

$$\psi \to [\textstyle\int_K |\psi(kMN)|^2 dk]^{1/2},$$

dk being Haar measure on K. Each conical distribution lies in one of the spaces $\mathfrak{D}_\lambda'$.

THEOREM 3.3. *Suppose* $s\lambda \neq \lambda$ *for all* $s \neq e$ *in* W *and assume the function* $\boldsymbol{e}_{s*}$ *is not zero at* λ. *Then the linear combinations*

$$\sum_{s \in W} a_s \Psi_{\lambda, s} \qquad a_s \in \boldsymbol{C}$$

constitute all the conical distributions in $\mathfrak{D}_\lambda'$.

The following lemma plays a crucial role in the proof of Theorem 3.2. If $g \in G$, let $H(g) \in \mathfrak{a}$, $k(g) \in K$ be determined by $g = k(g) \exp H(g)n$ $(n \in N)$, and write m^* for m_{s*}. If $g \in \bar{N}MAN$, let $\bar{n}(g) \in \bar{N}$ be determined by $g = \bar{n}(g)man$ $(m \in M, a \in A, n \in N)$.

LEMMA 3.4. *Suppose* G/K *has rank one and put* $|Z|^2 = -\langle Z, \theta Z\rangle$ *for* $Z \in \mathfrak{g}$. *Then if* $\bar{n} = \exp X \exp Y$ $(X \in \mathfrak{g}_{-\alpha}, Y \in \mathfrak{g}_{-2\alpha})$, *we have*

$$e^{\rho(H(\bar{n}))} = [(1 + c|X|^2)^2 + 4c|Y|^2]^{1/4(m_\alpha + 2m_{2\alpha})},$$

$$e^{\rho(H(\bar{n}(m^*\bar{n})) - H(\bar{n}))} = [c^2|X|^4 + 4c|Y|^2]^{-1/4(m_\alpha + 2m_{2\alpha})}.$$

The constant c is given by $c^{-1} = 4(m_\alpha + 4m_{2\alpha})$.

Remark: The first formula provides a simplified way of computing Harish-Chandra's $\boldsymbol{c}$-function

$$\boldsymbol{c}(\lambda) = \int_{\bar{N}} e^{-(i\lambda+\rho)(H(\bar{n}))}\, d\bar{n}$$

in the rank-one case.[9] It also gives an explicit formula for the *Poisson kernel*

$$P(gK, kM) = e^{-2\rho(H(g^{-1}k))}$$

because

$$P(aK, k(\bar{n})M) = e^{2\rho(\log a)}\, e^{2\rho(H(\bar{n}) - H(a^{-1}\bar{n}a))}, \; a \in A, \; \bar{n} \in \bar{N}.$$

4. *Intertwining Operators, Highest Weight Vectors, and Irreducibility Criteria.*—Let τ_λ and $\bar{\tau}_\lambda$ denote the natural representations of G on $\mathcal{K}_\lambda$ and $\mathfrak{D}'_\lambda$, respectively. Then τ_λ is equivalent to the representation of G induced[10] by the character $man \to e^{i\lambda(\log a)}$ of MAN; moreover, τ_λ is irreducible if and only if $\bar{\tau}_\lambda$ is irreducible (no proper closed invariant subspace $\neq 0$). We refer to the set of

representations τ_λ as the *spherical principal series.* Let ψ_0 denote the "unique" vector in $\mathcal{K}_\lambda$ fixed under K. Let $f \to \hat{f}$ denote the Radon transform on X and $\Psi \to \check{\Psi}$ its dual given by $\check{\Psi}(f) = \Psi(\hat{f})$ ($f \in \mathfrak{D}(X)$, $\Psi \in \mathfrak{D}'(\Xi)$). We define

λ *irreducible* if τ_λ is irreducible;

λ *cyclic* if ψ_0 is a cyclic vector for τ_λ;

λ *simple* if $\Psi \to \check{\Psi}$ is injective on $\mathfrak{D}_\lambda'$.

LEMMA 4.1. *Let $\lambda \in \mathfrak{a}_c^*$ be arbitrary. Then*
(i) *λ is cyclic $\langle = \rangle$ $-\lambda$ is simple;*
(ii) *λ is irreducible $\langle = \rangle$ λ and $-\lambda$ are simple.*

LEMMA 4.2. *For each $\lambda \in \mathfrak{a}_c^*$ and each $s \in W$, $(\Psi_{\lambda,s})\check{}$ is a function on X and*

$$(\Psi_{\lambda,s})\check{}(naK) = c\boldsymbol{e}_s(\lambda)e^{(is\lambda+\rho)(\log a)}, \quad n \in N, \ a \in A, \tag{3}$$

where c is a constant.

On the basis of this relation, we find that λ is not simple if $\boldsymbol{e}_{s^*}(\lambda) = 0$. It also implies that if $\boldsymbol{e}_{s^*}(\lambda) \neq 0$ and λ not simple, then for each $s \in W$, $s\lambda$ is not simple. Now Kostant has proved[4] that

$$\lambda \text{ is cyclic if } Re(\langle i\lambda, \alpha\rangle) \leq 0 \quad \text{for} \quad \alpha \in \Sigma^+. \tag{4}$$

Hence we deduce from Lemma 4.1 that

$$\lambda \text{ is simple if and only if } \boldsymbol{e}_{s^*}(\lambda) \neq 0, \tag{5}$$

and we deduce

COROLLARY 4.3. *λ is irreducible if and only if*

$$\boldsymbol{e}_{s^*}(\lambda)\boldsymbol{e}_{s^*}(-\lambda) \neq 0.$$

Remark: The "if" part required (4), which follows from Lemma 4.1 (i) except for the justification of certain interchange of limit and integration. This justification has only been carried out for the strict inequality in (4) and also for rank $(X) = 1$, so the reasoning indicated above does not constitute an independent proof of the irreducibility criterion.

A vector $\Psi \in \mathfrak{D}_\lambda'$ is called a *highest weight vector* if it is an eigenvector of $\tilde{\tau}_\lambda(A)$ and fixed under $\tilde{\tau}_\lambda(N)$. From Theorem 3.3 and Lemma 4.2, we deduce

COROLLARY 4.4. *Let λ be as in Theorem 3.3. Then the scalar multiples of the vectors $\Psi_{\lambda,s}$ constitute precisely the highest weight vectors in $\mathfrak{D}_\lambda'$.*

With the natural pairing $\langle , \rangle$ between $\mathcal{K}_\lambda$ and $\mathfrak{D}_{-\lambda}'$ whereby the representation $\tilde{\tau}_{-\lambda}$ becomes contragredient to τ_λ, we have

$$\Psi_{\lambda,e}(\phi) = \langle \phi \times \Psi_{\lambda,s}, \Psi_{s\lambda,s^{-1}}\rangle, \quad \phi \in \mathfrak{D}(\Xi), \ s \in W, \tag{6}$$

where $\times$ denotes convolution[11] on Ξ. These formulas are infinite-dimensional analogs of the formula[12] relating conical functions and finite-dimensional conical representations.

We now transfer the representation space of $\tilde{\tau}_\lambda$ from $\mathfrak{D}_\lambda'$ to $\mathfrak{D}'(B)$. Let $\bar{N}_s = \bar{N} \cap (m_s^{-1}Nm_s)$ and let $d\bar{n}_s$ and da denote Haar measures on $\bar{N}_s$ and A, respectively.

LEMMA 4.5. *The mapping $S \to \Psi$ given by*

$$\Psi(\phi) = \int_{K/M}\left(\int_A \phi(kaMN)e^{(i\lambda+\rho)(\log a)}\, da\right)dS(kM), \ \phi \in \mathfrak{D}(\Xi)$$

is a bijection of $\mathfrak{D}'(B)$ onto $\mathfrak{D}_\lambda'$ mapping $L^2(B)$ onto $\mathcal{K}_\lambda$. Under this map $\Psi_{\lambda,s}$ corresponds to the distribution

$$S_{\lambda,s}\colon F \to \boldsymbol{d}_s(\lambda)^{-1} \int_{\bar{N}_s} F(m_s k(\bar{n}_s)M)e^{-(i\lambda+\rho)(H(\bar{n}_s))}\, d\bar{n}_s, \ F \in \mathcal{E}(B),$$

defined for all $\lambda \in \mathfrak{a}_c^*$ by analytic continuation.

The distributions $S_{\lambda,s}$ now give the construction of the interwining operators. Let $S_1 \times S_2$ denote the convolution product of distributions on B induced by convolution on K. Although in general this product only commutes with the action of K, we see now that for $S_2 = S_{\lambda,s}$ we get commutation with the action of G.

THEOREM 4.6. *Assume λ irreducible. The convolution operator $S \to S \times S_{s\lambda,s^{-1}}$ on $\mathfrak{D}'(B)$ induces a bijection $A_{\lambda,s}$: $\mathfrak{D}_\lambda' \to \mathfrak{D}_{s\lambda}'$ which intertwines the representations $\bar{\tau}_\lambda$ and $\bar{\tau}_{s\lambda}$. If λ is real, then $A_{\lambda,s}$ gives a unitary operator $\mathcal{K}_\lambda$ onto $\mathcal{K}_{s\lambda}$.*

Note added in proof: Since the completion of this paper, G. Schiffman has kindly sent me a copy of his thesis. It contains a statement of Lemma 3.4, the last part of Theorem 4.6, and several other results on intertwining operators.

* This work was supported in part by the National Science Foundation (NSF GP 9003).

[1] Helgason, S., "Radon-Fourier transforms on symmetric spaces and related group representations," *Bull. Am. Math. Soc.*, **71**, 757–763 (1965).

[2] For complex G the problem is settled by R. A. Kunze and E. M. Stein ("Uniformly bounded representations, III," *Am. J. Math.* **89**, 385–442 (1967)), and G. Schiffmann ("Intégrales d'entrelacement," *Compt. Rend.*, **266**, 47–49 (1968)). Schiffmann also treats the case of G which split over R and for general G reduces the problem to the rank-one case which, however, is left open; see also Gelfand, I. M., M. I. Graev, and I. I. Pyatetskii-Shapiro, *Representation Theory and Automorphic Functions* (Philadelphia: W. B. Saunders Co., 1969), p. 377.

[3] Parthasarthy, K. R., R. Ranga Rao, and V. S. Varadarajan, "Representations of complex semisimple Lie groups and Lie algebras," *Ann. Math.*, **85**, 383–429 (1967).

[4] Kostant, B., "On the existence and irreducibility of certain series of representations," *Bull. Am. Math. Soc.*, to appear.

[5] Schwartz, L., *Théorie des Distributions* (Paris: Hermann, 1966).

[6] Helgason, S., "Duality and Radon transform for symmetric spaces," *Am. J. Math.*, **85**, 667–692 (1963).

[7] Cf. Harish-Chandra, "Spherical functions on a semisimple Lie group I," *Am. J. Math.*, **80**, 303 (1958); Gindikin, S. G., and F. I. Karpelevič, "Plancherel measure for Riemannian symmetric spaces of nonpositive curvature," *Soviet Math.*, **3**, 962–965 (1962).

[8] See reference in ref. 6, §4.

[9] See Harish-Chandra reference in ref. 7.

[10] Here, "induced" is meant in the sense of Mackey, G. W., "Induced representations of locally compact groups I," *Ann. Math.*, **55**, 101–139 (1952), and Bruhat, F., "Sur les représentations induites des groupes de Lie," *Bull. Soc. Math. France*, **84**, 97–205 (1956).

[11] Helgason, S., "A duality in integral geometry on symmetric spaces," *Proc. U.S.-Japan Seminar in Differential Geometry, Kyoto* (1965), p. 53.

[12] Cf. reference in ref. 1, Theorem 2.6.

Reprinted from ADVANCES IN MATHEMATICS
All Rights Reserved by Academic Press, New York and London
Vol. 5, No. 1, August 1970
Printed in Belgium

A Duality for Symmetric Spaces with Applications to Group Representations

SIGURDUR HELGASON[1]

Massachusetts Institute of Technology

Table of Contents

[1] Present address: Institut Mittag-Leffler, Djursholm, Sweden.

607/5/1-1

Preface

En aldrei sá neinn
þann sem augað gaf.
E.B.

Let X be a symmetric space of the noncompact type, Ξ the (dual) space of horocycles in X. The largest connected group G of isometries of X acts transitively both on X and on Ξ. The present paper deals with various topics in analysis on X and on Ξ and relates these to the representation theory of G. We give now a very rough indication of the contents of the paper; Chapter I contains a detailed summary of the main results. The proofs are given in the subsequent chapters where the relationship to past and current work in representation theory is also explained.

The analogies between the spaces X and Ξ are first studied in some detail and function theory on Ξ developed from the point of view of these analogies. The principal concepts here are the conical functions and conical distributions on Ξ; they are the counterparts to the spherical functions on X. The joint eigenspaces of the G-invariant differential operators on Ξ give models for irreducible representation of G; similarly for X. This provides two distinct models for the Gelfand–Naimark principal series of class one representations of G, the connection between the two models being provided by the Radon transform (mapping functions on X into functions on Ξ) and its dual (mapping distributions on Ξ into distributions on X). On the other hand when the class one principal series is realized by means of the models over Ξ, certain equivalences take place and the corresponding intertwining operators are constructed by means of the conical distributions. These distributions also furnish extreme weight vectors for the representations in the class one principal series. The two dual models of the class one principal series occur when the natural representations of G on $L^2(X)$ and on $L^2(\Xi)$ are decomposed into a direct integral of irreducible representations. The decomposition is most explicitly exhibited by means of a suitable Fourier transform on X and on Ξ.

The Plancherel measure on X is given by Harish-Chandra's function $\boldsymbol{c}(\lambda)$ which is given by a ratio of certain Gamma factors. Whereas the spherical functions on X involve the $\boldsymbol{c}$-function alone we find that the numerator itself enters significantly in the construction of the conical distributions. The denumerator on the other hand turns out to be intimately connected with the principal series of representations mentioned above.

This paper is fairly self-contained but uses many results from Harish-Chandra's papers [21(i), (j)] on spherical functions and a few results from [22(c)] where the Radon transform on X is applied to differential equations. Many of the results have been given in research announcements [22(d), (g), (h), (k)]. The material in Chapter II was presented at the Scandinavian Mathematical Congress, Oslo, 1968 and the material in Chapter III in lectures at Eidgenössische Technische Hochschule, Zürich whose hospitality I enjoyed during the summer 1969. I wish also to acknowledge support by the National Science Foundation during the preparation of this paper.

Finally, I wish to express my appreciation to G. Schiffmann for showing me his proof of his formula reducing intertwining operators to the rank one case [40(a)] and to B. Kostant for enlightenment about his work [28(b)].

Notation and Preliminaries

§1. *General Notation.* We shall use the standard notation $\boldsymbol{Z}$, $\boldsymbol{R}$, $\boldsymbol{C}$ for the ring of integers, the field of real numbers and the field of complex numbers, respectively; $\boldsymbol{Z}^+$ is the set of nonnegative integers, $\boldsymbol{R}^+$ the set of nonnegative real numbers. If S is a set, T a subset and f a function on S, the restriction of f to T is denoted $f \mid T$. If S is a topological space $Cl(T)$ denotes the closure of T in S. The space of continuous functions on S is denoted by $C(S)$, $C_c(S)$ the set of those of compact support. Composition of functions and operators will often be denoted by $\circ$.

§2. *Manifolds.* If M is a manifold (satisfying the second countability axiom) and $m \in M$ the tangent space to M at m is denoted M_m. Following Schwartz [41] we write $\mathscr{D}(M)$ for the space of C^∞ functions on M of compact support, topologized by means of uniform convergence of functions along with their derivatives; $\mathscr{D}'(M)$ denotes the dual space of all distributions on M. The space $\mathscr{E}(M)$ denotes the space of all C^∞ functions on M topologized in a similar way as $\mathscr{D}(M)$ and $\mathscr{E}'(M)$ denotes the dual space of distributions on M of compact support. If V is a vector space over $\boldsymbol{R}$, $\mathscr{S}(V)$ denotes the space of rapidly decreasing functions on V (Schwartz [41]). Let τ be a diffeomorphism of M onto itself, and let $f \in \mathscr{E}(M)$, $T \in \mathscr{D}'(M)$ and D a differential operator on M. We put

$$f^\tau(m) = f(\tau^{-1}(m)), \qquad m \in M$$

$$T^\tau(f) = T(f^{\tau^{-1}}), \qquad f \in \mathscr{D}(M)$$

$$D^\tau(f) = (Df^{\tau^{-1}})^\tau, \qquad f \in \mathscr{E}(M)$$

Then $f^\tau \in \mathscr{E}(M)$, $T^\tau \in \mathscr{D}'(M)$ and D^τ is another differential operator on M. The value of Df at a point m will usually be denoted by $(Df)(m)$ but sometimes it is convenient to write $D_m(f(m))$. If Φ is a differentiable mapping from a manifold M into another manifold $d\Phi_m$ (and sometimes $d\Phi$) denotes the differential of Φ at m.

§3. *Lie Groups.* If A is a group and $a \in A$, $L(a)$ denotes the left translation $x \to ax$ and $R(a)$ denotes the right translation $x \to xa$ on A. If $B \subset A$ is a subset we write $B^a = aBa^{-1}$; if B is a subgroup A/B denotes the set of left cosets aB, $a \in A$. The transformation $xB \to axB$ of A/B will always be denoted by $\tau(a)$.

Lie groups will be denoted by Latin capital letters and their Lie algebras by corresponding lower case German letters. If G is a Lie group and $\mathfrak{g}$ its Lie algebra the adjoint representation of G is denoted by Ad (or Ad_G) and the adjoint representation of $\mathfrak{g}$ by ad (or $\mathrm{ad}_{\mathfrak{g}}$).

We shall now list some standard notation concerning semisimple Lie groups which will be utilized throughout the paper. Let G be a connected semisimple Lie group with finite center, $\mathfrak{g}$ the Lie algebra of G, and $\langle\ ,\ \rangle$ (sometimes B) the Killing form of $\mathfrak{g}$. Let θ be a *Cartan involution* of $\mathfrak{g}$, that is an involutive automorphism such that the form $(X, Y) \to -\langle X, \theta Y\rangle$ is strictly positive definite on $\mathfrak{g} \times \mathfrak{g}$. Let $\mathfrak{g} = \mathfrak{k} + \mathfrak{p}$ be the decomposition of $\mathfrak{g}$ into eigenspaces of θ (a *Cartan decomposition*) and K the analytic subgroup of G with Lie algebra $\mathfrak{k}$. Let $\mathfrak{a} \subset \mathfrak{p}$ be a maximal abelian subspace, $\mathfrak{a}^*$ its dual, $\mathfrak{a}_c^*$ the complexification of $\mathfrak{a}^*$, i.e. the space of $\boldsymbol{R}$-linear maps of $\mathfrak{a}$ into $\boldsymbol{C}$. Let $A = \exp \mathfrak{a}$ and log the inverse of the map exp: $\mathfrak{a} \to A$. For $\lambda \in \mathfrak{a}^*$ put

$$\mathfrak{g}_\lambda = \{X \in \mathfrak{g} \mid [H, X] = \lambda(H)X, \qquad \text{for all} \qquad H \in \mathfrak{a}\}.$$

If $\lambda \neq 0$ and $\mathfrak{g}_\lambda \neq \{0\}$ then λ is called a (*restricted*) *root* and $m_\lambda = \dim(\mathfrak{g}_\lambda)$ is called its *multiplicity*. Let $\mathfrak{g}_c$ denote the complexification of $\mathfrak{g}$ and if $\mathfrak{s}$ is any subspace of $\mathfrak{g}$ let $\mathfrak{s}_c$ denote the complex subspace of $\mathfrak{g}_c$ spanned by $\mathfrak{s}$. If $\lambda, \mu \in \mathfrak{a}_c^*$ let $H_\lambda \in \mathfrak{a}_c$ be determined by $\lambda(H) = \langle H_\lambda, H\rangle$ for $H \in \mathfrak{a}$ and put $\langle \lambda, \mu\rangle = \langle H_\lambda, H_\mu\rangle$. Since $\langle\ ,\ \rangle$ is positive definite on $\mathfrak{p}$ we put $|\lambda| = \langle\lambda, \lambda\rangle^{1/2}$ for $\lambda \in \mathfrak{a}^*$ and $|X| = \langle X, X\rangle^{1/2}$ for $X \in \mathfrak{p}$. Let $\mathfrak{a}'$ be the open subset of $\mathfrak{a}$ where all restricted roots are $\neq 0$. The components of $\mathfrak{a}'$ are called *Weyl chambers*. Fix a Weyl chamber $\mathfrak{a}^+$ and call a (restricted) root α *positive* if it is positive on $\mathfrak{a}^+$. Let $\mathfrak{a}_+^*$ denote the corresponding Weyl chamber in $\mathfrak{a}^*$, that is the preimage of $\mathfrak{a}^+$ under the mapping $\lambda \to H_\lambda$. Let Σ denote the set of restricted roots, Σ^+ the set of positive roots and Σ^- the set of negatives of the members in Σ^+. Let

$\Sigma_0 = \{\alpha \in \Sigma \mid \frac{1}{2}\alpha \notin \Sigma\}$, and put $\Sigma_0^+ = \Sigma^+ \cap \Sigma_0$, $\Sigma_0^- = \Sigma^- \cap \Sigma_0$. A root $\alpha \in \Sigma^+$ is called *simple* if it is not a sum of two positive roots. The walls of the Weyl chamber $\mathfrak{a}^+$ lie on the hyperplanes $\alpha_1 = 0, \ldots, \alpha_l = 0$, $\alpha_1, \ldots, \alpha_l$ being the simple roots. Let ρ denote half the sum of the positive roots with multiplicity, i.e., $\rho = \frac{1}{2}\Sigma_{\alpha \in \Sigma^+} m_\alpha \alpha$. Let $\mathfrak{n} = \Sigma_{\alpha > 0} \mathfrak{g}_\alpha$, $\bar{\mathfrak{n}} = \theta\mathfrak{n}$ and let N and $\bar{N}$ denote the corresponding analytic subgroups of G. Let M denote the centralizer of A in K, M' the normalizer of A in K, W the (finite) factor group M'/M, the *Weyl group*. The group W acts as a group of linear transformations of $\mathfrak{a}$ and also on $\mathfrak{a}_c^*$ by $(s\lambda)(H) = \lambda(s^{-1}H)$ for $H \in \mathfrak{a}$, $\lambda \in \mathfrak{a}_c^*$ and $s \in W$. Let w denote the order of W, and let $m_1', \ldots, m_w'$ be a complete set of representatives in M' (mod M). Let $A^+ = \exp \mathfrak{a}^+$, $B = K/M$, $P = MAN$; then we have the decompositions

$$G = KCl(A^+)K \qquad \text{(Cartan decomposition)}, \tag{1}$$

$$G = KAN \qquad \text{(Iwasawa decomposition)}, \tag{2}$$

$$G = \bigcup_{i=1}^{w} Pm_i'P \qquad \text{(Bruhat decomposition)}. \tag{3}$$

Here (1) means that each $g \in G$ can be written $g = k_1 A(g) k_2$ where $k_1, k_2 \in K$ and $A(g) \in Cl(A^+)$; here $A(g)$ is actually unique. In (2) each $g \in G$ can be uniquely written

$$g = k(g) \exp H(g)\, n(g), \qquad k(g) \in K, \quad H(g) \in \mathfrak{a}, \quad n(g) \in N.$$

In (3) the union is a disjoint union; exactly one of the summands, namely Pm^*P (where $m^* \in M'$ satisfies $Ad(m^*)\,\mathfrak{a}^+ = -\mathfrak{a}^+$) is open in G. Thus the set $\bar{N}MAN$ is open in G; if $g \in \bar{N}MAN$ is written $g = \bar{n}man$, $\bar{n} \in \bar{N}$ is uniquely determined by g and will be denoted $\bar{n} = \bar{n}(g)$. Let s^* denote the coset m^*M in W.

The number l which equals $\dim \mathfrak{a}$ is called the *real rank* of G and the *rank* of the symmetric space $X = G/K$.

§4. *Normalization of Measures.* It is convenient to make some conventions concerning the normalization of certain invariant measures. Let $l = \dim \mathfrak{a}$. The Killing form induces Euclidean measures on A, $\mathfrak{a}$ and $\mathfrak{a}^*$; multiplying these by the factor $(2\pi)^{-(1/2)l}$ we obtain invariant measures da, dH and $d\lambda$, and the inversion formula for the Fourier transform

$$f^*(\lambda) = \int_A f(a)\, e^{-i\lambda(\log a)}\, da, \qquad \lambda \in \mathfrak{a}^*, \tag{1}$$

holds without any multiplicative constant,

$$f(a) = \int_{\mathfrak{a}^*} f^*(\lambda)\, e^{i\lambda(\log a)}\, d\lambda, \qquad f \in \mathscr{S}(A). \tag{2}$$

We normalize the Haar measures dk and dm on the compact groups K and M, respectively, such that the total measure is 1. The Haar measures of the nilpotent groups N, $\bar{N}$ are normalized such that

$$\theta(dn) = d\bar{n}, \qquad \int_{\bar{N}} e^{-2\rho(H(\bar{n}))}\, d\bar{n} = 1. \tag{3}$$

The Haar measure dg on G can be normalized such that

$$\int_G f(g)\, dg = \int_{KAN} f(kan)\, e^{2\rho(\log a)}\, dk\, da\, dn, \qquad f \in \mathscr{D}(G). \tag{4}$$

Since each automorphism $n \to mnm^{-1}$ of N by an element $m \in M$ preserves the measure dn the measure $dmdn$ is a bi-invariant measure $d(mn)$ on MN.

Suppose U is a locally compact group and P a closed subgroup, both unimodular. Suppose du and dp are Haar measures on U and P, respectively. Let du_P denote the U-invariant measure on U/P given by

$$\int_U f(u)\, du = \int_{U/P} \Big(\int_P f(up)\, dp\Big)\, du_P\,, \qquad f \in C_c(U). \tag{5}$$

In particular we have a K-invariant measure dk_M on $B = K/M$ of total measure one. For the coset spaces $X = G/K$, $\Xi = G/MN$ which play the main roles in this paper we write for simplicity

$$dx = dg_K\,, \qquad d\xi = dg_{MN}\,. \tag{6}$$

If S is a locally compact space with a measure μ and $p \geqslant 1$, then $L^p(S)$ denotes the set of measurable functions f such that $|f|^p$ is μ-integrable.

§5. *Representations.* Let Q be a locally compact group, V a locally convex topological vector space, Aut(V) the group of linear homeomorphisms of V onto itself. A *representation* π of Q on V is a homomorphism of Q into Aut(V) such that the map $(q, v) \to \pi(q)v$ of $Q \times V$ into V is continuous; π is called *irreducible* if the only closed subspaces of V which are invariant under each $q \in Q$ are $\{0\}$ and V. If π' is a representation of Q on another locally convex topological vector space V' then π

and π' are said to be *equivalent* if there exists a linear homeomorphism A of V onto V' such that $A\pi(q) = \pi'(q)A$ for all $q \in Q$. The operator A is called an *intertwining operator*.

§6. *Spherical Functions.* A *spherical function* (often called zonal spherical function) is a continuous function $\varphi \not\equiv 0$ on G satisfying the functional equation

$$\int_K \varphi(xky)\, dk = \varphi(x)\, \varphi(y), \qquad x, y \in G. \tag{7}$$

Such a function is necessarily bi-invariant under K. In Chapter II we make extensive use of Harish-Chandra's profound and basic results on spherical functions [21(i), (j), (k)] so we recall these briefly. First, the spherical functions are precisely the functions of the form

$$\varphi_\lambda(g) = \int_K e^{(i\lambda-\rho)(H(gk))}\, dk, \qquad g \in G, \tag{8}$$

where $\lambda \in \mathfrak{a}_c{}^*$ is arbitrary; moreover, $\varphi_\lambda \equiv \varphi_\mu$ if and only if $\lambda = s\mu$ for some $s \in W$.

If $F \in \mathscr{D}(G)$ is bi-invariant under K, i.e., $F(kgk') \equiv F(g)$, its *spherical Fourier transform* is defined by (conj = complex-conjugate)

$$\tilde{F}(\lambda) = \int_G F(g) \operatorname{conj} \varphi_\lambda(g)\, dg, \qquad \lambda \in \mathfrak{a}^*. \tag{9}$$

Then

$$F(g) = w^{-1} \int_{\mathfrak{a}*} \tilde{F}(\lambda)\, \varphi_\lambda(g)\, |\, \boldsymbol{c}(\lambda)|^{-2}\, d\lambda, \qquad g \in G \tag{10}$$

and

$$\int_G |\, F(g)|^2\, dg = w^{-1} \int_{\mathfrak{a}*} |\, \tilde{F}(\lambda)|^2\, |\, \boldsymbol{c}(\lambda)|^{-2}\, d\lambda, \tag{11}$$

where $|\, \boldsymbol{c}(\lambda)|^{-2} = \boldsymbol{c}(\lambda)\, \boldsymbol{c}(-\lambda)$ for $\lambda \in \mathfrak{a}^*$ and

$$\boldsymbol{c}(\lambda) = \int_{\bar{N}} e^{-(i\lambda+\rho)(H(\bar{n}))}\, d\bar{n} \tag{12}$$

the function $\boldsymbol{c}(\lambda)^{-1}$ having an analytic continuation to all of $\mathfrak{a}^*$. An important contribution to the proof was given by Gindikin–Karpelevič [19(a)].

§7. *Differential operators.* If A is a Lie group $\boldsymbol{D}(A)$ denotes the algebra of all left invariant differential operators on A. If $B \subset A$ is a closed subgroup, $\boldsymbol{D}(A/B)$ denotes the algebra of A-invariant differential operators on A/B. The notation being as in §3 let $\boldsymbol{D}_0(G)$ denote the set of $D \in \boldsymbol{D}(G)$ which are invariant under all right translations from K. There is a homomorphism μ of $\boldsymbol{D}_0(G)$ onto $\boldsymbol{D}(G/K)$ such that

$$(\mu(D)f) \circ \pi = D(f \circ \pi) \qquad \text{for} \qquad D \in \boldsymbol{D}_0(G), \qquad f \in \mathscr{E}(X),$$

π denoting the natural mapping of G onto G/K. The spherical functions on G can be characterized as the joint eigenfunctions of all $D \in \boldsymbol{D}_0(G)$, bi-invariant under K, normalized by $\varphi(e) = 1$.

Chapter I. Summary

Since the notation which is used throughout the paper has already been explained, it seems feasible to give in this chapter a fairly detailed summary of the main results of the paper. We have also found it practical to insert proofs here of some of the simplest results.

§1. The Space of Horocycles in a Symmetric Space

Let X be a symmetric space of the noncompact type, that is a coset space $X = G/K$ where G is a connected semisimple Lie group with finite center and K a maximal compact subgroup. Let $G = KAN$ be an Iwasawa decomposition (Notation, §3). A *horocycle* in X is by definition an orbit in X of a subgroup of G conjugate to N. The horocycles are all closed submanifolds of X. Let Ξ denote the set of all horocycles. The natural action of G on Ξ is transitive and if o denotes the origin in X and if $\xi_o = N \cdot o$ then the subgroup of G which maps ξ_o into itself equals MN so we make the identification

$$\Xi = G/MN. \tag{1}$$

(Note that in [22(c), Theorem 3.4] this identification is proved only when G has center $\{e\}$; for general G one must verify the simple fact, that since the center of G lies in K, the group $Ad_G(M)$ is the centralizer of $Ad_G(A)$ in $Ad_G(K)$).

The spaces X and Ξ have the same dimension and show many analogies reminiscent of the duality between points and hyperplanes in $\boldsymbol{R}^n$. Thus we call Ξ *the dual space* of X. We have the following two propositions, parts (i) being well known, parts (ii) proved in [22(c)].

PROPOSITION 1.1.

(i) *The mapping* $(kM, a) \to kaK$ *is a differentiable mapping of* $K/M \times A$ *onto* X *and if* $A' = \exp \mathfrak{a}'$ *the restriction to* $K/M \times A'$ *is a covering of order* w *of the regular set* X' *in* X.

(ii) *The mapping* $(kM, a) \to kaMN$ *is a diffeomorphism of* $K/M \times A$ *onto* Ξ.

PROPOSITION 1.2. *The following natural identifications hold*:

(i) $K\backslash G/K = A/W$;

(ii) $MN\backslash G/MN = A \times W$.

Here A/W is the space obtained from A by identifying points which are conjugate under W; thus (i) is essentially equivalent to the Cartan decomposition of G. On the other hand, (ii) amounts to the fact that each $g \in G$ can be written (with obvious notation) $g = m_1 n_1 m' a m_2 n_2$ where $a \in A$ is unique and $m' \in M'$ is unique mod M. This is the Bruhat decomposition [21(g)] except for the uniqueness of a which is proved in [22(c), p. 674] in the case when G has center $\{e\}$. In order for the proof to generalize to the present case we must establish the relation

$$N(\theta N) \cap K = \{e\}. \tag{2}$$

But if $n_1(\theta n_2) = k$ we obtain by applying $Ad = Ad_G$, $Ad(n_1)\, Ad(\theta n_2) = Ad(k)$ whence by Lemme 3.2 in [22(c)], $Ad(\theta n_2) = Ad(n_1^{-1})$. But in terms of a suitable basis of $\mathfrak{g}_c$, $Ad(N)$ is given by a group of upper unipotent matrices, $Ad(\theta N)$ by a group of lower unipotent matrices. Hence $Ad(n_1) = Ad(n_2) = e$ so (2) follows.

By a *Weyl chamber* in $\mathfrak{p}$ we understand a Weyl chamber in some maximal abelian subspace of $\mathfrak{p}$. Then K acts transitively on the set of all Weyl chambers in $\mathfrak{p}$ and the subgroup which leaves $\mathfrak{a}^+$ invariant equals M; thus $B = K/M$ is naturally identified with this set. Now by Proposition 1.1 $\xi \in \Xi$ can be written $\xi = ka \cdot \xi_o$ where $kM \in K/M$ and $a \in A$ are unique. We say the Weyl chamber kM is *normal* to ξ and call a the *complex distance* from o to ξ. The following result is then immediate.

PROPOSITION 1.3. *Given* $b \in B$, $x \in X$ *there exists a unique horocycle passing through* x *with normal* b.

Definition. Let $\xi(x, b)$ denote the horocycle in Proposition 1.3, let $a(x, b) \in A$ denote its complex distance from o and put $A(x, b) = \log a(x, b)$.

Each Weyl chamber $\mathfrak{a}_i^+ \subset \mathfrak{a}$ gives rise to a group N_i for which the Iwasawa decomposition $G = KAN_i$ holds. Correspondingly we have

Proposition 1.4. *Given a horocycle ξ and a point $x \in \xi$ there exist exactly w distinct horocycles passing through x with tangent space at x equal to the tangent space ξ_x.*

To see this we may assume $x = o$, $\xi = N \cdot o$. If ξ' is any horocycle through o then $\xi' = kN \cdot o$ $(k \in K)$. If in addition ξ and ξ' have the same tangent space at o then k maps this tangent space, and hence its orthogonal complement, into itself. Hence by [22(c), Lemma 3.3], $Ad(k)\mathfrak{a} \subset \mathfrak{a}$ so $k \in M'$, and $\xi' = N_i \cdot o$ for some N_i above. To see that two different N_i give different orbits of o suppose $N_i \cdot o = N_j \cdot o$. This means $N_iK = N_jK$ which in view of the relations $N_j \subset N_i\theta N_i$ [22(c), p. 674] and $K \cap (N_i\theta N_i) = \{e\}$ implies $N_i = N_j$.

Proposition 1.5. *Let $\xi_1, \xi_2 \in \Xi$. The following conditions are equivalent*:

(i) *ξ_1 and ξ_2 have the same normal.*

(ii) *There exists a geodesic γ perpendicular to both ξ_1 and ξ_2 such that $T\xi_1 = \xi_2$ for a transvection* [22(c), p. 672] *T along γ.*

(iii) *ξ_1 and ξ_2 are orbits of the same group gNg^{-1}.*

Proof. (i) $\Rightarrow$ (ii). We have $\xi_1 = k_1a_1 \cdot \xi_o$, $\xi_2 = k_2a_2 \cdot \xi_o$ and $k_1 = k_2$. Since $a_1 \cdot \xi_o$ and $a_2 \cdot \xi_0$ have the property in (ii) so do ξ_1 and ξ_2. For (ii) $\Rightarrow$ (iii) select $g \in G$ such that $g \cdot \xi_2 = \xi_o$. Then $g \cdot \xi_1 = T \cdot \xi_o$ for a transvection T along a geodesic γ perpendicular to $g \cdot \xi_1$ and ξ_o; let us say γ intersects ξ_o at the point $n \cdot o$. Thus $n^{-1} \cdot \gamma$ is a geodesic in $A \cdot o$ so for some $a \in A$, $n^{-1}g \cdot \xi_1 = a \cdot \xi_o$. Thus $\xi_1 = g^{-1}Ng \cdot (g^{-1}a \cdot o)$ and $\xi_2 = g^{-1}Ng \cdot (g^{-1} \cdot o)$.

Finally for (iii) $\Rightarrow$ (i) we may assume that ξ_1 and ξ_2 are orbits of the group $kNk^{-1} = N^k$, $(k \in K)$. Since $G = N^kA^kK$ each orbit of the group N^k intersects $A^k \cdot o$. Thus for suitable $a_1, a_2 \in A$ we have $\xi_i = N^kka_i \cdot o$ $(i = 1, 2)$ so ξ_1 and ξ_2 have the same normal kM.

Although the coset space G/MN is not in general reductive it does have something analogous to geodesics in X. Let $t \to \gamma(t)$ be a geodesic in X which is *regular*, that is its stabilizer has minimum dimension. The tangent vector $\dot{\gamma}(t)$ at a point $x = \gamma(t)$ determines by its centralizer a maximal abelian subspace in the tangent space $X_x \subset \mathfrak{g}$ and $\gamma(t)$ lies in a unique Weyl chamber in this abelian subspace. This Weyl chamber defines a nilpotent group N_γ, depending only on γ, not on x (cf. Chap. III §1). The curve $t \to \xi(t) = N_\gamma \cdot \gamma(t)$ will be called a *geodesic in* Ξ. A submanifold $\Xi' \subset \Xi$ is called *totally geodesic* of each geodesic in Ξ tangential to Ξ' lies in Ξ'.

PROPOSITION 1.6. *Let X and X' be symmetric space of the noncompact type, X' a totally geodesic submanifold of X. Then the horocycles in X' are all of the form $\xi' = \xi \cap X'$ where ξ is a horocycle in X.*

In the case when X' is a *boundary component of* X (see definition in Chap. III, §1) this result can be given a more precise form. Let $G = I_o(X)$, the largest connected group of isometries of X in the compact open topology and let G' denote the closed subgroup of $I_o(X)$ generated by the transvections (in X) along the geodesics in X'. Then $X = G/K$, $X' = G'/K'$ where $K' = G' \cap K$. The next result identifies the G'-orbit of ξ_o in Ξ with the dual space Ξ' of X'.

PROPOSITION 1.7. *Let X' be a boundary component of X and put*

$$\Xi_o' = G' \cdot \xi_o \subset \Xi.$$

Then Ξ_o' is totally geodesic in Ξ and the mapping $\xi \to \xi \cap X'$ is a bijection of Ξ_o' onto Ξ'.

A particularly useful case of the situation in Proposition 1.6 arises via the following reduction principle. We recall that $\boldsymbol{SU}(2, 1)$ denotes the group of complex 3×3 matrices with determinant 1 which leave invariant the form $z_1\bar{z}_1 + z_2\bar{z}_2 - z_3\bar{z}_3$.

THEOREM 1.8. *Let $\mathfrak{g}$ be a noncompact semisimple Lie algebra. Assume α and 2α are restricted roots. Select any $X \neq 0$ in $\mathfrak{g}_\alpha$, $Y \neq 0$ in $\mathfrak{g}_{2\alpha}$. Then the subalgebra $\mathfrak{g}(\alpha)$ of $\mathfrak{g}$ generated by X, Y, θX and θY is isomorphic to* $\mathfrak{su}(2, 1)$, *the Lie algebra of* $\boldsymbol{SU}(2, 1)$.

This theorem is particularly useful because analytic groups $G(\alpha) \subset G$ corresponding to $\mathfrak{g}(\alpha) \subset \mathfrak{g}$ have compatible Cartan-, Iwasawa-, and Bruhat decompositions. Thus if a group element $g \in G$ is embedded in a group of the form $G(\alpha)$ its components $A(g)$, $k(g)$, $H(g)$, $n(g)$, $\bar{n}(g)$

can be found by direct matrix calculation within $\boldsymbol{SU}(2, 1)$. In this way we compute in Chap. III, §1, formulas for $e^{\rho(H(\bar{n}))}$, $e^{\rho(H(\bar{n}(m^*\bar{n})))}$ and the Poisson kernel on $X \times B$ for X of rank one.

§2. Function Theory on X and on Ξ

Let $\boldsymbol{D}(X)$ and $\boldsymbol{D}(\Xi)$ denote the algebras of G-invariant differential operators on X and Ξ, respectively. Let $\boldsymbol{D}(A)$ denote the algebra of translations—invariant differential operators on A (i.e., the differential operators on A with constant coefficients), and let $\boldsymbol{I}(A)$ denote the set of W-invariant elements in $\boldsymbol{D}(A)$. The following result is well-known ([21(i), p. 260]; cf. also [22(b), p. 432]).

Theorem 2.1. *$\boldsymbol{D}(X)$ is canonically isomorphic to $\boldsymbol{I}(A)$.*

In order to describe this isomorphism Γ: $\boldsymbol{D}(X) \to \boldsymbol{I}(A)$ note that by Proposition 1.2 the K-invariant functions on X can be regarded as the W-invariant functions on A. When an operator $D \in \boldsymbol{D}(X)$ is applied to such functions it becomes a singular differential operator on A. The principal part of this operator has constant coefficients and this coincides with the highest order term in $\Gamma(D)$.

Theorem 2.2. *$\boldsymbol{D}(\Xi)$ is canonically isomorphic to $\boldsymbol{D}(A)$.*

In order to describe this isomorphism let us consider the action of G on Ξ. In view of Proposition 1.1, Ξ can be viewed as a trivial fiber bundle with base K/M and fiber over kM given by $\{kaMN \mid a \in A\}$. Now if $g \in G$ let $k(g)$, $a(g)$, $n(g)$ denote the components in the decomposition $G = KAN$. Since

$$gkaMN = k(gk)\,a(gk)\,aMN$$

we see that the action of g on Ξ is fiber preserving and induces a translation on each fiber. Thus if $D \in \boldsymbol{D}(A)$, the operator $\hat{\Gamma}(D)$ on $\mathscr{E}(\Xi)$ given by

$$(\hat{\Gamma}(D)f)(kaMN) = D_a(f(kaMN)) \tag{1}$$

lies in $\boldsymbol{D}(\Xi)$. The mapping $D \to \hat{\Gamma}(D)$ is the isomorphism in Theorem 2.2, [22(c), p. 676].

The mapping $n \to n \cdot o$ is a diffeomorphism of N onto ξ_o and since dn is MN-invariant there is induced a measure ds on each horocycle ξ;

ds is proportional to the volume element corresponding to the Riemannian structure on ξ induced by that of X.

If f is a function on X we define $\hat{f}$ by

$$\hat{f}(\xi) = \int_\xi f(x)\,ds(x) \tag{2}$$

for all $\xi \in \Xi$ for which this integral exists. The function $\hat{f}$ is called the *Radon transform* of f. In Chap. II we prove

THEOREM 2.3. *If $f \in L^1(X)$ then $\hat{f}(\xi)$ exists for almost all $\xi \in \Xi$. If $\hat{f} = 0$ almost everywhere then $f = 0$ almost everywhere.*

Since the spherical function φ_λ satisfies conj $\varphi_\lambda = \varphi_{-\lambda}$ for $\lambda \in \mathfrak{a}^*$ we extend the definition of the *spherical Fourier transform* as follows: If F is a function on G bi-invariant under K we define

$$\tilde{F}(\lambda) = \int_G F(g)\,\varphi_{-\lambda}(g)\,dg$$

for all $\lambda \in \mathfrak{a}_c{}^*$ for which this integral makes sense.

As a simple consequence of this theorem we obtain

COROLLARY 2.4. *Let $L^\natural(G)$ denote the space of functions on G which are integrable on G and bi-invariant under K. If $F \in L^\natural(G)$ and $\tilde{F} = 0$ on $\mathfrak{a}^*$ then $F = 0$.*

It is an interesting problem to characterize the space of functions $\tilde{F}$ as F runs through naturally defined spaces of functions on G which are bi-invariant under K. Theorem 2.5 below gives some results of this nature. Let the function δ on $\mathfrak{p}$ be determined by

$$\int_X f(x)\,dx = \int_{\mathfrak{p}} f((\exp X)K)\,\delta(X)\,dX$$

where dX is the Euclidean measure on $\mathfrak{p}$ (times a constant factor determined by $\delta(0) = 1$). If $g = k \exp X$ ($k \in K$, $X \in \mathfrak{p}$) we put $|g| = |X|$ and for each $q \in \boldsymbol{Z}^+$,

$$\delta_q(g) = \delta(X)(1 + |g|^q).$$

For $p \geqslant 1$ let $I^p(G)$ denote the set of all C^∞ functions F on G satisfying the following two conditions: (a) F is bi-invariant under K; (b) for each left invariant differential operator D on G and each $q \in \boldsymbol{Z}^+$,

$$\sup_g |\,\delta_q(g)^{1/p}(DF)(g)| < \infty.$$

The space $I^2(G)$ is the space $I(G)$ introduced by Harish-Chandra in [21(j), p. 585]. Note that $I^1(G) \subset L^1(G)$, $I^2(G) \subset L^2(G)$ and $I^1(G) \subset I^2(G)$. Let $I_c(G) = \mathscr{D}(G) \cap I^1(G)$. In order to describe the spherical Fourier transforms of the functions in these spaces let us call a holomorphic function ψ on $\mathfrak{a}_c{}^*$ a *rapidly decreasing holomorphic function of exponential type* if it satisfies the condition:

There exists a constant $A \geqslant 0$ such that for each holomorphic differential operator D on $\mathfrak{a}_c{}^*$ with polynomial coefficients

$$\sup_\lambda e^{-A|\mathrm{Im}\lambda|}\, |(D\psi)(\lambda)| < \infty,$$

where Im denotes imaginary part, ($\lambda = \mathrm{Re}\,\lambda + i\,\mathrm{Im}\,\lambda$ where $\mathrm{Re}\,\lambda$, $\mathrm{Im}\,\lambda \in \mathfrak{a}^*$).

Finally we let C_ρ denote the convex hull of the points $s\rho$ ($s \in W$) and $\mathfrak{a}_\rho{}^*$ the interior of the tube $\mathfrak{a}^* + iC_\rho$. We call a holomorphic function ψ on $\mathfrak{a}_\rho{}^*$ a *rapidly decreasing holomorphic function in the tube* $\mathfrak{a}_\rho{}^*$ if it satisfies the condition:

For each holomorphic differential operator D on $\mathfrak{a}_c{}^*$ with polynomial coefficients

$$\sup_{\lambda \in \mathfrak{a}_\rho{}^*} |(D\psi)(\lambda)| < \infty.$$

For the spherical Fourier transform of the spaces $I^1(G)$, $I^2(G)$ and $I_c(G)$ the following results hold.

THEOREM 2.5.

(i) *The map $F \to \tilde{F}$ is a bijection of $I^2(G)$ onto the space of W-invariant elements in the Schwartz' space $\mathscr{S}(\mathfrak{a}^*)$.*

(ii) (*The Paley-Wiener theorem*). *The map $F \to \tilde{F}$ is a bijection of $I_c(G)$ onto the space of W-invariant rapidly decreasing holomorphic functions on $\mathfrak{a}_c{}^*$ of exponential type.*

(iii) *Assume that either rank $(G/K) = 1$ or that G is complex. Then the map $F \to \tilde{F}$ is a bijection of $I^1(G)$ onto the space of W-invariant rapidly decreasing holomorphic functions in the tube $\mathfrak{a}_\rho{}^*$.*

Remarks. 1. The injectivity in Part (i) was proved by Harish-Chandra [21(k), p. 48], the surjectivity observed in [22(e), p. 572]. As noted in [22(i)], Part (ii) gives a simplified proof of Part (i).

2. Part (ii) comes from a combination of [22(i)] and recent work of R. Gangolli; see Ch. II, §2 for details.

3. The proof of Part (iii) which is based on the same idea as Part (ii) and is similar in details was found in 1966; it will be given in Ch. II, §2. I understand that another proof was found recently by P. Trombi for the case rank $(G/K) = 1$.

4. For $G = \boldsymbol{SL}(2, \boldsymbol{R})$ results similar to (ii) and (iii) were proved by Ehrenpreis–Mautner [9, pp. 417, 434]. For $G = \boldsymbol{SO}_0(1, n)$, (ii) is proved in Takahashi [47, p. 330].

5. The basic tool for proving Parts (ii) and (iii) is Harish–Chandra's series expansion for the spherical function.

The spherical Fourier transform is defined for functions on G, bi-invariant under K, or equivalently for K-invariant functions on X. This can however be extended to arbitrary functions on X as follows [22(g)]. Since $A(x, b)$ (in §1) can be viewed as the symmetric space analog to the Euclidean scalar product (x, ω) $(x, \omega \in \boldsymbol{R}^n$, ω unit vector) we define the *Fourier transform* on X by

$$\tilde{f}(\lambda, b) = \int_X f(x)\, e^{(-i\lambda+\rho)(A(x,b))}\, dx, \qquad \lambda \in \mathfrak{a}^*, \quad b \in B \tag{3}$$

for any $f \in \mathscr{D}(X)$. Then we have ($\mathfrak{a}_+{}^*$ being defined as in Notations, §3).

THEOREM 2.6. *For each* $f \in \mathscr{D}(X)$

$$f(x) = w^{-1} \int_{\mathfrak{a}^* \times B} \tilde{f}(\lambda, b)\, e^{(i\lambda+\rho)(A(x,b))}\, |\, \boldsymbol{c}(\lambda)|^{-2}\, d\lambda\, db$$

$$\int_X |f(x)|^2\, dx = w^{-1} \int_{\mathfrak{a}^* \times B} |\tilde{f}(\lambda, b)|^2\, \boldsymbol{c}(\lambda)|^{-2}\, d\lambda\, db$$

The mapping $f(x) \to \tilde{f}(\lambda, b)$ *extends to an isometry of* $L^2(X)$ *onto* $L^2(\mathfrak{a}_+{}^* \times B)$ *(with the measure* $|\, \boldsymbol{c}(\lambda)|^{-2}\, d\lambda\, db$ *on* $\mathfrak{a}_+{}^* \times B$*).*

The last statement of this theorem will be proved in Chap. III, §5; the rest is proved in [22(j)] by a reduction to Harish-Chandra's inversion formula for the spherical Fourier transform.

Remark. Since (by Proposition 1.1) the regular set X' in X can be identified with $\mathfrak{a}^+ \times B$ the symmetric space appears self-dual under the Fourier transform.

We shall now discuss the Radon transform in more detail. If $x \in X$ the isotropy subgroup K_x at x acts transitively on the set $\check{x}$ of horocycles which pass through x. Thus $\check{x}$ has a unique K_x-invariant measure $d\mu$ such that $\check{x}$ has total measure 1. If φ is a function on Ξ we put

$$\check{\varphi}(x) = \int_{\check{x}} \varphi(\xi)\, d\mu(\xi) \tag{4}$$

for all $x \in X$ for which this integral exists. The function $\check{\varphi}$ is called the *dual Radon transform of* φ. In group theoretic terms (2) and (4) can be written

$$\hat{f}(gMN) = \int_N f(gnK)\, dn \tag{5}$$

and

$$\check{\varphi}(gK) = \int_K \varphi(gkMN)\, dk. \tag{6}$$

For the G-invariant measures dx on X, $d\xi$ on Ξ (cf., Notation §4) we have the relation

$$\int_X f(x)\, \check{\varphi}(x)\, dx = \int_{\Xi} \hat{f}(\xi)\, \varphi(\xi)\, d\xi \tag{7}$$

for f and φ of compact support. We digress briefly to indicate how (5) and (6) can be extended to and (7) proved for general homogeneous spaces. For the moment let G be any unimodular locally compact group, X and Ξ two left coset spaces of G,

$$X = G/H_X \qquad \text{and} \qquad \Xi = G/H_\Xi,$$

where the subgroups H_X and H_Ξ are assumed to satisfy the conditions:

(i) $H_X H_\Xi$ closed, H_X, H_Ξ and $H_X \cap H_\Xi$ closed and unimodular.

(ii) If $h_X \in H_X$, $h_X H_\Xi \subset H_\Xi H_X$ then $h_X \in H_\Xi$; if $h_\Xi \in H_\Xi$, $h_\Xi H_X \subset H_X H_\Xi$ then $h_\Xi \in H_X$.

Two elements in X and Ξ, respectively, are called *incident*[1] if as cosets in G they have a point in common. If $x \in X$ put

$$\check{x} = \{\xi \in \Xi \mid x, \xi \text{ incident}\},$$

[1] This notion from [8] was kindly suggested by Chern in this context.

and if $\xi \in \Xi$, put

$$\hat{\xi} = \{x \in X \mid x, \xi \text{ incident}\}.$$

If x_o is the origin in X, ξ_o the origin in Ξ we have the obvious identifications

$$\check{x}_o = H_X/H_X \cap H_\Xi \qquad \text{and} \qquad \hat{\xi}_o = H_\Xi/H_X \cap H_\Xi \,;$$

and more generally, if $x = gH_X$, $\xi = \gamma H_\Xi$,

$$\check{x} = (H_X)^g/(H_X \cap H_\Xi)^g \qquad \text{and} \qquad \hat{\xi} = (H_\Xi)^\gamma/(H_X \cap H_\Xi)^\gamma,$$

the superscript denoting conjugation. On the basis of (ii) it is easily verified that the maps $x \to \check{x}$ $(x \in X)$, $\xi \to \hat{\xi}$ $(\xi \in \Xi)$ are injective; hence the elements of Ξ can be regarded as subsets of X, the elements of X as subsets of Ξ, both systems of subsets being permuted transitively by G.

Put $H = H_X \cap H_\Xi$ and fix Haar measures dg, dh_X, dh_Ξ and dh. Then we obtain canonically invariant measures dm on each $\check{x}$ and $d\mu$ on each $\hat{\xi}$ as follows: If g is any element in G such that $g \cdot \check{x}_o = \check{x}$ then the invariant measure $d(h_X)_H$ on $\check{x}_o$ can be carried over (by g) to a measure dm on $\check{x}$. If $g_1 \in G$ is another element satisfying $g_1 \cdot \check{x}_o = \check{x}$ then $(g \cdot x_o)^\vee = (g_1 \cdot x_o)^\vee$ so by the observed injectivity, $g \cdot x_o = g_1 \cdot x_o$. Hence $g^{-1}g_1 \in H_X$ so the measure given $\check{x}$ is independent of the choice of g. Similarly we obtain a canonical measure $d\mu$ on each $\hat{\xi}$.

In view of (i) the subsets $\check{x}$ and $\hat{\xi}$ are closed. Thus if f (resp. φ) is a continuous function of compact support on X (resp. Ξ) we can define the functions $\hat{f}$ on Ξ, $\check{\varphi}$ on X by

$$\hat{f}(\xi) = \int_{\hat{\xi}} f(x)\, dm(x), \qquad \check{\varphi}(x) = \int_{\check{x}} \varphi(\xi)\, d\mu(\xi).$$

Denoting the invariant measures dg_{H_X} on X, dg_{H_Ξ} on Ξ by dx and $d\xi$, respectively, formula (7) is proved as follows. Consider the natural projections

$$p_X : G/H \to X \qquad \text{and} \qquad p_\Xi : G/H \to \Xi,$$

and the function

$$P = (f \circ p_X)(\varphi \circ p_\Xi) \qquad \text{on} \quad G/H.$$

607/5/1-2

Since the space G/H has a double fibering, namely over X and over Ξ, the integral of P can be evaluated in two ways:

$$\int_{G/H} P(gH)\,dg_H = \int_{G/H_X} dg_{H_X} \int_{H_X/H} P(gh_XH)\,d(h_X)_H\,,$$

$$\int_{G/H} P(gH)\,dg_H = \int_{G/H_\Xi} dg_{H_\Xi} \int_{H_\Xi/H} P(gh_\Xi H)\,d(h_\Xi)_H\,.$$

The identity of the right sides is just formula (7).

The above abstraction of the Radon transform was done in [22(h)]; a further interesting generalization is given in [16].

A case of interest (for the theory of Eisenstein series) is $H_X = \Gamma$ (suitable discrete subgroup of the semisimple group G) and $H_\Xi = N$. Here the transforms $f \to \hat{f}$, $\varphi \to \check{\varphi}$ are integration over $N/N \cap \Gamma$ and summation over $\Gamma/\Gamma \cap N$.

We return now to the situation in (5) and (6) where $H_X = K$, $H_\Xi = MN$. Formula (7) suggests how the Radon transform and its dual should be defined for distributions: If $S \in \mathscr{E}'(X)$, $\Sigma \in \mathscr{D}'(\Xi)$ we define $\hat{S}$ and $\check{\Sigma}$ by

$$\hat{S}(\varphi) = S(\check{\varphi}) \qquad \text{and} \qquad \check{\Sigma}(f) = \Sigma(\hat{f}), \tag{8}$$

for $\varphi \in \mathscr{E}(\Xi)$, $f \in \mathscr{D}(X)$.

In Chap. II we prove the following inversion formula and Plancherel formula for the Radon transform.

THEOREM 2.6. *For a certain integro-differential operator Λ on Ξ, Λ^* its adjoint, w order of W,*

$$f = w^{-1}(\Lambda\Lambda^*\hat{f})^{\vee}, \tag{9}$$

$$\int_X |f(x)|^2\,dx = w^{-1}\int_\Xi |\Lambda\hat{f}(\xi)|^2\,d\xi, \qquad f \in \mathscr{D}(X). \tag{10}$$

In case all Cartan subalgebras of $\mathfrak{g}$ are conjugate we have

$$f = \Box((\hat{f})^{\vee}), \qquad f \in \mathscr{D}(X), \tag{11}$$

where $\Box$ is a certain differential operator on X.

The proof is based on Harish-Chandra's inversion formula for the spherical Fourier transform on G (Notation §6). The operators Λ and $\Box$

are constructed from the function $c(\lambda)$ by means of the Fourier transform; they reduce to differential operators exactly when c^{-1} reduces to a polynomial which in view of the formula ([21(i), p. 303] and [19(a)])

$$c(\lambda) = \prod_{\alpha \in \Sigma^+} B\left(\frac{1}{2} m_\alpha, \frac{1}{4} m_{\alpha/2} + \frac{\langle i\lambda, \alpha \rangle}{\langle \alpha, \alpha \rangle}\right) \tag{12}$$

(B = beta function) happens precisely when m_α is even for each α (and consequently $m_{\alpha/2} = 0$), or equivalently, when all Cartan subalgebras are conjugate (cf. Ch. II, Prop. 3.6).

We recall that a *spherical function* on X is by definition a K-invariant eigenfunction φ of all the operators in $D(X)$ normalized by $\varphi(o) = 1$. If φ is a spherical function on X then $g \to \varphi(gK)$ is a spherical function on G and vice versa. In an analogous manner one could define a spherical distribution on X; however since $D(X)$ contains an elliptic operator, the Laplace Beltrami operator on X, it is clear from the regularity theorems for elliptic operators that a spherical distribution would necessarily be a spherical function. The following definition is therefore a precise analogy.

DEFINITION. A distribution (resp. a C^∞ function) on $\Xi = G/MN$ is called *conical* if it is (1) MN-invariant; (2) eigendistribution (resp. eigenfunction) of each $D \in D(\Xi)$.

Remark. It was noted earlier that Ξ depends only on the Lie algebras $\mathfrak{g}$, $\mathfrak{k}$, and $\mathfrak{n}$ but not on the choice of G with Lie algebra $\mathfrak{g}$. Similarly, the transformation group $\tau(MN)$ of Ξ is independent of this choice of G and so is, therefore, the concept of conical distribution.

Since the set of spherical functions is parametrized by $\mathfrak{a}_c{}^*/W$ (cf., Notations §6) one would in view of Propositions 1.1 and 1.2 expect a parametrization of the set of conical distributions by $\mathfrak{a}_c{}^* \times W$. We shall now explain how this expectation is essentially verified.

For each $s \in W$ fix $m_s \in M'$ in the coset s and put $\xi_s = m_s MN$. If Ξ_s denotes the orbit of ξ_s in Ξ under the group MAN the Bruhat decomposition implies $\Xi = \bigcup_{s \in W} \Xi_s$ (disjoint union). Also if $\xi \in \Xi_s$ there exists a unique $a(\xi) \in A$ such that $\xi \in MNa(\xi) \cdot \xi_s$. The group MAN gives a natural measure dv on Ξ_s and for each $\lambda \in \mathfrak{a}_c{}^*$ we consider the functional

$$\Psi'_{\lambda,s} : \varphi \to \int_{\Xi_s} \varphi(\xi)\, e^{(is\lambda + s\rho)(\log a(\xi))}\, dv(\xi).$$

The integral converges absolutely for all $\varphi \in \mathscr{D}(\Xi)$ if and only if λ belongs to a certain open tube in $\mathfrak{a}_c^*$ (Lemma 4.2, Chap. III). For such λ, $\Psi'_{\lambda,s}$ is a conical distribution. The obvious problem is to continue the mapping $\lambda \to \Psi'_{\lambda,s}$ to a meromorphic distribution-valued function on $\mathfrak{a}_c^*$. The values obtained by analytic continuation are still conical distributions and so are the „residues" at the singularities. Remarkably enough it turns out that the singularities of $\Psi'_{\lambda,s}$ are the same as those of the *numerator* $\mathbf{d}_s(\lambda)$ of the partial $\boldsymbol{c}$-function (cf. (12))

$$\prod_{\alpha\in\Sigma^+\cap s^{-1}\Sigma^-} B\left(\frac{1}{2}m_\alpha, \frac{1}{4}m_{\alpha/2} + \frac{\langle i\lambda, \alpha\rangle}{\langle\alpha, \alpha\rangle}\right) \tag{13}$$

(we shall see that the *denumerator* $\boldsymbol{e}_s(\lambda)^{-1}$ has its significance too, namely in representation theory). After dividing $\Psi'_{\lambda,s}$ by this numerator $\mathbf{d}_s(\lambda)$ we obtain a distribution $\Psi_{\lambda,s}$ and find

THEOREM 2.7. *For each $s \in W$ the mapping $\lambda \to \mathbf{d}_s(\lambda)^{-1}\Psi'_{\lambda,s} = \Psi_{\lambda,s}$ extends to an entire function on $\mathfrak{a}_c^*$. Its values are conical distributions on Ξ.*

The values of $\Psi_{\lambda,s}$ at the "removable singularities" are "residues" of $\Psi'_{\lambda,s}$; geometrically they turn out to be certain transversal derivatives of other $\Psi'_{\lambda,s'}$ (corresponding to other MNA-orbits in the closure of Ξ_s).

For each $D \in \boldsymbol{D}(\Xi)$ the distributions $\Psi_{\lambda,s}$ ($s \in W$) all have the same eigenvalue, say $\lambda(D)$. Defining for each $\lambda \in \mathfrak{a}_c^*$ the vector space

$$\mathscr{D}_\lambda'(\Xi) = \{\Psi \in \mathscr{D}'(\Xi) \mid D\Psi = \lambda(D)\Psi \quad \text{for all} \quad D \in \boldsymbol{D}(\Xi)\}, \tag{14}$$

we prove in Chap. III, Theorem 4.9 that each conical distribution lies in some $\mathscr{D}_\lambda'(\Xi)$ and that for a generic λ the linear combinations $\sum_{s\in W} a_s\Psi_{\lambda,s}$ constitute all the conical distributions in $\mathscr{D}_\lambda'(\Xi)$. For the hyperbolic spaces X of dimension >2, this result holds for *all* $\lambda \in \mathfrak{a}_c^*$.

We can now describe all the conical *functions* on Ξ. Let as usual s^* denote the element in the Weyl group which interchanges Σ^+ and Σ^- and put $\xi^* = \xi_{s^*}$. The orbit $\Xi^* = MNA \cdot \xi^*$ is an open dense submanifold of Ξ and there exists a function $\psi_\lambda \in \mathscr{E}(\Xi^*)$ such that

$$\Psi_{\lambda,s^*}(\varphi) = \int_{\Xi^*} \psi_\lambda(\xi)\, \varphi(\xi)\, d\xi, \qquad \varphi \in \mathscr{D}(\Xi^*).$$

THEOREM 2.8. *The function ψ_λ is on Ξ^* given by*

$$\psi_\lambda(mna \cdot \xi^*) = \psi_\lambda(\xi^*)e^{(is^*\lambda+\rho)(\log a)} \qquad m \in M,\ n \in N,\ a \in A. \tag{15}$$

It extends to a C^∞ function on Ξ if and only if

$$\frac{\langle i\lambda - \rho, \alpha\rangle}{\langle \alpha, \alpha\rangle} \in \boldsymbol{Z}^+ \qquad \text{for each} \quad \alpha \in \Sigma^+. \tag{16}$$

The functions (15) (*with $\psi_\lambda(\xi^*)$ arbitrary and λ satisfying* (16)) *constitute all the conical functions on Ξ.*

In the next section we describe the role of conical functions and conical distributions in representation theory.

§3. Two Dual Models for the Spherical Representations

Preserving the earlier notation we make the following

Definition. A representation π of G on a vector space E will be called (1) *spherical* if there exists a nonzero vector in E fixed under $\pi(K)$; (2) *conical* if there exists a nonzero vector in E fixed under $\pi(MN)$.

A. *Representation models over X*

There is a well-known relationship between positive definite spherical functions and irreducible unitary spherical representations. Dropping the positive definiteness condition we still have a simple relationship.

Proposition 3.1. *Let $\lambda \in \mathfrak{a}_c^*$ and $\mathscr{E}_{(\lambda)}$ the closure in $\mathscr{E}(X)$ of the linear combinations of all G-translates of the spherical function φ_λ. Let T_λ denote the natural representation of G on $\mathscr{E}_{(\lambda)}$, $T_\lambda{}^*$ the contragredient representation on the dual $\mathscr{E}^*_{(\lambda)}$ (strong topology). Then T_λ and $T_\lambda{}^*$ are irreducible spherical representations. The fixed vectors of $T_\lambda(K)$ and of $T_\lambda{}^*(K)$ are unique up to a scalar multiple and, with a proper normalization, these vectors φ_λ and δ_λ, satisfy*

$$\varphi_\lambda(g \cdot o) = \langle T_\lambda(g^{-1})\, \varphi_\lambda\, , \delta_\lambda\rangle, \qquad g \in G.$$

In terms of the isomorphism Γ: $\boldsymbol{D}(X) \to \boldsymbol{I}(A)$ from Theorem 2.1 we have by [22(i), Lemma 18]

$$D\varphi_\lambda = \Gamma(D)(i\lambda)\, \varphi_\lambda\, .$$

Hence we define for each $\lambda \in \mathfrak{a}_c^*$,

$$\mathscr{E}_\lambda(X) = \{f \in \mathscr{E}(X) \mid Df = \Gamma(D)(i\lambda)f \quad \text{for} \quad D \in \boldsymbol{D}(X)\} \tag{1}$$

and give this space the relative topology of $\mathscr{E}(X)$. Of course, $\mathscr{E}_{(\lambda)}(X) \subset \mathscr{E}_\lambda(X)$. A function on a space on which K acts is called *K-finite* if its transforms under K span a finite-dimensional space.

THEOREM 3.2. *Suppose $X = G/K$ has rank one. Then for each $\lambda \in \mathfrak{a}_c^*$ the K-finite functions in $\mathscr{E}_\lambda(X)$ are precisely the functions*

$$f(x) = \int_B e^{(i\lambda+\rho)(A(x,b))} F(b)\, db, \tag{2}$$

where F is a K-finite function on B.

Since the rank is one $\boldsymbol{D}(X)$ consists of the polynomials in the Laplacian Δ so the functions (2) for varying F and varying $\lambda \in \mathfrak{a}_c^*$ constitute all the K-finite eigenfunctions of Δ. As a result of Theorem 3.2 we find that *all* eigenfunctions of Δ can be written in the form

$$x \to \int_B e^{(i\lambda+\rho)(A(x,b))}\, dT(b), \tag{3}$$

where $\lambda \in \mathfrak{a}_c^*$ and T is a "functional" on B (cf. Corollary 1.6 in Chap. IV). For example if X is the non-Euclidean disk, $i\lambda \in \mathfrak{a}^*$, and T an arbitrary analytic functional on B the functions (3) constitute all eigenfunctions of Δ with eigenvalue $\geqslant - \langle \rho, \rho \rangle$. (Theorem 1.8, Chap. IV).

Let $\boldsymbol{e}_{s^*}(\lambda)^{-1}$ denote the denominator of the $\boldsymbol{c}$-function ((12) §2; see (31), Chap. III §4 for the precise formula). Using Theorem 3.2 one can prove the following criterion relating irreducibility of G on $\mathscr{E}_\lambda(X)$ with the Plancherel measure $\boldsymbol{c}(\lambda)\, \boldsymbol{c}(-\lambda)\, d\lambda$.

THEOREM 3.3. *Assume $X = G/K$ has rank one. Let $\lambda \in \mathfrak{a}_c^*$. Then the natural representation of G on the eigenspace $\mathscr{E}_\lambda(X)$ is irreducible if and only if $\boldsymbol{e}_{s^*}(\lambda)\, \boldsymbol{e}_{s^*}(-\lambda) \neq 0$.*

Remark. The proofs of Theorems 3.2 and 3.3 work for X of arbitrary rank except for a result on partial differential equations (a generalization of Lemma 1.5, Chap. IV to X of arbitrary rank) which does not seem to be available in the literature.

B. *Representation models over Ξ*

We begin with the relationship between conical functions and conical representations. If φ is a function on Ξ let E_φ denote the vector space spanned by all the G-translates of φ and let π_φ denote the natural representation of G on E_φ.

THEOREM 3.4. *The mapping* $\psi \to \pi_\psi$ *is a bijection of the set of conical functions on* Ξ *onto the set of finite-dimensional irreducible conical representations of* G (*here we identify proportional conical functions and identify equivalent representations*). *Moreover,*

$$\psi(g \cdot \xi_o) = \langle \pi_\psi(g^{-1})\, e, e' \rangle, \qquad g \in G, \tag{4}$$

where e *and* e', *respectively, are contained in the* (*one-dimensional*) *highest weight spaces of* π_ψ *and of its contragredient representation.*

The proof gives the following

COROLLARY 3.5. *A finite-dimensional representation of* G *is spherical if and only if it is conical.*

We consider now the eigenspace $\mathscr{D}_\lambda' = \mathscr{D}_\lambda'(\Xi)$ defined in (14) §2. In addition we consider for each $\lambda \in \mathfrak{a}_c^*$ the space $\mathscr{K}_\lambda$ of measurable functions on Ξ satisfying the conditions,

(i) $\psi(ga \cdot \xi_o) = e^{(i\lambda-\rho)(\log a)}\psi(g \cdot \xi_0)$,

(ii) $(|\, \psi\, |_\lambda)^2 = \int_K |\, \psi(k \cdot \xi_o)|^2\, dk < \infty$.

Then $\mathscr{K}_\lambda$ is a Hilbert space with norm $\psi \to |\, \psi\, |_\lambda$. Viewing its elements as distributions on Ξ, $\mathscr{K}_\lambda$ becomes a dense subspace of $\mathscr{D}_\lambda'$ (strong topology). The group G acts in a natural way on $\mathscr{K}_\lambda$, giving a representation τ_λ and on $\mathscr{D}_\lambda'$, giving a representation $\tilde{\tau}_\lambda$. The representation τ_λ is equivalent to the representation of G *induced* (in the sense of [6(a)], [33]) by the one-dimensional representation $man \to e^{i\lambda(\log a)}$ of MAN; moreover, τ_λ is irreducible if and only if $\tilde{\tau}_\lambda$ is irreducible. The vector $\psi_0 \in \mathscr{K}_\lambda$ given by $\psi_0(ka \cdot \xi_o) = e^{(i\lambda-\rho)(\log a)}$ is a K-fixed vector.

PROPOSITION 3.6. *The vector* $\Psi_{\lambda,e}$ (cf., Theorem 2.7) *is a cyclic vector for the representation* $\tilde{\tau}_\lambda$ *of* G *on* $\mathscr{D}_\lambda'$.

The following result has many applications in our work.

THEOREM 3.7. *The dual Radon transform* $\Psi \to \check{\Psi}$ *maps* $\mathscr{D}_\lambda'(\Xi)$ *into* $\mathscr{E}_\lambda(X)$; *moreover,*

$$\check{\Psi}_{\lambda,s}(na \cdot \xi_o) = a_s e_s(\lambda)\, e^{(is\lambda+\rho)(\log a)}, \qquad \lambda \in \mathfrak{a}_c^*, \quad s \in W, \tag{5}$$

where a_s *is a constant and* $e_s(\lambda)^{-1}$ *is the denominator of the partial* **c**-*function* ((13) §2 *and* (31) Chap. III §4).

Since the N-invariant functions in $\mathscr{E}_\lambda(X)$ can be determined completely, (5) plays a crucial role in proving that for a generic λ the conical distributions in $\mathscr{D}_\lambda'$ are just the linear combinations of the $\Psi_{\lambda,s}$ ($s \in W$). Theorem 3.7 also suggests the following

DEFINITION. An element $\lambda \in \mathfrak{a}_c^*$ is called *simple* if the map $\Psi \to \check{\Psi}$ of $\mathscr{D}_\lambda'(\Xi)$ into $\mathscr{E}_\lambda(X)$ is injective.

LEMMA 3.8. *The representation τ_λ is irreducible if and only if λ and $-\lambda$ are both simple.*

Now Theorem 3.7 explains the relation between irreducibility of τ_λ and the denominator of the $\mathbf{c}$-function (cf., Theorem 1.1 in Chap. IV).

For $\lambda \in \mathfrak{a}^*$ it is well known that the unitary representation τ_λ is equivalent to each $\tau_{s\lambda}$ ($s \in W$). Passing from the representation τ_λ on the Hilbert space $\mathscr{K}_\lambda$ to the representation $\tilde{\tau}_\lambda$ on the distribution space $\mathscr{D}_\lambda'$ we extend this equivalence to all (complex) irreducible $\lambda \in \mathfrak{a}_c^*$; the equivalence is realized by means of an intertwining operator $A_{\lambda,s}$ constructed by means of the conical distribution $\Psi_{\lambda,s}$. For this it is convenient to transfer the representation space from $\mathscr{D}_\lambda'$ to $\mathscr{D}'(B)$. For each $s \in W$ select a representative $m_s \in M'$, put $\bar{N}_s = \bar{N} \cap m_s^{-1} N m_s$, and let da and $d\bar{n}_s$ denote Haar measures on A and $\bar{N}_s$, respectively.

LEMMA 3.9. *The map $S \to \Psi$ given by*

$$\Psi(\varphi) = \int_{K/M} \left(\int_A \varphi(ka \cdot \xi_o)\, e^{(i\lambda+\rho)(\log a)}\, da \right) dS(k_M), \qquad \varphi \in \mathscr{D}(\Xi),$$

is a bijection of $\mathscr{D}'(B)$ onto $\mathscr{D}_\lambda'(\Xi)$ mapping $L^2(B)$ onto $\mathscr{K}_\lambda$. Under this map $\Psi_{\lambda,s}$ (in Theorem 2.7) *corresponds to the distribution on B:*

$$S_{\lambda,s} : F \to \mathbf{d}_s(\lambda)^{-1} \int_{\bar{N}_s} F(m_s k(\bar{n}_s) M) e^{-(i\lambda+\rho)(H\bar{n}_s))}\, d\bar{n}_s$$

defined for all $\lambda \in \mathfrak{a}_c^*$ by analytic continuation.

THEOREM 3.10. *Assume $\tilde{\tau}_\lambda$ irreducible. Then the convolution operator $S \to S \times S_{s\lambda,s^{-1}}$ on $\mathscr{D}'(B)$ induces a bijection $A_{\lambda,s}$ of $\mathscr{D}_\lambda'(\Xi)$ onto $\mathscr{D}'_{s\lambda}(\Xi)$, which intertwines $\tilde{\tau}_\lambda$ and $\tilde{\tau}_{s\lambda}$; in particular $\tilde{\tau}_\lambda$ and $\tilde{\tau}_{s\lambda}$ are equivalent.*

If $\lambda \in \mathfrak{a}^*$ then $A_{\lambda,s}$ (suitably normalized) gives a unitary operator of $\mathscr{K}_\lambda$ onto $\mathscr{K}_{s\lambda}$.

There are other advantages in passing from the Hilbert space $\mathscr{H}_\lambda$ to the distribution space $\mathscr{D}_\lambda'$. A vector in $\mathscr{D}_\lambda'$ is called an *extreme weight vector* if it is an eigenvector of the operators $\tilde{\tau}_\lambda(A)$ and fixed under $\tilde{\tau}_\lambda(N)$.

PROPOSITION 3.11. *Let λ be as in Theorem* 3.10, *and in addition regular. Then the scalar multiples of the vectors $\Psi_{\lambda,s}$ $(s \in W)$ constitute precisely the extreme weight vectors in $\mathscr{D}_\lambda'(\Xi)$.*

With this interpretation of the vectors $\Psi_{\lambda,s}$ there are infinite-dimensional analogs of formula (4) for the conical function. Since the subgroup MN of G is noncompact the convolution of distributions on G does not immediately give a convolution product $\times$ of distributions on Ξ. However, if one of the factors is MN-invariant this can be done in a natural way (Chap. III, §5).

THEOREM 3.12. *Let $\tilde{\tau}_\lambda$ be irreducible, and $s \in W$. For a suitable constant $a(\lambda, s) \in \mathbf{C}$ we have*

$$\Psi_{-s\lambda,e}(\varphi) = a(\lambda, s)\langle \varphi \times \Psi_{\lambda,s}, \Psi_{-\lambda,s}\rangle \qquad \varphi \in \mathscr{D}(\Xi).$$

Here $\langle\ ,\ \rangle$ denotes the bilinear form on the product of the spaces on which the contragredient representations τ_λ and $\tilde{\tau}_{-\lambda}$ operate.

In the case when G is a complex semisimple Lie group M is a torus and more general results are obtained by considering G/N instead of G/MN. Here the full principal series of irreducible representations of G is realized on joint eigenspaces of the G-invariant differential operators on G/N (cf., Chap. IV, §4, Remark 2).

C. *Equivalence Between the Two Models* [22(h), §4]

Let T_X denote the natural representation of G on $L^2(X)$ and T_Ξ the natural representation of G on $L^2(\Xi)$. If $\lambda \in \mathfrak{a}_c^*$ is *simple* let $\mathscr{H}_\lambda$ denote the Hilbert space

$$\mathscr{H}_\lambda = \left\{ g(x) = \int_B e^{(i\lambda+\rho)(A(x,b))} F(b)\, db \mid F \in L^2(B) \right\}, \tag{6}$$

the norm of g being by definition the L^2 norm of F. Let T_λ denote the natural representation of G on $\mathscr{H}_\lambda$. Then Theorem 2.6 implies that in terms of direct integral theory

$$L^2(X) = \int_{\mathfrak{a}^*/W} \mathscr{H}_\lambda \mid \mathbf{c}(\lambda)|^{-2}\, d\lambda, \qquad T_X = \int_{\mathfrak{a}^*/W} T_\lambda \mid \mathbf{c}(\lambda)|^{-2}\, d\lambda. \tag{7}$$

In his papers [21(b)], [21(j), p. 612], Harish-Chandra gives a Plancherel formula for the reduction of T_X (cf. also [20(b)], [35(a)]). However (7) is slightly more complete in the sense that the constituents $\mathscr{H}_\lambda$ are explicitly given and the "component" of $f \in \mathscr{D}(X)$ in $\mathscr{H}_\lambda$ is given by

$$f_\lambda(x) = \int_B \tilde{f}(\lambda, b)\, e^{(i\lambda+\rho)(A(x,b))}\, db.$$

The reduction of T_Ξ is essentially just elementary Fourier analysis on A (compare a similar problem for G/N, G complex, treated in [14]). In analogy to the definition (3) §2 of Fourier transform on X we define for $\varphi \in \mathscr{D}(\Xi)$

$$\tilde{\varphi}(\lambda, kM) = \int_A \varphi(ka \cdot \xi_o)\, e^{(-i\lambda+\rho)(\log a)}\, da. \tag{8}$$

Then

$$\varphi(ka \cdot \xi_o) = \int_{\mathfrak{a}^*} \tilde{\varphi}(\lambda, kM)\, e^{(i\lambda-\rho)(\log a)}\, d\lambda. \tag{9}$$

The functions φ_λ on Ξ given by

$$\varphi_\lambda(ka \cdot \xi_o) \equiv \tilde{\varphi}(\lambda, kM)\, e^{(i\lambda-\rho)(\log a)} \tag{10}$$

form a dense subspace of the Hilbert space $\mathscr{K}_\lambda$ (as φ runs through $\mathscr{D}(\Xi)$). The invariant measure $d\xi$ on Ξ is given by $d\xi = e^{2\rho(\log a)}\, da\, dk_M$ (Lemma 3.1, Chap. II) so (9) implies an analog of Theorem 2.6,

$$\int_\Xi |\varphi(\xi)|^2\, d\xi = \int_{\mathfrak{a}^*} |\varphi_\lambda|_\lambda^2\, d\lambda. \tag{11}$$

Now (9) and (11) imply the direct integral decompositions

$$L^2(\Xi) = \int_{\mathfrak{a}^*} \mathscr{K}_\lambda\, d\lambda, \qquad T_\Xi = \int_{\mathfrak{a}^*} \tau_\lambda\, d\lambda. \tag{12}$$

In Chap. III, Proposition 5.7 we shall see how the decompositions (7) and (12) correspond under the Radon transform and its dual.

Chapter II. The Radon Transform and Its Dual

§1. The Injectivity

As in Chap I, let X be a symmetric space of the noncompact type and Ξ its dual; as in Chap. I, §2 the Radon transform of a function f on X will be denoted by $\hat{f}$.

THEOREM 1.1. *If* $f \in L^1(X)$ *then* $\hat{f}(\xi)$ *exists for almost all* $\xi \in \Xi$. *If* $\hat{f}(\xi) = 0$ *for almost all* ξ *then* $f(x) = 0$ *for almost all* x.

Proof. If $f \in L^1(X)$ then the function F on G given by $F(g) = f(g \cdot o)$ lies in $L^1(G)$ and the relation

$$\begin{aligned} \infty > \int_G |F(g)|\, dg &= \int_{KAN} |F(kan)|\, e^{2\rho(\log a)}\, dk\, da\, dn \\ &= \int_{KA} e^{2\rho(\log a)} \int_N |f(kan \cdot o)|\, dn\, dk\, da \\ &= \int_{(K/M)\times A} e^{2\rho(\log a)} \int_N |f(kan \cdot o)|\, dn\, dk_M\, da, \end{aligned}$$

shows that for almost all $(kM, a) \in (K/M) \times A$ the integral

$$\int_N |f(kan \cdot o)|\, dn < \infty.$$

Since the map $(kM, a) \to ka \cdot \xi_o$ is a diffeomorphism of $(K/M) \times A$ onto Ξ this proves that $\hat{f}(\xi)$ exists for almost all $\xi \in \Xi$.

Now let $L^\natural(G)$ denote the space of integrable functions on G which are bi-invariant under K. Let $\lambda \in \mathfrak{a}_c{}^*$ be such that φ_λ is a *bounded* spherical function. Let $\lambda = \xi + i\eta$ with $\xi, \eta \in \mathfrak{a}^*$. By Theorem 2.1 in [23], $\varphi_{i\eta}$ is also bounded so if $F \in L^\natural(G)$,

$$\int_K \int_G |F(g)\, e^{(i\lambda-\rho)(H(gk))}|\, dg\, dk = \int_G |F(g)|\, \varphi_{i\eta}(g)\, dg < \infty. \tag{1}$$

Thus the inner integral is finite for almost all k, hence for all k because of the bi-invariance of F under K. Thus, using the Fubini theorem on $G = KAN$ we get since $dg = e^{2\rho(\log a)}\, dk\, da\, dn$,

$$\int_G F(g)\, \varphi_\lambda(g)\, dg = \int_G F(g)\, e^{(i\lambda-\rho)(H(g))}\, dg = \int_A \varphi_F(a)\, e^{i\lambda(\log a)}\, da, \tag{2}$$

whenever φ_λ is bounded, where

$$\varphi_F(a) = e^{\rho(\log a)} \int_N F(an)\, dn.$$

Now suppose $f \in L^1(X)$ satisfies $\hat{f}(\xi) = 0$ for almost all $\xi \in \Xi$. Let $\varphi \in L^\natural(G) \cap \mathscr{D}(G)$ and consider the function

$$F_0(g) = \int_G \varphi(h) f(h^{-1}gK)\, dh$$

which clearly belongs to $L^\natural(G)$ and satisfies

$$\int_N F_0(an)\,dn = \int_G \varphi(h)\hat{f}(h^{-1}a \cdot \xi_o)\,dh = 0 \tag{3}$$

for all $a \in A$. Now under convolution and L^1-norm on G, $L^\natural(G)$ is a commutative Banach algebra which is semisimple and has all its maximal ideals of the form

$$\left\{F \,\middle|\, \int_G F(g)\,\varphi_\lambda(g)\,dg = 0\right\},$$

where φ_λ is a bounded spherical function on G (see, e.g. [22(b), pp. 410, 453]). Thus (3) and (2) imply that $F_0 \equiv 0$. But since $\varphi \in L^\natural(G) \cap \mathscr{D}(G)$ is arbitrary it follows that $f(x) = 0$ for almost all $x \in X$ as desired.

COROLLARY 1.2. *Suppose $F \in L^\natural(G)$ such that the spherical Fourier transform $\tilde{F}$ satisfies $\tilde{F}(\lambda) = 0$ for all $\lambda \in \mathfrak{a}^*$. Then $F = 0$.*

In fact define $f \in L^1(X)$ by $f(g \cdot o) = F(g)$. Replacing in (2) F by $|F|$ and putting $\lambda = 0$ we deduce that $\varphi_F \in L^1(A)$. Thus $\tilde{F}(\lambda) = 0$ for $\lambda \in \mathfrak{a}^*$ implies by (2) that $\varphi_F = 0$. But then Theorem 1.1 implies $f = 0$.

Remark. Corollary 1.2 occurs in [12 (a), p. 218] but the proof there seems to me to have a gap because the crucial formula (2) above (which we derived from Theorem 2.1 in [23]) is taken for granted for all $F \in L^\natural(G)$ and all $\lambda \in \mathfrak{a}_c^*$ for which φ_λ is bounded.

§2. DIGRESSION ON THE SPHERICAL FOURIER TRANSFORM

We shall now prove Theorem 2.5 in Chap. I characterizing the spherical Fourier transforms of various function spaces on G. Let $\mathscr{H}(\mathfrak{a}_\rho^*)$ denote the space of W-invariant rapidly decreasing holomorphic functions in the tube $\mathfrak{a}_\rho^*$ (in the sense of Chap. I, §2).

THEOREM 2.1. *Suppose that either G is complex or of real rank one. Then the spherical Fourier transform $F \to \tilde{F}$ is a bijection of $I^1(G)$ onto $\mathscr{H}(\mathfrak{a}_\rho^*)$.*

Let us first verify (for G arbitrary) that $\tilde{F} \in \mathscr{H}(\mathfrak{a}_\rho^*)$ if $F \in I^1(G)$. The Weyl group invariance is obvious from $\varphi_\lambda = \varphi_{s\lambda}$ ($\lambda \in \mathfrak{a}_c^*$, $s \in W$). Let

$\alpha_1, \ldots, \alpha_l$ denote the simple restricted roots which then form a basis of $\mathfrak{a}^*$ and determine $\epsilon_1, \ldots, \epsilon_l \in \mathfrak{a}^*$ by $\langle \alpha_i, \epsilon_j \rangle = \delta_{ij}$. For each $\lambda \in \mathfrak{a}_c^*$ let $\lambda = \xi + i\eta$ ($\xi, \eta \in \mathfrak{a}^*$), $\lambda = \sum_j \lambda_j \epsilon_j$ ($\lambda_j \in \boldsymbol{C}$) and $\lambda_j = \xi_j + i\eta_j$ ($\xi_j, \eta_j \in \boldsymbol{R}$), We shall express the holomorphic differential operators on $\mathfrak{a}_c^*$ by means of the coordinates $\lambda_1, \ldots, \lambda_l$. Now by definition

$$\tilde{F}(\lambda) = \int_G F(g)\, \varphi_{-\lambda}(g)\, dg \tag{1}$$

and

$$\frac{\partial}{\partial \lambda_j} \int_K e^{-(i\lambda+\rho)(H(gk))}\, dk = \int_K e^{-(i\lambda+\rho)(H(gk))}(-i)\, \epsilon_j(H(gk))\, dk. \tag{2}$$

Now put $A^+ = \exp \mathfrak{a}^+$. If $a \in A^+$ we have by [21(i), p. 279] for all $H \in \mathfrak{a}^+$,

$$\langle \log a - H(ak), H \rangle \geqslant 0 \tag{3}$$

and

$$\langle -s^* \log a + H(ak), H \rangle \geqslant 0, \tag{4}$$

where $s^* \in W$ is determined by $s^*\rho = -\rho$. For each $\lambda \in \mathfrak{a}_c^*$ let $H_\lambda \in \mathfrak{a}_c$ be determined by $\langle H_\lambda, H \rangle = \lambda(H)$ for all $H \in \mathfrak{a}_c$. Then $H_{\epsilon_i} \in \mathrm{Cl}(\mathfrak{a}^+)$ so by (3) and (4)

$$\epsilon_j(s^* \log a) \leqslant \epsilon_j(H(ak)) \leqslant \epsilon_j(\log a)$$

so

$$|\, \epsilon_j(H(ak))| \leqslant |\, \epsilon_j \,|\, |\log a\,|, \qquad a \in A^+, \quad k \in K. \tag{5}$$

Thus by (2),

$$\left| \frac{\partial}{\partial \lambda_j} \varphi_{-\lambda}(a) \right| \leqslant \varphi_{-i\eta}(a)\, |\, \epsilon_j\,|\, |\log a\,|, \qquad (a \in A^+). \tag{6}$$

For

$$\Delta(\exp H) = \prod_{\alpha \in \Sigma^+} (e^{\alpha(H)} - e^{-\alpha(H)})^{m_\alpha} \qquad (H \in \mathfrak{A})$$

and a suitable multiple da^* of da

$$\int_G F(g)\, dg = \int_{A^+} F(a)\, \Delta(a)\, da^* \qquad F \in L^\natural(G). \tag{6'}$$

If $F \in I^1(G)$ we have for each $m \in \boldsymbol{Z}^+$,

$$\int_{A^+} |F(a)|\, |\log a\,|^m\, \Delta(a)\, da^* < \infty. \tag{7}$$

If λ belongs to the tube $\mathfrak{a}_\rho{}^*$ it is clear from the boundedness of $\varphi_{-i\eta}$ together with (6) and (7) that (1) can be differentiated under the integral sign so $\tilde{F}$ is holomorphic in $\mathfrak{a}_\rho{}^*$. More generally, iterating (2) and (6), we see by (7) that for each polynomial P in l variables the integral

$$\int_G F(g)\, P\left(\frac{\partial}{\partial\lambda_1}, \ldots, \frac{\partial}{\partial\lambda_l}\right) \varphi_{-\lambda}(g)\, dg$$

converges uniformly for $\lambda \in \mathfrak{a}_\rho{}^*$. Thus (1) above can be differentiated under the integral sign and we conclude that

$$P\left(\frac{\partial}{\partial\lambda_1}, \ldots, \frac{\partial}{\partial\lambda_l}\right) \tilde{F} \tag{8}$$

is bounded for $\lambda \in \mathfrak{a}_\rho{}^*$.

Let D be a left invariant differential operator on G. For each $F \in I^1(G)$, DF is integrable on G and by [21(i), p. 294]

$$|(D\varphi_\lambda)(g)| \leqslant a(1 + |\xi|^2 + |\eta|^2)^d\, \varphi_{i\eta}(g),$$

where a and d are constants. Thus if $\lambda \in \mathfrak{a}_\rho{}^*$, $D(F\varphi_\lambda)$ is integrable. Now if X, X' are left invariant vector fields on G and Q and XQ integrable C^∞ functions on G then

$$\int_G |(XQ)(g \exp tX)|\, dg$$

converges uniformly for small t so

$$\int_G (XQ)(g)\, dg = \left\{\frac{d}{dt} \int_G Q(g \exp tX)\, dg\right\}_{t=0} = 0.$$

Thus if $\lambda \in \mathfrak{a}_\rho{}^*$

$$\int_G (XX'F)(g)\, \varphi_\lambda(g)\, dg = -\int_G (X'F)(g)(X\varphi_\lambda)(g)\, dg$$
$$= \int_G F(g)(X'X\varphi_\lambda)(g)\, dg.$$

Hence if ω is the Casimir operator on G which satisfies $\omega\varphi_\lambda = -(\langle\lambda, \lambda\rangle + \langle\rho, \rho\rangle)\, \varphi_\lambda$ [21(i), p. 271] we deduce (since ω leaves $I^1(G)$ invariant)

$$(\omega^n F)^\sim(\lambda) = (-\langle\lambda, \lambda\rangle - \langle\rho, \rho\rangle)^n\, \tilde{F}(\lambda), \qquad (\lambda \in \mathfrak{a}_\rho{}^*).$$

Consequently, for each polynomial P and each $n \in \boldsymbol{Z}^+$

$$\sup_{\lambda \in \mathfrak{a}_\rho^*} \left| P\left(\frac{\partial}{\partial \lambda_1}, \ldots, \frac{\partial}{\partial \lambda_l}\right) \{(\langle \lambda, \lambda \rangle + \langle \rho, \rho \rangle)^n \tilde{F}(\lambda)\} \right| < \infty.$$

Now an easy induction shows that $\tilde{F} \in \mathscr{H}(\mathfrak{a}_\rho^*)$.

Next let $\psi \in \mathscr{H}(\mathfrak{a}_\rho^*)$. Under the assumption of the theorem we shall construct $F \in I^1(G)$ such that $\tilde{F} = \psi$. Of course we define F by Harish-Chandra's inversion formula

$$F(g) = w^{-1} \int_{\mathfrak{a}^*} \psi(\xi)\, \varphi_\xi(g) \mid \boldsymbol{c}(\xi)|^{-2}\, d\xi \tag{9}$$

(cf., Notation §6). We shall prove $F \in I^1(G)$. By the Cauchy–Riemann equations for ψ it is clear that each of its derivatives with respect to the variables $\xi_1, \ldots, \xi_l, \eta_1, \ldots, \eta_l$ is bounded in $\mathfrak{a}_\rho^*$ so extends by uniform continuity to the closure $\mathfrak{a}^* + iC_\rho$ (C_ρ = convex hull of the points $s\rho$, $s \in W$). Consequently, if D is any differential operator in the variables $\xi_1, \ldots, \xi_l$ with polynomial coefficients,

$$\sup_{\xi + i\eta \in \mathfrak{a}^* + iC_\rho} |(D\psi)(\xi + i\eta)| < \infty. \tag{10}$$

In order to prove $F \in I^1(G)$ we use Harish-Chandra's series expansion for $\varphi_\lambda(\exp H)$ valid for $H \in \mathfrak{a}^+$ ([21(i), Lemma 37]; see also [22(i), §4]). Let L denote the set of linear combinations $\mu = m_1\alpha_1 + \cdots + m_l\alpha_l$ ($m_i \in \boldsymbol{Z}^+$) and put $m(\mu) = m_1 + \cdots + m_l$. Let the rational functions ${}^{`}\Gamma_\mu$ on $\mathfrak{a}_c^*$ be defined by induction on $m(\mu)$ as follows: Put ${}^{`}\Gamma_0 = 1$ and

$$\{\langle \mu, \mu - 2\rho \rangle - 2\mu\}\, {}^{`}\Gamma_\mu = 2 \sum_{\alpha \in \Sigma^+} m_\alpha \sum_{k \geqslant 1} {}^{`}\Gamma_{\mu - 2k\alpha} \{\langle \mu - 2k\alpha, \alpha \rangle - \alpha\}. \tag{11}$$

The expansion in question is (since $\boldsymbol{c}(\lambda)\, \boldsymbol{c}(-\lambda) = \boldsymbol{c}(s\lambda)\, \boldsymbol{c}(-s\lambda)$)

$$\boldsymbol{c}(\lambda)^{-1}\, \boldsymbol{c}(-\lambda)^{-1}\, \varphi_\lambda(\exp H) = \sum_{\mu \in L} \zeta_\mu(\lambda, H), \qquad (H \in \mathfrak{a}^+), \tag{12}$$

where

$$\zeta_\mu(\lambda, H) = \sum_{s \in W} \boldsymbol{c}(-s\lambda)^{-1}\, {}^{`}\Gamma_\mu(is\lambda - \rho)\, e^{(is\lambda - \rho - \mu)(H)}, \qquad (H \in \mathfrak{a}_c). \tag{13}$$

In both (12) and (13) λ runs through $\mathfrak{a}_c^*$ with certain complex hyperplanes excepted, namely those which form the singularities of

$$\boldsymbol{c}(s\lambda)^{-1}, \qquad \boldsymbol{c}(-s\lambda)^{-1}, \qquad {}^{`}\Gamma_\mu(is\lambda - \rho).$$

In particular (12) and (13) hold for all $\lambda \in \mathfrak{a}^*$. We shall also use the relation $\boldsymbol{c}(\lambda)\, \boldsymbol{c}(-\lambda) = \mid \boldsymbol{c}(\lambda)|^2$ for $\lambda \in \mathfrak{a}^*$.

Now let us consider the case rank $(G/K) = 1$. Then we have by [22(i), p. 304]

$$|\,{}'\Gamma_\mu(i\lambda - \rho))| \leqslant c(1 + m(\mu)^d)(1 + |\,\lambda\,|^e), \tag{14}$$

where c, d, and e are constants, $\mu \in L$ and $\lambda \in \mathfrak{a}^*$. In the proof we replace λ by $\xi + i\eta$ (η fixed) and find that it remains valid provided $\langle\alpha_1, \alpha_1 + \eta\rangle > 0$ (α_1 = unique simple root) and $|\,\lambda\,|$ is replaced by $(|\,\xi\,|^2 + |\,\eta\,|^2)^{1/2}$. In particular (14) remains valid for λ in a strip around the line $\eta = \rho$. Hence we can use Cauchy's formula to estimate the derivatives of the function $\lambda \to {}'\Gamma_\mu(i\lambda - \rho)$ for points on the line; specifically, for each $m \in \boldsymbol{Z}^+$ there exists a constant c_m such that

$$\left|\frac{d^m}{d\xi^m}\,{}'\Gamma_\mu(i\xi - 2\rho)\right| \leqslant c_m(1 + m(\mu)^d)(1 + |\,\xi\,|^e). \tag{15}$$

The functions $\boldsymbol{c}(\xi)^{-1}$ and $\boldsymbol{c}(\xi - i\rho)^{-1}$ are products of Gamma factors $\Gamma(a + i\xi)/\Gamma(b + i\xi)$ where $a, b > 0$ ([21(i), p. 303] or [19(a)]) so by [22(e), Lemma 3.6] $\boldsymbol{c}(\xi)^{-1}$ and $\boldsymbol{c}(\xi - i\rho)^{-1}$ have each derivative bounded by a polynomial.

Let $H \in \mathfrak{a}^+$; the function

$$\psi(\xi)\, \boldsymbol{c}(-s\xi)^{-1}\,{}'\Gamma_\mu(is\xi - \rho)\, e^{(is\xi - \rho - \mu)(H)}, \qquad \xi \in \mathfrak{a}^*,$$

is clearly integrable and since

$$\sum_{\mu \in L} e^{-\mu(H)}(1 + m(\mu)^d) < \infty \tag{16}$$

the following interchange of summation and integration is legitimate: If $H \in \mathfrak{a}^+$, then

$$\begin{aligned} F(\exp H) &= w^{-1} \int_{\mathfrak{a}^*} \psi(\xi) \sum_{\mu \in L} \sum_{s \in W} \boldsymbol{c}(-s\xi)^{-1}\,{}'\Gamma_\mu(is\xi - \rho)\, e^{(is\xi - \rho - \mu)(H)}\, d\xi \\ &= e^{-\rho(H)} \sum_{\mu \in L} e^{-\mu(H)} \int_{\mathfrak{a}^*} \psi(-\xi)\, \boldsymbol{c}(\xi)^{-1}\,{}'\Gamma_\mu(-i\xi - \rho)\, e^{-i\xi(H)}\, d\xi. \end{aligned}$$

Here we also used the invariance of ψ under W. In this last integral we shall use Cauchy's theorem to shift the integration from $\mathfrak{a}^*$ to $\mathfrak{a}^* - i\rho$, that is, we claim

$$\begin{aligned} &\int_{\mathfrak{a}^*} \psi(-\xi)\, \boldsymbol{c}(\xi)^{-1}\,{}'\Gamma_\mu(-i\xi - \rho)\, e^{-i\xi(H)}\, d\xi \\ &\quad = \int_{\mathfrak{a}^*} \psi(-\xi + i\rho)\, \boldsymbol{c}(\xi - i\rho)^{-1}\,{}'\Gamma_\mu(-i\xi - 2\rho)\, e^{-\rho(H)}\, e^{-i\xi(H)}\, d\xi. \end{aligned} \tag{17}$$

This shift is permissible because if $0 < t < 1$ the integrand is a holomorphic function of ξ on the closed strip bounded by $\mathfrak{a}^*$ and $\mathfrak{a}^* - it\rho$ and the integral behaves suitably at ∞ because of the rapid decrease of ψ and the mentioned estimates for $\boldsymbol{c}$ and ${}^\backprime\Gamma_\mu$. Letting $t \to 1$, the claimed relation follows. Hence

$$e^{2\rho(H)}F(\exp H) = \sum_{\mu \in L} e^{-\mu(H)}F_\mu(H), \qquad (H \in \mathfrak{a}^+), \tag{18}$$

where

$$F_\mu(H) = \int_{\mathfrak{a}^*} \psi(-\xi + i\rho)\, \boldsymbol{c}(\xi - i\rho)^{-1}\, {}^\backprime\Gamma_\mu(-i\xi - 2\rho)\, e^{-i\xi(H)}, \qquad (H \in \mathfrak{a}). \tag{19}$$

Thus (10), (15), and the indicated estimates for $\boldsymbol{c}(\xi - i\rho)^{-1}$ show that $F_\mu \in \mathscr{S}(\mathfrak{a})$, more precisely, for each $m, n \in \boldsymbol{Z}^+$.

$$\sup_{t \in \boldsymbol{R}} \left| (1+t)^m \frac{d^n}{dt^n} F_\mu(\exp tH_{\alpha_1}) \right| \leqslant c_{m,n}(1 + m(\mu)^d), \qquad (\mu \in L).$$

Thus by (16) and (18), for each differential operator E on $\mathfrak{a}$ with polynomial coefficients,

$$\sup_{H \in \mathfrak{a}^+} | e^{2\rho(H)}(EF)(H)| < \infty. \tag{20}$$

The function δ used in the definition of $I'(G)$ is given by [see [21(j), §12] or [21(b), p. 379])

$$\delta(H) = \prod_{\alpha \in \Sigma^+} \left(\frac{\sinh \alpha(H)}{\alpha(H)} \right)^{m_\alpha}$$

and clearly $\delta(H)\, e^{-2\rho}$ remains bounded on $\mathfrak{a}^+$. Now let D be a left invariant differential operator on G. Then, according to [21(i), Lemma 23] there exists a differential operator $\delta'(D)$ on $\mathfrak{a}^+$ such that

$$(DF)(\exp H) = [\delta'(D)((F \circ \exp)^-)]\,(H)$$

for each $F \in \mathscr{E}(G)$, bi-invariant under K, the bar denoting restriction to $\mathfrak{a}^+$. Moreover the coefficients of $\delta'(D)$ belong to the ring of functions on $\mathfrak{a}^+$ generated by the functions

$$H \to (e^{2\alpha(H)} - 1)^{-1},$$

607/5/1-3

where α is a positive restricted root. But since we are in the rank one case these coefficients are bounded except near the origin. Thus (20) implies that for each $q \in \boldsymbol{Z}^+$

$$\sup_{H \in \mathfrak{a}^+} |(1 + | H |^q)\, \delta(H)(DF)(\exp H)| < \infty.$$

If $k_1, k_2 \in K$ then

$$(DF)(k_1 \exp Hk_2) = (DF)(k_2^{-1} \exp Hk_2) = [Ad(k_2)\, DF](\exp H)$$

(where the action of $Ad(k_2)$ on $\mathfrak{g}$ is extended to the universal enveloping algebra). The function δ_q (Chap. I, §2) being bi-invariant under K we thus conclude the last inequality that $F \in I^1(G)$. Now by Theorem 5, p. 610 of [21(j)] we have

$$\tilde{F}(\xi) = \psi(\xi), \qquad \xi \in \mathfrak{a}^*.$$

(See [22(i), p. 301] for a fuller explanation.) But since $\tilde{F}$ and ψ extend to holomorphic functions on $\mathfrak{a}_\rho{}^*$ they coincide there too. This proves Theorem 2.1 for the case rank $(G/K) = 1$.

In order to prove the theorem for G complex it is more convenient to use the formula ([(21(c)]; [18] for $G = \boldsymbol{SL}(n, \boldsymbol{C})$),

$$\pi(\lambda)\, \varphi_\lambda(a) \sum_{s \in W} \epsilon(s)\, e^{s\rho(\log a)} = i^{-d}\pi(\rho) \sum_{s \in W} \epsilon(s)\, e^{is\lambda(\log a)}, \qquad (a \in A), \qquad (21)$$

for the spherical function rather that the expansion (12). In this formula π denotes the product of the positive restricted roots, $d =$ degree π and we write $\pi(\lambda)$ instead of $\pi(H_\lambda)$; $\epsilon(s)$ denotes the determinant of the linear transformation s of $\mathfrak{a}$.

Going back to (9) we rewrite it using (21) and the fact that for G complex, $\boldsymbol{c}(\xi) = i^{-d}\pi(\rho)/\pi(\xi)$, [21(i), §14]. Then a trivial calculation gives

$$\sum_{s \in W} \epsilon(s)\, e^{s\rho(H)} F(\exp H) = i^d \pi(\rho)^{-1} \int_{\mathfrak{a}^*} \psi(\xi)\, \pi(\xi)\, e^{i\xi(H)}\, d\xi, \qquad (H \in \mathfrak{a}). \quad (22)$$

Again we can use Cauchy's theorem to shift the integration to the extremes of the tube $\mathfrak{a}^* + iC_\rho$. Thus for each $\sigma \in W$,

$$\int_{\mathfrak{a}^*} \psi(\xi)\, \pi(\xi)\, e^{i\xi(H)}\, d\xi = \int_{\mathfrak{a}^*} \psi(\xi + i\sigma\rho)\, \pi(\xi + i\sigma\rho)\, e^{(i\xi - \sigma\rho)(H)}\, d\xi;$$

so by (22),

$$\Big(\sum_{s\in W} \epsilon(s)\, e^{s\rho(H)}\Big)\Big(\sum_{\sigma\in W} e^{\sigma\rho(H)}\Big) F(\exp H) = i^d \pi(\rho)^{-1} \int_{\mathfrak{a}^*} \psi^*(\xi)\, \pi(\xi)\, e^{i\xi(H)}\, d\xi, \quad (23)$$

where

$$\psi^*(\xi) = \pi(\xi)^{-1} \sum_{\sigma\in W} \psi(\xi + i\sigma\rho)\, \pi(\xi + i\sigma\rho).$$

The sum here is rapidly decreasing in ξ and is skew (that is the substitution $\xi \to \sigma\xi$ results in multiplication by $\epsilon(\sigma)$). By Lemma 3.1 in [22(e)], $\psi^* \in \mathscr{S}(\mathfrak{a}^*)$. Let F^* denote the unique function on G which is bi-invariant under K and such that

$$F^*(\exp H) = \sum_{\sigma\in W} e^{\sigma\rho(H)} F(\exp H), \qquad (H \in \mathfrak{a}).$$

Then (23) implies

$$F^*(g) = w^{-1} \int_{\mathfrak{a}^*} \psi^*(\xi)\, \varphi_\xi(g)\, |\, \mathbf{c}(\xi)|^{-2}\, d\xi$$

so by [21(j), p. 586], $F^* \in I^2(G)$, that is

$$\sup_g |\, \delta_q(g)^{1/2}\, (DF^*)(g)| < \infty \quad (24)$$

for each $q \in \mathbf{Z}^+$ and each left invariant differential operator D on G.

Let Δ_0 denote the extension of the function

$$a \to \sum_{s\in W} e^{s\rho(\log a)}, \qquad a \in A,$$

to a function in $\mathscr{E}(G)$, bi-invariant under K. We must now estimate the derivatives of this function.

LEMMA 2.2. *For each left invariant differential operator D on G*

$$|(D\Delta_0)(g)| \leqslant c(1 + |\, g\,|^d)\, \Delta_0(g), \qquad g \in G,$$

where c is a constant and d = degree(π).

Proof. The function $\Delta_0(a)$ on A is invariant under the Weyl group and by simple divisibility arguments (see Exposé 19 in [44] by Cartier) the product

$$\Big(\sum_{s\in W} \epsilon(s)\, e^{s\rho}\Big)\Big(\sum_{\sigma\in W} e^{\sigma\rho}\Big)$$

can be written as a linear combination

$$\sum_{\mu} a_{\mu} \sum_{s \in W} \epsilon(s)\, e^{s\mu}, \qquad a_{\mu} \in \boldsymbol{R} - \{0\},$$

where each μ lies in $\mathfrak{a}^*$ and satisfies $H_\mu \in \mathfrak{a}^+$. Comparing with formula (21) for the spherical function we see that $\sum_{s \in W} e^{s\rho}$ is a linear combination of spherical functions

$$\varphi_{-i\mu}, \qquad H_\mu \in \mathfrak{a}^+, \qquad \mu = s\rho + \sigma\rho \qquad (s, \sigma \in W).$$

Now we use the inequalities

$$| D\varphi_\lambda(g)| \leqslant c_D(1 + | \xi |^2 + | \eta |^2)^{d_D} \varphi_{i\eta}(g), \quad g \in G$$

and

$$| \varphi_0(a)| \leqslant c_0(1 + | \log a |^d)\, e^{-\rho(\log a)}, \qquad a \in A^+.$$

proved in [21(a), pp. 279, 294]; (c_D, d_D, and c_0 are constants). Now using (3) we deduce for the $\varphi_{-i\mu}$ above and $a \in A^+$,

$$\begin{aligned} |(D\varphi_{-i\mu})(a)| &\leqslant \text{const}\, \varphi_{-i\mu} = \text{const} \int_K e^{(\mu-\rho)(H(ak))}\, dk \\ &\leqslant \text{const} \int_K e^{-\rho(H(ak))}\, dk\; e^{\mu(\log a)} \\ &\leqslant \text{const}(1 + | \log a |^d)\, e^{-\rho(\log a)}\, e^{\mu(\log a)}. \end{aligned}$$

But $\mu = s\rho + \sigma\rho$ and $\rho - s\rho$, $\rho - \sigma\rho$ are positive on $\mathfrak{a}^+$. Hence

$$|(D\Delta_0)(a)| \leqslant \text{const}(1 + | \log a |^d)\, e^{\rho(\log a)}, \qquad (a \in A^+). \tag{25}$$

If $k_1, k_2 \in K$ then

$$(D\Delta_0)(k_1 a k_2) = ((Ad(k_2)D)\Delta_0)(a)$$

and since the operators $Ad(k)D$ ($k \in K$) lie in a finite-dimensional vector space the lemma follows from (25). Now

$$\delta(H) = \left\{ \sum_s \epsilon(s)\, e^{(s\rho)(H)} \Big/ \pi(H) \right\}^2$$

and D is a linear combination of products $X_1 \cdots X_r$ where X_i is a left invariant vector field on G. Using the lemma and (24) we deduce by induction on the order of D

$$\sup_g |\delta_q(g)(DF)(g)| < \infty,$$

so $F \in I^1(G)$. Thus $\tilde{F}$ is holomorphic on $\mathfrak{a}_\rho{}^*$ and since we have seen $\tilde{F} = \psi$ on $\mathfrak{a}^*$ we deduce $\tilde{F} = \psi$ on $\mathfrak{a}_\rho{}^*$. This concludes the proof of Theorem 2.1.

Remark. In order to extend Theorem 2.1 to arbitrary G it seems necessary not just to use the quoted result of Harish–Chandra that $F \in I^2(G)$ if ψ is rapidly decreasing on $\mathfrak{a}^*$ but to generalize its proof. More specifically one needs Lemma 37 in [21(j)] with ρ replaced by 2ρ provided the function a lies in $\mathscr{H}(\mathfrak{a}_\rho{}^*)$. For this one must extend somewhat the analysis in [21(j)] relating the spherical functions on G/K to the corresponding spherical functions on the boundary component $M/M \cap K$. While Harish–Chandra's estimates are given in a sufficiently small tube around $\mathfrak{a}^*$ it would now be necessary to allow λ to vary in the entire tube $\mathfrak{a}_\rho{}^*$. With the $\boldsymbol{c}$-functions for G and M explicitly known this does not seem to present serious obstacles. Thus there is little doubt that Theorem 2.1 remains true for any G.

The following fact was established during the proof.

COROLLARY 2.3. *If the function $F \in I^1(G)$ has spherical Fourier transform $\tilde{F}$ then the function $\Delta_0 F$ lies in $I^2(G)$ and has spherical Fourier transform given by*

$$(\Delta_0 F)^{\sim}(\xi) = \sum_{s \in W} \tilde{F}(\xi + is\rho)\, \pi(\xi + is\rho)\big/\pi(\xi), \qquad (\xi \in \mathfrak{a}^*).$$

THEOREM 2.4 (The Paley-Wiener Theorem). *The spherical Fourier transform $F \to \tilde{F}$ is a bijection of $I_c(G)$ onto the space of W-invariant rapidly decreasing holomorphic functions on $\mathfrak{a}_c{}^*$ of exponential type.*

This theorem was proved in [22(i)] except for the full justification of a certain term-by-term integration (condition (Γ) p. 304); this justification is now provided by recent work of R. Gangolli. We give below a simplified justification.

Lemma 4.1 in [22(i)] shows that for each $H \in \mathfrak{a}^+$

$$|\,{}'\Gamma_\mu(i\lambda - \rho)| \leqslant K e^{\mu(H)}, \qquad \mu \in L, \tag{26}$$

where $K = K_{\lambda,H}$ is a constant. It suffices to prove that (26) holds for all $\lambda \in \mathfrak{a}^*$ with $K = K_H$ (i.e., independent of λ) because the cited condition (Γ) is then clearly satisfied.

Let E denote the Laplacian on A. According to [21(i), pp. 270–272] the function

$$\Phi_\lambda(\exp H) = \sum_{\mu \in L} {}'\Gamma_\mu(i\lambda - \rho)\, e^{(i\lambda - \rho - \mu)(H)}, \qquad H \in \mathfrak{a}^+, \tag{27}$$

and the radial part $\delta'(\omega)$ of ω satisfy

$$\delta'(\omega)\, \Phi_\lambda = (-\langle \lambda, \lambda \rangle - \langle \rho, \rho \rangle)\, \Phi_\lambda \tag{28}$$

and

$$\delta'(\omega) = E + \text{lower order terms}. \tag{29}$$

Since ω is symmetric, $\delta'(\omega)$ is symmetric with respect to $\Delta(a)\, da$; the same is true of the operator $\Delta^{-1/2} E \circ \Delta^{1/2}$ which equals E up to lower order terms. Since no first order operator on A is symmetric (29) implies

$$\delta'(\omega) = \Delta^{-1/2} E \circ \Delta^{1/2} - d \tag{30}$$

where d is a function; clearly $d = \Delta^{-1/2} E(\Delta^{1/2})$. The functions $\Delta^{1/2}$, $\Delta^{-1/2}$ and d have expansions on $\mathfrak{a}^+$,

$$\Delta^{-1/2}(\exp H) = e^{-\rho(H)} \sum_{\mu \in L} b_\mu e^{-\mu(H)},$$

$$\Delta^{1/2}(\exp H) = e^{\rho(H)} \sum_{\mu \in L} c_\mu e^{-\mu(H)}, \qquad d = \sum_{\mu \in L} d_\mu e^{-\mu(H)}, \qquad d_0 = \langle \rho, \rho \rangle$$

where the coefficients b_μ, c_μ, and d_μ have at most polynomial growth in μ. Now (28) and (30) imply that the function

$$\Psi_\lambda(\exp H) = \Delta^{1/2}(\exp H)\, \Phi_\lambda(\exp H) = \sum_{\mu \in L} a_\mu(\lambda)\, e^{(i\lambda - \mu)(H)}, \tag{31}$$

where

$$a_\mu(\lambda) = \sum_{\nu,\, \mu - \nu \in L} {}'\Gamma_{\mu - \nu}(i\lambda - \rho)\, c_\nu$$

satisfies the simpler differential equation

$$E\Psi_\lambda = (-\langle\lambda, \lambda\rangle - \langle\rho, \rho\rangle + d)\,\Psi_\lambda\,. \tag{32}$$

This gives the recursion formula

$$a_\mu(\lambda)(\langle\mu, \mu\rangle - 2i\langle\lambda, \mu\rangle) = \sum_{0\neq\nu,\,\mu-\nu\in L} d_\nu a_{\mu-\nu}(\lambda),$$

so for a constant c

$$|\, a_\mu \,| \leqslant cm(\mu)^{-1} \sum_{0\neq\nu,\,\mu-\nu\in L} d_\nu \,|\, a_{\mu-\nu} \,|, \qquad (\lambda \in \mathfrak{a}^*). \tag{33}$$

Let $H \in \mathfrak{a}^+$; repeating (with $N_0 \geqslant c \sum d_\nu e^{-\nu(H)}$) the proof of Lemma 4.1 in [22(i)] we deduce from (33)

$$|\, a_\mu(\lambda)| \leqslant C_H e^{\mu(H)}, \qquad \mu \in L, \quad \lambda \in \mathfrak{a}^*,$$

where $C_H \in \boldsymbol{R}$ only depends on H. But

$$`\Gamma_\mu(i\lambda - \rho) = \sum_{\nu,\,\mu-\nu\in L} a_{\mu-\nu}(\lambda)\, b_\nu$$

so since $H \in \mathfrak{a}^+$ is arbitrary the same kind of estimate for $`\Gamma_\mu(i\lambda - \rho)$ follows. Q.E.D.

§3. The Inversion Formula and the Plancherel Formula for the Radon Transform

To begin with we have to express the measure $d\xi$ on Ξ by means of the diffeomorphism $\psi : (kM, a) \to kaMN$ of $(K/M) \times A$ onto Ξ (Proposition 1.1, Chap. I). The invariant measures dk on K and dm on M, normalized by $\int dk = \int dm = 1$ determine a K-invariant measure dk_M on K/M. Let da be normalized as in Notations, §4.

Lemma 3.1. *For each $\varphi \in \mathcal{D}(\Xi)$,*

$$\int_\Xi \varphi(\xi)\, d\xi = \int_{K/M\times A} \varphi(ka \cdot \xi_o)\, e^{2\rho(\log a)}\, da\, dk_M\,.$$

Proof. It is clear, since ψ is a diffeomorphism, that there exists a function $D \in \mathscr{E}(K/M \times A)$ such that

$$\int_{\Xi} \varphi(\xi)\, d\xi = \int_{K/M} \int_A \varphi(ka \cdot \xi_o)\, D(kM, a)\, da\, dk_M \tag{1}$$

Since $d\xi$ is K-invariant we see, replacing φ by $\varphi^{\tau(k)}$, that D is constant in the first variable so we write $D(a)$ instead of $D(kM, a)$.

To calculate $D(a)$ let $\mathfrak{l}$ denote the orthogonal complement of $\mathfrak{m}$ in $\mathfrak{k}$ (with respect to the Killing form on $\mathfrak{g}$). Select open neighborhoods U of 0 in $\mathfrak{l}$ and V of e in A such that the map $(T, a) \to (\exp T)a$ is a diffeomorphism of $U \times V$ onto a local cross section in G over a neighborhood of ξ_o in Ξ. Select $F \in \mathscr{D}(G)$ with support contained in $(\exp U)\, V(MN)$ such that

$$F((\exp T)\, amn) = \alpha(a)\, \beta(n), \qquad (T \in U, a \in V, m \in M, n \in N),$$

where $\alpha \in \mathscr{D}(A)$ and $\beta \in \mathscr{D}(N)$ has integral over N equal to 1. Let

$$\varphi(gMN) = \int_{MN} F(gmn)\, dm\, dn.$$

Then

$$\int_{\Xi} \varphi(\xi)\, d\xi = \int_G F(g)\, dg = \int_{KAN} F(kan)\, e^{2\rho(\log a)}\, dk\, da\, dn = \int_A \alpha(a)\, e^{2\rho(\log a)}\, da$$

since F is K-invariant. On the other hand, using (1) we find by the K-invariance

$$\int_{\Xi} \varphi(\xi)\, d\xi = \int_A \alpha(a)\, D(a)\, da.$$

Hence $D(a) = e^{2\rho(\log a)}$ for $a \in V$, hence by analyticity for $a \in G$.

If F_1 and F_2 are two continuous functions on G with compact support we define the function $F_1 \# F_2$ by

$$(F_1 \# F_2)(g) = \int_G F_1(h)\, F_2(hg)\, dh.$$

If f_1 and f_2 are continuous functions of compact support on X we write $f_1 \# f_2$ for the function on X defined by

$$(f_1 \# f_2)(g \cdot o) = (F_1 \# F_2)(g)$$

if $F_i(g) = f_i(g \cdot o)$, $(i = 1, 2, g \in G)$.

LEMMA 3.2. *Let $f_1, f_2 \in C_c(X)$. Then $f_1 \# f_2$ is K-invariant and the Radon transform satisfies*

$$(f_1 \# f_2)^\wedge (a \cdot \xi_o) = \int_K \int_A e^{2\rho(\log b)} \hat{f}_1(kb \cdot \xi_o) \hat{f}_2(kba \cdot \xi_o)\, dk\, db.$$

Proof. The K-invariance is obvious and

$$(f_1 \# f_2)^\wedge (a \cdot \xi_o) = \int_N (F_1 \# F_2)(an)\, dn = \int_G F_1(g) \hat{f}_2(ga \cdot \xi_o)\, dg$$

$$= \int_{KAN} F_1(kbn) \hat{f}_2(kbna \cdot \xi_o)\, e^{2\rho(\log b)}\, dk\, db\, dn.$$

Since $\hat{f}_2(kbna \cdot \xi_o) = \hat{f}_2(kba \cdot \xi_0)$ the lemma follows.

For any vector space V over $\boldsymbol{R}$ let $\mathscr{S}(V)$ denote the Schwartz space of rapidly decreasing functions on V ([41]). Considering any K-invariant Riemannian structure on K/M we can define rapidly decreasing functions on the product manifold as in [22(f)] and thus obtain a well-defined space $\mathscr{S}(\Xi)$. Consider now the function $\boldsymbol{c}(\lambda)$. As observed in [22(e)] the multiplication by $\boldsymbol{c}(\lambda)^{-1}$ is a mapping of $\mathscr{S}(\mathfrak{a}^*)$ into itself. Hence there exist linear operators j and $\bar{j}$ on $\mathscr{S}(A)$ such that for $\lambda \in \mathfrak{a}^*$

$$(jF)^* (\lambda) = \boldsymbol{c}(\lambda)^{-1} F^*(\lambda) \tag{2}$$

and

$$(\bar{j}F)^* (\lambda) = \boldsymbol{c}(-\lambda)^{-1} F^*(\lambda), \tag{3}$$

where $*$ denotes the Euclidean Fourier transform

$$F^*(\lambda) = \int_A F(h)\, e^{-i\lambda(\log h)}\, dh.$$

We can then define the endomorphisms J, $\bar{J}$ of $\mathscr{S}(\Xi)$ by

$$(J\varphi)(kaMN) = j_a(\varphi(kaMN))$$

$$(\bar{J}\varphi)(kaMN) = \bar{j}_a(\varphi(kaMN))$$

for $\varphi \in \mathscr{S}(\Xi)$. Let e^ρ denote the function on $\Xi = (K/M) \times A$ given by $e^\rho(kM, a) = e^{\rho(\log a)}$ for $a \in A$, $k \in K$, and define the space $\mathscr{S}_\rho(\Xi)$ by

$$\mathscr{S}_\rho(\Xi) = \{\varphi \in \mathscr{E}(\Xi) \mid e^\rho \varphi \in \mathscr{S}(\Xi)\}.$$

Then the operators $\Lambda = e^{-\rho} J \circ e^{\rho}$ and $\bar{\Lambda} = e^{-\rho} \bar{J} \circ e^{\rho}$ map the space $\mathscr{S}_\rho(\Xi)$ into itself.

LEMMA 3.3. *The spaces $\mathscr{S}(\Xi)$ and $\mathscr{S}_\rho(\Xi)$ as well as the operators J, $\bar{J}$, Λ, and $\bar{\Lambda}$ are all invariant under the action of G on Ξ.*

This is immediate from the fact that if Ξ is viewed as a trivial fibre bundle over K/M the action of each $g \in G$ on Ξ is fibre-preserving (Chap. I, §2) and induces a translation on each fibre, combined with the fact that j is invariant under translations on A.

THEOREM 3.4.[1] *For each $f \in \mathscr{D}(X)$ we have*

$$w \int_X |f(x)|^2 \, dx = \int_\Xi |(\Lambda \hat{f})(\xi)|^2 \, d\xi.$$

Proof. Let us consider two functions $f_1, f_2 \in \mathscr{D}(X)$ and put $F_i(g) = f_i(g \cdot o)$ $(g \in G,\ i = 1, 2)$. Since the function $F_1 \# F_2$ is bi-invariant under K we have by Harish-Chandra's inversion formula for the spherical Fourier transform (Notations, §6),

$$(F_1 \# F_2)(e) = w^{-1} \int_{\mathfrak{a}^*} |\, \boldsymbol{c}(\lambda)|^{-2} (F_1 \# F_2)^{\sim}(\lambda) \, d\lambda. \tag{4}$$

On the other hand, we have, using (2) in §1 and $\boldsymbol{c}(-\lambda) = \operatorname{conj} \boldsymbol{c}(\lambda)$ $(\lambda \in \mathfrak{a}^*)$,

$$(j\bar{j}\varphi_F)^* (\lambda) = |\, \boldsymbol{c}(\lambda)|^{-2} \tilde{F}(\lambda), \tag{5}$$

if $F \in \mathscr{D}(G)$ is bi-invariant under K. Thus by the Euclidean Fourier inversion formula

$$F(e) = w^{-1}(j\bar{j}\varphi_F)(e). \tag{6}$$

Now using Lemma 3.2 we have

$$\begin{aligned}\varphi_{F_1 \# F_2}(a) &= e^{\rho(\log a)}(f_1 \# f_2)^{\wedge}(a \cdot \xi_o)\\ &= \int_K \int_A e^{\rho(\log b)} \hat{f}_1(kb \cdot \xi_o)\, e^{\rho(\log ab)} \hat{f}_2(kba \cdot \xi_o) \, dk \, db;\end{aligned}$$

so, viewing $\varphi_i = e^{\rho} \hat{f}_i$ as a function on $(K/M) \times A$,

$$\varphi_{F_1 \# F_2}(a) = \int_K \int_A \varphi_1(kM, b)\, \varphi_2(kM, ab) \, db \, dk.$$

[1] Cf. [22(d), pp. 443–444].

As a function on $\mathfrak{a}^*$, c^{-1} has each of its derivatives bounded by a polynomial so the operators $j, \bar{j}$, and $j\bar{j}$ are convolutions by tempered distributions. Hence

$$(j\varphi_{F_1 \# F_2})(a) = \int_K \int_A \varphi_1(kM, b)(J\varphi_2)(kM, ab)\, db\, dk$$

$$= \int_K \int_A \varphi_1(kM, a^{-1}b)(J\varphi_2)(kM, b)\, db\, dk.$$

Now if $F \in \mathscr{S}(A)$ and F' denotes the function $a \to F(a^{-1})$ it is easy to prove $\bar{j}F' = (jF)'$ so

$$\bar{j}_a(F(a^{-1}b)) = [\bar{j}((F')^{L(b)}](a) = [(\bar{j}F')^{L(b)}](a)$$
$$= (jF)'(ab^{-1}) = (jF)(a^{-1}b)$$

whence

$$[j\bar{j}\varphi_{F_1 \# F_2}](a) = \int_K \left(\int_A (J\varphi_1)(kM, a^{-1}b)(J\varphi_2)(kM, b)\, db\right) dk.$$

Here we put $a = e$, introduce the operator Λ, and combine this with (6) and Lemma 3.1. Then we get

$$w \int_X f_1(x) f_2(x)\, dx = \int_\Xi (\Lambda\hat{f}_1)(\xi)(\Lambda\hat{f}_2)(\xi)\, d\xi.$$

Now $F \in \mathscr{S}(A)$ is real if and only if $F^*(-\lambda) = \operatorname{conj} F^*(\lambda)$ $(\lambda \in \mathfrak{a}^*)$; it follows that j maps real functions into real functions and

$$\Lambda(\operatorname{conj} \varphi) = \operatorname{conj}(\Lambda\varphi), \qquad \varphi \in \mathscr{D}(\Xi).$$

Now Theorem 3.4 follows.

THEOREM 3.5. (i) *Let Λ^* denote the adjoint of Λ with respect to the volume element $d\xi$. Then*

$$f = w^{-1}(\Lambda\Lambda^*\hat{f})^{\vee}, \qquad f \in \mathscr{D}(X). \tag{7}$$

(ii) *Suppose all Cartan subgroups of G are conjugate. Then Λ is a differential operator and for a certain differential operator $\square$ on X,*

$$f = \square((\hat{f})^{\vee}), \qquad f \in \mathscr{D}(X). \tag{8}$$

Proof. Given $f \in \mathscr{D}(X)$ define the function $F_0 \in \mathscr{D}(G)$ by

$$F_0(g) = \int_K f(kg \cdot o)\, dk.$$

Then F_0 is bi-invariant under K and

$$\varphi_{F_0}(a) = e^{\rho(\log a)} \int_{N \times K} f(kan \cdot o)\, dk\, dn,$$

so by (6)

$$f(o) = w^{-1} \left[(j\bar{j})_a \left\{ e^{\rho(\log a)} \int_{K \times N} f(kan \cdot o)\, dk\, dn \right\} \right]_{a=e}. \tag{9}$$

Now we put

$$F_1(g) = e^{\rho(H(g))} \int_{K \times N} f(kgn \cdot o)\, dk\, dn.$$

Then F_1 is right invariant under MN so $F_1(g) = \varphi_1(g \cdot \xi_o)$ where $\varphi_1 \in \mathscr{D}(\Xi)$. From (9) we have

$$f(o) = w^{-1}[J\bar{J}\varphi_1](\xi_o). \tag{10}$$

But with our definition of e^ρ on Ξ

$$\varphi_1(\xi) = e^{\rho}(\xi) \int_K \hat{f}(k \cdot \xi)\, dk$$

so since $\Lambda\bar{\Lambda} = e^{-\rho} J\bar{J} \circ e^{\rho}$, (10) implies

$$\begin{aligned} f(o) &= w^{-1} \left[(\Lambda\bar{\Lambda})_\xi \left(\int_K \hat{f}(k \cdot \xi)\, dk \right) \right]_{\xi=\xi_o} \\ &= w^{-1} \left[\int_K (\Lambda\bar{\Lambda})_\xi (\hat{f}(k \cdot \xi))\, dk \right]_{\xi=\xi_o} = w^{-1} \int_K (\Lambda\bar{\Lambda}\hat{f})(k \cdot \xi_o)\, dk, \end{aligned}$$

the last relation coming from Lemma 3.3. Thus the relation

$$f = w^{-1}(\Lambda\bar{\Lambda}\hat{f})^{\vee}$$

holds at the point $o \in X$, hence everywhere on X by the G-invariance of the operations $\wedge$ and $\vee$.

Now consider φ_1, φ_2 arbitrary in $\mathscr{D}(\Xi)$ and put

$$h(a) = \int_K \int_A \varphi_1(kM, b)\, \varphi_2(kM, ab)\, dk\, db.$$

Then, as before,

$$(jh)(a) = \int_K \int_A \varphi_1(kM, b)(J\varphi_2)(kM, ab)\, dk\, db$$

and

$$h'(a) = h(a^{-1}) = \int_K \int_A \varphi_1(kM, ab)\, \varphi_2(kM, b)\, db,$$

so

$$(jh')(a) = \int_{K\times A} (\bar{J}\varphi_1)(kM, ab)\, \varphi_2(kM, b)\, db;$$

so, since $jh' = (jh)'$,

$$\int_{K\times A} \varphi_1(kM, b)(J\varphi_2)(kM, b)\, db = \int_{K\times A} (\bar{J}\varphi_1)(kM, b)\, \varphi_2(kM, b)\, db$$

or equivalently,

$$\int_{\Xi} \varphi_1(\xi)(J\varphi_2)(\xi)\, e^{-2\rho}(\xi)\, d\xi = \int_{\Xi} (\bar{J}\varphi_1)(\xi)\, \varphi_2(\xi)\, e^{-2\rho}(\xi)\, d\xi.$$

Putting $\psi_i = e^{-\rho}\varphi_i$ $(i = 1, 2)$ this becomes

$$\int_{\Xi} \psi_1(\xi)(\Lambda\psi_2)(\xi)\, d\xi = \int_{\Xi} (\bar{\Lambda}\psi_1)(\xi)\, \psi_2(\xi)\, d\xi$$

which shows that Λ has an adjoint Λ^*, namely $\bar{\Lambda}$. This proves (i).

For (ii) we start with a general result about Cartan subalgebras.

PROPOSITION 3.6. *Let* $\mathfrak{g}$ *be any semisimple Lie algebra. The following properties are equivalent.*

(i) *All Cartan subalgebras of* $\mathfrak{g}$ *are conjugate under the adjoint group.*

(ii) *All restricted roots have even multiplicity.*

Proof. Extend $\mathfrak{a}$ to a maximal abelian subalgebra $\mathfrak{h}$ of $\mathfrak{g}$, let $\mathfrak{g}_c$ denote the complexification of $\mathfrak{g}$ and $\mathfrak{h}_c$ the complex subalgebra of $\mathfrak{g}_c$ generated by $\mathfrak{h}$. Let Δ denote the set of roots of $\mathfrak{g}_c$ with respect to $\mathfrak{h}_c$, and if $\beta \in \Delta$ let $H_\beta \in \mathfrak{h}_c$ be determined by $B_c(H, H_\beta) = \beta(H)$ for $H \in \mathfrak{h}_c$, B_c denoting the Killing form of $\mathfrak{g}_c$. Then

$$\sum_{\beta\in\Delta} \mathbf{R}H_\beta = i(\mathfrak{h} \cap \mathfrak{k}) + \mathfrak{a}$$

and this subspace $\mathfrak{h}_\Delta$ is invariant under the conjugation η of $\mathfrak{g}_c$ with respect to $\mathfrak{g}$. The transpose $\sigma = {}^t\eta$ of the map $\eta \mid \mathfrak{h}_\Delta$ permutes the set Δ.

The set Σ consists of the restrictions of Δ to $\mathfrak{a}$. If $\alpha \in \Sigma$ let Δ_α denote the set of $\beta \in \Delta$ such that $\beta \mid \mathfrak{a} = \alpha$. Since $\eta = 1$ on $\mathfrak{a}$, σ induces a permutation of Δ_α of order 2. There is at most one fixed point $\beta \in \Delta_\alpha$, in which case $H_\beta \in \mathfrak{a}$; thus $H_\beta \in \mathfrak{a}$ if and only if the restriction $\alpha = \beta \mid \mathfrak{a}$ has odd multiplicity (cf., Araki [2, Proposition 2.2]).

Now it is well-known [21(f), §2], [28(a)] that any Cartan subalgebra of $\mathfrak{g}$ is conjugate to a Cartan subalgebra $\mathfrak{t}$ satisfying $\mathfrak{t} = \mathfrak{t} \cap \mathfrak{k} + \mathfrak{t} \cap \mathfrak{p}$, and two such subalgebras are conjugate if and only if their intersections with $\mathfrak{p}$ (their "vector parts") are conjugate.

To prove Proposition 3.6 suppose (i) holds. We have $\mathfrak{h} = \mathfrak{h} \cap \mathfrak{k} + \mathfrak{a}$ and now extend $\mathfrak{h} \cap \mathfrak{k}$ to a maximal abelian subalgebra $\mathfrak{t}_\mathfrak{k}$ of $\mathfrak{k}$ which we then extend to a maximal abelian subalgebra $\mathfrak{t}$ of $\mathfrak{g}$. If $H \in \mathfrak{t}_\mathfrak{k}$, $Z \in \mathfrak{t}$ then $[H, Z + \theta Z] = 0$ so by $Z + \theta Z \in \mathfrak{k}$, we have $Z + \theta Z \in \mathfrak{t}_\mathfrak{k}$ so $\theta\mathfrak{t} \subset \mathfrak{t}$, i.e., $\mathfrak{t} = \mathfrak{t}_\mathfrak{k} + \mathfrak{t} \cap \mathfrak{p}$. Hence $\mathfrak{t}$ is a Cartan subalgebra of $\mathfrak{g}$ so by (i) and the remarks above $\mathfrak{h} \cap \mathfrak{k} = \mathfrak{t}_\mathfrak{k}$. Now if $\beta \in \Delta$ such that $\alpha = \beta \mid \mathfrak{a}$ has odd multiplicity then by the above $H_\beta \in \mathfrak{a}$. Select $X_\beta \neq 0$ in $\mathfrak{g}_c$ such that $[H, X_\beta] = \beta(H)\, X_\beta$ for $H \in \mathfrak{h}$. Since $\beta(\mathfrak{h} \cap \mathfrak{k}) = 0$ this X_β can be chosen in $\mathfrak{g}$. Writing $X_\beta = X_\mathfrak{k} + X_\mathfrak{p}$ according to $\mathfrak{g} = \mathfrak{k} + \mathfrak{p}$ we deduce that both $X_\mathfrak{k}$ and $X_\mathfrak{p}$ are $\neq 0$ and that $[H, X_\mathfrak{k}] = 0$ for $H \in \mathfrak{h} \cap \mathfrak{k}$ contradicting the maximality of $\mathfrak{t}_\mathfrak{k}$.

On the other hand, suppose the multiplicity m_α even for each $\alpha \in \Sigma$. By the above, $H_\beta \notin \mathfrak{a}$ for all $\beta \in \Delta$. If X_β is chosen as above then the complexification $\mathfrak{k}_c$ satisfies

$$\mathfrak{k}_c = (\mathfrak{h} \cap \mathfrak{k})_c + \sum_{\beta \in \Delta} \boldsymbol{C}(X_\beta + \theta X_\beta).$$

But $H_\beta \notin \mathfrak{a}$ for all $\beta \in \Delta$ shows that $(\mathfrak{h} \cap \mathfrak{k})_c$ is maximal abelian in $\mathfrak{k}_c$, i.e., $\mathfrak{h} \cap \mathfrak{k}$ is maximal abelian in $\mathfrak{k}$. Now if $\mathfrak{t}$ is any Cartan subalgebra of $\mathfrak{g}$ we may by the above assume $\mathfrak{t} = \mathfrak{t} \cap \mathfrak{k} + \mathfrak{t} \cap \mathfrak{p}$. Here $\mathfrak{t} \cap \mathfrak{p}$ is conjugate under K to a subspace of $\mathfrak{a}$ so we may assume $\mathfrak{t} \cap \mathfrak{p} \subset \mathfrak{a}$. But then $\mathfrak{t} \cap \mathfrak{p} = \mathfrak{a}$ because otherwise $\dim(\mathfrak{t} \cap \mathfrak{k}) > \dim(\mathfrak{h} \cap \mathfrak{k})$ would contradict the maximality of $\mathfrak{h} \cap \mathfrak{k}$. Thus $\mathfrak{t} \cap \mathfrak{k} \subset \mathfrak{m}$, so $\mathfrak{t} \cap \mathfrak{k}$ and $\mathfrak{h} \cap \mathfrak{k}$ are conjugate under an inner automorphism of $\mathfrak{m}$. This leaves $\mathfrak{a}$ pointwise fixed, so we have proved that $\mathfrak{t}$ and $\mathfrak{h}$ are conjugate.

We shall now prove the latter part of Theorem 3.5. According to Proposition 3.6 all multiplicities m_α ($\alpha \in \Sigma$) are now even so by Araki [2, Proposition 2.4], Σ is a proper root system, that is twice a restricted root is never a restricted root. In view of the formula $\Gamma(x + 1) = x\Gamma(x)$

the $\boldsymbol{c}$-function (cf. (12), §2) reduces to the reciprocal of a polynomial. Hence j and Λ are differential operators and $j\bar{j}$ is invariant under the action of W on A. Let $\square \in \boldsymbol{D}(X)$ correspond to $w^{-1}j\bar{j}$ according to Theorem 2.1, Chap. I and let $\tilde{\square} \in \boldsymbol{D}_0(G)$ be chosen such that $(\tilde{\square}F)(g) = (\square f)(g \cdot o)$ for all $f \in \mathscr{E}(X)$, $F(g) = f(g \cdot o)$. Then by Theorem 3.3 in [22(e)],

$$\varphi_{\tilde{\square}F_0} = w^{-1}(j\bar{j})\,\varphi_{F_0}\,. \tag{11}$$

Combining this with (9) we get

$$f(o) = (\varphi_{\tilde{\square}F_0})(e), \tag{12}$$

i.e.,

$$f(o) = \int_{K\times N} (\square f)(kn \cdot o)\,dk\,dn. \tag{13}$$

Now fix $g \in G$ and replace in (13) f by $f^{\tau(g^{-1})}$. Then since $\square$ is invariant under g,

$$f(g \cdot o) = \int_{K\times N} (\square f)(gkn \cdot o)\,dk\,dn,$$

and by the Darboux equation [22(a), p. 277]

$$\int_K (\square f)(gk \cdot x)\,dk = \tilde{\square}_g\left(\int_K f(gk \cdot x)\,dk\right), \qquad x \in X,$$

whence

$$\begin{aligned} f(g \cdot o) &= \int_N \tilde{\square}_g\left(\int_K f(gkn \cdot o)\,dk\right)dn \\ &= \tilde{\square}_g\left(\int_K\int_N f(gkn \cdot o)\,dn\,dk\right) = \tilde{\square}_g((\hat{f})^{\vee}(g \cdot o)), \end{aligned}$$

which finishes the proof of Theorem 3.5.

Remarks. Theorem 3.5 was given in [22(c), (d), (g)]. The inversion formula for the Radon transform on $\boldsymbol{R}^n$ had been generalized in a different way to odd-dimensional hyperbolic spaces in [22(a)] and later in [42] for any dimension. A formula closely related to (7) was given in [14] §5 for the case G complex and in [13], [17, Chap. V] for the Lobachevski spaces. For G complex, (7) is essentially a special case of the representation of a function on G by means of its integrals over all conjugacy classes and their translates (cf. [18, §26], [21(d), p. 514]). The inversion problem for $f \to \hat{f}$ was also discussed in [19(b)].

Chapter III. Conical Functions, Conical Distributions and Representation Theory

§1. Totally Geodesic Submanifolds of X and of Ξ. Reduction to the Hermitian Hyperbolic Case

Let X be a symmetric space of the noncompact type, $I(X)$ the group of isometries of X in the compact open topology and $G = I_0(X)$ the identity of $I(X)$. We recall that a totally geodesic submanifold X' of X is also a symmetric space in the metric induced by that of X. Suppose the symmetric space X' is also of the noncompact type and let G' denote the closed subgroup of $I(X)$ generated by the transvections (in X) along the geodesics in X'. If $o \in X'$ and $\mathfrak{p}'$ denotes the tangent space X_o' considered as a subspace of the Lie algebra $\mathfrak{g}$ of $I(X)$ then $I(X')$ has Lie algebra $\mathfrak{p}' + [\mathfrak{p}', \mathfrak{p}']$, which thus is semisimple and must therefore coincide with the Lie algebra of G'. Hence $I_0(X')$ is isomorphic to $Ad(G')$. On the other hand, suppose G' is a connected semisimple Lie subgroup of $I(X)$ without compact components. According to Mostow [37] if $\mathfrak{g}' = \mathfrak{k}' + \mathfrak{p}'$ is a Cartan decomposition of the Lie algebra of G' there exists a Cartan decomposition $\mathfrak{g} = \mathfrak{k} + \mathfrak{p}$ of $\mathfrak{g}$ such that $\mathfrak{k}' \subset \mathfrak{k}$, $\mathfrak{p}' \subset \mathfrak{p}$. If $K \subset G$ is the analytic subgroup corresponding to $\mathfrak{k}$ then K, acting on X, has a unique fixed point, say o; the orbit $X' = G' \cdot o$ is a totally geodesic submanifold of X of the noncompact type. Since the orthogonal complement of $[\mathfrak{p}', \mathfrak{p}']$ in $\mathfrak{k}'$ is an ideal in $\mathfrak{g}'$ we have (since G' has no compact components) $\mathfrak{g}' = \mathfrak{p}' + [\mathfrak{p}', \mathfrak{p}']$; hence G' coincides with the closed subgroup of $I(X)$ generated by the transvections (in X) along the geodesics in X'.

Proposition 1.1. *Let X and X' be symmetric spaces of the noncompact type, X' totally geodesic in X. Then each horocycle in X' has the form $\xi' = \xi \cap X'$ where ξ is a horocycle in X.*

Proof. Preserving the notation above we have compatible Cartan decompositions $\mathfrak{g} = \mathfrak{k} + \mathfrak{p}$, $\mathfrak{g}' = \mathfrak{k}' + \mathfrak{p}'$ of the Lie algebras of $I(X)$ and $I(X')$, respectively. Let $\mathfrak{a}' \subset \mathfrak{p}'$ be a maximal abelian subspace and extend $\mathfrak{a}'$ to a maximal abelian subspace $\mathfrak{a}$ of $\mathfrak{p}$. We then have corresponding restricted root space decompositions

$$\mathfrak{g} = \sum_{\lambda} \mathfrak{g}_{\lambda} \qquad \text{and} \qquad \mathfrak{g}' = \sum_{\mu} \mathfrak{g}_{\mu}',$$

where $\lambda \in \mathfrak{a}^*$, $\mu \in (\mathfrak{a}')^*$, and

$$\mathfrak{g}_{\lambda} = \{X \in \mathfrak{g} \mid [H, X] = \lambda(H)\, X \text{ for } H \in \mathfrak{a}\}$$

and

$$\mathfrak{g}_{\mu}' = \{X \in \mathfrak{g}' \mid [H, X] = \mu(H)\, X \text{ for } H \in \mathfrak{a}'\}.$$

Denoting the restriction of λ to $\mathfrak{a}'$ by $\bar{\lambda}$ it is obvious that for each $\mu \in (\mathfrak{a}')^*$

$$\mathfrak{g}_{\mu}' = \mathfrak{g}' \cap \sum_{\bar{\lambda}=\mu} \mathfrak{g}_{\lambda}\,. \tag{1}$$

Let $\Sigma \subset \mathfrak{a}^*$ and $\Sigma' \subset (\mathfrak{a}')^*$ denote the sets of restricted roots. We turn $\mathfrak{a}^*$ and $(\mathfrak{a}')^*$ into ordered vector spaces with *compatible orderings* that is orderings such that $\lambda \in \mathfrak{a}^*$, $\lambda > 0 \Rightarrow \bar{\lambda} \geqslant 0$. Let $\Sigma^+ \subset \Sigma$, $(\Sigma')^+$ be the subsets of positive elements, let

$$\mathfrak{n} = \sum_{\lambda \in \Sigma^+} \mathfrak{g}_{\lambda}\,, \qquad \mathfrak{n}' = \sum_{\mu \in (\Sigma')^+} \mathfrak{g}_{\mu}'$$

and let N and N' be the corresponding analytic subgroups of G and G', respectively. By (1) and the compatibility,

$$(\Sigma')^+ \subset \{\bar{\alpha} \mid \alpha \in \Sigma^+\}, \tag{2}$$

whence $N' \subset N$; since $K' \subset K$, $A' \subset A$ and $G' = K'A'N'$, $G = KAN$ we have $K' = G' \cap K$, $A' = G' \cap A$, and $N' = G' \cap N$. Hence if $\xi_0' = N' \cdot o$, $\xi_o = N \cdot o$ we have $\xi_o' \subset \xi_o \cap X'$. But if $x \in \xi_o \cap X'$ then $x = g' \cdot o = n \cdot o$ with $g' \in G'$, $n \in N$. Hence $g' = nk$ $(k \in K)$ but then $G' = N'A'K'$ implies $n \in N'$ so $x \in \xi_o'$. Thus $\xi_o' = X' \cap \xi_o$ and the proposition follows.

607/5/1-4

As usual let $\mathfrak{m}$ denote the centralizer of $\mathfrak{a}$ in $\mathfrak{k}$, and θ the Cartan involution of $\mathfrak{g}$ (corresponding to the decomposition $\mathfrak{g} = \mathfrak{k} + \mathfrak{p}$). In the present case G acts effectively on G/K so $[\mathfrak{p}, \mathfrak{p}] = \mathfrak{k}$ and we have a convenient description of $\mathfrak{m}$ by means of the restricted root subspaces.

LEMMA 1.2.

$$\mathfrak{m} + \mathfrak{a} = \sum_{\alpha \neq 0} [\mathfrak{g}_\alpha, \mathfrak{g}_{-\alpha}].$$

Proof. Since $\mathfrak{g}_o = \mathfrak{m} + \mathfrak{a}$ each $[\mathfrak{g}_\alpha, \mathfrak{g}_{-\alpha}]$ is contained in $\mathfrak{m} + \mathfrak{a}$. On the other hand, $\mathfrak{a} \subset \sum_{\alpha > 0} [\mathfrak{g}_\alpha, \mathfrak{g}_{-\alpha}]$ so if the lemma were not true there would exist a $T \neq 0$ in $\mathfrak{m}$ such that

$$\langle T, [\mathfrak{g}_\alpha, \mathfrak{g}_{-\alpha}]\rangle = 0 \qquad \text{for all} \quad \alpha > 0, \tag{3}$$

$\langle\ ,\ \rangle$ denoting as usual the Killing form of $\mathfrak{g}$. We imbed T in a maximal abelian subalgebra $\mathfrak{h}_k$ of $\mathfrak{m}$; then $\mathfrak{h} = \mathfrak{h}_k + \mathfrak{a}$ is a Cartan subalgebra of $\mathfrak{g}$. We consider the complexifications $\mathfrak{h}_c \subset \mathfrak{g}_c$ and the corresponding root space decomposition, $\mathfrak{g}_c = \sum_\beta (\mathfrak{g}_c)^\beta$. We order the roots in a way compatible with the ordering of the restricted roots. Let P be the set of positive roots, P_+ the set of positive roots which do not vanish identically on $\mathfrak{a}$. Then we have

$$\mathfrak{n} = \mathfrak{g} \cap \sum_{\beta \in P_+} (\mathfrak{g}_c)^\beta. \tag{4}$$

Let $\mathfrak{q}$ denote the orthogonal complement of $\mathfrak{a}$ in $\mathfrak{p}$; then the map $X \to X - \theta X$ is a bijection of $\mathfrak{n}$ onto $\mathfrak{q}$. Since $[\mathfrak{p}, \mathfrak{p}] = \mathfrak{k}$ (the orthogonal complement of $[\mathfrak{p}, \mathfrak{p}]$ in $\mathfrak{k}$ is an ideal in $\mathfrak{g}$) it is clear from (4) that there must exist a root $\beta \in P_+$ such that $\beta(T) \neq 0$ (otherwise T would lie in the center of $\mathfrak{g}$). Select $X_\beta \neq 0$ in $(\mathfrak{g}_c)^\beta$ and write $X_\beta = Y + iZ$, $(Y, Z \in \mathfrak{g})$. If $\alpha = \beta \mid \mathfrak{a}$ then $Y, Z \in \mathfrak{g}_\alpha$ and at least one of the vectors $[T, Y]$, $[T, Z]$ is $\neq 0$. Thus by (3) we get a nonzero vector in $\mathfrak{g}_\alpha$ which is perpendicular to $\mathfrak{g}_{-\alpha}$. This contradiction proves the lemma.

Although the coset space G/MN is not in general reductive we can nevertheless define *geodesics* in Ξ as follows. Let $t \to \gamma(t)$ $(t \in \boldsymbol{R})$ be an oriented geodesic in X which is *regular* (that is a geodesic in X whose stabilizer has minimum dimension). Pick any point x on the geodesic, and consider the isotropy subgroup K_x of G at x, its Lie algebra $\mathfrak{k}_x \subset \mathfrak{g}$, the orthogonal complement $\mathfrak{p}_x$ of $\mathfrak{k}_x$ in $\mathfrak{g}$ and consider the isomorphism $d\pi_x : \mathfrak{p}_x \to X_x$ induced by the projection $\pi_x : g \to g \cdot x$ of G onto X. Let $d_x \in \mathfrak{p}_x$ be the vector such that $(d\pi_x)_e(d_x)$ is the tangent vector to the geodesic at x. The Iwasawa group corresponding to *the* Weyl chamber (in the centralizer of d_x in $\mathfrak{p}_x$) which contains d_x is given by

$$N_x = \{z \in G \mid \lim_{t \to +\infty} \exp(-td_x)\, z \exp(td_x) = e\}.$$

Let us verify that N_x is independent of the choice of x on γ. So let y be another point on γ. To prove $N_x = N_y$ we may assume $x = o$, $d_x = d_o \in \mathfrak{a}^+$ so $N_x = N$. Then $y = \exp sd_o \cdot o$ ($s \in \boldsymbol{R}$), $K_y = \exp sd_o\, K \exp(-sd_0)$ and $\mathfrak{p}_y = \mathrm{Ad}(\exp sd_o)\mathfrak{p}$. In particular, $d_o \in \mathfrak{p}_y$. Secondly,

$$\pi_y(\exp td_o) = \exp(t + s)\, d_o \cdot o$$

so $(d\pi_y)_e(d_o) =$ tangent vector to γ at y. Since $d_o \in \mathfrak{p}_y$ the definition of d_y implies $d_o = d_y$, whence $N = N_y$. Thus we write N_γ instead of N_x. Putting

$$\Gamma_\gamma(t) = N_\gamma \cdot \gamma(t) \qquad (t \in \boldsymbol{R}) \tag{5}$$

we call the curve Γ_γ a *geodesic in* Ξ.

LEMMA 1.3. *The action of G on Ξ permutes the geodesics; more precisely* $g \cdot \Gamma_\gamma(t) = \Gamma_{g \cdot \gamma}(t)$ ($t \in \boldsymbol{R}, g \in G$).

Proof. We have for $g \in G$, $t \in \boldsymbol{R}$,

$$g \cdot \Gamma_\gamma(t) = gN_\gamma \cdot \gamma(t) = gN_\gamma g^{-1} \cdot (g \cdot \gamma)(t). \tag{6}$$

If $x_1 = g \cdot x$ then $K_{x_1} = gK_x g^{-1}$, $\mathfrak{k}_{x_1} = \mathrm{Ad}(g)\,\mathfrak{k}_x$, $\mathfrak{p}_{x_1} = Ad(g)\,\mathfrak{p}_x$ and $\pi_{x_1}(\exp t\, Ad(g)\, d_x) = g \exp td_x \cdot x$ so $(d\pi_{x_1})_e(Ad(g)\, d_x)$ equals the tangent vector to $g \cdot \gamma$ at x_1. Thus

$$N_{g \cdot \gamma} = gN_\gamma g^{-1}$$

which by (6) proves the lemma.

We also note the following result which is clear from the fact that each transvection of X along γ normalizes N_γ.

LEMMA 1.4. *Any two horocycles $\Gamma_\gamma(t_1)$ and $\Gamma_\gamma(t_2)$ on the curve Γ_γ correspond under a transvection of X along γ.*

A submanifold Ξ_1 of Ξ will be called *totally geodesic* if each geodesic in Ξ which is tangential to Ξ_1 at some point lies completely in Ξ_1.

We shall now take for the submanifold $X' \subset X$ a *boundary component* (cf., e.g. [21(j)], [25]) of the space X. We recall the construction. Let γ be a nonregular geodesic in X and let Z_γ denote the set of isometries in $I_0(X)$ which commute with the transvections along γ. The group Z_γ is reductive and if Z_γ' denotes the semisimple part of the identity component of Z_γ an orbit of a point o of γ under Z_γ' is called a boundary component of X. This orbit $X' = Z_\gamma' \cdot o$ is then a symmetric space of the noncompact type, totally geodesic in X. Let G' be constructed as before. Let ξ_γ be a horocycle in X through o such that the Weyl chamber giving the normal to ξ_γ contains the tangent vector to γ at o in its closure.

THEOREM 1.5. *Let X' be a boundary component of X, Ξ' and Ξ their respective duals and Ξ_γ the orbit $G' \cdot \xi_\gamma$ in Ξ. Then Ξ_γ is totally geodesic in Ξ and the map $\xi \to \xi \cap X'$ is a bijection of Ξ_γ onto Ξ'.*

Let K denote the isotropy subgroup of G at o and $\mathfrak{k}$ its Lie algebra, $\mathfrak{g} = \mathfrak{k} + \mathfrak{p}$ the Cartan decomposition. Then $\mathfrak{g}' = \mathfrak{k}' + \mathfrak{p}'$ where $\mathfrak{k}' = \mathfrak{k} \cap \mathfrak{g}'$, $\mathfrak{p}' = \mathfrak{g}' \cap \mathfrak{p}$. The orthogonal complement in $\mathfrak{p}$ of the tangent space $(\xi_\gamma)_o$ is a maximal abelian subspace $\mathfrak{a}$ of $\mathfrak{p}$ [22(c), Lemma 3.3] and we have $\gamma(t) = \exp tH_o \cdot o$ where $H_o \in \mathfrak{a}$. Let Σ denote the set of restricted roots of $\mathfrak{g}$ with respect to $\mathfrak{a}$ and let Σ^0 denote the set of $\alpha \in \Sigma$ such that $\alpha(H_o) = 0$. Then the Lie algebra of Z_γ is given by

$$\mathfrak{z}_\gamma = \mathfrak{m} + \mathfrak{a} + \sum_{\alpha \in \Sigma^0} \mathfrak{g}_\alpha \tag{7}$$

and $[\mathfrak{z}_\gamma, \mathfrak{z}_\gamma]$ is the Lie algebra of Z_γ'. Since $\mathfrak{p}' = [\mathfrak{z}_\gamma, \mathfrak{z}_\gamma] \cap \mathfrak{p}$ it is clear that the subspace

$$\sum_{\alpha \in \Sigma^0} \boldsymbol{R} H_\alpha$$

of $\mathfrak{a}$ is a maximal abelian subspace $\mathfrak{a}'$ of $\mathfrak{p}'$. Then $\mathfrak{a} = \mathfrak{a}' + \mathfrak{a}''$ (direct sum) where $\mathfrak{a}'' = \{H \in \mathfrak{a} \mid \alpha(H) = 0 \text{ for } \alpha \in \Sigma^0\}$. Thus $[\mathfrak{z}_\gamma, \mathfrak{a}''] = 0$ so in particular the centralizer $\mathfrak{m}'$ of $\mathfrak{a}'$ in $\mathfrak{k}'$ satisfies

$$\mathfrak{m}' = \mathfrak{m} \cap \mathfrak{k}'. \tag{8}$$

Similarly since each $Ad_{G'}(g)$ $(g \in G')$ is the identity on $\mathfrak{a}''$ we have

$$M' = M \cap K'. \tag{9}$$

Now let $\mathfrak{a}^+$ denote the Weyl chamber in $\mathfrak{a}$ which is normal to the horocycle ξ_γ, and let $\Sigma^+ \subset \Sigma$ be the corresponding set of positive roots. The restrictions $\bar{\alpha} = \alpha \mid \mathfrak{a}'$, as α runs through Σ^0, form the set Σ' of restricted roots of $\mathfrak{g}'$ with respect to $\mathfrak{a}'$. The subset

$$\{H \in \mathfrak{a}' \mid \alpha(H) > 0 \text{ for } \alpha \in \Sigma^0 \cap \Sigma^+\}$$

is a Weyl chamber in $\mathfrak{a}'$; let $(\Sigma')^+$ denote the corresponding set of positive roots, so

$$(\Sigma')^+ = \{\bar{\alpha} \mid \alpha \in \Sigma^0 \cap \Sigma^+\}.$$

Thus we have as in the beginning of this section

$$K' = K \cap G', \qquad A' = A \cap G', \qquad N' = N \cap G'. \tag{10}$$

Since $\xi_\gamma = N \cdot o$ (9) and (10) imply that the map $g' \cdot \xi_\gamma \to g'M'N'$ is a well-defined bijection of Ξ_γ onto Ξ'. Since $\xi_\gamma \cap X' = N' \cdot o$ this proves the last statement of the theorem. For the first statement we consider first the special case that Γ_δ is a geodesic in Ξ tangential to Ξ_γ at the point ξ_γ. By Lemma 1.4, there exists a $Z \in \mathfrak{p}$ such that $\Gamma_\delta(t) = \exp tZ \cdot \xi_\gamma$ for $t \in \boldsymbol{R}$. The tangent space to $\Xi = G/MN$ at ξ_γ is canonically identified with the factor space $\mathfrak{g}/(\mathfrak{m} + \mathfrak{n})$; the assumption about Γ_δ therefore means that there exists a vector $Y \in \mathfrak{g}'$ such that $Z - Y \in \mathfrak{m} + \mathfrak{n}$. Since $\mathfrak{g}' = \mathfrak{n}' + \mathfrak{m}' + \mathfrak{a}' + \theta\mathfrak{n}'$ we may (by (8)) assume $Y \in \mathfrak{a}' + \theta\mathfrak{n}'$. Let $\mathfrak{q}$ (resp. $\mathfrak{q}'$) denote the orthogonal complement of $\mathfrak{a}$ in $\mathfrak{p}$ (resp. of $\mathfrak{a}'$ in $\mathfrak{p}'$). Then the map $X \to X - \theta X$ is a bijection of $\mathfrak{n}$ onto $\mathfrak{q}$ mapping $\mathfrak{n}'$ onto $\mathfrak{q}'$. Let $X_1, ..., X_d$ be a basis of $\mathfrak{n}$ such that $X_1, ..., X_c$ span $\mathfrak{n}'$. Then for certain $H \in \mathfrak{a}$, $H' \in \mathfrak{a}'$ we have

$$Z = H + \sum_1^d c_i(X_i - \theta X_i)$$

and

$$Y = H' + \sum_1^c b_j \theta X_j, \qquad (c_i, b_j \in \boldsymbol{R}),$$

so the relation $Z - Y \in \mathfrak{m} + \mathfrak{n}$ implies $Z \in \mathfrak{p}'$ so Γ_δ lies in Ξ_γ.

Now Lemma 1.3 reduces the general case to the special one considered, so the theorem is proved.

We shall now for an arbitrary X construct a whole family of totally geodesic submanifolds $X' \subset X$ which are isometric to the Hermitian hyperbolic space of two complex dimensions. This will be particularly useful to us because it turns out that certain general computations in X can be done entirely within the spaces X'. As usual, $\boldsymbol{SU}(2, 1)$ denotes the group of complex 3×3 matrices of determinant 1 which leave invariant the form $z_1\bar{z}_1 + z_2\bar{z}_2 - z_3\bar{z}_3$.

THEOREM 1.6. *Let* $\mathfrak{g}$ *be an arbitrary noncompact semisimple Lie algebra,* θ *a Cartan involution of* $\mathfrak{g}$, $\mathfrak{g} = \mathfrak{k} + \mathfrak{p}$ *the corresponding Cartan decomposition. Let* $\mathfrak{a}$ *be a maximal abelian subspace of* $\mathfrak{p}$. *Assume* α *and* 2α *are restricted roots of* $\mathfrak{g}$ *with respect to* $\mathfrak{a}$. *Select* $X_\alpha \in \mathfrak{g}_\alpha$, $X_{2\alpha} \in \mathfrak{g}_{2\alpha}$ *both* $\neq 0$. *Then the subalgebra* $\mathfrak{g}^*$ *of* $\mathfrak{g}$ *generated by* X_α, $X_{2\alpha}$, θX_α, $\theta X_{2\alpha}$ *is isomorphic to* $\mathfrak{su}(2, 1)$, *the Lie algebra of* $SU(2, 1)$.

The proof requires a few easy lemmas. We put $Y_\alpha = [\theta X_\alpha, X_{2\alpha}]$, and use now B for the Killing form.

LEMMA 1.7. $[X_\alpha, Y_\alpha] = cX_{2\alpha}$ *where* $c = 2\alpha(H_\alpha)\, B(X_\alpha, \theta X_\alpha)$.

Proof. By Jacobi identity

$$[X_\alpha, [\theta X_\alpha, X_{2\alpha}]] + [\theta X_\alpha, [X_{2\alpha}, X_\alpha]] + [X_{2\alpha}, [X_\alpha, \theta X_\alpha]] = 0.$$

The middle term is 0 so the lemma follows from

$$[X_\alpha, \theta X_\alpha] = B(X_\alpha, \theta X_\alpha)\, H_\alpha \tag{11}$$

LEMMA 1.8. *The vector* $[X_\alpha, \theta Y_\alpha]$ *is* $\neq 0$ *and lies in* $\mathfrak{m}$.

Proof. We have $[X_\alpha, \theta Y_\alpha] \in \mathfrak{g}_0 = \mathfrak{m} + \mathfrak{a}$ and if $H \in \mathfrak{a}$,

$$B(H, [X_\alpha, \theta Y_\alpha]) = B([H, X_\alpha], \theta Y_\alpha) = \alpha(H)\, B(X_\alpha, [X_\alpha, \theta X_{2\alpha}]) = 0.$$

Thus $[X_\alpha, \theta Y_\alpha] \in \mathfrak{m}$. To see that this vector is $\neq 0$ we consider the vector $[[X_\alpha, \theta Y_\alpha], X_\alpha]$. By the first part of the proof it equals

$$\begin{aligned}[[\theta X_\alpha, Y_\alpha], X_\alpha] &= -[[Y_\alpha, X_\alpha], \theta X_\alpha] - [[X_\alpha, \theta X_\alpha], Y_\alpha] \\ &= c[X_{2\alpha}, \theta X_\alpha] - B(X_\alpha, \theta X_\alpha)\, \alpha(H_\alpha)\, Y_\alpha\end{aligned}$$

so by Lemma 1.7,

$$[[X_\alpha, \theta Y_\alpha], X_\alpha] = -3\alpha(H_\alpha)\, B(X_\alpha, \theta X_\alpha)\, Y_\alpha \neq 0. \tag{12}$$

LEMMA 1.9. *The linear transformation* $ad_{\mathfrak{g}}([X_\alpha, \theta Y_\alpha])$ *leaves the plane* $\boldsymbol{R}X_\alpha + \boldsymbol{R}Y_\alpha$ *invariant.*

Proof. We have

$$\begin{aligned}[[X_\alpha, \theta Y_\alpha], Y_\alpha] &= -[[\theta Y_\alpha, Y_\alpha], X_\alpha] - [[Y_\alpha, X_\alpha], \theta Y_\alpha] \\ &= +B(Y_\alpha, \theta Y_\alpha)\,\alpha(H_\alpha)\,X_\alpha + c[X_{2\alpha}, [X_\alpha, \theta X_{2\alpha}]\end{aligned}$$

which by the Jacobi identity equals

$$= +B(Y_\alpha, \theta Y_\alpha)\,\alpha(H_\alpha)\,X_\alpha - c\alpha(H_{2\alpha})\,B(X_{2\alpha}, \theta X_{2\alpha})\,X_\alpha,$$

which together with (12) proves the lemma.

LEMMA 1.10. $[[X_\alpha, \theta Y_\alpha], X_{2\alpha}] = 0$.

Proof. Putting $T = [X_\alpha, \theta Y_\alpha]$ we have seen above that $[T, X_\alpha] \subset \boldsymbol{R} Y_\alpha$, $[T, Y_\alpha] \subset \boldsymbol{R} X_\alpha$ so $[T, [X_\alpha, Y_\alpha]] = 0$. Now use Lemma 1.7.

LEMMA 1.11. $[Y_\alpha, \theta X_{2\alpha}] \subset \boldsymbol{R}\theta X_\alpha$.

Proof. We have by (11)

$$\begin{aligned}[Y_\alpha, \theta X_{2\alpha}] &= [[\theta X_\alpha, X_{2\alpha}], \theta X_{2\alpha}] \\ &= -[[X_{2\alpha}, \theta X_{2\alpha}], \theta X_\alpha] - [\theta X_{2\alpha}, \theta X_\alpha], X_{2\alpha}] \in \boldsymbol{R}\theta X_\alpha .\end{aligned}$$

These lemmas show that

$$\begin{aligned}\mathfrak{g}^* = (\boldsymbol{R}[X_\alpha, \theta Y_\alpha] + \boldsymbol{R}H_\alpha) + (\boldsymbol{R}X_\alpha + \boldsymbol{R}Y_\alpha) + \boldsymbol{R}X_{2\alpha} \\ + (\boldsymbol{R}\theta X_\alpha + \boldsymbol{R}\theta Y_\alpha) + \boldsymbol{R}\theta X_{2\alpha} .\end{aligned}$$

Since $\mathfrak{g}^*$ is a θ-invariant subalgebra of $\mathfrak{g}$ it is reductive. The first parenthesis above is the centralizer of H_α in $\mathfrak{g}^*$ so (12) shows that $\mathfrak{g}^*$ has center $\{0\}$, so $\mathfrak{g}^*$ is semisimple; the decomposition above is the restricted root space decomposition.

Now it will be convenient to give all general Lie algebra concepts connected with $\mathfrak{g}^*$ the superscript *. (Dual spaces will not be considered so no confusion with our customary use of * should result.)

Thus $H_\alpha^* \in \mathfrak{a}^*$ is defined by

$$B^*(H, H_\alpha^*) = \alpha(H), \qquad H \in \mathfrak{a}^*.$$

But the formula for $\mathfrak{g}^*$ shows that

$$B^*(H, H) = \mathrm{Tr}(ad_{\mathfrak{g}^*}(H)\,ad_{\mathfrak{g}^*}(H)) = 4\alpha(H)^2 + 2(2\alpha(H))^2 = 12\alpha(H)^2,$$

so

$$\alpha(H_\alpha{}^*) = \frac{1}{12}. \tag{13}$$

Put

$$X_\alpha{}^* = c_\alpha X_\alpha, \qquad X_{2\alpha}^* = c_{2\alpha} X_{2\alpha}$$

the constants c_α and $c_{2\alpha}$ being determined such that

$$-B^*(X_\alpha{}^*, \theta X_\alpha{}^*) = -B^*(X_{2\alpha}^*, \theta X_{2\alpha}^*) = 24, \tag{14}$$

and finally put

$$Y_\alpha{}^* = [\theta X_\alpha{}^*, X_{2\alpha}^*].$$

Then the lemmas above hold also for the starred vectors in $\mathfrak{g}^*$. Combining with (13) and (14) we derive the relations

$$[X_\alpha{}^*, Y_\alpha{}^*] = -4X_{2\alpha}^*, \qquad [X_\alpha{}^*, [X_\alpha{}^*, \theta Y_\alpha{}^*]] = -6Y_\alpha{}^*,$$
$$[Y_\alpha{}^*, \theta X_{2\alpha}^*] = -4\theta X_\alpha{}^*, \qquad [Y_\alpha{}^*, \theta Y_\alpha{}^*] = -96H_\alpha{}^*,$$
$$[[X_\alpha{}^*, \theta Y_\alpha{}^*], Y_\alpha{}^*] = -24X_\alpha{}^*.$$

Now we consider the Lie algebra $\mathfrak{su}(2, 1)$ and add the subscript 0 to all general Lie algebra concepts connected with it. The Killing form is

$$B_0(X, Y) = 6\,\mathrm{Tr}(XY)$$

and the automorphism $X \to I_{2,1} X I_{2,1}$ of $\mathfrak{su}(2, 1)$ where

$$I_{2,1} = \begin{vmatrix} -1 & 0 & 0 \\ 0 & -1 & 0 \\ 0 & 0 & 1 \end{vmatrix}$$

is a Cartan involution θ of $\mathfrak{su}(2, 1)$ [22(b), p. 348]. Moreover, $\mathfrak{k}_0$ is the set of matrices

$$\left|\begin{array}{c|c} A & \begin{matrix} 0 \\ 0 \end{matrix} \\ \hline 0 \quad 0 & -\mathrm{Tr}(A) \end{array}\right|, \qquad A \text{ skew-hermitian},$$

and $\mathfrak{p}_0$ is the set of matrices

$$\left|\begin{array}{cc|c} 0 & 0 & z_1 \\ 0 & 0 & z_2 \\ \hline \bar{z}_1 & \bar{z}_2 & 0 \end{array}\right|, \qquad z_1, z_2 \in \boldsymbol{C}.$$

For $\mathfrak{a}_0$ and $\mathfrak{a}_0^+$ we can take $\mathfrak{a}_0 = \boldsymbol{R}H_0$, $\mathfrak{a}_0^+ = \{tH_0 \mid t > 0\}$ where

$$H_0 = \begin{vmatrix} 0 & 0 & 1 \\ 0 & 0 & 0 \\ 1 & 0 & 0 \end{vmatrix}.$$

Now K_0 and M_0 are defined by the convention above. The restricted roots are α_0, $2\alpha_0$, $-\alpha_0$, $-2\alpha_0$ where

$$\alpha_0(H_0) = 1, \qquad \text{so} \quad H_{\alpha_0} = H_0/12.$$

Also $\rho_0 = 2\alpha_0$ and $\mathfrak{m}_0 = \boldsymbol{R}T_0$ where

$$T_0 = \begin{vmatrix} i & 0 & 0 \\ 0 & -2i & 0 \\ 0 & 0 & i \end{vmatrix}.$$

The root spaces $\mathfrak{g}_{\alpha_0}$ and $\mathfrak{g}_{-\alpha_0}$ are given by the set of matrices

$$\begin{vmatrix} 0 & z & 0 \\ -\bar{z} & 0 & \bar{z} \\ 0 & z & 0 \end{vmatrix} \quad \text{for } \mathfrak{g}_{\alpha_0}, \qquad \begin{vmatrix} 0 & z & 0 \\ -\bar{z} & 0 & -\bar{z} \\ 0 & -z & 0 \end{vmatrix} \quad \text{for } \mathfrak{g}_{-\alpha_0},$$

where $z \in \boldsymbol{C}$. The root spaces $\mathfrak{g}_{2\alpha_0}$ and $\mathfrak{g}_{-2\alpha_0}$ are given by the set of matrices

$$\begin{vmatrix} it & 0 & -it \\ 0 & 0 & 0 \\ it & 0 & -it \end{vmatrix} \quad \text{for } \mathfrak{g}_{2\alpha_0}, \qquad \begin{vmatrix} it & 0 & it \\ 0 & 0 & 0 \\ -it & 0 & -it \end{vmatrix} \quad \text{for } \mathfrak{g}_{-2\alpha_0}$$

where $t \in \boldsymbol{R}$. Consider now the linear mapping τ of $\mathfrak{g}^*$ into $\mathfrak{su}(2, 1)$ given by

$$X_\alpha^* \to \begin{vmatrix} 0 & 1 & 0 \\ -1 & 0 & 1 \\ 0 & 1 & 0 \end{vmatrix}, \qquad X_{2\alpha}^* \to \begin{vmatrix} i & 0 & -i \\ 0 & 0 & 0 \\ i & 0 & -i \end{vmatrix},$$

$$\theta X_\alpha^* \to \begin{vmatrix} 0 & 1 & 0 \\ -1 & 0 & -1 \\ 0 & -1 & 0 \end{vmatrix}, \qquad \theta X_{2\alpha}^* \to \begin{vmatrix} i & 0 & i \\ 0 & 0 & 0 \\ -i & 0 & -i \end{vmatrix},$$

$$Y_\alpha^* \to -2\begin{vmatrix} 0 & i & 0 \\ i & 0 & -i \\ 0 & i & 0 \end{vmatrix}, \qquad \theta Y_\alpha^* \to -2\begin{vmatrix} 0 & i & 0 \\ i & 0 & i \\ 0 & -i & 0 \end{vmatrix},$$

$$[X_\alpha^*, \theta Y_\alpha^*] \to -4\begin{vmatrix} i & 0 & 0 \\ 0 & -2i & 0 \\ 0 & 0 & i \end{vmatrix}, \qquad H_\alpha^* \to \frac{1}{12}\begin{vmatrix} 0 & 0 & 1 \\ 0 & 0 & 0 \\ 1 & 0 & 0 \end{vmatrix}.$$

The commutation relations in $\mathfrak{g}^*$ show quickly that τ is an isomorphism whereby Theorem 1.6 is proved.

Preserving the notation above let G be a connected Lie group with finite center and Lie algebra $\mathfrak{g}$. We order the restricted roots (of $\mathfrak{g}$ with respect to $\mathfrak{a}$) in some way such that $\alpha > 0$. Let G^*, K, K^*, N, N^*, A, A^*, respectively, denote the analytic subgroups of G which correspond to the subalgebras $\mathfrak{g}^*$, $\mathfrak{k}$, $\mathfrak{k}^*$, $\mathfrak{n} = \sum_{\lambda>0} \mathfrak{g}_\lambda$, $\mathfrak{n}^* = \boldsymbol{R}X_\alpha + \boldsymbol{R}Y_\alpha + \boldsymbol{R}X_{2\alpha}$, $\mathfrak{a}$ and $\mathfrak{a}^* = \boldsymbol{R}H_\alpha$. As usual let $\mathfrak{m}$(resp. M) denote the centralizer of $\mathfrak{a}$ in $\mathfrak{k}$ (resp. $\mathfrak{a}$ in K) and define $\mathfrak{m}^*$ and M^* similarly.

LEMMA 1.12. *$K^* = G^* \cap K$, $N^* = G^* \cap N$ and $M^* = G^* \cap M$.*

Proof. The first two statements are contained in the proof of Proposition 1.1. Since $\mathfrak{m}^* = \boldsymbol{R}[X_\alpha, \theta Y_\alpha]$ it is clear that $\mathfrak{m}^* = \mathfrak{g}^* \cap \mathfrak{m}$,and of course $G^* \cap M \subset M^*$. On the other hand, the symmetric space G^*/K^* has rank one so K^*/M^* is a 3-sphere hence simply connected whence M^* is connected. Thus $M^* \subset G^* \cap M$ so the lemma is proved.

Repeating the proof of Theorem 1.5 we obtain

COROLLARY 1.13. *Let $X = G/K$, $\Xi = G/MN$, $o = \{K\}$, $\xi_0 = \{MN\}$. The Hermitian hyperbolic space $X^* = G^*/K^*$ and its dual space $\Xi^* = G^*/M^*N^*$ are totally geodesic in X and Ξ, respectively, the imbeddings given as the orbits $G^* \cdot o \subset X$, and $G^* \cdot \xi_o \subset \Xi$.*

We recall that m_α denotes the multiplicity of a restricted root α and ρ denotes half of the sum of the positive roots with multiplicity. Putting $\bar{N} = \theta N$ the Bruhat decomposition of G shows that $\bar{N}MAN$ is a dense subset of G. Moreover, if $g \in \bar{N}MAN$ then there exists a unique element $\bar{n}(g) \in \bar{N}$ such that $g \in \bar{n}(g)\, MAN$ [22(c), p. 674]. Select $m^* \in K$ such that $Ad_G(m^*)$ interchanges the Weyl chambers $\mathfrak{a}^+$ and $-\mathfrak{a}^+$.

THEOREM 1.14. *Let G/K be a symmetric space of rank one and put $|Z|^2 = -B(Z, \theta Z)$ for $Z \in \mathfrak{g}$.*

Then if $\bar{n} = \exp X \exp Y$ $(X \in \mathfrak{g}_{-\alpha}, Y \in \mathfrak{g}_{-2\alpha})$ *we have*

$$e^{\rho(H(\bar{n}))} = [(1 + c \mid X \mid^2)^2 + 4c \mid Y \mid^2]^{\frac{1}{4}(m_\alpha + 2m_{2\alpha})}$$

$$e^{\rho(H(\bar{n}(m^*\bar{n})) - H(\bar{n}))} = [c^2 \mid X \mid^4 + 4c \mid Y \mid^2]^{-\frac{1}{4}(m_\alpha + 2m_{2\alpha})}$$

The constant c *is given by* $c^{-1} = 4(m_\alpha + 4m_{2\alpha})$.

We use Theorem 1.6 to reduce Theorem 1.14 to the case $\boldsymbol{SU}(2, 1)$ for which explicit computations are easy. We make use of the notation introduced during the proof of Theorem 1.6.

LEMMA 1.15. *Let* $X = G/K$ *be a symmetric space of rank one,* $G = I_0(X)$. *Then an element* $m^* \in K$, *for which* $Ad_G(m^*)$ *induces the nontrivial Weyl group element, can be chosen such that* $(m^*)^2 = e$.

Proof. For the group $\boldsymbol{SU}(2, 1)$ we can take for m^* the element

$$\begin{vmatrix} -1 & 0 & 0 \\ 0 & -1 & 0 \\ 0 & 0 & 1 \end{vmatrix}. \tag{15}$$

Consider the universal covering group $\boldsymbol{SU}(2, 1)^\sim$, and $\tilde{K}_0$ the analytic subgroup with Lie algebra $\mathfrak{k}_0$ and $\tilde{M}_0$ the centralizer of $\mathfrak{a}_0$ in $\tilde{K}_0$. Again $\tilde{K}_0/\tilde{M}_0$ is a 3-sphere so $\tilde{M}_0$ is connected; since $\boldsymbol{SU}(2, 1)^\sim$ has infinite center, $\tilde{K}_0$ is non compact and so is $\tilde{M}_0$; hence $\tilde{M}_0$ is isomorphic to $\boldsymbol{R}$. Select any $\tilde{m}^* \in \tilde{K}_0$ which induces the non trivial Weyl group element on $\mathfrak{a}_0$. The conjugation by $\tilde{m}^*$ induces an automorphism of $\tilde{M}_0$; thus there exists a $c \in \boldsymbol{R}$ such that

$$\tilde{m}^* m (\tilde{m}^*)^{-1} = m^c \qquad \text{for} \quad m \in \tilde{M}_0 .$$

Here we write m^c for $\exp(cX)$ if $m = \exp X$ where $X \in \mathfrak{m}_0$. This c is independent of the choice of $\tilde{m}^*$.

Now $(\tilde{m}^*)^2$ is an element in $\tilde{M}_0$, say m_1. For $t \in \boldsymbol{R}$

$$(\tilde{m}^* m_1{}^t)^2 = (m_1{}^t)^c \, (\tilde{m}^*)^2 \, m_1{}^t = (m_1)^{ct+t+1}.$$

If c were -1 then the conjugation by the element (15) would induce the automorphism $m \rightarrow m^{-1}$ of M_0 which is impossible; in fact (15) commutes elementwise with M_0. Thus $c \neq -1$ so putting $t = -(c + 1)^{-1}$ we get $(\tilde{m}^* m_1{}^t)^2 = e$.

Since the subgroup $G^* \subset G$ is covered by $\boldsymbol{SU}(2, 1)^{\sim}$ with $\tilde{K}_0$ being mapped onto K^* by the covering map, we can take for m^* the image of $\tilde{m}^* m_1{}^t$ under this map. This proves the lemma in the case when X is not a hyperbolic space; but for this case we see from the model used in the proof of Theorem 4.10 below that m^* can be chosen of square e.

Definition. The element m^* being as in Lemma 1.15 we call the map $\bar{n} \to \bar{n}(m^*\bar{n})$ the *inversion* of $\bar{N}$.

It is clear that this is an involutive mapping of $\bar{N} - \{e\}$ onto itself; for the hyperbolic spaces it reduces to the ordinary inversion. Note that under the bijection $\bar{n} \to k(\bar{n})M$ of $\bar{N}$ onto $K/M - \{m^*M\}$ the inversion corresponds to the map $kM \to m^* \, k \, M$ of K/M onto itself which interchanges the points M and m^*M. Now we need the explicit formula for this inversion in the case of a Hermitian hyperbolic space.

Lemma 1.16. *Let $G = \boldsymbol{SU}(2, 1)$ and m^* the element* (15). *Then the inversion $\bar{n} \to \bar{n}(m^*\bar{n})$ of $\bar{N}$ is given as follows. Let for $\bar{n} \neq e$,*

$$m^*\bar{n} = \bar{n}(m^*\bar{n})\, man', \qquad m \in M, \qquad a \in A, \qquad n' \in N. \tag{16}$$

If (in accordance with the formulas for $\mathfrak{g}_{-\alpha_0}$ and $\mathfrak{g}_{-2\alpha_0}$)

$$\bar{n} = \exp \begin{vmatrix} it & z & it \\ -\bar{z} & 0 & -\bar{z} \\ -it & -z & -it \end{vmatrix}, \qquad \bar{n}(m^*\bar{n}) = \exp \begin{vmatrix} iu & w & iu \\ -\bar{w} & 0 & -\bar{w} \\ -iu & -w & -iu \end{vmatrix}$$

then

$$w = -z/(|z|^2 + 2it) \qquad \text{and} \qquad u = -t/(|z|^4 + 4t^2). \tag{17}$$

Also

$$e^{\rho_0(\log a)} = |z|^4 + 4t^2 \qquad \text{and} \qquad e^{\rho_0(H(\bar{n}))} = (1 + |z|^2)^2 + 4t^2.$$

Proof. We write down the equation (16) where

$$m = \begin{vmatrix} e^{i\alpha} & 0 & 0 \\ 0 & e^{-2i\alpha} & 0 \\ 0 & 0 & e^{i\alpha} \end{vmatrix}, \qquad a = \begin{vmatrix} \cosh r & 0 & \sinh r \\ 0 & 1 & 0 \\ \sinh r & 0 & \cosh r \end{vmatrix},$$

and

$$n' = \exp \begin{vmatrix} iu' & w' & -iu' \\ -\bar{w}' & 0 & \bar{w}' \\ iu' & w' & -iu' \end{vmatrix}.$$

Then (16) represents 9 equations in the variables u, w, α, r, u', w' which can be solved in a routine fashion. We indicate the steps. Let the matrix (16) be denoted by (a_{ij}). The we find

$$\tfrac{1}{2}(a_{21} + a_{23}) = \bar{z} = -\bar{w}e^{i\alpha}e^{r}, \tag{18}$$

$$\tfrac{1}{2}(a_{21} - a_{23}) = 0 = -\bar{w}'e^{-2i\alpha} + \bar{w}e^{i\alpha}e^{r}(|\, w' \,|^2 - 2iu'), \tag{19}$$

$$\tfrac{1}{2}(a_{12} + a_{32}) = -z = w'e^{i\alpha}e^{r}, \tag{20}$$

$$\tfrac{1}{2}(a_{12} - a_{32}) = 0 = we^{-2i\alpha} - w'e^{i\alpha}e^{r}(|\, w \,|^2 - 2iu). \tag{21}$$

From (18) and (20) we derive

$$\bar{w}w'e^{2i\alpha} = |\, z \,|^2 e^{-2r} = w\bar{w}'e^{-2i\alpha}, \tag{22}$$

$$|\, w \,|^2 = |\, w' \,|^2 = |\, z \,|^2\, e^{-2r}, \tag{23}$$

and invoking also (21),

$$e^{i\alpha}e^{r} = (|\, w \,|^2 - 2iu)^{-1}. \tag{24}$$

But (17) can also be written

$$m^*\bar{n}(m^*\bar{n}) = \bar{n}m^{-1}a^{-1}n'', \qquad (n'' \in N), \tag{25}$$

so (18) and (24) give by symmetry

$$e^{-i\alpha}e^{-r} = (|\, z \,|^2 - 2it)^{-1}, \qquad \bar{w} = -\bar{z}e^{-i\alpha}e^{-r} \tag{26}$$

Now (24) and (26) give the formulas for the inversion on $\bar{N}$. Secondly, $\rho_0 = \tfrac{1}{2}(2\alpha_0 + 2\alpha_0)$ so

$$e^{\rho_0(\log a)} = e^{2r} = |\, z \,|^4 + 4t^2.$$

Finally, suppose $\bar{n} = k \exp H(\bar{n})u$ where $k \in K$, $u \in N$. Suppose $s \in \boldsymbol{R}$ such that $\alpha_0(H(\bar{n})) = s$ and write

$$u = \exp \begin{vmatrix} i\tau & \zeta & -i\tau \\ -\bar{\zeta} & 0 & \bar{\zeta} \\ i\tau & \zeta & -i\tau \end{vmatrix}, \qquad \tau \in \boldsymbol{R}, \quad \zeta \in \boldsymbol{C}.$$

Then we have

$$\theta(\bar{n})^{-1}\bar{n} = \theta u^{-1} \exp 2H(\bar{n})u \tag{27}$$

and again we substitute matrices on both sides. Denoting the matrix $\theta(\bar{n})^{-1}\bar{n}$ by (b_{ij}) we get

$$b_{22} = 1 + 2 \mid z \mid^2 = 1 + 2 \mid \zeta \mid^2 e^{2s}, \tag{28}$$

$$b_{21} = 2it\bar{z} - \bar{z} \mid z \mid^2 = -\bar{\zeta} + \bar{\zeta} e^{2s}(1 + 2i\tau - \mid \zeta \mid^2), \tag{29}$$

$$b_{32} = -2z - 2itz - z \mid z \mid^2 = \zeta + \zeta e^{2s}(1 + 2i\tau + \mid \zeta \mid^2). \tag{30}$$

We conjugate (30) and add to (29); this gives

$$-\bar{z}(1 - 2it + \mid z \mid^2) = \bar{\zeta} e^{2s}. \tag{31}$$

Taking the modulus and using (28) we get

$$e^{2s} = (1 + \mid z \mid^2)^2 + 4t^2,$$

which finishes the proof of the lemma.

We turn now to the proof of Theorem 1.14. Assume first both $\mathfrak{g}_{-\alpha}$ and $\mathfrak{g}_{-2\alpha} \neq \{0\}$; take $X = \theta X_\alpha$, $Y = \theta X_{2\alpha}$ in Theorem 1.6 and construct $\mathfrak{g}^*$ etc. In view of Lemma 1.12, G and G^* have compatible Iwasawa decompositions as well as compatible Bruhat decompositions.

For the Killing forms we have

$$B(H, H) = \mathrm{Tr}(\mathrm{ad}\, H \,\mathrm{ad}\, H) = 2m_\alpha \alpha(H)^2 + 2m_{2\alpha}(2\alpha(H))^2$$

so

$$B(H, H) = 2(m_\alpha + 4m_{2\alpha})(\alpha(H))^2,$$

whereas

$$B^*(H, H) = 12\alpha(H)^2$$

so

$$H_\alpha^* = \tfrac{1}{6}(m_\alpha + 4m_{2\alpha})\, H_\alpha\,,$$

whence by (11)

$$B(X_\alpha\,, \theta X_\alpha) = \tfrac{1}{6}(m_\alpha + 4m_{2\alpha})\, B^*(X_\alpha\,, \theta X_\alpha) \tag{32}$$

and

$$B(X_{2\alpha}\,, \theta X_{2\alpha}) = \tfrac{1}{6}(m_\alpha + 4m_{2\alpha})\, B^*(X_{2\alpha}\,, \theta X_{2\alpha}). \tag{33}$$

Now since $\mathfrak{g}_{-\alpha}$ and $\mathfrak{g}_{-2\alpha}$ commute elementwise, $\bar{n} = \exp X \exp Y = \exp(X + Y)$ and by the compatibility of the Iwasawa decompositions $H(\bar{n})$ is the same whether computed in $\mathfrak{g}$ or in $\mathfrak{g}^*$. We have

$$e^{\rho(H(\bar{n}))} = (e^{\rho^*(H(\bar{n}))})^{\frac{1}{4}(m_\alpha+2m_{2\alpha})}. \tag{34}$$

Now under the isomorphism $\tau : \mathfrak{g}^* \to \mathfrak{su}(2, 1)$ let $X + Y$ correspond to

$$\begin{vmatrix} it & z & it \\ -\bar{z} & 0 & -\bar{z} \\ -it & -z & -it \end{vmatrix}.$$

Then z is real and

$$B^*(X, \theta X) = |z|^2 B^*(\theta X_\alpha{}^*, X_\alpha{}^*) = -24 |z|^2$$

which together with (32) gives

$$|X|^2 = 4(m_\alpha + 4m_{2\alpha}) |z|^2 \tag{35}$$

and

$$B^*(Y, \theta Y) = t^2 B^*(\theta X_{2\alpha}^*, X_{2\alpha}^*) = -24t^2,$$

which by (33) gives

$$|Y|^2 = 4(m_\alpha + 4m_{2\alpha})t^2. \tag{36}$$

But by Lemma 1.16

$$e^{\rho_*(H(\bar{n}))} = (1 + |z|^2)^2 + 4t^2$$

and now the desired formula for $e^{\rho(H(\bar{n}))}$ follows from (34), (35), and (36).

For the second formula in Theorem 1.14 we note first that $H(\bar{n}(m^*\bar{n}))$ is independent of the choice of m^*. Thus we choose m^* in the group K^*. Then the decomposition within G^*,

$$m^*\bar{n} = \bar{n}(m^*\bar{n})\, man' \tag{37}$$

can by Lemma 1.12 also be viewed as an analogous decomposition in G. Now (37) implies that

$$H(\bar{n}(m^*\bar{n})) + \log a = H(m^*\bar{n}) = H(\bar{n})$$

and now the last formula of Theorem 1.14 follows from the formula for $e^{\rho_0(\log a)}$ in Lemma 1.16.

Finally if $\mathfrak{g}_{2\alpha} = \{0\}$ then G/K is a hyperbolic space. In this case Theorem 1.14 is much easier, either by direct computation for $G = \boldsymbol{SO}_0(n, 1)$ or by reduction to the group $\boldsymbol{SO}_0(2, 1)$ (whose Lie algebra is generated by X and θX).

Remarks. Theorem 1.14 which is Lemma 3.4 in [22(k)] was found independently (by classification) by Schiffmann whose thesis contains an a priori proof found later by Godement. With the knowledge of the first formula of Theorem 1.14, Knapp and Stein [26(c)] subsequently proved the latter one. For $G = \boldsymbol{SL}(n, \boldsymbol{R})$ and $G = \boldsymbol{Sp}(n, \boldsymbol{R})$ explicit formulas for $H(\bar{n})$ itself were given by Bhanu-Murthy [3(a), (b)].

As a consequence of Theorem 1.14 we obtain a fairly direct evaluation of Harish-Chandra's function

$$\boldsymbol{c}(\lambda) = \int_{\bar{N}} e^{-(i\lambda+\rho)(H(\bar{n}))}\, d\bar{n}$$

in the rank one case. Recall that $d\bar{n}$ has been normalized by $\boldsymbol{c}(-i\rho) = 1$.

COROLLARY 1.17. (*X of rank one*).

$$\boldsymbol{c}(\lambda) = c_0 \frac{2^{-\langle i\lambda, \alpha_0\rangle}\Gamma\langle i\lambda, \alpha_0\rangle}{\Gamma(\frac{1}{2}(\frac{1}{2}m_\alpha + 1 + \langle i\lambda, \alpha_0\rangle))\, \Gamma(\frac{1}{2}(\frac{1}{2}m_\alpha + m_{2\alpha} + \langle i\lambda, \alpha_0\rangle))} \tag{38}$$

where $\alpha_0 = \alpha/\langle\alpha, \alpha\rangle$ *and*

$$c_0 = \Gamma(\tfrac{1}{2}(m_\alpha + m_{2\alpha} + 1))\, 2^{\frac{1}{2}m_\alpha + m_{2\alpha}}$$

Proof. For $\lambda \in \mathfrak{a}_c^*$ let $z \in \boldsymbol{C}$ be determined by $\lambda = z\rho$. Then if dX and dY denote measures on $\mathfrak{g}_{-\alpha}$ and $\mathfrak{g}_{-2\alpha}$, respectively (corresponding to the norm $|\ |$), we have for $\operatorname{Im}(z) < 0$, $P = \frac{1}{4}(m_\alpha + 2m_{2\alpha})$, $c_1 =$ constant,

$$\begin{aligned}\boldsymbol{c}(\lambda) &= c_1 \int_{\mathfrak{g}_{-\alpha}} \int_{\mathfrak{g}_{-2\alpha}} [(1 + c\,|\,X\,|^2)^2 + 4c\,|\,Y\,|^2]^{-P(iz+1)}\, dX\, dY \\ &= c_2 \int_0^\infty \int_0^\infty [(1 + r)^2 + s]^{-P(iz+1)}\, r^{\frac{1}{2}m_\alpha - 1} s^{\frac{1}{2}m_{2\alpha}-1}\, dr\, ds,\end{aligned}$$

where c_2 is a constant. Here we first evaluate the integral over s by introducing the variable t by $s = (1 + r)^2 t$. Then we get

$$\boldsymbol{c}(\lambda) = c_2 \int_0^\infty r^{\frac{1}{2}m_\alpha - 1}(1 + r)^{-2P(iz+1)+m_{2\alpha}}\, dr \int_0^\infty t^{\frac{1}{2}m_{2\alpha}-1}(1 + t)^{-P(iz+1)} dt.$$

Using the formula (see, e.g., Erdèlyi [10], p. 9)

$$\int_0^\infty v^{x-1}(1 + v)^{-x-y}\, dv = \frac{\Gamma(x)\,\Gamma(y)}{\Gamma(x + y)}$$

we obtain (with a new constant c_3)

$$c(\lambda) = c_3 \frac{\Gamma(\frac{1}{2}(\langle i\lambda + \rho, \alpha_0\rangle - m_{2\alpha}))\, \Gamma(\langle i\lambda, \alpha_0\rangle)}{\Gamma(\frac{1}{2}\langle i\lambda + \rho, \alpha_0\rangle)\, \Gamma(\langle i\lambda + \rho, \alpha_0\rangle - m_{2\alpha})},$$

so by the duplication formula

$$\Gamma(2z) = 2^{2z-1}\pi^{-\frac{1}{2}}\Gamma(z)\, \Gamma(z + \tfrac{1}{2})$$

and the fact that $c(-i\rho) = 1$, Corollary 1.17 follows.

Remark. Formula (38) had be derived quite differently by Harish-Chandra (with the powers of 2 missing) in [21(i), p. 303] by means of expressing the spherical function φ_λ in terms of the hypergeometric function.

Another consequence of Theorem 1.10 is the following

COROLLARY 1.18. *The Poisson kernel on $X \times B$*

$$P(gk, kM) = e^{-2\rho(H(g^{-1}k))}, \qquad (g \in G,\, k \in K),$$

is in the rank one case given by

$$P(a_tK, k(\bar{n})M) = \left[\frac{e^{2t}((1 + c\mid X\mid^2)^2 + 4c\mid Y\mid^2)}{(1 + c\mid e^tX\mid^2)^2 + 4c\mid e^{2t}Y\mid^2}\right]^{\langle \rho, \alpha_0\rangle}$$

if $a_t = \exp tH$, $\alpha(H) = 1$, $\bar{n} = \exp X \exp Y (X \in \mathfrak{g}_{-\alpha}, Y \in \mathfrak{g}_{-2\alpha})$ *and* $\alpha_0 = \alpha/\langle\alpha, \alpha\rangle$.

In fact, if $a \in A$, $\bar{n} \in \bar{N}$ then

$$P(ak, k(\bar{n})M) = e^{2\rho(\log a)}e^{2\rho(H(\bar{n}))}e^{-2\rho(H(a^{-1}\bar{n}a))}$$

so the formula follows immediately from Theorem 1.14.

Remark. For the hyperbolic spaces, real and Hermitian, the Poisson kernel was well known.

§2. CONICAL DISTRIBUTIONS. ELEMENTARY PROPERTIES

As usual let G be a connected semisimple Lie group with finite center, $X = G/K$ the associated symmetric space and $\Xi = G/MN$ its dual. A distribution (resp. a C^∞-function) on Ξ is called *conical* if it is

(i) MN-invariant;

(ii) eigendistribution (resp. eigenfunction) of each G-invariant differential operator on Ξ.

607/5/1-5

In this section we shall prove a preliminary structure theorem about these distributions. Following our conventions let s^* denote the Weyl group element which interchanges Σ^+ and Σ^-, let $m^* \in M'$ be such that $Ad_G(m^*)$ restricted to $\mathfrak{a}$ equals s^*, let ξ^* denote the point m^*MN in Ξ and Ξ^* the orbit $MNA \cdot \xi^*$. From the Bruhat lemma we know that $\bar{N}MAN$ is an open subset of G whose complement has Haar measure 0. Thus $\Xi - \Xi^*$ has measure 0 for the G-invariant measure $d\xi$ on Ξ. If $\xi \in \Xi^*$ we know (cf., Theorem 3.8 in [22(c)]) that there exists a unique element $a(\xi) \in A$ such that $\xi = mna(\xi) \cdot \xi^*$ for some $m \in M$, $n \in N$.

Let $U \to D_U$ denote the isomorphism of $\boldsymbol{D}(A)$ onto $\boldsymbol{D}(\Xi)$ (Chap. I, §2) which satisfies

$$(D_U\varphi)(ka \cdot \xi_0) = U_a(\varphi(ka \cdot \xi_0)), \qquad \varphi \in \mathscr{D}(\Xi). \tag{1}$$

If $\mu \in \mathfrak{a}_c^*$, extend μ to a homomorphism of $\boldsymbol{D}(A)$ into $\boldsymbol{C}$; in view of the above isomorphism of $\boldsymbol{D}(A)$ onto $\boldsymbol{D}(\Xi)$ we obtain a homomorphism $\chi_u : \boldsymbol{D}(\Xi) \to \boldsymbol{C}$. Every homomorphism of $\boldsymbol{D}(\Xi)$ into $\boldsymbol{C}$ is of this form.

The main result of this section is the following theorem.

THEOREM 2.1. *Let Ψ be a conical distribution on Ξ and let $\lambda \in \mathfrak{a}_c^*$ be determined such that*

$$D\Psi = \chi_{i\lambda-\rho}(D)\, \Psi, \qquad D \in \boldsymbol{D}(\Xi).$$

Then there exists a function $\psi_\lambda \in \mathscr{E}(\Xi^*)$ such that $\Psi = \psi_\lambda$ on Ξ^*, that is

$$\Psi(\varphi) = \int_{\Xi^*} \psi_\lambda(\xi)\, \varphi(\xi)\, d\xi \qquad \textit{for} \quad \varphi \in \mathscr{D}(\Xi^*). \tag{2}$$

This function ψ_λ is given by

$$\psi_\lambda(\xi) = e^{(is^*\lambda+\rho)(\log a(\xi))}\psi_\lambda(\xi^*), \qquad (\xi \in \Xi^*). \tag{3}$$

On the other hand, let $\lambda \in \mathfrak{a}_c^*$ and let $\psi_\lambda \in \mathscr{E}(\Xi^*)$ be defined by (3) *and $\psi_\lambda(\xi^*) \neq 0$. Then ψ_λ is a locally integrable function on Ξ if and only if*

$$\operatorname{Re}(\langle i\lambda, \alpha\rangle) > 0 \qquad \textit{for all} \quad \alpha \in \Sigma^+, \tag{4}$$

(Re *denoting real part*). *If* (4) *is satisfied then ψ_λ, as a distribution on Ξ, is a conical distribution.*

We begin with a simple lemma concerning the G-invariant measure $d\xi$ on Ξ (with the normalizations in Notation, §4). The last relation of the lemma was already proved in Chap. II, §3.

Each $a \in A$ normalizes MN and therefore defines a diffeomorphism $\sigma(a)\colon gMN \to gaMN$ of Ξ onto itself.

LEMMA 2.2. *For any $\varphi \in \mathscr{D}(\Xi)$, $a \in A$ we have*

$$\int_\Xi \varphi(\xi)\, d\xi = \int_{N\times A} \varphi(na \cdot \xi^*)\, e^{-2\rho(\log a)}\, dn\, da, \tag{5}$$

$$\int_\Xi \varphi(\xi)\, d\xi = e^{2\rho(\log a)} \int_\Xi \varphi(\sigma(a) \cdot \xi)\, d\xi, \tag{6}$$

$$\int_\Xi \varphi(\xi)\, d\xi = \int_{(K/M)\times A} \varphi(kaMN)\, e^{2\rho(\log a)}\, dk_M\, da. \tag{7}$$

Proof. Let S denote the group NA. Then as noted before the mapping $s \to s \cdot \xi^*$ is a bijection of S onto Ξ^*. The measure on S which hereby corresponds to $d\xi$ is a left invariant Haar measure on S. But the measure

$$F \to \int_{NA} F(na)\, e^{-2\rho(\log a)}\, dn\, da$$

is a left invariant measure on S, so (5) holds with a constant factor, say c, on the right. To evaluate c we compute (putting $bgb^{-1} = g^b$)

$$\int_\Xi \varphi(\xi)\, d\xi = \int_\Xi \varphi(m^* \cdot \xi)\, d\xi = c \int_{NA} \varphi(n^{m^*}a^{m^*}MN)\, e^{-2\rho(\log a)}\, da\, dn$$

$$= c \int_{\bar{N}A} \varphi(\bar{n}aMN)\, e^{+2\rho(\log a)}\, d\bar{n}\, da.$$

$$= c \int_{\bar{N}A} \varphi(k(\bar{n})\, aMN)\, e^{2\rho(\log a)} e^{-2\rho(H(\bar{n}))}\, d\bar{n}\, da$$

Since $d\bar{n}$ is normalized by $\int_{\bar{N}} e^{-2\rho(H(\bar{n}))}\, d\bar{n} = 1$ we obtain by comparing with (7), $c = 1$. We also get Harish-Chandra's formula [21(i), p. 287]

$$\int_{K/M} \varphi(kM)\, dk_M = \int_{\bar{N}} \varphi(k(\bar{n})M)\, e^{-2\rho(H(\bar{n}))}\, d\bar{n}. \tag{8}$$

Formula (6) follows immediately by using (5) or (7) in the function $\xi \to \varphi(\sigma(a) \cdot \xi)$.

If $D \in \boldsymbol{D}(\Xi)$ the adjoint $D^* \in \boldsymbol{D}(\Xi)$ is defined by

$$\int \varphi(\xi)(D\psi)(\xi)\, d\xi = \int (D^*\varphi)(\xi)\, \psi(\xi)\, d\xi, \qquad \varphi, \psi \in \mathscr{D}(\Xi).$$

The operation $U \to U^*$ on $\boldsymbol{D}(A)$ is defined similarly by means of da. Let $D_1 \circ D_2$ denote composition of two differential operators D_1 and D_2.

LEMMA 2.3. *If* $U \in \boldsymbol{D}(A)$ *then*

$$(D_U)^* = D_{e^{-2\rho}U^* \circ e^{2\rho}}$$

Proof. Let $U' \in \boldsymbol{D}(A)$ be determined by $(D_U)^* = D_{U'}$. Then using (7) and (1),

$$\int_{K/M \times A} U_a(\varphi(ka \cdot \xi_0))\, \psi(ka \cdot \xi_0)\, e^{2\rho(\log a)}\, dk_M\, da$$

$$= \int_{K/M \times A} \varphi(ka \cdot \xi_0)\, e^{2\rho(\log a)}\, U_a'(\psi(ka \cdot \xi_0))\, dk_M\, da.$$

Let $\alpha, \beta \in \mathscr{D}(A)$ and put

$$\varphi(ka \cdot \xi_0) \equiv \alpha(a) \qquad \text{and} \qquad \psi(ka \cdot \xi_0) \equiv e^{-2\rho(\log a)}\beta(a).$$

Then the above relation gives $U^* = e^{2\rho} U' \circ e^{-2\rho}$ as desired.

Let $U \in \boldsymbol{D}(A)$ and put $D = D_U$. Let $\varphi \in \mathscr{D}(\Xi)$, $g \in G$ and $a \in A$. Then by (1)

$$(D\varphi)(ga \cdot \xi_0) = (D\varphi^{\tau(g^{-1})})(a \cdot \xi_0) = U_a(\varphi^{\tau(g^{-1})}(a \cdot \xi_0))$$

so

$$(D\varphi)(\xi) = [U_a(\varphi^{\sigma(a^{-1})}(\xi))]_{a=e}\,. \tag{9}$$

Now we need a lemma giving the analog of (9) for distributions.

LEMMA 2.4. *Let* Ψ *be an arbitrary distribution on* Ξ *and for each* $a \in A$ *define the functional* $U_a(\Psi^{\sigma(a)})$ *by*

$$U_a(\Psi^{\sigma(a)}) : \varphi \to U_a(\Psi^{\sigma(a)}(\varphi)), \qquad \varphi \in \mathscr{D}(\Xi).$$

Then $U_a(\Psi^{\sigma(a)})$ *is a distribution on* Ξ *and*

$$D^*\Psi = [U_a(\Psi^{\sigma(a)})]_{a=e}\,. \tag{10}$$

Proof. We define the function $\Phi \in \mathscr{E}(A \times \Xi)$ by

$$\Phi(a, \xi) = \varphi(\sigma(a) \cdot \xi),$$

and for $a_0 \in A$ let u denote the distribution given by

$$u(F) = (UF)(a_0), \qquad F \in \mathscr{D}(A).$$

If Ψ is any distribution on Ξ the tensor product $u \otimes \Psi$ is a distribution on $A \times \Xi$ which satisfies

$$(u \otimes \Psi)_{(a,\xi)}(f(a, \xi)) = u_a(\Psi_\xi(f(a, \xi))) = \Psi_\xi(u_a(f(a, \xi))) \tag{11}$$

for all $f \in \mathscr{D}(A \times \Xi)$, [41, Ch. IV]. Now the support of $u \otimes \Psi$ is contained in $\{a_0\} \times \Xi$ which intersects the support of Φ in a compact set. Hence $(u \otimes \Psi)(\Phi)$ can be defined as $(u \otimes \Psi)(\alpha\Phi)$ where α is an arbitrary function in $\mathscr{D}(A \times \Xi)$ which is identically 1 on a neighborhood of the compact set indicated, [41, p. 90]. Then by (11),

$$u_a(\Psi_\xi(\alpha(a, \xi)\, \varphi(\sigma(a) \cdot \xi))) = \Psi_\xi(u_a(\alpha(a, \xi)\, \varphi(\sigma(a) \cdot \xi))). \tag{12}$$

But if a is sufficiently close to a_0 then $\alpha(a, \xi) = 1$ on a neighborhood of the support of the function $\xi \to \varphi(\sigma(a) \cdot \xi))$. For such a we have

$$\Psi_\xi(\alpha(a, \xi)\, \varphi(\sigma(a) \cdot \xi)) = \Psi_\xi(\varphi(\sigma(a) \cdot \xi))$$

so by (12), since u has support in $\{a_0\}$

$$u_a(\Psi_\xi(\varphi^{\sigma(a)^{-1}}(\xi))) = \Psi_\xi(u_a(\varphi(\sigma(a) \cdot \xi)),$$

that is

$$[U_a(\Psi(\varphi^{\sigma(a)^{-1}})]_{a=a_0} = \Psi_\xi([U_a(\varphi^{\sigma(a)^{-1}}(\xi))]_{a=a_0}). \tag{13}$$

Combining this formula with (9) we obtain

$$\Psi(D\varphi) = [U_a(\Psi(\varphi^{\sigma(a)^{-1}}))]_{a=e}\,. \tag{14}$$

Consequently, $[U_a(\Psi^{\sigma(a)})]_{a=e}$ is a distribution on Ξ, and in fact,

$$D^*\Psi = [U_a(\Psi^{\sigma(a)})]_{a=e}\,. \tag{15}$$

That $U_b(\Psi^{\sigma(b)})$ for each $b \in A$ is a distribution on Ξ follows by using (14) with Ψ replaced by $\Psi^{\sigma(b)}$. This proves the lemma.

Using this lemma we shall prove that condition (ii) in the definition of a conical distribution can be put in a more convenient form.

PROPOSITION 2.5. *A distribution Ψ on Ξ is an eigendistribution of each*

$D \in \boldsymbol{D}(\Xi)$ if and only if for each $a \in A$, $\Psi^{\sigma(a)}$ is a constant multiple of Ψ. More precisely, let $\lambda \in \mathfrak{a}_c^*$; then for any distribution Ψ on Ξ, the relation

$$D\Psi = \chi_{i\lambda-\rho}(D)\Psi \qquad \textit{for all} \quad D \in \boldsymbol{D}(\Xi) \tag{16}$$

is equivalent to

$$\Psi^{\sigma(a)} = e^{-(i\lambda+\rho)(\log a)}\Psi \qquad \textit{for all} \quad a \in A. \tag{17}$$

Proof. Suppose Ψ is an eigendistribution of each $D \in \boldsymbol{D}(\Xi)$. Then there exists a unique $\lambda \in \mathfrak{a}_c^*$ such that (16) holds. Hence (15) implies

$$[U_a(\Psi^{\sigma(a)}]_{a=e} = \chi_{i\lambda-\rho}(D^*)\Psi. \tag{18}$$

Fix $b \in A$, $\varphi \in \mathscr{D}(\Xi)$ and apply both sides to $\varphi^{\sigma(b)^{-1}}$. Since

$$[U_a(\Psi^{\sigma(a)})]_{a=e}\,(\varphi^{\sigma(b)^{-1}}) = [U_a(\Psi^{\sigma(a)}(\varphi^{\sigma(b)^{-1}}))]_{a=e}$$

we obtain

$$[U_a(\Psi^{\sigma(ab)}(\varphi))]_{a=e} = \chi_{i\lambda-\rho}(D^*)\,\Psi^{\sigma(b)}(\varphi).$$

Thus the function $t_\varphi : b \to \Psi^{\sigma(b)}(\varphi)$ on A satifies

$$Ut_\varphi = \chi_{i\lambda-\rho}(D^*)t_\varphi \tag{19}$$

for all $U \in \boldsymbol{D}(A)$. It follows that

$$\Psi^{\sigma(b)}(\varphi) = \Psi(\varphi)\,e^{\mu(\log b)} \tag{20}$$

where $\mu \in \mathfrak{a}_c^*$ satisfies $\chi_{i\lambda-\rho}(D^*) = \chi_\mu(D)$ for all $D \in \boldsymbol{D}(\Xi)$. In particular μ is independent of the choice of φ. Now the automorphism $U \to e^{-2\rho}U \circ e^{2\rho}$ of $\boldsymbol{D}(A)$ is the same as the automorphism induced by the mapping $H \to H + 2\rho(H)I$, $(H \in \mathfrak{a})$, I being the identity operator. Also if $\nu \in \mathfrak{a}_c^*$

$$\chi_\nu(H + 2\rho(H)I) = \chi_{\nu+2\rho}(H), \qquad H^* = -H,$$

so by Lemma 2.3,

$$\chi_{i\lambda-\rho}((D_U)^*) = \chi_{i\lambda+\rho}(D_{U^*}) = \chi_{-i\lambda-\rho}(D_U)$$

so $\mu = -(i\lambda + \rho)$. This proves the implication (16) $\Rightarrow$ (17) in Proposition 2.5. The converse follows by reversing the above steps.

We can now prove Theorem 2.1. Let Ψ be a conical distribution and select $\lambda \in \mathfrak{a}_c^*$ such that

$$D\Psi = \chi_{i\lambda-\rho}(D)\Psi, \qquad D \in \boldsymbol{D}(\Xi).$$

We define a C^∞ function α on Ξ^* by the formula

$$\alpha(\xi) = e^{(-is*\lambda+\rho)(\log a(\xi))}: \qquad \xi \in \Xi^*.$$

Then

$$\alpha^{\sigma(a)} = e^{(i\lambda+\rho)(\log a)}\alpha.$$

Combining this with (17) and the N-invariance of Ψ we obtain for the restriction $\Psi^* = \Psi \mid \mathscr{D}(\Xi^*)$,

$$(\alpha\Psi^*)^{\tau(n)\sigma(a)} = \alpha\Psi^*, \qquad n \in N, \quad a \in A. \tag{21}$$

Consider now the *direct* product $N \times A$. The mapping $(n, a) \to nm^*a \cdot \xi_0$ is a diffeomorphism of $N \times A$ onto Ξ^*. Under this diffeomorphism $\alpha\Psi^*$ corresponds to a distribution on $N \times A$ which by (21) is invariant under all left translations (since A is abelian), and therefore is a constant multiple of Haar measure on the product group $N \times A$. Thus

$$(\alpha\Psi^*)(\varphi) = c \int_{N\times A} \varphi(nm^*a \cdot \xi_0)\, dn\, da,$$

where c is a constant. Since s^* preserves da we deduce from Lemma 2.2,

$$(\alpha\Psi^*)(\varphi) = c \int_{\Xi} \varphi(\xi)\, \beta(\xi)\, d\xi,$$

where $\beta(\xi) = e^{2\rho(\log a(\xi))}$. Since $1/\alpha \in \mathscr{E}(\Xi^*)$ this proves that Ψ^* is a C^∞ function and in fact given by (3).

On the other hand, let ψ_λ be defined by (3), taking for simplicity $\psi_\lambda(\xi^*) = 1$. The local integrability of ψ_λ is equivalent to

$$\int_{\Xi} \mid \psi_\lambda(\xi)\, \varphi(\xi) \mid d\xi < \infty \qquad \text{for each} \quad \varphi \in \mathscr{D}(\Xi). \tag{22}$$

Now we observe that the set

$$\{kM \mid k \cdot \xi_0 \in \Xi - \Xi^*\} \tag{23}$$

is a subset of K/M of lower dimension. In fact, under the diffeomorphism $kM \to kMAN$ of K/M onto G/MAN the complement of (23) in K/M corresponds to the unique maximum dimensional orbit of MAN on G/MAN. Thus the integral

$$\int_{K/M} |\psi_\lambda(k \cdot \xi_0)|\, dk_M \tag{24}$$

makes sense and since

$$\psi_\lambda(ka \cdot \xi_0) = e^{(i\lambda-\rho)(\log a)}\psi_\lambda(k \cdot \xi_0)$$

it is clear from Lemma 2.2, (7), that ψ_λ is locally integrable if and only if the integral (24) is finite or equivalently by (8), if and only if

$$\int_{\bar{N}} |\psi_\lambda(m^*k(\bar{n}) \cdot \xi_0)|\, e^{-2\rho(H(\bar{n}))}\, d\bar{n} < \infty. \tag{25}$$

But

$$\psi_\lambda(m^*k(\bar{n}) \cdot \xi_0) = e^{(-i\lambda+\rho)(H(\bar{n}))}$$

so the condition amounts to the integral

$$\int_{\bar{N}} e^{-(i\lambda+\rho)(H(\bar{n}))}\, dn$$

being absolutely convergent. But this is the integral representing Harish-Chandra's **c**-function and the absolute convergence is well-known ([21(i), p. 291] and [19(i)]; cf. (30) in §4) to be equivalent to

$$\operatorname{Re}(\langle i\lambda, \alpha\rangle) > 0 \qquad \text{for each} \quad \alpha \in \Sigma^+.$$

Finally, if ψ_λ as given by (3) is locally integrable on Ξ, then it is obvious that the distribution

$$\varphi \to \int_{\Xi} \psi_\lambda(\xi)\, \varphi(\xi)\, d\xi, \qquad (\varphi \in \mathscr{D}(\Xi))$$

is MN-invariant; using Proposition 2.5 we see finally that it is a conical distribution. Theorem 2.1 is now completely proved.

The following result will be useful later. Let $\mathscr{D}_\lambda'$ denote the set of distributions Ψ satisfying (16).

PROPOSITION 2.6. *Let $\Psi \in \mathscr{D}_\lambda'$ be a conical distribution and $\Phi \in \mathscr{D}_\mu'$ a conical function. Then $\Phi\Psi$ is a conical distribution in $\mathscr{D}'_{\mu+\lambda+i\rho}$.*

Proof. We just have to verify the appropriate homogeneity under $\sigma(a)$, $(a \in A)$. The distribution $\Psi \in \mathscr{D}_\lambda'$ satisfies $\Psi^{\sigma(a)} = e^{-(i\lambda+\rho)(\log a)}\Psi$ whereas Φ *as a function* satisfies $\Phi^{\sigma(a)} = e^{-(i\mu-\rho)(\log a)}\Phi$. Thus if $\varphi \in \mathscr{D}(\Xi)$,

$$(\Phi\Psi)^{\sigma(a)}(\varphi) = \Psi(\Phi\varphi^{\sigma(a)^{-1}}) = \Psi^{\sigma(a)}(\Phi^{\sigma(a)}\varphi) = e^{-i(\lambda+\mu)(\log a)}(\Phi\Psi)(\varphi)$$

proving the proposition.

§3. Conical Functions and Finite-Dimensional Representations

In this section we shall determine all the conical functions on G/MN and establish their close connection with representation theory. We first state the main results.

Theorem 3.1. *Let ψ be a conical function on Ξ. Then*

$$\psi(\xi) = e^{-\mu(\log a(\xi))}\psi(\xi^*), \qquad (\xi \in \Xi^*), \tag{1}$$

where $\mu \in \mathfrak{a}_c^*$ satisfies

$$\frac{\langle \mu, \alpha \rangle}{\langle \alpha, \alpha \rangle} \in \boldsymbol{Z}^+ \qquad \text{for} \quad \alpha \in \Sigma^+. \tag{2}$$

Conversely, if $\mu \in \mathfrak{a}_c^*$ satisfies (2) *then the function ψ defined by* (1) *and $\psi(\xi^*)$ arbitrary extends to a conical function on Ξ.*

A representation π of G on a vector space E is called *conical* (resp. *spherical*) if there exists a nonzero vector in E fixed under $\pi(MN)$ (resp. under $\pi(K)$). Let ζ be an arbitrary function on Ξ, E_ζ the smallest vector space containing all the G-translates of ζ and π_ζ the natural representation of G on E_ζ, i.e. $(\pi_\zeta(g)\psi)(\xi) = \psi(g^{-1} \cdot \xi)$ for $g \in G$, $\xi \in \Xi$, $\psi \in E_\zeta$.

Theorem 3.2. *The mapping $\psi \to \pi_\psi$ is a bijection of the set of conical functions $\not\equiv 0$ on Ξ (proportional ψ identified) onto the set of finite-dimensional irreducible conical representations of G (equivalent representations identified). Also*

$$\psi(g \cdot \xi_o) = \langle \pi_\psi(g^{-1})\, \boldsymbol{e}, \boldsymbol{e}^* \rangle, \qquad (g \in G). \tag{3}$$

where e *and* e^*, *respectively, are contained in the* (*one-dimensional*) *highest weight spaces of* π_ψ *and of its contragredient representation. Finally,* μ *in* (1) *is the restriction to* $\mathfrak{a}$ *of the highest weight of* π_ψ.

It is convenient to prove these two theorems together. The idea is as follows. First we determine the conical functions in the case $G = \boldsymbol{SL}(2, \boldsymbol{R})$ For a general G and Ξ one constructs for each restricted root $\alpha > 0$ a subgroup $G^\alpha \subset G$ locally isomorphic to $\boldsymbol{SL}(2, \boldsymbol{R})$ and a corresponding horocycle space $\Xi^\alpha \subset \Xi$. If ψ is a conical function on Ξ a suitable modification of it has a restriction to Ξ^α which is a nonzero conical function (ψ may vanish on $\Xi - \Xi^*$) and one arrives at the integrality condition (2). On the other hand, if (2) is satisfied, one can construct a certain finite-dimensional representation π_μ of G whose highest weight $\tilde{\mu}$ satisfies $\tilde{\mu} \mid \mathfrak{a} = \mu$. Using the structure of the group of components of M one finds that the function $g \to \langle \pi_\mu(g^{-1})e, e^* \rangle$ defines a conical function on Ξ which is proportional to the function (1).

Now let $G = \boldsymbol{SL}(2, \boldsymbol{R})$ and consider the standard Iwasawa decomposition $G = KAN$ where $K = \boldsymbol{SO}(2)$ and

$$A = \left\{ a = \begin{pmatrix} d & 0 \\ 0 & d^{-1} \end{pmatrix} \middle| \, d > 0 \text{ in } \boldsymbol{R} \right\}, \qquad N = \left\{ n = \begin{pmatrix} 1 & x \\ 0 & 1 \end{pmatrix} \middle| \, x \in \boldsymbol{R} \right\}.$$

LEMMA 3.3. *The conical functions on* G/MN ($G = \boldsymbol{SL}(2, \boldsymbol{R})$) *are the constant multiples of the functions*

$$\psi \left(\begin{vmatrix} x_{11} & x_{12} \\ x_{21} & x_{22} \end{vmatrix} \cdot \xi_0 \right) = (x_{21})^{2n}, \qquad n \in \boldsymbol{Z}^+. \tag{4}$$

Proof. The element $H_0 = \left| \begin{smallmatrix} 1 & 0 \\ 0 & -1 \end{smallmatrix} \right|$ forms a basis of $\mathfrak{a}$ and $\rho(H_0) = 1$. Here M consists of $\left| \begin{smallmatrix} \epsilon & 0 \\ 0 & \epsilon \end{smallmatrix} \right|$ with $\epsilon = \pm 1$ and for m^* we can use $\left| \begin{smallmatrix} 0 & 1 \\ -1 & 0 \end{smallmatrix} \right|$. Suppose ψ is a conical function $\not\equiv 0$. Then by Theorem 2.1, there exists a $\mu \in \mathfrak{a}_c^*$ such that

$$\psi(\xi) = e^{-\mu(\log a(\xi))} \psi(\xi^*), \qquad (\xi \in \Xi^*).$$

Let $z \in \boldsymbol{C}$ be such that $\mu = z\rho$. Then

$$\psi \left(\begin{vmatrix} \epsilon & 0 \\ 0 & \epsilon \end{vmatrix} \begin{vmatrix} 1 & x \\ 0 & 1 \end{vmatrix} \begin{vmatrix} d & 0 \\ 0 & d^{-1} \end{vmatrix} \begin{vmatrix} 0 & 1 \\ -1 & 0 \end{vmatrix} \cdot \xi_0 \right) = \psi(\xi^*) \, e^{-z \log d}$$

so that on the set Ξ^*, ψ is given by

$$\psi \left(\begin{vmatrix} x_{11} & x_{12} \\ x_{21} & x_{22} \end{vmatrix} \cdot \xi_0 \right) = \psi(\xi^*) \, | \, x_{21} \, |^z.$$

Since $\psi \in \mathscr{E}(\Xi)$ we see that z must be an even integer $\geqslant 0$, that is (2) holds. Conversely, if $n \in \boldsymbol{Z}^+$ formula (4) gives a well-defined C^∞ function on G/MN. Also

$$\psi(mn_1 ga \cdot \xi_0) = \psi(g \cdot \xi_0)\, e^{2n\rho(\log a)}$$

so by Proposition 2.5, ψ is a conical function.

Now we need a simple lemma from Satake [39, p. 93]. Let $\mathfrak{g}_c$ denote the complexification of $\mathfrak{g}$ and G_c an analytic group with Lie algebra $\mathfrak{g}_c$.

LEMMA 3.4. *Suppose the group G is a Lie subgroup of G_c. Let M^0 denote the identity component of M. Then*

$$M = M^0(\exp(i\mathfrak{a}) \cap K).$$

We sketch a proof. Let $U \subset G_c$ denote the (compact) analytic subgroup with Lie algebra $\mathfrak{k} + i\mathfrak{p}$. Let $k \in M$. Then there exists a maximal torus $T \subset U$ containing k and $\exp(i\mathfrak{a})$. Also $\mathfrak{t}$, the Lie algebra of T, satisfies $\mathfrak{t} = \mathfrak{k} \cap \mathfrak{t} + i\mathfrak{a}$. Thus $k \in \exp \mathfrak{t} \subset \exp(\mathfrak{k} \cap \mathfrak{t}) \exp(i\mathfrak{a}) \subset M^0 \exp i\mathfrak{a}$.

We can now prove the first half of Theorem 3.1. Let $\alpha > 0$ be a restricted root of $\mathfrak{g}$ with respect to $\mathfrak{a}$. Select $X_\alpha \neq 0$ in $\mathfrak{g}_\alpha$. Then the space

$$\mathfrak{g}^\alpha = \boldsymbol{R}H_\alpha + \boldsymbol{R}X_\alpha + \boldsymbol{R}\theta X_\alpha$$

is a subalgebra of $\mathfrak{g}$, isomorphic to $\mathfrak{sl}(2, \boldsymbol{R})$. Let G^α denote the corresponding analytic subgroup of G. If $\mathfrak{k}^\alpha = \mathfrak{g}^\alpha \cap \mathfrak{k}$, $\mathfrak{p}^\alpha = \mathfrak{g}^\alpha \cap \mathfrak{p}$, the decomposition $\mathfrak{g}^\alpha = \mathfrak{k}^\alpha + \mathfrak{p}^\alpha$ is a Cartan decomposition of $\mathfrak{g}^\alpha$. The space $\mathfrak{a}^\alpha = \boldsymbol{R}H_\alpha$ is maximal abelian in $\mathfrak{p}^\alpha$, and the positive multiples of H_α form a Weyl chamber $(\mathfrak{a}^\alpha)^+$ in $\mathfrak{a}^\alpha$ and in the usual manner we form the subgroups K^α, A^α, M^α, N^α, $\bar{N}^\alpha$ of G^α and the subalgebras $\mathfrak{m}^\alpha$, $\mathfrak{n}^\alpha$ and $\bar{\mathfrak{n}}^\alpha$. Considering the Iwasawa decompositions we have $K^\alpha = G^\alpha \cap K$, $N^\alpha = G^\alpha \cap N$, $A^\alpha = G^\alpha \cap A$. The relation $M \cap G^\alpha \subset M^\alpha$ is obvious, the converse follows since G^α centralizes the orthogonal complement of $\mathfrak{a}^\alpha$ in $\mathfrak{a}$. Thus $M \cap G^\alpha = M^\alpha$. This implies

LEMMA 3.5. *The mapping $gM^\alpha N^\alpha \to gMN$ is an injection of Ξ^α $(= G^\alpha/M^\alpha N^\alpha)$ into Ξ.*

Let the Killing form of $\mathfrak{g}$ and $\mathfrak{g}^\alpha$ be denoted by $\langle\ ,\ \rangle$ and $\langle\ .\ \rangle_\alpha$, respectively. If $\mu \in \mathfrak{a}_c{}^*$ let $\bar{\mu}$ denote the restriction $\mu \mid \mathfrak{a}^\alpha$. Since $\bar{\mu} = c\bar{\alpha}$ $(c \in \boldsymbol{C})$ we have

$$\frac{\langle \mu, \alpha \rangle}{\langle \alpha, \alpha \rangle} = \frac{\langle \bar{\mu}, \bar{\alpha} \rangle_\alpha}{\langle \bar{\alpha}, \bar{\alpha} \rangle_\alpha}. \tag{5}$$

Suppose now $\psi \not\equiv 0$ is a conical function on Ξ. Then by Theorem 2.1 there exists a $\mu \in \mathfrak{a}_c{}^*$ such that (1) holds, say with $\psi(\xi^*) = 1$. Let $m_\alpha{}^* \in K^\alpha$ such that $Ad_{G^\alpha}(m_\alpha{}^*)$ induces the nontrivial Weyl group element on $\mathfrak{a}^\alpha$. Viewing Ξ^α as a subset of Ξ (Lemma 3.5) we consider the function on Ξ^α given by

$$\psi_\alpha = (\psi^{\tau(m^*)}/\Xi^\alpha)^{\tau(m_\alpha{}^*)}.$$

Then in the usual notation

$$\psi_\alpha(m_\alpha n_\alpha a_\alpha m_\alpha{}^* M^\alpha N^\alpha) = \psi(m^* m_\alpha{}^* m_\alpha n_\alpha a_\alpha m_\alpha{}^* MN) = \psi(m^* a_\alpha^{-1} MN) = e^{(s^*\mu)(\log a_\alpha)}$$

so ψ_α is a conical function on Ξ^α. Using (5) and Lemma 3.3 (cf., Remark at the beginning of §4) we deduce that

$$\frac{\langle -s^*\mu, \alpha\rangle}{\langle \alpha, \alpha\rangle} \in \boldsymbol{Z}^+.$$

Since $-s^*$ induces a permutation of the positive restricted roots the first part of Theorem 3.1 is proved.

For the converse we first need a preliminary lemma. Extend $\mathfrak{a}$ to a maximal abelian subalgebra $\mathfrak{h}$ of $\mathfrak{g}$. Then the complexification $\mathfrak{h}_c$ is a Cartan subalgebra of $\mathfrak{g}_c$. If β is a function on $\mathfrak{h}_c$ we write $\bar{\beta}$ for the restriction $\beta \mid \mathfrak{a}$. We order the roots of $\mathfrak{g}_c$ with respect to $\mathfrak{h}_c$ in such a way that $\beta > 0 \Rightarrow \bar{\beta} \geqslant 0$.

Lemma 3.6. *Let β be a root of $\mathfrak{g}_c$ with respect to $\mathfrak{h}_c$ such that $\bar{\beta} \neq 0$ and determine m by*

$$\langle \beta, \beta\rangle = m\langle \bar{\beta}, \bar{\beta}\rangle.$$

Then

$$m = \begin{cases} 1 \quad \text{or } 2 \text{ if } 2\bar{\beta} \text{ is not a restricted root} \\ 1,2 \text{ or } 4 \text{ if } 2\bar{\beta} \text{ is a restricted root.} \end{cases}$$

This lemma is of course contained in Araki's extensive study [2] of the relation between roots and restricted roots. We shall however give a direct elementary proof. Extend θ to an automorphism of $\mathfrak{g}_c$ and write $\theta\beta$ for the root given by $H_{\theta\beta} = \theta H_\beta$. Then $\bar{\beta} = 1/2\,(\beta - \theta\beta)$; if $\beta = -\theta\beta$ then $m = 1$ so we assume $\beta + \theta\beta \neq 0$. If $\beta + \theta\beta$ were a root the roots subspace $(\mathfrak{g}_c)_{\beta+\theta\beta}$ would be contained in $\mathfrak{m}_c$, whereas clearly $[X_\beta, \theta X_\beta] \in \mathfrak{p}_c$; hence $\beta + \theta\beta$ is not a root whence by elementary root theory $2\langle \beta, \theta\beta\rangle/\langle \beta, \beta\rangle$ is an integer $\geqslant 0$, hence 0, 1, or 2. Since $\bar{\beta} \neq 0$

the value 2 is excluded so $m = 2$ or 4. If $m = 4$ then $\langle \beta, \theta\beta \rangle > 0$ so by root theory, $\beta - \theta\beta$ is a root, so $2\bar{\beta}$ is a restricted root. Q.E.D.

Now let $\mu \in \mathfrak{a}_c^*$ satisfy condition (2). Since $\mathfrak{h}_c$ is invariant under θ we can extend μ to a complex linear function on $\mathfrak{h}_c$ (also denoted μ) by putting $\mu(\mathfrak{h}_c \cap \mathfrak{k}_c) = 0$. Then if β is any positive root we can write

$$2\,\frac{\langle \mu, \beta \rangle}{\langle \beta, \beta \rangle} = \begin{cases} 2m^{-1}\langle \mu, \bar{\beta} \rangle / \langle \bar{\beta}, \bar{\beta} \rangle & \text{if } \; 2\bar{\beta} \notin \Sigma^+, \\ 4m^{-1}\langle \mu, 2\bar{\beta} \rangle / \langle 2\bar{\beta}, 2\bar{\beta} \rangle & \text{if } \; 2\bar{\beta} \in \Sigma^+. \end{cases}$$

According to (2) and Lemma 3.6 the left side is in $\boldsymbol{Z}^+$. Thus if $\tilde{G}_c$ is the simply connected Lie group with Lie algebra $\mathfrak{g}_c$ there exists a finite-dimensional irreducible representation π of $\tilde{G}_c$ (and $\mathfrak{g}_c$) on a complex vector space E such that π has highest weight μ. Let $\boldsymbol{e} \in E$ be a highest weight vector $\neq 0$. Let $\tilde{G}$, $\tilde{K}$, $\tilde{A}$, $\tilde{N}$ denote the analytic subgroups of $\tilde{G}_c$ corresponding to $\mathfrak{g}$, $\mathfrak{k}$, $\mathfrak{a}$, $\mathfrak{n}$, respectively and let $\tilde{M}$ denote the centralizer of $\tilde{A}$ in $\tilde{K}$, and $\tilde{M}^0$ its identity component.

We shall now prove that π restricted to $\tilde{G}$ is a conical representation. The algebra $\mathfrak{m}_c$ is spanned by $\mathfrak{h}_c \cap \mathfrak{k}_c$ and the root vectors X_β, $X_{-\beta}$ where $\beta > 0$ and $\bar{\beta} = 0$. If $H \in \mathfrak{h}_c \cap \mathfrak{k}_c$ then $\pi(\exp H)\boldsymbol{e} = \mu(H)\boldsymbol{e} = 0$; since $\mu + \beta$ is not a weight of π, $\pi(X_\beta)\boldsymbol{e} = 0$. Also $\pi(X_{-\beta})\boldsymbol{e} = 0$; in fact since $\langle \mu, \beta \rangle = 0$, the linear form $\mu - \beta$ is the image of $\mu + \beta$ under the Weyl reflexion s_β and thus is not a weight. This proves $\pi(m)\boldsymbol{e} = \boldsymbol{e}$ for all $m \in \tilde{M}^0$; in view of Lemma 3.4 we still must prove $\pi(\exp H)\boldsymbol{e} = \boldsymbol{e}$ (or equivalently $\mu(H) \in 2\pi i\boldsymbol{Z}$) for each $H \in i\mathfrak{a}$ satisfying $\exp H \in \tilde{K}$. But this is an immediate consequence of the following well-known lemma (for a proof see Loos [32, p. 77].)

LEMMA 3.7. *With* exp *denoting the exponential mapping of* $\mathfrak{g}_c$ *into* $\tilde{G}_c$ *the set*

$$\{H \in i\mathfrak{a} \mid \exp H \in \tilde{K}\}$$

consists of the integral linear combinations of the vectors

$$2\pi i H_\alpha / \langle \alpha, \alpha \rangle, \qquad \alpha \in \Sigma. \tag{6}$$

We have now proved that π gives an irreducible conical representation of $\tilde{G}$. By Schur's lemma the center of $\tilde{G}$ is mapped into scalars by π; but this center is contained in $\tilde{M}$ and $\pi(m)\boldsymbol{e} = \boldsymbol{e}$ for $m \in \tilde{M}$. Thus π is identity on the center so we obtain an irreducible representation π_μ of G on E by putting

$$\pi_\mu(g) = \pi(\tilde{g})$$

if $g \in G$, $\tilde{g} \in \tilde{G}$ are related by $Ad_G(g) = Ad_{\tilde{G}}(\tilde{g})$. Since M and $\tilde{M}$ have the same images in the adjoint group of $\mathfrak{g}$ it is clear that $\pi_\mu(mn)\boldsymbol{e} = \boldsymbol{e}$ for $m \in M$, $n \in N$, so π_μ is conical.

Let $\pi_\mu{}^*$ denote the contragredient representation of G on the dual space E^* of E, and let $e^* \in E^*$ be a nonzero highest weight vector. Consider now the function

$$F(g) = \langle \pi_\mu(g^{-1})\, \boldsymbol{e}, \boldsymbol{e}^* \rangle, \qquad g \in G.$$

It is clear that $F(mngn') \equiv F(g)$. Hence we have $F(gmn) = F(g)$ for all $m \in M$, $n \in N$, and all $g \in MNAm^*N$. Since this last subset is dense in G we can define $\psi \in \mathscr{E}(\Xi)$ such that

$$F(g) = \psi(g \cdot \xi_0), \qquad g \in G,$$

and then

$$\psi(mna \cdot \xi^*) = F(am^*) = e^{-\mu(\log a)}\psi(\xi^*)$$

so ψ is conical. Since the elements $\pi_\mu(g^{-1})\boldsymbol{e}$ span E, the function F, and therefore ψ, is $\not\equiv 0$. Thus Theorem 3.1 is proved.

As for Theorem 3.2 we have, starting with μ satisfying (2), constructed the irreducible conical representation π_μ and shown that the function ψ defined by (1) satisfies

$$\psi(g \cdot \xi_0) = \langle \pi_\mu(g^{-1})\, \boldsymbol{e}, \boldsymbol{e}^* \rangle, \qquad g \in G.$$

This relation shows that the space E_ψ is finite-dimensional; the correspondence

$$\sum a_i \psi^{\tau(g_i)} \to \sum a_i \pi_\mu(g_i)\boldsymbol{e}, \qquad a_i \in \boldsymbol{C}, \quad g_i \in G,$$

is a well-defined bijection of E_ψ onto E, setting up an equivalence between π_ψ and π_μ. Next we prove the "onto" statement in Theorem 3.2. So let π_0 be any finite-dimensional, irreducible conical representation of G on a vector space E_0. Let $\boldsymbol{e}_0 \neq 0$ be a vector under $\pi_0(MN)$ and $\boldsymbol{e}_0{}^* \neq 0$ a vector in the highest weight space of the representation contragredient to π_0. Again we consider the function

$$F^*(g) = \langle \pi_0(g^{-1})\boldsymbol{e}_0 , \boldsymbol{e}_0{}^* \rangle$$

and show just as before that $F^*(g) = \psi^*(g \cdot \xi_0)$ where ψ^* is a conical function on Ξ and that the representation π_{ψ^*} is equivalent to π_0. Of Theorem 3.2 there just remains to prove that if ψ_1 and ψ_2 are conical functions such that π_{ψ_1} and π_{ψ_2} are equivalent then ψ_1 and ψ_2 are proportional. This, however, is clear from (3), the highest weight spaces being one-dimensional.

Corollary 3.8. *Let π be an irreducible finite-dimensional representation of G. Then π is spherical if and only if it is conical.*

Suppose first π is spherical, $v \neq 0$ a vector fixed under K, φ the spherical function on G/K given by

$$\varphi(gk) = (\pi(g^{-1})v, v), \qquad g \in G,$$

where (,) is a K-invariant inner product in the representation space such that $(v, v) = 1$. Then the G-translates of φ span a finite-dimensional function space E_φ on G/K and the natural representation π_φ of G on this space is equivalent to π. Let $\psi \in E_\varphi$ be in the highest weight space, $\psi \neq 0$. Then $\pi_\varphi(n)\psi = \psi$ so by $G = NAK$ we have

$$\psi(nmg \cdot o) = \psi(nmn_1a_1k_1 \cdot o) = \psi(g \cdot o)$$

so π_φ and hence π is conical. On the other hand suppose π conical. By Theorem 3.2, π is equivalent to π_ψ where ψ is a conical function on Ξ given by (1). It remains to verify that the vector

$$\psi_0 = \int_K \pi_\psi(k)\psi \, dk$$

is $\neq 0$. But using (8) in §2 we have

$$\begin{aligned}\psi_0(\xi_0) &= \int_{K/M} \psi(m^*k \cdot \xi_0) \, dk_M \\ &= \int_{\bar{N}} \psi(m^*k(\bar{n}) \cdot \xi_0) \, e^{-2\rho(H(\bar{n}))} \, d\bar{n} = \psi(\xi^*) \int_{\bar{N}} e^{(s^*\mu - 2\rho)(H(\bar{n}))} \, d\bar{n}\end{aligned}$$

which is $\neq 0$.

§4. The Determination of the Conical Distributions

Having in §2 and §3 dealt with elementary properties of conical distributions and conical functions we shall now start on the problem of finding all the conical distributions. As explained in Chapter I, §2, one would on the basis of the analogy with spherical functions expect the set of all conical distributions parametrized by $\mathfrak{a}_c^* \times W$.

Denoting, as always, the space of distributions on the horocycle space Ξ by $\mathscr{D}'(\Xi)$, we define for $\lambda \in \mathfrak{a}_c^*$,

$$\mathscr{D}_\lambda'(\Xi) = \{\Psi \in \mathscr{D}'(\Xi) \mid D\Psi = \chi_{i\lambda-\rho}(D)\Psi \text{ for each } D \in \boldsymbol{D}(\Xi)\}.$$

Here $\boldsymbol{D}(\Xi)$ and $\chi_{i\lambda-\rho}$ has of course the same meaning as in §2. We shall often write $\mathscr{D}_\lambda'$ instead of $\mathscr{D}_\lambda'(\Xi)$.

Of course each conical distribution $\Psi \neq 0$ lies in exactly one of the spaces $\mathscr{D}_\lambda'$. It is useful to remark, on the basis of Theorem 3.1, that $\mathscr{D}_\lambda'$ contains a conical function if and only if

$$\frac{\langle i\lambda - \rho, \alpha\rangle}{\langle \alpha, \alpha\rangle} \in \boldsymbol{Z}^+ \qquad \text{for} \quad \alpha \in \Sigma^+.$$

Proposition 4.1. *The space $\mathscr{D}_\lambda'(\Xi)$ consists precisely of the distributions Ψ on Ξ given by*

$$\Psi(\varphi) = \int_{K/M} \left(\int_A \varphi(kaMN)\, e^{(i\lambda+\rho)(\log a)}\, da\right) dS(k_M), \tag{1}$$

where S is an arbitrary distribution on K/M. The mapping $S \to \Psi$ given by (1) *is one-to-one.*

Proof. Since by Proposition 2.5 the elements of $\mathscr{D}_\lambda'$ are characterized by

$$\Psi^{\sigma(a)} = e^{-(i\lambda+\rho)(\log a)}\Psi, \qquad a \in A \tag{2}$$

it is clear that Ψ as defined by (1) is a member of $\mathscr{D}_\lambda'$. On the other hand suppose $\Phi \in \mathscr{D}_\lambda'$. Fix $\varphi \in \mathscr{E}(K/M)$ and if $\alpha \in \mathscr{D}(A)$ let us denote the function $(kM, a) \to \varphi(kM)\,\alpha(a)$ on $(K/M) \times A$ by $\varphi \otimes \alpha$. By Proposition 1.1 in Chap. I we can regard $\varphi \otimes \alpha$ as a function on Ξ. The functional τ_φ given by

$$\tau_\varphi : \alpha \to \Phi(\varphi \otimes \alpha), \qquad \alpha \in \mathscr{D}(A)$$

is clearly a distribution on A; if $R(b)$ denotes the right translation $a \to ab$ on A we find

$$(\tau_\varphi)^{R(b)}(\alpha) = \Phi^{\sigma(b)}(\varphi \otimes \alpha)$$

so in view of (2),

$$(\tau_\varphi)^{R(b)} = e^{-(i\lambda+\rho)(\log b)}\tau_\varphi\,.$$

This implies that τ_φ is a constant multiple of a character of A, more precisely

$$\Phi(\varphi \otimes \alpha) = \tau_\varphi(\alpha) = c_\varphi \int_A e^{(i\lambda+\rho)(\log a)}\alpha(a)\,da \tag{3}$$

for all $\alpha \in \mathscr{D}(A)$, c_φ being a constant. Fix $\alpha_0 \in \mathscr{D}(A)$ such that the integral of $e^{(i\lambda+\rho)(\log a)}\alpha_0(a)$ over A is 1; then we see that the map $\varphi \to c_\varphi = \Phi(\varphi \otimes \alpha_0)$ is a distribution S on K/M and by (3),

$$\Phi(\varphi \otimes \alpha) = \int_{K/M}\left(\int_A (\varphi \otimes \alpha)(kaMN)\, e^{(i\lambda+\rho)(\log a)} da\right) dS(k_M)$$

for all $\varphi \in \mathscr{E}(K/M)$, $\alpha \in \mathscr{D}(A)$. The corresponding functions $\varphi \otimes \alpha$ span a dense subspace of $\mathscr{D}(K/M \times A)$ (see [41], Chap. IV) so the proposition follows.

In view of the characterizing condition (2) the problem of determining the conical distributions on Ξ amounts to finding the MN-invariants in $\mathscr{D}'(\Xi)$ which are "homogeneous" under the transformations $\sigma(a)$, $a \in A$. This leads to considering the orbits of the group MAN acting on Ξ. For each $s \in W$ fix $m_s \in M'$ in the coset $s \in M'/M$ and put $\xi_s = m_sMN$. Let Ξ_s denote the orbit $MNA \cdot \xi_s$; then the Bruhat decomposition implies

$$\Xi = \bigcup_{s \in W} \Xi_s \qquad \text{(disjoint union).} \tag{4}$$

Moreover, Proposition 1.2 (ii) in Chap I shows that if $\xi \in \Xi_s$ there exists a unique $a(\xi) \in A$ such that $\xi = mna(\xi) \cdot \xi_s$ with $m \in M$, $n \in N$. Now fix $s \in W$ and consider the groups

$$N^s = m_sNm_s^{-1}, \qquad \bar{N}^s = m_s\bar{N}m_s^{-1}$$

$$N_s' = N \cap \bar{N}^s, \qquad \bar{N}_s = m_s^{-1}N_s'm_s = \bar{N} \cap N^{s^{-1}}$$

whose Lie algebras, respectively, are given by

$$\mathfrak{n}^s = \sum_{s^{-1}\beta>0} \mathfrak{g}_\beta, \qquad \bar{\mathfrak{n}}^s = \sum_{s^{-1}\beta<0} \mathfrak{g}_\beta$$

$$\mathfrak{n}_s' = \sum_{\beta>0,\, s^{-1}\beta<0} \mathfrak{g}_\beta, \qquad \bar{\mathfrak{n}}_s = \sum_{\beta<0,\, s\beta>0} \mathfrak{g}_\beta$$

We have $\mathfrak{n} = (\mathfrak{n} \cap \mathfrak{n}^s) + \mathfrak{n}_s'$ and as is well-known,

$$N = (N \cap N^s)N_s' = N_s'(N \cap N^s) \tag{5}$$

607/5/1-6

in the sense that the product maps are diffeomorphisms. Furthermore, Haar measures on the factors correspond to dn under the product maps; in particular if we choose a Haar measure dn_s' on N_s' it corresponds via the identification $N/(N \cap N^s) = N_s'$ to an N-invariant measure on $N/(N \cap N^s)$. The subgroup of MNA which leaves ξ_s fixed equals $M(N \cap N^s)$ so the map

$$(n_s', a) \to n_s'a \cdot \xi_s$$

is a diffeomorphism of $N_s' \times A$ onto Ξ_s; in view of the remarks above the measure $dn_s'da$ is hereby carried into an MN-invariant measure $d\nu$ on Ξ_s.

Ignoring convergence questions for the time being we ask what properties a function ψ on Ξ_s should have in order that the functional

$$\varphi \to \int_{\Xi_s} \varphi(\xi)\, \psi(\xi)\, d\nu(\xi)$$

should be MN-invariant and satisfy (2). These requirements lead to the relations

$$\psi(n_s'a \cdot \xi_s) \equiv \psi(a \cdot \xi_s) \equiv e^{(is\lambda+s\rho)(\log a)}\psi(\xi_s),$$

so we define a functional $\Psi'_{\lambda,s}$ by

$$\Psi'_{\lambda,s}(\varphi) = \int_{\Xi_s} \varphi(\xi)\, e^{(is\lambda+s\rho)(\log a(\xi))}\, d\nu(\xi) \tag{6}$$

on the set of $\varphi \in \mathscr{D}(\Xi)$ for which the integral converges absolutely. If $s = e$ then $\Xi_s = A \cdot \xi_0$ which is closed in Ξ so $\Psi'_{\lambda,e}$ is for each $\lambda \in \mathfrak{a}_c^*$ a distribution on Ξ, in fact a conical distribution on the space $\mathscr{D}_\lambda'(\Xi)$. For the other orbits Ξ_s we have the following criterion for absolute convergence.

Lemma 4.2. *The integral defining $\Psi'_{\lambda,s}$ converges absolutely for all $\varphi \in \mathscr{D}(\Xi)$ if and only if*

$$\mathrm{Re}(\langle \alpha, i\lambda \rangle) > 0 \qquad \text{for} \quad \alpha \in \Sigma^+ \cap s^{-1}\Sigma^-, \tag{7}$$

Re *denoting real part. If* (7) *is satisfied $\Psi'_{\lambda,s}$ is a conical distribution in $\mathscr{D}_\lambda'(\Xi)$.*

Proof. Let $\mu = is\lambda + s\rho$. Then, assuming absolute convergence,

$$\Psi'_{\lambda,s}(\varphi) = \int_{N_s' \times A} \varphi(n_s' a m_s \cdot \xi_0)\, e^{\mu(\log a)}\, dn_s'\, da.$$

The image of dn_s' under the map $n_s' \to m_s^{-1}n_s'm_s$ is a Haar measure $d\bar{n}_s$ on $\bar{N}_s$. Writing $a^{s^{-1}} = m_s^{-1}am_s$ the integral above becomes

$$\int_{\bar{N}_s\times A} \varphi(m_s\bar{n}_s a^{s^{-1}} \cdot \xi_0)\, e^{\mu(\log a)}\, d\bar{n}_s\, da$$

$$= \int_{\bar{N}_s\times A} \varphi(m_s k(\bar{n}_s) \exp H(\bar{n}_s)\, a \cdot \xi_0)\, e^{(s^{-1}\mu)(\log a)}\, d\bar{n}_s\, da$$

$$= \int_{\bar{N}_s} e^{-(s^{-1}\mu)(H(\bar{n}_s))}\, d\bar{n}_s \int_A \varphi(m_s k(\bar{n}_s)\, a \cdot \xi_0)\, e^{(s^{-1}\mu)(\log a)}\, da.$$

Here the integral over A is absolute convergent for any $\varphi \in \mathscr{D}(\Xi)$; taking φ K-invariant the computation above shows that absolute convergence of (6) for all $\varphi \in \mathscr{D}(\Xi)$ holds if and only if the integral

$$\int_{\bar{N}_s} e^{-(i\lambda+\rho)(H(\bar{n}_s))}\, d\bar{n}_s \tag{8}$$

is absolutely convergent. But according to [19(a), Theorem 1], this happens if and only if $\operatorname{Re}\langle(i\lambda, \alpha)\rangle > 0$ for $\alpha \in \Sigma^+ \cap s^{-1}\Sigma^-$. (This is a consequence of (30) derived below.) Assuming now this to be the case we consider the functional $S'_{\lambda,s}$ on $\mathscr{E}(K/M)$ defined by

$$S'_{\lambda,s}(F) = \int_{\bar{N}_s} F(m_s k(\bar{n}_s)M)\, e^{-(i\lambda+\rho)(H(\bar{n}_i))}\, d\bar{n}_s\,. \tag{9}$$

Thus $S'_{\lambda,s}$ is a distribution on K/M (even extends to a measure), the computations above show that

$$\Psi'_{\lambda,s}(\varphi) = \int_{K/M} \left(\int_A \varphi(kaMN)\, e^{(i\lambda+\rho)(\log a)}\, da\right) dS'_{\lambda,s}(k_M) \tag{10}$$

so by Proposition 4.1, $\Psi'_{\lambda,s} \in \mathscr{D}_\lambda'$. The MN-invariance was already remarked so the lemma is proved.

The problem is now to extend the function $\lambda \to \Psi'_{\lambda,s}$ on the tube (7) to a meromorphic distribution-valued function $\mathfrak{a}_c{}^*$ and build all conical distributions up from these.

The convolution $*$ of the distributions on K induces a convolution $\times$ of distributions on K/M as follows. If T_1, $T_2 \in \mathscr{D}'(K)$ we have as usual

$$(T_1 * T_2)(f) = \int_{K\times K} f(k_1k_2)\, dT_1(k_1)\, dT_2(k_2), \qquad f \in \mathscr{D}(K).$$

Let π denote the natural mapping of G onto G/K and if $f \in \mathscr{D}(K)$ let $f_\pi \in \mathscr{D}(K/M)$ be defined by

$$f_\pi(kM) = \int_M f(km)\, dm.$$

Given $S \in \mathscr{D}'(K/M)$ we define $\tilde{S} \in \mathscr{D}'(K)$ by $\tilde{S}(f) = S(f_\pi)$ $(f \in \mathscr{D}(K))$. In particular, if $F \in \mathscr{D}(K/M)$ then $\tilde{F} = F \circ \pi$. Now if S_1, S_2 are two distributions on K/M their convolution $S_1 \times S_2$ is the distribution on K/M given by

$$(S_1 \times S_2)(F) = (\tilde{S}_1 * \tilde{S}_2)(\tilde{F}), \qquad F \in \mathscr{D}(K/M).$$

LEMMA 4.3. *Let S_1, $S_2 \in \mathscr{D}'(K/M)$ and assume that S_2 is M-invariant. Then*

$$(S_1 \times S_2)(F) = \iint F(k_1k_2M)\, dS_1(k_1M)\, dS_2(k_2M).$$

Proof. We use the notation above and put

$$g(k_1) = \int_K \tilde{F}(k_1k_2)\, d\tilde{S}_2(k_2).$$

Then

$$g(k_1) = \int_{K/M} F(k_1k_2M)\, dS_2(k_2M)$$

so since S_2 is M-invariant, $g(k_1m) \equiv g(k_1)$. Hence

$$(S_1 \times S_2)(F) = \int_K g(k_1)\, d\tilde{S}_1(k_1) = \int_{K/M} g(k_1)\, dS_1(k_1M)$$

as desired.

Each $s \in W$ can be written $s = s_{\alpha_1} \cdots s_{\alpha_p}$ where each α_i is a simple restricted root and s_{α_i} the corresponding Weyl reflexion. This expression for s is called *reduced* if p is as small as possible. If α is a simple root then s_α permutes $\Sigma^+ - \{\alpha, 2\alpha\}$ so $\bar{\mathfrak{n}}_{s_\alpha} = \mathfrak{g}_{-\alpha} + \mathfrak{g}_{-2\alpha}$.

THEOREM 4.4. *Let $s \in W$ be arbitrary and $s = s_{\alpha_1} \cdots s_{\alpha_p}$ any reduced expression of s. Put $s^{(q)} = s_{\alpha_{q+1}} \cdots s_{\alpha_p}$ $(1 \leqslant q \leqslant p-1)$, $s^{(p)} = e$. Then for a suitable normalization of the Haar measures on $\bar{N}_{s_{\alpha_i}}$ we have*

$$S'_{\lambda,s} = S'_{s'\lambda, s_{\alpha_1}} \times \cdots \times S'_{s^{(p)}\lambda, s_{\alpha_p}}$$

for all λ in the tube

$$\mathrm{Re}(\langle\alpha, i\lambda\rangle) > 0 \qquad \textit{for} \quad \alpha \in \Sigma^+ \cap s^{-1}\Sigma^-.$$

Proof. We assume λ lies in the tube above so (9) is absolutely convergent. It is known [24, Lemma 2.3] that since the expression for s is reduced, $\beta = -s^{-1}\alpha_1$ is > 0. Let (α_1) denote the set of positive roots proportional to α_1 (so $(\alpha_1) = \Sigma^+ \cap \{\alpha_1, 2\alpha_1\}$); define (β) and $(-\beta)$ similarly. Since s_{α_1} permutes $\Sigma^+ - (\alpha_1)$ and since $s^{-1} = (s')^{-1}s_{\alpha_1}$ we have

$$s^{-1}(\Sigma^+ - (\alpha_1)) = (s')^{-1}(\Sigma^+ - (\alpha_1))$$

that is

$$s^{-1}\Sigma^+ - (-\beta) = (s')^{-1}\,\Sigma^+ - (\beta)$$

whence we obtain the disjoint union

$$\Sigma^- \cap s^{-1}(\Sigma^+) = (\Sigma^- \cap (s')^{-1}\,\Sigma^+) \cup (-\beta). \tag{11}$$

As in [19(a)] we consider for each $\alpha \in \Sigma_0{}^+$ the subalgebra $\mathfrak{g}_{(\alpha)}$ of $\mathfrak{g}$ generated by the root spaces $\mathfrak{g}_\alpha$ and $\mathfrak{g}_{-\alpha}$ and let G_α denote the analytic subgroup of G corresponding to $\mathfrak{g}_{(\alpha)}$. Then G_α is semisimple and if $\mathfrak{k}_\alpha = \mathfrak{k} \cap \mathfrak{g}_{(\alpha)}$, $\mathfrak{p}_\alpha = \mathfrak{p} \cap \mathfrak{g}_{(\alpha)}$ then $\mathfrak{g}_{(\alpha)} = \mathfrak{k}_\alpha + \mathfrak{p}_\alpha$ is a Cartan decomposition of $\mathfrak{g}_{(\alpha)}$. The space $\mathfrak{a}_\alpha = \boldsymbol{R}H_\alpha$ is a maximal abelian subspace of $\mathfrak{p}_\alpha$; let $\mathfrak{a}_\alpha{}^+$ be the Weyl chamber containing H_α and $G_\alpha = K_\alpha A_\alpha N_\alpha$ the corresponding Iwasawa decomposition of G_α. Since $[\mathfrak{g}_\alpha, \mathfrak{g}_\alpha] = \mathfrak{g}_{2\alpha}$ the group N_α has Lie algebra $\mathfrak{g}_\alpha + \mathfrak{g}_{2\alpha}$. Put $\bar{N}_\alpha = \theta N_\alpha$; if α is simple $\bar{N}_\alpha = \bar{N}_{s_\alpha}$. If M_α denotes the centralizer of A_α in K_α then

$$M_\alpha = M \cap K_\alpha \tag{12}$$

because G_α leaves fixed each point on the hyperplane $\alpha = 0$ in $\mathfrak{a}$. Since $P = MAN$ we put $P_\alpha = M_\alpha A_\alpha N_\alpha$.

Now by (11) we have

$$\bar{\mathfrak{n}}_s = \bar{\mathfrak{n}}_{s'} + \bar{\mathfrak{n}}_\beta$$

which, as is well-known implies the global decomposition

$$\bar{N}_s = \bar{N}_{s'}\bar{N}_\beta = \bar{N}_\beta\bar{N}_{s'}, \tag{13}$$

in the sense that the product maps are diffeomorphisms and Haar measures on the factors correspond to $d\bar{n}_s$ under the product maps. Identifying K/M and G/P we have, therefore,

$$S'_{\lambda,s}(F) = \int_{\bar{N}_s} F(m_s\bar{n}_s P)\, e^{-(i\lambda+\rho)(H(\bar{n}_s))}\, d\bar{n}_s$$

$$= \int_{\bar{N}_\beta \bar{N}_{s'}} F(m_s\bar{n}_\beta\bar{n}_{s'}P)\, e^{-(i\lambda+\rho)(H(\bar{n}_\beta\bar{n}_{s'}))}\, d\bar{n}_\beta\, d\bar{n}_{s'},$$

both sides being absolutely convergent. Now according to $G_\beta = K_\beta A_\beta N_\beta$ we can write $\bar{n}_\beta = k_\beta a_\beta n_\beta$ so the integral equals

$$\int_{\bar{N}_\beta \bar{N}_{s'}} F(m_s k_\beta a_\beta n_\beta \bar{n}_{s'}P)\, e^{-(i\lambda+\rho)(H(a_\beta n_\beta \bar{n}_{s'}))}\, d\bar{n}_\beta\, d\bar{n}_{s'}\,.$$

Now since $\bar{N}_{s'} = N^{(s')^{-1}} \cap \bar{N}$ we have (cf., Eq. (5) of this section)

$$N^{(s')^{-1}} = \bar{N}_{s'}(N^{(s')^{-1}} \cap N) \tag{14}$$

in the sense that the product map is a measure-preserving diffeomorphism. Thus we have the identification

$$\bar{N}_{s'} = N^{(s')^{-1}}/(N^{(s')^{-1}} \cap N), \tag{15}$$

also in the sense that the invariant measures correspond. Now N_β obviously normalizes $N^{(s')^{-1}}$ so we have by (14) a unique decomposition

$$n_\beta \bar{n}_{s'} n_\beta^{-1} = (\bar{n}_{s'})^* n.$$

The mapping $\bar{n}_{s'} \to (\bar{n}_{s'})^*$ of $\bar{N}_{s'}$ onto itself corresponds via (15) to conjugation by n_β which by the nilpotency of N_β preserves the invariant measure on the coset space. Thus n_β can be omitted in the last integral. Next we use the substitution $\bar{n}_{s'} \to a_\beta \bar{n}_{s'} a_\beta^{-1}$; it is clear that $d(a_\beta \bar{n}_{s'} a_\beta^{-1}) = e^{2\vartheta(\log a_\beta)}\, d\bar{n}_{s'}$ where $2\vartheta = \sum_{\alpha<0,s'\alpha>0} \alpha$. Since

$$-2\rho = \sum_{\alpha<0,s'\alpha>0} \alpha + \sum_{\alpha<0,s'\alpha<0} \alpha,$$

$$2(s')^{-1}\rho = \sum_{\alpha<0,s'\alpha>0} \alpha + \sum_{\alpha>0,s'\alpha>0} \alpha,$$

we find by addition $2\vartheta = (s')^{-1}\rho - \rho$. Hence the integral above reduces to

$$\int_{\bar{N}_\beta \bar{N}_{s'}} F(m_s k_\beta \bar{n}_{s'} P)\, e^{-(i\lambda+\rho)(H(\bar{n}_{s'}))}\, e^{-(i\lambda+(s')^{-1}\rho)(\log a_\beta)}\, d\bar{n}_\beta\, d\bar{n}_{s'}\,. \tag{16}$$

Now we construct the group G_{α_1} and the appropriate subgroups K_{α_1}, A_{α_1}, N_{α_1}, $\bar{N}_{\alpha_1}$, and M_{α_1}. These groups are obtained by conjugating the previous subgroups by $m_{s'}$ because $s'\beta = \alpha_1$. Thus if $\bar{n}_{\alpha_1} = m_{s'}\bar{n}_\beta m_{s'}^{-1}$ is decomposed

$$\bar{n}_{\alpha_1} = k_{\alpha_1} a_{\alpha_1} n_{\alpha_1} \qquad (k_{\alpha_1} \in K_{\alpha_1}\,,\, a_{\alpha_1} \in A_{\alpha_1}\,,\, n_{\alpha_1} \in N_{\alpha_1})$$

we have

$$k_{\alpha_1} = m_{s'} k_\beta m_{s'}^{-1}, \qquad a_{\alpha_1} = m_{s'} a_\beta m_{s'}^{-1}, \qquad n_{\alpha_1} = m_{s'} n_\beta m_{s'}^{-1}.$$

Since $m_s m_{s'}^{-1} = m m_{s_{\alpha_1}}$ $(m \in M)$ our integral becomes (with $d\bar{n}_{\alpha_1}$ corresponding to $d\bar{n}_\beta$)

$$\int_{\bar{N}_{\alpha_1}\bar{N}_{s'}} F(m_{s_{\alpha_1}} k_{\alpha_1}(\bar{n}_{\alpha_1})\, m_{s'}\bar{n}_{s'} P)\, e^{-(i\lambda+\rho)(H(\bar{n}_{s'}))} e^{-(is'\lambda+\rho)(\log a_{\alpha_1})}\, d\bar{n}_{\alpha_1}\, d\bar{n}_{s'}\,.$$

We have written $m_{s_{\alpha_1}}$ instead of $mm_{s_{\alpha_1}}$ since the distribution $S'_{\lambda,s}$ is M-invariant. Putting $k = m_{s_{\alpha_1}} k_{\alpha_1}(\bar{n}_{\alpha_1})$ this integral equals ($\tau(k)$ as in Notation §2),

$$\int_{\bar{N}_{\alpha_1}} S'_{\lambda,s'}(F^{\tau(k^{-1})})\, e^{-(is'\lambda+\rho)(H(\bar{n}_{\alpha_1}))}\, d\bar{n}_{\alpha_1}$$

which in view of (9) and Lemma 4.3 equals

$$(S'_{s'\lambda, s_{\alpha_1}} \times S'_{\lambda,s'})(F).$$

The theorem now follows by iteration, because the expression for $s^{(q)}$ is necessarily reduced.

Remark. This proof is an adaptation of Schiffmann's proof of Theorem 1 in [40(a)]. It is an extension of the proof in the paper [19(a)] by Gindikin–Karpelevič, where in fact the integral (9) is reduced to (16) in the case when F is constant. See also H. Jacquet: Fonctions de Whittaker associées aux groupes de Chevalley, *Bull. Soc. Math. France* **95** (1967) 243–309; in particular §2.

Now let $\alpha \in \Sigma^+$ be simple and $\mu \in \mathfrak{a}_c^*$. Since s_α permutes $\Sigma^+ - (\alpha)$, Lemma 4.2 implies that

$$S'_{\mu,s_\alpha}(F) = \int_{\bar{N}_\alpha} F(m_{s_\alpha}\bar{n}_\alpha P)\, e^{-(i\mu+\rho)(H(\bar{n}_\alpha))}\, d\bar{n}_\alpha \tag{17}$$

converges absolutely for $F \in \mathscr{E}(G/P)$ if and only if $\mathrm{Re}(\langle \alpha, i\mu \rangle) > 0$. Select $m_\alpha^* \in K_\alpha$ such that $Ad_{G_\alpha}(m_\alpha^*) = -1$ on $\mathfrak{a}_\alpha$. Then $m_{s_\alpha}^{-1} m_\alpha^* \in M$. Note that by (12) G_α/P_α is canonically imbedded as a submanifold in G/P. Let $\bar{\mu}$ denote the restriction of μ to $\mathfrak{a}_\alpha$ and consider the distribution $T_{\bar{\mu},s_\alpha}$ on G_α/P_α given by

$$T_{\bar{\mu},s_\alpha}(f) = \int_{\bar{N}_\alpha} f(m_\alpha^* \bar{n}_\alpha P_\alpha)\, e^{-(i\mu+\rho)(H(\bar{n}_\alpha))}\, d\bar{n}_\alpha \tag{18}$$

for $f \in \mathscr{E}(G_\alpha/P_\alpha)$. According to Lemma 4.2 for the rank-one space G_α/K_α, absolute convergence of (18) is equivalent to $\mathrm{Re}(\langle \bar{\alpha}, i\bar{\mu} + \bar{\rho} - \rho_\alpha \rangle) > 0$ where ρ_α is the ρ-function for G_α/K_α. But comparing with (17) we deduce since S'_{μ,s_α} is M-invariant that

$$S'_{\mu,s_\alpha}(F) = T_{\bar{\mu},s_\alpha}(\bar{F}), \qquad F \in \mathscr{E}(G/P), \tag{19}$$

$\bar{F}$ denoting restriction of F to G_α/P_α. Comparing the convergence criteria we have also $\langle \alpha, \rho \rangle = \langle \alpha, \rho_\alpha \rangle$, a fact which is easily derived directly.

Theorem 4.4 and (19) reduce the problem of finding a meromorphic continuation of the functions $\lambda \to S'_{\lambda,s}$ to a problem in a single complex variable. For this problem we use Theorem 1.14.

THEOREM 4.5. *Let $s \in W$ be arbitrary and define the function $\boldsymbol{d}_s$ on $\mathfrak{a}_c^*$ by*

$$\boldsymbol{d}_s(\lambda) = \prod_{\alpha \in \Sigma_0^+ \cap s^{-1}\Sigma_0^-} \Gamma(\langle i\lambda, \alpha_0 \rangle)\, 2^{-\langle i\lambda, \alpha_0 \rangle},$$

where for any $\alpha \in \Sigma$ we put $\alpha_0 = \alpha/\langle \alpha, \alpha \rangle$. Then the map

$$\lambda \to \boldsymbol{d}_s^{-1}(\lambda)\, \Psi'_{\lambda,s}$$

extends from the tube (7) *to a holomorphic function, denoted $\Psi_{\lambda,s}$, on $\mathfrak{a}_c^*$ with values in $\mathscr{D}'(\Xi)$. For each $\lambda \in \mathfrak{a}_c^*$, $\Psi_{\lambda,s}$ is a conical distribution in $\mathscr{D}_\lambda'(\Xi)$. Let $S_{\lambda,s}$ correspond to $\Psi_{\lambda,s}$ in the sense of Proposition* 4.1.

Proof. First let us consider the case when G/K has rank one. In this case W has order 2. If $s = e$ the theorem is obvious as remarked earlier; we assume therefore $s \neq e$ in W. Then α being the unique element in Σ_0^+,

$$\boldsymbol{d}_s(\lambda) = \Gamma(\langle i\lambda, \alpha_0 \rangle)\, 2^{-\langle i\lambda, \alpha_0 \rangle}$$

is meromorphic on $\mathfrak{a}_c^*$; the poles are all simple and are given by

$$-\langle i\lambda, \alpha_0\rangle \in \boldsymbol{Z}^+. \tag{20}$$

Now m^* being as in Lemma 1.15 we have for $\mathrm{Re}(\langle\alpha, i\lambda\rangle) > 0$,

$$S'_{\lambda,s}(F) = \int_{\bar{N}} F(m^*k(\bar{n})M)\, e^{-(i\lambda+\rho)(H(\bar{n}))}\, d\bar{n}. \tag{21}$$

Let $z \in \boldsymbol{C}$ be determined by $\lambda = z\rho$ and let ψ denote the function on $K/M - \{m^*M\}$ given by $\psi(k(\bar{n})M) = e^{\rho(H(\bar{n}))}$. Then by (8) §2 the right side of (21) equals

$$\int_{K/M} F(m^*kM)\, \psi(kM)^{-iz+1}\, dk_M = \int_{K/M} F(kM)\, \psi(m^*kM)^{-iz+1}\, dk_M$$

so

$$S'_{\lambda,s}(F) = \int_{\bar{N}} e^{-(i\lambda-\rho)(H(\bar{n}(m^*\bar{n})))} F(\bar{n}P)\, e^{-2\rho(H(\bar{n}))}\, d\bar{n}$$

and now we use Theorem 1.14. Using the same notation as in the proof of Corollary 1.17 ($4p = m_\alpha + 2m_{2\alpha}$, $\lambda = z\rho$) we find that the integral becomes

$$c_1 \int_{\mathfrak{g}_{-\alpha}} \int_{\mathfrak{g}_{-2\alpha}} F^*(X, Y) \left[1 + \frac{1 + 2c\,|X|^2}{c^2\,|X|^4 + 4c\,|Y|^2}\right]^{p(1-iz)} dX\, dY,$$

where F^* is a C^∞ integrable function. In order to extend this to a meromorphic function of z we may assume F^* of compact support; in fact if we write $F^* = F' + F''$, where $F' = 0$ near 0, $F'' \in \mathscr{D}(\mathfrak{g}_{-\alpha} \times \mathfrak{g}_{-2\alpha})$ the contribution from F' will be holomorphic for all $z \in \boldsymbol{C}$.

Now (assuming $\mathfrak{g}_{-2\alpha} \neq \{0\}$) we use polar coordinates in each of the spaces $\mathfrak{g}_{-\alpha}$, $\mathfrak{g}_{-2\alpha}$ and write $r = c^{1/2}\,|X|$, $s = 2c^{1/2}\,|Y|$. Then the integral becomes

$$\int_0^\infty \int_0^\infty F_1(r, s) \left(1 + \frac{1 + 2r^2}{r^4 + s^2}\right)^{p(1-iz)} r^{m_\alpha - 1} s^{m_{2\alpha}-1}\, dr ds, \tag{22}$$

where F_1 is (essentially) the average of F^* over $S^r \times S^s$ (product of spheres). Next we put $x = r^2$, $y = s$ and then use polar coordinates (ρ, θ) in the quadrant $x \geqslant 0$, $y \geqslant 0$. Our expression then becomes

$$\int_0^\infty \rho^{2piz-1} \varphi_z(\rho)\, d\rho, \qquad \mathrm{Re}(iz) > 0, \tag{23}$$

where

$$\varphi_z(\rho) = \int_0^{\pi/2} F_1((\rho \cos \theta)^{1/2}, \rho \sin \theta)(1 + 2\rho \cos \theta + \rho^2)^{p(1-iz)}$$
$$\times (\cos \theta)^{\frac{1}{2}m_\alpha - 1} (\sin \theta)^{m_{2\alpha}-1}\, d\theta. \tag{24}$$

The function $F_1(r, s)$ is an even C^∞ function of r and s; thus $\varphi_z \in \mathscr{D}(\boldsymbol{R})$, and $z \to \varphi_z(\rho)$ extends to an analytic function on $\boldsymbol{C}$. But as is well-known an integral of the form (23) extends to a meromorphic function of $z \in \boldsymbol{C}$ with at most simple poles when

$$-2piz \in \boldsymbol{Z}^+. \tag{25}$$

In the case when $\mathfrak{g}_{-2\alpha} = \{0\}$ we still get the expression (23) but this time $\varphi_z(\rho)$ is the simpler function

$$\tfrac{1}{2}F_1(\rho^{1/2})(1 + \rho)^{2p(1-iz)},$$

so the possible poles for (23) are still given by (25). Since this condition coincides with (20) we have obtained the desired extension of $\lambda \to \boldsymbol{d}_s(\lambda)^{-1}\Psi'_{\lambda,s}$ to a holomorphic mapping from $\mathfrak{a}_c{}^*$ to $\mathscr{D}'(\Xi)$ in the rank-one case.

For the case of arbitrary rank we appeal to Theorem 4.4. Consider the j-th factor $S'_{s^{(j)}\lambda, s_{\alpha_j}}$ in the decomposition of $S'_{\lambda,s}$ and use (19). Consider the rank-one symmetric space $G_{\alpha_j}/K_{\alpha_j}$ and let $\langle\ ,\ \rangle_j$ denote the associated Killing form. It was proved above that for each $f \in \mathscr{E}(G_{\alpha_j}/P_{\alpha_j})$

$$\nu \to \Gamma(\langle i\nu, (\bar{\alpha}_j)_0\rangle_j)^{-1}\, T_{\nu, s_{\alpha_j}}(f)$$

is holomorphic on $(\mathfrak{a}_{\alpha_j})_c{}^*$. Hence by (19) and Theorem 4.4,

$$\lambda \to \prod_{1 \leqslant j \leqslant p} \Gamma(\langle i(s^{(j)}\lambda)^-, (\bar{\alpha}_j)_0\rangle)^{-1}\, S'_{\lambda,s} \tag{26}$$

is holomorphic on the entire space $\mathfrak{a}_c{}^*$. But since $\mathfrak{a}_{\alpha_j}$ is one-dimensional we have just as in (5), §3, that the argument for Γ above equals

$$\langle is^{(j)}\lambda, (\alpha_j)_0\rangle. \tag{27}$$

Now we have by (11) if $\beta_1 = -s^{-1}\alpha_1$,

$$\Sigma_0{}^+ \cap s^{-1}\Sigma_0{}^- = (\Sigma_0{}^+ \cap (s')^{-1}\, \Sigma_0{}^-) \cup \{\beta_1\}. \tag{28}$$

But for each q the expression $s^{(q)} = s_{\alpha_{q+1}} \cdots s_{\alpha_p}$ is reduced so as used earlier $\beta_q = -s^{(q)^{-1}}\alpha_{q+1}$ is in Σ_0^+; by iteration of (28) we find

$$\Sigma_0^+ \cap s^{-1}\Sigma_0^- = \{\beta_1, \ldots, \beta_p\}. \tag{29}$$

Since

$$s^{(j)^{-1}}\alpha_j = -s_{\alpha_p} \cdots s_{\alpha_{j+1}} s_{\alpha_j} \alpha_j = -s^{(j-1)^{-1}}\alpha_j = \beta_j$$

the expression (27) equals $\langle i\lambda, (\beta_j)_0\rangle$ so by (29) the product of the factors in (26) equals the Γ-factor in $\boldsymbol{d}_s(\lambda)^{-1}$. This proves Theorem 4.5.

Remark. If α is a simple root then $\Psi_{\lambda, s_\alpha} \neq 0$ for all $\lambda \in \mathfrak{a}_c^*$.

In fact, because of (19) we may assume that G/K has rank one so one just has to verify that if $F \in \mathscr{E}(K/M)$ is suitably chosen the integral (23) as a function of z has a nonvanishing residue at the points $2piz = -m$ ($m \in \boldsymbol{Z}^+$), or equivalently, that

$$\left(\frac{d^m \varphi_z}{d\rho^m}\right)_{\rho=0} \neq 0.$$

But this is immediate if we choose F such that for $X \in \mathfrak{g}_{-\alpha}$, $Y \in \mathfrak{g}_{-2\alpha}$ sufficiently small, $F^*(X, Y) = |X|^{2m}$.

Since

$$\begin{aligned}(i\lambda + (s')^{-1}\rho)(\log a_\beta) &= (is'\lambda + \rho)(\log a_{\alpha_1})\\ &= (is'\lambda + \rho_{\alpha_1})(\log a_{\alpha_1}) = (i\lambda + \rho_\beta)(\log a_\beta),\end{aligned}$$

Theorem 4.4 gives for $F \equiv 1$ the product formula of Gindikin–Karpelevič [19(a)],

$$\int_{\bar{N}_s} e^{-(i\lambda+\rho)(H(\bar{n}_s))}\, d\bar{n}_s = a_s' \prod_{1 \leqslant j \leqslant p} \boldsymbol{c}_j(\lambda_{\beta_j})$$

where $\boldsymbol{c}_j$ is the $\boldsymbol{c}$-function for the space G_{β_j}/K_{β_j}, λ_{β_j} the restriction of λ to $\mathfrak{a}_{\beta_j}$ and a_s' a constant. Denoting the left side of this formula by $\boldsymbol{c}_s$ we have by Corollary 1.17

$$\int_{\bar{N}_s} e^{-(i\lambda+\rho)(H(\bar{n}_s))}\, d\bar{n}_s = \boldsymbol{c}_s(\lambda) = a_s \boldsymbol{d}_s(\lambda)\, \boldsymbol{e}_s(\lambda) \tag{30}$$

where $d_s(\lambda)$ is as above,

$$e_s(\lambda)^{-1} = \prod_{\alpha\in\Sigma_0^+\cap s^{-1}\Sigma_0^-} \Gamma\left(\frac{1}{2}\left(\frac{m_\alpha}{2}+1+\langle i\lambda,\alpha_0\rangle\right)\right)\Gamma\left(\frac{1}{2}\left(\frac{m_\alpha}{2}+m_{2\alpha}+\langle i\lambda,\alpha_0\rangle\right)\right) \tag{31}$$

and a_s is a constant.

We consider now the dual transform $\Psi \to \check{\Psi}$ (Chap. I, §2) on $\mathscr{D}'(\Xi)$ given by

$$\check{\Psi}(f) = \Psi(\hat{f}), \qquad f \in \mathscr{D}(X).$$

Then $\check{\Psi}$ is a distribution on X. We recall the isomorphisms

$$\Gamma : \boldsymbol{D}(X) \to \boldsymbol{I}(A)$$

and

$$\hat{\Gamma} : \boldsymbol{D}(\Xi) \to \boldsymbol{D}(A)$$

from §2, Chap. I, and the vector spaces $\mathscr{E}_\lambda(X)$, $\mathscr{D}_\lambda'(\Xi)$ defined for each $\lambda \in \mathfrak{a}_c{}^*$ by

$$\mathscr{E}_\lambda(X) = \{f \in \mathscr{E}(X) \mid Df = \Gamma(D)(i\lambda)f \text{ for } D \in \boldsymbol{D}(X)\},$$
$$\mathscr{D}_\lambda'(\Xi) = \{\Psi \in \mathscr{D}'(\Xi) \mid D\Psi = \hat{\Gamma}(D)(i\lambda - \rho)\Psi \text{ for } D \in \boldsymbol{D}(\Xi)\}.$$

These spaces are eigenspaces of families of differential operators corresponding to a given system of eigenvalues. Note that $\hat{\Gamma}(D)(i\lambda - \rho) = \chi_{i\lambda-\rho}(D)$ so the space $\mathscr{D}_\lambda'(\Xi)$ above is the one defined in the beginning of this section. We recall also that if Δ is the Laplace–Beltrami operator corresponding to the Riemannian structure on X given by the Killing form of $\mathfrak{g}$ then by [21(i), p. 271], (cf. also [22(b) p. 451]),

$$\Gamma(\Delta)(i\lambda) = -\langle\lambda,\lambda\rangle - \langle\rho,\rho\rangle.$$

The space $\mathscr{E}(X)$ is given the usual topology corresponding to uniform convergence of sequences of functions and of their derivatives. The space $\mathscr{D}'(\Xi)$ is given the strong dual topology. The spaces $\mathscr{E}_\lambda(X)$ and $\mathscr{D}_\lambda'(\Xi)$ are closed subspaces and are given the relative topologies.

PROPOSITION 4.6. *Let $\lambda \in \mathfrak{a}_c{}^*$. Then*

$$\check{\Psi} \in \mathscr{E}_\lambda(X) \qquad \textit{for each} \quad \Psi \in \mathscr{D}_\lambda'(\Xi).$$

More precisely, if according to Proposition 4.1 *we represent* Ψ *in the form*

$$\Psi(\varphi) = \int_{K/M} \left(\int_A \varphi(ka \cdot \xi_0) e^{(i\lambda+\rho)(\log a)} \, da \right) dS(k_M)$$

then

$$\check{\Psi}(x) = \int_B e^{(i\lambda+\rho)(A(x,b))} \, dS(b),$$

where $A(x, b) \in \mathfrak{a}$ *is the log of the complex distance from* o *to the horocycle through* x *with normal* b (*cf. Proposition* 1.4, *Chap. I*).

Proof. Let $f \in \mathscr{D}(X)$. Then

$$\hat{f}(ka \cdot \xi_o) = \int_N f(kan \cdot o) \, dn$$

and

$$\int_X f(x) \, dx = \int_{N \times A} f(an \cdot o) \, dn \, da.$$

Hence

$$\begin{aligned} \int_A \hat{f}(a \cdot \xi_o) \, e^{(i\lambda+\rho)(\log a)} \, da &= \int_A e^{(i\lambda+\rho)(\log a)} \int_N f(an \cdot o) \, dn \, da \\ &= \int_X e^{(i\lambda+\rho)(A(x,eM))} f(x) \, dx. \end{aligned}$$

Replacing here f by $f^{\tau(k^{-1})}$ if $b = kM$, we find

$$\begin{aligned} \check{\Psi}(f) = \Psi(\hat{f}) &= \int_{K/M} \left(\int_A \hat{f}(ka \cdot \xi_o) \, e^{(i\lambda+\rho)(\log a)} \, da \right) dS(k_M) \\ &= \int_B \left(\int_X e^{(i\lambda+\rho)(A(x,b))} f(x) \, dx \right) dS(b). \end{aligned}$$

If $\alpha \in \mathscr{D}(X)$ is identically 1 on a neighborhood of the support of f we can conclude from the Fubini theorem for distributions [41, p. 109], that

$$\begin{aligned} \check{\Psi}(f) &= \int_B \left(\int_X \alpha(x) \, e^{(i\lambda+\rho)(A(x,b))} f(x) \, dx \right) dS(b) \\ &= \int_X f(x) \left(\int_B e^{(i\lambda+\rho)(A(x,b))} \, dS(b) \right) dx \end{aligned}$$

which proves the stated formula for $\check{\Psi}(x)$.

Since [22(j), pp. 60–61],

$$A(gk \cdot x, k_1M) = A(x, k^{-1}k(g^{-1}k_1)M) + A(g \cdot o, k_1M)$$

we see that the function $h(x) = e^{(i\lambda+\rho)(A(x,b))}$ satisfies

$$\int_K h(gk \cdot x)\, dk = \Big(\int_K h(k \cdot x)\, dk\Big)\, h(g \cdot o)$$

and thus by [22(b), p. 439] is an eigenfunction of each $D \in \mathbf{D}(X)$. Since $A(k \cdot x, k_1M) = A(x, k^{-1}k_1M)$ the eigenvalues for h are independent of b, and are therefore the same as for the function

$$\int_B e^{(i\lambda+\rho)(A(x,b))}\, db.$$

Here $db = dk_M$ if $b = kM$. However since $A(gK, kM) = -H(g^{-1}k)$ [22(j), p. 61], this function equals the spherical function $\varphi_{-\lambda}(g^{-1})$ which by [21(i), p. 295], equals $\varphi_\lambda(g)$. Thus $h \in \mathscr{E}_\lambda(X)$ (cf. [22(b), p. 432]) as stated.

Definition. *Let* $\lambda \in \mathfrak{a}_c^*$. We shall call λ *simple* if the mapping $\Psi \to \tilde{\Psi}$ of $\mathscr{D}_\lambda'$ into $\mathscr{E}_\lambda$ is injective.

Lemma 4.7. *Let* $\lambda \in \mathfrak{a}_c^*$. *Then* λ *is simple if and only if the mapping*

$$F \in \mathscr{E}(B) \to f \in \mathscr{E}_\lambda(X) \tag{32}$$

given by

$$f(x) = \int_B e^{(i\lambda+\rho)(A(x,b))} F(b)\, db$$

is injective.

Proof. It suffices to prove the "if" part. Assume therefore $S \in \mathscr{D}'(B)$ satisfies

$$\int_B e^{(i\lambda+\rho)(A(x,b))} dS(b) = 0 \qquad \text{for} \quad x \in X.$$

If $F \in \mathscr{E}(K)$ it follows that

$$\int_K \Big(\int_B e^{(i\lambda+\rho)(A(k^{-1}\cdot x,b))}\, dS(b)\Big) F(k)\, dk \equiv 0.$$

This can be written

$$\int_K \Big(\int_K e^{(i\lambda+\rho)(A(x,kk_0M))}\, d\tilde{S}(k_0) \Big)\, F(k)\, dk \equiv 0$$

so the *function* $F_1 \in \mathscr{E}(B)$ defined by

$$F_1(kM) = \int_K F(kk_0^{-1})\, d\tilde{S}(k_0)$$

satisfies

$$\int_B e^{(i\lambda+\rho)(A(x,b))} F_1(b)\, db = 0.$$

Since the map (32) is assumed injective, $F_1 \equiv 0$, so $S = 0$, F being arbitrary.

LEMMA 4.8. *For each* $\lambda \in \mathfrak{a}_c^*$ *and each* $s \in W$, *the function* $(\Psi_{\lambda,s})^{\vee}$ *on* X *is given by*

$$(\Psi_{\lambda,s})^{\vee}\,(na \cdot o) = a_s e_s(\lambda)\, e^{(is\lambda+\rho)(\log a)}, \qquad n \in N, \quad a \in A,$$

or equivalently

$$\int_B e^{(i\lambda+\rho)(A(x,b))}\, dS_{\lambda,s}(b) \equiv a_s e_s(\lambda) \int_B e^{(is\lambda+\rho)(A(x,b))}\, d\delta_0(b).$$

Here $a_s \neq 0$ *is the constant from* (30) *and* δ_o *is the delta-distribution on* B *at the origin* eM.

Proof. We have by the N-invariance of $\Psi_{\lambda,s}$

$$(\Psi'_{\lambda,s})^{\vee}\,(na \cdot o) = (\Psi'_{\lambda,s})^{\vee}\,(a \cdot o)$$

and since $A(gK, kM) = -H(g^{-1}k)$ we get from (9)

$$(\Psi'_{\lambda,s})^{\vee}\,(a \cdot o) = \int_{\bar{N}_s} e^{-(i\lambda+\rho)(H(a^{-1}m_s k(\bar{n}_s))} e^{-(i\lambda+\rho)(H(\bar{n}_s))}\, d\bar{n}_s\,.$$

We put $c = m_s^{-1}\, a^{-1} m_s$ and we use the relation

$$H(ck(\bar{n})) = \log c + H(c\bar{n}c^{-1}) - H(\bar{n})$$

so

$$(\Psi'_{\lambda,s})^{\vee}(a \cdot o) = e^{-(i\lambda+\rho)(\log c)} \int_{\bar{N}_s} e^{-(i\lambda+\rho)(H(c\bar{n}_s c^{-1}))}\, d\bar{n}_s\,,$$

which (cf., proof of Theorem 4.4) equals

$$e^{-(i\lambda+\rho)(\log c)} e^{(\rho - s^{-1}\rho)(\log c)} \int_{\bar{N}_s} e^{-(i\lambda+\rho)(H(\bar{n}_s))}\, d\bar{n}_s\,.$$

In view of (30) this proves the lemma.

We recall now that each conical distribution on Ξ lies in $\mathscr{D}_\lambda'$ for some $\lambda \in \mathfrak{a}_c^*$.

THEOREM 4.9. *Assume* $\lambda \in \mathfrak{a}_c^*$ *satisfies the conditions*[1]

(i) λ *is regular* ($s\lambda \neq \lambda$ *for each* $s \neq e$ *in* W);

(ii) λ *is simple*;

(iii) $\mathbf{e}_{s^*}(\lambda) \neq 0 \qquad (s^* \in W$ *mapping* $\mathfrak{a}^+$ *onto* $-\mathfrak{a}^+)$.

Then the linear combinations

$$\sum_{s \in W} c_s \Psi_{\lambda,s}\,, \qquad c_s \in \mathbf{C}$$

constitute all the conical distributions in $\mathscr{D}_\lambda'$.

Proof. The distributions $\Psi_{\lambda,s} \in \mathscr{D}_\lambda'$ being conical, suppose $\Psi \in \mathscr{D}_\lambda'$ is a conical distribution. Then $\check{\Psi}$ is an N-invariant function in $\mathscr{E}_\lambda(X)$. Let $f \in \mathscr{E}(G)$ be given by $f(g) = \check{\Psi}(g \cdot o)$. Now let $\mathbf{D}(G)$ denote the set of left invariant differential operators on G. Given $D \in \mathbf{D}(G)$ let $D_{\mathfrak{a}}$ denote the unique element in $\mathbf{D}(A)$ such that

$$D - D_{\mathfrak{a}} \in \mathfrak{n}\mathbf{D}(G) + \mathbf{D}(G)\mathfrak{k}.$$

It is clear that $(Df)(a) = (D_{\mathfrak{a}}f)(a)$ for $a \in A$. Moreover if $\mathbf{D}_o(G)$ denotes the set of elements in $\mathbf{D}(G)$ which are invariant under right translations from K, the mapping $D \to e^{-\rho} \circ D_{\mathfrak{a}} \circ e^{\rho}$ is a homomorphism of $\mathbf{D}_o(G)$ onto $\mathbf{I}(A)$, the set of W-invariants in $\mathbf{D}(A)$ with kernel $\mathbf{D}_o(G) \cap \mathbf{D}(G)\mathfrak{k}$. [21(i), Theorem 1]. It follows that for each $L \in \mathbf{I}(A)$

$$L(e^{-\rho}f) = L(i\lambda)(e^{-\rho}f) \tag{33}$$

[1] For the equivalence of (ii) and (iii) see [22(k), (5)]; for proof in the rank one case see Theorem 1.1 in the following Chapter IV, §1.

where on the right L is viewed as a polynomial function on $\mathfrak{a}_c^*$. As proved by Steinberg [45] and Harish-Chandra (unpublished) the dimension of the space of solutions to (33) equals the order $|W|$ of W. In our case where λ is regular this is easier: The functions $a \to e^{is\lambda(\log a)}$ constitute $|W|$ linearly independent solutions of (33); on the other hand, there exist $|W|$ elements $u_j \in \boldsymbol{D}(A)$ such that $\boldsymbol{D}(A) = \sum_j \boldsymbol{I}(A)\, u_j$ [21(h), Lemma 11] from which follows easily that a solution φ to (33) is determined by the values $(u_j \cdot \varphi)(e)$ $(1 \leqslant j \leqslant |W|)$.

Under the regularity assumption on λ we have now proved that

$$\check{\Psi}(na \cdot o) = \sum_{s \in W} c_s e^{(is\lambda+\rho)(\log a)} \qquad c_s \in \boldsymbol{C}.$$

Since λ is simple and each $\boldsymbol{e}_s(\lambda) \neq 0$ it now follows from Lemma 4.8 that Ψ is a linear combination of the $\Psi_{\lambda,s}$.

Remark 1. If $\lambda = 0$ then all the distributions $\Psi_{\lambda,s}(s \in W)$ are proportional.

In fact, the residue of the expression (23) at the pole $z = 0$ is $\varphi_0(0)$ so we deduce from Theorem 4.4 that for $\lambda = 0$ each $S_{\lambda,s}$ is a constant multiple of the delta function on B at eM.

Remark 2. Theorem 4.9 shows that our guess that the set of conical distributions is parametrized by $\mathfrak{a}_c^* \times W$ is to a large extent true. But Remark 1 above raises an interesting question: Are the conical distributions in $\mathscr{D}_\lambda'$ for $\lambda = 0$ all proportional? As pointed out in the above proof the system (33) has $|W|$ independent solutions even for $\lambda = 0$. Are there conical distributions Ψ whose transforms $\check{\Psi}$ correspond to these solutions? According to Remark 1 these would not arise by our method of construction. The next theorem shows in particular that at least when G is the identity component of the orthogonal group for signature $(1, n)$ these extra Ψ do not exist.

THEOREM 4.10. *Suppose* $G = \boldsymbol{SO}_0(1, n)$, $n > 2$. *For each* $\lambda \in \mathfrak{a}_c^*$ *the linear combinations*

$$\sum_{s \in W} c_s \Psi_{\lambda,s}, \qquad c_s \in \boldsymbol{C}$$

constitute all the conical distributions in $\mathscr{D}_\lambda'$.

We first describe the main steps in the proof.

(i) If $\lambda \neq 0$ and $e_s(\lambda) \neq 0$ the statement follows from Theorem 4.9 and Theorem 1.1, Chap. IV.

(ii) Let ξ_o denote the origin in G/MN. We determine those $\lambda \in \mathfrak{a}_c^*$ for which $\mathscr{D}_\lambda'$ contains a conical distribution Φ with support contained in the orbit $\Xi_o = MNA \cdot \xi_o$ (and not proportional to $\Psi_{\lambda,e}$). It turns out that these λ are just the nonzero poles of the map $\lambda \to \Psi'_{\lambda,s}$ and the distributions Φ are just the corresponding residues; these are certain transversal derivatives of $\Psi_{\lambda,e}$.

(iii) If $\lambda = 0$ all conical distributions in $\mathscr{D}_\lambda'$ are proportional. This is the principal step in the proof; a similar phenomenon occurs in the theory of Eisenstein series.

(iv) Finally suppose λ not simple and $\Psi \in \mathscr{D}_\lambda'$. If λ is not a pole of $\Psi'_{\lambda,s}$, the remark following Theorem 4.5 shows that there exists a constant c such that $\Psi - c\Psi'_{\lambda,s}$ has support contained in Ξ_o and thus by (ii) is a multiple of $\Psi_{\lambda,e}$. If on the other hand λ is a pole of $\Psi'_{\lambda,s}$ one can, using Proposition 2.6 and Theorem 3.1, construct a conical function φ such that $\varphi\Psi \in \mathscr{D}_0'$. By (iii) $\varphi\Psi$ is proportional to $\Psi_{0,e}$, which shows in particular that Ψ has support contained in Ξ_o and is therefore by (ii) a linear combination of $\Psi_{\lambda,e}$ and $\Psi_{\lambda,s}$.

Passing now to the proof of the theorem we must describe the Lie algebras $\mathfrak{k}$, $\mathfrak{a}$, $\mathfrak{n}$, $\mathfrak{m}$ and the orthogonal complement $\mathfrak{l}$ of $\mathfrak{m}$ in $\mathfrak{k}$ (cf., Takahashi [47]. Now K is the set of matrices

$$\begin{vmatrix} 1 & 0 \\ 0 & u \end{vmatrix}, \qquad u \in \boldsymbol{SO}(n)$$

and if E_{ij} $(1 \leqslant i, j \leqslant n+1)$ denotes the usual basis of the space of $(n+1) \times (n+1)$ matrices, $E_{ij} - E_{ji}$ $(2 \leqslant i < j \leqslant n+1)$ is a basis for $\mathfrak{k}$. For $\mathfrak{a}$ we take

$$\mathfrak{a} = \boldsymbol{R}H \qquad \text{and} \qquad H = E_{12} + E_{21},$$

so M is the set of matrices

$$\begin{vmatrix} 1 & 0 & 0 \\ 0 & 1 & 0 \\ 0 & 0 & u \end{vmatrix}, \qquad u \in \boldsymbol{SO}(n-1)$$

and the vectors $X_{jk} = E_{jk} - E_{kj}$ $(3 \leqslant j < k \leqslant n+1)$ form a basis of $\mathfrak{m}$. Then the vectors

$$T_3 = E_{23} - E_{32}, \ldots, T_{n+1} = E_{2,n+1} - E_{n+1,2}$$

form a basis of $\mathfrak{l}$. We put

$$Y_3 = E_{13} + E_{31}, \ldots, Y_{n+1} = E_{1,n+1} + E_{n+1,1}$$

and

$$X_3 = Y_3 + T_3, \ldots, X_{n+1} = Y_{n+1} + T_{n+1},$$

so if we take $\mathfrak{a}^+$ to be the Weyl chamber in $\mathfrak{a}$ which contains H then

$$\mathfrak{n} = \sum_{j=3}^{n+1} \boldsymbol{R} x_j .$$

The unique positive restricted root α satisfies $\alpha(H) = 1$. For $Z \in \mathfrak{g}$ let Z^+ (or just Z) denote the vector field on given by

$$(Z^+ f)(\xi) = \left\{\frac{d}{dt} f(\exp(-tZ) \cdot \xi)\right\}_{t=0}, \qquad f \in \mathscr{E}(\Xi).$$

Then $Z \to Z^+$ is a Lie algebra homomorphism. The adjoint $(Z^+)^*$ of Z^+, as defined in §2, equals $-Z^+$ (since $d\xi$ is G-invariant) so if $\Psi \in \mathscr{D}'(\Xi)$

$$(Z^+\Psi)(\varphi) = \Psi(-Z^+\varphi), \qquad \varphi \in \mathscr{D}(\Xi).$$

When the representation $m \to Ad_G(m) \mid \mathfrak{l}$ is expressed in terms of the basis (T_j) of $\mathfrak{l}$ it becomes the ordinary representation of $\boldsymbol{SO}(n-1)$ on $\boldsymbol{R}^{n-1}$. Hence the differential operator

$$\Delta^+ = \sum_{j=3}^{n+1} (T_j{}^+)^2$$

on Ξ is M-invariant.

LEMMA 4.11. *Let $\Psi \in \mathscr{D}_\lambda'$ be a conical distribution with support in Ξ_0. Then there exists a polynomial P such that*

$$\Psi = P(\Delta^+)\Psi_{\lambda,e} .$$

Proof. For each $\xi \in \Xi_0$ the linearly independent vectors $(T_j^+)_\xi$ $(3 \leqslant j \leqslant n+1)$ together with the tangent space $(\Xi_0)_\xi$ span the whole tangent space Ξ_ξ. By the general structure of distributions with support contained in a submanifold (Schwartz [41], Bruhat [6(a)]) we can state that there exist unique $\Psi_{p_3,\ldots,p_{n+1}} \subset \mathscr{D}'(\Xi_0)$ such that

$$\Psi = \sum_{(p)} (T_3^+)^{p_3} \cdots (T_{n+1}^+)^{p_{n+1}} \tilde{\Psi}_{p_3,\ldots,p_{n+1}} \tag{34}$$

where the tilda denotes the canonical extension of a distribution on Ξ_0 to a distribution on Ξ. Now if $a \in A$ the diffeomorphism $\sigma(a) : gMN \to gaMN$ of Ξ leaves each Z^+ invariant so by the uniqueness of the representation (34) each $\tilde{\Psi}_{p_3,\ldots,p_{n+1}}$ belongs to $\mathscr{D}_\lambda'$ and is therefore a constant multiple of $\Psi_{\lambda,e}$. Thus there exists a polynomial Q such that

$$\Psi = Q(T_3^+,\ldots, T_{n+1}^+)\, \Psi_{\lambda,e}\,. \tag{35}$$

This means that there exists a differential operator D on K/M such that if $\tilde{D}$ denotes its canonical extension to a differential operator on $(K/M) \times A = \Xi$ then

$$\Psi = \tilde{D}\Psi_{\lambda,e}\,. \tag{36}$$

From Lemma 2.2 (7) it is clear that $\sim$ preserves adjoints, i.e., $(D^*)^\sim = (\tilde{D})^*$. Since both Ψ and $\Psi_{\lambda,e}$ are M-invariant we can take $\tilde{D}$ (and therefore D and D^*) invariant under M. But then there exists a unique K-invariant differential operator D^0 on K/M such that for all $F \in \mathscr{E}(K/M)$

$$(D^0F)(eM) = (D^*F)(eM). \tag{37}$$

Then if $\varphi \in \mathscr{D}((K/M) \times A)$,

$$((D^0)^\sim\varphi)(eM, a) = ((D^*)^\sim\varphi)(eM, a) = ((D^\sim)^*\varphi)(eM, a);$$

so since $\Psi_{\lambda,e}(\varphi)$ is determined by the restriction of φ to Ξ_0 we have

$$\Psi(\varphi) = \Psi_{\lambda,e}((D^0)^\sim\varphi). \tag{38}$$

But D^0 is a polynomial P in the Laplacian Δ on K/M [22(a), p. 270] which is self-adjoint so

$$\Psi = P(\tilde{\Delta})\, \Psi_{\lambda,e}\,. \tag{39}$$

Now if $Z \in \mathfrak{k}$ we let temporarily Z also denote the corresponding left invariant vector field on K. Then

$$\omega = \sum_{3 \leqslant j \leqslant n+1} T_j^2 + \sum_{3 \leqslant j < k \leqslant n+1} X_{jk}^2$$

is proportional to the Casimir operator on K, and if $\pi : K \to K/M$ is the natural projection we have (with appropriate normalization of the metric on K/M) $\omega(F \circ \pi) = \Delta F \circ \pi$ [22(b), p. 451]. On the other hand, ω lies in the center of the universal enveloping algebra of $\mathfrak{k}$ so using Corollary 7.3 in [22(e)] we deduce from $(\Delta F) \circ \pi = \omega(F \circ \pi)$ that $\omega^+ = \tilde{\Delta}$ where ω^+ is the image of ω under the homomorphism $Z \to Z^+$ $(Z \in \mathfrak{g})$. But Δ^+ and $\Psi_{\lambda,e}$ are both M-invariant so we have by induction

$$(\omega^+)^p \Psi_{\lambda,e} = (\Delta^+)^p \Psi_{\lambda,e}, \qquad (p \in \boldsymbol{Z}^+)$$

which concludes the proof of the lemma.

Lemma 4.12. *Let* $\lambda \in \mathfrak{a}_c^*$ *and suppose* P *is a polynomial of degree* $N > 0$ *such that* $\Psi = P(\Delta^+) \Psi_{\lambda,e}$ *is a conical distribution. Then* P *is determined up to a constant factor and*

$$\frac{\langle i\lambda, \alpha\rangle}{\langle \alpha, \alpha\rangle} = -N$$

Proof. For this we shall need the commutation relations between the basis vectors of $\mathfrak{l}$, $\mathfrak{a}$, $\mathfrak{n}$ and $\mathfrak{m}$. The algebra $\mathfrak{n}$ is abelian and

$$[H, T_p] = X_p - T_p, \quad [Y_p, T_k] = -\delta_{pk}H, \quad [X_j, T_k] = -\delta_{jk}H - X_{jk}$$
$$[X_{jk}, T_m] = \delta_{km}T_j - \delta_{jm}T_k, \qquad [T_j, T_k] = -X_{jk},$$
$$[X_j, [X_j, T_k]] = -X_k + 2\delta_{jk}X_j.$$

The distribution $\Psi = P(\Delta^+) \Psi_{\lambda,e}$ is (for any polynomial P) invariant under M and belongs to $\mathscr{D}_\lambda'$. It is conical if and only if

$$\Psi^{\tau(\exp tX_j)} = \Psi \qquad \text{for} \quad t \in \boldsymbol{R}, \quad 3 \leqslant j \leqslant n+1. \tag{40}$$

Writing $P(\Delta^+) = \sum_0^N a_i(\Delta^+)^i$ $(a_N = 1)$ and using

$$Ad(\exp tX_j)\, T_k = T_k - t(\delta_{jk}H + X_{jk}) + \cdots$$

condition (40) amounts to

$$\Big[\sum_i a_i \Big\{\sum_k (T_k - t(\delta_{jk}H + X_{jk}) + \cdots)^2\Big\}^i\Big](\Psi_{\lambda,e})$$

$$= \Big[\sum_i a_i \Big(\sum_k T_k^2\Big)^i\Big](\Psi_{\lambda,e}), \qquad t \in \boldsymbol{R}, \quad 3 \leqslant j \leqslant n+1,$$

where we have written Z instead of Z^+ ($Z \in \mathfrak{g}$). An equivalent condition is obtained by equating the coefficient of t to 0. Thus (40) is equivalent to

$$\sum_i a_i \mathfrak{S}\Big[\Big(\sum T_k^2\Big)^{i-1} \Big\{T_jH + HT_j + \sum_k (T_kX_{jk} + X_{jk}T_k)\Big\}\Big](\Psi_{\lambda,e}) = 0 \tag{41}$$

where

$$\mathfrak{S}(A^jB) = \sum_{0 \leqslant i \leqslant j} A^{j-i}BA^i.$$

Using the commutation relations we can replace the expression in {,} in (41) by

$$2T_jH + X_j + (n-3)\,T_j + 2\sum_k T_kX_{jk}\,. \tag{42}$$

But the M-invariance of Δ^+ and $\Psi_{\lambda,e}$ implies

$$\Big[X_{jk}\,, \sum_p T_p^2\Big] = 0 \qquad \text{and} \qquad X_{jk}(\Psi_{\lambda,e}) = 0, \tag{43}$$

so (41) can be written

$$\sum_{i=1}^N a_i \mathfrak{S}\Big[\Big(\sum_k T_k^2\Big)^{i-1} \{2T_jH + X_j + (n-3)\,T_j\}\Big](\Psi_{\lambda,e}) = 0. \tag{44}$$

Using the commutation relations, (43), and

$$X_j(\Psi_{\lambda,e}) = 0, \qquad H(\Psi_{\lambda,e}) = -(i\lambda + \rho)(H)\,\Psi_{\lambda,e} \tag{45}$$

we can reduce (44) to a relation of the form

$$\Big(\sum_{(p)} b_{p_3\cdots p_{n+1}} T_3^{p_3} \cdots T_{n+1}^{p_{n+1}}\Big)(\Psi_{\lambda,e}) = 0. \tag{46}$$

Applying this to a function $\varphi \in \mathscr{D}(\Xi)$ of the form $\varphi(kM, a) = \alpha(kM)\beta(a)$ we see that in (46) each coefficient b is 0. This means that if we write the left side of (44) as a top degree term

$$c\left(\sum_k T_k{}^2\right)^{N-1} T_j(\Psi_{\lambda,e}) \tag{47}$$

plus a linear combination of lower degree terms

$$T_{i_1}\cdots T_{i_q}(\Psi_{\lambda,e}), \qquad q < 2N-1, \tag{48}$$

we can conclude $c = 0$. Since

$$\begin{aligned}
T_j\left(\sum_p T_p{}^2\right)\left(\sum_q T_q{}^2\right)^k(\Psi_{\lambda,e}) &= \sum_p (T_pT_jT_p - X_{jp}T_p)\left(\sum_q T_q{}^2\right)^k(\Psi_{\lambda,e})\\
&= \sum_p (T_p{}^2T_j - 2T_pX_{jp} - T_j + \delta_{jp}T_p)\left(\sum_q T_q{}^2\right)^k(\Psi_{\lambda,e})\\
&= \left(\left(\sum_p T_p{}^2\right)T_j - (n-2)\,T_j\right)\left(\sum_q T_q{}^2\right)^k(\Psi_{\lambda,e})
\end{aligned}$$

we have by iteration,

$$T_j\left(\sum_p T_p{}^2\right)^{k+1}(\Psi_{\lambda,e}) = \left\{\left(\sum_p T_p{}^2\right) - (n-2)\right\}^{k+1} T_j(\Psi_{\lambda,e}). \tag{49}$$

A similar computation gives

$$\begin{aligned}
T_jH\left(\sum_p T_p{}^2\right)\left(\sum_p T_p{}^2\right)^k(\Psi_{\lambda,e}) &= T_j\left(\sum_p T_p{}^2\right)(H-2)\left(\sum_p T_p{}^2\right)^k(\Psi_{\lambda,e})\\
&\quad + T_j\left(\sum_p T_pX_p + X_pT_p\right)\left(\sum_p T_p{}^2\right)^k(\Psi_{\lambda,e}).
\end{aligned}$$

The last term on the right can be omitted because $k+1 \leqslant N-1$ so a term like

$$T_jX_pT_pT_r^{2k}(\Psi_{\lambda,e})$$

is congruent to

$$T_jT_pT_r^{2k}X_p(\Psi_{\lambda,e})$$

modulo a sum of terms (48) with

$$q \leqslant 2k+1 \leqslant 2N-3.$$

Similarly the X_j term in (44) makes no contribution to the top degree term (47). Thus, modulo a sum of terms (48), the left side of (44) is congruent to

$$\sum_{0 \leqslant i \leqslant N-1} \left(\left(\sum_k T_k^2 \right)^i (2T_j H + (n-3)\, T_j) \left(\sum_k T_k^2 \right)^{N-i-1} \right) (\Psi_{\lambda,e})$$

which again is congruent to

$$\sum_{0 \leqslant i \leqslant N-1} \left(\sum_k T_k^2 \right)^i 2T_j \left(\sum_k T_k^2 \right)^{N-i-1} (H - 2(N-i-1))\, (\Psi_{\lambda,e})$$
$$+ \sum_{0 \leqslant i \leqslant N-1} (n-3) \left(\sum_k T_k^2 \right)^i \left\{ \left(\sum_k T_k^2 \right) - (n-2) \right\}^{N-i-1} T_j(\Psi_{\lambda,e}).$$

Here the first sum equals (by (45) and (49))

$$\sum_{0 \leqslant i \leqslant N-1} 2 \left(\sum_k T_k^2 \right)^i \left\{ \left(\sum_k T_k^2 \right) - (n-2) \right\}^{N-i-1} T_j(d_i \Psi_{\lambda,e}),$$

where the scalar d_p equals

$$-(i\lambda + \rho)(H) - 2(N - p - 1).$$

The constant c in (47) therefore equals

$$\sum_{0 \leqslant p \leqslant N-1} \{(n-3) - 2(i\lambda + \rho)(H) - 4(N - p - 1)\}.$$

Since $\rho = \frac{1}{2}(n-1)\alpha$, the relation $c = 0$ gives $i\lambda(H) = -N$ as desired. If P' is another polynomial with leading coefficient 1 with the same properties as required of P we get a contradiction by applying to $P - P'$ the result already obtained.

During the proof of Theorem 4.5 we saw that the poles of $\lambda \to \Psi'_{\lambda,s}$ occur precisely for $\lambda(H) = im$ ($m \in \boldsymbol{Z}^+$) and that the residues at these points are conical distributions with support contained in Ξ_0. Lemmas 4.11 and 4.12 therefore supply step (ii) in the proof of Theorem 4.10. The following lemma now gives step (iii).

LEMMA 4.13. *If $\lambda = 0$ then all the conical distributions in $\mathcal{D}_\lambda'$ are proportional.*

Proof. Let Ψ be any conical distribution in $\mathscr{D}_0'(\Xi)$. According to Theorem 2.1, the restriction of Ψ to Ξ^* is a function given by

$$\psi(\xi) = ce^{\rho(\log a(\xi))}, \qquad (\xi \in \Xi^*), \tag{50}$$

where c is a constant. We wish to prove $c = 0$. Suppose this were not the case. Let $S \in \mathscr{D}'(K/M)$ correspond to Ψ by means of Proposition 4.1. Then S restricted to $(K/M - eM)$ is a C^∞ function, say s. The N-invariance of Ψ implies that if $\varphi \in \mathscr{D}(\Xi)$, $n \in N$,

$$\begin{aligned}\Psi(\varphi) &= \int_{K/M} \left(\int_A \varphi(nka \cdot \xi_o)\, e^{\rho(\log a)}\, da \right) dS(kM) \\ &= \int_{K/M} \left(\int_A \varphi(k(nk) \exp H(nk)\, a \cdot \xi_o)\, e^{\rho(\log a)}\, da \right) dS(kM).\end{aligned}$$

It follows that

$$\int_{K/M} \beta(k(nk)M)\, e^{-\rho(H(nk))}\, dS(kM) = \int_{K/M} \beta(kM)\, dS(kM) \tag{51}$$

for each $\beta \in \mathscr{D}(K/M)$. In order to obtain more precise information about S (and Ψ) we multiply Ψ by a suitable conical function resulting in a conical distribution on which we can use Theorem 4.9.

We know from Theorem 3.1 that the function

$$\Phi(\xi) = e^{-\alpha(\log a(\xi))}, \qquad \xi \in \Xi^*,$$

extends to a conical function on Ξ, and in fact $\Phi \in \mathscr{D}'_{-i(\rho+\alpha)}$. Thus by Proposition 2.6,

$$\Phi\Psi \in \mathscr{D}'_{-i\alpha}. \tag{52}$$

Now by Theorem 1.1, Chap. IV, λ is not simple if and only if

$$-\frac{\langle i\lambda + \rho, \alpha\rangle}{\langle \alpha, \alpha\rangle} \in \boldsymbol{Z}^+. \tag{53}$$

In particular $-i\alpha$ is simple so by Theorem 4.9 and (52) we obtain a sharpening of (50)

$$\Phi\Psi = c_1\Psi_{-i\alpha,s} + c_2\Psi_{-i\alpha,e}, \qquad c_1, c_2 \in \boldsymbol{C}. \tag{54}$$

With the vectors H and T_3 introduced earlier, put

$$u_\theta = \exp \theta T_3, \qquad a_t = \exp tH, \qquad (\theta, t \in \boldsymbol{R}).$$

Then the function Φ above is completely determined by the values $\Phi(u_\theta a_t \cdot \xi_o)$. For m^* we can take $m^* = u_\pi$; then the Bruhat decomposition

$$u_\theta a_t = mn_1 a_r m^* n_2 , \qquad m \in M, \quad r \in \boldsymbol{R}, \quad n_1 , n_2 \in N \tag{55}$$

in the group $\boldsymbol{SO}_0(1, n)$ actually takes place within the subgroup $\boldsymbol{SO}_0(1, 2)$. Writing therefore

$$n_1 = \exp \xi X_3 , \qquad n_2 = \exp \eta X_3 ,$$

the equation (55) can be used to determine r in terms of θ and t. Computing the last column and last row in (55) we obtain the equations

$$\begin{aligned} 0 &= \eta e^{-r} + \xi^2 \eta e^{-r} - \xi, \\ \sin \theta &= -\eta e^{-r} + \xi^2 \eta e^{-r} - \xi, \\ \cos \theta &= 2\xi\eta e^{-r} - 1, \\ -\sinh t \sin \theta &= \xi e^{-r} + \xi\eta^2 e^{-r} - \eta, \\ -\cosh t \sin \theta &= \xi e^{-r} - \xi\eta^2 e^{-r} + \eta, \end{aligned}$$

which gives

$$e^r = e^{-t}(\sin \tfrac{1}{2}\theta)^{-2} \tag{56}$$

and hence

$$\Phi(u_\theta a_t \cdot \xi_o) = e^t (\sin \tfrac{1}{2}\theta)^2. \tag{57}$$

The conical distribution $\Psi_{-i\alpha,s}$ is according to Theorem 2.1 given by the locally integrable function

$$e^{(\rho-\alpha)(\log a(\xi))}$$

up to a constant factor. Each $kM \in K/M$ can be written $kM = mu_\theta M$ $(0 \leqslant \theta = \theta(kM) \leqslant \pi)$ and

$$\int_{K/M} F(kM)\, dk_M = c_3 \int_0^\pi \Big(\int_M F(mu_\theta M)\, dm \Big) \sin^{n-2} \theta \, d\theta,$$

since we are in the case $n > 2$; c_3 is a constant. If $f \in \mathscr{E}(\boldsymbol{R})$ is an even function of period 2π and $g \in \mathscr{D}(A)$ there exists a $\varphi \in \mathscr{D}(\Xi)$ such that

$$\varphi(mu_\theta a \cdot \xi_o) = f(\theta)\, g(a), \qquad m \in M, \quad \theta \in \boldsymbol{R}, \quad a \in A.$$

Applying (54) to such a function φ we obtain (writing $b = kM$, $B = K/M$)

$$\begin{aligned} &\int_B (\sin \tfrac{1}{2}\theta(b))^2 f(\theta(b))\, dS(b) \\ &\qquad = d_1 \int_0^\pi (\sin \tfrac{1}{2}\theta)(\cos \tfrac{1}{2}\theta)^{n-2} f(\theta)\, d\theta + d_2 f(0) \end{aligned} \tag{58}$$

for certain constants d_1, d_2. On the other hand comparing Φ with Ψ we see that s is given by

$$s(kM) = c\,|\sin \tfrac{1}{2}\theta(kM)|^{-(n-1)},$$

so by the structure of distributions with point supports

$$S(\beta) = c \int_B |\sin \tfrac{1}{2}\,\theta(b)|^{-(n-1)}(\beta(b) - \beta(o))\,db + (D\beta)(o), \tag{59}$$

where $o = eM$ and D is a differential operator on B, which we can take to be M-invariant. As in (37) we can therefore take D to be a polynomial Q in the Laplacian Δ on K/M. The mapping $\varphi \to f$ where $\varphi(u_\theta M) = f(\theta)$ is a bijection of the space of M-invariant functions in $\mathscr{D}(K/M)$ onto the space of even functions in $\mathscr{E}(\boldsymbol{R})$ with period 2π. Under this mapping the functional $\varphi \to (Q(\Delta)\varphi)(o)$ corresponds to a distribution on $\boldsymbol{R}$ with support 0 and invariant under $\theta \to -\theta$; thus there exists a polynomial Q_0 such that

$$(Q(\Delta)\varphi)(o) = \left(Q_0\left(\frac{d^2}{d\theta^2}\right)f\right)(0) \tag{60}$$

for all φ and f as above. Clearly Q_0 and Q have the same degree. Comparing (58) with (59) for the function

$$\beta(b) = (\sin \tfrac{1}{2}\theta(b))^2 f(\theta(b))$$

we deduce that

$$\left[Q_0\left(\frac{d^2}{d\theta^2}\right)(\sin^2 \tfrac{1}{2}\theta f(\theta))\right]_{\theta=0} = d_2 f(0) \tag{61}$$

for every even function $f \in \mathscr{E}(\boldsymbol{R})$. The distribution

$$F \to \left(Q_0\left(\frac{d^2}{d\theta^2}\right)F\right)(0)$$

on $\boldsymbol{R}$ can of course also be expressed by means of the coordinate function $\tau = \sin \frac{1}{2}\theta$. For a new polynomial Q_* we then get from (61),

$$\left[Q_*\left(\frac{d^2}{d\tau^2}\right)(\tau^2 f(\theta))\right]_{\tau=0} = d_2 f(0).$$

Taking $f(\theta) = (\sin \frac{1}{2}\theta)^{2j}$ $(j \in \boldsymbol{Z}^+)$ we see that Q_*, and therefore also Q_0 and Q, is a first degree polynomial. Returning to (59) we have obtained for the structure of S,

$$S(\beta) = \int_B s(b)(\beta(b) - \beta(o))\, db + a_0\beta(o) + a_1(\Delta\beta)(o), \tag{62}$$

where a_0, $a_1 \in \boldsymbol{C}$. We shall now use (51) in the special case $\beta \equiv 1$. Then we get

$$\int_B s(b)(t_x(b) - 1)\, db + a_1(\Delta t_x)(o) = 0, \tag{63}$$

where $t_x(kM) = e^{-\rho(H(xk))}$, $x \in N$. We shall now see that the two terms do not show the same behavior near $x = e$ and thus must both vanish. We emphasize that (63) is an equation for the hyperbolic space $\boldsymbol{H}^n = \boldsymbol{SO}_0(1, n)/\boldsymbol{SO}(n)$ but does not depend in any way on the model chosen for the group of isometries. The Laplacian Δ was defined by means of a specific model but it is determined up to a constant factor by its K-invariance. Thus we can view (63) in terms of the unit ball model for $\boldsymbol{H}^n$, i.e., the unit ball $|y| < 1$ in $\boldsymbol{R}^n$ with the Riemannian structure

$$ds^2 = (1 - (y_1^2 + \cdots + y_n^2))^{-2}\,(dy_1^2 + \cdots + dy_n^2).$$

As is well-known the Poisson kernel for this Riemannian manifold $\boldsymbol{H}^n$ is both given by[2]

$$P(y, b) = \left(\frac{1 - |y|^2}{|y - b|^2}\right)^{n-1}$$

and

$$P(gK, kM) = e^{-2\rho(H(g^{-1}k))},$$

the last formula being valid in any symmetric space of the noncompact type. The function t_x is therefore given by

$$t_x(b) = \left(\frac{1 - |x^{-1}\cdot 0|^2}{|x^{-1}\cdot 0 - b|^2}\right)^{\frac{1}{2}(n-1)}$$

The group K fixes the origin 0 and we choose for M the subgroup fixing the point $o = (1, 0,..., 0)$. Then

$$s(b) = c\,|\sin \tfrac{1}{2}\theta(b)|^{-(n-1)} = c2^{n-1}\,|o - y|^{-(n-1)},$$

[2] This formula was first communicated to me by L. Ahlfors.

where again $\theta(b)$ is the distance on the unit sphere between $(1, 0, \ldots, 0)$ and b. The horocycles in this model are the spheres tangential to the unit sphere $|y| = 1$. Select for $t \in \boldsymbol{R}$ an element $x \in N$ such that

$$x^{-1} \cdot 0 = (\cos^2 t, \sin t \cos t, 0, \ldots, 0).$$

In order to compute $(\Delta t_x)(o)$ we express Δ in geodesic polar coordinates around the point $o_t = (\cos t, \sin t, 0, \ldots, 0)$ because $t_x(b)$ only depends on the spherical distance

$$r = r(b) = (o_t b).$$

Now

$$| x^{-1} \cdot o - b |^2 = 1 - 2 \cos r \cos t + \cos^2 t$$

and since Δ restricted to functions depending only on r has the form

$$\Delta_r = \frac{d^2}{dr^2} + (n-2) \cot r \frac{d}{dr}$$

we find that

$$(\Delta t_x)(o) = | \sin t |^{n-1} \{\Delta_r (1 - 2 \cos r \cos t + \cos^2 t)^{-\frac{1}{2}(n-1)}\}_{r=|t|}$$

which gives

$$(\Delta t_x)(o) = 2(n-1) \cot^2 t. \tag{64}$$

On the other hand, the integral in (63) is a constant multiple of

$$\int_{|y|=1} (1 - y_1)^{-\frac{1}{2}(n-1)} \{(1 - 2y_2 | \cot t | - 2(1 - y_1) \cot^2 t)^{-\frac{1}{2}(n-1)} - 1\} \, d\omega(y)$$

where $d\omega$ is the surface element on $|y| = 1$. For $\tau = \cot t$ small we can expand the expression inside $\{\ \}$ in a power series in τ. The coefficient to τ^4 is easily seen to be > 0; comparing with (64) and (63) we deduce that $a_1 = 0$ and $s \equiv 0$. This contradicts our assumption that $c \neq 0$ so Lemma 4.13 is proved.

Step (iv) was already outlined in sufficient detail so Theorem 4.10 is now completely proved.

§5. The Spherical Representations and the Fourier Transform on X

We recall that a representation π of the group G on a vector space is called spherical if the vector space contains a vector fixed under all of $\pi(K)$. We shall now describe a family of spherical representations.

For each $\lambda \in \mathfrak{a}_c^*$ let $\mathscr{H}_\lambda$ denote the space of complex-valued measurable functions ψ on Ξ satisfying the conditions

(i) $\psi(gaMN) = e^{(i\lambda-\rho)(\log a)}\psi(gMN)$,

(ii) $\int_K |\psi(kMN)|^2 \, dk < \infty$.

Then $\mathscr{H}_\lambda$ is a Hilbert space, the norm $\| \psi \|_\lambda$ being the square root of the expression in (ii). Each $\psi \in \mathscr{H}_\lambda$ determines a distribution

$$\varphi \to \int_\Xi \varphi(\xi)\, \psi(\xi)\, d\xi, \qquad \varphi \in \mathscr{D}(\Xi),$$

which will be denoted by the same letter. Using Lemma 2.2 and Proposition 2.5 we see that $\mathscr{H}_\lambda \subset \mathscr{D}_\lambda'$. We put $\mathscr{E}_\lambda(\Xi) = \mathscr{E}(\Xi) \cap \mathscr{H}_\lambda$.

The dual space $\mathscr{D}'(\Xi)$ of $\mathscr{D}(\Xi)$ is given the strong topology and $\mathscr{D}_\lambda'$ is given the relative topology. Let $\tilde{\tau}_\lambda$ denote the natural representation of G on $\mathscr{D}_\lambda'$, i.e., $\tilde{\tau}_\lambda(g)\Psi = \Psi^{\tau(g)}$ for $g \in G$, $\Psi \in \mathscr{D}_\lambda'$. Restricting each $\tilde{\tau}_\lambda(g)$ to the subspace $\mathscr{H}_\lambda$ we obtain a representation τ_λ of G on $\mathscr{H}_\lambda$. If we view $\mathscr{H}_\lambda$ as a space of functions on G, τ_λ becomes the representation of G induced by the one-dimensional representation $man \to e^{i\lambda(\log a)}$ of MAN [33], [6(a)]. If $\lambda \in \mathfrak{a}^*$ then τ_λ is unitary [21(a), p. 242]. The vector $\psi_0 \in \mathscr{H}_\lambda$ given by $\psi_0(kaMN) \equiv e^{(i\lambda-\rho)(\log a)}$ is fixed under $\tau_\lambda(K)$ so each τ_λ is spherical. The family $\{\tau_\lambda \mid \lambda \in \mathfrak{a}_c^*\}$ will be referred to as the *spherical principal series* (since it consists of the spherical representations of the principal series). We shall now relate some simple properties of these representations to the Radon transform.

Since MN is noncompact one cannot use the convolution product $*$ on G to define a convolution product on $\mathscr{E}'(\Xi)$. However, if one of the factors is MN-invariant this can be done as follows. The mapping $F \to \bar{F}$ where

$$\bar{F}(gMN) = \int_{MN} F(gmn)\, dm\, dn$$

maps $\mathscr{D}(G)$ onto $\mathscr{D}(\Xi)$; the dual map $\Psi \to \tilde{\Psi}$ is an injection of $\mathscr{D}'(\Xi)$ into $\mathscr{D}'(G)$.

If $\varphi \in \mathscr{D}(\Xi)$ and if $\eta \in \mathscr{D}'(\Xi)$is MN-invariant we put

$$(\varphi \times \eta)(gMN) = \int_G F(gh^{-1})\, d\tilde{\eta}(h) \tag{1}$$

if $F \in \mathscr{D}(G)$ is any function satisfying $\bar{F} = \varphi$. To show that this definition is valid we must verify

$$\int_G F(gh^{-1})\, d\tilde{\eta}(h) \equiv 0 \qquad \text{if} \quad \bar{F} \equiv 0. \tag{2}$$

But $\tilde{\eta}$ is left invariant under MN and if $\tilde{\eta}'$ denotes the image of $\tilde{\eta}$ under the diffeomorphism $g \to g^{-1}$ of G then $\tilde{\eta}'$ is right invariant under MN. If $\tilde{\eta}'$ were a function on G right invariant under MN, (2) would be obvious. But in any case $\tilde{\eta}'$ can be approximated by functions $\alpha * \tilde{\eta}'$ ($\alpha \in \mathscr{D}(G)$) which are right invariant under MN so (2) follows. Relation (2) shows also that $\eta^* \in \mathscr{D}'(\Xi)$ can be defined by

$$\eta^*(\bar{F}) = \eta(\bar{F}'), \qquad F \in \mathscr{D}(G), \tag{3}$$

where $F'(g) = F(g^{-1})$. We can then generalize (1) and define for $\Psi \in \mathscr{E}'(\Xi)$, $\eta \in \mathscr{D}'(\Xi)$, η MN-invariant,

$$(\Psi \times \eta)(\varphi) = \Psi(\varphi \times \eta^*), \qquad \varphi \in \mathscr{D}(\Xi). \tag{4}$$

Then $\Psi \times \eta$ is a distribution on Ξ and if $\Psi = \psi \in \mathscr{D}(\Xi)$ a routine computation shows that $\Psi \times \eta = \psi \times \eta$.

Lemma 5.1. *For each $\lambda \in \mathfrak{a}_c^*$ define $\Psi_{\lambda,e} \in \mathscr{D}_\lambda'$ as in Theorem* 4.5,

$$\Psi_{\lambda,e} : \varphi \to \int_A \varphi(aMN)\, e^{(i\lambda+\rho)(\log a)}\, da, \qquad \varphi \in \mathscr{D}(\Xi).$$

Then

(i) $\mathscr{D}(\Xi) \times \Psi_{\lambda,e}$ *is dense in* $\mathscr{K}_\lambda$.
(ii) $\mathscr{D}(\Xi) \times \Psi_{\lambda,e}$ *is dense in* $\mathscr{D}_\lambda'$.
(iii) $\Psi_{\lambda,e}$ *is a cyclic vector for the representation* $\tilde{\tau}_\lambda$ *of* G *on* $\mathscr{D}_\lambda'$.

Proof. Formula (1) implies that

$$(\varphi \times \eta)(gMN) = \eta^*(\varphi \circ \tau(g)).$$

But if $\eta = \Psi_{\lambda,e}$ then (3) shows by a simple computation that $\eta^* = \Psi_{-\lambda,e}$. Thus

$$(\varphi \times \Psi_{\lambda,e})(kaMN) = \Psi_{-\lambda,e}(\varphi \circ \tau(ka))$$
$$= e^{(i\lambda-\rho)(\log a)} \int_A \varphi(kcMN)\, e^{(-i\lambda+\rho)(\log c)}\, dc. \tag{5}$$

Taking φ of the form $\varphi(kaMN) = \alpha(kM)\,\beta(a)$ statement (i) follows.

For (ii) let L be any linear functional on $\mathscr{D}_\lambda'$. Extending L to $\mathscr{D}'(\Xi)$ and using the reflexivity of $\mathscr{D}(\Xi)$ we can find $l \in \mathscr{D}(\Xi)$ such that

$$L(\Psi) = \Psi(l), \qquad (\Psi \in \mathscr{D}_\lambda'). \tag{6}$$

Let $\varphi(kaMN) = \alpha(kM)\,\beta(a)$ where $\alpha \in \mathscr{D}(K/M)$ is arbitrary and

$$\int_A \beta(a)\, e^{(-i\lambda+\rho)(\log a)}\, da = 1.$$

Putting $\Psi = \varphi \times \Psi_{\lambda,e}$ we obtain from Lemma 2.2,

$$L(\varphi \times \Psi_{\lambda,e}) = \int_{K/M} \alpha(kM) \left(\int_A l(kaMN)\, e^{(i\lambda+\rho)(\log a)}\, da \right) dk_M .$$

If L vanishes on $\mathscr{D}(\Xi) \times \Psi_{\lambda,e}$ we deduce that

$$\int_A l(kaMN)\, e^{(i\lambda+\rho)(\log a)}\, da = 0, \qquad k \in K. \tag{7}$$

But then Proposition 4.1 shows that $\Psi(l) = 0$ for $\Psi \in \mathscr{D}_\lambda'$ so $L = 0$.

For (iii) suppose L in (6) vanishes on all the translates of $\Psi_{\lambda,e}$. Then (7) holds so again $L = 0$.

PROPOSITION 5.2. *Let $\lambda \in \mathfrak{a}_c^*$. Then the representation τ_λ is irreducible if and only if $\tilde{\tau}_\lambda$ is irreducible. (Here irreducibility means as usual absence of nontrivial closed invariant subspaces.)*

Proof. Suppose τ_λ is irreducible and let $E \subset \mathscr{D}_\lambda'$ be a proper closed invariant subspace. Let (Ψ_i) be a sequence in $\mathscr{K}_\lambda$ converging (in $\mathscr{K}_\lambda$) to $\Psi \in \mathscr{K}_\lambda$. Let $\varphi \in \mathscr{D}(\Xi)$ and select $\beta \in \mathscr{D}(A)$ such that $\varphi(kaMN) \equiv \varphi(kaMN)\,\beta(a)$. Then by Lemma 2.2,

$$\int_\Xi (\Psi - \Psi_i)(\xi)\, \varphi(\xi)\, d\xi$$
$$= \int_A e^{(i\lambda+\rho)(\log a)} \beta(a) \left(\int_K (\Psi - \Psi_i)(kMN)\, \varphi(kaMN)\, dk \right) da$$

So by the Schwarz' inequality, $\Psi_i(\varphi) \to \Psi(\varphi)$. But since $\mathscr{D}'(\Xi)$ is a Montel space, weakly convergent sequences are strongly convergent [41, p. 74] so $\Psi_i \to \Psi$ in the topology of $\mathscr{D}'(\Xi)$. Taking $\Psi_i \in E$ we see that $E \cap \mathscr{K}_\lambda$ is closed in $\mathscr{K}_\lambda$ so by the irreducibility of τ_λ, $E \cap \mathscr{K}_\lambda = \{0\}$ or $\mathscr{K}_\lambda$. The latter possibility implies by Lemma 5.1 that $E = \mathscr{D}_\lambda'$ so $E \cap \mathscr{K}_\lambda = \{0\}$. In order to conclude $E = \{0\}$ we consider for $f \in \mathscr{D}(G)$, $\Psi \in \mathscr{D}'(\Xi)$ the distribution on Ξ given by

$$\int_G \Psi^{\tau(g)} f(g)\, dg : \varphi \to \int_G \Psi^{\tau(g)}(\varphi) f(g)\, dg.$$

If we approximate f by step functions the property of sequences in $\mathscr{D}'(\Xi)$ mentioned shows that E is invariant under the map

$$\Psi \to \int_G \Psi^{\tau(g)} f(g)\, dg.$$

Now a trivial computation shows that

$$\left(\int_G \Psi^{\tau(g)} f(g)\, dg\right)^{\sim} = f * \tilde{\Psi}.$$

In particular if $\Psi \in E$ the distribution $\int_G \Psi^{\tau(g)} f(g)\, dg$ is a function in E, hence belongs to $E \cap \mathscr{K}_\lambda = \{0\}$. Thus $f * \tilde{\Psi} = 0$ for all $f \in \mathscr{D}(G)$ so $\Psi = 0$. This proves that $\tilde{\tau}_\lambda$ is irreducible.

On the other hand, if $\tilde{\tau}_\lambda$ is irreducible let $E \subset \mathscr{K}_\lambda$ be a proper closed invariant subspace. Then there exists a function $\alpha \neq 0$ in $L^2(K/M)$ such that

$$\int_{K/M} \varphi(kMN)\, \alpha(kM)\, dk_M = 0, \qquad \text{for} \quad \varphi \in E.$$

Since E is invariant under convolution by functions in $\mathscr{D}(K/M)$ we may take α in $\mathscr{D}(K/M)$. The continuous linear functional on $\mathscr{D}_\lambda'$ given by

$$L_\alpha : \Psi \to S(\alpha),$$

where $S \in \mathscr{D}'(K/M)$ is as in Proposition 4.1, vanishes on E hence also on its closure $\bar{E}$ in $\mathscr{D}_\lambda'$. But $\bar{E}$ is G-invariant, $\tilde{\tau}_\lambda$ irreducible so $E = 0$ as desired.

Let $\psi_0 \in \mathscr{K}_\lambda$ denote the function $\psi_0(kaMN) = e^{(i\lambda-\rho)(\log a)}$; up to a scalar multiple ψ_0 is the unique K-fixed vector in $\mathscr{K}_\lambda$.

DEFINITION. Let $\lambda \in \mathfrak{a}_c^*$. We call

(i) λ *irreducible* if τ_λ is irreducible;
(ii) λ *cyclic* if ψ_0 is a cyclic vector for τ_λ;
(iii) λ *simple* if the map $\Psi \to \tilde{\Psi}$ is injective on $\mathscr{D}_\lambda'$ (cf. §4).

PROPOSITION 5.3. *Let* $\lambda \in \mathfrak{a}_c^*$. *Then*

(i) λ *is cyclic* $\Leftrightarrow -\lambda$ *is simple*
(ii) λ *is irreducible* $\Leftrightarrow \lambda$ *and* $-\lambda$ *are simple.*

Proof. We have if $g \in G$,

$$\psi_0^{\tau(g^{-1})}(kaMN) = \psi_0(\exp H(gk)\, aMN) = e^{(i\lambda-\rho)(H(gk)+\log a)}$$

so if $\langle\ ,\ \rangle_\lambda$ denotes the scalar product on $\mathscr{H}_\lambda$,

$$\langle\psi_0^{\tau(g^{-1})}, F\rangle_\lambda = \int_K e^{(i\lambda-\rho)(H(gk))}\mathrm{conj}(F(kMN))\, dk$$
$$= \int_B e^{(-i\lambda+\rho)(A(g^{-1}K,b))}\mathrm{conj}(F(b))\, db$$

so (i) follows. Let $\bar{\lambda}$ denote the complex conjugate of λ (so $\bar{\lambda}(H) = \mathrm{conj}(\lambda(H))$ for $H \in \mathfrak{a}$) and consider the sesquilinear form $\{\ ,\ \}$ on $\mathscr{H}_\lambda \times \mathscr{H}_{\bar{\lambda}}$ given by

$$\{\psi_1, \psi_2\} = \int_K \psi_1(kMN)\, \mathrm{conj}(\psi_2(kMN))\, dk. \tag{8}$$

Since the mapping $k \to k(gk)$ of K onto itself is a diffeomorphism with Jacobian $e^{-2\rho(H(gk))}$ [21(a), p. 241] it is easy to see that G leaves $\{\ ,\ \}$ invariant. Hence if $V \subset \mathscr{H}_\lambda$ is a closed invariant subspace the annihilator

$$V' = \{\psi \in \mathscr{H}_{\bar{\lambda}} \mid \{v, \psi\} = 0 \text{ for } v \in V\}$$

is also a closed invariant subspace. Hence λ is irreducible if and only if $\bar{\lambda}$ is irreducible.

Now for each $\lambda \in \mathfrak{a}_c^*$ let $\mathscr{N}_\lambda$ denote the kernel of the map $\psi \to \tilde{\psi}$ of $\mathscr{H}_\lambda$ into $\mathscr{E}_\lambda(X)$. Then $\mathscr{N}_\lambda \neq \mathscr{H}_\lambda$, G-invariant, and, by the Schwarz inequality on $L^2(B)$, closed in $\mathscr{H}_\lambda$. Hence if λ is not simple, so $\mathscr{N}_\lambda \neq \{0\}$, then λ is not irreducible. But obviously, λ is simple if and only if $-\bar{\lambda}$ is simple; thus we have proved that if λ is irreducible then λ and $-\lambda$ are simple.

For the converse we define for each simple $\lambda \in \mathfrak{a}_c^*$ the Hilbert space $\mathscr{H}_\lambda$ by

$$\mathscr{H}_\lambda = \left\{ h(x) = \int_B e^{(i\lambda+\rho)(A(x,b))} F(b)\, db \mid F \in L^2(B) \right\} \tag{9}$$

the norm of h being the L^2-norm of F. Assuming λ simple the map $\psi \to \tilde{\psi}$ is an isometry of $\mathscr{K}_\lambda$ onto $\mathscr{H}_\lambda$. Suppose $V \subset \mathscr{H}_\lambda$ is a nonzero, G-invariant closed subspace, and let V_0 be the corresponding subspace of $\mathscr{K}_\lambda$. There exists $h \in V$ such that $h(o) \neq 0$. Consider the function

$$h^\natural(x) = \int_K h(k \cdot x)\, dk = \int_B e^{(i\lambda+\rho)(A(x,b))} F^\natural(b)\, db,$$

where

$$F^\natural(b) = \int_K F(k \cdot b)\, dk.$$

The space $V_0 \subset \mathscr{K}_\lambda$ is closed, K-invariant and therefore invariant under convolution

$$F(b) \to \int_K f(k) F(k \cdot b)\, dk, \qquad f \in L^2(K).$$

In particular the constant function $F^\natural$ lies in V_0 so $h^\natural \in V$ and $h^\natural(o) = h(o) \neq 0$. If $-\lambda$ is also simple then by (i) ψ_0 is a cyclic vector for τ_λ so the vector $\tilde{\psi}_0$ (which is a constant multiple of $h^\natural$) has its G-translates dense in $\mathscr{H}_\lambda$. This proves $V = \mathscr{H}_\lambda$ so λ is irreducible.

Consider now the spherical function $\varphi_\lambda(g) = \varphi_\lambda(gK)$ which is given by

$$\varphi_\lambda(x) = \int_B e^{(i\lambda+\rho)(A(x,b))}\, db. \tag{10}$$

It is well-known that if φ_λ is positive definite on G the linear combinations of the translates of φ_λ form a vector space V_λ with a G-invariant inner product

$$\left\langle \sum \alpha_i \varphi_\lambda^{L(g_i)}, \sum \beta_j \varphi_\lambda^{L(h_j)} \right\rangle = \sum \alpha_i \bar{\beta}_j \varphi_\lambda(g_i^{-1} h_j)$$

($\alpha_i, \beta_i \in \mathbf{C}$, $g_i, h_j \in G$); the corresponding representation T_λ of G on the completion of V_λ is irreducible and unitary and

$$\varphi_\lambda(g) = \langle \varphi_\lambda, T_\lambda(g)\varphi_\lambda \rangle. \tag{11}$$

In the case when λ is real (so φ_λ is automatically positive definite) we now express this in more concrete terms and identify $\mathscr{H}_\lambda$ with the completion of V_λ. Let $\times$ denote the operation on functions on X induced by the convolution product $*$ on G.

PROPOSITION 5.4. *Suppose $\lambda \in \mathfrak{a}^*$. Then the expression*

$$|f \times \varphi_\lambda|_\lambda = \left[\int_X (f \times \varphi_\lambda)(x) \operatorname{conj}(f(x))\, dx\right]^{1/2}, \qquad f \in \mathscr{D}(X) \tag{12}$$

is a well-defined norm on the space $\mathscr{D}(X) \times \varphi_\lambda$. If λ is simple the identity map is an isometry of $\mathscr{D}(X) \times \varphi_\lambda$ onto a dense subspace of $\mathscr{H}_\lambda$.

Proof. We first prove that the norm (12) is well-defined. Writing $g(kM) = k(gk)M$ we have as mentioned before

$$A(g^{-1} \cdot x, b) = A(x, g(b)) - A(g \cdot o, g(b)). \tag{13}$$

The Jacobian of $b \to g(b)$ being $e^{-2\rho(A(g\cdot o, g(b)))}$ we see that

$$\varphi_\lambda(g^{-1} \cdot x) = \int_B e^{(i\lambda+\rho)(A(x,b))} e^{(-i\lambda+\rho)(A(g\cdot o, b))}\, db.$$

Since

$$f \times \varphi_\lambda(x) = \int_G f(g \cdot o)\, \varphi_\lambda(g^{-1} \cdot x)\, dg$$

we obtain

$$(f \times \varphi_\lambda)(x) = \int_B e^{(i\lambda+\rho)(A(x,b))} \tilde{f}(\lambda, b)\, db, \tag{14}$$

where $\tilde{f}$ is the Fourier transform [22(g)]

$$\tilde{f}(\lambda, b) = \int_X f(x)\, e^{(-i\lambda+\rho)(A(x,b))}\, dx. \tag{15}$$

Since λ is real it follows that

$$(|f \times \varphi_\lambda|_\lambda)^2 = \int_B |\tilde{f}(\lambda, b)|^2\, db. \tag{16}$$

Now (14) and (16) show that if $|f \times \varphi_\lambda|_\lambda = 0$ then $f \times \varphi_\lambda \equiv 0$; moreover the identity map of $\mathscr{D}(X) \times \varphi_\lambda$ (with the norm (12)) into $\mathscr{H}_\lambda$ is an isometry. The space $\mathscr{D}(X) \times \varphi_\lambda$ being G-invariant and $\mathscr{H}_\lambda$ irreducible the proposition follows.

COROLLARY 5.5. *If $\lambda \in \mathfrak{a}^*$ is simple then the space of functions on B,*

$$\{\tilde{f}(\lambda, b) \mid f \in \mathscr{D}(X)\} \tag{17}$$

is dense in $L^2(B)$.

In analogy with Theorem 3.2 and the subsequent Theorem 6.4 we extend (11) to the case when φ_λ is not necessarily positive definite.

PROPOSITION 5.6. *Let $\lambda \in \mathfrak{a}_c^*$ and $\mathscr{E}_{(\lambda)}$ the closed subspace of $\mathscr{E}(X)$ generated by the translates of φ_λ under G. Let T_λ denote the natural representation of G on $\mathscr{E}_{(\lambda)}$ and $T_\lambda{}^*$ the contragredient representation of G on the dual space $\mathscr{E}^*_{(\lambda)}$ (strong topology). Then $T_\lambda{}^*(K)$ admits a fixed vector δ_λ (unique up to a factor), T_λ and $T_\lambda{}^*$ are both irreducible and with a suitable normalization*

$$\varphi_\lambda(gK) = \langle T_\lambda(g^{-1})\, \varphi_\lambda\, , \delta_\lambda \rangle, \qquad g \in G. \tag{18}$$

Proof. That $T_\lambda{}^*$ has the continuity properties required of a representation is clear since $\mathscr{E}_{(\lambda)}$ is semireflexive (Bruhat [6(a), p. 114]). By the Hahn–Banach theorem each $\sigma \in \mathscr{E}^*_{(\lambda)}$ can be extended to a continuous linear functional on $\mathscr{E}(X)$, i.e., to a distribution T_σ on X of compact support. If σ is K-invariant we can also take T_σ to be K-invariant so

$$\begin{aligned} T_\sigma(\varphi_\lambda^{\tau(g)}) &= \int_X \varphi_\lambda(g^{-1} \cdot x)\, dT_\sigma(x) = \int_X \varphi_\lambda(g^{-1}k \cdot x)\, dT_\sigma(x) \\ &= \varphi_\lambda(g^{-1} \cdot o)\, T_\sigma(\varphi_\lambda) = T_\sigma(\varphi_\lambda)\, \delta(\varphi_\lambda^{\tau(g)}), \end{aligned}$$

where δ is the delta-distribution on X at o. Thus $\sigma(f) = \sigma(\varphi_\lambda)\, \delta(f)$ for $f \in \mathscr{E}_{(\lambda)}$ which proves the existence and uniqueness of δ_λ as well as (18). For the irreducibility of T_λ let $V \subset \mathscr{E}_{(\lambda)}$ be a nonzero closed invariant subspace. Then there exists an $f \in V$ such that $f(o) \neq 0$. The integral of the V-valued function $k \to f^{\tau(k)}$, i.e. $\int_K f^{\tau(k)}\, dk$, is an element $f^\natural$ of V, since V is complete [5, p. 82]. Evaluation at a point is a continuous linear functional on V so (by definition) $f^\natural(x) = \int f^{\tau(k)}(x)\, dk$ whence $f^\natural \neq 0$ and $f^\natural$ is a nonzero multiple of φ_λ. Thus $V = \mathscr{E}_{(\lambda)}$.

Finally, let $W \subset \mathscr{E}^*_{(\lambda)}$ be a closed invariant subspace and W^0 its annihilator in $\mathscr{E}_{(\lambda)}$. By the irreducibility of T_λ, $W^0 = \mathscr{E}_{(\lambda)}$ or $W^0 = \{0\}$. In the first case, $W = \{0\}$ and in the second case the semireflexivity of $\mathscr{E}_{(\lambda)}$ implies $W = \mathscr{E}_{(\lambda)}$, as desired.

Harish-Chandra's inversion formula for the spherical Fourier transform can be generalized as follows [22(a); (b), Theorem 4.5]. If $f \in \mathscr{D}(X)$ then

$$f(x) = \frac{1}{w} \int_{\mathfrak{a}^* \times B} \tilde{f}(\lambda, b)\, e^{(i\lambda+\rho)(A(x,b))} \mid \mathbf{c}(\lambda)|^{-2}\, d\lambda\, db, \tag{19}$$

$$\int_X |f(x)|^2\, dx = \frac{1}{w} \int_{\mathfrak{a}^* \times B} |\tilde{f}(\lambda, b)|^2 \mid \mathbf{c}(\lambda)|^{-2}\, d\lambda\, db. \tag{20}$$

Here $\tilde{f}(\lambda, b)$ is the Fourier transform (15), w is the order of the Weyl group. If T_X denotes the natural representation of G on $L^2(X)$, formulas (19) and (20) (together with Corollary 5.5) imply that in terms of direct integral theory,

$$L^2(X) = \int_{\mathfrak{a}^*/W} \mathscr{H}_\lambda \mid \mathbf{c}(\lambda)|^{-2}\, d\lambda, \qquad T_X = \int_{\mathfrak{a}^*/W} T_\lambda \mid \mathbf{c}(\lambda)|^{-2}\, d\lambda. \tag{21}$$

Formula (19) can also be written (cf. (14))

$$f(x) = \int_{\mathfrak{a}^*} (f \times \varphi_\lambda)(x) \mid \mathbf{c}(\lambda)|^{-2}\, d\lambda, \qquad f \in \mathscr{D}(X). \tag{22}$$

On the other hand, we have by (5)

$$\varphi(\xi) = \int_{\mathfrak{a}^*} (\varphi \times \Psi_{\lambda,e})(\xi)\, d\lambda \tag{23}$$

and

$$\int_\Xi |\varphi(\xi)|^2\, d\xi = \int_{\mathfrak{a}^*} \| \varphi \times \Psi_{\lambda,e} \|_\lambda\, d\lambda, \qquad \varphi \in \mathscr{D}(\Xi). \tag{24}$$

This means (by Lemma 5.1 (i)) that the natural representation T_Ξ of G on $L^2(\Xi)$ has the direct integral decomposition

$$L^2(\Xi) = \int_{\mathfrak{a}^*} \mathscr{K}_\lambda\, d\lambda, \qquad T_\Xi = \int_{\mathfrak{a}^*} \tau_\lambda\, d\lambda. \tag{25}$$

The decompositions (22) and (23) correspond under the Radon transform and its dual:

PROPOSITION 5.7. *For $f \in \mathscr{D}(X)$ the function $| \mathbf{c}(\lambda)|^{-2} f \times \varphi_\lambda$ has Radon transform $\hat{f} \times \Psi_{\lambda,e}$ in the following weak sense:*

$$\left[\frac{1}{w} \int_{\mathfrak{a}^*} A(\lambda) \mid \mathbf{c}(\lambda)|^{-2} f \times \varphi_\lambda\, d\lambda\right]^\wedge = \int_{\mathfrak{a}^*} A(\lambda)(\hat{f} \times \Psi_{\lambda,e})\, d\lambda \tag{26}$$

for all $A \in \mathscr{D}(\mathfrak{a}^*)$ *invariant under* W. *Moreover,*

$$f \times \varphi_\lambda = (\hat{f} \times \Psi_{\lambda,e})^{\vee}, \qquad \lambda \in \mathfrak{a}_c^*, \tag{27}$$

and

$$|f \times \varphi_\lambda|_\lambda = \|\hat{f} \times \Psi_{\lambda,e}\|_\lambda, \qquad \lambda \in \mathfrak{a}^*.$$

Proof. As recalled in Chap. I (Theorem 2.5(i)) the function $A(\lambda)$ can be written

$$A(\lambda) = \int_X a(x)\, \mathrm{conj}(\varphi_\lambda(x))\, dx,$$

where the function $g \to a(g \cdot o)$ on G lies in the space $I^2(G)$ and is given by the inversion formula

$$a(x) = w^{-1} \int_{\mathfrak{a}^*} A(\lambda)\, \varphi_\lambda(x)\, |\, c(\lambda)|^{-2}\, d\lambda.$$

Hence

$$(f \times a)(x) = w^{-1} \int_{\mathfrak{a}^*} A(\lambda)(f \times \varphi_\lambda)(x)\, |\, c(\lambda)|^{-2}\, d\lambda. \tag{28}$$

Now according to [21(j), p. 593], the integral

$$\int_N a(gn \cdot o)\, dn$$

converges absolutely for each $g \in G$. Putting $f_1(x) = f(k \cdot x)$ we have (assuming all integrals absolutely convergent) for $b \in A$

$$\begin{aligned}
(f \times a)^{\wedge}(kbMN) &= \int_N (f_1 \times a)(bn \cdot o)\, dn \\
&= \int_G f_1(g \cdot o)\, \hat{a}(g^{-1}bMN)\, dg \\
&= \int_{AN} f_1(hn \cdot o)\, \hat{a}(n^{-1}h^{-1}b \cdot \xi_o)\, dh\, dn \\
&= \int_A \hat{f}_1(hMN)\, \hat{a}(h^{-1}bMN)\, dh.
\end{aligned}$$

Since the computations hold with $|f|$ and $|a|$ replacing f and a giving a finite result the computations above are valid. But the function $A(\lambda)$

is the Fourier transform of the function $h \to e^{\rho(\log h)}\hat{a}(hMN)$ on A [21(i), p. 262] so we obtain

$$(f \times a)^{\wedge}(kbMN) = \int_{\mathfrak{a}^*} F(\lambda)\, A(\lambda)\, d\lambda$$

where

$$F(\lambda) = \int_A \hat{f}(kbhMN)\, e^{\rho(\log h)} e^{-i\lambda(\log h)}\, dh,$$

which we have seen equals $(\hat{f} \times \Psi_{\lambda,e})(kbMN)$. In view of (28) this proves (26).

Next we have

$$(\hat{f} \times \Psi_{\lambda,e})(kMN) = \int_A \left(\int_N f(kan \cdot o)\, dn\right) e^{(-i\lambda+\rho)(\log a)}\, da$$

which equals $\tilde{f}(\lambda, kM)$ since

$$\int_X f(x)\, dx = \int_{AN} f(an \cdot o)\, da\, dn.$$

Comparing with (14) and (16) the rest of Proposition 5.7 follows.

Let $\mathfrak{a}_+{}^*$ denote the positive Weyl chamber in $\mathfrak{a}^*$, that is the set of $\lambda \in \mathfrak{a}^*$ such that the vector $H_\lambda \in \mathfrak{a}$ determined by $B(H_\lambda, H) = \lambda(H)$ $(H \in \mathfrak{a})$ belongs to the positive Weyl chamber $\mathfrak{a}^+$.

THEOREM 5.8. *The Fourier transform $f(x) \to \tilde{f}(\lambda, b)$ defined by* (15) *extends to an isometry of $L^2(X)$ onto $L^2(\mathfrak{a}_+{}^* \times B)$ (with the measure $|\boldsymbol{c}(\lambda)|^{-2}\, d\lambda\, db$ on $\mathfrak{a}_+{}^* \times B$).*

Proof. Let $\mathscr{K}(X)$ denote the set of continuous functions on X of compact support; if φ_n denotes the characteristic function of a closed ball in X with center 0 and radius $n \in \boldsymbol{Z}^+$ put $\mathscr{K}^* = \bigcup_{n=0}^{\infty} \varphi_n \mathscr{K}(X)$. Let $\varphi, f \in \mathscr{K}^*$ and assume φ is K-invariant. Then $\tilde{\varphi}(\lambda, b)$ equals the spherical Fourier transform $\tilde{\varphi}(\lambda)$ of the function $g \to \varphi(g \cdot 0)$. Also $f \times \varphi \in \mathscr{K}^*$ and using $A(g \cdot x, b) = A(x, g^{-1}(b)) + A(g \cdot o, b)$ we find

$$(f \times \varphi)^{\sim}(\lambda) = \tilde{\varphi}(\lambda) \tilde{f}(\lambda, b). \tag{29}$$

Suppose now $F \in L^2(\mathfrak{a}_+{}^* \times B)$ is orthogonal to the range of the map $f \to \tilde{f}$ $(f \in \mathscr{K}^*)$. Then

$$\int_{\mathfrak{a}_+{}^* \times B} F(\lambda, b)\, \tilde{\varphi}(\lambda) \tilde{f}(\lambda, b)\, |\boldsymbol{c}(\lambda)|^{-2}\, d\lambda\, db = 0, \tag{30}$$

for all φ and f as above. Since the functions $\tilde{\varphi}$ (φ K-invariant in $\mathscr{D}(X)$) form a dense subspace of $L^2(\mathfrak{a}_+^*)$ we deduce that there exists a null set N_f (dependent on f) such that if $\lambda \notin N_f$ then $b \to F(\lambda, b)$ lies in $L^2(B)$ and

$$\int_B F(\lambda, b) \tilde{f}(\lambda, b)\, db = 0, \qquad (f \in \mathscr{K}^*). \tag{31}$$

Now fix a global coordinate system on X, and consider the set $\mathfrak{M}$ of all functions on X of the form $\varphi_n(x)\, p(x)$ where $n \in \mathbf{Z}^+$ and p is a polynomial in the coordinates with rational coefficients. Then $\mathfrak{M}$ is a countable subset of $\mathscr{K}^*$ and dense in the uniform topology. Now let $N = \bigcup N_f$ as f runs through $\mathfrak{M}$, so (31) holds for all $f \in \mathfrak{M}$ and all $\lambda \notin N$. But if $f \in \mathscr{K}^*$ and n_0 such that $f = f\varphi_{n_0}$, then a suitable sequence $f_k = \varphi_{n_0} p_k$ ($k = 1, 2, \ldots$) from $\mathfrak{M}$ converges uniformly to f and for each $\lambda \in \mathfrak{a}^*$, $\tilde{f}_k(\lambda, b)$ converges to $\tilde{f}(\lambda, b)$ uniformly on B. Thus (31) holds for all $f \in \mathscr{K}^*$ and all λ outside the fixed null set N. Now by Bruhat [6(a), p. 193] each $\lambda \in \mathfrak{a}_+^*$ is irreducible, hence by Proposition 5.3 simple, so using Corollary 5.5 we deduce that if $\lambda \notin N$, then

$$\int_B F(\lambda, b)\, P(b)\, db = 0, \qquad \text{for all} \quad P \in L^2(B).$$

Now the Fubini theorem shows that $F = 0$ almost everywhere on $\mathfrak{a}_+^* \times B$.

§6. Construction of the Intertwining Operators

Suppose now λ belongs to the real space $\mathfrak{a}^*$. Then as mentioned earlier, the representation τ_λ of G on the Hilbert space $\mathscr{K}_\lambda$ is unitary. Furthermore τ_λ is irreducible and $\tau_{s\lambda}$ is equivalent to τ_λ for each $s \in W$ (Bruhat [6(a)], Kostant [28(b)]. Thus there exists an operator $A_{\lambda,s}$: $\mathscr{K}_\lambda \to \mathscr{K}_{s\lambda}$ unique up to a scalar multiple, such that $A_{\lambda,s}\tau_\lambda(g) = \tau_{s\lambda}(g)A_{\lambda,s}$ for $g \in G$; the operator $A_{\lambda,s}$ is called an *intertwining operator* for τ_λ and $\tau_{s\lambda}$, [33], [6(a)].

We shall now construct the operators $A_{\lambda,s}$ explicitly by means of the conical distributions; by passing from the Hilbert space $\mathscr{K}_\lambda$ to the distribution space $\mathscr{D}_\lambda'$ we generalize the equivalence $\tau_\lambda \sim \tau_{s\lambda}$ to complex λ. We recall that the mapping $S \to \Psi$ where

$$\Psi(\varphi) = \int_{K/M} \left(\int_A \varphi(ka \cdot \xi_o)\, e^{(i\lambda+\rho)(\log a)}\, da \right) dS(kM) \tag{1}$$

is a bijection of $\mathscr{D}'(B)$ onto $\mathscr{D}_\lambda'(\Xi)$, mapping $L^2(B)$ onto $\mathscr{H}_\lambda$; hereby $\mathscr{D}'(B)$ and $L^2(B)$ become representation spaces for $\tilde{\tau}_\lambda$ and τ_λ, respectively. Under the map (1) the distribution $\Psi_{\lambda,s}$ corresponds to the distribution

$$S_{\lambda,s} : F \to \boldsymbol{d}_s(\lambda)^{-1} \int_{\bar{N}_s} F(m_s k(\bar{n}_s)M)\, e^{-(i\lambda+\rho)(H(\bar{n}_s))}\, d\bar{n}_s\,, \qquad F \in \mathscr{E}(B), \qquad (2)$$

defined for all $\lambda \in \mathfrak{a}_c{}^*$ by analytic continuation. Let $S_1 \times S_2$ denote the convolution product of distributions on B as defines in §4. Although in general this product only commutes with the action of K we see now that for $S_2 = S_{\lambda,s}$ we get commutation with the action of G.

Theorem 6.1. *Assume $\lambda \in \mathfrak{a}_c{}^*$ irreducible*[3] *and that $\boldsymbol{e}_{s^*}(\lambda)\, \boldsymbol{e}_{s^*}(-\lambda) \neq 0$. Let $s \in W$. Then the convolution operator $S \to S \times S_{s\lambda,s^{-1}}$ on $\mathscr{D}'(B)$ induces* (via (1)) *a bijection $A_{\lambda,s}$ of $\mathscr{D}_\lambda'$ onto $\mathscr{D}_{s\lambda}'$ which intertwines $\tilde{\tau}_\lambda$ and $\tilde{\tau}_{s\lambda}$*, i.e.,

$$A_{\lambda,s}\tilde{\tau}_\lambda(g) = \tilde{\tau}_{s\lambda}(g)\, A_{\lambda,s}\,, \qquad g \in G. \qquad (3)$$

In particular, $\tilde{\tau}_\lambda$ and $\tilde{\tau}_{s\lambda}$ are equivalent.

Proof. By definition, if $\varphi \in \mathscr{D}(\Xi)$, then

$$(A_{\lambda,s}(\Psi))(\varphi) = \int_{K/M} \left(\int_A \varphi(ka \cdot \xi_o)\, e^{(is\lambda+\rho)(\log a)}\, da \right) d(S \times S_{s\lambda,s^{-1}})(kM). \qquad (4)$$

Using Lemma 4.3 and Proposition 4.6 we get

$$(A_{\lambda,s}\Psi)^\vee(x) = \iint e^{(is\lambda+\rho)(A(x,k_1k_2M))}\, dS(k_1M)\, dS_{s\lambda,s^{-1}}(k_2M). \qquad (5)$$

Writing $A(x, k_1k_2M) = A(k_1^{-1} \cdot x, k_2M)$, Lemma 4.8 transforms this expression into

$$b_s \int e^{(i\lambda+\rho)(A(k^{-1}\cdot x, eM))}\, dS(kM), \qquad (6)$$

where $b_s = a_{s^{-1}}\boldsymbol{e}_{s^{-1}}(s\lambda)$ so

$$(A_{\lambda,s}\Psi)^\vee(x) = b_s\check{\Psi}(x). \qquad (7)$$

[3] See Chap. IV, Theorem 1.1 and the subsequent Remark.

Since the dual transform $\Psi \to \check{\Psi}$ commutes with the action of G and since $\tilde{\tau}_\lambda(g)\Psi = \Psi^{\tau(g)}$ we see that

$$\begin{aligned}(A_{\lambda,s}\tilde{\tau}_\lambda(g)\Psi)^\vee &= b_s(\tilde{\tau}_\lambda(g)\Psi)^\vee = b_s(\Psi^{\tau(g)})^\vee = b_s(\check{\Psi})^{\tau(g)} \\ &= ((A_{\lambda,s}\Psi)^\vee)^{\tau(g)} = (\tilde{\tau}_{s\lambda}(g)\, A_{\lambda,s}\Psi)^\vee.\end{aligned}$$

Since $s\lambda$ is simple (Proposition 5.3) it follows that

$$A_{\lambda,s}\tilde{\tau}_\lambda(g) = \tilde{\tau}_{s\lambda}(g)\, A_{\lambda,s}\,. \tag{8}$$

To see that $A_{\lambda,s}$ is surjective it suffices to prove that the mapping $S \to S \times S_{s\lambda,s^{-1}}$ maps $\mathscr{D}'(B)$ onto itself. But using Lemma 4.8 on (5) and (6) for $S = S_{\lambda,s}$ we see that

$$\begin{aligned}&\int_B e^{(is\lambda+\rho)(A(x,b))}\, d(S_{\lambda,s} \times S_{s\lambda,s^{-1}})(b) \\ &\qquad = \boldsymbol{a}_{s^{-1}}\boldsymbol{e}_{s^{-1}}(s\lambda) \int_B e^{(i\lambda+\rho)(A(x,b))}\, dS_{\lambda,s}(b) \\ &\qquad = \boldsymbol{a}_{s^{-1}}\boldsymbol{a}_s\boldsymbol{e}_{s^{-1}}(s\lambda)\, \boldsymbol{e}_s(\lambda) \int_B e^{(is\lambda+\rho)(A(x,b))}\, d\delta_o(b),\end{aligned}$$

where δ_o is the delta-distribution on B at eM. Again since $s\lambda$ is simple it follows that

$$S_{\lambda,s} \times S_{s\lambda,s^{-1}} = \boldsymbol{a}_{s^{-1}}\boldsymbol{a}_s\boldsymbol{e}_{s^{-1}}(s\lambda)\, \boldsymbol{e}_s(\lambda)\, \delta_o\,.$$

But using for example Lemma 1.2, Chap. IV we see that this formula holds not only when $s\lambda$ is simple but on the entire space $\mathfrak{a}_c{}^*$ by analytic continuation. Moreover $\boldsymbol{e}_{s^{-1}}(s\lambda) = \boldsymbol{e}_s(-\lambda)$ so we obtain

$$S_{\lambda,s} \times S_{s\lambda,s^{-1}} = \boldsymbol{a}_{s^{-1}}\boldsymbol{a}_s\boldsymbol{e}_s(\lambda)\, \boldsymbol{e}_s(-\lambda)\, \delta_0\,, \qquad (\lambda \in \mathfrak{a}_c{}^*). \tag{9}$$

Taking convolution with S on the left the surjectivity follows. The injectivity is clear from (7).

THEOREM 6.2. *Suppose $\lambda \in \mathfrak{a}^*$. Then a suitable constant multiple of the operator $A_{\lambda,s}$ is an isometry of $\mathscr{H}_\lambda$ onto $\mathscr{H}_{s\lambda}$.*

Proof. The statement will follow by proving that the operator

$$B_{\lambda,s} : F \to b_s F \times S_{s\lambda,s^{-1}}$$

is an isometry of L^2 onto itself, b_s being the constant above. The expressions (5) and (6) being equal, the isometry statement amounts to that if $F \in L^2$ and $S_1 \in \mathscr{D}'(B)$ such that

$$\int_B e^{(i\lambda+\rho)(A(x,b))} F(b)\, db = \int_B e^{(is\lambda+\rho)(A(x,b))}\, dS_1(b) \tag{10}$$

for all $x \in X$, then $S_1 \in L^2(B)$ and F and S_1 have the same L^2-norm. But if $f \in \mathscr{D}(X)$ we have (§5)

$$f \times \varphi_\lambda(x) = \int_B e^{(i\lambda+\rho)(A(x,b))}\, \tilde{f}(\lambda, b)\, db,$$

so $\varphi_\lambda = \varphi_{s\lambda}$ implies

$$\int_B e^{(i\lambda+\rho)(A(x,b))} \tilde{f}(\lambda, b)\, db = \int_B e^{(is\lambda+\rho)(A(x,b))} \tilde{f}(s\lambda, b)\, db. \tag{11}$$

Since $s\lambda$ is simple we find from (10) and (11) that if $F(b) = \tilde{f}(\lambda, b)$ then S_1 is the L^2-function $b \to \tilde{f}(s\lambda, b)$; by the proof of Proposition 5.4

$$\int_X (f \times \varphi_\lambda)(x)\, \mathrm{conj}(f(x))\, dx = \int_B |\tilde{f}(\lambda, b)|^2\, db$$

so F and S_1 have the same norm. Now if $F \in L^2(B)$ is arbitrary there exists by Corollary 5.5 a sequence $(f_n) \in \mathscr{D}(X)$ such that $\tilde{f}_n(\lambda, \cdot)$ converges to F in the L^2 norm. Let $F_s \in L^2(B)$ denote the limit of the sequence $\tilde{f}_n(s\lambda, \cdot)$. Replacing in (11) f by f_n we deduce by Schwarz' inequality that S_1 in (10) equals F_s which of course has the same norm as F. That $B_{\lambda,s}$ is surjective is clear by taking left convolution in (9) with an arbitrary $F_0 \in L^2(B)$; the first part of the proof having shown that $F_0 \times S_{\lambda,s} \in L^2(B)$.

Remark 1. The proof shows that on the subspace $\{\tilde{f}(\lambda, \cdot) \mid f \in \mathscr{D}(X)\}$ of $L^2(B)$ the intertwining operator $B_{\lambda,s}$ is given by

$$B_{\lambda,s} : \tilde{f}(\lambda, b) \to \tilde{f}(s\lambda, b).$$

Remark 2. Theorems 6.1 and 6.2 were announced in [22(k)]; subsequently I received preprints of Schiffmann [40(b)] and Knapp–Stein [26(c)] kindly sent by the authors. Here Schiffmann gives a different proof of Theorem 6.2; he also proves the analyticity of the integral (2) in the case when F is a K-finite function on B and thereby obtains an analog of (3) for the action of the Lie algebra $\mathfrak{g}$ on the space of K-finite vectors in $\mathscr{H}_\lambda$. This amounts to *infinitesimal equivalence* [22(a), p. 230] of the representations τ_λ and $\tau_{s\lambda}$. Knapp and Stein deal with G of real rank one and for this case indicate a meromorphic continuation of the integral (2) but the isometry of the intertwining operators is not discussed.

For complex G the intertwining operators for the principal series were constructed by Kunze–Stein [30] (cf. also, Zhelobenko [49(b)]) and in a more precise form by Schiffmann [40(a)] where Theorem 6.2 is also reduced to the rank one case.

DEFINITION. Let $\lambda \in \mathfrak{a}_c^*$. A vector $\Psi \neq 0$ in $\mathscr{D}_\lambda'$ will be called an *extreme weight vector* if there exists a $\mu \in \mathfrak{a}_c^*$ such that

$$\tilde{\tau}_\lambda(an)\Psi = e^{\mu(\log a)}\Psi \qquad \text{for} \quad a \in A, \quad n \in N. \tag{12}$$

The corresponding linear forms μ in (12) will be called the *extreme weights* of the representation $\tilde{\tau}_\lambda$.

THEOREM 6.3. *Let λ be as in Theorem 4.9. Then $\Psi_{\lambda,s} \neq 0$ for $s \in W$, the nonzero scalar multiples of the $\Psi_{\lambda,s}$ constitute precisely the extreme weight vectors in $\mathscr{D}_\lambda'$ and*

$$\tilde{\tau}_\lambda(an)\, \Psi_{\lambda,s} = e^{-(is\lambda+\rho)(\log a)}\Psi_{\lambda,s} \qquad a \in A, \quad n \in N, \tag{13}$$

so the linear forms $-(is\lambda + \rho)$, $s \in W$, constitute the extreme weights of $\tilde{\tau}_\lambda$.

Proof. The first statement and (13) follow from Lemma 4.8, λ being simple. Now if $\Psi \in \mathscr{D}_\lambda'$ is an extreme vector, $\check{\Psi}$ is an N-invariant element in the eigenspace $\mathscr{E}_\lambda(X)$. By the proof of Theorem 4.9, $\check{\Psi}(na \cdot o)$ is a linear combination of the exponentials $e^{(is\lambda+\rho)(\log a)}$, $(s \in W)$. Since the $s\lambda$ are distinct we conclude from (12) that $\check{\Psi}(na \cdot o)$ is proportional to one of these exponentials, so by the simplicity of λ, Ψ is proportional to one of the vectors $\Psi_{\lambda,s}$.

Remark. Extreme weights of infinite-dimensional representations have for example been considered by Harish–Chandra [21(e), p. 751] (where two irreducible representation of a complex semisimple $\mathfrak{g}$ with the same extreme weight are shown equivalent) and Zhelobenko [49(c)] (where for complex G the set of extreme weights of a given irreducible representation is related to the Weyl group).

We have seen that if $\langle i\lambda - \rho, \alpha\rangle/\langle\alpha, \alpha\rangle \in \boldsymbol{Z}^+$ for each $\alpha \in \Sigma^+$ then $\mathscr{D}_\lambda'$ contains a conical function ψ, and if E_ψ denotes the G-invariant subspace generated by ψ then E_ψ is finite-dimensional, irreducible under the natural representation π_ψ of G on E_ψ and

$$\psi(g \cdot \xi_o) = \langle \pi_\psi(g^{-1})e, e^*\rangle, \tag{14}$$

where $\boldsymbol{e}$ and $\boldsymbol{e}^*$, respectively, are highest weight vectors for π_ψ and its contragredient representation. We shall now obtain analogs of (14) for the infinite-dimensional representation τ_λ in terms of the extreme weight vectors $\Psi_{\lambda,s}$ from Theorem 6.3.

Consider now the pairing $\langle\ ,\ \rangle$ between the spaces $\mathscr{E}_\lambda(\Xi) = \mathscr{E}(\Xi) \cap \mathscr{K}_\lambda$ and $\mathscr{D}'_{-\lambda}$ given by

$$\langle \psi, \Psi \rangle = \int_{K/M} \psi(k \cdot \xi_o)\, dS(kM) \tag{15}$$

if

$$\Psi(\varphi) = \int_{K/M} \left(\int_A \varphi(ka \cdot \xi_o)\, e^{(-i\lambda+\rho)(\log a)}\, da \right) dS(kM).$$

THEOREM 6.4. *With the pairing* (15) *the representation τ_λ on $\mathscr{E}_\lambda$ and $\tilde{\tau}_{-\lambda}$ on $\mathscr{D}'_{-\lambda}$ are contragredient. Moreover, with λ as in Theorem* 6.1, *we have for each $s \in W$ the following analog of* (14): $\mathscr{D}(\Xi) \times \Psi_{\lambda,s} \subset \mathscr{E}_\lambda$ *and there exists a constant $a(\lambda, s) \in \boldsymbol{C}$ such that*

$$\Psi_{-s\lambda,e}(\varphi) = a(\lambda, s)\langle \varphi \times \Psi_{\lambda,s}, \Psi_{-\lambda,s}\rangle, \qquad \varphi \in \mathscr{D}(\Xi). \tag{16}$$

Proof. We must first prove the invariance of $\langle\ ,\ \rangle$,

$$\langle \tau_\lambda(g)\psi, \tilde{\tau}_{-\lambda}(g)\Psi\rangle = \langle \psi, \Psi\rangle, \qquad g \in G. \tag{17}$$

If S in (15) is a function, (17) follows just as the invariance of the form $\{\ ,\ \}$ in the proof of Proposition 5.3. For arbitrary S, the invariance follows easiest by approximation; this proves the first part of the theorem.

Turning now to (16) we recall first that by (1) and (3) in §5, the distribution $\Psi^*_{\lambda,s}$ satisfies

$$(\varphi \times \Psi_{\lambda,s})(g \cdot \xi_o) = \Psi^*_{\lambda,s}(\varphi \circ \tau(g)) \qquad g \in G, \tag{18}$$

so we must compute $\Psi^*_{\lambda,s}$.

LEMMA 6.5. *With λ as in Theorem* 6.4,

$$\Psi^*_{\lambda,s} = c(s, \lambda)\, \Psi_{-s\lambda, s^{-1}}$$

where $c(s, \lambda) \in \boldsymbol{C}$.

Proof. For simplicity put $\eta = \Psi_{\lambda,s}$ and let $\tilde{\eta} \in \mathscr{D}'(G)$ be as in §5. Then (18) implies that η^* is MN-invariant. We have now for $a \in A$,

$$(\eta^*)^{\sigma(a)} = e^{(is\lambda-\rho)(\log a)}\eta^* \tag{19}$$

and

$$(\eta^*)^{\tau(a)} = e^{(i\lambda-\rho)(\log a)}\eta^*. \tag{20}$$

We indicate the proof of the first relation, the proof for the second is similar and will be left to the reader. We first have

$$(\eta^*)^{\sigma(a)}(\varphi \circ \tau(g)) = \eta^*(\varphi^{\sigma(a)^{-1}} \circ \tau(g)) \tag{21}$$

and if as in §5, $\varphi = \bar{F}$, $(F \in \mathscr{D}(G))$, then the formula

$$\int_N F(na)\, dn = e^{2\rho(\log a)} \int_N F(an)\, dn$$

gives in terms of the right translation $R(a) : g \to ga$ on G,

$$\varphi^{\sigma(a)} = e^{2\rho(\log a)}(F^{R(a)})^-, \qquad a \in A. \tag{22}$$

Thus by (18)

$$(\eta^*)^{\sigma(a)}(\varphi \circ \tau(g)) = e^{-2\rho(\log a)} \int_G F(gh^{-1}a)\, d\tilde{\eta}(h).$$

Now $\tilde{\eta}'$ being as in §5 we find easily

$$(\tilde{\eta}')^{R(a)} = (\tilde{\eta}^{L(a^{-1})})'$$

and

$$\tilde{\eta}^{L(a)} = (\eta^{\tau(a)})^\sim$$

and by Lemma 4.8,

$$\eta^{\tau(a)} = e^{-(is\lambda+\rho)(\log a)}\eta.$$

Thus

$$\int_G F(gh^{-1}a)\, d\tilde{\eta}(h) = \int_G F(gha)\, d\tilde{\eta}'(h) = \int_G F(gh)\, d(\tilde{\eta}')^{R(a)}(h);$$

so invoking the relations above we find

$$\begin{aligned}(\eta^*)^{\sigma(a)}(\varphi \circ \tau(g)) &= e^{(is\lambda-\rho)(\log a)} \int_G F(gh)\, d\tilde{\eta}'(h)\\ &= e^{(is\lambda-\rho)(\log a)}(\varphi \times \eta)(g \cdot \xi_o),\end{aligned}$$

so by (18), formula (19) follows. We have thus proved that $\Psi^*_{\lambda,s}$ is a conical distribution in $\mathscr{D}'_{-s\lambda}$. If λ is regular we deduce from (20) and Theorem 6.3 that

$$\Psi^*_{\lambda,s} = c(\lambda, s)\, \Psi_{-s\lambda, s^{-1}}, \qquad \text{where} \quad c(\lambda, s) \in \mathbf{C}. \tag{23}$$

If λ is singular we approximate it with regular λ_n (all λ_n and λ satisfying the assumptions of the lemma). For each $\varphi \in \mathscr{D}(\Xi)$ satisfying $\Psi_{-s\lambda,s^{-1}}(\varphi) \neq 0$ we have for n sufficiently large,

$$c(\lambda_n\,, s) = \Psi^*_{\lambda_n, s}\,(\varphi)/\Psi_{-s\lambda_n, s^{-1}}(\varphi).$$

The right side tends to a limit as $\lambda_n \to \lambda$ and this limit is independent of the choice of the sequence (λ_n). The left side shows that the limit is also independent of φ; hence we denote the limit by $c(\lambda, s)$. Now by (23) we have for *all* $\varphi \in \mathscr{D}(\Xi)$

$$\Psi^*_{\lambda_n, s}\,(\varphi) = c(\lambda_n\,, s)\,\Psi_{-{\circ}\lambda_n, s^{-1}}(\varphi).$$

Letting $\lambda_n \to \lambda$ we obtain (23) also for the singular point λ so the lemma is proved.

We can now complete the proof of Theorem 6.4. Using (18) and (20) we see that $\varphi \times \Psi_{\lambda,s} \in \mathscr{K}_\lambda$. Now it remains to compute

$$\int_{K/M} \Psi_{-s\lambda, s^{-1}}(\varphi \circ \tau(k))\, dS_{-\lambda, s}(kM).$$

This expression equals

$$\int_A e^{(-is\lambda+\rho)(\log a)} \Big(\int_{K/M}\int_{K/M} \varphi(kk_1 a \cdot \xi_o)\, dS_{-s\lambda, s^{-1}}(k_1 M)\, dS_{-\lambda, s}(kM)\Big)\, da$$

which by Lemma 4.3 is

$$\int_A \int_{K/M} \varphi(ka \cdot \xi_o)\, e^{(-is\lambda+\rho)(\log a)}\, d(S_{-\lambda, s} \times S_{-s\lambda, s^{-1}})\,(kM)\, da$$

and by (9) this reduces to

$$\boldsymbol{a}_{s^{-1}}\boldsymbol{a}_s \boldsymbol{e}_{s^{-1}}(-s\lambda)\, \boldsymbol{e}_s(-\lambda)\, \Psi_{-s\lambda, e}(\varphi).$$

This concludes the proof of Theorem 6.4.

Remark. Formula (16) looks slightly different from formula (6) in [22(h)]; the reason is that the convolution product on Ξ is not defined in [22(h)] in exactly the same way as here in the beginning of §5.

Chapter IV. Eigenspaces of Invariant Differential Operators as Models for Irreducible Representations

Let U be a Lie group, P a closed subgroup, $\boldsymbol{D}(U/P)$ the set of U-invariant differential operators on the manifold U/P. The joint eigenspaces of the operators in $\boldsymbol{D}(U/P)$ give natural representation spaces for the group U. For $U = \boldsymbol{SO}(n+1)$ acting on the sphere $U/P = \boldsymbol{S}^n$, $\boldsymbol{D}(U/P)$ consists of the polynomials in the Laplacian and the eigenspaces are well-known to be irreducible under the action of U. For noncompact spaces this is no longer true in general as the space of constants within the space of harmonic functions shows.

§1. The Eigenfunctions of the Laplacian Δ on a Symmetric Space of Rank One

In this section we assume that the symmetric space $X = G/K$ has rank one, i.e., $\dim A = 1$, although many of our arguments (e.g., Lemma 1.2) hold for arbitrary rank. We preserve the notation from previous chapters. Now the Weyl group has two elements, e and s. The following result was used in the proof of Lemma 4.13 in Chap. III.

Theorem 1.1. *An element $\lambda \in \mathfrak{a}_c{}^*$ is simple if and only if $\boldsymbol{e}_s(\lambda) \neq 0$, hence λ is irreducible if and only if $\boldsymbol{e}_s(\lambda)\, \boldsymbol{e}_s(-\lambda) \neq 0$.*

Proof. Suppose $\boldsymbol{e}_s(\lambda) = 0$. By the remark following Theorem 4.5, Chap. III we have $\Psi_{\lambda,s} \neq 0$ whereas by Lemma 4.8, $(\Psi_{\lambda,s})^\vee = 0$. Hence λ is not simple. The converse requires some preparation.

Lemma 1.2. (*X of arbitrary rank*). *Suppose $\lambda \in \mathfrak{a}_c{}^*$ satisfies*

$$\operatorname{Re}(\langle i\lambda, \alpha\rangle) > 0 \qquad \text{for} \quad \alpha \in \Sigma^+. \tag{1}$$

Then λ is simple.

Proof. If λ is not simple choose $F \neq 0$ in $\mathscr{E}(B)$ such that

$$\int_B e^{(i\lambda+\rho)(A(x,b))} F(b)\, db = 0, \qquad x \in X. \tag{2}$$

Here we can assume $F(eM) \neq 0$, replacing F by $F^{\tau(k)}$ if necessary. Take $x = a \cdot o$ $(a \in A)$ and use the formula (8) in Chap. III, §2 to transfer the integration in (2) from B to $\overline{N}$. Since

$$A(a \cdot o, k(\bar{n})M) = -H(a^{-1}k(\bar{n})) = H(\bar{n}) - H(a^{-1}\bar{n}a) + \log a$$

formula (2) implies

$$\int_{\bar{N}} e^{-(i\lambda+\rho)(H(a^{-1}\bar{n}a)}e^{(i\lambda-\rho)(H(\bar{n}))}F(k(\bar{n})M)\,d\bar{n} = 0. \tag{3}$$

Using the substitution $\bar{n} \to a\bar{n}a^{-1}$ we obtain

$$\int_{\bar{N}} e^{(i\lambda-\rho)(H(a\bar{n}a^{-1})}e^{-(i\lambda+\rho)(H(\bar{n})}F(k(a\bar{n}a^{-1})M)\,d\bar{n} = 0 \tag{4}$$

for all $a \in A$. Let $\lambda = \xi + i\eta$ $(\xi, \eta \in \mathfrak{a}^*)$. Our assumption (1) amounts to $-H_\eta \in \mathfrak{a}^+$. Let $H_0 \in \mathfrak{a}^+$ and put $a_t = \exp tH_0$ $(t \geqslant 0)$. Then by [22(i), Lemma 43],

$$\langle H, H(\bar{n})\rangle \geqslant 0, \langle H, H(\bar{n}) - H(a_t\bar{n}a_t^{-1})\rangle \geqslant 0 \qquad \text{for} \quad H \in \mathfrak{a}^+,$$

so if we choose ϵ such that $0 < \epsilon < 1$, $H_\rho + \epsilon H_\eta \in \mathfrak{a}^+$ and put $\kappa = \sup |F|$, the integrand above is for $a = a_t$ majorized by

$$e^{-(\eta+\rho)(H(a_t\bar{n}a_t^{-1}))}e^{(\eta-\rho)(H(\bar{n}))}\kappa \leqslant e^{(\eta+\epsilon\eta)(H(a_t\bar{n}a_t^{-1}))}e^{(\eta-\rho)(H(\bar{n}))}\kappa$$

$$\leqslant e^{(-\eta+\epsilon\eta)(H(\bar{n}))}e^{(\eta-\rho)(H(\bar{n}))}\kappa = e^{(\epsilon\eta-\rho)(H(\bar{n}))}\kappa,$$

which is integrable. Thus we can in (4) put $a = a_t$ and let $t \to +\infty$ under the integral sign. Since $a_t\bar{n}a_t^{-1} \to e$ as $t \to \infty$ we obtain

$$F(eM)\int_{\bar{N}} e^{-(i\lambda+\rho)(H(\bar{n}))}\,d\bar{n} = 0,$$

so by the choice of F, $c(\lambda) = 0$ which contradicts the assumption about λ (cf. (30) §4, Chap. III).

Lemma 1.3. *The point $\lambda = 0$ is simple.*

Proof. If $\lambda = 0$ is not simple there exists a $F \in \mathscr{E}(B)$, $F(eM) \neq 0$ such that

$$\int_B e^{\rho(A(x,b))}F(b)\,db = 0, \qquad x \in X,$$

which, as we saw in the last proof, implies

$$\int_{\bar{N}} e^{-\rho(H(a\bar{n}a^{-1})+H(\bar{n}))}F(k(\bar{n})M)\,d\bar{n} = 0, \qquad a \in A. \tag{5}$$

Now we use the expression for $e^{\rho(H(\bar{n}))}$ from Theorem 1.14, Chap. III. Put $f(X, Y) = F(k(\bar{n})M)$ if $\bar{n} = \exp X \exp Y$, where $X \in \mathfrak{g}_{-\alpha}$, $Y \in \mathfrak{g}_{-2\alpha}$. If $H \in \mathfrak{a}$ and we put $a = \exp tH$ in (5) we obtain for $s = e^{\alpha(H)t}$

$$\int_{\mathfrak{g}_{-\alpha}\times\mathfrak{g}_{-2\alpha}} K_s(X, Y) f(X, Y)\, dX\, dY = 0, \qquad (s > 0), \tag{6}$$

where

$$K_s(X, Y) = [((s^2 + c \mid X \mid^2)^2 + 4c \mid Y \mid^2)((1 + c \mid X \mid^2)^2 + 4c \mid Y \mid^2)]^{-p},$$

p denoting $(m_\alpha + 2m_{2\alpha})/4$. Let $k(s)$ denote the integral of $K_s(X, Y)$ over $\mathfrak{g}_{-\alpha} \times \mathfrak{g}_{-2\alpha}$ and put $L_s(X, Y) = k(s)^{-1}K_s(X, Y)$. Then

$$\int L_s(X, Y)\, dX\, dY = 1. \tag{7}$$

Furthermore, for any $\delta > 0$

$$\lim_{s\to 0} \int_{|X|+|Y|\geqslant\delta} L_s(X, Y)\, dX\, dY = 0. \tag{8}$$

In fact for each fixed $\delta > 0$

$$\int_{|X|+|Y|\geqslant\delta} K_s(X, Y)\, dX\, dY \leqslant \int_{|X|+|Y|\geqslant\delta} K_0(X, Y)\, dX\, dY \leqslant C_\delta,$$

where C_δ is a constant; secondly the monotone convergence theorem gives

$$\lim_{s\to 0} k(s) = \int K_0(X, Y)\, dX\, dY = \infty$$

because repeated polar coordinate transformation shows

$$\int_{|X|\leqslant 1, |Y|\leqslant 1} (\mid X \mid^4 + \mid Y \mid^2)^{-p}\, dX\, dY = \infty.$$

By Eq. (7)

$$\begin{aligned}\int L_s(X, Y) f(X, Y)\, dX\, dY - f(0, 0) &= \int_{|X|+|Y|<\delta} L_s(X, Y)(f(X, Y) - f(0, 0))\, dX\, dY \\ &+ \int_{|X|+|Y|\geqslant\delta} L_s(X, Y)(f(X, Y) - f(0, 0))\, dX\, dY\end{aligned}$$

607/5/1-9*

so by (8) this expression converges to 0 as $s \to 0$. Hence by (6), $f(0, 0) = 0$ which is a contradiction.

Remark. This proof of Lemma 1.3 was suggested by R. Coifman. My original proof was more complicated but did give the lemma without any restriction on the rank of X.

Now according to Bruhat [6(a), p. 193], each real $\lambda \neq 0$ in $\mathfrak{a}^*$ is irreducible, hence by Proposition 5.3 Chap. III, simple. Combining this with Lemmas 1.2 and 1.3 we see that for each $\lambda \in \mathfrak{a}_c^*$ at least one of the elements λ, $-\lambda$ is simple.

To finish the proof of Theorem 1.1, suppose λ is not simple. Assume $\boldsymbol{e}_s(\lambda) \neq 0$. By the above, $-\lambda$ is simple so the first part of the proof gives $\boldsymbol{e}_s(-\lambda) \neq 0$. Now we use the basic formula (9) in Chap. III, §6 which here becomes

$$S_{-\lambda,s} \times S_{\lambda,s} = \boldsymbol{a}_s^2 \boldsymbol{e}_s(\lambda)\, \boldsymbol{e}_s(-\lambda)\, \delta_0 .$$

Taking convolution with a $F \in \mathscr{E}(B)$ on the left we obtain

$$F_1 \times S_{\lambda,s} = \boldsymbol{a}_s^2 \boldsymbol{e}_s(\lambda)\, \boldsymbol{e}_s(-\lambda)\, F, \tag{9}$$

where $F_1 = F \times S_{-\lambda,s}$. On the other hand Lemma 4.8 in Chap. III implies

$$\int_B e^{(i\lambda+\rho)(A(x,b))}(F_1 \times S_{\lambda,s})(b)\, db = \boldsymbol{a}_s \boldsymbol{e}_s(\lambda) \int_B e^{(-i\lambda+\rho)(A(x,b))} F_1(b)\, db.$$

Choosing $F \in \mathscr{E}(B)$ such that $F \not\equiv 0$ and

$$\int_B e^{(i\lambda+\rho)(A(x,b))} F(b)\, db \equiv 0$$

the last formula implies

$$\int_B e^{(-i\lambda+\rho)(A(x,b))} F_1(b)\, db \equiv 0.$$

By (9), $F_1 \not\equiv 0$ so $-\lambda$ is not simple. This contradiction proves $\boldsymbol{e}_s(\lambda) = 0$ so Theorem 1.1 is proved.

Remarks. The irreducible λ had been determined by Parthasarathy, Rao, and Varadarajan [38] for complex G and by Kostant [28(b)] for real G. The methods are algebraic and deal with algebraic irreducibility for the Lie algebra; the results for G come via Harish-Chandra's Theorem 5 in [21(a)] which relates algebraic irreducibility for the Lie algebra to irreducibility for the group.

Formally (that is ignoring some convergence difficulties) the proof above of Theorem 1.1 extends to X of arbitrary rank (cf. [22(k), §4]) giving

$$e_{s^*}(\lambda)\, e_{s^*}(-\lambda) \neq 0$$

as a necessary and sufficient condition for the irreducibility of λ. This criterion had been suspected on the basis of Proposition 5.3 (ii) and special cases of Lemma 4.8, Chap. III. Using the poles of the Gamma function, its duplication formula and the fact that if α and 2α are both restricted roots then their respective multiplicities are even and odd [2], one can verify that the irreducibility criterion above is equivalent to that in [28(b), Theorem 2].

The connection between the irreducibility of the induced representation and the singularities[1] of the Plancherel measure $c(\lambda)^{-1}c(-\lambda)\, d\lambda$ as expressed in [22(k), Corollary 4.3] had to my knowledge not been observed before, even in special cases. It would be interesting to see if such a connection persists in the $\mathfrak{p}$-adic case cf. [6(b)], [36], [34], [35(b)]. It is also natural to conjecture that the relationship will generalize to the other series of representations which enter into Harish-Chandra's Plancherel formula for G [21 (l), Corollary of Theorem 19]. The irreducibility criterion given in [26(a), Theorem 3] for X of rank one, $\lambda = 0$, is in [26(b)] interpreted in these terms.

We recall that if H is a transformation group of a topological space S, a function h on S is called *H-finite* if the transforms of the function under H span a finite-dimensional vector space V_h. If in addition the natural representation of H on V_h decomposes up to equivalence into finitely many copies of the same irreducible representation δ the function h is called *H-finite of type* δ.

THEOREM 1.4. *Suppose the symmetric space* $X = G/K$ *has rank one and let* Δ *denote the Laplace–Beltrami operator on* X. *Then the K-finite eigenfunctions of* Δ *on* X *are precisely the functions*

$$f(x) = \int_B e^{(i\lambda+\rho)(A(x,b))} F(b)\, db, \tag{10}$$

where $\lambda \in \mathfrak{a}_c^*$ *and* F *is a K-finite function on* B.

[1] Including those which occur both in $d_{s^*}(\lambda)$ and $e_{s^*}(\lambda)^{-1}$ but can be cancelled out in $c(\lambda)^{-1}$.

Proof. We know from Proposition 4.6, Chap. III that each function (10) is an eigenfunction of Δ; more precisely it lies in the eigenspace $\mathscr{E}_\lambda(X)$ and is obviously K-finite. For the converse we need a uniqueness lemma concerning the eigenfunctions of Δ.

LEMMA 1.5. *For each $\lambda \in \mathfrak{a}_c{}^*$ there exists a K-orbit O_λ in X such that whenever $f \in \mathscr{E}_\lambda(X)$ vanishes identically on O_λ it vanishes identically on X.*

Proof. Although this lemma is at least implicity in the literature as a consequence of the Gårding inequality (see e.g., Agmon [1], p. 102) we give a direct proof. Let $\langle\ ,\ \rangle$ denote the Riemannian structure on X induced by the Killing form of $\mathfrak{g}$, Δ the corresponding Laplace–Beltrami operator and dx_0 the corresponding volume element. As usual grad u denotes the gradient of a function u. Let $R > 0$ and let S^R and B^R, respectively, denote the sphere and open ball in X with center o and radius R and put

$$\| u \| = \left[\int_{B^R} | u(x)|^2 \, dx_0 + \int_{B^R} \langle \text{grad } u, \text{grad } u\rangle(x) \, dx_0\right]^{1/2}. \tag{11}$$

Let $u \in \mathscr{E}(X)$ such that $u \equiv 0$ on S^R. Let us prove that given $\epsilon > 0$ there exists a $\varphi \in \mathscr{D}(B^R)$ such that

$$\| u - \varphi \| < \epsilon. \tag{12}$$

For this let $(r, \theta_1, \ldots, \theta_{n-1})$ be geodesic polar coordinates around o, let for $\delta > 0$, α_δ be the continuous function on $[0, R]$ defined by: 1) $\alpha_\delta = 1$ on $[0, R - 2\delta]$; 2) $\alpha_\delta = 0$ on $[R - \delta, R]$; 3) α_δ is linear on $[R - 2\delta, R - \delta]$. Let α be defined on B^R by

$$\alpha(r, \theta_1, \ldots, \theta_{n-1}) = \alpha_\delta(r). \tag{13}$$

We first prove that if δ is small enough then

$$\| \alpha u - u \| < \tfrac{1}{2}\epsilon \tag{14}$$

Since

$$\text{grad } u = \sum_{i,j=1}^{n} g^{ij} \frac{\partial u}{\partial \theta_i} \frac{\partial}{\partial \theta_j}, \qquad (r = \theta_n),$$

where

$$g_{ij} = \left\langle \frac{\partial}{\partial \theta_i}, \frac{\partial}{\partial \theta_j} \right\rangle, \qquad g^{ij} = (g_{ij})^{-1},$$

and since

$$\frac{\partial}{\partial\theta_i}(\alpha u - u) = (\alpha - 1)\frac{\partial u}{\partial\theta_i} + \frac{\partial\alpha}{\partial\theta_i} u$$

it would be sufficient to prove

$$\int_{B^R} \left| \frac{\partial\alpha}{\partial r} u \right|^2 dx_0 \to 0 \qquad \text{as} \quad \delta \to 0. \tag{15}$$

But if $g(r, \theta_1, ..., \theta_{n-1}) = \det(g_{ij})$ we have

$$\int_{B^R} \left| \frac{\partial\alpha}{\partial r} u \right|^2 dx_0 = \frac{1}{\delta^2} \int_{R-2\delta}^{R-\delta} dr \int_{(\theta)} | u |^2 g^{1/2} \, d\theta_1 \cdots d\theta_{n-1} .$$

Using now $u \equiv 0$ on S^R we have

$$| u(r, \theta_1, ..., \theta_{n-1})|^2 = \left| \int_r^R \frac{\partial u}{\partial p} dp \right|^2 \leqslant \int_r^R dp \int_r^R \left| \frac{\partial u}{\partial p} \right|^2 dp,$$

so we obtain

$$| u(r, \theta_1, ..., \theta_{n-1})|^2 \leqslant 2\delta \int_{R-2\delta}^R \left| \frac{\partial u}{\partial p} \right|^2 dp, \qquad R - 2\delta \leqslant r \leqslant R.$$

Hence if $m = \max | \partial u/\partial p |^2$ on B^R we have

$$\int_{B^R} \left| \frac{\partial\alpha}{\partial r} u \right|^2 dx_0 \leqslant 4m(\mathrm{vol}(B^{R-\delta}) - \mathrm{vol}(B^{R-2\delta})).$$

Now (15) and therefore (14) follow. Since αu has compact support contained in B^R there exists a $\varphi \in \mathscr{D}(B^R)$ such that $\| \alpha u - \varphi \| < \frac{1}{2}\epsilon$ so (12) is proved.

Now we choose an orthonormal basis of the tangent space X_o and let $(x_1, ..., x_n)$ be a system of normal coordinates with respect to this basis. Let $\varphi \in \mathscr{D}(B^R)$. Using Schwarz's inequality on

$$\varphi(x) = \int_{-R}^{x_j} \frac{\partial\varphi}{\partial x_j}(x_1, ..., t, ..., x_n) \, dt$$

we derive the Poincaré inequality

$$\int_{B^R} | \varphi(x)|^2 \, dx_1 \cdots dx_n \leqslant 4R^2 \int_{B^R} \left| \frac{\partial\varphi}{\partial x_j} \right|^2 dx_1 \cdots dx_n ,$$

by integrating over slices of the ball $x_1^2 + \cdots + x_n^2 < R^2$ perpendicular to the x_j axis. Now let

$$c_R = \inf_{i, x \in B^R} g^{ii}(x), \qquad d_R = \sup_{i \neq j, x \in B^R} g^{ij}(x),$$

$$e_R = \sup_{x \in B^R} g^{1/2}(x), \qquad f_R = \inf_{x \in B^R} g^{1/2}(x).$$

Then if φ is real

$$\sum_{i,j} g^{ij} \frac{\partial \varphi}{\partial x_i} \frac{\partial \varphi}{\partial x_j} \geqslant c_R \sum_i \left(\frac{\partial \varphi}{\partial x_i} \right)^2 - d_R \sum_{i \neq j} \left| \frac{\partial \varphi}{\partial x_i} \frac{\partial \varphi}{\partial x_j} \right|$$

$$\geqslant (c_R - (n-1)\, d_R) \sum_i \left(\frac{\partial \varphi}{\partial x_i} \right)^2.$$

We assume R so small that $c_R - (n-1)\, d_R > 0$. Then

$$\int_{B^R} |\varphi(x)|^2\, dx_0 \leqslant e_R 4R^2 \int_{B^R} \sum_i \left(\frac{\partial \varphi}{\partial x_i} \right)^2 dx_1 \cdots dx_n$$

$$\leqslant 4 e_R R^2 (c_R - (n-1)\, d_R)^{-1} \int_{B^R} \sum_{i,j} g^{ij} \frac{\partial \varphi}{\partial x_i} \frac{\partial \varphi}{\partial x_j}\, dx_1 \cdots dx_n$$

$$\leqslant 4 e_R f_R^{-1} R^2 (c_R - (n-1)\, d_R)^{-1} \int_{B^R} \sum_{i,j} g^{ij} \frac{\partial \varphi}{\partial x_i} \frac{\partial \varphi}{\partial x_j}\, dx_0 .$$

Hence we have for small R,

$$\int_{B^R} |\varphi(x)|^2\, dx_0 \leqslant g_R 4R^2 \int_{B^R} \langle \operatorname{grad} \varphi, \operatorname{grad} \varphi \rangle\, dx_0 , \tag{16}$$

where g_R is a constant depending only on the Riemannian structure of X and $g_R \to 1$ as $R \to 0$.

Now to prove Lemma 1.5 suppose $\Delta u = cu$ where $c \in \boldsymbol{C}$ and $u \in \mathscr{E}(X)$, $u \equiv 0$ on S^R. Since we have for $\varphi, \psi \in \mathscr{D}(B^R)$

$$\int_{B^R} \varphi \Delta \psi\, dx_0 = - \int_{B^R} \langle \operatorname{grad} \varphi, \operatorname{grad} \psi \rangle\, dx_0 ,$$

we obtain from (12)

$$\int_{B^R} u \Delta \psi\, dx_0 = - \int_{B^R} \langle \operatorname{grad} u, \operatorname{grad} \psi \rangle\, dx_0$$

and by the symmetry of Δ and $\Delta u = cu$

$$c \int_{B^R} u\psi \, dx_0 = -\int_{B^R} \langle \operatorname{grad} u, \operatorname{grad} \psi \rangle \, dx_0 .$$

Again by (12)

$$c \int_{B^R} | u(x)|^2 \, dx_0 = -\int_{B^R} \langle \operatorname{grad} u, \operatorname{grad} \bar{u} \rangle \, dx_0 . \tag{17}$$

This relation shows c real so we may assume u real. But using (12) we may replace φ by u in (16) so if R is small enough (17) implies $u = 0$ on B^R, hence on X by analyticity.

For each irreducible finite-dimensional representation δ of K let $\mathscr{E}_{\lambda,\delta}(X)$ denote the space of functions in $\mathscr{E}_\lambda(X)$ which are K-finite of type δ. Similarly we consider the subspace $\mathscr{E}_\delta(B)$ of $\mathscr{E}(B)$ and the supspace $\mathscr{E}_\delta(O)$ of $\mathscr{E}(O)$, O being any K-orbit in X, which is not the single point o.

To conclude the proof of Theorem 1.4 suppose $\lambda \in \mathfrak{a}_c{}^*$ is simple. Select an orbit O_λ with the property of Lemma 1.5. The mapping $F \to f$ given by (10) induces a one-to-one linear mapping of $\mathscr{E}_\delta(B)$ into $\mathscr{E}_{\lambda,\delta}(X)$. On the other hand the restriction to O_λ induces a one-to-one linear mapping of $\mathscr{E}_{\lambda,\delta}(X)$ into $\mathscr{E}_\delta(O_\lambda)$. Since $\mathscr{E}_\delta(B)$ and $\mathscr{E}_\delta(O_\lambda)$ have the same finite dimension we deduce that both of these maps are surjective. Hence the map $F \to f$ given by (10) maps the space of K-finite functions on B onto the space of K-finite functions in $\mathscr{E}_\lambda(X)$.

Finally, let $\lambda \in \mathfrak{a}_c{}^*$ be arbitrary. As remarked during the proof of Theorem 1.1, at least one of the elements λ, $s\lambda$ is simple. But by its definition, $\mathscr{E}_\lambda(X) = \mathscr{E}_{s\lambda}(X)$, so Theorem 1.4 is proved.

We shall now use Theorem 1.4 to get some idea about *arbitrary* eigenfunctions of Δ. By a *functional* T on B we shall mean a linear form on some linear space of functions on B. We write $\int_B F(b)\, dT(b)$ instead of $T(F)$, F being a function on B.

COROLLARY 1.6. *Let f be an arbitrary eigenfunction of Δ on X. Then there exists a $\lambda \in \mathfrak{a}_c{}^*$ and a functional T on B such that*

$$f(x) = \int_B e^{(i\lambda+\rho)(A(x,b))} \, dT(b), \qquad x \in X. \tag{18}$$

Proof. For each irreducible representation δ of K we fix a basis $F_{\delta,j}$ of $\mathscr{E}_\delta(B)$, orthonormal with respect to the usual inner product on $L^2(B)$.

Put

$$f_{\delta,j}(x) = \int_B e^{(i\lambda+\rho)(A(x,b))} F_{\delta,j}(b)\, db.$$

Let χ_δ denote the character of δ, $d(\delta)$ the dimension of δ and put $\alpha_\delta = d(\delta)\operatorname{conj}(\chi_\delta)$. Putting for $\alpha \in \mathscr{E}(K)$, $f \in \mathscr{E}(X)$

$$(\alpha * f)(x) = \int_K \alpha(k) f(k^{-1} \cdot x)\, dk$$

we have

$$f(x) = \sum_\delta (\alpha_\delta * f)(x), \tag{19}$$

the series being absolutely convergent, [21(k), p. 13].

Now suppose f is an eigenfunction of Δ. Select $\lambda \in \mathfrak{a}_c{}^*$ simple such that $f \in \mathscr{E}_\lambda(X)$. Then clearly $\alpha_\delta * f \in \mathscr{E}_{\lambda,\delta}(X)$. By virtue of Theorem 1.4

$$\alpha_\delta * f = \sum_j a_{\delta,j} f_{\delta,j} \tag{20}$$

for certain constants $a_{\delta,j}$. Now define a functional T on B by

$$T(F) = \sum_\delta \Big(\sum_j a_{\delta,j} \int_B F_{\delta,j}(b) F(b)\, db\Big), \tag{21}$$

for all $F \in \mathscr{E}(B)$ for which the sum over δ converges. In particular, $T(F)$ is defined for all K-finite F. Now given $x \in X$ put $F(b) = e^{(i\lambda+\rho)(A(x,b))}$ in (21). Then (20) and (19) give the desired formula (18) for f.

Of course Corollary 1.6 is hardly satisfactory until the functionals T are specified further. In the next theorem we do this for the special case $G = \boldsymbol{SU}(1, 1)$. Let K denote the subgroup of matrices $\left|\begin{smallmatrix} t & 0 \\ 0 & t^{-1}\end{smallmatrix}\right|$, $|\,t\,| = 1$. If we to each

$$g = \begin{vmatrix} a & b \\ \bar{b} & \bar{a} \end{vmatrix} \in G \qquad (|\,a\,|^2 - |\,b\,|^2 = 1)$$

associate the mapping $T_g : z \to (az + b)/(\bar{b}z + \bar{a})$ of the unit disk $|\,z\,| < 1$ onto itself then the mapping $gK \to T_g(0)$ is a diffeomorphism of the symmetric space $X = G/K$ onto the disk. A G-invariant Riemannian metric on the disk is given by

$$ds_0{}^2 = [1 - (x^2 + y^2)]^{-2}\,(dx^2 + dy^2), \tag{22}$$

and the operator

$$\Delta_0 = [1 - (x^2 + y^2)]^2 \left(\frac{\partial^2}{\partial x^2} + \frac{\partial^2}{\partial y^2}\right) \tag{23}$$

is the corresponding Laplace–Beltrami operator.

Let $\mathfrak{A}(B)$ denote the space of analytic functions on the boundary B of X considered as an analytic manifold. The space $\mathfrak{A}(B)$ carries a natural topology (Köthe [29]) which can be briefly described as follows. Let U be an open annulus containing B, $\mathscr{H}(U)$ the space of holomorphic functions on U topologized by uniform convergence on compact subsets. Then $\mathfrak{A}(B)$ can be identified with the union $\bigcup_U \mathscr{H}(U)$ and its topology is the inductive limit topology. The elements of the dual space $\mathfrak{A}'(B)$ of $\mathfrak{A}(B)$ are called *analytic functionals*. To each $T \in \mathfrak{A}'(B)$ one can associate a Fourier series

$$T \sim \sum a_n e^{in\theta} \tag{24}$$

where

$$a_n = \int_0^{2\pi} e^{-in\theta}\, dT(\theta).$$

Using elementary criteria for convergence of power series Köthe proves in [29], p. 23 that the series in (24) represents an analytic functional if and only if

$$\sum_{n \in \boldsymbol{Z}} r^{|n|} \mid a_n \mid < \infty \qquad \text{for each} \quad 0 \leqslant r < 1. \tag{25}$$

Let $P(z, b)$ denote the usual Poisson kernel

$$P(z, b) = \frac{1 - |z|^2}{1 - 2|z|\cos(\theta - \varphi) + |z|^2}, \qquad (z = |z|\, e^{i\theta},\ b = e^{i\varphi}).$$

We can then state the indicated sharpening of Corollary 1.6 (cf. [22(j), p. 14]; concerning the case $\mu = 1$ see also [29] and [31]).

THEOREM 1.7. *The functions*

$$f(z) = \int_B P(z, b)^\mu\, dT(b) \tag{26}$$

where $\mu \in \boldsymbol{R}$ and T is an analytic functional on B constitute precisely the eigenfunctions of Δ_0 with[1] *eigenvalue $\geqslant -1$.*

[1] This restriction has now been removed: *The functions* (26) *with μ complex constitute all the eigenfunctions of Δ_0.*

Proof. A direct computation gives

$$(\Delta_0)_z\,(P(z,b)^\mu) = 4\mu(\mu-1)\,P(z,b)^\mu$$

so the eigenvalue is $\geqslant -1$ and is independent of b. Thus f in (26) is an eigenfunction with eigenvalue $\geqslant -1$. For the converse we need a lemma which is a special case of Theorem 1.4 but which can easily be proved directly.

LEMMA 1.8. *Let $c \in \boldsymbol{C}$ and $n \in \boldsymbol{Z}$. The eigenfunctions of Δ_0 satisfying*

$$\Delta_0 = cf, \qquad f(e^{i\theta}z) \equiv e^{in\theta}f(z), \tag{27}$$

are all proportional to the function

$$f_n(z) = \frac{1}{2\pi}\int_0^{2\pi} P(z, e^{i\varphi})^\mu\, e^{in\varphi}\, d\varphi, \tag{28}$$

where $4\mu(\mu-1) = c$, $\operatorname{Re}(\mu) \geqslant 1/2$.

In fact, since $P(z,b) = e^{2\rho(A(z,b))}$, the inequality $\operatorname{Re}(\mu) \geqslant 1/2$ amounts to the fact that λ in Theorem 1.4 can be taken such that $\operatorname{Re}(\langle i\lambda, \alpha\rangle) \geqslant 0$ for $\alpha \in \Sigma^+$. This guarantees that λ is simple so (28) does not vanish identically.

To complete the proof of Theorem 1.7 we express $f_n(z)$ in (28) by means of the hypergeometric function $F(a,b;c;z)$. Using [10], p. 81 we obtain

$$f_n(r) = (1-r^2)^\mu r^{|n|}\frac{\Gamma(|n|+\mu)}{\Gamma(\mu)(|n|)!}F(\mu, |n|+\mu; |n|+1; r^2). \tag{29}$$

Let $f \in \mathscr{E}(X)$ satisfy

$$\Delta_0 f = cf, \qquad c \geqslant -1.$$

We select $\mu \in \boldsymbol{R}$ such that $\mu \geqslant 1/2$, $4\mu(\mu-1) = c$ and for a fixed z, $|z| < 1$, expand the function $\theta \to f(e^{i\theta}z)$ in an absolutely convergent Fourier series

$$f(e^{i\theta}z) = \sum_n c_n(z)\, e^{in\theta}. \tag{30}$$

Since

$$c_n(z) = \frac{1}{2\pi}\int f(e^{i\theta}z)\, e^{-in\theta}\, d\theta,$$

c_n satisfies the relations (27) so by the lemma

$$c_n(z) = a_n f_n(z),$$

where a_n is a constant. Thus by (30)

$$f(z) = \sum_n a_n f_n(z), \tag{31}$$

the series converging absolutely. Now μ being positive, the power series expansion

$$F(\mu, |n| + \mu; |n| + 1; r^2) = \sum_{m \geqslant 0} b_{m,n} r^{2m}$$

has all coefficients $b_{m,n}$ nonnegative and $b_{o,n} = 1$. The absolute convergence of (31) therefore implies

$$\sum_{n \in Z} |a_n| r^{|n|} \gamma_n < \infty \qquad 0 \leqslant r < 1, \tag{32}$$

where

$$\gamma_n = \frac{\Gamma(|n| + \mu)}{\Gamma(\mu)(|n|)!}.$$

Since $(\gamma_n)^{1/n} \to 1$, we deduce from (32) that

$$\sum_{n \in Z} |a_n| r^{|n|} < \infty$$

and consequently there exists an analytic functional T on B with Fourier series $\sum a_n e^{in\theta}$. It remains to prove (26). For each z, $|z| < 1$ the function

$$\zeta \to (1 - z\bar{z})^\mu / (\zeta - z)^\mu (\zeta^{-1} - \bar{z})^\mu$$

is holomorphic in an annulus U containing B and can be expanded there in a Laurent series. When we restrict this Laurent series to B we obtain the Fourier series of the function $b \to P(z, b)^\mu$ which by (28) is

$$P(z, e^{i\varphi})^\mu = \sum_n f_n(z) e^{-in\varphi}. \tag{33}$$

Since the Laurent series converges uniformly on compact subsets and since by definition of the topology of $\mathfrak{A}(B)$, the injection $\mathscr{H}(U) \to \mathfrak{A}(B)$ is continuous, the series (33) converges in the topology of $\mathfrak{A}(B)$. Hence we can apply T to (33) termwise; using (31) the desired representation of f follows so Theorem 1.7 is proved.

Remark. Using formulas (23) p. 64 and (46) p. 104 in [10] it is readily seen that formally

$$c_\mu(1 - |z|^2)^{\mu 1-} f(z) \to T \qquad \text{as} \qquad |z| \to 1$$

c_μ denoting the constant $\Gamma(\mu)^2/\Gamma(2\mu - 1)$. This remark is suggested by recent work of John Lewis on Eisenstein series for which such a limit was actually shown to exist.

With possible generalizations of Theorem 1.7 in mind we restate it in general terms.

THEOREM 1.8. *Let $G = SU(1, 1)$ and $X = G/K$ the associated symmetric space with Riemannian structure induced by the Killing form of $\mathfrak{g}$, Δ the corresponding Laplace–Beltrami operator. Then the functions*

$$f(x) = \int_B e^{(i\lambda+\rho)(A(x,b))}\, dT(b),$$

where $i\lambda \in \mathfrak{a}^*$, $T \in \mathfrak{A}'(B)$, constitute precisely the eigenfunctions of Δ with[1] *eigenvalue $c \geqslant -\langle\rho, \rho\rangle$.*

Since the Killing form on $\mathfrak{g}$ is given by $\langle X, X\rangle = 4\,\mathrm{Tr}(XX)$ it is not hard to see that the new Riemannian structure is eight times the old one so $\Delta_0 = 8\Delta$. Also $\langle\rho, \rho\rangle = 1/8$ so Theorems 1.7 and 1.8 coincide.

Remark. The *positive* eigenfunctions of Δ have been determined for arbitrary symmetric spaces by Karpelevič [25], §17 (see also Furstenberg [11, Theorem 5.1] for the positive joint eigenfunctions of all $D \in \boldsymbol{D}(G/K)$).

§2. THE IRREDUCIBILITY OF THE EIGENSPACES OF Δ

We still assume that $X = G/K$ has rank one and are now going to investigate the irreducibility question for the eigenspaces of the Laplacian Δ. In the present case these are just the spaces $\mathscr{E}_\lambda(\lambda \in \mathfrak{a}_c^*)$ which now can be described by

$$\mathscr{E}_\lambda(X) = \{f \in \mathscr{E}(X) \mid \Delta f = -(\langle\lambda, \lambda\rangle + \langle\rho, \rho\rangle)f\}. \tag{1}$$

These spaces are topologized as explained in Chap. III, §4, and in accordance with the notation in Chap. III, §5, we let T_λ denote the natural representation of G on $\mathscr{E}_\lambda$, i.e., $T_\lambda(g)f = f^{\tau(g)}$ for $f \in \mathscr{E}_\lambda$, $g \in G$.

[1] See footnote on p. 139.

THEOREM 2.1. *Let $\lambda \in \mathfrak{a}_c^*$. The natural representation of G on the eigenspace $\mathscr{E}_\lambda$ of Δ is irreducible if and only if*

$$\boldsymbol{e}_s(\lambda)\, \boldsymbol{e}_s(-\lambda) \neq 0. \tag{2}$$

We shall derive this result on the basis of Theorems 1.1 and 1.4. By "irreducible" we mean as usual absence of nontrivial closed invariant subspaces.

LEMMA 2.2. *Let $\lambda \in \mathfrak{a}_c^*$ be simple and as in* Chap. III, §5 *let $\mathscr{H}_\lambda$ denote the Hilbert space of functions f of the form*

$$f(x) = \int_B e^{(i\lambda+\rho)(A(x,b))} F(b), \qquad F \in L^2(B), \tag{3}$$

the norm $|f|_\lambda$ defined as the L^2 norm of F. Then $\mathscr{H}_\lambda$ is dense in $\mathscr{E}_\lambda(X)$.

Proof. As a result of Theorem 1.4, $\mathscr{H}_\lambda$ contains each of the spaces $\mathscr{E}_{\lambda,\delta}(X)$. Thus by (19) §1, $\mathscr{H}_\lambda$ is dense in $\mathscr{E}_\lambda(X)$.

Suppose now (2) holds. Then λ is simple. Let $V \subset \mathscr{E}_\lambda(X)$ be a closed invariant subspace. In order to prove that the invariant subspace $V \cap \mathscr{H}_\lambda$ of $\mathscr{H}_\lambda$ is closed let (v_n) be a sequence in $V \cap \mathscr{H}_\lambda$ converging to $f \in \mathscr{H}_\lambda$. Writing

$$v_n(x) = \int_B e^{(i\lambda+\rho)(A(x,b))} V_n(b)\, db \qquad V_n \in L^2(B)$$

we see by (3) and Schwarz's inequality that if D is any differential operator on X,

$$|(Dv_n)(x) - (Df)(x)|^2 \leqslant \int_B |\, D_x(e^{(i\lambda+\rho)(A(x,b))})|^2\, db \,|\, v_n - f\,|_\lambda^2.$$

This shows that the sequence (v_n) converges to f in $\mathscr{E}(X)$. Hence $f \in V$ so $V \cap \mathscr{H}_\lambda$ is closed in $\mathscr{H}_\lambda$. But the map $\psi \to \bar{\psi}$ of $\mathscr{H}_\lambda$ onto $\mathscr{H}_\lambda$ is an isometry so by Proposition 5.3, Chap. III, $V \cap \mathscr{H}_\lambda = \{0\}$ or $\mathscr{H}_\lambda$. The second possibility implies $V = \mathscr{E}_\lambda(X)$ in view of Lemma 2.2. If on the other hand $V \cap \mathscr{H}_\lambda = \{0\}$ consider for each $f \in V$ the function $\alpha_\delta * f$ in (19). By Theorem 1.4, $\alpha_\delta * f \in \mathscr{H}_\lambda$. To see that $\alpha_\delta * f \in V$ consider the dual space $\mathscr{E}'(X)$ of $\mathscr{E}(X)$ which consists of the distributions on X of compact support. The annihilator V^0 of V is G-invariant and if $T \in V^0$ we have

$$\int_X (\alpha_\delta * f)(x)\, dT(x) = \int_K \alpha_\delta(k) \int_X f(x)\, dT^{\tau(k^{-1})}(x) = 0$$

so $\alpha_\delta * f$ belongs to double annihilator $(V^0)^0$ which however equals V,

V being closed. Thus $\alpha_\delta * f \in V \cap \mathscr{H}_\lambda = \{0\}$ so using (19) $V = \{0\}$. This proves that the action of G on $\mathscr{E}_\lambda$ is irreducible.

On the other hand, suppose the representation T_λ of G on $\mathscr{E}_\lambda$ is irreducible. Since $\mathscr{E}_\lambda = \mathscr{E}_{s\lambda}$ and since at least one of the elements λ, $s\lambda$ is simple we may assume λ simple. Let $\mathscr{E}_\lambda{}^\infty$ denote the space of K-finite vectors in $\mathscr{E}_\lambda$. Then the action of the universal enveloping algebra $U(\mathfrak{g})$ on $\mathscr{E}_\lambda{}^\infty$ is algebraically irreducible. In fact, if $\mathscr{E}_\lambda$ were a Banach space this would be contained in Harish-Chandra's Theorem 5 in [21(a), p. 228]; but Godement's proof of the algebraic irreducibility in [20(a), p. 545] is valid for $\mathscr{E}_\lambda$, the essential point being the finite-dimensionality of each space $\mathscr{E}_{\lambda,\mathfrak{s}}$ which is insured, for example by Theorem 1.4. Since the bijection $\psi \to \tilde{\psi}$ of $\mathscr{K}_\lambda$ onto $\mathscr{H}_\lambda$ commutes with the action of G it maps the space $\mathscr{K}_\lambda{}^\infty$ of K-finite vectors in $\mathscr{K}_\lambda$ onto the space $\mathscr{H}_\lambda{}^\infty$ of K-finite vectors in $\mathscr{H}_\lambda$ in such a way that $(u\psi)^\vee = u\tilde{\psi}$ for $u \in U(\mathfrak{g})$, $\psi \in \mathscr{K}_\lambda{}^\infty$. Since by Theorem 1.4 $\mathscr{H}_\lambda{}^\infty = \mathscr{E}_\lambda{}^\infty$, it follows that the action of $U(\mathfrak{g})$ on $\mathscr{K}_\lambda{}^\infty$ is algebraically irreducible so by the quoted result of Harish-Chandra G acts irreducibly on $\mathscr{K}_\lambda$; thus (2) holds by Theorem 1.1.

§3. The Eigenspaces of the Invariant Differential Operators on Ξ

According to Proposition 2.5, Chap. III the joint eigenspaces for the operators in $\boldsymbol{D}(\Xi)$ acting on $\mathscr{D}'(\Xi)$ are the spaces $\mathscr{D}_\lambda'(\Xi)$, $\lambda \in \mathfrak{a}_c{}^*$, and the eigenspace $\mathscr{D}_\lambda'$ is irreducible under the action of G if and only if $\boldsymbol{e}_{s^*}(\lambda)\, \boldsymbol{e}_{s^*}(-\lambda) \neq 0$ (cf., Chap. III, Proposition 5.2 and remarks following Theorem 1.1). In this context, Theorem 3.2 of Chap. III can be somewhat sharpened.

Proposition 3.1. *Let $\lambda \in \mathfrak{a}_c{}^*$ and let $\mathscr{D}'_{\lambda,G}$ denote the space of G-finite vectors in $\mathscr{D}_\lambda'$. Then $\mathscr{D}'_{\lambda,G} \neq \{0\}$ if and only if*

$$\frac{\langle i\lambda - \rho, \alpha\rangle}{\langle \alpha, \alpha\rangle} \in \boldsymbol{Z}^+ \quad \textit{for} \quad \alpha \in \Sigma^+. \tag{1}$$

In this case $\mathscr{D}'_{\lambda,G}$ is finite-dimensional and irreducible under G and as λ runs through the elements of $\mathfrak{a}_c{}^*$ satisfying (1) *the spaces $\mathscr{D}'_{\lambda,G}$ are models for all the finite-dimensional irreducible conical representations of G.*

Proof. Suppose Ψ is a nonzero element in $\mathscr{D}'_{\lambda,G}$. Let $S \in \mathscr{D}'(B)$ be determined as in Proposition 4.1, Chap. III. Since Ψ is G-finite, it is K-finite and so is S. But a K-finite distribution on B is necessarily a C^∞ function so $\Psi \in \mathscr{E}(\Xi)$. Let $\mathscr{D}_\Psi$ denote the (finite-dimensional) subspace of $\mathscr{D}_\lambda'$ spanned by all $\Psi^{\tau(g)}$, $g \in G$. By weight theory there exists a nonzero vector $\Psi_0 \in \mathscr{D}_\Psi$ which is N-invariant. But in view of the Bruhat lemma an N-invariant function on Ξ is also M-invariant so Ψ_0 is a conical function in $\mathscr{D}_\lambda'$ and (1) follows, as well as the finite-dimensionality of $\mathscr{D}'_{\lambda,G}$.

Now assume (1). Then $\mathscr{D}_\lambda'$ contains a conical function, say ψ_λ, unique up to a constant multiple, so $\mathscr{D}'_{\lambda,G} \neq \{0\}$. By the above argument each N-invariant in $\mathscr{D}'_{\lambda,G}$ is proportional to ψ_λ. Hence the irreducibility.

§4. The Eigenspaces of the Invariant Differential Operators on G/N, (G Complex)

In this section we assume G to be a simply connected complex semisimple Lie group. In this case we can give a sharper form of Proposition 3.1 and thereby include *all* the holomorphic irreducible finite-dimensional representations of G. (A finite-dimensional representation π of G is called *holomorphic* if its differential $d\pi$ is a complex-linear representation of $\mathfrak{g}$, i.e., $d\pi(JX) = i\, d\pi(X)$, where J is the complex structure of $\mathfrak{g}$ and $X \in \mathfrak{g}$). Now we have $\mathfrak{m} = J\mathfrak{a}$ so M is the centralizer of the torus $\exp(J\mathfrak{a})$ in the compact group K, hence M is connected. The subalgebra $\mathfrak{h} = \mathfrak{a} + J\mathfrak{a}$ is a Cartan subalgebra of $\mathfrak{g}$. Let Δ denote the set of roots of $\mathfrak{g}$ with respect to $\mathfrak{h}$, Δ^+ the set of $\beta \in \Delta$ whose restriction to $\mathfrak{a}$ belongs to Σ^+. Then Δ^+ is the set of positive roots for an ordering of Δ. Let H denote the analytic subgroup of G corresponding to $\mathfrak{h}$; H, N and $\bar{N}$ are now complex analytic subgroups of G. Let o denote the origin in G/N.

Theorem 4.1. *Let G be a simply connected complex semisimple Lie group, $\boldsymbol{D}(G/N)$ the algebra of G-invariant differential operators on G/N (as a real analytic manifold). Then*

(i) *$\boldsymbol{D}(G/N)$ is commutative;*

(ii) *For each algebra homomorphism*

$$\chi : \boldsymbol{D}(G/N) \to \boldsymbol{C}$$

let V_χ denote the space of holomorphic functions f on G/N satisfying

$$Df = \chi(D)f \qquad D \in \boldsymbol{D}(G/N)$$

and let π_χ denote the natural representation of G on V_χ. As χ varies, π_χ runs through all the irreducible finite-dimensional holomorphic representations of G.

For the proof we first determine the structure of the algebra $\boldsymbol{D}(G/N)$. Let $\boldsymbol{D}(H)$ denote the algebra of left invariant differential operators on H, considered as a real Lie group.

PROPOSITION 4.2. *The mapping $U \to D_U$ where*

$$(D_U f)(gN) = \{U_h(f(ghN))\}_{h=e}, \qquad f \in \mathscr{E}(G/N), \tag{1}$$

is an isomorphism of $\boldsymbol{D}(H)$ onto $\boldsymbol{D}(G/N)$.

In order to prove this we must first determine the polynomials on the tangent space $(G/N)_o$ which are invariant under the action of N. Corresponding to the direct decomposition $\mathfrak{g} = (\bar{\mathfrak{n}} + \mathfrak{h}) + \mathfrak{n}$ we have a projection

$$\sigma : \mathfrak{g} \to \bar{\mathfrak{n}} + \mathfrak{h}.$$

Consider also the natural projection $\pi : G \to G/N$ whose differential gives a bijection

$$(d\pi)_e : \bar{\mathfrak{n}} + \mathfrak{h} \to (G/N)_o .$$

Since

$$(d\pi)_e (\sigma \circ Ad_G(n)X) = d\tau(n)((d\pi)_e (X)) \qquad X \in \bar{\mathfrak{n}} + \mathfrak{h}$$

cf. [22(b), p. 368], our problem calls for determining the invariants of the group $\sigma \circ Ad_G(N)$ in the symmetric algebra $S(\bar{\mathfrak{n}} + \mathfrak{h})$; here $\bar{\mathfrak{n}}$ and $\mathfrak{h}$ are considered as vector spaces over $\boldsymbol{R}$.

LEMMA 4.3. *Let $I(\bar{\mathfrak{n}} + \mathfrak{h})$ denote the set of elements in $S(\bar{\mathfrak{n}} + \mathfrak{h})$ which are invariant under the group $\sigma \circ Ad_G(N)$. Then*

$$I(\bar{\mathfrak{n}} + \mathfrak{h}) = S(\mathfrak{h}).$$

Proof. Since the restriction of each $\sigma \circ Ad_G(n)$ to $\mathfrak{h}$ is the identity map we have $S(\mathfrak{h}) \subset I(\bar{\mathfrak{n}} + \mathfrak{h})$. For the converse let $\alpha_1 > \cdots > \alpha_p$ denote the positive restricted roots in decreasing order and put $\beta_i = -\alpha_i$. Now each root space $\mathfrak{g}_{\beta_i}$ has dimension 2 over $\boldsymbol{R}$ and is invariant under J so let E_{β_i}, F_{β_i} ($F_{\beta_i} = JE_{\beta_i}$) be a basis. For $X \in \mathfrak{n}$ let $d(X)$ denote the derivation of $S(\bar{\mathfrak{n}} + \mathfrak{h})$ which extends the endomorphism $\sigma \circ ad_{\mathfrak{g}}(X)$ of $\bar{\mathfrak{n}} + \mathfrak{h}$. The group $\sigma \circ Ad_G(N)$ being connected it is clear that $P \in I(\bar{\mathfrak{n}} + \mathfrak{h})$ if and only if $d(X)P = 0$ for all $X \in \mathfrak{n}$. Let $P \in I(\bar{\mathfrak{n}} + \mathfrak{h})$. Then

$$P = \sum_{(m)(n)} r_{(m),(n)} E_{\beta_1}^{m_1} F_{\beta_1}^{n_1} \cdots E_{\beta_p}^{m_p} F_{\beta_p}^{n_p},$$

where $r_{(m),(n)} \in S(\mathfrak{h})$. Let $X_\alpha \neq 0$ in $\mathfrak{g}_\alpha$, $X_\beta \neq 0$ in $\mathfrak{g}_\beta$ ($\alpha > 0$, $\beta < 0$). Then

$$d(X_\alpha)\, X_\beta = 0, \qquad \text{if } \alpha + \beta > 0,$$

and

$$d(X_\alpha)\, X_\beta \in \mathfrak{h} - \{0\}, \qquad \text{if } \alpha + \beta = 0.$$

Hence $d(X_{\alpha_1})$ annihilates $\mathfrak{h}$ and all $\mathfrak{g}_{\beta_i}$ except for $i = 1$. Now we write P in the form

$$P = \sum_{m_1, n_1} a(m_1, n_1)\, E_{\beta_1}^{m_1} F_{\beta_1}^{n_1},$$

where $a(m_1, n_1)$ is a polynomial in the vectors E_{β_i}, F_{β_i} ($i > 1$) and coefficients in $S(\mathfrak{h})$. Since

$$d(X_{\alpha_1})(E_{\beta_1}^{m_1} F_{\beta_1}^{n_1}) = m_1 E_{\beta_1}^{m_1 - 1} F_{\beta_1}^{n_1} A_1 + n_1 E_{\beta_1}^{m_1} F_{\beta_1}^{n_1 - 1} J A_1,$$

where $A_1 = [X_{\alpha_1}, E_{\beta_1}]$, the relation $d(X_{\alpha_1})P = 0$ implies

$$(m+1)\, a(m+1, n)\, A_1 + (n+1)\, a(m, n+1)\, J A_1 = 0 \qquad (m \geqslant 0, n \geqslant 0).$$

Here we can put $X_{\alpha_1} = E_{\alpha_1}$ and also $X_{\alpha_1} = F_{\alpha_1}$. The two resulting relations imply $a(m+1, n) = a(m, n+1) = 0$. Thus the expression for P contains neither E_{β_1} nor F_{β_1}. Applying $d(X_{\alpha_2}), \ldots, d(X_{\alpha_p})$ successively we see that $P \in S(\mathfrak{h})$ and the lemma follows.

Turning now to Proposition 4.2 we shall prove the surjectivity of the map $U \rightarrow D_U$, the other properties stated being obvious. Let $X_1, \ldots, X_q$ be a basis of $\bar{\mathfrak{n}} + \mathfrak{h}$, composed of a basis of $\bar{\mathfrak{n}}$ and a basis of $\mathfrak{h}$. If $a > 0$ is sufficiently small the map

$$(t_1, \ldots, t_q) \rightarrow \exp\left(\sum_1^q t_i X_i\right) N$$

is a diffeomorphism of the cube $|\, t_i \,| < a$ onto an open neighborhood of o in G/N. The inverse map is a coordinate system on V and the coordinate vector fields $\partial/\partial t_i$ satisfy

$$(\partial/\partial t_i)_o = (d\pi)_e\, (X_i), \qquad 1 \leqslant i \leqslant q.$$

Let $D \in \boldsymbol{D}(G/N)$. Then there exists a unique polynomial P in q variables such that for all $f \in \mathscr{E}(G/N)$

$$(Df)(o) = \left\{ P\left(\frac{\partial}{\partial t_1}, \ldots, \frac{\partial}{\partial t_q}\right) f\left(\exp\left(\sum t_i X_i\right) N\right) \right\}_{t=0}. \tag{2}$$

In the decomposition of P into monomials, let P^* denote the sum of the components of highest total degree. The invariance of D implies

$$(Df^{\tau(n)})(o) = (Df)(o), \qquad n \in N. \tag{3}$$

In this relation we substitute the expression (2). Although the transformation $\tau(n)$, when expressed in terms of the $(t_1, \ldots t_q)$, is not necessarily linear, the highest order derivatives in $P(\partial/\partial t_1, \ldots, \partial/\partial t_q)$ do transform according to the *linear* transformation $d\tau(n)_o$. Hence (3) implies that P^* is invariant under $d\tau(n)$, and thus by Lemma 4.3, $P^* \in S(\mathfrak{h})$. Let $D_{P^*} \in \boldsymbol{D}(G/N)$ be determined by Proposition 4.2. Then the differential operator $D - D_{P^*}$ has by (1) and (2) lower order than D at the point o; this holds at any point because of the invariance under G. It is now clear by induction that D has the form $D = D_U$ for some $U \in S(\mathfrak{h})$.

The following consequenee of Proposition 4.2 is obvious.

COROLLARY 4.4. *A function $f \in \mathscr{E}(G/N)$ is an eigenfunction of each $D \in \boldsymbol{D}(G/N)$ if and only if there exists a C^∞ homomorphism ω of H into $\boldsymbol{C}^*$, the multiplicative group of $\boldsymbol{C}^* - \{0\}$, such that*

$$f(ghN) = f(gN)\,\omega(h) \qquad g \in G, \quad h \in H. \tag{4}$$

LEMMA 4.5. *Let $\omega : H \to \boldsymbol{C}^*$ be a holomorphic homomorphism and V_ω the space of holomorphic functions f on G/N satisfying* (4). *Then V_ω has finite dimension.*

Proof. We first consider the case $G = \boldsymbol{SL}(2, \boldsymbol{C})$. Writing $\varphi(g) = f(gN)$ we have for some $n \in \boldsymbol{Z}$

$$\varphi\left(\begin{vmatrix} 1 & 0 \\ y & 1 \end{vmatrix}\begin{vmatrix} a & 0 \\ 0 & a^{-1} \end{vmatrix}\begin{vmatrix} 1 & x \\ 0 & 1 \end{vmatrix}\right) = a^n\varphi\left(\begin{vmatrix} 1 & 0 \\ y & 1 \end{vmatrix}\right), \qquad x, y \in \boldsymbol{C}, \quad a \in \boldsymbol{C}^*,$$

or explicitly

$$\varphi\begin{pmatrix} a & ax \\ ay & axy + a^{-1} \end{pmatrix} = a^n\varphi\begin{pmatrix} 1 & 0 \\ y & 1 \end{pmatrix}. \tag{5}$$

Putting here $x = a^{-1}$, $y = -a^{-1}$ we have by the continuity of φ

$$\lim_{a \to 0} a^n\varphi\begin{pmatrix} 1 & 0 \\ -a^{-1} & 1 \end{pmatrix} = \varphi\begin{pmatrix} 0 & 1 \\ -1 & 0 \end{pmatrix}. \tag{6}$$

The function

$$F(z) = \varphi\begin{pmatrix} 1 & 0 \\ z & 1 \end{pmatrix}, \qquad z \in \boldsymbol{C},$$

is an entire function which by (6) satisfies

$$\lim_{z\to\infty} z^{-n}F(z) \qquad \text{exists.} \tag{7}$$

If $n < 0$, (7) implies via the Liouville theorem that $F \equiv 0$ so $\varphi \equiv 0$. Suppose therefore $n \geqslant 0$. Then $z^{-n}F(z)$ has a pole of order $\leqslant n$ at $z = 0$. Hence

$$z^{-n}F(z) = p(z^{-1}) + h(z),$$

where p is a polynomial of degree $\leqslant n$ and $h(z)$ regular at $z = 0$; by (7) h is bounded, hence a constant. Thus F is a polynomial P of degree $\leqslant n$, and (5) amounts to

$$\varphi\begin{pmatrix} a & b \\ c & d \end{pmatrix} = a^n P(ca^{-1}) \tag{8}$$

holding on the dense subset $\bar{N}HN$ of G. Hence φ is a polynomial in the a, b, c, d of degree $\leqslant n$ so $\dim V_\omega < \infty$. The lemma also holds for the adjoint group of $\boldsymbol{SL}(2, \boldsymbol{C})$.

Next we consider general G and put $\varphi(g) = f(gN)$. Let $X_1, \ldots, X_p$ be a basis of $\bar{\mathfrak{n}}$ over $\boldsymbol{C}$ consisting of root vectors and $\tilde{X}_i$ the left invariant vector field on G generated by X_i ($1 \leqslant i \leqslant p$). Since φ is holomorphic on $\bar{N}$ we have a Taylor series

$$\varphi(\exp z_1X_1 \cdots \exp z_pX_p) = \sum z_1^{m_1} \cdots z_p^{m_p}(\tilde{X}_1^{m_1} \cdots \tilde{X}_p^{m_p}\varphi)(e). \tag{9}$$

Fix j and suppose $X_j \in \mathfrak{g}_{\beta_j}$. Let G^{β_j} denote the analytic subgroup of G corresponding to the subalgebra

$$\mathfrak{g}_{\beta_j} + \mathfrak{g}_{-\beta_j} + \boldsymbol{C}H_{\beta_j}.$$

Then G^{β_j} is locally isomorphic to $\boldsymbol{SL}(2, \boldsymbol{C})$ so by the special case considered there exists an integer $m \geqslant 0$ depending only on ω such that

$$(\tilde{X}_j{}^q\varphi)(e) = 0, \qquad 1 \leqslant j \leqslant p, \quad q \geqslant m.$$

Replacing in (4) f by the function $gN \to f(\bar{n}gN)$ we have by the left invariance of $\tilde{X}_j$,

$$(\tilde{X}_j{}^q\varphi)(\bar{n}) = 0, \qquad 1 \leqslant j \leqslant p, \quad q \geqslant m. \tag{10}$$

In particular, the right side of (9) is a polynomial in z_p

$$\varphi(\exp z_1X_1 \cdots \exp z_pX_p) = \sum_{j=1}^{m} P_j(z_1, \ldots, z_{p-1})\, z_p^{\,j} \tag{11}$$

and

$$P_j(z_1, \ldots, z_{p-1}) = \left\{\frac{\partial^j}{\partial t^j}\varphi(\exp z_1X_1 \cdots \exp z_{p-1}X_{p-1} \exp tX_p)\right\}_{t=0}$$
$$= \left\{\frac{\partial^j}{\partial t^j}\varphi(\exp(t\, Ad(g)\, X_p)g)\right\}_{t=0},$$

where

$$g = \exp z_1X_1 \cdots \exp z_{p-1}X_{p-1}.$$

If $X \in \mathfrak{g}$ let $\bar{X}$ denote the corresponding *right* invariant vector field on G. Then the last expression gives

$$P_j(z_1, \ldots, z_{p-1}) = ((\overline{Ad(g)\, X_p})^j\, \varphi)(g).$$

Since $\bar{\mathfrak{n}}$ is nilpotent this is a finite sum of terms

$$q(z_1, \ldots, z_{p-1})(Q(\bar{X}_1, \ldots, \bar{X}_p)\varphi)(\exp z_1X_1 \cdots \exp z_{p-1}X_{p-1}) \tag{12}$$

where q is a polynomial in $p - 1$ variables and Q is a polynomial (noncommutative). Now we replace in (10) φ by $Q(\bar{X}_1, \ldots, \bar{X}_p)\varphi$ and take $z_p = 0$. Since $\tilde{X}_{p-1}$ commutes with $Q(\bar{X}_1, \ldots, \bar{X}_p)$ we deduce from (10) that the expression (12), and therefore P_j, is a polynomial in z_{p-1}. By iteration, the left side of (11) is a polynomial in $z_1, \ldots, z_p$, whose degree has a bound determined by ω alone. But since $\bar{N}HN$ is dense in G, f is determined by the restriction of φ to $\bar{N}$; hence $\dim V_\omega < \infty$.

LEMMA 4.6. *The space V_ω is irreducible under G.*

Proof. By complete reducibility, V_ω decomposes into finitely many irreducible subspaces. A lowest weight vector in each of these is determined by a function φ on G satisfying

$$\varphi(\bar{n}g) = \varphi(g) \qquad \text{and} \qquad \varphi(ghn) = \varphi(g)\,\omega(h). \tag{13}$$

Again since $\bar{N}HN$ is dense in G (13) determines φ up to a constant factor so E_f is irreducible.

It is now clear that each π_χ in Theorem 4.1 is irreducible. To complete the proof let π be an irreducible finite-dimensional holomorphic representation of G. Let v be a lowest weight vector and let $\omega : H \to C^*$

be the homomorphism determined by $\pi(h)v = \omega(h)v$. Then ω is holomorphic and the natural representation of G on V_ω has the same lowest weight as π, hence is equivalent to it. This proves Theorem 4.1.

Remark 1. The models described in Theorem 4.1 are related to known models for the irreducible finite-dimensional representations of G. The space V_ω in Lemma 4.5 can be viewed as the space of holomorphic sections of a complex line bundle over G/HN so since G/HN is compact Lemma 4.5 becomes a special case of the theorem that the space of holomorphic sections of a complex vector bundle over a compact, manifold is finite-dimensional[2] ([27], [7]). With this interpretation of V_ω, Lemma 4.6 becomes the theorem of Borel–Weil, cf. [43]. For other proofs see, e.g., [4] and [48]. The models of the irreducible representations given by Zhelobenko [49(a), Theorem 5], are null spaces of certain differential operators on N, and nonholomorphic representations of G are also considered.

A global treatment of the weight theory was already given by Godement [20(a), Appendix], at least for $\boldsymbol{SL}(n, \boldsymbol{C})$; models for the representations were constructed by means of functions φ satisfying (13).

Remark 2. For any homomorphism χ of $\boldsymbol{D}(G/N)$ into $\boldsymbol{C}$ let

$$\mathscr{E}_\chi(G/N) = \{f \in \mathscr{E}(G/N) \mid Df = \chi(D)f \text{ for } D \in \boldsymbol{D}(G/N)\}$$

with the topology induced by that of $\mathscr{E}(G/N)$. According to Proposition 4.2 the homomorphism $U \to \chi(D_U)$ $(U \in \boldsymbol{D}(H))$ determines a point $\lambda_\chi \in \mathfrak{h}_c{}^*$ (the complex dual of $\mathfrak{h}$ as a vector space over $\boldsymbol{R}$) such that $\lambda_\chi(U) = \chi(D_U)$ $(U \in \mathfrak{h})$. By Corollary 4.4, the space $\mathscr{E}_\chi$ is $\{0\}$ unless the map

$$\lambda_\chi : \mathfrak{h} \to \boldsymbol{C}$$

is the differential $d\omega$ of a homomorphism $\omega : H \to \boldsymbol{C}^*$. In his paper [49(d)] Zhelobenko determines explicitly the homomorphisms ω, say ω_0, for which the space of functions f satisfying (4) is not irreducible. Excepting the corresponding χ_0 which are given by

$$\chi_0(D_U) = d\omega_0(U), \qquad U \in \mathfrak{h},$$

the eigenspace $\mathscr{E}_\chi(G/N)$ is always irreducible under the action of G. Assuming now the results of [50] *it follows that these irreducible spaces $\mathscr{E}_\chi$ together with a specific subspace of each non-irreducible $\mathscr{E}_\chi$, give all completely irreducible representations of G up to weak equivalence.*

[2] A very short proof of this fact has been shown to me by J.-P. Serre.

BIBLIOGRAPHY

1. S. AGMON, "Lectures on Elliptic Boundary Value Problems," Van Nostrand Co., Inc. Princeton, N. J., 1965.
2. S. ARAKI, On root systems and an infinitesimal classification of irreducible symmetric spaces, *J. Math. Osaka Univ.* **13** (1962), 1–34.
3. T. S. BHANU-MURTHY, (a) Plancherel's measure for the factor space $\mathbf{SL}(n, \mathbf{R})/\mathbf{SO}(n, \mathbf{R})$, *Dokl. Akad. Nauk. SSSR* **133** (1960), 503–506; (b) The asymptotic behaviour of zonal spherical functions on the Siegel upper half plane, *Dokl. Akad. Nauk. SSSR* **135** (1960), 1027–1030.
4. R. BLATTNER, On induced representations II: Infinitesimal induction, *Amer. J. Math.* **83** (1961), 499–512.
5. N. BOURBAKI, "Éléments de Mathématique," Vol. VI, "Intégration," Chaps. I–V, Hermann, Paris, 1952.
6. F. BRUHAT, (a) Sur les représentations induites des groupes de Lie, *Bull. Soc. Math. France* **84** (1956), 97–205; (b) Sur les représentations des groupes $\mathfrak{p}$-adiques, *Amer. J. Math.* **83** (1961), 321–338, 343–368.
7. H. CARTAN AND J.-P. SERRE, Un théorème de finitude concernant les variétés analytiques compactes, *C.R. Acad. Sci. Paris* **237** (1953), 128–130.
8. S. S. CHERN, On integral geometry in Klein spaces, *Ann. Math.* **43** (1942), 178–189.
9. L. EHRENPREIS AND F. MAUTNER, Some properties of the Fourier transform on semisimple Lie groups, I, *Ann. of Math.* **61** (1955), 406–439.
10. A. ERDÉLYI ET AL., "Higher Transcendental Functions," (Bateman manuscript project) Vol. I, McGraw-Hill Book Co., New York, 1953.
11. H. FURSTENBERG, Translation-invariant cones of functions on semisimple Lie groups, *Bull. Amer. Math. Soc.* **71** (1965), 271–326.
12. R. GANGOLLI, (a) Isotropic infinitely divisible measures on symmetric spaces, *Acta Math.* **111** (1964), 213–246; (b) On the Plancherel formula and the Paley-Wiener theorem for spherical functions on semisimple Lie groups (to appear).
13. I. M. GELFAND, Integral geometry and its relation to group representations, *Russian Math. Surveys* (1960), 143–151.
14. I. M. GELFAND AND M. I. GRAEV, The geometry of homogeneous spaces, group representations in homogeneous spaces and questions in integral geometry related to them, *Amer. Math. Soc. Transl.* **37** (1964), 351–429.
15. I. M. GELFAND, M. I. GRAEV, AND I. I. PYATETSKII-SHAPIRO, "Representation Theory and Automorphic Functions," W. B. Saunders Co., Philadelphia, Pa., 1969.
16. I. M. GELFAND, M. I. GRAEV, AND S. J. SHAPIRO, Differential forms and integral geometry, *Functional Anal. Appl.* **3** (1969), 24–40.
17. I. M. GELFAND, M. I. GRAEV, AND N. VILENKIN, "Integral Geometry and Representation Theory," "Generalized Functions," Vol. 5, Academic Press, Inc., New York, 1966.
18. I. M. GELFAND AND M. A. NAIMARK, "Unitäre Darstellungen der klassischen Gruppen," Akademie Verlag, Berlin, Germany, 1957.
19. S. G. GINDIKIN AND F. I. KARPELEVIČ, (a) Plancherel measure of Riemannian symmetric spaces of nonpositive curvature, *Sov. Math.* **3** (1962), 962–965; (b) "On a Problem in Integral Geometry," Chebotarev Memorial Vol., Kazan University, 1964.
20. R. GODEMENT, (a) A theory of spherical functions, *Trans. Amer. Math. Soc.* **73** (1952), 496–556; (b) "Introduction aux Travaux de Selberg," Séminaire Bourbaki, 1957.

21. HARISH-CHANDRA, (a) Representations of a semisimple Lie group on a Banach space I, *Trans. Amer. Math. Soc.* **75** (1953), 185–243; (b) On the Plancherel formula for the right-invariant functions on a semisimple Lie group, *Proc. Nat. Acad. Sci. USA* **40** (1954), 200–204; (c) Representations of semisimple Lie groups III, *Trans. Amer. Math. Soc.* **76** (1954), 234–253; (d) The Plancherel formula for complex semisimple Lie groups, *Trans. Amer. Math. Soc.* **76** (1954), 485–528; (e) Representations of semisimple Lie groups IV, *Amer. J. Math.* **77** (1955), 743–777; (f) The characters of semisimple Lie groups, *Trans. Amer. Math. Soc.* **83** (1956), 98–163; (g) On a lemma of Bruhat, *J. Math. Pures Appl.* **35** (1956), 203–210; (h) Differential operators on a semisimple Lie algebra, *Amer. J. Math.* **79** (1957), 87–120; (i) Spherical functions on a semisimple Lie group I, *Amer. J. Math.* **80** (1958), 241–310; (j) Spherical functions on a semisimple Lie group II, *Amer. J. Math.* **80** (1958), 553–613; (k) Discrete series for semisimple Lie groups II, *Acta Math.* **116** (1966), 1–111; (l) Harmonic analysis on semisimple Lie groups, *Amer. Math. Soc. Colloquium Lectures* (1969).

22. S. HELGASON, (a) Differential operators on homogeneous spaces, *Acta Math.* **102** (1959), 239–299; (b) "Differential Geometry and Symmetric Spaces," Academic Press, Inc., New York, 1962; (c) Duality and Radon transform for symmetric spaces, *Amer. J. Math.* **85** (1963), 667–692; (d) A duality in integral geometry; some generalizations of the Radon transform, *Bull. Amer. Math. Soc.* **70** (1964), 435–446; (e) Fundamental solutions of invariant differential operators on symmetric spaces, *Amer. J. Math.* **86** (1964), 565–601; (f) The Radon transform on Euclidean spaces, compact two-point homogeneous spaces and Grassmann manifolds, *Acta Math.* **113** (1965), 153–180; (g) Radon–Fourier transforms on symmetric spaces and related group representations, *Bull. Amer. Math. Soc.* **71** (1965), 757–763; (h) "A Duality in Integral Geometry on Symmetric Spaces," pp. 37–56, Proc. U.S.–Japan Seminar in Differential Geometry, Kyoto, Japan, 1965; (i) An analog of the Paley–Wiener theorem for the Fourier transform on certain symmetric spaces, *Math. Ann.* **165** (1966), 297–308; (j) "Lie Groups and Symmetric Spaces. Battelle Rencontres 1967," pp. 1–71, W. A. Benjamin, Inc., New York, 1968; (k) Applications of the Radon transform to representations of semisimple Lie groups, *Proc. Nat. Acad. Sci. USA* **63** (1969), 643–647.

23. S. HELGASON AND K. JOHNSON, The bounded spherical functions on symmetric spaces, *Advan. Math.* **3** (1969), 586–593.

24. N. IWAHORI, On the structure of a Hecke ring of a Chevalley group over a finite field, *J. Fac. Sci. Univ. Tokyo* **10** (1964), 215–236.

25. F. I. KARPELEVIČ, The geometry of geodesics and the eigenfunctions of the Beltrami–Laplace operator on symmetric spaces, *Trans. Moscow Math. Soc.* **14** (1965), 48–185.

26. A. W. KNAPP AND E. M. STEIN, (a) Singular integrals and the principal series, *Proc. Nat. Acad. Sci. USA* **63** (1969), 281–284; (b) Singular integrals and the principal series II, *Proc. Nat. Acad. Sci.*, to appear; (c) Existence of Complementary Series (to appear).

27. K. KODAIRA, On a differential-geometric method in the theory of analytic stacks, *Proc. Nat. Acad. Sci. USA* **39** (1953), 865–868.

28. B. KOSTANT, (a) On the conjugacy of real Cartan subalgebras, *Proc. Nat. Acad. Sci. USA* **41** (1955), 967–970; (b) On the existence and irreducibility of certain series of representations, *Bull. Amer. Math. Soc.* **75** (1969), 627–642.

29. G. KÖTHE, Die Randverteilungen analytischer Funktionen, *Math. Z.* **57** (1952), 13–33.

30. R. A. Kunze and E. M. Stein, Uniformly bounded representations III, *Amer. J. Math.* **89** (1967), 385–442.
31. J. L. Lions and E. Magenes, Problèmes aux limits non homogènes VII, *Ann. Math. Pura Appl.* (4) **63** (1963), 201–224.
32. O. Loos, "Symmetric Spaces," Vol. II, W. A. Benjamin, Inc., New York, 1969.
33. G. W. Mackey, Induced representations of locally compact groups I, *Ann. Math.* **55** (1952), 101–140.
34. H. Matsumoto, Fonctions sphériques sur un groupe semisimple $\mathfrak{p}$-adique, *C.R. Acad. Sci. Paris* **269** (1969), 829–832.
35. F. I. Mautner, (a) Fourier analysis and symmetric spaces, *Proc. Nat. Acad. Sci. USA* **37** (1951), 529–533; (b) Spherical functions over $\mathfrak{p}$-adic fields I, *Amer. J. Math.* **80** (1958), 441–457.
36. I. MacDonald, Spherical functions on a $\mathfrak{p}$-adic Chevalley group, *Bull. Amer. Math. Soc.* **74** (1968), 520–525.
37. G. D. Mostow, Some new decomposition theorems for semisimple Lie groups, *Mem. Amer. Math. Soc.* **14** (1955), 31–54.
38. K. R. Parthasarathy, R. Ranga Rao, and V. S. Varadarajan, Representations of complex semisimple Lie groups and Lie algebras, *Ann. Math.* **85** (1967), 383–429.
39. I. Satake, On representations and compactifications of symmetric Riemann spaces, *Ann. Math.* **71** (1960), 77–110.
40. G. Schiffmann, (a) Intégrales d'entrelacement, *C.R. Acad. Sci. Paris* **266** (1968), 47–49; (b) "Sur les Intégrales d'entrelacement de R. A. Kunze et E. M. Stein," Thése, Paris, 1969.
41. L. Schwartz, "Théorie des Distributions," Hermann and Co., Paris, 1966.
42. V. I. Semyanistyi, Homogeneous functions and some problems in integral geometry in spaces of constant curvature, *Soviet Math.* **2** (1961), 59–62.
43. J.-P. Serre, "Représentations Linéaires et Espaces Homogènes Kähleriens des Groupes de Lie Compacts," Sém. Bourbaki, 1954.
44. "Séminaire Sophus Lie, Théorie des algèbres de Lie, Topologie des groupes de Lie," Paris, 1955.
45. R. Steinberg, Differential equations invariant under finite reflection groups, *Trans. Amer. Math. Soc.* **112** (1964), 392–400.
46. A. I. Stern, Completely irreducible class I representations of real semisimple Lie groups, *Soviet Math.* **20** (1969), 1254–1257.
47. R. Tahahashi, Sur les représentations unitaires des groupes de Lorentz généralisées, *Bull. Soc. Math. France* **91** (1963), 289–433.
48. N. Wallach, Induced representations of Lie algebras and a theorem of Borel–Weil, *Trans. Amer. Math. Soc.* **136** (1969), 181–187.
49. D. P. Zhelobenko, (a) The theory of linear representations of complex and real Lie groups, *Trans. Moscow Math. Soc.* **12** (1963), 57–110; (b) Symmetry in the class of elementary representations of a semisimple Lie group, *Functional Anal. Appl.* **1** (1967), 15–38; (c) Analogue of the Cartan–Weyl theory for infinite-dimensional representations of semisimple complex Lie group, *Soviet Math.* **8** (1967), 798–801; (d) The analysis of irreducibility in the class of elementary representations of a complex semisimple Lie group, *Izv. Akad. Nauk SSSR* **32** (1968), 105–128.
50. D. P. Zhelobenko and M. A. Naimark, A characterization of completely irreducible representations of a semisimple complex Lie group, *Soviet Math.* **7** (1966), 1403–1406.

PRINTED IN BELGIUM BY THE ST. CATHERINE PRESS, TEMPELHOF 37, BRUGES, LTD.

Actes, Congrès intern. Math., 1970. Tome 2, p. 313 à 319.

GROUP REPRESENTATIONS AND SYMMETRIC SPACES

by Sigurdur HELGASON

1. Introduction.

In this lecture I shall discuss some special instances of the following three general problems concerning a homogeneous space G/H, H being a closed subgroup of a Lie group G.

(A) Determine the algebra $\mathbf{D}(G/H)$ of all differential operators on G/H which are invariant under G.

(B) Determine the functions on G/H which are eigenfunctions of each $D \in \mathbf{D}(G/H)$.

(C) For each joint eigenspace for the operators in $\mathbf{D}(G/H)$ study the natural representation of G on this eigenspace ; in particular, when is it irreducible and what representations of G are so obtained ?

Here we shall deal with the case of a symmetric space X of the noncompact type and with the case of the space Ξ of horocycles in X. We refer to [6] for proofs of most of the results reported here.

2. The eigenfunctions of the Laplacian on the non-Euclidean disk.

Let X denote the open unit disk in the plane equipped with the Riemannian metric

$$ds^2 = \frac{dx^2 + dy^2}{[1 - (x^2 + y^2)]^2} .$$

The corresponding Laplace-Beltrami operator is given by

$$\Delta = [1 - (x^2 + y^2)]^2 \left(\frac{\partial^2}{\partial x^2} + \frac{\partial^2}{\partial y^2} \right).$$

We shall begin by stating some recent results about the eigenfunctions of Δ. Let B denote the boundary of X and $P(z, b)$ the Poisson kernel

$$P(z, b) = \frac{1 - |z|^2}{|z - b|^2} \qquad z \in X, \ b \in B .$$

It is then easily verified that if $\mu \in \mathbf{C}$ then $\Delta_z(P(z, b)^\mu) = 4\mu(\mu - 1)P(z, b)^\mu$ so for any measure m on B the function $z \to \int_B P(z, b)^\mu \, dm(b)$ is an eigenfunction of Δ. If $\mu \in \mathbf{R}$ and $m \geqslant 0$ this gives all the positive eigenfunctions of Δ (cf. [1],

[7]). More generally one can take m to be a distribution on B and even more generally, an *analytic functional* on B, that is a continuous linear functional on the space of analytic functions on the boundary B with the customary topology.

THEOREM 1. – *The eigenfunctions of the Laplace-Beltrami operator on the non-Euclidean disk are precisely the functions*

$$f(z) = \int_B P(z, b)^\mu \, dT(b) \tag{1}$$

where $\mu \in \mathbf{C}$ and T is an analytic functional on B.

The functional T is related to the boundary behaviour of f. Assuming, as we can, that μ in (1) satisfies $\operatorname{Re} \mu \geqslant 1/2$ we have as $|z| \to 1$

$$c_\mu (1 - |z|^2)^{\mu - 1} f(z) \to T \qquad c_\mu = \Gamma(\mu)^2 / \Gamma(2\mu - 1) \tag{2}$$

in the sense that the Fourier series of the left hand side converge formally for $|z| \to 1$ to the Fourier series of T. (For $\operatorname{Re} \mu = 1/2$ a minor modification of (2) is necessary).

The case $\mu = 1$ in Theorem 1 is closely related to Köthe's Cauchy kernel representation of holomorphic functions by analytic functionals, [9]. For Eisenstein series a result analogous to (2) was proved by John Lewis in his thesis.

It is well known that the eigenspaces of the Laplacian on a sphere are irreducible under the action of the rotation group. The analogous statement for X is in general false : The largest connected group G of isometries of X does not act irreducibly on the space of harmonic functions ($\mu = 1$). In fact, the constants form an invariant subspace. However we have the following result.

THEOREM 2. – *For $\mu \in \mathbf{C}$ let V_μ denote the space of eigenfunctions of Δ for the eigenvalue $4\mu(\mu - 1)$ with the topology induced by that of $C^\infty(X)$. Then G acts irreducibly on V_μ if and only if μ is not an integer.*

3. The Fourier transform on a symmetric space X. Spherical functions.

In order to motivate the definition I restate the Fourier inversion formula for $\mathbf{R}^n$ in a suggestive form. If $f \in L^1(\mathbf{R}^n)$ and $(\ , \ )$ denotes the inner product on $\mathbf{R}^n$ the Fourier transform $\tilde{f}$ is defined by

$$\tilde{f}(\lambda\omega) = \int_{\mathbf{R}^n} f(x)\, e^{-i\lambda(x,\omega)}\, dx \qquad \lambda \geqslant 0 , |\omega| = 1 ,$$

and if for example $f \in C_c^\infty(\mathbf{R}^n)$ we have

$$f(x) = (2\pi)^{-n} \iint_{\mathbf{R}^+ \times \mathbf{S}^{n-1}} \tilde{f}(\lambda\omega)\, e^{i\lambda(x,\omega)}\, \lambda^{n-1}\, d\lambda\, d\omega \tag{3}$$

where $\mathbf{R}^+$ denotes the set of nonnegative reals and $d\omega$ is the surface element on $\mathbf{S}^{n-1}$.

Now consider a symmetric space X of the noncompact type, that is a coset space $X = G/K$ where G is a connected semisimple Lie group with finite center and K a maximal compact subgroup. We fix an Iwasawa decomposition $G = KAN$

of G, A and N being abelian and nilpotent, respectively. The horocycles in X are the orbits in X of the subgroups of G conjugate to N ; the group G permutes the horocycles transitively and the set Ξ of all horocycles is naturally identified with the coset space G/MN where M is the centralizer of A in K. Let $\mathfrak{g}$, $\mathfrak{k}$, $\mathfrak{a}$, $\mathfrak{n}$, $\mathfrak{m}$ denote the respective Lie algebras of the groups introduced and log the inverse of the map exp : $\mathfrak{a} \to A$. It is clear from the above that each $\xi \in \Xi$ can be written $\xi = kaMN$, where $kM \in K/M$ and $a \in A$ are unique. Here the coset kM is called the *normal* to ξ and a the *complex distance* from the origin o in X to ξ. If $x \in X$, $b \in B (= K/M)$ there exists exactly one horocycle, denoted $\xi(x, b)$, through x with normal b. Let $a(x, b) \in A$ denote the complex distance from o to $\xi(x, b)$ and put $A(x, b) = \log a(x, b)$. This element of $\mathfrak{a}$ is the symmetric space analog of the inner product (x, ω) in $\mathbf{R}^n$. Denoting by $\mathfrak{a}^*$ the dual space of $\mathfrak{a}$ and defining $\rho \in \mathfrak{a}^*$ by $\rho(H) = \frac{1}{2}\operatorname{Tr}(\operatorname{ad} H \mid \mathfrak{n})$, where ad is adjoint representation and | restriction, we can define the *Fourier transform* $\tilde{f}$ of a function $f \in C_c^\infty(X)$ by

$$\tilde{f}(\lambda, b) = \int_X f(x)\, e^{(-i\lambda+\rho)(A(x,b))}\, dx, \qquad \lambda \in \mathfrak{a}^*, \ b \in B, \tag{4}$$

dx denoting the volume element on X, suitably normalized. The inversion formula for this Fourier transform is

$$f(x) = w^{-1} \int_{\mathfrak{a}^*} \int_B \tilde{f}(\lambda, b)\, e^{(i\lambda+\rho)(A(x,b))}\, |\mathbf{c}(\lambda)|^{-2}\, d\lambda\, db, \tag{5}$$

where w is the order of the Weyl group W of X, db the normalized K-invariant measure on B and $\mathbf{c}(\lambda)$ Harish-Chandra's function which can be expressed explicitly in terms of Γ-functions as we shall explain later in more detail.

A *spherical function* on X is by definition a K-invariant eigenfunction φ of each G-invariant differential operator on X, normalized by $\varphi(o) = 1$. By a simple reformulation of a theorem of Harish-Chandra the spherical functions are just the functions

$$\varphi_\lambda(x) = \int_B e^{(i\lambda+\rho)(A(x,b))}\, db \tag{6}$$

λ being arbitrary in the complex dual $\mathfrak{a}_c^*$; also $\varphi_\lambda = \varphi_\mu$ if and only if $\lambda = s\mu$ for some $s \in W$. The c-function arises in Harish-Chandra's work from a study of the behaviour of $\varphi_\lambda(x)$ for large x ; roughly speaking, $\varphi_\lambda(a)$ behaves for large a in the Weyl chamber A^+ as $\Sigma_{s\in W}\, \mathbf{c}(s\lambda)\, e^{(is\lambda-\rho)(\log a)}$ if $\lambda \in \mathfrak{a}^*$.

If f in (4) is K-invariant, then $\tilde{f}$ is independent of b and by use of (6) formula (5) reduces to Harish-Chandra's inversion formula for the spherical Fourier transform. On the other hand the general formula (5) can be derived quite easily from this special case, [5].

It is of course of interest to characterize the images of various function spaces on X under the Fourier transform $f \to \tilde{f}$. In this regard we have the following result (where $\mathfrak{a}_+^*$ denotes the positive Weyl chamber in $\mathfrak{a}^*$).

THEOREM 3. – *The Fourier transform $f \to \tilde{f}$ extends to an isometry of $L^2(X)$ onto $L^2(\mathfrak{a}_+^* \times B)$ (with the measure $|\mathbf{c}(\lambda)|^{-2}\, d\lambda\, db$).*

A point $x \in X$ is called *regular* if the geodesic (ox) has stabilizer of minimum dimension. Since $(K/M) \times A^+$ is by the "polar coordinate representation" identified with the set X' of all regular points in X, Theorem 3 shows that "*X is self-dual under the Fourier transform*". The c-function is given by

$$\mathbf{c}(\lambda) = c_0 \prod_{\alpha \in P^+} \frac{\Gamma(\langle i\lambda, \alpha_0 \rangle)\, 2^{-i\langle \lambda, \alpha_0 \rangle}}{\Gamma\left(\frac{1}{2}\left(\frac{1}{2} m_\alpha + 1 + \langle i\lambda, \alpha_0 \rangle\right)\right) \Gamma\left(\frac{1}{2}\left(\frac{1}{2} m_\alpha + m_{2\alpha} + \langle i\lambda, \alpha_0 \rangle\right)\right)}$$

where c_0 is a constant. Here P^+ is the set of positive roots which are not integral multiples of other positive roots, m_α and $m_{2\alpha}$ are the multiplicities, $\langle\ ,\ \rangle$ the inner product on $\mathfrak{a}_c^*$ and $\alpha_0 = \alpha / \langle \alpha, \alpha \rangle$. This formula was proved by Harish-Chandra and Bhanu-Murthy in special cases and by Gindikin and Karpelevič [2] in general. Every detail in this formula has turned out to be conceptually significant : the location of the singularities for the Paley-Wiener theorem, the asymptotic behaviour for the Fourier transform of rapidly decreasing functions, the Radon transform on X is inverted by a differential operator (not just a pseudo-differential operator) if and only if $\mathbf{c}^{-1}$ is a polynomial ; using the formula for $\mathbf{c}$ one shows [6] that this happens exactly when all Cartan subgroups of G are conjugate. Finally, we shall now see that the numerator and the denominator have their individual importance. We have in fact the following generalization of Theorem 2.

Let X have rank 1, i.e. dim $A = 1$, let Δ denote the Laplace-Beltrami operator on X and for $\lambda \in \mathfrak{a}_c^*$ $c_\lambda \in \mathbf{C}$ the eigenvalue given by $\Delta \varphi_\lambda = c_\lambda \varphi_\lambda$. Let $\mathcal{E}_\lambda$ be the eigenspace of Δ for the eigenvalue c_λ, this space taken with the topology induced by the usual topology of $C^\infty(X)$.

THEOREM 4. – *Let* $\mathbf{e}(\lambda)^{-1}$ *denote the denominator in the expression for* $\mathbf{c}(\lambda)$. *Then the natural representation of G on* $\mathcal{E}_\lambda$ *is irreducible if and only if*

$$\mathbf{e}(\lambda)\, \mathbf{e}(-\lambda) \neq 0 .$$

4. The conical distribution on Ξ.

The spaces
$$X = G/K, \ \Xi = G/MN$$

have many analogies reminiscent of the duality between points and hyperplanes in $\mathbf{R}^n$. For example, we have the following natural identifications for the orbit spaces of K on X, MN on Ξ,

$$K \backslash G/K = A/W, \quad MN \backslash G/MN = A \times W . \tag{7}$$

In the spirit of this analogy we define the counterparts to the spherical functions.

DEFINITION. – A distribution on Ξ is called a *conical distribution* if it is an MN-invariant eigendistribution of each G-invariant differential operator on Ξ.

Since by (6) the set of spherical functions is parametrized by $\mathfrak{a}_c^*/W$, the identifications (7) suggest that the set of conical distributions should somehow be parametrized by $\mathfrak{a}_c^* \times W$. We shall now explain how this turns out to be essentially so. The Bruhat decomposition for G implies that Ξ decomposes into finitely many disjoint orbits under MNA

$$\Xi = \bigcup_{s \in W} \Xi_s , \quad \Xi_s = MNA \cdot \xi_s .$$

There is a natural measure $d\nu$ on the orbit Ξ_s and if $\lambda \in \mathfrak{a}_c^*$ we consider the functional

$$\Phi_{\lambda,s} : \varphi \to \int_{\Xi_s} \varphi(\xi)\, e^{(is\lambda+s\rho)\,(\log a(\xi))}\, d\nu(\xi) , \quad \varphi \in C_c^\infty(\Xi) ,$$

where $a(\xi)$ denotes the A-component of $\xi \in \Xi_s$. Since Ξ_s is not in general closed there is no guarantee of convergence. However one does have absolute convergence for all $\varphi \in C_c^\infty(\Xi)$ if and only if $\mathrm{Re}(\langle i\lambda , \alpha\rangle) > 0$ for all

$$\alpha \in P^+ \cap s^{-1}P^- \qquad (P^- = -P^+) .$$

If this is the case, $\Phi_{\lambda,s}$ is a conical distribution. One would now like to obtain a meromorphic continuation of the distribution-valued function $\lambda \to \Phi_{\lambda,s}$ because then all the values and "residues" of this extension would still be conical distributions. Remarkably enough it turns out that the singularities are the same as those in the numerator for the c-function except that one restricts the product to $P^+ \cap s^{-1} P^-$. Thus we have

THEOREM 5. – *Let* $s \in W$, $\alpha_0 = \alpha/\langle\alpha , \alpha\rangle$, $\mathbf{d}_s(\lambda) = \prod_{\alpha \in P^+ \cap s^{-1} P^-} \Gamma(\langle i\lambda , \alpha_0\rangle)$. *Then the mapping*

$$\lambda \to \Psi_{\lambda,s} = \frac{1}{\mathbf{d}_s(\lambda)}\, \Phi_{\lambda,s}$$

extends to an entire function on $\mathfrak{a}_c^*$.

The residues of $\Phi_{\lambda,s}$, that is the values of $\Psi_{\lambda,s}$ at the removable singularities λ_0, have the following geometric interpretation. The closure of Ξ_s in Ξ is a union of Ξ_s and some other orbits $\Xi_{s'}$. Then the residue $\mathrm{Res}_{\lambda=\lambda_0} \Phi_{\lambda,s}$ is a linear combination of certain transversal derivatives of the various $\Psi_{\lambda,s'}$ constructed from the other orbits in the closure.

We have now to each $(\lambda , s) \in \mathfrak{a}_c^* \times W$ associated a conical distribution $\Psi_{\lambda,s}$. One can now prove ([6], Ch. III) that essentially all the conical distributions arise in this manner. This is done by transferring the differential equations for the conical distributions to differential equations on X by means of the dual to the Radon transform. Under this transform a conical distribution is sent into a C^∞ function so the differential equations become easier to handle. Thus our preliminary guess that the set of conical distributions can be parametrized by $\mathfrak{a}_c^* \times W$ is essentially verified.

5. Eigenspaces of invariant differential operators on Ξ.

Let $\mathbf{D}(\Xi)$ denote the algebra of all G-invariant differential operators on Ξ. For $\lambda \in \mathfrak{a}_c^*$ let the eigenvalue $\gamma_\lambda(D)$ be determined by $D\Psi_{\lambda,s} = \gamma_\lambda(D)\,\Psi_{\lambda,s}$ for all $D \in \mathbf{D}(\Xi)$. As indicated by the notation, the eigenvalue is independent of $s \in W$. Let $\mathcal{D}'(\Xi)$ denote the set of all distributions on Ξ and put

$$\mathcal{D}'_\lambda = \{\Psi \in \mathcal{D}'(\Xi) \mid D\Psi = \gamma_\lambda(D)\,\Psi \quad \text{for} \quad D \in \mathbf{D}(\Xi)\} .$$

Each eigendistribution of the operators in $\mathbf{D}(\Xi)$ lies in one of the spaces $\mathcal{D}'_\lambda$. Let τ_λ denote the natural representation of G on the distribution space $\mathcal{D}'_\lambda$ (strong distribution topology).

THEOREM 6. — *The representation τ_λ is irreducible if and only if* $\mathbf{e}(\lambda)\,\mathbf{e}(-\lambda) \neq 0$.

This is proved by relating the representation τ_λ to the Hilbert space representation π_λ of G induced by the one-dimensional representation $man \to e^{i\lambda(\log a)}$ of MAN. According to Harish-Chandra [3], Theorem 5, irreducibility of π_λ is equivalent to the (algebraic) irreducibility of the representation $d\pi_\lambda$ of $\mathfrak{g}$ on the space of K-finite vectors in the Hilbert space and a criterion for the algebraic irreducibility of $d\pi_\lambda$ is given by Kostant [8], p. 63. For X of rank one an entirely different proof of Theorem 6 is given in [6].

The distributions $\Psi_{\lambda,s}$ all belong to $\mathcal{D}'_\lambda$ and play the role of *extreme weight vectors* for the infinite-dimensional representation τ_λ.

If the irreducibility condition for τ_λ is satisfied the representations τ_λ and $\tau_{s\lambda}$ are equivalent for each $s \in W$. The intertwining operator realizing this equivalence is a kind of a convolution operator by means of the conical distribution $\Psi_{s\lambda,s^{-1}}$. More precisely, the map $S \to \Psi$ given by

$$\Psi(\varphi) = \int_{K/M} \left(\int_A e^{(i\lambda+\rho)(\log a)}\,\varphi(kaMN)\,da\right) dS(kM)$$

is a bijection of $\mathcal{D}'(B)$ onto $\mathcal{D}'_\lambda$, Thus τ_λ can be regarded as a representation of G on $\mathcal{D}'(B)$. If $S_{s\lambda,s^{-1}} \in \mathcal{D}'(B)$ corresponds to $\Psi_{s\lambda,s^{-1}}$ then the convolution operator $S \to S \times S_{s\lambda,s^{-1}}$ sets up the equivalence between τ_λ and τ_{s_λ}. The relationship of these results with the work of Knapp, Kunze, Schiffmann, Stein and Zhelobenko on intertwining operators is explained in [6], Ch. III, § 6.

For $\lambda \in \mathfrak{a}^*$ the intertwining operator is very simply described in terms of the Fourier transform $\tilde{f}(\lambda, b)$. In fact, the space $\{\tilde{f}(\lambda, .) \mid f \in C_c^\infty(X)\}$ is dense in $L^2(B)$ as well as in $\mathcal{D}'(B)$ and the operator $\tilde{f}(\lambda, .) \to \tilde{f}(s\lambda, .)$ extends to an isometry of $L^2(B)$ onto itself and induces the operator which intertwines τ_λ and $\tau_{s\lambda}$.

REFERENCES

[1] FURSTENBERG. — Translation-invarient cones of functions on semisimple Lie groups, *Bull. Amer. Math. Soc.*, 71, 1965, p. 271-326.

[2] GINDIKIN S.G. and KARPELEVIČ F.I. — Plancherel measure of Riemannian symmetric spaces of nonpositive curvature, *Soviet Math.*, 3, 1962, p. 962-965.

[3] HARISH-CHANDRA. — Representations of a semisimple Lie group on a Banach space I, *Trans. Amer. Math. Soc.*, 75, 1953, p. 185-243.

[4] HARISH-CHANDRA. — Spherical functions on a semisimple Lie group I, II, *Amer. J. Math.*, 80, 1958, p. 241-310, p. 553-613.

[5] HELGASON S. — Lie groups and symmetric spaces, *Battelle Rencontres*, 1967, 1-71. W.A. Benjamin, New York, 1968.

[6] HELGASON S. — A duality for symmetric spaces with applications to group representations, *Advances in Math.*, 5, 1970, p. 1-154.
[7] KARPELEVIČ. — The geometry of geodesics and the eigenfunctions of the Beltrami-Laplace operator on symmetric spaces, *Trans. Moscow Math. Soc.*, 14, 1965, p. 48-185.
[8] KOSTANT B. — On the existence and irreducibility of certain series of representations, *Bull. Amer. Math. Soc.*, 75, 1969, p. 627-642.
[9] KÖTHE G. — Die Randverteilungen analytischer Funktionen, *Math. Zeitschr.*, 57, 1952, p. 13-33.

Massachusetts Institute of Technology
Dept. of Mathematics 2-182
Cambridge
Massachusetts 02 139 (USA)

J. DIFFERENTIAL GEOMETRY
6 (1972) 411-419

A FORMULA FOR THE RADIAL PART OF THE LAPLACE-BELTRAMI OPERATOR

SIGURDUR HELGASON

Let V be a manifold and H a Lie transformation group of V. Suppose $Du = 0$ is a differential equation on V, both the differential operator D and the function u assumed invariant under H. Then the differential equation will involve several inessential variables, a fact which may render general results about differential operators rather ineffective for the differential equation at hand. Thus although D may not be an elliptic operator it might become one after the inessential variables are eliminated (cf. [3, p. 99]).

This viewpoint leads to the general definition (cf. [7]) of the transversal part and radial part of a differential operator on V given in §§ 2 and 3. The radial part has been constructed for many special differential operators in the literature; see for example [1], [3], [4], [5], [8] for Lie groups, Lie algebras and symmetric spaces, [9], [6] for some Lorentzian manifolds. Our main result, formula (3.3) in Theorem 3.2, includes various known examples worked out by computations suited for each individual case. See Harish-Chandra [4, p. 99] for the Laplacian on a semisimple Lie algebra, Berezin [1] and Harish-Chandra [3, § 8] for the Laplacian on a semisimple Lie group, and Harish-Chandra [5, § 7] and Karpelevič [8, § 15] for the Laplacian on a symmetric space. The author is indebted to J. Lepowsky for useful critical remarks.

Notation. If V is a manifold and $v \in V$, then the tangent space to V at v will be denoted V_v; the differential of a differentiable mapping φ of one manifold into another is denoted $d\varphi$. We shall use Schwartz' notation $\mathscr{E}(V)$ (resp. $\mathscr{D}(V)$) for the space of complex-valued C^∞ functions (resp. C^∞ functions of compact support) on V. Composition of differential operators D_1, D_2 is denoted $D_1 \circ D_2$.

2. The transversal part of a differential operator

Let V be a manifold satisfying the second axiom of countability, and H a Lie transformation group of V. If $h \in H$, $v \in V$, let $h \cdot v$ denote the image of v under H and let H^v denote the isotropy subgroup of H at v. Let $\mathfrak{h}$ denote the Lie algebra of H. If $X \in \mathfrak{h}$, let X^+ denote the vector field on V *induced by* X, i.e.,

Received November 3, 1971. Supported in part by National Science Foundation Grant GP-22928.

$$(2.1)\qquad (X^+f)(v) = \left\{\frac{d}{dt} f(\exp tX\cdot v)\right\}_{t=0}, \qquad f\in\mathscr{D}(V), \quad v\in V.$$

A C^∞ function f on an open subset of V is said to be *locally invariant* if $X^+f = 0$.

Lemma 2.1. *Suppose $W \subset V$ is a submanifold such that for each $w \in W$ the tangent spaces at w satisfy the condition:*

$$(2.2)\qquad V_w = W_w + (H\cdot w)_w \qquad (\textit{direct sum}).$$

Let $w_0 \in W$. Then there exists an open relatively compact neighborhood W_0 of w_0 in W and a relatively compact submanifold $B \subset H, e \in B$ such that the natural projection $\pi\colon H \to H/H^{w_0}$ is a diffeomorphism of B onto an open neighborhood U_0 of $\pi(e)$ in H/H^{w_0} and such that the mapping $\eta\colon (b, w) \to b\cdot w$ is a diffeomorphism of $B \times W_0$ onto an open neighborhood V_0 of w_0 in V.

Proof. Let $\mathfrak{h}^0$ denote the Lie algebra of H^{w_0}, and $\mathfrak{n} \subset \mathfrak{h}$ any subspace complementary to $\mathfrak{h}^0$. Then the mapping $\varphi\colon (X, w) \to \exp X\cdot w$ of $\mathfrak{n} \times W$ into V is regular at $(0, w_0)$. In fact, since $(d\varphi)_{(0,w_0)}$ fixes W_{w_0}, it suffices to prove

$$(2.3)\qquad (d\varphi)_{(0,w_0)}(\mathfrak{n}\times 0) = (H\cdot w_0)_{w_0}.$$

This however is clear from dimensionality considerations. Now the lemma follows from the standard fact that if $\mathfrak{n}_0$ is a sufficiently small neighborhood of 0 in $\mathfrak{n}$, then exp is a diffeomorphism of $\mathfrak{n}_0$ onto a submanifold $B \subset H$ diffeomorphic under π to an open neighborhood of w_0 in H/H^{w_0}.

It was pointed out to me by R. Palais that the local integration of involutive distributions (Chevalley [3, p. 89]) shows that a submanifold W satisfying (2.2) always exists.

Now let us assume that V has a Riemannian structure g invariant under the action of H. Assuming furthermore that all the orbits of H have the same dimension, we shall with each differential operator D on V associate a new differential operator D_T on V which acts "transversally to the orbits".

Fix $s_0 \in V$ and let S denote the orbit $H\cdot s_0$. For each $s \in S$ consider the geodesics in V starting at s, perpendicular to S. If we take sufficiently short pieces of these geodesics, their union is a submanifold $S_s^\perp$ of V. Shrinking $S_{s_0}^\perp$ if necessary we may assume that it satisfies transversality condition (2.2) for W. Take w_0 as s_0, and let W_0, B and V_0 be as in the lemma. For $f \in \mathscr{E}(V)$ (or even for functions defined on V_0) we define a new function f_{s_0} on V_0 by

$$f_{s_0}(b\cdot w) = f(w), \quad b\in B, \quad w \in W_0.$$

We then define D_T by

$$(2.4)\qquad (D_Tf)(s_0) = (Df_{s_0})(s_0), \qquad s_0 \in V.$$

Since $B \cdot w$ is a neighborhood of w in the orbit $H \cdot w$, and since D decreases supports, the choice of B above is immaterial, and (2.4) is indeed a valid defini tion; the operator D_T decreases supports and is therefore a differential operator, which we call *the transversal part* of D.

Theorem 2.2. *Let V be a Riemannian manifold, H a Lie transformation group of isometries of V, all orbits assumed to have the same dimension. Let S be any H-orbit and let $\bar{f}$ denote restriction of a function f to S. Then the Laplace-Beltrami operators $L = L_V$ and L_S on V and S, respectively, satisfy*

$$(2.5) \qquad (Lf)^- = L_S\bar{f} + (L_T f)^- \qquad f \in \mathscr{E}(V) .$$

Proof. Let $(y_1, \cdots, y_r)$ be any coordinate system on B such that $y_1(e) = \cdots = y_r(e) = 0$, and let $w \to (z_{r+1}(w), \cdots, z_n(w))$ be a coordinate system on W_0 such that the geodesics forming $S_{s_0}^{\perp}$ correspond to the straight lines through 0. Then we define a coordinate system $(x_1, \cdots, x_n)$ on V_0 by

$$\begin{aligned}(x_1(b \cdot w), \cdots, x_r(b \cdot w), x_{r+1}(b \cdot w), \cdots, x_n(b \cdot w)) \\ = (y_1(b), \cdots, y_r(b), z_{r+1}(w), \cdots, z_n(w)) .\end{aligned}$$

The Laplace-Beltrami operator is given by

$$L = \sum_{p,q=1}^{n} g^{pq}(\partial_{pq} - \sum_t \Gamma^t_{pq}\partial_t) ,$$

where $\partial_p = \partial/\partial x_p$, $\partial_{pq} = \partial^2/\partial x_p \partial x_q$, g^{pq} is the inverse of the matrix $g_{pq} = g(\partial_p, \partial_q)$, and Γ^t_{pq} is the Christoffel symbol

$$\Gamma^r_{pq} = \tfrac{1}{2} \sum_s g^{rs}(\partial_q g_{ps} + \partial_p g_{qs} - \partial_s g_{pq}) .$$

Suppose $\psi \in \mathscr{E}(V_0)$ satisfies the condition

$$(2.6) \qquad \psi(x_1, \cdots, x_n) \equiv \psi(0, \cdots, 0, x_{r+1}, \cdots, x_n) ,$$

or equivalently

$$\psi(b \cdot w) = \psi(w) , \quad b \in B , \quad w \in W_0 .$$

Then

$$(2.7) \qquad \psi = \psi_{s_0} , \qquad (L\psi)(s_0) = (L_T\psi)(s_0) .$$

On the other hand, suppose $\varphi \in \mathscr{E}(V_0)$ satisfies

$$(2.8) \qquad \varphi(x_1, \cdots, x_n) \equiv \varphi(x_1, \cdots, x_r, 0, \cdots, 0) ,$$

or equivalently

$$\varphi(b\cdot w) = \varphi(b\cdot s_0)\,, \quad b\in B\,, \quad w\in W_0\,.$$

For each set of real numbers $a_{r+1}, \cdots, a_n$, not all 0, the curve

$$t \to (x_1(s_0), \cdots, x_r(s_0), a_{r+1}t, \cdots, a_nt)$$

is a geodesic in V. The differential equation for geodesics

$$\ddot{x}_i + \sum_{p,q} \Gamma^i_{pq}\dot{x}_p\dot{x}_q = 0$$

(dot denoting differentiation with respect to t) therefore shows that

$$\Gamma^i_{\alpha\beta}(s_0) = 0\,, \quad 1 \leq i \leq n\,, \quad r+1 \leq \alpha, \beta \leq n\,.$$

Since the geodesic is perpendicular to S at s_0,

(2.9) $$g_{i\alpha}(s_0) = g^{i\alpha}(s_0) = 0\,, \quad \text{for} \quad 1 \leq i \leq r\,, \quad r+1 \leq \alpha \leq n\,.$$

It follows that

$$(L\varphi)(s_0) = \sum_{1\leq i,j\leq r} g^{ij}(\partial_{ij}\varphi - \sum_{1\leq k\leq r} \Gamma^k_{ij}\partial_k\varphi)(s_0)\,.$$

But by (2.9), $\Gamma^k_{ij}(s_0)$ is the same for S and for V, so

(2.10) $$(L\varphi)(s_0) = (L_S\bar{\varphi})(s_0)\,.$$

But

$$L(\varphi\psi) = \varphi L\psi + 2g\,(\text{grad}\,\varphi, \text{grad}\,\psi) + \psi L\varphi\,,$$

where for any $f \in \mathscr{E}(V_0)$,

$$\text{grad}\,f = \sum_{p,q} g^{pq}(\partial_p f)\partial_q\,.$$

Hence (2.9) implies

(2.11) $$L(\varphi\psi)(s_0) = \varphi(s_0)(L\psi)(s_0) + \psi(s_0)(L\varphi)(s_0)\,.$$

But φ_{s_0} is a constant function, so by (2.4) and (2.7)

$$\varphi_{s_0}(L\psi)(s_0) = L((\varphi\psi)_{s_0})(s_0) = (L_T(\varphi\psi))(s_0)\,.$$

Similarly, since $\bar{\psi}$ is a constant function, (2.10) implies

$$\psi(s_0)(L\varphi)(s_0) = L_S(\bar{\varphi}\bar{\psi})(s_0)\,.$$

This gives formula (2.5) for the function $f = \varphi\psi$, and since the linear combinations of such products form a dense subspace of $\mathscr{D}(V_0)$ the theorem follows by approximation.

Remark. The theorem remains true with the same proof if V is a manifold with a pseudo-Riemannian structure g provided g is nonsingular on S.

3. The radial part of a differential operator

Again let V be a manifold satisfying the second axiom of countability, and H a Lie transformation group of V. Suppose $W \subset V$ is a submanifold satisfying transversality condition (2.2) in Lemma 2.1.

Lemma 3.1. *Let D be a differential operator on V. Then there exists a unique differential operator $\Delta(D)$ on W such that*

$$(3.1) \qquad (Df)^{-} = \Delta(D)\bar{f}$$

for each locally invariant function f on an open subset of V, the bar denoting restriction to W.

Proof. Let $w_0 \in W$ and select W_0, B and V_0 as in Lemma 2.1. If $\varphi \in \mathscr{E}(W_0)$, we define f on V_0 by

$$f(b \cdot w) = \varphi(w) \,, \quad b \in B \,, \quad w \in W_0 \,.$$

The mapping $\varphi \longrightarrow (Df)^{-}$ gives an operator $D_{w_0, W_0, B}$ of $\mathscr{E}(W_0)$ into itself. It is now an easy matter to verify that the linear transformation $\Delta(D)$ given by

$$(\Delta(D)\psi)(w_0) = (D_{w_0, W_0, B}\psi)(w_0)$$

is a well-defined differential operator on $\mathscr{E}(W)$, with the properties stated in the lemma.

The operator $\Delta(D)$ is called the *radial part* of D. We shall now give a formula for the radial part of the Laplace-Beltrami operator on V under a strengthening of transversality assumption (2.2); in fact we assume that each H-orbit intersects W just once and orthogonally.

Theorem 3.2. *Suppose V is a Riemannian manifold, H a closed unimodular subgroup of the Lie group of all isometries of V (with the compact open topology). Let $W \subset V$ be a submanifold satisfying the condition: For each $w \in W$,*

$$(3.2) \qquad (H \cdot w) \cap W = \{w\}; \; V_w = (H \cdot w)_w \oplus W_w \,,$$

where $\oplus$ denotes orthogonal direct sum. Let L_V and L_W denote the Laplace-Beltrami operators on V and W, respectively. Then

$$(3.3) \qquad \Delta(L_V) = \delta^{-\frac{1}{2}} L_W \circ \delta^{\frac{1}{2}} - \delta^{-\frac{1}{2}} L_W(\delta^{\frac{1}{2}}) \,,$$

where the function δ is the volume element ratio in (3.8) *below.*

Proof. Let V^* denote the subset $H \cdot W$ of V. Since the mapping $(h, w) \longrightarrow h \cdot w$ of $H \times W$ into V has (by (3.2)) a surjective differential at each point, V^* is an open subset of V. Since H is closed, the isotropy subgroup H^w at each point $w \in W$ is compact and the orbit $H \cdot w$ is closed; if we fix a left invariant Haar measure on H and a Haar measure on H^w (with total measure 1), we obtain in a standard way an H-invariant measure $d\dot{h}$ on each orbit $H \cdot w = H/H^w$. Denoting by dv and dw the Riemannian measures on V and W, respectively, we shall prove that there exists a function $\delta \in \mathscr{E}(W)$ such that

$$(3.4) \qquad \int_{V^*} F(v)dv = \int_W \delta(w) \left(\int_{H \cdot w} F(h \cdot w) d\dot{h} \right) dw \,, \qquad F \in \mathscr{D}(V^*) \,.$$

Let $w_0 \in W$. Because of the second part of (3.2) there exist a coordinate neighborhood W_0 of w_0 in W, a vector subspace $\mathfrak{m} \subset \mathfrak{h}$ of dimension $\dim V - \dim W$ and a neighborhood $\mathfrak{m}_0$ of 0 in $\mathfrak{m}$ such that the map

$$\eta : (X, w) \to \exp X \cdot w$$

is a diffeomorphism of $\mathfrak{m}_0 \times W_0$ onto an open neighborhood V_0 of w_0 in V. Let $(x_1, \cdots, x_r)$ be a Cartesian coordinate system on $\mathfrak{m}$, and $(x_{r+1}, \cdots, x_n)$ an arbitrary coordinate system on W_0. In the formulas below let $1 \leq i, j \leq r, r + 1 \leq \alpha, \beta \leq n$. Let the coordinate system $(x_1, \cdots, x_n)$ on V_0 be determined by

$$x_i\,(\exp X \cdot w) = x_i(X) \,, \qquad x_\alpha\,(\exp X \cdot w) = x_\alpha(w) \,.$$

Let g denote the Riemannian structure of V, and put $g_{pq} = g(\partial_p, \partial_q)$ as usual, so that

$$dv = \bar{g}^{\frac{1}{2}} dx_1 \cdots dx_n \,, \qquad dw = \bar{\gamma}^{\frac{1}{2}} dx_{r+1} \cdots dx_n \,,$$

where

$$(3.5) \qquad \bar{g} = |\det ((g_{pq})_{1 \leq p,q \leq n})| \,, \qquad \bar{\gamma} = |\det (g_{\alpha\beta})| \,.$$

Because of the orthogonality in (3.2) we have

$$(3.6) \qquad g_{i\alpha}(w) = 0 \,, \qquad w \in W_0 \,.$$

But if $h = \exp X$ ($X \in \mathfrak{m}_0$) then our choice of coordinates implies for the differential dh,

$$dh\left(\frac{\partial}{\partial x_\alpha}\right)_w = \left(\frac{\partial}{\partial x_\alpha}\right)_{h \cdot w} \,, \qquad dh\left(\frac{\partial}{\partial x_i}\right)_w = \sum_{j=1}^r a_{ij} \left(\frac{\partial}{\partial x_j}\right)_{h \cdot w} \,,$$

where $a_{ij} \in \boldsymbol{R}$. Hence $g_{\alpha\beta}(h \cdot w) = g_{\alpha\beta}(w)$ and using (6), $g_{i\alpha}(h \cdot w) = 0$; consequently

$$(3.7) \qquad \bar{g}(h \cdot w) = \det (g_{ij})(h \cdot w) \bar{\gamma}(w) \,.$$

However

$$|\{\det (g_{ij})\}^{\frac{1}{2}}(h\cdot w)|\, dx_1\cdots dx_r(h\cdot w)$$

is just the Riemannian volume element $d\sigma_w$ on the orbit $H\cdot w$. Thus, if $F\in\mathscr{D}(V_0)$ we obtain from the Fubini theorem and (3.7) that

$$\int_V F(v)dv = \int_W \bar{\gamma}^{\frac{1}{2}}(w)\left(\int_{H\cdot w} F(p)d\sigma_w(p)\right)dx_{r+1}\cdots dx_n(w)\ .$$

But $d\sigma_w$ is invariant under H, so it must be a scalar multiple of $d\dot{h}$,

$$d\sigma_w = \delta(w)d\dot{h}\ . \tag{3.8}$$

This proves (3.4) for all $F\in\mathscr{D}(V_0)$; then it holds also if F has support inside $h\cdot V_0$ for some $h\in H$. But as w_0 runs through W, the sets $h\cdot V_0$ form a covering of V^*. Passing to a locally finite refinement and a corresponding partition of unity, (3.4) follows for all $F\in\mathscr{D}(V^*)$.

Let $\dot{F}(w)$ denote the inner integral in (3.4), so that

$$\dot{F}(w) = \int_{H\cdot w} F(h\cdot w)d\dot{h}\ . \tag{3.9}$$

It is a routine matter to verify that the mapping $F\to\dot{F}$ is surjective, i.e.,

$$\mathscr{D}(V^*)^{\cdot} = \mathscr{D}(W)\ . \tag{3.10}$$

For the determination of $\Delta(L_V)$ we first observe that

$$\Delta(L_V) = L_W + \text{lower order terms.} \tag{3.11}$$

This is clear from the coordinate expression for L_V together with (3.6) if we also note that the vector fields $\partial/\partial x_i$ are tangential to the H-orbits. Next we recall that L_V is symmetric with respect to dv, i.e.,

$$\int_V (L_Vf_1)(v)f_2(v)dv = \int_V f_1(v)(L_Vf_2)(v)dv \tag{3.12}$$

for all $f_1, f_2\in\mathscr{D}(V^*)$. But then this relation holds for all $f_2\in\mathscr{E}(V^*)$. In particular we can use it on f_2 invariant under H. Applying (3.4) to the left hand side of (3.12) we obtain

$$\int_W \delta(w)f_2(w)\left(\int_{H\cdot w}(L_Vf_1)(h\cdot w)d\dot{h}\right)dw\ . \tag{3.13}$$

But for each $v\in V$ the isotropy subgroup H^v is compact, so by invariance of L_V

$$(L_V)_v\left(\int_H f_1(h\cdot v)\,dh\right) = \int_H (L_Vf_1)(h\cdot v)dh\ .$$

Now putting here $v = w$ we get the inner integral in (3.13) equal to $(\Delta(L_V)\dot{f}_1)(w)$; thus the left hand side of (3.12) is

$$\int_W (\Delta(L_V)\dot{f}_1)(w)\bar{f}_2(w)\delta(w)dw ,$$

the bar denoting restriction to W. But using the H-invariance of $L_V f_2$, formula (3.4) and the definition of radial part, the right hand side of (3.12) reduces to

$$\int_W \dot{f}_1(w)(\Delta(L_V)\bar{f}_2)(w)\delta(w)dw .$$

But in view of (3.10) the functions $\dot{f}_1$ (and of course the $\bar{f}_2$) fill up $\mathscr{D}(W)$, so the equality of the two last expressions implies that $\Delta(L_V)$ is symmetric with respect to $\delta(w)dw$. Now since L_W is symmetric with respect to dw, a simple computation shows that the composition $\delta^{-\frac{1}{2}}L_W \circ \delta^{\frac{1}{2}}$ is symmetric with respect to $\delta(w)dw$ and it clearly agrees with L_W up to lower order terms. Thus by (3.11) the symmetric operators $\Delta(L_V)$ and $\delta^{-\frac{1}{2}}L_W \circ \delta^{\frac{1}{2}}$ agree up to an operator of order ≤ 1. But this operator, being symmetric, must be a function, and now (3.3) follows by applying the operators to the function 1.

It is of interest to generalize Theorem 3.2 to pseudo-Riemannian manifolds V. If V has a pseudo-Riemannian structure g, which for each $w \in W$ is non-degenerate on the closed orbit $H \cdot w$, and if each H^w $(w \in W)$ is compact, then Theorem 3.2 remains valid. In fact, the isotropy group H^v is then compact for each $v \in V^*$, so no change is necessary in the proof.

When a semisimple Lie group H acts on its Lie algebra by the adjoint representation, the regular elements of a Cartan subalgebra constitute a transversal submanifold W where the isotropy subgroup H^w is the same for all $w \in W$. This then provides an example for the following variation of Theorem 3.2.

Theorem 3.3. *Let the assumptions be as in Theorem* 3.2 *except that V has only a pseudo-Riemannian structure g. Then formula* (3.3) *remains valid if we further assume that*

(i) *for each $w \in W$ the orbit $H \cdot w$ is closed and g is non-degenerate on it,*

(ii) *H^w is the same for all $w \in W$, and its Lie algebra is its own normalizer in the Lie algebra of H.*

Proof. Put $H^0 = H^w$ $(w \in W)$ and $\dot{h} = hH^0$, and fix an H-invariant measure $d\dot{h}$ on the coset space H/H^0. Such a measure exists since each orbit $H \cdot w$ has an H-invariant measure $d\sigma_w$ defined as above. If γ is a geodesic in V tangential to W at w then γ is left fixed by each $h \in H_0$. Thus (ii) implies $\gamma \subset W$ so W is a totally geodesic submanifold of V. Defining δ by (3.8) the only part of the proof above which requires change is the justification of the formula

$$\int_{H \cdot w} (L_V f_1)(h \cdot w)d\dot{h} = (\Delta(L_V)\dot{f}_1)(w) . \tag{3.14}$$

For this we use Theorem 2.2 and the subsequent remark to split L_V into its "orbital part" and transversal part. The orbital part gives integral 0 over $H\cdot w$, so in the integral (3.14) we can replace L_V by its transversal part $L_{V,T}$. Putting $f_1^h(w) = f_1(h\cdot w)$ for $h \in H$, we have, by the H-invariance of $L_{V,T}$,

$$(L_{V,T}f_1)(h\cdot w) = (L_{V,T}(f_1^h))(w) ,$$

which, by the definition of transversal part and radial part, equals $\Delta(L_V)(\overline{f_1^h})(w)$, W being totally geodesic. But then the left hand side of (3.14) equals

$$\int_{H/H^0} \Delta(L_V)_w(f_1(\dot{h}\cdot w))d\dot{h} ,$$

which equals $(\Delta(L_V)\dot{f}_1)(w)$ because now $\dot{h}$ and w are independent variables. This proves (3.14) and therefore also Theorem 3.3.

References

[1] F. A. Berezin, *Laplace operators on semisimple Lie groups,* Trudy Moskov. Mat. Obšč. **6** (1957) 371–463.
[2] C. Chevalley, *Theory of Lie groups.* I, Princeton University Press, Princeton, 1946.
[3] Harish-Chandra, *The characters of semisimple Lie groups,* Trans. Amer. Math. Soc. **83** (1956) 98–163.
[4] ——, *Differential operators on a semisimple Lie algebra,* Amer. J. Math. **79** (1957) 87–120.
[5] ——, *Spherical functions on a semisimple Lie group.* I, Amer. J. Math. **80** (1958) 241–310.
[6] S. Helgason, *Differential operators on homogeneous spaces,* Acta Math. **102** (1959) 239–299.
[7] ——, *Radon-Fourier transforms on symmetric spaces and related group representations,* Bull. Amer. Math. Soc. **71** (1965) 757–763.
[8] F. I. Karpelevič, *The Geometry of geodesics and the eigenfunctions of the Beltrami-Laplace operator on symmetric spaces,* Trudy Moskov. Mat. Obšč. **14** (1963) 48–185.
[9] P. D. Methée, *Sur les distributions invariantes dans le groupe des rotations de Lorentz,* Comment. Math. Helv. **28** (1954) 1–49.

MASSACHUSETTS INSTITUTE OF TECHNOLOGY

BULLETIN OF THE
AMERICAN MATHEMATICAL SOCIETY
Volume 79, Number 1, January 1973

PALEY-WIENER THEOREMS AND SURJECTIVITY OF INVARIANT DIFFERENTIAL OPERATORS ON SYMMETRIC SPACES AND LIE GROUPS

BY SIGURDUR HELGASON[1]

Communicated by Robert T. Seeley, June 14, 1972

1. **Introduction.** The principal result of this paper is that if D is an invariant differential operator on a symmetric space X of the noncompact type then, for each function $f \in C^\infty(X)$, the differential equation $Du = f$ has a solution $u \in C^\infty(X)$. This is proved by means of a Paley-Wiener type theorem for the Radon transform on X. As a consequence we also obtain a Paley-Wiener theorem for the Fourier transform on X, that is an intrinsic characterization of the Fourier transforms of the functions in $C_c^\infty(X)$. In [**2**], Eguchi and Okamoto characterized the Fourier transforms of the Schwartz space on X. Invoking in addition the division theorem of Hörmander [**16**] and Lojasiewicz [**18**] we obtain by the method of [**11**] the surjectivity of D on the space of tempered distributions on X.

Finally, as a consequence of a structure theorem of Harish-Chandra [**8**] for the bi-invariant differential operators on a noncompact semisimple Lie group G, we obtain a local solvability theorem for each such operator.

2. **The range of invariant differential operators.** Let X be a symmetric space of the noncompact type, that is a coset space G/K where G is a connected, noncompact semisimple Lie group with finite center and K a maximal compact subgroup. Let $\boldsymbol{D}(X)$ denote the set of differential operators on X, invariant under G and let $C^\infty(X)$ denote the set of all C^∞ functions on X and $C_c^\infty(X)$ the set of $f \in C^\infty(X)$ of compact support.

THEOREM 2.1. *Let $D \neq 0$ in $\boldsymbol{D}(X)$. Then*

$$DC^\infty(X) = C^\infty(X).$$

As in Malgrange's proof of an analogous theorem for constant coefficient operators on $\boldsymbol{R}^n$ ([**3**], [**20**]) our proof proceeds by proving that if V is a closed ball in X then

$$f \in C_c^\infty(X), \operatorname{supp}(Df) \subset V \quad \textit{implies} \quad \operatorname{supp}(f) \subset V,$$

supp denoting support. This is proved by means of Theorem 2.2 below

---

AMS (MOS) subject classification (1970). Primary 22E30, 43A85, 58G99, 35A05.

Key words and phrases. Symmetric spaces, Lie groups, invariant differential operators, Radon transform, Fourier transform.

[1] Supported in part by the National Science Foundation NSF GP-22928.

for the Radon transform [**10**] $f \to \hat{f}$ on X. If ξ is a horocycle in X then $\hat{f}(\xi)$ is the integral of f over ξ. The following Paley-Wiener type theorem for the Radon transform is the analog for X of Theorem 2.1 in [**12**]. The proof is however quite different and is in part based on Harish-Chandra's expansion for general Eisenstein integrals on G [**7**]. I am indebted to Harish-Chandra for communicating to me this expansion which has not been published, but will appear in [**22**]. It is a generalization of the asymptotic expansion for the spherical functions in [**6**].

THEOREM 2.2. *Let $L \in C_c^\infty(X)$ and let V be a closed ball in X. Assume $\hat{f}(\xi) = 0$ whenever the horocycle ξ in X is disjoint from V. Then $f(x) = 0$ for $x \notin V$.*

REMARK. Instead of assuming $f \in C_c^\infty(X)$ it suffices to assume that the function $g \to f(gK)$ belongs to the Schwartz space on G in the sense of [**9**, p. 19].

The analog of Theorem 2.1 for left invariant differential operators D on a Lie group L is in general false. In fact, it was proved to me by Hörmander in 1964 (independently proved in Cerèzo-Rouvière [**1**]) that if for a given L one assumes local solvability for every D then either L is abelian or has an abelian normal subgroup of codimension 1. However for each bi-invariant (i.e., left and right invariant) operator on the semisimple group G we have the following local solvability result.

THEOREM 2.3. *There exists an open neighborhood V of e in G with the following property: For each bi-invariant differential operator $D \neq 0$ on G,*

$$DC^\infty(V) \supset C_c^\infty(V).$$

The proof is easily deduced from a structure theorem for D (Harish-Chandra [**8** p. 477]) combined with Proposition 1.4 in Raïs [**21**] which deals with nilpotent groups.

3. **The Fourier transform on** X. Let $G = KAN$ be an Iwasawa decomposition of G, A and N being abelian and nilpotent, respectively. Let $\mathfrak{a}$ denote the Lie algebra of A, $\mathfrak{a}^*$ its dual and $\mathfrak{a}_c^*$ the complexification of $\mathfrak{a}^*$. If $\lambda \in \mathfrak{a}_c^*$ let $\operatorname{Im} \lambda$ denote its imaginary part. Let $|\lambda|$ denote the norm on $\mathfrak{a}^*$ given by the Killing form of the Lie algebra of G. If $H \in \mathfrak{a}$ the map $X \to [H, X]$ is an endomorphism of the Lie algebra $\mathfrak{n}$ of N whose trace we denote $2\rho(H)$. Let M be the centralizer of A in K, put $B = K/M$ and let db be the K-invariant measure on B with total measure 1. For $x \in X$, $b = kM \in B$, let $A(x, b) \in \mathfrak{a}$ be determined by $n \in N$, $x = kn \exp A(x, b)K$. Fixing a G-invariant measure dx on X the Fourier transform $\tilde{f}$ of a function f on X is defined by

$$\tilde{f}(\lambda, b) = \int_X e^{(-i\lambda+\rho)(A(x,b))} f(x)\, dx$$

for all $\lambda \in \mathfrak{a}_c^*$, $b \in B$, for which this integral converges absolutely [**13**]. It satisfies

$$(1) \qquad \int_B e^{(is\lambda+\rho)(A(x,b))} \tilde{f}(s\lambda, b)\, db \equiv \int_B e^{(i\lambda+\rho)(A(x,b))} \tilde{f}(\lambda, b)\, db$$

for $f \in C_c^\infty(X)$, and every element s in the Weyl group W of X, and the mapping $f \to \tilde{f}$ extends to an isometry of $L^2(X, dx)$ onto

$$L^2(\mathfrak{a}_+^* \times B, |\mathbf{c}(\lambda)|^{-2}\, d\lambda\, db)$$

[**15**, pp. 120, 124]. Here $\mathfrak{a}_+^*$ is the positive Weyl chamber in $\mathfrak{a}^*$, $\mathbf{c}(\lambda)$ is Harish-Chandra's $\mathbf{c}$-function and $d\lambda$ is a suitably normalized Euclidean measure on $\mathfrak{a}^*$. Combining this characterization of $L^2(X)^\sim$ with Theorem 2.2, we obtain a characterization of the Fourier transforms of $C_c^\infty(X)$.

DEFINITION. A C^∞ function $\psi(\lambda, b)$ on $\mathfrak{a}_c^* \times B$, holomorphic in λ, will be called a *holomorphic function of uniform exponential type* if there exists a constant $A \geqq 0$ such that, for each polynomial $P(\lambda)$ on $\mathfrak{a}_c^*$,

$$\sup_{\lambda \in \mathfrak{a}_c^*, b \in B} e^{-A|\mathrm{Im}\lambda|} |P(\lambda)\, \psi(\lambda, b)| < \infty.$$

THEOREM 3.1. *The mapping $f \to \tilde{f}$ is a bijection of $C_c^\infty(X)$ onto the space of holomorphic functions of uniform exponential type satisfying* (1).

For the case when f is assumed K-invariant this reduces to a known result ([**4**, p. 434], for $\mathbf{SL}(2, \mathbf{R})$, [**14**], [**5**], [**15**, p. 37]). Finally, let $\mathscr{S}'(X)$ denote the dual space of the Schwartz space $\mathscr{S}(X)$. Its elements are distributions on X, the tempered distributions. In the manner indicated in the introduction we obtain an extension of Theorem 4.2 in [**11**].

THEOREM 3.2. *Let $D \neq 0$ in $\mathbf{D}(X)$. Then*

$$D\mathscr{S}'(X) = \mathscr{S}'(X).$$

REFERENCES

1. A. Cerèzo and F. Rouvière, *Résolubilité locale d'un opérateur différentiel invariant du premier ordre*, Ann. Sci. École Norm. Sup. **4** (1971), 21–30.

2. M. Eguchi and K. Okamoto, *The Fourier transform of the Schwartz space of a symmetric space*, 1972 (preprint).

3. L. Ehrenpreis, *Solution of some problems of division*. I. *Division by a polynomial of derivation*, Amer. J. Math. **76** (1954), 883–903. MR **16**, 834.

4. L. Ehrenpreis and F. I. Mautner, *Some properties of the Fourier transform on semi-simple Lie groups*. I, Ann. of Math. (2) **61** (1955), 406–439. MR **16**, 1017.

5. R. Gangolli, *On the Plancherel formula and the Paley-Wiener theorem for spherical functions on semisimple Lie groups*, Ann. of Math. (2) **93** (1971), 150–165.

6. Harish-Chandra, *Spherical functions on a semisimple Lie group*. I, II, Amer. J. Math.

80 (1958), 241–310, 553–613. MR **20** #925; MR **21** #92.
7. ———, *Differential equations and semisimple Lie groups*, 1960 (unpublished).
8. ———, *Invariant eigendistributions on a semisimple Lie group*, Trans. Amer. Math. Soc. **119** (1965), 457–508. MR **31** #4862d.
9. ———, *Discrete series for semisimple Lie groups*. II. *Explicit determination of the characters*, Acta Math. **116** (1966), 1–111. MR **36** #2745.
10. S. Helgason, *Duality and Radon transform for symmetric spaces*, Amer. J. Math. **85** (1963), 667–692. MR **28** #1632.
11. ———, *Fundamental solutions of invariant differential operators on symmetric spaces*, Amer. J. Math. **86** (1964), 565–601. MR **29** #2323.
12. ———, *The Radon transform on Euclidean spaces, compact two-point homogeneous spaces and Grassmann manifolds*, Acta Math. **113** (1965), 153–180. MR **30** #2530.
13. ———, *Radon-Fourier transforms on symmetric spaces and related group representations*, Bull. Amer. Math. Soc. **71** (1965), 757–763. MR **31** #3543.
14. ———, *An analog of the Paley-Wiener theorem for the Fourier transform on certain symmetric spaces*, Math. Ann. **165** (1966), 297–308. MR **36** #6545.
15. ———, *A duality for symmetric spaces with applications to group representations*, Advances in Math. **5** (1970), 1–154. MR **41** #8587.
16. L. Hörmander, *On the division of distributions by polynomials*, Ark. Mat. **3** (1958), 555–568. MR **23** #A2044.
17. P. D. Lax and R. S. Phillips, *The Paley-Wiener theorem for the Radon transform*, Comm. Pure Appl. Math. **23** (1970), 409–424; Correction, ibid. **24** (1971), 279–288. MR **42** #8189; #8190.
18. S. Lojasiewicz, *Sur le problème de la division*, Studia Math. **18** (1959), 87–136. MR **21** #5893.
19. D. Ludwig, *The Radon transform on euclidean space*, Comm. Pure Appl. Math. **19** (1966), 49–81. MR **32** #8064.
20. B. Malgrange, *Existence et approximation des solutions des équations aux dérivées partielles et des équations de convolution*, Ann. Inst. Fourier (Grenoble) **6** (1955/56), 271–355. MR **19**, 280.
21. M. Raïs, *Solutions élémentaires des opérateurs différentiels bi-invariants sur un groupe de Lie nilpotent*, C. R. Acad. Sci. Paris **273** (1971), 495–498.
22. G. Warner, *Harmonic analysis on semisimple Lie groups*, Springer-Verlag, Berlin and New York (to appear).

DEPARTMENT OF MATHEMATICS, MASSACHUSETTS INSTITUTE OF TECHNOLOGY, CAMBRIDGE, MASSACHUSETTS 02139

The surjectivity of invariant differential operators on symmetric spaces I

Dedicated to Leifur Ásgeirsson on the occasion of his 70th birthday.

By SIGURDUR HELGASON*

1. Introduction

The principal result of this paper is that if D is an invariant differential operator on a symmetric space X of the noncompact type then for each function $f \in C^\infty(X)$ the differential equation $Du = f$ has a solution $u \in C^\infty(X)$ (Theorem 8.2). The proof which is given in §§ 3–8 can be outlined as follows.

In § 3 we show (Theorem 3.3) that D is surjective on $C^\infty(X)$ if for each closed ball $V \subset X$ the implication

$$f \in C_c^\infty(X), \qquad \operatorname{supp}(Df) \subset V \Longrightarrow \operatorname{supp}(f) \subset V \tag{1}$$

holds. (Here supp = support and subscript c means compact support.) This is based on the existence of a fundamental solution for D ([10(d)]) and on the elementary Corollary 3.2, which depends on the symmetry of $X = G/K$. In § 4 we combine the Fourier inversion formula for $f \in C_c^\infty(X)$ with the decomposition of f in a Fourier series according to the irreducible representations δ of the isotropy group K. This gives a Fourier inversion formula (Theorem 4.4) for each component f^δ in the decomposition, the kernel being an Eisenstein integral in the sense of [9(f)], of a rather special type. We use an expansion of such integrals due to Harish-Chandra and in §§ 6–7 establish some identities and estimates for the terms in the expansion. As a result of these identities it is found that, in contrast to the Fourier inversion formula (10, § 4) for f, the expansion for each f^δ (5, § 8) consists of integrals of functions holomorphic in a tube over a Weyl chamber. We can then use Cauchy's theorem to shift the path of integration into the tube and derive Lemma 8.1. In order to conclude that the Implication (1) holds we use the fact that the Radon transform $(Df)^\wedge$ equals a constant coefficient operator $\hat{D}$ acting on $\hat{f}$. The Lions-Titchmarsh convolution theorem therefore gives precise information about the support of $\hat{f}$ and now (1) follows from Lemma 8.1.

As a by-product of the proof we obtain a Paley-Wiener theorem for the Fourier transform $f \rightarrow \tilde{f}$ on X (Theorem 8.3), a Paley-Wiener theorem for

* Supported in part by the National Science Foundation NSP GP-22928.

the Radon transform $f \to \hat{f}$ on X (Theorem 8.4), and a criterion for solving the differential equation $DS = T$ in the space of distributions on X of compact support (Theorem 8.5).

Finally in § 9 we prove local solvability for each bi-invariant differential operator on the noncompact semi-simple Lie group. The proof is an immediate consequence of Harish-Chandra's representation of these operators together with an invariance property of fundamental solutions in Raïs [18].

The main results of this paper were announced in [10(j)]. I am grateful to Harish-Chandra for permission to report (in § 5) on some of his results from [9(c)]. Their proof appears in Warner's book [21, Ch. 9].

2. Notation

As usual **R**, **C**, **Z** denote the sets of real numbers, complex numbers and integers, respectively. The sets of nonnegative elements in **R** and **Z** are denoted by $\mathbf{R}^+$ and $\mathbf{Z}^+$, respectively. If $\alpha = \beta + i\gamma$, $\beta, \gamma \in \mathbf{R}$ we write $\beta = \operatorname{Re} \alpha$, $\gamma = \operatorname{Im} \alpha$ and $\bar{\alpha}$ (or $\operatorname{conj}(\alpha)$) for $\beta - i\gamma$. If A is a group and $a, b \in A$ we put $a^b = bab^{-1}$. If $B \subset A$ is a subgroup the set of left cosets aB will be denoted A/B and if $a_0 \in A$, the transformation $aB \to a_0aB$ of A/B onto itself is denoted $\tau(a_0)$.

Manifolds will be assumed to satisfy the 2nd axiom of countability. If M is a manifold and $m \in M$, then M_m denotes the tangent space to M at m. If f is a C^∞ function of M and D a differential operator we sometimes allow ourselves to write $D_m(f(m))$ for the value of Df at m. The closure of a subset $S \subset M$ is denoted $\operatorname{Cl}(S)$ and the support of a function f is denoted $\operatorname{supp}(f)$.

If L is a Lie group with Lie algebra $\mathfrak{l}$, $\exp: \mathfrak{l} \to L$ denotes the exponential mapping and ad (resp. Ad) denotes the adjoint representation of $\mathfrak{l}$ (resp. L).

3. Distributions on Lie groups and symmetric spaces

Let M be a C^∞ manifold, $\mathcal{E}(M)$ the space of complex-valued C^∞ functions on M and $\mathbf{E}(M)$ the set of differential operators on M with coefficients in $\mathcal{E}(M)$. For a compact subset K of M let $\mathcal{D}_K(M)$ denote the set of elements in $\mathcal{E}(M)$ with support contained in K and $\mathcal{D}(M)$ their union. We give $\mathcal{E}(M)$ the usual Fréchet space topology, $\mathcal{D}_K(M)$ the topology induced from $\mathcal{E}(M)$ and $\mathcal{D}(M)$ the inductive limit topology of $\mathcal{D}_{K_i}(M)$, the (K_i) being an increasing family with union M. The duals $\mathcal{D}'(M)$ and $\mathcal{E}'(M)$ consist of the set of all distributions on M and of the subspace of distributions of compact support, respectively.

Let μ be a measure on M, equivalent to Lebesgue measure. Let D^*

denote the adjoint of D with respect to μ. If $T \in \mathfrak{D}'(M)$ DT denotes the distribution $f \to T(D^*f)$. If τ is a diffeomorphism of M and $f \in \mathfrak{E}(M)$, $T \in \mathfrak{D}'(M)$, $D \in \mathbf{E}(M)$ we define as usual $f^\tau \in \mathfrak{E}(M)$, $T^\tau \in \mathfrak{D}'(M)$ and D^τ by $f^\tau = f \circ \tau^{-1}$, $T^\tau(f) = T(f^{\tau^{-1}})$, $D^\tau(f) = (Df^{\tau^{-1}})^\tau$; D (and similarly f and T) is called invariant under τ if $D^\tau = D$. If τ leaves μ invariant then $(D^*)^\tau = (D^\tau)^*$ and $(DT)^\tau = D^\tau(T^\tau)$.

Now let G be a separable unimodular Lie group, dg a fixed Haar measure on G and $D \to D^*$ the corresponding adjoint operation on $\mathbf{E}(G)$. Let $\mathbf{D}(G)$ be the set of $D \in \mathbf{E}(G)$ invariant under left translations. If t_1, $t_2 \in \mathfrak{D}'(G)$, one of compact support, their convolution is denoted by $t_1 * t_2$. Then

$$(1) \qquad (f*t)(x) = \int_G f(xy^{-1})dt(y) , \qquad (t*f)(x) = \int_G f(y^{-1}x)dt(y)$$

if $t \in \mathfrak{D}'(G)$, $f \in \mathfrak{D}(G)$. The mapping $'$: $x \to x^{-1}$ is a diffeomorphism of G and if for $D \in \mathbf{D}(G)$ we let ε_D denote the distribution $f \to (Df')(e)$ we have

$$(2) \qquad \varepsilon_{D^*} = (\varepsilon_D)' , \qquad Dt = t*\varepsilon_D .$$

Let $K \subset G$ be a compact subgroup with Haar measure dk normalized by $\int_K dk = 1$. Let π be the natural mapping of G onto G/K and fix a G-invariant measure dx on G/K by

$$(3) \qquad \int_{G/K} F(x)dx = \int_G (F\circ\pi)(g)dg , \qquad F \in \mathfrak{D}(G/K) .$$

Let $\mathfrak{k}$ and $\mathfrak{g}$ denote the Lie algebras of K and G, respectively. Let $\mathbf{D}_0(G)$ be the set of $D \in \mathbf{D}(G)$ which are invariant under all right translations from K and $\mathbf{D}(G/K)$ the set of all G-invariant differential operators on G/K. Then there is a homomorphism μ: $\mathbf{D}_0(G) \to \mathbf{D}(G/K)$ given by $(\mu(D)F)\circ\pi = D(F\circ\pi)$ which induces an isomorphism of $\mathbf{D}_0(G)/(\mathbf{D}_0(G) \cap \mathbf{D}(G)\mathfrak{k})$ onto $\mathbf{D}(G/K)$, [10(a), p. 265]. Also by (2), $\mu(D^*) = \mu(D)^*$.

Let $o = \pi(e)$, δ the distribution $F \to F(o)$ on G/K, and δ_K the distribution $f \to \int_K f(k)dk$ on G. For $f \in \mathfrak{D}(G)$ let $f_\pi \in \mathfrak{D}(G/K)$ be determined by $f_\pi\circ\pi = f*\delta_K$ and if $T \in \mathfrak{D}'(G/K)$ let $\tilde{T} \in \mathfrak{D}'(G)$ be determined by

$$(4) \qquad \tilde{T}(f) = T(f_\pi) , \qquad f \in \mathfrak{D}(G) .$$

If T_1 and T_2 are distributions on G/K, one of compact support, their "convolution" $T_1 \times T_2$ is the distribution defined by

$$(5) \qquad (T_1 \times T_2)(F) = (\tilde{T}_1 * \tilde{T}_2)(\tilde{F}) .$$

Then $(T_1 \times T_2)^\sim = \tilde{T}_1 * \tilde{T}_2$ so $\times$ satisfies the associative law. Also $T \times \delta = T$. If $D \in \mathbf{D}(G/K)$ we have

(6) $$D(T_1 \times T_2) = T_1 \times DT_2 .$$

This follows easily by selecting $E \in \mathbf{D}_0(G)$ such that $\mu(E) = D$ and using the relation

(7) $$(DT)^{\sim} = E\tilde{T} , \qquad T \in \mathfrak{D}'(G/K) .$$

We shall now see that in the case when G/K is symmetric there is an analog of (6) for D operating on the first factor. So in addition to our assumption that G is a separable unimodular group and K compact we assume now that (G, K) is a *symmetric pair*, that is, there exists an involutive automorphism θ of G such that $(K_\theta)_0 \subset K \subset K_\theta$ where K_θ is the fixed point set of θ and $(K_\theta)_0$ its identity component. Let s denote the symmetry $gK \to \theta(g)K$ of G/K onto itself.

LEMMA 3.1. *Let* (G, K) *be a symmetric pair and* $D \in \mathbf{D}(G/K)$. *Then if* $J \in \mathfrak{D}'(G/K)$ *is* K*-invariant and* $T \in \mathcal{E}'(G/K)$ *we have*

$$D(J \times T) = DJ \times T ,$$
$$D(T \times J) = \big((D^*)^s T\big) \times J .$$

Proof. Select $E \in \mathbf{D}_0(G)$ such that $\mu(E) = D$ and put $j = \tilde{J}$, $t = \tilde{T}$. By the symmetry, $\theta(g) \subset Kg^{-1}K$ $(g \in G)$ so putting

$$f^\natural(g) = \iint f(kgk')dkdk' , \qquad f \in \mathfrak{D}(G)$$

we have $(f^\theta)^\natural = (f')^\natural$. Since $j(f) = j(f^\natural)$ we deduce

(8) $$j^\theta = j' .$$

In particular, since $j*t$ is bi-invariant under K, $(j*t)^\theta = (j*t)'$. Now θ is involutive and therefore leaves dg invariant so using (2) we obtain

$$\begin{aligned}\big(E(j*t)\big)^\theta &= E^\theta(j*t)^\theta = E^\theta(j*t)' = E^\theta(t'*j^\theta)\\ &= t'*(Ej)^\theta = (Ej*t)' = (Ej*t)^\theta .\end{aligned}$$

Combining this with (7) we get the first relation of the lemma.

Since j is bi-invariant under K and ε_E invariant by conjugation by each $k \in K$ it is clear that $j*\varepsilon_E$ is bi-invariant under K. Thus, by (2) and (8),

$$\varepsilon_E*j = (j^\theta*\varepsilon_{E^*})' = (j^\theta*\varepsilon_{E^*})^\theta = j*(\varepsilon_{E^*})^\theta .$$

But $(\varepsilon_{E^*})^\theta = \varepsilon_{(E^*)^\theta}$ so $\varepsilon_E*j = j*\varepsilon_{(E^*)^\theta}$. Taking left convolution with t and using (2) we get $(Et)*j = (E^*)^\theta(t*j)$. Also $\mu\big((E^*)^\theta\big) = \big(\mu(E^*)\big)^s = (D^*)^s$ and now (7) implies the second relation of the Lemma.

COROLLARY 3.2. *Let* (G, K) *be a symmetric pair and* $D \in \mathbf{D}(G/K)$. *Then*

$$D(S \times T) = DS \times T = S \times DT$$

for all $S,\ T \in \mathfrak{D}'(G/K)$, *at least one of compact support. Moreover*

$$D^* = D^s .$$

If $g \in G$, the mapping $\tau(g)$ of X, extended to distributions, commutes with D as well as with the right convolution $S \to S \times T$. Hence

(9) $$D(J^{\tau(g)} \times T) = D(J^{\tau(g)}) \times T .$$

As g runs through G, and J runs through the K-invariant elements in $\mathfrak{D}'(X)$, the linear span of $J^{\tau(g)}$ forms a dense subspace of $\mathfrak{D}'(X)$ because the dual space is $\mathfrak{D}(X)$. An analogous statement holds for $\mathfrak{D}'(X)$ replaced by $\mathfrak{E}'(X)$. Thus (9) gives the first relation of the corollary by continuity. The second relation comes from taking $J = \delta$ in Lemma 3.1.

Remark. The relation $D^* = D^s$ had been stated by Selberg (J. Ind. Math. Soc. 20 (1956) p. 53).

Suppose now G is a connected noncompact semisimple Lie group with finite center, $K \subset G$ a maximal compact subgroup. Then (G, K) is a symmetric pair and the space $X = G/K$ is the most general symmetric space of the non-compact type. It has a canonical metric d defined by means of the Killing form of $\mathfrak{g}$.

THEOREM 3.3. *Let* $X = G/K$ *be a symmetric space of the noncompact type and* D *any* G*-invariant differential operator on* X*. Assume that for each closed ball* $V \subset X$,

(10) $$f \in \mathfrak{D}(X) , \qquad \operatorname{supp}(Df) \subset V \Longrightarrow \operatorname{supp}(f) \subset V .$$

Then

$$D\mathfrak{E}(X) = \mathfrak{E}(X) .$$

We begin by modifying condition (10).

LEMMA 3.4. *If the operator* D *satisfies* (10) *then*

(11) $$T \in \mathfrak{E}'(X) , \qquad \operatorname{supp}(D^*T) \subset V \Longrightarrow \operatorname{supp}(T) \subset V .$$

Proof. In general let $V^r(x)$ denote the closed ball of radius r centered at x. We may assume V in (11) centered at $o = \pi(e)$. Let R denote its radius. For $0 < \varepsilon < 1$ let ψ_ε be a non-negative radial function in $\mathfrak{D}(X)$ with support in $V^\varepsilon(o)$ and total integral 1.

Now suppose $T \in \mathfrak{E}'(X)$, $\operatorname{supp}(D^*T) \subset V$. Consider the function $T^s \times \psi_\varepsilon$. By Lemma 3.1

(12) $$D(T^s \times \psi_\varepsilon) = (D^*T)^s \times \psi_\varepsilon .$$

So if we put $L_r = \pi^{-1}(V^r(o))$ we have, since $V = V^R(o)$,

$$\operatorname{supp}(D(T^s \times \psi_\varepsilon)) \subset \pi(L_R L_\varepsilon) .$$

If $R(\varepsilon) = \sup d(o, q)$ as q varies in $\pi(L_R L_\varepsilon)$ we conclude from (10) (since $\psi_\varepsilon^s = \psi_\varepsilon$),

$$\operatorname{supp}(T \times \psi_\varepsilon) \subset V^{R(\varepsilon)}(o) .$$

Now fix η, $(0 < \eta < 1)$. Then of course

$$\operatorname{supp}(T \times \psi_\varepsilon) \subset V^{R(\eta)}(o) , \qquad 0 < \varepsilon \leqq \eta . \tag{13}$$

On the other hand let $F \in \mathfrak{D}(X)$; since $(\tilde{\psi}_\varepsilon)' = \tilde{\psi}_\varepsilon$ we have

$$(T \times \psi_\varepsilon)(F) = T(F \times \psi_\varepsilon) , \tag{14}$$

and by standard arguments,

$$\lim_{\varepsilon \to 0} F \times \psi_\varepsilon = F$$

in $\mathfrak{D}(X)$. Thus by (13) and (14), $\operatorname{supp}(T) \subset V^{R(\eta)}(o)$ so taking intersection over η, $\operatorname{supp}(T) \subset V$.

LEMMA 3.5. *Let* $D \neq 0$ *in* $\mathbf{D}(G/K)$. *Then the map* $D: \mathcal{E}'(G/K) \to \mathcal{E}'(G/K)$ *is injective.*

Proof. Using Theorem 4.2 in [10(d)] let J be a fundamental solution for D, that is, a distribution on X such that $DJ = \delta$. (Although in the cited theorem D was assumed to have real coefficients, the proof is valid without this restriction.) Then by Corollary 3.2

$$T = DT \times J , \qquad T \in \mathcal{E}'(X) , \tag{15}$$

so the lemma follows.

We can now prove Theorem 3.3 invoking the same functional analysis tools as Malgrange does in [17] for the constant coefficient case (cf. also [4]). Since $D: \mathcal{E}(X) \to \mathcal{E}(X)$ is a continuous linear mapping of a Fréchet space into itself its surjectivity is by a theorem of Banach [1] equivalent to D^* being one-to-one on $\mathcal{E}'$ and $D^*\mathcal{E}'$ being closed in $\mathcal{E}'$ (for the weak topology $\sigma(\mathcal{E}', \mathcal{E})$). Thus by Lemma 3.5 it remains to prove that $D^*\mathcal{E}'$ is closed or equivalently, by another theorem of Banach [1], that whenever $B' \subset \mathcal{E}'$ is bounded (for $\sigma(\mathcal{E}', \mathcal{E})$) then $(D^*\mathcal{E}') \cap B'$ is closed in B'. But just as in Schwartz [20, p. 90], B' is contained in a set $\mathcal{E}'_C$ of elements in $\mathcal{E}'$ with support in a compact subset $C \subset X$. Thus it suffices to prove that if $C \subset X$ is a compact subset then

$$(D^*\mathcal{E}') \cap \mathcal{E}'_C \text{ is closed in } \mathcal{E}' . \tag{16}$$

Let $T_j = D^*S_j$ $(S_j \in \mathcal{E}')$ be a net in $(D^*\mathcal{E}') \cap \mathcal{E}'_C$ converging to a limit $T \in \mathcal{E}'$. Then $T \in \mathcal{E}'_C$. Select r such that $C \subset V^r(o)$. Then $\operatorname{supp}(T_j) \subset V^r(o)$ so by Lemma 3.4, $\operatorname{supp}(S_j) \subset V^r(o)$. If J^* is a K-invariant fundamental solution for D^* we have by (15),

$$S_j = T_j \times J^* \tag{17}$$

which converges (in $\sigma(\mathcal{D}', \mathcal{D})$) to $T \times J^*$. In particular $\operatorname{supp}(T \times J^*) \subset V^r(o)$. By (17)

$$T_j = D^*(T_j \times J^*)$$

so taking limits $T = D^*(T \times J^*) \in D^*\mathcal{E}'$. This proves (16) and the theorem.

4. The Radon transform and the Fourier transform on X

As in Theorem 3.3 let $X = G/K$ be a symmetric space of the non-compact type. We list briefly the customary notation associated with X and refer for example to [10(h), p. 4–10] for more explicit definitions. Let $\mathfrak{g}$ and $\mathfrak{k}$ denote the Lie algebras of G and K, respectively, $\mathfrak{g}_c$ the complexification of $\mathfrak{g}$, and $\mathfrak{p}$ the orthogonal complement of $\mathfrak{k}$ in $\mathfrak{g}$ with respect to the Killing form $\langle\, ,\rangle$ of $\mathfrak{g}_c$. Let $\mathfrak{a} \subset \mathfrak{p}$ be a maximal abelian subspace, Σ the corresponding set of (restricted) roots. Let $\mathfrak{a}^*$ be the dual of $\mathfrak{a}$, $\mathfrak{a}^*_c$ and $\mathfrak{a}_c$ their respective complexifications. If $\lambda, \mu \in \mathfrak{a}^*_c$ let $H_\lambda \in \mathfrak{a}_c$ be determined by $\langle H, H_\lambda\rangle = \lambda(H)$ $(H \in \mathfrak{a})$ and put $\langle \lambda, \mu\rangle = \langle H_\lambda, H_\mu\rangle$. If $\lambda \in \mathfrak{a}^*$, $X \in \mathfrak{p}$, put $|\lambda| = \langle\lambda, \lambda\rangle^{1/2}$, $|X| = \langle X, X\rangle^{1/2}$. Fix an open Weyl chamber $\mathfrak{a}^+ \subset \mathfrak{a}$ and let $\mathfrak{a}^*_+$ denote its preimage in $\mathfrak{a}^*$ under the map $\lambda \to H_\lambda$. Let Σ^+ (resp. Σ^-) denote the corresponding set of positive (resp. negative) roots and $\mathfrak{n}$ (resp. $\bar{\mathfrak{n}}$) the direct sum of the corresponding positive (resp. negative) root subspaces. Then we have the Iwasawa decompositions $\mathfrak{g} = \mathfrak{k} + \mathfrak{a} + \mathfrak{n}$, and, for the corresponding groups, $G = KAN$. Let $A^+ = \exp \mathfrak{a}^+$. Then we have the Cartan decomposition $G = K\operatorname{Cl}(A^+)K$. If $g \in G$ we write the Iwasawa and Cartan decompositions, $g = k(g) \exp H(g) n(g)$, $g = k_1 \exp A(g) k_2$ where $k(g) \in K$, $n(g) \in N$, $H(g) \in \mathfrak{a}$, $A(g) \in \operatorname{Cl}(\mathfrak{a}^+)$ are uniquely determined. Let log denote the inverse of the map $\exp\colon \mathfrak{a} \to A$. Let M (resp. M') denote the centralizer (resp. normalizer) of A in K, W the Weyl group M'/M, which acts as a group of linear transformations of $\mathfrak{a}$ and on $\mathfrak{a}^*_c$. If $\lambda \in \mathfrak{a}^*_c$, $s \in W$ we write $s\lambda$ instead of λ^s. If $\lambda \in \Sigma$ let m_λ denote its multiplicity and put $\rho = (1/2)\sum_{\alpha>0} m_\alpha \alpha$. Let $\Sigma_0 = \{\alpha \in \Sigma \mid (1/2)\alpha \notin \Sigma\}$, put $\Sigma_0^\pm = \Sigma^\pm \cap \Sigma_0$, and let π denote the polynomial function $\prod_{\alpha\in\Sigma_0^+} \alpha$.

Let $B = K/M = G/MAN$ and denote the actions of G on X and on B by $(g, x) \to g\cdot x$ and $(g, b) \to g(b)$. We put

$$A(gK, kM) = -H(g^{-1}k) . \tag{1}$$

Then $\exp A(x, b)$ is the complex distance from o to the horocycle in X through x with normal b and

$$A(g\cdot x, g(b)) = A(x, b) + A(g\cdot o, g(b)) . \tag{2}$$

For each $\mu \in \mathfrak{a}_c^*$, $b \in B$ the function $x \to e^{\mu(A(x,b))}$ is an eigenfunction of each $D \in \mathbf{D}(G/K)$. In fact

$$D_x(e^{(i\lambda+\rho)(A(x,b))}) = \Gamma(D)(i\lambda)e^{(i\lambda+\rho)(A(x,b))} \tag{3}$$

where $D \to \Gamma(D)$ is an isomorphism of the algebra $\mathbf{D}(G/K)$ onto the algebra of W-invariant polynomials on $\mathfrak{a}$ with coefficients in $\mathbf{C}$ (cf. [10(h), pp. 10, 94]).

The Killing form $\langle , \rangle$ induces Euclidean measures on A, $\mathfrak{a}$ and $\mathfrak{a}^*$; multiplying these by the factor $(2\pi)^{-(1/2)l}$ ($l = \dim A$) we obtain measures da, dH, and $d\lambda$ and the Fourier transform

$$f^*(\lambda) = \int_A f(a)e^{-i\lambda(\log a)}da$$

is inverted without any multiplicative constant. The Haar measures dm on M, $d\bar{n}$ on $\bar{N}$ are normalized such that 1 and $e^{-2\rho(H(\bar{n}))}$ have integrals 1 over M and $\bar{N}$ respectively. We then use $dn = \theta(d\bar{n})$ as a Haar measure on N. Then the measure dg on G is normalized such that

$$\int_G f(g)dg = \int_{KAN} f(kan)e^{2\rho(\log a)}dk\,da\,dn , \tag{4}$$

$$\int_G f(g)dg = \int_K \int_K \int_{A^+} f(k_1ak_2)\Delta(a)da\,dk_1dk_2 \tag{5}$$

where, for a suitable constant c,

$$\Delta(\exp H) = c\prod_{\alpha\in\Sigma^+}(\sinh \alpha(H))^{m_\alpha} \qquad (H \in \mathfrak{a}) .$$

Let $dx = dg_K$ and $db = dk_M$, respectively, be the invariant measures on $X = G/K$, $B = K/M$, induced by dg, dk, and dm. The measure dn induces a measure $d\sigma$ on the horocycle $\xi_0 = N\cdot o$ in X and by translation on any horocycle ξ in X. If f is a function on X its Radon transform $\hat{f}$ and its Fourier transform $\tilde{f}$ are defined by

$$\hat{f}(\xi) = \int_\xi f(x)d\sigma(x) , \qquad \tilde{f}(\lambda, b) = \int_X f(x)e^{(-i\lambda+\rho)(A(x,b))}dx \tag{6}$$

for all ξ and all $(\lambda, b) \in \mathfrak{a}_c^* \times B$ for which these integrals converge absolutely. They are related by

$$\tilde{f}(\lambda, kM) = \int_A \hat{f}(ka\cdot\xi_0)e^{(-i\lambda+\rho)(\log a)}da \tag{7}$$

because by (4)

$$\int_X f(x)dx = \int_{AN} f(an\cdot o)dadn\ . \tag{8}$$

The map $f \to \tilde{f}$ which for each $s \in W$ satisfies the identities

$$\int_B e^{(is\lambda+\rho)(A(x,b))}\tilde{f}(s\lambda, b)db = \int_B e^{(i\lambda+\rho)(A(x,b))}\tilde{f}(\lambda, b)db \tag{9}$$

extends to an isometry of $L^2(X, dx)$ onto $L^2(\mathfrak{a}_+^* \times B, |c(\lambda)|^{-2}d\lambda db)$ and is inverted by

$$f(x) = w^{-1}\int_{\mathfrak{a}^*\times B} \tilde{f}(\lambda, b)e^{(i\lambda+\rho)(A(x,b))}\,|\,c(\lambda)\,|^{-2}d\lambda db\ , \qquad f \in \mathfrak{D}(X) \tag{10}$$

(cf. [10(h), pp. 15, 120, 124]). Here w is the order of W and $c(\lambda)$ is Harish-Chandra's c-function [9(a), pp. 291, 612].

Later on it will be useful to decompose $f \in \mathfrak{D}(X)$ according to the representations of K and we therefore write down the inversion formula for the individual components.

Let δ be a unitary representation of K on a vector space V_δ of dimension $d(\delta)$. For $f \in \mathfrak{E}(X)$ we put

$$f^\delta(x) = d(\delta)\int_K f(k\cdot x)\delta(k^{-1})dk\ . \tag{11}$$

Then f^δ is a C^∞ function on G with values in Hom (V_δ, V_δ). Moreover

$$f^\delta(k\cdot x) = \delta(k)f^\delta(x)\ . \tag{12}$$

LEMMA 4.1. *If $f^\delta \neq 0$ then $\delta(M)$ has a fixed vector $\neq 0$.*

In fact by (12) there exists $a \in A$ such that $f^\delta(a\cdot o) \neq 0$ so there exists a $v \in V_\delta$ such that $f^\delta(a\cdot o)v \neq 0$. But then for $m \in M$,

$$\delta(m)f^\delta(a\cdot o)v = f^\delta(ma\cdot o)v = f^\delta(a\cdot o)v\ .$$

COROLLARY 4.2. *If $f \neq 0$ in $\mathfrak{E}(X)$ is K-finite, that is, its K-translates span a finite-dimensional space E_f then E_f contains a non-zero fixed vector under M.*

In fact E_f contains a function φ which is not identically 0 on $A\cdot o$. Then $\int_M \varphi^{\tau(m)}dm$ has the desired property.

LEMMA 4.3. *Let $v \in V_\delta$ be a unit vector fixed under $\delta(M)$. Each $F \not\equiv 0$ in $\mathfrak{D}(A^+\cdot o)$ can be extended to $f \in \mathfrak{D}(X)$ such that $f^\delta \not\equiv 0$.*

In fact it suffices to define f by $f(ka\cdot o) = \langle\delta(k)v, v\rangle F(a)$ where $\langle\,,\rangle$ is the inner product on V.

In view of Lemma 4.1 we consider representations δ for which the fixed point space under M, denoted V_δ^M, is $\neq 0$. In analogy with $\hat{f}$ and $\tilde{f}$ we define

$$\hat{f}^\delta(\xi) = \int_\xi f^\delta(x)d\sigma(x) ,$$

$$\tilde{f}^\delta(\lambda, b) = \int_X f^\delta(x)e^{(-i\lambda+\rho)(A(x,b))}dx .$$

Then

$$\tilde{f}^\delta(\lambda, kM) = \delta(k)\tilde{f}^\delta(\lambda) , \qquad \hat{f}^\delta(ka\cdot\xi_0) = \delta(k)\hat{f}^\delta(a) , \tag{13}$$

where

$$\tilde{f}^\delta(\lambda) = \tilde{f}^\delta(\lambda, eM) , \qquad \hat{f}^\delta(a) = \hat{f}^\delta(a\cdot\xi_0) .$$

THEOREM 4.4. *Let $f \in \mathfrak{D}(X)$ and let δ be any finite-dimensional unitary representation of K. Then*

$$\tilde{f}^\delta(\lambda) = \int_{A^+}\left(\int_K e^{(-i\lambda+\rho)(A(a\cdot o,kM))}\delta(k^{-1})dk\right)f^\delta(a\cdot o)\Delta(a)da$$

and for $a \in A$,

$$f^\delta(a\cdot o) = w^{-1}\int_{\mathfrak{a}^*}\left(\int_K e^{(i\lambda+\rho)(A(a\cdot o,kM))}\delta(k)dk\right)\tilde{f}^\delta(\lambda)\,|\,\mathbf{c}(\lambda)\,|^{-2}d\lambda .$$

Here $\Delta(a)$ has the same meaning as in (5).

If we use the invariance $A(k\cdot x, k(b)) = A(x, b)$ this is a simple consequence of definition (6) and the inversion formula (10).

The functions f^δ determine f in the following precise sense. Let $\hat{K}$ denote the set of equivalence classes of irreducible unitary representations of K. Then

$$f = \sum_{\delta\in\hat{K}} \text{Trace}\,(f^\delta) , \qquad f \in \mathcal{E}(X) \tag{14}$$

in the topology of $\mathcal{E}(X)$ (Harish-Chandra [9(e), p. 13]). Now it is immediate that all the linear transformations $f^\delta(a\cdot o)$ $(f \in \mathcal{E}(X))$, $\tilde{f}^\delta(\lambda)$ $(f \in \mathfrak{D}(X))$ and

$$\int_K e^{(i\lambda+\rho)(A(a\cdot o,kM))}\delta(k)dk$$

map the space V_δ into V_δ^M. Thus the orthogonal projection $E_\delta: V_\delta \to V_\delta^M$ $\left(\text{which equals } \int \delta(m)dm\right)$ satisfies $f^\delta(a\cdot o) = E_\delta f^\delta(a\cdot o)$ so by (14)

$$f(a\cdot o) = \sum_\delta \text{Trace}\,(f^\delta(a\cdot o)E_\delta) . \tag{15}$$

Thus we conclude

LEMMA 4.5. *Let* | *denote restriction. Then*

$$f(a\cdot o) = \sum_{\delta\in\hat{K}} \text{Trace}\left(f^{\delta}(a\cdot o) \mid V_{\delta}^{M}\right)$$

for all $f \in \mathfrak{S}(X)$.

5. Harish-Chandra's series expansion

Let V be a finite dimensional vector space over C and $\sigma = (\sigma_1, \sigma_2)$ a double unitary representation of K on V, σ_1 operating on the left, σ_2 on the right. Let $\lambda \in \mathfrak{a}_c^*$ and consider the function

$$\varphi(g\colon v) = \int_K \sigma_1\big(k(gk)\big) v \sigma_2(k^{-1}) e^{(i\lambda-\rho)(H(gk))} dk \tag{1}$$

for $g \in G$, $v \in V$. This can be viewed as a function on G with values in Hom (V, V) and we have $\varphi(k_1 g k_2\colon v) = \sigma_1(k_1)\varphi(g\colon v)\sigma_2(k_2)$. Let

$$V_\sigma^M = \{v \in V \mid \sigma_1(m)v = v\sigma_2(m) \text{ for } m \in M\} .$$

In his paper [9(c)] (cf. also [9(b)] and [21, Ch. 9]), Harish-Chandra gives the following expansion of $\varphi(a\colon v)$ $(a \in A^+, v \in V_\sigma^M)$ in terms of exponential functions on A. In the case when both σ_1 and σ_2 are the identity representation of K this reduces to his expansion for the spherical function φ_λ in [9(a), Lemma 37].

Let $\alpha_1, \cdots, \alpha_l$ be the simple restricted roots, L the set of integral linear combinations $n_1\alpha_1 + \cdots + n_l\alpha_l$ $(n_i \in \mathbf{Z}^+)$, and $L' = L - \{0\}$. The indicated result of [9(c)] is the following

THEOREM 5.1. *There exist certain meromorphic functions* $\mathbf{C}_s$ $(s \in W)$ *on* $\mathfrak{a}_c^*$ *and rational functions* Γ_μ $(\mu \in L)$ *on* $\mathfrak{a}_c^*$ *all with values in* Hom (V_σ^M, V_σ^M) *such that for* $a \in A^+$, $v \in V_\sigma^M$

$$e^{\rho(\log a)} \int_K \sigma_1\big(k(ak)\big) v \sigma_2(k^{-1}) e^{(i\lambda-\rho)(H(ak))} dk = \sum_{s\in W} \Phi(s\lambda\colon a)\mathbf{C}_s(\lambda)v , \tag{2}$$

where

$$\Phi(\lambda\colon a) = e^{i\lambda(\log a)} \sum_{\mu\in L} \Gamma_\mu(\lambda) e^{-\mu(\log a)} . \tag{3}$$

Here λ *varies in a certain open dense subset* ${}^*\mathfrak{a}_c'$ *of* $\mathfrak{a}_c^*$, *the functions* Γ_μ *are given by certain explicit recursion formulas, depending on* σ (cf. (4) below).

We shall need this theorem just for the case $\sigma_2 =$ identity representation and give ${}^*\mathfrak{a}_c'$ and the recursion formulas for this case.

Extend $\mathfrak{a}_c$ to a Cartan subalgebra $\mathfrak{h}_c$ of $\mathfrak{g}_c$ and let P_+ be the set of roots of $\mathfrak{g}_c$ with respect to $\mathfrak{h}_c$ which restrict to members of Σ^+. For each $\beta \in P_+$ select $X_\beta \in \mathfrak{g}_c$ such that $[H, X_\beta] = \beta(H)X_\beta$ and $\langle X_\beta, X_{-\beta}\rangle = 1$ and put $Z_\beta = (1/2)(X_\beta + \theta X_\beta)$. Extending σ to the universal enveloping algebra of $\mathfrak{k}_c$ the functions Γ_μ are given by the following recursion formula (where $\bar{\beta} = \beta \mid \mathfrak{a}$):

$$(4)\qquad \begin{aligned}&(\langle\mu,\mu\rangle - 2i\langle\mu,\lambda\rangle)\Gamma_\mu(\lambda)\\ &= 2\sum_{\beta\in P_+}\sum_{k\geq 1}\{\langle\bar\beta,\mu-2k\bar\beta+\rho\rangle - i\langle\bar\beta,\lambda\rangle\}\Gamma_{\mu-2k\bar\beta}(\lambda)\\ &\quad + 8\sum_{\beta\in P_+}\sum_{k\geq 1}k\sigma(Z_\beta Z_{-\beta})\Gamma_{\mu-2k\bar\beta}(\lambda)\,, \qquad \Gamma_0\equiv 1\,.\end{aligned}$$

Here k runs over the set of integers $\geqq 1$ for which $\mu - 2k\bar\beta \in L$. The set ${}^*\mathfrak{a}_c'$ is the set of regular elements in $\mathfrak{a}_c^*$ where all $\Gamma_\mu(s\lambda)$ $(\mu\in L,\ s\in W)$ are defined, that is,

$$(5)\qquad {}^*\mathfrak{a}_c' = \mathfrak{a}_c^* - \bigcup_{s\in W}\bigcup_{\mu\in L'} s\cdot\tau_\mu - \bigcup_{\alpha\in\Sigma}\pi_\alpha\,,$$

where τ_μ is the hyperplane $2i\langle\lambda,\mu\rangle = \langle\mu,\mu\rangle$ and π_α is the hyperplane $\langle\lambda,\alpha\rangle = 0$ in $\mathfrak{a}_c^*$. The functions $\mathbf{C}_s$ are holomorphic on ${}^*\mathfrak{a}_c'$.

The background for the recursion formula (4) is as follows. Put $\sigma_2 = 1$, $\sigma_1 = \sigma$ in (1). Since $H(gk) = -H(g^{-1}k(gk))$ and since $k\to k(gk)$ is a homeomorphism of K onto itself with Jacobian $e^{-2\rho(H(gk))}$ (that is $d(k(gk)) = e^{-2\rho(H(gk))}dk$) (cf. [9(a), p. 294]) we see by (1) § 4 that

$$\varphi(g\colon v) = \int_K e^{(i\lambda+\rho)(A(g\cdot o,kM))}\sigma(k)v\,dk\,.$$

Writing $\varphi(gK\colon v)$ instead of $\varphi(g\colon v)$ we get from (3) § 4 for $D\in\mathbf{D}(G/K)$,

$$(D\varphi)(x\colon v) = \Gamma(D)(i\lambda)\varphi(x\colon v)\,.$$

In particular, if L is the Laplace-Beltrami operator on X then

$$(6)\qquad (L\varphi)(x\colon v) = (-\langle\lambda,\lambda\rangle - \langle\rho,\rho\rangle)\varphi(x\colon v)\,.$$

In [9(c)] L is calculated in terms of the Cartan decomposition $G = KAK$. The result can be stated

$$(7)\qquad \begin{aligned}(Lf)(ka\cdot o) &= [(\Delta^{-1/2}L_A\circ\Delta^{1/2} - d)(f\circ\pi)](ka)\\ &\quad -2\sum_{\beta\in P_+}(\sinh\beta(\log a))^{-2}[\mathrm{Ad}\,(a^{-1})(Z_\beta Z_{-\beta})(f\circ\pi)](ka)\,,\end{aligned}$$

where $d = \Delta^{-1/2}L_A(\Delta^{1/2})$; it can also be derived from Theorem 2.2 in [10(i)], which gives a general decomposition of the Laplace-Beltrami operator into radial part and orbital part. When L is applied to a V-valued function f satisfying $f(k\cdot x) = \sigma(k)f(x)$, it is easily seen that

$$(8)\qquad (Lf)(a\cdot o) = (\delta'(L)\bar f)(a\cdot o)\,,$$

where $\bar f$ is the restriction of f to $A^+\cdot o$ and

$$(9)\qquad \delta'(L) = (\Delta^{-1/2}L_A\circ\Delta^{1/2} - d) - 2\sum_{\beta\in P_+}(\sinh\bar\beta)^{-2}\sigma(Z_\beta Z_{-\beta})\,.$$

Recursion formula (4) is equivalent to the differential equation

$$(10)\qquad \delta'(L)\Phi_\lambda = (-\langle\lambda,\lambda\rangle - \langle\rho,\rho\rangle)\Phi_\lambda$$

for the function $\Phi_\lambda(a) = e^{-\rho(\log a)}\Phi(\lambda\colon a)$, which in turn corresponds to (6) via (8).

In his announcement [9(f)] Harish-Chandra considers integrals even more general than (2) and states without proof a number of properties of the associated C-functions, in particular their remarkable relationship to the Plancherel measure on G. In the next section we prove some other results about the C-functions for the case $\sigma_2 = 1$, which is the one of particular use to us.

6. The C-functions

We have seen in the last section that if δ is any finite-dimensional unitary representation of K then for $v \in V_\delta^M$, $a \in A^+$,

$$e^{\rho(\log a)}\int_K e^{(i\lambda+\rho)(A(a\cdot o,kM))}\delta(k)v\,dk = \sum\nolimits_{t\in W}\Phi(t\lambda\colon a)\mathrm{C}_t(\lambda)v\,. \tag{1}$$

If in addition $\mathrm{Re}\,(\langle i\lambda, \alpha\rangle) > 0$ for $\alpha \in \Sigma^+$ then as is proved in [10(h), p. 130], if F is continuous on B,

$$e^{(-i\lambda+\rho)(\log a)}\int_B e^{(i\lambda+\rho)(A(a\cdot o,b))}F(b)\,db \longrightarrow F(eM)\boldsymbol{c}(\lambda) \tag{2}$$

as $a \to \infty$ along a line in A^+. Using (3) § 5 formula (1) above therefore gives

$$\boldsymbol{c}(\lambda)v = \mathrm{C}_e(\lambda)v\,, \qquad v \in V_\delta^M\,. \tag{3}$$

For $s \in W$ fix a representative $m_s \in M'$ and let

$$\bar{N}_s = \bar{N} \cap (m_s^{-1}Nm_s)\,.$$

Fix a Haar measure $d\bar{n}_s$ on the nilpotent group $\bar{N}_s$ and as in [10(h), p. 91], put

$$\boldsymbol{c}_s(\lambda) = \int_{\bar{N}_s} e^{-(i\lambda+\rho)(H(\bar{n}_s))}d\bar{n}_s = a_s\boldsymbol{d}_s(\lambda)\boldsymbol{e}_s(\lambda)\,,$$

where a_s is a constant and by [9(a), p. 303] and [7],

$$\boldsymbol{d}_s(\lambda) = \prod\nolimits_{\alpha\in\Sigma_0^+\cap s^{-1}\Sigma_0^-}\Gamma(\langle i\lambda, \alpha_0\rangle)2^{-\langle i\lambda,\alpha_0\rangle}\,,$$

$$\boldsymbol{e}_s(\lambda)^{-1} = \prod\nolimits_{\alpha\in\Sigma_0^+\cap s^{-1}\Sigma_0^-}\Gamma\Big(\frac{1}{2}\Big(\frac{m_\alpha}{2} + 1 + \langle i\lambda, \alpha_0\rangle\Big)\Big)$$
$$\times\,\Gamma\Big(\frac{1}{2}\Big(\frac{m_\alpha}{2} + m_{2\alpha} + \langle i\lambda, \alpha_0\rangle\Big)\Big)\,,$$

where $\alpha_0 = \alpha/\langle\alpha, \alpha\rangle$. In particular, $\boldsymbol{c}_s(\lambda)$ is a meromorphic function on $\mathfrak{a}_c^*$.

LEMMA 6.1. *For* $v \in V_\delta^M$, $\lambda \in {}^*\mathfrak{a}_c'$,

$$\mathrm{C}_{s^{-1}}(s\lambda)v = \boldsymbol{c}(\lambda)\boldsymbol{c}_s(\lambda)^{-1}\int_{\bar{N}_s} e^{-(i\lambda+\rho)(H(\bar{n}_s))}\delta\big(k(\bar{n}_s)^{-1}m_s^{-1}\big)v\,d\bar{n}_s\,.$$

Proof. In formula (1) we replace λ by $s\lambda$ and use the identity

$$e^{(is\lambda+\rho)(A(a\cdot o,kM))} = a_s^{-1}e_s(\lambda)^{-1}\int_B e^{(i\lambda+\rho)(A(a\cdot o,k\cdot b))}dS_{\lambda,s}(b) ,$$

from [10(h), p. 95] where $S_{\lambda,s}$ is the distribution on B given by

$$S_{\lambda,s}\colon F \to d_s(\lambda)^{-1}\int_{\bar{N}_s} F\big(m_s k(\bar{n}_s)M\big)e^{-(i\lambda+\rho)(H(\bar{n}_s))}d\bar{n}_s , \qquad F \in \mathcal{E}(B) .$$

The result is

$$e^{\rho(\log a)}a_s^{-1}e_s(\lambda)^{-1}\int_K\int_K e^{(i\lambda+\rho)(A(a\cdot o,klM))}\delta(k)v\,dk\,d\tilde{S}_{\lambda,s}(l)$$
$$= \sum\nolimits_{t\in W} \Phi(t\lambda\colon a)C_{ts^{-1}}(s\lambda)v ,$$

where $\tilde{S}_{\lambda,s} \in \mathcal{D}'(K)$ is defined by

$$\int_K F(k)d\tilde{S}_{\lambda,s}(k) = \int_{K/M}\Big(\int_M F(km)dm\Big)dS_{\lambda,s}(kM)$$

(cf. (4) § 3). But then if $\chi(k) = \delta(k)v$,

$$e^{(-i\lambda+\rho)(\log a)}a_s^{-1}e_s(\lambda)^{-1}\int_K e^{(i\lambda+\rho)(A(a\cdot o,kM))}(\chi*\tilde{S}_{\lambda,s})(k)dk$$
$$= e^{-i\lambda(\log a)}\sum\nolimits_{t\in W}\Phi(t\lambda\colon a)C_{ts^{-1}}(s\lambda)v .$$

But $\chi*\tilde{S}_{\lambda,s}$ is right invariant under M so if $\operatorname{Re}(\langle i\lambda, \alpha\rangle) > 0$ for all $\alpha \in \Sigma^+$, $\lambda \in {}^*\mathfrak{a}_c'$ then by (2)

$$a_s^{-1}e_s(\lambda)^{-1}(\chi*\tilde{S}_{\lambda,s})(e)c(\lambda) = C_{s^{-1}}(s\lambda)v .$$

Using the formulas for $c_s(\lambda)$ and $S_{\lambda,s}$ this gives the lemma by analytic continuation.

LEMMA 6.2. *For* $v \in V_\delta^M$,

$$\int_K e^{(is\lambda+\rho)(A(x,kM))}\delta(k)v dk = \int_K e^{(i\lambda+\rho)(A(x,kM))}\delta(k)dk\, c(\lambda)^{-1}C_{s^{-1}}(s\lambda)v .$$

Proof. Consider $\Gamma_{s,\lambda} \in \operatorname{Hom}(V_\delta, V_\delta)$ given by

$$\Gamma_{s,\lambda} = c_s(\lambda)^{-1}\int_{\bar{N}_s}\delta\big(k(\bar{n}_s)^{-1}m_s^{-1}\big)e^{-(i\lambda+\rho)(H(\bar{n}_s))}d\bar{n}_s .$$

Then again by [10(h), p. 95]

$$\int_K e^{(is\lambda+\rho)(A(x,kM))}\delta(k)dk$$
$$= a_s^{-1}e_s(\lambda)^{-1}\int_K\int_K e^{(i\lambda+\rho)(A(x,klM))}\delta(k)dk\,d\tilde{S}_{\lambda,s}(l)$$
$$= a_s^{-1}e_s(\lambda)^{-1}\int_K e^{(i\lambda+\rho)(A(x,kM))}\delta(k)\delta(l^{-1})dk\,d\tilde{S}_{\lambda,s}(l)$$
$$= \Big(\int_K e^{(i\lambda+\rho)(A(x,kM))}\delta(k)dk\Big)\Gamma_{s,\lambda} .$$

This gives the lemma via Lemma 6.1.

LEMMA 6.3. *Let* $f \in \mathfrak{D}(X)$ *and let* $\tilde{f}^{\delta}(\lambda)$ *be as defined in* § 4. *Then if* $s \in W$, $v \in V_{\delta}^{M}$

$$\tilde{f}^{\delta}(s\lambda)v = \mathbf{c}(-\lambda)^{-1}\mathrm{C}_{s^{-1}}(s\bar{\lambda})^{*}\tilde{f}^{\delta}(\lambda)v ,$$

where asterisk denotes adjoint.

Proof. If $v, w \in V_{\delta}^{M}$ and if $\langle , \rangle$ denotes the scalar product on V_{δ} then

$$\begin{aligned}\langle \tilde{f}^{\delta}(s\lambda)v, w\rangle &= d(\delta)\int_{X}\Big\langle f(x)\Big(\int_{K} e^{(-is\lambda+\rho)(A(x,kM))}\delta(k^{-1})vdk,\ w\Big)\Big\rangle dx\\ &= d(\delta)\int_{X} f(x)\Big(\Big\langle v,\ \int_{K} e^{+(is\bar{\lambda}+\rho)A(x,kM)}\delta(k)w\Big\rangle dk\Big)dx\\ &= d(\delta)\int_{X} f(x)\Big\langle v,\ \int_{K} e^{(i\bar{\lambda}+\rho)(A(x,kM))}\delta(k)dk\mathbf{c}(\bar{\lambda})^{-1}\mathrm{C}_{s^{-1}}(s\bar{\lambda})w\Big\rangle dx\\ &= d(\delta)\int_{X} f(x)\Big\langle\int_{K} e^{(-i\lambda+\rho)(A(x,kM))}\delta(k^{-1})dkv,\ \mathbf{c}(\bar{\lambda})^{-1}\mathrm{C}_{s^{-1}}(s\bar{\lambda})w\Big\rangle dx\\ &= \langle f^{\delta}(\lambda)v,\ \mathbf{c}(\bar{\lambda})^{-1}\mathrm{C}_{s^{-1}}(s\bar{\lambda})w\rangle ,\end{aligned}$$

which proves the result since $f^{\delta}(\lambda)$ maps V_{δ}^{M} into itself and since $\mathbf{c}(\bar{\lambda}) = \operatorname{conj}(\mathbf{c}(-\lambda))$.

Let s^* denote the Weyl group element which interchanges $\mathfrak{a}^+$ and $-\mathfrak{a}^+$ and put $m^* = m_{s^*}$.

LEMMA 6.4. *Suppose rank* $X = 1$. *Given* $\bar{n} \in \bar{N}$ *there exists a unique* $\bar{n}_0 \in \bar{N}$ *such that*

$$m^*k(\bar{n}_0)M = k(\bar{n})^{-1}m^*M . \tag{4}$$

It satisfies

$$H(\bar{n}_0) = H(\bar{n}) . \tag{5}$$

Proof. The existence and uniqueness of $\bar{n}_0$ is clear since the map $\bar{n} \to k(\bar{n})M$ is a bijection of $\bar{N}$ onto $K/M - \{m^*M\}$. Next we observe that the validity of (5) is independent of how m^* is chosen mod M. We shall now prove (5) by the reduction technique of [10(h), p. 54] which in fact enables us to reduce the question to the case when G is SU(2,1). Let α (and possibly 2α) be the positive restricted roots so we have the root space decomposition

$$\bar{\mathfrak{n}} = \mathfrak{g}_{-\alpha} + \mathfrak{g}_{-2\alpha} .$$

Then $\bar{n} = \exp X \exp Y$ ($X \in \mathfrak{g}_{-\alpha}$, $Y \in \mathfrak{g}_{-2\alpha}$). Suppose $Y \neq 0$. Then the analytic subgroup G^* of G whose Lie algebra $\mathfrak{g}^*$ is generated by X, Y, θX, θY is locally isomorphic to SU(2, 1) [10(h), p. 54]. By the remark above we may take m^* inside G^* and for convenience we take it such that $(m^*)^2 = e$ [10(h), p. 59]. Since the groups G^* and G have compatible Iwasawa and Bruhat decompositions it suffices to prove (5) for the case $G = \mathbf{SU}(2,1)$. For this we use the computations is [10(h), p. 60–63]. We have

$$\bar{n} = k(\bar{n}) \exp H(\bar{n})u \;, \qquad u \in N \;,$$

and if

$$\bar{n} = \exp\begin{vmatrix} it & z & it \\ -\bar{z} & 0 & -\bar{z} \\ -it & -z & -it \end{vmatrix}, \qquad u = \exp\begin{vmatrix} i\tau & \zeta & -i\tau \\ -\bar{\zeta} & 0 & \bar{\zeta} \\ i\tau & \zeta & -i\tau \end{vmatrix}$$

$z, \zeta \in \mathbf{C}$, $t, \tau \in \mathbf{R}$, then by [10(h), p. 62],

$$\zeta = -z/(1 + |z|^2 - 2it) \;, \qquad \tau = -t/\big((1 + |z|^2)^2 + 4t^2\big)$$

and

$$e^{\rho(H(\bar{n}))} = (1 + |z|^2)^2 + 4t^2 \;. \tag{6}$$

We can choose m^* such that $g \to m^*gm^*$ is the automorphism θ of $\mathbf{SU}(2,1)$ [10(h), p. 59] and then

$$k(\bar{n})^{-1}m^* = m^*m^*k(\bar{n})^{-1}m^* = m^* \exp\big(-H(\bar{n})\big)\theta(u)\theta(\bar{n})^{-1} \;,$$

from which follows that

$$k(\bar{n})^{-1}m^* = m^*k\big(\exp\big(-H(\bar{n})\big)\theta(u) \exp\big(H(\bar{n})\big)\big) \;.$$

Thus $\bar{n}_0 = \theta(u)^{\exp(-H(\bar{n}))}$ and writing

$$\bar{n}_0 = \exp\begin{vmatrix} it_0 & z_0 & it_0 \\ -\bar{z}_0 & 0 & -\bar{z}_0 \\ -it_0 & -z_0 & -it_0 \end{vmatrix}, \qquad t_0 \in \mathbf{R},\; z_0 \in \mathbf{C} \;,$$

we obtain

$$\begin{aligned} t_0 &= -t \;, \\ z_0 &= -z(1 + |z|^2 - 2it)^{-1}\big((1 + |z|^2)^2 + 4t^2\big)^{1/2} \;. \end{aligned}$$

In particular $|z_0| = |z|$ so (5) follows from (6).

The case $Y = 0$ is even simpler because then we just have to compute in $\mathbf{SL}(2, \mathbf{R})$.

LEMMA 6.5. *Suppose rank* $X = 1$ *and* $v \in V_\delta^M$. *Then*

$$\begin{aligned} &\int_{\bar{N}} \langle \delta\big(k(\bar{n})^{-1}m^*\big)v, v\rangle e^{-(i\lambda+\rho)(H(\bar{n}))}d\bar{n} \\ &= \int_{\bar{N}} \langle \delta\big(m^*k(\bar{n})\big)v, v\rangle e^{-(i\lambda+\rho)(H(\bar{n}))}d\bar{n} \;, \end{aligned}$$

for $\operatorname{Re}(\langle i\lambda, \alpha\rangle) > 0$.

Proof. Since we did not prove that the mapping $\bar{n} \to \bar{n}_0$ was measure-preserving (except for the case $G = SU(2,1)$) this lemma is not quite an immediate consequence of Lemma 6.4. Let $F(kM) = \langle\delta(k)v, v\rangle$ and let the function ψ on $K/M - \{m^*M\}$ be given by $\psi(k(\bar{n})M) = e^{(-i\lambda+\rho)(H(\bar{n}))}$. Since by [9(a), p. 287], $d(k(n)M) = e^{-2\rho(H(\bar{n}))}d\bar{n}$, the last integral above is

$$\int_{K/M} F(m^*kM)\psi(kM)dk_M = \int_K \tilde{F}(k)\tilde{\psi}(m^*k)dk ,$$

where $\tilde{F}(k) = F(kM)$ etc. and we used $m^{*2} = e$, [10(h), p. 59]. Here we substitute $k \to k^{-1}$ and use the bi-invariance of $\tilde{F}$ and $\tilde{\psi}$ under M. The result is

$$\int_{K/M} F(k^{-1}m^*M)\psi(m^*k^{-1}m^*)dk_M$$

which we transfer back to $\bar{N}$, obtaining

$$\int_{\bar{N}} \langle\delta(k(\bar{n})^{-1}m^*)v, v\rangle\psi(m^*k(\bar{n})^{-1}m^*)e^{-2\rho(H(\bar{n}))}d\bar{n} .$$

Since $\psi(m^*k(\bar{n})^{-1}m^*) = \psi(k(\bar{n}_0))$, Lemma 6.4 now gives the desired result.

We are now ready to prove a result which plays a key role in the proof of the main theorems. I understand that it has also been proved by Harish-Chandra.

THEOREM 6.6. *Let δ be a unitary irreducible representation of K on V_δ and assume that V_δ^M, the set of fixed vectors under $\delta(M)$, is $\neq 0$. Then if I denotes the identity operator,*

$$\frac{\mathbf{C}_{s^{-1}}(-s\lambda)}{\mathbf{c}(-\lambda)}\,\frac{\mathbf{C}_{s^{-1}}(-s\bar{\lambda})^*}{\mathbf{c}(\lambda)} = I \qquad \text{on } V_\delta^M . \tag{7}$$

Proof. In Lemma 6.3 replace λ by $t\lambda$ $(t \in W)$ and s by t^{-1}. Then we obtain

$$\begin{aligned}\tilde{f}^\delta(\lambda)v &= \mathbf{c}(-t\lambda)^{-1}\mathbf{C}_t(\bar{\lambda})^*\tilde{f}^\delta(t\lambda)v \\ &= \mathbf{c}(-t\lambda)^{-1}\mathbf{C}_t(\bar{\lambda})^*\mathbf{c}(-\lambda)^{-1}\mathbf{C}_{t^{-1}}(t\bar{\lambda})^*\tilde{f}^\delta(\lambda)v .\end{aligned} \tag{8}$$

Now assume G/K of rank one. Then by Kostant [13, Theorem 6], $\dim V_\delta^M = 1$, so $\tilde{f}^\delta(\lambda)$ is a scalar on V_δ^M. In view of Lemma 4.3 and Theorem 4.4 we can choose $f \in \mathfrak{D}(X)$ such that this scalar is a holomorphic function on $\mathfrak{a}_c^*$ not identically 0. Hence (8) implies, if t is the nontrivial element in the Weyl group,

$$\mathbf{c}(\lambda)^{-1}\mathbf{C}_t(\bar{\lambda})^*\mathbf{c}(-\lambda)^{-1}\mathbf{C}_t(-\bar{\lambda})^* = 1 . \tag{9}$$

But V_δ^M is here one-dimensional so that adjoint operation amounts to complex conjugation. In addition

$$\mathbf{C}_t(-\bar{\lambda})^* = \mathbf{C}_t(\lambda)$$

as a consequence of Lemmas 6.1 and 6.5. This combined with (9) gives the theorem for G/K of rank one.

For the general case consider the endomorphism $A(\lambda, s)$ of V_δ^M given by

$$A(\lambda, s)v = \int_{\bar{N}_s} e^{-(i\lambda+\rho)(H(\bar{n}_s))}\delta(m_s k(\bar{n}_s))v\, d\bar{n}_s \,. \tag{10}$$

Let $w \in V_\delta$ and put $F(kM) = \langle \delta(k)v, w\rangle$. Then

$$\langle A(\lambda, s)v, w\rangle = S'_{\lambda,s}(F)\ , \tag{11}$$

where $S'_{\lambda,s}$ is the distribution $\boldsymbol{d}_s(\lambda)S_{\lambda,s}$ on K/M. Let $s = s_{\alpha_1}\cdots s_{\alpha_p}$ be a reduced expression of s, that is, s_{α_i} is a reflection in the plane $\alpha_i = 0$ where α_i is a simple root and p is as small as possible. Put $s^{(q)} = s_{\alpha_{q+1}}\cdots s_{\alpha_p}$ $(1 \leqq q \leqq p-1)$, $s^{(p)} = e$. We normalize the Haar measures on each $\bar{N}_{s_{\alpha_i}}$ such that $\exp\left(-2\rho(H(\bar{n}_{s_{\alpha_i}}))\right)$ has integral 1. Adjusting $d\bar{n}_s$ by a constant factor if necessary we have for the convolution $\times$ on K/M,

$$S'_{\lambda,s}(F) = S'_{s'\lambda,s_{\alpha_1}} \times \cdots \times S'_{s^{(p)}\lambda,s_{\alpha_p}}(F)\ , \tag{12}$$

by Theorem 4.4, p. 84 in [10(h)], whose proof follows a method of Schiffmann [19]. This implies by (11)

$$A(s'\lambda, m_{s_{\alpha_1}})A(s''\lambda, m_{s_{\alpha_2}}) \cdots A(s^{(p)}\lambda, m_{s_{\alpha_p}}) = A(\lambda, s) \tag{13}$$

(cf. [19], p. 18). For each simple root α consider the rank-one symmetric space G_α/K_α where G_α is the analytic subgroup of G whose Lie algebra is generated by the root spaces $\mathfrak{g}_\alpha$ and $\mathfrak{g}_{-\alpha}$, and $K_\alpha = G_\alpha \cap K$. If $\mathfrak{a}_\alpha = \mathbf{R}H_\alpha$ and $A_\alpha = \exp \mathfrak{a}_\alpha$, $\bar{N}_\alpha = \bar{N}_{s_\alpha}$, then $G_\alpha = K_\alpha A_\alpha \bar{N}_\alpha$ is an Iwasawa decomposition of G_α and if $\boldsymbol{c}_\alpha$ is the c-function for G_α/K_α we have by (13) (for δ trivial)

$$\boldsymbol{c}_s(\lambda) = \textstyle\prod_{j=1}^{p} \boldsymbol{c}_{\alpha_j}(\lambda_j)\ , \tag{14}$$

where $\lambda_j = (s^{(j)}\lambda) \mid \mathfrak{a}_{\alpha_j}$ (cf. [7]). The adjoint of $A(\bar{\lambda}, s)$ is given by

$$A(\bar{\lambda}, s)^*v = \int_{\bar{N}_s} e^{-(-i\lambda+\rho)(H(\bar{n}_s))}\delta\left(k(\bar{n}_s)^{-1}m_s^{-1}\right)v\, d\bar{n}_s \qquad (v \in V_\delta^M)\ ,$$

since, by $k(\bar{n}_s^m) = k(\bar{n}_s)^m$, the right hand side belongs to V_δ^M. Since

$$A(\bar{\lambda}, s)^* = A(s^{(p)}\bar{\lambda}, m_{s_{\alpha_p}})^* \cdots A(s'\bar{\lambda}, m_{s_{\alpha_1}})^*\ ,$$

Theorem 6.6 will follow if we prove for each j,

$$\frac{A(s^{(j)}\bar{\lambda}, m_{s_{\alpha_j}})^*}{\boldsymbol{c}_{\alpha_j}(-\lambda_j)}\frac{A(s^{(j)}\lambda, m_{s_{\alpha_j}})}{\boldsymbol{c}_{\alpha_j}(\lambda_j)} = I \tag{15}$$

on the space V_δ^M. Let V denote the K_{α_j}-invariant subspace of V_δ generated by V_δ^M and let $V = \sum_i V_i$ be its decomposition into K_{α_j}-irreducible subspaces. If in the above sum we omit those V_i which have no M-invariant vector the remaining sum would still be a K_j-invariant subspace containing V_δ^M. It follows that $V_\delta^M \cap V_i \neq 0$ for each i so by Kostant's theorem quoted, $\dim(V_\delta^M \cap V_i) = 1$. Choosing a unit vector v_i in each $V_\delta^M \cap V_i$ we have therefore $V_\delta^M = \sum_i \mathbf{C}v_i$. Let δ_i denote the (irreducible) representation of K_{α_j} on V_i given by δ. On the subspace $V_\delta^M \cap V_i$, the operator $\boldsymbol{c}_{\alpha_j}(\lambda_j)^{-1} \times A(s^{(j)}\lambda, m_{s_{\alpha_j}})$ is equal to the scalar

$$(16) \qquad \boldsymbol{c}_{\alpha_j}(\lambda_j)^{-1}\int_{\bar N_{\alpha_j}} \langle \delta_i(m_{s_{\alpha_j}}k(\bar n_{\alpha_j}))v_i, v_i\rangle e^{-(i\lambda_j+\rho_{\alpha_j})(H(\bar n_{\alpha_j}))}d\bar n_{\alpha_j} ,$$

where we used the fact that $\rho \,|\, \mathfrak{a}_{\alpha_j}$ is the ρ-function ρ_{α_j} for the space $G_{\alpha_j}/K_{\alpha_j}$. On the same space the operator $\boldsymbol{c}_{\alpha_j}(-\lambda_j)^{-1}A(s^{(j)}\bar\lambda, m_{s_{\alpha_j}})^*$ is equal to the scalar

$$(17) \qquad \boldsymbol{c}_{\alpha_j}(-\lambda_j)^{-1}\int_{\bar N_{\alpha_j}} \langle \delta_i(k(\bar n_{\alpha_j})^{-1}m_{s_{\alpha_j}})v_i, v_i\rangle e^{-(-i\lambda_j+\rho_{\alpha_j})(H(\bar n_{\alpha_j}))}d\bar n_{\alpha_j} .$$

Since we had proved the theorem for the rank-one case the product of these two scalars is 1. This proves (15) so Theorem 6.6 is completely proved.

COROLLARY 6.7. *Let* $\|A\|$ *denote the Hilbert-Schmidt norm* $(\mathrm{Tr}\,(AA^*))^{1/2}$. *Then*

$$\|\mathbf{C}_s(\lambda)\|^2 = |\boldsymbol{c}(\lambda)|^2 \dim V_\delta^M \quad \textit{for } \lambda \in \mathfrak{a}^* .$$

In fact, if $\lambda \in \mathfrak{a}^*$ then $|\boldsymbol{c}(\lambda)|^2 = \boldsymbol{c}(\lambda)\boldsymbol{c}(-\lambda) = \boldsymbol{c}(s\lambda)\boldsymbol{c}(-s\lambda)$ so the relation follows from Theorem 6.6.

Remark. In their paper [12], Johnson and Wallach show that (16) is a certain rational function of λ. Since the basis $\{v_i\}$ diagonalizes $A(s^{(j)}\lambda, m_{s_{\alpha_j}})$, (13) and (14) imply that $\mathbf{C}_{s^{-1}}(s\lambda)/\boldsymbol{c}(\lambda)$ is a matrix whose entries are rational functions of λ.

7. Estimates of the coefficients Γ_μ

In order to use expansion (1) § 6 in the inversion formula of Theorem 4.4 it is necessary to have, in addition to the established properties of $\mathbf{C}_s$, some estimates of the functions Γ_μ. For this we use differential equations (10) § 5, generalizing the methods of [6] and [10(h), p. 38] for the case $\sigma = 1$.

In formula (9) § 5 for the operator $\delta'(L)$ we put for $\alpha \in \Sigma^+$,

$$B_\alpha v = \sum\nolimits_{\beta \in P_+, \bar\beta=\alpha} \sigma(Z_\beta Z_{-\beta})v , \qquad v \in V_\sigma^M .$$

If $m \in M$, $\mathrm{Ad}_G(m)$ permutes P_+ so B_α maps V_σ^M into itself and the function $\Psi_\lambda = \Delta^{1/2}\Phi_\lambda$ defined on A^+ with values in $\mathrm{Hom}\,(V_\sigma^M, V_\sigma^M)$ satisfies the differential equation

(1) $$L_A\Psi_\lambda - d\Psi_\lambda - 2\sum_{\alpha\in\Sigma^+}(\sinh\alpha)^{-2}B_\alpha\Psi_\lambda = (-\langle\lambda,\lambda\rangle - \langle\rho,\rho\rangle)\Psi_\lambda .$$

Here $d = \Delta^{-1/2}L_A(\Delta^{1/2})$ and

$$\Delta(\exp H) = \text{const. } e^{2\rho(H)}\prod_{\alpha\in\Sigma^+}(1 - e^{-2\alpha(H)})^{m_\alpha} ,$$

so we have expansions on $\mathfrak{a}^+$

$$\Delta^{-1/2}(\exp H) = e^{-\rho(H)}\sum_{\mu\in L} b_\mu e^{-\mu(H)} ,$$
$$\Delta^{1/2}(\exp H) = e^{\rho(H)}\sum_{\mu\in L} c_\mu e^{-\mu(H)} ,$$
$$d(\exp H) = \langle\rho,\rho\rangle + \sum_{\mu\in L'} d_\mu e^{-\mu(H)} ,$$
$$(\sinh\alpha(H))^{-2} = \sum_{\mu\in L'} e_\mu^\alpha e^{-\mu(H)} ,$$

where the coefficients b_μ, c_μ, d_μ, and e_μ^α have at most polynomial growth in μ. Hence

$$\Psi_\lambda(\exp H) = \sum_{\mu\in L} A_\mu(\lambda)e^{(i\lambda-\mu)(H)} ,$$

where $A_\mu(\lambda)\in\operatorname{Hom}(V_\sigma^M, V_\sigma^M)$ is given by

$$A_\mu(\lambda) = \sum_{\nu,\mu-\nu\in L}\Gamma_{\mu-\nu}(\lambda)c_\nu .$$

We have also if $F_\mu = 2\sum_\alpha e_\mu^\alpha B_\alpha$,

$$2\sum_{\alpha\in\Sigma^+}(\sinh\alpha)^{-2}(H)B_\alpha = \sum_{\mu\in L'} F_\mu e^{-\mu(H)} .$$

We now substitute the formula for Ψ_λ into (1) and use

$$(L_A\Psi_\lambda)(\exp H) = \sum_{\mu\in L}\langle i\lambda-\mu, i\lambda-\mu\rangle A_\mu(\lambda)e^{-\mu(H)} .$$

Equating the coefficients to $e^{-\mu(H)}$ we obtain the recursion formula

(2) $$\{\langle\mu,\mu\rangle - 2i\langle\lambda,\mu\rangle\}A_\mu(\lambda) = \sum_{0\neq\nu,\,\mu-\nu\in L}(d_\nu + F_\nu)A_{\mu-\nu}(\lambda) .$$

For an endomorphism A of V_σ^M let $\|A\|$ denote its Hilbert-Schmidt norm $(\operatorname{Tr}(AA^*))^{1/2}$. Then $\|AB\|\leqq\|A\|\,\|B\|$.

LEMMA 7.1. *Let $H\in\mathfrak{a}^+$. Then there exists a constant K_H such that*

$$\|\Gamma_\mu(\lambda)\|\leqq K_H e^{\mu(H)} , \qquad \mu\in L,\ \lambda\in\mathfrak{a}^* + i\mathfrak{a}_+^* .$$

Proof (cf. Lemma 4.1 in [10(g)]). If $\mu = m_1\alpha_1 + \cdots + m_l\alpha_l$ put $m(\mu) = m_1 + \cdots + m_l$. Then (2) implies

(3) $$\|A_\mu(\lambda)\|\leqq c\,m(\mu)^{-1}\sum_{0\neq\nu,\,\mu-\nu\in L}(|d_\nu| + \|F_\nu\|)\,\|A_{\mu-\nu}(\lambda)\|$$

where c is a constant. Since $|d_\nu|$ and $\|F_\nu\|$ have at most polynomial growth in ν there exists an integer N_0 such that

(4) $$c\sum_{\nu\in L}(|d_\nu| + \|F_\nu\|)e^{-\nu(H)}\leqq N_0 .$$

Then we select K_H' such that

$$\| A_\nu(\lambda) \| \leqq K'_H e^{\nu(H)} \tag{5}$$

for $\nu \in L$, $m(\nu) \leqq N_0$, $\lambda \in \mathfrak{a}^* + i\mathfrak{a}^*_+$. This can be done since, by induction, each $A_\nu(\lambda)$ is bounded for this range of λ. We shall now prove by induction that (5) holds for all $\nu \in L$. Let $N > N_0$ be an integer and suppose (5) holds for all $\nu \in L$ with $m(\nu) < N$. Let $\mu \in L$ with $m(\mu) = N$. Then by (3), (4), and (5),

$$\begin{aligned} \| A_\mu(\lambda) \| &\leqq N^{-1} c \textstyle\sum_{\nu \in L} \big(| d_\nu | + \| F_\nu \|\big) K'_H e^{\mu-\nu(H)} \\ &\leqq K'_H e^{\mu(H)} , \end{aligned}$$

proving (5). But

$$\Gamma_\mu(\lambda) = \textstyle\sum_{\nu,\, \mu-\nu \in L} A_{\mu-\nu}(\lambda) b_\nu$$

so since b_ν has at most polynomial growth in ν the estimate (5) implies a similar one for $\Gamma_\mu(\lambda)$.

8. The main theorems

We are now ready to prove that the assumption (10) in Theorem 3.3 is indeed satisfied for the symmetric space X.

LEMMA 8.1. *Let $f \in \mathfrak{D}(X)$ and V be a closed ball in X. Assume that the Radon transform $\hat{f}(\xi)$ is 0 whenever the horocycle ξ is disjoint from V. Then $f(x) = 0$ for $x \notin V$.*

Proof. Since the Radon transform commutes with the action of G we assume that V is centered at o. Let ξ_0 denote the horocycle $N \cdot o$. Any other horocycle can be written $\xi = ka \cdot \xi_0$ $(k \in K,\ a \in A)$ and by Lemma 5.4 in [10(g)] the point $ka \cdot o$ is the only point on ξ which realizes the distance $d(o, \xi)$ from o to ξ. Our assumption can therefore be written

$$\hat{f}(ka \cdot \xi_0) = 0 \qquad \text{if } k \in K,\ |\log a| > R , \tag{1}$$

if R denotes the radius of V. Since $\tilde{f}(\lambda, kM)$ is the Euclidean Fourier transform of the C^∞ function $a \longrightarrow e^{\rho(\log a)} \hat{f}(ka \cdot \xi_0)$ (cf. (7) § 4) it follows from the classical Paley-Wiener theorem that for each $N \in \mathbf{Z}^+$ there exists a constant C_N such that

$$|\tilde{f}(\xi + i\eta, kM)| \leqq C_N(1 + |\xi + i\eta|)^{-N} e^{R|\eta|} \qquad (k \in K) , \tag{2}$$

where $\xi, \eta \in \mathfrak{a}^*$ and $|\xi + i\eta| = \big(|\xi|^2 + |\eta|^2\big)^{1/2}$.

Now we wish to conclude $f(a \cdot o) = 0$ for $|\log a| > R$ by means of Theorem 4.4 and Lemma 4.5. Let $\delta \in \hat{K}$ be arbitrary. By Theorem 4.4 and (1) § 6, if $H \in \mathfrak{a}^+$, $v \in V_\delta^M$,

$$e^{\rho(H)}f^{\delta}(\exp H\cdot o)v = w^{-1}\int_{\mathfrak{a}^*}\sum_{s\in W}\Phi(s\lambda\colon \exp H)\mathbf{C}_s(\lambda)\tilde{f}^{\delta}(\lambda)\,|\,\boldsymbol{c}(\lambda)\,|^{-2}d\lambda\, v$$
$$= w^{-1}\sum_{s\in W}\int_{\mathfrak{a}^*}\sum_{\mu\in L}e^{-\mu(H)}\Gamma_{\mu}(s\lambda)\mathbf{C}_s(\lambda)e^{is\lambda(H)}\tilde{f}^{\delta}(\lambda)\,|\,\boldsymbol{c}(\lambda)\,|^{-2}d\lambda\, v\ .$$

Here we would like to interchange the integration over $\mathfrak{a}^*$ and the summation over L. Since

$$\tilde{f}^{\delta}(\lambda) = d(\delta)\int_K \tilde{f}(\lambda, kM)\delta(k^{-1})dk \tag{3}$$

it is clear from (2) that $||\tilde{f}^{\delta}(\lambda)||$ satisfies the same inequality with possibly a different constant. Since $\boldsymbol{c}(\lambda)^{-1}$ is bounded by a polynomial on $\mathfrak{a}^*$ [10(d), p. 574], Corollary 6.7 shows that $\mathbf{C}_s(\lambda)\tilde{f}^{\delta}(\lambda)\,|\,\boldsymbol{c}(\lambda)\,|^{-2}$ is integrable on $\mathfrak{a}^*$. Now using Lemma 7.1 (with $(1/2)H$ in place of H) we conclude from the Fubini theorem that

$$\begin{aligned} &e^{\rho(H)}f^{\delta}(\exp H\cdot o)v \\ &= w^{-1}\sum_{s\in W}\sum_{\mu\in L}e^{-\mu(H)}\int_{\mathfrak{a}^*}\Gamma_{\mu}(s\lambda)\mathbf{C}_s(\lambda)e^{is\lambda(H)}\tilde{f}^{\delta}(\lambda)\,|\,\boldsymbol{c}(\lambda)\,|^{-2}d\lambda\, v\ . \end{aligned}\tag{4}$$

Since for $\lambda\in\mathfrak{a}^*$, $|\,\boldsymbol{c}(\lambda)\,|^2 = \boldsymbol{c}(\lambda)\boldsymbol{c}(-\lambda) = \boldsymbol{c}(s\lambda)\boldsymbol{c}(-s\lambda)$ the integral can be rewritten as

$$\int_{\mathfrak{a}^*}e^{-i\lambda(H)}\Gamma_{\mu}(-\lambda)\boldsymbol{c}(\lambda)^{-1}\mathbf{C}_s(-s^{-1}\lambda)\boldsymbol{c}(-\lambda)^{-1}\tilde{f}^{\delta}(-s^{-1}\lambda)d\lambda\, v\ ,$$

which by Lemma 6.3 equals

$$\int_{\mathfrak{a}^*}e^{-i\lambda(H)}\Gamma_{\mu}(-\lambda)\boldsymbol{c}(\lambda)^{-1}\Big\{\frac{\mathbf{C}_s(-s^{-1}\lambda)}{\boldsymbol{c}(-\lambda)}\cdot\frac{\mathbf{C}_s(-s^{-1}\bar{\lambda})^*}{\boldsymbol{c}(\bar{\lambda})}\Big\}\tilde{f}^{\delta}(-\lambda)d\lambda\, v\ .$$

Now $\tilde{f}^{\delta}(-\lambda)$ is holomorphic in $\mathfrak{a}_c^*$ and $\Gamma_{\mu}(-\lambda)\boldsymbol{c}(\lambda)^{-1}$ is holomorphic in the tube $\mathfrak{a}^* + i(-\mathfrak{a}_+^*)$. We would now like to use Cauchy's theorem and shift the integration from $\mathfrak{a}^*$ to a plane $\mathfrak{a}^* + i\eta_0$ in this tube imitating the proof of Proposition 5.1 in Helgason [10(g)]. But for this we must make sure that the expression inside the braces is holomorphic in the tube. But now Theorem 6.6 shows that the expression is in fact equal to 1! This being so, formula (4) can be simplified and we obtain

$$\begin{aligned} &e^{\rho(H)}f^{\delta}(\exp H\cdot o)v \\ &= \sum_{\mu\in L}e^{-\mu(H)}\int_{\mathfrak{a}^*}e^{-i\lambda(H)}\Gamma_{\mu}(-\lambda)\boldsymbol{c}(\lambda)^{-1}\tilde{f}^{\delta}(-\lambda)d\lambda\, v\ , \end{aligned}\tag{5}$$

where the functions $\mathbf{C}_s$ have dropped out! Although the functions Γ_{μ} and $\tilde{f}^{\delta}$ are matrix-valued the proof of the cited Proposition 5.1 applies virtually without change and we conclude that the integral above vanishes if $H\in\mathfrak{a}^+$

and $|H| > R$. Thus $f^\delta(\exp H\cdot o)v = 0$ for these H, all $\delta \in \hat{K}$ and all $v \in V_\delta^M$. Hence by Lemma 4.5, $f(a\cdot o) = 0$ for $a \in A^+$, $|\log a| > R$. Applying this to the rotated function $x \to f(k\cdot x)$ on X we conclude $f(x) = 0$ for $x \notin V$ so the lemma is proved.

Remark. The assumption $f \in \mathfrak{D}(X)$ in Lemma 8.1 can be weakened by examining the proof. Since Theorem 4.4 is based on Harish-Chandra's inversion formula for the spherical Fourier transform for the functions in the Schwartz space $\mathcal{C}(G)$ [9(e), p. 19] which are bi-invariant under K, Lemma 8.1 remains valid for all $f \in \mathfrak{E}(X)$ for which $f\circ\pi \in \mathcal{C}(G)$.

THEOREM 8.2. *Let $D \neq 0$ in $\mathbf{D}(G/K)$. Then*

$$D\mathfrak{E}(G/K) = \mathfrak{E}(G/K) .$$

Proof. We shall conclude this from Theorem 3.3 by verifying that if $V \subset X$ is a closed ball and if $f \in \mathfrak{D}(X)$, $\operatorname{supp}(Df) \subset V$ then $\operatorname{supp}(f) \subset V$. Since D is invariant under G we may assume that V is centered at o. Now $\operatorname{supp}(Df) \subset V$ implies for the Radon transform,

$$(Df)^\wedge(\xi) = 0 \tag{6}$$

if $\xi \cap V = \emptyset$. But by Theorem 5.3 in [10(c)], $(Df)^\wedge = \hat{D}\hat{f}$ where $\hat{D}$ is a G-invariant differential operator on the horocycle space $\Xi = G/MN$. But under the diffeomorphism $(kM, a) \to ka\cdot\xi_0$ of $K/M \times A$ onto Ξ the operator $\hat{D}$ is given by a constant coefficient operator Δ on the Euclidean space A [10(c), p. 676]. Thus if V has radius R, (6) means that

$$\Delta_a(\hat{f}(ka\cdot\xi_0)) = 0 , \qquad k \in K,\ |\log a| > R . \tag{7}$$

Since the function $a \to \hat{f}(ka\cdot\xi_0)$ has compact support, (7) implies by a special case of the Lions-Titchmarsh convexity theorem [15] that $\hat{f}(ka\cdot\xi_0) = 0$ for $|\log a| > R$. But then by Lemma 8.1 we conclude $\operatorname{supp}(f) \subset V$ and Theorem 8.2 is completely proved.

We can also use Lemma 8.1 and its proof to establish a Paley-Wiener theorem for the Fourier transform on X, that is an intrinsic characterization of the Fourier transforms $\tilde{f}(\lambda, b)$ as f runs through $\mathfrak{D}(X)$.

Definition. A C^∞ function $\psi(\lambda, b)$ on $\mathfrak{a}_c^* \times B$, holomorphic in λ, will be called a *holomorphic function of uniform exponential type* if there exists a constant $R \geqq 0$ such that for each integer $N \geqq 0$

$$\sup_{\lambda \in \mathfrak{a}_c^*, b \in B} e^{-R|\operatorname{Im}\lambda|}\left(1 + |\lambda|\right)^N |\psi(\lambda, b)| < \infty , \tag{8}$$

where $\operatorname{Im}\lambda = \eta$, if $\lambda = \xi + i\eta$ ($\xi, \eta \in \mathfrak{a}^*$).

THEOREM 8.3. *The Fourier transform* $f(x) \to \tilde{f}(\lambda, b)$ *is a bijection of* $\mathfrak{D}(X)$ *onto the set of holomorphic functions* $\psi(\lambda, b)$ *of uniform exponential type satisfying the identities*

$$\int_B e^{(is\lambda+\rho)(A(x,b))}\psi(s\lambda, b)db = \int_B e^{(i\lambda+\rho)(A(x,b))}\psi(\lambda, b)db \tag{9}$$

for $s \in W$, $\lambda \in \mathfrak{a}_c^*$, $x \in X$. *Moreover* $\tilde{f}$ *satisfies* (8) *if and only if* f *has support in the closed ball* $V^R(o)$.

Proof. Because of formulas (7) and (9) in § 4 only the "onto" part remains to be verified. In other words we must show that if $\psi \in \mathfrak{E}(\mathfrak{a}_c^* \times B)$ is holomorphic in the first variable and satisfies (8) and (9) above, then there exists $f \in \mathfrak{D}(X)$ with support in $V^R(o)$ such that $\tilde{f} = \psi$. Because of the Fourier inversion formula (10) § 4 we define f on X by

$$f(x) = w^{-1}\int_{\mathfrak{a}^* \times B} \psi(\lambda, b)e^{(i\lambda+\rho)(A(x,b))}\,|\,\boldsymbol{c}(\lambda)\,|^{-2}d\lambda db\,. \tag{10}$$

Then $f \in \mathfrak{E}(X)$. The next thing to prove is that $f(x) = 0$ for $d(o, x) > R$. For this we use the method of Lemma 8.1. For $\delta \in \hat{K}$ we derive from (10)

$$f^\delta(x) = w^{-1}\int_{\mathfrak{a}^*}\Big(\int_K e^{(i\lambda+\rho)(A(x,kM))}\delta(k)dk\Big)\psi^\delta(\lambda)\,|\,\boldsymbol{c}(\lambda)\,|^{-2}d\lambda\,, \tag{11}$$

where

$$\psi^\delta(\lambda, b) = d(\delta)\int_K \psi(\lambda, k\cdot b)\delta(k^{-1})dk\,, \qquad \psi^\delta(\lambda) = \psi^\delta(\lambda, eM)\,.$$

In (9) we replace x by $k\cdot x$, use $A(k\cdot x, b) = A(x, k^{-1}(b))$, then make the substitution $b \to k(b)$, multiply by $\delta(k^{-1})$ and integrate over K. The result is

$$\int_B e^{(is\lambda+\rho)(A(x,b))}\psi^\delta(s\lambda, b)db = \int_B e^{(i\lambda+\rho)(A(x,b))}\psi^\delta(\lambda, b)db\,.$$

But $\psi^\delta(\lambda, kM) = \delta(k)\psi^\delta(\lambda)$ so

$$\begin{aligned}&\int_K e^{(is\lambda+\rho)(A(x,kM))}\delta(k)dk\,\psi^\delta(s\lambda)\\ &= \int_K e^{(i\lambda+\rho)(A(x,kM))}\delta(k)dk\,\psi^\delta(\lambda)\,.\end{aligned}$$

Now $\psi^\delta(\lambda)$ maps V_δ into V_δ^M and using Lemma 6.2 we obtain

$$\int_K e^{(i\lambda+\rho)(A(x,kM))}\delta(k)dk\frac{\mathbf{C}_{s^{-1}}(s\lambda)}{\boldsymbol{c}(\lambda)}\psi^\delta(s\lambda) = \int_K e^{(i\lambda+\rho)(A(x,kM))}\delta(k)dk\,\psi^\delta(\lambda)\,.$$

Now if $\operatorname{Re}(\langle i\lambda, \alpha\rangle) > 0$ for $\alpha \in \Sigma^+$ the limit relation (2) § 6 implies that

$$\frac{\mathbf{C}_{s^{-1}}(s\lambda)}{\mathbf{c}(\lambda)}\psi^\delta(s\lambda) = \psi^\delta(\lambda)$$

so by Theorem 6.6,

$$\psi^\delta(s\lambda) = \frac{\mathbf{C}_{s^{-1}}(s\bar{\lambda})^*}{\mathbf{c}(-\lambda)}\psi^\delta(\lambda) \qquad \text{on } V_\delta^M \tag{12}$$

and this holds as an identity between meromorphic functions on $\mathfrak{a}_c^*$. Now using the expansion (1) § 6 we see that formula (4) in the present section holds with $\tilde{f}^\delta(\lambda)$ replaced by $\psi^\delta(\lambda)$. Lemma 6.3 is now replaced by (12) and since each matrix element in $\psi^\delta(\lambda)$ satisfies growth condition (8) we can repeat the proof of Lemma 8.1 and conclude $f^\delta(a\cdot o)v = 0$ if $\log a \in \mathfrak{a}^+$, $|\log a| > R$, and $v \in V_\delta^M$. This implies $f(a\cdot o) = 0$ for these a by Lemma 4.5 and finally, since the assumptions about ψ are rotation invariant, $f(ka\cdot o) = 0$ for $k \in K$, $a \in A^+$, $|\log a| > R$ so $\operatorname{supp}(f) \subset V^R(o)$.

It remains to prove $\tilde{f} = \psi$. By the inversion formula for f we have if $\varphi = \tilde{f} - \psi$

$$\int_{\mathfrak{a}^*}\left(\int_B \varphi(\lambda, b)e^{(i\lambda+\rho)(A(x,b))}\,|\mathbf{c}(\lambda)|^{-2}db\right)d\lambda = 0\,. \tag{13}$$

But by the compatibility conditions (9) the inner integral is invariant under each transformation $\lambda \to s\lambda$ $(s \in W)$. So in the last formula we can replace $\mathfrak{a}^*$ by the Weyl chamber $s\mathfrak{a}_+^*$. We integrate the resulting relation against an arbitrary $F(x) \in \mathcal{D}(X)$. This gives (after the substitution $\lambda \to -\lambda$)

$$\int_{-s\mathfrak{a}_+^* \times B} \varphi(-\lambda, b)\tilde{F}(\lambda, b)\,|\mathbf{c}(\lambda)|^{-2}d\lambda db = 0\,.$$

According to [10(h), p. 120] the functions $\tilde{F}(\lambda, b)$ fill up a dense subset of $L^2(\mathfrak{a}_+^* \times B)$ as F runs through $\mathcal{D}(X)$. But the proof gives the same statement for the space $L^2(-s\mathfrak{a}_+^* \times B)$. The last relation therefore shows $\varphi(-\lambda, b) = 0$ for almost all $(\lambda, b) \in -s\mathfrak{a}_+^* \times B$. Since $s \in W$ is arbitrary and φ continuous we conclude $\varphi \equiv 0$ and the proof is finished.

Remark. In the case when $f(x)$ is K-invariant, $\tilde{f}(\lambda, b)$ is independent of $b \in B$ so (9) amounts to W-invariance in λ; then Theorem 8.3 reduces to a known result (cf. [5, p. 434] for $\mathbf{SL}(2, \mathbf{R})$, [10(g)], [6], [10(h)]).

Since

$$e^{\rho(\log a)}\hat{f}(ka\cdot\xi_0) = \int_{\mathfrak{a}^*}\tilde{f}(\lambda, kM)e^{i\lambda(\log a)}d\lambda \tag{14}$$

Theorem 8.3 has as a consequence a Paley-Wiener theorem for the Radon transform on X, that is a characterization of the space $\mathcal{D}(X)^\wedge$. If $D \in \mathbf{D}(A)$ (cf. definition in § 3) then the function

$$P_D(b, \lambda) = \{D_a(e^{(i\lambda+\rho)(A(a\cdot o, b))})\}_{a=e}$$

is a polynomial in λ with coefficients in $\mathcal{E}(B)$. Let $L(b, D)$ be the differential operator on A with coefficients in $\mathcal{E}(B)$ which under the Fourier transform on $\mathfrak{a}^*$ corresponds to the multiplication by $P_D(b, \lambda)$; i.e., if

$$F(a) = \int_{\mathfrak{a}^*} F^*(\lambda) e^{i\lambda(\log a)} d\lambda$$

then

$$(L(b, D)F)(a) = \int_{\mathfrak{a}^*} P_D(b, \lambda) F^*(\lambda) e^{i\lambda(\log a)} d\lambda\ .$$

THEOREM 8.4. *The Radon transform $f \to \hat{f}$ is a bijection of $\mathcal{D}(X)$ onto the set of functions $\varphi \in \mathcal{D}(\Xi)$ which satisfy the condition: For each $u \in K$ and each $D \in \mathbf{D}(A)$ the function*

$$\int_K L(ukM, D)_a(e^{\rho(\log a)}\varphi(ka\cdot\xi_0))dk \tag{15}$$

is W-invariant, as a function of $a \in A$.

Proof. Because of (14) we just have to characterize the Euclidean Fourier transforms of the functions $\psi(\lambda, b)$ which are holomorphic of uniform exponential type and satisfy (9). Writing $x = k^{-1}a\cdot o$, condition (9) is, by analyticity, equivalent to

$$\int_B P_D(b, \lambda)\psi(\lambda, kb)db\ ,$$

being W-invariant in λ for each $k \in K$. Now the theorem follows by taking the inverse Fourier transform (14).

Remark. For the case of a Euclidean space a more satisfactory Paley-Wiener theorem for the Radon transform was given in [10(e), Theorem 3.2], [10(f), Corollary 4.3]. Different proofs were later given in [16, p. 56] and [14].

In [3], Eguchi and Okamoto consider the Schwartz space $\mathcal{C}(X)$ on X and state a characterization of its Fourier transform. Using their result and repeating the proof of Theorem 8.4 one gets a characterization of $\mathcal{C}(X)^\wedge$ as the set of functions φ in the Schwartz space on $\Xi = A \times K/M$ for which (15) is W-invariant for each $u \in K$, $D \in \mathbf{D}(A)$.

Generalizing the Fourier transform in an obvious way to $T \in \mathcal{E}'(X)$ one can prove the following solvability criterion in $\mathcal{E}'(X)$.

THEOREM 8.5. *If* $D \in \mathbf{D}(G/K)$ *and* $T \in \mathcal{E}'(X)$ *the differential equation*

$$DS = T$$

has a solution $S \in \mathcal{E}'(X)$ *if and only if for each* $b \in B$ *the function* $\lambda \to \tilde{T}(\lambda, b)/\Gamma(D)(i\lambda)$ *is an entire function on* $\mathfrak{a}_c^*$.

The proof uses Theorem 8.3 above combined with [3] and the method of Malgrange for the constant coefficient case ([11], Ch. III, § 4).

9. Local solvability on Lie groups

Let L be a connected Lie group. Since the elements of $\mathbf{D}(L)$, that is, the left-invariant differential operators on L, form the analogs of the constant coefficient differential operators on $\mathbf{R}^n$ it is natural to inquire whether for each such $D \in \mathbf{D}(L)$ the equation $Du = f$ $(f \in \mathcal{E}(L))$ has a solution $u \in \mathcal{E}(L)$, at least locally. It was shown to me by Hörmander in 1964, as a consequence of his solvability condition (Theorem 6.1.1 in [11]), that local solvability for every $D \in \mathbf{D}(L)$ implies that either L is abelian or has an abelian normal subgroup of codimension 1. This was proved independently by Cerèso-Rouvière [2]. However for each bi-invariant (i.e. left and right invariant) differential operator on our semisimple group G (in §§ 4–8) we have the following local solvability result.

THEOREM 9.1. *There exists an open neighborhood* V *of* e *in* G *with the following property: For each bi-invariant differential operator* $D \neq 0$ *on* G,

$$D\mathcal{E}(V) \supset \mathcal{D}(V) .$$

Proof. Let $U_{\mathfrak{g}}$ be a connected open neighborhood of 0 in $\mathfrak{g}$ such that exp is a diffeomorphism of $U_{\mathfrak{g}}$ onto an open neighborhood U_G of e in G and such that, in addition, $U_{\mathfrak{g}}$ is completely invariant, that is, for each compact subset $C \subset U_{\mathfrak{g}}$ the closure $\mathrm{Cl}\,(\mathrm{Ad}(G)\cdot C) \subset U_{\mathfrak{g}}$. Such a neighborhood $U_{\mathfrak{g}}$ exists [9(d), p. 461]. We shall first prove that D has a fundamental solution on U_G, that is, there exists a $T \in \mathcal{D}'(U_G)$ such that $DT = \delta_e$, the delta-distribution at e. Let

$$\eta(X) = \det\left(\frac{1 - e^{-\mathrm{ad}\,X}}{\mathrm{ad}X}\right), \qquad X \in U_{\mathfrak{g}} .$$

Then $\eta^{-1} \in \mathcal{E}(U_{\mathfrak{g}})$ (cf. e.g. [10(b), p. 95]). According to Harish-Chandra [9(d), p. 476] there is a specific $\mathrm{Ad}\,(G)$-invariant differential operator p_D on $\mathfrak{g}$ with constant coefficients such that

$$(1) \qquad (Df)\circ\exp = \eta^{-1/2} p_D(\eta^{1/2} f\circ\exp) \qquad \text{on } U_{\mathfrak{g}}$$

whenever f is a C^∞ function on U_G, locally invariant under inner automorphisms. Let S be a fundamental solution of p_D on $\mathfrak{g}$, i.e., $p_D S = \delta_0$, the delta-distribution at 0. According to Rais [18], Proposition 1.4, S can be taken as invariant under $\mathrm{Ad}\,(G)$. We can then define a distribution T on U_G by

$$T(f) = S(\eta^{1/2} f \circ \exp)\ , \qquad f \in \mathfrak{D}(U_G) \tag{2}$$

and T is invariant. This relation shows that S coincides with the distribution τ_T in [9(d), p. 477] which satisfies $\tau_{DT} = p_D(\tau_T)$. We conclude that $(DT)(f) = \tau_{DT}(\eta^{1/2} f \circ \exp) = f(e)$ as desired.

Now let V be an open neighborhood of e in G such that $V = V^{-1}$ and $VV \subset U_G$. Let $f \in \mathfrak{D}(V)$. Then a solution to the differential equation $Du = f$ on V is given by

$$u(x) = \int f(xy) dT(y)\ , \qquad x \in V$$

provided $T \in \mathfrak{D}'(U_G)$ is determined by $D^* T = \delta_e$. In fact, $u \in \mathfrak{E}(V)$ and

$$(Du)(x) = \int (Df)(xy) dT(y) = \int [D(f^{L(x^{-1})})](y) dT(y) = (D^* T)(f^{L(x^{-1})})$$

so $(Du)(x) = f(x)$.

MASSACHUSETTS INSTITUTE OF TECHNOLOGY

REFERENCES

[1] N. BOURBAKI, *Espaces vectoriels topologiques*, Ch. III-V, Hermann, Paris. 1955.

[2] A. CERÈSO and F. ROUVIÈRE, Résolubilité locale d'un opérateur différentiel invariant du premier ordre. Ann. Scient. Éc. Norm. Sup. **4** (1971), 21-30.

[3] M. EGUCHI and K. OKAMOTO, The Fourier transform of the Schwartz space of a symmetric space, preprint, 1972.

[4] L. EHRENPREIS, Solutions of some problems of division I, Amer. J. Math. **76** (1954), 883-903.

[5] ——— and F. MAUTNER, Some properties of the Fourier transform on semisimple Lie groups, I. Ann. of Math. **61** (1955), 406-439.

[6] R. GANGOLLI, On the Plancherel formula and the Paley-Wiener theorem for spherical functions on semisimple Lie groups. Ann. of Math. **93** (1971), 150-165.

[7] S. G. GINDIKIN and F. I. KARPELEVIČ, Plancherel measure of Riemannian symmetric spaces of nonpositive curvature, Sov. Math. **3** (1962), 962-965.

[8] R. GODEMENT, A theory of spherical functions I, Trans. Amer. Math. Soc. **73** (1952), 496-556.

[9] HARISH-CHANDRA,

(a) Spherical functions on a semisimple Lie group I, II, Amer. J. Math. **80** (1958), 241-310, 553-613.

(b) Some results on differential equations and their applications, Proc. Nat. Acad. Sci. USA **45** (1959), 1763-1764.

(c) Differential equations and semisimple Lie groups (1960) (unpublished).

(d) Invariant eigendistributions on a semisimple Lie group, Trans. Amer. Math. Soc. **119** (1965), 457-508.

(e) Discrete series for semisimple Lie groups II, Acta Math. **116** (1966), 1-111.

f) On the theory of the Eisenstein integral, Proc. Int. Conf. on Harm. Anal. Univ. of Maryland, 1971, lecture notes in Math. No. 266, Springer-Verlag, 1972.

[10] S. HELGASON,

(a) Differential operators on homogeneous spaces, Acta Math. **102** (1959), 239–299.

(b) *Differential Geometry and Symmetric Spaces*, Academic Press, New York, 1962.

(c) Duality and Radon transform for symmetric spaces, Amer. J. Math. **85** (1963), 667–692.

(d) Fundamental solutions of invariant differential operators on symmetric spaces, Amer. J. Math. **86** (1964), 565–601.

(e) A duality in integral geometry; some generalizations of the Radon transform, Bull. Amer. Math. Soc. **70** (1964), 435–446.

(f) The Radon transform on Euclidean spaces, compact two-point homogeneous spaces and Grassman manifolds, Acta Math. **113** (1965), 153–180.

(g) An analog of the Paley-Wiener theorem for the Fourier transform on certain symmetric spaces, Math. Ann. **165** (1966), 297–308.

(h) A duality for symmetric spaces, with applications to group representations, Advan. Math. **5** (1970) 1–154.

(i) A formula for the radial part of the Laplace-Beltrami operator, J. Diff. Geom. **6** (1972), 411–419.

(j) Paley-Wiener theorems and surjectivity of invariant differential operators on symmetric spaces and Lie groups, Bull. Amer. Math. Soc. **79** (1973), 129–132.

[11] L. HÖRMANDER, *Linear partial differential operators*, Springer-Verlag, 1963.

[12] K. Johnson, and N. R. WALLACH, Composition series and interwining operator for the spherical principal series, Bull. Amer. Math. Soc. **78** (1972), 1053–1059.

[13] B. KOSTANT, On the existence and irreducibility of certain series of representations, Bull. Amer. Math. Soc. **75** (1969), 627–642.

[14] P. D. LAX and R. S. PHILLIPS, The Paley-Wiener theorem for the Radon transform, Comm. Pure Appl. Math. **23** (1970), 409–424, Correction ibid. 1971, p. 279.

[15] J. L. LIONS, Supports dans la transformation de Laplace, J. Analyse Math. **2** (1952–53), 369–380.

[16] D. LUDWIG, The Radon transform in Euclidean space, Comm. Pure Appl. Math. **19** (1966), 49–81.

[17] B. MALGRANGE, Équations aux dérivées partielles à coéfficients constants. 2. Équations avec second membre, C. R. Acad. Sci. Paris **238** (1954), 196–198.

[18] M. RAÏS, Solutions élémentaires des opérateurs différentiels bi-invariants sur un groupe de Lie nilpotent, C. R. Acad. Sci. Paris, **273** (1971), 495–498.

[19] G. SCHIFFMANN, Intégrales d'entrelacement et fonctions de Whittaker, Bull. Soc. Math. France **99** (1971), 3–72.

[20] L. Schwartz, *Théorie des distributions*, Hermann, Paris, 1966.

[21] G. Warner, *Harmonic analysis on semisimple Lie groups*, Springer-Verlag, 1972.

(Received July 10, 1972)

Reprinted from JOURNAL OF FUNCTIONAL ANALYSIS
All Rights Reserved by Academic Press, New York and London

Vol. 17, No. 3, November 1974
Printed in Belgium

Eigenspaces of the Laplacian; Integral Representations and Irreducibility*

SIGURDUR HELGASON

Massachusetts Institute of Technology, Department of Mathematics, Cambridge, Massachusetts 02139

Communicated by I. Segal

Received March 15, 1974

An integral representation is given for eigenfunctions of the Laplacian on a noncompact two-point homogeneous space. For the flat case an irreducibility criterion for the eigenspace representations is proved, complementing an earlier result [6a, p. 143] for the nonflat cases.

1. INTRODUCTION

The following result was stated in [6a, p. 139].

THEOREM 1.1. *The eigenfunctions of the Laplace–Beltrami operator on the hyperbolic plane D are precisely the functions*

$$z \to \int_B P(z, b)^\mu \, dT(b) \qquad (z \in D)$$

where P is the usual Poisson kernel, μ any complex number and T any analytic functional on the boundary B of D.

We give here in Section 6 the proof which was sketched in [6b], and which generalizes to higher dimensions. We also prove an extension of Theorem 1.1 (for μ real) to noncompact two-point homogeneous spaces along the lines of the proof in [6a, p. 139], sharpening an earlier result [6a, p. 137]. A summary appears in [6e]. Independently, various extensions of Theorem 1.1 have been given by Hashizume, Kowata, Minemura and Okamoto, (cf. [4], [5] and [12a,b,c]), the last two papers giving, respectively, the generalization of Theorem 1.1 to higher-dimensional real hyperbolic spaces, and our Theorem 7.1 for the case of a hermitian hyperbolic space.

* Dedicated to Salomon Bochner on the occasion of his 75th birthday.

The group of isometries of X has a natural representation on each eigenspace of the Laplace–Beltrami operator L_X on X. For $X \neq \mathbf{R}^n$ the irreducibility criterion for these representations was established in [6a, p. 143]; here we prove a much simpler result for $X = \mathbf{R}^n$: there is irreducibility if and only if the eigenvalue is $\neq 0$.

2. Notation

As usual $\mathbf{R}$ and $\mathbf{C}$ denote the field of real numbers and the field of complex numbers, respectively, $\mathbf{Z}$ the ring of integers, $\mathbf{R}^+$ the set of nonnegative real numbers and $\mathbf{Z}^+ = \mathbf{R}^+ \cap \mathbf{Z}$. If $\lambda \in \mathbf{C}$, $\lambda = \alpha + i\beta$ where $\alpha, \beta \in \mathbf{R}$ we put $\alpha = \mathrm{Re}(\lambda)$, $\lambda^* = \alpha - i\beta$.

If M is a C^∞ manifold, M_m denotes its tangent space at a point $m \in M$. As usual, $\mathscr{E}(M)$ (resp. $\mathscr{D}(M)$) stands for the space of complex-valued C^∞ functions (resp. the C^∞ functions of compact support) on M with the customary Schwartz topologies. Then their duals $\mathscr{D}'(M)$ (resp. $\mathscr{E}'(M)$) consist of the space of distributions (resp. distributions of compact support) on M. The composition of two maps f and g is denoted $f \circ g$.

Lie groups will be denoted by capital letters L, M, etc. and their Lie algebras by the corresponding lower case German letters, $\mathfrak{l}$, $\mathfrak{m}$, etc. The adjoint representations of $\mathfrak{l}$ and L are respectively denoted by ad and Ad. Each $\mathrm{Ad}(x)$ $(x \in L)$ extends to an automorphism of the universal enveloping algebra of $\mathfrak{l}$, denoted by the same symbol. For a coset space L/H the translation $xH \to lxH$ is denoted $\tau(l)$.

3. The Laplacian on X in Polar Coordinates (X of any Rank)

Let $X = G/K$ be a symmetric space of the noncompact type, G being a connected semisimple Lie group with finite center and K a maximal compact subgroup. Let $\mathfrak{g} = \mathfrak{k} + \mathfrak{p}$ be the corresponding Cartan decomposition of the Lie algebra $\mathfrak{g}$ of G, $\mathfrak{p}$ being the orthogonal complement to $\mathfrak{k}$ with respect to the Killing form $\langle\ ,\ \rangle$ of $\mathfrak{g}_c$, the complexification of $\mathfrak{g}$. Let $\mathfrak{g} = \mathfrak{k} + \mathfrak{a} + \mathfrak{n}$, $G = KAN$ be Iwasawa decompositions with $\mathfrak{a} \subset \mathfrak{p}$, let $\Sigma^+(\Sigma^-)$ be the corresponding sets of positive (negative) restricted roots and put $\Sigma = \Sigma^+ \cup \Sigma^-$. Let $\mathfrak{a}^+$ be the Weyl chamber in $\mathfrak{a}$ where all members of Σ^+ are positive and put $A^+ = \exp \mathfrak{a}^+$. Let m_α denote the multiplicity of a root α. Let $\mathfrak{a}^*$ be the dual space of $\mathfrak{a}$, $\mathfrak{a}_c^*$ and $\mathfrak{a}_c$ their respective complexifications and for each $\lambda \in \mathfrak{a}_c^*$ let $H_\lambda \in \mathfrak{a}_c$ be determined by $\lambda(H) = \langle H, H_\lambda \rangle$ for $H \in \mathfrak{a}$.

Let M (with Lie algebra $\mathfrak{m}$) denote the centralizer of A in K. The mapping $(kM, a) \to kaK$ is then a diffeomorphism of $(K/M) \times A^+$ onto an open dense set in G/K, the polar-coordinate decomposition of X. In this section we write out the Laplace–Beltrami operator L_X according to this decomposition (Theor. 3.1), the Riemannian structure on X being defined by $\langle \, , \, \rangle$ restricted to $\mathfrak{p}$. While this could be done by means of Harish–Chandra's derivation of the generalized radial parts of the Casimir operator relative to the Cartan decomposition $G = KAK$ [16, p. 277], we shall give a more geometric derivation.

Let o denote the origin in X and $\bar{f}$ the restriction to $A \cdot o$ of a function f on X. Then if $a \in A^+$ we have by Theorem 1.4 and Lemma 2.8, Chapter I in [6c],

$$(L_X f(a \cdot o) = (L_{Ka \cdot o} f^*)(a \cdot o) + (\Delta(L_X)\bar{f})(a \cdot o), \tag{1}$$

where $\Delta(L_X)$ is the radial part of L_X (for K acting on X), $L_{Ka \cdot o}$ is the Laplace–Beltrami operator on the orbit $Ka \cdot o$ and f^* denotes the restriction of f to this orbit. The operator $\Delta(L_X)$ is given by

$$\Delta(L_X) = L_{A \cdot o} + \sum_{\alpha \in \Sigma^+} m_\alpha (\coth \alpha) \, H_\alpha \, , \tag{2}$$

[3, p. 269]. We shall now determine the first term on the right in (1) more explicitly. Let $\mathfrak{l}$ be the orthogonal complement of $\mathfrak{m}$ in $\mathfrak{k}$; then

$$\mathfrak{l} = \sum_{\alpha \in \Sigma^+} \mathfrak{l}_\alpha \qquad \text{(direct sum)},$$

where

$$\mathfrak{l}_\alpha = \{T \in \mathfrak{k} \mid (\operatorname{ad} H)^2 T = \alpha(H)^2 T \text{ for } H \in \mathfrak{a}\}.$$

Then $\dim \mathfrak{l}_\alpha = m_\alpha$; let $T_1^\alpha, \ldots, T_{m_\alpha}^\alpha$ be a basis of $\mathfrak{l}_\alpha$, orthonormal with respect to $-\langle \, , \, \rangle$ and put $\omega_\alpha = \sum_1^{m_\alpha} T_i^\alpha \cdot T_i^\alpha$ where the T_i^α are viewed as left-invariant differential operators on G. If π denotes the natural mapping of G onto G/K and log denotes the inverse of the mapping exp: $\mathfrak{a} \to A$ we shall prove the following result.

THEOREM 3.1. *The Laplace–Beltrami operator L_X has the following polar-coordinate decomposition,*

$$(L_X f)(ka \cdot o) = \Big[\Big(L_{A\cdot o} + \sum_{\alpha\in\Sigma^+} m_\alpha \coth \alpha H_\alpha\Big) f\Big] (ka \cdot o)$$
$$+ \sum_{\alpha\in\Sigma^+} (\sinh \alpha(\log a))^{-2} [\mathrm{Ad}(a^{-1})\, \omega_\alpha(f \circ \pi)](ka),$$

for $k \in K$, $a \in A^+$ and $f \in \mathscr{E}(X)$.

For the proof, fix $a \in A^+$, put $O = a \cdot o$ and let ϕ: $k \to k \cdot O$ denote the natural mapping of K onto the orbit $K \cdot O$ (which can be identified with K/M).

LEMMA 3.2. *Let $X_i^\alpha = (\sinh \alpha(\log a))^{-1} T_i^\alpha$. Then the vectors*

$$d\phi(X_i^\alpha) \qquad 1 \leqslant i \leqslant m_\alpha, \qquad \alpha \in \Sigma^+,$$

form an orthonormal basis of the tangent space $(K \cdot O)_O$.

Proof. The differential $d\phi$ maps $\mathfrak{l}$ onto the tangent space and since for $t \in \mathbf{R}$

$$\phi(\exp tT_i^\alpha) = a \exp \mathrm{Ad}(a^{-1})\, tT_i^\alpha \cdot o,$$

we see that

$$d\phi(T_i^\alpha) = d\tau(a)(\mathrm{Ad}(a^{-1})\, T_i^\alpha)_{\mathfrak{p}},$$

where $\tau(a)$ is the translation $gk \to agK$ of G/K, $\mathfrak{p}$ is identified with $(G/K)_0$ and the subscript $\mathfrak{p}$ refers to the component for the decomposition $\mathfrak{g} = \mathfrak{k} + \mathfrak{p}$. But

$$(\mathrm{Ad}(a^{-1})\, T_i^\alpha)_{\mathfrak{p}} = (1/2)[\mathrm{Ad}(a^{-1})\, T_i^\alpha - \mathrm{Ad}(a)\, T_i^\alpha],$$

so

$$\langle d\phi(T_i^\alpha), d\phi(T_j^\beta)\rangle = (1/4)\langle T_i^\alpha - \mathrm{Ad}(a^2)\, T_i^\alpha, T_j^\beta - \mathrm{Ad}(a^2)\, T_j^\beta\rangle. \tag{3}$$

Since

$$\mathrm{Ad}(a^2)\, T_i^\alpha \equiv \cosh 2\alpha(\log a)\, T_i^\alpha \qquad (\mathrm{mod}\ \mathfrak{p}),$$

the result follows from (3).

We consider now a differential operator D on the orbit $K \cdot O$ defined by

$$(Df)(k \cdot O) = \Big\{\sum_{i,\alpha} \partial_{i,\alpha}^2 f(k \exp(\Sigma x_{i,\alpha} X_i^\alpha) \cdot O)\Big\}\, x_{i,\alpha} = 0, \tag{4}$$

where $\partial_{i,\alpha} = \partial/\partial x_{i,\alpha}$ etc. This operator is indeed well defined since for each $m \in M$, $\mathrm{Ad}(m)$ gives an orthogonal transformation of each $\mathfrak{l}_\alpha$. Moreover D is invariant under the action of K.

For the local coordinates

$$\exp(\Sigma x_{i,\alpha} X_i^\alpha) \cdot O \to (x_{i,\alpha}),$$

near O on the orbit $K \cdot O$ the tangent vectors $(\partial_{i,\alpha})_O$ are orthonormal. This follows from Lemma 3.2, since

$$(\partial_{i,\alpha})_O(f) = \{(d/dt) f(\exp tX_i^\alpha \cdot O)\}_{t=0} = d\phi(X_i^\alpha)(f).$$

It follows that the Laplacian $L' = L_{K \cdot O}$ satisfies

$$(L'f)(O) = \sum_{i,\alpha} (\partial_{i,\alpha}^2 f)(O) + \text{lower order terms},$$

so $D - L'$ has order <2 at O, hence everywhere, by the K-invariance. But K/M has no first order invariant operator ($\mathrm{Ad}_K(M)$ leaves no vector in $\mathfrak{l}$ fixed) except when K/M is one-dimensional. In any case, since D and L' both annihilate constants, we conclude $D = L'$.

Now Theorem 3.1 follows from (1) if we put in formula (4) for D, the definition of X_i^α and $O = a \cdot o$.

4. The K-Finite Eigenfunctions and the Generalized C-Function

A function ϕ on a space on which K acts is called K-finite if the transforms of ϕ under K span a finite-dimensional space $\{K \cdot \phi\}$. Let δ be a unitary irreducible representation of K on a vector space V_δ with inner product $\langle\ ,\ \rangle$. The function ϕ is called K-finite of type δ if the natural representation of K on $\{K \cdot \phi\}$ decomposes into finitely many copies of δ. If X has rank one, then $\dim V_\delta^M$ is 0 or 1 (Kostant [9]). Let $d(\delta) = \dim V_\delta$.

Proposition 4.1. *Let V_δ^M be the space of vectors in V_δ fixed under $\delta(M)$. Let $v_1, \ldots, v_{d(\delta)}$ be an orthonormal basis of V_δ such that the first m vectors form a basis of V_δ^M. Then the function*

$$kM \to \langle v_j, \delta(k) v_i\rangle \qquad 1 \leqslant j \leqslant d(\delta), \quad 1 \leqslant i \leqslant m, \tag{1}$$

form a basis of the space of K-finite functions on K/M of type δ.

In fact, the Peter–Weyl theory generalized to K/M (cf. Weil

[17, p. 81]) shows that the space of K-finite functions of type δ has dimension md(δ). But the functions (1) are of this type and are linearly independent.

Suppose now X has rank one. Then Σ^+ consists of α and possibly 2α. Let m^* be an element in the normalizer of A in K, inducing the nontrivial Weyl group element on $\mathfrak{a}$. For each $\lambda \in \mathfrak{a}_c{}^*$ let

$$\mathscr{E}_\lambda(X) = \{f \in \mathscr{E}(X) \mid L_X f = -(\langle \lambda, \lambda \rangle + \langle \rho, \rho \rangle) f\}$$

and let $\mathscr{E}_{\lambda,\delta}(X)$ be the set of K-finite functions in $\mathscr{E}_\lambda(X)$ of type δ. It is clear by considering restriction from X to $A \cdot o$ that if $\mathscr{E}_{\lambda,\delta}(X) \neq \{0\}$ then $V_\delta{}^M \neq \{0\}$ and is therefore of dimension one. Fix a unit vector $v = v_1 \in V_\delta{}^M$ and put

$$\Phi_{\lambda,\delta}(x) = \int_K e^{(i\lambda+\rho)(A(x,kM))} \langle v, \delta(k)v \rangle \, dk, \tag{2}$$

where, as in [6a], p. 10, $A(x, b) \in \mathfrak{a}$ is the log of the complex distance from o to the horocycle through x with normal b. (In terms of the Iwasawa decomposition $g = k(g) \exp H(g)\, n(g)$ of $g \in G$ we have $A(gK, kM) = -H(g^{-1}k)$). More generally, we consider the functions

$$f_{\delta,j}(x) = d(\delta)^{1/2} \int_K e^{(i\lambda+\rho)(A(x,kM))} \langle v_j\,, \delta(k)v \rangle \, dk \qquad 1 \leqslant j \leqslant d(\delta), \tag{3}$$

which by Theorem 1.4, p. 133 in [6a] and Prop. 4.1 above form a basis of $\mathscr{E}_{\lambda,\delta}(X)$ if $\operatorname{Re}\langle i\lambda, \alpha \rangle \geqslant 0$.

Lemma 4.2. *With the notation above,*

$$f_{\delta,j}(ka \cdot o) = d(\delta)^{1/2} \langle v_j\,, \delta(k)v \rangle\, \Phi_{\lambda,\delta}(a \cdot o) \qquad 1 \leqslant j \leqslant d(\delta)$$

for $k \in K$, $a \in A$.

Proof. Write λ^* for the complex conjugate conj(λ) and define $F\colon X \to V_\delta$ by

$$F(x) = \int_K e^{(-i\lambda^*+\rho)(A(x,kM))} \delta(k) v \, dk. \tag{4}$$

Then

$$f_{\delta,j}(x) = d(\delta)^{1/2} \langle v_j\,, F(x) \rangle, \tag{5}$$

and since $F(k \cdot x) = \delta(k) F(x)$, we have $F(a \cdot o) \in V_\delta{}^M$ and consequently,

$$F(a \cdot o) = \operatorname{conj}(\Phi_{\lambda,\delta}(a \cdot o)) v. \tag{6}$$

Now the lemma follows from (5) and (6).

Thus the space $\mathscr{E}_{\lambda,\delta}(X)$ is essentially determined by the single function $\Phi_{\lambda,\delta}$. We shall now use Theorem 3.1 to determine $\Phi_{\lambda,\delta}$ quite explicitly. For this we use (6).

If $T \in \mathfrak{k}$ and $a \in A$, we have

$$[(\mathrm{Ad}(a^{-1})T)(F \circ \pi)](a) = \{(d/dt)F(\exp tTa \cdot o)\}_{t=0} = \delta(T)(F(a \cdot o)),$$

where we use δ to denote the representation induced on $\mathfrak{k}$. Extending it to the universal enveloping algebra of $\mathfrak{k}$ (without change of notation) we deduce from Theorem 3.1

$$(L_X F)(a \cdot o) = (\Delta(L_X)\bar{F})(a \cdot o) + \sum_{\alpha \in \Sigma^+} (\sinh \alpha(\log a))^{-2}\, \delta(\omega_\alpha)(F(a \cdot o)) \qquad (7$$

for $a \in A^+$, $\bar{F}$ denoting the restriction of F to $A \cdot o$. Also

$$L_X F = -(\langle \lambda^*, \lambda^* \rangle + \langle \rho, \rho \rangle)F. \qquad (8)$$

But $\mathrm{Ad}(m)\omega_\alpha = \omega_\alpha$ so $\delta(\omega_\alpha)(F(a \cdot o))$ lies in V_δ^M. Let $c_\alpha \in \mathbf{C}$ be determined by

$$\delta(\omega_\alpha)(F(a \cdot o)) = c_\alpha F(a \cdot o).$$

LEMMA 4.4. *Let* $H \in \mathfrak{a}^+$ *be determined by* $\alpha(H) = 1$ *and put* $a_t = \exp tH$, $d_\alpha = 2(m_\alpha + 4m_{2\alpha})c_\alpha$, $d_{2\alpha} = 2(m_\alpha + 4m_{2\alpha})c_{2\alpha}$. *Then the function*

$$\phi(t) = \Phi_{\lambda,\delta}(a_t \cdot o)$$

satisfies the differential equation

$$(d^2\phi/dt^2) + (m_\alpha \coth t + 2m_{2\alpha} \coth 2t)(d\phi/dt)$$
$$+ [d_\alpha \sinh^{-2} t + d_{2\alpha} \sinh^{-2} 2t + \lambda(H)^2 + \rho(H)^2]\,\phi(t) = 0.$$

This lemma follows immediately from (6)–(8) if we take into account formula (2) Section 3 for the radial part and the relations

$$\langle H, H \rangle = 2(m_\alpha + 4m_{2\alpha}), \qquad \lambda(H)^2 = 2(m_\alpha + 4m_{2\alpha})\langle \lambda, \lambda \rangle,$$

(cf. [3, p. 302]) which follow readily from the definition of the Killing form.

We shall now use this lemma to determine $\Phi_{\lambda,\delta}$ explicitly in terms of the hypergeometric function. Let $l = (i\lambda - \rho)(H)$ so

$$l(l + m_\alpha + 2m_{2\alpha}) = -[\lambda(H)^2 + \rho(H)^2]. \qquad (9)$$

The constants d_α and $d_{2\alpha}$ being $\leqslant 0$, we can uniquely determine r and s by

$$r(r + m_{2\alpha} - 1) = -\tfrac{1}{4}d_{2\alpha}, \qquad r \geqslant 0 \tag{10}$$

$$s(s + m_\alpha + m_{2\alpha} - 1) = -d_\alpha - \tfrac{1}{4}d_{2\alpha}, \qquad s \geqslant 0. \tag{11}$$

Comparing the differential equation in Lemma 4.4 with [15], p. 543, we have

$$\begin{aligned}\phi(t) = c_{\lambda,\delta} \tanh^s t \cosh^l t\, F(((s + r - l)/2), ((s - r - l + 1 - m_{2\alpha})/2),\\ s + ((m_\alpha + m_{2\alpha} + 1)/2), \tanh^2 t)\end{aligned} \tag{12}$$

where $c_{\lambda,\delta}$ is constant. Here we used the fact that $\phi(t)$ is smooth near $t = 0$. On the other hand, we know from [6a], p. 130 that if $\mathrm{Re}(\langle i\lambda, \alpha\rangle) > 0$, then

$$e^{(-i\lambda+\rho)(tH)}\phi(t) \to \mathbf{c}(\lambda) \qquad \text{as} \quad t \to +\infty \tag{13}$$

where $\mathbf{c}(\lambda)$ is Harish–Chandra's $\mathbf{c}$-function. Using the expression for this function in terms of the Γ-function ([3], p. 303; see also [6a], p. 64) and the value of the hypergeometric function $F(a, b, c, x)$ for $x = 1$, we obtain

$$c_{\lambda,\delta} = \frac{\left(\begin{aligned}\Gamma(\tfrac{1}{2}[(i\lambda + \rho)(H) + s + r])\, \Gamma(\tfrac{1}{2}[(i\lambda + \rho)(H) + s - r + 1 - m_{2\alpha}])\\ \times \Gamma(\tfrac{1}{2}[m_\alpha + m_{2\alpha} + 1])\end{aligned}\right)}{\Gamma(\tfrac{1}{2}(i\lambda + \rho)(H))\, \Gamma(\tfrac{1}{2}[(i\lambda + \rho)(H) + 1 - m_{2\alpha}])\Gamma(s + \tfrac{1}{2}[m_\alpha + m_{2\alpha} + 1])}.$$

The formula holds for all λ by meromorphic continuation.

On the other hand, we have by some transformation formulas for Eisenstein integrals (Sect. 6 in [6a]),

$$\Phi_{\lambda,\delta}(x) = \mathbf{c}(-\lambda)^{-1}\, \mathbf{C}(\lambda)\, \Phi_{-\lambda,\delta}(x), \tag{15}$$

where

$$\mathbf{C}(-\lambda) = \int_{\bar{N}} \langle \delta(m^* k(\bar{n}) v, v\rangle\, e^{-(i\lambda+\rho)(H(\bar{n}))}\, d\bar{n}. \tag{16}$$

Here $\bar{N} = m^* N (m^*)^{-1}$ and $d\bar{n}$ is the Haar measure normalized by $\int_{\bar{N}} e^{-2\rho(H(\bar{n}))}\, d\bar{n} = 1$. Now (12), (14) and (15) imply, by letting $t \to 0$,

$$C(-\lambda) = \mathbf{c}(\lambda)(p_{r,s}(-\lambda)/p_{r,s}(\lambda))(q_{r,s}(-\lambda)/q_{r,s}(\lambda)), \tag{17}$$

$$p_{r,s}(\lambda) = \Gamma(\tfrac{1}{2}[(i\lambda + \rho)(H) + s + r])/\Gamma(\tfrac{1}{2}(i\lambda + \rho)(H)), \tag{18}$$

$$q_{r,s}(\lambda) = \frac{\Gamma(\tfrac{1}{2}[(i\lambda + \rho)(H) + 1 - m_{2\alpha} + s - r])}{\Gamma(\tfrac{1}{2}[(i\lambda + \rho)(H) + 1 - m_{2\alpha}])}. \tag{19}$$

Now we must relate r and s to the representation δ. Let $\hat{K}_0$ denote the set of equivalence classes of unitary irreducible representations δ of K with a fixed unit vector under $\delta(M)$. The set $\hat{K}_0$ was parametrized in Kostant [9] by a pair of integers; we shall use the related parametrization from Johnson–Wallach [8a,b] and Johnson [7]. It has the following properties.

(1) If $m_{2\alpha} = 0$, $\hat{K}_0$ corresponds to the set of pairs (p, q) with $p \in \mathbf{Z}^+$, $q = 0$.

(2) If $m_{2\alpha} = 1$, $\hat{K}_0$ corresponds to the set of pairs (p, q) with $(p, q) \in \mathbf{Z}^+ \times \mathbf{Z}$, where $p \pm q \in 2\mathbf{Z}^+$.

(3) If $m_{2\alpha} = 3$ or 7, $\hat{K}_0$ corresponds to the set of pairs $(p, q) \in \mathbf{Z}^+ \times \mathbf{Z}^+$ where $p \pm q \in 2\mathbf{Z}^+$.

We indicate how this parametrization arises. For $k \in \mathbf{Z}^+$, let $\mathscr{H}^k$ denote the space of homogeneous kth degree harmonic polynomials on $\mathfrak{p}$. Viewing K/M as the unit sphere in $\mathfrak{p}$ and identifying the elements of $\mathscr{H}^k$ with the restriction to K/M we have

$$L^2(K/M) = \sum_{k=0}^{\infty} \mathscr{H}^k \tag{20}$$

$$L^2(K/M) = \sum_{\delta \in \hat{K}_0} V_\delta\,, \tag{21}$$

each space $\mathscr{H}^k$ decomposing under K into finitely many inequivalent components. If $m_{2\alpha} = 0$, $\mathscr{H}^k$ is irreducible under K giving the parametrization (1). If $m_{2\alpha} = 1$, the complex structure on $\mathfrak{p}$ gives a decomposition

$$\mathscr{H}^p = \sum_q \mathscr{H}^{1/2(p-q),1/2(p+q)},$$

where for $r, s \in \mathbf{Z}^+$,

$$\mathscr{H}^{r,s} = \{f \in \mathscr{H}^{r+s} \mid f(\alpha Z) = \alpha^r(\alpha^*)^s f(Z)\},$$

and q runs through the integers for which $p \pm q \in 2\mathbf{Z}^+$. This gives (2). The parametrization in (3) arises in a somewhat analogous manner from the quaternions and the Cayley numbers.

If we now compare our formula (17) with Theorem 3.1 in Johnson–Wallach [8a] (using (p, q) in place of their (m, k)) we find the following relationships between (s, r) and δ.

(1) If $m_{2\alpha} = 0$, then using the duplication formula for the Γ function we find

$$(s, r) = (p, 0).$$

(2) If $m_{2\alpha} = 1$ we find since $r \geqslant 0$ that

$$\tfrac{1}{2}(s + r) = \tfrac{1}{2}(p + q), \qquad \tfrac{1}{2}(s - r) = \tfrac{1}{2}(p - q) \qquad \text{if} \quad q \geqslant 0,$$
$$\tfrac{1}{2}(s + r) = \tfrac{1}{2}(p - q), \qquad \tfrac{1}{2}(s - r) = \tfrac{1}{2}(p + q) \qquad \text{if} \quad q < 0.$$

Thus we have in both cases

$$(s, r) = (p, | q |).$$

(3) If $m_{2\alpha} = 3$ or 7 we find similarly

$$(s, r) = (p, q).$$

In all cases the normalized eigenvalues d_α and $d_{2\alpha}$ are given by

$$d_\alpha = -p(p + m_\alpha + m_{2\alpha} - 1) + q(q + m_{2\alpha} - 1)$$
$$d_{2\alpha} = -4q(q + m_{2\alpha} - 1).$$

Summarizing the results above we have proved the following.

THEOREM 4.5. *Select $H \in \mathfrak{a}$ such that $\alpha(H) = 1$ and put $a_t = \exp tH$, $l = (i\lambda - \rho)(H)$. Then if $\delta \in \hat{K}_0$ corresponds to (p, q) in the above parametrization, we have*

$$\Phi_{\lambda,\delta}(a_t \cdot o)$$
$$= c_{\lambda,\delta} \tanh^p t \cosh^l t \, F(((-l + p + q)/2), ((-l + p - q + 1 - m_{2\alpha})/2),$$
$$p + ((m_\alpha + m_{2\alpha} + 1)/2), \tanh^2 t), \tag{22}$$

where

$$c_{\lambda,\delta} = \frac{\begin{pmatrix} \Gamma(\frac{1}{2}[(i\lambda + \rho)(H) + p + q]) \, \Gamma(\frac{1}{2}[(i\lambda + \rho)(H) + p - q + 1 - m_{2\alpha}]) \\ \times \, \Gamma(\frac{1}{2}[m_\alpha + m_{2\alpha} + 1]) \end{pmatrix}}{\Gamma(\frac{1}{2}(i\lambda + \rho)(H)\, \Gamma(\frac{1}{2}[(i\lambda + \rho)(H) + 1 - m_{2\alpha}])\, \Gamma(p + \frac{1}{2}[m_\alpha + m_{2\alpha} + 1])}.$$

Note that for $m_{2\alpha} = 1$, we should have put $r = | q |$ in (12) and (14); however, the absolute value can be dropped because the hypergeometric function is symmetric in the two first arguments.

5. General Eigenfunctions

It was shown in [6a, p. 137] that each eigenfunction f of L_X could be written

$$f(x) = \int_B e^{(i\lambda+\rho)(A(x,b))}\, dT(b)$$

where $B = K/M$, $\lambda \in \mathfrak{a}_c{}^*$ and T is a "functional." In order to sharpen this we consider the space $\mathcal{A}(B)$ of analytic functions on B with its customary topology (Lions–Magenes [11]) which can be defined as follows.

For $M > 0$ put

$$|F|_M = \sup_{k\in Z^+} (1/(2k)!\, M^k)\, \|L_B{}^k F\|,$$

where $\| \ \|$ is the L^2 norm on the sphere B with respect to a normalized measure db and put

$$\mathcal{A}_M(B) = \{F \in \mathcal{E}(B) \mid |F|_M < \infty\}.$$

Then $\mathcal{A}_M$ is a Banach space, $\mathcal{A}$ is the union of the spaces $\mathcal{A}_M$ and is accordingly given the inductive limit topology. The dual space $\mathcal{A}'(B)$ consists, by definition, of the *analytic functionals* on B. According to Roumieu [13] each $T \in \mathcal{A}'(B)$ can be written

$$T(\phi) = \sum_0^\infty \int_B (L_B{}^k\phi)(b)\, d\mu_k(b) \qquad \phi \in \mathcal{A}(B) \tag{1}$$

where the μ_k are measures on B whose total variations $|\mu_k|$ satisfy

$$\sum_0^\infty M^k(2k)!\, |\mu_k| < \infty \qquad \text{for each} \quad M > O. \tag{2}$$

Conversely measures μ_k satisfying (2) give by (1) an analytic functional.

Lemma 5.1. *For each $\lambda \in \mathfrak{a}_c{}^*$ and each $T \in \mathcal{A}'(B)$ the function*

$$f(x) = \int_B e^{(i\lambda+\rho)(A(x,b))}\, dT(b) \tag{3}$$

is an eigenfunction of L_X and $f \in \mathcal{E}_\lambda(X)$.

Proof. For each $x \in X$, the function $b \to e^{(i\lambda+\rho)(A(x,b))}$ is analytic so (3) is well defined. As a consequence of (1) (cf. Roumieu [13, p. 167]) $f(x)$ is analytic and its derivatives can be computed by differentiating under the "integral sign." But then $f \in \mathscr{E}_\lambda(X)$ (cf. [6a, p. 94]).

If $n = \dim B$ the eigenvalues of L_B are well known to be $-i(i + n - 1)$, $i \in \mathbf{Z}^+$. Let S_{ij} $(1 \leqslant j \leqslant d(i))$ be a basis of the corresponding eigenspace, orthonormal with respect to the inner product on $L^2(B, db)$. To each $T \in \mathfrak{A}'(B)$ we associate an expansion into spherical harmonics

$$T \sim \sum_{0 \leqslant i} \sum_{1 \leqslant j \leqslant d(i)} \alpha_{ij} S_{ij}\,, \tag{4}$$

where

$$\alpha_{ij} = \int_B S_{ij}(b)^* \, dT(b).$$

Proposition 5.2. (i) *The series*

$$\sum_{0 \leqslant i, 1 \leqslant j \leqslant d(i)} a_{ij} S_{ij} \tag{5}$$

is the spherical harmonic expansion for a function $F \in \mathfrak{A}(B)$ if and only if

$$\sum_{0}^{\infty} r^i \sum_{j=1}^{d(i)} |\, a_{ij} \,|^2 < \infty \qquad \textit{for some} \quad r > 1. \tag{6}$$

(ii) *If (6) holds then (5) converges to F in the topology of $\mathfrak{A}(B)$.*

(iii) *A series (5) represents an analytic functional on B if and only if*

$$\sum_{0}^{\infty} r^i \sum_{j=1}^{d(i)} |\, a_{ij} \,|^2 < \infty \qquad \text{or} \quad 0 \leqslant r < 1. \tag{7}$$

Part (i) is contained in Seeley's paper [14] and originally I had used it to derive (ii) and (iii) (cf. [10] and [6a] for $n = 1$). We can omit the proof since (ii) and (iii) are proved in [5] even for compact analytic Riemannian manifolds.

Proposition 5.3. *Let $f \in \mathscr{E}_\lambda(X)$ be arbitrary. Then f has an expansion*

$$f(x) = \sum_{\delta \in \hat{K}_0} \left(\sum_{j=1}^{d(\delta)} a_{\delta,j} f_{\delta,j}(x) \right) \tag{8}$$

and

$$\int_K |f(ka\cdot o)|^2\,dk = \sum_{\delta\in \hat{K}_0} |\Phi_{\lambda,\delta}(a\cdot o)|^2 \sum_{j=1}^{d(\delta)} |a_{\delta,j}|^2 < \infty. \tag{9}$$

Proof. Formula (8) is in [6a], p. 138 and (9) is immediate from Lemma 4.2 and the Schur orthogonality relations.

6. The Eigenspace of L_X on the Real Hyperbolic Spaces

We shall now prove the converse of Lemma 5.1 for the case of a real hyperbolic space (cf. the indication in [6b] for the hyperbolic plane).

Theorem 6.1. *The eigenfunctions of L_X for the hyperbolic space X are precisely*

$$f(x) = \int_B e^{(i\lambda+\rho)(A(x,b))}\,dT(b),$$

where $\lambda \in \mathfrak{a}_c^*$ and $T \in \mathcal{O}\!\mathcal{l}'(B)$.

Proof. Let f be an eigenfunction of L_X, select $\lambda \in \mathfrak{a}_c^*$ with $\mathrm{Re}(i\lambda(H)) \geqslant 0$ such that $f \in \mathcal{E}_\lambda(X)$. We expand f according to Proposition 5.3. We shall use (9) Section 5 and the formula for $\Phi_{\lambda,\delta}$ (Theor. 4.5) to conclude that the criterion of Proposition 5.2(iii) is satisfied. In Theorem 4.5 we have $m_{2\alpha} = 0$, $q = 0$ and we are now going to use successively the transformation formulas

$$F(a, a+\tfrac{1}{2}, c, (4z/(1+z)^2)) = (1+z)^{2a} F(2a, 2a+1-c, c, z) \tag{1}$$

$$F(a, b, c, z) = (1-z)^{-b} F(c-a, b, c, (z/(z-1))) \tag{2}$$

so that in the expression for $\Phi_{\lambda,\delta}$ the "variable" p enters in F only in the third argument. Combining (1) and (2) we have

$$F\left(a, a+\frac{1}{2}, c, \frac{4z}{(1+z)^2}\right) = \left(\frac{1+z}{1-z}\right)^{2a} (1-z)^{c-1} F\left(c-2a, 2a-c+1, c, \frac{z}{z-1}\right). \tag{3}$$

To simplify the expression for $c_{\lambda,\delta}$ we use the duplication formula for Γ. We note also that

$$\tanh t = \frac{2\tanh t/2}{1+\tanh^2 t/2}, \qquad \cosh t = \frac{1+\tanh^2 t/2}{1-\tanh^2 t/2}$$

and write $r = \tanh t/2$, $x = i\lambda(H)$, $\rho(H) = (1/2)m_\alpha$ whence $l = x - (1/2)m_\alpha$. The result is

$$\Phi_{\lambda,\delta}(a_t \cdot o) = c_{x,p}(1 - r^2)^{(1/2)(m_\alpha - 1)}$$
$$\times\, r^p F\left(x + \frac{1}{2}, -x + \frac{1}{2}, p + \frac{1}{2}(m_\alpha + 1), \frac{r^2}{r^2 - 1}\right), \tag{4}$$

where

$$c_{x,p} = (\Gamma(p + x + \tfrac{1}{2}m_\alpha)/\Gamma(x + \tfrac{1}{2}m_\alpha))\,(\Gamma(\tfrac{1}{2}(m_\alpha + 1)))/\Gamma(p + \tfrac{1}{2}(m_\alpha + 1))).$$

We substitute this result into (9) Section 5. For fixed x we have $c_{x,p} = \text{const. } p^{x-(1/2)} + 0(p^{-1})$ ([2], p. 47), so putting

$$a_p = \left\{\sum_1^{d(\delta)} | a_{\delta,j} |^2\right\}^{1/2}$$

we derive

$$\sum_0^\infty r^{2p} \left| F\left(x + \frac{1}{2}, -x + \frac{1}{2}, p + \frac{1}{2}(m_\alpha + 1), \frac{r^2}{r^2 - 1}\right)\right|^2 a_p^2 < \infty \tag{5}$$

for each $0 \leqslant r < 1$. Here we put $y = r^2/(r^2 - 1)$ and expand F in a finite Taylor series

$$F(x + \tfrac{1}{2}, -x + \tfrac{1}{2}, p + \tfrac{1}{2}(m_\alpha + 1), y) = q_n(p, y) + \rho_{n+1}(p, y),$$

where $q_n(p, y)$ is an nth degree polynomial in y and the remainder term $\rho_{n+1}(p, y)$ satisfies the estimate

$$| \rho_{n+1}(p, y)| \leqslant C_n \, | y |^{n+1} \, (p + \tfrac{1}{2}(m_\alpha + 1)^{\mathrm{Re}(2x)-2-n}$$

(cf. [2], p. 76). We put

$$p_1 = p + \tfrac{1}{2}(m_\alpha + 1), \qquad n_1 = n + 2 - 2\,\mathrm{Re}(x).$$

Then (5) can be written

$$\sum_0^\infty a_p^2 r^{2p} C_n^2 \, | y |^{2n+2} p_1^{-2n_1} \, | p_1^{n_1} C_n^{-1} \, | y |^{-n-1} q_n(p, y) + \epsilon_{n+1}(p, y)|^2 < \infty \tag{6}$$

where $| \epsilon_{n+1}(p, y)| \leqslant 1$. Fix n_0 such that $n_0 + 2 - 2\,\mathrm{Re}(x) > 0$, let r_0 be arbitrary $(0 < r_0 < 1)$ and put $y_0 = r_0^2/(r_0^2 - 1)$. Then choose P_0 so large that

$$\left| \frac{p_1^{n_0+2-2\mathrm{Re}(x)}}{C_{n_0} y_0^{n_0+1}} \, q_{n_0}(p, y_0) \right| \geqslant 2 \qquad \text{for } p \geqslant P_0 .$$

This can be done since the Taylor polynomial $q_{n_0}(p, y_0) \to 1$ for $p \to \infty$. Then (6) implies

$$\sum_0^\infty a_p{}^2 r_0^{2p} C_{n_0}^2 \mid y_0 \mid^{2n_0+2} p_1^{4\mathrm{Re}(x)-2n_0-4} < \infty.$$

Since r_0 is arbitrary we obtain the desired conclusion $\sum a_p{}^2 r^{2p} < \infty$. Now Proposition 5.2 shows that there exists an analytic functional $T \in \mathfrak{A}'(B)$ such that

$$a_{\delta,j} = \int_B S_{pj}(b)^* \, dT(b) \qquad (1 \leqslant j \leqslant d(\delta)) \tag{7}$$

the representation $\delta \in \hat{K}_0$ corresponding to $(p, 0)$. For S_{pj} we can take the functions

$$S_{pj}(kM) = d(\delta)^{1/2} \langle v_j \, , \delta(k)v\rangle \qquad 1 \leqslant j \leqslant d(\delta).$$

For a fixed $x \in X$ the function $e^{(-i\lambda^*+\rho)(A(x,b))}$ has an expansion in spherical harmonics

$$e^{(-i\lambda^*+\rho)(A(x,b))} = \sum_{\delta\in\hat{K}_0} \sum_{1\leqslant j\leqslant d(\delta)} b_{\delta,j} \, d(\delta)^{1/2} \langle v_j \, , \delta(k)v\rangle, \tag{8}$$

where

$$b_{\delta,j} = \int_K e^{(-i\lambda^*+\rho)(A(x,kM))} \, d(\delta)^{1/2} \langle v_j \, , \delta(k)v\rangle^* \, dk.$$

By Proposition 5.2 (8) converges in the topology of $\mathfrak{A}(B)$ so applying T to (8) term-by-term (after complex conjugation) we get

$$\int e^{(i\lambda+\rho)(A(x,b))} \, dT = \sum_{\delta,j} a_{\delta,j} b_{\delta\ j}^* = \sum_{\delta,j} a_{\delta,j} f_{\delta,j}(x) = f(x),$$

proving Theorem 6.1.

7. The Eigenfunctions of L_X on General Two-Point Homogeneous Spaces

We shall now prove a weaker analog of Theorem 6.1 for the other symmetric spaces of rank one.

THEOREM 7.1. *Let X be a symmetric space of rank one and assume*

$m_{2\alpha} \geqslant 1$. *The eigenfunction of* L_X *with eigenvalue* $\geqslant -\langle \rho, \rho \rangle$ *are precisely*

$$f(x) = \int_B e^{(i\lambda+\rho)(A(x,b))}\, dT(b),$$

where $i\lambda \in \mathfrak{a}^*$ *and* $T \in \mathcal{Ol}'(B)$.

Proof. The restriction on the eigenvalue $-\langle \lambda, \lambda \rangle - \langle \rho, \rho \rangle$ is equivalent to $i\lambda \in \mathfrak{a}^*$, i.e. $l \in \mathbf{R}$. We write again $x = i\lambda(H)$ which we may assume $\geqslant 0$. After using the formula

$$F(a, b, c, z) = (1 - z)^{c-a-b}\, F(c - a, c - b, c, z) \tag{1}$$

(which incidentally corresponds to (15) Sect. 4) we have by Theor. 4.5,

$$\begin{aligned}
&\Phi_{\lambda,\delta}(a_t \cdot o) \\
&\quad = \frac{\Gamma(\frac{1}{2}[m_\alpha + m_{2\alpha} + 1])}{\Gamma(p + \frac{1}{2}[m_\alpha + m_{2\alpha} + 1])} \tanh^p t \cosh^{-x-\rho(H)} t \\
&\quad\quad \times \sum_{k=0}^{\infty} \frac{\Gamma(k + \frac{1}{2}[p - q + x + \frac{1}{2}m_\alpha + 1])}{\Gamma(\frac{1}{2}[x + \frac{1}{2}m_\alpha + 1])} \\
&\quad\quad \times \frac{\Gamma(k + \frac{1}{2}[p + q + x + \rho(H)])}{\Gamma(\frac{1}{2}[x + \rho(H)])} \\
&\quad\quad \times \frac{\Gamma(p + \frac{1}{2}(m_\alpha + m_{2\alpha} + 1))}{\Gamma(k + p + \frac{1}{2}(m_\alpha + m_{2\alpha} + 1))} \tanh^2 t
\end{aligned} \tag{2}$$

We now estimate this from below as q varies, in order to use (9) Section 5.

LEMMA 7.2. *If* $k \geqslant 0$, $m_{2\alpha} \geqslant 1$ *and* $p \geqslant 3$ *we have*

$$\begin{aligned}
&\Gamma(\tfrac{1}{2}[x + \tfrac{1}{2}m_\alpha + 1] + \tfrac{1}{2}(p - q) + k)\, \Gamma(\tfrac{1}{2}[x + \tfrac{1}{2}m_\alpha + m_{2\alpha}] + \tfrac{1}{2}(p + q) + k) \\
&\quad \geqslant \Gamma(\tfrac{1}{2}[x + \tfrac{1}{2}(m_\alpha + m_{2\alpha} + 1)] + \tfrac{1}{2}(p - 1) + k) \\
&\quad\quad \times \Gamma\left(\tfrac{1}{2}[x + \tfrac{1}{2}(m_\alpha + m_{2\alpha} + 1)] + \frac{p}{2} + k\right).
\end{aligned}$$

Proof. The integer q is $\geqslant 0$ except in the case $m_{2\alpha} = 1$ but for that case both factors on the left are equal so we can assume $q \geqslant 0$. We write the left hand side as

$$\begin{aligned}\Gamma(a+k)\,\Gamma(a+q+\tfrac{1}{2}(m_{2\alpha}-1)+k) \\ = \Gamma(a+k)(a+q-1+\tfrac{1}{2}(m_{2\alpha}-1)+k) \\ \times\,\Gamma(a+q-1+\tfrac{1}{2}(m_{2\alpha}-1)+k) \\ \geqslant \Gamma(a+k+1)\,\Gamma(a+q-1+\tfrac{1}{2}(m_{2\alpha}-1)+k),\end{aligned}$$

so by iteration, the product $\Gamma(a+k)\,\Gamma(a+q+\frac{1}{2}(m_{2\alpha}-1)+k)$ majorizes $\Gamma(a+\frac{1}{2}[q+\frac{1}{2}(m_{2\alpha}-1)]+k)^2$ if $q+\frac{1}{2}(m_{2\alpha}-1)$ is even, and majorizes

$$\Gamma(a+\tfrac{1}{2}[q+\tfrac{1}{2}(m_{2\alpha}-1)-1]+k)\,\Gamma(a+\tfrac{1}{2}[q+\tfrac{1}{2}(m_{2\alpha}-1)+1]+k),$$

if $q+\frac{1}{2}(m_{2\alpha}-1)$ is odd. Now putting

$$a = \tfrac{1}{2}[x+\tfrac{1}{2}m_\alpha+1]+\tfrac{1}{2}(p-q),$$

and using the fact that $\Gamma(y)$ is increasing for $y \geqslant 2$ the statement of the lemma follows.

Now using Lemma 7.1 on (2) we get for $p \geqslant 3$,

$$\begin{aligned}&\Phi_{\lambda,\delta}(a_t\cdot o) \\ &\quad\geqslant d_{p,x}\tanh^{p}t\,\cosh^{-(x+\rho(H))}t\,F(\tfrac{1}{2}[x+\tfrac{1}{2}(m_\alpha+m_{2\alpha}+1)+p-1], \\ &\quad\tfrac{1}{2}[x+\tfrac{1}{2}(m_\alpha+m_{2\alpha}+1)+p],\,p+\tfrac{1}{2}(m_\alpha+m_{2\alpha}+1),\,\tanh^2 t),\qquad (3)\end{aligned}$$

where $d_{p,x}$ is a certain expression in terms of Γ functions. Using the duplication formula we find easily

$$d_{p,x} = d_x\,\frac{\Gamma(x+\frac{1}{2}[m_\alpha+m_{2\alpha}+1]+p-1)}{2^p\Gamma(\frac{1}{2}[m_\alpha+m_{2\alpha}+1]+p)},$$

where d_x only depends on x. Now we use the quadratic transformation formula (1) Section 6 on the right hand side in the inequality above. After a routine calculation we obtain the following.

LEMMA 7.3. *For $x = i\lambda(H) \geqslant 0$, $m_{2\alpha} \geqslant 1$, $p \geqslant 3$ we have*

$$\begin{aligned}\Phi_{\lambda,\delta}(a_t\cdot o) \geqslant c(x,t)\,\frac{\Gamma(p+x-1+\frac{1}{2}[m_\alpha+m_{2\alpha}+1])}{\Gamma(p+\frac{1}{2}[m_\alpha+m_{2\alpha}+1])}\tanh^p\frac{t}{2} \\ \times F\Big(x+p-1+\frac{1}{2}[m_\alpha+m_{2\alpha}+1],\,x, \\ p+\frac{1}{2}[m_\alpha+m_{2\alpha}+1],\,\tanh^2\frac{t}{2}\Big),\end{aligned}$$

where $c(x,t)$ is independent of p.

Suppose now $f \in \mathscr{E}_\lambda(X)$. We write the expansion in Proposition 5.3 as

$$f(x) = \sum_{(p,q)} \sum_j a_{p,q,j} f_{p,q,j}(x).$$

For fixed x, the Γ fraction in Lemma 7.3 behaves for large p as $(p + \frac{1}{2}(m_\alpha + m_{2\alpha} + 1))^{x-1}$ and F is $\geqslant 1$, all its arguments being positive. Writing $r = \tanh^2 t/2$ we therefore deduce from Proposition 5.3,

$$\sum_{p=0}^{\infty} r^p \sum_q \sum_{j=1}^{d(\delta)} | a_{p,q,j} |^2 < \infty, \qquad 0 \leqslant r < 1, \tag{4}$$

where, for a fixed p, q runs over those integers for which $(p, q) = \delta \in \hat{K}_0$. Now recall the decomposition

$$\mathscr{H}^\delta = \sum_{(p,q)=\delta\in\hat{K}_0} V^\delta \qquad \text{on } \mathfrak{p} \tag{5}$$

indicated in Section 4, which, by restriction to K/M, gives a decomposition of the space of spherical harmonics of order p. Thus (4) implies that the condition in Proposition 5.2(iii) is satisfied so there exists a $T \in \mathfrak{A}'(B)$ such that

$$a_{p,q,j} = \int_{K/M} d(\delta)^{1/2} \langle v_j , \delta(k)v\rangle^* \, dT(kM), \qquad 1 \leqslant j \leqslant d(\delta).$$

Now the proof is completed just as that of Theorem 6.1.

8. Eigenspaces of the Laplacian on $\mathbf{R}^n$

Let L denote the usual Laplacian on $\mathbf{R}^n$ and for each $\lambda \in \mathbf{C}$ let $\mathscr{E}_\lambda(\mathbf{R}^n)$ denote the eigenspace

$$\mathscr{E}_\lambda(\mathbf{R}^n) = \{f \in \mathscr{E}(\mathbf{R}^n) \mid Lf = -\lambda^2 f\}.$$

with the topology induced by that of $\mathscr{E}(\mathbf{R}^n)$. Let G denote the group of all isometries of $\mathbf{R}^n$, and K the group of rotations $\mathbf{O}(n)$.

Theorem 8.1. *The natural action of G on $\mathscr{E}_\lambda(\mathbf{R}^n)$ is irreducible if and only if $\lambda \neq 0$.*

It is clear that each function

$$f(x) = \int_{\mathbf{S}^{n-1}} e^{i\lambda(x,w)} F(w) \, dw, \qquad F \in L^2(\mathbf{S}^{n-1}), \tag{1}$$

lies in $\mathscr{E}_\lambda(\mathbf{R}^n)$; here (,) denotes the usual inner product on $\mathbf{R}^n$ and dw the normalized volume element.

LEMMA 8.2. *Let $\lambda \neq 0$. Then the mapping $F \to f$ defined by* (1) *is one-to-one.*

Let $p(\zeta) = p(\zeta_1, \ldots, \zeta_n)$ be a polynomial and D the corresponding constant coefficient differential operator on $\mathbf{R}^n$ such that

$$\int_{\mathbf{R}^n} e^{i(x,\zeta)} D_x(e^{-(1/2)|x|^2}) = p(\zeta)\, e^{-(1/2)(\zeta_1^2+\cdots+\zeta_n^2)} \tag{2}$$

for $\zeta \in \mathbf{C}^n$. If $f \equiv 0$ in (1) we deduce from (2) that

$$\int_{\mathbf{S}^{n-1}} p(\lambda w_1, \ldots, \lambda w_n)\, F(w)\, dw = 0.$$

Since $\lambda \neq 0$, this implies $F \equiv 0$.

LEMMA 8.3. *Let $\lambda \neq 0$. The K-finite solutions f of the equation $Lf = -\lambda^2 f$ are precisely*

$$f(x) = \int_{\mathbf{S}^{n-1}} e^{i\lambda(x,w)} F(w)\, dw$$

where F is a K-finite function on $\mathbf{S}^{n-1}$.

Proof. Let δ be an irreducible representation of K and if Σ is any sphere in $\mathbf{R}^n$ with center at 0 let $\mathscr{E}_\delta(\Sigma)$ denote the space of K-finite functions in $\mathscr{E}(\Sigma)$ of type δ. We know from Lemma 1.5 p. 134 in [6a] that if Σ is suitably chosen each function f in $\mathscr{E}_\lambda(\mathbf{R}^n)$ is determined by its restriction $f \mid \Sigma$ to Σ. With F and f as in (1) it follows that the maps

$$F \to f \mid \Sigma, \qquad F \to f \qquad F \in L^2(\mathbf{S}^{n-1})$$

are one-to-one and commute with the action of K. For reasons of dimensionality, the first must therefore map $\mathscr{E}_\delta(\mathbf{S}^{n-1})$ onto $\mathscr{E}_\delta(\Sigma)$. The lemma now follows.

For $\lambda \neq 0$ let $\mathscr{H}_\lambda$ denote the space of functions f as defined in (1); $\mathscr{H}_\lambda$ is a Hilbert space if the norm of f is the L^2 norm of F on $\mathbf{S}^{n-1}$.

LEMMA 8.4. *Let $\lambda \neq 0$. Then the space $\mathscr{H}_\lambda$ is dense in $\mathscr{E}_\lambda(\mathbf{R}^n)$.*

Proof. Each eigenfunction of L can be expanded in a convergent series of K-finite eigenfunctions (cf. Sect. 5) so the lemma follows from Lemma 8.3.

We can now prove Theorem 8.1. We first prove that G acts irreducibly on $\mathscr{H}_\lambda$. Let $V \neq 0$ be a closed invariant subspace of $\mathscr{H}_\lambda$. Then there exists an $h \in V$ such that $h(0) = 1$. We write

$$h(x) = \int_{\mathbf{S}^{n-1}} e^{i\lambda(x,w)} H(w)\, dw,$$

and the average $h^\natural(x) = \int_K h(k \cdot x)\, dk$ is then

$$h^\natural(x) = \varphi_\lambda(x) = \int_{\mathbf{S}^{n-1}} e^{i\lambda(x,w)}\, dw.$$

If f in (1) lies in the annihilator V^0 of V the functions F and H are orthogonal on $\mathbf{S}^{n-1}$. Since V^0 is K-invariant this remains true for H replaced by its integral over K, in other words φ_λ belongs to the double annihilator $(V^0)^0 = V$. Now, since V is invariant under translations it follows that for each $t \in \mathbf{R}^n$ the function

$$x \to \int_{\mathbf{S}^{n-1}} e^{i\lambda(x,w)} e^{i\lambda(t,w)}\, dw$$

belongs to V. But then Lemma 8.2 shows that the annihilator of V in $\mathscr{H}_\lambda$ is $\{0\}$, whence the irreducibility of G on $\mathscr{H}_\lambda$.

Passing now to $\mathscr{E}_\lambda$ let $V \subset \mathscr{E}_\lambda$ be a closed invariant subspace. Then $V \cap \mathscr{H}_\lambda$ is an invariant subspace of $\mathscr{H}_\lambda$; Schwartz' inequality shows easily that it is closed. Thus, by the above, $V \cap \mathscr{H}_\lambda$ is $\{0\}$ or $\mathscr{H}_\lambda$. In the second case $V = \mathscr{E}_\lambda$ by Lemma 8.4. In the first case consider for each $f \in V$ the convergent expansion

$$f = \sum_{\delta \in \hat{K}} \alpha_\delta * f, \tag{2}$$

where $\alpha_\delta = d(\delta)\, \chi_\delta{}^*$ and

$$(\alpha_\delta * f)(x) = \int_K \alpha_\delta(k)\, f(k^{-1} \cdot x)\, dk,$$

χ_δ being the character of δ. Then $\alpha_\delta * f \in \mathscr{H}_\lambda$ by Lemma 8.3. Let $V^0 \subset \mathscr{E}'(\mathbf{R}^n)$ be the annihilator of V. Then V^0 is G-invariant and if $T \in V^0$,

$$\int_{\mathbf{R}^n} (\alpha_\delta * f)(x)\, dT(x) = \int_K \alpha_\delta(k) \int_{\mathbf{R}^n} f(x)\, dT(k \cdot x)\, dk$$

so $\alpha_\delta * f$ belongs to the double annihilator $(V^0)^0 = V$. Thus $\alpha_\delta * f \in V \cap \mathscr{H}_\lambda = \{0\}$ so, by (2), $f = 0$. Thus $V = \{0\}$ so the proof is finished.

We shall now give an integral representation similar to Theorem 1.1, of all the eigenfunctions of L on $\mathbf{R}^2$.

Given $a, b > 0$ let $E_{a,b}$ denote the space of holomorphic functions f on $\mathbf{C} - \{0\}$ satisfying

$$\|f\|_{a,b} = \sup_z |f(z)|\, e^{-a|z|-b|z|^{-1}} < \infty. \tag{3}$$

Then $E_{a,b}$ is a Banach space with norm the left hand side of (3). Also $E_{a,b} \subset E_{a',b'}$ if and only if $a \leqslant a'$, $b \leqslant b'$ and in this case $\|f\|_{a,b} \geqslant \|f\|_{a',b'}$ so the injection of $E_{a,b}$ into $E_{a',b'}$ is continuous. We can therefore give the union

$$E = \bigcup_{a,b} E_{a,b}\,,$$

the inductive limit topology. We identify the members of E with their restrictions to the unit circle $\mathbf{S}^1$ and call the members of the dual space E' *entire functionals* on $\mathbf{S}^1$.

THEOREM 8.5. *The eigenfunctions of the Laplacian on* $\mathbf{R}^2$ *are precisely the harmonic functions and the functions*

$$f(x) = \int_{\mathbf{S}^1} e^{i\lambda(x,w)}\, dT(w) \tag{4}$$

where $\lambda \in \mathbf{C} - \{0\}$ *and* T *is an entire functional on* $\mathbf{S}^1$.

We note that the right hand side of (4) is indeed defined because if $x = (x_1, x_2)$ the integrand is the restriction to $\mathbf{S}^1$ of the function

$$z \to \exp((1/2)i\lambda x_1(z + z^{-1}) + (1/2)\lambda x_2(z - z^{-1})),$$

which certainly belongs to E.

The well-known characterization of the Taylor coefficients for entire functions of exponential type (see Boas [1], p. 73) generalizes easily to E as follows.

LEMMA 8.6. *Let* f *be holomorphic in* $\mathbf{C} - \{0\}$ *and let*

$$f(z) = \sum_n \alpha_n z^n \tag{5}$$

be its Laurent expansion. Then

$$f \in E_{a+\epsilon,b+\epsilon} \quad \textit{for all} \quad \epsilon > 0 \quad \textit{if and only if}$$

$$\alpha_n = 0\left(\frac{(a+\epsilon)^n}{n!}\right) \qquad \alpha_{-n} = 0\left(\frac{(b+\epsilon)^n}{n!}\right) \qquad n \geqslant 0$$

for each $\epsilon > 0$.

Proof. It is clear by the decomposition

$$f(z) = \sum_0^\infty \alpha_n z^n + \sum_{-\infty}^{-1} \alpha_n z^n = f_+(z) + f_-(z) \tag{6}$$

that $E_{a,b} = E_{a,0} + E_{0,b}$. Now the lemma reduces to the case where f is an entire function.

Given $T \in E'$ we associate with it a Fourier series

$$T \sim \sum a_n e^{in\theta}, \tag{7}$$

where, by definition, $a_n = T(e^{-in\theta})$ $(n \in \mathbf{Z})$. Now we prove the following result which, compared to Proposition 5.2, shows the difference between analytic and entire functionals on $\mathbf{S}^1$.

PROPOSITION 8.7. *A series* (7) *is the Fourier series for an entire functional if and only if*

$$\sum_n |a_n| (r^{|n|}/|n|!) < \infty \qquad \text{for} \quad r \geqslant 0. \tag{8}$$

Proof. Let $T \in E'$ have the Fourier series (7). Then for each $a, b > 0$ and each $\epsilon > 0$, T has a continuous restriction to $E_{a+\epsilon,0}$ and $E_{0,b+\epsilon}$. The series $\sum_0^\infty a^n z^n/n!$ converges to e^{az} in the topology of $E_{a+\epsilon,0}$ so

$$T(e^{az}) = \sum_0^\infty a^n a_n/n!$$

Using a similar argument for $E_{0,b+\epsilon}$ the convergence (8) follows.

On the other hand suppose (8) holds. Then we can define a linear form T on E by

$$T(f) = \sum a_n \alpha_{-n} \qquad f \in E, \tag{9}$$

the (α_n) being as in (5). By a standard property of the inductive limit, T is continuous provided that for each $a, b > 0$ the restriction

$T \mid E_{a,b}$ is continuous. Now it is easily seen by means of the closed graph theorem that with the notation in (6), the map $f \in E_{a,b} \to f_+ \in E_{a,0}$ is continuous. Thus it suffices to prove that for each $a > 0$ the restriction $T \mid E_{a,0}$ is continuous. For $f \in E_{a,0}$ let $f(z) = \sum_0^\infty \alpha_n z^n$ denote its Taylor series and

$$F(z) = \int_0^\infty f(t)\, e^{-zt}\, dt \qquad (\mathrm{Re} > a),$$

its Fourier–Borel transform whose Laurent series is clearly given by

$$F(z) = \sum_0^\infty n!\, \alpha_n z^{-n-1}.$$

If a sequence $(f_k) \subset E_{a,0}$ converges to 0 we have

$$|f_k(z)| \leqslant A_k e^{a|z|}, \tag{10}$$

where $A_k \to 0$. The Fourier–Borel transforms $F_k(z)$ then satisfy for $x > a$

$$|F_k(x)| \leqslant A_k \int_0^\infty e^{(a-x)t} = A_k(x-a)^{-1}. \tag{11}$$

But since the right hand side of (10) is rotation-invariant and since the function $z \to f(e^{i\theta}z)$ has Fourier–Borel transform $e^{-i\theta}F(e^{-i\theta}z)$ the estimate (11) implies

$$|F_k(z)| \leqslant A_k(|z| - a)^{-1} \qquad \text{for} \quad |z| > a. \tag{12}$$

But if $\alpha_n^{(k)}$ are the Taylor coefficients of $f_k(z)$ we have

$$n!\, |\alpha_n^{(k)}| = (2\pi i)^{-1} \int F_k(z)\, z^n\, dz$$

with the integration over the circle $|z| = a + 1$, say. But then by (12)

$$n!\, |\alpha_n^{(k)}| \leqslant A_k(a+1)^{n+1}$$

so

$$|T(f_k)| \leqslant \left| \sum_0^\infty a_{-n} \alpha_n^{(k)} \right| \leqslant A_k \sum |a_{-n}|\, (a+1)^{n+1}\, (n!)^{-1}$$

so $T(f_k) \to 0$, proving the continuity.

We can now prove Theorem 8.5. Let f satisfy $Lf = -\lambda^2 f$ where $\lambda \in \mathbf{C} - \{0\}$. We expand in an absolutely convergent Fourier series

$$f(e^{i\theta}x) = \sum_n c_n(x)\, e^{in\theta} \qquad x \in \mathbf{C}, \tag{13}$$

where

$$c_n(x) = (2\pi)^{-1} \int_0^{2\pi} f(e^{i\theta}x)\, e^{-in\theta}\, d\theta.$$

By Lemma 8.3, $c_n(x) = a_n f_n(x)$ where

$$f_n(x) = \frac{1}{2\pi} \int_0^{2\pi} \exp[i\lambda(x, e^{i\theta})]\, e^{in\theta}\, d\theta.$$

Thus

$$f_n(r) = i^n J_n(\lambda r) = i^n (2\pi)^{-1} \int_0^{2\pi} e^{i\lambda r \sin\theta} e^{-in\theta}\, d\theta.$$

The Bessel function $J_n(z)$ satisfies

$$J_{-n}(z) = J_n(-z) = (-1)^n J_n(z)$$
$$\lim_{n\to+\infty} n!\, ((1/2)z)^{-n} J_n(z) = 1$$
$$n!\, |\, J_n(z)| \leqslant ((1/2)|\, z\, |)^n e^y \qquad \text{if} \quad z = x + iy,$$

so the relation

$$\sum_n |\, a_n J_n(\lambda r)| < \infty,$$

which follows from (13), implies

$$\sum_n |\, a_n\, |\, r^{|n|}/|\, n\, |! < \infty.$$

Thus, by Proposition 8.7 there exists a $T \in E'$ such that

$$T \sim \sum a_n e^{in\theta}.$$

The formula for f_n shows that

$$\exp[i\lambda(x, e^{i\theta})] = \sum_n f_n(x)\, e^{-in\theta}$$

which if $x = (x_1, x_2)$ gives the Laurent expansion

$$\exp[\tfrac{1}{2} i\lambda x_1(z + z^{-1}) + \tfrac{1}{2}\lambda x_2(z - z^{-1})] = \sum_n f_n(x)\, z^{-n}, \tag{14}$$

which converges in the topology of E. Applying T we get the desired formula

$$\int_{S^1} e^{i\lambda(x,w)}\, dT(w) = \sum f_n(x)\, a_n = f(x).$$

On the other hand if $T \in E'$ has Fourier series (7), Proposition 8.7 shows that the function (4) which equals

$$\sum a_n f_n(x),$$

is indeed differentiable and its derivatives can be obtained by termwise differentiation. This concludes the proof.

References

1. R. P. Boas, Jr., Entire Functions, Academic Press, New York, 1954.
2. A. Erdélyi *et al.*, Higher Transcendental Functions (Bateman manuscript project) Vol. **I**, McGraw-Hill Book Co., New York, 1953.
3. Harish-Chandra, Spherical functions on a semisimple Lie group I, *Amer. J. Math.* **80** (1958), 241–310.
4. M. Hashizume, A. Kowata, K. Minemura, and K. Okamoto, An integral representation of an eigenfunction of the Laplacian on the Euclidean space, *Hiroshima Math. J.* **2** (1972), 535–545.
5. M. Hashizume, K. Minemura, and K. Okamoto, Harmonic functions on hermitian hyperbolic spaces, *Hiroshima Math. J.* **3** (1973), 81–108.
6. S. Helgason, (a) A duality for symmetric spaces with applications to group representations, *Advan. Math.* **5** (1970), 1–154. (b) Harmonic analysis in the nonEuclidean disk. *Proc. Int. Conf. on Harmonic analysis.* Univ. of Maryland 1971. Lect. Notes in Math. No. 266, Springer-Verlag 1972. (c) Analysis on Lie groups and homogeneous spaces, *Conf. Board of the Math. Sci. Regional Conf. Ser. Math.*, no. **14**. *Amer. Math. Soc.*, Providence, R.I. 1972. (d) The surjectivity of invariant differential operators on symmetric spaces I, *Ann. of Math.* **98** (1973), 451–479. (e) The eigenfunctions of the Laplacian on a two-point homogeneous space; integral representations and irreducibility, *Amer. Math. Soc. Symposium*, Stanford, 1973.
7. K. D. Johnson, Composition series and intertwining operators for the spherical principal series II (to appear).
8. K. D. Johnson and N. Wallach, (a) Composition series and intertwining operators for the spherical principal series, *Bull. Amer. Math. Soc.* **78** (1972), 1053–1059. (b) Composition series and intertwining operators for the spherical principal series (preprint).
9. B. Kostant, On the existence and irreducibility of certain series of representations, *Bull. Amer. Math. Soc.* **75** (1969), 627–642.
10. G. Köthe, Die Randverteilungen analytischer Funktionen, *Math. Z.* **57** (1952), 13–33.
11. J. L. Lions and E. Magenes, Problèmes aux limites non homogènes (VII), *Ann. Math. Pura Appl.* 4, **63** (1963), 201–224.

12. K. Minemura, (a) Harmonic functions on real hyperbolic spaces, *Hiroshima Math. J.* **3** (1973), 121–151. (b) Eigenfunctions of the Laplacian on a real hyperbolic space *J. Math. Soc. Japan* (to appear). (c) Eigenfunctions of the Laplacian on a hermitian hyperbolic space (preprint).
13. C. Roumieu, Ultra-distributions définies sur $\mathbf{R}^n$ et sur certaines classes de variétés differentiables, *J. Analyse Math.* **X** (1962–63), 153–192.
14. R. Seeley, Eigenfunction expansion of analytic functions, *Proc. Amer. Math. Soc.* **21** (1969), 734–738.
15. N. J. Vilenkin, Special functions and the theory of group representations, *Amer. Math. Soc.*, Providence, R.I., 1968.
16. G. Warner, Harmonic analysis on semisimple Lie groups, Vol. **II**, Springer-Verlag, (1972.
17. A. Weil, L'intégration dans les groupes topologiques et les applications, Hermann, Paris, 1940.

Printed by the St Catherine Press Ltd., Tempelhof 37, Bruges, Belgium.

Reprinted from ADVANCES IN MATHEMATICS Vol. 22, No. 2, November 1976
All Rights Reserved by Academic Press, New York and London Printed in Belgium

A Duality for Symmetric Spaces with Applications to Group Representations, II. Differential Equations and Eigenspace Representations*

SIGURDUR HELGASON

Department of Mathematics, Massachusetts Institute of Technology, Cambridge, Massachusetts 02139

IN MEMORY OF NORMAN LEVINSON

1. INTRODUCTION

Let $X = G/K$ be a symmetric space of the noncompact type, where G is a connected semisimple Lie group with finite center and K a maximal compact subgroup. Let $\mathscr{D}(X)$ denote the space of complex-valued C^∞-functions on X of compact support. In earlier papers we have defined the Fourier transform $\tilde{f}$ of functions f on X and characterized the image $\mathscr{D}(X)^\sim$. For the subspace $\mathscr{D}_\delta(X) \subset \mathscr{D}(X)$ of functions transforming under K according to δ we prove in the present paper a sharper, topological Paley–Wiener theorem (Theorem 7.1). As a consequence, if D is an invariant differential operator on X and T a K-finite distribution on X there exists another such distribution S such that $DS = T$. Another consequence is an integral representation of the joint K-finite eigenfunctions of all D (Corollary 7.4).

We also use Theorem 7.1 to settle an instance of the following general problem about ***eigenspace representations*** defined in [8(d)]. Let L/H be a homogeneous space, H being a closed subgroup of the Lie group L, $\boldsymbol{D}(L/H)$ the algebra of all differential operators on L/H which are invariant under L. For each homomorphism $\chi: \boldsymbol{D}(L/H) \to \boldsymbol{C}$ consider the joint eigenspace

$$E_\chi = \{f \in C^\infty(L/H) \mid Df = \chi(D)f \quad \text{for} \quad D \in \boldsymbol{D}(L/H)\}.$$

* Supported in part by the National Science Foundation Grant MCS 75-21415. Some of the results (Sect. 7) were presented at the Harmonic Analysis conference in Oberwolfach, August 1975.

Let T_χ denote the natural representation of G on this eigenspace, $(T_\chi(g)f)(x) = f(g^{-1} \cdot x)$ for $g \in G$, $f \in E_\chi$, $x \in L/H$. The problem is: When is T_χ irreducible and what representations are so obtained? The problem has an obvious generalization to invariant differential operators on vector bundles considered by Bott [2]. These ***eigenspace representations*** T_χ of L which arise from homomorphisms χ: $\boldsymbol{D}(L/H) \to \boldsymbol{C}$ are counterparts to the *induced representations* of L which are defined by means of representations of H.

For the case G/K above we prove that the representation T_λ of G on the joint eigenspace $\mathscr{E}_\lambda(X)$ containing the spherical function ϕ_λ is irreducible if and only if $\mathbf{e}(\lambda)\,\mathbf{e}(-\lambda) \neq 0$, $\mathbf{e}(\lambda)^{-1}$ denoting the denominator of Harish-Chandra's $\mathbf{c}$-function (Sect. 4). This was proved in [8(e)] for X of rank 1. Since $T_\lambda = T_{s\lambda}$ for each s in the Weyl group W, the family $\{T_\lambda\}$ is not to be confused with the spherical principal series of representations τ_λ of G induced by characters χ_λ of a minimal parabolic subgroup (here τ_λ and $\tau_{s\lambda}$ need not even be equivalent). Nevertheless, our irreducibility criterion for T_λ shows *a posteriori* that when T_λ is irreducible it is infinitesimally equivalent (under the dual to the Radon transform) to τ_λ.

In addition to the Paley–Wiener theorem, our approach uses some important results about harmonic polynomials from Kostant–Rallis [15] and Kostant [14b] and the expression for the intertwining operator on K-finite functions found by Johnson and Wallach [12]. Along the way we obtain an analytic form of Kostant's irreducibility criterion for τ_λ [14(a)] which for complex G had been proved in [19]. As shown in [8(e)] this implies an irreducibility criterion for the eigenspace representations associated with the horocycle space $\Xi = G/MN$. Other examples of irreducibility criteria for eigenspace representations are G/N (G complex semisimple, N maximal nilpotent) in [8(e), p. 145], coset spaces of nilpotent groups [9], the Euclidean space $\boldsymbol{R}^n = \boldsymbol{O}(n)\,\boldsymbol{R}^n/\boldsymbol{O}(n)$ [8(h)] and compact symmetric spaces U/K[8(a), p. 454; 21].

In a similar spirit the discrete series for G is by Hotta [10], Wallach [23] (and Takahashi [22] for the de Sitter group) realized on certain eigenspaces of the Casimir operator on suitable vector bundles over G/K and by Parthasarathy [25] and Schmid [26] on null spaces of the Dirac operator on certain vector bundles over G/K; we mention also the unitary representations of the conformal group on the space of solutions of the mass O, spin s equations in flat Minkowski space [27].

In Section 4 we prove that the generalized spherical function is a derivative of the (zonal) spherical function ϕ_λ; this leads to a simplified proof of Harish-Chandra's expansion for the generalized spherical function which enters significantly in the proof of the Paley–Wiener theorem.

2. Notation and Preliminaries

We shall use the standard notation $\boldsymbol{Z}$, $\boldsymbol{R}$, $\boldsymbol{C}$ for the integers, real numbers and complex numbers, respectively, $\boldsymbol{Z}^+$ the set of nonnegative integers. For a manifold M we use the notation $\mathscr{D}(M)$ and $\mathscr{E}(M)$ for the spaces $C_c^\infty(M)$, $C^\infty(M)$ with the customary topologies. The duals $\mathscr{D}'(M)$ and $\mathscr{E}'(M)$ consist of the space of all distributions on M and of the subspace of distributions of compact support, respectively. Let τ be a diffeomorphism of M onto M and let $f \in \mathscr{E}(M)$, $T \in \mathscr{D}'(M)$, and D a differential operator on M. We put

$$f^\tau(m) = f(\tau^{-1}(m)) \qquad m \in M,$$

$$T^\tau(f) = T(f^{\tau^{-1}}) \qquad f \in \mathscr{D}(M),$$

$$D^\tau f = (Df^{\tau^{-1}})^\tau \qquad f \in \mathscr{E}(M).$$

The value of Df at $m \in M$ will usually be denoted $(Df)(m)$ but sometimes it is more convenient to write $D_m(f(m))$.

The invariant measure on a homogeneous space P will be denoted dp. If P is compact we normalize dp by $\int dp = 1$.

If A is a group and $a \in A$, $L(a)$ denotes the left translation $x \to ax$, $R(a)$ the right translation $x \to xa$; if $B \subset A$ is a subgroup the translation $xB \to axB$ of A/B is denoted $\tau(a)$. The adjoint representation of a Lie group will be denoted Ad.

We now list some standard notation from the theory of semisimple Lie groups which will be used throughout the paper. Let G be a connected semisimple Lie group with finite center, $\mathfrak{g}$ its Lie algebra and $\langle\ ,\ \rangle$ the Killing form of $\mathfrak{g}$. Let K be a maximal compact subgroup of G, $\mathfrak{k}$ its Lie algebra, θ and $\mathfrak{g} = \mathfrak{k} + \mathfrak{p}$ the associated Cartan involution and Cartan decomposition. Let $\mathfrak{a} \subset \mathfrak{p}$ be a maximal abelian subspace, $\mathfrak{a}^*$ its dual and $\mathfrak{a}_c^*$ the space of $\boldsymbol{R}$-linear maps of $\mathfrak{a}$ into $\boldsymbol{C}$. Let Σ denote the system of (restricted) roots on $\mathfrak{a}$, Σ_0 the system of indivisible roots, Σ^+ the set of $\alpha \in \Sigma$ which are positive on some fixed Weyl chamber $\mathfrak{a}^+ \subset \mathfrak{a}$ and put $\Sigma_0^+ = \Sigma^+ \cap \Sigma_0$, $\rho = \frac{1}{2} \sum_{\alpha \in \Sigma^+} m_\alpha \alpha$, m_α being the multiplicity of α. For $\alpha \in \Sigma$, let $\mathfrak{g}_\alpha$ denote the corresponding rootspace and let $\mathfrak{n} = \sum_{\alpha \in \Sigma^+} \mathfrak{g}_\alpha$. Put $N = \exp \mathfrak{n}$, $A = \exp \mathfrak{a}$, $A^+ = \exp \mathfrak{a}^+$ and let M and M', respectively, denote the centralizer and normalizer of A in K. The Weyl group W is M'/M and we put $B = K/M$. Let $g = k(g) \exp H(g)\, n(g)$ denote the factoring of an element $g \in G$ according to the Iwasawa decomposition $G = KAN$ and for $x \in X$, $b \in B$ let $A(x, b) \in \mathfrak{a}$ be defined by $A(gK, kM) = -H(g^{-1}k)$.

Let $\boldsymbol{D}(G)$ denote the algebra of left invariant differential operators on G and $\boldsymbol{D}(X) = \boldsymbol{D}(G/K)$ the algebra of G-invariant differential operators on $X = G/K$. The natural mapping of G onto G/K will be denoted by π and we often put $\tilde{f} = f \circ \pi$.

3. K-Finite Functions on X

In this section X denotes an arbitrary complete Riemannian manifold on which a compact Lie transformation group K acts. Let dk denote the Haar measure on K normalized by $\int_K = 1$. If f and ϕ are continuous functions on X and K, respectively, we write $\phi * f(x) = \int_K \phi(k) f(k^{-1} \cdot x)\, dk$.

Let δ be a unitary irreducible representation of K on a (finite-dimensional) Hilbert space V_δ. Let $d(\delta) = \dim V_\delta$, χ_δ the character of δ and $\check{\delta}$ the representation of K contragradient to δ.

We consider the space

$$\mathscr{D}(X, \operatorname{Hom}(V_\delta, V_\delta)) \tag{1}$$

of C^∞ functions on X with compact support, having values in $\operatorname{Hom}(V_\delta, V_\delta)$. Let

$$\mathscr{D}^\delta(X) = \{F \in \mathscr{D}(X, \operatorname{Hom}(V_\delta, V_\delta)) \mid F(k \cdot x) \equiv \delta(k) F(x)\}. \tag{2}$$

The space (1) carries a natural topology [21] in which it is an LF-space, in fact the strict inductive limit of the Fréchet spaces $\mathscr{D}_R(X, \operatorname{Hom}(V_\delta, V_\delta))$ $(R = 0, 1, \ldots)$ of functions in (1) with support contained in the ball of radius R around some fixed point in X. We consider also the subspace $\mathscr{D}_\delta(X) \subset \mathscr{D}(X)$ consisting of all K-finite functions of type δ. The spaces $\mathscr{D}^\delta(X)$ and $\mathscr{D}_\delta(X)$ are given the topologies induced from (1) and $\mathscr{D}(X)$, respectively.

For any continuous function f on X we put

$$f^{\delta}(x) = d(\delta) \int_K f(k^{-1} \cdot x)\, \delta(k)\, dk. \tag{3}$$

PROPOSITION 3.1. *The mapping* $Q\colon F(x) \to \mathrm{Tr}(F(x))$ *is a homeomorphism of* $\mathscr{D}^{\delta}(X)$ *onto* $\mathscr{D}_{\check{\delta}}(X)$. *The inverse is given by* $f \to f^{\delta}$.

Furthermore, the maps

$$p\colon f \in \mathscr{D}(X) \to d(\delta)\, \chi_{\delta} * f \in \mathscr{D}_{\check{\delta}}(X),$$

$$q\colon f \in \mathscr{D}(X) \to f^{\delta} \in \mathscr{D}^{\delta}(X)$$

are continuous open surjections, and the subspace $\mathscr{D}_{\check{\delta}}(X) \subset \mathscr{D}(X)$ *is closed.*

Proof. If $F \in \mathscr{D}^{\delta}(X)$ is written in matrix form along with $\delta(k)$ we find the relation

$$F_{ij}(k^{-1} \cdot x) = \sum_l \bar{\delta}_{li}(k)\, F_{lj}(x)$$

(where the bar denotes complex conjugate), which proves that each F_{ij}, and therefore $\mathrm{Tr}(F)$, belongs to $\mathscr{D}_{\check{\delta}}(X)$. Also if $f(x) = \mathrm{Tr}(F(x))$ we have from (3) and the orthogonality relations,

$$f^{\delta}(x) = d(\delta) \int_K \mathrm{Tr}(\delta(k)\, F(x))\, \delta(k^{-1})\, dk = F(x).$$

In particular, Q is one-to-one. Also

$$f(x) = \mathrm{Tr}(d(\delta) \int_K \delta(k)\, f(k^{-1} \cdot x)\, dk) = d(\delta)\, \chi_{\delta} * f(x).$$

If $g \in \mathscr{D}_{\check{\delta}}(X)$ then $g^{\delta} \in \mathscr{D}^{\delta}(X)$ and $\mathrm{Tr}(g^{\delta}) = d(\delta)\, \chi_{\delta} * g = g$ so Q is surjective. The continuity of Q and Q^{-1} being obvious, Q is a homeomorphism.

The remarks above show that p is a continuous projection of $\mathscr{D}(X)$ onto $\mathscr{D}_{\check{\delta}}(X)$. This implies that p is open. In fact, let V be a neighborhood of O in $\mathscr{D}(X)$, $U \subset V$ another such, satisfying $p(U) \subset V$. Putting $V_0 = U \cup p(U)$ we have

$$p(V_0) = \{f \in V_0 \mid pf = f\} = V_0 \cap \mathscr{D}_{\check{\delta}}(X)$$

so $p(V) \supset p(V_0)$ is a neighborhood of O in $\mathscr{D}_{\check{\delta}}(X)$. Also $p^2 = p$ implies that $\mathscr{D}_{\check{\delta}}(X)$ is closed. Since $Q \circ q = p$ the theorem follows.

LEMMA 3.2. *The spaces $\mathscr{D}^\delta(X)$ and $\mathscr{D}_\delta(X)$ are LF-spaces.*

This is a consequence of Proposition 3.1 and the following general lemma (which is necessary since a closed subspace of an LF-space is not necessarily an LF-space).

LEMMA 3.3. *Let $E = \bigcup_{i=1}^{\infty} E_i$ be a strict inductive limit of Fréchet spaces $E_1 \subset E_2 \subset \cdots$, $M \subset E$ a closed topological subspace and $M_i = M \cap E_i$. Assume $p: E \to M$ is a continuous open surjection which for each i maps E_i onto M_i. Then M is the strict inductive limit of $(M_i)_{i=1,2,\ldots}$, each M_i having the topology induced by E.*

By general theory E_i carries the relative topology of E so by completeness is closed. The restriction $p_i = p \mid E_i$ gives a continuous surjection $p_i: E_i \to M_i$. We must prove that a convex subset $V \subset M$ is a neighborhood of O if and only if for each i, $V \cap M_i$ is a neighborhood of O in M_i. The "only if" being obvious we assume $V \cap M_i$ is a neighborhood of O in M_i ($i = 1, 2,\ldots$). Then $p_i^{-1}(V \cap M_i) = W_i$ is a convex neighborhood of O in E_i and $W_i \subset W_{i+1}$ so $W = \bigcup_{i=1}^{\infty} W_i$ is convex. Also $E_i \cap W_{i+1} = W_i$ and by induction, $E_i \cap W_{i+j} = W_i$ ($j \geqslant 1$), whence $E_i \cap W = W_i$. Thus by definition of inductive limit W is a neighborhood of O in E. Finally, $p(W) = V$ so since p is open, V is a neighborhood of O in M as desired.

4. THE GENERALIZED SPHERICAL FUNCTION

Following now the notation of Section 2 let $\hat{K}_0$ denote the set of equivalence classes of irreducible representations δ of K such that $\delta(M)$ has a fixed nonzero vector. Let V_δ be a representation space of δ (with inner product $\langle\ ,\ \rangle$) and $V_\delta{}^M$ the space of fixed vectors under $\delta(M)$. Let $d(\delta) = \dim V_\delta$, $l(\delta) = \dim V_\delta{}^M$. For $\delta \in \hat{K}_0$ let $\Phi_{\lambda,\delta}$ denote the generalized spherical function

$$\Phi_{\lambda,\delta}(x) = \int_K e^{(i\lambda+\rho)(A(x,kM))}\,\delta(k)\,dk, \qquad x \in X,\ \lambda \in \mathfrak{a}_c{}^*,$$

which when δ is the unit representation reduces to the zonal spherical function φ_λ. We recall that $\lambda \in \mathfrak{a}_c{}^*$ is called *simple* if the mapping

$$F \in \mathscr{E}(B) \to f \in \mathscr{E}(X)$$

given by

$$f(x) = \int_B e^{(i\lambda+\rho)(A(x,b))} F(b)\, db \tag{1}$$

is injective, [8(e), p. 94]. We know [8(e), p. 129] that if $\mathrm{Re}(\langle i\lambda, \alpha\rangle) > 0$ for $\alpha \in \Sigma^+$ then λ is simple.

THEOREM 4.1. *Fix $v, w \in V_\delta$ and assume $-\lambda$ simple. Then there exists a right-invariant differential operator L on G such that*

$$\langle \Phi_{\lambda,\delta}(gK)v, w\rangle = (L\tilde{\varphi}_\lambda)(g).$$

Proof. First we note that [8(e), p. 116] if o denotes the origin in X

$$\varphi_\lambda(g^{-1}\cdot x) = \int_B e^{(i\lambda+\rho)(A(x,b))} e^{(-i\lambda+\rho)(A(g\cdot o,b))}\, db.$$

Since $-\lambda$ is simple the functions

$$b \to \sum_j a_j e^{(-i\lambda+\rho)(A(g_j\cdot o,b))}, \qquad a_j \in \boldsymbol{C},\ g_j \in G,$$

form a dense subspace of $L^2(B)$. With F and f as in (1) it is clear from Schwarz's inequality that the mapping $F \to f$ is continuous from $L^2(B)$ into $\mathscr{E}(X)$. Thus we conclude that the closed subspace $\mathscr{E}_{(\lambda)}(X) \subset \mathscr{E}(X)$ generated by the G-translates of φ_λ contains all the functions f in (1) with $F \in L^2(B)$. On the other hand (cf. [8(f), p. 33]), the natural representation T of G on $\mathscr{E}_{(\lambda)}(X)$ is irreducible, and if dT denotes the differential of T, representing $\mathfrak{g}$ on the space $\mathscr{E}^\infty_{(\lambda)}$ of K-finite vectors in $\mathscr{E}_{(\lambda)}$, then $dT(\boldsymbol{D}(G))\,\varphi_\lambda = \mathscr{E}^\infty_{(\lambda)}$ and $dT(\boldsymbol{D}_\delta(G))\,\varphi_\lambda = \mathscr{E}_{(\lambda),\delta}$. Here $\mathscr{E}_{(\lambda),\delta}$ denotes the space of K-finite vectors in $\mathscr{E}_{(\lambda)}$ of type δ and $\boldsymbol{D}_\delta(G)$ is the set of elements in $\boldsymbol{D}(G)$ which under $Ad_G(K)$ transform according to δ. But viewed as differential operators on G, the members of $dT(\boldsymbol{D}(G))$ are right-invariant so the proposition follows with some L from $dT(\boldsymbol{D}_\delta(G))$.

For $\alpha \in \Sigma^+$ let

$$\mathfrak{l}_\alpha = \{T \in \mathfrak{k} \mid (\mathrm{ad}\, H)^2\, T = \alpha(H)^2\, T \text{ for } H \in \mathfrak{a}\},$$

let $T_1^\alpha, \ldots, T^\alpha_{m_\alpha}$ be a basis of $\mathfrak{l}_\alpha$, orthonormal with respect to $-\langle\ ,\ \rangle$, and put

$$\omega_\alpha = \sum_1^{m_\alpha} T_i^\alpha \cdot T_i^\alpha,$$

where the T_i^α are viewed as left-invariant differential operators on G. Let $\alpha_1, ..., \alpha_l$ be the simple restricted roots, and Λ the set of integral linear combinations $n_1\alpha_1 + \cdots + n_l\alpha_l$ ($n_i \in \boldsymbol{Z}^+$). We can now state a useful expansion for the generalized spherical function $\Phi_{\lambda,\delta}$.

THEOREM 4.2. *There exist certain meromorphic functions* $\mathbf{C}_s$ ($s \in W$) *on* $\mathfrak{a}_c^*$ *and rational functions* Γ_μ ($\mu \in \Lambda$) *on* $\mathfrak{a}_c^*$ *all with values in* $\mathrm{Hom}(V_\delta^M, V_\delta^M)$ *such that for* $H \in \mathfrak{a}^+$, $v \in V_\delta^M$,

$$\Phi_{\lambda,\delta}(\exp H \cdot o)v = \sum_{s\in W, \mu\in\Lambda} e^{(is\lambda-\rho-\mu)(H)}\Gamma_\mu(s\lambda)\, \mathbf{C}_s(\lambda)v. \tag{2}$$

Here the functions Γ_μ *are given by the recursion formulas*

$$\begin{aligned}(\langle\mu, \mu\rangle - 2i\langle\mu, \lambda\rangle)\, \Gamma_\mu(\lambda) &\\ = 2 \sum_{\alpha\in\Sigma^+} m_\alpha \sum_{k\geq 1} \{\langle\alpha, \mu - 2k\alpha + \rho\rangle - i\langle\alpha, \lambda\rangle\}\, \Gamma_{\mu-2k\alpha}(\lambda) &\\ - 4 \sum_{\alpha\in\Sigma^+} \sum_{k\geq 1} k\, \delta(\omega_\alpha)\, \Gamma_{\mu-2k\alpha}(\lambda), \qquad \Gamma_0 \equiv 1,&\end{aligned}$$

k running over the set of integers ≥ 1 *for which* $\mu - 2k\alpha \in \Lambda$.

The expansion holds for all $\lambda \in \mathfrak{a}_c^*$ with the exception of a countable family of hyperplanes, more precisely on the set of regular elements in $\mathfrak{a}_c^*$, where all $\Gamma_\mu(s\lambda)$ are defined (cf. [8(g), p. 462]).

This expansion was proved by Harish-Chandra in [7(c)]; this proof is given in Warner [24, pp. 264–318, 426–449]. Since the proof is so long (it applies to a double representation δ of K) we sketch a simple proof based on Theorem 4.1.

For the case when δ is the unit representation the expansion for φ_λ was given by Harish-Chandra [7(b), pp. 263–283]; it is the expansion (2) with $\mathbf{C}_s(\lambda) = \mathbf{c}(s\lambda)$. Now let $\Delta(L)$ be the radial part of the operator L for the action of $K \times K$ on G ((k_1, k_2): $g \to k_1 g k_2^{-1}$) with A^+ as transversal submanifold [8(f), p. 10] or [7(b), p. 265]. Then by [7(b), p. 267],

$$\Delta(L) = E_0 + \sum_i g_i E_i,$$

where E_0, E_i are constant coefficient differential operators on A and each g_i has an absolutely convergent series

$$g_i(\exp H) = \sum_{\mu\in\Lambda-O} \gamma_{i\mu} e^{-\mu(H)} \qquad \gamma_{i\mu} \in \boldsymbol{R}.$$

From the expansion of φ_λ and Proposition 4.1 (remembering that L depends on λ) we deduce an expansion,

$$\Phi_{\lambda,\delta}(\exp H \cdot o)v = \sum_{s\in W, \mu\in\Lambda} e^{(is\lambda-\rho-\mu)(H)} A_{s,\mu}(\lambda)v$$

for $H \in \mathfrak{a}^+$, $v \in V_\delta{}^M$, where $A_{s,\mu}$ is a meromorphic function on $\mathfrak{a}_c{}^*$ with values in $\mathrm{Hom}(V_\delta{}^M, V_\delta{}^M)$. Now that we know that an expansion of the stated form exists, its coefficients are computed by using only the fact that $\Phi_{\lambda,\delta}$ is an eigenfunction of the Laplacian L_X of X,

$$L_X\Phi_{\lambda,\delta} = -(\langle\lambda, \lambda\rangle + \langle\rho, \rho\rangle)\, \Phi_{\lambda,\delta}\,.$$

In fact (cf. [8(h), p. 331]) if $H_\alpha \in \mathfrak{a}$ satisfies $\langle H_\alpha\,, H\rangle = \alpha(H)$

$$(L_Xf)(ka \cdot o) = \Big[\Big(L_{A\cdot o} + \sum_{\alpha\in\Sigma^+} m_\alpha(\coth \alpha)\, H_\alpha\Big) f\Big]\, (ka \cdot o)$$

$$+ \sum_{\alpha\in\Sigma^+} (\sinh \alpha\, (\log a))^{-2}\, [Ad(a^{-1})\, \omega_\alpha(f \circ \pi)](ka),$$

where $L_{A\cdot o}$ is the Laplacian on $A \cdot o$ so the restriction $\bar{\Phi}_{\lambda,\delta}$ of $\Phi_{\lambda,\delta}$ to $A^+ \cdot o$ satisfies

$$\Big\{L_{A\cdot o} + \sum_{\alpha\in\Sigma^+} m_\alpha(\coth \alpha)\, H_\alpha + \sum_{\alpha\in\Sigma^+} (\sinh \alpha)^{-2}\, \delta(\omega_\alpha) + \langle\rho, \rho\rangle + \langle\lambda, \lambda\rangle\Big\}\, \bar{\Phi}_{\lambda,\delta} = 0.$$

This implies that for each $s \in W$, $A_{s,\mu}(\lambda)$ satisfies the recursion relations stated for $\Gamma_\mu(s\lambda)$; thus

$$A_{s,\mu}(\lambda) = \Gamma_\mu(s\lambda)\, A_{s,0}(\lambda)$$

so putting $\mathbf{C}_s(\lambda) = A_{s,0}(\lambda)$ we get (2).

Finally, the functions $\mathbf{C}_s$ are given by (cf. [8(g), p. 463])

$$\mathbf{C}_{s^{-1}}(s\lambda)\, \mathbf{c}(\lambda)^{-1} = \int_{\bar{N}_s} \delta(k(\bar{n}_s)^{-1}\, m_s^{-1})\, e^{-(i\lambda+\rho)(H(\bar{n}_s))}\, d\bar{n}_s \Big/ \int_{\bar{N}_s} e^{-(i\lambda+\rho)(H(\bar{n}_s))}\, d\bar{n}_s\,,$$

where $m_s \in M'$ is a representative of s, $\bar{N} = \theta N$, $\bar{N}_s = \bar{N} \cap (m^{-1}Nm_s)$ and $d\bar{n}_s$ is any Haar measure on $\bar{N}_s$. Here $\mathbf{c}(\lambda)$ is Harish-Chandra's $\mathbf{c}$-function which, as in [8(e), p. 92] we write in the form

$$\mathbf{c}(\lambda) = c_0\mathbf{d}(\lambda)\, \mathbf{e}(\lambda),$$

where

$$\mathbf{d}(\lambda) = \sum_{\alpha \in \Sigma_0^+} \Gamma(\langle i\lambda, \alpha_0\rangle)\, 2^{-\langle i\lambda, \alpha_0\rangle}$$

$$\mathbf{e}(\lambda)^{-1} = \prod_{\alpha \in \Sigma_0^+} \Gamma(\tfrac{1}{2}(\tfrac{1}{2}m_\alpha + 1 + \langle i\lambda, \alpha_0\rangle))\, \Gamma(\tfrac{1}{2}(\tfrac{1}{2}m_\alpha + m_{2\alpha} + \langle i\lambda, \alpha_0\rangle)),$$

$\alpha_0 \alpha / \langle \alpha, \alpha\rangle$ and c_0 is a constant such that $\mathbf{c}(-i\rho) = 1$.

5. Harmonic Polynomials

Let B be a nondegenerate quadratic form on a vector space E over $\boldsymbol{R}$ and let U be a subgroup of the orthogonal group $\boldsymbol{O}(B)$. Let S denote the space of complex-valued polynomial functions on E, $I \subset S$ the set of U-invariants in S and I_+ the set of polynomials in I without constant term. Let $H \subset S$ denote the set of U-harmonic polynomials, that is, polynomials annihilated by the constant coefficient differential operators on E coming from I_+. Let N_U denote the variety in the complexification E_c defined by the joint zeros of the elements in I_+. Then identifying E_c and its dual E_c^* by means of B we have easily, $X^p \in H$ for each $X \in N_U$ and each $p \in \boldsymbol{Z}^+$. Let $H_1 \subset H$ be the span of the functions X^p ($p \in \boldsymbol{Z}^+$, $X \in N_U$) and let $H_2 \subset H$ consist of the harmonic polynomials which vanish identically on N_U. The following result is proved in [8(b)].

Proposition 5.1. *Assume that either*

(1) $U \subset \boldsymbol{O}(B)$ *is connected and semisimple*

or

(2) $U \subset \boldsymbol{O}(B)$ *is compact and* B *positive definite.*

Then

(i) $S = IH$,

(ii) $H = H_1 + H_2$ *(direct sum).*

For the case of K acting on $\mathfrak{p}$ we shall use the following result from [15].

Theorem 5.2. *If* $X \in \mathfrak{a}$ *is regular then each* $h \in H$ *is determined by its values on the orbit* $Ad(K) \cdot X$.

For $X \in \mathfrak{a}$ regular consider the imbedding $K/M = Ad(K) \cdot X \subset \mathfrak{p}$.

Then as well known [8(b), p. 250] each K-finite function on K/M is the restriction of a polynomial $p \in S$ which by Proposition 5.1(i) can be taken harmonic. Thus, by Theorem 5.2 we see that the restriction mapping $h \to h \mid Ad(K) \cdot X$ is a bijection of H onto the space $\mathscr{E}^{\infty}(K/M)$ of K-finite functions on K/M, and for each $\delta \in \hat{K}_0$ it maps the subspace $H_\delta \subset H$ onto the subspace $\mathscr{E}_\delta(K/M) \subset \mathscr{E}^{\infty}(K/M)$, the subscript δ indicating the space of K-finite elements of type δ. Let $v_1, \ldots, v_{d(\delta)}$ be a basis of V_δ such that $v_1, \ldots, v_{l(\delta)}$ span $V_\delta{}^M$. Then the functions $kM \to \langle v_j, \delta(k) v_i \rangle$ $1 \leqslant j \leqslant d(\delta)$, $1 \leqslant i \leqslant l(\delta)$ form a basis of $\mathscr{E}_\delta(K/M)$; in particular $\dim H_\delta{}^M = l(\delta)$, where $H_\delta{}^M$ is the space of M-fixed vectors in H_δ and H_δ decomposes into $l(\delta)$ K-irreducible subspaces

$$H_\delta = \sum_{j=1}^{l(\delta)} H_{\delta,j}\,, \tag{1}$$

where each $H_{\delta,j}$ can be taken so as to consist of homogeneous polynomials of degree, say $d_j(\delta)$. We shall need the following result [14(b), p. 279].

THEOREM 5.3. *If X has rank* 1

$$\dim H_\delta{}^M = 1 \qquad \textit{for each} \quad \delta \in \hat{K}_0\,.$$

Again let X have arbitrary rank. For each $\alpha \in \Sigma_0{}^+$ let $\mathfrak{g}^\alpha \subset \mathfrak{g}$ be the Lie algebra generated by the root spaces $\mathfrak{g}_\alpha$ and $\mathfrak{g}_{-\alpha}$ and $G^\alpha \subset G$ the corresponding analytic subgroup. Let $K^\alpha = G^\alpha \cap K$, $\mathfrak{p}^\alpha = \mathfrak{g}^\alpha \cap \mathfrak{p}$, $M^\alpha = G^\alpha \cap M$. Then (cf. [6]) G^α is semisimple and the symmetric space $X^\alpha = G^\alpha/K^\alpha$ of noncompact type has rank 1. Now let $\delta \in \hat{K}_0$ and let V denote the K^α-invariant subspace of V_δ generated by $V_\delta{}^M$ and let $V = \Sigma_i V_i$ be a decomposition into K^α-irreducible subspaces. If in this sum we omit those V_i which have no M-invariant vector the remaining sum would still be a K^α-invariant subspace of V_δ containing $V_\delta{}^M$; thus $V_\delta{}^M \cap V_i \neq 0$ for each i so by Theorem 5.3, $\dim V_\delta{}^M \cap V_i = 1$. Thus $V = \Sigma_i V_i$ has $l(\delta)$ summands so we write $V = \sum_{i=1}^{l(\delta)} V_i$. Let $\delta_i{}^\alpha$ denote the irreducible representation of K^α on V_i induced by δ as described above and let $D(\delta_i{}^\alpha)$ denote the degree of the K^α-harmonic polynomials in $H(\mathfrak{p}^\alpha)_{\delta_i{}^\alpha}$. Then we shall need the following result of Kostant [14(b), p. 287],

$$\sum_{i=1}^{l(\delta)} d_i(\delta) = \sum_{\alpha \in \Sigma_0{}^+} \sum_{i=1}^{l(\delta)} D(\delta_i{}^\alpha). \tag{2}$$

6. A Criterion for Simplicity

We recall that $\lambda \in \mathfrak{a}_c^*$ is not simple if and only if there exists a function $F \neq 0$ in $\mathscr{E}(B)$ such that

$$f(x) = \int_B e^{(i\lambda+\rho)(A(x,b))} F(b)\, db = 0 \qquad \text{for} \quad x \in X.$$

We shall now analyze this condition by means of the equivalent condition that all derivatives of f vanish at the point $x = 0$.

For the Lorentz group $SO(n, 1)$ the nonsimple λ are given by $\lambda = i\rho(1 + 2k(n-1)^{-1})$ $(k \in Z^+)$ (cf. Takahashi [22, p. 322]). For X of rank 1 it was proved in [8(b)] that the nonsimple λ are precisely the singularities of the denominator $\mathbf{e}(\lambda)^{-1}$ of the $\mathbf{c}$-function (cf. Sect. 4). In this section we deal with the general case.

Theorem 6.1. *An element $\lambda \in \mathfrak{a}_c^*$ is simple if and only if $\mathbf{e}(\lambda) \neq 0$.*

Let $U(\mathfrak{g})$ and $U(\mathfrak{a})$ denote the universal enveloping algebras of the complexifications of $\mathfrak{g}$ and $\mathfrak{a}$, respectively, and consider the map $u \to q^u$ of $U(\mathfrak{g})$ into $U(\mathfrak{a})$ given by

$$u - q^u \in U(\mathfrak{g})\mathfrak{k} + \mathfrak{n}U(\mathfrak{g})$$

(cf. [7(b), p. 247; 14(a), p. 631]). Let $S(\mathfrak{g})$ and $S(\mathfrak{p})$ denote the complex symmetric algebra over $\mathfrak{g}$ and $\mathfrak{p}$, respectively, and $p \to p^*$ the symmetrization map of $S(\mathfrak{g})$ onto $U(\mathfrak{g})$. Let $H(\mathfrak{p}) \subset S(\mathfrak{p})$ denote the subspace of harmonic polynomials; identifying $H(\mathfrak{p})$ with H via the Killing form, let H^*, H_δ^*, $H_{\delta,i}^*$ be the images of H, H_δ and $H_{\delta,i}$ under the symmetrization. Fix $\delta \in \hat{K}_0$ and as before let v_i $(1 \leqslant i \leqslant l(\delta))$ be a basis of V_δ^M; in accordance with (1) in Section 5 we fix a basis ϵ_j $(1 \leqslant j \leqslant l(\delta))$ of the space of linear maps $V_\delta \to H_\delta^*$ commuting with the action of K. For $\lambda \in \mathfrak{a}_c^*$ we consider the $l(\delta) \times l(\delta)$-matrix

$$Q^\delta(\lambda)_{ij} = q^{\epsilon_j(v_i)}(\rho - i\lambda) \qquad (1 \leqslant i, j \leqslant l(\delta))$$

(cf. [14(a), p. 633]). We have then the following counterpart to [14(a), Proposition 4].

Lemma 6.2. *Let $\lambda \in \mathfrak{a}_c^*$. Then $-\lambda$ is simple if and only if* $\det(Q^\delta(\lambda)) \neq 0$ *for all $\delta \in \hat{K}_0$.*

Proof. By definition $-\lambda$ is not simple if and only if there exists a $F \not\equiv 0$ in $\mathscr{E}(B)$ such that

$$\int_B e^{(-i\lambda+\rho)(A(x,b))} F(b)\, db = 0, \qquad x \in X. \tag{1}$$

Here we can of course assume $F(o) \neq 0$ and therefore also that F is invariant under M. Condition (1) is equivalent to

$$\int_K \psi_0(gk)\, F(kM)\, dk \qquad \text{for all} \quad g \in G, \tag{2}$$

where $\psi_0(g) = e^{(i\lambda-\rho)(H(g))}$. Now under the left regular representation of G on $\mathscr{E}(G)$ each $u \in U(\mathfrak{g})$ is represented by a right-invariant differential operator $\tau(u)$ such that if $X_1, \ldots, X_r \in \mathfrak{g}$ then

$$(\tau(X_1, \ldots, X_r))f(g) = \left\{ \frac{\partial^r}{\partial t_1, \ldots, \partial t_r} f(\exp(-t_r X_r) \cdots \exp(-t_1 X_1) g) \right\}_{t=0}.$$

Then, since $\psi_0(kgn) \equiv \psi_0(g)$ we have

$$(\tau(u)\, \psi_0)(e) = \tau(q^u)\, \psi_0)(e) = q^u(\rho - i\lambda). \tag{3}$$

By analyticity, (2) is equivalent to

$$\left\{ \tau(u)_g \left(\int_K \psi_0(gk)\, F(kM)\, dk \right) \right\}_{g=e} = 0, \qquad u \in U(\mathfrak{g}). \tag{4}$$

But by [8(a), p. 394], $U(\mathfrak{g}) = U(\mathfrak{g})\mathfrak{k} \oplus S(\mathfrak{p})^*$ so by Proposition 5.1(i) and standard properties of the symmetrization mapping, $U(\mathfrak{g}) = U(\mathfrak{g})\mathfrak{k} + H^* I^*$ (cf. [14(b), p. 243]). But $\tau(U(\mathfrak{g})\mathfrak{k})$ annihilates ψ_0 and if $u \in I^*$ then $(\tau(u)\, \psi_0)(k) = (\tau(u)\, \psi_0)(e) = (\tau(u)\, \phi_\lambda)(e) = c_\lambda \psi_0(\kappa)$ so that it suffices to look at (4) for $u \in H^*$. Thus by (1) in Section 5, Eq. (4) is equivalent to

$$\left\{ \tau(\epsilon_j(v_i))_g \left(\int_K \psi_0(gk)\, F(kM)\, dk \right) \right\}_{g=e} = 0 \tag{5}$$

for all $\delta \in \hat{K}_0$ and all $1 \leqslant i \leqslant d(\delta)$, $1 \leqslant j \leqslant l(\delta)$. But

$$\begin{aligned}
&[\tau(\epsilon_j(v_i))\, \psi_0^{R(k^{-1})}](e) \\
&\quad = (\tau(\epsilon_j(v_i))\, \psi_0)^{L(k^{-1})}\, (e) \\
&\quad = (\tau(\epsilon_j(v_i))^{L(k^{-1})}\, \psi_0)(e) = [\tau(Ad(k^{-1})(\epsilon_j(v_i)))\, \psi_0](e) \\
&\quad = [\tau(\epsilon_j(\delta(k^{-1})\, v_i))\, \psi_0](e) = \sum_{p=1}^{d(\delta)} \langle v_i\, , \delta(k)\, v_p \rangle (\tau(\epsilon_j(v_p))\, \psi_0)(e)
\end{aligned}$$

if the basis $v_1, \dots, v_{d(\delta)}$ is orthonormal. But the M-invariance of F implies that

$$\int_K \langle v_i , \delta(k) v_p\rangle F(kM)\, dk = 0$$

unless $1 \leqslant i, p \leqslant l(\delta)$. Taking (3) into account we see that (5) amounts to

$$\sum_{p=1}^{l(\delta)} Q^\delta(\lambda)_{pj} \int_K \langle v, \delta(k) v_p\rangle F(kM)\, dk = 0 \tag{6}$$

for all $1 \leqslant j \leqslant l(\delta)$ and all $\delta \in \hat{K}_0$, $v \in V_\delta$. But the functions $\langle v, \delta(k) v_p\rangle$ generate $\mathscr{E}(B)$ so (6) amounts to $\det Q^\delta(\lambda) = 0$ for some δ so the lemma is proved.

Consider now, as in Section 4, the operator on V_δ given by

$$B(s, \lambda) \doteq \int_{\bar{N}_s} \delta(m_s k(\bar{n}_s))\, e^{-(i\lambda+\rho)(H(\bar{n}_s))}\, d\bar{n}_s \Big/ \int_{\bar{N}_s} e^{-(i\lambda+\rho)(H(\bar{n}_s))}\, d\bar{n}_s$$

which is essentially the intertwining operator for the spherical principal series restricted to K-finite functions of type δ [20, p. 18; 8(e), p. 122]. Using this intertwining property, the simplicity of the point $\lambda = -i\rho$, and the bijection $H \to \mathscr{E}^\infty(K/M)$ (Sect. 4), Johnson and Wallach proved [12],

THEOREM 6.3. *The operator $B(\lambda, s)$ is on $V_\delta{}^M$ given by*

$$B(\lambda, s) \mid V_\delta{}^M = Q^{\check{\delta}}(-s\lambda)\, Q^{\check{\delta}}(-\lambda)^{-1}$$

provided the bases (v_i) of $V_{\check{\delta}}{}^M$ and (ϵ_j) of $\mathrm{Hom}_K(V_{\check{\delta}}, H_{\check{\delta}}{}^*)$ *have been suitably chosen.*

Let $s = s_{\alpha_1}, \dots, s_{\alpha_p}$ be a reduced expression of s, that is, s_{α_i} is a reflection in the plane $\alpha_i = 0$, where α_i is a simple root and p is as small as possible. Put $s^{(q)} = s_{\alpha_{q+1}} \cdots s_{\alpha_p}$ $(1 \leqslant q \leqslant p-1)$, $s^{(p)} = e$. Then by Schiffmann's product formula [20, p. 18],

$$B(s^1\lambda, s_{\alpha_1})\, B(s^{(2)}\lambda, s_{\alpha_2}) \cdots B(s^{(p)}\lambda, s_{\alpha_p}) = B(\lambda, s) \tag{7}$$

and for a suitable basis of $V_\delta{}^M$,

$$B(s^{(q)}\lambda, s_{\alpha_q}) \mid V_\delta{}^M$$

is a diagonal matrix with diagonal elements

$$Q_{\alpha_q}^{\delta(q,i)^\vee}(\lambda_q)\, Q_{\alpha_q}^{\delta(q,i)^\vee}(-\lambda_q)^{-1}. \tag{8}$$

(cf. [8(g), pp. 468–469]). Here the Q-polynomial is computed for the symmetric space X^{α_q}, $\delta(q, i)$ stands for the representation $\delta_i^{\alpha_q}$ (Sect. 4) and $\lambda_q = s^{(q)}\lambda \mid RH_{\alpha_q}$. But putting $\beta_q = s^{(q)^{-1}}\alpha_q$ we have by [8(e), p. 91]

$$\lambda_q(H_{\alpha_q}) = \langle\lambda, \beta_q\rangle = \lambda \mid \mathfrak{a}_{\beta_q},$$

and

$$\Sigma_0^+ \cap s^{-1}\Sigma_0^- = \{\beta_1, \ldots, \beta_p\}.$$

Now let s be the Weyl group element s^* which interchanges $\mathfrak{a}^+$ and $-\mathfrak{a}^+$. Then we get from Theorem 6.3 and (7)

$$\det Q^\delta(s^*\lambda) \det Q^\delta(\lambda)^{-1} = \prod_{\beta\in\Sigma_0^+, 1\leqslant i\leqslant l(\delta)} Q_\beta^{\delta(\beta,i)}(-\lambda_\beta)\, Q_\beta^{\delta(\beta,i)}(\lambda_\beta)^{-1}, \tag{9}$$

where $\delta(\beta, i) = \delta_i^\beta$ and $\lambda_\beta = \lambda \mid \mathfrak{a}_\beta$.

In his paper [14(b)] Kostant determined the polynomials $Q^\delta(\lambda)$ in the rank 1 case by a study of the harmonic polynomials, Johnson and Wallach [12] and Johnson [11] determined them by classification. We shall now show how they can be obtained from the differential equation for the generalized spherical function studied in [8(h)].

THEOREM 6.4. *Suppose X has rank 1 and select $H \in \mathfrak{a}$ such that $\alpha \in \Sigma_0^+$ satisfies $\alpha(H) = 1$. Then up to a constant factor,*

$$Q^\delta(\lambda) = p_{r,s}(-\lambda)\, q_{r,s}(-\lambda),$$

where $p_{r,s}$ and $q_{r,s}$ are polynomials given by

$$p_{r,s}(\lambda) = \frac{\Gamma(\frac{1}{2}[(i\lambda + \rho)(H) + s + r])}{\Gamma(\frac{1}{2}(i\lambda + \rho)(H))},$$

$$q_{r,s}(\lambda) = \frac{\Gamma(\frac{1}{2}[(i\lambda + \rho)(H) + 1 - m_{2\alpha} + s - r])}{\Gamma(\frac{1}{2}[(i\lambda + \rho)(H) + 1 - m_{2\alpha}])}.$$

Here r and s are nonnegative integers satisfying

$$\delta(\omega_\alpha) \mid V_\delta^M = \{r(r + m_{2\alpha} - 1) - s(s + m_\alpha + m_{2\alpha} - 1)\}(2(m_\alpha + 4m_{2\alpha}))^{-1}, \tag{10}$$

$$\delta(\omega_{2\alpha}) \mid V_\delta^M = -2r(r + m_{2\alpha} - 1)(m_\alpha + 4m_{2\alpha})^{-1}. \tag{11}$$

Proof. Because the left-hand sides of (10) and (11) are negative these equations determine unique real solutions $r \geqslant 0$ and $s \geqslant 0$. Using the transformation formula

$$\Phi_{s,\lambda,\delta} = \Phi_{\lambda,\delta} \mathbf{C}_{s^{-1}}(s\lambda)/\mathbf{c}(\lambda)$$

as well as the formula $B(\lambda, s) \mid V_\delta{}^M = \mathbf{C}_{s^{-1}}(s\lambda)/\mathbf{c}(\lambda)$ for the rank 1 case (cf. [8(g), pp. 464–466]) it is proved in [8(h), p. 335] that

$$Q^\delta(\lambda)/Q^\delta(-\lambda) = p_{r,s}(-\lambda)\, q_{r,s}(-\lambda)/(p_{r,s}(\lambda).q_{r,s}(\lambda)), \qquad s \geqslant r \geqslant 0. \quad (12)$$

Now we observe that $\frac{1}{2}(1 - m_{2\alpha})$ is an integer except when $m_{2\alpha} = 0$ (in which case $r = 0$). Since the right-hand side has only finitely many poles we conclude that both $\frac{1}{2}(s + r)$ and $\frac{1}{2}(s - r)$ must be integers, so $r, s \in \boldsymbol{Z}^+$, and $p_{r,s}$ and $q_{r,s}$ are polynomials. Now

$$p_{r,s}(\lambda) = 0 \quad \text{if} \quad i\lambda(H) = -\rho(H), -\rho(H) - 2, \ldots,$$

$$q_{r,s}(\lambda) = 0 \quad \text{if} \quad i\lambda(H) = -\tfrac{1}{2}(m_\alpha + 1), -\tfrac{1}{2}(m_\alpha + 1) - 2, \ldots$$

so $p_{r,s}(\lambda)\, q_{r,s}(\lambda)$ and $p_{r,s}(-\lambda)\, q_{r,s}(-\lambda)$ are relatively prime. Hence by (12)

$$Q^\delta(\lambda) = p_{r,s}(-\lambda)\, q_{r,s}(-\lambda)\, h(\lambda), \quad (13)$$

where $h(\lambda)$ is an even polynomial.

We claim now that h is a constant. If $d =$ degree $Q^\delta(\lambda)$ we can, by Section 5, use a subspace $H_{\delta,d} \subset H_d$ (the homogeneous harmonic polynomials of degree d) as a model for V_δ. By (13) we have

$$d \geqslant \tfrac{1}{2}(s + r) + \tfrac{1}{2}(s - r) = s.$$

On the other hand we clearly have the following Fourier expansion

$$e^{-(i\lambda+\rho)(H(g^{-1}))} = e^{(i\lambda+\rho)(A(gK,eM))} = \sum_{\sigma \in \hat{K}_0} d(\sigma) \langle \Phi_{\lambda,\sigma}(gK)v, v\rangle \quad (14)$$

$v \in V_\delta{}^M$ being a unit vector. Applying $u \in (H_{\delta,d})^*$ to the left-hand side and evaluating at $g = e$ we get a dth-degree polynomial in λ. Now the function $f_\sigma(g) = \langle \Phi_{\lambda,\sigma}(gK)v, v\rangle$ is K-finite of type $\check{\sigma}$ and u is K-finite of type δ. If $k \in K$ then

$$(\tau(u)\, f_\sigma)(e) = (\tau(u)\, f_\sigma)(k^{-1}) = \tau(Ad(k)u)\, f_\sigma^{L(k)}(e)$$

and integrating this over K we see that it vanishes unless $\sigma = \delta$. Also

$$\begin{aligned}\langle \Phi_{\lambda,\delta}(kaK)v, v\rangle &= \langle \Phi_{\lambda,\delta}(aK)v, \delta(k^{-1})v\rangle \\ &= \langle \Phi_{\lambda,\delta}(aK)v, v\rangle\langle \delta(k)v, v\rangle\end{aligned}$$

and by [8(h), Theorem 4.5]

$$\langle \Phi_{\lambda,\delta}(aK)v, v\rangle = p_s(\lambda)\, \rho(\log a)^s + \text{higher powers of } \rho(\log a),$$

p_s being an sth degree polynomial.

Thus we conclude from (14) that $s \geqslant d$, whence $d = s$ so $h(\lambda)$ is a constant. Q.E.D.

Since for the rank 1 case the polynomials $Q^\delta(\lambda)$ and $Q^\delta(-\lambda)$ are relatively prime we conclude from (9) that

$$\det Q^\delta(\lambda) = P(\lambda) \prod_{\beta \in \Sigma_0^+, 1 \leqslant i \leqslant l(\delta)} Q_\beta^{\delta(\beta,i)}(\lambda_\beta),$$

where $P(\lambda)$ is a polynomial. But $\det Q^\delta(\lambda)$ has degree $\sum_{i=1}^{l(\delta)} d_i(\delta)$ so by (2) in Section 5, P is a constant. Thus we have obtained the following result [14(b), p. 292].

THEOREM 6.5. *For a nonzero constant c*

$$\det Q^\delta(\lambda) = c \prod_{\beta \in \Sigma_0^+, 1 \leqslant i \leqslant l(\delta)} Q_\beta^{\delta(\beta,i)}(\lambda_\beta).$$

We can now complete the proof of Theorem 6.1. First we recall (Sect. 5) that the representations $\delta \in \hat{K}_0$ can always be realized on spaces of harmonic polynomials. Given $\alpha \in \Sigma_0^+$, $\exp[\mathfrak{p}^\alpha, \mathfrak{p}^\alpha]$ generates the subgroup $K^\alpha \subset K$ so by a direct argument $H(\mathfrak{p}^\alpha) \subset H(\mathfrak{p})$. It follows that as δ runs through $\hat{K}_0$ the representations δ_i^α ($\alpha \in \Sigma_0^+$, $1 \leqslant i \leqslant l(\delta)$) run through $(K^\alpha)_{\hat{0}}$ (this was remarked in [14(b), p. 293]). Thus we can conclude from Lemma 6.2 and Theorem 6.5 that $\lambda \in \mathfrak{a}_c^*$ is simple if and only if for each $\beta \in \Sigma_0^+$ the restriction λ_β is simple. But by [8(e), p. 129], λ_β is simple if and only if $\mathbf{e}^\beta(\lambda_\beta) \neq 0$, where $\mathbf{e}^\beta$ is the $\mathbf{e}$-function for the symmetric space G^β/K^β. Since

$$\mathbf{e}(\lambda) = \prod_{\beta \in \Sigma_0^+} \mathbf{e}^\beta(\lambda_\beta)$$

Theorem 6.1 follows.

7. A Paley–Wiener Theorem

If f is a function on X its Fourier transform $\tilde{f}$ is defined by

$$\tilde{f}(\lambda, b) = \int_X f(x)\, e^{(-i\lambda+\rho)(A(x,b))}\, dx, \qquad \lambda \in \mathfrak{a}_\alpha{}^*,\ b \in B.$$

We intend now to determine the space of Fourier transforms $\tilde{f}(\lambda, b)$ of the functions $f \in \mathscr{D}_{\check{\delta}}(X)$ (cf. Sect. 3) more explicitly than in [8(g)]. Since $f = d(\delta)\, \chi_\delta * f$ we have

$$\tilde{f}(\lambda, kM) = d(\delta) \int_X f(x) \left(\int_K e^{(-i\lambda+\rho)(A(x,vM))} \chi_\delta(kv^{-1})\, dv\right) dx$$

so we define for $f \in \mathscr{D}_{\check{\delta}}(X)$

$$\tilde{f}(\lambda) = d(\delta) \int_X f(x)\, \Phi_{\bar{\lambda},\delta}(x)^*\, dx,$$

where $\bar{\lambda}$ denotes the linear form $H \to \overline{\lambda(H)}$ on $\mathfrak{a}$ and $*$ denotes the adjoint of a linear transformation of V_δ with respect to the inner product $\langle\ ,\ \rangle$ on V_δ.

Let $\mathscr{H}(\mathfrak{a}^*, \mathrm{Hom}(V_\delta, V_\delta{}^M))$ denote the space of holomorphic functions on $\mathfrak{a}_c{}^*$ of exponential type with values in $\mathrm{Hom}(V_\delta, V_\delta{}^M)$. The space $\mathscr{H}(\mathfrak{a}^*, \mathrm{Hom}(V_\delta, V_\delta{}^M))$ is given the topology such that the Fourier transform is a homeomorphism of the space $\mathscr{D}(A, \mathrm{Hom}(V_\delta, V_\delta{}^M))$ (of C^∞ functions on A of compact support with values in $\mathrm{Hom}(V_\delta, V_\delta{}^M)$) onto $\mathscr{H}(\mathfrak{a}^*, \mathrm{Hom}(V_\delta, V_\delta{}^M))$. For this topology the multiplication by a fixed nonzero polynomial $P(\lambda)$ is a homeomorphism of $\mathscr{H}(\mathfrak{a}^*, \mathrm{Hom}(V_\delta, V_\delta{}^M))$ onto the subspace $P\mathscr{H}(\mathfrak{a}^*, \mathrm{Hom}(V_\delta, V_\delta{}^M))$ ([5, Chap. V]; another description of the topology has been shown to me by Hörmander (cf. also Dostal [4])). We denote by $\mathscr{H}^\delta(\mathfrak{a}^*)$ the topological subspace

$$\{F \in \mathscr{H}(\mathfrak{a}^*, \mathrm{Hom}(V_\delta, V_\delta{}^M) \mid (Q^{\check{\delta}})^{-1} F^{\check{\delta}}\ W\text{-invariant}\}.$$

We can now state the main result of this section.

Theorem 7.1. *The mapping $f \to \tilde{f}$, where*

$$\tilde{f}(\lambda) = d(\delta) \int_X f(x)\, \Phi_{\bar{\lambda},\delta}(x)^*\, dx \tag{1}$$

is a homeomorphism of $\mathscr{D}_{\check{\delta}}(X)$ onto $\mathscr{H}^\delta(\mathfrak{a}^*)$.

For $s \in W$, $\lambda \in \mathfrak{a}_c^*$ consider again the operator

$$\Gamma_{s,\lambda} = \int_{\bar{N}_s} \delta(k(\bar{n}_s)^{-1} m_s^{-1}) e^{-(i\lambda+\rho)(H(\bar{n}_s))} d\bar{n}_s \Big/ \int_{\bar{N}_s} e^{-(i\lambda+\rho)(H(\bar{n}_s))} d\bar{n}_s .$$

Then using the expansion in Theorem 4.2 we have [8(g), p. 464],

$$\Phi_{s\lambda,\delta}(x) = \Phi_{\lambda,\delta}(x) \Gamma_{s,\lambda} \tag{2}$$

so

$$\tilde{f}(s\lambda) = (\Gamma_{s,\bar{\lambda}})^* \tilde{f}(\lambda). \tag{3}$$

Since

$$\tilde{f}(\lambda) = d(\delta) \int_X f(x) \left(\int_K e^{(-i\lambda+\rho)(A(x,kM))} \delta(k^{-1}) dk\right) dx$$

it is obvious that $\tilde{f}(\lambda) \in \mathrm{Hom}(V_\delta, V_\delta^M)$, so by (3) and Theorem 6.3

$$Q^{\check{\delta}}(\lambda)^{-1} \tilde{f}(\lambda) \text{ is } W\text{-invariant.}$$

Also if $\hat{f}$ denotes the Radon transform of f we have if ξ_0 is the origin in the horocycle space $\Xi = G/MN$,

$$\tilde{f}(\lambda, kM) = \int_A \hat{f}(ka \cdot \xi_0) e^{(-i\lambda+\rho)(\log a)} da. \tag{4}$$

Here log is the inverse of the map $\exp: \mathfrak{a} \to A$, and da is a suitable Haar measure on A. But if f has support in a closed ball $B^R(o)$ around the origin then since the distance d from o on X satisfies $d(kan \cdot o) \geqslant d(a \cdot o)$, we have $\hat{f}(ka \cdot o) = 0$ for $d(a \cdot o) \geqslant R$. Thus $\tilde{f}(\lambda, kM)$ has, for k fixed, exponential type $\leqslant R$ so by

$$\tilde{f}(\lambda) = d(\delta) \int_K \tilde{f}(\lambda, kM) \delta(k^{-1}) dk,$$

it follows that the function $\lambda \to \tilde{f}(\lambda)$ belongs to $\mathscr{H}^\delta(\mathfrak{a}^*)$.

For the surjectivity let $\psi \in \mathscr{H}^\delta(\mathfrak{a}^*)$. Then the function

$$\Psi(\lambda, kM) = \mathrm{Tr}(\delta(k) \psi(\lambda))$$

on $\mathfrak{a}_c^* \times B$ is a holomorphic function of uniform exponential type ([8(g), p. 473]. Let us check that it also satisfies the compatibility relations

$$\int_K \Psi(\lambda, kM) e^{(i\lambda+\rho)(A(x,kM))} dk = \int_K \Psi(s\lambda, kM) e^{(is\lambda+\rho)(A(x,kM))} dk. \tag{5}$$

For this it suffices to see that

$$\Phi_{s\lambda,\delta}(x)\,\psi(s\lambda) = \Phi_{\lambda,\delta}(x)\,\psi(\lambda). \tag{6}$$

But $\Gamma_{s,\bar{\lambda}}$ and $(\Gamma_{s,\bar{\lambda}})^*$ both leave $V_\delta{}^M$ invariant so

$$(\Gamma_{s,\bar{\lambda}})^* \mid V_\delta{}^M = (\Gamma_{s,\bar{\lambda}} \mid V_\delta{}^M)^*$$

and by [8(g), Theorem 6.6]

$$(\Gamma_{s,\lambda} \mid V_\delta{}^M)(\Gamma_{s,\bar{\lambda}} \mid V_\delta{}^M)^* = I \tag{7}$$

so if $\lambda \in \mathfrak{a}^*$, the restriction of $\Gamma_{s,\lambda}$ to $V_\delta{}^M$ is unitary. But by Theorem 6.3, (10), and the fact that $\psi(\lambda)$ has range in $V_\delta{}^M$, we deduce

$$\Gamma_{s,\lambda}\psi(s\lambda) = \psi(\lambda),$$

which by (2) implies (6). Thus by [8(g), Theorem 3.8] there exists a unique $F \in \mathscr{D}(X)$ such that $\tilde{F}(\lambda, kM) \equiv \Psi(\lambda, kM)$. Then the function $f(x) = d(\delta)\,\chi_\delta * F$ belongs to $\mathscr{D}_{\check{\delta}}(X)$ and

$$\begin{aligned}\tilde{f}(\lambda) &= d(\delta)^2 \int_K \int_K \tilde{F}(\lambda, u^{-1}kM)\,\chi_\delta(u)\,\delta(k^{-1})\,dk\,du \\ &= d(\delta)^2 \int_K \int_K \operatorname{Tr}(\delta(k)\,\psi(\lambda))\,\chi_\delta(u)\,\delta(k^{-1}u^{-1})\,dk\,du \\ &= \psi(\lambda)\end{aligned}$$

as a result of the orthogonality relations. Also $\tilde{f}(\lambda, kM) = \operatorname{Tr}(\delta(k)\,\psi(\lambda))$ so the mapping $f(x) \to \tilde{f}(x)$ is one-to-one.

It remains to prove the topological statement in the theorem. First we remark that as a result of Theorem 6.4, det $Q^\delta(\lambda)$ and det $Q^\delta(s^*\lambda)$ are relatively prime. From this we shall deduce that each element in $\mathscr{H}^\delta(\mathfrak{a}^*)$ is divisible by $Q^{\check{\delta}}(\lambda)$. More precisely, if $\mathscr{I}^\delta(\mathfrak{a}^*)$ denotes the set of Weyl group invariants in $\mathscr{H}(\mathfrak{a}^*, \operatorname{Hom}(V_\delta, V_\delta{}^M))$ we have the following result.

Lemma 7.2. *The mapping*

$$\psi(\lambda) \to Q^{\check{\delta}}(\lambda)\,\psi(\lambda)$$

maps $\mathscr{I}^\delta(\mathfrak{a}^*)$ *onto* $\mathscr{H}^\delta(\mathfrak{a}^*)$.

For this let $F \in \mathscr{H}^\delta(\mathfrak{a}^*)$. Then

$$Q^{\check{\delta}}(s^*\lambda)^{-1}\,F(s^*\lambda) = Q^{\check{\delta}}(\lambda)^{-1}\,F(\lambda). \tag{8}$$

Let $(Q_c(\lambda)_{ij})$ denote the matrix formed by the cofactors in $Q^{\check{\delta}}(\lambda)_{ji}$. Then $Q_c(\lambda)$ is another $l(\delta) \times l(\delta)$-matrix with polynomial coefficients and it satisfies

$$Q_c(\lambda)\, Q^{\check{\delta}}(\lambda) = Q^{\check{\delta}}(\lambda)\, Q_c(\lambda) = \det Q^{\check{\delta}}(\lambda) I,$$

where I is the unit matrix. Then by (8)

$$\det Q^{\check{\delta}}(\lambda)\, Q_c(s^*\lambda)\, F(s^*\lambda) = \det Q^{\check{\delta}}(s^*\lambda)\, Q_c(\lambda)\, F(\lambda). \tag{9}$$

We know from Theorems 6.3 and 6.4 that $\det Q^{\check{\delta}}(\lambda)$ decomposes into linear factors none of which occurs in the similar decomposition of $\det Q^{\check{\delta}}(s^*\lambda)$. Thus $Q_c(\lambda)\, F(\lambda)$ vanishes on the zero set for each of these linear factors and, by the power series expansion, must be divisible by it. Thus $\det Q^{\check{\delta}}(\lambda)^{-1}\, Q_c(\lambda)\, F(\lambda)$ is holomorphic on $\mathfrak{a}_c{}^*$ and therefore, by a result of Malgrange [17, p. 306], of exponential type. Consequently,

$$F(\lambda) = Q^{\check{\delta}}(\lambda)\, \psi(\lambda),$$

where $\psi(\lambda)$ is entire of exponential type and since $F \in \mathscr{H}^{\delta}(\mathfrak{a}^*)$ we have $\psi \in \mathscr{I}^{\delta}(\mathfrak{a}^*)$.

Corollary 7.3. *The space* $\mathscr{H}^{\delta}(\mathfrak{a}^*)$ *with the topology induced by* $\mathscr{H}(\mathfrak{a}^*, \operatorname{Hom}(V_\delta, V_\delta{}^M))$ *is an LF-space.*

Applying Lemma 3.3 to the map $F(\lambda) \to \sum_{s \in W} F(s\lambda)$ we see that $\mathscr{I}^{\delta}(\mathfrak{a}^*)$ is an LF-space. The mapping $\psi(\lambda) \to Q^{\check{\delta}}(\lambda)\, \psi(\lambda)$ of $\mathscr{I}^{\delta}(\mathfrak{a}^*)$ onto $\mathscr{H}^{\delta}(\mathfrak{a}^*)$ is obviously continuous. Its inverse can be composed of two maps

$$F(\lambda) \to (\det Q^{\check{\delta}}(\lambda))^{-1} F(\lambda) \to Q_c(\lambda) \det Q^{\check{\delta}}(\lambda)^{-1} F(\lambda).$$

both of which are continuous. Thus $\mathscr{H}^{\delta}(\mathfrak{a}^*)$ is homeomorphic to $\mathscr{I}^{\delta}(\mathfrak{a}^*)$.

Thus $f \to \tilde{f}$ is a continuous map of the LF-space $\mathscr{D}_{\check{\delta}}(X)$ onto the LF-space $\mathscr{H}^{\delta}(\mathfrak{a}^*)$, hence a homeomorphism (cf. [3]). Thus Theorem 7.1 is completely proved.

Corollary 7.4. *The K-finite eigenfunctions of all the operators* $D \in \boldsymbol{D}(X)$ *are precisely*

$$f(x) = \int_B e^{(i\lambda+\rho)(A(x,b))}\, F(b)\, db, \tag{10}$$

where $\lambda \in \mathfrak{a}_c{}^*$ *and F is a K-finite function on B.*

Let $\lambda \in \mathfrak{a}_c^*$ and as usual let $\Gamma(D)$ be the W-invariant polynomial on $\mathfrak{a}_c^*$ such that

$$D\phi_\lambda = \Gamma(D)(i\lambda)\,\phi_\lambda\,, \qquad D \in \mathbf{D}(X).$$

Let $\mathscr{E}_\lambda(X)$ denote the joint eigenspace

$$\mathscr{E}_\lambda(X) = \{f \in \mathscr{E}(X) \mid Df = \Gamma(D)(i\lambda)f \quad \text{for} \quad D \in \mathbf{D}(X)\}.$$

For $\delta \in \hat{K}_0$ let $\mathscr{E}_{\lambda,\delta}(X)$ and $\mathscr{E}_\delta(B)$, respectively, denote the spaces of K-finite elements of type δ in $\mathscr{E}_\lambda(X)$ and $\mathscr{E}(B)$. Each joint eigenfunction of $\mathbf{D}(X)$ lies in $\mathscr{E}_\lambda(X)$ for some $\lambda \in \mathfrak{a}_c^*$; since $\mathscr{E}_{s\lambda}(X) = \mathscr{E}_\lambda(X)$ we may by Theorem 6.1 choose $s \in W$ such that $s\lambda$ is simple. So we can assume λ simple. Since (cf. Sect. 4)

$$\dim \mathscr{E}_\delta(B) = \dim(V_\delta)\dim(V_\delta{}^M)$$

it is clear that

$$\dim \mathscr{E}_{\lambda,\delta}(X) \geqslant \dim(V_\delta)\dim(V_\delta{}^M), \tag{11}$$

the map $F \to f$ defined by (10) being an injection of $\mathscr{E}(B)$ into $\mathscr{E}_\lambda(X)$, commuting with K.

Now let $h \in \mathscr{E}_{\lambda,\delta}(X)$; viewing h as a distribution we define its Fourier transform $\tilde{h}$ as a linear functional on $\mathscr{H}^\delta(\mathfrak{a}^*)$:

$$\tilde{h}(\tilde{f}) = h(f) = \int_X h(x)\,f(x)\,dx, \qquad f \in \mathscr{D}_{\check{\delta}}(X).$$

Then the mapping

$$\psi \in \mathscr{I}^\delta(\mathfrak{a}^*) \to \tilde{h}(Q^{\check{\delta}}\psi)$$

is in the dual space of $\mathscr{I}^\delta(\mathfrak{a}^*)$ so by Theorem 7.1 (for δ the identity representation) this mapping is the spherical transform of a K-invariant $\mathrm{Hom}(V_\delta\,,\,V_\delta{}^M)$-valued distribution j on X, that is,

$$\tilde{j}(\tilde{\phi}) = \tilde{h}(Q^{\check{\delta}}\tilde{\phi}) = j(\phi) \tag{12}$$

for all K-invariant ϕ in $\mathscr{D}(X, \mathrm{Hom}(V_\delta\,,\,V_\delta{}^M))$.

Now putting $p_D(\mu) = \Gamma(D)(-i\mu)$ for $D \in \mathbf{D}(X)$, $\mu \in \mathfrak{a}_c^*$ the relations $Dh = \Gamma(D)(i\lambda)h$ imply that

$$p_D\tilde{h} = p_D(-\lambda)\tilde{h}, \qquad D \in \mathbf{D}(X).$$

But by (12), $\tilde{j}$ satisfies these relations too so j is an eigendistribution of all $D \in \boldsymbol{D}(X)$. But then $j = \phi_\lambda A$, where $A \in \operatorname{Hom}(V_\delta, V_\delta{}^M)$. Consequently,

$$\dim \mathscr{E}_{\lambda,\delta}(X) \leqslant \dim V_\delta \dim V_\delta{}^M$$

so by (11) we are done.

Remark. As in [8(e), p. 133–138] where Corollary 7.4 was proved for X of rank 1 we can deduce that each eigenfunction $f \in \mathscr{E}_\lambda(X)$ has the form

$$f(x) = \int_B e^{(i\lambda+\zeta)(A(x,b))}\, dT(b),$$

where T is a "functional" on B. The evidence so far (cf. [13], [8(h)] and the references there) suggests that the functionals T are precisely the analytic functionals. For X of rank 1 this is indeed so as a result of [8(h)] and [13].

We also state the inversion formula and the Plancherel formula for the transform (1). Let w denote the order of W. If dx and $d\lambda$ are normalized as in [8(e)] we have for $\phi \in \mathscr{D}(X)$

$$\phi(x) = w^{-1} \int_{\mathfrak{a}^*} \int_K \tilde{\phi}(\lambda, b)\, e^{(i\lambda+\rho)(A(x,kM))}\, |\mathbf{c}(\lambda)|^{-2}\, d\lambda\, dk. \tag{13}$$

COROLLARY 7.5. *The transform* (1) *is inverted by*

$$f(x) = \operatorname{Tr}\left[w^{-1} \int_{\mathfrak{a}^*} \Phi_{\lambda,\delta}(x)\, \tilde{f}(\lambda)\, |\mathbf{c}(\lambda)|^{-2}\, d\lambda\right], \qquad f \in \mathscr{D}_{\check{\delta}}(X). \tag{14}$$

Moreover

$$\int_X |f(x)|^2\, dx = w^{-1} d(\delta)^{-1} \int_{\mathfrak{a}^*} \operatorname{Tr}(\tilde{f}(\lambda)\, \tilde{f}(\lambda)^*)\, |\mathbf{c}(\lambda)|^{-2}\, d\lambda. \tag{15}$$

Here Tr *and* $*$ *denote trace and adjoint on* V_δ.

In fact the proof above showed that $\tilde{f}(\lambda, kM) = \operatorname{Tr}(\delta(k)\tilde{f}(\lambda))$ so (14) follows from (13). Similarly (15) follows from the Plancherel formula for (13) (cf. [8(e)]).

8. Application to Invariant Differential Equations

For $\delta \in \hat{K}_0$, let $\mathscr{D}_\delta'(X)$ denote the space of K-finite distributions on X which under K transform according to δ; then the sum $\bigoplus_{\delta \in \hat{K}_0} \mathscr{D}_\delta'(X) = \mathscr{D}_0'(X)$ consists of all K-finite distributions on X.

THEOREM 8.1. *Let $D \neq 0$ be any G-invariant differential operator on X. Then*

$$D\,\mathscr{D}_0'(X) = \mathscr{D}_0'(X).$$

We derive this result from Theorem 7.1. If $T \in \mathscr{D}'(X)$ we denote by $\chi_\delta * T$ the distribution $\int_K \chi_\delta(k)\, T^{\tau(k)}\, dk$, where $T^{\tau(k)}$ is the distribution $T^{\tau(k)}(f) = \int_X f(k\cdot x)\, dT(x)$ $(f \in \mathscr{D}(X))$. Then $(\chi_{\check\delta} * T)(f) = T(\chi_\delta * f)$. This identifies $\mathscr{D}_\delta'(X)$ with the dual space of $\mathscr{D}_{\check\delta}(X)$. Now let $P(\lambda)$ be the polynomial on $\mathfrak{a}_c^*$ which satisfies $(Df)^\sim(\lambda, b) \equiv P(\lambda)\tilde f(\lambda, b)$ $(f \in \mathscr{D}(X))$. Since multiplication by $P(\lambda)$ is a homeomorphism of $\mathscr{H}^\delta(\mathfrak{a}^*)$ onto $P\mathscr{H}^\delta(\mathfrak{a}^*)$, Theorem 7.1 implies that the mapping $\phi \to D\phi$ is a homeomorphism of $\mathscr{D}_{\check\delta}(X)$ onto $D\mathscr{D}_{\check\delta}(X)$.

Now if $S \in \mathscr{D}_\delta'(X)$ and D^* denotes the adjoint of D, the linear form $D^*f \to S(f)$ on $D^*\mathscr{D}_{\check\delta}(X)$ is continuous. By the Hahn–Banach theorem there exists an element in the dual of $\mathscr{D}_{\check\delta}(X)$, i.e., a $T \in \mathscr{D}_\delta'(X)$ such that $T(D^*f) = S(f)$ for $f \in \mathscr{D}_{\check\delta}(X)$. This implies that $DT = S$ on $\mathscr{D}_{\check\delta}(X)$, hence on all of $\mathscr{D}(X)$.

9. EIGENSPACE REPRESENTATIONS

We consider again the symmetric space $X = G/K$ and for each $\lambda \in \mathfrak{a}_c^*$ the joint eigenspace $\mathscr{E}_\lambda(X)$. Let T_λ denote the associated eigenspace representation, i.e., $(T_\lambda(g)f)(x) = f(g^{-1}\cdot x)$. Using Theorem 6.1, Corollary 7.4, and the Radon transform method applied to the rank 1 case in [8(e), p. 143] we can answer completely the irreducibility question for T_λ.

THEOREM 9.1. *The eigenspace representation T_λ is irreducible if and only if*

$$\mathbf{e}(\lambda)\,\mathbf{e}(-\lambda) \neq 0. \tag{1}$$

Proof. Consider the space $\mathscr{H}_\lambda$ of functions

$$f(x) = \int_B e^{(i\lambda+\rho)(A(x,b))} F(b)\, db, \qquad F \in L^2(B).$$

Assume (1). Let $0 \neq V \subset \mathscr{E}_\lambda(X)$ be a closed invariant subspace. Then $\varphi_\lambda \in V$. Since $-\lambda$ is simple the proof of Theorem 4.1 shows $\mathscr{H}_\lambda \subset V$. Since λ is simple the proof of Corollary 7.4 shows that $\mathscr{H}_\lambda$ contains each

$\mathscr{E}_{\lambda,\delta}(X)$ whence by [8(e), p. 138] $\mathscr{H}_\lambda$ is dense in $\mathscr{E}_\lambda(X)$. Thus $V = \mathscr{E}_\lambda(X)$ so T_λ is irreducible.

Conversely, suppose T_λ is irreducible. Since $\mathscr{E}_\lambda = \mathscr{E}_{s\lambda}$ $(s \in W)$, since condition (1) is Weyl group invariant, and since at least one of the elements $s\lambda$ $(s \in W)$ is simple we may assume λ simple. Then we can give $\mathscr{H}_\lambda$ the L^2-norm on B. Let $0 \neq E \subset \mathscr{H}_\lambda$ be a closed invariant subspace. Since T_λ is irreducible and $\mathscr{H}_\lambda \subset \mathscr{E}_\lambda(X)$ dense, $E \cap \mathscr{E}_{\lambda,\delta}$ is dense in $\mathscr{E}_{\lambda,\delta}$ for each δ. Thus by Corollary 7.4, $E \cap \mathscr{E}_{\lambda,\delta} = \mathscr{E}_{\lambda,\delta}$, $E = \mathscr{H}_\lambda$ and G irreducible on $\mathscr{H}_\lambda$. Thus the translates $\varphi_\lambda(g \cdot x)$ generate $\mathscr{H}_\lambda$ so by the proof of Theorem 4.1, $-\lambda$ is simple. Now (1) follows from Theorem 6.1.

10. The Rank 1 Case

In this section we assume rank $G/K = 1$. Then for each $\delta \in \hat{K}_0$, the polynomial $Q^\delta(\lambda)$ is a scalar polynomial on the one-dimensional space $\mathfrak{a}_c{}^*$. Let $p(\delta)$ denote its degree. Let $\mathscr{D}^\delta(A)$ denote the space of functions $\Phi \in \mathscr{D}(A, \operatorname{Hom}(V_\delta, V_\delta))$ satisfying the conditions

(i) $\Phi(a^k) = \delta(k)\,\Phi(a)$ if a, $a^k = kak^{-1} \in A$, $k \in K$.

(ii) $\Phi(a)/\rho(\log a)^{p(\delta)}$ is smooth and even on A, log denoting the inverse of the map exp: $\mathfrak{a} \to A$.

It is clear from (i) that $\Phi(a)$ maps V_δ into $V_\delta{}^M$. We identify A with the submanifold $A \cdot o$ of X.

Lemma 10.1. *The restriction $F \to F \mid A$ is a bijection of $\mathscr{D}^\delta(X)$ onto $\mathscr{D}^\delta(A)$.*

If $F \in \mathscr{D}^\delta(X)$ then $F(ka) = \delta(k)F(a)$ so (i) is satisfied and the restriction map is one-to-one. Put $f(x) = \operatorname{Tr}(F(x))$. By Proposition 3.1, $f \in \mathscr{D}_{\check\delta}(X)$, $F = f^\delta$ and by [8(g), Theorem 4.4],

$$F(a) = w^{-1} \int_{\mathfrak{a}^*} \Phi_\delta(\lambda, a)\,\vec{f}(\lambda)\,|\,\mathbf{c}(\lambda)|^{-2}\,d\lambda. \tag{1}$$

Fix a unit vector $v_0 \in V_\delta{}^M$; then for each $v \in V_\delta$, $\vec{f}(\lambda)v = \langle \vec{f}(\lambda)v, v_0\rangle\, v_0$ and it follows from [8(h), Lemma 4.2, Theorem 4.5] that the function $\langle \Phi_\delta(\lambda, a)\, v_0, v_0\rangle\, \rho(\log a)^{-p(\delta)}$ is smooth and even on A. Then (1) implies that condition (ii) is verified.

Conversely, let $\Phi \in \mathscr{D}^\delta(A)$ and define F on X by

$$F(ka \cdot o) = \delta(k)\, \Phi(a), \qquad k \in K,\ a \in A.$$

Condition (i) implies that F is well defined. In fact if $a \neq e$ and $ka \cdot o = k_1 a_1 \cdot o$ then $ka^2k^{-1} = k_1 a_1{}^2 k_1^{-1}$ so $k_1^{-1}k = k_0$ lies in the Weyl group and $a_1 = a^{k_0}$. Then

$$\delta(k_1)\, \Phi(a_1) = \delta(k)\, \delta(k_0^{-1})\, \Phi(a_1) = \delta(k)\, \Phi(a).$$

It remains to prove that F is smooth. Let S be the unit sphere in $\mathfrak{p}$ and fix $H_0 \in \mathfrak{a} \cap S$. The map $kM \to Ad(k)\, H_0$, where Ad denotes the adjoint representation identifies K/M with S so we have the Hilbert space decomposition

$$L^2(S) = \sum_{\delta \in \hat{K}_0} \mathscr{E}(S)_\delta,$$

where $\mathscr{E}(S)_\delta$ is the set of K-finite functions on S of type δ. Then $\dim \mathscr{E}(S)_\delta = d(\delta) \dim V_\delta{}^M = d(\delta)$ so $\mathscr{E}(S)_\delta$ is irreducible under K. If H denotes the space of harmonic polynomial functions on $\mathfrak{p}$ the restriction $p \to p \mid S$ is a one-to-one map of H into $\mathscr{E}(S)$ and $\mathscr{E}(S)_\delta$ is the image under this map of a subspace of the space $H^{p(\delta)}$ of homogeneous harmonic polynomials of degree $p(\delta)$. We take this subspace of $H^{p(\delta)}$ as our space V_δ. The maps

$$v \in V_\delta \to v \mid S \in \mathscr{E}(S)_\delta,$$

$$v \in V_\delta \to h \in \mathscr{E}(S)_\delta \qquad (\text{where } h(Ad(k)\, H_0) = \langle v, \delta(k)\, v_0 \rangle)$$

both commute with the action of K so must by Schur's lemma be proportional; thus for a constant c_δ

$$v(Ad(k)\, H_0) = c_\delta \langle v, \delta(k)\, v_0 \rangle, \qquad v \in V_\delta,\ k \in K.$$

If $v' \in V_\delta$ we have

$$\begin{aligned} \langle v', F(ka \cdot o) v \rangle &= \langle v', \delta(k)\, \Phi(a) v \rangle \\ &= \langle v_0, \Phi(a) v \rangle \langle v', \delta(k)\, v_0 \rangle = c_\delta^{-1} \langle v_0, \Phi(a) v \rangle\, v'(Ad(k)\, H_0) \\ &= c \langle v_0, \rho(\log a)^{-p(\delta)}\, \Phi(a) v \rangle\, v'(Ad(k) \log a), \end{aligned}$$

where c is another constant. The scalar function $\langle v_0, \rho(\log a)^{-p(\delta)}\, \Phi(a) v \rangle$ on A is by (ii) W-invariant so extends to a smooth K-invariant function

ϕ_v on X (this holds for X of any rank by a Taylor series argument). It follows that if Exp: $\mathfrak{p} \to X$ is the exponential mapping,

$$\langle v', F(\text{Exp } X)v\rangle = c\phi_v(\text{Exp } X)\, v'(X), \qquad X \in \mathfrak{p}$$

so F is smooth and the lemma is proved.

Putting, as in [8(h), Sect. 4],

$$\Psi_{\lambda,\delta}(x) = \int_K e^{(i\lambda+\rho)(A(x,kM))}\langle v_0\,,\, \delta(k)\, v_0\rangle\, dk$$

we have the following consequence of [8(g), Theorems 3.1, 4.4, Lemma 5.1] ($dx = \Delta(a)\, da\, dk$ if $x = ka \cdot o$).

Corollary 10.2. *The mapping* $F \to \tilde{F}$ *given by*

$$\tilde{F}(\lambda) = w^{-1} \int_A \Psi_{-\lambda,\delta}(a)\, F(a)\, \Delta(a)\, da$$

is a bijection of $\mathscr{D}^{\delta}(A)$ *onto* $\mathscr{H}^{\delta}(\mathfrak{a}^*)$ *with inverse*

$$F(a) = w^{-1} \int_{\mathfrak{a}^*} \text{conj } \Psi_{-\lambda,\delta}(a)\, \tilde{F}(\lambda)\, |\, \mathbf{c}(\lambda)|^{-2}\, d\lambda.$$

Here conj denotes complex conjugation. Taking traces on both sides we obtain an integral transform in one variable with a hypergeometric function as kernel. Let

$$K_\delta(\lambda, a) = \tfrac{1}{2}\Psi_{-\lambda,\delta}(a \cdot o)\, \rho(\log a)^{-p(\delta)}\, Q^{\delta}(\lambda)^{-1},$$

$$\phi(a) = \text{Tr}(F(a))\, \rho(\log a)^{-p(\delta)}, \qquad \tilde{\phi}(\lambda) = \text{Tr}(\tilde{F}(\lambda))\, Q^{\delta}(\lambda)^{-1}$$

$$\Delta^{\delta}(a) = \rho(\log a)^{2p(\delta)}\, \Delta(a), \qquad \mathbf{c}^{\delta}(\lambda) = \mathbf{c}(\lambda)\, Q^{\delta}(-\lambda)^{-1}.$$

Here $\Psi_{-\lambda,\delta}$ is explicitly expressible by means of the hypergeometric function [8(h), Theorem 4.5]. The functions $\Delta(a)$ and $\mathbf{c}(\lambda)$ are also explicitly known in terms of the multiplicities of the restricted roots α and 2α of X.

Now Corollary 10.2 can be restated:

Corollary 10.3. *The mapping* $\phi(a) \to \tilde{\phi}(\lambda)$, *where*

$$\tilde{\phi}(\lambda) = \int_A K_\delta(\lambda, a)\, \phi(a)\, \Delta^{\delta}(a)\, da$$

is a bijection of the space of even functions in $\mathscr{D}(A)$ onto the space of even functions on $\mathfrak{a}^*$ which extend to entire functions of exponential type on $\mathfrak{a}_c^*$. The inverse is given by

$$\phi(a) = \int_{\mathfrak{a}^*} K_\delta(\lambda, a)^- \tilde{\phi}(\lambda) \mid \mathbf{c}^\delta(\lambda)|^{-2} \, d\lambda.$$

If δ is the identity representation of K, this is of course just the spherical transform.

11. The Euclidean Case

While the classical Paley–Wiener theorem characterizes the Fourier transforms $\tilde{f}(\zeta_1, ..., \zeta_n)$ of the functions $f \in \mathscr{D}(\boldsymbol{R}^n)$ as entire functions of exponential type in the n complex variables ζ_i we might in analogy with [8(g), Theorem 8.3] ask for a characterization in terms of the variables $\lambda \in \boldsymbol{R}$, $\omega \in \boldsymbol{R}^n$, $|\omega| = 1$ when we write the Fourier transform of $f \in L^1(\boldsymbol{R}^n)$ in polar coordinate form,

$$\tilde{f}(\lambda\omega) = \int_{\boldsymbol{R}^n} f(x) \, e^{-i\lambda(x,\omega)} dx. \tag{1}$$

Here (x, ω) denotes the inner product on $\boldsymbol{R}^n$.

Theorem 11.1. *The mapping $f \to \tilde{f}$ maps $\mathscr{D}(\boldsymbol{R}^n)$ onto the set of functions $\phi(\lambda, \omega) \in C^\infty(\boldsymbol{R} \times \boldsymbol{S}^{n-1})$ satisfying*

(i) *For each ω, the function $\lambda \to \phi(\lambda, \omega)$ extends to a holomorphic function on $\mathbf{C}$ such that there exists a constant $A > 0$ for which*

$$\sup_{\lambda \in \boldsymbol{C}, \omega} |\phi(\lambda, \omega)| \, (1 + |\lambda|)^N \, e^{-A|\operatorname{Im} \lambda|} < \infty$$

for each integer $N \in \boldsymbol{Z}$.

(ii) *If P_k is a homogeneous polynomial of degree $k \geqslant 0$ then the function*

$$\phi_k(\lambda) = \int_{\boldsymbol{S}^{n-1}} \phi(\lambda, \omega) \, P_k(\omega_1, ..., \omega_n) \, d\omega$$

has the property that $\phi_k(\lambda) \, \lambda^{-k}$ is even and holomorphic on $\mathbf{C}$.

Proof. If $f \in \mathscr{D}(\boldsymbol{R}^n)$ then (i) obviously holds for the function $\phi = \tilde{f}$.

Also $d^k\tilde{f}/d\lambda^k$ at $\lambda = 0$ is a homogeneous kth degree polynomial in $\omega_1, \ldots, \omega_n$ so (ii) is obvious.

Conversely, suppose ϕ satisfies (i) and (ii). We define f by

$$f(x) = (2\pi)^{-n} \int_{\boldsymbol{R}^+ \times \boldsymbol{S}^{n-1}} \phi(\lambda, \omega)\, e^{i\lambda(x,\omega)} \lambda^{n-1}\, d\lambda\, d\omega, \tag{2}$$

where $\boldsymbol{R}^+$ denotes the set of positive real numbers. Now we expand $\phi(\lambda, \omega)$ in spherical harmonics

$$\phi(\lambda, \omega) = \sum_{0 \leqslant l} \sum_{1 \leqslant m \leqslant d(l)} \alpha_{lm}(\lambda)\, S_{lm}(\omega).$$

Here S_{lm} $(1 \leqslant m \leqslant d(l))$ is an orthonormal basis of the eigenspace of the Laplacian on $\boldsymbol{S}^{n-1}$ for eigenvalue $-l(l + n - 2)$; the S_{lm} are restrictions to $\boldsymbol{S}^{n-1}$ of homogeneous polynomials on $\boldsymbol{R}^n$ of degree l. Clearly $f \in C^\infty(\boldsymbol{R}^n)$ so since the assumptions for ϕ are invariant under rotation of ω it suffices to prove that $f(r, 0,\ldots, 0) = 0$ for $r \in \boldsymbol{R}$ sufficiently large. It is well known (cf. Bochner [1, p. 37]) that

$$\int_{\boldsymbol{S}^{n-1}} e^{i\lambda r\omega_1} S_{lm}(\omega)\, d\omega = c_{l,m,n}[J_{l+(n/2)-1}(\lambda r)/(\lambda r)^{(n/2)-1}], \tag{3}$$

where $c_{l,m,n}$ is a constant. The theorem will therefore follow if we prove that

$$\int_0^\infty \alpha_{lm}(\lambda)(\lambda r)^{1-(n/2)}\, J_{l+(n/2)-1}(\lambda r)\, \lambda^{n-1}\, d\lambda = 0, \qquad r > A. \tag{4}$$

Assuming first that f is a radial function, $f(x) = F(|\, x\,|)$, (1) takes the form (with $\tilde{f}(\xi) = \tilde{F}(|\, \xi\,|)$),

$$\tilde{F}(\lambda) = (2\pi)^{-n/2}\, \lambda^{-(n/2)+1} \int_0^\infty F(r)\, J_{(n/2)-1}(\lambda r)\, r^{n/2}\, dr.$$

Suppose $\tilde{F}$ extends to an entire function on $\boldsymbol{C}$ satisfying for each $N \in \boldsymbol{Z}$

$$\sup_{\lambda \in \boldsymbol{C}} |\, \tilde{F}(\lambda)|\, (1 + |\, \lambda\,|)^N\, e^{-A|\operatorname{Im} \lambda|} < \infty. \tag{5}$$

Then

$$F(r) = 0 \qquad \text{for} \quad r > A. \tag{6}$$

For this we follow a method of Ludwig [16, p. 58]. Since $\tilde{F}$ is even,

$\tilde{F}(\lambda) = g(\lambda^2)$, where g is entire and we can define an entire function G on C^n by

$$G(\zeta) = G(\zeta_1, \ldots, \zeta_n) = g(\zeta_1^2 + \cdots + \zeta_n^2). \tag{7}$$

If $\lambda^2 = \zeta_1^2 + \cdots + \zeta_n^2$ it is easy to see that

$$|\operatorname{Im} \lambda| \leqslant (|\operatorname{Im} \zeta_1|^2 + \cdots + (\operatorname{Im} \zeta_n|^2)^{1/2}.$$

Denoting the right-hand side by $|\operatorname{Im} \zeta|$ we see from (5) and (7) that

$$|G(\zeta)| \leqslant C_1 e^{A|\operatorname{Im} \zeta|} \tag{8}$$

and for $\xi \in R^n$,

$$|G(\xi)| = |\tilde{F}(|\xi|)| \leqslant C_2(1 + |\xi|)^{-N}, \tag{9}$$

where C_1 and C_2 are constants. Since $\tilde{f}(\xi) = G(\xi)$, (8) and (9) imply by the Paley–Wiener theorem [18], that $f(x) = F(|x|) = 0$ for $|x| > A$.

Now we drop the assumption that f is radial. Because of (ii), the coefficient $\alpha_{lm}(\lambda)$ has the property that $\alpha_{lm}(\lambda)\,\lambda^{-l}$ is even and holomorphic on C and by (i) it satisfies condition (5) for $\tilde{F}$. Now write the expression in (4) as r^l times

$$r^{-((n/2)+l)+1} \int_0^\infty (\alpha_{lm}(\lambda)\,\lambda^{-l})\, J_{l+(n/2)-1}(\lambda r)\, \lambda^{(n/2)+l}\, d\lambda. \tag{10}$$

But up to a constant factor this is the Fourier transform in R^{n+2l} of the radial function α given by

$$\alpha(\xi) = \alpha_{lm}(|\xi|)\,|\xi|^{-l} \qquad (\xi \in R^{n+2l}).$$

But since (5) $\Rightarrow$ (6) we conclude that (4) holds so the theorem is proved.

12. The Spherical Principal Series

Following the notation of Section 6 we discuss briefly the spherical principal series of G and the irreducibility question for the eigenspace representations on the horocycle space $\Xi = G/MN$.

For $\lambda \in \mathfrak{a}_c^*$ let $\mathscr{H}_\lambda$ denote the space of complex-valued measurable functions ψ on G satisfying the conditions

(i) $\psi(gam\, n) = e^{(i\lambda-\rho)(\log a)}\psi(gmn)$,

(ii) $\int_K |\psi(k)|^2\, dk < \infty$.

Then $\mathscr{H}_\lambda$ is a Hilbert space, the norm being the square root of the expression in (ii). Putting $(\tau_\lambda(g)\,\psi)(x) = \psi(g^{-1}x)$ we have a representation τ_λ of G on $\mathscr{H}_\lambda$. The family τ_λ $(\lambda \in \mathfrak{a}_c{}^*)$ is called the *spherical principal series*. According to [8(e), p. 114], τ_λ is irreducible if and only if λ and $-\lambda$ are both simple. Thus we have

THEOREM 12.1. *The representation τ_λ of the spherical principal series is irreducible if and only if*

$$\mathbf{e}(\lambda)\,\mathbf{e}(-\lambda) \neq 0.$$

We now explain how this result is equivalent to the irreducibility criterion of Kostant [14(b), p. 631] or [14(a), p. 321], which for complex G had been proved by Parthasarathy, Ranga Rao and Varadarajan [19]. Kostant considers the space $\mathscr{H}_\lambda{}^\infty$ of K-finite functions in $\mathscr{H}_\lambda$ and the associated representation $d\tau_\lambda$ of $U(\mathfrak{g})$ on $\mathscr{H}_\lambda{}^\infty$. For each $\alpha \in \Sigma_0{}^+$ let $I_\alpha \subset \boldsymbol{R}$ be the interval $|t| < m_\alpha/2$ if $2\alpha \notin \Sigma^+$ and $|t| < m_\alpha/2 + 1$ if $2\alpha \in \Sigma^+$. Also define

$$l_\alpha = \tfrac{1}{2}m_\alpha, \qquad n_\alpha = 1 \qquad \text{if } 2\alpha \notin \Sigma^+$$

$$l_\alpha = \tfrac{1}{2}m_\alpha + 1, \qquad n_\alpha = 2 \qquad \text{if } 2\alpha \in \Sigma^+.$$

Kostant then proves that $d\tau_\lambda$ is not (algebraically) irreducible if and only if there exists a root $\alpha \in \Sigma_0{}^+$ such that

(1) $\langle i\lambda, \alpha_0\rangle \notin I_\alpha$,

(2) $\langle i\lambda, \alpha_0\rangle - l_\alpha \in n_\alpha \boldsymbol{Z}$.

Using the poles of the Gamma function, its duplication formula, and the fact that if α and 2α are both restricted roots then their respective multiplicities are even and odd, one verifies that the λ satisfying conditions (1) and (2) are the solutions to the equation $\mathbf{e}(\lambda)\,\mathbf{e}(-\lambda) = 0$. For complete agreement with Theorem 12.1 one now recalls Harish-Chandra's basic result [7(a), p. 228] which insures *a priori* that $d\tau_\lambda$ is algebraically irreducible if and only if τ_λ is irreducible. Our method, being analytic, does not require this result nor does it require our weight–theoretic parametrization of the finite-dimensional spherical representations [8(c), (e)], whereas Kostant's method uses the analogous parametrization for the complexification G_c. However, our proof of Theorem 6.1 for the rank 1 case [8(e), p. 132] did require Bruhat's irreducibility criterion. A direct proof (using the explicit formula for $e^{\rho(H(\bar{n}))}$) has been found by J. Dadok.

For each $\lambda \in \mathfrak{a}_c{}^*$ let $\mathscr{D}_\lambda'(\Xi)$ denote the joint distribution eigenspace of the operators $D \in \boldsymbol{D}(G/MN)$ containing the function $gMN \to e^{(i\lambda-\zeta)(H(g))}$, and $\tilde{\tau}_\lambda$ the natural representation of G on $\mathscr{D}_\lambda'(\Xi)$. From [8(e), p. 112] we deduce,

COROLLARY 12.2. *The eigenspace representation $\tilde{\tau}_\lambda$ is irreducible if and only if*

$$\mathbf{e}(\lambda)\,\mathbf{e}(-\lambda) \neq 0.$$

REFERENCES

1. S. BOCHNER, "Harmonic Analysis and the Theory of Probability," *Univ. of California Press*, Berkeley, 1955.
2. R. BOTT, The index theorem for homogeneous differential operators, *in* "Differential and Combinatorial Topology" (S. S. Cairns, Ed.), pp. 167–186, Princeton Univ. Press, Princeton, N.J., 1965.
3. J. DIEUDONNÉ AND SCHWARTZ, La dualité dans les espaces F et LF, *Ann. Inst. Fourier* **1** (1949), 61–101.
4. M. A. DOSTAL, A complex characterization of the Schwartz space $\mathscr{D}(\Omega)$, *Math. Ann.* **195** (1972), 175–191.
5. L. EHRENPREIS, "Fourier Analysis in Several Complex Variables," Interscience, New York, 1970.
6. S. G. GINDIKIN AND F. I. KARPELEVIČ, Plancherel measure of Riemannian symmetric spaces of nonpositive curvature, *Soviet Math. Dokl.* **3** (1962), 962–965.
7. HARISH-CHANDRA, (a) Representations of a semisimple Lie group on a Banach space, I, *Trans. Amer. Math. Soc.* **75** (1953), 185–243. (b) Spherical functions on a semisimple Lie group I, *Amer. J. Math.* **80** (1958), 241–310. (c) Differential equations and semisimple Lie groups (1960), unpublished.
8. S. HELGASON, (a) "Differential Geometry and Symmetric Spaces," Academic Press, New York, 1962. (b) Invariants and fundamental functions, *Acta Math.* **109** (1963), 241–258. (c) Radon–Fourier transforms on symmetric spaces and related group representations, *Bull. Amer. Math. Soc.* **71** (1965), 757–763. (d) Group representations and symmetric spaces, *Actes Congr. Intern. Math.* **2** (1970), 313–319. (e) A duality for symmetric spaces with applications to group representations, *Advances in Math.* **5** (1970), 1–154. (f) Analysis on Lie Groups and Homogeneous Spaces, Regional conference series, No. 14, American Mathematical Society, Providence, R.I., 1972. (g) The surjectivity of invariant differential operators on symmetric spaces I, *Ann. of Math.* **98** (1973), 451–479. (h) Eigenspaces of the Laplacian; integral representations and irreducibility, *J. Functional Analysis* **17** (1974), 328–353.
9. A. HOLE, Representations of nilpotent Lie groups on eigenspaces, preprint.

10. R. Hotta, On the realization of the discrete series for semisimple Lie groups, *J. Math. Soc. Japan* **23** (1971), 384–407.
11. K. D. Johnson, Composition series and intertwining operators for the spherical principal series II, *Trans. Amer. Math. Soc.* **215** (1976), 269–283.
12. K. D. Johnson and N. Wallach, Composition series and intertwining operators for the spherical principal series, *Trans. Amer. Math. Soc.*, to appear.
13. M. Kashiwara, A. Kowata, K. Minemura, K. Okamoto, T. Oshima, and M. Tanaka, Eigenfunctions of invariant differential operators on a symmetric space, preprint.
14. B. Kostant (a), On the existence and irreducibility of certain series of representations, *Bull. Amer. Math. Soc.* **75** (1969), 627–642. (b) On the existence and irreducibility of certain series of representations, *in* "Lie Groups and Their Representations" (I. M. Gelfand, Ed.), pp. 231–329, Halsted Press, New York, 1975.
15. B. Kostant and Rallis, Orbits and Lie group representations associated to symmetric spaces, *Amer. J. Math.* **93** (1971), 753–809.
16. D. Ludwig, The Radon transform on Euclidean space, *Comm. Pure Appl. Math.* **19** (1966), 49–81.
17. B. Malgrange, Existence et approximation des solutions des équations aux derivées partielles et des équations de convolution, *Ann. Inst. Fourier* **4** (1955–1956), 271–354.
18. R. E. A. C. Paley and N. Wiener, "Fourier Transforms in the Complex Domain," American Mathematical Society Colloquium Publications, American Mathematical Society, Providence, R. I., 1934.
19. K. R. Parthasarathy, R. Ranga Rao, and V. S. Varadarajan, Representations of complex semisimple Lie groups and Lie algebras, *Ann. of Math.* **85** (1967), 383–429.
20. G. Schiffmann, Intégrales d'entrelacement et fonctions de Whittaker, *Bull. Soc. Math. France* **99** (1971), 3–72.
21. M. Sugiura, Spherical functions and representation theory of compact Lie groups, *Sci. Papers College Gen. Ed. Univ. Tokyo* **10** (1960), 187–193.
22. R. Takahashi, Sur les représentations unitaires des groupes de Lorentz généralisées, *Bull. Soc. Math. France* **91** (1963), 289–433.
23. N. Wallach, On the Enright–Varadarajan modules. A construction of the discrete series, *Ann. Sci. École Norm. Sup.* **9** (1976), 81–101.
24. G. Warner, "Harmonic Analysis on Semisimple Lie Groups," Vol. II, Springer–Verlag, New York/Berlin, 1972.
25. R. Parthasarathy, Dirac operator and the discrete series, *Ann. of Math.* **96** (1972), 1–30.
26. W. Schmid, Some properties of square-integrable representations of semisimple Lie groups, *Ann. Math.* **102** (1975), 535–564.
27. H. P. Jakobsen and M. Vergne, Wave and Dirac operators and representations of the conformal group, preprint.

Printed by the St Catherine Press Ltd., Tempelhof 37, Bruges, Belgium.

Reprinted from ADVANCES IN MATHEMATICS Vol. 36, No. 3, June 1980
All Rights Reserved by Academic Press, New York and London *Printed in Belgium*

A Duality for Symmetric Spaces with Applications to Group Representations. III. Tangent Space Analysis*

SIGURDUR HELGASON

Department of Mathematics, Massachusetts Institute of Technology, Cambridge, Massachusetts 02139

1. INTRODUCTION

Let $X = G/K$ be a symmetric space of the noncompact type, G being a connected semisimple Lie group with finite center and K a maximal compact subgroup. Let X_0 be the tangent space to X at the fixed pont o of K and let G_0 denote the group of affine transformations of X_0 generated by the translations and the natural action of K on X_0. The group G_0 which is the semi-direct product $K \times_s X_0$ is often called the Cartan motion group. While the irreducible unitary representations of G_0 are described by Mackey's general semi-direct product theorem [13, p. 131], and their characters given by Gindikin [4], our purpose here is to develop G_0-invariant analysis on the space X_0. More specifically, let $\mathbf{D}(G_0/K)$ denote the algebra of G_0-invariant differential operators on $G_0/K = X_0$. We are interested in the joint eigenfunctions of $\mathbf{D}(G_0/K)$ and in the representations of G_0 on the joint eigenspaces, ([8(d)] Ch. IV).

As tools for this study we prove Paley–Wiener type theorems for the spherical transform and for the generalized Bessel transform on X_0. The first of these comes from Theorem 2.1 which is an analog for entire functions of exponential type of Chevalley's restriction theorem for polynomials. An important property of the generalized Bessel transform comes from Theorem 4.6 which shows that the generalized Bessel function is divisible by the $\check{J}^\delta$-matrix. Another useful tool is Theorem 4.8 which represents the generalized Bessel function as a certain derivative of the zonal spherical function.

These results are used in Section 6 to derive an integral formula for the K-finite joint eigenfunctions of $\mathbf{D}(G_0/K)$. This in turn is used to give (in Theorem 6.6) an explicit irreducibility criterion for the associated eigenspace representation of G_0. This is the principal application of our results.

The results for the flat space $X_0 = G_0/K$ are to a large extent analogous to those proved in [8(g)] for the curved space $X = G/K$; most of the proofs are

* Supported in part by the National Science Foundation, Grant MCS 76-21044.

0001-8708/80/060297-27$05.00/0

however quite different because Harish–Chandra's expansion for the zonal spherical function on G/K, which is a basic tool in analysis on G/K, does not seem to have a natural analog for the zonal spherical function ψ on the flat space $G_0/K = X_0$. In fact, if $\beta_1, \ldots, \beta_p$ are the indivisible restricted roots of G/K, Harish–Chandra expands ϕ in a series of exponentials $e^{-n_1\beta_1 - \cdots - n_p\beta_p}$; for X_0 the natural analog would be an expansion of ψ in terms of the monomials $\beta_1^{-n_1} \cdots \beta_p^{-n_p}$. If G/K has rank one such an expansion is given by the asymptotic series of the Bessel function $J_n(z)$ (see e.g. [19, p. 368]) and if G is complex such an expansion is a part of Harish–Chandra's formula for the invariant integral $\int_K \exp\langle X, k \cdot Y\rangle \, dk$ [7, Theorem 2] where $\langle , \rangle$ is the Killing form of the Lie algebra $\mathfrak{k}$ of K and $X, Y \in \mathfrak{k}$. For general G however, such an analog of Harish–Chandra's expansion[1] fails; the monomials above are not even linearly independent. For example $\sum_{i \neq j} \langle \beta_i, \beta_j \rangle \beta_i^{-1} \beta_j^{-1} = 0$ for $G = \mathbf{SL}(3, \mathbf{R})$.

The analogy between X and X_0 is nevertheless quite helpful, particularly for heuristic guesses. But in some respects X_0 is more subtle. An example is the divisibility in Lemma 4.4 and Theorem 4.6. For X the divisibility by $\check{Q}^{\delta}(\lambda)$ is a consequence of the Weyl group invariance in (8) Section 4; for X_0 on the other hand the Weyl group invariance in Theorem 4.6 is quite easy to prove directly, but it does not imply the divisibility by $\check{J}^{\delta}(\lambda)$. This divisibility is obtained from the curved space case by a suitable limit argument; I do not know of a proof which works directly within the flat space.

Some general notation: As usual $\mathbf{Z}$, $\mathbf{R}$ and $\mathbf{C}$ denote the rings of integers, real numbers, and complex numbers, respectively. If L is a Lie group with Lie algebra $\mathfrak{l}$, Ad and ad denote the adjoint representation of L and of $\mathfrak{l}$, respectively. Sometimes we write Ad_L and $\mathrm{ad}_{\mathfrak{l}}$ for emphasis.

If M is a manifold and $\phi: M \to M$ a diffeomorphism we write $f^{\phi} = f \circ \phi$ if f is a function on M. If D is a differential operator on M we write D^{ϕ} for the differential operator given by $D^{\phi}(f) = (D(f^{\phi^{-1}}))^{\phi}$. If V is a finite-dimensional vector space $S(V)$ denotes the symmetric algebra over V, and $S_d(V)$ the set of homogeneous elements in $S(V)$ of degree d.

2. Invariant Entire Functions of Exponential Type

Let G be a connected semisimple Lie group with finite center, $K \subset G$ a maximal compact subgroup. Let $\mathfrak{g} = \mathfrak{k} + \mathfrak{p}$ be the corresponding Cartan decomposition of the Lie algebra $\mathfrak{g}$ of G, $\mathfrak{k}$ being the Lie algebra of K and $\mathfrak{p}$ the orthogonal complement of $\mathfrak{k}$ in $\mathfrak{g}$ with respect to the Killing form B of $\mathfrak{g}$. Let θ be the corresponding Cartan involution and put $|X| = B(X, X)^{1/2}$ for $X \in \mathfrak{p}$.

Let $\mathfrak{a} \subset \mathfrak{p}$ be a maximal abelian subspace, W the Weyl group acting on $\mathfrak{a}$. This action extends to the dual $\mathfrak{a}^*$ and to the complexifications $\mathfrak{a}_c$ and $\mathfrak{a}_c^*$.

[1] A substitute has, however, been shown to me by J-L. Clerc.

Let R be a positive real number. A function f on the complex space $\mathbf{C}^n$ is called an entire function of *exponential type R* if f is holomorphic on $\mathbf{C}^n$ and if for each integer $N \geqslant 0$ there exists a constant C_N such that

$$|f(\zeta)| \leqslant C_N(1 + |\zeta|)^{-N} e^{R|\operatorname{Im} \zeta|}, \qquad \zeta \in \mathbf{C}^n.$$

Here $\zeta = \xi + i\eta$; $\xi, \eta \in \mathbf{R}^n$, $\eta = \operatorname{Im} \zeta$, and $|\zeta| = (|\xi|^2 + |\eta|^2)^{1/2}$. The space of such functions will be denoted by $\mathscr{H}^R(\mathbf{C}^n)$ and we put $\mathscr{H}(\mathbf{C}^n) = \bigcup_{R>0} \mathscr{H}^R(\mathbf{C}^n)$.

THEOREM 2.1. *Let ϕ be a W-invariant entire function of exponential type R on $\mathfrak{a}_c$. Then ϕ extends uniquely to a K-invariant entire function Φ of exponential type R on the complexification $\mathfrak{p}_c$ of $\mathfrak{p}$.*

We start with a simple lemma, cf. [3, p. 130].

LEMMA 2.2. *Each W-invariant holomorphic function ϕ on $\mathfrak{a}_c$ extends uniquely to a K-invariant holomorphic function Φ on $\mathfrak{p}_c$.*

Proof. Let $l = \dim \mathfrak{a}$ and let $p_1, \ldots, p_l$ be homogeneous algebraically independent generators of $I(\mathfrak{a}^*)$, the algebra of W-invariants in the algebra $S(\mathfrak{a}_c^*)$ of polynomial functions on $\mathfrak{a}_c$ [1]. Let

$$\phi(H) = \sum_{(\alpha)} a_{\alpha_1 \cdots \alpha_l} h_1^{\alpha_1} \cdots h_l^{\alpha_l}, \qquad H = (h_1, \ldots, h_l) \in \mathfrak{a}_c, \tag{1}$$

be the Taylor series expansion of ϕ, which we write as

$$\phi(H) = \sum_0^\infty A_m(H), \tag{2}$$

where

$$A_m(H) = \sum_{\alpha_1 + \cdots + \alpha_l = m} a_{\alpha_1, \ldots, \alpha_l} h_1^{\alpha_1} \cdots h_l^{\alpha_l}.$$

Each A_m is W-invariant and can therefore be expressed

$$A_m = \sum b_{\beta_1 \cdots \beta_l} p_1^{\beta_1} \cdots p_l^{\beta_l},$$

where $\sum_1^l \beta_i (\deg p_i) = m$. It is well-known (e.g., [11(a), p. 355], [3, p. 129]) that the mapping

$$p\colon H \to (p_1(H), \ldots, p_l(H))$$

maps $\mathfrak{a}_c$ onto $\mathbf{C}^l$. In fact, if $(\xi_1, \ldots, \xi_l) \in \mathbf{C}^l$, the mapping $p_i \to \xi_i$ $(1 \leqslant i \leqslant l)$ gives, by the algebraic independence of the p_i, a homomorphism of $I(\mathfrak{a}_c^*)$ into $\mathbf{C}$, which, since $S(\mathfrak{a}_c^*)$ is integral over $I(\mathfrak{a}_c^*)$, is given by point evaluation ([8(a), Ch. X], Lemma 6.9 and Theorem 5.5(i)).

Thus the series

$$\phi^*(\xi) = \sum_0^\infty B_m(\xi), \tag{3}$$

where $\xi = (\xi_1, ..., \xi_l)$ and

$$B_m(\xi) = \sum_\beta b_{\beta_1 \cdots \beta_l} \xi_1^{\beta_1} \cdots \xi_l^{\beta_l}, \qquad \left(\sum_i \beta_i \deg(p_i) = m\right)$$

converges absolutely. To see that (3) converges uniformly on compacts let $C \subset \mathbf{C}^l$ be a compact set. Since the map p is not only surjective but proper ([11(a), p. 356]) there exists a compact set $\tilde{C} \subset \mathfrak{a}_c$ such that $p(\tilde{C}) = C$. The series (1), and therefore (2), converges uniformly for $H \in \tilde{C}$. Since $B_m(p(H)) = A_m(H)$ it follows that (3) converges uniformly for $\xi \in C$. This proves that ϕ^* is holomorphic on $\mathbf{C}^l$. By a theorem of Chevalley ([8(a), p. 430]) each p_i extends to a K-invariant polynomial P_i on $\mathfrak{p}_c$. The function

$$\Phi(Z) = \phi^*(P_1(Z), ..., P_l(Z))$$

has the property of Lemma 2.2. The uniqueness is obvious so the lemma is proved.

In order to prove Theorem 2.1, let $\mathfrak{g}_c = \mathfrak{k}_c + \mathfrak{p}_c$ be the complexification of the Cartan decomposition $\mathfrak{g} = \mathfrak{k} + \mathfrak{p}$. Let G_c be the simply connected Lie group with Lie algebra $\mathfrak{g}_c$, K_c and G the analytic subgroups corresponding to $\mathfrak{k}_c$ and $\mathfrak{g}$. We can assume this G coincides with the one above. For $X \in \mathfrak{p}_c$ write $X = X' + iX''$ $(X', X'' \in \mathfrak{p})$. Let $\phi \in \mathscr{H}^R(\mathfrak{a}_c)$ be W-invariant and Φ the extension to $\mathfrak{p}_c$ given by Lemma 2.2. The function Φ being holomorphic, its K-invariance implies its K_c-invariance. In order to deduce from the inequality

$$|\phi(H' + iH'')| \leqslant Ce^{R|H''|}, \qquad H', H'' \in \mathfrak{a},$$

an inequality

$$|\Phi(X' + iX'')| \leqslant Ce^{R|X''|}, \qquad X', X'' \in \mathfrak{p},$$

by means of the K_c-invariance of Φ it would suffice to show that the infimum of $|Z''|$ as Z runs through the orbit $K_c \cdot (X' + iX'')$ is reached for some $Z \in \mathfrak{a}_c$. Here $g \cdot Z = \mathrm{Ad}(g)Z$ $(g \in G_c, Z \in \mathfrak{g}_c)$.

Let $X \in \mathfrak{p}_c$ be regular and semisimple. Since the orbit $K_c \cdot X$ is closed ([12, p. 783]) the function $k \to |(\mathrm{Ad}(k)X)''|^2$ has a minimum[2] on K_c. Let $\mathfrak{g}_c{}^R$ denote the Lie algebra $\mathfrak{g}_c$ considered as a real Lie algebra and B^R its Killing form. Then $B^R = 2B$ on $\mathfrak{g} \times \mathfrak{g}$. We have the Cartan decomposition

$$\mathfrak{g}_c{}^R = \mathfrak{u} + i\mathfrak{u}, \qquad \text{where} \quad \mathfrak{u} = \mathfrak{k} + i\mathfrak{p},$$

[2] As remarked by Nicole Bopp one should here take note of (3a) below.

and we denote by η the corresponding Cartan involution of $\mathfrak{g}_c{}^R$. If $Z = Z' + iZ''$ where $Z', Z'' \in \mathfrak{p}$ then $\eta(Z) = -Z' + iZ''$ so $Z'' = -(i/2)(Z + \eta(Z))$. Using the invariance of B^R under $\mathrm{Ad}(K_c)$ and η, and the relation $B^R(iZ, iZ) = -B^R(Z, Z)$ we obtain

$$|(\mathrm{Ad}(k)X)''|^2 = -\tfrac{1}{4}\{B^R(X, X) + B^R(\eta\,\mathrm{Ad}(k)X, \mathrm{Ad}(k)X)\}. \tag{3a}$$

Thus if $k_c \in K_c$ is a minimum point for our function we would have

$$\left\{\frac{d}{dt} B^R(\eta\,\mathrm{Ad}(\exp tT)\,\mathrm{Ad}(k_c)X, \mathrm{Ad}(\exp tT)\,\mathrm{Ad}(k_c)X)\right\}_{t=0} = 0 \tag{4}$$

for each $T \in \mathfrak{k}_c$. Using $\eta^2 = 1$ it follows that

$$B^R(\eta\,\mathrm{Ad}(k_c)X, [\mathfrak{k}_c, \mathrm{Ad}(k_c)X]) = 0. \tag{5}$$

But B^R is invariant and is nondegenerate on $\mathfrak{k}_c$ so this implies

$$[\eta\,\mathrm{Ad}(k_c)X, \mathrm{Ad}(k_c)X] = 0. \tag{6}$$

But $\mathrm{Ad}(k_c)X$ is a regular semisimple element of $\mathfrak{p}_c$ and therefore lies in a unique Cartan subspace $\mathfrak{b}_c$ of $\mathfrak{p}_c$. Since $\eta\mathfrak{p}_c \subset \mathfrak{p}_c$, (6) implies $\eta(\mathrm{Ad}(k_c)X) \subset \mathfrak{b}_c$ so by the uniqueness we deduce $\eta\mathfrak{b}_c = \mathfrak{b}_c$. Hence

$$\mathfrak{b}_c = (\mathfrak{b}_c \cap \mathfrak{u}) + \mathfrak{b}_c \cap (i\mathfrak{u}) \tag{7}$$

is the decomposition of $\mathfrak{b}_c$ into a toral part and a vector part. We have also

$$\mathfrak{a}_c = \mathfrak{a}_c \cap \mathfrak{u} + \mathfrak{a}_c \cap (i\mathfrak{u}) \tag{8}$$

and $\mathfrak{a}_c \cap \mathfrak{u} = i\mathfrak{a}$, $\mathfrak{a}_c \cap (i\mathfrak{u}) = \mathfrak{a}$. The Cartan subspaces $\mathfrak{b}_c$ and $\mathfrak{a}_c$ being conjugate [12, p. 764] we select $k_1 \in K_c$ such that $\mathrm{Ad}(k_1)\mathfrak{b}_c = \mathfrak{a}_c$. The eigenvalues being preserved we deduce

$$\mathrm{Ad}(k_1)(\mathfrak{b}_c \cap \mathfrak{u}) = i\mathfrak{a}, \qquad \mathrm{Ad}(k_1)(\mathfrak{b}_c \cap i\mathfrak{u}) = \mathfrak{a}.$$

Let $Z \in \mathfrak{b}_c \cap \mathfrak{u}$ and put $H = \mathrm{Ad}(k_1)Z \in i\mathfrak{a}$. Now by the Cartan decomposition $K_c = \exp(i\mathfrak{k})K$ we decompose $k_1 = (\exp T)k_0$, where $T \in i\mathfrak{k}$, $k_0 \in K$. Applying the Cartan involution η of G_c to the equation

$$k_1 \exp Zk_1^{-1} = \exp H$$

we get

$$\exp(-T)\, k_0 \exp Zk_0^{-1} \exp T = \exp H \tag{9}$$

whence

$$\exp(-H) \exp 2T \exp H = \exp 2T.$$

But diagonalizing Ad(exp T) we deduce from this that Ad(exp H) and Ad(exp T) commute so by (9)

$$e^{\operatorname{ad}(\operatorname{Ad}(k_0) Z)} = \operatorname{Ad}(k_0 \exp Z k_0^{-1}) = \operatorname{Ad}(\exp H) = e^{\operatorname{ad} H}.$$

Since Z is arbitrary, this proves that $\operatorname{Ad}(k_0)(\mathfrak{b}_c \cap \mathfrak{u}) = i\mathfrak{a}$ and $\operatorname{Ad}(k_0)\mathfrak{b}_c = \mathfrak{a}_c$. But then the vector

$$H = \operatorname{Ad}(k_0) \operatorname{Ad}(k_c) X \in \mathfrak{a}_c$$

satisfies

$$H'' = (\operatorname{Ad}(k_0) \operatorname{Ad}(k_c)X)'' = \operatorname{Ad}(k_0)((\operatorname{Ad}(k_c)X)'')$$

so

$$| H'' | = \min_{k \in K_c} |(\operatorname{Ad}(k)X)'' |.$$

This proves that if $\phi \in \mathscr{H}^R(\mathfrak{a}_c)$ is W-invariant its K-invariant holomorphic extension Φ satisfies

$$| \Phi(X)| \leqslant C e^{R|X''|} \tag{10}$$

for all $X \in \mathfrak{p}_c$ which are regular and semisimple, C being a constant. But such X are dense in $\mathfrak{p}_c$ so (10) holds on all of $\mathfrak{p}_c$ by continuity. We have also by the K-invariance of Φ on $\mathfrak{p}$ that for each integer $N \geqslant 0$ there exists a constant C_N such that

$$| \Phi(X')| \leqslant C_N(1 + | X' |)^{-N}, \qquad X' \in \mathfrak{p}. \tag{11}$$

Conditions (10) and (11) imply that $\Phi \in \mathscr{H}^R(\mathfrak{p}_c)$. For this let $X_1, \ldots, X_n$ be an orthonormal basis of $\mathfrak{p}$ and put

$$X = \sum_1^n z_j X_j = \sum_1^n (x_j + iy_j) X_j .$$

Then

$$| X'' | = (y_1^2 + \cdots + y_n^2)^{1/2} \leqslant | z_1 | + \cdots + | z_n |$$

so (10) and (11) imply by the classical Paley–Wiener theorem that Φ is the Fourier transform of an L^2-function on $\mathfrak{p}$ vanishing outside the ball of radius R around 0. But then (11) implies that this L^2-function is smooth so $\Phi \in \mathscr{H}^R(\mathfrak{p}_c)$ as stated. This completes the proof of Theorem 2.1.

Remark. For the case when G/K has rank one and for the case when G is complex, Theorem 2.1 has a considerably simpler proof (cf. [8(g), p. 216] and [3, sect. 3]).

We next prove an analog of Theorem 2.1 for the Schwartz spaces $\mathscr{S}(\mathfrak{p})$ and $\mathscr{S}(\mathfrak{a})$ of rapidly decreasing functions. For the case when G is complex it was proved by Harish–Chandra [7], Theorem 4, by a method special to the complex case.

PROPOSITION 2.3. *Each W-invariant $\phi \in \mathscr{S}(\mathfrak{a})$ extends to a K-invariant $\Phi \in \mathscr{S}(\mathfrak{p})$.*

Proof. By the theorem of Glaeser [5] or Schwarz [17], ϕ has the form $\phi(H) = \phi^*(p_1, ..., p_l)$ where $\phi^* \in C^\infty(\mathbf{R}^l)$ and the $p_1, ..., p_l$ are algebraically independent homogeneous generators of $I(\mathfrak{a}^*)$. Thus the function $\phi^*(P_1, ..., P_l)$ gives an extension of ϕ to a K-invariant C^∞ function[3] Φ on $\mathfrak{p}$. We next prove that for all integers $k, l \geqslant 0$

$$\sup_X |(1 + |X|^l)(L_{\mathfrak{p}}^k \Phi)(X)| < \infty, \tag{12}$$

$L_{\mathfrak{p}}$ denoting the Laplacian on $\mathfrak{p}$. Then $\Phi \in \mathscr{S}(\mathfrak{p})$ would follow using the Fourier transform on $\mathfrak{p}$.

We prove (12) using the radial part of $L_{\mathfrak{p}}$. For this we recall the formula of Theorem 3.2 in [8(f)] for the radial part $\Delta(L_V)$ of the Laplace Beltrami operator L_V on a Riemannian manifold V under the action of an isometry group with orthogonal transversal submanifold $S \subset V$. If L_S is the Laplace Beltrami operator on S and ω the density function on S for the group action we have

$$\Delta(L_V) = \omega^{-1/2} L_S \circ \omega^{1/2} - \omega^{-1/2} L_S(\omega^{1/2}). \tag{13}$$

Using the formula (the dot $\cdot$ denoting inner product)

$$L_S(uv) = uL_S(v) + vL_S(u) + 2 \operatorname{grad} u \cdot \operatorname{grad} v$$

and the formula $\omega^{-1/2} \operatorname{grad}(\omega^{1/2}) = \frac{1}{2}\omega^{-1} \operatorname{grad}(\omega)$, (13) can be written

$$\Delta(L_V) f = L_S f + \omega^{-1} \operatorname{grad} \omega \cdot \operatorname{grad} f, \qquad f \in C^\infty(S). \tag{14}$$

To make this explicit for $V = \mathfrak{p}$, let Σ denote the set of roots of $\mathfrak{g}$ with respect to $\mathfrak{a}$, fix a Weyl chamber $\mathfrak{a}^+ \subset \mathfrak{a}$ and let Σ^+ be the corresponding set of positive roots. Then $\mathfrak{a}^+$ is an orthogonal transversal submanifold for the action of $\mathrm{Ad}_G(K)$ on $\mathfrak{p}$ and the density function $\omega(H)$ which satisfies

$$\int_{\mathfrak{p}} F(X)\, dX = \int_{\mathfrak{a}^+} \left(\int_K F(\mathrm{Ad}(k)H)\, dk \right) \omega(H)\, dH$$

[3] A more direct proof of this has been found by J. Dadok.

($F \in C_c^\infty(\mathfrak{p})$, dX, dH Euclidean measures, dk Haar measure) is given by

$$\omega(H) = c \left| \prod_{\alpha \in \Sigma^+} \alpha(H)^{m_\alpha} \right| \qquad H \in \mathfrak{a}. \tag{15}$$

Here m_α is the multiplicity of α and c is a constant. If $A_\alpha \in \mathfrak{a}$ is determined by $B(A_\alpha, H) = \alpha(H)$ ($H \in \mathfrak{a}$) then formula (14) takes the form

$$\Delta(L_\mathfrak{p})f = L_\mathfrak{a} f + \sum_{\alpha \in \Sigma^+} m_\alpha \alpha^{-1} A_\alpha f \tag{16}$$

for $f \in C^\infty(\mathfrak{a}^+)$. Since $\alpha^{-1}A_\alpha$ is unchanged under $\alpha \to -\alpha$ we can in (16) sum over all $\alpha \in \Sigma$; thus the choice of Weyl chamber is immaterial and our functions ϕ and Φ satisfy

$$(L_\mathfrak{p}\Phi)(H) = (\Delta(L_\mathfrak{p})\phi)(H) = (L_\mathfrak{a}\phi)(H) + \tfrac{1}{2} \sum_{\alpha \in \Sigma} m_\alpha \alpha(H)^{-1}(A_\alpha \phi)(H) \tag{17}$$

for $H \in \mathfrak{a}$ regular. The function $\phi \in \mathscr{S}(\mathfrak{a})$ being invariant under the symmetry s_α the function $\psi = A_\alpha \phi$ satisfies $\psi^{s_\alpha} = -\psi$ so the function $\alpha^{-1}A_\alpha\phi$ extends from $\mathfrak{a} - \alpha^{-1}(0)$ to a function in $\mathscr{S}(\mathfrak{a})$. Thus (17) holds for all $H \in \mathfrak{a}$ and shows that $L_\mathfrak{p}\Phi$ is the K-invariant extension of a W-invariant $\phi_1 \in \mathscr{S}(\alpha)$. Now (12) is clear by iteration in k.

The next and final variation will not be needed in this paper but we include it since we feel it might lead to a proof of Chevalley's extension theorem for polynomials, already used. Let us call a function f on $\mathbf{R}^n$ Lipschitzean if for each $R > 0$ there exists a constant C_R such that

$$|f(X) - f(Y)| \leqslant C_R |X - Y| \qquad \text{for} \qquad X, Y \in \mathbf{R}^n, \ |X| < R, \ |Y| < R.$$

PROPOSITION 2.4. *Each W-invariant Lipschitzean function ϕ on $\mathfrak{a}$ extends to a K-invariant Lipschitzean function Φ on $\mathfrak{p}$.*

Proof. For $R > 0$ select C_R such that

$$|\phi(H) - \phi(H')| \leqslant C_R |H - H'| \qquad \text{for} \quad |H| < R, \quad |H'| < R. \tag{18}$$

Since all maximal abelian subspaces $\mathfrak{b}$ of $\mathfrak{p}$ are K-conjugate and since Φ is K-invariant it is clear that for each such $\mathfrak{b}$ the restriction $\psi = \Phi \mid \mathfrak{b}$ satisfies (18) (with the same constant C_R). Given $X, Y \in \mathfrak{p}$ let $\mathrm{Ad}(k_0)Y$ be the point on the orbit $\mathrm{Ad}(K)Y$ which minimizes the square distance $F(k) = |X - \mathrm{Ad}(k)Y|^2$. The relation

$$\left\{\frac{d}{dt} F((\exp tT)\, k_0)\right\}_{t=0} = 0, \qquad T \in \mathfrak{k},$$

leads, by simple calculation, to $[X, \mathrm{Ad}(k_0)Y] = 0$. Thus X and $\mathrm{Ad}(k_0)Y$ lie in a maximal abelian subspace $\mathfrak{b}$ and if $|X| < R$, $|Y| < R$,

$$\begin{aligned}|\Phi(X) - \Phi(Y)| &= |\psi(X) - \psi(\mathrm{Ad}(k_0)Y)| \leqslant C_R\, |X - \mathrm{Ad}(k_0)Y| \\ &\leqslant C_R\, |X - Y|,\end{aligned}$$

finishing the proof.

3. The Spherical Transform

Let $S(\mathfrak{p})$ denote the symmetric algebra over $\mathfrak{p}$ (with complex coefficients) and if $p \in S(\mathfrak{p})$ let $\partial(p)$ denote the corresponding constant coefficient differential operator on $\mathfrak{p}$. As p runs through the set $I(\mathfrak{p})$ of K-invariants in $S(\mathfrak{p})$ the operators $\partial(p)$ run through the algebra $\mathbf{D}(G_0/K)$ of G_0-invariant differential operators on $\mathfrak{p} = G_0/K$. For $\lambda \in \mathfrak{a}_c^*$ let $A_\lambda \in \mathfrak{a}_c$ be determined by $B(H, A_\lambda) = \lambda(H)$ $(H \in \mathfrak{a})$. The following analog of Harish–Chandra's integral formula for spherical functions on G/K is well-known (e.g. Gindikin [4], Korányi [10, sect. 1]) and can be proved in the same way (e.g. [8(a), p. 432]). Related results are given by Godement [5(a), Sect. 17].

Proposition 3.1. *The K-invariant joint eigenfunctions of the operators in $\mathbf{D}(G_0/K)$ are the constant multiples of the functions*

$$\psi_\lambda(X) = \int_K e^{iB(kA_\lambda, X)}\, dk \qquad (X \in \mathfrak{p})$$

as $\lambda \in \mathfrak{a}_c^*$; moreover $\psi_\lambda \equiv \psi_\mu$ if and only if λ and μ are W-conjugate.

Given a K-invariant function f on $\mathfrak{p}$ its *spherical transform* is given by

$$\tilde{f}(\lambda) = \int_{\mathfrak{p}} f(X)\, \psi_{-\lambda}(X)\, dX,$$

whenever the integral converges. Let $\mathscr{D}_K(\mathfrak{p})$ denote the set of K-invariant functions in $\mathscr{D}(\mathfrak{p})$ $(=C^\infty(\mathfrak{p}))$. Similarly, we define the subspace $\mathscr{S}_K(\mathfrak{p}) \subset \mathscr{S}(\mathfrak{p})$, $\mathscr{H}_K(\mathfrak{p}_c^*) \subset \mathscr{H}(\mathfrak{p}_c^*)$, $\mathscr{H}_W(\mathfrak{a}_c^*) \subset \mathscr{H}(\mathfrak{a}_c^*)$, $\mathscr{S}_W(\mathfrak{a}^*) \subset \mathscr{S}(\mathfrak{a}^*)$.

Theorem 3.2. *The spherical transform $f \to \tilde{f}$ is a bijection of $\mathscr{D}_K(\mathfrak{p})$ onto $\mathscr{H}_W(\mathfrak{a}^*)$.*

Proof. If $f \in \mathscr{D}_K(\mathfrak{p})$ we have by the K-invariance

$$\tilde{f}(\lambda) = \int_{\mathfrak{p}} f(X)\, e^{-iB(A_\lambda, X)}\, dX \tag{1}$$

Let $\mathfrak{q}$ be the orthogonal complement of $\mathfrak{a}$ in $\mathfrak{p}$ and let us call the translates of $\mathfrak{q}$ by elements of G_0, *planes* in $\mathfrak{p}$. Defining the *Radon transform* of f by

$$\hat{f}(\xi) = \int_{\xi} f(Y)\, dY,$$

where dY is the Euclidean volume element on the plane ξ we have by (1)

$$\tilde{f}(\lambda) = \int_{\mathfrak{a}} e^{i\lambda(H)} \hat{f}(H + \mathfrak{g})\, dH. \tag{2}$$

As a function of H, $\hat{f}(H + \mathfrak{q})$ is W-invariant and has compact support. Hence $\tilde{f} \in \mathscr{H}_W(\mathfrak{a}_c^*)$. Conversely, let $\phi \in \mathscr{H}_W(\mathfrak{a}_c^*)$. By Theorem 2.1, ϕ extends to a function $\Phi \in \mathscr{H}_K(\mathfrak{p}_c^*)$ which by the classical Paley–Wiener theorem is the usual Fourier transform of some $f \in \mathscr{D}_K(\mathfrak{p})$. But then $\phi = \tilde{f}$ as desired.

The following result follows in a similar manner from Proposition 2.3.

THEOREM 3.3. *The spherical transform is a bijection of $\mathscr{S}_K(\mathfrak{p})$ onto $\mathscr{S}_W(\mathfrak{a}^*)$.*

Let $\mathscr{E}'(\mathfrak{p})$ denote the space of distributions on $\mathfrak{p}$ of compact support and $\mathscr{E}'_K(\mathfrak{p})$ the subspace of K-invariants. For $T \in \mathscr{E}'_K(\mathfrak{p})$ we define the spherical transform $\tilde{T}$ by

$$\tilde{T}(\lambda) = \int_{\mathfrak{p}} \psi_{-\lambda}(X)\, dT(X), \qquad \lambda \in \mathfrak{a}_c^*.$$

In analogy with Schwartz [16, Ch. VII, sect. 8] and Hörmander [9, Ch. I, sect. 7], we can use Theorems 3.2 and 3.3 to characterize the function $\tilde{T}$.

COROLLARY 3.4. *The spherical transform $T \to \tilde{T}$ is a bijection of $\mathscr{E}'_K(\mathfrak{p})$ onto the space of W-invariant holomorphic functions f on $\mathfrak{a}_c^*$ satisfying an inequality of the form*

$$\sup_{\zeta}(1 + |\zeta|)^{-N} e^{-R|\mathrm{Im}\zeta|} |f(\zeta)| < \infty$$

for some $R, N \geqslant 0$.

4. The J-Polynomials

The decompositions $\mathfrak{p} = \mathfrak{a} + \mathfrak{q}$, $\mathfrak{g} = \mathfrak{k} + \mathfrak{p}$, $S(\mathfrak{g}) = S(\mathfrak{p}) + S(\mathfrak{g})\mathfrak{k}$ give rise to a direct decomposition $S(\mathfrak{g}) = S(\mathfrak{a}) \oplus (S(\mathfrak{g})\mathfrak{k} + \mathfrak{q}S(\mathfrak{g}))$. Let $p \to j^p$ denote the corresponding projection of $S(\mathfrak{g})$ onto $S(\mathfrak{a})$. Let $H(\mathfrak{p}^*) \subset S(\mathfrak{p}^*)$ and $H(\mathfrak{p}) \subset S(\mathfrak{p})$ denote the subspaces of harmonic polynomials, that is $q \in H(\mathfrak{p}^*)$ if $\partial(p)q = 0$ for all $p \in I(\mathfrak{p})$ without constant term. Let M denote the centralizer of $A = \exp \mathfrak{a}$

in K and let $\hat{K}$ (resp. $\hat{K}_M$) denote the set of equivalence classes of irreducible unitary representations δ of K (resp. such that $\delta(M)$ has a fixed nonzero vector). Let V_δ be a representation space of δ (with inner product $\langle,\rangle$) and $V_\delta^M \subset V_\delta$ the subspace of fixed vectors under $\delta(M)$. Let $v_1, \ldots, v_{d(\delta)}$ be a basis of V_δ such that $v_1, \ldots, v_{l(\delta)}$ is a basis of V_δ^M. Under the adjoint action of K on $\mathfrak{p}$ (resp. coadjoint action of K on $\mathfrak{p}^*$) let $S_\delta(\mathfrak{p})$, $H_\delta(\mathfrak{p})$ (resp. $S_\delta(\mathfrak{p}^*)$, $H_\delta(\mathfrak{p}^*)$) be the subspaces of elements of type δ. Under the mapping of $\mathfrak{p}$ onto $\mathfrak{p}^*$ whereby $X \in \mathfrak{p}$ is mapped into the linear form $Y \to B(X, Y)$, $S_\delta(\mathfrak{p})$ gets mapped onto $S_\delta(\mathfrak{p}^*)$. Note also that if $p \in S(\mathfrak{p})$ then the differential operator $\partial(p)$ satisfies $\partial(\mathrm{Ad}(k) \cdot p) = \partial(p)^{\mathrm{Ad}(k)}$ $(k \in K)$. Let $\epsilon_1, \ldots, \epsilon_{l'(\delta)}$ be a basis of the space $E_\delta = \mathrm{Hom}_K(V_\delta, H_\delta(\mathfrak{p}))$ of linear maps $V_\delta \to H_\delta(\mathfrak{p})$ commuting with the actions of K. Thus $\epsilon_j(\delta(k)v) = \mathrm{Ad}(k)(\epsilon_j(v))$, $v \in V_\delta$. For $\lambda \in \mathfrak{a}_c^*$ we consider the $l(\delta) \times l'(\delta)$ matrix

$$J^\delta(\lambda)_{ij} = j^{\epsilon_j(v_i)}(-i\lambda). \tag{1}$$

If the bases (v_i), (ϵ_j) are replaced by new bases (v_i'), (ϵ_j') by means of linear transformations A and B respectively, the new matrix $J'^\delta(\lambda)$ is given by

$$J'^\delta(\lambda) = {}^tAJ^\delta(\lambda)B, \tag{2}$$

${}^tA$ being the transpose of A. As proved by Kostant and Rallis [12, p. 759], $H_\delta(\mathfrak{p})$ decomposes into $l(\delta)$ irreducible subspaces

$$H_\delta(\mathfrak{p}) = \sum_{j=1}^{l(\delta)} H_{\delta,j} \tag{3}$$

and each $H_{\delta,j}$ can be assumed to consist of homogeneous polynomials of degree, say $d_j(\delta)$. Thus $l'(\delta) = l(\delta)$. If the basis (ϵ_j) had been chosen such that, by (3), $\epsilon_j(V_\delta) = H_{\delta,j}$ $(1 \leqslant j \leqslant l(\delta))$ we see that the matrix entry $J^\delta(\lambda)_{ij}$ is a homogeneous polynomial of degree $d_j(\delta)$.

PROPOSITION 4.1. *For any choice of bases (v_i) of V_δ^M, (ϵ_j) of E_δ, the polynomial* $\det(J^\delta(\lambda))$ *is a homogeneous polynomial of degree* $\sum_{j=1}^{l(\delta)} d_j(\delta)$.

This is clear from the above remarks combined with (2).

We shall now relate the matrices J^δ to their curved space analogs introduced in [15] and [11(b)]. Let $\mathfrak{g} = \mathfrak{k} + \mathfrak{a} + \mathfrak{n}$ be the Iwasawa decomposition of $\mathfrak{g}$ corresponding to the Weyl chamber $\mathfrak{a}^+$. Let $p \to p^*$ denote the symmetrization map of $S(\mathfrak{g})$ onto the universal enveloping algebra $U(\mathfrak{g})$ and for $u \in U(\mathfrak{g})$ let $q^u \in S(\mathfrak{a})$ be determined by $u - q^u \in U(\mathfrak{g})\mathfrak{k} + \mathfrak{n}U(\mathfrak{g})$. For $\lambda \in \mathfrak{a}_c^*$ we consider as in [8(g), p. 198] the $l(\delta) \times l(\delta)$ matrix

$$Q^\delta(\lambda)_{ij} = q^{\epsilon_j(v_i)^*}(\rho - i\lambda), \qquad 1 \leqslant i, j \leqslant l(\delta), \tag{4}$$

where ρ is half the sum of the elements in Σ^+, counted with multiplicity.

LEMMA 4.2. *Let the basis* (ϵ_j) *be chosen such that* $\epsilon_j(V_\delta) = H_{\delta,j} \subset S_{d_j(\delta)}(\mathfrak{p})$. *Then*

$$\lim_{R\to\infty} R^{-d_j(\delta)} Q^\delta(R\lambda)_{ij} = J^\delta(\lambda)_{ij},$$

uniformly as λ *varies in any compact set.*

Proof. Let $p \in S_d(\mathfrak{g})$. Then for certain $T_i \in \mathfrak{k}$, $Z_j \in \mathfrak{n}$, $p_i, q_j \in S_{d-1}(\mathfrak{g})$, $h \in S_d(\mathfrak{a})$ we have

$$p = \sum_i p_i T_i + \sum_j Z_j q_j + h. \tag{5}$$

By symmetrization

$$p^* = \sum_i p_i^* T_i^* + \sum_j Z_j^* q_j^* + h^* + \text{term of degree} < d$$

so that

$$q^{p^*} = h^* + \text{term of degree} < d. \tag{6}$$

On the other hand, (5) implies

$$p = \sum_i p_i T_i + \sum_j \tfrac{1}{2}(Z_j + \theta Z_j)\, q_j + \sum_j \tfrac{1}{2}(Z_j - \theta Z_j)\, q_j + h$$

so since $Z_j - \theta Z_j \in \mathfrak{q}$ we have

$$j^p = h. \tag{7}$$

Since we can identify $S(\mathfrak{a})$ and $S(\mathfrak{a})^*$ ($\mathfrak{a}$ being commutative) we deduce from (6) and (7)

$$q^{\epsilon_j(v_i)^*}(-i\lambda) = j^{\epsilon_j(v_i)}(-i\lambda) + r(\lambda),$$

where r has degree $< d_j(\delta)$. Since $q^u(-i\lambda)$ and $q^u(\rho - i\lambda)$ have the same terms of highest degree in λ the lemma follows.

If $g \in G$ let $A(g) \in \mathfrak{a}$ be determined by $g \in N \exp A(g) K$ and if $k \in K$ define the element $A(gK, kM) \in \mathfrak{a}$ by $A(gK, kM) = A(k^{-1}g)$. Geometrically, $\exp A(g)$ is the "composite distance" from the origin o in G/K to the horocycle through $g \cdot o$ parallel to $N \cdot o$ ([8(d), p. 9]). Since for $H \in \mathfrak{a}$ the plane $H + \mathfrak{q}$ in $\mathfrak{p}$ is the tangent space analog to the horocycle $(\exp H)N \cdot o$ in G/K the following lemma is geometrically plausible [4, sect. 6]. Let $A\colon \mathfrak{p} \to \mathfrak{a}$ denote the orthogonal projection.

LEMMA 4.3. *Let* $H \in \mathfrak{a}$. *Then*

$$\left\{\frac{d}{dt} A(k \exp tH)\right\}_{t=0} = A(k \cdot H)$$

uniformly for $k \in K$.

Proof. We have

$$A(k \exp tH) = A(k \exp tHk^{-1}) = A(\exp tk \cdot H) = A(\exp t(A(k \cdot H) + X))$$

where $X \in \mathfrak{q}$. But $X = Z - \theta Z$ where $Z \in \mathfrak{n}$ so $X = 2Z - (Z + \theta Z)$. Thus by standard properties of exp,

$$\begin{aligned} tA(k \cdot H) &= A(\exp tA(k \cdot H)) = A(\exp t2Z \exp tA(k \cdot H) \exp t(-Z - \theta Z)) \\ &= A(\exp[t(2Z + A(k \cdot H) - (Z + \theta Z)) + O(t^2)]) \\ &= A(\exp[tk \cdot H + O(t^2)]), \end{aligned}$$

where $O(t^2)/t^2$ is bounded, uniformly in k, for small t. The lemma follows.

For $\delta \in \hat{K}_M$, $\lambda \in \mathfrak{a}_c^*$ we consider as in [8(g), sect. 4] the generalized spherical function on X given by

$$\Phi_{\lambda,\delta}(x) = \int_K e^{(i\lambda+\rho)(A(x,kM))} \delta(k)\, dk, \qquad x \in X.$$

The analogous concept for $\mathfrak{p}$ is the function

$$\Psi_{\lambda,\delta}(X) = \int_K e^{iB(k \cdot A_\lambda, X)} \delta(k)\, dk, \qquad X \in \mathfrak{p},$$

which we refer to as the *generalized Bessel function* in accordance with the terminology of Gross–Kunze [6], where such functions are considered for compact automorphism groups of locally compact abelian groups.

We now recall an invariance property of the generalized spherical function $\Phi_{\lambda,\delta}$ proved in [8(e), 8(g), p. 205]. Let $\check{\delta}$ denote the representation of K, contragredient to δ operating on the dual space $V_\delta^* = V_{\check{\delta}}$. Let $v \to v'$ denote the semilinear mapping of V_δ onto $V_{\check{\delta}}$ defined by $v'(u) = \langle u, v\rangle$ for $u, v \in V_\delta$ and define the inner product $\langle , \rangle$ on $V_{\check{\delta}}$ by $\langle v', w'\rangle = \langle w, v\rangle$. For a linear transformation L of V_δ let L' denote the linear transformation $L'(v') = (Lv)'$, $v' \in V_{\check{\delta}}$. If L has matrix expression (l_{ij}) in a basis $v_1, \ldots, v_{d(\delta)}$, then L' has the expression $(\overline{l_{ij}})$ in the basis (v_i'). Since $\check{\delta}(k) = {}^t(\delta(k^{-1}))$, we have $\check{\delta}(k) = \delta(k)'$, $\check{\delta}$ is unitary, and in terms of the bases (v_i) and (v_i'), $\check{\delta}(k)_{ij} = \overline{\delta(k)}_{ij}$.

Lemma 4.4. *Let $v_1, \ldots, v_{d(\delta)}$ be an orthonormal basis of V_δ, $v_1, \ldots, v_{l(\delta)}$ spanning $V_\delta{}^M$, (v_i') the dual basis of $V_{\check{\delta}}$. Let the matrix $Q^{\check{\delta}}(\lambda)$ be calculated by means of the basis $v_1', \ldots, v_{l(\delta)}'$ of $V_{\check{\delta}}^M$ and an arbitrary basis (ϵ_j') of $E_{\check{\delta}}$. Expressing δ in terms of (v_i) we have the matrix identities*

$$\begin{aligned} &Q^{\check{\delta}}(\lambda)^{-1} \int_K e^{(-i\lambda+\rho)(A(k \exp H))} \delta(k)\, dk \\ &\quad = Q^{\check{\delta}}(s\lambda)^{-1} \int_K e^{(-is\lambda+\rho)(A(k \exp H))} \delta(k)\, dk \qquad (8) \end{aligned}$$

for $s \in W$, $H \in \mathfrak{a}$ and both sides are holomorphic in λ on $\mathfrak{a}_c^*$.

Proof. Formula (8) amounts to (2) Section 7 in [8(*g*)] taking into account Theorem 6.3 which assumes the bases (v_i) and (v_i') chosen as above and in addition (ϵ_j') chosen such that $Q^{\check{\delta}}(i\rho) = I$ (the identity). But passing to a new basis of $E_{\check{\delta}}$ means (as in (2)) changing $Q^{\check{\delta}}(\lambda)$ to $Q^{\check{\delta}}(\lambda)B$, where B is a constant nonsingular matrix which would cancel out in (8). That both sides of (8) are holomorphic on all of $\mathfrak{a}_c^*$ follows from the Weyl group invariance by the argument in [8(g), p. 207].

Remark. If we change $v_1, \ldots, v_{l(\delta)}$ to another orthonormal basis (w_i) where $w_i = Av_i = \sum_j a_{ji} v_j$ then $w_i' = \sum_j \bar{a}_{ji} v_j'$. This changes $Q^{\check{\delta}}(\lambda)$ to ${}^t\bar{A}Q^{\check{\delta}}(\lambda)$ (cf. (2)) and changes $Q^{\check{\delta}}(\lambda)^{-1}$ to $Q^{\check{\delta}}(\lambda)^{-1}A$, A being unitary. On the other hand the matrix expression, say C, of the integral in (8) restricted to $V_\delta{}^M$ is by the basis change replaced by $A^{-1}CA$. Thus the entire expression is just changed by multiplication on the right by A.

Next we need an expression for the determinant of $J^\delta(\lambda)$. Let $\Sigma_0{}^+ \subset \Sigma^+$ denote the set of indivisible positive roots and for $\alpha \in \Sigma_0{}^+$ consider the subgroup $G^\alpha \subset G$ of real rank one with Lie algebra $\mathfrak{g}^\alpha$ generated by the root spaces $\mathfrak{g}_\alpha$ and $\mathfrak{g}_{-\alpha}$. With $K^\alpha = G^\alpha \cap K$, $\mathfrak{p}^\alpha = \mathfrak{g}^\alpha \cap \mathfrak{p}$ consider as in [8(g), p. 197] the representations $\delta_i{}^\alpha$ $(1 \leqslant i \leqslant l(\delta))$ of K^α and the degree $D(\delta_i{}^\alpha)$ of the K^α-harmonic polynomials in the K^α-irreducible space $H(\mathfrak{p}^\alpha)_{\delta_i{}^\alpha}$. Let $D(\alpha) = \sum_{i=1}^{l(\delta)} D(\delta_i{}^\alpha)$.

LEMMA 4.5. *For a certain nonzero constant c,*

$$\det J^\delta(\lambda) = c \prod_{\alpha \in \Sigma_0{}^+} \langle \lambda, \alpha \rangle^{D(\alpha)}.$$

Proof. If G/K has rank one this is clear from Proposition 4.1. The general case follows from Kostant's product formula for $\det Q^\delta(\lambda)$ ([11, p. 292] or [8(g), p. 203]) if we take Lemma 4.2 into account.

We can now prove the main result of this section which shows that the generalized Bessel function is divisible by the J-matrix.

THEOREM 4.6. *With the bases of $V_\delta{}^M$, $V_{\check{\delta}}{}^M$ as in Lemma 4.4 and the basis (ϵ_j') of $E_{\check{\delta}}$ arbitrary we have the matrix identities*

$$J^{\check{\delta}}(\lambda)^{-1} \int_K e^{-i\lambda(A(k\cdot H))}\, \delta(k)\, dk$$
$$= J^{\check{\delta}}(s\lambda)^{-1} \int_K e^{-is\lambda(A(k\cdot H))}\, \delta(k)\, dk \tag{9}$$

for $s \in W$, $H \in \mathfrak{a}$ and both sides are holomorphic in λ on all of $\mathfrak{a}_c^*$.

Proof. Because of (2) we can assume the basis (ϵ_j') chosen such that $\epsilon_j'(V_{\check{\delta}}) = H_{\check{\delta},j}$.

Let $(Q_c(\lambda)_{ij})$ be the matrix formed by the cofactors in $(Q^{\check{\delta}}(\lambda)_{ji})$ so that

$$Q^{\check{\delta}}(\lambda)^{-1} = (\det Q^{\check{\delta}}(\lambda))^{-1} Q_c(\lambda).$$

We define $(J_c(\lambda)_{ij})$ similarly. Then the ijth matrix element in the left hand side of (8) is

$$\det Q^{\check{\delta}}(\lambda)^{-1} \left[\sum_{p=1}^{l(\delta)} Q_c(\lambda)_{ip} \int_K e^{(-i\lambda+\rho)(A(k\exp H))} \delta_{pj}(k)\, dk \right]. \tag{10}$$

Let $\alpha_1, ..., \alpha_l$ be the simple roots in Σ^+, let $e_1, ..., e_l$ be the dual basis of $\mathfrak{a}$ and write $m_i(\alpha)$ for the positive integers determined by $\alpha = \sum_1^l m_i(\alpha)\alpha_i$, $\alpha \in \Sigma^+$. Let $\mathfrak{a}_c'$ denote the set of $\lambda \in \mathfrak{a}_c^*$ where $\det J^{\check{\delta}}(\lambda) \neq 0$. Writing $A_\lambda = \sum_1^l \lambda_i e_i$, $\mathfrak{a}_c'$ is the space $\mathfrak{a}_c^*$ with the finitely many hyperplanes $\sum_1^l m_i(\alpha)\lambda_i = 0$ removed $(\alpha \in \sum_0{}^+)$. Now Lemma 4.2 implies that

$$\lim_{R\to\infty} R^{d_i(\check{\delta})} \det Q^{\check{\delta}}(R\lambda)^{-1} Q_c(R\lambda)_{ip} = \det J^{\check{\delta}}(\lambda)^{-1} J_c(\lambda)_{ip}$$

uniformly in each compact subset of the region $\mathfrak{a}_c'$. Replacing in (10). λ by $R\lambda$, H by H/R we deduce from Lemmas 4.2–4.3 that as $R \to \infty$ the left hand side of (10) multiplied by $R^{d_i(\check{\delta})}$ converges to

$$\det J^{\check{\delta}}(\lambda)^{-1} \left[\sum_1^{l(\delta)} J_c(\lambda)_{ip} \int_K e^{-i\lambda(A(k\cdot H))} \delta_{pj}(k)\, dk \right], \tag{11}$$

uniformly on each compact subset of $\mathfrak{a}_c'$. Since Σ^+ is a finite set we can select numbers $d_1 > 0, ..., d_l > 0$ such that the set

$$\mathfrak{a}_d = \{\lambda \in \mathfrak{a}_c^* \mid |\lambda_1| = d_1, ..., |\lambda_l| = d_l\}$$

belongs to $\mathfrak{a}_c'$. (For example take $d_i > m(d_{i+1} + \cdots + d_l)$ for $1 \leqslant i < l$ where $m = \operatorname{Max}_{i,\alpha}(m_i(\alpha))$). Cauchy's theorem for the polydisk $|\lambda_1| \leqslant d_1, ..., |\lambda_l| \leqslant d_l$ then implies that the above convergence to (11) is uniform in each polydisk $|\lambda_i| \leqslant d_i' < d_i$ $(1 \leqslant i \leqslant l)$. Since $\mathfrak{a}_{td} \subset \mathfrak{a}_c'$ for each $t > 0$ we can conclude that (11) is holomorphic on all of $\mathfrak{a}_c^*$. The Weyl group invariance follows by means of this limit argument as well, but can also be verified by a direct computation.

Finally we show that the generalized Bessel function can be expressed by means of certain derivatives of the zonal spherical function ([8(g), Theorem 4.1], [2]). First we note an obvious lemma.

Lemma 4.7. *Let $k \in K$, $\lambda \in \mathfrak{a}_c^*$, $p \in S(\mathfrak{p})$ and put*

$$f_k(X) = e^{iB(kA_\lambda, X)}, \qquad X \in \mathfrak{p}.$$

Then

$$\partial(p)\, f_k = j^{(k^{-1}\cdot p)}(i\lambda)\, f_k\,.$$

For notational simplicity we shall now write p instead of $\partial(p)$, $p \in S(\mathfrak{p})$. Let $p \to p^t$ denote the complex linear automorphism of $S(\mathfrak{p})$ induced by the mapping $X \to -X$ of $\mathfrak{p}$. With bases (v_i), (v_i') and (ϵ_j') as in Lemma 4.4 we define the matrix differential operator $L^{\check{\delta}}$ (with $d(\delta)$ rows, $l(\delta)$ columns) by

$$(L^{\check{\delta}} f)_{ij}(X) = (\epsilon_j'(v_i')^t\, f)(X), \qquad f \in C^\infty(\mathfrak{p}).$$

THEOREM 4.8. *The generalized Bessel function can be written*

$$\Psi_{\lambda,\delta}(X) = (L^{\check{\delta}}\psi_\lambda)(X)\; J^{\check{\delta}}(\lambda)^{-1} \qquad \textit{on} \quad V_\delta^{\,M}.$$

Proof. We may assume the basis (ϵ_j') chosen such that $\epsilon_j'(V_{\check{\delta}}) = H_{\check{\delta},j} \subset S_{d_j(\check{\delta})}$, (cf. Lemma 4.2). In fact if we go to a new basis $B\epsilon_j'$ then $L^{\check{\delta}} f$ and $J^{\check{\delta}}(\lambda)$ are replaced by $(L^{\check{\delta}} f)B$ and $J^{\check{\delta}}(\lambda)B$ respectively. Since

$$k^{-1}\cdot \epsilon_j'(v_i') = \epsilon_j'(\check{\delta}(k^{-1})\, v_i'),\ \check{\delta}(k^{-1})\, v_i' = \sum_q \langle \delta(k)\, v_q\,,\, v_i\rangle\, v_q'$$

we obtain

$$\begin{aligned}(\epsilon_j'(v_i')\,\psi_\lambda)(X) &= \int_K j^{\epsilon_j'(\check{\delta}(k^{-1})v_i')}(i\lambda)\; e^{iB(kA_\lambda, X)}\, dk \\ &= \sum_{q=1}^{d(\delta)} j^{\epsilon_j'(v_q')}(i\lambda) \int_K \langle \delta(k)\, v_q\,,\, v_i\rangle\, e^{iB(k\cdot A_\lambda, X)}\, dk.\end{aligned}$$

Replacing k by km and integrating over M we see that the integral vanishes for $q > l(\delta)$. Noting also that if $p \in S_k(\mathfrak{p})$ then

$$p^t = (-1)^k\, p, \qquad j^{p^t}(i\lambda) = j^p(-i\lambda)$$

we obtain

$$(\epsilon_j'(v_i')^t\, \psi_\lambda)(X) = \sum_{q=1}^{l(\delta)} \Psi_{\lambda,\delta}(X)_{iq}\; J^{\check{\delta}}(\lambda)_{qj}\,,$$

proving the theorem.

5. The Fourier Transform of K-Finite Functions

If f is a K-finite function on $\mathfrak{p}$ of type $\check{\delta}$ we define its *generalized Bessel transform* by

$$\tilde{f}(\lambda) = d(\delta) \int_{\mathfrak{p}} f(X)\, \Psi_{\bar{\lambda},\delta}(X)^*\, dX = d(\delta) \int_{\mathfrak{p}} f(X) \int_K e^{-iB(kA_\lambda, X)} \delta(k^{-1})\, dk\, dX \tag{1}$$

for all $\lambda \in \mathfrak{a}_c^*$ for which the integral converges. Here $*$ denotes adjoint on $\operatorname{Hom}(V_\delta, V_\delta)$. This definition is the flat analog of the one in [8(e), sect. 4], [8(g), sect. 7] and resembles the one in [6, sect. 3].

Let $\mathscr{H} = \mathscr{H}(\mathfrak{a}^*, \operatorname{Hom}(V_\delta, V_\delta^M))$ denote the space of holomorphic functions on $\mathfrak{a}_c^*$ of exponential type with values in $\operatorname{Hom}(V_\delta, V_\delta^M)$; this space is provided with its usual topology such that the Euclidean Fourier transform is a homeomorphism of the space $\mathscr{D}(A, \operatorname{Hom}(V_\delta, V_\delta^M))$ (of C^∞ compactly supported functions on A with values in $\operatorname{Hom}(V_\delta, V_\delta^M)$) onto $\mathscr{H}(\mathfrak{a}^*, \operatorname{Hom}(V_\delta, V_\delta^M))$. We denote by $\mathscr{H}^\delta(\mathfrak{a}^*)$ the topological subspace

$$\mathscr{H}^\delta(\mathfrak{a}^*) = \{F \in \mathscr{H} \mid (J^{\check{\delta}})^{-1} F \ W\text{-invariant and holomorphic}\},$$

the bases used being as in Lemma 4.4. This space is analogous to the one defined in [8(g), p. 204]. But while there the Weyl group invariance of the function $Q^{\check{\delta}}(\lambda)^{-1}F(\lambda)$ implied (loc. cit. p. 206) that it was holomorphic the analogous statement for $J^{\check{\delta}}$ is false and our present definition of $\mathscr{H}^\delta(\mathfrak{a}^*)$ takes this into account.

Theorem 5.1. *The mapping $f \to \tilde{f}$ where*

$$\tilde{f}(\lambda) = d(\delta) \int_{\mathfrak{p}} f(X)\, \Psi_{\bar{\lambda},\delta}(X)^*\, dX$$

is a homeomorphism of $\mathscr{D}_{\check{\delta}}(\mathfrak{p})$ onto $\mathscr{H}^\delta(\mathfrak{a}^*)$. The inverse is given by

$$f(X) = c \operatorname{Tr} \left[\int_{\mathfrak{a}^*} \Psi_{\lambda,\delta}(X)\, \tilde{f}(\lambda)\, \omega(\lambda)\, d\lambda \right]$$

where c is a constant, independent of f, Tr *means trace on V_δ, and the density ω is given by* (15), sect. 2.

Proof. From Theorem 4.6 we know that $J^{\check{\delta}}(\lambda)^{-1}\Psi_{\bar{\lambda},\delta}(X)^*$ is W-invariant and holomorphic on $\mathfrak{a}_c^*$. Thus $f \to \tilde{f}$ maps $\mathscr{D}_{\check{\delta}}(\mathfrak{p})$ into $\mathscr{H}^\delta(\mathfrak{a}^*)$.

For the surjectivity, let $F \in \mathscr{H}^\delta(\mathfrak{a}^*)$ and define f by

$$f(X) = \operatorname{Tr} \left[\int_{\mathfrak{a}^*} \Psi_{\lambda,\delta}(X)\, F(\lambda)\, \omega(\lambda)\, d\lambda \right]. \tag{2}$$

Here we substitute the formula for $\Psi_{\lambda,\delta}$ from Theorem 4.8 and write $F = J^{\check{\delta}}\Phi$, where Φ is W-invariant and holomorphic. But writing

$$\Phi(\lambda) = \det J^{\check{\delta}}(\lambda)^{-1}(J_c(\lambda)\, F(\lambda))$$

we see from Malgrange's theorem [14, p. 306] that Φ is of exponential type. Thus Φ belongs to the space $\mathscr{I}^{\delta}(\mathfrak{a}^*)$ of W-invariants in $\mathscr{H}(\mathfrak{a}^*, \operatorname{Hom}(V_\delta, V_\delta{}^M))$. Then if ϕ denotes the inverse spherical transform

$$\phi(X) = \int_{\mathfrak{a}^*} \psi_\lambda(X)\, \Phi(\lambda)\, \omega(\lambda)\, d\lambda$$

we obtain by (2)

$$f = c_0 \operatorname{Tr}(L^{\check{\delta}}\phi), \tag{3}$$

where c_0 is a constant. Note that $L^{\check{\delta}}$ has $d(\delta)$ rows, $l(\delta)$ columns, ϕ has $l(\delta)$ rows, $d(\delta)$ columns. More explicitly, if $\phi(X)v_i = \sum_1^{l(\delta)} \phi_{ji}(X)v_j$, $1 \leqslant i \leqslant d(\delta)$, then

$$f = c_0 \sum_{\substack{1\leqslant i\leqslant d \\ 1\leqslant j\leqslant l}} \epsilon_j'(v_i')^t\, \phi_{ji}\,.$$

By Theorem 3.2, ϕ, and therefore f, has compact support so by (2) $f \in \mathscr{D}_{\check{\delta}}(\mathfrak{p})$. The generalized Bessel transform is then

$$\tilde{f}(\lambda) = c_0\, d(\delta) \int_K \delta(k)\, dk \sum_{i,j} \int_{\mathfrak{p}} (\epsilon_j'(v_i')^t\, \phi_{ji})(X)\, e^{-iB(k^{-1}\cdot A_\lambda, X)}\, dX.$$

The integral over $\mathfrak{p}$ is by Lemma 4.7

$$\int_{\mathfrak{p}} \phi_{ji}(X)\, j^{\epsilon_j'(\check{\delta}(k)v_i')}(-i\lambda)\, e^{-iB(k^{-1}A_\lambda, X)}\, dX$$

so, since

$$\check{\delta}(k)\, v_i' = \sum_1^{d(\delta)} \overline{\delta(k)}_{qi}\, v_q' = \sum_{q=1}^{d(\delta)} \langle v_q\,, \delta(k)\, v_i\rangle\, v_q'\,,$$

we obtain by the K-invariance of ϕ_{ji},

$$\tilde{f}(\lambda) = c_0 d(\delta) \sum_{i,j,q} \int_K \delta(k)\langle v_q\,, \delta(k)\, v_i\rangle\, dk\, j^{\epsilon_j'(v_q')}(-i\lambda)\, \widetilde{\phi_{ji}}(\lambda),$$

where $1 \leqslant i, q \leqslant d(\delta)$, $1 \leqslant j \leqslant l(\delta)$. But by its definition $\tilde{f}(\lambda) = \delta(m)\tilde{f}(\lambda)$, $m \in M$, so multiplying the equation above by $\delta(m)$ we have on the right the integral

$$\int_K \delta(mk)\langle v_q\,, \delta(k)\, v_i\rangle\, dk = \int_K \delta(k)\langle \delta(m)\, v_q\,, \delta(k)\, v_i\rangle\, dk.$$

Integrating this over M we see that it vanishes for $q > l(\delta)$ and for $1 \leqslant q \leqslant l(\delta)$ it equals $d(\delta)^{-1}E_{qi}$ by the orthogonality relations. Thus

$$\begin{aligned} \tilde{f}(\lambda) &= c_0 \sum_{\substack{1\leqslant j,q\leqslant l(\delta)\\ 1\leqslant i\leqslant d(\delta)}} E_{qi} J^{\check{\delta}}(\lambda)_{qj}\, \widetilde{\phi_{ji}}(\lambda) \\ &= c J^{\check{\delta}}(\lambda)\, \Phi(\lambda) = cF(\lambda), \end{aligned}$$

where $c \neq 0$ is another constant. This proves the surjectivity.

The injectivity follows from the formula

$$e^{i\lambda(A(k\cdot X))} = \sum_{i,j,\delta} d(\delta) \langle \Psi_{\lambda,\delta}(X)\, v_j\,,\, v_i\rangle \langle \delta(k)\, v_i\,,\, v_j\rangle,$$

which is obtained by Peter–Weyl expansion along K.

Finally, we must prove the topological statement of the theorem. The space $\mathscr{D}_{\check{\delta}}(\mathfrak{p})$ is an LF-space (strict inductive limit of Fréchet spaces) as proved in [8(g), p. 192]. Moreover, $\mathscr{I}^{\delta}(\mathfrak{a}^*)$ is an LF-space [8(g), p. 207]. The mapping $\Phi(\lambda) \to J^{\check{\delta}}(\lambda)\, \Phi(\lambda)$ of $\mathscr{I}^{\delta}(\mathfrak{a}^*)$ onto $\mathscr{H}^{\delta}(\mathfrak{a}^*)$ is obviously continuous; its inverse is the composite $I_1^{-1} \circ I_2$ where the maps I_1 and I_2 are given by

$$I_2: F(\lambda) \in \mathscr{H}^{\delta}(\mathfrak{a}^*) \to J_c(\lambda)\, F(\lambda) \in J_c \mathscr{H}^{\delta}(\mathfrak{a}^*)$$

$$I_1: \Phi(\lambda) \in \mathscr{I}^{\delta}(\mathfrak{a}^*) \to \det(J^{\delta}(\lambda))\, \Phi(\lambda) \in (\det J^{\delta}) \mathscr{I}^{\delta}(\mathfrak{a}^*).$$

Since I_2 is continuous and I_1 a homeomorphism, $I_1^{-1} \circ I_2$ is continuous. Thus $\mathscr{H}^{\delta}(\mathfrak{a}^*)$ is an LF-space and $f \to \tilde{f}$ is a one-to-one a continuous map of one LF-space onto another, hence a homeomorphism.

From (3) we now obtain the following description of the K-finite elements of $C_c^\infty(\mathfrak{p})$. The analog for G/K was obtained by Torasso [18] in his alternative proof of the Paley–Wiener theorem in [8(e)].

COROLLARY 5.2. *Each K-finite function $f \in C_c^\infty(\mathfrak{p})$ is a linear combination of constant coefficient derivatives of K-invariant functions in $C_c^\infty(\mathfrak{p})$.*

Because of Theorem 3.3 the proof of Theorem 5.1 also yields its analog for the spaces of rapidly decreasing functions. Let $\mathscr{S}_{\check{\delta}}(\mathfrak{p})$ denote the space of rapidly decreasing C^∞ functions on $\mathfrak{p}$ of type $\check{\delta}$ and $\mathscr{S}^{\delta}(\mathfrak{a}^*)$ the space of functions $F \in \mathscr{S}(\mathfrak{a}^*, \mathrm{Hom}(V_\delta\,, V_\delta^M))$ such that $J^{\check{\delta}}(\lambda)^{-1}F(\lambda)$ is W-invariant and of class C^∞ on $\mathfrak{a}^*$. Writing

$$J^{\check{\delta}}(\lambda)^{-1}\, F(\lambda) = \det\, J^{\check{\delta}}(\lambda)^{-1}\, J_c(\lambda)\, F(\lambda), \tag{4}$$

and noting that by Lemma 4.5, $\det J^{\check{\delta}}(\lambda)$ is a product of functions of a single variable, we see from [8(b), Lemma 3.1] that the function (4) is rapidly decreasing on $\mathfrak{a}^*$. We therefore obtain the following results in the manner indicated.

THEOREM 5.3. *The mapping $f \to \tilde{f}$ is a homeomorphism of $\mathscr{S}_{\delta}^{\vee}(\mathfrak{p})$ onto $\mathscr{S}^{\delta}(\mathfrak{a}^*)$.*

The topological statement is now trivial because the spaces are closed subspaces of Fréchet spaces, hence are also Fréchet.

COROLLARY 5.4. *Each K-finite $f \in \mathscr{S}(\mathfrak{p})$ is a linear combination of constant coefficient derivatives of functions in $\mathscr{S}_K(\mathfrak{p})$.*

6. EIGENFUNCTIONS AND EIGENSPACE REPRESENTATIONS

For $\lambda \in \mathfrak{a}_c^*$ let $\mathscr{E}_\lambda(\mathfrak{p})$ denote the joint eigenspace

$$\mathscr{E}_\lambda(\mathfrak{p}) = \{f \in C^\infty(\mathfrak{p}) \mid \partial(p)f = p(i\lambda)f \text{ for all } p \in I(\mathfrak{p})\}.$$

This space is given the topology induced by that of $C^\infty(\mathfrak{p})$. Each joint eigenspace of the operators $\partial(I(\mathfrak{p})) = \mathbf{D}(G_0/K)$ is of this form for a suitable λ and $\mathscr{E}_{s\lambda}(\mathfrak{p}) = \mathscr{E}_\lambda(\mathfrak{p})$ for all $s \in W$. If F is a continuous function on K/M then its "Poisson transform"

$$f(X) = \int_{K/M} e^{iB(k \cdot A_\lambda, X)} F(KM)\, dk_M \tag{1}$$

belongs to $\mathscr{E}_\lambda(\mathfrak{p})$. Here dk_M is the normalized K-invariant measure on K/M.

DEFINITION [8(d), p. 94]. The point $\lambda \in \mathfrak{a}_c^*$ is said to be *simple* if the mapping $F \to f$ defined by (1) is injective.

We recall that $\lambda \in \mathfrak{a}_c^*$ is said to be *regular* if $\alpha(A_\lambda) \neq 0$ for all α in Σ, the set of restricted roots; otherwise λ is said to be *singular*. Let K_λ denote the stabilizer of λ in K.

PROPOSITION 6.1. *A point $\lambda \in \mathfrak{a}_c^*$ is regular if and only if $K_\lambda = M$.*

Proof. If λ is singular K_λ has a larger Lie algebra than M so $K_\lambda \neq M$. If λ is regular then K_λ and M have the same Lie algebra. If $k \in K_\lambda$ and $\lambda = \xi + i\eta$ ($\xi, \eta \in \mathfrak{a}^*$) then $k\xi = \xi$, $k\eta = \eta$. For each $\alpha \in \Sigma$ either $\alpha(A_\xi)$ or $\alpha(A_\eta)$ is $\neq 0$. Thus the centralizers $\mathfrak{z}_{A_\xi}$ and $\mathfrak{z}_{A_\eta}$ of ξ and η, respectively, in $\mathfrak{p}$, have intersection $\mathfrak{a}$. But k leaves each of these centralizers invariant so $k \in M'$, the normalizer of $\mathfrak{a}$ in K. Then the Weyl group element $s = \mathrm{Ad}(k) \mid \mathfrak{a}$ fixes the set $\mathbf{R}A_\xi + \mathbf{R}A_\eta$ pointwise. But $|\alpha(A_\xi)| + |\alpha(A_\eta)| > 0$ for all $\alpha \in \Sigma$ so this set contains a regular element. Hence $s = e$ and $k \in M$.

THEOREM 6.2. *A point $\lambda \in \mathfrak{a}_c^*$ is simple if and only if it is regular.*

Proof. We write the integral (1) in the form

$$\int_{K/M} e^{iB(kA_\lambda, X)} F(kM)\, dk_M$$
$$= \int_{K/K_\lambda} e^{iB(kA_\lambda, X)}\, dk_{K_\lambda} \int_{K_\lambda/M} F(klM)\, dl_M .$$

If λ is singular, Proposition 6.1 shows $K_\lambda \neq M$ and we shall find an $F \not\equiv 0$ such that the last integral is 0 for all k. Select a continuous function $g \not\equiv 0$ on K_λ/M such that $\int_{K_\lambda/M} g(lM)\, dl_M = 0$. If ϕ is an arbitrary function on K/M put

$$F(kM) = \int_{K_\lambda} \tilde{\phi}(kl^{-1})\, \tilde{g}(l)\, dl \tag{2}$$

where $\tilde{\phi}$, $\tilde{g}$ are the "lifts" of ϕ and g to K and K_λ, respectively, and dl is the normalized Haar measure on K_λ. Then

$$\int_{K_\lambda/M} F(kuM)\, du_M = \int_{K_\lambda} \tilde{g}(l)\, dl \int_{K_\lambda} \tilde{\phi}(kul^{-1})\, du$$

which vanishes because l^{-1} can, by the invariance, be removed in the last integral. But clearly ϕ can be determined such that the right hand side of (2) is $\not\equiv 0$. This proves that λ is not simple.

The converse of Theorem 6.2 does not seem to come so easily. By differentiation we see that f in (1) is $\equiv 0$ if and only if

$$\int_{K/M} p(ik \cdot \lambda)\, F(kM)\, dk_M = 0 \qquad \text{for} \quad p \in S(\mathfrak{p}). \tag{3}$$

If λ is regular the algebra of functions $kM \to p(ik \cdot \lambda)$ separates points on K/M (Proposition 6.1) but the lack of self-adjointness makes the Stone–Weierstrass theorem inapplicable. Instead, we use the method of [8(g), sect. 6].

So we suppose λ regular and that f in (1) is $\equiv 0$. Using (3) for $p = \epsilon_j(v_i)$ $(1 \leqslant i \leqslant d(\delta),\ 1 \leqslant j \leqslant l(\delta))$ we obtain from Lemma 4.7,

$$\int_K j^{k^{-1}\cdot \epsilon_j(v_i)}(i\lambda)\, F(kM)\, dk = 0.$$

But

$$k^{-1} \cdot \epsilon_j(v_i) = \epsilon_j(\delta(k^{-1})\, v_i) = \sum_{p=1}^{d(\delta)} \langle v_i\, , \delta(k)\, v_p\rangle\, \epsilon_j(v_p)$$

so

$$\sum_{p=1}^{d(\delta)} \int_K \langle v_i\, , \delta(k)\, v_p\rangle\, j^{\epsilon_j(v_p)}(i\lambda)\, F(kM)\, dk = 0.$$

If $p > l(\delta)$ the integral is 0 as we see by replacing k by km and integrating over M.

Hence we have for $1 \leqslant i \leqslant d(\delta)$, $1 \leqslant j \leqslant l(\delta)$,

$$\sum_{p=1}^{l(\delta)} J_{pj}^{\delta}(-\lambda) \int_K \langle v_i , \delta(k) v_p \rangle F(kM) \, dk_M = 0. \tag{4}$$

If F were $\not\equiv 0$ this integral would be $\neq 0$ for some δ, i and p. But the regularity of λ guarantees by Lemma 4.5 that $\det J^{\delta}(\lambda) \neq 0$ so (4) gives a contradiction. Thus $F \equiv 0$ so we have proved that λ is simple.

Remark. Theorem 6.2 shows strong contrast to the curved case [8(g), p. 208] where each Weyl group orbit contains a simple point.

We can now determine the K-finite joint eigenfunctions of $\mathbf{D}(G_0/K)$ for "regular" eigenvalues.

THEOREM 6.3. *Assume $\lambda \in \mathfrak{a}_c^*$ regular. Then the K-finite elements in $\mathscr{E}_\lambda(\mathfrak{p})$ are the functions*

$$f(X) = \int_{K/M} e^{iB(k\cdot A_\lambda, X)} F(kM) \, dk_M ,$$

where F is a K-finite continuous function on K/M.

Since we know by Theorem 6.2 that λ is simple the proof is almost identical to that of Corollary 7.4 in [8(g)] so we omit it.

For $\lambda \in \mathfrak{a}_c^*$ let I_λ denote the set of K_λ-invariants in $S(\mathfrak{p})$ and $H_\lambda \subset S(\mathfrak{p}^*)$ the corresponding set of K_λ-harmonic polynomials.

LEMMA 6.4. *Let $\lambda \in \mathfrak{a}_c^*$ and let h be a K_λ-harmonic polynomial on $\mathfrak{p}$. Then the function*

$$f(X) = h(X) \, e^{iB(A_\lambda, X)}, \qquad X \in \mathfrak{p},$$

belongs to the eigenspace $\mathscr{E}_\lambda(\mathfrak{p})$.

Proof. We shall prove a bit more, namely that f is an eigenfunction of each $\partial(p)$ $(p \in I_\lambda)$. Consider the automorphism $p \to p^\lambda$ of $S(\mathfrak{p})$ given by $p^\lambda(\nu) = p(\nu + i\lambda)$, $(\nu \in \mathfrak{a}_c^*)$, and let $e^{i\lambda}$ denote the function $X \to e^{iB(A_\lambda, X)}$ on $\mathfrak{p}$. Then

$$\begin{aligned} \partial(p) f &= \partial(p) \, e^{i\lambda} h = e^{i\lambda} \, e^{-i\lambda} \, \partial(p)(e^{i\lambda} h) \\ &= e^{i\lambda} \, \partial(p^\lambda) h = e^{i\lambda}(p^\lambda(0) h) + e^{i\lambda} \, \partial(p^\lambda - p^\lambda(0)) h. \end{aligned}$$

But if $p \in I_\lambda$ then $p^\lambda \in I_\lambda$ so $\partial(p^\lambda - p^\lambda(0))h = 0$. Hence $\partial(p) f = p(i\lambda) f$ as claimed. This argument is similar to one by Harish–Chandra, cf. Warner [20, p. 316].

COROLLARY 6.5. *Let $h \in S(\mathfrak{p})$ be an M-harmonic polynomial. Then if $\lambda \in \mathfrak{a}_c^*$*

$$\int_K h(k \cdot X)\, e^{iB(A_\lambda, k\cdot X)}\, dk = h(0)\, \psi_\lambda(X) \tag{5}$$

In fact, since $M \subset K_\lambda$, h is K_λ-harmonic so (5) follows from Lemma 6.4 and Proposition 3.1.

We can now decide the irreducibility question for the eigenspace representations of $\mathbf{D}(G_0/K)$. We recall that each joint eigenspace of these operators has the form $\mathscr{E}_\lambda(\mathfrak{p})$ for some $\lambda \in \mathfrak{a}_c^*$.

THEOREM 6.6. *Let $\lambda \in \mathfrak{a}_c^*$ and let T_λ denote the natural representation of G_0 on the joint eigenspace $\mathscr{E}_\lambda(\mathfrak{p})$. Then T_λ is irreducible if and only if λ is regular.*

Proof. Suppose λ is regular. Let $0 \neq V \subset \mathscr{E}_\lambda(\mathfrak{p})$ be a closed invariant subspace. Then V contains the zonal spherical function ψ_λ. Since λ is simple we can define the Hilbert space

$$\mathscr{H}_\lambda = \left\{ f(X) = \int_{K/M} e^{iB(k\cdot A_\lambda, X)} F(kM)\, dk_M \mid F \in L^2(K/M) \right\},$$

giving f the L^2-norm of F on K/M. Since $-\lambda$ is simple the functions

$$e_Y\colon kM \to e^{-iB(k\cdot A_\lambda, Y)}$$

as Y runs through $\mathfrak{p}$ span a dense subspace of $L^2(K/M)$. Hence the Poisson transforms of the functions e_Y span a dense subspace of $\mathscr{H}_\lambda$. But these Poisson transforms are just the translates $X \to \psi_\lambda(X - Y)$ of ψ_λ, which belong to V. The imbedding $\mathscr{H}_\lambda \to \mathscr{E}_\lambda(\mathfrak{p})$ being continuous (by Schwartz inequality) we deduce $\mathscr{H}_\lambda \subset V$. Therefore, by Theorem 6.3, V contains all the K-finite elements in $\mathscr{E}_\lambda(\mathfrak{p})$ so, by Fourier expansion along K, $V = \mathscr{E}_\lambda(\mathfrak{p})$. This proves the irreducibility of T_λ.

Conversely, suppose λ is singular. Let V be the closed invariant subspace of $\mathscr{E}_\lambda(\mathfrak{p})$ generated by the eigenfunction $e^{iB(A_\lambda, X)}$. Let $T \in \mathscr{E}'(\mathfrak{p})$ (the space of distributions of compact support) belong to the annihilator of V, i.e.,

$$\int_{\mathfrak{p}} e^{iB(A_\lambda, kX+Y)}\, dT(X) = 0 \qquad k \in K, \quad Y \in \mathfrak{p}.$$

This amounts to

$$\tilde{T}(kA_{-\lambda}) = 0 \qquad \text{for} \quad k \in K, \tag{6}$$

where $\tilde{T}$ denotes the Fourier transform $\tilde{T}(Z) = T(e^{-iB(Z,\cdot)})$ for $Z \in \mathfrak{p}_c$. We would like to determine T such that in addition to (6) we have for some h K_λ harmonic and homogeneous,

$$\int_{\mathfrak{p}} h(X)\, e^{iB(A_\lambda, X)}\, dT(X) \neq 0. \tag{7}$$

In view of Lemma 6.4 this would prove $V \neq \mathscr{E}_\lambda(\mathfrak{p})$, that is non-irreducibility of T_λ. Using the homogeneity of h (7) can be written

$$(\partial(h)\tilde{T})(A_{-\lambda}) \neq 0. \tag{8}$$

Our problem is thus to satisfy (6) and (8).

Let Σ_λ denote the set of $\alpha \in \Sigma$ such that $\alpha(A_\lambda) = 0$ and let Σ' denote the complement $\Sigma' = \Sigma - \Sigma_\lambda$. Consider the root space decompositions ([8(c), sect. 2]),

$$\mathfrak{k} = \mathfrak{m} + \sum_{\mu \in \Sigma} \mathfrak{k}_\mu, \qquad \mathfrak{p} = \mathfrak{a} + \sum_{\mu \in \Sigma} \mathfrak{p}_\mu,$$

where $\mathfrak{k}_\mu$ and $\mathfrak{p}_\mu$ are the joint eigenspaces (in $\mathfrak{k}$ and $\mathfrak{p}$, respectively) of the operators $(\mathrm{ad}\, H)^2$ $(H \in \mathfrak{a})$ with eigenvalues $\mu(H)^2$. We put

$$\mathfrak{a}(\lambda) = \sum_{\alpha \in \Sigma_\lambda} \mathbf{R} A_\alpha, \qquad \mathfrak{k}(\lambda) = \sum_{\mu \in \Sigma_\lambda} \mathfrak{k}_\mu, \qquad \mathfrak{p}(\lambda) = \sum_{\mu \in \Sigma_\lambda} \mathfrak{p}_\mu,$$

$$\mathfrak{k}' = \sum_{\mu \in \Sigma'} \mathfrak{k}_\mu, \qquad \mathfrak{p}' = \sum_{\mu \in \Sigma'} \mathfrak{p}_\mu.$$

Then K_λ has Lie algebra

$$\mathscr{L}(K_\lambda) = \mathfrak{m} + \mathfrak{k}(\lambda)$$

and by the bracket relations

$$[\mathfrak{k}_\mu, \mathfrak{p}_\nu] \subset \mathfrak{p}_{\nu+\mu} + \mathfrak{p}_{\nu-\mu}, \qquad [\mathfrak{k}_\mu, \mathfrak{p}_\mu] \subset \mathfrak{p}_{2\mu} + \mathbf{R} A_\mu \tag{9}$$

([8(c), sect. 2]) the identity component $(K_\lambda)^0$ of K_λ leaves $\mathfrak{a}(\lambda) + \mathfrak{p}(\lambda)$ invariant. Since $\mathfrak{m} + \mathfrak{k}(\lambda)$ does not commute elementwise with $\mathfrak{p}(\lambda)$ we can pick a nonzero vector $X \in \mathfrak{p}(\lambda)$ orthogonal to the fixed point set of $(K_\lambda)^0$ in $\mathfrak{a}(\lambda) + \mathfrak{p}(\lambda)$. Then the corresponding function X^*: $Y \to B(X, Y)$ on $\mathfrak{p}$ is a $(K_\lambda)^0$-harmonic function, and *a fortiori*, a K_λ-harmonic function. Condition (8) is clearly satisfied by taking $h = \tilde{T} = X^*$.

On the other hand, since $[\mathfrak{k}, A_\lambda] \subset \mathfrak{p}'$ it is clear that the orbit $K \cdot A_{-\lambda}$ lies in the translated space $-A_\lambda + (\mathfrak{p}')_c$. But the function $\tilde{T} = X^*$ vanishes on this space so (6) and (8) are simulteneously satisfied with the linear function $\tilde{T} = X^*$. This concludes the proof.

In contrast to the situation for the curved case, the conclusion of Theorem 6.3 fails to hold if the regularity assumption is dropped. We conclude this paper with a variation of Theorem 6.3 which deals with the singular $\lambda \in \mathfrak{a}_c^*$ as well.

THEOREM 6.7. *Let $\lambda \in \mathfrak{a}_c^*$ and let $f \in \mathscr{E}_\lambda(\mathfrak{p})$ be K-finite. Then f is a finite sum*

$$f(X) = \sum_j q_j(X) \int_{K/M} e^{iB(kA_\lambda, X)} F_j(kM)\, dk_M$$

where q_j is a K-harmonic polynomial on $\mathfrak{p}$ and F_j is a K-finite function on K/M.

Proof. Fix a representation $\delta \in \hat{K}_M$ and let $v_1, \ldots, v_{d(\delta)}$ be an orthonormal basis of V_δ such that $v_1, \ldots, v_{l(\delta)}$ span $V_\delta{}^M$. Let $\mathscr{E}_{\lambda,\delta}(\mathfrak{p})$ denote the set of elements $\mathscr{E}_\lambda(\mathfrak{p})$ of type δ. Select a $\mu \in \mathfrak{a}_c^*$ such that $\lambda + z\mu$ is regular for all sufficiently small $z \in \mathbf{C} - \{0\}$. For such z the functions

$$f_{ij}^z(X) = \int_{K/M} e^{iB(kA_{\lambda+z\mu}, X)} \langle v_j, \delta(k)\, v_i\rangle\, dk_M,$$

$(1 \leqslant i \leqslant l(\delta), 1 \leqslant j \leqslant d(\delta))$ form a basis of $\mathscr{E}_{\lambda+z\mu,\delta}(\mathfrak{p})$ as we know from Theorem 6.3. Using[4] Proposition 2.21 in Oshima–Sekiguchi [21] we can find functions $c_{ij,rs}(z)$ holomorphic near $z = 0$ and integers n_{rs} such that the functions

$$g_{rs}^z(X) = z^{-n_{rs}} \sum_{i,j} c_{ij,rs}(z)\, f_{ij}^z(X)$$

$(1 \leqslant r \leqslant d(\delta), 1 \leqslant s \leqslant l(\delta))$ are, for each z in a suitable disk $|z| < \epsilon$, linearly independent. But our proof of Corollary 7.4 in [8(g)] shows that $\dim \mathscr{E}_{\nu,\delta}(\mathfrak{p}) \leqslant \dim V_\delta \dim V_\delta{}^M$ for each $\nu \in \mathfrak{a}_c^*$, regular or singular. Thus we conclude that the functions g_{rs}^0 form a basis of $\mathscr{E}_{\lambda,\delta}(\mathfrak{p})$. Writing

$$e^{iB(kA_{\lambda+z\mu}, X)} = e^{iB(kA_\lambda, X)}\{1 + z(iB(kA_\mu, X)) + \cdots\}$$

we see that g_{rs}^0 is a linear combination of functions

$$\int_{K/M} e^{iB(k\cdot A_\lambda, X)} Q(B(A_\mu, k^{-1}X)) \langle v_j, \delta(k)\, v_i\rangle\, dk_M,$$

where Q is a polynomial. Expanding Q in a sum of products of K-invariants and K-harmonics the desired representation of f follows.

The proof has the following consequence.

COROLLARY 6.8. *For $\delta \in \hat{K}_M$ and $\lambda \in \mathfrak{a}_c^*$ let $\mathscr{E}_{\lambda,\delta}(\mathfrak{p})$ denote the space of K-finite elements of type δ in the joint eigenspace $\mathscr{E}_\lambda(\mathfrak{p})$. Then*

$$\dim \mathscr{E}_{\lambda,\delta}(\mathfrak{p}) = \dim V_\delta \dim V_\delta{}^M.$$

[4] I am indebted to T. Oshima for informing me of this result.

References

1. C. Chevalley, Invariants of finite groups generated by reflections, *Amer. J. Math.* **77** (1955), 778–782.
2. J. Dadok, "Fourier Transforms of Distributions on Symmetric Spaces," thesis, MIT, 1976.
3. M. Flensted-Jensen, Spherical functions on a real semisimple Lie group. A method of reduction to the complex case, *J. Functional Analysis* **30** (1978), 106–146.
4. S. G. Gindikin, Unitary representations of groups of automorphisms of Riemann symmetric spaces of null curvature, *Functional Anal. Appl.* **1** (1967), 28–32.
5. G. Glaeser, Fonctions composées différentiables, *Ann. of Math.* **77** (1963), 193–209.

5(a). R. Godement, A theory of spherical functions, *Trans. Amer. Math. Soc.* **73** (1952), 496–556.

6. K. I. Gross and R. A. Kunze, Bessel functions and representation theory, *J. Functional Anal.* **22** (1976), 73–105.
7. Harish-Chandra, Differential operators on a semisimple Lie algebra, *Amer. J. Math.* **79** (1957), 87–120.

8(a). S. Helgason, "Differential Geometry and Symmetric Spaces," Academic Press, New York, 1962.

8(b). S. Helgason, Fundamental solutions of invariant differential operators on symmetric spaces, *Amer. J. Math.* **86** (1964), 565–601.

8(c). S. Helgason, Totally geodesic spheres in compact symmetric spaces, *Math. Ann.* **165** (1966), 309–317.

8(d). S. Helgason, A duality for symmetric spaces with applications to group representations, *Advances in Math.* **5** (1970), 1–154.

8(e). S. Helgason, The surjectivity of invariant differential operators on symmetric spaces, I, *Ann. of Math.* **98** (1973), 451–479.

8(f). S. Helgason, A formula for the radial part of the Laplace Beltrami operator, *J. Differential Geometry* **6** (1972), 411–419.

8(g). S. Helgason, A duality for symmetric spaces with applications to group representations. II. Differential equations and eigenspace representations, *Advances in Math.* **22** (1976), 187–219.

9. L. Hörmander, "Linear Partial Differential Operators," Springer, Berlin, 1963.
10. A. Korányi, A survey of harmonic functions on symmetric spaces, *in* "Proc. Symp. Pure Math. Vol. XXXV," Amer. Math. Soc., Providence, R.I., 1979.

11(a). B. Kostant, Lie group representations on polynomial rings, *Amer. J. Math.* **86** (1963), 327–402.

11(b). B. Kostant, On the existence and irreducibility of certain series of representations, *in* "Lie Groups and Their Representations," pp. 231–329, Halsted Press, New York, 1975.

12. B. Kostant and S. Rallis, Orbits and Lie group representations associated to symmetric spaces, *Amer. J. Math.* **93** (1971), 753–809.
13. G. W. Mackey, Induced representations of locally compact groups, I, *Ann. of Math.* **55** (1952), 101–140.
14. B. Malgrange, Existence et approximation des solutions des équations aux dérivées partielles et des équations de convolution, *Ann. Inst. Fourier Grenoble* **4** (1955–1956), 271–354.
15. K. R. Parthasarathy, R. Ranga Rao, and V. S. Varadarajan, Representations of complex semisimple Lie groups and Lie algebras, *Ann. of Math.* **85** (1967), 383–429.
16. L. Schwartz, "Théorie des distributions", Hermann et Cie, Paris, 1966.

17. G. W. SCHWARZ, Smooth functions invariant under the action of a compact Lie group, *Topology* **14** (1975), 63–68.
18. P. TORASSO, Le théorème de Paley–Wiener pour l'espace des fonctions indéfinement différentiables et à support compact sur un espace symmétrique de type noncompact, *J. Functional Analysis* **26** (1977), 201–213.
19. E. T. WHITTAKER AND G. N. WATSON, "A Course of Modern Analysis," 4th ed. Cambridge Univ. Press, London, 1927.
20. G. WARNER, "Harmonic Analysis on Semi-Simple Lie Groups, II," Springer, New York, 1972.
21. T. OSHIMA AND J. SEKIGUCHI, Eigenspaces of invariant differential operators on an affine symmetric space, *Invent. Math.* (1980).

Printed by the St. Catherine Press Ltd., Tempelhof 37, Bruges, Belgium

Reprinted from ADVANCES IN MATHEMATICS
All Rigts Reserved by Academic Press, New York and London
Vol. 60, No. 1, April 1986
Printed in Belgium

Invariant Differential Operators on Grassmann Manifolds

F. GONZALEZ AND S. HELGASON*

Department of Mathematics, Case Western Reserve University, Cleveland, Ohio, 44106, and Department of Mathematics, Massachusetts Institute of Technology, Cambridge, Massachusetts 02139

1. INTRODUCTION

For a Lie group G and a closed subgroup $H \subset G$ let $\boldsymbol{D}(G/H)$ denote the algebra of differential operators D on the manifold G/H which are invariant under the action of G. In this note we determine this algebra in the case when G is the group $\boldsymbol{M}(n)$ of rigid motions of the Euclidean space $\boldsymbol{R}^n$ and H is the subgroup of $\boldsymbol{M}(n)$ leaving a certain p-dimensional plane in $\boldsymbol{R}^n$ invariant. Here $0 \leqslant p \leqslant n-1$ and

$$G/H = \boldsymbol{M}(n)/(\boldsymbol{M}(p) \times \boldsymbol{O}(n-p))$$

is the space $\boldsymbol{G}(p, n)$ of p-planes in $\boldsymbol{R}^n$. For $p=0$ we have $G/H = \boldsymbol{R}^n$ and $\boldsymbol{D}(G/H)$ consists of the polynomials in the Laplacian. For $p=n-1$, G/H is the space of hyperplanes $(x, \omega) = r$ (ω a fixed unit vector and $r \in \boldsymbol{R}$) and $\boldsymbol{D}(G/H)$ is generated by the operator $\square$: $\phi(\omega, r) \to d^2\phi(\omega, r)/dr^2$. This is proved in [4, pp. 173–175]; however, as kindly pointed out to us by R. Sulanke, the general description of $\boldsymbol{D}(G/H)$ in [4, Lemma 7.1(ii)] and [5, Part (ii) of Exercise C.2, p. 337] is only valid for the above cases $p=0$ and $p=n-1$. For the case $p=1$, $n=3$, where G/H consists of all lines in $\boldsymbol{R}^3$, F. Richter has found that $\boldsymbol{D}(G/H)$ has two generators of orders 2 and 4, respectively.

In this note we deal with the case of arbitary p and n and show that $\boldsymbol{D}(G/H)$ is an algebra with $\min(p+1, n-p)$ algebraically independent generators which have orders 2, 4, 6,.... We also compare this with the simpler description of the algebra $\boldsymbol{D}(\boldsymbol{O}(n)/\boldsymbol{O}(p) \times \boldsymbol{O}(n-p))$ of invariant differential operators on the manifold $\boldsymbol{G}_{p,n}$ of p-dimensional subspaces of $\boldsymbol{R}^n$.

We thank Professor R. Sulanke for pointing out the gap indicated above and for other informative remarks.

* Supported in part by NSF Grant DMS-8504029.

0001-8708/86 $7.50

2. The Linear Isotropy Group

The group $G = \boldsymbol{M}(n)$ is isomorphic to the $(n+1)\times(n+1)$ matrix group

$$\left\{\left(\begin{array}{c|c} k & V \\ \hline 0 & 1 \end{array}\right) : k \in \boldsymbol{O}(n),\ V \in \boldsymbol{R}^n\right\}, \tag{2.1}$$

and it acts on $\boldsymbol{R}^n$ by

$$\left(\begin{array}{c|c} k & V \\ \hline 0 & 1 \end{array}\right)\cdot\begin{pmatrix} Y \\ 1 \end{pmatrix} = \begin{pmatrix} k\cdot Y + V \\ 1 \end{pmatrix}, \qquad Y \in \boldsymbol{R}^n.$$

The Lie algebra $\mathfrak{g}$ of G is given by the set of matrices

$$\left(\begin{array}{c|c} T & Z \\ \hline 0 & 0 \end{array}\right); \qquad T \in \mathfrak{o}(n),\ Z \in \boldsymbol{R}^n,$$

$\mathfrak{o}(n)$ being the Lie algebra of $\boldsymbol{O}(n)$ and the adjoint representation $\mathrm{Ad}_G = \mathrm{Ad}$ satisfies

$$\mathrm{Ad}\left(\begin{array}{c|c} k & V \\ \hline 0 & 1 \end{array}\right)\left(\begin{array}{c|c} T & Z \\ \hline 0 & 0 \end{array}\right) = \left(\begin{array}{c|c} kTk^{-1} & k\cdot Z - kTk^{-1}\cdot V \\ 0 & 0 \end{array}\right). \tag{2.2}$$

The subgroup H of G leaving invariant the p-plane spanned by the p first coordinate axes is given by $H = \boldsymbol{M}(p)\times\boldsymbol{O}(n-p)$ and consists of the matrices

$$h = \left(\begin{array}{c|c|c} a & 0 & V \\ \hline 0 & b & 0 \\ \hline 0 & 0 & 1 \end{array}\right) : a \in \boldsymbol{O}(p),\ b \in \boldsymbol{O}(n-p),\ V \in \boldsymbol{R}^p. \tag{2.3}$$

Let $\mathfrak{h}$ denote the Lie algebra of H and $\mathfrak{m} \subset \mathfrak{g}$ the subspace of matrices

$$S = \begin{pmatrix} 0 & Y & 0 \\ -{}^tY & 0 & W \\ 0 & 0 & 0 \end{pmatrix} : \begin{array}{l} Y \text{ any } p\times(n-p) \text{ matrix}, \\ {}^tY \text{ its transpose}, \\ W \in \boldsymbol{R}^{n-p}. \end{array} \tag{2.4}$$

Then $\mathfrak{g}=\mathfrak{h}+\mathfrak{m}$ (direct sum) and

$$\mathrm{Ad}(h)\,S=\left(\begin{array}{c|c|c} 0 & aYb^{-1} & 0 \\ \hline -b^{t}Ya^{-1} & 0 & b\cdot W+b^{t}Ya^{-1}\cdot V \\ \hline 0 & 0 & 0 \end{array}\right) \tag{2.5}$$

so $\mathfrak{m}$ is $\mathrm{Ad}_G(H)$-invariant. If J is the $(n+1)\times(n+1)$ diagonal matrix with diagonal elements $(d_1,\dots,d_p,\ d_{p+1},\dots,\ d_n,1)$, $d_i=1$ $(1\leqslant i\leqslant p)$, $d_i=-1$ $(p+1\leqslant i\leqslant n)$, H is the subgroup of fixed points of the involution $g\to JgJ$ of G. Thus G/H is a symmetric space and consequently ([6], [8], [1]) $\boldsymbol{D}(G/H)$ is a commutative algebra.

As usual let E_{ij} denote the matrix $(\varepsilon_{rs})_{1\leqslant r,s\leqslant n+1}$ with $\varepsilon_{rs}=\delta_{ri}\delta_{sj}$ and put

$$X_{ij}=E_{ip+j}-E_{p+ji} \qquad (1\leqslant i\leqslant p,\ 1\leqslant j\leqslant n-p), \tag{2.6}$$

$$U_k=E_{p+k,n+1} \qquad (1\leqslant k\leqslant n-p). \tag{2.7}$$

These vectors form a basis of $\mathfrak{m}$.

3. The Algebra $\boldsymbol{D}(G/H)$

Let $I(\mathfrak{m})$ denote the algebra of $\mathrm{Ad}_G(H)$-invariant elements in the symmetric algebra $S(\mathfrak{m})$ (consisting of all polynomials in $\{X_{ij},U_k\}$ with complex coefficients). As proved in [3] there is a canonical linear bijection

$$P\to D_P \tag{3.1}$$

of $I(\mathfrak{m})$ onto $\boldsymbol{D}(G/H)$. We recall that if (Z_i) is any basis of $\mathfrak{m}$ then

$$(D_Pf)(gH)=\left\{P\left(\frac{\partial}{\partial t_1},\dots,\frac{\partial}{\partial t_s}\right)f\left(g\exp\left(\sum t_iZ_i\right)H\right)\right\}_{(t)=0} \tag{3.2}$$

Because of Lemma 4.2 we know that if $P_1,\dots,P_l$ are algebraically independent generators of $I(\mathfrak{m})$ then $D_{P_1},\dots,D_{P_l}$ are algebraically independent generators of $\boldsymbol{D}(G/H)$. Our problem is therefore to describe explicitly a set of algebraically independent generators of $I(\mathfrak{m})$.

THEOREM 3.1. *Consider the* $(p+1)\times(n-p)$*-matrix with vector entries*

$$A=\begin{pmatrix} U_1,\dots,U_{n-p} \\ X_{11},\dots,X_{1n-p} \\ \vdots \quad\quad \vdots \\ X_{p1},\dots,X_{pn-p} \end{pmatrix}.$$

For $1 \leqslant k \leqslant \min(p+1, n-p)$ *let* $Q_k \in S(\mathfrak{m})$ *be the sum of the squares of the* $k \times k$*-minors of* A *having vectors* U_j *in the first row*:

$$Q_k = \sum_{\substack{1 \leqslant i_1 < \cdots < i_k \leqslant n-p \\ 1 \leqslant j_1 < \cdots < j_{k-1} \leqslant p}} \det{}^2 \begin{pmatrix} U_{i_1} & ,\ldots, & U_{i_k} \\ X_{j_1 i_1} & ,\ldots, & X_{j_1 i_k} \\ \vdots & & \\ X_{j_{k-1} i_1} & ,\ldots, & X_{j_{k-1} i_k} \end{pmatrix}.$$

Then the polynomials Q_k *are algebraically independent generators of* $I(\mathfrak{m})$.

Remarks. (i) $Q_1 = U_1^2 + \cdots + U_{n-p}^2$ so D_{Q_1} is the Laplacian $\Box$ on the fibers of the vector bundle $\boldsymbol{G}(p, n)$ over $\boldsymbol{G}_{p,n} = \boldsymbol{O}(n)/\boldsymbol{O}(p) \times \boldsymbol{O}(n-p)$ defined in [4].

(ii) If $p+q+1=n$ then for the two dual homogeneous spaces, $\boldsymbol{G}(p, n)$ and $\boldsymbol{G}(q, n)$ ([4, Sect. 8], [2]), the algebras of invariant differential operators have the same number of generators.

(iii) The number $\min(p+1, n-p)$ is also suggested in [9] as the natural one to expect from another point of view.[1]

For the proof of this theorem we view $S(\mathfrak{m})$ as the algebra of polynomial functions on the dual space $\mathfrak{m}^*$ whereby $I(\mathfrak{m})$ is identified with the algebra $I_H(\mathfrak{m}^*)$ of polynomial functions on $\mathfrak{m}^*$ invariant under the coisotropy action of H on $\mathfrak{m}^*$. Thus our task is to determine $I_H(\mathfrak{m}^*)$.

We start with a general remark about the vector space $M_{m,n}$ of real $m \times n$ matrices. Let E_{ij} $(1 \leqslant i \leqslant m,\ 1 \leqslant j \leqslant n)$ be the canonical basis of $M_{m,n}$ and $\{\varepsilon_{ij}\}$ the dual basis of $M_{m,n}^*$ (the entry functions) such that $\varepsilon_{ij}(E_{kl}) = \delta_{ik}\delta_{jl}$. For $A \in M_{m,m}$, $B \in M_{n,n}$ consider the linear transformation $\varphi: X \to AXB$ of $M_{m,n}$ into itself. It is then easy to verify that the transpose map ${}^t\varphi: M_{m,n}^* \to M_{m,n}^*$ satisfies

$${}^t\varphi(\varepsilon_{ij}) = \sum_{q \leqslant m, r \leqslant n} a_{iq} \varepsilon_{qr} b_{rj}$$

so that

$${}^t\varphi: \begin{pmatrix} \varepsilon_{11} & ,\ldots, & \varepsilon_{1n} \\ \vdots & & \vdots \\ \varepsilon_{m1} & ,\ldots, & \varepsilon_{mn} \end{pmatrix} \to A \begin{pmatrix} \varepsilon_{11} & ,\ldots, & \varepsilon_{1n} \\ \vdots & & \vdots \\ \varepsilon_{m1} & ,\ldots, & \varepsilon_{mn} \end{pmatrix} B. \tag{3.3}$$

Coming back to our particular situation we write $M_{p+1,n-p}$ as the set of matrices

$$\begin{pmatrix} U \\ X \end{pmatrix} = \begin{pmatrix} u_1 & ,\ldots, & u_{n-p} \\ x_{11} & ,\ldots, & x_{1n-p} \\ \vdots & & \vdots \\ x_{p1} & ,\ldots, & x_{pn-p} \end{pmatrix}.$$

[1] *Note added in proof.* We understand from Prof. R. Sulanke that another method for determining $\boldsymbol{D}(G/H)$ has been found by F. Richter.

We consider the bijection of $M_{p+1,n-p}$ onto $\mathfrak{m}^*$ given by $J\binom{U}{X}=\lambda_{U,X}$, where, with S as in (2.4),

$$\lambda_{U,X}(S)=\operatorname{Tr}[\binom{U}{X}(W\,|\,{}^tY)]=UW+\operatorname{Tr}(X^tY).$$

Let h be as in (2.3) so the map $h\to h^{-1}$ is $(a, b, V)\to(a^{-1}, b^{-1}, -a^{-1}V)$. The co-isotropy representation of H on $\mathfrak{m}^*$ is given by

$$(\mathrm{Ad}^*(h)\,\lambda_{U,X})(S)=\lambda_{U,X}(\mathrm{Ad}(h^{-1})\,S) \tag{3.4}$$

and by (2.5) this last expression equals

$$Ub^{-1}W-Ub^{-1}\,{}^tYV+\operatorname{Tr}(Xb^{-1}\,{}^tYa). \tag{3.5}$$

Next we claim that on $M_{p+1,n-p}$

$$\mathrm{Ad}^*(h)\circ J=J\circ\tilde{h}, \tag{3.6}$$

where $\tilde{h}$: $M_{p+1,n-p}\to M_{p+1,n-p}$ is the map

$$\tilde{h}\colon\begin{pmatrix}U\\X\end{pmatrix}\to\begin{pmatrix}1&0\\-V&I_p\end{pmatrix}\begin{pmatrix}1&0\\0&a\end{pmatrix}\begin{pmatrix}U\\X\end{pmatrix}(b^{-1}), \tag{3.7}$$

I_p denoting the $p\times p$ identity matrix. In fact, the right-hand side of (3.7) equals

$$\begin{pmatrix}U'\\X'\end{pmatrix}=\begin{pmatrix}Ub^{-1}\\aXb^{-1}-VUb^{-1}\end{pmatrix}$$

and

$$\lambda_{U',X'}(S)=Ub^{-1}W+\operatorname{Tr}[(aXb^{-1}-VUb^{-1})\,{}^tY],$$

which is readily seen to coincide with (3.5). Thus by (3.8) we can (via J) regard $I(\mathfrak{m})=I_H(\mathfrak{m}^*)$ as the algebra of polynomial functions on $M_{p+1,n-p}$ invariant under

(i) left multiplication by $\left(\begin{smallmatrix}1&0\\0&a\end{smallmatrix}\right)$, $a\in\boldsymbol{O}(p)$;

(ii) left multiplication by $\left(\begin{smallmatrix}1&0\\V&I_p\end{smallmatrix}\right)$, $V\in\boldsymbol{R}^p$;

(iii) right multiplication by $b\in\boldsymbol{O}(n-p)$.

Next observe that under the transpose ${}^tJ$: $\mathfrak{m}\to M^*_{p+1,n-p}$ we have ${}^tJ(X_{ij})=\xi_{ij}$, ${}^tJ(U_k)=\mu_k$, where ξ_{ij}, μ_k are the entry functions

$$\xi_{ij}\begin{pmatrix}U\\X\end{pmatrix}=x_{ij},\qquad \mu_k\begin{pmatrix}U\\X\end{pmatrix}=u_k.$$

Because of (3.6) and (3.4) we have

$${}^tJ\circ\mathrm{Ad}(h^{-1})={}^t\tilde{h}\circ{}^tJ\quad\text{on}\quad\mathfrak{m}, \tag{3.8}$$

and by (3.3) the transpose ${}^t\tilde{h}$ is given by

$${}^t\tilde{h} = h_{-V,a,b^{-1}}: \begin{pmatrix} \mu \\ \xi \end{pmatrix} \to \begin{pmatrix} 1 & 0 \\ -V & 0 \end{pmatrix} \begin{pmatrix} 1 & 0 \\ 0 & a \end{pmatrix} \begin{pmatrix} \mu \\ \xi \end{pmatrix} b^{-1},$$

with

$$\begin{pmatrix} \mu \\ \xi \end{pmatrix} = \begin{pmatrix} \mu_1, \dots, \mu_{n-p} \\ \xi_{11}, \dots, \xi_{1n-p} \\ \vdots \quad\quad \vdots \\ \xi_{p1}, \dots, \xi_{pn-p} \end{pmatrix},$$

a matrix with entries in $M^*_{p+1,n-p}$. Let $Q_k(\mu, \xi)$ be the polynomial function on $M_{p+1,n-p}$, obtained by replacing in Q_k the U_k by μ_k and the X_{ij} by ξ_{ij}. Since (ii) is an elementary row transformation it is clear that $Q_k(\mu, \xi)$ is invariant under $h_{-V,I_p,I_{n-p}}$. To see that $Q_k(\mu, \xi)$ is also invariant under $h_{0,a,b^{-1}}$ we use Lemma 4.1. In fact, $Q_k(\mu, \xi) = P_k(\mu, \xi) - P_k(\xi)$, where $P_k(\mu, \xi)$ (respectively $P_k(\xi)$) is the sum of the squares of the $k \times k$ minors of $\binom{\mu}{\xi}$ (respectively ξ). Both of these are by Lemma 4.1 invariant under $h_{0,a,b^{-1}}$ so we conclude that $Q_k(\mu, \xi)$ is invariant under all ${}^t\tilde{h}$ $(h \in H)$. However, $Q_k(\mu, \xi)$ is just the function $Q_k \circ J$ so (3.6) shows that Q_k does indeed belong to $I(\mathfrak{m})$.

Next let $(\mathfrak{m}^*)'$ be the subset of $\mathfrak{m}^* = M_{p+1,n-p}$ consisting of the matrices $\binom{U}{X}$ for which $|U|^2 = u_1^2 + \cdots + u_{n-p}^2 \neq 0$. Then let $\mathfrak{m}_0^* \subset \mathfrak{m}^*$ be the subspace of matrices

$$\begin{pmatrix} u_1 & 0 \\ 0 & x \end{pmatrix} = \begin{pmatrix} u_1, 0, \dots, 0 \\ 0, x_{12}, \dots, x_{1n-p} \\ \vdots \\ 0, x_{p2}, \dots, x_{pn-p} \end{pmatrix}, \quad u_1 \in \boldsymbol{R},\ X \in M_{p,n-p-1}.$$

Applying the transformations (iii) and (ii) above we see that the $\mathrm{Ad}^*(H)$-orbit of each point in $(\mathfrak{m}^*)'$ intersects $\mathfrak{m}_0^*$. Consider now the subgroup $H_0 \subset H$ of elements h in (2.3) with $V = 0$, $a \in \boldsymbol{O}(p)$ and b of the form

$$b = \begin{pmatrix} \pm 1 & 0 \\ 0 & b_1 \end{pmatrix}, \qquad b_1 \in \boldsymbol{O}(n-p-1).$$

Then the action of $\mathrm{Ad}^*(H_0)$ on $\mathfrak{m}_0^*$ is given by

$$\begin{pmatrix} u_1 & 0 \\ 0 & X \end{pmatrix} \to \begin{pmatrix} \pm u_1 & 0 \\ 0 & aXb_1^{-1} \end{pmatrix} \quad a \in \boldsymbol{O}(p),\ b_1 \in \boldsymbol{O}(n-p-1). \tag{3.9}$$

Let $I_{H_0}(\mathfrak{m}_0^*)$ denote the corresponding algebra of $\mathrm{Ad}^*(H_0)$-invariant polynomial functions on $\mathfrak{m}_0^*$. The restriction mapping $Q \to \bar{Q} = Q \mid \mathfrak{m}_0^*$ then maps $I_H(\mathfrak{m}^*)$ into $I_{H_0}(\mathfrak{m}_0^*)$. Since $\mathrm{Ad}^*(H)\,\mathfrak{m}_0^*$ contains $(\mathfrak{m}^*)'$ which is dense in $\mathfrak{m}^*$ this map is injective. Because of Lemma 4.1 in the Appendix, $I_{H_0}(\mathfrak{m}_0^*)$ is generated by u_1^2 and the algebraically independent polynomials $P_k(X)$ $(1 \leqslant k \leqslant l = \min(p, n-p-1))$, where $P_k(X)$ is the sum of the squares of the $k \times k$ minors in X. It follows that u_1^2, $u_1^2 P_1, \ldots, u_1^2 P_l$ (which coincide with $\bar{Q}_1, \bar{Q}_2, \ldots, \bar{Q}_{l+1}$) are algebraically independent so by the injectivity of the map $Q \to \bar{Q}$ the polynomials $Q_1, \ldots, Q_{l+1}$ are algebraically independent over $\boldsymbol{C}$.

It remains to prove that the algebra I generated by the $Q_1, \ldots, Q_{l+1}$ equals $I_H(\mathfrak{m}^*)$. Suppose there exists a $Q \in I_H(\mathfrak{m}^*)$ not in I. Then $\bar{Q}$ is a polynomial

$$S(u_1^2, P_1, \ldots, P_l) = S(\bar{Q}_1, \bar{Q}_2/\bar{Q}_1, \ldots, \bar{Q}_{l+1}/\bar{Q}_1).$$

By the injectivity,

$$Q = S(Q_1, Q_2/Q_1, \ldots, Q_{l+1}/Q_1) = \frac{S_1(Q_1, \ldots, Q_{l+1})}{Q_1^k}, \tag{3.10}$$

where S_1 is another polynomial. Since $Q \notin I$, we have $k \geqslant 1$. By the algebraic independence of the Q_i we may assume that the variable t_1 does not divide $S_1(t_1, \ldots, t_{l+1})$. Now write

$$S_1(t_1, \ldots, t_{l+1}) = S'(t_2, \ldots, t_{l+1}) + t_1 S''(t_1, \ldots, t_{l+1}).$$

Then $S'(Q_2, \ldots, Q_{l+1}) \neq 0$. We shall now show that there exists a complex matrix $\zeta_0 \in M_{p+1, n-p}^{C}$ such that

$$Q_1(\zeta_0) = 0, \quad S'(Q_2(\zeta_0), \ldots, Q_{l+1}(\zeta_0)) \neq 0 \tag{3.11}$$

For this consider the complex matrices of the form

$$\zeta = \begin{pmatrix} 1, i, & 0 \;, \ldots, 0 \\ 0, z_{12}, z_{13}, \ldots, z_{1n-p} \\ \vdots \\ 0, z_{p2}, z_{p3}, \ldots, z_{pn-p} \end{pmatrix} \tag{3.12}$$

and put $Z = (z_{ij})$ $(1 \leqslant i \leqslant p, 2 \leqslant j \leqslant n-p)$. Also for each $1 \leqslant k \leqslant l$, let $Q_k'(Z)$ denote the sum of the squares of the $k \times k$ minors of Z with entries from $z_{12}, \ldots, z_{p2}$ in the first column. Then it is easy to see (because of a pairwise cancellation of minors outside the first column of Z) that

$$Q_1(\zeta) = 0, \; Q_k(\zeta) = Q_{k-1}'(Z) \quad (2 \leqslant k \leqslant l+1).$$

Thus

$$S'(Q_2(\zeta),\ldots, Q_{l+1}(\zeta)) = S'(Q'_1(Z),\ldots, Q'_l(Z)). \tag{3.13}$$

However, the polynomial functions $Q'_1,\ldots, Q'_l$ were already seen to be algebraically independent over $\boldsymbol{C}$ so by (3.13) there exists a ζ_0 of the form (3.12) such that (3.11) holds. This, however, contradicts identity (3.10). Thus $I = I_H(\mathfrak{m}^*)$ and the theorem is proved.

Remark. It is of some interest to compare this description of $\boldsymbol{D}(\boldsymbol{G}(p, n))$ with the simpler problem of determining the algebra

$$\boldsymbol{D}(\boldsymbol{G}_{p,n}) = \boldsymbol{D}(\boldsymbol{O}(n)/\boldsymbol{O}(p) \times \boldsymbol{O}(n-p))$$

of invariant differential operators on the manifold $\boldsymbol{G}_{p,n}$ of p-dimensional subspaces of $\boldsymbol{R}^n$. Here $\mathfrak{g} = \mathfrak{o}(n)$, $\mathfrak{h} = \mathfrak{o}(p) \oplus \mathfrak{o}(n-p)$ and $\mathfrak{m}$ consists of the matrices

$$\begin{pmatrix} 0 & X \\ -{}^tX & 0 \end{pmatrix} : X \in M_{p,n-p}.$$

The adjoint action of H on $\mathfrak{m}$ is given by

$$\mathrm{Ad}_G \begin{pmatrix} a & 0 \\ 0 & b \end{pmatrix} \begin{pmatrix} 0 & X \\ -{}^tX & 0 \end{pmatrix} = \begin{pmatrix} 0 & aXb^{-1} \\ -b\,{}^tXa^{-1} & 0 \end{pmatrix},$$

$a \in \boldsymbol{O}(p)$, $b \in \boldsymbol{O}(n-p)$. By Lemma 4.1 we conclude:

PROPOSITION 3.2. *$I(\mathfrak{m})$ is generated by the polynomials $P_1,\ldots, P_l$ ($l = \min(p, n-p)$), where P_k is the sum of the squares of the $k \times k$ minors of the vector-valued matrix*

$$\begin{pmatrix} X_{11},\ldots, & X_{1n-p} \\ \vdots & \vdots \\ X_{p1},\ldots, & X_{pn-p} \end{pmatrix}$$

with X_{ij} as in (2.6). *The algebra $\boldsymbol{D}(\boldsymbol{G}_{p,n})$ is generated by $D_{P_1},\ldots, D_{P_l}$ given by* (3.2).

4. APPENDIX

The following lemma has been used in the proofs; while it presumably is known we could not find an explicit statement of it in the literature. For $r = s$, a related result is given in [7, Lemma II].

LEMMA 4.1. *Let r and s be positive integers and let J be the algebra of polynomial functions on $M_{r,s}$ invariant under the action of $\boldsymbol{O}(r) \times \boldsymbol{O}(s)$ given by $X \to aXb^{-1}$ ($a \in \boldsymbol{O}(r)$, $B \in \boldsymbol{O}(s)$). Then J is generated by the polynomials*

P_k $(1 \leqslant k \leqslant \min(r, s))$, *where* $P_k(X)$ *is the sum of squares of the* $k \times k$ *minors of* X. *Moreover, the* P_k *are algebraically independent over* $\boldsymbol{C}$.

Proof. First we observe that $P_k \in J$. It suffices to verify this for a and b separately, say for a. With the standard bases (e_j) of $\boldsymbol{R}^s$, (f_i) of $\boldsymbol{R}^r$ we define the linear transformations $\xi\colon \boldsymbol{R}^s \to \boldsymbol{R}^r$, $\alpha\colon \boldsymbol{R}^r \to \boldsymbol{R}^r$ by the matrices $X = (X_{ij})$, $a = (a_{lm})$:

$$\xi e_j = \sum_i X_{ij} f_i, \qquad \alpha f_i = \sum_l a_{li} f_l.$$

We also put $|\xi|^2 = \operatorname{Tr}(X^t X)$. With the standard orthonormal bases of spaces of k-vectors,

$$\{e_{j1} \wedge \cdots \wedge e_{jk}\} \text{ of } \Lambda^k(\boldsymbol{R}^s), \quad \{f_{i_1} \wedge \cdots \wedge f_{i_k}\} \text{ of } \Lambda^k(\boldsymbol{R}^r)$$

the linear transformation $\Lambda^k \xi\colon \Lambda^k(\boldsymbol{R}^s) \to \Lambda^k(\boldsymbol{R}^r)$ is represented by the $k \times k$ minors of X so

$$P_k(X) = |\Lambda^k \xi|^2. \tag{4.1}$$

Now

$$\alpha \xi e_j = \sum_i X_{ij} \alpha f_i = \sum_i X_{ij} \sum_l a_{li} f_l = \sum_l (aX)_{lj} f_l \tag{4.2}$$

so

$$P_k(aX) = |\Lambda^k(\alpha\xi)|^2.$$

On the other hand, if we use the orthonormal basis (αf_i) to form a basis of $\Lambda^k(\boldsymbol{R}^r)$, (4.2) shows that

$$|\Lambda^k(\alpha\xi)|^2 = P_k(X),$$

so $P_k \in J$ as claimed.

For the rest we may assume $r \leqslant s$. We then observe that each $X \in M_{r,s}$ is conjugate under the $\boldsymbol{O}(r) \times \boldsymbol{O}(s)$-action (in fact under the $\boldsymbol{O}(s)$-action) to an element $(Y|0)$, where $Y \in M_{r,r}$, $0 =$ zero $r \times (s\text{-}r)$-matrix. Let $M'_{r,r}$ denote the set of nonsingular elements of $M_{r,r}$ and $M'_{r,s}$ the set of elements in $M_{r,s}$ which are $\boldsymbol{O}(r) \times \boldsymbol{O}(s)$-conjugate to an element in $M'_{r,r}$. Decomposing a $Y \in M'_{r,r}$ into an orthogonal matrix times a symmetric matrix we see that each $X \in M'_{r,s}$ is conjugate to an element of the set D of matrices

$$\begin{pmatrix} d_1 & & 0 & 0 \cdots 0 \\ & \ddots & & \\ 0 & & d_r & 0 \cdots 0 \end{pmatrix}, \qquad d_i \in \boldsymbol{R}.$$

Let $Q \in J$ and $\bar{Q}$ the restriction $Q \mid D$. Since $M'_{r,s}$ is dense in $M_{r,s}$, $Q \to \bar{Q}$ is injective. Also $\bar{Q}$ is invariant under the transformations $d_i \to \varepsilon_i d_{\sigma(i)}$ $(1 \leqslant i \leqslant r)$ of D, where $\varepsilon_i = \pm 1$ and σ is any permutation (the Weyl group of $\boldsymbol{O}(r)$). Thus $\bar{Q}$ is a polynomial in the algebraically independent elementary symmetric functions of $d_1^2, ..., d_R^2$. However, these functions are just the restrictions to D of the polynomials P_k. By the injectivity mentioned, the P_k are algebraically independent and Q is a polynomial in them.

LEMMA 4.2. *Let $P_1, ..., P_l$ be algebraically independent generators of $I(\mathfrak{m})$ and $D_i = D_{P_i}$ $(1 \leqslant i \leqslant l)$ the elements of $\mathbf{D}(G/H)$ given by* (3.1). *Then $D_1, ..., D_l$ are algebraically independent generators of $\mathbf{D}(G/H)$.*

Proof. We know from [3] that the D_i are generators. It remains to prove that they are algebraically independent. Suppose not; let

$$P = \sum a_{n_1, ..., n_l} x_1^{n_1} \cdots x_l^{n_l}$$

be a nonzero polynomial such that $P(D_1, ..., D_l) = 0$. Let $d_i = \text{degree}(P_i)$ and let $N = \max(\sum d_i n_i)$, the maximum taken over the set of l-tuples $(n_1, ..., n_l)$ for which $a_{n_1, ..., n_l} \neq 0$. We write the polynomial

$$S = \sum a_{n_1, ..., n_l} P_1^{n_1} \cdots P_l^{n_l}$$

as the sum $S = Q + R$, where

$$Q = \sum_{\Sigma d_i n_i = N} a_{n_1, ..., n_l} P_1^{n_1} \cdots P_l^{n_l}$$

and degree $(R) < N$. Also $Q \neq 0$ by assumption. Consider now the operator

$$\sum a_{n_1, ..., n_l} D_1^{n_1} \cdots D_l^{n_l} - D_S,$$

which by [3, Remark p. 269] has order $< N$. This operator equals $0 - D_Q - D_R$, which by (3.2) has order N. This contradiction proves the lemma.

Note added in proof. Another proof of Lemma 4.1 was indicated to us by M. Raïs.

REFERENCES

1. M. DUFLO, Opérateurs invariants sur un espace symétrique, *C. R. Acad. Sci. Paris Sér. A* **289** (1979), 135–137.
2. F. GONZALEZ, "Radon Transforms on Grassmann Manifolds," Thesis, M.I.T., 1984.
3. S. HELGASON, Differential operators on homogeneous spaces, *Acta. Math.* **102** (1959), 239–299.

4. S. Helgason, The Radon transform on Euclidean spaces, two-point homogeneous spaces and Grassmann manifolds, *Acta Math.* **113** (1965), 153–180.
5. S. Helgason, "Groups and Geometric Analysis," Academic Press, Orlanda, Flo., 1984.
6. A. Lichnèrowicz, Opérateurs differentiels invariants sur un espace symmétrique, *C. R. Acad. Sci.* **257** (1963), 3548–3550.
7. M. Raïs, Distributions homogènes sur des espaces de matrices, *Bull. Soc. Math. France, Suppl. Mémoire* **30** (1972).
8. W. Smoke, Commutativity of the invariant differential operators on a symmetric space, *Proc. Amer. Math. Soc.* **19** (1968), 222–224.
9. R. S. Strichartz, Harmonic analysis on Grassmann bundles, preprint.
10. S. J. Takiff, Invariant polynomials on Lie algebras of inhomogeneous unitary and special orthogonal groups, *Trans. Amer. Math. Soc.* **170** (1972), 221–230.

Printed by the St. Catherine Press Ltd., Tempelhof 41, Bruges, Belgium

Value-distribution Theory for Holomorphic Almost Periodic Functions

By SIGURÐUR HELGASON

1. *Introduction*

I am deeply indebted to the organizers for inviting me to participate in this conference on the occasion of the 100th anniversary of Harald Bohr.

My student years here in Copenhagen happened to coincide with the last 5 years of Bohr's life. Although many years have passed I still have a vivid memory of his inspiring lectures and of his personal kindness.

Today I am going to talk about some work of my own from this time; this was a response to a prize question posed by the University for 1950 concerning holomorphic almost periodic functions. This work was inspired by papers by Bohr and by Jessen, together with the works of the Finnish mathematical school (primarily Rolf Nevanlinna and Lars Ahlfors) on value-distribution theory of meromorphic functions. The new results are described in §§ 4-5.

First I will describe some background material. Here I am indebted to Prof. B. Fuglede and Prof. H. Tornehave for some informative references. Bohr's early work on Dirichlet series and the Riemann zeta-function led him to the theory of almost periodic functions. While the principal results of his theory of almost periodic functions on **R** have to some extent been absorbed in the theory of continuous functions on compact abelian groups, his theory of holomorphic almost periodic functions [2] has retained its independence and its charm.

A holomorphic function $f(s)$ in a vertical strip (α, β): $\alpha < \operatorname{Re} s < \beta$ is said to be *almost periodic* if to each $\epsilon > 0$ there exists a number $l = l(\epsilon)$ such that each interval $t_0 < t < t_0 + l$ of length l contains a number τ such that

$$|f(s+i\tau) - f(s)| \leq \epsilon$$

for all s in the strip. (Here α and β are allowed to be infinite.) In other words, if $s = \sigma + it$, $\alpha < \sigma < \beta$, the function $t \to f(\sigma + it)$ is almost periodic on **R** and uniformly so for $\alpha < \sigma < \beta$.

To each such function f one can associate its Dirichlet series

$$f(s) \sim \sum_n A_n e^{\Lambda_n s} \qquad \Lambda_n \in \mathbf{R}, \tag{1}$$

which determines it uniquely. Here

$$A_n = \mathcal{M}_t\,(f(\sigma + it)e^{-\Lambda_n(\sigma + it)}) \tag{2}$$

where $\mathcal{M}$ is the mean value

$$\mathcal{M}(\varphi) = \lim_{S-R\to\infty} \frac{1}{S-R} \int_R^S \varphi(t)dt.$$

For φ almost periodic this limit does indeed exist and then the holomorphy of f implies that A_n is indeed independent of σ. A uniformly convergent Dirichlet series (say $\varsigma(s) = \Sigma_1^\infty e^{-(\log n)s}$ for Re $s > 1 + \epsilon$) is almost periodic; on the other hand, to an almost periodic function $f(s)$ in a strip can be associated a sequence $f_p(s)$ of exponential polynomials $\Sigma\, A_n^{(p)} \exp(\Lambda_n^{(p)} s)$ which converge to $f(s)$ uniformly in any closed substrip ($\alpha_1 \leq$ Re $s \leq \beta_1$, where $\alpha < \alpha_1 < \beta_1 < \beta$).

The original Dirichlet series

$$\sum_1^\infty \frac{a_n}{n^s} = \sum_1^\infty a_n e^{-(\log n)s}$$

were generalized to series of the form

$$\sum_n a_n e^{\Lambda_n s} \qquad \Lambda_1 > \Lambda_2 > \ldots \tag{3}$$

and both at the beginning and the end of his career Bohr investigated problems of convergence, summability etc. for such series (3). It is therefore worth stressing that in (1) the order of the exponents is unrestricted.

2. *Result of Jessen. The Jensen Function*

With the Riemann zeta-function as motivation it becomes a problem of interest to study the distribution of zeros of a function $f(s)$ almost periodic in a strip (α, β). For such functions f the basic general results were obtained by Jessen [6]. He showed the existence of the limit

$$\varphi_f(\sigma) = \mathcal{M}_t(\log |f(\sigma + it)|) = \lim_{S-R\to\infty} \frac{1}{S-R} \int_R^S \log |f(\sigma + it)|\, dt \tag{1}$$

(in spite of the fact that f may have zeros) and proved that it is a *convex* function of σ. Jessen's principal result is the following theorem. If $\alpha < \alpha' < \beta' < \beta$ let $n(\alpha', \beta'; R, S)$ denote the number of zeros of f in the rectangle $\alpha' < \sigma < \beta'$, $R < t < S$ counted with multiplicity.

THEOREM 2.1. *If φ_f is differentiable at α' and β' then the density of zeros*

$$H(\alpha', \beta') = \lim_{S-R\to\infty} \frac{n(\alpha', \beta'; R, S)}{S-R}$$

exists and

$$H(\alpha', \beta') = \frac{1}{2\pi}\{\varphi_f'(\beta') - \varphi_f'(\alpha')\}.$$

If the function $t \to f(\sigma + it)$ has a fixed period p it turns out that this is equivalent to the classical Jensen formula in complex function theory; for this case the function φ_f is a piecewise linear function. Jessen called φ_f *the Jensen function* for f.

Indication of proof. First we assume that the boundary of the rectangle $\alpha' < \sigma < \beta'$, $R < t < S$ contains no zero of $f(s)$. Then by standard complex variable theory.

$$2\pi\, n(\alpha', \beta'; R, S) =$$
$$\int_R^S \frac{f'(\beta' + it)}{f(\beta' + it)}dt - \int_R^S \frac{f'(\alpha' + it)}{f(\alpha' + it)}dt - i\int_{\alpha'}^{\beta'} \frac{f'(\sigma + iR)}{f(\sigma + iR)}d\sigma + i\int_{\alpha'}^{\beta'} \frac{f'(\sigma + iS)}{f(\sigma + iS)}d\sigma.$$

Consider the vertical segments $\alpha' + it$ $(R \le t \le S)$ and $\beta' + it$ $(R \le t \le S)$. We can find a simply connected region Ω containing both of these segments and no zeros for f. We can then define the logarithm $\log f(s)$ in Ω, divide the relation above by $S-R$ and let it tend to ∞. We can restrict the R and S in such a way that the two last terms above give no contribution in the limit. The identity in Theorem 2.1 follows by taking real parts. The restriction on α' and β' is then removed by a continuity argument.

It is now an interesting problem to characterize the convex functions $\varphi(\sigma)$ which arise as Jensen functions φ_f for suitable almost periodic $f(s)$. This question was investigated by Buch [4] whose results imply for example that any convex function which is not linear on any interval arises in this fashion. A complete characterization of the φ_f was given by Jessen and Tornehave [7], § 112. It implies for example that a convex function $\varphi(\sigma)$, $\alpha < \sigma < \beta$, having infinitely many intervals of linearity in a compact subinterval of (α, β) cannot be a Jensen function φ_f if the slopes $\varphi'(\sigma)$ in these intervals are linearly independent over the rational numbers.

3. *Normal Almost Periodic Functions*

Already in his original paper [2] Bohr made a special investigation of almost periodic functions $f(s)$ in a half-plane $(-\infty, \beta)$ and expressed their behaviour near $\sigma = -\infty$ in terms of their Dirichlet expansion

$$f(s) \sim \sum_n A_n e^{\Lambda_n s} \qquad (A_n \neq 0). \tag{1}$$

In [3] he singled out the so-called *normal almost periodic functions* $f(s)$ for which among the nonzero exponents Λ_n the smallest one exists. These have the following property:

Given any $a \in \mathbf{C}$, there exists a half-plane $(-\infty, \sigma_a)$ which contains no a-point for $f(s)$ (i.e. a zero of $f(s) - a$).

Let us for a moment view such a function $f(s)$ via the substitution $s = \log z$ as a function $\varphi(z)$ on a piece $0 < \rho < \rho_0$ of the Riemann surface of $\log z$. The series (1) then becomes a generalized Laurent series

$$\varphi(z) \sim \Sigma\, A_n z^{\Lambda_n} \qquad A_n \neq 0. \tag{2}$$

Let us for simplicity assume the normalizing property that the lowest nonzero exponent, say Λ_0, is > 0. Bohr showed in [3] that the inverse function is also normal almost periodic. I have proved in [5] that a similar statement can be made about the composition of two normal almost periodic functions (having the above normalizing property).

4. *Value-distribution Theory. The first Fundamental Theorem*

Consider a fixed $\beta \leq \infty$ and let $z = f(s)$ be normal almost periodic in $\{-\infty, \beta\}$ (that is normal almost periodic in any substrip $(-K, \beta_1)$ where $\beta_1 < \beta$). We apply stereographic projection of the z-plane $\mathbf{C} \cup \{\infty\}$ onto the Riemann sphere S with diameter 1, tangential to the z-plane at $z = 0$. Given $z_1, z_2 \in \mathbf{C} \cup \{\infty\}$ the (chordal) distance of the corresponding points on S is given by

$$k(z_1,z_2) = \frac{|z_1 - z_2|}{\sqrt{(1+|z_1|^2)\,(1+|z_2|^2)}}$$

so the arc-length element $d\sigma$ on S is

$$d\sigma = \frac{|dz|}{1+|z|^2}.$$

For any $a \in \mathbf{C} \cup \{\infty\}$ let Λ_a denote the smallest exponent in the Dirichlet expansion of $f(s) - a$. (Here $f(s) - \infty$ is to be understood as $1/f(s)$ which is also normal almost periodic). Let A_a denote the corresponding coefficient. We now introduce a quantity which measures how well the function $f(s)$ approximates the value a on the line Re $s = \sigma$. Put

$$M(\sigma,a) = \mathcal{M}_t\{-\log(k(f(\sigma+it), a))\} + [\log k(f(-\infty),a)], \tag{1}$$

where the remainder term is

$$[\log k(f(-\infty),a)] = \begin{cases} \log k(f(-\infty),a), & f(-\infty) \neq a \\ \log \dfrac{|A_a|}{1+|a|^2} & f(-\infty) = a \neq \infty \\ \log \ |A_\infty| & f(-\infty) = a = \infty. \end{cases}$$

The existence of the integral in (1) is clear from the existence of (1) § 2. Next we put $n(\sigma,a;R,S)$ = the number of a-points (with multiplicity) of $f(s)$ in the rectangle $-\infty < \tau < \sigma$, $R < t < S$ (with $s = \tau + it$).

$$N(\sigma,a) = \lim_{S-R\to\infty} \frac{2\pi}{S-R}\int_{-\infty}^{\sigma} n(\tau,a;R,S)\, d\tau + n(-\infty,a;-\infty,\infty)\sigma, \tag{2}$$

where

$$n(-\infty,a;-\infty,\infty) = \operatorname{Max}(\Lambda_a,0).$$

The existence of the last limit is easily established by means of tools used in the proof of Theorem 2.1. The function $N(\sigma,a)$ is taken as a measure for the number of a-points of f in the half-plane Re $s < \sigma$. Note that the remainder term in (2) appears only if $a = \lim_{\sigma\to-\infty} f(s)$.

THEOREM 4.1. *If $f(s)$ is normal almost periodic in $\{-\infty, \beta\}$ then the sum*

$$M(\sigma,a) + N(\sigma,a) = T(\sigma) \tag{3}$$

is independent of a. Also

$$T(\sigma) = \lim_{S-R\to\infty} \frac{2\pi}{S-R}\int_{-\infty}^{\sigma} A_{RS}(\tau)d\tau, \tag{4}$$

where $A_{RS}(\tau)$ *is the area of the Riemann surface* $V_{R,S}(\tau)$ *over the Riemann sphere onto which the function* f *maps the rectangle* $-\infty < \rho < \tau$, $R < t < S$.

The fact that the sum $M(\sigma, a) + N(\sigma, a)$ is independent of a is an analog of Nevanlinna's first fundamental theorem for meromorphic functions. It implies that if $N(\sigma,a)$ is small that is, if $f(s)$ has few a-points, then the approximation term $M(\sigma,a)$ is large and vice-versa. The function $T(\sigma)$ is called *the characteristic function*. The geometric interpretation (4) of $T(\sigma)$ is an analog of a similar interpretation for the classical (periodic) case given by Ahlfors [1] and Shimizu [9].

The proof of Theorem 4.1 proceeds along lines similar to the classical theory (Nevanlinna [8], VI, § 3) but requires in addition some tools utilized in the proof of Theorem 2.1. A brief indication follows. Let $A \in \mathbf{C}$ and put

$$w(s) = A + \frac{1}{f(s)}, \qquad v(s) = \log(1 + |w(s)|^2).$$

We use Gauss' formula

$$\int_{\Gamma} \frac{\partial v}{\partial n}\, dl = \iint_{\Omega} \Delta v \; d\tau dt \tag{5}$$

on a region Ω which is the rectangle $\sigma_0 < \tau < \sigma$, $R < t < S$ with small disks removed having the zeros of f as the centers. Here Γ is the boundary of Ω (with the appropriate orientation), dl the arc element on Γ, Δ the Laplacian in the (τ,t) variables and $\partial/\partial n$ the outgoing normal derivative.

The proof now proceeds along the following steps.

(i) We use the Laurent series of $1/f(s)$ around each zero of $f(s)$ to estimate the contributions to the left hand side of (5) of the circular parts of Γ. Then we let the radii of the disks considered tend to 0.

(ii) By direct computation

$$(\Delta v)(s) = 4\frac{|w'(s)|^2}{(1+|w(s)|^2)^2}.$$

Viewing $w(s)$ as a map from the s-plane to the Riemann sphere lying on the w-plane we have

$$\frac{d\sigma}{|ds|} = \frac{1}{1+|w(s)|^2}|w'(s)|.$$

Thus if $A_{RS}(\tau)$ is the area function (for the function $w(s)$) we have (with $s = \tau + it$)

$$A_{RS}(\sigma) = \int_R^S dt \int_{-\infty}^{\sigma} \frac{|w'(s)|^2}{(1+|w(s)|^2)^2} d\tau.$$

(iii) In (5) we divide by $S-R$ and let $S-R \to \infty$ through special values of S and R, such that on the corresponding horizontal lines $\tau + iR$, $\tau + iS$ $(\sigma_0 < \tau < \sigma)$ $f(s)$ is bounded away from 0. Then the horizontal pieces of the boundary Γ in (5) give no contribution to the limit. The normal derivatives in (5) can be pulled outside the integral as $\partial/\partial\sigma$.

(iv) Now let $\sigma_0 \to -\infty$ in (5) and then integrate with respect to σ from $-\infty$ to σ. Considering the behaviour of $w(s)$ as $\sigma \to -\infty$ we obtain after some manipulation the formula

$$\mathcal{M}_t(\log \sqrt{1+|w(\sigma+it)|^2}) + N(\sigma,0) = \tag{6}$$

$$= \lim_{S-R\to\infty} \frac{2}{S-R}\int_{-\infty}^{\sigma} A_{RS}(\tau)d\tau + \log\sqrt{1+|w(-\infty)|^2}.$$

The last term should be replaced by $\log(1/|A_0|)$ in case $w(-\infty) = \infty$.

(v) Consider a fixed $a \in \mathbf{C}$ and the function

$$w_1(s) = \frac{1+\bar{a}f(s)}{f(s)-a} = \bar{a} + \frac{1}{(f(s)-a)\,(1+|a|^2)^{-1}}.$$

The values of $w_1(s)$ are obtained from the values of $f(s)$ by rotation of the sphere so A_{RS} is the same for w_1 and for f. Also

$$(1 + |w_1(s)|^2)^{-1} = k(w_1(s), \infty) = k(f(s),a)$$

so when (6) is used on w_1 we do obtain Theorem 4.1.

5. *The Second Fundamental Theorem. Applications*

While the first fundamental theorem expresses the constancy of the *total affinity* $M(\sigma,a) + N(\sigma,a)$ of $f(s)$ to the value a the second fundamental theorem will show that for most a $N(\sigma,a)$ is the principal component. This is based on an estimate of the sum $\sum_{v=1}^{p} M(\sigma,a_v)$ for arbitrary distinct $a_1, \dots a_p$.

Motivated by the classical (periodic) theory we consider the number $n_1(\sigma,a;R,S)$ of multiple roots in the equation $f(s)-a=0$ in $-\infty<\tau<\sigma$, $R<t<S$, such that a k-fold root is only counted $(k-1)$ times. We also put $n_1(-\infty,a;-\infty,\infty)=\mathrm{Max}(\Lambda_a',0)$, where Λ_a' is the smallest exponent in the expansion of $f'(s)-a$. Clearly $\sum_a n_1(\sigma,a;R,S)$ is the number of zeros for $f'(s)$ in the rectangle indicated. Then we put

$$N_1(\sigma,a)=\lim_{S-R\to\infty}\frac{2\pi}{S-R}\int_{-\infty}^{\sigma} n_1(\tau,a;R,S)d\tau+n_1(-\infty,a;-\infty,\infty)\sigma\ , \tag{7}$$

and by the remark above, $\sum_a N_1(\sigma,a)$ is bounded by the function $N(\sigma,0)$ taken for the derivative f'. In the next theorem (the analog of Nevanlinna's second fundamental theorem) we distinguish between the two cases: β finite and $\beta=\infty$.

THEOREM 5.1. I. *Let $f(s)$ be normal almost periodic in $\{-\infty,\infty\}$ and $a_1,\dots,a_p$ arbitrary distinct complex numbers. The inequality*

$$\sum_{\nu=1}^{p} M(\sigma,a_\nu)+\sum_a N_1(\sigma,a)\le T(\sigma)+O(\log T(\sigma))+O(\log|\sigma|) \tag{8}$$

holds for all σ except on a set of σ of finite measure.
II. *Let $f(s)$ be normal almost periodic in $\{-\infty,0\}$ and $a_1,\dots,a_p$ any distinct complex numbers. Then inequality* (8) *holds with* $\log|\sigma|$ *replaced by* $\log(1/(1-e^{\sigma}))$ *and the inequality holds for all $\sigma<0$ except for at set of σ over which the integral of $e^{\sigma}(1-e^{\sigma})^{-1}$ is finite.*

In the proof of this theorem the passage from periodic functions to almost periodic functions gives rise to certain technical difficulties. The proof is therefore too complicated to describe here in detail. Instead, I will show how the theorem implies the analog of Nevanlinna's defect relation.

Application. Let f be normal almost periodic in $\{-\infty,\beta\}$. For each $a\in\mathbf{C}$ we define the *defect* by

$$\delta(a)=1-\overline{\lim_{\sigma\to\beta}}\frac{N(\sigma,a)}{T(\sigma)}$$

and the *ramification index*

$$\nu(a)=\underline{\lim_{\sigma\to\beta}}\frac{N_1(\sigma,a)}{T(\sigma)}\ .$$

In the case when f is a nonconstant normal almost periodic function in $\{-\infty,\infty\}$ it is easily deduced from Theorem 4.1 that $\underline{\lim}_{\sigma\to\infty} T(\sigma)/\sigma>0$. From Theorem 5.1 we can therefore deduce the following result.

THEOREM 5.2. Let $f(s)$ *be nonconstant and normal almost periodic in* $\{-\infty, \infty\}$. *Then the defect* $\delta(a)$ *and the ramification index* $\nu(a)$ *are strictly positive for at most countably many a and*

$$\sum_a \delta(a) + \sum_a \nu(a) \leq 1.$$

The defect $\delta(a)$ is a measure for how rarely f takes the value a. If a is omitted by $f(s)$ altogether in $-\infty \leq \operatorname{Re} s < \infty$ then $\delta(a) = 1$ so we deduce from $\sum_a \delta(a) \leq 1$ that there can be at most one such value a.

For the case $\beta = 0$ we obtain similarly the following result.

THEOREM 5.3. *Let* $f(s)$ *be normal almost periodic in* $\{-\infty, 0\}$ *and assume*

$$\lim_{\sigma \to 0} \frac{\log\left(\frac{1}{1-e^{\sigma}}\right)}{T(\sigma)} = 0.$$

Then

$$\sum_a \delta(a) \leq 1.$$

Again this implies that f omits at most one value in the strip $-\infty \leq \operatorname{Re} s < 0$.

References

[1] Ahlfors, L. Beiträge zur Theorie der meromorphen Funktionen. *7. Congr. Math. Scand. Oslo 1929.*
[2] Bohr, H. Zur Theorie der fastperiodischen Funktionen, III Teil. Dirichletentwicklung analytischer Funktionen. *Acta Math. 47* (1926), 237-281.
[3] ———— On the inverse function of an analytic almost periodic function. *Ann. of Math.* 32 (1931), 247-260.
[4] Buch, K. R. Über die Nullstellenverteilung einer analytischen grenzperiodischen Funktion. Mat. Fys. Medd. Danske Vid. Selsk. 16 no. 4 1938.
[5] Helgason, S. *Værdifordelingsteori for analytiske næstenperiodiske functioner.* Prisopgave, Københavns Universitet, 1950.
[6] Jessen, B. Über die Nullstellen einer analytischen fastperiodischen Funktion. Eine Verallgemeinerung der Jensenschen Formel. *Math. Ann.* 108 (1933), 485-516.
[7] Jessen, B. and H. Tornehave, Mean motions and zeros of almost periodic functions. *Acta Math.* 77 (1945), 137-279.
[8] Nevanlinna, R. *Eindeutige analytische Funktionen.* Springer, Berlin 1936.
[9] Shimizu, T. On the theory of meromorphic functions. *Japanese J. Math.* 6 (1929), 119-171.

Department of Mathematics
MIT 2-182
Cambridge, MA 02139
U.S.A.

Contemporary Mathematics
Volume **113**, 1990

The Totally-Geodesic Radon Transform on Constant Curvature Spaces

SIGURDUR HELGASON

1. Introduction. Let X be an n-dimensional simply connected Riemannian manifold of constant curvature $\kappa \neq 0$. We normalize the metric such that $\kappa = \pm 1$ so X is either the sphere $\mathbf{S}^n$ with the standard metric or the hyperbolic space $\mathbf{H}^n$, that is the unit ball $|x| < 1$ in $\mathbf{R}^n$ with Riemannian structure

$$ds^2 = 4\,\frac{dx_1^2 + \cdots + dx_n^2}{(1 - x_1^2 - \cdots - x_n^2)^2} \tag{1.1}$$

of constant curvature -1. For a fixed k $(1 \leq k \leq n-1)$ let ξ be an arbitrary totally geodesic submanifold of X of dimension k. The *totally geodesic Radon transform* $f \longrightarrow \hat{f}$ is defined by

$$\hat{f}(\xi) = \int_\xi f(x)\,dm(x), \tag{1.2}$$

where dm is the surface element ξ induced by (1.1) and f is a continuous function on X integrable on each ξ. Denoting by Ξ the space of all ξ we consider also the *dual transform* $\phi \longrightarrow \check{\phi}$ given by

$$\check{\phi}(x) = \int_{x\in\xi} \phi(\xi)\,d\mu(\xi), \qquad x \in X. \tag{1.3}$$

Here ϕ is a continuous function on Ξ and μ the measure on the set of ξ passing through x, invariant under all rotations around x and having total measure 1.

In this paper we present general inversion formulas and a support theorem for the transform (1.2).

1980 *Mathematics Subject Classification* (1985 *Revision*). Primary 44A05; Secondary 43A85, 53C65, 58C35.

Supported in part by NSF Grant DMS 8805665.

Remarks. The transform (1.2) was first considered in [5a] and there the following inversion formulas are given for k even:

(1.4)
$$cf = [L+(k-1)(n-k)][L+(k-3)(n-k+2)]\dots[L+1(n-2)](\hat{f})^{\vee}, \quad \kappa = -1,$$

(1.5)
$$2cf = [L-(k-1)(n-k)][L-(k-3)(n-k+2)]\dots[L-1\cdot(n-2)](\hat{f})^{\vee}, \quad \kappa = +1,$$

the latter formula requiring f to be a symmetric function on $\mathbf{S}^n$. Here L is the Laplacian on X and the constant c is given by

$$c = \frac{\Gamma(\frac{n}{2})}{\Gamma(\frac{n-k}{2})}(-4\pi)^{\frac{k}{2}}.$$

At that time (1958) formula (3.6) below was also proved but was not included in [5a] because the formula for k odd seemed unreasonably complicated compared to (1.4) for k even, and since the case $k = 1$, when (1.2) is the X-ray transform, had not acquired its later distinction through tomography. However, the method of proof of (3.6) was shortly afterwards used to prove the general support theorem for the transform (1.2) on the hyperbolic space (see [5b], Theorem 5.2).

In response to their question, a proof of the inversion formula (3.6) was communicated to Berenstein and Tarabusi; I understand that they have subsequently found another formulation of this inversion [1]. In [7] a different formula is given for the case $k = n - 1$.

Recently, Ingi Bjarnason called my attention to the potential use in geophysics of a support theorem for the X-ray transform on a sphere. Such a theorem is proved in §4. Because of the necessary symmetry assumption on the function there are some difficulties in extending the proof for the hyperbolic space in **[5d]** to the sphere, so our proof below of Theorem 4.1 follows different lines. We remark that this result was proved by Quinto [**6**] p. 520, however under the additional assumption, that f vanish identically on some belt around the equator.

2. The Generalized Transforms. Let G denote the group $I(X)$ of isometries of X, K the subgroup fixing o. Fix a $\xi_o \in \Xi$ passing through o and let H_o denote the subgroup of G leaving ξ_o stable. Then

(2.1)
$$X = G/K, \qquad \Xi = G/H_o$$

and the transforms $f \longrightarrow \hat{f}$, $\phi \longrightarrow \check{\phi}$ in (1.2) and (1.3) are the Radon transforms for the double fibration

$$\begin{array}{ccc} & G/(H_o \cap K) & \\ \swarrow & & \searrow \\ G/K & & G/H_o \end{array}$$

as defined in [5c]:

$$\hat{f}(\gamma H_o) = \int_{gK \cap \gamma H_o \neq \emptyset} f(gK)\,dm(gK), \tag{2.2}$$

$$\check{\phi}(gK) = \int_{\gamma H_o \cap gK \neq \emptyset} \phi(\gamma H_o)\,d\mu(\gamma H_o).$$

However, we can also fix a $\xi_p \in \Xi$ at a distance $p \geq 0$ from o and let H_p be the subgroup leaving ξ_p stable. By Chern's definition [2], $x \in X$ and $\xi \in \Xi$ are *incident* if as cosets $x = gK$, $\xi = \gamma H_p$ have a point in common. It is then readily seen that

$$x \text{ and } \xi \text{ are incident} \iff d(x, \xi) = p, \tag{2.3}$$

d denoting distance. The group-theoretic viewpoint therefore leads us to the definitions

$$(\hat{f}_p)(\xi) = \int_{d(x,\xi)=p} f(x)\,ds(x), \qquad (\check{\phi}_p)(x) = \int_{d(x,\xi)=p} \phi(\xi)\,d\sigma(\xi), \tag{2.4}$$

where ds the Riemannian measure on the set in question and $d\sigma$ is again the average over the set of ξ at distance p from x.

The transforms (2.4) are in reality the double-fibration Radon transforms from [5c] when specialized to the double fibration

$$\begin{array}{ccc} & G/(K \cap H_p) & \\ / & & \backslash \\ G/K & & G/H_p \end{array}$$

but here they have also an intuitive geometric interpretation. For $\mathbf{H}^2$, $\hat{f}_p$ is the integral of f over an "equidistant curve" in the sense of non-Euclidean geometry; for $\mathbf{S}^2$, $\hat{f}_p$ is the integral of f over a circle on $\mathbf{S}^2$ of length $2\pi \cos p$. Note that

$$\hat{f}_o = \hat{f}, \qquad \check{\phi}_o = \check{\phi}.$$

3. The Inversion Formulas. Consider now a pair $x \in X$, $\xi \in \Xi$ at distance p, i.e. $d(x, \xi) = p$. Choose $g \in G$ such that $g \cdot o = x$. Then if dk is the normalized Haar measure on K,

$$(\hat{f})_p^\vee(g \cdot o) = \int_K \hat{f}(gkg^{-1} \cdot \xi)\,dk = \int_K dk \int_\xi f(gkg^{-1} \cdot y)\,dm(y)$$

$$= \int_\xi \left(\int_K f(gkg^{-1} \cdot y)\,dk \right) dm(y) = \int_\xi \left(M^{d(x,y)} f \right)(g \cdot o)\,dm(y),$$

where, in general, $(M^r f)(z)$ is the average of f on the sphere $S_r(z)$ with radius r and center z. Now let x_o be *a* point on ξ at minimum distance(that is p) from x and let

$$r = d(x_o, y), \qquad q = d(x, y), \tag{3.1}$$

d denoting the distance function on X. Since ξ is totally geodesic, $d(x_o, y)$ equals the distance between x_o and y in ξ. The last integrand $(M^{d(x,y)} f)(x)$ is constant in y on each sphere in ξ with center x_o.

Consider first the case $X = \mathbf{H}^n$. Using geodesic polar coordinates in ξ with center x_o we then obtain

$$(\hat{f})_p^\vee)(x) = \Omega_k \int_0^\infty (M^q f)(x) \sinh^{k-1} r\, dr, \tag{3.2}$$

where Ω_k is the area of the unit sphere in $\mathbf{R}^k$.

Next we use the cosine formula on the non-Euclidean *right-angled* triangle (xx_oy). By (3.1) and $p = d(x, x_o)$ we then have (curvature being -1),

$$\cosh q = \cosh p \cosh r. \tag{3.3}$$

Now we fix x and put

$$F(\cosh q) = (M^q f)(x), \quad \hat{F}(\cosh p) = (\hat{f})_p^\vee(x). \tag{3.4}$$

Then by (3.2) and (3.3),

$$\hat{F}(\cosh p) = \Omega_k \int_0^\infty F(\cosh p \cosh r)(\sinh r)^{k-1} dr.$$

As shown in [5d], p. 95 (where $(k-1)/2$ should be $k/2 - 1$) this is inverted by

$$F(u^{-1})u^{-k} = c\,u \left(\frac{d}{d(u^2)}\right)^k \int_0^u (u^2 - v^2)^{\frac{k}{2}-1} \hat{F}(v^{-1})\, dv, \tag{3.5}$$

where $c^{-1} = (k-1)!\,\Omega_{k+1} \big/ 2^{k+1}$. Let $\operatorname{lm} v$ denote the positive solution q to the equation $\cosh q = v^{-1}$ $(v \leq 1)$. Putting $q = 0$, $u = 1$ we deduce the following result from (3.4) and (3.5).

Theorem 3.1. *The k-dimensional totally geodesic Radon transform $f \longrightarrow \hat{f}$ on the hyperbolic space $\mathbf{H}^n$ is inverted by*

$$f(x) = c \left[\left(\frac{d}{d(u^2)}\right)^k \int_0^u ((\hat{f})_{\ell m\, v}^\vee))(x)(u^2 - v^2)^{\frac{k}{2}-1} dv\right]_{u=1}, \tag{3.6}$$

where

$$c = \frac{2^{k+1}}{(k-1)!\Omega_{k+1}}, \quad \operatorname{lm} v = \cosh^{-1}(v^{-1}).$$

We consider now the remaining case $X = \mathbf{S}^n$ and assume f symmetric, i.e. $f(-s) = f(s)$. (For f odd, $\hat{f} \equiv 0$). Again, we start with the formula

$$(\hat{f})_p^\vee(x) = \int_\xi (M^{d(x,y)} f)(x) dm(y)$$

and use geodesic polar coordinates in ξ with center x_o. Then we get in place of (3.2), using now the symmetry of f,

$$(\hat{f})_p^\vee(x) = 2\Omega_k \int_0^{\pi/2} (M^q f)(x) \sin^{k-1} r dr. \tag{3.7}$$

Now we have by spherical trigonometry used on the right-angled triangle (xx_oy),

$$\cos q = \cos p \cos r.$$

Fix x and put

$$F(\cos q) = (M^q f)(x), \quad \hat{F}(\cos p) = (\hat{f})_p^\vee(x). \tag{3.8}$$

We put $v = \cos p$ and $u = v \cos r$. Then

$$v^{k-1} \hat{F}(v) = 2\Omega_k \int_0^v F(u)(v^2 - u^2)^{\frac{k}{2}-1} du. \tag{3 .10}$$

Again, this is inverted by

$$F(u) = \frac{c}{2} u \left(\frac{d}{d(u^2)} \right)^k \int_0^u (u^2 - v^2)^{\frac{k}{2}-1} v^k \hat{F}(v) dv, \tag{3.11}$$

where c is the same as in (3.5). Thus we get the following counterpart to Theorem 3.1.

Theorem 3.2. *The k-dimensional totally geodesic Radon transform $f \longrightarrow \hat{f}$ on $\mathbf{S}^n$ is for f symmetric inverted by*

$$f(x) = \frac{c}{2} \left[\left(\frac{d}{d(u^2)} \right)^k \int_0^u \left((\hat{f})^\vee_{\cos^{-1}(v)} \right) (x) v^k (u^2 - v^2)^{\frac{k}{2}-1} dv \right]_{u=1} \tag{3.12}$$

Formulas (3.6) and (3.12) have a geometric interpretation: $(\hat{f})_q^\vee(x)$ is the average of the integrals of f over the k-dimensional totally geodesic manifolds tangent to $S_q(x)$.

For the case when k is even one can carry out the differentiations in (3.6) and (3.12) and the integral sign disappears. Thus the inversion formula is local in that the value of f near x only depends on the values of $\hat{f}$ for ξ close to x. This we of course knew on the basis of (1.4) and (1.5).

Example. The simplest case of the inversion formulas (1.4) and (3.6) is of course when $k = 2$, $n = 3$. Since $d/d(u^2) = (2u)^{-1} d/du$ formula (3.6) becomes after carrying out the differentiations,

$$f(x) = -\frac{1}{2\pi}\left\{(\hat{f})^{\vee}(x) - \left[\frac{d}{du}\left((\hat{f})^{\vee}_{\ell m\, u}(x)\right)\right]_{u=1}\right\},$$

whereas (1.4) becomes

$$f = -\frac{1}{2\pi}(L+1)(\hat{f})^{\vee}.$$

Thus we obtain the amusing formula

$$\left[\frac{d}{du}(\hat{f})^{\vee}_{\ell m\, u}\right]_{u=1} = -L\left((\hat{f})^{\vee}\right).$$

4. The Support Theorem for the X-ray transform on $\mathbf{S}^n$. The corollary of Theorem 3.2 that a symmetric function on $\mathbf{S}^2$ is determined by its great circle integrals was proved by Funk **[3]**. He used the result to prove that the standard metric on the projective plane cannot be deformed in such a way that the geodesics remain closed and of the same length. However, it was proved by Guillemin **[4]** that for $\mathbf{S}^2$ such deformations are possible.

In this section we shall prove a "support theorem" for the X-ray transform on $\mathbf{S}^n$. Since the transform $\phi \longrightarrow \check{\phi}_p$ requires the consideration of great circles with unrestricted tangents this theorem cannot be derived from Theorem 3.2.

Theorem 4.1. *Let B be a closed spherical cap on $\mathbf{S}^n$, B' the cap symmetric to B with respect to O. Let f be a symmetric C^1 function satisfying*

$$\int_{\gamma} f(s)\,dm(s) = 0 \tag{4.1}$$

for every great circle γ on $\mathbf{S}^n$ which does not enter the "arctic zones" B and B'. Then if f is 0 on the equator *

$$f \equiv 0 \quad \textit{outside} \quad B \cup B'. \tag{4.2}$$

Remarks. The desirability of such a result for tomographical applications in geophysics was pointed out to me by Ingi Bjarnason. If in addition to (4.1) f is assumed to vanish identically in some belt around the equator the conclusion (4.2) was proved by Quinto **[6]**, p. 520.

For the proof of Theorem 4.1 we may, replacing f by a suitable convolution

$$f_1(x) = \int_{\mathbf{O}(n+1)} \phi(g) f(g^{-1}\cdot x)\,dg, \qquad \phi \in C^{\infty}(\mathbf{O}(n+1)),$$

assume f to be smooth. Considering the restriction $f|(\mathbf{S}^n \cap E)$ where $E \subset \mathbf{R}^{n+1}$ is an arbitrary 3-dimensional vector subspace (so $\mathbf{S}^n \cap E = \mathbf{S}^2$) is it sufficient to consider the case $n = 2$ in Theorem 4.1. Suppose B and

*For the case $n = 2$ all derivatives are assumed zero on the equator, cf. Helgason [B4] (from "The Selected Works Bibliography", p. xxxv).

B' have the north pole N and the south pole S as centers. Consider the spherical coordinates of a point $(x_1, x_2, x_3) \in \mathbf{S}^2$

$$x_1 = \cos\phi \sin\theta \quad x_2 = \sin\phi \sin\theta, \quad x_3 = \cos\theta \tag{4.3}$$

and let k_ϕ denote the rotation by the angle ϕ around the x_3-axis. Then we decompose f into a Fourier expansion

$$f(x) = \sum_{n\in\mathbf{Z}} f_n(x), \qquad f_n(x) = \frac{1}{2\pi}\int_0^{2\pi} f(k_\phi \cdot x)e^{-in\phi}d\phi \tag{4.4}$$

relative to the group K of rotations k_ϕ. Since both the assumption (4.1) and the conclusion (4.2) are K-invariant, and since each f_n is symmetric we can by (4.4) assume that f satisfies a homogeneity condition

$$f(k_\phi \cdot x) = e^{in\phi} f(x). \tag{4.5}$$

But then the X-ray transform $\hat{f}$ satisfies this condition too:

$$\hat{f}(k_\phi \cdot \gamma) = e^{in\phi}\hat{f}(\gamma). \tag{4.6}$$

In particular, (4.5) shows that f is determined by its values on the intersection of $\mathbf{S}^2$ with the plane $x_1 = 0$. We put

$$g(\cos\theta) = f(0, \sin\theta, \cos\theta).$$

Then (4.5) implies, since f is even, that $g((\cos(\pi - \theta)) = (-1)^n g(\cos\theta)$, i.e.,

$$g(-u) = (-1)^n g(u) \tag{4.7}$$

Let Γ denote the set of great circles which are disjoint from $B \cup B'$ and whose normal lies in the plane $x_1 = 0$. For $\gamma \in \Gamma$ let x_γ be the intersection of γ with the half-plane $x_1 = 0, x_2 > 0$ and let α denote the spherical distance from N to x_γ. Now by the symmetry of f, if $\gamma \in \Gamma$ then

$$\hat{f}(\gamma) = 2\int_0^{\pi} f(x_s)ds, \tag{4.8}$$

where x_s is the point on γ at distance s from x_γ with $x_1(x_s) > 0$. Let ϕ and θ be the coordinates of x_s according to (4.3). Considering the right angled triangle $x_s N x_\gamma$ we have

$$\cos\theta = \cos s \cos\alpha \tag{4.9}$$

and since the angle at N is $\pi/2 - \phi$, (4.5) implies

$$f(x_s) = e^{-in(\pi/2-\phi)} g(\cos\theta).$$

Writing

$$\hat{g}(\cos\alpha) = \hat{f}(\gamma), \qquad \gamma \in \Gamma,$$

equation (4.8) becomes

$$\hat{g}(\cos\alpha) = 2(-i)^n \int_0^\pi e^{in\phi} g(\cos\theta)\,ds$$

$$= 2(-i)^n \int_0^\pi e^{in\phi} g(\cos s\cos\alpha)\,ds.$$

Put $v = \cos\alpha$ and $u = v\cos s$ so $du = v(-\sin s)ds = -(v^2-u^2)^{\frac{1}{2}}ds$. Then

$$\hat{g}(v) = 2(-i)^n \int_{-v}^{v} e^{in\phi(u,v)} g(u)(v^2-u^2)^{-\frac{1}{2}}\,du, \tag{4.10}$$

where the dependence of ϕ on u and v is indicated (for $v \neq 0$).

Next we observe from the geometry or from (4.3) that $\phi(-u,v) = -\phi(u,v)$. Thus by (4.7) we can rewrite (4.10) as

$$\hat{g}(v) = 4(-1)^{\frac{n}{2}} \int_0^v \cos(n\phi(u,v))g(u)(v^2-u^2)^{-\frac{1}{2}}\,du, \qquad n \text{ even};\tag{4.11}$$

$$\hat{g}(v) = 4(-1)^{\frac{n-1}{2}} \int_0^v \sin(n\phi(u,v))g(u)(v^2-u^2)^{-\frac{1}{2}}\,du, \qquad n \text{ odd}.\tag{4.12}$$

From the theory of Abel-type Volterra integral equations

$$\Phi(v) = \int_0^v K(u,v)(v^2-u^2)^{-\frac{1}{2}}F(u)\,du$$

it is known **[8]** that if $K(u,u) \neq 0$ and F of $0 = 0$, then F is uniquely determined by Φ. Since $\phi(u,u) = \pi/2$ we see that for each of the equations (4.11), (4.12) is this condition $K(u,u) \neq 0$ luckily satisfied.

If the cap B has spherical radius β then our assumption (4.1) amounts to $\hat{g}(\cos\alpha) = 0$ for $\frac{\pi}{2} > \alpha > \beta$, i.e. $\hat{g}(v) = 0$ for $0 < v < \cos\beta$. But then the uniqueness mentioned implies $g(u) = 0$ for $0 < u < \cos\beta$ so

$$f(o, \sin\theta, \cos\theta) = g(\cos\theta) = 0 \quad \text{for } \frac{\pi}{2} > \theta > \beta.$$

By (4.5) this proves the theorem.

BIBLIOGRAPHY

1. C. Berenstein and E.C. Tarabusi, (to appear).
2. S.S. Chern, On integral geometry in Klein spaces. Ann. of Math. **43** (1942), 178–189.
3. P. Funk, Über eine geometrische Anwendung der Abelschen Integral gleichung, Math. Ann. **77** (1916), 129–135.
4. V. Guillemin, Radon transform on Zoll surfaces. Advan. Math. 22 (1976), 85–199.

5. S. Helgason, a) Differential operators on homogeneous spaces. Acta Math. **102** (1959), 239–299.
b) A duality in integral geometry; some generalizations of the Radon transform. Bull. Amer. Math. Soc. **70** (1964), 435–446.
c) A duality in integral geometry on symmetric spaces. Proc. U.S.-Japan Seminar in Differential Geometry Kyoto (1965), Nippon Hyronsha, Tokyo (1966).
d) Support of Radon transforms. Advan. Math. **38** (1980), 91–100.
6. E.T. Quinto, The invertibility of rotation invariant Radon transforms. J. Math. Anal. Appl. **91** (1983), 510–522.
7. V.I. Semyanisty, Homogeneous functions and some problems of integral geometry in spaces of constant curvature. Sov. Math. Dokl. **2** (1961), 59–62.
8. K. Yosida, Lectures on Differential and Integral Equations, Interscience, New York, 1960.

Department of Mathematics, Massachusetts
Institute of Technology, Cambridge, Ma 02139

SOME RESULTS ON INVARIANT DIFFERENTIAL OPERATORS ON SYMMETRIC SPACES

By Sigurdur Helgason

1. Introduction. Let G be a connected, noncompact semisimple Lie group with finite center, K a maximal compact subgroup and $X = G/K$ the associated symmetric space. Let $\boldsymbol{D}(G/K)$ denote the algebra of differential operators on X which are invariant under the transformations $\tau(g) : hK \to ghK$ of X, $g \in G$ being arbitrary. In this note we shall investigate connections between this algebra $\boldsymbol{D}(G/K)$ and the algebra $\boldsymbol{Z}(G)$ of differential operators on G which are invariant under all left and right translations.

To state the results let $\boldsymbol{D}(G)$ denote the algebra of all left-invariant differential operators on G, $\boldsymbol{D}_K(G)$ the subalgebra of $\boldsymbol{D}(G)$ consisting of those operators which are also invariant under right translations from K. Let $\mu : \boldsymbol{D}_K(G) \to \boldsymbol{D}(G/K)$ denote the homomorphism given by $(\mu(D)f) \circ \pi = D(f \circ \pi)$, where $\pi : G \to G/K$ is the natural map and $f \in C^\infty(X)$. Since G/K is reductive, μ is surjective [7(a)].

We consider another homomorphism ν of $\boldsymbol{D}(G)$ into the algebra $\boldsymbol{E}(X)$ of all differential operators on X. The map which to $g \in G$ assigns the linear transformation $f \to f \circ \tau(g^{-1})$ of $C^\infty(X)$ is a representation of G whose differential is the indicated homomorphism $\nu : \boldsymbol{D}(G) \to \boldsymbol{E}(X)$. It is known [7(b)] that

$$\nu(D) = \mu(D^*) \qquad D \in \boldsymbol{Z}(G),$$

the asterisk denoting adjoint. The image of $\boldsymbol{Z}(G)$ under μ (and ν) will be denoted $\boldsymbol{Z}(G/K)$. We shall prove that each $D \in \boldsymbol{D}(G/K)$ is a "quotient" of two members $Z_1, Z_2 \in \boldsymbol{Z}(G/K)$, that is

$$DZ_1 = Z_2, \qquad (Z_1 \neq 0).$$

Manuscript received 21 June 1989.
Supported by the National Science Foundation DMS-8805665.
American Journal of Mathematics 114 (1992), 789–811.

For G classical (real or complex) we have actually $\boldsymbol{D}(G/K) = \boldsymbol{Z}(G/K)$ [7(b)] and we shall see in the present paper that this remains true for all noncompact irreducible symmetric spaces with exactly four exceptions, namely the symmetric pairs

$$(1.1) \quad (\mathfrak{e}_6, \mathfrak{so}(10) + \boldsymbol{R}), (\mathfrak{e}_6, \mathfrak{f}_4), (\mathfrak{e}_7, \mathfrak{e}_6 + \boldsymbol{R}), (\mathfrak{e}_8, \mathfrak{e}_7 + \mathfrak{su}(2)).$$

The proofs are based on relating these differential operators to Weyl group invariants.

I am very grateful to G. Shimura and D. Vogan for their invaluable help as explained in more detail late in Section 3. I also thank F. Bien and T. Oshima as well as H. Ochiai and T. Kurokawa for their illuminating comments (cf. remark in Section 4).

As customary, Lie groups will be denoted by capital letters and their Lie algebras by the corresponding lower case German letters.

2. Elements of $\boldsymbol{D}(G/K)$ as fractions. Let

$$(2.1) \qquad \mathfrak{g} = \mathfrak{k} + \mathfrak{p}$$

be the Cartan decomposition corresponding to G/K and θ the associated Cartan involution. Fix a maximal abelian subspace $\mathfrak{a} \subset \mathfrak{p}$, let $\Sigma = \Sigma(\mathfrak{g}, \mathfrak{a})$ be the set of roots of $\mathfrak{g}$ with respect to $\mathfrak{a}$ and extend $\mathfrak{a}$ to a Cartan subalgebra $\mathfrak{h}$ of $\mathfrak{g}$. Denoting complexification by superscript $\mathbf{C}$ let $\Delta = \Delta(\mathfrak{g}^{\mathbf{C}}, \mathfrak{h}^{\mathbf{C}})$ denote the set of roots of $\mathfrak{g}^{\mathbf{C}}$ with respect to $\mathfrak{h}^{\mathbf{C}}$. Let $\overline{F}$ denote the restriction to $\mathfrak{a}$ (or $\mathfrak{a}^{\mathbf{C}}$) of a function F on $\mathfrak{h}$ (or $\mathfrak{h}^{\mathbf{C}}$). The space $\mathfrak{h}_{\boldsymbol{R}} = i(\mathfrak{h} \cap \mathfrak{k}) + \mathfrak{a}$ is the real form of $\mathfrak{h}^{\mathbf{C}}$ where each $\alpha \in \Delta$ is real-valued. We select orderings in the duals $\mathfrak{a}^*$, $(\mathfrak{h}_{\boldsymbol{R}})^*$ which are compatible in the sense that $\lambda \in (\mathfrak{h}_{\boldsymbol{R}})^*$ is >0 whenever $\overline{\lambda} > 0$. The set Δ^+ of positive roots is divided into two classes P and P_0:

$$P = \{\alpha \in \Delta^+ : \overline{\alpha} \neq 0\}, \qquad P_0 = \{\alpha \in \Delta^+ : \overline{\alpha} = 0\}.$$

Then $\Sigma^+ = \{\overline{\alpha} : \alpha \in P\}$. Let

$$\rho^* = \frac{1}{2} \sum_{\alpha \in \Delta^+} \alpha, \qquad \rho_0 = \frac{1}{2} \sum_{\alpha \in P_0} \alpha, \qquad \rho = \frac{1}{2} \sum_{\alpha \in P} \alpha.$$

Since $\alpha \to -\theta\alpha$ is a permutation of P we see that

$$(2.2)\qquad \rho(H) = 0\ (H \in \mathfrak{h} \cap \mathfrak{k}), \qquad \rho_0(H) = 0,\ (H \in \mathfrak{a}).$$

Let $\mathfrak{g} = \mathfrak{k} + \mathfrak{a} + \mathfrak{n}$, $G = KAN$ be the Iwasawa decompositions corresponding to the positive system Σ^+, let M denote the centralizer of A in K and $\mathfrak{m}$ its Lie algebra. For the complexified Lie algebras we have

$$(2.3)\qquad \mathfrak{m}^{\mathbf{C}} = (\mathfrak{h} \cap \mathfrak{k})^{\mathbf{C}} + \sum_{\alpha \in P_0} ((\mathfrak{g}^{\mathbf{C}})^{\alpha} + (\mathfrak{g}^{\mathbf{C}})^{-\alpha})$$

$$(2.4)\qquad \mathfrak{n}^{\mathbf{C}} = \sum_{\alpha \in P} (\mathfrak{g}^{\mathbf{C}})^{\alpha}.$$

For $\mu \in (\mathfrak{h}^{\mathbf{C}})^*$, $\nu \in (\mathfrak{a}^{\mathbf{C}})^*$ we define $H_\mu \in \mathfrak{h}^{\mathbf{C}}$, $A_\nu \in \mathfrak{a}^{\mathbf{C}}$ by the condition

$$B(H_\mu, H) = \mu(H),\ (H \in \mathfrak{h}), \qquad B(A_\nu, H) = \nu(H),\ (H \in \mathfrak{a}),$$

B denoting the Killing form. Let W denote the Weyl group of the root system Σ; it acts on $\mathfrak{a}$, $\mathfrak{a}^{\mathbf{C}}$, the duals $\mathfrak{a}^*$, $(\mathfrak{a}^{\mathbf{C}})^*$ and the symmetric algebras $S(\mathfrak{a}), \ldots, S((\mathfrak{a}^{\mathbf{C}})^*)$, resulting in corresponding algebras $I(\mathfrak{a}), \ldots, I((\mathfrak{a}^{\mathbf{C}})^*)$ of Weyl group invariants. Similar notions pertain to the Weyl group $\tilde{W}$ of the root system Δ. We recall that W is the set of linear transformations of $\mathfrak{a}^{\mathbf{C}}$ induced by the subgroup of $\tilde{W}$ leaving $\mathfrak{a}^{\mathbf{C}}$ invariant. Moreover, if two elements in $\mathfrak{a}^{\mathbf{C}}$ are conjugate under $\tilde{W}$ they are conjugate under W. We have also to consider the translation of $\tilde{W}$ by ρ_0, that is the group $\tilde{W}_0$ of transformations

$$\mu \to s\mu + s\rho_0 - \rho_0, \qquad \mu \in (\mathfrak{h}^{\mathbf{C}})^*,$$

as s runs through $\tilde{W}$. Then the corresponding algebra $I_0(\mathfrak{h}^{\mathbf{C}})$ of invariants consists of the functions $\mu \to P(\rho_0 + \mu)$ on $(\mathfrak{h}^{\mathbf{C}})^*$ as P runs through $I(\mathfrak{h}^{\mathbf{C}})$. We view $\mathfrak{a}^*$ as a subset of $\mathfrak{h}_R^*$ taking each $\lambda \in \mathfrak{a}^*$ to be $\equiv 0$ on $i(\mathfrak{h} \cap \mathfrak{k})$.

PROPOSITION 2.1. *The following properties of the symmetric space G/K are equivalent*:

(i) $\boldsymbol{Z}(G/K) = \boldsymbol{D}(G/K)$;
(ii) *The restriction from $\mathfrak{h}_R^*$ to $\mathfrak{a}^*$ maps $I_0(\mathfrak{h}^{\mathbf{C}})$ onto $I(\mathfrak{a}^{\mathbf{C}})$.*

Proof. In the Iwasawa decomposition $G = KAN$ let $\exp H(g)$ ($H(g) \in \mathfrak{a}$) denote the A-component of $g \in G$ and consider the function

$$f(g) = e^{(i\lambda - \rho)(H(g))}, \tag{2.5}$$

λ being fixed in $(\mathfrak{a}^*)^{\mathbf{C}}$. By Harish-Chandra [6(a)] Lemma 18 we have an isomorphism $\gamma' : \boldsymbol{Z}(G) \to S(\mathfrak{h}^{\mathbf{C}})$ given by

$$Z - \gamma'(Z) \in \sum_{\alpha \in \Delta^+} \boldsymbol{D}(G)(\mathfrak{g}^{\mathbf{C}})^{\alpha},$$

where elements of $\mathfrak{g}^{\mathbf{C}}$ are viewed as left-invariant differential operators on G. Since f is right invariant under MN we thus conclude from (2.3) and (2.4),

$$Zf = \gamma'(Z)f. \tag{2.6}$$

In addition, $f(g \exp H) = e^{(i\lambda - \rho)(H_1)}f(g)$ for $H \in \mathfrak{h}$ where H_1 is the $\mathfrak{a}$-component of H. We have

$$(i\lambda - \rho)(H_1) = (i\lambda - \rho)(H) = (i\lambda - \rho^* + \rho_0)(H)$$

so

$$(Hf)(g) = (i\lambda - \rho^* + \rho_0)(H)f(g). \tag{2.7}$$

On the other hand, the Harish-Chandra isomorphism γ of $\boldsymbol{Z}(G)$ onto $I(\mathfrak{h}^{\mathbf{C}})$ is given by $\gamma(Z) = (\gamma'(Z))^0$ where $P \to P^0$ is the automorphism of $S(\mathfrak{h}^{\mathbf{C}})$ given by $H^0 = H - \rho^*(H)$ (loc. cit. Lemma 19). Viewing $P \in S(\mathfrak{h}^{\mathbf{C}})$ as a function on $(\mathfrak{h}^{\mathbf{C}})^*$ we have $P^0 = P \circ T(-\rho^*)$ where $T(\mu)$ denotes translation by μ. Thus by (2.6) and (2.7)

$$(Zf)(g) = \gamma'(Z)(i\lambda - \rho^* + \rho_0)f(g) = \gamma(Z)(i\lambda + \rho_0)f(g). \tag{2.8}$$

Since Z is invariant under right translations, (2.8) implies that the spherical function

$$\phi_\lambda(g) = \phi_\lambda(gK) = \int_K e^{(i\lambda - \rho)(H(gk))}dk$$

satisfies

$$(2.9) \qquad Z\phi_\lambda = \gamma(Z)(\rho_0 + i\lambda)\phi_\lambda.$$

On the other hand, the isomorphism Γ of $\boldsymbol{D}(G/K)$ onto $I(\mathfrak{a}^{\mathbf{c}})$ satisfies

$$(2.10) \qquad D\phi_\lambda = \Gamma(D)(i\lambda)\phi_\lambda.$$

Thus we deduce that if $\boldsymbol{Z}(G/K) = \boldsymbol{D}(G/K)$ then $(I_0(\mathfrak{h}^{\mathbf{c}})^- = I(\mathfrak{a}^{\mathbf{c}})$, the bar denoting restriction.

Conversely, let us assume (ii) and let $D \in \boldsymbol{D}(G/K)$. Then there exists a $Q \in I_0(\mathfrak{h}^{\mathbf{c}})$ such that $Q(i\lambda) = \Gamma(D)(i\lambda)$ for all $\lambda \in \mathfrak{a}^*$. Thus $P(\rho_0 + i\lambda) = \Gamma(D)(i\lambda)$ for some $P \in I(\mathfrak{h}^{\mathbf{c}})$ so by the surjectivity of γ there exists a $Z \in \boldsymbol{Z}(G/K)$ such that $Z\phi_\lambda = D\phi_\lambda$ for all $\lambda \in (\mathfrak{a}^*)^{\mathbf{c}}$. By the inversion formula for the spherical transform, $Zf = Df$ for $f \in C_c^\infty(G/K)$ invariant under $\tau(K)$. Since the $\tau(G)$-translates of such functions span a dense subset of $C_c^\infty(G/K)$ ([7(d)], Ch. II, Lemma 5.6) Z and D must in fact coincide.

Next we prove that each $D \in \boldsymbol{D}(G/K)$ is a "fraction" of two members of $\boldsymbol{Z}(G/K)$.

Theorem 2.2. *Given $D \in \boldsymbol{D}(G/K)$ there exist $Z_1 \neq 0$, $Z_2 \in \boldsymbol{Z}(G/K)$ such that*

$$(2.11) \qquad DZ_1 = Z_2.$$

Proof. Denoting by $T(\rho_0)$ the translation of $(\mathfrak{h}^{\mathbf{c}})^*$ by ρ_0 the mapping $\tau : Q \to Q \circ T(-\rho_0)$ is an automorphism of $S(\mathfrak{h}^{\mathbf{c}})$ mapping $I_0(\mathfrak{h}^{\mathbf{c}})$ onto $I(\mathfrak{h}^{\mathbf{c}})$. Consider the commutative diagram

$$\begin{array}{ccc} I_0(\mathfrak{h}^{\mathbf{c}}) & \xrightarrow{\ \tau\ } & I(\mathfrak{h}^{\mathbf{c}}) \\ \sigma \big\downarrow & & \big\downarrow \zeta \\ I(\mathfrak{a}^{\mathbf{c}}) & \xrightarrow{\ Id\ } & I(\mathfrak{a}^{\mathbf{c}}), \end{array}$$

where $\sigma(Q) = Q|\mathfrak{a}^*$, $\zeta(P) = (P \circ T(\rho_0))|\mathfrak{a}^*$. The surjectivity of γ and formula (2.9) show that σ and ζ do indeed map the respective spaces $I_0(\mathfrak{h}^{\mathbf{c}})$, $I(\mathfrak{h}^{\mathbf{c}})$ into $I(\mathfrak{a}^{\mathbf{c}})$. Using again the spherical transform we see that the operators DZ_1 and Z_2 in (2.11) coincide if and only if they have the same effect on the spherical functions ϕ_λ. Thus (2.11) is equivalent to the polynomial identity

$$\Gamma(D)(i\lambda)\gamma(Z_1)(\rho_0 + i\lambda) = \gamma(Z_2)(\rho_0 + i\lambda),$$

in other words

$$\Gamma(D)\zeta(\gamma(Z_1)) = \zeta(\gamma(Z_2)).$$

Thus the problem is to prove that the image $\zeta(I(\mathfrak{h}^{\mathbf{C}}))$ has quotient field $\boldsymbol{C}(\zeta(I(\mathfrak{h}^{\mathbf{C}})))$ equal to $\boldsymbol{C}(I(\mathfrak{a}^{\mathbf{C}}))$. By the commutativity of the diagram above, this amounts to proving that if J_0 denotes the image of $I_0(\mathfrak{h}^{\mathbf{C}})$ under the restriction $\sigma : Q \to Q|\mathfrak{a}^*$ then the quotient field $\boldsymbol{C}(J_0)$ equals $\boldsymbol{C}(I(\mathfrak{a}^{\mathbf{C}}))$.

First note that $S(\mathfrak{h}^{\mathbf{C}})$ is integral over $I(\mathfrak{h}^{\mathbf{C}})$ so since τ^{-1} is an automorphism, $S(\mathfrak{h}^{\mathbf{C}})$ is also integral over $I_0(\mathfrak{h}^{\mathbf{C}})$. Hence $S(\mathfrak{a}^{\mathbf{C}})$ is integral over J_0.

Suppose $\boldsymbol{C}(J_0)$ were a proper subset of $\boldsymbol{C}(I(\mathfrak{a}^{\mathbf{C}}))$. Then we can select $R \in I(\mathfrak{a}^{\mathbf{C}})$, $R \notin \boldsymbol{C}(J_0)$. By the above, R is algebraic over $\boldsymbol{C}(J_0)$ and as the homomorphic image of $I_0(\mathfrak{h}^{\mathbf{C}})$, J_0 has finitely many generators, say $\xi_1, \ldots, \xi_r$. Let

$$p_R(x) = x^n + f_1(\xi)x^{n-1} + \cdots + f_n(\xi)$$

be a polynomial with coefficients in $\boldsymbol{C}(J_0) = \mathbf{C}(\xi_1, \ldots, \xi_r)$ of lowest degree and leading coefficient 1 having R as a root. By assumption, $n \geq 2$. Let $f(\xi)$ denote the product of the denominators of the $f_i(\xi)$ and let $q(\xi)$ denote the product of $f(\xi)$ with the discriminant of the polynomial $f(\xi)p_R(x)$. Then $q(\xi)$ is a polynomial function on $(\mathfrak{a}^{\mathbf{C}})^*$. It is also $\not\equiv 0$ because if the equation $p_R(x) = 0$ had some multiple roots, the relation $p_R'(R) = 0$ would contradict the minimality of p_R.

Select $\mu^* \in \mathfrak{a}^*$ such that $\lambda^* = i\mu^*$ satisfies

$$q(\lambda^*) \neq 0, \tag{2.12}$$

and consider the homomorphism

$$\chi : j \to j(\lambda^*) \quad \text{of} \quad J_0 \to \boldsymbol{C}. \tag{2.13}$$

The image of $p_R(x)$ under χ is a polynomial with coefficients in $\boldsymbol{C}$ and discriminant $\neq 0$ having n distinct roots $\alpha_1, \ldots, \alpha_n \in \boldsymbol{C}$. For each i $(1 \leq i \leq n)$ we wish to extend χ to a homomorphism

$$\chi_i : \boldsymbol{C}[\xi_1, \ldots, \xi_r, R] \to \boldsymbol{C} \text{ by putting } \chi_i(R) = \alpha_i.$$

To see that this is possible suppose $p(\xi, x)$ is a polynomial in x with coefficients in $J_0 = \boldsymbol{C}[\xi_1, \ldots, \xi_r]$ having R as a root. Then $p_R(x)$ divides $p(\xi, x) : p(x, \xi) = p_R(x)q(\xi, x)$. The coefficients of $q(\xi, x)$ are found by long division so they appear as rational expressions in $\xi_1, \ldots, \xi_r$ with denominators dividing $f(\xi)$. By (2.12), $\chi(f(\xi)) \neq 0$ so the image of $p(x, \xi)$ under χ has α_i as a zero. Thus χ_i can be defined by

$$\chi_i(t(\xi_1, \ldots, \xi_r, R)) = t(\chi(\xi_1), \ldots, \chi(\xi_r), \alpha_i)$$

t being an arbitrary polynomial in $r + 1$ variables. Since $S(\mathfrak{a}^{\mathbf{C}})$ is integral over J_0 and hence over $\boldsymbol{C}[\xi_1, \ldots, \xi_r, R]$ there is a homomorphism $\Lambda_i : S(\mathfrak{a}^{\mathbf{C}}) \to \boldsymbol{C}$ extending χ_i. This is given by evaluation at a point $\lambda_i \in (\mathfrak{a}^{\mathbf{C}})^*$, i.e. $\Lambda_i(P) = P(\lambda_i)$ for $P \in S(\mathfrak{a}^{\mathbf{C}})$. In particular, if $Q \in I_0(\mathfrak{h}^{\mathbf{C}})$ then $\chi(\sigma(Q)) = Q(\lambda^*)$ equals both $Q(\lambda_1)$ and $Q(\lambda_2)$ so

$$(2.14) \qquad P(\rho_0 + i\mu^*) = P(\rho_0 + \lambda_1) = P(\rho_0 + \lambda_2) \quad \text{for} \quad P \in I(\mathfrak{h}^{\mathbf{C}}).$$

This implies that the elements $\rho_0 + i\mu^*$, $\rho_0 + \lambda_1$, $\rho_0 + \lambda_2$ are conjugate under $\tilde{W}$. Thus writing $\lambda_j = \nu_j + i\mu_j$ $(\mu_j, \nu_j \in \mathfrak{a}^*, j = 1, 2)$

$$(2.15) \qquad s_1(\rho_0 + i\mu^*) = \rho_0 + \nu_1 + i\mu_1 \quad s_2(\rho_0 + i\mu^*) = \rho_0 + \nu_2 + i\mu_2$$

for some $s_1, s_2 \in \tilde{W}$. Since $\tilde{W}$ preserves $(\mathfrak{h}_{\boldsymbol{R}})^*$, (2.15) implies $s_j\rho_0 = \rho_0 + \nu_j$, $s_j\mu^* = \mu_j$ $(j = 1, 2)$ so $\nu_1 = \nu_2 = 0$ and $s_2 s_1^{-1}\lambda_1 = \lambda_2$. Hence by [7(d)], Ch. VII, Corollary 8.9, λ_1 and λ_2 are conjugate under the Weyl group W of Σ. In particular,

$$\alpha_1 = \chi_1(R) = R(\lambda_1) = R(\lambda_2) = \chi_2(R) = \alpha_2$$

contradicting $\alpha_1 \neq \alpha_2$. This proves the theorem.

3. The modified criterion. We shall now modify condition (ii) in

Proposition 2.1 so that it becomes a more effective criterion. Now we assume that G is simple.

PROPOSITION 3.1. *Denoting again restriction from* $(\mathfrak{h}^{\mathbf{C}})^*$ *to* $(\mathfrak{a}^{\mathbf{C}})^*$ *by bar the following conditions are equivalent*:

(i) $I_0(\mathfrak{h}^{\mathbf{C}})^- = I(\mathfrak{a}^{\mathbf{C}})$;
(ii) $I(\mathfrak{h}^{\mathbf{C}})^- = I(\mathfrak{a}^{\mathbf{C}})$.

We first prove by induction on the degrees that (ii) $\Rightarrow$ (i). Let $I_0^d(\mathfrak{h})$, $I^d(\mathfrak{h})$ and $I^d(\mathfrak{a})$ denote the sets of elements in $I_0(\mathfrak{h}^{\mathbf{C}})$, $I(\mathfrak{h}^{\mathbf{C}})$, and $I(\mathfrak{a}^{\mathbf{C}})$, respectively, of degree $\leq d$. For induction we assume $I_0^e(\mathfrak{h})^- = I^e(\mathfrak{a})$ holds for $e < d$. Let $R \in I^d(\mathfrak{a})$ and select (by (ii)) a $P \in I^d(\mathfrak{h})$ such that $\overline{P} = R$. Then $\mathbf{Q} = P \circ T(\rho_0) \in I_0^d(\mathfrak{h})$ and for $\lambda \in (\mathfrak{a}^{\mathbf{C}})^*$,

$$\mathbf{Q}(\lambda) = P(\rho_0 + \lambda) = P(\lambda) + S(\lambda),$$

where $S \in I^{d-1}(\mathfrak{a})$. By induction there exists a $Q_1 \in I_o^{d-1}(\mathfrak{h})$ such that $\overline{Q}_1 = S$. Then $(Q - Q_1)^- = R$ (cf. [3] Theorem 6.4).

For the converse (i) $\Rightarrow$ (ii) we must unfortunately resort to the use of the classification. First we recall [7(b)] that if G is classical (real or complex) then $I(\mathfrak{h}^{\mathbf{C}})^- = I(\mathfrak{a}^{\mathbf{C}})$. We therefore assume for the rest of the proof of Proposition 3.1 that G is simple and exceptional. We have also $I(\mathfrak{h}^{\mathbf{C}})^- = I(\mathfrak{a}^{\mathbf{C}})$ in the cases when G is complex, a normal real form or of real rank one. For these cases the proposition is contained in the above proof. This leaves us with the following six cases.

TABLE

Symmetric pair	Basis of Δ	Basis of Σ
$EII(\mathfrak{e}_6, \mathfrak{su}(6) + \mathfrak{su}(2))$	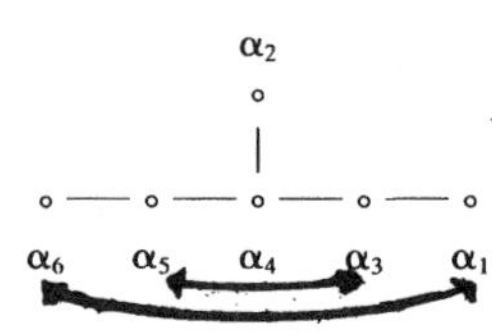	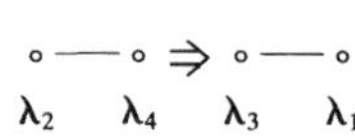
Degrees:	2,5,6,8,9,12	2,6,8,12
$EIII(\mathfrak{e}_6, \mathfrak{su}(10) + \mathbf{R}$	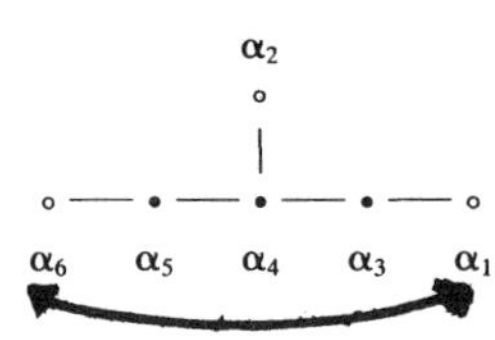	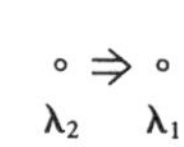

Symmetric pair	Basis of Δ	Basis of Σ
Degrees:	2,5,6,8,9,12	2,4
$EIV(\mathfrak{e}_6, \mathfrak{f}_4)$	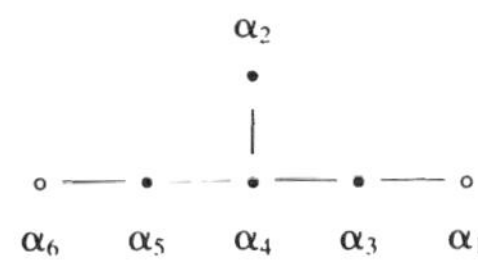	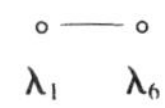
Degrees:	2,5,6,8,9,12	2,3
$EVI(\mathfrak{e}_7, \mathfrak{so}(12) + \mathfrak{su}(2))$	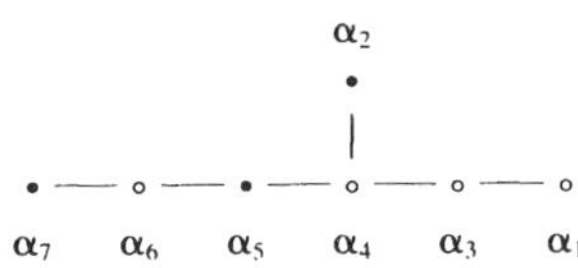	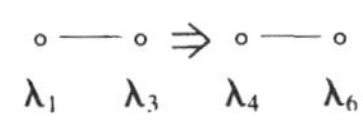
Degrees:	2,6,8,10,12,14,18	2,6,8,12
$EVII(\mathfrak{e}_7, \mathfrak{e}_6 + \boldsymbol{R})$	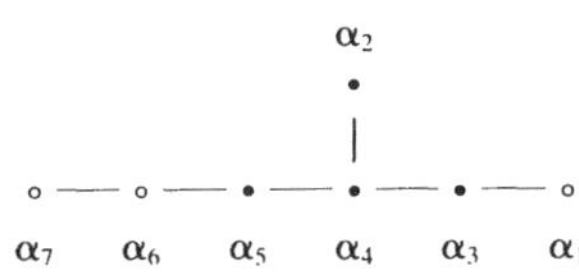	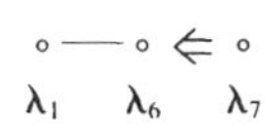
Degrees:	2,6,8,10,12,14,18	2,4,6
$EIX(\mathfrak{e}_8, \mathfrak{e}_7 + \mathfrak{su}(2))$	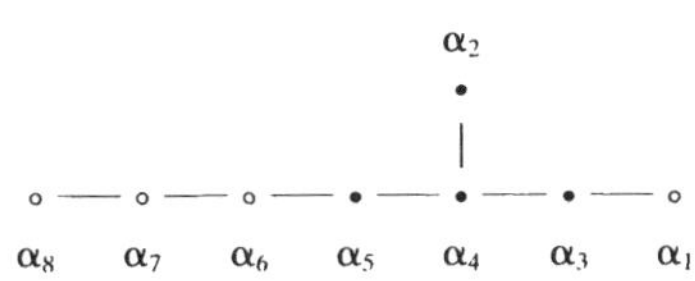	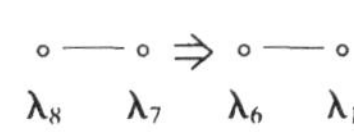
Degrees:	2,8,12,14,18,20,24,30	2,6,8,12

In the table above we list the pair $(\mathfrak{g}, \mathfrak{k})$, the Dynkin diagram for a basis of Σ and a Satake-Tits diagram for a basis of Δ, the black dots representing a basis of P_0. The curved arrows refer to two $\alpha, \beta \in \Sigma$ such that $\bar{\alpha} = \bar{\beta}$ (see Araki [1]). We also list the degrees of the uniquely determined homogeneous algebraically independent generators of $I(\mathfrak{h}^{\mathbf{C}})$ and $I(\mathfrak{a}^{\mathbf{C}})$. Looking first at *EIII* the algebraic independence implies that each homogeneous element in $I(\mathfrak{h}^{\mathbf{C}})$ has degree of the form $2n_1 + 5n_2 + 6n_3 + 8n_4 + 9n_5 + 12n_6$ $(n_i \geq 0)$. Clearly the generator of $I(\mathfrak{a}^{\mathbf{C}})^-$ of degree 4 cannot belong to the restriction $I(\mathfrak{h}^{\mathbf{C}})^-$. Thus

$$I(\mathfrak{h}^{\mathbf{C}})^- \neq I(\mathfrak{a}^{\mathbf{C}}) \quad \text{for } EIII. \tag{3.1}$$

The same degree considerations imply (3.1) for the cases EIV, EVII and EIX. We shall now, following Vogan, show that $I_0(\mathfrak{h}^{\mathbf{C}})^- \neq I(\mathfrak{a}^{\mathbf{C}})$ for these four cases. Finally we shall prove that $I(\mathfrak{h}^{\mathbf{C}})^- = I(\mathfrak{a}^{\mathbf{C}})$ for the cases EII, EVI as one would expect by considering the degrees. This would prove Proposition 3.1.

LEMMA 3.2. (Vogan). *For EIII, EIV, EVII and EIX there exist two dominant integral functions* $\lambda \neq \mu$ *on* $\mathfrak{h}_{\boldsymbol{R}}$ *such that* $\lambda, \mu \in \mathfrak{a}^*$ *(i.e. they vanish on* $\mathfrak{h} \cap \mathfrak{k}$*) and such that* $\rho_0 + \lambda$ *and* $\rho_0 + \mu$ *are conjugate by* $\tilde{W}$.

This proves $I_0(\mathfrak{h}^{\mathbf{C}})^- \neq I(\mathfrak{a}^{\mathbf{C}})$ for these cases. In fact, being distinct highest weights, λ and μ are nonconjugate under $\tilde{W}$, hence nonconjugate under W, thus separated by a W-invariant polynomial $P \in I(\mathfrak{a}^{\mathbf{C}})$. Since each $Q \in I_0(\mathfrak{h}^{\mathbf{C}})$ satisfies $Q(\lambda) = Q(\mu)$ we have $P \notin I_0(\mathfrak{h}^{\mathbf{C}})^-$.

Remark. A more complete study of all "anomalies" as in Lemma 3.2 has been carried out by Bien [2] who had independently proved Lemma 3.2 for the case EIII.

We use the inner product $\langle \ , \ \rangle$ on $\mathfrak{h}_{\boldsymbol{R}}^*$ for which each $\alpha \in \Delta$ satisfies $\langle \alpha, \alpha \rangle = 2$ ([7(c)], Ch. X, Section 3) and write each $\lambda \in \mathfrak{h}_{\boldsymbol{R}}^*$ in terms of the dual basis (ω_i), $\langle \omega_i, \alpha_j \rangle = \delta_{ij}$, $\lambda = \Sigma_i \lambda_i \omega_i$. (Since all roots have the same length the ω_i are a fixed multiple of the fundamental weights.) Then $\lambda \in \mathfrak{a}^*$ if and only if $\theta\lambda = -\lambda$, i.e. $\lambda_i = 0$ if $\theta\alpha_i = \alpha_i$ (i.e. α_i a black root) and $\lambda_k = \lambda_l$ whenever $\alpha_k - \alpha_l = \theta\alpha_k - \theta\alpha_l$ (i.e. α_k and α_l joined by a curved arrow). Since all α_i have the same length we have for the Weyl reflection s_{α_i},

$$(3.2) \qquad \alpha_i \neq \alpha_j \Rightarrow \begin{cases} s_{\alpha_i}(\alpha_j) = \alpha_i + \alpha_j & \alpha_i, \alpha_j \text{ adjacent} \\ s_{\alpha_i}(\alpha_j) = \alpha_j & \text{otherwise.} \end{cases}$$

Consider first the case EIII. We have the correspondence

$$(3.3) \qquad \lambda \in \mathfrak{h}_{\boldsymbol{R}}^* \leftrightarrow \begin{matrix} & & & & \lambda_2 & & & & \\ & & & & | & & & & \\ \lambda_6 & - & \lambda_5 & - & \lambda_4 & - & \lambda_3 & - & \lambda_1, \end{matrix} \qquad \lambda_i = \langle \lambda, \alpha_i \rangle,$$

and by the remark above

$$\lambda \in \mathfrak{a}^* \Leftrightarrow \lambda_6 = \lambda_1, \qquad \lambda_3 = \lambda_4 = \lambda_5 = 0. \tag{3.4}$$

For the action of $\tilde{W}$ we have for example by (3.2)–(3.3)

$$\begin{array}{c} \qquad\qquad\qquad\qquad\qquad\qquad\qquad\qquad\qquad\qquad\quad \lambda_2 \\ \qquad\qquad\qquad\qquad\qquad\qquad\qquad\qquad\qquad\qquad\quad | \\ s_{\alpha_5}(\lambda) \leftrightarrow (\lambda_5 + \lambda_6) - (-\lambda_5) - (\lambda_4 + \lambda_5) - \lambda_3 - \lambda_1. \end{array}$$

The black roots form a basis of the root system for $\mathfrak{a}_3$ so in terms of (3.3)

$$\begin{array}{c} \qquad\qquad\qquad\qquad\qquad\qquad\qquad\qquad\qquad\qquad (-2) \\ \qquad\qquad\qquad\qquad\qquad\qquad\qquad\qquad\qquad\qquad | \\ \rho_0 = \tfrac{3}{2}\alpha_5 + 2\alpha_4 + \tfrac{3}{2}\alpha_3 \leftrightarrow (-\tfrac{3}{2}) - (1) - (1) - (1) - (-\tfrac{3}{2}) \end{array} \tag{3.5}$$

Now take in (3.4)

$$\begin{array}{c} \qquad\qquad\qquad 0 \\ \qquad\qquad\qquad | \\ \lambda \leftrightarrow \tfrac{5}{2} - 0 - 0 - 0 - (\tfrac{5}{2}), \end{array} \qquad \begin{array}{c} \qquad\qquad\qquad 3 \\ \qquad\qquad\qquad | \\ \mu \leftrightarrow \tfrac{1}{2} - 0 - 0 - 0 - \tfrac{1}{2} \end{array}$$

so

$$\begin{array}{c} \qquad\qquad\qquad (-2) \\ \qquad\qquad\qquad | \\ \lambda + \rho_0 \leftrightarrow 1 - 1 - 1 - 1 - 1, \end{array}$$

$$\begin{array}{c} \qquad\qquad\qquad 1 \\ \qquad\qquad\qquad | \\ \mu + \rho_0 \leftrightarrow (-1) - 1 - 1 - 1 - (-1) \end{array}$$

Then

$$s_{\alpha_4}s_{\alpha_2}(\lambda + \rho_0) = s_{\alpha_4}\left(\begin{array}{c}\\ \\ 1 - 1 - \end{array}\begin{array}{c}2\\ |\\ (-1)\end{array}\begin{array}{c}\\ \\ - 1 - 1\end{array}\right)$$

$$= \begin{array}{c}\\ \\ 1 - 0 - \end{array}\begin{array}{c}1\\ |\\ 1\end{array}\begin{array}{c}\\ \\ - 0 - 1\end{array}$$

$$s_{\alpha_1}s_{\alpha_6}(\mu + \rho_0) = s_{\alpha_1}\left(\begin{array}{c}\\ \\ 1 - 0 - \end{array}\begin{array}{c}1\\ |\\ 1\end{array}\begin{array}{c}\\ \\ - 1 - (-1)\end{array}\right)$$

$$= \begin{array}{c}\\ \\ 1 - 0 - \end{array}\begin{array}{c}1\\ |\\ 1\end{array}\begin{array}{c}\\ \\ - 0 - 1\end{array}$$

so $\lambda + \rho_0$ and $\mu + \rho_0$ are $\tilde{W}$-conjugate.

For EIV, EVII and EIX we proceed similarly. The black roots form a basis of the root system for $\mathfrak{d}_4$ so for all the cases

$$\rho_0 = 3(\alpha_2 + \alpha_3 + \alpha_5) + 5\alpha_4 \leftrightarrow \begin{array}{c}\\ \\ \cdots (-3) - 1 - \end{array}\begin{array}{c}1\\ |\\ 1\end{array}\begin{array}{c}\\ \\ - 1 - (-3)\end{array}$$

where $\cdots$ represents one entry (zero) for EVII and two entries (both zeros) for EIX. Now take

$$\lambda \leftrightarrow \begin{array}{c}\\ \\ \cdots 4 - 0 - \end{array}\begin{array}{c}0\\ |\\ 0\end{array}\begin{array}{c}\\ \\ - 0 - 1,\end{array}$$

$$
\begin{array}{cccccccccccccc}
 & & & & & & & & 0 & & & & & \\
 & & & & & & & & | & & & & & \\
\mu \leftrightarrow & \cdots 2 & - & 1 & - & 0 & - & & 0 & - & 0 & - & 4 &
\end{array}
$$

dots representing 0 and the 2 appearing only for EVII and EIX. Then

$$
\begin{array}{cccccccccc}
 & & & & & 1 & & & & \\
 & & & & & | & & & & \\
\lambda + \rho_0 \leftrightarrow & \cdots 1 & - & 1 & - & 1 & - & 1 & - & (-2),
\end{array}
$$

$$
\begin{array}{cccccccccccc}
 & & & & & & & 1 & & & & \\
 & & & & & & & | & & & & \\
\mu + \rho_0 \leftrightarrow & \cdots 2 & - & (-2) & - & 1 & - & 1 & - & 1 & - & 1
\end{array}
$$

$$
\begin{array}{cccccccccc}
 & & & & & 1 & & & & \\
 & & & & & | & & & & \\
s_{\alpha_3} s_{\lambda_1}(\lambda + \rho_0) = s_{\alpha_3}(& \cdots 1 & - & 1 & - & 1 & - & (-1) & - & 2)
\end{array}
$$

$$
\begin{array}{cccccccccc}
 & & & & & 0 & & & & \\
 & & & & & | & & & & \\
= (& \cdots 1 & - & 1 & - & 0 & - & 1 & - & 1)
\end{array}
$$

Note: the label above the node in the preceding diagram is 1:

$$
\begin{array}{cccccccccc}
 & & & & & 1 & & & & \\
 & & & & & | & & & & \\
= (& \cdots 1 & - & 1 & - & 0 & - & 1 & - & 1)
\end{array}
$$

$$
\begin{array}{cccccccc}
 & & & & 1 & & & \\
 & & & & | & & & \\
s_{\alpha_5} s_{\alpha_6}(\mu + \rho_0) = s_{\alpha_5}(& \cdots 0, 2, (-1) & - & & 1 & - & 1 & - 1)
\end{array}
$$

$$
\begin{array}{cccccccccc}
 & & & & & 1 & & & & \\
 & & & & & | & & & & \\
= (& \cdots 1 & - & 1 & - & 0 & - & 1 & - & 1)
\end{array}
$$

so in all three cases $\lambda + \rho_0$ and $\mu + \rho_0$ are conjugate under $\tilde{W}$. This proves the lemma.

The proof of Proposition 3.1 is now concluded with the following result.

LEMMA 3.3. *For the symmetric pairs*

$$EVI : (\mathfrak{e}_7, \mathfrak{so}(12) + \mathfrak{su}(2)), \qquad EII : (\mathfrak{e}_6, \mathfrak{su}(6) + \mathfrak{su}(2))$$

we have $I(\mathfrak{h}^{\mathbf{C}})^- = I(\mathfrak{a}^{\mathbf{C}})$.

Proof. Here we use the determination of the invariants in Lee [9]. We shall also use a different description of the root systems in order to get a simpler description of the Weyl group. Consider the orthonormal basis $e_1, \ldots, e_8$ in $\boldsymbol{R}^8$, and put $e^{(8)} = \Sigma_1^8 e_i$. Then the system of vectors

$$\Delta = \{e_i - e_j, \tfrac{1}{2}e^{(8)} - e_i - e_j - e_k - e_l : i, j, k, l = 1, \ldots, 8\}$$

is a root system in the seven dimensional space

$$\left\{(r_1, \ldots, r_8) : \sum_1^8 r_i = 0\right\}.$$

As a basis for this root system we can choose

$$\beta_i = e_i - e_{i+1}\ (1 \leq i \leq 6), \qquad \beta_7 = \tfrac{1}{2}e^{(8)} - e_1 - e_2 - e_3 - e_4$$

(cf. [10], Section 66). In this notation our diagrams for EVI are

(3.6)

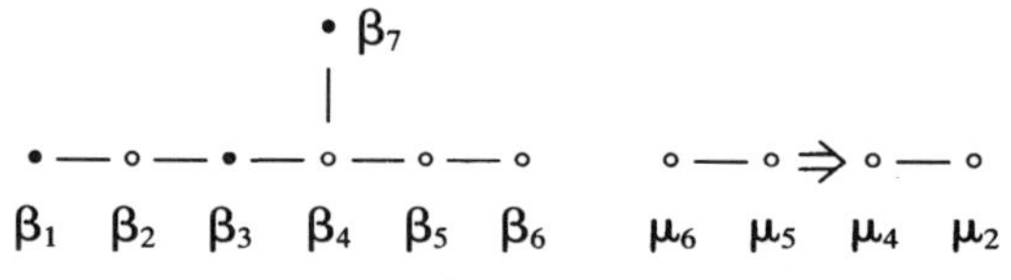

where $\mu_i = \bar{\beta}_i$. The reflections s_{β_i} $(1 \leq i \leq 6)$ are the transpositions $(r_1 r_2), \ldots, (r_6 r_7)$ and s_{β_7} is

$$(r_1, \ldots, r_8) \rightarrow (r_1, \ldots, r_8) + \tfrac{1}{2}r^{(4)}(-1, -1, -1, -1, 1, 1, 1, 1),$$

where $r^{(4)} = r_1 + r_2 + r_3 + r_4$. As shown in [9] the algebra $I(\mathfrak{h}^C)$ is generated by the polynomials

$$(3.7) \qquad j_m = \sum_{i<j} (Y_i + Y_j)^m \qquad m = 2, 6, 8, 10, 12, 14, 18$$

where

$$Y_1 = 3H_1 + 2H_2 + H_3 - H_7$$

$$Y_2 = -H_1 + 2H_2 + H_3 - H_7$$

$$Y_3 = -H_1 - 2H_2 + H_3 - H_7$$

$$Y_4 = -H_1 - 2H_2 - 3H_3 - H_7$$

$$Y_5 = -H_1 - 2H_2 - 3H_3 - 4H_4 - H_7$$

$$Y_6 = -H_1 - 2H_2 - 3H_3 - 4H_4 - 4H_5 - H_7$$

$$Y_7 = -H_1 - 2H_2 - 3H_3 - 4H_4 - 4H_5 - 4H_6 - H_7$$

$$Y_8 = -Y_1 - Y_2 - \cdots - Y_7.$$

Here $H_i = H_{\beta_i}$ in our notation of Section 2. Also $H_\alpha|\mathfrak{a}^* = A_{\bar{\alpha}}$ in our identification of $\mathfrak{a}^*$ with a subset of $\mathfrak{h}_R^*$. From our diagram (3.6) we see that

$$H_i|\mathfrak{a}^* = \begin{cases} 0 & \text{if } i = 1, 3, 7; \\ A_i = A_{\mu_i} & \text{if } i = 2, 4, 5, 6. \end{cases}$$

This gives

$$\frac{1}{2\cdot 4^m} j_m|\mathfrak{a}^* = A_2^m + A_4^m + (A_2 + A_4)^m + (A_4 + A_5)^m$$

$$+ (A_2 + A_4 + A_5)^m + (A_4 + A_5 + A_6)^4$$

$$+ (A_2 + 2A_4 + A_5)^m + (A_2 + A_4 + A_5 + A_6)^m$$

$$+ (A_2 + 2A_4 + A_5 + A_6)^m + (2A_2 + 3A_4 + 2A_5 + A_6)^m$$

$$+ (A_2 + 3A_4 + 2A_5 + A_6)^m + (A_2 + 2A_4 + 2A_5 + A_6)^m.$$

On the other hand, the generators of the invariants for the root system $\mathfrak{f}_4$:

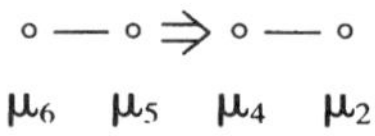

under the group generated by the reflections

$$s_{\mu_i} A_{\mu_j} = A_{\mu_j} - 2\frac{\langle \mu_i, \mu_j \rangle}{\langle \mu_i, \mu_i \rangle} A_{\mu_i}$$

are by Lee [9] given by

$$i_m = \sum_{i<j} (Z_i + Z_j)^m + (Z_i - Z_j)^m, \qquad m = 2, 6, 8, 12,$$

where

$$Z_1 = 2A_2 + 4A_4 + 3A_5 + 2A_6$$

$$Z_2 = 2A_2 + 2A_4 + A_5 \qquad Z_3 = 2A_4 + A_5 \qquad Z_4 = A_5$$

Then it is readily seen that up to a constant factor, i_m is the restriction of j_m. This proves the first case of Lemma 3.3.

For the case EII we use the diagram

(3.8)

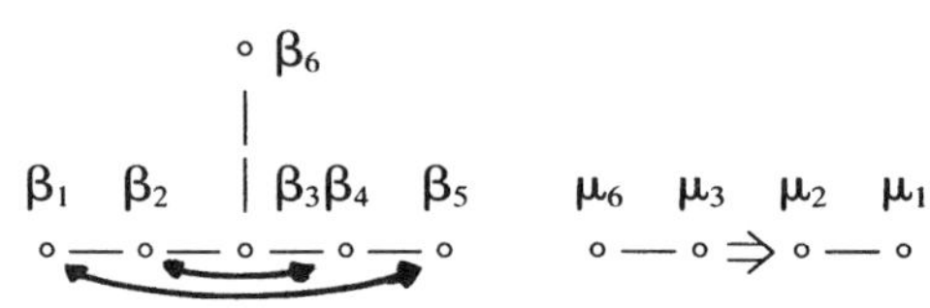

where $\mu_i = \overline{\beta}_i$. As shown in [9], and earlier by Coxeter, the algebra

$I(\mathfrak{h}^c)$ is generated by the polynomials

$$(3.9) \quad j_m = \sum_{1 \le i \le 6} a_i^m + \sum_{1 \le i \le 6} b_i^m + \sum_{1 \le i < j \le 6} c_{ij}^m, \qquad m = 2, 5, 6, 8, 9, 12,$$

where

$$a_i = Y_i + Y, \qquad b_i = Y_i - Y \qquad c_{ij} = -Y_i - Y_j \ (i < j)$$

with

$$Y_1 = 5H_1 + 4H_2 + 3H_3 + 2H_4 + H_5$$

$$Y_2 = -H_1 + 4H_2 + 3H_3 + 2H_4 + H_5$$

$$Y_3 = -H_1 - 2H_2 + 3H_3 + 2H_4 + H_5$$

$$Y_4 = -H_1 - 2H_2 - 3H_3 + 2H_4 + H_5$$

$$Y_5 = -H_1 - 2H_2 - 3H_3 - 4H_4 + H_5$$

$$Y_6 = -Y_1 - \cdots - Y_5$$

$$Y = -3(H_1 + 2H_2 + 3H_3 + 2H_4 + H_5 + 2H_6), \qquad H_i = H_{\beta_i}.$$

In the diagram (3.8) we have

$$H_1|\mathfrak{a}^* = H_5|\mathfrak{a}^* = A_1 = A_{\mu_1}, \qquad H_2|\mathfrak{a}^* = H_4|\mathfrak{a}^* = A_2 = A_{\mu_2}$$

and it follows that the restrictions $y_i = Y_i|\mathfrak{a}^*$ are given by

$$y_1 = -y_6 = 6A_1 + 6A_2 + 3A_3$$

$$y_2 = -y_5 = 6A_2 + 3A_3$$

$$y_3 = -y_4 = 3A_3.$$

We then find that the restriction $\bar{j}_m = j_m|\mathfrak{a}^*$ is given by

$$\begin{aligned}3^{-m}\bar{j}_m = (2^m + (-2)^m)[A_1^m + A_2^m + (A_1 + A_2)^m + (A_2 + A_3)^m\\ + (A_1 + A_2 + A_3)^m + (A_2 + A_3 + A_6)^m\\ + (A_1 + 2A_2 + A_3)^m + (A_1 + A_2 + A_3 + A_6)^m\\ + (A_1 + 2A_2 + A_3 + A_6)^m + (2A_1 + 3A_2 + 2A_3 + A_6)^m\\ + (A_1 + 3A_2 + 2A_3 + A_6)^m + (A_1 + 2A_2 + 2A_3 + A_6)^m].\end{aligned}$$

Again, this shows that $I(\mathfrak{h}^{\mathbf{C}})$ restricts onto $I(\mathfrak{a}^{\mathbf{C}})$ and explains, in passing, why $I(\mathfrak{a}^{\mathbf{C}})$ has no generators of degree 5 and 9.

Remark. I am indebted to G. Shimura for pointing out a gap in the proof of Proposition 7.4 in [7(b)] (which is Proposition 5.32, Chapter II in [7(d)]). This result is nevertheless correct because of Propositions 2.1 and 3.1 in the present paper. Lemma 3.2 plays a crucial role here and I am deeply grateful to my colleague David Vogan for communicating it to me.

THEOREM 3.4. *Assume G/K is irreducible. Then $\mathbf{Z}(G/K) = \mathbf{D}(G/K)$ if and only if $(\mathfrak{g}, \mathfrak{k})$ is not one of the pairs* (1.1).

This is a consequence of Propositions 2.1, 3.1, and Lemmas 3.2, 3.3, combined with the fact that (ii) in Proposition 3.2 is satisfied for G classical.

Examples. (i) The case EIV : $G/K = E_6/F_4$. Here the restricted root system is that of $\mathfrak{a}_2$. In the notation of (3.8) we have

$$H_{\beta_i}|\mathfrak{a}^* = 0 \quad \text{for} \quad i = 2, 3, 4, 6. \tag{3.10}$$

From (3.9) we thus obtain (with $A_j = H_j|\mathfrak{a}^*$ $(j = 1, 2, \ldots, 6)$)

$$\begin{aligned}\bar{j}_m = (2^{m+3} + (-4)^m)[(A_1 - A_5)^m\\ + (A_1 + 2A_5)^m + (-1)^m(2A_1 + A_5)^m].\end{aligned}$$

This vanishes for $m = 3$, explaining why $I(\mathfrak{h}^{\mathbf{C}})^-$ has no invariant of degree 3. For $m = 5$ we know *a priori* by the algebraic independence of the generators of $I(\mathfrak{a}^{\mathbf{C}})$ that $\bar{j}_5$ must be a product $i_2 \cdot i_3$ where $i_k \in I(\mathfrak{a}^{\mathbf{C}})$ is a generator of degree k. We see also directly that

$$(A_1 - A_5)^5 + (A_1 + 2A_5)^5 - (2A_1 + A_5)^5$$

$$= (-15)(A_1^2 + A_1A_2 + A_2^2)(A_1 - A_5)(2A_1 + A_5)(A_1 + 2A_5)$$

(ii) The case $EVII : G/K = E_7/E_6 \times \boldsymbol{R}$. Here the restricted root system is $\mathfrak{c}_3$. The algebra $I(\mathfrak{h}^{\mathbf{C}})$ is generated by the polynomials (3.7). The Satake-Tits diagram in the Table shows that

$$H_{\beta_i}|\mathfrak{a}^* = 0 \quad \text{for} \quad i = 3, 4, 5, 7$$

so with $A_i = H_{\beta_i}|\mathfrak{a}^*$ $(i = 1, 2, 6)$ we obtain for $m = 2, 6, 8, 10, 12, 14, 18$,

$$2^{-m}\bar{j}_m = 8A_1^m + 8(A_1 + 2A_2)^m + (A_1 - 2A_6)^m + (A_1 + 2A_6)^m$$
$$+ (3A_1 + 4A_2 + 2A_6)^m + (A_1 + 4A_2 + 2A_6)^m$$
$$+ 8(A_1 + 2A_2 + 2A_6)^m.$$

The generators of $I(\mathfrak{a}^{\mathbf{C}})$ are

$$i_2 = A_1^2 + (A_1 + 2A_2)^2 + (A_1 + 2A_2 + 2A_6)^2,$$

$$i_4 = A_1^2(A_1 + 2A_2)^2 + (A_1 + 2A_2)^2(A_1 + 2A_2 + 2A_6)^2$$
$$+ A_1^2(A_1 + 2A_2 + 2A_6)^2,$$

$$i_6 = A_1^2(A_1 + 2A_2)^2(A_1 + 2A_2 + 2A_6)^2.$$

One then verifies that

$$\text{(3.11)} \qquad \bar{j}_4 = ci_2^2 \qquad (c \doteq \text{constant}).$$

Of course we know from Theorem 7.6 in [7(b)] that i_4 is a rational function of the $\bar{j}_k$. Here however the situation is more complicated than for EIV, in fact,

$$\bar{j}_6 = a(i_2^3 + 2i_2i_4 + 18i_6)$$

$$\bar{j}_8 = b(3i_2^4 + 16i_2^2i_4 + 20i_4^2 + 264i_2i_6) \tag{3.12}$$

$$\bar{j}_{10} = c(i_2^5 + 10i_2^3i_4 + 30i_2i_4^2 + 650i_2^2i_6 - 930i_4i_6)$$

with suitable nonzero constants a, b, c.

A natural question is whether each $i \in I(\mathfrak{a}^{\mathbf{C}})$ might be representable as a rational function of the $\bar{j}_k$ with a power of $\bar{j}_2$ as denominator; the following general result may be of some interest.

PROPOSITION 3.5. *Let Ω denote the Casimir operator on G. Then*

$$\gamma(\Omega)(\rho_0 + i\lambda) = L_A(i\lambda) - \langle\rho, \rho\rangle, \qquad \lambda \in \mathfrak{a}^*$$

where L_A is the Laplace operator on A.

Proof. Let $H_1, \ldots, H_l, \ldots, H_m$ be an orthonormal basis of $\mathfrak{h}_{\boldsymbol{R}}$ compatible with the direct decomposition $\mathfrak{h}_{\boldsymbol{R}} = \mathfrak{a} + i(\mathfrak{h} \cap \mathfrak{k})$. If $X_\alpha \in (\mathfrak{g}^{\mathbf{C}})^\alpha$ is for each $\alpha \in \Delta$ chosen such that $B(X_\alpha, X_{-\alpha}) = 1$ then as shown in [6(b)],

$$\Omega = \sum_1^m H_i^2 + \sum_{\alpha \in \Delta^+} (X_\alpha X_{-\alpha} + X_{-\alpha}X_\alpha) \tag{3.13}$$

so by $[X_\alpha, X_{-\alpha}] = H_\alpha$ we have

$$\gamma'(\Omega) = \sum_1^m H_i^2 + 2H_{\rho^*}. \tag{3.14}$$

Hence

$$\gamma(\Omega)(\rho_0 + i\lambda) = \gamma'(\Omega)(\rho_0 + i\lambda - \rho^*) = \gamma'(\Omega)(i\lambda - \rho)$$

$$= \left(\sum_1^l H_i^2\right)(i\lambda - \rho) + 2H_{\rho^*}(i\lambda - \rho)$$

$$= L_A(i\lambda) - 2\langle i\lambda, \rho\rangle + \langle \rho, \rho\rangle + 2\langle H_{\rho^*}, A_{i\lambda-\rho}\rangle.$$

Since

$$\langle H_{\rho^*}, A_{i\lambda-\rho}\rangle = \langle \rho, i\lambda\rangle - \langle \rho, \rho\rangle$$

the result follows.

Added in Proof. Another verification of the second case in Lemma 3.3 is given in [11].

4. Joint eigenfunctions. Because of Theorem 3.4, the joint eigenfunctions are the same for the algebra $\boldsymbol{Z}(G/K)$ and for the algebra $\boldsymbol{D}(G/K)$ except possibly when $(\mathfrak{g}, \mathfrak{k})$ is one of the four exceptional pairs (1.1). For these let $D_1, \ldots, D_l$ be generators of $\boldsymbol{D}(G/K)$ and fix $Z_1, \ldots, Z_l, W_1, \ldots, W_l$ in $\boldsymbol{Z}(G/K)$ such that $Z_iD_i = W_i$, $Z_i \neq 0$, for $1 \leq i \leq l$. From Theorem 2.2 we have the following consequence.

COROLLARY 4.1. *Suppose ψ is a joint eigenfunction of $\boldsymbol{Z}(G/K)$, i.e.*

$$Z\psi = c(Z)\psi \qquad Z \in \boldsymbol{Z}(G/K)$$

with $c(Z) \in \boldsymbol{C}$. Assume

$$c(Z_i) \neq 0 \quad \textit{for} \quad 1 \leq i \leq l. \tag{4.1}$$

Then ψ is a joint eigenfunction of $\boldsymbol{D}(G/K)$.

Remarks. Because of the indications in the discussion in Section 3 of the examples EIV, EVII there is reason to believe that it may be sufficient to impose condition (4.1) for the Casimir operator Ω alone. In connection with Proposition 3.5 it is therefore of interest to recall a consequence of Karpelevič's Theorem 17.2.1 in [8]: *A nonnegative eigenfunction of the Laplace-Beltrami operator L_X with eigenvalue $-\langle \rho, \rho\rangle$ is a joint eigenfunction of $\boldsymbol{D}(G/K)$.*

A similar statement holds also for the bounded solutions of $L_Xu = 0$ ([8], [4]). A simplified proof of these results is given in [5].

On the other hand, as remarked by Bien, condition (4.1) cannot be omitted because with λ, μ as in Lemma 3.2 any linear combination $a\phi_\lambda + b\phi_\mu$ $(ab \neq 0)$ is a joint eigenfunction of $\boldsymbol{Z}(G/K)$ but not of $\boldsymbol{D}(G/K)$.

We conclude with a simple positive result about the case of a compact symmetric space U/K. Let $\boldsymbol{D}(U/K)$ denote the algebra of U-invariant differential operators on U/K and $\boldsymbol{Z}(U/K)$ the subalgebra induced by the algebra $\boldsymbol{Z}(U)$ of bi-invariant differential operators on U.

PROPOSITION 4.2. *Any joint eigenfunction $f \in C^\infty(U/K)$ of $\boldsymbol{Z}(U/K)$ is a joint eigenfunction of $\boldsymbol{D}(U/K)$.*

Proof. Let $\pi : U \to U/K$ denote the natural map and put $\tilde{f} = f \circ \pi$. The Peter-Weyl theorem implies

$$\tilde{f} = \sum_{\delta \in \hat{U}} d(\delta)\tilde{f} * \chi_\delta, \tag{4.2}$$

where $\hat{U}$ is the unitary dual of U, $d(\delta)$ and χ_δ denote the degree and character of δ and $*$ denotes convolution on U. We have $Z\tilde{f} = c(Z)\tilde{f}$ and $Z\chi_\delta = c_\delta(Z)\chi_\delta$ for each $Z \in \boldsymbol{Z}(U)$ with $c(Z), c_\delta(Z) \in \boldsymbol{C}$. Applying Z to (4.2) we thus conclude that

$$\sum_{\delta \in \hat{U}} (c(Z) - c_\delta(Z))\tilde{f} * \chi_\delta = 0 \tag{4.3}$$

for $Z \in \boldsymbol{Z}(U)$ and consequently

$$c(Z) = c_\delta(Z) \quad \text{for} \quad Z \in \boldsymbol{Z}(U) \quad \text{if} \quad \tilde{f} * \chi_\delta \neq 0. \tag{4.4}$$

On the other hand, $c_{\delta_1}(Z) = c_{\delta_2}(Z)$ for all $Z \in \boldsymbol{Z}(U)$ implies that δ_1 is equivalent to δ_2 so by (4.3) and (4.4),

$$\tilde{f} = \tilde{f} * \chi_\delta \tag{4.5}$$

for a single δ. Then the spherical function $\phi_\delta(u) = \int \chi_\delta(uk)dk$ satisfies $\tilde{f} = \tilde{f} * \phi_\delta$ and as a result f is a joint eigenfunction of $\boldsymbol{D}(U/K)$.

MIT

REFERENCES

[1] S. I. Araki, On root systems and an infinitesimal classification of irreducible symmetric spaces, *J. Math. Osaka City Univ.*, **13** (1964), 1–34.

[2] F. Bien, A *D*-module Study of Spherical Representations, *Mathematical Notes*, Princeton University Press, (1990).

[3] M. Flensted-Jensen, Spherical functions on a real semisimple Lie group. A method of reduction to the complex case, *J. Funct. Anal.*, **30** (1978), 106–146.

[4] H. Furstenberg, A Poisson formula for semisimple Lie groups, *Ann. of Math.*, **77** (1963), 335–386.

[5] Y. Guivarc'h, Sur la representation integrale des fonctions harmoniques et des fonctions propres positives dans un espace Riemannien symétrique, *Bull. Sci. Math.*, **108** (1984), 373–392.

[6] Harish-Chandra, (a) The characters of semi-simple Lie groups, *Trans. Amer. Math. Soc.*, **83** (1956), 98–163; (b) Spherical functions on a semisimple Lie group I, *Amer. J. Math.*, **80** (1958), 241–310.

[7] S. Helgason, (a) Differential operators on homogeneous spaces, *Acta Math.*, **102** (1959), 239–299; (b) Fundamental solutions of invariant differential operators on a symmetric space, *Amer. J. Math.*, **86** (1964), 565–601; (c) *Differential Geometry, Lie Groups and Symmetric Spaces*, Academic Press, 1978; (d) *Groups and Geometric Analysis*, Academic Press, 1984.

[8] F. I. Karpelevič, The geometry of geodesics and the eigenfunctions of the Beltrami-Laplace operator on symmetric spaces, *Trans. Moscow Math. Soc.*, **14** (1965), 48–185.

[9] C. Y. Lee, Invariant polynomials of Weyl groups and applications to the centers of universal enveloping algebras, *Can. J. Math.*, **XXVI** (1974), 583–592.

[10] L. Pontrjagin, *Topological Groups* German transl. Teubner, Leipzig 1957, 1958.

[11] L. Clozel et P. Delorme, Le théorème de Paley-Wiener invariant pour les groupes de Lie réductifs II, *Ann. Sci. École Norm. Sup.*, **23** (1990), 193–228.

Reprinted from ADVANCES IN MATHEMATICS
All Rights Reserved by Academic Press, New York and London

Vol. 91, No. 2, February 1992
Printed in Belgium

The Flat Horocycle Transform for a Symmetric Space

SIGURDUR HELGASON*

Department of Mathematics, Massachusetts Institute of Technology, Cambridge, Massachusetts 02139

1. INTRODUCTION

Let $X = G/K$ be a symmetric space of the noncompact type of dimension n and rank l. In the tangent space X_o to X at the origin $o = \{K\}$ there is singled out an n-dimensional family Ξ_o of $(n-l)$-dimensional affine planes, the *horocycle planes*. They are permuted transitively by the Cartan motion group G_o which is generated by the isotropy action of K on X_o and the translations in X_o. The *Radon transform* for the pair (X_o, Ξ_o) (*the flat horocycle transform*) associates to a function f on X_o the function $\hat{f}$ on Ξ_o given by

$$\hat{f}(\xi) = \int_\xi f(Y)\, dm(Y), \qquad \xi \in \Xi_o, \tag{1.1}$$

dm being the Euclidean measure on ξ. Using the Fourier transform on X_o the transform (1.1) can readily be inverted as shown by Gindikin in [1].

In this paper we prove a support theorem for (1.1) and determine ranges for (1.1) and its dual transform. The principal tools are the tangent space analysis in [7] and some new results on the Fourier transform on X_o established in Section 3.

A transform generalizing (1.1) (replacing dm by a rotation-invariant measure) has been considered by Orloff [13]. He obtained support and invertibility results for the case when the rank l is 1.

If V is a finite-dimensional vector space, V^* denotes its dual and $S(V)$, $S(V^*)$ the corresponding symmetric algebras, $\mathscr{S}(V)$ the space of rapidly decreasing functions on V, and $\mathscr{S}'(V)$ the corresponding space of tempered distributions. If X is a manifold we write $\mathscr{D}(X) = C_c^\infty(X)$, $\mathscr{E}(X) = C^\infty(X)$ (complex-valued functions), and $\mathscr{D}'(X)$ (resp. $\mathscr{E}'(X)$) for the space of distributions on X (resp. distributions of compact support). If $C \subset X$ is a compact subset, $\mathscr{D}_C(X)$ denotes the set of $f \in \mathscr{D}(X)$ with support contained in C.

* Supported by the National Science Foundation, DMS-8805665.

0001-8708/92 $7.50

If g is a diffeomorphism of X we put for $f \in \mathscr{E}(X)$, $T \in \mathscr{D}'(X)$, and D an operator on $\mathscr{E}(X)$:

$$f^g = f \circ g^{-1}, \qquad T^g(\phi) = T(\phi^{g^{-1}}), \qquad D^g f = (D(f \circ g)) \circ g^{-1}.$$

If X is a metric space $B_r(p)$ denotes the open ball in X with center p and radius r. The action of g on X will generally be denoted $g \cdot x$ $(x \in X)$. The adjoint representation of a Lie group will generally be denoted Ad, but we shall often write $g \cdot Y$ for $\mathrm{Ad}(g)(Y)$.

2. The Flat Horocycle Transform

Let $X = G/K$ be a symmetric space of the noncompact type, G being a connected noncompact semisimple Lie group with finite center and K a maximal compact subgroup. Let o denote the origin $\{K\}$ and X_o the tangent space to X at o. For $g \in G$ let $\tau(g)$ denote the translation $hK \to ghK$. Then if $k \in K$, the differential $d\tau(k)_o$ maps X_o onto itself. Let G_o denote the group of affine transformations of X_o generated by all the translations $T_Y \colon Z \to Z + Y$ of X_o $(Y \in X_o)$ and the linear transformations $d\tau(k)_o$, $k \in K$. Then $X_o = G_o/d\tau(K)$, which we write as $X_o = G_o/K$ for simplicity.

Let $G = KAN$, $\mathfrak{g} = \mathfrak{k} + \mathfrak{a} + \mathfrak{n}$ be Iwasawa decompositions of G and of its Lie algebra $\mathfrak{g}$. Let $\mathfrak{p}$ be the orthogonal complement of $\mathfrak{k}$ in $\mathfrak{g}$, relative to the Killing form B of $\mathfrak{g}$. Then $\mathfrak{a}$ is a maximal abelian subspace of $\mathfrak{p}$ and we denote by $\mathfrak{q}$ its orthogonal complement in $\mathfrak{p}$. Thus we have the orthogonal decompositions

$$\mathfrak{g} = \mathfrak{k} + \mathfrak{p}, \qquad \mathfrak{p} = \mathfrak{a} + \mathfrak{q}. \tag{2.1}$$

Under the natural isomorphisms

$$X_o = (G/K)_o = (G \cdot o)_o \approx \mathfrak{g}/\mathfrak{k} \approx \mathfrak{p} \tag{2.2}$$

we have, by restriction,

$$(A \cdot o)_o \approx \mathfrak{a}, \qquad (N \cdot o)_o \approx \mathfrak{q} \tag{2.3}$$

because the manifolds $A \cdot o$ and $N \cdot o$ are orthogonal in o. Since $\mathfrak{q}$ is thus identified with the tangent space to the horocycle $N \cdot o$ in o and since the horocycles in X are permuted transitively by G we are led to the following definition.

Definition. A *horocycle plane* in $\mathfrak{p}$ is the image $g \cdot \mathfrak{q}$ of $\mathfrak{q}$ under some $g \in G_o$.

Let Ξ_o denote the manifold of all horocycle planes. Each $\xi \in \Xi_o$ has a natural measure dm induced by B. Also if $g \in G_o$ is such that $g \cdot \mathfrak{q}$ contains 0 then if $Z \in \mathfrak{q}$ is such that $g \cdot Z = 0$ so $gT_Z = k \in K$ we have $g \cdot \mathfrak{q} = k\mathfrak{q}$ where

$k \in K$. Thus K acts transitively on the set $\check{0}$ of horocycle planes through 0. The Haar measure on K thus induces a normalized invariant measure on $\check{0}$ and more generally such a measure $d\mu$ on each set

$$\check{Y} = \{\xi \in \Xi_o : Y \in \xi\}, \qquad Y \in \mathfrak{p}. \tag{2.4}$$

For the spaces X_o and Ξ_o both of which are homogeneous spaces of G_o, we can now define the Radon transform $f \to \hat{f}$ and its dual $\phi \to \check{\phi}$,

$$\hat{f}(\xi) = \int_\xi f(Y)\, dm(Y), \tag{2.5}$$

$$\check{\phi}(Y) = \int_{\check{Y}} \phi(\xi)\, d\mu(\xi). \tag{2.6}$$

In this paper we prove an inversion formula for (2.5) (Theorem 5.1), a support theorem (Theorem 6.1), and range theorems for (2.5) and (2.6) as well as for the extension of the former to compactly supported distributions on X_o (Theorems 7.3, 7.4, and Corollary 7.5).

The principal tool in the proofs is a detailed study of the Fourier transform on $\mathfrak{p}$ when expressed according to the polar coordinate decomposition of $\mathfrak{p}$,

$$(H, kM) \to \mathrm{Ad}(k)H \qquad (H \in \mathfrak{a}^+, k \in K). \tag{2.7}$$

Here M denotes the centralizer of $\mathfrak{a}$ in K, and $\mathfrak{a}^+$ is the positive Weyl chamber corresponding to the ordering given by $\mathfrak{n}$.

Also let M' denote the normalizer of A in K and $W = M'/M$ the Weyl group which acts on A, $\mathfrak{a}$, the dual $\mathfrak{a}^*$, and its complexification $\mathfrak{a}^*_{\mathbf{C}}$. We sometimes write B for the left coset space K/M.

The Killing form of $\mathfrak{g}$ is also denoted by B; it includes a norm $|\ |$ on $\mathfrak{p}$, $\mathfrak{a}$, and $\mathfrak{a}^*$ and a distance d on $\mathfrak{p}$ given by $d(X, Y) = |X - Y|$. It also induces Euclidean measures on A, $\mathfrak{a}$, and $\mathfrak{a}^*$. Multiplying these by $(2\pi)^{-l/2}$ we get measures da, dH, and $d\lambda$ for which the Fourier inversion formula holds without multiplicative constants.

3. A Paley–Wiener Theorem for the Tangent Space Fourier Transform

Definition. A C^∞ function $\psi(\lambda, b)$ on $\mathfrak{a}^*_{\mathbf{C}} \times B$, holomorphic in λ, is called a *holomorphic function of uniform exponential type* if there exists a constant $A \geqslant 0$ such that for each $N \in \mathbf{Z}^+$,

$$\sup_{\lambda \in \mathfrak{a}^*_{\mathbf{C}},\, b \in B} e^{-A|\mathrm{Im}\,\lambda|}(1 + |\lambda|^N)\, |\psi(\lambda, b)| < \infty.$$

Here $\mathrm{Im}\,\lambda = \eta$ if $\lambda = \xi + i\eta$ $(\xi, \eta \in \mathfrak{a}^*)$ and $|\lambda| = (|\xi|^2 + |\eta|^2)^{1/2}$.

Let $\mathscr{H}(\mathfrak{a}^* \times K/M)$ denote the space of all such functions. In [6] we proved that as f runs through $\mathscr{D}(\mathbf{R}^n)$ the Fourier transform in polar coordinate form

$$\tilde{f}(\lambda\omega) = \phi(\lambda, \omega) = \int_{\mathbf{R}^n} f(x)\, e^{-i\lambda(x,\omega)}\, dx$$

runs through the space of functions $\phi \in \mathscr{H}(\mathbf{R} \times \mathbf{S}^{n-1})$ satisfying the condition:

For each $k \in \mathbf{Z}^+$ and each isotropic vector $a \in \mathbf{C}^n$ the function

$$\lambda \to \lambda^{-k} \int_{\mathbf{S}^{n-1}} \phi(\lambda, \omega)(a, \omega)^k\, d\omega$$

is even and holomorphic on $\mathbf{C}$.

Here $a \in \mathbf{C}^n$ is *isotropic* if $a_1^2 + \cdots a_n^2 = 0$. We shall now extend this theorem to $\mathscr{D}(\mathfrak{p})$ relative to the "polar coordinate decomposition" $(H, kM) \to \mathrm{Ad}(k)H$ $(H \in \mathfrak{a}^+, k \in K)$.

For $\lambda \in \mathfrak{a}_{\mathbf{C}}^*$ let $A_\lambda \in \mathfrak{a}_{\mathbf{C}}$ be determined by $B(H, A_\lambda) = \lambda(H)$ $(H \in \mathfrak{a})$ and write the Fourier transform

$$\hat{f}(Y) = \int_{\mathfrak{p}} f(X)\, e^{-iB(X, Y)}\, dX$$

in the form

$$\tilde{f}(k \cdot A_\lambda) = \int_{\mathfrak{p}} f(X)\, e^{-B(k \cdot A_\lambda, X)}\, dX. \tag{3.1}$$

For the applications to the Radon transform (2.4) we wish to characterize the space of functions $\phi(\lambda, kM) = \tilde{f}(k \cdot A_\lambda)$ on $\mathfrak{a}_{\mathbf{C}}^* \times K/M$ as f runs through $\mathscr{D}(\mathfrak{p})$. We require some notation.

Let $\hat{K}$ denote the set of equivalence classes of irreducible unitary representations δ of K and $\hat{K}_M$ the set of $\delta \in \hat{K}$ such that $\delta(M)$ has a nonzero fixed vector. Let V_δ be the representation space of δ (with inner product $\langle\, ,\, \rangle$) and $V_\delta^M \subset V_\delta$ the set of vectors fixed under $\delta(M)$. Let $v_1, \ldots, v_{d(\delta)}$ be a basis of V_δ such that $v_1, \ldots, v_{l(\delta)}$ is a basis of V_δ^M. Corresponding to (2.1) we have the direct decomposition of the symmetric algebra,

$$S(\mathfrak{g}) = S(\mathfrak{a}) \oplus (S(\mathfrak{g})\mathfrak{k} + \mathfrak{q}S(\mathfrak{g}))$$

and we let $j \to j^p$ denote the corresponding projection of $S(\mathfrak{g})$ onto $S(\mathfrak{a})$. Let $H(\mathfrak{p}^*) \subset S(\mathfrak{p}^*)$ and $H(\mathfrak{p}) \subset S(\mathfrak{p})$ denote the subspaces of K-harmonic polynomials. Under the adjoint action of K on $\mathfrak{p}$ (resp. coadjoint action of K on $\mathfrak{p}^*$) let $S_\delta(\mathfrak{p}), H_\delta(\mathfrak{p})$ (resp. $S_\delta(\mathfrak{p}^*), H_\delta(\mathfrak{p}^*)$) be the subspaces of elements of type δ. The bijection of $\mathfrak{p}$ onto $\mathfrak{p}^*$ (by means of B) extends to an isomorphism of $S_\delta(\mathfrak{p})$ onto $S_\delta(\mathfrak{p}^*)$. If $p \in S(\mathfrak{p})$ then the differential operator $\partial(p)$ satisfies $\partial(\mathrm{Ad}(k) \cdot p) = \partial(p)^{\mathrm{Ad}(k)}$ $(k \in K)$. As shown by Kostant and Rallis [12] the space $H_\delta(\mathfrak{p})$ decomposes into $l(\delta)$ irreducible subspaces on which the representation Ad is equivalent to δ. Thus the space $E_\delta = \mathrm{Hom}_K(V_\delta, H_\delta(\mathfrak{p}))$ of linear K-maps of V_δ into $H_\delta(\mathfrak{p})$ has dimension $l(\delta)$. Let $\varepsilon_1, \ldots, \varepsilon_{l(\delta)}$ be a basis of E_δ and for $\lambda \in \mathfrak{a}_{\mathbf{C}}^*$ consider the $l(\delta) \times l(\delta)$ matrix

$$J^\delta(\lambda)_{ij} = j^{\varepsilon_j(v_i)}(-i\lambda). \tag{3.2}$$

Let $\check{\delta}$ denote the representation of K which is contragredient to δ.

THEOREM 3.1. *The range $\mathscr{D}(\mathfrak{p})^\sim$ of $\mathscr{D}(\mathfrak{p})$ under the Fourier transform* (3.1) *is the set of functions $\phi(\lambda, kM)$ on $\mathfrak{a}_{\mathbf{C}}^* \times K/M$ satisfying*:

(i) $\phi \in \mathscr{H}(\mathfrak{a}^* \times K/M)$.

(ii) *For each $\delta \in \hat{K}_M$, the function*

$$\lambda \to J^{\check{\delta}}(\lambda)^{-1} \int_K \phi(\lambda, kM)\, \delta(k^{-1})\, dk$$

is W-invariant and holomorphic on $\mathfrak{a}_{\mathbf{C}}^*$.

First we prove a simple lemma relating the Radon and the Fourier transform.

LEMMA 3.1. *Let $f \in L^1(\mathfrak{p})$ and $k \in K$. Then for each $\lambda \in \mathfrak{a}^*$*

$$\tilde{f}(k \cdot A_\lambda) = c_o \int_{\mathfrak{a}} \hat{f}(k \cdot (H + \mathfrak{q}))\, e^{-i\lambda(H)}\, dH \qquad (c_o = \mathrm{const.}).$$

Proof. We have

$$\begin{aligned}\tilde{f}(k \cdot A_\lambda) &= \int_{\mathfrak{p}} f(X)\, e^{-iB(k \cdot A_\lambda, X)}\, dX \\ &= \int_{\mathfrak{p}} f(k \cdot Z)\, e^{-iB(A_\lambda, Z)}\, dZ \\ &= c_o \int_{\mathfrak{a}} \int_{\mathfrak{q}} f(k \cdot (H + Q))\, e^{-iB(A_\lambda, H)}\, dH\, dQ,\end{aligned}$$

proving the lemma.

The distance from 0 to the horocycle plane $k\cdot(H+\mathfrak{q})$ is $|H|$ so if $f\in\mathscr{D}(\mathfrak{p})$ then the classical Paley–Wiener theorem implies that the function $(\lambda, kM)\to\tilde{f}(k\cdot A_\lambda)$ belongs to $\mathscr{H}(\mathfrak{a}^*\times K/M)$. To verify condition (ii) consider first the polar coordinate integration formula

$$\int_{\mathfrak{p}} F(X)\,dX=\int_{\mathfrak{a}^+}\left(\int_K F(\mathrm{Ad}(k)\,H)\,dk\right)\omega(H)\,dH, \qquad F\in C_c(\mathfrak{p}), \quad (3.3)$$

where the density ω is given by

$$\omega(H)=c\prod_{\alpha\in\Sigma^+}\alpha(H)^{m_\alpha} \qquad (c=\text{const.}), \quad (3.4)$$

where Σ^+ is the set of positive roots of $\mathfrak{g}$ with respect to $\mathfrak{a}$, positivity referring to the chosen $\mathfrak{n}$ above. Next we consider the *generalized Bessel function*

$$\Psi_{\lambda,\delta}(X)=\int_K e^{iB(kA_\lambda,X)}\,\delta(k)\,dk. \quad (3.5)$$

Then, using (3.1) we have

$$\int_K \tilde{f}(k\cdot A_\lambda)\,\delta(k^{-1})\,dk$$

$$=\int_{\mathfrak{p}} f(X)\,\Psi_{\bar{\lambda},\delta}(X)^*\,dX$$

$$=\int_{\mathfrak{a}^+}\left(\int_K f(k\cdot H)\,\Psi_{\bar{\lambda},\delta}(H)^*\,\delta(k^{-1})\,dk\right)\omega(H)\,dH. \quad (3.6)$$

In [7, Theorem 4.6] we proved that

$$\lambda\to J^\delta(\lambda)^{-1}\,(\Psi_{\bar{\lambda},\delta}(H))^*$$

is holomorphic on $\mathfrak{a}_{\mathbf{C}}^*$ and W-invariant. By (3.5) this verifies condition (ii).

On the other hand, suppose ϕ satisfies (i) and (ii). We define f by

$$f(X)=\int_{K/M} dk_M\int_{\mathfrak{a}_+^*}\phi(\lambda, kM)\,e^{iB(k\cdot A_\lambda,X)}\omega(\lambda)\,d\lambda, \quad (3.7)$$

where $\omega(\lambda)=\omega(A_\lambda)$ and $\mathfrak{a}_+^*$ corresponds to $\mathfrak{a}^+$ under the map $\lambda\to A_\lambda$. Here we decompose by the Peter–Weyl theorem

$$\phi(\lambda, kM)=\sum_{\delta\in\hat{K}_M} d(\delta)\,\mathrm{Tr}(\phi_\delta(\lambda)\,\delta(k)), \quad (3.8)$$

where

$$\phi_\delta(\lambda) = \int_K \phi(\lambda, kM)\, \delta(k^{-1})\, dk.$$

We substitute (3.8) into (3.7) and invoke (3.5). Then we obtain

$$f(X) = \sum_{\delta \in \hat{K}_M} d(\delta) \int_{\mathfrak{a}_+^*} \operatorname{Tr}(\phi_\delta(\lambda)\, \Psi_{\lambda,\delta}(X))\, \omega(\lambda)\, d\lambda. \tag{3.9}$$

For $s \in W$ let $m_s \in M'$ be a representative. A routine computation involving the formulas

$$j^{m_s \cdot p} = m_s \cdot (j^p), \qquad \varepsilon_l(\delta(k)v) = k \cdot \varepsilon_l(v)$$

proves that

$$J^\delta(s\lambda) = \delta(m_s)\, J^\delta(\lambda). \tag{3.10}$$

Condition (ii) implies therefore

$$\phi(s^{-1}\lambda, km_sM) = \phi(\lambda, kM) \tag{3.11}$$

so

$$\phi_\delta(s\lambda) = \delta(m_s)\, \phi_\delta(\lambda). \tag{3.12}$$

Also $\Psi_{s\lambda,\delta}(X) = \Psi_{\lambda,\delta}(X)\, \delta(m_s^{-1})$ and (3.4) shows that $|\omega(\lambda)|$ is Weyl group invariant. Thus (3.9) can also be written

$$f(X) = |W|^{-1} \sum_{\delta \in \hat{K}_M} d(\delta) \int_{\mathfrak{a}^*} \operatorname{Tr}(\phi_\delta(\lambda)\, \Psi_{\lambda,\delta}(X))\, |\omega(\lambda)|\, d\lambda, \tag{3.13}$$

where $|W|$ is the order of the Weyl group. By (ii) $J^\delta(\lambda)^{-1}\phi_\delta(\lambda)$ is W-invariant and holomorphic of exponential type R (the exponential type of ϕ) so by Theorem 5.1 in [7], $\phi_\delta(\lambda)$ is the generalized Bessel transform $\tilde{F}_\delta(\lambda)$ of the function

$$F_\delta(X) = c \operatorname{Tr}\left(\int_{\mathfrak{a}^*} \Psi_{\lambda,\delta}(X)\, \phi_\delta(\lambda)\, |\omega(\lambda)|\, d\lambda \right) \tag{3.14}$$

of support in the ball $|X| < R$ in $\mathfrak{p}$. Thus by (3.13), $f \in \mathcal{D}(\mathfrak{p})$. The inversion formula and (3.7) now show that $\phi(\lambda, kM)$ is a constant multiple of $\tilde{f}(k \cdot A_\lambda)$. This concludes the proof.

4. The Flat Horocycle Space

By definition, each $g \in G_o$ has the form $g \cdot X = \mathrm{Ad}(k_o)(X_o + X)$ $(X \in \mathfrak{p})$ for some fixed $X_o \in \mathfrak{p}$, $k_o \in K$. By (2.1) each $\xi \in \Xi_o$ can thus be written

$$\xi = \mathrm{Ad}(k)(H + \mathfrak{q}), \qquad k \in K,\, H \in \mathfrak{a}. \tag{4.1}$$

Let $M'\mathfrak{q}$ denote the group of affine transformations $X \to \mathrm{Ad}(m')(Y + X)$ of $\mathfrak{p}$ with $m' \in M'$ and $Y \in \mathfrak{q}$. Since $\mathrm{Ad}(m')\mathfrak{q} \subset \mathfrak{q}$, $M'\mathfrak{q}$ is indeed a group.

Let $(K/M \times \mathfrak{a})_W$ denote the product $K/M \times \mathfrak{a}$ with the equivalence relation

$$(k_1 M, H_1) \sim (k_2 M, H_2)$$

if $H_2 = s^{-1} H_1$, $k_2 = k_1 m_s$ where $s \in W$ and $m_s \in M'$ is a representative of s. We often write $k \cdot X = \mathrm{Ad}(k) \cdot X$ for short. The following lemma is essentially given in Orloff [13].

Lemma 4.1. (i) *$M'\mathfrak{q}$ is the subgroup of G_o leaving $\mathfrak{q}$ invariant so*

$$\Xi_o = G_o / M'\mathfrak{q}.$$

(ii) *The mapping $(kM, H) \to \mathrm{Ad}(k)(H + \mathfrak{q})$ induces a bijection*

$$(K/M \times \mathfrak{a})_W \sim \Xi_o.$$

Proof. (i) If $g \cdot \mathfrak{q} = \mathfrak{q}$ then $g \cdot 0 = Y_o \in \mathfrak{q}$ so the transformation $X \to -Y_o + g \cdot X$ $(X \in \mathfrak{p})$ is an element $k \in K$ and $k \cdot \mathfrak{q} = \mathfrak{q}$. Hence $k \cdot \mathfrak{a} = \mathfrak{a}$ so $k \in M'$ whence $g \in M'\mathfrak{q}$.

(ii) By (4.1) it is clear that $(kM, H) \to \mathrm{Ad}(k)(H + \mathfrak{q})$ induces a well-defined mapping of $(K/M \times \mathfrak{a})_W$ onto Ξ_o. For injectivity suppose $\mathrm{Ad}(k_1)(H_1 + \mathfrak{q}) = \mathrm{Ad}(k_2)(H_2 + \mathfrak{q})$. Since H_i is the point in $H_i + \mathfrak{q}$ closest to the origin we deduce $\mathrm{Ad}(k_1)\, H_1 = \mathrm{Ad}(k_2)\, H_2$ so $\mathrm{Ad}(k_1)\mathfrak{q} = \mathrm{Ad}(k_2)\mathfrak{q}$ so by (i) $k_2 = k_1 m_s$ for some $s \in W$. This proves (ii).

The transforms (2.5), (2.6) are thus the Radon transforms (in the sense of [4]) for the double fibration,

$$\begin{array}{ccc} & G_o/M' & \\ \swarrow & & \searrow \\ X_o = G_o/K & & \Xi_o = G_o/M'\mathfrak{q} \end{array} \tag{4.2}$$

which is a flat analog of the double fibration

$$\begin{array}{ccc} & G/M & \\ \swarrow & & \searrow \\ X = G/K & & \Xi = G/MN \end{array} \tag{4.3}$$

and our Theorems 6.1, 7.1, 7.2, 7.4, 7.5 are analogous to the corresponding results for (4.3) [5; 8; 9; 10, Prop. 4.4]. The proofs are, however, quite different because the main results of the spherical function theory so crucial for (4.3) are not available for (4.2) (cf. [7, Introduction]).

Because of the identification $(K/M\times\mathfrak{a})_W=\Xi_o$ it is convenient to view functions ϕ on Ξ_o as functions on $K/M\times\mathfrak{a}$ satisfying $\phi(s^{-1}H, km_sK)\equiv\phi(H, kK)$ $(s\in W)$. The functions

$$(H, kM)\to\hat{f}(k\cdot(H+\mathfrak{q})),\qquad (\lambda, kM)\to\tilde{f}(k\cdot A_\lambda)\tag{4.4}$$

satisfy these invariance conditions.

We consider now the signum function for Σ^+ defined by

$$\operatorname{sgn}_\Sigma(\lambda)=\omega(\lambda)/|\omega(\lambda)|,\qquad \lambda\in\mathfrak{a}^*,\tag{4.5}$$

where, by definition, the value is 1 if $\omega(\lambda)=0$. Then $\operatorname{sgn}_\Sigma$ is a tempered distribution on $\mathfrak{a}^*$. We then define the "Hilbert transform" $\mathscr{H}_\Sigma$ as the convolution operator with the Fourier transform of $\operatorname{sgn}_\Sigma$ on $\mathfrak{a}$, i.e.,

$$(\mathscr{H}_\Sigma F)^*(\lambda)=\operatorname{sgn}_\Sigma(\lambda)\,F^*(\lambda),\qquad F\in\mathscr{S}(\mathfrak{a}).\tag{4.6}$$

Viewing the vector $A_\alpha\in\mathfrak{a}$ $(\alpha\in\Sigma^+)$ as a differential operator on $\mathfrak{a}$ we define $\Lambda_o\phi$ for $\phi\in\mathscr{S}(\Xi_o)$ by

$$(\Lambda_o\phi)(H, kM)=\Big(\mathscr{H}_\Sigma\prod_{\alpha\in\Sigma^+}A_\alpha^{m_\alpha}\Big)_H(\phi(H, kM)),\qquad H\in\mathfrak{a},\tag{4.7}$$

where the subscript H indicates the argument on which the operator operates. Then by (4.5) and (4.6),

$$(\Lambda_o\phi)^*(\lambda, kM)=|\omega(\lambda)|\,\phi^*(\lambda, kM)\,i^p\tag{4.8}$$

(with $p=\sum_{\alpha\in\Sigma^+}m_\alpha$). Since the invariance condition $\psi(s^{-1}H, km_sM)\equiv\psi(H, kM)$ implies $\psi^*(s^{-1}\lambda, km_sM)\equiv\psi^*(\lambda, kM)$ we deduce from (4.8) that

$$(\Lambda_o\phi)(s^{-1}H, km_sH)\equiv(\Lambda_o\phi)(H, kM)\tag{4.9}$$

so $\Lambda_o\phi$ is a function on Ξ_o.

5. The Inversion Formula

By combining Lemma 3.1 with the Fourier inversion formula on $\mathfrak{p}$ it is easy to invert the Radon transform (2.5).

THEOREM 5.1. *The Radon transform*

$$\hat{f}(\xi)=\int_\xi f(Y)\,dm(Y),\qquad \xi\in\Xi_o$$

is inverted by the formula

$$f = c(\Lambda_o \hat{f})^\vee, \qquad f \in \mathcal{S}(\mathfrak{p}), \tag{5.1}$$

where c is a constant independent of f.

Proof. We have by classical Fourier transform theory

$$f(0) = c_1 \int_{\mathfrak{p}} \tilde{f}(Y)\, dY, \qquad (c_1 = (2\pi)^{-\dim(\mathfrak{p})}),$$

which by (3.3) can be written

$$f(0) = c_2 \int_{K/M} dk_M \int_{\mathfrak{a}^*_+} \tilde{f}(kA_\lambda)\, \omega(\lambda)\, d\lambda \qquad (c_2 = \text{const.})$$

so by the Weyl group invariance of $|\omega(\lambda)|$ and (4.4)

$$|W|\, f(0) = c_2 \int_{K/M} dk_M \int_{\mathfrak{a}^*} \tilde{f}(kA_\lambda)\, |\omega(\lambda)|\, d\lambda.$$

Using Lemma 3.1 we thus obtain

$$f(0) = c_3 \int_{K/M} dk_M \int_{\mathfrak{a}^*} |\omega(\lambda)|\, d\lambda \int_{\mathfrak{a}} \hat{f}(k \cdot (H + \mathfrak{q}))\, e^{-i\lambda(H)}\, dH$$

so by (4.8),

$$f(0) = c_4 \int_{K/M} dk_M \int_{\mathfrak{a}^*} d\lambda \int_{\mathfrak{a}} (\Lambda_o \hat{f})(H, kM)\, e^{-i\lambda(H)}\, dH.$$

Using the Fourier inversion formula on $\mathfrak{a}$ we thus find

$$f(0) = c \int_{K/M} (\Lambda_o \hat{f})(0, kM)\, dk_M.$$

Here the c's are constants independent of f. Bringing in the dual transform (2.6) the last formula is

$$f(0) = c(\Lambda_o \hat{f})^\vee\, (0). \tag{5.2}$$

Since the operators $f \to \hat{f}$ and Λ_o are translation-invariant, (5.2) implies the theorem.

Remark. A formula similar to (5.1) but differing in some details is given by Gindikin in [1].

6. The Support Theorem

Let $f \in \mathscr{S}(\mathbf{R}^n)$ and $0 < d < n$. It is known [3] that if f has integral 0 over each d-dimensional plane outside a fixed ball B then f vanishes identically outside B. This amounts to an assumption on an $(n-d)(d+1)$-dimensional family of d-planes.

We shall now prove a stronger version of this result in the case of horocycle planes in $\mathfrak{p}$. Here the assumption is made for just an n-dimensional family of d-planes ($n = \dim \mathfrak{p}$, $d = \dim \mathfrak{q}$).

THEOREM 6.1. *Let $f \in \mathscr{S}(\mathfrak{p})$ and assume*

$$\hat{f}(\xi) = 0 \quad \textit{for} \quad \xi \in \Xi_o, \quad \textit{with} \quad d(0, \xi) > R. \tag{6.1}$$

Then

$$f(Y) = 0 \quad \textit{for} \quad Y \in \mathfrak{p}, \ |Y| > R.$$

Proof. We shall use Fourier transform theory of K-finite functions on $\mathfrak{p}$ as developed in [7]. If $\delta \in \hat{K}_M$ we put for $f \in \mathscr{E}(\mathfrak{p})$, $\phi \in \mathscr{E}(\Xi_o)$

$$\begin{aligned} f^\delta(X) &= d(\delta) \int_K f(k \cdot X)\, \delta(k^{-1})\, dk, \\ \phi^\delta(\xi) &= d(\delta) \int_K \phi(k \cdot \xi)\, \delta(k^{-1})\, dk \end{aligned} \tag{6.2}$$

and

$$(f^\delta)^\wedge\ (\xi) = \int_\xi f^\delta(X)\, dm(X), \qquad f \in \mathscr{S}(\mathfrak{p}). \tag{6.3}$$

Then by (6.2)

$$(f^\delta)^\wedge = (\hat{f})^\delta, \qquad f \in \mathscr{S}(\mathfrak{p}). \tag{6.4}$$

Let $\check{\delta}$ denote the representation of K, contragredient to δ, and let $\mathscr{D}_{\check{\delta}}(\mathfrak{p})$ denote the space of K-finite functions in $\mathscr{D}(\mathfrak{p})$ of type $\check{\delta}$. Also let

$$\mathscr{D}^\delta(\mathfrak{p}) = \{F \in \mathscr{D}(\mathfrak{p}, \operatorname{Hom}(V_\delta, V_\delta)) : F(k \cdot X) \equiv \delta(k)\, F(X)\} \tag{6.5}$$

and consider $\mathscr{D}_{\check{\delta}}(\mathfrak{p})$ and $\mathscr{D}^\delta(\mathfrak{p})$ with the relative topologies. Then the mapping

$$F(X) \to \operatorname{Tr}(F(X)) \tag{6.6}$$

is a homeomorphism of $\mathscr{D}^{\delta}(\mathfrak{p})$ onto $\mathscr{D}_{\delta}(\mathfrak{p})$ and the inverse if given by $f \to f^{\delta}$ [6, Sect. 3]. We define the spaces $\mathscr{S}_{\delta}(\mathfrak{p})$ and $\mathscr{S}^{\delta}(\mathfrak{p})$ in the same way. The *generalized Bessel transform* of $f \in \mathscr{S}_{\delta}(\mathfrak{p})$ is defined by

$$\tilde{f}(\lambda) = d(\delta) \int_{\mathfrak{p}} f(X)\, \Psi_{\bar{\lambda},\delta}(X)^*\, dX$$

$$= d(\delta) \int_{\mathfrak{p}} f(X) \int_K e^{-iB(k \cdot A_{\lambda}, X)} \delta(k^{-1})\, dk\, dX \tag{6.7}$$

for all $\lambda \in \mathfrak{a}_{\mathbf{C}}^*$ for which the integral converges. Then we have from Lemma 3.1,

$$\tilde{f}(\lambda) = c_o \int_{\mathfrak{a}} (f^{\delta})^{\wedge}\, (H + \mathfrak{q})\, e^{-i\lambda(H)}\, dH. \tag{6.8}$$

Suppose $f \in \mathscr{S}(\mathfrak{p})$ satisfies (6.1). Note that if $\xi = k \cdot (H + \mathfrak{q})$ then $d(0, \xi) = |H|$. By the vector-valued Peter–Weyl theorem,

$$f = \sum_{\delta \in \hat{K}_M} \mathrm{Tr}(f^{\delta}) \tag{6.9}$$

and by (6.6), the function $F_{\delta} = \mathrm{Tr}(f^{\delta})$ belongs to $\mathscr{S}_{\delta}(\mathfrak{p})$ and $(F_{\delta})^{\delta} = f^{\delta}$. Also

$$(f^{\delta})^{\wedge}\, (\xi) = d(\delta) \int_K \hat{f}(k \cdot \xi)\, \delta(k^{-1})\, dk = 0 \qquad \text{for} \quad d(0, \xi) > R \tag{6.10}$$

and by (6.8)

$$(F_{\delta})^{\sim}\, (\lambda) = c_o \int_{\mathfrak{a}} (f^{\delta})^{\wedge}\, (H + \mathfrak{q})\, e^{-i\lambda(H)}\, dH. \tag{6.11}$$

Thus $\tilde{F}_{\delta}$ belongs to the space $\mathscr{H}(\mathfrak{a}^*, \mathrm{Hom}(V_{\delta}, V_{\delta}^M))$ of holomorphic functions on $\mathfrak{a}_{\mathbf{C}}^*$ of exponential type with values in $\mathrm{Hom}(V_{\delta}, V_{\delta}^M)$. By (6.10), (6.11), $\tilde{F}_{\delta}$ has exponential type R. In addition we know from [7, Theorem 5.3] that the function $\lambda \to J^{\delta}(\lambda)^{-1}\, \tilde{F}_{\delta}(\lambda)$ is smooth on the real space $\mathfrak{a}^*$ and W-invariant. We now write

$$J^{\delta}(\lambda)^{-1} = \det(J^{\delta}(\lambda))^{-1}\, J_c(\lambda), \tag{6.12}$$

where $(J_c(\lambda)_{ij})$ is the matrix formed by the cofactors in $(J^{\delta}(\lambda)_{ji})$. The entries of $J_c(\lambda)$ being polynomials in λ the function $\lambda \to J_c(\lambda)\, \tilde{F}_{\delta}(\lambda)$ is holomorphic of exponential type on all of $\mathfrak{a}_c^*$. Also by Lemma 4.5 in [7], $\det(J^{\delta}(\lambda))$ is a product of linear factors. Now it is obvious that if $F(z)$ is holomorphic on $\mathbf{C}$ and $x^{-1}F(x)$ is smooth on $\mathbf{R}$ then $z^{-1}F(z)$ is holomorphic on $\mathbf{C}$. Thus by the above $\lambda \to J^\delta(\lambda)^{-1}\,\tilde{F}_\delta(\lambda)$ is holomorphic on $\mathfrak{a}_{\mathbf{C}}^*$ and of exponential type R. Thus by Theorem 5.1 in [7] F_δ vanishes outside the ball $|Y| \leqslant R$ and by (6.9) the same holds for f. This proves the theorem.

7. The Ranges

Let $\delta \in \hat{K}_M$ and fix an orthonormal basis $v_1, ..., v_{d(\delta)}$ of V^δ such that $v_1, ..., v_{l(\delta)}$ span V_δ^M. Assigning to $v \in V_\delta$ the linear form $v'\colon w \to \langle w, v\rangle$ on V_δ we get a conjugate-linear bijection $v \to v'$ of V_δ onto $V_{\check\delta}$. We define the inner product $\langle\ ,\ \rangle$ on $V_{\check\delta}$ by $\langle v', w'\rangle = \langle w, v\rangle$. Then v_i' $(1 \leqslant i \leqslant d(\delta))$ is an orthonormal basis of $V_{\check\delta}$ with v_j' $(1 \leqslant j \leqslant l(\delta))$ spanning $V_{\check\delta}^M$.

We define now the matrix $J^{\check\delta}(D)$ with constant-coefficient differential operator entries on $\mathfrak{a}$ by

$$(J^\delta(D)_{ij}F)(H) = \int_{\mathfrak{a}^*} J^\delta(\lambda)_{ij}F^*(\lambda)\, e^{i\lambda(H)}\, d\lambda. \tag{7.1}$$

Viewing a function $\Phi \in \mathscr{D}(\mathfrak{a}, \operatorname{Hom}(V_\delta, V_\delta^M))$ as $l(\delta) \times d(\delta)$ matrix-valued we have

$$(J^{\check\delta}(D)\Phi)(H) = \int_{\mathfrak{a}^*} J^{\check\delta}(\lambda)\,\Phi^*(\lambda)\, e^{i\lambda(H)}\, d\lambda. \tag{7.2}$$

Theorem 7.1. *The range $\mathscr{D}_{\check\delta}(\mathfrak{p})^\wedge$ consists of the functions*

$$\phi(k\cdot(H+\mathfrak{q})) = \operatorname{Tr}(\delta(k)(J^{\check\delta}(D)\Psi)(H)), \tag{7.3}$$

where $\Psi \in \mathscr{D}(\mathfrak{a}, \operatorname{Hom}(V_\delta, V_\delta^M))$ is W-invariant.

Proof. From Theorem 5.1 in [7], already used, we know that as f runs through $\mathscr{D}_{\check\delta}(\mathfrak{p})$ the functions $\tilde{f}(\lambda)$ in (6.8) exhaust the space $J^{\check\delta}\mathscr{T}^\delta(\mathfrak{a}^*)$ where $\mathscr{T}^\delta(\mathfrak{a}^*)$ is the space of W-invariants in $\mathscr{H}(\mathfrak{a}^*, \operatorname{Hom}(V_\delta, V_\delta^M))$. Thus by (6.8) the functions $(f^\delta)^\wedge\,|\,\mathfrak{a}$ run through the space of functions $J^{\check\delta}(D)\Psi$, where $\Psi \in \mathscr{D}(\mathfrak{a}, \operatorname{Hom}(V_\delta, V_\delta^M))$ is W-invariant. On the other hand, by the analog of (6.6) for Ξ_o, combined with (6.4), we have

$$\hat{f} = \operatorname{Tr}(\hat{f}^\delta) = \operatorname{Tr}((f^\delta)^\wedge)$$

so

$$\begin{aligned}\hat{f}(k\cdot(H+\mathfrak{q})) &= \operatorname{Tr}((f^\delta)^\wedge\,(k\cdot(H+\mathfrak{q})))\\ &= \operatorname{Tr}(\delta(k)(f^\delta)^\wedge\,(H+\mathfrak{q})),\end{aligned}$$

and this proves the result.

Next we describe the kernel $\mathcal{N}$ of the dual transform $\phi \to \check{\phi}$ of $\mathcal{E}(\Xi_0)$ into $\mathcal{E}(\mathfrak{p})$. Let $\mathcal{N}_\delta$ denote the set of $\phi \in \mathcal{N}$ of type δ. A vector-valued function ζ on $\mathfrak{a}$ will be called *W-odd* if $\sum_{s \in W} \zeta^s = 0$. We have clearly

$$\zeta \text{ is } W\text{-odd} \Leftrightarrow \int_{\mathfrak{a}} \zeta(H)\, \psi(H)\, dH = 0 \tag{7.4}$$

for all W-invariant $\psi \in \mathcal{D}(\mathfrak{a})$.

THEOREM 7.2. *Let $\delta \in \hat{K}_M$. The space $\mathcal{N}_\delta$ then consists of the functions*

$$\zeta(k \cdot (H + \mathfrak{q})) = \mathrm{Tr}(\check{\delta}(k)\, Z(H)),$$

where $Z \in \mathcal{E}(\mathfrak{a}, \mathrm{Hom}(V_\delta, V_\delta^M))$ and the function

$$^t J^\delta(D)^* Z \qquad \text{on } \mathfrak{a}$$

is W-odd.

Here ^tA denotes the transpose of a matrix A and if A has differential operator entries, $(A^*)_{ij} = (A_{ij})^*$, star denoting adjoint.

Proof. By general theory the transforms (2.5), (2.6) are adjoint of each other, i.e.,

$$\int_{\Xi_o} \hat{f}(\xi)\, \phi(\xi)\, d\xi = \int_{\mathfrak{p}} f(Y)\, \check{\phi}(Y)\, dY, \tag{7.5}$$

where $d\xi$ is a G_o-invariant measure on Ξ_o. In the parametrization of Lemma 4.1(ii) $d\xi$ is a constant multiple of $dH\, dk_M$. In the bilinear form $\langle\, ,\, \rangle$ on $\mathcal{D}(\Xi_o) \times \mathcal{E}(\Xi_o)$ given by

$$\langle \phi_1, \phi_2 \rangle = \int_{\Xi_o} \phi_1(\xi)\, \phi_2(\xi)\, d\xi \tag{7.6}$$

we see from (7.5) that, $\perp$ denoting annihilator,

$$\mathcal{N} = (\mathcal{D}(\mathfrak{p})^\wedge)^\perp. \tag{7.7}$$

On the other hand, the orthogonality relations on K show that

$$\langle \mathcal{D}_\delta(\Xi_o), \mathcal{E}_\pi(\Xi_o) \rangle = 0 \tag{7.8}$$

for all $\pi \in \hat{K}_M$ not equivalent to δ. Consequently, by (7.6)

$$\mathcal{N}_\delta = \left\{ \zeta \in \mathcal{E}_\delta(\Xi_o) : \int_{\Xi_o} \zeta(\xi)\, \phi(\xi)\, d\xi = 0 \text{ for } \phi \in \mathcal{D}_\delta(\mathfrak{p})^\wedge \right\}. \tag{7.9}$$

Writing (7.3) out in matrix form we have

$$\phi(k\cdot(H+\mathfrak{q}))=\sum_{1\leqslant i\leqslant d(\delta),1\leqslant j\leqslant l(\delta)}\delta_{ij}(k)\,\psi_{ji}(H), \tag{7.10}$$

where we have taken into account the fact that $\Psi(H)$ maps V_δ into V_δ^M. Writing $\zeta\in\mathcal{N}_\delta$ in the form

$$\zeta(k\cdot(H+\mathfrak{q}))=\mathrm{Tr}(\check{\delta}(k)\,Z(H)) \tag{7.11}$$

we have $\mathrm{Tr}(\check{\delta}(k)\,\check{\delta}(m)\,Z(H))=\mathrm{Tr}(\check{\delta}(k)\,Z(H))$ for $m\in M$ so by the irreducibility of $\check{\delta}$, $\check{\delta}(m)\,Z(H)=Z(H)$, in other words

$$Z\in\mathscr{E}(\mathfrak{a},\mathrm{Hom}(V_{\check{\delta}},V_{\check{\delta}}^M)).$$

Since $\check{\delta}(k)_{ij}=(\delta(k)_{ij})^-$ expression (7.11) equals

$$\zeta(k\cdot(H+\mathfrak{q}))=\sum_{1\leqslant p\leqslant d(\delta),1\leqslant q\leqslant l(\delta)}(\delta_{pq}(k)^-)\,Z_{qp}(H).$$

Hence, by (7.3) and (7.8) and the orthogonality relations we get ($1\leqslant i\leqslant d(\delta)$, $1\leqslant j\leqslant l(\delta)$)

$$\sum_{i,j}\int_{\mathfrak{a}}\sum_{k=1}^{l(\delta)}(J_{jk}^{\delta}(D)\,\Psi_{ki})(H)\,Z_{ji}(H)\,dH=0.$$

This amounts to

$$\sum_{i,k}\int_{\mathfrak{a}}\Psi_{ki}(H)({}^{t}J^{\delta}(D)^*\,Z)_{ki}\,(H)\,dH=0 \tag{7.12}$$

and the Ψ_{ki} are arbitrary W-invariants in $\mathscr{D}(\mathfrak{a})$. In view of (7.4) this implies the statement of the theorem.

THEOREM 7.3. *In terms of the bilinear form $\langle\ ,\ \rangle$ on $\mathscr{D}(\Xi_o)\times\mathscr{E}(\Xi_o)$,*

$$\mathscr{D}(\mathfrak{p})^\wedge=\mathcal{N}^\perp\qquad(\textit{annihilator in }\mathscr{D}(\Xi_o)). \tag{7.13}$$

In particular, $\mathscr{D}(\mathfrak{p})^\wedge$ is a closed subspace of $\mathscr{D}(\Xi_o)$.

Proof. Relation (7.9) can be written

$$\mathcal{N}_{\check{\delta}}=(\mathscr{D}_{\delta}(\mathfrak{p})^\wedge)^\perp\qquad(\text{annihilator in }\mathscr{E}_{\check{\delta}}(\Xi_o)). \tag{7.14}$$

If L is a constant-coefficient differential operator on $\mathbf{R}^n$ then as a well-known consequence of the existence of a fundamental solution for L we have that the subspace $L\mathscr{D}(\mathbf{R}^n)$ is closed in $\mathscr{D}(\mathbf{R}^n)$. Thus by Theorem 7.1,

$\mathscr{D}_\delta(\mathfrak{p})^\wedge$ is a closed subspace of $\mathscr{D}_\delta(\Xi_o)$ and therefore equals its double annihilator in $\mathscr{E}_\delta(\Xi_o)$. Hence by (7.14),

$$\mathscr{D}_\delta(\mathfrak{p})^\wedge = (\mathscr{N}_\delta)^\perp \qquad \text{(annihilator in } \mathscr{D}_\delta(\Xi_o)). \tag{7.15}$$

The inclusion $\subset$ in (7.13) being obvious, let $\phi \in \mathscr{N}$. Consider the decomposition

$$\phi = \sum_{\delta \in \hat{K}_M} d(\delta)(\chi_\delta * \phi), \tag{7.16}$$

where

$$d(\delta)(\chi_\delta * \phi)(H, kM) = d(\delta) \int_K \phi(H, ukM)\, \chi_\delta(u^{-1})\, du$$
$$= \mathrm{Tr}(\phi^\delta(H, kM))$$

with ϕ^δ as in (6.2). The function

$$d(\delta)\, \chi_\delta * \phi \tag{7.17}$$

belongs to $(\mathscr{N}_\delta)^\perp = \mathscr{D}_\delta(\mathfrak{p})^\wedge$ so by Theorem 7.1,

$$d(\delta)(\chi_\delta * \phi)(H, kM) = \mathrm{Tr}(\delta(k)(J^\delta(D)\Psi)(H)),$$

where $\Psi \in \mathscr{D}(\mathfrak{a}, \mathrm{Hom}(V_\delta, V_\delta^M))$ is W-invariant. This means that

$$\mathrm{Tr}(\delta(k)\, d(\delta) \int_K \phi(H, uM)\, \delta(u^{-1})\, du) = \mathrm{Tr}(\delta(k)\, J^\delta(D)\Psi)$$

so $\phi^\delta(H, eM) = (J^\delta(D)\Psi)(H)$ and

$$\phi^\delta(H, kM) = \delta(k)(J^\delta(D)\Psi)(H). \tag{7.18}$$

Putting

$$\phi^*(\lambda, kM) = \int_{\mathfrak{a}} \phi(H, kM)\, e^{-i\lambda(H)}\, dH \tag{7.19}$$

we obtain from (7.18)

$$d(\delta) \int_K \phi^*(\lambda, kM)\, \delta(k^{-1})\, dk = \int (J^\delta(D)\Psi)(H)\, e^{-i\lambda(H)}\, dH$$
$$= J^\delta(\lambda)\, \tilde{\Psi}(\lambda).$$

Now Theorem 3.1 shows that

$$\phi^*(\lambda, kM) = \tilde{f}(k \cdot A_\lambda)$$

for some $f \in \mathscr{D}(\mathfrak{p})$. Then (7.19) and Lemma 3.1 show that $\hat{f}$ is a constant multiple of ϕ. Hence $\phi \in \mathscr{D}(\mathfrak{p})^\wedge$ and the theorem is proved.

Next we have to extend this theorem to distributions. If $T \in \mathscr{E}'(\mathfrak{p})$ its Radon transform $\hat{T}$ is defined by

$$\hat{T}(\phi) = T(\check{\phi}), \qquad \phi \in \mathscr{D}(\Xi_o). \tag{7.20}$$

This definition is motivated by (7.5). Suppose T has support inside the ball $B_R(0)$. If $\phi(H, kM) = 0$ for $|H| \leqslant R$ and all $k \in K$ then $\check{\phi}(Y) = 0$ for $|Y| \leqslant R$ so $\hat{T}(\phi) = 0$. Thus $\hat{T} \in \mathscr{E}'(\Xi_o)$.

Theorem 7.4. *The range $\mathscr{E}'(\mathfrak{p})^\wedge$ is given by*

$$\mathscr{E}'(\mathfrak{p})^\wedge = \{\Sigma \in \mathscr{E}'(\Xi_o) : \Sigma(\mathscr{N}) = 0\}. \tag{7.21}$$

Proof. If we combine the inversion formula (5.1) with (7.5) we obtain the Plancherel formula

$$\int_{\mathfrak{p}} f(Y)\, g(Y)\, dY = c \int_{\Xi_o} (\Lambda_o \hat{f})(\xi)\, \hat{g}(\xi)\, d\xi. \tag{7.22}$$

We shall use this, the support theorem, and (7.13) to prove Theorem 7.4. The inclusion $\subset$ in (7.21) being obvious suppose now $\Sigma \in \mathscr{E}'(\Xi_o)$ satisfies $\Sigma(\mathscr{N}) = 0$. Motivated by (7.22) we define the distribution S on $\mathfrak{p}$ by

$$S(f) = c\Sigma(\Lambda_o \hat{f}), \qquad f \in \mathscr{D}(\mathfrak{p}). \tag{7.23}$$

We shall show that S has compact support and that $\hat{S} = \Sigma$. For $F \in \mathscr{D}(G_o)$ we define $\sigma \in \mathscr{D}'(\Xi_o)$ by

$$\sigma(\phi) = \int_{G_o} F(g)\, \Sigma^{g^{-1}}(\phi)\, dg, \tag{7.24}$$

where dg is a Haar measure on G_o. Clearly σ has compact support. Consider the map $F \in \mathscr{D}(G_o) \to \dot{F} \in \mathscr{D}(\Xi_o)$ where

$$\dot{F}(gH) = \int_H F(gh)\, dh$$

with $H = M'\mathfrak{q}$. Let $\Psi \in \mathscr{D}'(\Xi_o) \to \tilde{\Psi} \in \mathscr{D}(G_o)$ denote the adjoint. Select $\Phi \in \mathscr{D}(G_o)$ such that $\dot{\Phi} = \phi$. Then $\Sigma(\phi^g) = \tilde{\Sigma}(\Phi^{L(g)})$ so

$$\sigma(\phi) = \int_{G_o} \left(\int_{G_o} F(g)\, \Phi(g^{-1}h)\, dg \right) d\tilde{\Sigma}(h) = (F' * \tilde{\Sigma})(\Phi), \tag{7.25}$$

where $F'(g)=F(g^{-1})$ and $*$ denotes convolution

$$(F * T)(g)=\int_{G_o} F(gh^{-1})\, dT(h).$$

Now (7.25) shows that σ is a function so by the above, $\sigma \in \mathscr{D}(\Xi_o)$. Of course $\mathscr{N}$ is G_o-invariant so $\sigma(\mathscr{N})=0$, whence by Theorem 7.3, $\sigma=\hat{s}$ where $s \in \mathscr{D}(\mathfrak{p})$.

Since the Radon transform $f \rightarrow \hat{f}$ and the operator Λ_o commute with the action of G_o we have

$$S^g(f)=c\Sigma^g(\Lambda_o \hat{f}). \tag{7.26}$$

Invoking the Plancherel formula (7.22) we get by (7.24)

$$\int_{\mathfrak{p}} s(Y) f(Y)\, dY=c\sigma(\Lambda_o \hat{f})=\int_{G_o} F(g)\, S^{g^{-1}}(f)\, dg \tag{7.27}$$

so

$$s(gK)=(F' * \tilde{S})(g). \tag{7.28}$$

Since $F \in \mathscr{D}(G)$ was arbitrary, (7.28) proves that

$$\mathscr{D}(G_o) * \tilde{S} \subset \mathscr{D}(G_o).$$

Actually, (7.28) gives the following more precise statement since the supports of s and $\sigma=\hat{s}$ are directly related by Theorem 6.1: For each compact neighborhood C of e in G, there exists another compact $C' \subset G$ such that $C_1 \subset C_2 \Rightarrow C_1' \subset C_2'$ and

$$\mathscr{D}_C(G_o) * \tilde{S} \subset \mathscr{D}_{C'}(G_o). \tag{7.29}$$

Letting C shrink to $\{e\}$ we deduce from (7.29) that $\tilde{S}$, and hence also S, has compact support. This being so, (7.27) remains valid for $f \in \mathscr{E}(\mathfrak{p})$, in particular for $f=\check{\phi}$. Then

$$\int_{G_o} F(g)\, S^{g^{-1}}(\check{\phi})\, dg=\int_{\mathfrak{p}} s(Y)\, \check{\phi}(Y)\, dY=\sigma(\phi)=(F' * \tilde{\Sigma})(\Phi).$$

Letting F approximate the delta function at $\{e\}$ we deduce

$$S(\check{\phi})=\tilde{\Sigma}(\Phi)=\Sigma(\phi)$$

so $\hat{S}=\Sigma$ as desired.

COROLLARY 7.5. *The dual transform $\phi \to \check{\phi}$ is surjective*:

$$\mathscr{E}(\Xi_o)^\vee = \mathscr{E}(X_o).$$

In fact $\phi \in \mathscr{E}(\Xi_o) \to \check{\phi} \in \mathscr{E}(X_o)$ is a continuous map of Fréchet spaces. By a standard theorem on Fréchet spaces it must be surjective because the adjoint map $T \in \mathscr{E}'(X_o) \to \hat{T} \in \mathscr{E}'(\Xi_o)$ is one-to-one and (by Theorem 7.4) has closed range.

Remark. It is known from Gonzalez [2] that every smooth point function on X_o has the form

$$f(Y) = \int_{\mathbf{O}(n)} \psi(Y + u \cdot \xi_o)\, du, \qquad Y \in X_o,$$

where ψ is a smooth function on the space $\mathbf{G}(d, n)$ of all affine d-planes in X_o. For $d = n-1$ this was first proved by A. Hertle [11]. Corollary 7.5 is a strengthening of this result (for $d = \dim \mathfrak{q}$, $n = \dim \mathfrak{p}$) in that only a certain lower dimensional family of d-planes is required.

Correction. At the end of the proof of Theorem 6.6 in [7] a correction is needed. The choice $\tilde{T} = X^*$ should be replaced by: For $\tilde{T}$ we choose a K-invariant polynomial such that $(X\tilde{T})(A_{-\lambda}) \neq 0$. Since $X \in \mathfrak{p}(\lambda)$ is transversal to the tangent space $(k \cdot A_{-\lambda})_{A_{-\lambda}}$ this choice is possible. Thus both conditions (6) and (8) are satisfied.

REFERENCES

1. S. G. GINDIKIN, Unitary representations of groups of automorphisms of Riemann symmetric spaces of null curvature, *Functional Anal. Appl.* **1** (1967), 28–32.
2. F. GONZALEZ, Invariant differential operators and the range of the Radon d-plane transform, *Trans. Amer. Math. Soc.*, in press.
3. S. HELGASON, The Radon transform on Euclidean spaces, compact two-point homogeneous spaces and Grassmann manifolds, *Acta Math.* **113** (1965), 153–180.
4. S. HELGASON, A duality in integral geometry on symmetric spaces, *in* "Proc. U.S.–Japan Seminar in Differential Geometry, Kyoto 1965," pp. 37–56, Nippon Hyronsha, Tokyo, 1966.
5. S. HELGASON, The surjectivity of invariant differential operators on symmetric spaces, *Ann. of Math.* (2) **98** (1973), 451–480.
6. S. HELGASON, A duality for symmetric spaces with applications to group representations, II, *Adv. Math.* **22** (1976), 187–219.
7. S. HELGASON, A duality for symmetric spaces with applications to group representations. III. Tangent space analysis, *Adv. Math.* **30** (1980), 297–323.
8. S. HELGASON, The range of the Radon transform on a symmetric space, *in* "Proc. Conf. Representation Theory of Reductive Groups, Utah 1982" (P. Trombi, Ed.), pp. 145–151.

9. S. Helgason, Operational properties of the Radon transform with applications, *in* "Proc. Conf. Differential Geometry with Appl. Nové Město, 1983," pp. 59–75.
10. S. Helgason, Some results on Radon transforms, Huygens' principle and X-ray transforms, *Contemp. Math.* **63** (1987), 151–177.
11. A. Hertle, On the range of the Radon transform and its dual, *Math. Ann.* **267** (1984), 91–99.
12. B. Kostant and S. Rallis, Orbits and Lie group representations associated to symmetric spaces, *Amer. J. Math.* **93** (1971), 753–809.
13. J. Orloff, Invariant Radon transform for a symmetric space, *Trans. Amer. Math. Soc.* **318** (1990), 581–660.

Printed by Catherine Press, Ltd., Tempelhof 41, B-8000 Brugge, Belgium

Reprinted from JOURNAL OF FUNCTIONAL ANALYSIS
All Rights Reserved by Academic Press, New York and London

Vol. 107, No. 2, August 1, 1992
Printed in Belgium

Huygens' Principle for Wave Equations on Symmetric Spaces

SIGURDUR HELGASON*

Department of Mathematics, Room 2-182,
Massachusetts Institute of Technology, Cambridge, Massachusetts 02139

Communicated by the Editors

Received August 7, 1991

Let $X = G/K$ be a symmetric space, L_X the Laplace–Betrami operator on X, and 2ρ the sum of the positive restricted roots of X (with multiplicity). Using the Fourier transform on X we write down the solution of the modified wave equation $\partial^2 u/\partial t^2 = (L_X + |\rho|^2)u$ on X. It is proved, using the Paley–Wiener theorem for this Fourier transform (S. HELGASON, *Ann. of Math.* **98** (1973), 451–480) that the equation satisfies Huygens' principle if dim X is odd and all Cartan subgroups of G are conjugate.

1. INTRODUCTION

In his work [5], Hadamard suggested the problem of deciding for which Riemannian manifolds X the wave equation satisfies Huygens' principle. If $X = \mathbf{R}^{2n+1}$ it is well known that the wave equation satisfies this principle. Hadamard proved that the condition dim $X =$ odd is a necessary condition. For a very readable report on the problem, see Günther's article [4a].

In their paper [10], Ólafsson and Schlichtkrull show that Huygens' principle holds for the wave equation on an odd-dimensional symmetric space X of the noncompact type for which the isometry group has all Cartan subgroups conjugate. The proof uses the theory of the Radon transform on X. Such a result, using similar methods, had been indicated without proof by Solomatina [11].

In the present note we give another independent proof of essentially the same result (Theorem 4.1) using the Fourier transform on X. This proof was inspired by the paper [1] of Branson and Ólafsson on the energy partition problem.

The case G complex had been settled in [6g] and [6h]; these papers also establish Huygens' principle for the modified wave equation on a

* Supported by the National Science Foundation, DMS-8805665.

0022-1236/92 $5.00

compact semisimple Lie group of odd dimension. For the case $G = \mathbf{SO}_o(2n+1, 1)$ explicit solution formulas are known [4b, 6g, 8, 9] implying, in particular, Huygens' principle.

I am greatly indebted to Ólafsson and Schlichtkrull for discussions about the problem and for sending me a copy of their paper [10] before publication.

2. Huygens' Principle

Let X be a Riemannian manifold with distance function d. As usual we denote the open ball and sphere, respectively, by

$$B_r(x) = \{y \in X : d(x, y) < r\}, \qquad S_r(x) = \{y \in X : d(x, y) = r\}.$$

Let L_X be the Laplace–Beltrami operator on X and $P(d/dt)$ a differential operator on $\mathbf{R}$. Given $f \in \mathscr{D}(X) = \mathscr{C}_c^\infty(X)$ we consider the Cauchy problem

$$P\left(\frac{d}{dt}\right) u = L_X u, \qquad u(x, 0) = 0, \; u_t(x, 0) = f(x). \tag{1}$$

Huygens' principle is said to hold for (1) if for each $\varepsilon > 0$

$$u(x, t) \text{ depends only on the restriction } f \mid V_{t,\varepsilon}(x), \tag{2}$$

$V_{t,\varepsilon}(x)$ denoting the shell $B_{t+\varepsilon}(x) - B_{t-\varepsilon}(x)$. If $X = \mathbf{R}^n$ and $P(d/dt) = d^2/dt^2$, (1) is the standard wave equation and the solution u is given by the Poisson–Tedone formula (for $n \geqslant 2$)

$$u(x, t) = \frac{1}{(n-2)!} \frac{\partial^{n-2}}{\partial t^{n-2}} \int_0^t (M^r f)(x)\, r(t^2 - r^2)^{(n-3)/2}\, dr \tag{3}$$

(cf. [7, p. 33]), $(M^r f)(x)$ denoting the average of f on $S_r(x)$. If n is odd the right hand side of (3) reduces to an expression

$$u(x, t) = \sum_{k=0}^{(n-3)/2} a_k t^{k+1} \frac{d^k}{dt^k} (M^t f)(x),$$

so Huygens' principle does indeed hold.

3. Radon and Fourier Transforms on Riemannian Symmetric Spaces

Let $X = G/K$ be a symmetric space of the noncompact type, i.e., G is a connected semisimple Lie group with finite center and K is a maximal compact subgroup. Consider an Iwasawa decomposition $G=NAK$ where N is nilpotent, A abelian, and let $\mathfrak{g}=\mathfrak{n}+\mathfrak{a}+\mathfrak{k}$ be the corresponding Lie algebra decomposition. Let M be the centralizer of A in K. If $g\in G$ we determine $A(g)\in\mathfrak{a}$ by $g\in N\exp A(g)K$ and put $A(gK, kM)=A(k^{-1}g)$. Let $\rho\in\mathfrak{a}^*$ be determined in terms of the Jacobian of the map $n\to ana^{-1}$ by

$$\rho(\log a)=\frac{1}{2}\log\left(\frac{d(ana^{-1})}{dn}\right), \qquad a\in A,$$

and let $B=K/M$.

The *Fourier transform* of a function f on X is defined by (cf. [6b])

$$\tilde{f}(\lambda, b)=\int_X f(x)\, e^{(-i\lambda+\rho)(A(x,b))}\, dx \tag{1}$$

for all $\lambda\in\mathfrak{a}_c^*$, $b\in B$ for which the integral converges. Here dx is the volume element on X. A *horocycle* in X is by definition an orbit of a subgroup of G conjugate to N. The group G permutes the horocycles transitively; more precisely, if $o=\{K\}$ and $\xi_0=N\cdot o$ the mapping

$$(kM, a)\to ka\cdot\xi_o$$

is a bijection of $K/M\times A$ onto the space Ξ of horocycles. The Radon transform of a function f on X is defined by (cf. [6a])

$$\hat{f}(\xi)=\int_\xi f(x)\, dm(x) \tag{2}$$

for all $\xi\in\Xi$ for which the integral exists, dm being the volume element on the submanifold ξ. The two transforms are related by

$$\tilde{f}(\lambda, kM)=\int_A \hat{f}(ka\cdot\xi_o)\, e^{(-i\lambda+\rho)(\log a)}\, da, \tag{3}$$

da being a suitably normalized Haar measure on A.

If ϕ is a continuous function on Ξ its dual transform $\check{\phi}$ is defined by

$$\check{\phi}(gK)=\int_K \phi(gk\cdot\xi_o)\, dk, \tag{4}$$

which is the average of ϕ over the set of horocycles passing through the point gK in X. We then have for a fixed G-invariant measure $d\xi$ on Ξ,

$$\int_X f(x)\,\check{\phi}(x)\, dx=\int_\Xi \hat{f}(\xi)\,\phi(\xi)\, d\xi. \tag{5}$$

The Radon transform is inverted by the following formula: for a certain pseudodifferential operator L on Ξ,

$$f = (L\hat{f})^{\vee} \qquad f \in \mathscr{D}(X) \tag{6}$$

(cf. [6b, c]). Furthermore, L is a differential operator exactly in the case when all Cartan subgroups of G are conjugate. As proved in [6d], the range of $\mathscr{D}(X)$ under the Fourier transform consists of the functions $\psi \in \mathscr{C}^{\infty}(\mathfrak{a}_c^* \times B)$ satisfying the following two conditions: (i) $\lambda \to \psi(\lambda, b)$ is a holomorphic function on $\mathfrak{a}_c^*$ satisfying

$$\sup_{(\lambda, b) \in \mathfrak{a}_c^* \times B} (1 + |\lambda|)^N e^{-R\,|\mathrm{Im}\,\lambda|}\, |\psi(\lambda, b)| < \infty \tag{7}$$

for some fixed $R > 0$ and all $N \geqslant 0$. (Here $\mathrm{Im}\,\lambda$ denotes the imaginary part of λ.) (ii) If W denotes the Weyl group of $(\mathfrak{g}, \mathfrak{a})$,

$$\int_B e^{(is\lambda + \rho)(A(x,b))} \psi(s\lambda, b)\, db = \int_B e^{(i\lambda + \rho)(A(x,b))} \psi(\lambda, b)\, db \tag{8}$$

for all $s \in W$, $x \in X$.

The proof also gives the following support theorem for the Radon transform [6d, Lemma 8.1]. Let

$$\beta_r = \{\xi \in \Xi : d(o, \xi) < R\}, \qquad \sigma_r = \{\xi \in \Xi : d(o, \xi) = r\}.$$

Then we have

$$f \in \mathscr{D}(X), \qquad \mathrm{supp}(\hat{f}) \subset \beta_r \Leftrightarrow \mathrm{supp}(f) \subset \overline{B_r(o)}. \tag{9}$$

We also need the definition of the Fourier transform $\tilde{T}$ and Radon transform $\hat{T}$ for all T in the space $\mathscr{E}'(X)$ of distributions of compact support. The analog of (1) is obviously

$$\tilde{T}(\lambda, b) = \int_X e^{(-i\lambda + \rho)(A(x,b))}\, dT(x), \tag{10}$$

and in accordance with (5) we define $\hat{T}(\phi) = T(\check{\phi})$. The range $\mathscr{E}'(X)^{\sim}$ is described similarly to $\mathscr{D}(X)^{\sim}$ above except that condition (7) holds with *one* $N < 0$ (cf. [6d, Theorem 8.5] or [2]). We now need the extension of (9) to distributions.

LEMMA 3.1. *Let $T \in \mathscr{E}'(X)$ satisfy the condition*

$$\mathrm{supp}(\hat{T}) \subset \beta_A.$$

Then

$$\operatorname{supp}(T) \subset \overline{B_A(o)}.$$

Proof. The proof is analogous to the one for the Euclidean case [6f, Chap. I, Theorem 2.20]. For $F \in \mathscr{D}(G)$, $\phi \in \mathscr{D}(\Xi)$, and $\sigma \in \mathscr{D}'(\Xi)$ consider the mixed convolution

$$(F \times \phi)(\xi) = \int_G F(g)\, \phi(g^{-1} \cdot \xi)\, dg, \; (F \times \sigma)(\phi) = \int_G F(g)\, \sigma(\phi^{g^{-1}})\, dg,$$

where $\phi^g(\xi) = \phi(g^{-1} \cdot \xi)$. Then we have

$$(F \times \phi)^{\vee}(x) = \int_G F(g)\, \check{\phi}(g^{-1} \cdot x)\, dg. \tag{11}$$

By the definition of $\hat{T}$ the support assumption on $\hat{T}$ implies

$$T(\check{\phi}) = 0 \tag{12}$$

for all $\phi \in \mathscr{D}(\Xi)$ with $\operatorname{supp}(\phi) \cap \beta_A = \varnothing$. Let $\varepsilon > 0$ and let $\phi \in \mathscr{D}(\Xi)$ satisfy $\operatorname{supp}(\phi) \cap \overline{\beta_{A+\varepsilon}} = \varnothing$. Let N_ε be a symmetric neighborhood of e in G such that $g \cdot \beta_A \subset \beta_{A+\varepsilon}$ for $g \in N_\varepsilon$. Then

$$g \cdot \operatorname{supp}(\phi) \cap \beta_A = \varnothing \qquad \text{for} \quad g \in N_\varepsilon.$$

If F is symmetric and $\operatorname{supp}(F) \subset N_\varepsilon$ we have

$$\operatorname{supp}(F \times \phi) \cap \beta_A = \varnothing. \tag{13}$$

Thus the distribution $F \times T$ defined by

$$(F \times T)(f) = \int_G F(g^{-1})\, T(f^g)\, dg$$

(where $f^g(x) = f(g^{-1}x)$) satisfies, since $F(g^{-1}) = F(g)$,

$$\begin{aligned}(F \times T)(\check{\phi}) &= \int_G F(g)\, T((\check{\phi})^g)\, dg \\ &= \int_G F(g) \left(\int_X \check{\phi}(g^{-1}x)\, dT(x) \right) dg = T((F \times \phi)^{\vee}) = 0\end{aligned}$$

by (12)–(13). This means that

$$(F \times T)^{\wedge}(\phi) = (F \times T)(\check{\phi}) = 0$$

so $(F \times T)^\wedge$ has support in $\overline{\beta_{A+\varepsilon}}$. On the other hand, $F \times T$ is a function, as we see by lifting T to a distribution on G. Thus the support theorem (9) implies

$$\operatorname{supp}(F \times T) \subset \overline{B_{A+\varepsilon}(o)}.$$

Letting $\varepsilon \to 0$ we get the desired conclusion.

Remark. The converse of Lemma 3.1 holds also as a result of (9) and the relation $(F \times T)^\wedge = F \times \hat{T}$.

4. The Wave Equation on X

With X as in Section 3 we consider the Cauchy problem for the modified wave equation on X,

$$\frac{\partial^2 u}{\partial t^2} = (L_X + |\rho|^2)u, \qquad u(x, 0) = 0, \; u_t(x, 0) = f(x), \tag{1}$$

and prove the following result.

THEOREM 4.1. *Let X be an odd-dimensional symmetric space of the non-compact type and suppose G has all its Cartan subgroups conjugate. Then the modified wave equation* (1) *satisfies Huygens principle and the solution is given by* (8) *below.*

Proof. We take f in (1) in $\mathscr{D}(X)$ and apply the Fourier transform to the equation above. Then

$$\tilde{u}_{tt}(t, \lambda, b) + |\lambda|^2 \, \tilde{u}(t, \lambda, b) = 0 \tag{2}$$

so, taking the initial data into account,

$$\tilde{u}(t, \lambda, b) = \tilde{f}(\lambda, b) \frac{\sin |\lambda| \, t}{|\lambda|} \tag{3}$$

(cf. [1, Sect. 2]). Now $|\lambda|^{-1} \sin |\lambda| \, t$ is the Euclidean Fourier transform of a distribution T_t on A of compact support:

$$\frac{\sin |\lambda| \, t}{|\lambda|} = \int_A e^{-i\lambda(\log a)} \, dT_t(a). \tag{4}$$

For a quick proof of this in the present context consider the wave equation on A (as a Euclidean space):

$$L_A v = \frac{\partial^2 v}{\partial t^2}, \qquad v(a, 0) = 0, \qquad v_t(a, 0) = F(a), \qquad F \in \mathscr{D}(A).$$

Taking Fourier transform on A we obtain (compare (3))

$$\tilde{v}(\lambda, t) = \tilde{F}(\lambda) \frac{\sin |\lambda| t}{|\lambda|}. \tag{4'}$$

For t fixed the map $F \to v(\cdot, t)$ commutes with the translations on A so

$$v(a, t) = (F * T_t)(a), \qquad \text{where} \quad T_t \in \mathscr{D}'(A). \tag{4''}$$

From (3) Section 1 we deduce that

$$\operatorname{supp}(T_t) \subset \overline{B_t(e)} \qquad \text{if} \quad \dim A \geqslant 2$$

$$\operatorname{supp}(T_t) \subset \overline{S_t(e)} \qquad \text{if} \quad \dim A \geqslant 2, \text{ odd}.$$

Now (4) follows from (4′) and (4″). Note also that if $\dim A = 1$, (4) holds with T_t as half the characteristic function of $B_t(e)$. On the other hand, the Paley–Wiener theorem for $\mathscr{E}'(X)$ mentioned in Section 3 implies that

$$\frac{\sin |\lambda| t}{|\lambda|} = \int_X e^{(-i\lambda + \rho)(A(x, b))} \, d\tau_t(x), \tag{5}$$

where $\tau_t \in \mathscr{E}'(X)$. Integrating over b we get

$$\frac{\sin |\lambda| t}{|\lambda|} = \int_X \phi_{-\lambda}(x) \, d\tau_t(x),$$

where $\phi_{-\lambda}(x)$ is the spherical function. By the uniqueness, τ_t is K-invariant. So is its Radon transform $\hat{\tau}_t$, and we now relate it to T_t.

Lemma 4.2. *In terms of the product decomposition $\Xi = K/M \times A$ we have*

$$\hat{\tau}_t = 1 \otimes e^{\rho} T_t.$$

Proof. If $g = kan$ relative to the decomposition $G = KAN$ we put $a = \exp H(g)$. Let $\alpha \in \mathscr{C}^\infty(K/M)$, $\beta(a) = e^{(-i\lambda - \rho)(\log a)}$, and $\alpha^\natural$ the integral of α over K/M. Then by K-invariance,

$$\hat{\tau}_t(\alpha \otimes \beta) = \alpha^\natural \hat{\tau}_t(1 \otimes \beta) = \alpha^\natural \tau_t((1 \otimes \beta)^\vee)$$

$$(1 \otimes \beta)^\vee (g \cdot o) = \int_K (1 \otimes \beta)(gk \cdot \xi_o) \, dk = \int_K \beta(\exp H(gk)) \, dk = \phi_{-\lambda}(g)$$

so

$$\hat{\tau}_t(\alpha \otimes \beta) = \alpha^\natural \tau_t(\phi_{-\lambda}) = \alpha^\natural T_t(e^{-i\lambda}) = (1 \otimes e^\rho T_t)(\alpha \otimes \beta).$$

Since these $\alpha \otimes \beta$ span a dense subspace of $\mathscr{C}^\infty(\Xi)$ the lemma is proved.

LEMMA 4.3. *Suppose G has all its Cartan subgroups conjugate, and let $A > 0$. Then if $T \in \mathscr{E}'(X)$,*

$$\operatorname{supp}(\hat{T}) \subset \sigma_A \Rightarrow \operatorname{supp}(T) \subset S_A(o). \tag{6}$$

Proof. Let $\varepsilon > 0$ and suppose f satisfies $\operatorname{supp}(f) \subset B_{A-\varepsilon}(o)$. Then $\operatorname{supp}(\hat{f}) \subset \overline{\beta_{A-\varepsilon}}$ and since L in (6) Section 3 is now a differential operator, $\operatorname{supp}(L\hat{f}) \subset \overline{\beta_{A-\varepsilon}}$. Hence by (6) Section 3

$$T(f) = T((L\hat{f})^\vee) = \hat{T}(L\hat{f}) = 0$$

so $\operatorname{supp}(T) \cap B_{A-\varepsilon} = \varnothing$. Combining this with Lemma 3.1, (6) follows.

Remark. This lemma is a distribution counterpart to Proposition 1 of [10].

Now if dim A is odd and >1 we have as already noted $\operatorname{supp}(T_t) \subset S_t(e)$. Hence, by Lemma 4.3 we deduce for dim X odd and rank $X > 1$,

$$\operatorname{supp}(\tau_t) \subset S_t(o). \tag{7}$$

Now if $f \in \mathscr{D}(X)$ and the distribution $T \in \mathscr{E}'(X)$ is K-invariant we have for the convolution $f \times T$ on X

$$(f \times T)^\sim(\lambda, b) = \tilde{f}(\lambda, b)\, \tilde{T}(\lambda, b).$$

Hence by (3) $u(x, t) = (f \times \tau_t)(x)$ so by the symmetry of τ_t,

$$u(g \cdot o, t) = \int_X f(g \cdot x)\, d\tau_t(x) \tag{8}$$

and by (7) this proves Theorem 4.1 for rank $X > 1$.

If rank $X = 1$, then by the classification, X is a hyperbolic space and the result known as mentioned in the introduction.

While the above proof of Theorem 4.1 required the classification at the last point, the following variation does not require the classification.

THEOREM 4.4. *Let X and G satisfy the assumptions of Theorem* 4.1. *Then the modified wave equation*

$$\frac{\partial^2 v}{\partial t^2} = (L_X + |\rho|^2)v, \qquad v(x,0) = f(x), \qquad v_t(x,0) = 0 \tag{9}$$

satisfies Huygens' principle.

In fact, if u satisfies (1) then the function

$$v(g \cdot o, t) = u_t(g \cdot o, t) = \int_X f(g \cdot x)\, d\tau_t'(x) \tag{10}$$

satisfies (9). By Lemma 4.2

$$(\tau_t')^\wedge = 1 \otimes e^\rho T_t'$$

and from the properties recalled about T_t we have

$$\operatorname{supp}(T_t') \subset S_t(e),$$

both when $\dim A = 2k+1 > 1$ and when $\dim A = 1$. Thus by Lemma 4.3, $\operatorname{supp}(\tau_t') \subset S_t(0)$ and now (10) implies the theorem.

REFERENCES

1. T. P. BRANSON AND G. ÓLAFSSON, Equipartition of energy for waves in symmetric spaces, *J. Funct. Anal.* **97** (1991), 403–416.
2. M. EGUCHI, M. HASHIZUME, AND K. OKAMOTO, The Paley–Wiener theorem for distributions on symmetric spaces, *Hiroshima Math. J.* **3** (1973), 109–120.
3. I. M. GELFAND AND G. E. SHILOV, "Generalized Functions," Vol. I, Academic Press, New York, 1964.

4a. P. GÜNTHER, Huygens' principle and Hadamard's conjecture, *Math. Intelligencer* **13** (1991), 56–63.

4b. P. GÜNTHER, Über einige spezielle Probleme aus der Theorie der linearen partiellen Differentialgleichungen zweiter Ordnung, *Ber. Verh. Sächs. Akad. Wiss. Leipzig* **102** (1957), 1–50.

4c. P. GÜNTHER, "Huygens' Principle and Hyperbolic Equations," Perspectives in Math., Vol. 5, Academic Press, New York, 1988.

5. J. HADAMARD, "Lectures on Cauchy's Problem," New Haven, CT, Yale Univ. Press, 1923.

6a. S. HELGASON, Duality and Radon Transform for symmetric spaces, *Amer. J. Math.* **85** (1963), 667–692.

6b. S. HELGASON, Radon–Fourier transforms on symmetric spaces and related group representations, *Bull. Amer. Math. Soc.* **71** (1965), 757–763.

6c. S. HELGASON, A duality for symmetric spaces with applications to group representations, *Adv. Math.* **5** (1970), 1–154.

6d. S. HELGASON, The surjectivity of invariant differential operators on symmetric spaces, *Ann. of Math.* **98** (1973), 451–480.

6e. S. HELGASON, "Differential Geometry, Lie Groups, and Symmetric Spaces," Academic Press, New York, 1978.
6f. S. HELGASON, "Groups and Geometric Analysis. Integral Geometry, Invariant Differential Operators, and Spherical Functions," Academic Press, New York, 1984.
6g. S. HELGASON, Wave equations on homogeneous spaces, *in* "Lie group representations III" (R. Herb *et al.*, Eds.), pp. 254–287, Lecture Notes in Math., Vol. 1077, Springer-Verlag, Berlin, 1984.
6h. S. HELGASON, Solvability questions for invariant differential operators, *in* "Group Theoretical Methods in Physics, Montreal," Academic Press, New York, 1977.
7. F. JOHN, "Plane Waves and Spherical Means," Wiley (Interscience), New York, 1955.
8. I. A. KIPRIJANOV AND L. A. IVANOV, The Euler–Poisson–Darboux equation in a Riemannian space, *Soviet Math. Dokl.* **24** (1981), 331–335.
9. P. D. LAX AND R. S. PHILLIPS, Translation representation for the solution of the non-Euclidean wave equation, *Comm. Pure Appl. Math.* **32** (1979), 617–667.
10. G. ÓLAFSSON AND H. SCHLICHTKRULL, Wave propagation on Riemannian symmetric spaces, *J. Funct. Anal.*, to appear.
11. L. E. SOLOMATINA, Translation representation and Huygens principle for an invariant wave equation in a Riemannian symmetric space, *Soviet Math. Izv.* **30** (1986), 108–111.

Printed by Catherine Press, Ltd., Tempelhof 41, B-8000 Brugge, Belgium

Radon Transforms for Double Fibrations. Examples and Viewpoints.

SIGURDUR HELGASON
Massachusetts Institute of Technology
Cambridge, MA 02139, USA

1 Introduction

The title of Radon's pioneering paper *[25]* is (in translation): "On the determination of a function by its integrals over certain manifolds". Examples of problems suggested by this title were successfully considered by Fritz John in the thirties *[21, 22]* (lines and spheres in $\boldsymbol{R}^3$)and by Gelfand, Naimark and Graev in the fifties (*[6]*, *[8]*; conjugacy classes and their translates in the complex classical groups).

My interest in Radon's general problem goes back to the mid-fifties when I got acquainted with John's book "Plane Waves and Spherical Means". The idea was to look at Radon's problem from the point of view of group theory and thereby take advantage of Lie theory technics. While the term "group" does not appear in John's book, several of its themes are particularly inviting for group-theoretic variations.

Already in Funk's paper *[5]* it is shown that an even function on $\boldsymbol{S}^2$ is determined by its integrals over the great circles. In Radon's paper it is proved that a function on $\boldsymbol{R}^2$ is by an explicit formula determined by its line integrals. We have the following homogeneous space representations,

$$\boldsymbol{S}^2 = \boldsymbol{O}(3)/\boldsymbol{O}(2), \quad \{\text{great circles}\} = \boldsymbol{O}(3)/\boldsymbol{O}(2)\boldsymbol{Z}_2, \qquad (I)$$

$$\boldsymbol{R}^2 = \boldsymbol{M}(2)/\boldsymbol{O}(2), \quad \{\text{lines}\} = \boldsymbol{M}(2)/\boldsymbol{M}(1)\boldsymbol{Z}_2, \qquad (II)$$

where $\boldsymbol{O}(n)$ is the orthogonal group in $\boldsymbol{R}^n$ and $\boldsymbol{M}(n)$ the motion group.

In each of these two cases we are dealing with homogeneous spaces of the same group, namely $\boldsymbol{O}(3)$ and $\boldsymbol{M}(2)$, respectively. This serves as a basis for our generalization of the Radon transform to double fibrations, proposed in *[13]*.

Notation: We use the standard notation $\boldsymbol{R}$ and $\boldsymbol{C}$ for the real and complex numbers, respectively. If X is a topological space, $C(X)$ (resp. $C_c(X)$) denotes

Proceeding of the Conference
75 YEARS OF RADON TRANSFORM
 pages 175– 188

the space of complex-valued continuous (resp. compactly supported continuous) functions on X. If X is a manifold, $\mathcal{E}(X)$ (resp. $\mathcal{D}(X)$) denotes the space of complex-valued C^∞ functions (resp. compactly supported C^∞ functions) on X.

2 The Radon Transform for a Double Fibration.

Let G be a topological group, K and H closed subgroups. Consider the homogeneous spaces

$$X = G/K, \qquad \Xi = G/H \tag{1}$$

and put $L = K \cap H$. Now it is no longer so that the elements $\xi \in \Xi$ are subsets of X. However, we can use Chern's notion of *incidence [4]*: $x \in X$ and $\xi \in \Xi$ are incident if as cosets in G they intersect.

If $x \in X$, $\xi \in \Xi$ we put

$$\check{x} = \{\eta \in \Xi : \ x, \eta \text{ incident}\}, \quad \widehat{\xi} = \{y \in X : \ y, \xi \text{ incident}\}.$$

Then if $x = gK$, $\xi = \gamma H$ we have

$$\check{x} = \{gkH : \ k \in K\}, \qquad \widehat{\xi} = \{\gamma hK : \ h \in H\}. \tag{2}$$

Thus $\check{x}$ is the g translate of the orbit $K \cdot \{eH\} = K/L$ in Ξ and $\widehat{\xi}$ the γ translate of the orbit $H \cdot \{eK\}$ in X. This leads to the integral transforms

$$\widehat{f}(\gamma H) = \int_{H/L} f(\gamma hK)\,dh_L, \qquad f \in C_c(X), \tag{3}$$

$$\check{\phi}(gK) = \int_{K/L} \phi(gkH)\,dk_L, \qquad \phi \in C_c(\Xi), \tag{4}$$

dh_L and dk_L being fixed invariant measures on H/L and K/L, respectively. (In case of compactness, normalization is such that the total measure is 1).

Principal Problems. *Find kernels and ranges for the transforms $f \longrightarrow \widehat{f}$, $\phi \longrightarrow \check{\phi}$ and, if possible, find explicit inversion formulas.*

The following simple result is a group-theoretic condition under which the members of Ξ can be viewed as subsets of X and vice-versa.

Proposition 2.1: *Let*

$$K_H = \{k \in K : kH \cup k^{-1}H \subset HK\},\, H_K = \{h \in H : hK \cup h^{-1}K \subset KH\}.$$

Then the following properties are equivalent:

a) $K_H = H_K = K \cap H$.

b) The maps $x \longrightarrow \check{x}$ $(x \in X)$, $\xi \longrightarrow \widehat{\xi}$ $(\xi \in \Xi)$ *are one-to-one.*

c) The functions in $C_c(X)^\wedge$ *separate points on* Ξ *and the functions in* $C_c(\Xi)^\vee$ *separate points on* X.

Other simple general properties of the incidence are contained in the following propositions.

Proposition 2.2: *The following conditions are equivalent:*

a) $KH \cap HK = K \cup H$.

b) For any $x_1 \neq x_2$ *in* X *there is at most one* $\xi \in \Xi$ *incident to both.*

c) For any $\xi_1 \neq \xi_2$ *in* Ξ *there is at most one* $x \in X$ *incident to both.*

Proposition 2.3: *The following conditions are equivalent:*

a) $KHK = G$.

b) Any two $x_1, x_2 \in X$ *are incident to some* $\xi \in \Xi$.

Similar statement for $HKH = G$.

Assuming the groups L, K, H and G all unimodular so we in particular have invariant measures $dx = dg_K$, $d\xi = d\gamma_H$ on X and Ξ, respectively, the transforms (3) and (4) are adjoint operators (*[13]*):

$$\int_X f(x)\check{\phi}(x)dx = \int_\Xi \widehat{f}(\xi)\phi(\xi)d\xi. \tag{5}$$

We stress that even when the ξ can be viewed as subsets of X the incidence definition above is more general than just the inclusion $x \in \xi$. We shall now see that this increased generality is essential for an inversion formula even in the simple case of Funk's problem above. As in (I) we write

$$X = \boldsymbol{S}^2 = G/K, \quad \Xi = \{\text{great circles}\} = G/H,$$

where $G = \boldsymbol{O}(3)$, K is the subgroup fixing the North Pole x_0 and H is the subgroup of G mapping a certain great circle ξ_0 into itself. If $d(x_0, \xi_0) = p$ ($d =$ distance) then we find easily

$$x \text{ incident to } \xi \Longleftrightarrow d(x, \xi) = p. \tag{6}$$

Fixing p we write (3) and (4) in the form

$$\widehat{f}_p(\xi) = \int_{d(x,\xi)=p} f(x)dm(x), \qquad \check{\phi}_p(x) = \int_{d(x,\xi)=p} \phi(\xi)d\mu(\xi). \tag{7}$$

While Funk did not in *[5]* have a general inversion formula for f in terms of $\widehat{f}$ this can be given in terms of (7) (cf. *[20]*) as follows:

$$f(x) = \frac{1}{2\pi}\left[\frac{d}{du}\int_0^u (\widehat{f})^\vee_{\cos^{-1}(v)}(x)v(u^2 - v^2)^{-\frac{1}{2}}dv\right]_{u=1} \tag{8}$$

Note that $(\widehat{f})_p^\vee(x)$ is the average of the integrals of f over the great circles which have distance p along S^2 from x.

3 Examples with $G=SU(1,1)$.

We shall now consider several examples of (1) §1 with G the group $\boldsymbol{SU}(1,1)$ of matrices

$$\begin{pmatrix} a & b \\ \bar{b} & \bar{a} \end{pmatrix}, \quad |a|^2 - |b|^2 = 1.$$

operating on the unit disk $|z| < 1$ by

$$\begin{pmatrix} a & b \\ \bar{b} & \bar{a} \end{pmatrix} \cdot z = \frac{az+b}{\bar{b}z+\bar{a}}.$$

Consider the subgroups K, M, M', A, N, Γ given as follows:

$$K = \left\{ k_\theta = \begin{pmatrix} e^{i\theta} & 0 \\ 0 & e^{-i\theta} \end{pmatrix} : 0 \leq \theta < 2\pi \right\},$$

$$M = \{k_o, k_\pi\}, \qquad M' = \{k_o, k_\pi, k_{-\frac{\pi}{2}}, k_{\frac{\pi}{2}}\},$$

$$A = \left\{ \begin{pmatrix} ch\, t & sh\, t \\ sh\, t & ch\, t \end{pmatrix} : t \in \boldsymbol{R} \right\}, N = \left\{ \begin{pmatrix} 1+ix & -ix \\ ix & 1-ix \end{pmatrix} : x \in \boldsymbol{R} \right\},$$

$$\Gamma = C\boldsymbol{SL}(2, \boldsymbol{Z})C^{-1},$$

where C is the transformation $w \longrightarrow (w-i)(w+i)^{-1}$ mapping the upper-half plane onto the unit disk. The group G preserves the Riemannian structure

$$ds^2 = (1-x^2-y^2)^{-2}(dx^2+dy^2) \tag{1}$$

on the disk permuting the geodesics (the arcs orthogonal to $|z|=1$) transitively and permuting the horocycles (the circles inside the disk tangential to the boundary) transitively. Let d denote the distance on X.

Example I.

$$X = G/K, \qquad \Xi = G/M'A. \tag{2}$$

Here X is the disk, Ξ is the set of geodesics and the incidence reduces to $x \in \xi$. Here $f \longrightarrow \widehat{f}$ is the so called *X-ray transform.* The inversion is given by (cf. *[20]*)

$$f(x) = \frac{1}{\pi} \left[\frac{d}{du} \int_0^u (\widehat{f})^\vee_{\operatorname{lm} v}(x)(u^2-v^2)^{-\frac{1}{2}} dv \right]_{u=1} \tag{3}$$

where $\operatorname{lm} v = ch^{-1}(v^{-1})$. Note that the inversion involves the general incidence in analogy with (7) §2.

Example II. Here we take

$$X = G/K, \quad \Xi = G/MN. \tag{4}$$

Here Ξ is the set of horocycles in the disk X. Again incidence reduces to $x \in \xi$ and the transform $f \longrightarrow \widehat{f}$ is the *horocycle transform*. Let $\xi_{t,\theta}$ denote the horocycle through $e^{i\theta}$ having distance t (with sign) from the origin. Then by *[18]* p. 42 and *[14]*, p. 43

$$f = (S\widehat{f})^{\vee}, \tag{5}$$

where S is an operator in the variable t alone and is a convolution

$$S\phi = \frac{1}{\pi}(T' - 1) * \phi, \quad \phi \in \mathcal{D}(\boldsymbol{R}), \tag{6}$$

where T is the distribution

$$T\phi = \frac{1}{2}\int_{\boldsymbol{R}} (sht)^{-1}\phi(t)dt \quad \text{(principal value)} \tag{7}$$

and T' its derivative. Another inversion of $f \longrightarrow \widehat{f}$ is given in *[7]*, Ch. V, §2.

Example III. *A curious coincidence.*

Examples I and II have analogs for the hyperbolic space $X = \boldsymbol{H}^n$ of n-dimensions. We take the customary model of $\boldsymbol{H}^n$ as the ball $|x| < 1$ in $\boldsymbol{R}^n$ with the Riemannian structure

$$ds^2 = \frac{dx_1^2 + \cdots + dx_n^2}{(1 - x_1^2 - \cdots - x_n^2)^2}.$$

The totally geodesic hypersurfaces are the spherical caps perpendicular to the sphere $|x| = 1$ whereas the horocycles are the spheres tangential to $|x| = 1$. In both cases the isometry group $I(X)$ permutes these submanifolds transitively so we get two dual spaces

$$\Xi_1 = I(X)/H_1, \quad \Xi_2 = I(X)/H_2.$$

Let us denote the corresponding Radon transforms by $f \longrightarrow \widehat{f}$ and $f \longrightarrow \mathcal{R}f$, respectively. We also get two dual transforms

$$\phi \longrightarrow \check{\phi}, \quad \psi \longrightarrow \mathcal{R}^*\psi$$

by averaging a function ϕ on Ξ_1 (resp. ψ on Ξ_2) over the set of $\xi \in \Xi_1$ (resp. $\xi \in \Xi_2$) passing through a point in X.

In general $(\widehat{f})^{\vee}$ and $\mathcal{R}^*\mathcal{R}f$ are different point functions. However, in the case $n = 3$ they happen to coincide:

$$(\widehat{f})^{\vee} = \mathcal{R}^*\mathcal{R}f. \tag{8}$$

The inversion formula *[11]*

$$f = -\frac{1}{2\pi}(L+1)((\widehat{f})^{\vee}),$$

where L is the Laplace-Beltrami operator thus implies the same formula for $f \longrightarrow \mathcal{R}f$.

Example IV. Now we take

$$X = G/MAN, \quad \Xi = G/K. \tag{9}$$

Then Ξ is the disk and X its boundary. In this case each x is incident to each ξ. Now (3) takes the form

$$\widehat{f}(gK) = \int_{K/L} f(gkMAN)dk_L = \int_B f(g \cdot b)db, \tag{10}$$

where db is the K-invariant measure on the boundary B and $b \longrightarrow g \cdot b$ is the action of G on B. As is well known, the Poisson kernel

$$P(z,b) = \frac{1-|z|^2}{|z-b|^2}$$

is also given by the Jacobian,

$$P(g \cdot o, b) = \frac{d(g^{-1} \cdot b)}{db}.$$

Thus we deduce from (10)

$$\widehat{f}(z) = \int_B \frac{1-|z|^2}{|z-b|^2} f(b)db, \tag{11}$$

so the abstract Radon transform simply becomes the *Poisson integral.* The inversion formula is now just the classical Schwarz theorem from potential theory,

$$f(b) = \lim_{z \to b} \widehat{f}(z). \tag{12}$$

On the other hand, the transform $\phi \longrightarrow \check{\phi}$ is less interesting. Since the invariant measure on $MAN/M = AN$ is given by

$$\phi \longrightarrow \int_{AN} \phi(anK)da\,dn,$$

which happens to coincide with the non-Euclidean measure on the disk Ξ, the transform $\phi \longrightarrow \check{\phi}$ is simply

$$\check{\phi}(b) = \int_{\Xi} \phi(z)dz, \quad b \in B.$$

Example V. Here we take

$$X = G/K, \quad \Xi = G/H, \tag{13}$$

where H is the subgroup of G fixing a certain point $x_0 \neq o$. If $r = d(o, x_0)$ and ξ_0 is the circle $\{x \in X : d(x_0, x) = r\}$ we can identify Ξ with the set of non Euclidean circles in X of radius r. Here

$$\widehat{f}(gH) = \int_{H/L} f(ghK)dh_L = \int_{g\cdot\xi_0} f(x)dm(x)$$

the integral of f over the circle $g \cdot \xi_0$. The injectivity question for $f \longrightarrow \widehat{f}$ thus becomes the so-called *Pompeiu problem.* While the answer is positive for f sufficiently small at ∞ this is not so for $f \in C(X)$. However, if we take two $r_1, r_2 > 0$ (which do not satisfy a certain transcendental relation) the mapping

$$f \longrightarrow (\widehat{f}_{r_1}, \widehat{f}_{r_2})$$

is injective. For this see Berenstein-Zalcman *[2]*.

Example VI. Here we take

$$X = G/N, \quad \Xi = G/\Gamma. \tag{14}$$

Under the isomorphism

$$g \longrightarrow CgC^{-1} \text{ of } \boldsymbol{SL}(2, \boldsymbol{R}) \text{ onto } G$$

the group

$$N^* = \left\{ \begin{pmatrix} 1 & x \\ 0 & 1 \end{pmatrix} : x \in \boldsymbol{R} \right\}$$

corresponds to N. Since $\boldsymbol{SL}(2, \boldsymbol{R})/N^*$ is naturally identified with the punctured plane $\boldsymbol{R}^2 - \{0\}$ we could identify X with $\boldsymbol{R}^2 - \{0\}$. Here we put as usual $\Gamma_\infty = \Gamma \cap N$ and then (3) (4) in §2 take the form

$$\widehat{f}(g\Gamma) = \sum_{\Gamma/\Gamma_\infty} f(g\gamma N), \quad \check{\phi}(gN) = \int_{N/\Gamma_\infty} \phi(gn\Gamma)dn_{\Gamma_\infty}. \tag{15}$$

The series in (15) is called *theta series* and the null space of $\phi \longrightarrow \check{\phi}$ consists by definition of the *cusp forms.* Because of (5),

$$\{\text{Cusp forms}\} = (C_c(X)^\wedge)^\perp, \tag{16}$$

with orthogonality $\perp$ referring to $L^2(\Xi)$.

These examples show that the Radon transform $f \longrightarrow \widehat{f}$ and its dual $\phi \longrightarrow \check{\phi}$ for the double fibration

$$\begin{array}{ccc} & G/L & \\ \swarrow & & \searrow \\ G/K & & G/H \end{array}$$

gives rise to a multitude of questions, even when we restrict G to the simple case $\boldsymbol{SU}(1,1)$. This establishes a higher viewpoint for a variety of problems. The advantage of this is that *one* result for just *one* example automatically suggests problems for all the others. What for instance is the range and kernel for the map $f \longrightarrow \widehat{f}$ given by the theta series (15)?

At the same time the detailed results stated for the examples show sufficient variety that they could hardly be fully captured by a general theory for the transforms (3) and (4) in §2. Thus it seems more profitable to consider our abstract setup in §2 as a framework for examples rather than as axioms for a general theory.

Example VII. We shall now discuss the X-ray transform in $\boldsymbol{R}^3$ and point out an interesting analogy with Example IV. Denoting by $\boldsymbol{M}(3)$ the group of isometrics of $\boldsymbol{R}^3$, we have

$$X = \boldsymbol{R}^3 = \boldsymbol{M}(3)/\boldsymbol{O}(3), \quad \Xi = \{\text{lines}\} = \boldsymbol{M}(3)/H, \tag{17}$$

where H is the subgroup mapping a line, say the x_1-axis, into itself. Writing a line $\xi \in \Xi$ in the form

$$\xi: \ x_1 = \alpha_1 t + \beta_1, \quad x_2 = \alpha_2 t + \beta_2, \quad x_3 = t,$$

the X-ray transform $f \longrightarrow \widehat{f}$ is given by

$$\widehat{f}(\xi) = (\alpha_1^2 + \alpha_2^2 + 1)^{\frac{1}{2}} \int_{\boldsymbol{R}} f(\alpha_1 t + \beta_1, \alpha_2 t + \beta_2, t)\, dt. \tag{18}$$

Then by differentiating, we obtain

$$\left(\frac{\partial^2}{\partial\alpha_1 \partial\beta_2} - \frac{\partial^2}{\partial\alpha_2 \partial\beta_1} \right) \left((\alpha_1^2 + \alpha_2^2 + 1)^{-\frac{1}{2}} \widehat{f} \right) = 0. \tag{19}$$

John proved in *[22]* that, roughly speaking, this differential equation characterizes the range. An improved version was given by Richter *[26]* as follows (and extended by Gonzalez *[9]* to k-planes in $\boldsymbol{R}^n$). In analogy with the space $\mathcal{S}(\boldsymbol{R}^3)$ of rapidly decreasing functions on $\boldsymbol{R}^3$ one can define the space $\mathcal{S}(\Xi)$. Then for a certain explicit fourth degree $\boldsymbol{M}(3)$-invariant differential operator $\Box$ on Ξ the following result holds.

Theorem 3.1:

$$\mathcal{S}(\boldsymbol{R}^3)^\wedge = \{\phi \in \mathcal{S}(\Xi) : \quad \Box\phi = 0\}. \tag{20}$$

We compare this now to the situation in Example IV. By a standard theorem about the Poisson integral we have (Ξ being now the unit disk with Laplacian L) :

$$L^\infty(B)^\wedge = \{\phi \in L^\infty(\Xi) : \ L\phi = 0\}. \tag{21}$$

This striking analogy between (20) and (21) goes further if we ask for the analog for line functions of Gauss' mean value theorem for harmonic functions:

$$L\phi = 0 \Longleftrightarrow \int_S f(s)dw(s) = f(s_o)$$

if S is a circle with center s_o and dw the normalized invariant measure. This analog is another result from John's paper *[22]*:

Theorem 3.2: *A line function ϕ satisfies $\Box\phi = 0$ if and only if it has the following mean value property. For each hyperboloid of revolution H of one sheet the mean values of ϕ over the two families of generating lines of H are equal. (The variable of integration is the polar angle in the equatorial plane of H).*

John proves this by transforming the differential equation (19) to the ultrahyperbolic equation

$$\frac{\partial^2 u}{\partial x_1^2} + \frac{\partial^2 u}{\partial x_2^2} - \frac{\partial^2 u}{\partial x_3^2} - \frac{\partial^2 u}{\partial x_4^2} = 0 \tag{22}$$

on which he then uses Ásgeirsson's mean value theorem *[1]*.

4 Horocycles in a Symmetric Space.

In this section we take X as a symmetric space and Ξ the space of horocycles in X. This means

$$X = G/K, \quad \Xi = G/MN, \tag{1}$$

where G is a connected semisimple Lie group with finite center, K a maximal compact subgroup, $G = KAN$ an Iwasawa decomposition of G (N nilpotent, A abelian) and M the centralizer of A in K. The horocycles are the orbits in X of a group of the form gNg^{-1} ($g \in G$). If $o = eK$ is the origin in X and ξ_0

the horocycle $N \cdot o$, MN is the isotropy subgroup of ξ_0. If $g \in G$, let $k(g) \in K$, $n(g) \in N$ and $H(g) \in \mathfrak{a}$ (the Lie algebra of A) be determined by

$$g = k(g) \exp H(g) n(g). \tag{2}$$

The Radon transform and its dual are defined by (3) (4) in §2 with $H = MN$. Concerning the principal problems stated in §2 the range $\mathcal{D}(X)^\wedge$ and inversion formula for $f \longrightarrow \widehat{f}$ is given in *[15]* and *[13]*, respectively; the range $\mathcal{E}(\Xi)^\vee$ and the kernel $\mathcal{N}$ of the dual transform $\phi \longrightarrow \check{\phi}$ are given in *[19]* and *[17]*, respectively. Actually,

$$\mathcal{E}(\Xi)^\vee = \mathcal{E}(X). \tag{3}$$

While the description of the range $\mathcal{D}(X)^\wedge$ is a bit complicated, crisper results are obtained in *[16]* by considering the range $\mathcal{D}_\delta(X)^\wedge$ of the subspace of $\mathcal{D}(X)$ of functions of type δ, where δ is a fixed irreducible representation of K. In particular, if δ is the identity representation we get the Abel transform

$$(\mathcal{A}f)(a) = e^{\rho(\log a)} \int_N f(an \cdot o) dn, \tag{4}$$

where 2ρ is the sum of the positive restricted roots with multiplicity. This $\mathcal{A}$ is a homeomorphism of $\mathcal{D}_K(X)$ (the space of K-invariant functions in $\mathcal{D}(X)$) onto $\mathcal{D}_W(A)$ (the space of W-invariant functions in $\mathcal{D}(A)$, W being the Weyl group). The transpose $\mathcal{A}^*$ is then a bijection of the distribution space $\mathcal{D}'_W(A)$ onto $\mathcal{D}'_K(X)$.

The following result follows fairly simply from (3).

Theorem 4.1: *The restriction $\mathcal{A}^*|\mathcal{E}_W(A)$ is given by*

$$(\mathcal{A}^*\phi)(gK) = \int_K \phi(\exp H(gk)) e^{-\rho(H(gk))} dk$$

and it is a bijection of $\mathcal{E}_W(A)$ onto $\mathcal{E}_K(X)$.

For X of rank one, a part of this result is proved by Bray and Solmon *[3]*.

5 Conical Distributions and Intertwining Operators.

Let $\boldsymbol{D}(X)$ and $\boldsymbol{D}(\Xi)$ denote the algebras of G-invariant differential operators on X and Ξ, respectively. While the *spherical functions* on X are, by definition, the K-invariant joint eigenfunctions of $\boldsymbol{D}(X)$, the *conical distributions* on Ξ are defined as the MN-invariant joint eigen-distributions of $\boldsymbol{D}(\Xi)$. While the set of spherical functions is parametrized by $\mathfrak{a}_c^*/W$ (*[10]*) it is proved in *[14]*, Ch. III, §4 that the set of conical distributions is, roughly speaking, parametrized by $\mathfrak{a}_c^* \times W$. We recall briefly the construction.

Let $\overline{N} = \theta N$ where θ is the Cartan involution of G with fixed point group K. For $s \in W$ let m_s denote a representative in K. Put

$$\overline{N}_s = \overline{N} \cap m_s^{-1} N m_s \tag{1}$$

and define the distribution $S'_{\lambda,s}$ on $B = K/M$ by

$$S'_{\lambda,s}(F) = \int_{\overline{N}_s} F(m_s k(\bar{n}_s)M) e^{-(i\lambda+\rho)(H(\bar{n}_s))} d\bar{n}_s \tag{2}$$

for those $\lambda \in \mathfrak{a}_{\mathbf{C}}^*$ for which the integral converges. If $\boldsymbol{d}_s(\lambda)$ is the numerator in the formula for the partial $\boldsymbol{c}$-function $\boldsymbol{c}_s(\lambda)$ then the mapping

$$\lambda \longrightarrow S_{\lambda,s} = \boldsymbol{d}_s(\lambda)^{-1} S'_{\lambda,s} \tag{3}$$

extends to an entire function on $\mathfrak{a}_{\mathbf{C}}^*$ with values in $\mathcal{D}'(B)$. This is proved in *[14]*, Ch. III, §4; part of the proof is based on a method of Schiffmann *[26]* where he gives an important generalization of the Gindikin-Karpelevič product formula for the $\boldsymbol{c}$-function.

In accordance with the product decomposition $\Xi = K/M \times A$, in which $d\xi = e^{2\rho(\log a)} dk_M\, da$, the tensor product

$$\Psi_{\lambda,s} = S_{\lambda,s} \otimes e^{i\lambda-\rho} \tag{4}$$

is given by

$$\Psi_{\lambda,s}(\phi) = \int_{K/M\times A} \phi(ka\cdot\xi_0) e^{(i\lambda+\rho)(\log a)} dS_{\lambda,s}(k_M) da \tag{5}$$

and $\Psi_{\lambda,s}$ is a conical distribution on Ξ. Let $c_D(\lambda) \in \boldsymbol{C}$ be determined by

$$D\Psi_{\lambda,s} = c_D(\lambda)\Psi_{\lambda,s}, \quad D \in \boldsymbol{D}(\Xi).$$

Then $c_D(\lambda)$ is independent of s and as λ runs through $\mathfrak{a}_{\mathbf{C}}^*$ the spaces

$$\mathcal{D}'_\lambda(\Xi) = \{\Psi \in \mathcal{D}'(\Xi) : \ D\Psi = c_D(\lambda)\Psi \text{ for } D \in \boldsymbol{D}(\Xi)\}$$

constitute all the joint eigenspaces of $\boldsymbol{D}(\Xi)$. Actually,

$$\mathcal{D}'_\lambda(\Xi) = \mathcal{D}'(B) \otimes e^{i\lambda-\rho} \tag{6}$$

in the sense of (4), (5). The functions in the subspace

$$\mathcal{K}_\lambda(\Xi) = L^2(B) \otimes e^{i\lambda-\rho} \tag{7}$$

are characterized by

$$\psi(ga\cdot\xi_0) = e^{(i\lambda-\rho)(\log a)} \psi(g\cdot\xi_0), \tag{8}$$

$$\int_K |\psi(k \cdot \xi_0)|^2 dk < \infty. \tag{9}$$

Denoting by τ_λ and $\tilde{\tau}_\lambda$ the natural representations of G on $\mathcal{K}_\lambda(\Xi)$ and on $\mathcal{D}'_\lambda(\Xi)$, respectively, the family $\{\tau_\lambda : \lambda \in \mathfrak{a}_c^*\}$ as well as the family $\{\tilde{\tau}_\lambda : \lambda \in \mathfrak{a}_c^*\}$ is called the *spherical principal series.*

Let $\times$ denote the convolution product on K/M. The following result is proved in *[14]*, Ch. III, §6 giving the construction of the intertwining operators for the spherical principal series.

Theorem 5.1: *Assume $\tilde{\tau}_\lambda$ irreducible. Then the mapping*

$$A_{\lambda,s} : \ S \otimes e^{i\lambda-\rho} \longrightarrow (S \times S_{s\lambda,s^{-1}}) \otimes e^{is\lambda-\rho} \tag{10}$$

is a bijection of $\mathcal{D}'_\lambda(\Xi)$ onto $\mathcal{D}'_{s\lambda}(\Xi)$ commuting with the action of G :

$$A_{\lambda,s}\tilde{\tau}_\lambda(g) = \tilde{\tau}_{s\lambda}(g)A_{\lambda,s}, \quad g \in G. \tag{11}$$

A somewhat different construction of the intertwining operators is given by Schiffmann *[26]* (and Knapp-Stein *[24]*). Schiffmann considers the intertwining integral

$$(I'_{\lambda,s}\psi)(g) = \int\limits_{N/(N\cap N^s)} \psi(gnm_s)d\dot{n}, \quad \psi \in \mathcal{K}_\lambda(\Xi), \tag{12}$$

where $N^s = m_s N m_s^{-1}$ and $d\dot{n}$ the invariant measure. Since the quotient can be identified with $\overline{N}^s = N \cap (m_s \overline{N} m_s^{-1})$ whereby $d\dot{n}$ goes over in $d\bar{n}^s$, and since the conjugation by m_s^{-1} sends the pair $(\overline{N}^s, d\bar{n}^s)$ to $(\overline{N}_s, d\bar{n}_s)$ we see that

$$(I'_{\lambda,s}\psi)(g) = \int\limits_{\overline{N}_s} \psi(gm_s\bar{n}_s \cdot \xi_0)d\bar{n}_s,$$

which is a function on Ξ. Moreover, it can be proved without difficulty that

$$I'_{\lambda,s}(F \otimes e^{i\lambda-\rho}) = (F \times (S'_{-\lambda,s})^*) \otimes e^{is\lambda-\rho}, \quad F \in \mathcal{E}(B), \tag{13}$$

where $S \longrightarrow S^*$ is the (well-defined) operation on *M-invariant* distributions on K/M induced by the diffeomorphism $k \longrightarrow k^{-1}$ of K. Since intertwining operators between irreducible representations are unique up to a scalar, (10) and (13) suggest a connection between $(S_{-\lambda,s})^*$ and $S_{s\lambda,s^{-1}}$. And indeed we have the following result.

Proposition 5.2: *For all $\lambda \in \mathfrak{a}_c^*$ and all $s \in W$,*

$$(S_{-\lambda,s})^* = S_{s\lambda,s^{-1}}.$$

For the case of $\boldsymbol{SU}(2,1)$ this rests on the fact that the involution $\bar{n} \longrightarrow \bar{n}_0$ of $\overline{N}$ given by

$$t_0 = -t, \quad z_0 = -t\frac{1+|z|^2+2it}{[(1+|z|^2)^2+4t^2]^{\frac{1}{2}}},$$

with

$$\bar{n} = \exp\begin{pmatrix} it & z & it \\ -\bar{z} & 0 & -\bar{z} \\ -it & -z & -it \end{pmatrix}, \quad t \in \boldsymbol{R},\ z \in \boldsymbol{C},$$

has the property: $H(\bar{n}_0) = H(\bar{n})$.

Proposition 5.2 now establishes the identity

$$A_{\lambda,s} = \boldsymbol{d}_s(-\lambda)^{-1} I'_{\lambda,s} \tag{14}$$

between the two constructions of intertwining operators.

References

[1] Àsgeirsson, L., *Über eine Mittelwertseigenschaft von Lösungen homogener linearer partieller Differentialgleichungen 2. Ordnung mit konstanten Koefficienten*, 321–346, 1937, **113**, Math. Ann.

[2] Berenstein, C. and L. Zalcman, *Pompeiu's problem on spaces of constant curvature*, 113–130, 1976, **30**, J. d'Analyse Math.

[3] Bray, W.O. and D.C. Solmon, *Paley-Wiener theorems on rank-one symmetric spaces of non- compact type*, 17–29, 1990, **113**, Contemp. Math.

[4] Chern, S.S., *On integral geometry in Klein spaces*, 178–189, 1942, **43**, Ann. of Math.

[5] Funk, P., *Über eine geometrische Anwendung der Abelschen Integralglerchung*, 129–135, 1916, **77**, Math. Ann.

[6] Gelfand, I.M. and Graev, M.I., *Analogue of the Plancherel formula for the classical groups*, 375–404, 1955, **4**, Trudy Moscow Mat. Obš č.

[7] Gelfand, I.M. Graev, M.I. and Vilenkin, N., *Generalized Functions*, 1966, **5**, Academic Press New York

[8] Gelfand, I.M. and M.A. Naimark, *Unitäre Darstellungen der Klassischen Gruppen*, 1957, Akademie- Verlag, Berlin

[9] Gonzalez, F., *Invariant differential operators and the range of the Radon d-plane transform*, 627–635, 1990, **287**, Math. Ann.

[10] Harish-Chandra, *Spherical functions on a semisimple Lie group I*, 241–310, 1958, **80**, Amer. J. Math.

[11] Helgason, S., *Differential operators on homogeneous spaces*, 239–299, 1959, **102**, Acta Math.

[12] Helgason, S., *Radon-Fourier transforms on symmetric spaces and related group representations* , 451–480, 1973, **98**, Ann. of Math.

[13] Helgason, S., *A duality in integral geometry on symmetric spaces*, Proc. U.S. - Japan Seminar in Differential Geometry, Kyoto, 1965, Nippon Hironsha, Tokyo, 1966.

[14] Helgason, S., *A duality for symmetric spaces with applications to group representations* ,1–154, 1970, **5**, Advan. Math.

[15] Helgason, S., *The surjectivity of invariant differential operators on symmetric spaces* , 451–480, 1973, **98**, Ann. of Math.

[16] Helgason, S., *The range of the Radon transform on a symmetric space*, 145–151, 1982, in Proc. Conf. Representation Theory of Reductive Groups, Utah

[17] Helgason, S., *Operational properties of the Radon transform with applications*, 59–75, 1983, in Proc. Conf. Differential Geometry with Appl. Nové Město

[18] Helgason, S., *Groups and Geometric Analysis* , 1984, Academic Press, Orlando

[19] Helgason, S., *Some results on Radon transforms, Huygens' principle and X-ray transforms* , 151–177, 1987, **63**, Contemp. Math.

[20] Helgason, S., *The totally-geodesic Radon transform on constant curvature spaces* , 141–149, 1990, **113**, Contemp. Math.

[21] John, F., *Abhängigkeiten zwischen den Flächenintegralen einer stetiger Funktion* , 541–559, 1935, **111**, Math. Ann.

[22] John, F., *The ultrahyperbolic differential equation with* 4 *independent variables*, 300–322, 1938, **4**, Duke Math. J.

[23] John, F., *Plane Waves and Spherical Means, Applied to partial Differential Equations* , 1955, Wiley (Interscience), New York

[24] Knapp, A.W. and Stein, E, *Intertwining operators for semisimple groups*, 489–578, 1971, **93**, Ann. of Math.

[25] Radon, J., *Über die Bestimmung von Funktionen durch ihre Integralwerte längs gewisser Mannigfaltigkeiten*, 262–277, 1917, **69**, Ber. Verh. Sächs. Akad.

[26] Schiffmann, G., *Integrales d'entrelacement et fonctions de Whittaker*, 3–72, 1971, **99**, Bull. Soc. Math. France

Integral Geometry and Multitemporal Wave Equations

SIGURDUR HELGASON
Massachusetts Institute of Technology

In Memory of Fritz John

Introduction

While the wave equation $(L - \partial^2/\partial t^2)u = 0$ on a Euclidean space has an analogue for any Riemannian manifold X, the case when X is a symmetric space offers another natural analogue where the operator $L - \partial^2/\partial t^2$ is replaced by a system (see (2.7) and (2.8)) and the time variable is now multidimensional. This was introduced by Semenov-Tjan-Šanskiĭ in [17] and was further developed by Shahshahani [18] and Phillips and Shahshahani [16] .

In the present paper we study the above system by means of Fourier transform methods described in Section 3. In particular, we show (Theorem 6.3) that the spectral representations $\mathcal{E}^\sigma$ are directly related to the Fourier transform on X. This is the basis of the proof of the spectral representation theorem (Theorem 6.4).

In Section 4 we prove a uniqueness theorem for the system and obtain some new formulas for the solution. In Section 5 we prove a characterization of incoming waves in terms of their sources. This is an analogue of results by Lax and Phillips for the flat case. This proof follows ideas of Phillips and Shahshahani for similar results but because of the available tools is a bit simpler. I am indebted to Shahshahani for useful discussions and to Schlichtkrull for some concrete suggestions, specified later.

The following notation will be used. $\mathbb{R}^n$ and $\mathbb{S}^n$ denote the Euclidean n-space and n-sphere, respectively. $\mathcal{S}(\mathbb{R}^n)$ denotes the space of rapidly decreasing functions on $\mathbb{R}^n$. If X is a manifold, $\mathcal{E}(X)$ and $\mathcal{D}(X)$, respectively, constitute the space of smooth functions on X and smooth functions of compact support. Their duals $\mathcal{D}'(X)$ and $\mathcal{E}'(X)$ constitute the space of distributions on X and distributions of compact support, respectively. Support is denoted supp. If σ is a diffeomorphism of X, $f \in \mathcal{E}(X)$, and J an operator on $\mathcal{E}(X)$, then we put

$$f^\sigma = f \circ \sigma^{-1}, \quad J^\sigma(f) = (J f^{\sigma^{-1}})^\sigma,$$

for $f \in \mathcal{E}(X)$. If X is a metric space, $B_r(x)$ denotes the open ball of radius r and center x and $S_r(x)$ denotes the corresponding sphere. If G is a group with Haar measure, then $*$ denotes the convolution on G; if $K \subset G$ is a compact

Communications on Pure and Applied Mathematics, Vol. LI, 1035–1071 (1998)
 CCC 0010–3640/98/091035-37

subgroup, then $\times$ denotes the convolution on G/K that is derived from $*$ on G. The translation of $gK \to hgK$ on G/K is denoted $\tau(h)$. The adjoint representation of a Lie group is denoted by Ad and ${}^{\mathsf{T}}A$ denotes the transpose of a linear transformation A.

1 The Wave Equation in $\mathbb{R}^n$

Let f be a function on $\mathbb{R}^n$ integrable over each hyperplane. Let $\widehat{f}(\xi)$ denote the integral

$$\widehat{f}(\xi) = \int_\xi f(x)dm(x) \tag{1.1}$$

of f over the hyperplane ξ, dm being the Euclidean measure. The map $f \to \widehat{f}$ is called the *Radon transform*. We can write the hyperplane ξ in the form $\langle x, \omega\rangle = p$ where $\langle\cdot,\cdot\rangle$ is the scalar product, ω a unit vector, and $p \in \mathbb{R}$. Defining $\widehat{f}(\omega, p) = \widehat{f}(\xi)$, $\widehat{f}$ becomes a function on $\mathbb{S}^{n-1} \times \mathbb{R}$ satisfying the symmetry condition $\widehat{f}(-\omega, -p) = \widehat{f}(\omega, p)$. If φ is a smooth function on $\mathbb{S}^{n-1} \times \mathbb{R}$ that satisfies this symmetry condition $\varphi(-\omega, -p) = \varphi(\omega, p)$ and in addition is rapidly decreasing in p, we define the operator Λ by

$$(\Lambda\varphi)(\omega, p) = \begin{cases} \frac{d^{n-1}}{dp^{n-1}}\varphi(\omega, p) & n \text{ odd}\,, \\ \mathcal{H}_p \frac{d^{n-1}}{dp^{n-1}}\varphi(\omega, p) & n \text{ even}\,, \end{cases}$$

where $\mathcal{H}$ is the Hilbert transform

$$(\mathcal{H}F)(t) = \frac{i}{\pi}\int_{\mathbb{R}} \frac{F(p)}{p - t}\, dp\,.$$

The function $\Lambda\varphi$ also satisfies the above symmetry condition and thus can be viewed as a function on the space Ξ of hyperplanes.

The Radon transform is then inverted by the formula

$$c_0 f(x) = (-i)^{n-1} \int_{\mathbb{S}^{n-1}} \left\{(\Lambda\widehat{f})(\omega, p)\right\}_{p=\langle\omega,x\rangle} d\omega\,, \tag{1.2}$$

where $c_0 = (2\pi)^n/\pi$.

We consider now the Cauchy problem for the wave equation

$$Lu = \frac{\partial^2 u}{\partial t^2}\,, \qquad u(x, 0) = f_0(x)\,, \quad u_t(x, 0) = f_1(x)\,, \tag{1.3}$$

f_0 and f_1 being given functions in $\mathcal{D}(\mathbb{R}^n)$.

LEMMA 1.1 *If $h \in C^2(\mathbb{R})$ and $\omega \in \mathbb{S}^{n-1}$, then the function*

$$v(x,t) = h(\langle x, \omega\rangle + t) \tag{1.4}$$

satisfies $Lv = (\partial^2/\partial t^2)v$.

The proof is obvious. It is now easy on the basis of (1.2) and (1.4) to write down the solution to the Cauchy problem (1.3). For a different method in the case n odd, see Lax and Phillips [15, IV, sec. 2].

THEOREM 1.2 *The solution to* (1.3) *is given by*

$$u(x,t) = \int_{\mathbb{S}^{n-1}} (Sf)(\omega, \langle x, \omega\rangle + t)d\omega\,, \tag{1.5}$$

where

$$Sf = \begin{cases} c(\partial^{n-1}\widehat{f}_0 + \partial^{n-2}\widehat{f}_1)\,, & n \text{ odd}\,, > 1\,. \\ c(\mathcal{H}(\partial^{n-1}\widehat{f}_0 + \partial^{n-2}\widehat{f}_1)\,, & n \text{ even}\,. \end{cases} \tag{1.6}$$

Here $\partial = \partial/\partial p$, and the constant c equals

$$c = \frac{1}{2}(2\pi i)^{1-n}\,.$$

PROOF: Because of Lemma 1.1 we just have to check the initial conditions in (1.3).

(i) If $n > 1$ is odd, then $\omega \longrightarrow (\partial^{n-1}\widehat{f}_0)(\omega, \langle x, \omega\rangle)$ is an even function on $\mathbb{S}^{n-1}$ but $\omega \longrightarrow (\partial^{n-2}\widehat{f}_1)(\omega, \langle x, \omega\rangle)$ is odd. Thus $u(x,0)) = f_0(x)$ by (1.2). Applying $\partial/\partial t$ to the right-hand side of (1.5) and putting $t = 0$ gives $u_t(x,0) = f_1(x)$, this time because the function

$$\omega \longrightarrow (\partial^n \widehat{f}_0)(\omega, \langle x, \omega\rangle)$$

is odd and

$$\omega \longrightarrow (\partial^{n-1}\widehat{f}_1)(\omega, \langle, x, \omega\rangle)$$

is even.

(ii) n even. Here the proof is the same if one remarks that $\mathcal{H}$ interchanges even and odd functions on $\mathbb{R}$.

∎

DEFINITION For the initial data $f = \{f_0, f_1\}$ we shall refer to the function Sf in (1.6) on $\mathbb{S}^{n-1} \times \mathbb{R}$ as the *source*.

Note that for n odd, Lax and Phillips refer to $-\partial^{(n+1)/2}\widehat{f}_0 + \partial^{(n-1)/2}\widehat{f}_1$ as the *translation representation* [15, IV, theorem 2.2]. They also call the wave $u(x,t)$ *outgoing* if $u(x,t) = 0$ in the *forward cone* $|x| < t$ and *incoming* if $u(x,t) = 0$ in the *backward cone* $|x| < -t$.

COROLLARY 1.3 *The solution $u(x,t)$ to* (1.3) *is*

(i) *outgoing if and only if $(Sf)(\omega, s) = 0$ for $s > 0$.*

(ii) *incoming if and only if $(Sf)(\omega, s) = 0$ for $s < 0$.*

PROOF (adapted from [15, IV, sec. 2] for n odd):
For (i) suppose $(Sf)(\omega, s) = 0$ for $s > 0$. Then if $|x| < t$ we have $\langle x, \omega\rangle + t > -|x| + t > 0$ so by (1.5) u is outgoing. Conversely, suppose $u(x,t) = 0$ for $|x| < t$. Let $t_0 > 0$ be arbitrary and $\varphi \in \mathcal{D}(t_0, \infty)$. Then if $|x| < t_0$ we have

$$0 = \int_{\mathbb{R}} u(x,t)\varphi(t)dt = \int_{\mathbb{S}^{n-1}} d\omega \int_{\mathbb{R}} (Sf)(\omega, \langle x, \omega\rangle + t)\varphi(t)dt$$
$$= \int_{\mathbb{S}^{n-1}} d\omega \int_{\mathbb{R}} (Sf)(\omega, p)\varphi(p - \langle x, \omega\rangle)dp = 0\,.$$

Taking $\partial^{|k|}/\partial x_{i_1}, \ldots, \partial x_{i_k}$ at $x = 0$, we deduce

$$\int_{\mathbb{R}} \left(\int_{\mathbb{S}^{n-1}} (Sf)(\omega, p)\omega_{i_1}, \ldots, \omega_{i_k}\, d\omega\right) \left(\partial^{|k|}\varphi\right)(p)dp = 0$$

for each k and each $\varphi \in \mathcal{D}(t_0, \infty)$. Integrating by parts in the $\mathbb{R}$ integral, we deduce that the function

$$p \to \int_{\mathbb{S}^{n-1}} (Sf)(\omega, p)\omega_{i_1}, \ldots, \omega_{i_k}\, d\omega \tag{1.7}$$

has its kth derivative $\equiv 0$ for $p > t_0$. Having compact support if n is odd and being $o(1)$ if n is even, we deduce that the function (1.7) vanishes identically for $p > t_0$. Now varying k, we see that $(Sf)(\omega, p)$ vanishes for $p > t_0$. Since $t_0 > 0$ was arbitrary, this proves (i). Part (ii) is proved in the same way. ∎

Remark. The solution to (1.3) can also be written in the well-known form

$$u(x,t) = (f_0 * T_t')(x) + (f_1 * T_t)(x)\,, \quad T_t' = \frac{dT_t}{dt}\,, \tag{1.8}$$

where $*$ denotes convolution on $\mathbb{R}^n$ and T_t and T_t' in $\mathcal{E}'(\mathbb{R}^n)$ are given by their Fourier transforms,

$$\frac{\sin(|\lambda|\,t)}{|\lambda|} = \int_{\mathbb{R}^n} e^{-i(\lambda,x)}\, dT_t(x)\,, \quad \cos(|\lambda|\,t) = \int_{\mathbb{R}^n} e^{-i(\lambda,x)}\, dT_t'(x)\,. \tag{1.9}$$

Since

$$LT_t = \frac{d^2}{dt^2}T_t\,, \quad LT'_t = \frac{d^2}{dt^2}T'_t\,, \tag{1.10}$$

and

$$(T_t)_{t=0} = 0\,, \quad (T'_t)_{t=0} = \delta_0\,, \tag{1.11}$$

one refers to the pair $\{T_t, T'_t\}$ as the *fundamental solution* to the wave equation.

For n odd, (1.5) immediately implies the classical Huygens' principle.

COROLLARY 1.4 *Let $n > 1$ be odd. Assume f_0 and f_1 have support in $B_R(0)$. Then u has support in the conical shell*

$$|t| - R \leq |x| \leq |t| + R\,. \tag{1.12}$$

In fact, for n odd, (1.5) implies

$$u(0,t) = 0 \quad \text{for } |t| \geq R\,. \tag{1.13}$$

If $y \in \mathbb{R}^n$ the translated initial data f_0^{-y}, f_1^{-y} give the solution $x \to u(x+y,t)$. Thus by (1.13) since $\operatorname{supp}(f_i^{-y}) \subset B_{R+|y|}(0)$,

$$u(y,t) = 0 \quad \text{for } t \geq R + |y|\ .$$

On the other hand, since T_t in (1.8) has support in $B_{|t|}(0)$, (1.8) implies that u has support in the region $|x| \leq |t| + R$. This proves (1.12).

Remark. The shell (1.12) is the union

$$\bigcup_{|y| \leq R} C_y$$

where C_y is the light cone $\{(x,t) : |x-y| = |t|\}$.

For $n > 1$ odd one also has the following limit theorem of Friedlander:

THEOREM 1.5 *For n odd the solution u to* (1.3) *satisfies*

$$\lim_{|t| \to \infty} t^{\frac{n-1}{2}} u_t((t+p)\omega, t) = \left(-\partial^{(n+1)/2} \widehat{f}_0 + \partial^{(n-1)/2} \widehat{f}_1 \right)(\omega, p)\,.$$

For a proof, see Lax and Phillips [15, p. 108].

2 Multitemporal Wave Equations on Symmetric Spaces

Since the Laplacian in (1.3) generalizes to Riemannian manifolds, the analogue of the Cauchy problem (1.3) exists for any Riemannian manifold X. In the case when X is a Riemannian symmetric space, there is another natural analogue of (1.3), defined by Semenov-Tjan-Šanskiĭ in [17]; it is a system of equations and the "time variable" becomes multidimensional.

In order to describe the problem, we need to recall some notions from symmetric space theory. Let $X = G/K$ be a symmetric space of the noncompact type; that is, G is a connected, noncompact, semisimple Lie group with finite center and K a maximal compact subgroup. Let $\mathfrak{g}$ and $\mathfrak{k}$ denote the corresponding Lie algebras, and θ the Cartan involution of $\mathfrak{g}$ with fixed point set $\mathfrak{k}$. Then $\mathfrak{g} = \mathfrak{k} + \mathfrak{p}$ where $\mathfrak{p}$, the -1 eigenspace of θ, is the orthogonal complement of $\mathfrak{k}$ in $\mathfrak{g}$ relative to the Killing form B of $\mathfrak{g}$. For $X, Y \in \mathfrak{p}$, we put $\langle X, Y\rangle = B(X, Y)$ and $|X| = \langle X, X\rangle^{1/2}$. We fix a maximal abelian subspace $\mathfrak{a} \subset \mathfrak{p}$, $\mathfrak{a}^*$ its dual, and let $\Sigma = \Sigma(\mathfrak{g}, \mathfrak{a})$ be the system of roots of $\mathfrak{g}$ relative to $\mathfrak{a}$. Let Σ_0 denote the set of indivisible roots. The Weyl group W of Σ acts on $\mathfrak{a}$ and $\mathfrak{a}^*$ by

$$(\sigma \cdot \lambda)(H) = \lambda(\sigma^{-1}H) = \lambda^\sigma(H), \quad \sigma \in W, \lambda \in \mathfrak{a}^*, H \in \mathfrak{a}.$$

This action extends to automorphisms of the symmetric algebras $S(\mathfrak{a})$ and $S(\mathfrak{a}^*)$.

Fix a Weyl chamber $\mathfrak{a}^+$ in $\mathfrak{a}$, and let $\mathfrak{a}^*_+$ correspond to $\mathfrak{a}^+$ under the map $\lambda \to H_\lambda$ where $\langle H, H_\lambda\rangle = \lambda(H)$ $(H \in \mathfrak{a})$. Let Σ^+ be the corresponding set of positive roots, let 2ρ denote their sum with multiplicity, and put $\Sigma_0^+ = \Sigma^+ \cap \Sigma_0$. Let

$$G = NAK, \qquad \mathfrak{g} = \mathfrak{n} + \mathfrak{a} + \mathfrak{k}, \tag{2.1}$$

be the Iwasawa decompositions corresponding to this ordering, and let $\mathfrak{n}$ be the (nilpotent) Lie algebra of N. Let M be the centralizer of A in K and put $B = K/M, \overline{N} = \theta N$. If $H \in \mathfrak{a}$, let $\partial(H)$ denote the corresponding directional derivative. The map $H \to \partial(H)$ extends to an isomorphism of $S(\mathfrak{a})$ (respectively, $S^C(\mathfrak{a}) = \mathbf{C} \otimes S(\mathfrak{a})$) onto the algebra of all differential operators on $\mathfrak{a}$ with constant real (respectively, complex) coefficients. The map $\lambda \to H_\lambda$ extends to an isomorphism of $S^C(\mathfrak{a}^*)$ onto $S^C(\mathfrak{a})$.

Let $I(\mathfrak{a})$ denote the algebra of W-invariants in $S(\mathfrak{a})$ and $I_+(\mathfrak{a})$ the set of $q \in I(\mathfrak{a})$ without constant term. A polynomial $p \in S(\mathfrak{a}^*)$ is said to be *W-harmonic* if $\partial(q)p = 0$ for all $q \in I_+(\mathfrak{a})$. Let $H(\mathfrak{a}^*)$ be the space of W-harmonic polynomials and let $H(\mathfrak{a}) \subset S(\mathfrak{a})$ correspond to $H(\mathfrak{a}^*)$ under the extension of $\lambda \to H_\lambda$. It is well-known that

$$S(\mathfrak{a}) = S(\mathfrak{a})I_+(\mathfrak{a}) \oplus H(\mathfrak{a}) \tag{2.2}$$

and

$$\dim H(\mathfrak{a}) = w \quad \text{(the order of } W\text{)}\,. \tag{2.3}$$

For the adjoint action of K on $\mathfrak{p}$, we have analogous notions, $S(\mathfrak{p})$, $S(\mathfrak{p}^*)$, $I(\mathfrak{p})$, $I_+(\mathfrak{p})$, $H(\mathfrak{p})$, and $H(\mathfrak{p}^*)$. The analogue to (2.2) is also valid [10, III, theorem 1.2].

Let $\mathbf{D}(G)$ denote the algebra of left-invariant differential operators on G and $\mathbf{D}_K(G)$ the subalgebra of $D \in \mathbf{D}(G)$ that are right-invariant under K. If $X \in \mathfrak{g}$, let $\widetilde{X} \in \mathbf{D}(G)$ denote the corresponding left-invariant vector field. The *symmetrization* [10, II, sec. 4]

$$Y_1 \cdots Y_p \to \frac{1}{p} \sum_{\sigma \in \mathfrak{S}_p} \widetilde{Y}_{\sigma(1)} \cdots \widetilde{Y}_{\sigma(p)}$$

($\mathfrak{S}_p =$ the symmetric group) is a linear bijection of $S(\mathfrak{g})$ onto $\mathbf{D}(G)$. Let H^* denote the image of $H(\mathfrak{p})$ under this map.

According to (2.1) we have the direct decomposition

$$\mathbf{D}\,(G) = \mathbf{D}(A) \oplus \left(\tilde{\mathfrak{n}}\,\mathbf{D}(G) + \mathbf{D}(G)\,\tilde{\mathfrak{k}}\right). \tag{2.4}$$

Let $D \to q^D$ denote the corresponding projection of $\mathbf{D}(G)$ onto $\mathbf{D}(A) = S(\mathfrak{a})$. Then the map

$$\gamma : D \to e^{-\rho} q^D \circ e^{\rho}$$

is a homomorphism of $\mathbf{D}_K(G)$ onto $I(\mathfrak{a})$ with kernel $\mathbf{D}_K(G) \cap \mathbf{D}(G)\tilde{\mathfrak{k}}$ (see [4, sec. 3] or [10, III, theorem 5.17]). On the other hand, because G/K is reductive, the algebra $\mathbf{D}_K(G)/\mathbf{D}_K(G) \cap \mathbf{D}(G)\mathfrak{k}$ is naturally isomorphic to the algebra $\mathbf{D}(G/K)$ of G-invariant differential operators on G/K [5]. This gives a surjective isomorphism

$$\Gamma : \mathbf{D}(G/K) \to I(\mathfrak{a})\,. \tag{2.5}$$

We now fix a real homogeneous basis

$$p_1 = 1\,, p_2\,, \ldots, p_w\,, \tag{2.6}$$

of $H(\mathfrak{a})$. For fixed $f_i \in \mathcal{D}(X)$ $(1 \le i \le w)$, we now consider the "Cauchy problem" for $u \in \mathcal{E}(X \times \mathfrak{a})$,

$$Du = \partial(\Gamma(D))u\,, \quad D \in \mathbf{D}(G/K)\,, \tag{2.7}$$

with initial conditions

$$(\partial(p_i)u)(x,0) = f_i(x)\,, \quad 1 \le i \le w\,. \tag{2.8}$$

In (2.7) D operates on the first variable and $\partial(\Gamma(D))$ operates on the second variable. As remarked by Schlichtkrull, the system (2.8) is equivalent to

$$(\partial(p)u)(x,0) = F(p)\,, \quad p \in H(\mathfrak{a})\,,$$

where

$$F \in \mathrm{Hom}_{\mathbb{R}}(H(\mathfrak{a}), \mathcal{D}(X)) \quad (F(p_i) = f_i)\,.$$

The system (2.7)–(2.8) was first considered by Semenov-Tjan-Šanskiĭ [17] and further studied by Shahshahani and Phillips and Shahshahani ([18] and [16]). Here we shall present some additional results, some generalizations of older results, and some new and simpler proofs.

3 The Fourier Transform

To study the system (2.7)–(2.8), we use the Fourier transform theory on X developed in [6]–[9] and [11]. We now indicate the results needed. Let us normalize the invariant measures on G, N, A, K, M, $\mathfrak{a}$, $\mathfrak{a}^*$, $X = G/K$, $B = K/M$, as in [11, II, sec. 3.1]. According to (2.1) each $g \in G$ is decomposed $g = n \exp A(g) k$ ($A(g) \in \mathfrak{a}$), and we can define the "inner product"

$$A(gK, kM) = A(k^{-1}g)\,, \quad g \in G\,,\ k \in K\,. \tag{3.1}$$

Then for each $b \in B$, the function $x \to e^{(i\lambda+\rho)(A(x,b))}$ is an eigenfunction of each $D \in \mathbf{D}(G/K)$ and

$$D_x\left(e^{(i\lambda+\rho)(A(x,b))}\right) = \Gamma(D)(i\lambda)e^{(i\lambda+\rho)(A(x,b))} \tag{3.2}$$

For a function f on X, we define the *Fourier transform* $\widetilde{f}$ by

$$\widetilde{f}(\lambda, b) = \int_X f(x) e^{(-i\lambda+\rho)(A(x,b))}\, dx \tag{3.3}$$

for all $b \in B$ and $\lambda \in \mathfrak{a}_c^* = \mathbf{C} \otimes \mathfrak{a}^*$ for which the integral exists. For $f \in \mathcal{D}(X)$ we have the *inversion formula* [6]

$$f(x) = \frac{1}{w}\int_{\mathfrak{a}^* \times B} \widetilde{f}(\lambda, b) e^{(i\lambda+\rho)(A(x,b))}\, d\mu(\lambda, b)\,, \tag{3.4}$$

where $d\mu(\lambda, b) = |\mathbf{c}(\lambda)|^{-2}\, d\lambda\, db$, $\mathbf{c}(\lambda)$ being Harish-Chandra's $\mathbf{c}$-function. The *Plancherel theorem* [7, III, theorem 5.8] asserts that $f \to \widetilde{f}$ extends to an isometry of $L^2(X)$ onto $L^2(\mathfrak{a}_+^* \times B, \mu)$. The *Paley-Wiener theorem* [8, theorem 8.3] describes the range $\mathcal{D}(X)^\sim$ as a certain subspace of the space $\mathcal{H}(\mathfrak{a}_c^* \times B)$ of smooth functions $\varphi(\lambda, b)$ holomorphic in $\lambda \in \mathfrak{a}_C^*$ of exponential type (uniform in b). That subspace is specified by the identities

$$\int_B \varphi(\lambda, b) e^{(i\lambda+\rho)(A(x,b))}\, db \quad \text{is } W\text{-invariant in } \lambda\,. \tag{3.5}$$

We now relate this theory to representations of K. Let $\widehat{K}_M$ be the set of unitary, irreducible representations δ of K with fixed vector under $\delta(M)$. Let V_δ be the representation space of δ and $V_{\check{\delta}}^M$ the space of fixed vectors under $\delta(M)$. We choose an orthonormal basis $v_1, \dots, v_{d(\delta)}$ of V_δ such that $v_1, \dots, v_{\ell(\delta)}$ span V_δ^M. With H^* defined as before, let $E_\delta = \mathrm{Hom}_K(V_\delta, H^*)$ denote the space of linear maps $V_\delta \to H^*$ intertwining $\delta(k)$ and $\mathrm{Ad}(k)\, (k \in K)$. By [14], $\dim E_\delta = \ell(\delta)$. Let $\check{\delta}$ be the contragredient to δ operating on the dual space $V_\delta' = V_{\check{\delta}}$. For $v \in V_\delta$ let $v' \in V_\delta'$ be determined by $v'(u) = \langle u, v\rangle$ for $u \in V_\delta$. For $\lambda \in \mathfrak{a}_C^*$, consider the linear map

$$Q^\delta(\lambda) : E_\delta \to V_{\check{\delta}}^M \tag{3.6}$$

given by

$$(Q^\delta(\lambda)(e))(v) = q^{e(v)}(\rho - i\lambda)\,, \quad e \in E_\delta\,,\ v \in V_\delta\,, \tag{3.7}$$

q^D being defined after (3.4). If $\varepsilon_1, \dots, \varepsilon_{\ell(\delta)}$ is any basis of E_δ, we define

$$Q^\delta(\lambda)_{ij} = q^{\varepsilon_j(v_i)}(\rho - i\lambda)\,, \quad 1 \le i, j \le \ell(\delta)\,.$$

The matrix $Q^\delta(\lambda)_{ij}$ whose entries are polynomials in λ are related to the P_{ij}^δ in [13, sec. 4] by a simple variable shift. If $C_\delta : V_{\check{\delta}}^M \to E_\delta$ is the linear transformation determined by $C_\delta v_j' = \varepsilon_j$, then

$$Q^\delta(\lambda) C_\delta v_j' = \sum_{i=1}^{\ell(\delta)} Q^\delta(\lambda)_{ij} v_i'\,. \tag{3.8}$$

If (3.3) is specialized to $f \in \mathcal{D}_{\check{\delta}}(X)$, the space of K-finite f of type $\check{\delta}$, then [9, sec. 7],

$$\widetilde{f}(\lambda, kM) = \mathrm{Tr}_{V_\delta}(\widetilde{f}(\lambda)\delta(k))\,, \quad \mathrm{Tr} = \text{trace}\,, \tag{3.9}$$

where $\widetilde{f}(\lambda)$ is the *δ-spherical transform*

$$\widetilde{f}(\lambda) = d(\delta) \int_X f(x) \left(\int_K e^{(-i\lambda+\rho)(A(x,kM))} \delta(k^{-1}) dk \right) dx \,. \tag{3.10}$$

Then $\widetilde{f}(\lambda)$ maps V_δ into $V_{\check{\delta}}^M$ and has the invariance property

$$Q^{\check{\delta}}(\lambda)^{-1} \widetilde{f}(\lambda) \quad \text{is } W\text{-invariant in } \lambda \,. \tag{3.11}$$

Also, as f runs through $\mathcal{D}_{\check{\delta}}(X)$, the functions (3.11) fill up the space $\mathcal{I}^\delta(\mathfrak{a}^*)$ of W-invariants in the space $\mathcal{H}(\mathfrak{a}_c^*, \mathrm{Hom}(V_\delta, V_{\check{\delta}}^M))$ of $\mathrm{Hom}(V_\delta, E_{\check{\delta}})$-valued holomorphic functions of exponential type on $\mathfrak{a}_C^*$ [9, sec. 7].

4 Properties of Solutions of the Multitemporal Wave Equations

It is known from [18] and [17] that the system (2.7)–(2.8) has a solution that for each $H \in \mathfrak{a}$ has compact support in the first variable. We now prove a uniqueness statement for such solutions.

PROPOSITION 4.1 *Suppose u' and u'' are two solutions to* (2.7)–(2.8) *and that for each $H \in \mathfrak{a}$, the functions $x \to u'(x, H)$ and $x \to u''(x, H)$ have compact support. Then $u' = u''$.*

PROOF: Let $u = u' - u''$. Then u satisfies (2.7)–(2.8) with $f_j \equiv 0$ for all j. Consider the Fourier transform $\widetilde{u}(\lambda, b; H)$ of u in the x-variable. Then by (3.2) and (3.3), if $D \in \mathbf{D}(G/K)$ and D^* its adjoint,

$$\begin{aligned} \partial(\Gamma(D))_H(\widetilde{u}(\lambda, b; H)) &= \Gamma(D^*)(-i\lambda)\widetilde{u}(\lambda, b; H) \\ &= \Gamma(D)(i\lambda)\widetilde{u}(\lambda, b; H) \end{aligned} \tag{4.1}$$

by corollary 5.3 and lemma 5.21 in [10, II]. Also,

$$\{\partial(p_i)_H(u(x, H))\}_{H=0} \equiv 0 \,. \tag{4.2}$$

Because of (2.2)–(2.3) each $p \in S(\mathfrak{a})$ can be written $p = \sum_{j=1}^{w} q_j p_j$ $(q_j \in I(\mathfrak{a}))$. Then using (2.7) and (4.1) we get

$$\begin{aligned} \partial(p)_H(\widetilde{u}(\lambda, b; H)) &= \sum_j \partial(p_j)\partial(q_j)_H(\widetilde{u}(\lambda, b; H)) \\ &= \sum_j \partial(p_j)_H(q_j(i\lambda)\widetilde{u}(\lambda, b; H)) \end{aligned}$$

∎

Using the inversion formula for $u(x, H)$ and (4.2), we deduce

$$\{\partial(p)_H(u(x,H))\}_{H=0} = 0\,, \quad p \in S(\mathfrak{a})\,. \tag{4.3}$$

On the other hand, if $\lambda \in \mathfrak{a}^*$ is regular, (4.1) implies by a result of Steinberg and Harish-Chandra [10, III, theorem 3.13] that for constants C_s

$$\widetilde{u}(\lambda, b; H) = \sum_{s\in W} C_s(\lambda, b) e^{is\lambda(H)}\,.$$

Thus the equation $\{\partial(p)_H(\widetilde{u}(\lambda, b\,; H))\}_{H=0} = 0$ for all p implies

$$\sum_{s\in W} p(is\lambda) C_s(\lambda, b) = 0\,, \quad p \in S(\mathfrak{a})\,.$$

Choosing p such that $p(is\lambda) = 0$ for all $s \neq s_0$ in W and $p(is_0\lambda) = 1$, we get $C_{s_0}(\lambda) = 0$. Thus $\widetilde{u}(\lambda, b; H) = 0$ for all $\lambda \in \mathfrak{a}^*$ regular and $b \in B$, so by the inversion formula, $u(x, H) \equiv 0$, as desired.

Consider now the quotient fields $C(S(\mathfrak{a}))$ and $C(I(\mathfrak{a}))$ and the bilinear form $(\cdot,\cdot)$ on $C(S(\mathfrak{a})) \times C(S(\mathfrak{a}))$ with values in $C(I(\mathfrak{a}))$ given by

$$(a, b) = \sum_{\sigma\in W} a^\sigma b^\sigma\,. \tag{4.4}$$

This form being nondegenerate, we can determine $q^j \in C(S(\mathfrak{a}))$ such that

$$(q^j, p_i) = \delta_{ij}\,. \tag{4.5}$$

THEOREM 4.2 *Given $f_1, \dots, f_w \in \mathcal{E}(X)$, the system*

$$Du = \partial(\Gamma(D))u\,, \quad D \in \mathbf{D}(G/K)\,, \tag{4.6}$$

with initial data

$$(\partial(p_j)u)(x, 0) = f_j(x) \quad (1 \le j \le w) \tag{4.7}$$

has a solution given by the convolution

$$u(x, H) = \sum_{j=1}^{w} (f_j \times S_H^j)(x) \tag{4.8}$$

where for each j, $H \in \mathfrak{a}$, S_H^j is in $\mathcal{E}'(X)$ and is for all b given by

$$\sum_{\sigma\in W} q^j(i\sigma\lambda) e^{(i\sigma\lambda)(H)} = \int_X e^{(-i\lambda+\rho)(A(x,b))}\, dS_H^j(x)\,. \tag{4.9}$$

Remark. Because of Proposition 4.1, we shall refer to (4.8) as *the* solution to (2.7)–(2.8).

We need some known results about p_i and q^j from [4, sec. 3] for which simplified proofs (due to Steinberg) can be found in [2, sec. 5.5]. Let $\pi \in S(\mathfrak{a})$ be given by $\pi \in \prod_{\alpha \in \sum_0^+} H_\alpha$.

LEMMA 4.3 *We have* $\pi q^j \in S(\mathfrak{a})$ *and, more precisely,*

(i) $S(\mathfrak{a}) = \bigoplus_{i=1}^{w} I(\mathfrak{a}) p_i$,

(ii) $S(\mathfrak{a}) = \bigoplus_{j=1}^{w} I(\mathfrak{a})(\pi q^j)$,

(iii) $\frac{1}{\pi} S(\mathfrak{a}) = \{x \in C(S(\mathfrak{a})) : (x, y) \in I(\mathfrak{a})$ *for all* $y \in S(\mathfrak{a})\}$.

If $s_1, \ldots, s_w$ runs through W, (4.5) implies that the matrices

$$e_{ij} = p_j(s_i\lambda) \quad \text{and} \quad f_{k\ell} = q^k(s_\ell \lambda)$$

are inverses. The determinant of e_{ij} being homogeneous in λ, it follows from Cramer's rule that each q^j is homogeneous in λ. Because of (ii) above, $q^j = h^j/\pi$ where $h^j \in S(\mathfrak{a})$, so the left-hand side of (4.9) equals

$$\frac{1}{\pi(i\lambda)} \sum_\sigma h^j(i\sigma\lambda)\varepsilon(\sigma) e^{i\sigma\lambda(H)} \tag{4.10}$$

where $\pi(\sigma\lambda) = \varepsilon(\sigma)\pi(\lambda)$ and $\varepsilon(\sigma) = \pm 1$. The sum in (4.10) is skew and is thus divisible by $\pi(i\lambda)$, so (4.10) is holomorphic in λ. Also, if $\lambda = \xi + i\eta$ $(\xi, \eta \in \mathfrak{a}^*)$, we have

$$\left| e^{i\sigma\lambda(H)} \right| \le e^{|H||\eta|} ,$$

so (4.10) is an entire function of exponential type of polynomial growth for λ real. By the Paley-Wiener theorem extended to $\mathcal{E}'(X)$ ([8, theorem 8.5], [1], or [11, III, corollary 5.9]), there exist unique $S_H^j \in \mathcal{E}'(X)$ satisfying (4.9) and

$$\operatorname{supp}(S_H^j) \subset B_{|H|}(o) . \tag{4.11}$$

Next we prove that S_H^j depends smoothly on $H \in \mathfrak{a}$; that is, for each $f \in \mathcal{D}(X)$, $S_H^j(f)$ is smooth in H. Since S_H^j is K-invariant, $(f \times S_H^j)^{\sim} = \tilde{f}(S_H^j)^{\sim}$ [11, III], so (3.4) and (4.9) imply

$$\int_X f(x)\, dS_H^j(x) = \int_{\mathfrak{a}^* \times B} \tilde{f}(\lambda, b) \sum_\sigma q^j(-i\sigma\lambda) e^{-i\sigma\lambda(H)}\, d\mu(\lambda, b) ,$$

so the smoothness in H follows.

Defining $u(x,H)$ by (4.8), we shall now verify (4.6)–(4.7). Since $D(f \times S) = f \times DS$ for $f \in \mathcal{E}(X)$, $S \in \mathcal{E}'(X)$, we consider DS^j_H and shall prove

$$DS^j_H = \partial(\Gamma(D))_H(S^j_H)\,. \tag{4.12}$$

For this we observe that by (4.9) and (2.3)

$$(DS^j_H)^{\sim}(\lambda, b) = \Gamma(D^*)(-i\lambda)(S^j_H)^{\sim}(\lambda, b) = \Gamma(D)(i\lambda)(S^j_H)^{\sim}(\lambda, b)\,.$$

On the other hand, applying $\partial(\Gamma(D))$ to (4.9), we see by the W-invariance of $\Gamma(D)$ that

$$(\partial(\Gamma(D))_H(S^j_H))^{\sim}(\lambda, b) = \Gamma(D)(i\lambda)(S^j_H)^{\sim}(\lambda, b)\,.$$

These formulas imply (4.12), so (4.6) follows.

Second, applying $\partial(p_k)_H$ to (4.9), we get

$$\sum_\sigma q^j(i\sigma\lambda)p_k(i\sigma\lambda)e^{i\sigma\lambda(H)} = (\partial(p_k)_H(S^j_H))^{\sim}(\lambda, b)\,.$$

Putting here $H = 0$ and using (4.5), we obtain

$$\left\{(\partial(p_k)_H(S^j_H))^{\sim}(\lambda, b)\right\}_{H=0} = \delta_{jk} \tag{4.13}$$

for all λ, b. By the injectivity of the Fourier transform, this means

$$\left\{\partial(p_k)S^j_H\right\}_{H=0} = \delta_{jk}\delta_o\,, \tag{4.14}$$

where δ_o the delta distribution of X at o. Now (4.7) follows immediately.

Remark. Relations (4.12)–(4.14) are the analogues to (1.10)–(1.11), and one can thus consider the family

$$\left\{S^1_H, \ldots, S^w_H\right\}$$

as the *fundamental solution* to (4.6)–(4.7).

Example. G/K of rank one. Let H_0 be the vector in $\mathfrak{a}^+$ of length 1. Then with $p_1 = 1$ and $p_2 = H_0$, we have $q^1 = \frac{1}{2}$, $q^2 = 1/(2H_0)$. Also, $\Gamma(L) = H_0^2 - |\rho|^2$. By writing $v(x,t) = u(x, tH_0)$, equations (4.6) and (4.7) become

$$(L + |\rho|^2)v = \frac{\partial^2}{\partial t^2}v\,, \quad v(x,0) = f_0(x)\,,\; v_t(x,0) = f_1(x)\,.$$

Writing S_t^j instead of $S_{tH_0}^j$, equation (4.9) becomes (since $\lambda(H_0) = \pm|\lambda|$)

$$\cos|\lambda|t = \int_X e^{(-i\lambda+\rho)(A(x,b))}\,dS_t^1(x)\,,$$
$$\frac{\sin|\lambda|t}{|\lambda|} = \int_X e^{(-i\lambda+\rho)(A(x,b))}\,dS_t^2(x)\,,$$

in exact analogy with (1.9).

From Theorem 4.2 we can now deduce the following result stating informally that the speed of propagation is ≤ 1. The result is established in [16, theorem 3.4] in a rather complicated fashion (see also [17, lemma 2]).

COROLLARY 4.4 *If* $\mathrm{supp}(f_j) \subset B_R(x_0)$ *for* $1 \leq j \leq w$, *then for each* $H \in \mathfrak{a}$,

$$u(x,H) = 0 \quad \textit{for } x \notin B_{R+|H|}(x_0)\,. \tag{4.15}$$

By the group invariance, we can take $x_0 = o$. Then the result is an immediate consequence of (4.8), (4.11), and the following simple lemma:

LEMMA 4.5 *Let* $f \in \mathcal{E}(X)$, $S \in \mathcal{E}'(X)$. *Then*

$$\mathrm{supp}(f) \subset \overline{B_r(o)}\,, \quad \mathrm{supp}(S) \subset \overline{B_s(o)} \Rightarrow \mathrm{supp}(f \times S) \subset \overline{B_{r+s}(o)}\,. \tag{4.16}$$

PROOF: By approximation we may assume S is a continuous function. If $\widetilde{S}$ denotes the lift of S to G ($\widetilde{S}(g) = S(g \cdot o)$), then

$$(f \times S)(g \cdot o) = \int_G f(gh^{-1} \cdot o)\widetilde{S}(h)dh\,.$$

If this is $\neq 0$, then, for some h, $d(gh^{-1} \cdot o, o) \leq r$ and $d(h \cdot o, o) \leq s$ whence,

$$d(o, g \cdot o) = d(g^{-1} \cdot o, o) \leq d(g^{-1} \cdot o, h^{-1} \cdot o) + d(h^{-1} \cdot o, o) \leq r + s\,,$$

proving (4.16). ∎

We can now reformulate the property in Corollary 4.4 as follows:

PROPOSITION 4.6 *Let* $f_j \in \mathcal{D}(X)(1 \leq j \leq w)$ *and* $u(x,H)$ *the corresponding solution to* (4.6)–(4.7). *Then if*

$$\overline{B_\rho(x_0)} \cap \bigcup_{j=1}^{w} \mathrm{supp}(f_j) = \varnothing\,, \tag{4.17}$$

we have for each $H \in \mathfrak{a}$,

$$u(x,H) = 0 \quad \textit{for } x \in B_{\rho-|H|}(x_0)\,. \tag{4.18}$$

PROOF: The compact set $C = \bigcup_{j=1}^{w} \operatorname{supp}(f_j)$ can be covered by finitely many balls $B_\varepsilon(x_i)$ (ε fixed) such that

$$\overline{B_\rho(x_0)} \cap \overline{B_\varepsilon(x_i)} = \varnothing \quad \text{for each } i\,. \tag{4.19}$$

Choose a corresponding partition of unity, that is, $\varphi_i \in \mathcal{D}(B_\varepsilon(x_i))$, $\varphi_i \geq 0$, $\sum_i \varphi_i \leq 1$, with equality in a neighborhood of C. Then $f_j = \sum_i f_j^i$ $(1 \leq j \leq w)$ where $f_j^i = \varphi_i f_j$ has support in $B_\varepsilon(x_i)$. Let $u^i(x,H)$ be the solution to (4.6)–(4.7) corresponding to $f_j^i (1 \leq j \leq w)$. By Corollary 4.4,

$$u^i(x,H) = 0 \quad \text{for } x \notin B_{\varepsilon+|H|}(x_i)\,. \tag{4.20}$$

However, $B_{\varepsilon+|H|}(x_i)$ is disjoint from $\overline{B_{\rho-|H|}(x_0)}$; in fact, a common point y would satisfy $d(x_i,y) \leq \varepsilon+|H|$, $d(x_0,y) \leq \rho-|H|$, whence $d(x_i,x_0) \leq \varepsilon+\rho$, contradicting (4.19). Thus by (4.20), $u^i(x,H) = 0$ for $x \in B_{\rho-|H|}(x_0)$ so (4.18) follows since $u = \sum_i u^i$.

∎

We shall now derive some other formulas for the solution. If $S \in \mathcal{E}'(X)$ is K-invariant and $f \in \mathcal{D}(X)$, we have

$$(f \times S)^{\sim} = \widetilde{f}\widetilde{S}$$

so by (4.8) and the inversion formula for the Fourier transform,

$$u(x,H) = \frac{1}{w}\int_{\mathfrak{a}^*} \sum_{j,\sigma} q^j(i\sigma\lambda)e^{i\sigma\lambda(H)}(\mathbf{c}(\lambda)\mathbf{c}(-\lambda))^{-1}\int_B \widetilde{f}_j(\lambda,b)e^{(i\lambda+\rho)(A(x,b))}\,db\,d\lambda\,.$$

Since $(\mathbf{c}(\lambda)\mathbf{c}(-\lambda))^{-1}$ as well as the integral over B are W-invariant in λ (cf. (3.5)), the terms in $\sum_\sigma$ all have the same integral over $\mathfrak{a}^*$. Thus we can replace the average $\frac{1}{w}\sum_\sigma$ by a single term. For each $\sigma \in W$, we thus obtain (as in [18, lemma 8],

$$u(x,H) = \int_{\mathfrak{a}^*\times B} \sum_k j_k^\sigma(\lambda)\widetilde{f}_k(\lambda,b)e^{i\lambda(A(x,b)+\sigma H)}e^{\rho(A(x,b))}\,d\lambda\,db\,, \tag{4.21}$$

where

$$j_k(\lambda) = \frac{q^k(i\lambda)}{\mathbf{c}(\lambda)\mathbf{c}(-\lambda)} \tag{4.22}$$

and as usual, $j_k^\sigma(\lambda) = j_k(\sigma^{-1}\lambda)$. Consider now the Euclidean Fourier transform

$$F^*(\lambda) = \int_A e^{-i\lambda(\log a)} F(a)da\,,$$

and let J_k denote the operator on $\mathcal{S}(A)$ determined by

$$(J_k F)^*(\lambda) = j_k(\lambda)F^*(\lambda)\,. \tag{4.23}$$

Since $q^k(\lambda)\pi(\lambda)$ is a polynomial and $(\pi(\lambda)\mathbf{c}(\lambda)\mathbf{c}(-\lambda))^{-1}$ has all its derivatives bounded by a polynomial, J_k does indeed map $\mathcal{S}(A)$ into itself. Also, by a simple computation,

$$(J_k^\sigma F)^*(\lambda) = j_k^\sigma(\lambda)F^*(\lambda)\,.$$

Consider now the Radon transform

$$\widehat{f}(\xi) = \int_\xi f(x)dm(x)\,,$$

where ξ is a horocycle $gNg^{-1}\cdot x$ and dm the measure on ξ derived from dn. More explicitly, if $\xi_0 = N\cdot o$,

$$\widehat{f}(g\cdot\xi_0) = \int_N f(gn\cdot o)dn\,.$$

Since each ξ equals $ka\cdot\xi_0$ where $kM \in K/M$ and $a \in A$ are unique, we can write $\widehat{f}(\xi)$ in the form $\widehat{f}(kM,a)$. Then J_k operates on functions on the horocycle space Ξ. Also by [11, p. 276],

$$\widetilde{f}(\lambda,b) = \int_A \widehat{f}(b,a)e^{(-i\lambda+\rho)(\log a)}\,da = (e^\rho\widehat{f})^*(\lambda,b)\,. \tag{4.24}$$

Thus by (4.21)

$$u(x,H) = \int_{\mathfrak{a}^*\times B}\sum_k (J_k^\sigma(e^\rho\widehat{f}_k))^*(\lambda,b)e^{i\lambda(A(x,b)+\sigma H)}e^{\rho(A(x,b))}\,d\lambda\,db$$

so by the inversion formula for the Euclidean Fourier transform,

$$u(x,H) = \int_B\sum_k J_k^\sigma(e^\rho\widehat{f}_k)(b,\exp(A(x,b)+\sigma H))e^{\rho(A(x,b))}\,db\,. \tag{4.25}$$

Here e^ρ is the function on Ξ given by $e^\rho(kM,a) = e^{\rho(\log a)}$. As in [11, II, sec. 3], we introduce the operators

$$\Lambda_{k,\sigma} = e^{-\rho}J_k^\sigma\circ e^\rho \quad (1\le k\le w)$$

on the space $\mathcal{S}_\rho(\Xi) = \{\varphi\in\mathcal{E}(\Xi) : e^\rho\varphi\in\mathcal{S}(\Xi)\}$, where $\mathcal{S}(\Xi) = \mathcal{S}(K/M\times A)$. Then the space $\mathcal{S}_\rho(\Xi)$ and the operators $\Lambda_{k,\sigma}$ are G-invariant [11, III, lemma 3.7].

THEOREM 4.7 *Fix $\sigma \in W$. The solution to* (4.16)–(4.17) *can be written*

$$u(g \cdot o, H) = e^{\rho(\sigma H)} \int_{K/M} \sum_j (\Lambda_{j,\sigma} \widehat{f}_j)(gk \exp \sigma H \cdot \xi_0) dk_M \,. \tag{4.26}$$

PROOF: If $g = e$ the right-hand side becomes

$$e^{\rho(\sigma H)} \int_B \sum_j (e^{-\rho} J_j^\sigma (e^\rho \widehat{f}_j))(b, \exp \sigma H) db = \int_B \sum_j (J_j^\sigma (e^\rho \widehat{f}_j))(b, \exp \sigma H) db \,,$$

which agrees with the right-hand side of (4.25) for $x = o$. By the G-invariance of (4.6)–(4.7) in the x-variable, the initial data $f_j^{\tau(g^{-1})}$ generate the solution $u^{\tau(g^{-1})}$. Since the Radon transform as well as the operators $\Lambda_{j,\sigma}$ commute with the G-action, we deduce from the above

$$\begin{aligned} u(g \cdot o, H) &= u^{\tau(g^{-1})}(o, H) \\ &= e^{\rho(\sigma H)} \int_{K/M} \sum_j (\Lambda_{j,\sigma} (\widehat{f_j^{\tau(g^{-1})}}))(k \exp \sigma H \cdot \xi_0) dk_M \\ &= e^{\rho(\sigma H)} \int_{K/M} \sum_j (\Lambda_{j,\sigma} \widehat{f}_j)(gk \exp \sigma H \cdot \xi_0) dk_M \end{aligned}$$

as claimed. ∎

DEFINITION Given $\sigma \in W$, the function

$$(S_\sigma f)(\xi) = \sum_j (\Lambda_{j,\sigma} \widehat{f}_j)(\xi) \,, \quad \xi \in \Xi \,,$$

will be called the *σ-source* of the solution u.

From (4.25) it is easy to see that

$$u(x, H) = e^{\rho(\sigma H)} \int_B (S_\sigma f)(b, \exp(A(x, b) + \sigma H)) e^{2\rho(A(x,b))} \, db \,, \tag{4.27}$$

and this formula is the symmetric space analogue to (1.5).

From (4.27) it is easy to deduce the analogue of Huygens' principle in Corollary 1.4. This is done in the same way in [18, sec. 4, cor. 2]; see also [17, prop. 18].

COROLLARY 4.8 *Assume G has all its Cartan subgroups conjugate. Then if* $\operatorname{supp}(f_j) \subset B_R(o)$ $(1 \leq j \leq w)$, *the solution $u(x, H)$ has support in the region*

$$|H| - R \leq d(x, o) \leq |H| + R \,. \tag{4.28}$$

In fact, the assumption on G is equivalent to all $\alpha \in \Sigma(\mathfrak{g}, \mathfrak{a})$ having even multiplicity [7, II, sec. 3]. This, in turn, implies by the Gindikin-Karpelevič formula for $\mathbf{c}(\lambda)$ that $\mathbf{c}(\lambda)^{-1}$ and even $(\pi(\lambda)\mathbf{c}(\lambda))^{-1}$ are polynomials [3]. In addition, $\pi(\lambda)q^j(i\lambda)$ is a polynomial so the operators $\Lambda_{j,\sigma}$ are differential operators. Since

$$d(o, a\cdot o) \leq d(o, na\cdot o)\,, \quad a\in A\,,\ n\in N\,, \tag{4.29}$$

we thus deduce from (4.27) that $u(x,H) = 0$ if $|H + A(x,b)| \geq R$ for $b \in B$. But if $x = gK$, $b = kM$, and $k^{-1}g = nak_0$, (4.29) implies

$$|A(x,b)| = \left|A(k^{-1}g)\right| = |\log a| \leq d(o, na\cdot o) \leq d(o,x)\,. \tag{4.30}$$

Thus if $|H| > R + d(o,x)$, we have for $b \in B$, $|H + A(x,b)| \geq |H| - |A(x,b)| \geq R$, so $u(x,H) = 0$ for $d(o,x) < |H| - R$. This, together with (4.15), implies the corollary.

5 Incoming Waves and Supports

In this section we shall prove the symmetric space analogue of Corollary 1.3, relating incoming waves to support properties of the sources S_σ. Results of this type are given in sections 4.1 and 4.2 in [17] and section 4 and 5 in [16], but we have not understood these papers well enough to make a detailed comparison. However, the proof of Theorem 5.1 below is mostly a variation, with some details added, of the proof of theorem 4.10 in [16].

As usual, we put

$${}^+\mathfrak{a} = \{H \in \mathfrak{a} : \langle H, H'\rangle > 0 \text{ for } H' \in \mathfrak{a}^+\}$$

and $\mathfrak{a}^- = -\mathfrak{a}^+$. Note that ${}^+\mathfrak{a} \cup \{0\}$ is closed [10, p. 438], and $\mathfrak{a}^+ \subset {}^+\mathfrak{a}$.

Definition Fix $\sigma \in W$. The wave $u(x,H)$ in (2.5) is said to be *σ-incoming* if

$$u(x,H) = 0 \quad \text{for } d(o,x) < |H|\,,\ \sigma H \in \mathfrak{a}^-\,. \tag{5.1}$$

To distinguish from balls in X, an open ball in $\mathfrak{a}$ of radius r and center H is denoted $V_r(H)$.

We shall now, in analogy with Corollary 1.3, characterize σ-incoming waves in terms of their σ-sources.

Theorem 5.1 *Fix $\sigma \in W$ and $f_j \in \mathcal{D}(X)$ $(1 \leq j \leq w)$. Then u is σ-incoming if and only if*

$$\operatorname{supp}(S_\sigma f)(b,\cdot) \subset \exp(\overline{{}^+\mathfrak{a}}) \quad \textit{for all } b \in B\,. \tag{5.2}$$

Naturally, formula (4.27) is basic to the proof. First we prove (5.2) $\Rightarrow$ (5.1). Suppose $H \in \mathfrak{a}$, $\sigma H \in \mathfrak{a}^-$, and $d(o,x) < |H|$. Then by (4.30), $A(x,b) \in V_{|H|}(0)$. Now let $H_0 \in {}^+\mathfrak{a} + (-\sigma H)$. Then $\langle H_0 + \sigma H, -\sigma H\rangle \geq 0$ so

$$- |H_0|\,|H| \leq \langle H_0, \sigma H\rangle \leq - |H|^2$$

so $H_0 \notin V_{|H|}(0)$. Hence $A(x,b) \notin {}^+\mathfrak{a} + (-\sigma H)$ so by (5.2) and (4.27), $u(x,H) = 0$. This proves (5.2) $\Rightarrow$ (5.1).

For the converse we assume (5.1). For $\varphi \in \mathcal{D}(\mathfrak{a})$ we consider the integral

$$v(x) = \int_{\mathfrak{a}} u(x,H)\varphi(H)dH\,. \tag{5.3}$$

Then by (4.21) we obtain

$$v(x) = \int_{\mathfrak{a}^*\times B} F_\lambda^\sigma(b)\psi^*(\lambda)e^{(i\lambda+\rho)(A(x,b))}\,d\lambda\,db\,, \tag{5.4}$$

where

$$\psi(H) = \varphi(-\sigma^{-1}H) \quad \text{and} \quad F_\lambda^\sigma(b) = \sum_k j_k^\sigma(\lambda)\widetilde{f}_k(\lambda,b)\,.$$

Note that

$$F_\lambda^\sigma(b) = h_\sigma(b,\,\cdot\,)^*(\lambda)\,,$$

the Fourier transform (in A) of the function

$$h_\sigma(b,a) = \sum_k J_k^\sigma(e^\rho\widehat{f}_k)(b,a)\,, \quad b\in B\,,\ a\in A\,. \tag{5.5}$$

Let $\sigma^* \in W$ be the element interchanging $\mathfrak{a}^+$ and $\mathfrak{a}^-$ and put $\tau = \sigma^{-1}\sigma^*$. Then $\sigma H \in \mathfrak{a}^- \iff H \in \tau\mathfrak{a}^+$, so by (5.1)

$$u(x,H) = 0 \quad \text{for } d(o,x) < |H|\,,\ H\in\tau\mathfrak{a}^+\,. \tag{5.6}$$

Now fix $H_0 \in \tau\mathfrak{a}^+$ and consider the solution $(x,H) \longrightarrow u(x,H+H_0)$ to (2.7) whose initial data are

$$\{\partial(p_i)_H(u(x,H+H_0))\}_{H=0} = (\partial(p_i)u)(x,H_0)\,.$$

We denote these by $F_i(x)$ $(1\leq i\leq w)$. Then (5.6) implies $F_i(x) = 0$ for $d(o,x) < |H_0|$, that is, $\operatorname{supp}(F_i)\cap B_{|H_0|}(o) = \varnothing$. Hence by Proposition 4.6,

$$u(x,H+H_0) = 0 \quad \text{for } x\in B_{|H_0|-|H|}(o)\,. \tag{5.7}$$

Of course, $H + H_0 \in V_{|H_0|-\,\varepsilon}(H_0) \iff |H| < |H_0| - \varepsilon$, which in turn implies

$$B_{|H_0|-|H|}(o) \supset B_\varepsilon(o)\,.$$

Thus (5.7) implies the following result:

LEMMA 5.2 *Let $\tau = \sigma^{-1}\sigma^*$, fix $H_0 \in \tau\mathfrak{a}^+$, and let $\varepsilon < |H_0|$. Then*

$$u(x, H + H_0) = 0 \quad \textit{for } d(o,x) < \varepsilon \quad \textit{if } H + H_0 \in V_{|H_0| - \varepsilon}(H_0)\,.$$

Now following [16, lemma 4.8], let

$$\mathfrak{a}_\varepsilon(\tau) = \{H \in \mathfrak{a} : \langle H, H' \rangle > \varepsilon\,|H'| \text{ for some } H' \in \tau\mathfrak{a}^+\}\,. \tag{5.8}$$

Then $\mathfrak{a}_\varepsilon(\tau) = \tau \cdot \mathfrak{a}_\varepsilon(e)$. Let H_0 in Lemma 5.2 run along a fixed ray $\{te_0 : t > 0,\ |e_0| = 1\}$ in $\tau\mathfrak{a}^+$. The balls $V_{|H_0|-\varepsilon}(H_0)$ will then fill up the half-space $\langle H, e_0 \rangle > \varepsilon$. As H_0 varies in $\tau\mathfrak{a}^+$ (and $|H_0| > \varepsilon$), these half-spaces have union $\mathfrak{a}_\varepsilon(\tau)$. Thus Lemma 5.2 implies the following result in [16, lemma 4.8]:

LEMMA 5.3 *Let $\tau = \sigma^{-1}\sigma^*$, $\varepsilon > 0$, and $H \in \mathfrak{a}_\varepsilon(\tau)$. Then*

$$u(x, H) = 0 \quad \textit{for } d(o,x) < \varepsilon\,.$$

Now let φ in (5.3) satisfy $\varphi \in \mathcal{D}(\mathfrak{a}_\varepsilon(\tau))$. Then

$$v(x) = 0 \quad \text{for } d(o,x) < \varepsilon\,, \tag{5.9}$$

which by (5.4) amounts to

$$v(x) = \int_{\mathfrak{a}^*} \psi^*(\lambda)\mathcal{P}_\lambda(F^\sigma_\lambda)(x)d\lambda = 0\,, \tag{5.10}$$

where $\mathcal{P}_\lambda$ is the Poisson transform defined by

$$(\mathcal{P}_\lambda F)(x) = \int_B e^{(i\lambda+\rho)(A(x,b))}F(b)db\,. \tag{5.11}$$

As in the proof in [9] of the injectivity criterion for the Poisson transform, we exploit (5.10) by means of the equation

$$\{D_g(v(g \cdot o))\}_{g=e} = 0 \tag{5.12}$$

for all $D \in \mathbf{D}(G)$, the algebra of left-invariant differential operators on G. The kernel in (5.11) is given by

$$e^{(i\lambda+\rho)(A(gK,kM))} = \zeta_{-\lambda}(k^{-1}g)\,,$$

where $\zeta_\lambda(g) = e^{(-i\lambda+\rho)(A(g))}$. Fix $\delta \in \widehat{K}_M$ (as in Section 3) and consider the operators $\varepsilon_j(v_i)$ in Section 3. By [11, p. 239], they satisfy

$$(\varepsilon_j(v_i)\zeta_{-\lambda})(k^{-1}) = \sum_{p=1}^{\ell(\delta)} \langle v_i, \delta(k)v_p \rangle Q^\delta(-\lambda)_{pj}\,. \tag{5.13}$$

Writing (5.4) and (5.10) in the form (with $dk_M = db$)

$$v(g \cdot o) = \int_{\mathfrak{a}^*} d\lambda \int_{K/M} \zeta_{-\lambda}(k^{-1}g) F(\lambda, kM) dk_M \,, \tag{5.14}$$

where $F(\lambda, b) = h_\sigma(b, \cdot)^*(\lambda)\psi^*(\lambda)$, we thus have

$$0 = \{\varepsilon_j(v_i)_g(v(g \cdot o))\}_{g=e} = \int_{\mathfrak{a}^*} \sum_{p=1}^{\ell(\delta)} Q^\delta(-\lambda)_{pj} F^\delta_{pi}(\lambda) d\lambda \,.$$

Here

$$F^\delta_{pi}(\lambda) = \int_{K/M} \langle v_i, \delta(k)v_p \rangle F(\lambda, kM) dk_M \,.$$

Now $F(\lambda, b) = (h_\sigma(b, \cdot) \times \psi)^*(\lambda)$, $\times$ denoting convolution on A (and $\mathfrak{a}$). Thus, defining

$$\zeta_{pi}(H) = \int_{K/M} h_\sigma(kM, \exp H) \langle v_i, \delta(k)v_p \rangle dk_M \quad (H \in \mathfrak{a})$$

we have for $1 \leq i, j \leq \ell(\delta)$,

$$\sum_{p=1}^{\ell(\delta)} \int_{\mathfrak{a}^*} Q^\delta(-\lambda)_{pj} (\zeta_{pi} \times \psi)^*(\lambda) d\lambda = 0 \,. \tag{5.15}$$

Define $Q^\delta_{pj} \in S(\mathfrak{a})$ by $Q^\delta_{pj}(\lambda) = Q^\delta(\lambda)_{pj}$ and $P^\delta_{pj} \in S(\mathfrak{a})$ by $P^\delta_{pj}(\lambda) = Q^\delta_{pj}(i\lambda)$. Then, since $P^\delta(i\lambda) = Q^\delta(-\lambda)$, (5.15) implies by the Fourier inversion formula,

$$\sum_{p=1}^{\ell(\delta)} \left[(\partial(P^\delta_{pj})\zeta_{pi}) \times \psi \right](0) = 0 \,,$$

so, if $\check{\psi}(H) = \psi(-H)$,

$$\sum_{p=1}^{\ell(\delta)} \int_{\mathfrak{a}} (\partial(P^\delta_{pj})\zeta_{pi})(H) \check{\psi}(H) dH = 0 \,. \tag{5.16}$$

This holds whenever $\varphi \in \mathcal{D}(\mathfrak{a}_\varepsilon(\tau))$ and $\varepsilon > 0$.

Let $\mathfrak{a}_0(\tau) = \lim_{\varepsilon \to 0} \mathfrak{a}_\varepsilon(\tau)$. Then $\mathfrak{a}_0(\tau) = \tau\mathfrak{a}_0(e)$, so the complement of $\mathfrak{a}_0(\tau)$ is $-\tau({}^+\mathfrak{a}) \cup \{0\} = \sigma^{-1}({}^+\mathfrak{a}) \cup \{0\}$. Since $\check{\psi}(H) = \varphi(\sigma^{-1}H)$, (5.16) implies

$$\sum_{p=1}^{\ell(\delta)} \partial\left(P^\delta_{pj}\right) \zeta_{pi} = 0 \quad \text{on } \mathfrak{a} \backslash {}^+\mathfrak{a} \,. \tag{5.17}$$

The matrix P^δ with entries $P^\delta_{pj} \in S(\mathfrak{a})$ satisfies

$$\det\left(P^\delta\right) = P_c P^\delta\,, \tag{5.18}$$

where P_c is a matrix whose entries $P^c_{pi} \in S(\mathfrak{a})$ are the cofactors of P^δ. Applying $\partial(P^c_{pj})$ to (5.17) and summing on j, we get by (5.18),

$$(\partial(\det P^\delta)\zeta_{pi})(H) = 0 \quad \text{for } H \in \mathfrak{a}\backslash{}^+\mathfrak{a}\,.$$

Now $\det P^\delta(\lambda) = \det Q^\delta(i\lambda)$, which by theorem 4.2 and corollary 11.3 in [11, III] is a product of factors $\langle\lambda,\alpha\rangle + c_{j,\alpha}$ where $c_{j,\alpha} > 0$. Here $\alpha \in \sum_0^+$ and j runs through a certain finite set. (Apart from variable shifts, this is contained in theorems 5 and 7 of [13].) With $H_\alpha \in \mathfrak{a}$ given by $\langle H_\alpha, H\rangle = \alpha(H)$, we thus have for $F = \zeta_{pi}$, which belongs to $\mathcal{S}(\mathfrak{a})$,

$$\prod_{\alpha,j}(\partial(H_\alpha) + c_{j,\alpha})F = 0 \quad \text{on } \mathfrak{a}\backslash{}^+\mathfrak{a}\,. \tag{5.19}$$

We pull out a single factor $\partial(H_\alpha) + c$ $(c > 0)$ and consider the equation

$$(\partial(H_\alpha) + c)f = 0 \quad \text{in } \mathfrak{a}\backslash{}^+\mathfrak{a}\,,\ f \in \mathcal{S}(\mathfrak{a})\,.$$

Then the function

$$g(H) = e^{c_0\alpha(H)}f(H)\,, \quad c_0 = c/\langle\alpha,\alpha\rangle \tag{5.20}$$

satisfies

$$\partial(H_\alpha)g = 0 \quad \text{in } \mathfrak{a}\backslash{}^+\mathfrak{a}\,. \tag{5.21}$$

Fix $H_0 \in \mathfrak{a}\backslash{}^+\mathfrak{a}$. Since $-H_\alpha \in \mathfrak{a}\backslash{}^+\mathfrak{a}$ and since $\mathfrak{a}\backslash{}^+\mathfrak{a}$ is star-shaped with respect to 0, we see that $H_0 - tH_\alpha = t(\frac{H_0}{t} - H_\alpha)$ belongs to $\mathfrak{a}\backslash{}^+\mathfrak{a}$ for $t > 0$ sufficiently large. Then (5.21) implies that the function $t \to g(H_0 - tH_\alpha)$ is a constant C for large $t > 0$, so by (5.20)

$$f(H_0 - tH_\alpha) = Ce^{-C_0\alpha(H_0)}e^{C_0\alpha(H_\alpha)t}\,. \tag{5.22}$$

This contradicts $f \in \mathcal{S}(\mathfrak{a})$ unless $f \equiv 0$ on $\mathfrak{a}\backslash{}^+\mathfrak{a}$. Iterating this argument for the factors in (5.19), we deduce

$$\zeta_{pi} \equiv 0 \quad \text{on } \mathfrak{a}\backslash{}^+\mathfrak{a}\,, \quad 1 \leq i\,,\ p \leq \ell(\delta)\,. \tag{5.23}$$

Observe now that for each $k_0 \in K$ assumption (5.1) is valid for the function $u_{k_0} : (x,H) \to u(k_0 \cdot x, H)$. Since $A(k_0 \cdot x, b) = A(x, k_0^{-1} \cdot b)$, it is clear from (4.25) that replacing u by u_{k_0} amounts to replacing h_σ by the function $(b, a) \to h_\sigma(k_0 \cdot b, a)$. Thus (5.23) implies

$$\int_{K/M} h_\sigma(k_0 kM, \exp H)\langle v_i, \delta(k)v_p\rangle \, dk_M = 0\,, \quad H \in \mathfrak{a}\backslash{}^+\mathfrak{a}\,, \tag{5.24}$$

and here we put $u = k_0 k$, $k = k_0^{-1}u$. Since the vectors $\delta(k_0)v_i$ ($k_0 \in K$, $1 \leq i \leq \ell(\delta)$) span V_δ, (5.24) shows that the function $uM \to h_\sigma(uM, \exp H)$ is orthogonal to all $\langle v, \delta(k)v_p\rangle (1 \leq p \leq \ell(\delta), v \in V_\delta)$. Since $\delta \in \widehat{K}_M$ is arbitrary this proves

$$h_\sigma(b, \exp H) = 0 \quad \text{for } b \in B\,,\ H \in \mathfrak{a}\backslash{}^+\mathfrak{a}\,,$$

so (5.2) is proved.

6 Energy and Spectral Representation

In [18, sec. 2], Shahshahani transfers the Cauchy problem (2.7)–(2.8) to a vector formulation and uses this to generalize the classical energy for the wave equation to the present context. We recall the definition.

Let $p \in S(\mathfrak{a})$. By Lemma 4.3(i) we have a matrix $L_p = (L_p)_{ij}$ with entries in $I(\mathfrak{a})$ such that

$$pp_j = \sum_1^w (L_p)_{ij} p_i\,, \quad 1 \leq j \leq w\,.$$

Let $D_p = (D_p)_{ij}$ denote the corresponding matrix with entries in $\mathbf{D}(G/K)$, that is, $\Gamma((D_p)_{ij}) = (L_p)_{ij}$ (cf. (2.5)). Let ${}^{\mathsf{T}}D_p$ denote the transpose of D_p.

LEMMA 6.1 *The Cauchy problem* (2.7)–(2.8) *is equivalent to the problem*

$${}^{\mathsf{T}}D_p\, \mu = \partial(p)\mu\,, \quad p \in S(\mathfrak{a})\,, \tag{6.1}$$

where the column vector μ is a smooth map from $X \times \mathfrak{a}$ to $\mathbf{C}^w$ and

$${}^{\mathsf{T}}\mu(x, 0) = (f_1(x), \dots, f_w(x))\,. \tag{6.2}$$

More precisely, if u satisfies (2.7)–(2.8), *then*

$${}^{\mathsf{T}}\mu(x, H) = ((\partial(p_1)u)(x, H), \dots, (\partial(p_w)u)(x, H)) \tag{6.3}$$

satisfies (6.1)–(6.2) *above, and if μ satisfies* (6.1)–(6.2), *then $u = \mu_1$ (first component) satisfies* (2.7)–(2.8).

PROOF: Since $p_k = p_k p_1 = \sum_i (L_{p_k})_{i1} p_i$, we have $(L_{p_k})_{i1} = \delta_{ik}$. Thus, if μ satisfies (6.1)–(6.2), then

$$\partial(p_k)\mu_1 = (\partial(p_k)\mu)_1 = ({}^t D_{p_k}\mu)_1 = \sum_i (D_{p_k})_{i1}\mu_i = \mu_k$$

so (2.8) holds. Also, if $D \in \mathbf{D}(G/K)$, $L_{\Gamma(D)}$ is a diagonal matrix, ${}^{\mathsf{T}}D_{\Gamma(D)} = DI_w$, so (6.1) implies $D\mu_1 = \partial(\Gamma(D))\mu_1$. Thus $u = \mu_1$ satisfies (2.7).

On the other hand, assume u satisfies (2.7)–(2.8). Define μ by (6.3). Then if $p \in S(\mathfrak{a})$,

$$\begin{aligned}(\partial(p)\mu)_j &= \partial(p)\mu_j = \partial(p)\partial(p_j)u = \partial(pp_j)u \\ &= \partial\Big(\sum_i (L_p)_{ij} p_i\Big)(u) = \sum_i \partial(p_i)\partial(\Gamma(D_p)_{ij})u \\ &= \sum_i \partial(p_i)(D_p)_{ij}u = \sum_i (D_p)_{ij}\partial(p_i)u = ({}^t D_p\mu)_j\end{aligned}$$

so (6.1) holds; also (6.2) is obvious. ∎

Remark. If (6.1) holds for $p \in \mathfrak{a}$, it holds for all $p \in S(\mathfrak{a})$. In fact, the map $p \to L_p$ is an isomorphism of $S(\mathfrak{a})$ onto an algebra of $w \times w$ matrices with entries in $I(\mathfrak{a})$. Thus $\partial(pq)\mu = {}^{\mathsf{T}}D_{pq}\mu$, justifying the remark.

Consider now the matrix $A = (A_{ij})$ with entries in $I(\mathfrak{a})$ given by

$$A_{ij} = (\pi q^j, \theta(\pi q^i))\,, \tag{6.4}$$

where $(\cdot,\cdot)$ is defined in (4.4). Let $\mathcal{A} = (\mathcal{A}_{ij})$ be the matrix with entries in $\mathbf{D}(G/K)$ given by $\Gamma(\mathcal{A}_{ij}) = A_{ij}$.

DEFINITION Given $u, v \in \mathcal{E}(X \times \mathfrak{a})$, the *energy form* is defined by

$$E(u, v; H) = \int_X ({}^{\mathsf{T}}\mu\mathcal{A}\bar{\nu})(x, H)\,dx\,, \tag{6.5}$$

the integral assumed convergent. Here ${}^{\mathsf{T}}\mu$ and ν are given by (6.3) from u and v, respectively.

It is a routine matter to verify that the function ${}^{\mathsf{T}}\mu\mathcal{A}\bar{\nu}$ is independent of the choice of basis of $H(\mathfrak{a})$. However, the homogeneity of each p_i plays a role in the proof of Theorem 6.2 and in Theorem 6.3.

In the rank-one example discussed before, this reduces to

$$E(u,v;tH_0)=\frac{\langle\pi,\pi\rangle}{2}\int_X\left(u_t\overline{v}_t-u\left(L+|\rho|^2\right)\overline{v}\right)dx\,,$$

which up to the factor $\langle\pi,\pi\rangle$ agrees with the usual formula [11, p. 501].

As used in the proof of (4.1), we have

$$\Gamma(D^*)=\theta\Gamma(D)\,,\quad D\in\mathbf{D}(G/K)\,.$$

Hence

$$\Gamma(\mathcal{A}_{ij}^*)(i\lambda)=\Gamma(\mathcal{A}_{ij})(-i\lambda)=A_{ij}(-i\lambda)=A_{ji}(i\lambda)=\Gamma(\mathcal{A}_{ji})(i\lambda)$$

so $\mathcal{A}_{ij}^*=\mathcal{A}_{ji}$. This implies the Hermitian symmetry,

$$E(u,v;H)=\overline{E(v,u;H)}\,.$$

While this relation is a simple formality, the positivity

$$E(u,u;H)\geq 0\,,$$

which is a consequence of (6.11) below in connection with Theorem 6.2, is subtler and seems to require the Fourier transform theory. Because of (6.11) we have a norm

$$\|F\|=\left\{\int_X\left({}^{\mathsf{T}}F\mathcal{A}\overline{F}\right)(x)\,dx\right\}^{1/2},\quad {}^{\mathsf{T}}F=(f_1,\ldots,f_w)$$

on $\mathcal{D}(X)\times\cdots\times\mathcal{D}(X)$ (w times), which by the injectivity in Theorem 6.4 is strictly positive for $F\not\equiv 0$.

THEOREM 6.2 *Let u and v be the solutions to* (2.7)–(2.8) *with initial data* $(f_i)\subset\mathcal{D}(X)$, $(g_j)\subset\mathcal{E}(X)$ *(Theorem* 4.2). *Then the energy form $E(u,v;H)$ is constant in the time $H\in\mathfrak{a}$.*

PROOF (Shahshahani [18]): Consider the matrix $B=(B_{ij})$ given by

$$B_{ij}=(\pi q^i,\pi q^j)\in I(\mathfrak{a})$$

and J the diagonal matrix whose i^{th} entry is $(-1)^{d_i}$ where $d_i=\deg p_i=-\deg(q^i)$. Then $A_{ij}=(\pi q^j,\theta(\pi q^i))=(-1)^{\deg(\pi)+d_i}B_{ij}$, so if $\mathcal{B}=(\mathcal{B}_{ij})$ with $\Gamma(\mathcal{B}_{ij})=B_{ij}$,

$$(-1)^{\deg(\pi)}JB=A\,,\quad(-1)^{\deg(\pi)}J\mathcal{B}=\mathcal{A}\,.$$

Taking the inner product with q^m and p_k, respectively, we derive the relations

$$\sum_i B_{ik}p_i = \pi^2 q^k\,, \quad pq^j = \sum_i (L_p)_{ji} q^i\,,\ p \in S(\mathfrak{a})\,,$$

which quickly imply that the matrix L_pB is symmetric. Thus

$$L_pB = B^{\mathsf{T}}L_p\,, \quad D_p\mathcal{B} = \mathcal{B}^{\mathsf{T}}D_p\,.$$

In particular, take $p = H \in \mathfrak{a}$ and observe that $\partial(p)\,{}^{\mathsf{T}}\mu = {}^{\mathsf{T}}(\partial(p)\mu)$. Then if u and v are solutions and μ and ν are determined by (6.3), we have (by (6.1))

$$\begin{aligned}
&\partial(p)_H(E(u,v;H))\\
&= \int_X \left[{}^{\mathsf{T}}\left({}^{\mathsf{T}}D_p\mu\right)A\overline{\nu} + {}^{\mathsf{T}}\mu A\partial(p)\overline{\nu}\right](x,H)dx\\
&= (-1)^{\deg(\pi)}\int_X \left[{}^{\mathsf{T}}\left({}^{\mathsf{T}}D_p\mu\right)J\mathcal{B}\overline{\nu} + {}^{\mathsf{T}}\mu J\mathcal{B}\,{}^{\mathsf{T}}D_p\overline{\nu}\right](x,H)dx\,.
\end{aligned}$$

In the last term we replace $\mathcal{B}^{\mathsf{T}}D_p$ by $D_p\mathcal{B}$ and put $\xi = \mathcal{B}\overline{\nu}$. The integrand is a sum of the terms

$$\begin{aligned}
\sum_j \left({}^{\mathsf{T}}\left({}^{\mathsf{T}}D_p\mu\right)\right)_j (J\xi)_j &= \sum_{i,j}(D_p)_{i,j}\mu_i(-1)^{d_j}\xi_j\,,\\
\sum_i ({}^{\mathsf{T}}\mu)_i(JD_p\xi)_i &= \sum_{i,j}\mu_i(-1)^{d_i}(D_p)_{ij}\xi_j\,,
\end{aligned}$$

so it would suffice to prove

$$\int_X \left[((D_p)_{ij}\mu_i)\xi_j + (-1)^{d_i-d_j}\mu_i(D_p)_{ij}\xi_j\right](x,H)\,dx = 0\,.$$

Since $pp_j = \sum_i (L_p)_{ij}p_i$, $(L_p)_{ij}$ is either 0 or homogeneous of degree $1+d_j-d_i$. Also $\Gamma(D^*) = \theta\Gamma(D)$ as used before, so

$$\Gamma((D_p)_{ij}{}^*) = \theta((L_p)_{ij}) = (-1)^{1+d_j-d_i}\Gamma((D_p)_{ij})\,,$$

whence

$$(D_p)_{ij}{}^* = (-1)^{1+d_j-d_i}(D_p)_{ij}\,.$$

Thus the last integral equals

$$\int_X [((D_p)_{ij}\mu_i)\xi_j - \mu_i(D_p)_{ij}{}^*\xi_j]\,(x,H) = 0\,.$$

Hence $\partial(p)_H(E(u,v;H)) = 0$ as claimed. ∎

Remark. In [17, prop. 1], a similar transfer of (2.7)–(2.8) to a vector formulation is described. The corresponding energy is given by (4) in [17, sec. 1.2]. It differs from formula (6.5) above in that the Cartan involution in (6.4) is absent. Thus the positivity of the energy stated in theorem 1 of [17] seems doubtful. The unproved statement in proposition 3 in [17] that this energy is time-independent seems to us not to be compatible with Theorem 6.2.

As in [17] and [18], we consider now for each $\sigma \in W$, $\lambda \in \mathfrak{a}_C^*$, and $b \in B$ the particular solution $\mu^\sigma(x, H; \lambda, b)$ of (6.1) given by (6.3) for

$$u^\sigma(x, H; \lambda, b) = e^{i\lambda(H)+(i\sigma\lambda+\rho)(A(x,b))}\,. \tag{6.6}$$

This is indeed a solution, for by (3.2),

$$(Du^\sigma) = \Gamma(i\sigma\lambda)u^\sigma = \Gamma(i\lambda)u^\sigma\,, \quad D \in \mathbf{D}(G/K)\,,$$

and

$$\partial(\Gamma(D))u^\sigma = \Gamma(i\lambda)u^\sigma\,.$$

We consider then the linear map

$$\mathcal{E}^\sigma : F(x) \to \int_X ({}^{\mathsf{T}}F\mathcal{A}\overline{\mu}^\sigma)(x, 0\,; \lambda, b)dx\,, \tag{6.7}$$

where ${}^{\mathsf{T}}F(x) = (f_1(x), \ldots, f_w(x))$, $f_i \in \mathcal{D}(X)$. Thus $\mathcal{E}^\sigma$ maps $\mathcal{D}(X) \times \cdots \times \mathcal{D}(X)$ into a function space on $\mathfrak{a}^* \times B$. The following result shows that $\mathcal{E}^\sigma$ is intimately related to the Fourier transform on X:

THEOREM 6.3 *For each $\sigma \in W$*

$$\mathcal{E}^\sigma(F)(\lambda, b) = \pi(\lambda)^2 \sum_k q^k(i\lambda)\widetilde{f}_k(\sigma\lambda, b)\,.$$

PROOF: For $\lambda \in \mathfrak{a}^*$ (real) we have

$$\mu_j^\sigma(x, 0\,; \lambda, b)^- = p_j(-i\lambda)e^{(-i\sigma\lambda+\rho)(A(x,b))}$$

so

$$\mathcal{E}^\sigma(F) = \int_X \sum_{i,j} f_i(x)A_{ij}(-i\sigma\lambda)p_j(-i\lambda)e^{(-i\sigma\lambda+\rho)(A(x,b))}\,dx\,, \tag{6.8}$$

which by the W-invariance of A_{ij} equals

$$\sum_k \widetilde{f}_k(\sigma\lambda, b) \sum_j A_{kj}(-i\lambda)p_j(-i\lambda)\,. \tag{6.9}$$

Here the sum $\sum_j$ equals

$$\sum_j \left[\sum_{\tau \in W} (\pi q^j)(-\tau i\lambda)(\pi q^k)(\tau i\lambda) \right] p_j(-i\lambda) = \\ \sum_{\tau \in W} (\pi q^k)(\tau i\lambda) \sum_j (\pi q^j)(-\tau i\lambda) p_j(-i\lambda) .$$

Now if d_j = degree p_j, then degree $q^j = -d_j$, so the last $\sum_j$ equals

$$(-1)^{\deg \pi} \sum_j (\pi q^j)(\tau i\lambda) p_j(i\lambda) .$$

Thus

$$\begin{aligned} \sum_j A_{kj}(-i\lambda) p_j(-i\lambda) &= (-1)^{\deg(\pi)} \sum_j p_j(i\lambda) \sum_\sigma (\pi q^k)(\sigma i\lambda)(\pi q^j)(\sigma i\lambda) \\ &= (-1)^{\deg(\pi)} \sum_j p_j(i\lambda)(\pi q^k, \pi q^j)(i\lambda) . \end{aligned}$$

However, $\sum_j p_j(\pi q^k, \pi q^j) = \pi^2 q^k$, as we see by taking the inner product with q^m and noting that $(\pi^2 q^k, q^m) = (\pi q^k, \pi q^m)$. Putting these formulas together, we see that expression (6.9) equals

$$\sum_k \widetilde{f}_k(\sigma\lambda, b)(-1)^{\deg \pi} \pi(i\lambda)^2 q^k(i\lambda) ,$$

and this proves the theorem. ∎

Given $H_0 \in \mathfrak{a}$, let U_{H_0} denote the map of $\mathcal{D}(X) \times \cdots \times \mathcal{D}(X)$ into itself given by

$$U_{H_0} : \mu(x, 0) \to \mu(x, H_0) , \tag{6.10}$$

μ satisfying (6.1). Since $\partial(p)$ is invariant under translations of $\mathfrak{a}$, it follows that for each $H \in \mathfrak{a}$, U_{H_0} maps the function $x \to \mu(x, H)$ into the function $x \to \mu(x, H + H_0)$.

In view of Theorem 6.3, the following result can be viewed as an enlargement of the Plancherel theorem for the Fourier transform on X. In fact, it will be proved by the tools described in Section 3.

THEOREM 6.4 *For each $\sigma \in W$ the map $\mathcal{E}^\sigma$ is an injective norm-preserving map of $\mathcal{D}(X) \times \cdots \times \mathcal{D}(X)$ onto a dense subspace of*

$$L^2(\mathfrak{a}^* \times B, \, d\lambda \, db / \, |\pi(\lambda)\mathbf{c}(\lambda)|^2)$$

Thus

$$\int_{\mathfrak{a}^*\times B} \mathcal{E}^\sigma(F)\overline{\mathcal{E}^\sigma(F)}(\lambda,b)\frac{d\lambda\, db}{|\pi(\lambda)\mathbf{c}(\lambda)|^2} = \int_X \left({}^T\!F\mathcal{A}\overline{F}\right)(x)dx\,. \tag{6.11}$$

Moreover, $\mathcal{E}^\sigma$ *intertwines* U_{H_0} *and the endomorphism*

$$e(H_0) : \varphi(\lambda,b) \to e^{i\lambda(H_0)}\varphi(\lambda,b) \quad \textit{of}\; L^2\Big(a^*\times B, \frac{d\lambda\, db}{|\pi(\lambda)\mathbf{c}(\lambda)|^2}\Big),$$

that is,

$$\mathcal{E}^\sigma U_{H_0} = e(H_0)\mathcal{E}^\sigma\,.$$

PROOF: We separate the various statements.

1. *The norm identity* (6.11). We have by Proposition 6.3,

$$\mathcal{E}^\sigma(F)\overline{\mathcal{E}^\sigma(F)} = \pi(\lambda)^4\sum_{k,\ell} q^k(i\lambda)q^\ell(-i\lambda)\widetilde{f}_k(\sigma\lambda,b)\overline{\widetilde{f}_\ell(\sigma\lambda,b)} \tag{6.12}$$

and by (12) in [11, III, sec. 1],

$$\int \widetilde{f}_k(\sigma\lambda,b)\overline{\widetilde{f}_\ell(\sigma\lambda,b)}db$$

is independent of σ. Thus the left-hand side of (6.11) equals

$$\frac{1}{w}\int_{\mathfrak{a}^*\times B}\sum_{k,\ell}\pi(\lambda)^2\widetilde{f}_k(\lambda,b)\overline{\widetilde{f}_\ell(\lambda,b)}\sum_{\sigma\in W}q^k(\sigma i\lambda)q^\ell(-\sigma i\lambda)\frac{d\lambda db}{|\mathbf{c}(\lambda)|^2}\,.$$

By writing $\pi(\lambda)^2 = \pi(\sigma i\lambda)\pi(-\sigma i\lambda)$, our expression becomes

$$\frac{1}{w}\int_{\mathfrak{a}^*\times B}\sum_{k,\ell}\widetilde{f}_k(\lambda,b)\overline{\widetilde{f}_\ell(\lambda,b)}A_{\ell k}(i\lambda)\frac{d\lambda\, db}{|\mathbf{c}(\lambda)|^2}\,. \tag{6.13}$$

On the other hand, if $g\in\mathcal{D}(X)$, we have the inversion formula

$$g(x) = \frac{1}{w}\int_{\mathfrak{a}^*\times B}\widetilde{g}(\lambda,b)e^{(i\lambda+\rho)(A(x,b))}\,|\mathbf{c}(\lambda)|^{-2}\,d\lambda\, db\,.$$

Also $A_{\ell k}$ is real, so $\mathcal{A}_{\ell k}$ has real coefficients and

$$(\mathcal{A}_{\ell k}g)^- = \mathcal{A}_{\ell k}\overline{g}\,,\quad \Gamma(\mathcal{A}_{\ell k})(i\lambda) = A_{\ell k}(i\lambda)\,. \tag{6.14}$$

By the Plancherel formula and (6.14),

$$\int_X f(x)(\mathcal{A}_{k\ell}\bar{g})(x)dx =$$

$$\frac{1}{w}\int_{\mathfrak{a}^*\times B}\widetilde{f}(\lambda,b)\Big(A_{k\ell}(i\lambda)\widetilde{g}(\lambda,b)\Big)^{-}|c(\lambda)|^2\,d\lambda\,db\,.$$

However, the definition (6.4) easily shows that for λ real, $\overline{A_{k\ell}(i\lambda)} = A_{\ell k}(i\lambda)$, and we introduce this in the last integral. Taking now $f = f_k$ and $g = f_\ell$ and summing on k and ℓ, we see that the right-hand side of (6.11) equals (6.13). This proves (6.11).

2. *The injectivity.* For this, suppose $\mathcal{E}^\sigma(F) = 0$. Then by Theorem 6.3

$$\sum_k q^k(i\lambda)\widetilde{f}_k(\sigma\lambda,b) = 0\,. \tag{6.15}$$

Here we multiply by $e^{(i\sigma\lambda+\rho)(A(x,b))}$ and integrate over B. Since this integral

$$\int_B \widetilde{f}_k(\lambda,b)e^{(i\lambda+\rho)(A(x,b))}\,db \tag{6.16}$$

is W-invariant in λ, we derive from (6.15) for each $s\in W$,

$$\sum_k q^k(si\lambda)\int_B \widetilde{f}_k(\lambda,b)e^{(i\lambda+\rho)(A(x,b))}\,db = 0\,.$$

We multiply by $p_\ell(si\lambda)$ and sum on s. This implies that (6.16) vanishes identically in x. By the injectivity (see [9]) of the Poisson transform for $\lambda\in\mathfrak{a}^*$, we have $\widetilde{f}_k(\lambda,b)\equiv 0$, whence by the injectivity of the Fourier transform, $f_k\equiv 0$.

The proof has the following consequence:

COROLLARY 6.5 (Schlichtkrull)

$$f_k(x) = \int_{\mathfrak{a}^*\times B} p_k(i\lambda)\mathcal{E}^\sigma(F)(\lambda,b)e^{(i\sigma\lambda+\rho)(A(x,b))}\frac{d\lambda\,db}{\pi(\lambda)^2\,|\mathbf{c}(\lambda)|^2}\,.$$

In fact, substitute for $\mathcal{E}^\sigma(F)$ from Theorem 6.3 and use the W-invariance of (6.16). Again, replacing λ by $s\lambda$ and summing on s the right-hand side reduces to $f_k(x)$ because of (3.4).

3. *The surjectivity.* Because of Theorem 6.3 this amounts to showing that if $g(\lambda, b)$ satisfies

$$\int_{\mathfrak{a}^* \times B} \widetilde{f}(\sigma\lambda, b)\pi(\lambda)^2 q^j(i\lambda)g(\lambda, b)\frac{d\lambda\, db}{|\pi(\lambda)\mathbf{c}(\lambda)|^2} = 0$$

for all $f \in \mathcal{D}(X)$ and all j $(1 \le j \le w)$, then $g \equiv 0$. Putting $g^j(\lambda, b) = q^j(\sigma^{-1}i\lambda)g(\sigma^{-1}\lambda, b)$, we then have

(6.17)
$$\int_{\mathfrak{a}^* \times B} \widetilde{f}(\lambda, b)g^j(\lambda, b)\, |\mathbf{c}(\lambda)|^{-2}\, d\lambda\, db = 0\,, \quad f \in \mathcal{D}(X)\,,\ 1 \le j \le w\,.$$

If we had this relation on $\mathfrak{a}^*_+ \times B$, the problem would be solved since by the Plancherel theorem (Section 3) the functions $\widetilde{f}(\lambda, b)$ are dense in $L^2(\mathfrak{a}^*_+ \times B)\,, |\mathbf{c}(\lambda)|^{-2}\, d\lambda\, db$. The fact that (6.17) holds for all j is decisive in passing from $\mathfrak{a}^*_+ \times B$ to $\mathfrak{a}^* \times B$.

Once again we invoke the representations $\delta \in \widehat{K}_M$ and the results for the δ-spherical transform described in Section 3. We specialize f in (6.17) to $f \in \mathcal{D}_{\check{\delta}}(X)$ and put

(6.18)
$$g^j_\delta(\lambda) = \int_K g^j(\lambda, kM)\delta(k)dk\,, \qquad g_\delta(\lambda) = \int_K g(\lambda, kM)\delta(k)dk\,.$$

If A, $B(k) \in \mathrm{Hom}(V_\delta, V_\delta)$, then

$$\int_K \mathrm{Tr}(AB(k))\, dk = \mathrm{Tr}\left(A\int_K B(k)\, dk\right).$$

Using in addition $\mathfrak{a}^* = \bigcup_{s \in W} s\overline{\mathfrak{a}^*_+}$ as well as formula (3.9), we derive from (6.17) the relation

$$\sum_{s\in W}\int_{\mathfrak{a}^*_+} \mathrm{Tr}\left(\widetilde{f}(s\lambda)g^j_\delta(s\lambda)\right)|\mathbf{c}(\lambda)|^{-2}\, d\lambda = 0\,. \tag{6.19}$$

Now according to (3.11) and (3.8)

$$\widetilde{f}(s\lambda) = Q^{\check{\delta}}(s\lambda)F_\delta(\lambda) = Q^{\check{\delta}}(s\lambda)C_{\check{\delta}}C_{\check{\delta}}^{-1}F_\delta(\lambda)\,, \tag{6.20}$$

where F_δ is a W-invariant holomorphic function on $\mathfrak{a}^*_C$, rapidly decreasing on $\mathfrak{a}^*$ and with values in $\mathrm{Hom}(V_\delta, E_{\check{\delta}})$. We insert the expression (6.20) into (6.19). The map $Q^{\check{\delta}}(s\lambda)C_{\check{\delta}}$ is a linear transformation of $V_{\check{\delta}}^M$ into itself, but we extend it to $V_{\check{\delta}}$ defining it to be 0 on the orthocomplement of $V_{\check{\delta}}^M$ in $V_{\check{\delta}}$. Then $g_{\delta}^j(s\lambda)$ can be shifted to the left in (6.19), and we obtain

$$\int_{\mathfrak{a}_+^*} \operatorname{Tr}(G_j(\lambda)F_\delta^o(\lambda))\,|\mathbf{c}(\lambda)|^{-2}\,d\lambda = 0\,, \tag{6.21}$$

where

$$G_j(\lambda) = \sum_{s\in W} g_\delta^j(s\lambda)Q^{\check{\delta}}(s\lambda)C_{\check{\delta}}\,, \quad \lambda \in \mathfrak{a}^*\,. \tag{6.22}$$

$$F_\delta^o(\lambda) = C_{\check{\delta}}^{-1}F_\delta(\lambda)\,, \qquad \lambda \in \mathfrak{a}^*\,. \tag{6.23}$$

Here $F_\delta^o(\lambda)$ maps V_δ into $V_{\check{\delta}}^M$. By the Paley-Wiener theorem for the δ-spherical transform (see remarks following (3.11)), we can, for each i and k $(1 \le i \le \ell(\delta))$, $1 \le k \le d(\delta))$, take $F_\delta^o(\lambda) = \varphi(\lambda)P^{ik}$ where $\varphi(\lambda)$ is an arbitrary W-invariant holomorphic scalar-valued function of exponential type on $\mathfrak{a}_C^*$, rapidly decreasing on $\mathfrak{a}^*$, and $P^{ik}(v_m) = \delta_{km}v_i$ $(1 \le m \le d(\delta))$. Then

$$\operatorname{Tr}(G_j(\lambda)F_\delta^o(\lambda)) = \varphi(\lambda)G_j(\lambda)_{ki}\,.$$

Thus by (6.21),

$$G_j(\lambda)_{ki} = 0\,, \quad 1 \le i \le \ell(\delta)\,,\ 1 \le k \le d(\delta)\,, \tag{6.24}$$

for almost all $\lambda \in \mathfrak{a}_+^*$ and hence by the W-invariance for almost all $\lambda \in \mathfrak{a}^*$. This means that $G_j(\lambda)$ maps $V_{\check{\delta}}^M$ into 0 and by definition it is 0 on $(V_{\check{\delta}}^M)^\perp$. Thus

$$G_j(\lambda) = 0 \quad \text{for almost all } \lambda \in \mathfrak{a}^*\,,\ 1 \le j \le w\,. \tag{6.25}$$

We now have by (6.18)

$$g_\delta^j(s\lambda) = q^j(\sigma^{-1}si\lambda)g_\delta(\sigma^{-1}s\lambda)\,, \quad 1 \le j \le w\,. \tag{6.26}$$

If $a_{mn}(\lambda)$ is an arbitrary matrix entry in $g_\delta(\sigma^{-1}\lambda)Q^{\check{\delta}}(\lambda)C_{\check{\delta}}$ and $s_1, \ldots,$ s_w are the elements in W, then by (6.25)–(6.26),

$$\sum_{r=1}^{w} q^j(\sigma^{-1}s_ri\lambda)a_{mn}(s_r\lambda) = 0\,, \quad 1 \le j \le w\,. \tag{6.27}$$

Because the matrix $\{q^j(s_r\nu)\}_{1\leq j,r\leq w}$ is nonsingular for regular ν [2, cor. 5.5.2], (6.27) implies

$$g_\delta(\sigma^{-1}\lambda)Q^{\check\delta}(\lambda)C_{\check\delta} = 0 \quad \text{for almost all } \lambda\,.$$

On $V_{\check\delta}^M$, $\det(Q^{\check\delta}(\lambda)C_{\check\delta}) \not\equiv 0$ so

$$g_\delta(\sigma^{-1}\lambda)v = 0 \quad \text{for } v \in V_{\check\delta}^M\,. \tag{6.28}$$

However, if $u \in V_\delta$ we have for $m \in M$

$$\begin{aligned} g_\delta(\lambda)u &= \int_K g(\lambda, kM)\delta(k)u\,dk \\ &= \int_K g(\lambda, kM)\delta(km)u\,dk \\ &= \int_M g(\lambda, kM)\delta(k)\,dk \int_M \delta(m)u\,dm \end{aligned}$$

which vanishes by (6.28). Thus $g_\delta(\lambda) = 0$ for all $\delta \in \widehat{K}_M$, whence $g(\lambda, b) = 0$ for almost all $\lambda \in \mathfrak{a}^*$ as desired.

4. *Intertwining.* Fix $H_0 \in \mathfrak{a}$. As remarked after (6.10), U_{H_0} maps the function $x \to \mu(x, H)$ into the function $x \to \mu(x, H+H_0)$, and the solution u equals μ_1. Passing from $u(x, H)$ in (4.21) to $u(x, H+H_0)$ amounts to replacing $\widetilde{f}_k(\lambda, b)$ by $\widetilde{f}_k(\lambda, b)e^{i\lambda(\sigma H_0)}$ (and $\widetilde{f}_k(\sigma\lambda, b)$ by $\widetilde{f}_k(\sigma\lambda, b)e^{i\lambda(H_0)}$). Now Theorem 6.3 shows that

$$\mathcal{E}^\sigma U_{H_0}(F) = e^{i\lambda(H_0)}\mathcal{E}^\sigma(F)\,. \qquad \blacksquare$$

Remarks.

1. While Theorem 6.3 is new, a proof of Theorem 6.4 for the case $\sigma = e$ appears in [18, pp. 214–224]. The proof is entirely different. In [17], theorem 1 is formally similar to Theorem 6.4 but differs from it in substance because the maps corresponding to $\mathcal{E}^\sigma$ are given in terms of the different definition of energy indicated in the remark following Theorem 6.2. We observe that in contrast to our proof of Theorem 6.4, the sketched proof of theorem 1 in [17] relies on the time invariance of the energy, which, as remarked before, seems at odds with Theorem 6.2.

2. In [12] the range of $\mathcal{E}^\sigma$ is determined more explicitly.

We conclude with the analogue of Friedlander's theorem 5.1. More general limit results appear in [17], and except for the formulation, the result with the proof below coincides with theorem 3.11 in [16].

In analogy with (4.23), let Q_k, J, and $\overline{J}$ denote the operators on $\mathcal{S}(A)$ that under the Fourier transform on A correspond to multiplication by $q^k(i\lambda)\pi(i\lambda)$, $c(\lambda)^{-1}$, and $c(-\lambda)^{-1}$, respectively.

THEOREM 6.6 *Let $\sigma \in W$ and $H, H_0 \in \mathfrak{a}^+$, and put $a_t = \exp(tH)$. Then if $k \in K$,*

$$\lim_{t\to+\infty} e^{t\rho(H)}(\partial(\pi)u)(ka_t \cdot o, -t\sigma H + \sigma H_0) = \left(\overline{J}\sum_k Q_k^{\sigma^{-1}}(e^\rho \widehat{f}_k)\right)(kM, H_0)\,.$$

PROOF: Because of (4.21),

$$(\partial(\pi)u)(x,H) = \int_{\mathfrak{a}^*\times B} F_\sigma(\lambda,b)e^{(i\lambda+\rho)(A(x,b))}e^{i\sigma\lambda(H)}\frac{d\lambda\,db}{|c(\lambda)|^2},$$

where

$$F_\sigma(\lambda,b) = \sum_k \widetilde{f}_k(\lambda,b)q^k(i\sigma\lambda)\pi(i\sigma\lambda)\,.$$

∎

Thus if $\ell \in K$

(6.29)
$$e^{t\rho(H)}(\partial(\pi)u)(\ell a_t\cdot o, -t\sigma H + \sigma H_o) = e^{t\rho(H)}\int_{\mathfrak{a}^*\times B} F_\sigma(\lambda,b)e^{(i\lambda+\rho)(A(a_t\cdot o,\ell^{-1}b))}e^{-it\lambda(H)+i\lambda(H_0)}\frac{d\lambda\,db}{|c(\lambda)|^2}.$$

By the K-invariance of db, we can replace

(6.30) $\quad F_\sigma(\lambda,b)$ by $F_\sigma(\lambda,\ell\cdot b)$, $\qquad A(a_t\cdot o,\ell^{-1}\cdot b)$ by $A(a_t\cdot o,b)$.

In the Iwasawa decomposition $G = KAN$, we write

(6.31) $$g = k(g)\exp H(g)n(g)\,,\quad H(g)\in\mathfrak{a}\,.$$

Then $A(gK, kM) = -H(g^{-1}k)$. If $b = k'M$ and we define $k \in K$ by $k' = k(a_t k)$, then $kM \to k'M$ is a diffeomorphism and

$$dk' = e^{-2\rho(H(a_t k))}\, dk\,. \tag{6.32}$$

Since $a_t k = k' \exp H(a_t k)n$ and A normalizes N, we have

$$H(a_{-t}k') = -H(a_t k)\,. \tag{6.33}$$

After the substitution (6.30) and $A(a_t \cdot o, k'M) = -H(a_{-t}k')$, relations (6.32)–(6.33) reduce the right-hand side of (6.29) to

$$e^{t\rho(H)} \int_{\mathfrak{a}^*} e^{-i\lambda(tH-H_0)} \int_B F_\sigma(\lambda, \ell k(a_t k)M) e^{(i\lambda-\rho)(H(a_t k))} \frac{d\lambda\, dk_M}{|c(\lambda)|^2}\,. \tag{6.34}$$

Now we use the map $\overline{n} \to k(\overline{n})M$ of $\overline{N}$ into K/M under which $dk_M = e^{-2\rho(H(\overline{n}))}\, d\overline{n}$ [4, p. 287] to transfer (6.34) to $\overline{N}$. Since A normalizes N, we have

$$a\overline{n} = ak(\overline{n}) \exp H(\overline{n})n_1 = k(ak(\overline{n})) \exp H(ak(\overline{n})) \exp H(\overline{n})n_2$$

so we have Harish-Chandra's relations

$$k(a\overline{n}) = k(ak(\overline{n})) = k(\overline{n}^a)\,, \qquad H(a\overline{n}) = H(ak(\overline{n})) + H(\overline{n}) \tag{6.35}$$

$$H(a\overline{n}) - \log a = H(\overline{n}^a)\,, \tag{6.36}$$

where $\overline{n}^a = a\overline{n}a^{-1}$. Thus (6.34) becomes

$$\int_{\mathfrak{a}^*} e^{-i\lambda(tH-H_0)} \int_{\overline{N}} F_\sigma(\lambda, \ell k(\overline{n}^{a_t})M) L(a_t, \overline{n}) \frac{d\lambda\, d\overline{n}}{c(\lambda)c(-\lambda)}\,, \tag{6.37}$$

where the kernel $L(a, \overline{n})$ equals

$$L(a, \overline{n}) = e^{(i\lambda-\rho)(H(\overline{n}^a))} e^{-(i\lambda+\rho)(H(\overline{n}))}\,. \tag{6.38}$$

Let $\lambda = \xi + i\eta (\xi, \eta \in \mathfrak{a}^*)$ and $-\eta \in \mathfrak{a}^*_+$ be so small that $(c(\lambda)c(-\lambda))^{-1}$ is holomorphic in $\mathfrak{a}^* + it\eta$ ($|t| \le 1$). Let $\varepsilon > 0$ be such that

$$0 < \varepsilon < 1\,, \quad \rho + \varepsilon\eta \in \mathfrak{a}^*_+\,.$$

Then by [10, p. 447, (41)],

$$|L(a_t, \overline{n})| \le e^{(\varepsilon\eta-\rho)(H(\overline{n}))}\,,$$

which is integrable over $\overline{N}$. Since F_σ is holomorphic in λ and for a fixed η is rapidly decreasing in ξ, the integral (6.37) can be shifted to $\mathfrak{a}^* + i\eta$. Letting then $t \to +\infty$, we get the expression

$$\int_{\mathfrak{a}^*+i\eta} e^{i\lambda(H_0)}(c(\lambda)c(-\lambda))^{-1}F_\sigma(\lambda, \ell M) \int_{\bar{N}} e^{-(i\lambda+\rho)(H(\bar{n}))}\,d\bar{n} = \int_{\mathfrak{a}^*+i\eta} e^{i\lambda(H_0)}F_\sigma(\lambda, \ell M)(c(-\lambda))^{-1}d\lambda\,.$$

Here we can shift the integration back and take $\eta = 0$. Using the definition of $\bar{J}$, Q_k, and relation (4.24), the theorem follows.

Acknowledgment. Supported by National Science Foundation Grant DMS-9402994.

Bibliography

[1] Eguchi, M., Hashizume, M., and Okamoto, K., *The Paley-Wiener theorem for distributions on symmetric spaces*, Hiroshima Math J. 3, 1973, pp. 109–120.

[2] Gangolli, R., and Varadarajan, V. S., *Harmonic Analysis of Spherical Functions on Real Reductive Groups*, Ergebnisse der Mathematik und ihrer Grenzgebiete [Results in Mathematics and Related Areas] No. 101, Springer-Verlag, Berlin–New York, 1988.

[3] Gindikin, S. G., and Karpelevič, F. I., *Plancherel measure for symmetric Riemannian spaces of non-positive curvature*, Dokl. Akad. Nauk SSSR 145, 1962, pp. 252–255.

[4] Harish-Chandra, *Spherical functions on a semisimple Lie group* I, Amer. J. Math. 80, 1958, pp. 241–310.

[5] Helgason, S., *Differential operators on homogenous spaces*, Acta Math. 102, 1959, pp. 239–299.

[6] Helgason, S., *Radon-Fourier transforms on symmetric spaces and related group representations*, Bull. Amer. Math. Soc. 71, 1965, pp. 757–763.

[7] Helgason, S., *A duality for symmetric spaces with applications to group representations*, Advances in Math. 5, 1970, pp. 1–154.

[8] Helgason, S., *The surjectivity of invariant differential operators on symmetric spaces* I , Ann. of Math. (2) 98, 1973, pp. 451–479.

[9] Helgason, S., *A duality for symmetric spaces with applications to group representations* II: *Differential equations and eigenspace representations*, Advances in Math. 22, 1976, pp. 187–219.

[10] Helgason, S., *Groups and Geometric Analysis. Integral Geometry, Invariant Differential Operators and Spherical Functions*, Pure and Applied Mathematics No. 113, Academic Press, Orlando, Fla., 1984.

[11] Helgason, S., *Geometric Analysis on Symmetric Spaces*, Mathematical Surveys and Monographs No. 39, American Mathematical Society, Providence, R.I., 1994.

[12] Helgason, S., and Schlichtkrull, H., *The Paley-Wiener space for the multitemporal wave equation*, Comm. Pure Appl. Math., to appear.

[13] Kostant, B., *On the existence and irreducibility of certain series of representations*, Bull. Amer. Math. Soc. 75, 1969, pp. 627–642.

[14] Kostant, B., and Rallis, S., *Orbits and representations associated with symmetric spaces*, Amer. J. Math. 93, 1971, pp. 753–809.

[15] Lax, P. D., and Phillips, R. S., *Scattering Theory*, Pure and Applied Mathematics No. 26, Academic Press, New York–London, 1967.

[16] Phillips, R. S., and Shahshahani, M. M., *Scattering theory for symmetric spaces of noncompact type*, Duke Math. J. 72, 1993, pp. 1–29.

[17] Semenov-Tjan-Šanskiĭ, M. A., *Harmonic analysis on Riemannian symmetric spaces of negative curvature, and scattering theory*, Izv. Akad. Nauk SSSR Ser. Mat. 40, 1976, pp. 562–592, 710.

[18] Shahshahani, M. M., *Invariant hyperbolic systems on symmetric spaces*, pp. 203–233 in: *Differential Geometry* (*College Park, Md., 1981/1982*), R. Brooks et al., eds., Birkhäuser, Boston, 1983.

SIGURDUR HELGASON
Massachusetts Institute of Technology
Department of Mathematics
77 Massachusetts Avenue
Cambridge, MA 02139-4307
E-mail: helgason@mit.edu

Received July 1997.
Revised December 1997.

The Paley-Wiener Space for the Multitemporal Wave Equation

S. HELGASON
Massachusetts Institute of Technology
AND
H. SCHLICHTKRULL
University of Copenhagen

Let X be a Riemannian symmetric space, and let $\mathcal{E}^\sigma : \mathcal{D}(X) \times \cdots \times \mathcal{D}(X) \to L^2(\mathfrak{a}^* \times B)$ be the spectral representation related to the multitemporal wave equation on X, as defined in [1, (6.7)]. In this short paper, we characterize the image of $\mathcal{E}^\sigma$. We use the notation from [1].

THEOREM 1 *The range of $\mathcal{D}(X) \times \cdots \times \mathcal{D}(X)$ under $\mathcal{E}^\sigma$ is the set of functions $\varphi(\lambda, kM)$ on $\mathfrak{a}_C^* \times B$ satisfying*

(i) $\varphi \in \mathcal{H}(\mathfrak{a}_C^* \times B)\pi$, *and*

(ii) *for each* $\delta \in \hat{K}_M$

$$\lambda \mapsto Q^{\check{\delta}}(\sigma\lambda)^{-1} \int_K \varphi(\lambda, kM)\delta(k^{-1})dk$$

is holomorphic on $\mathfrak{a}_C^*$.

In particular, we note that condition (ii) for the trivial K-type δ is a consequence of (i), since the polynomial $Q^{\check{\delta}}(\lambda)$ in that case is constant.

PROOF: We first prove the theorem with $\sigma = e$. Let $\mathcal{E} = \mathcal{E}^1$. In the proof we invoke the δ-spherical Fourier transform defined in [1, sec. 3]. We introduce the following notation:

For a function φ on $\mathfrak{a}_C^* \times B$ we define a $\mathrm{Hom}(V_\delta, V_\delta^M)$-valued function φ^δ on $\mathfrak{a}_C^*$ by

$$\varphi^\delta(\lambda) = d(\delta) \int_K \varphi(\lambda, kM)\delta(k^{-1})dk\,. \tag{1}$$

Likewise, for $f \in \mathcal{D}(X)$ we define a $\mathrm{Hom}(V_\delta, V_\delta)$-valued function f^δ on X by

$$f^\delta(x) = d(\delta) \int_K f(kx)\delta(k^{-1})dk\,.$$

Then $\mathrm{Tr}(f^\delta) = d(\delta)\chi_\delta * f \in \mathcal{D}_{\check{\delta}}(X)$ is the $\check{\delta}$-typical component of f. With the δ-spherical Fourier transform on $\mathcal{D}_{\check{\delta}}(X)$ defined as in [1, (3.10)], it is then easily seen that

$$\widetilde{\mathrm{Tr}(f^\delta)}(\lambda) = \tilde{f}^\delta(\lambda)\,, \tag{2}$$

where the expression on the right-hand side is defined by [1, (3.3)] and (1).

Communications on Pure and Applied Mathematics, Vol. LII, 0049–0052 (1999)

CCC 0010–3640/98/010049-04

Let $F = (f_1, \ldots, f_w) \in \mathcal{D}(X) \times \cdots \times \mathcal{D}(X)$ and put $\varphi = \mathcal{E}(F)$. Then (i) is obvious from [1, theorem 6.3]. Moreover, (ii) is satisfied, since as mentioned below [1, (3.11)], the function $Q^{\check{\delta}}(\lambda)^{-1}\widetilde{\mathrm{Tr}(f^{\delta})}(\lambda)$ is holomorphic, and by (2) the latter function equals $Q^{\check{\delta}}(\lambda)^{-1}\tilde{f}^{\delta}(\lambda)$.

Conversely, assume that a given function $\varphi(\lambda, kM)$ satisfies (i) and (ii). Define

$$f_j(x) = \int_{\mathfrak{a}_C^* \times B} p_j(i\lambda)\varphi(\lambda, b)e^{(i\lambda+\rho)A(x,b)} \frac{d\lambda\, db}{\pi(\lambda)^2|\mathbf{c}(\lambda)|^2} \tag{3}$$

for $x \in X$, $j = 1, \ldots, w$. We claim that $f_j \in \mathcal{D}(X)$ and $\varphi = \mathcal{E}(F)$ where $F = (f_1, \ldots, f_w)$.

It is well-known that $\lambda \mapsto \pi(\lambda)\mathbf{c}(\lambda)$ is nonzero on $\mathfrak{a}^*$ and that its reciprocal satisfies a polynomial estimate. It then follows from the uniform, rapid decrease of $\varphi(\,\cdot\,, b)$ that (3) defines a smooth function on X.

An easy computation shows that

$$f_j^{\delta}(x) = d(\delta)\int_K f_j(kx)\delta(k^{-1})dk = \int_{\mathfrak{a}^*} p_j(i\lambda)\Phi_{\lambda,\delta}(x)\varphi^{\delta}(\lambda)\frac{d\lambda}{\pi(\lambda)^2|\mathbf{c}(\lambda)|^2}$$

where $\Phi_{\lambda,\delta}(x)$ is the generalized spherical function $\int_K e^{(i\lambda+\rho)A(x,kM)}\delta(k)dk$. The function $\varphi^{\delta}(\lambda)$ is divisible by $Q^{\check{\delta}}(\lambda)$ because of our assumption (ii), and we can then write

$$f_j^{\delta}(x) = \int_{\mathfrak{a}^*} p_j(i\lambda)\Phi_{\lambda,\delta}(x)Q^{\check{\delta}}(\lambda)Q^{\check{\delta}}(\lambda)^{-1}\varphi^{\delta}(\lambda)\frac{d\lambda}{\pi(\lambda)^2|\mathbf{c}(\lambda)|^2}\,.$$

Averaging over W, and observing that $\lambda \mapsto \Phi_{\lambda,\delta}(x)Q^{\check{\delta}}(\lambda)$ is W-invariant (cf. [2, p. 289, theorem 5.15], we obtain

(4)
$$f_j^{\delta}(x) = \frac{1}{w}\int_{\mathfrak{a}^*} \Phi_{\lambda,\delta}(x)Q^{\check{\delta}}(\lambda)\left[\sum_{s\in W} p_j(is\lambda)Q^{\check{\delta}}(s\lambda)^{-1}\varphi^{\delta}(s\lambda)\right]\frac{d\lambda}{\pi(\lambda)^2|\mathbf{c}(\lambda)|^2}\,.$$

In order to prove that $\mathrm{Tr}(f_j^{\delta})$ is compactly supported, it now suffices, by [2, p. 289, cor. 5.14, and p. 290, thm. 5.15], to prove that the function

$$\lambda \mapsto \left[\sum_{s\in W} p_j(is\lambda)Q^{\check{\delta}}(s\lambda)^{-1}\varphi^{\delta}(s\lambda)\right]\frac{1}{\pi(\lambda)^2} \tag{5}$$

belongs to the space $\mathcal{J}^{\delta}(\mathfrak{a}^*)$ of W-invariant functions in $\mathcal{H}(\mathfrak{a}_C^*, \mathrm{Hom}(V_\delta, V_\delta^M))$. Clearly, (5) is W-invariant, and the expression in square brackets is holomorphic of exponential type, uniformly in δ. It remains to be seen that it is divisible by $\pi(\lambda)^2$. We rewrite (5) as follows:

$$\left[\sum_{s\in W} \varepsilon(s)p_j(is\lambda)Q^{\check{\delta}}(s\lambda)^{-1}\frac{\varphi^{\delta}(s\lambda)}{\pi(s\lambda)}\right]\frac{1}{\pi(\lambda)}\,.$$

Here the expression in square brackets is holomorphic by (i) and (ii), since $\pi(\lambda)$ and the determinant of $Q^{\check{\delta}}(\lambda)$ have no common zeroes, cf. [2, p. 267, theorem 4.2, and p. 348, corollary 11.3]. Moreover, being skew, it is divisible by $\pi(\lambda)$. Hence $\mathrm{Tr}(f_j^\delta) \in \mathcal{D}(X)$ with support uniform in δ, and so $f_j = \sum_\delta \mathrm{Tr}(f_j^\delta) \in \mathcal{D}(X)$.

It remains to be seen that $\mathcal{E}(F) = \varphi$. It follows from the above that the δ-spherical Fourier transform $\widetilde{\mathrm{Tr}(f_j^\delta)}$ of $\mathrm{Tr}(f_j^\delta)$ is given by $Q^{\check{\delta}}(\lambda)$ times (5). By (2) $\widetilde{\mathrm{Tr}(f_j^\delta)} = \widetilde{f_j}^{\delta}$ and hence

$$\begin{aligned}\pi(\lambda)^2 \sum_j q^j(i\lambda)\widetilde{f_j}^{\delta}(\lambda) &= \sum_j q^j(i\lambda)Q^{\check{\delta}}(\lambda) \sum_{s\in W} p_j(is\lambda)Q^{\check{\delta}}(s\lambda)^{-1}\varphi^\delta(s\lambda) \\ &= \sum_{s\in W}\left(\sum_j q^j(i\lambda)p_j(is\lambda)\right) Q^{\check{\delta}}(\lambda)Q^{\check{\delta}}(s\lambda)^{-1}\varphi^\delta(s\lambda).\end{aligned}$$

Now $\sum_j q^j(i\lambda)p_j(is\lambda) = 1$ for $s = e$, and otherwise it vanishes (use the remark just below lemma 4.3 in [1]); hence we obtain

$$\pi(\lambda)^2 \sum_j q^j(i\lambda)\widetilde{f_j}^{\delta}(\lambda) = \varphi^\delta(\lambda).$$

Since this expression holds for all δ, we conclude that

$$\pi(\lambda)^2 \sum_j q^j(i\lambda)\widetilde{f_j}(\lambda, b) = \varphi(\lambda, b)$$

for all $b \in B$, as claimed.

We now prove the theorem with arbitrary σ. We use the notation g^σ, φ^σ for $\lambda \mapsto g(\sigma^{-1}\lambda)$, $(\lambda, b) \mapsto \varphi(\sigma^{-1}\lambda, b)$. The proof will be based on the observation that conditions (i) and (ii) are independent of the basis $p = (p_1, \dots, p_w)$ of $H(\mathfrak{a})$ used.

Let $\mathcal{E}_p^\sigma$ denote the mapping $\mathcal{E}^\sigma$ with the basis p. Thus

$$\mathcal{E}_p^e(\mathcal{D} \times \cdots \times \mathcal{D}) = \mathcal{E}_{p'}^e(\mathcal{D} \times \cdots \times \mathcal{D})$$

if $p' = (p_1', \dots, p_w')$ is any homogeneous basis of $H(\mathfrak{a})$ with $p_1' = 1$. In particular,

$$\mathcal{E}_p^e(\mathcal{D} \times \cdots \times \mathcal{D}) = \mathcal{E}_{p^\sigma}^e(\mathcal{D} \times \cdots \times \mathcal{D}). \tag{6}$$

On the other hand,

$$(\mathcal{E}_p^\sigma F)(\lambda, b) = \sum_k (q^k)^\sigma(i\sigma\lambda)\widetilde{f_k}(\sigma\lambda, b) = (\mathcal{E}_{p^\sigma}^e)(\sigma\lambda, b)$$

so

$$\mathcal{E}_p^\sigma F = (\mathcal{E}_{p^\sigma}^e F)^{\sigma^{-1}}. \tag{7}$$

Using (6) and (7), the proof of (ii) is easily completed. □

Bibliography

[1] Helgason, Sigurdur. *Integral geometry and multitemporal wave equations*. Comm. Pure Appl. Math. **51** (1998), no. 9-10, 1035–1071.

[2] Helgason, Sigurdur. *Geometric analysis on symmetric spaces*. Mathematical Surveys and Monographs, 39. American Mathematical Society, Providence, RI, 1994.

SIGURDUR HELGASON
Massachusetts Institute of Technology
Department of Mathematics
77 Massachusetts Avenue
Cambridge, MA 02139-4307
E-mail: helgason@mit.edu

HENRIK SCHLICHTKRULL
University of Copenhagen
Department of Mathematics
Universitetsparken 5
DK 2100 Copenhagen Ø
DENMARK
E-mail: schlicht@math.ku.dk

Received September 1997.

Indag. Mathem., N.S., 16 (3–4), 531–551 December 19, 2005

The Abel, Fourier and Radon transforms on symmetric spaces

by Sigurdur Helgason

77 Massachusetts Avenue, Cambridge, MA 02139, USA

Dedicated to Gerrit van Dijk on the occasion of his 65th birthday

Communicated by Prof. J.J. Duistermaat at the meeting of June 20, 2005

1. INTRODUCTION

In this paper we prove some recent results on the three transforms in the title and discuss their relationships to older results. The spaces we deal with are symmetric spaces $X = G/K$ of the noncompact type, G being a connected noncompact semisimple Lie group with finite center and K a maximal compact subgroup.

For the two natural Radon transforms on X we prove a new inversion formula and a sharpening of an old support theorem; for the Abel transform we prove some new identities with some applications and for the Fourier transform a result for integrable functions which has a strong analog of the Riemann–Lebesgue lemma. These latter results are from a collaboration with Rawat, Sengupta and Sitaram.

Notation. Following Schwartz we use the notation $\mathcal{D}(X)$ for $C_c^\infty(X)$, $\mathcal{E}(X)$ for $C^\infty(X)$ and $\mathcal{S}(\mathbf{R}^n)$ for the space of rapidly decreasing functions on $\mathbf{R}^n$.

2. DIFFERENT RADON TRANSFORMS ON THE SYMMETRIC SPACE X

Radon's paper [40] suggested the general problem of determining a function on a manifold on the basis of its integrals over certain submanifolds. A natural case of

E-mail: helgason@mit.edu (S. Helgason).

this problem is the inversion of the X-ray transform on a Riemannian manifold. It is the transform $f \to \hat{f}$ defined by the arc-length integral

$$(2.1) \qquad \hat{f}(\gamma) = \int_{\gamma} f(x)\, dm(x),$$

f being an "arbitrary" function on the Riemannian manifold X and γ any complete geodesic in X.

In general this injectivity problem seems to be unresolved. For a Cartan symmetric space $X \neq \mathbf{S}^n$ the injectivity, however, holds. For a symmetric X of the noncompact the injectivity holds in the stronger form of the

Support theorem [26]. *If $\hat{f}(\gamma) = 0$ for all geodesics γ disjoint from a ball $B \subset X$ then $f(x) = 0$ for $x \notin B$.*

This last result requires stronger decay assumption at ∞ than the injectivity result does.

Here we shall prove an explicit inversion formula for the X-ray transform for rank $X > 1$. See Section 5 for the contact with Rouvière's different solution.

Funk [11] and Radon [40] inverted this transform for the sphere $\mathbf{S}^2$ and $\mathbf{R}^2$. Denoting the set of geodesics by Ξ we have the coset space representations

$$\mathbf{S}^2 = \mathbf{O}(3)/\mathbf{O}(2), \qquad \Xi = \mathbf{O}(3)/\mathbf{O}(2)\mathbf{Z}^2,$$
$$\mathbf{R}^2 = \mathbf{M}(2)/\mathbf{O}(2), \qquad \Xi = \mathbf{M}(2)/\mathbf{M}(1)\mathbf{Z}_2,$$

$\mathbf{M}(n)$ denoting the isometry group of $\mathbf{R}^n$.

This suggests the following generalization. Let $X = G/K$ and $\Xi = G/H$ be coset spaces of the same locally compact group G, K and H being closed subgroups. Here it will be convenient to assume all these groups as well as $L = K \cap H$ to be unimodular. We do not assume the elements $\xi \in \Xi$ to be subsets of X but instead use Chern's concept of incidence:

$$x = gK \text{ is incident to } \xi = \gamma H$$

if $gK \cap \gamma H \neq \emptyset$ as subsets of G. Given $x \in X$, $\xi \in \Xi$ define

$$\check{x} = \{\xi \in \Xi \colon x, \xi \text{ incident}\},$$
$$\hat{\xi} = \{x \in X \colon x, \xi \text{ incident}\}.$$

These are orbits of certain subgroups of G and have natural measures dm, $d\mu$ (up to factors) and we define the abstract Radon transform $f \to \hat{f}$ and its dual $\varphi \to \check{\varphi}$ by

$$(2.2) \qquad \hat{f}(\xi) = \int_{\hat{\xi}} f(x)\, dm(x), \qquad \check{\varphi}(x) = \int_{\check{x}} \varphi(\xi)\, d\mu(\xi).$$

The normalizations of dm and $d\mu$ are unified by taking $x_0 = eK$, $\xi_0 = eH$ and

$$(2.3) \qquad \hat{f}(\gamma H) = \int_{H/L} f(\gamma h \cdot x_0)\, dh_L, \qquad \check{\varphi}(gK) = \int_{K/L} \varphi(gk \cdot \xi_0)\, dk_L$$

the invariant measures dh_L, dk_L being fixed by Haar measures of H, K and L.

Main problems.

(i) Injectivity of $f \to \hat{f}$, $\varphi \to \check{\varphi}$.
(ii) Inversion formulas.
(iii) Range and kernel question for these transforms.
(iv) Applications elsewhere.

An easy general result relevant to problem (iii) is the following. For a suitable normalization of the measures $dx = dg_K$, $d\xi = d\gamma_H$ we have

$$(2.4) \qquad \int_X f(x)\check{\varphi}(x)\, dx = \int_\Xi \hat{f}(\xi)\varphi(\xi)\, d\xi,$$

a result which suggests the extension of (2.3) to distributions.

3. d-PLANES IN $\mathbf{R}^n$

Here we consider the space $X = \mathbf{R}^n$ and $\Xi = \mathbf{G}(d,n)$ the set of d-dimensional planes in $\mathbf{R}^n$. These are both homogeneous under the group $G = \mathbf{M}(n)$. Fix $x_0 \in \mathbf{R}^n$, $\xi_0 \in \mathbf{G}(d,n)$ at distance $d(x_0, \xi_0) = p$. Then we have

$$(3.1) \qquad X = \mathbf{R}^n = \mathbf{M}(n)/K_p, \qquad \Xi = \mathbf{G}(d,n) = \mathbf{M}(n)/H_p,$$

where K_p and H_p, respectively, are the stability groups of x_0 and ξ_0. Since various p will be considered the transforms (2.2) will be denoted by $\hat{f}_p$ and $\check{\varphi}_p$. Since the action of $\mathbf{M}(n)$ on X and Ξ is quite rich it turns out that for the coset space representation (3.1)

$$z \in X \text{ is incident to } \eta \in \Xi \quad \Leftrightarrow \quad d(z,\eta) = p.$$

Thus the transform $\hat{f}_p$ and $\check{\varphi}_p$ can be written

$$(3.2) \qquad \hat{f}_p(\xi) = \int_{d(x,\xi)=p} f(x)\, dm(x), \qquad \check{\varphi}_p(x) = \int_{d(x,\xi)=p} \varphi(\xi)\, d\mu(\xi).$$

In particular, $\hat{f}_0$ is the usual d-plane transform $\hat{f}$, but in order to invert it we need $\check{\varphi}_p$ for variable p. One of several versions of the inversion formula is the following (see [25,27]):

$$(3.3)\qquad f(x)=c(d)\left[\left(\frac{d}{d(r^2)}\right)^d \int_r^\infty p(p^2-r^2)^{d/2-1}(\hat{f})_p^\vee(x)\,dp\right]_{r=0}$$

with $c(d)$ and constant. Note that $(\hat{f})_p^\vee(x)$ is the average of the integrals of f over all d-planes at distance p from x.

For $d=1$ this formula reduces to

$$(3.4)\qquad f(x)=-\frac{1}{\pi}\int_0^\infty \frac{d}{dp}((\hat{f})_p^\vee(x))\frac{dp}{p},$$

which for $n=2$ coincides with Radon's original formula. Radon's proof is very elegant and is based on an exhaustion of the exterior $|x|>r$ by lines. As far as I know this proof has not been extended to higher dimensions n. Formula (3.4) for $n>2$ is crucial for the inversion of (2.1) given in Theorem 5.1.

4. d-DIMENSIONAL TOTALLY GEODESIC SUBMANIFOLDS IN HYPERBOLIC SPACE $\mathbf{H}^n$

A similar method works here and the analog of (3.3) is the formula

$$(4.1)\qquad f(x)=C(d)\left[\left(\frac{d}{d(r^2)}\right)^d \int_r^\infty (t^2-r^2)^{d/2-1}t^d(\hat{f})_{s(t)}^\vee(x)\,dt\right]_{r=1},$$

where $C(d)$ is a constant and $s(p)=\cosh^{-1}(p)$ (see [25,27]). Other versions of the inversion exist (e.g., [28] and [6]). For $d=1$ this reduces to

$$(4.2)\qquad f(x)=-\frac{1}{\pi}\int_0^\infty \frac{d}{dp}((\hat{f})_p^\vee(x))\frac{dp}{\sinh p}$$

a formula which for $n=2$ is stated without proof in Radon [40].

5. X-RAY INVERSION ON THE SYMMETRIC SPACE $X=G/K$

In communication from 2003, Rouvière proved an extension of formula (4.2) to symmetric spaces X of rank $\ell=1$. Inspired by his methods, I proved the inversion formula (5.2) for the X-ray transform for X of rank $\ell>1$. Then Rouvière [43] extended his formula to X of arbitrary rank ℓ. Actually he has several such formulas but they are all different from the formula (5.2) below.

Fix a flat totally geodesic submanifold E of X with $\dim E=\ell>1$ (ℓ the rank of X) passing through the origin $o=eK$ of X. Let $p>0$ and $S=S_p(o)$ be the sphere in E with radius p and center o. The geodesics γ in E tangent to

S are permuted transitively by the orthogonal group $\mathbf{O}(E)$. Let du and dk denote the normalized Haar measures on U and K. The spaces $k \cdot E$ as k runs through K constitute all flat totally geodesic subspaces of X through o of dimension ℓ. Thus the images $k \cdot \gamma$ ($k \in K$, γ tangent to S) constitute the set Γ_p of all geodesics γ in X lying in some flat ℓ-dimensional totally geodesic submanifold of X through o and $d(o, \gamma) = p$. The set Γ_p has a natural measure ω_p given by the functional

$$\omega_p : \varphi \to \int_K \left(\int_{\mathbf{O}(E)} \varphi\big(k(u \cdot \gamma)\big)\, du \right) dk. \tag{5.1}$$

Theorem 5.1. *The X-ray transform* (2.1) *on a symmetric space $X = G/K$ of rank $\ell > 1$ is inverted by the formula*

$$f(o) = -\frac{1}{\pi} \int_0^\infty \left(\frac{d}{dp} \int_{\Gamma_p} \hat{f}(\gamma) d\omega_p(\gamma) \right) \frac{dp}{p}, \qquad f \in \mathcal{D}(X). \tag{5.2}$$

Since Γ_p and $d\omega_p$ are K-invariant the formula holds at each point x by replacing f by $f \circ g$ where $g \in G$ is such that $g \cdot o = x$.

Proof. First assume f to be K-invariant and consider the restriction $f|E$. Fix an orthonormal frame $H_0, H \in E_0$, the tangent space to E at o, consider the one parameter subgroups $\exp tH_0$, $\exp tH$ and the geodesic $\gamma_0(t) = \exp tH_0 \cdot o$. Then the geodesic $\gamma(t) = \exp pH \cdot \gamma_o(t)$ lies in E and is tangent to $S_p(o)$. Because of (3.4) we have

$$f(o) = -\frac{1}{\pi} \int_0^\infty \frac{d}{dp} (\hat{f})_p^E(o) \frac{dp}{p}, \tag{5.3}$$

where the superscript E stands for the dual transform on geodesics in the space E. Thus

$$(\hat{f})_p^E(o) = \int_{\substack{\gamma \subset E \\ d(o,\gamma)=p}} (\hat{f})(\gamma)\, d\nu(\gamma) = \int_{\mathbf{O}(E)} (\hat{f})(u \cdot \gamma)\, du, \tag{5.4}$$

where ν stands for the average over the set of geodesics tangent to $S_p(o)$.

For $f \in \mathcal{D}(X)$ arbitrary we use (5.2) on the function

$$f^\natural(x) = \int_K f(k \cdot x)\, dk.$$

Taking into account the definition (5.1) the inversion formula (5.2) follows immediately. □

Remark. Note that the measure ω_p in (5.1) is a kind of convolution of the Haar measures dk and du. However it is not a strict convolution since the product ku is not defined.

6. THE HOROCYCLE TRANSFORM IN $X = G/K$

Consider the usual Iwasawa decomposition of G, $G = NAK$ where N and A are nilpotent and abelian, respectively. A *horocycle* is by definition [12] an orbit in X of a conjugate gNg^{-1} of N. The group G permutes the horocycles transitively and the space Ξ of horocycles can be written $\Xi = G/MN$ where M is the centralizer of A in K. In the double fibration

$$\begin{array}{ccc} & G/M & \\ \swarrow & & \searrow \\ X = G/K & & G/MN = \Xi \end{array}$$

it turns out that $x = gK$ is incident to $\xi = \gamma MN$ if and only if $x \in \xi$. The transforms (2.2) become

$$(6.1) \qquad \hat{f}(\gamma MN) = \int\limits_N f(\gamma n \cdot o)\,dn, \qquad \check{\varphi}(gK) = \int\limits_K \varphi(gk \cdot \xi_0)\,dk,$$

where $\xi_o = N \cdot o$. While the map $\varphi \to \check{\varphi}$ has a big kernel, the horocycle transform $f \to \hat{f}$ is injective (see [17] or [13]). The following result from [21] is considerably stronger.

Theorem 6.1 (Support theorem). *Let B be a closed ball in X. Then*

$$\begin{aligned} &\hat{f}(\xi) = 0 \quad \textit{for } \xi \cap B = \emptyset \quad \textit{implies} \\ &f(x) = 0 \quad \textit{for } x \notin B. \end{aligned}$$

Here one requires stronger decay conditions on f than for the injectivity. A different proof was given in [14]. We have also the following inversion and Plancherel formula for the Radon transform [18,19]. The pseudodifferential operator Λ and the differential operator $\Box$ below are constructed by means of the Harish-Chandra c-function, and w denotes the order of the Weyl group. For G complex a result similar to (6.2) appears in [12].

Theorem 6.2. *For $f \in \mathcal{D}(X)$ or sufficiently rapidly decreasing we have the inversion formula*

$$(6.2) \qquad f = \frac{1}{w}\left(\Lambda\bar{\Lambda}\hat{f}\right)^{\vee}.$$

If all Cartan subgroups of G are conjugate the formula has the improved version

$$f = \frac{1}{w}\Box((\hat{f})^{\vee}).$$

For G arbitrary

$$(6.3)\qquad w\int_X |f(x)|^2\,dx = \int_\Xi |\Lambda\hat{f}|^2(\xi)\,d\xi,$$

with a suitable normalization of the invariant measures dx and $d\xi$.

The range question (iii) for $f \to \hat{f}$ is more complicated. Consider first the hyperbolic plane $\mathbf{H}^2$ in the Poincaré unit disk model D. Here the horocycles are the circles in the disk tangential to the boundary $\{e^{i\theta}\colon\ \theta \in \mathbf{R}\}$. Let $\xi_{t,\theta}$ denote the horocycle through $e^{i\theta}$ with distance t (with sign) from the origin. Then we have the following result from [23].

Theorem 6.3. *The range $\mathcal{D}(D)\widehat{\ }$ consists of the functions $\psi \in \mathcal{D}(\Xi)$*

$$\psi(\xi_{t,\theta}) = \sum_n \psi_n(t)e^{in\theta}$$

where

$$(6.4)\qquad \psi_n(t) = e^{-t}\left(\frac{d}{dt}-1\right)\cdots\left(\frac{d}{dt}-2|n|+1\right)\varphi_n(t)$$

where $\varphi_n \in \mathcal{D}(D)$ is even.

This implies a relationship between $\psi(\xi_{t,\theta})$ and $\psi(\xi_{-t,\theta})$. More specifically, if $f'(t) = f(-t)$, $\Psi_n = e^t\psi_n$ then $*$ denoting convolution on $\mathbf{R}$

$$\Psi_n' = S_n * \Psi_n$$

where the distribution S_n on $\mathbf{R}$ has Fourier transform

$$\hat{S}_n = \frac{(i\lambda+1)\cdots(i\lambda+2|n|-1)}{(i\lambda-1)\cdots(i\lambda-2|n|+1)}, \qquad \lambda \in \mathbf{R}.$$

This relationship between $\psi(\xi_{-t,\theta})$ and $\psi(\xi_{t,\theta})$ implies that in (6.3) $f \to \Lambda\hat{f}$ does not map $L^2(X)$ onto $L^2(\Xi)$.

For the generalization of (6.4) to $X = G/K$ we need some additional notation. Let $\hat{K}$ be the unitary dual of K and $d(\delta)$ the degree of a $\delta \in \hat{K}$. Given δ acting on V_δ let

$$V_\delta^M = \{v \in V_\delta\colon\ \delta(m)v = v \text{ for } m \in M\}$$

and put $\ell(\delta) = \dim V_\delta^M$. Let

$$\hat{K}_M = \{\delta \in \hat{K} \colon \ell(\delta) > 0\}.$$

In the following theorem from [25] the expansion (6.5) is a generalization of (6.4). Put $\rho(H) = \frac{1}{2}\,\mathrm{Trace}(\mathrm{ad}\, H|\mathfrak{n})$.

Theorem 6.4. *The range $\mathcal{D}(X)\widehat{\ }$ consists of the functions $\psi \in \mathcal{D}(\Xi)$*

$$\psi(ka \cdot \xi_0) = \sum_{\delta \in \hat{K}_M} d(\delta)\,\mathrm{Tr}\big(\delta(k)\Psi_\delta(a)\big) \tag{6.5}$$

(Tr = Trace) *where Ψ_δ is a function on A with values in* $\mathrm{Hom}(V_\delta, V_\delta^M)$*, i.e., $\Psi_\delta \in \mathcal{D}(A, \mathrm{Hom}(V_\delta, V_\delta^M))$, given by*

$$\Psi_\delta(a) = e^{-\rho(\log a)}\, Q^\delta(D)_a\big(\Phi_\delta(a)\big), \quad a \in A, \tag{6.6}$$

where

$$\Phi_\delta \in \mathcal{D}\big(A, \mathrm{Hom}\big(V_\delta, V_\delta^M\big)\big) \tag{6.7}$$

is W-invariant and $Q^\delta(D)$ is a certain $\ell(\delta) \times \ell(\delta)$ matrix of constant coefficient differential operators on A.

From this result we can derive the following (unpublished) refinement of the support theorem above. Let A^+ be the Weyl chamber corresponding to the choice of the group N.

Theorem 6.5. *Suppose $f \in \mathcal{D}(X)$ satisfies*

$$\hat{f}(ka \cdot \xi_0) = 0 \quad \text{for } k \in K,\ a \in A^+,\ |\log a| > R.$$

Then

$$\hat{f}(ka \cdot \xi_0) = 0 \quad \text{for } k \in K,\ |\log a| > R,\ a \in A$$

so by Theorem 6.1

$$f(x) = 0 \quad \text{for } d(0, x) > R.$$

Proof. Let $Q_c(D)$ be the matrix of cofactors of $Q^\delta(D)$ so that

$$Q_c(D)Q^\delta(D) = \det Q^\delta(D) I. \tag{6.8}$$

Then (6.6) implies

$$Q_c(D)(e^\rho \Psi_\delta) = \det Q^\delta(D)\Phi_\delta. \tag{6.9}$$

Now it is known [34], [25, pp. 267, 348] that $\det Q^\delta(D)$ is a product of linear factors $\delta(H_i)+c$ where $H_i \in \mathfrak{a}$ and $\partial(H_i)$ the corresponding directional derivative.

Suppose the function $\psi = \hat{f}$ satisfies

$$\psi(ka\cdot\xi_0)=0 \quad \text{for } k\in K,\ a\in A^+,\ |\log a|>R.$$

Since

$$\Psi_\delta(a)=\int_K \psi(ka\cdot\xi_0)\delta(k^{-1})\,dk$$

we deduce from (6.8) and (6.9) that

$$\det Q^\delta(D)\Phi_\delta(a)=0 \quad \text{for } a\in A^+,\ |\log a|>R.$$

Consider this equation on a ray in A^+ starting at e. Because of the mentioned factorization of $\det Q^\delta(D)$ we deduce that on this ray Φ_δ satisfies an ordinary differential equation on the interval (R,∞). Having compact support we deduce that $\Phi_\delta(a)=0$ for $a\in A^+$, $|\log a|>R$. By its Weyl group invariance it vanishes for all $a\in A$, $|\log a|>R$ which by (6.5) proves the theorem. □

Consider the case rank $X=1$. Let $B_R(o)$ be a ball in X with radius R and center 0. Fix a unit vector H in the Lie algebra of A such that $\exp H\in A^+$. Put $a_t=\exp tH$. The *interior* of the horocycle $kNa_t\cdot o$ is the union $\bigcup_{\tau>t} kNa_\tau\cdot o$. A horocycle ξ is said to be *external* to $B_R(o)$ if its interior is disjoint from $B_R(o)$; ξ is said to *enclose* $B_R(o)$ if its interior contains $B_R(o)$.

Corollary 6.6. *Let X have rank one and $B_R(o)$ as above. Let $f\in\mathcal{D}(X)$. Then the following are equivalent*:

(i) $\hat{f}(\xi)=0$ *whenever ξ is external to $B_R(o)$.*
(ii) $\hat{f}(\xi)=0$ *whenever ξ encloses $B_R(o)$.*
(iii) $f\equiv 0$ *outside $B_R(o)$.*

For hyperbolic space this is clear from Theorem 6.3 and was proved in a different way by Lax and Phillips [35].

Problem (iii) for the dual transform $\varphi\to\check{\varphi}$ has a satisfactory answer (see [25, IV §§2 and 4]). The kernel can be described in the spirit of Theorem 6.5 and for the range one has the surjectivity

(6.10) $\quad \mathcal{E}(\Xi)^\vee=\mathcal{E}(X).$

Let $\mathcal{D}_K(X)$ denote the space of K-invariant functions in $\mathcal{D}(X)$. The *Abel transform* $f \to \mathcal{A}f$ is defined by

$$(7.1) \qquad (\mathcal{A}f)(a) = e^{\rho(\log a)} \int_N f(an \cdot o)\, dn, \qquad a \in A,\ f \in \mathcal{D}_K(X).$$

Except for the factor e^{ρ} it is the restriction of the Radon transform to K-invariant functions

$$(7.2) \qquad \mathcal{A}f = e^{\rho} \hat{f}.$$

Some of its properties are best analyzed by means of the spherical functions

$$(7.3) \qquad \varphi_\lambda(g) = \int_K e^{(i\lambda-\rho)(H(gk))}\, dk, \qquad g \in G,\ \lambda \in \mathfrak{a}_c^*,$$

where $H(g) \in \mathfrak{a}$ is determined by $g \in k \exp H(g) N$ and $\mathfrak{a}_c^*$ is the complex dual of $\mathfrak{a}$. The *spherical transform*

$$(7.4) \qquad \tilde{f}(\lambda) = \int_X f(x)\varphi_{-\lambda}(x)\, dx, \qquad f \in \mathcal{D}_K(X)$$

(where $\varphi_\lambda(g \cdot o) = \varphi_\lambda(g)$) is a homomorphism relative to convolution $\times$ on X:

$$(7.5) \qquad (f_1 \times f_2)^{\sim}(\lambda) = \tilde{f}_1(\lambda)\tilde{f}_2(\lambda).$$

As proved in [15], $\mathcal{A}$ intertwines the spherical transform and the Euclidean Fourier transform $F \to F^*$ on A so

$$(7.6) \qquad \int_A (\mathcal{A}f)(a) e^{-i\lambda(\log a)}\, da = \int_X \varphi_{-\lambda}(x) f(x)\, dx, \qquad (\mathcal{A}f)^* = \tilde{f}.$$

Thus $\mathcal{A}f$ is W-invariant and by (7.5)

$$(7.7) \qquad \mathcal{A}(f_1 \times f_2) = \mathcal{A}f_1 * \mathcal{A}f_2,$$

where $*$ is convolution on A. Let $\mathbf{D}(X)$ denote the algebra of G-invariant differential operators on X and $\Gamma : \mathbf{D}(X) \to \mathbf{D}_W(A)$ the isomorphism onto the W-invariant constant coefficient differential operators on A.

The Abel transform is a simultaneous transmutation operator between $\mathbf{D}(X)$ and $\mathbf{D}_W(A)$, i.e.,

$$(7.8) \qquad \mathcal{A}Df = \Gamma(D)\mathcal{A}f, \quad D \in \mathbf{D}(X),\ f \in \mathcal{D}_K(X)$$

as shown in [29] which for example can be used to prove that each D has a fundamental solution. By the Paley–Wiener theorem for (7.3) one has that

$\mathcal{A}: \mathcal{D}_K(X) \to \mathcal{D}_W(A)$ is a bijective homeomorphism. (Here the subscript W means W-invariance.) Hence we have a bijection

$$\text{(7.9)} \qquad \mathcal{A}^*: \mathcal{D}'_W(A) \to \mathcal{D}'_K(X)$$

between the corresponding distribution spaces. Also if $\varphi \in \mathcal{E}_W(A)$ we have easily (see [5] or [25, IV, §4])

$$\text{(7.10)} \qquad (\mathcal{A}^*\varphi)(gK) = \int_{K/M} \varphi(\exp H(gk))e^{-\rho(H(gk))}\,dk.$$

The Radon transform has the advantage over $\mathcal{A}$ that it commutes with the action of G. Thus we can deduce from (6.10) and (7.2) that as in [25]

$$\mathcal{A}^*\mathcal{E}_W(A) = \mathcal{E}_K(X).$$

We now add a few new results about $\mathcal{A}$ and $\mathcal{A}^*$ which will be useful later. Some are closely related to rank-one results in Bagchi and Sitaram in [3].

Because of the convolution property (7.7) one can ask how $\mathcal{A}^*$ behaves relative to convolution. Let L be the operator on $\mathcal{S}(A)$ given by

$$\text{(7.11)} \qquad (L\varphi)^*(\lambda) = |c(\lambda)|^{-2}\varphi^*(\lambda), \qquad \lambda \in \mathfrak{a}^*,$$

where $c(\lambda)$ is Harish-Chandra's c-function.

Theorem 7.1. *Let $\varphi \in \mathcal{D}_W(A)$, $\psi \in \mathcal{E}_W(A)$. Then*

$$\mathcal{A}^*(L\varphi) = w\mathcal{A}^{-1}(\varphi) \quad (w = \textit{order of } W)$$

and

$$\text{(7.12)} \qquad \mathcal{A}^*(\varphi * \psi) = \frac{1}{w}\mathcal{A}^*(L\varphi) \times \mathcal{A}^*\psi.$$

Proof. Using the inversion formula for the spherical transform we have

$$\begin{aligned} \mathcal{A}^*(L\varphi)(gK) &= \int_K (L\varphi)\big(\exp H(gk)\big)^{-\rho(H(gk))}\,dk \\ &= \int_K \left(\int_{\mathfrak{a}^*} (L\varphi)^*(\lambda)e^{i\lambda(H(gk))}\,d\lambda\right)e^{-\rho(H(gk))}\,dk \\ &= \int_{\mathfrak{a}^*} |c(\lambda)|^{-2}\varphi^*_\lambda(g)\,d\lambda = F(gK), \end{aligned}$$

where $\tilde{F}(\lambda) = \varphi^*(\lambda)w$. But $\tilde{F} = (\mathcal{A}F)^* = \varphi^* w$ so

$$\varphi = \frac{1}{w}\mathcal{A}F, \qquad F = w\mathcal{A}^{-1}(\varphi).$$

Thus $\mathcal{A}^*(L\varphi) = w\mathcal{A}^{-1}(\varphi)$. Consider now the average

$$\psi^\lambda(a) = \frac{1}{w} \sum_{s \in W} e^{is\lambda(\log a)}.$$

Then $\mathcal{A}^*\psi^\lambda = \varphi_\lambda$ and $\varphi * \psi^\lambda = \varphi^*(\lambda)\psi^\lambda$. But $\mathcal{A}^{-1}\varphi = \frac{1}{w}F \in \mathcal{D}_K(X)$ and

$$\mathcal{A}^{-1}\varphi \times \varphi_\lambda = \frac{1}{w}\tilde{F}(\lambda)\varphi_\lambda = \varphi^*(\lambda)\varphi_\lambda.$$

Combining these formulas we have

$$\mathcal{A}^*(\varphi * \psi^\lambda) = \mathcal{A}^{-1}\varphi \times \mathcal{A}^*(\psi^\lambda). \tag{7.13}$$

Now $\psi \in \mathcal{D}_W(A)$ is a superposition

$$\psi(a) = \int_{\mathfrak{a}^*} \psi^*(\lambda)\psi^\lambda(a)\, d\lambda$$

so the identity (7.12) follows from (7.13) for such ψ. For $\psi \in \mathcal{E}_W(A)$ the identity follows by an approximation because φ and $\mathcal{A}^*(L\varphi)$ have compact support and $\mathcal{A}^*$ is continuous on $\mathcal{E}_W(A)$. □

Theorem 7.1 implies the following inversion formula which in reality is a special case of (6.2). It appears also in [5].

Corollary 7.2. *The transform $f \to \mathcal{A}f$ has inversion*

$$f = \frac{1}{w}\mathcal{A}^*(L\mathcal{A}f), \qquad f \in \mathcal{D}_K(X).$$

The above results suggest various ways of defining $\mathcal{A}$ on the space $\mathcal{E}'_K(X)$ of K-invariant compactly supported distributions on X although formula (7.1) does not work.

Spherical transform method. If $T \in \mathcal{E}'_K(X)$, the spherical transform

$$\tilde{T}(\lambda) = \int_X \varphi_{-\lambda}(x)\, dT(x)$$

is a W-invariant entire function of exponential type on $\mathfrak{a}_c^*$ and of polynomial growth. (See [9] or [21, Theorem 8.5].) By the Euclidean Paley–Wiener theorem there exists an $S \in \mathcal{E}'_W(A)$ such that $\tilde{T} = S^*$. Thus in accordance with (7.6) we put

$$\mathcal{A}T = S. \tag{7.14}$$

Radon transform method. Because of (2.4) the Radon transform of a distribution $T \in \mathcal{E}'(X)$ is defined by

$$\hat{T}(\varphi) = T(\check{\varphi}), \quad \varphi \in \mathcal{E}(\Xi).$$

If T is K-invariant then so is $\hat{T}$ and since $\Xi = K/M \times A$ under the bijection $(kM, a) \to ka \cdot \xi_o$ we see that $\hat{T}$ has the form $\hat{T} = 1 \otimes \sigma$ where $\sigma \in \mathcal{E}'(A)$. Because of (7.2) we put

$$(7.15) \qquad \mathcal{A}T = e^{\rho}\sigma.$$

Functional analysis method. As remarked $\mathcal{A}^*$ is a bijection of $\mathcal{D}'_W(A)$ onto $\mathcal{D}'_K(X)$. The restriction of $\mathcal{A}^*$ to $\mathcal{E}_W(A)$ is a continuous bijection onto $\mathcal{E}_K(X)$ and in fact a homeomorphism since both spaces are Fréchet. Thus we have $(\mathcal{A}^*)^* : \mathcal{E}'_K(X) \to \mathcal{E}'_W(A)$ bijectively so we can define

$$(7.16) \qquad \mathcal{A}T = (\mathcal{A}^*)^*(T).$$

Proposition 7.3. *All the definitions* (7.14)–(7.16) *coincide.*

The convolution property in Theorem 7.1 extends readily to distributions so

$$\mathcal{A}^*(\mathcal{E}'_W(A) * \psi) = \mathcal{A}^{-1}\big(\mathcal{E}'_W(A)\big) \times \mathcal{A}^*\psi = \mathcal{E}'_K(X) \times \mathcal{A}^*\psi.$$

Thus putting

$$V_\psi = \mathcal{E}'_W(A) * \varphi, \quad \psi \in \mathcal{E}_W(A);$$
$$W_f = \mathcal{E}'_K(X) \times f, \quad f \in \mathcal{E}_K(X),$$

we conclude that

$$(7.17) \qquad \mathcal{A}^*(V_\psi) = W_{\mathcal{A}^*\psi}.$$

Theorem 7.4 (Bagchi–Sitaram). *If X has rank one and $f \in \mathcal{E}_K(X)$ then the closure of the space $W_f = \mathcal{E}'_K(X) \times f$ contains a spherical function.*

The authors use (7.17) to reduce the question to the analogous one for the one-dimensional space $A \sim \mathbf{R}$ where by Schwartz's theorem stated in Section 9 below some exponentials $e^{i\mu}$ and $e^{-i\mu}$ belong to the closure and $\mathcal{A}^*(e^{i\mu} + e^{-i\mu}) = 2\varphi_\mu$.

8. THE FOURIER TRANSFORM ON $X = G/K$

We now go to the notation of Section 6 with the Iwasawa decomposition $G = NAK$, and $\mathfrak{g} = \mathfrak{n} + \mathfrak{a} + \mathfrak{k}$ for the corresponding Lie algebras. For $g \in G$ let $A(g) \in \mathfrak{n}$ be determined by $g = n \exp A(g) k$ $(n \in N, k \in K)$. Given $x = gK$ in X, $b = kM$ in $B = K/M$ we put

$$A(x, b) = A(k^{-1}g)$$

and as usual we put $\rho(H) = \frac{1}{2}\,\mathrm{Trace}(\mathrm{ad}\,H|\mathfrak{n})$. Let $\mathfrak{a}_c^*$ denote the space of complex-valued linear forms on $\mathfrak{a}$.

Given a function f on X we define its *Fourier transform* by

$$(8.1) \qquad \tilde{f}(\lambda, b) = \int_X f(x) e^{(-i\lambda+\rho)(A(x,b))}\, dx$$

for those $(\lambda, b) \in \mathfrak{a}_c^* \times B$ for which the integral is defined. Many of the principal theorems for Fourier transforms on $\mathbf{R}^n$ have analogs for $X = G/K$.

Inversion formula [19]. For $f \in \mathcal{D}(X)$ we have

$$f(x) = \frac{1}{w} \int_{\mathfrak{a}^*} \int_B \tilde{f}(\lambda, b) e^{(i\lambda+\rho)(A(x,b))} |c(\lambda)|^{-2}\, d\lambda\, db$$

where $c(\lambda)$ is Harish-Chandra's c-function.

Plancherel formula [20]. The map $f \to \tilde{f}$ extends to an isometry of $L^2(X)$ onto $L^2(\mathfrak{a}_+^* \times B)$:

$$(8.2) \qquad \int_X |f(x)|^2\, dx = \int_{\mathfrak{a}_+^* \times B} |\tilde{f}(\lambda, b)|^2 |c(\lambda)|^{-2}\, d\lambda\, db.$$

Paley–Wiener theorem [21]. The map $f \to \tilde{f}$ maps the space $\mathcal{D}(X)$ onto the space of smooth $\varphi(\lambda, b)$ on $\mathfrak{a}_c^* \times B$ which are holomorphic on $\mathfrak{a}_c^*$ of exponential type (uniformly in B) satisfying the invariance condition

$$(8.3) \qquad \int_B \varphi(\lambda, b) e^{(i\lambda+\rho)(A(x,b))}\, db \text{ is } W\text{-invariant in } \lambda.$$

For the next result we refer to Eguchi's paper for full explanations of notation.

The Schwartz theorem [8]. Let $0 < p \leqslant 2$ and $\mathcal{S}^p(X) \subset L^p(X)$ the corresponding Schwartz space. Let $\varepsilon = 2/p - 1$ and $\mathcal{S}(\mathfrak{a}_\varepsilon^* \times B)$ the space of functions which are holomorphic in the "tube" $\mathfrak{a}_\varepsilon^* \times B$, are rapidly decreasing and satisfy (8.3). Then $f \to \tilde{f}$ is a bijection of $\mathcal{S}^p(X)$ onto $\mathcal{S}(\mathfrak{a}_\varepsilon^* \times B)$.

These results leave out the space $L^1(X)$ and one should think that a self-respecting Fourier transform should be defined here.

We shall now show (modifying a bit the proof of [31]) that this can be done and that a strong analog of the classical Riemann–Lebesgue lemma holds for $\tilde{f}$ in (8.1).

Let $C(\rho)$ denote the convex hull in $\mathfrak{a}^*$ of the set $\{s\rho\colon\ s \in W\}$ of Weyl group transforms of ρ.

Theorem 8.1. *Let $f \in L^1(B)$. Then there exists a subset $B' \subset B$ with $B - B'$ of measure 0 such that for each $b \in B'$*

(i) *$\tilde{f}(\lambda, b)$ is defined for λ in the tube $\mathfrak{a}^* + iC(\rho)$ and holomorphic in its interior.*
(ii) *$\lim_{\xi \to \infty} \tilde{f}(\xi + i\eta, b) = 0$ uniformly for $\eta \in C(\rho)$.*

Proof. Let $\lambda = \xi + i\eta$ where $\xi \in \mathfrak{a}^*$, $\eta \in C(\rho)$. Then

$$\int_B |\tilde{f}(\lambda, b)|\, db \leqslant \int_X |f(x)| \int_B e^{(\eta+\rho)(A(x,b))}\, db\, dx. \tag{8.4}$$

The integral over B is the spherical function $\varphi_{-i\eta}$ which is bounded by 1 [30]. Thus

$$\left\| \tilde{f}(\lambda, \cdot) \right\|_1 \leqslant \| f \|_1$$

and for each $\lambda \in \mathfrak{a}^* + iC(\rho)$, $\tilde{f}(\lambda, b)$ exists for all b in a subset $B_\lambda \subset B$ of full invariant measure. Let

$$B' = B'(f) = \bigcap_{s \in W} B_{is\rho}.$$

For the statements (i) and (ii) we may assume $f \geqslant 0$ in (8.1). Since $b \in B_{is\rho}$ for each $s \in W$ we have

$$\int_X f(x) e^{(s\rho+\rho)(A(x,b))}\, dx < \infty. \tag{8.5}$$

Fix $b \in B'$, $\eta \in C(\rho)$. Then

$$\int_X f(x) e^{(\rho+\eta)(A(x,b))}\, dx = \sum_{\sigma \in W} \int_{X_\sigma} f(x) e^{(\rho+\eta)(A(x,b))}\, dx \tag{8.6}$$

where

$$X_\sigma = \left\{ x \in X \colon \sigma\big(A(x,b)\big) \in \overline{\mathfrak{a}^+} \right\}.$$

Replace $\eta(A(x,b))$ by $(\sigma\eta)\ (\sigma(A(x,b)))$ and let $(\sigma\eta)^+$ be the element in $\overline{\mathfrak{a}^*_+}$, which is W-conjugate to $\sigma\eta$. Then since $(\sigma\eta)^+ - \sigma\eta \geqslant 0$ on $\mathfrak{a}^+$ we have

$$\int_{X_\sigma} f(x) e^{(\rho+\eta)(A(x,b))}\, dx \leqslant \int_{X_\sigma} f(x) e^{(\rho+(\sigma\eta)^+)(\sigma(A(x,b)))}\, dx.$$

Now by Lemma 8.3, Ch. IV in [24]

$$\overline{\mathfrak{a}^*_+} \cap C(\rho) = \overline{\mathfrak{a}^*_+} \cap (\rho + {}^-\mathfrak{a}^*),$$

where

$$^{-}\mathfrak{a}^* = \{\lambda \in \mathfrak{a}^* \mid \langle \lambda, \mu \rangle \leqslant 0 \text{ for } \mu \in \mathfrak{a}^*_+\}.$$

Thus

$$(\sigma\eta)^+ \in \overline{\mathfrak{a}^*_+} \cap (\rho + {}^{-}\mathfrak{a}^*),$$

whence

$$(8.7) \qquad (\sigma\eta)^+ - \rho \leqslant 0 \quad \text{on } \mathfrak{a}^+.$$

Thus the last integral is bounded by

$$\int_{X_\sigma} f(x) e^{(\rho+\sigma^{-1}\rho)(A(x,b))}\, dx < \infty$$

by (8.5). This shows by (8.6) that if $b \in B'$ and $\lambda \in \mathfrak{a}^* + iC(\rho)$ the integral (8.1) is absolutely convergent. The holomorphy statement follows by Morera's theorem. This proves (i).

For part (ii) we use the Radon transform (6.1). Since $f \in L^1(X)$, $\hat{f}(\xi)$ exist for almost all $\xi \in \Xi$ (see [20, II, §1]). Since $(kM, a) \to ka \cdot \xi_0$ is a diffeomorphism of K/M onto Ξ we write $\hat{f}(kM, a)$ for $\hat{f}(ka \cdot \xi_0)$. Enlarging B' to another subset of B of full invariant measure we may assume $\hat{f}(b, a)$ exists for $b \in B'$ and almost all a. Now we have

$$(8.8) \qquad \int_X f(x)\, dx = \int_{AN} f(anK)\, da\, dn$$

for suitable Haar measures on A and N. Applying this to the function $x \to f(k \cdot x)$ with $kM = b \in B'$ we get

$$(8.9) \qquad \int_X f(x)\, dx = \int_A \hat{f}(kM, a)\, da$$

so since $A(an \cdot o) = \log a$,

$$(8.10) \qquad \tilde{f}(\lambda, kM) = \int_A \hat{f}(kM, a) e^{(\rho+\eta)(\log a)} e^{-i\xi(\log a)}\, da$$
$$= \sum_{s \in W} \int_{s^{-1}A^+} \hat{f}(kM, a) e^{(\rho+\eta)(\log a)} e^{-i\xi(\log a)}\, da.$$

Now $a \in s^{-1}A^+$ implies $sa \in A^+$ and

$$\eta(\log a) = (s\eta)(s \log a) \leqslant (s\eta)^+(s \log a) \leqslant \rho(s \log a)$$

by (8.7). Thus on $s^{-1}A^+$,

$$(8.11)\qquad \hat{f}(kM,a)e^{(\rho+\eta)(\log a)} \leqslant \hat{f}(kM,a)e^{(\rho+s^{-1}\rho)(\log a)}.$$

For $b = kM \in B'$ the integral in (8.1) is absolutely convergent so by (8.9) the function

$$(8.12)\qquad a \to \hat{f}(kM,a)e^{(\rho+\eta)(\log a)}$$

belongs to $L^1(A)$. The first part of (8.10) combined with the Riemann–Lebesgue lemma for the Fourier transform on A shows that for each $\eta \in C(\rho)$

$$\lim_{\xi\to\infty} \tilde{f}(\lambda,b) = 0.$$

For the uniform convergence in (ii) we use the second part of (8.10). Let f_n be positive in $\mathcal{D}(X)$ such that $f_n \to f$ a.e. and $f_n(x) \leqslant f(x)$. In (8.10) and (8.11) we replace f by the function $g_n = f - f_n$. Then

$$\begin{aligned}|\tilde{g}_n(\lambda,kM)| &\leqslant \sum_{s\in W}\int_{s^{-1}A^+} \hat{g}_n(kM,a)e^{(\rho+\eta)(\log a)}\,da\\ &\leqslant \sum_{s\in W}\int_{s^{-1}A^+} \hat{g}_n(kM,a)e^{(\rho+s^{-1}\rho)(\log a)}\,da,\end{aligned}$$

which tends to 0 as $n \to \infty$ by (8.9), (8.12) and the dominated convergence theorem. Thus given $\varepsilon > 0$ we can fix N such that $|\tilde{g}_N(\lambda,kM)| < \varepsilon$ for all $\lambda \in \mathfrak{a}^* + iC_\rho$. By the Paley–Wiener theorem for $\mathcal{D}(X)$ there is an L such that $|\tilde{f}_N(\xi+i\eta,kM)| \leqslant \varepsilon$ for $|\xi| > L$ and $\eta \in C(\rho)$. Since $\tilde{g}_N = \tilde{f} - \tilde{f}_N$ this proves (ii). □

Remark. Another version of (ii) involving the L^1 norm over B is given in [45].

9. SPECTRAL ANALYSIS ON X

A theorem of Schwartz [48] states that if f is a function in $\mathcal{E}(\mathbf{R})$ ($f \not\equiv 0$) the closed subspace of $\mathcal{E}(X)$ (in its usual Fréchet space topology) generated by all the translates of f contains an exponential $e^{\mu x}$ for some $\mu \in \mathbf{C}$.

We shall now give the proof from [32] of the following analog of Schwartz's theorem.

Theorem 9.1. *Let $X = G/K$ have rank one and $f \neq 0$ a function in $\mathcal{E}(X)$. Then the closed subspace V_f of $\mathcal{E}(X)$ generated by the G-translates of f contains a function*

$$x \to e_{\mu,b}(x) = e^{\mu(A(x,b))}$$

for some $\mu \in \mathfrak{a}_c^*$.

For this we consider for $\lambda \in \mathfrak{a}_c^*$ the *Poisson transform*

$$\mathcal{P}_\lambda : F(b) \to f(x), \quad F \in L^1(B),$$

where

$$f(x) = \int_B e^{(i\lambda+\rho)(A(x,b))} F(b)\, db. \tag{9.1}$$

The element λ is said to be *simple* if $\mathcal{P}_\lambda$ is injective. The simplicity criterion for λ (see [22]) implies that for each $\lambda \in \mathfrak{a}_c^*$ one of the transforms $s\lambda$ $(s \in W)$ is simple. Consider now the spherical function φ_λ (7.3) which can also be written

$$\varphi_\lambda(x) = \int_B e^{(i\lambda+\rho)(A(x,b))}\, db.$$

We know from [25, III, Lemma 2.3] that if $-\lambda$ is simple then the closed space $\mathcal{E}_{(\lambda)}(X) \subset \mathcal{E}(X)$ generated by the G-translates of φ_λ contains the space $\mathcal{P}_\lambda(L^2(B))$.

Coming to the proof of the theorem we conclude from the Bagchi–Sitaram result (7.4) that the space V_f^K of K-invariants in V_f contains a spherical function φ_λ. By the simplicity result quoted, either λ or $-\lambda$ is simple so we can take $-\lambda$ simple. Thus by the conclusion above, V_f contains the space $\mathcal{P}_\lambda(L^2(B))$. Now by [25, III, Exercise B1, pp. 371 and 570],

$$e^{(i\lambda+\rho)A(x,eM)} = \sum_{\delta \in \hat{K}_M} d(\delta)\varphi_{\lambda,\delta}(x), \tag{9.2}$$

with δ and $\hat{K}_M$ as in (6.5) and

$$\varphi_{\lambda,\delta}(x) = \int_K e^{(i\lambda+\rho)(A(x,k))} \langle \delta(k)v, v\rangle\, dk.$$

Thus $\varphi_{\lambda,\delta} \in V_f$ so since (9.2) converges in the topology of $\mathcal{E}(X)$ the theorem follows.

Remark. Since Schwartz's theorem fails for $\mathbf{R}^n$ $(n > 1)$ the proof above via the Bagchi–Sitaram theorem is limited to the case of rank $X = 1$. However, this does not rule out the possibility that Theorem 9.1 might remain valid for X of higher rank.

10. Further results on the Fourier transform

A result of Hardy's [16] shows limitations on how fast a function on $\mathbf{R}^n$ and its Fourier transform can decay at ∞. Precisely, if

$$|f(x)| \leqslant Ae^{-\alpha|x|^2}, \quad |\tilde{f}(u)| \leqslant Be^{-\beta|u|^2}, \quad \alpha, \beta > 0,$$

and if $\alpha\beta > 1/4$ then $f = 0$. Sitaram and Sundari [53] proved an analog for a class of spaces X and Sengupta [49] extended this to all X. Many other variations of the result have been proved by Ray and Sarkar, Cowling, Sitaram and Sundari, Narayanan and Ray, Shimeno, Thangavelu. (See References.)

The following classical result is closely related to Wiener's Tauberian theorem. *Let $f \in L^1(\mathbf{R}^n)$ such that $\tilde{f}(u) \neq 0$ for all $u \in \mathbf{R}^n$. Then the translates of f span a dense subspace of $\mathbf{R}^n$*. Many papers deal with analogies of this result for semisimple Lie groups and symmetric spaces. See [10,46,51,52,45,36] for a sample.

The polar coordinate representation $(kM, a) \to kaK$ of X identifies X with $K/M \times A^+$ up to a null set. Thus one might interpret the Plancherel formula (8.2) as identifying X with its "dual". But in contrast to $\mathbf{R}^n$ where the Fourier transform is essentially equal to its inverse, the Fourier transform

$$(10.1) \qquad \tilde{f}(\lambda, b) = \int_X f(x) e^{(-i\lambda+\rho)(A(x,b))}\, dx,$$

and the inverse

$$(10.2) \qquad (\mathcal{F}^{-1}\varphi)(x) = \int_{\mathfrak{a}^* \times B} \varphi(\lambda, b) e^{(i\lambda+\rho)(A(x,b))} |c(\lambda)|^2\, d\lambda\, db$$

are quite different. Hence it is a natural problem to prove the analog of the Paley–Wiener theorem for $\mathcal{F}^{-1}$.

This was done by A. Pasquale [38] for the spherical transform for X of rank one or the case of G complex, and by N. Andersen [1] in general. Let L denote the Laplacian on X.

Theorem 10.1. *The image of $\mathcal{F}^{-1}(\mathcal{D}(\mathfrak{a}^* \times B))$ consists of the functions f on X satisfying*

$$\big(1 + d(o,x)\big)^m L^n f \in L^2(X) \quad \textit{for all } m, n \in \mathbf{Z}^+$$

and

$$\lim_{n \to \infty} \big\| \big(L + \langle \rho, \rho \rangle\big)^n \big\|_2^{1/2n} < \infty.$$

Another characterization was given by Pesenson [39], namely

$$\|L^\sigma f\|_2 \leqslant \big(\omega^2 + |\rho|^2\big)^\sigma \|f\|_2 \quad \text{for all } \sigma > 0.$$

REFERENCES

[1] Andersen N.B. – Real Paley–Wiener theorems for the inverse Fourier transform on a Riemannian symmetric space, Pacific J. Math. **213** (2004) 1–13.

[2] Anker J. – The spherical Fourier transform of rapidly decreasing functions—a simple proof of a characterization due to Harish-Chandra, Helgason, Trombi and Varadarajan, J. Funct. Anal. **96** (1991) 331–349.

[3] Bagchi S.C., Sitaram A. – Spherical mean-periodic functions on semisimple Lie groups, Pacific J. Math. **84** (1979) 241–250.

[4] van den Ban E.P., Schlichtkrull H. – Asymptotic expansions and boundary values of eigenfunctions on a Riemannian symmetric space, J. Reine Angew. Math. **380** (1987) 108–165.

[5] Beerends R.J. – The Fourier Transform of Harish-Chandra's c-function and inversion of the Abel transform, Math. Ann. **277** (1987) 1–23.

[6] Berenstein C.A., Casadio Tarabusi A.E. – Inversion formulas for k-dimensional Radon transform in real hyperbolic spaces, Duke Math. J. **62** (1991) 613–631.

[7] Cowling M., Sitaram A., Sundari M. – Hardy's uncertainty principle on semisimple Lie groups, Pacific J. Math. **192** (2000) 293–296.

[8] Eguchi M. – Asymptotic Expansions of Eisenstein Integrals and Fourier transform on symmetric spaces, J. Funct. Anal. **34** (1979) 167–216.

[9] Eguchi M., Hashizume M., Okamoto K. – The Paley–Wiener theorem for distributions on symmetric spaces, Hiroshima Math. J. **3** (1973) 109–120.

[10] Ehrenpreis L., Mautner F. – Some properties of the Fourier transform on semisimple Lie groups I, Ann. of Math. **61** (1955) 406–439.

[11] Funk P. – Über eine geometrische Anvendung der Abelschen Integralgleichung, Math. Ann. **77** (1916) 129–135.

[12] Gelfand I.M., Graev M.I. – The geometry of homogeneous spaces, group representations in homogeneous spaces and questions in integral geometry related to them, Amer. Math. Soc. Transl. **37** (1964).

[13] Gindikin S.G., Karpelevic F.I. – On a problem in integral geometry, Chebotarev Mem. Vol. Kazan Univ. 1964; Selecta Math. Sovietica **1** (1981) 169–184.

[14] Gonzalez F., Quinto E.T. – Support theorems for Radon transforms on higher rank symmetric spaces, Proc. Amer. Math. Soc. **122** (1994) 1045–1052.

[15] Harish-Chandra – Spherical functions on a semisimple Lie group II, Amer. J. Math. **80** (1958) 553–613.

[16] Hardy G.H. – A theorem concerning Fourier transforms, J. London Math. Soc. **8** (1933) 227–231.

[17] Helgason S. – Duality and Radon transforms for symmetric spaces, Amer. J. Math. **85** (1963) 667–692.

[18] Helgason S. – A duality in integral geometry. Some generalizations of the Radon transform, Bull. Amer. Math. Soc. **70** (1964) 435–446.

[19] Helgason S. – Radon–Fourier transforms on symmetric spaces and related group representations, Bull. Amer. Math. Soc. **71** (1965) 757–763.

[20] Helgason S. – A duality for symmetric spaces with applications to group representations, Adv. Math. **5** (1970) 1–154.

[21] Helgason S. – The surjectivity of invariant differential operators on symmetric spaces, Ann. of Math. **98** (1973) 451–480.

[22] Helgason S. – A duality for symmetric spaces with applications to group representations II. Differential equations and eigenspace representations, Adv. Math. **22** (1976) 187–219.

[23] Helgason S. – Ranges of Radon transforms, in: Proc. Symp. Appl. Math., vol. 27, Amer. Math. Soc., Providence, RI, 1982, pp. 63–70.

[24] Helgason S. – Groups and Geometric Analysis, Academic Press, San Diego, CA, 1984, Amer. Math. Soc., Providence, RI, 2000.

[25] Helgason S. – Geometric Analysis on Symmetric Spaces, Math. Surveys and Monographs, vol. 39, Amer. Math. Soc., Providence, RI, 1994.

[26] Helgason S. – The Radon Transform, Birkhäuser, Boston, 1999.

[27] Helgason S., Non-Euclidean analysis, in: Proc. Janos Bolyai Memorial Conference, Budapest, July 2002.

[28] Helgason S. – Differential operators on homogeneous spaces, Acta Math. **102** (1959) 239–299.

[29] Helgason S. – Fundamental solutions of invariant differential operators on symmetric spaces, Amer. J. Math. **86** (1964) 565–601.

[30] Helgason S., Johnson K. – The bounded spherical functions on symmetric spaces, Adv. Math. **3** (1969) 586–593.

[31] Helgason S., Rawat R., Sengupta J., Sitaram A. – Some remarks on the Fourier transform on a symmetric space, Tech. Report, Ind. Stat. Inst. Bangalore, 1998.

[32] Helgason S., Sengupta J. – Preprint, 1997.

[33] John F. – Plane Waves and Spherical Means, Wiley, New York, 1955.

[34] Kostant B. – On the existence and irreducibility of certain series of representations, in: Gelfand I.M. (Ed.), Lie Groups and Their Representations, Halsted, New York, 1975, p. 231, 329.

[35] Lax P.D., Phillips R.S. – Translation representation for the solution of the non-Euclidean wave equation, Comm. Pure Appl. Math. **32** (1979) 617–667.

[36] Mohanty P., Ray S.K., Sarkar R.P., Sitaram A. – Helgason Fourier transform for symmetric spaces II, J. Lie Theory **14** (2004) 227–242.

[37] Narayanan E.K., Ray S.K. – L^p version of Hardy's theorem on semisimple Lie groups, Proc. Amer. Math. Soc. **130** (2002) 1859–1866.

[38] Pasquale A. – A Paley–Wiener theorem for the inverse spherical transform, Pacific J. Math. **193** (2000) 143–176.

[39] Pesenson I., Deconvolution of band limited functions on non-compact symmetric spaces, Preprint.

[40] Radon J. – Über die Bestimmung von Funktionen durch ihre Integralwerte längs gewisser Mannigfaltigkeiten, Ber. Verth. Sächs. Akad. Wiss. Leipzig. Math. Nat. kl. **69** (1917) 262–277.

[41] Ray S.K., Sarkar R.P. – Cowling–Price theorem and characterization of heat kernel of symmetric spaces, Preprint.

[42] Rouvière F. – Inverting Radon transforms; the group-theoretic approach, Enseign. Math. **47** (2001) 205–252.

[43] Rouvière F. – Geodesic Radon transforms on symmetric spaces, Preprint, 2004.

[44] Rouvière F. – Sur la transformation d'Abel de groupes de Lie semisimple de rang un, Ann. Scuola. Norm. Sup. Pisa **10** (1983) 263–290.

[45] Sarkar R., Sitaram A. – The Helgason Fourier Transform for Symmetric Spaces. C.S. Seshadri Festschrift Volume, Birkhäuser, Boston, 2003.

[46] Sarkar R.P. – Wiener Tauberian theorem for rank one symmetric spaces, Pacific J. Math. **186** (1998) 349–358.

[47] Sawyer P. – The Abel transform on symmetric spaces of noncompact type, Amer. Math. Soc. Transl. (2) **210** (2003) 331–355.

[48] Schwartz L. – Théorie generale des fonctions moyenne-periodiques, Ann. of Math. **48** (1947) 857–929.

[49] Sengupta J. – The Uncertainty Principle on Riemannian symmetric spaces of the noncompact type, Proc. Amer. Math. Soc. **128** (2000) 2493–2499.

[50] Shimeno N. – An analog of Hardy's theorem for the Harish-Chandra transform, Hiroshima Math. J. **31** (2001) 383–390.

[51] Sitaram A. – An analog of the Wiener Tauberian theorem for spherical transforms on semisimple Lie groups, Pacific J. Math. **89** (1980) 439–445.

[52] Sitaram A. – On an analog of the Wiener Tauberian theorem for symmetric spaces of the noncompact type, Pacific J. Math. **133** (1988) 197–208.

[53] Sitaram A., Sundari M. – An analog of Hardy's theorem for very rapidly decreasing functions on semisimple Lie groups, Pacific J. Math. **177** (1997) 187–200.

[54] Thangavelu S. – An Introduction to the Uncertainty Principle, Birkhäuser, Basel, 2004.

[55] Thangavelu S. – Hardy's theorem for the Helgason Fourier transform on noncompact rank one symmetric spaces, Colloq. Math. **94** (2002) 263–280.

(Received February 2005)

Journal of Lie Theory
Volume **17** (2007) 307–315

The Inversion of the X-ray Transform on a Compact Symmetric Space

Sigurdur Helgason*

Communicated by G. Ólafsson

Abstract. The X-ray transform on a compact symmetric space M is here inverted by means of an explicit inversion formula. The proof uses the conjugacy of the minimal closed geodesics in M and of the maximally curved totally geodesic spheres in M, proved in [3].
Mathematics Subject Classification 2000 Primary: 22E30, 43A85, 44A12, 53C65; secondary: 22E46, 53C35.
Keywords and phrases: Compact symmetric spaces, Funk transform, Radon transform, X-ray transform.

1. Introduction

In his paper [1] Funk showed, using tools from a geometric paper by Minkowski, that an even function f on the sphere $\mathbf{S}^2$ is explicitly determined by its integrals $\widehat{f}(\xi)$ over the great circles ξ on the sphere. The evenness condition is clearly necessary since $\widehat{f} \equiv 0$ if f is odd.

The negative aspect of the result would suggest that Funk's theorem might not extend to geodesic integrals on a compact symmetric space since the concept of an even function is not present. However we shall see that in restated form the theorem generalizes to compact symmetric spaces.

Let $\mathbf{S}^+$ denote the top half $x_3 > 0$ of $\mathbf{S}^2$ and $f \in \mathcal{C}_c^\infty(\mathbf{S}^+)$. Then $g(x) = \frac{1}{2}(f(x) + f(-x))$ is even and $\widehat{f} = \widehat{g}$. The inversion formula for g (Corollary 2.1) thus gives an inversion formula for f on $\mathbf{S}^+$. In this form we extend Funk's injectivity result to compact symmetric spaces, even with an explicit inversion formula. See Corollary 3.3.

Let $M = U/K$ be an irreducible compact simply connected symmetric space, U being a compact semisimple Lie group. For this space we shall use results from our paper [3]; see also [4], VII, §11, whose notation we follow. Let $\mathfrak{u} = \mathfrak{k} + \mathfrak{p}_*$ be the eigenspace decomposition for the involution of the Lie algebra $\mathfrak{u}$ of U. If the metric on M is given by the negative of the Killing form B of $\mathfrak{u}$, the

* This paper is dedicated to François Rouvière whose paper [13] prompted me to reconsider the old problem dealt with here.

ISSN 0949–5932 / $2.50 © Heldermann Verlag

maximal sectional curvature on M equals $\|\bar{\delta}\|^2$ where $\bar{\delta}$ is the highest restricted root. We normalize the metric such that this maximal curvature is 1. We shall then use the following result from [3]. Here $m(\bar{\delta})$ denotes the multiplicity of $\bar{\delta}$.

Theorem 1.1.

(i) *The shortest closed geodesics in M have length 2π and they are permuted transitively by U.*

(ii) *M has totally geodesic spheres of curvature 1. Their maximum dimension is $1+m(\bar{\delta})$. All such spheres $\mathbf{S}^{1+m(\bar{\delta})}$ are conjugate under U.*

We need some further results for the geometry of M. Fix a maximal abelian subspace $\mathfrak{a}_* \subset \mathfrak{p}_*$. Let

$$\mathfrak{p}_\delta = \{X \in \mathfrak{p}_* : (\text{ad } H)^2 X = \delta(H)^2 X \text{ for } H \in \mathfrak{a}_*\}$$

fix $A(\bar{\delta}) \in \mathfrak{a}_*$ such that

$$B(H, A(\bar{\delta})) = \pi i \delta(H) \qquad H \in \mathfrak{a}_*,$$

and let $\mathfrak{a}_\delta$ denote the line $\mathbf{R}A(\bar{\delta})$ (notation of [4], VII, §11). If Exp denotes the map from $\mathfrak{p}_*$ to U/K given by $\text{Exp } X = (\exp X)K$ we have from [4], p. 343 that the set

$$M_\delta = \text{Exp}\,(\mathfrak{a}_\delta + \mathfrak{p}_\delta)$$

is a sphere, totally geodesic in M, of dimension $1+m(\delta)$ and curvature 1. The point Exp $A(\bar{\delta})$ is the point on M_δ antipodal to the point $o = eK$. Let $S \subset K$ be the subgroup fixing both o and Exp $A(\bar{\delta})$. From [4], p. 343 we have the following result.

Proposition 1.2. *The restriction of* $\text{Ad}_U(S)$ *to the tangent space* $(M_\delta)_0$ *contains* $\mathbf{SO}((M_\delta)_0)$.

Definition. Let $x \in M$. The *midpoint locus* A_x associated to x is the set of midpoints $m(\gamma)$ of all the closed minimal geodesics γ starting at x. Let $e_1(\gamma), e_2(\gamma)$ denote the midpoints of the arcs of γ which join x and $m(\gamma)$. Let E_x denote the set of these $e(\gamma)$. We call E_x the *equator* associated to x.

Theorem 1.3. *A_0 and E_0 are K-orbits and A_0 is a totally geodesic submanifold of M. Also*

$$A_0 = K/S.$$

Proof. A_0 is a K-orbit because of Theorem 1.1. For a similar statement for E_0 we must verify that the two midpoints $e(\gamma)$ on the same minimal γ are conjugate under K. This is obvious if we take Proposition 1.2 into account. For the rest see [4], VII, §11. ∎

Definition. The Funk transform for $M = U/K$ is the map $f \to \widehat{f}$ where

$$\widehat{f}(\xi) = \int_\xi dm(x)\,, \tag{1.1}$$

ξ being a closed geodesic in M of minimal length and dm the arc element.

Because of Theorem 1.1 we have a pair of homogeneous spaces:

$$M = U/K\,, \quad \Xi = \{\text{minimal geodesics}\} = U/H$$

where H is the stabilizer of a specific minimal geodesic ξ in M. We then have the corresponding dual transform $\varphi \to \check{\varphi}$ where

$$\check{\varphi}(gK) = \int_K \varphi(gk \cdot \xi)\, dk\,. \tag{1.2}$$

Notation: In a metric space $B_r(p)$ denotes the open ball with center p and radius r. $S_r(p)$ denotes the corresponding sphere.

2. Inversion on S^n

We consider now the sphere

$$X = \mathbf{S}^n = \mathbf{O}(n+1)/\mathbf{O}(n)\,,$$

where $L = \mathbf{O}(n)$ is the isotropy group of $o = (0, \ldots, 0, 1)$ and the space

$$\Xi = \{\text{totally geodesic } \mathbf{S}^k \subset \mathbf{S}^n\}$$

for k fixed, $1 \leq k \leq n-1$. We write

$$\Xi = \mathbf{O}(n+1)/H_p\,,$$

where H_p is the stability group of a k-sphere $\xi_p \subset X$ which has distance p from o. In addition to the Funk transform $f \to \widehat{f}$

$$\widehat{f}(\xi) = \int_\xi f(x)\, dm(x) \qquad \xi \in \Xi \tag{2.3}$$

we consider also the dual transform,

$$\check{\varphi}_p(gL) = \int_L \varphi(g\ell \cdot \xi_p)\, d\ell\,, \tag{2.4}$$

the average of φ over the set of $\mathbf{S}^k$ at distance p from $g \cdot o$. We write $\check{\varphi}$ for $\check{\varphi}_0$.

In [2] we inverted the transform $f \to \widehat{f}$ by the formula

$$f = P_k(\Delta)\left((\widehat{f})^\vee\right) \tag{2.5}$$

for k even, $P_k(\Delta)$ being an explicit polynomial in the Laplacian Δ of degree $k/2$. In the paper [16] this is augmented by the case $k = n-1$, k odd, and the transform $f \to \widehat{f}$ inverted by an integral which is then suitably regularized.

In [6] I published the inversion formula (2.6) below for $f \to \widehat{f}$, valid for all k and n; in comparison with (2.5) it seemed so unwieldy that I did not publish it in [2]; at that time the case $k = 1$ (the X-ray transform) had not gained the later prominence. Unexpectedly, the formula simplifies considerably for $k = 1$ and this version is the basis for the extension below to the compact space $M = U/K$. One more inversion of $f \to \widehat{f}$ on $\mathbf{S}^n$ with k arbitrary was given by Rubin [15].

From Theorem 3.2 in [6] we have the following inversion formula for (2.3). For $f \in \mathcal{C}^\infty(\mathbf{S}^n)$ even

$$f(x) = \frac{c}{2}\left[\left(\frac{d}{d(u^2)}\right)^k \int_0^u (\widehat{f})^\vee_{\cos^{-1} v}(x) v^k (u^2 - v^2)^{\frac{k}{2}-1}\, dv\right]_{u=1} \tag{2.6}$$

where

$$c = \frac{2^{k+1}}{(k-1)!\,\Omega_{k+1}},$$

and Ω_{k+1} is the area of the unit sphere in $\mathbf{R}^{k+1}$.

Remark 2.1. Let $(M^r f)(x)$ denote the average of f over a sphere in X with center x and radius r. The proof of (2.6) in [6] used the fact that the function $y \to (M^{d(x,y)} f)(x)$ on ξ_p is even, d denoting distance and $d(x, \xi_p) = p$. Indeed, if $g = \mathbf{O}(n+1)$ is such that $g \cdot o = x$ then it was shown that

$$(M^{d(x,y)} f)(x) = \int_L f(g\ell g^{-1} \cdot y)\, d\ell,$$

which is indeed even in y because of the linearity of $g\ell g^{-1}$.

For the case $k = 1$ we can derive a better version even without the evenness assumption. Note though that because of the integration over E_x formula (2.7) is not an exact inversion. The proof resembles that of Theorem 4.3 in [8].

Corollary 2.2. *The X-ray transform on $\mathbf{S}^n$ is inverted by the formula*

$$\tfrac{1}{2}(f(x) + f(-x)) = \int_{E_x} f(\omega)\, d\omega - \frac{1}{2\pi}\int_0^{\frac{\pi}{2}} \frac{d}{dp}(\widehat{f})^\vee_p(x)\frac{dp}{\sin p} \tag{2.7}$$

for every $f \in \mathcal{C}^\infty(\mathbf{S}^n)$. Here $d\omega$ is the normalized measure on the equator E_x.

Proof. Replacing f by $\frac{1}{2}(f(x) + f(-x))$ has no effect on $\widehat{f}$ so with $\widehat{F}(\cos p) = (\widehat{f})^\vee_p(x)$ we have for the right hand side of (2.6)

$$\frac{1}{2\pi}\left\{\frac{d}{du}\int_0^u (u^2 - v^2)^{-\frac{1}{2}} v \widehat{F}(v)\, dv\right\}_{u=1}$$
$$= -\frac{1}{2\pi}\left\{\frac{d}{du}\int_0^u \frac{d}{dv}(u^2 - v^2)^{\frac{1}{2}} \widehat{F}(v)\, dv\right\}_{u=1},$$

which by integration by parts becomes

$$-\frac{1}{2\pi}\left\{\frac{d}{du}\left[-u\widehat{F}(0)-\int_0^u (u^2-v^2)^{\frac{1}{2}}\frac{d}{dv}\widehat{F}(v)\,dv\right]\right\}_{u=1}$$

$$=\frac{1}{2\pi}\widehat{F}(0)+\frac{1}{2\pi}\int_0^1 (1-v^2)^{-\frac{1}{2}}\frac{d}{dv}\widehat{F}(v)\,dv$$

$$=\frac{1}{2\pi}(\widehat{f})^{\vee}_{\frac{\pi}{2}}(x)-\frac{1}{2\pi}\int_0^{\frac{\pi}{2}}\frac{d}{dp}(\widehat{f})^{\vee}_p(x)\frac{dp}{\sin p}.$$

The first term is an average of the integrals of f over geodesics at distance $\pi/2$ from x which thus lie in E_x. It represents a rotation–invariant functional on E_x hence a constant multiple of the integral over E_x. Taking $f \equiv 1$ the constant is 1 and the formula is proved. ∎

Corollary 2.3. *Suppose $f \in \mathcal{C}^\infty(\mathbf{S}^n)$ has support in the ball $B = \{x \in \mathbf{S}^n : d(o,x) < \frac{\pi}{4}\}$. Then*

$$f(x) = -\frac{1}{\pi}\int_0^{\frac{\pi}{2}}\frac{d}{dp}\left((\widehat{f})^{\vee}_p(x)\right)\frac{dp}{\sin p}, \qquad x \in B. \tag{2.8}$$

In fact if $x \in B$ then $f(-x) = 0$. If $x \in B$ any $y \in E_x$ then $d(o,y) \geq d(x,y) - d(o,x) \geq \frac{\pi}{2} - \frac{\pi}{4}$ so $f(y) = 0$.

3. The case of a compact symmetric space

Let $M = U/K$ be as in the Introduction. We shall now combine Theorem 1.1 and Corollary 2.2 to study the Funk transform (1.1). Note that this is the X-ray transform restricted to minimal geodesics.

Given $f \in \mathcal{C}^\infty(M)$ we consider its restriction $f|M_\delta$ to the sphere M_δ. For $0 \leq p \leq \frac{\pi}{2}$ we fix a geodesic $\xi_p \subset M_\delta$ at distance p from o. Let f_* denote the Funk transform $(f|M_\delta)\widehat{}$ and φ_p^* the dual transform (2.4). Note that φ_p^* is (for given p) independent of the choice of ξ_p. Then (2.7) implies

$$\tfrac{1}{2}\left(f(o)+f(\operatorname{Exp} A(\bar{\delta}))\right)=\int_{E_0'} f(\omega)\,d\omega-\frac{1}{2\pi}\int_0^{\frac{\pi}{2}}\frac{d}{dp}(f_*)^*_p(o)\frac{dp}{\sin p}, \tag{3.9}$$

where E_0' is the equator in M_δ associated to o. By Proposition 1.2, $E_0' = S \cdot \operatorname{Exp}(\frac{1}{2}A(\bar{\delta}))$. We now apply (3.9) to the function

$$f^{\natural}(x)=\int_K f(k\cdot x)\,dk.$$

Since $A_0 = K \cdot \operatorname{Exp} A(\bar{\delta})$ the left hand side becomes

$$\tfrac{1}{2}\Big(f(o)+\int_{A_0} f(\omega)\,d\omega\Big),$$

where $d\omega$ stands for average. The first term on the right becomes

$$\int_{K\cdot E_0'} f(k\cdot\omega)\,dk\,d\omega=\int_K f\Big(k\cdot\operatorname{Exp}\tfrac{1}{2}A(\bar{\delta})\Big)\,dk=\int_{E_0} f(\omega)\,d\omega,$$

where $E_0 = K \cdot \text{Exp}\ \frac{1}{2}A(\bar{\delta})$ which is contained in $S_{\frac{\pi}{2}}(o)$. For the second term on the right note that by the transitivity of the group S (Prop. 1.2)

$$\begin{aligned} ((f^\natural)_*)_p^*(o) &= \int_S (f^\natural)_*(s \cdot \xi_p)\, ds = \int_S (f^\natural)\widehat{\ }(s \cdot \xi_p)\, ds \\ &= \int_K \int_S \widehat{f}(ks \cdot \xi_p)\, ds\, dk = \int_K \widehat{f}(k \cdot \xi_p)\, dk = (\widehat{f})_p^\vee(o), \end{aligned}$$

where $\check{\varphi}_p$ is the dual transform (1.2) for $\xi = \xi_p$. Thus we have

$$\tfrac{1}{2}\Big(f(o) + \int_{A_0} f(\omega)\, d\omega\Big) = \int_{E_0} f(\omega)\, d\omega - \frac{1}{2\pi}\int_0^{\frac{\pi}{2}} \frac{d}{dp}\Big((\widehat{f})_p^\vee(o)\Big) \frac{dp}{\sin p}.$$

The set $\Xi_p = \{k \cdot \xi_p : k \in K\}$ constitutes the set of all minimal geodesics each lying in some totally geodesic sphere $S^{1+m(\delta)}$ through o having distance p from o. Let ω_p^0 denote the normalized K-invariant measure on this set. Thus

$$(\widehat{f})_p^\vee(o) = \int_{\Xi_p} \widehat{f}(\xi)\, d\omega_p^0(\xi).$$

Let $\Xi_p(x)$ be defined similarly for the point $x \in M$ and let ω_p denote the corresponding measure. Choose $g \in U$ such that $g \cdot o = x$. Then $A_x = gA_0$, $E_x = gE_0$ and $g \cdot \Xi_p = \Xi_p(x)$. Then we obtain the following result.

Theorem 3.1. *Let $f \in \mathcal{C}^\infty(M)$. Then*

$$\tfrac{1}{2}\Big(f(x) + \int_{A_x} f(\omega)\, d\omega\Big) = \int_{E_x} f(\omega)\, d\omega - \frac{1}{2\pi}\int_0^{\frac{\pi}{2}} \frac{d}{dp}\Big(\int_{\Xi_p(x)} \widehat{f}(\xi)\, d\omega_p(\xi)\Big) \frac{dp}{\sin p}.$$

Restricting the support of f we obtain the following result.

Corollary 3.2. *Let $f \in C_c^\infty(B_{\frac{\pi}{2}}(o))$. Then for $x \in B_{\frac{\pi}{2}}(o)$*

$$f(x) = 2\int_{E_x} f(\omega)\, d\omega - \frac{1}{\pi}\int_0^{\frac{\pi}{2}} \frac{d}{dp}\Big(\int_{\Xi_p(x)} \widehat{f}(\xi)\, d\omega_p(\xi)\Big) \frac{dp}{\sin p}.$$

For a full inversion formula we restrict the support of f further.

Corollary 3.3. *Let $f \in C_c^\infty(B_{\frac{\pi}{4}}(o))$. Then*

$$f(x) = -\frac{1}{\pi}\int_0^{\frac{\pi}{2}} \frac{d}{dp}\Big(\int_{\Xi_p(x)} \widehat{f}(\xi)\, d\omega_p(\xi)\Big) \frac{dp}{\sin p}, \quad x \in B_{\frac{\pi}{4}}(o).$$

For Corollary 3.2 we must show that

$$A_x \cap B_{\frac{\pi}{2}}(o) = \emptyset \text{ if } x \in B_{\frac{\pi}{2}}(o). \tag{3.10}$$

If $g \cdot o = x$ we have $A_x = g \cdot A_0$ and

$$\begin{aligned} d(o, gk \cdot \text{Exp}\ A(\bar{\delta})) &= d(g^{-1} \cdot o,\, k \cdot \text{Exp}\ A(\bar{\delta})) \\ &\geq d(o, k \cdot \text{Exp}\ A(\bar{\delta})) - d(o, g^{-1} \cdot o) \\ &\geq \pi - \frac{\pi}{2} = \frac{\pi}{2} \end{aligned}$$

so (3.10) follows. For Cor. 3.3 we must show

$$B_{\frac{\pi}{4}}(o) \cap E_x = \emptyset \quad \text{if} \quad x \in B_{\frac{\pi}{4}}(o)\,.$$

Better still we show that

$$B_{\frac{\pi}{4}}(o) \cap S_{\frac{\pi}{2}}(x) = \emptyset \quad \text{if} \quad x \in B_{\frac{\pi}{4}}(o)\,. \tag{3.11}$$

But if $z \in S_{\frac{\pi}{2}}(x)$ then

$$d(o,z) \geq d(x,z) - d(o,x) \geq \frac{\pi}{2} - \frac{\pi}{4} = \frac{\pi}{4}$$

proving (3.11).

Using Theorem 1.1 we can derive the following support theorem for the rank one case.

Theorem 3.4. *Assume M has rank one. Let $0 < \delta < \frac{\pi}{2}$. Suppose $f \in \mathcal{C}^\infty(B_{\frac{\pi}{2}}(o))$ satisfies*

(i) $\widehat{f}(\xi) = 0$ for $d(o,\xi) > \delta$.

(ii) For each $m > 0$

$$f(x) \cos d(o,x)^{-m} \text{ is bounded.}$$

Then

$$f(x) = 0 \quad \text{for} \quad d(o,x) > \delta\,. \tag{3.12}$$

Proof. Consider the restriction $f|M_\delta$. Because of (ii) $f|M_\delta$ can be extended to a symmetric function on M_δ so by the support theorem for the sphere ([12], [10] or [7]. I, § 3) (3.12) holds for $x \in M_\delta$. Since the rank is one the spheres kM_δ, $(k \in K)$ fill up M so (3.12) holds for $x \in M$. ∎

4. The non-compact case

Here we consider the case of an irreducible symmetric space $X = G/K$ of the non-compact type where G is simple, connected with finite center and K a maximal compact subgroup. As proved in [5] the X-ray transform is here injective. In his elegant paper [14] Rouvière proved an explicit inversion formula by a reduction to the hyperbolic plane. In my paper [9] another inversion formula is given which, however, requires rank $X > 1$.

In this section we present a third formula suggested by our method for the compact case. The proof is much simpler than in the compact case since there is no midpoint locus and no equator. Again we normalize the metric on X such that the maximal negative curvature is -1. A geodesic in X which lies in a totally geodesic hyperbolic space of curvature -1 will be called a *flexed geodesic*. From the duality for symmetric spaces ([4], Ch. V, §2) we know that the tangent spaces to G/K and U/K at o correspond under multiplication by i. This commutes with the action of K. Lie triple systems are mapped into Lie triple systems by this correspondence and sectional curvatures are turned into their negatives. Thus we have the following analog to Theorem 1.1.

Theorem 4.1.

(i) *X has hyperbolic totally geodesic submanifolds of curvature -1. Their maximum dimension is $1+m(\bar{\delta})$ and those $\mathbf{H}^{1+m(\bar{\delta})}$ are all conjugate under G.*

(ii) *The flexed geodesics in X are permuted transitively by G.*

We consider now the hyperbolic analog of M_δ, namely X_δ of curvature -1, dimension $1+m(\bar{\delta})$, passing through $o=eK$. We have then the following analog of Corollary 2.2 proved in [8], for $\mathbf{H}^n$ of all dimensions,

$$f(x) = -\frac{1}{\pi}\int_0^\infty \frac{d}{dp}\left((\widehat{f})_p^\vee(x)\right)\frac{dp}{\sinh p}, \quad x \in X_\delta . \tag{4.13}$$

For each $p \geq 0$ let $\Xi_p(x)$ denote the set of all flexed geodesics ξ in X, each lying in a totally geodesic $\mathbf{H}^{1+m(\bar{\delta})}$ passing through x with $d(x,\xi)=p$. Let ω_p denote the normalized measure on Ξ_p invariant under the isotropy group of x. The proof of Theorem 3.4 then yields the following result.

Theorem 4.2. *Let $f \in C_c^\infty(X)$. Then*

$$f(x) = -\frac{1}{\pi}\int_0^\infty \frac{d}{dp}\left(\int_{\Xi_p(x)} \widehat{f}(\xi)\, d\omega_p(\xi)\right)\frac{dp}{\sinh p}. \tag{4.14}$$

Remark 4.3. As kindly pointed out by Rouvière, (4.14) agrees with his formula in Theorem 1 in [14] which more generally holds for each root of $(\mathfrak{g},\mathfrak{a})$.

References

[1] Funk, P., *Über eine geometrische Anwendung der Abelschen Integralgleichung*, Math. Ann. **77** (1916), 129–135.

[2] Helgason, S., *Differential operators on homogeneous spaces*, Acta Math. **102** (1959), 239–299.

[3] —, *Totally geodesic spheres in compact symmetric spaces*, Math. Ann. **165** (1966), 309–317.

[4] —, "Differential Geometry, Lie Groups and Symmetric Spaces," Academic Press, New York, 1978. Reprinted by Amer. Math. Soc. 2001.

[5] —, *The X-ray transform on a symmetric space*, Lecture Notes in Math. No. 838, Springer–Verlag, New York, 1980, 145–148.

[6] —, *The totally geodesic Radon transform on constant curvature spaces*, Contemp. Math. **113** (1990), 141–149.

[7] —, "Geometric Analysis on Symmetric Spaces," Amer. Math. Soc. Mathematical Surveys and Monographs, 1994.

[8] —, *Non-Euclidean analysis*, in: A. Prékopa and E. Molnár, Eds., "Non-Euclidean Geometries," Mathematics and Its Applications **581**, Springer, New York, 2006, 367–384.

[9] —, *The Abel, Fourier and Radon transforms on symmetric spaces*, Indag. Math. N.S. **16** (2005), 531–551.

[10] Kurusa, A., *Support theorems for the totally geodesic Radon transform on constant curvature spaces*, Proc. Amer. Math. Soc. **122** (1994), 429—435.

[11] —, *The totally geodesic Radon transform on the Lorentz space of curvature* 1, Duke Math. J. **86** (1997), 565–583.

[12] Quinto, E.T., *The invertibility of rotation invariant Radon transforms*, J. Math. Anal. Appl. **91** (1983), 510–522.

[13] Rouvière. F., *Geodesic Radon transforms on symmetric spaces*, Preprint 2004.

[14] —, *Transformation aux rayons X sur un espace symétrique*, C. R. Acad. Sci. Paris Ser. I **342** (2006), 1–6.

[15] Rubin, B., *Inversion formulas for the spherical Radon transform and the generalised cosine transform*, Advan. Appl. Math. **29** (2002), 471–497.

[16] Semyanisty, V. I., *Homogeneous functions and some problems of integral geometry in spaces of constant curvature*, Soviet Math. Dokl. **2** (1961), 59–62.

Sigurdur Helgason
Department of Mathematics
Massachusetts Institute of Technology
Cambridge, MA 02139
helgason@mit.edu

Received Sept. 12, 2006
and in final form Nov. 14, 2006

Acknowledgments

The American Mathematical Society, the editors, and Sigurður Helgason gratefully acknowledge the kindness of these institutions and individuals in granting the following permissions:

Elsevier

(with K. Johnson) *The bounded spherical functions on symmetric spaces*, Adv. in Math. **3** (1969), 586-593. © Elsevier (1969).

A duality for symmetric spaces with applications to group representations, Adv. in Math. **5** (1970), 1-154. © Elsevier (1970).

Group representations and symmetric spaces, Proc. Internat. Congress Math. Nice 1970, Vol. 2. Gauthier-Villars, Paris 1971, 313-320. © Elsevier (1971).

Eigenspaces of the Laplacian; integral representations and irreducibility, J. Funct. Anal. **17** (1974), 328-353. © Elsevier (1974).

A duality for symmetric spaces with applications to group representations II. Differential equations and eigenspace representations, Adv. in Math. **22** (1976), 187-219. © Elsevier (1976).

A duality for symmetric spaces with applications to group representations III. Tangent space analysis, Adv. in Math. **36** (1980), 297-323. © Elsevier (1980).

(with F. Gonzalez) *Invariant differential operators on Grassmann manifolds*, Adv. in Math. **60** (1986), 81-91. © Elsevier (1986).

The flat horocycle transform for a symmetric space, Adv. in Math. **91** (1992), 232-251. © Elsevier (1992).

Huygens' principle for wave equations on symmetric spaces, J. Funct. Anal. **107** (1992) 279-288. © Elsevier (1992).

Heldermann Verlag

The inversion of the X-ray transform on a compact symmetric space, J. Lie Theory **17** (2007), 307-315.

Anna Loa Helgason for a photograph of Sigurður Helgason, page ii.

Artie Helgason for the pictures of Sigurður Helgason at the Lie Memorial Conference Oslo 1992 (p. xxxii) and at the Dr.h.c in Uppsala 1996 with Dennis A. Hejhal and C. Kieselman (p. xxxiv).

Institut Mittag-Leffler and the Royal Swedish Academy of Sciences

Differential operators on homogeneous spaces, Acta Math. **102** (1959), 239-299.

The Radon transform on Euclidean spaces, compact two-point homogeneous spaces and Grassmann manifolds, Acta Math. **113** (1965), 153-180.

International Press

Radon transforms for double fibrations. Examples and viewpoints, Proc. of Symposium *75 years of Radon Transform*, Vienna 1992. Int. Press, Boston 1994, 175-188. © 1994 International Press of Boston.

John Hopkins University Press

Duality and Radon transform for symmetric spaces, Amer. J. Math. **85** (1963), 667-692. The Johns Hopkins University Press. Reprinted with permission of The Johns Hopkins University Press.

Fundamental solutions of invariant differential operators on symmetric spaces, Amer. J. Math. **86** (1964), 565-601. The Johns Hopkins University Press. Reprinted with permission of The Johns Hopkins University Press.

Some results on invariant differential operators on symmetric spaces, Amer. J. Math. **114** (1992), 789-811. The Johns Hopkins University Press. Reprinted with permission of The Johns Hopkins University Press.

John Wiley and Sons

Integral geometry and multitemporal wave equations, Comm. Pure and Appl. Math. **51** (1998) 1035-1071.

(with H. Schlichtkrull) *The Paley-Wiener space for the multitemporal wave equation*, Comm. Pure and Appl. Math. **52** (1999) 49-52.

Lehigh University

A formula for the radial part of the Laplace-Beltrami operator, J. Diff. Geometry **6** (1972), 411-419.

Nippon Hyoronsha CO, LTD

A duality in integral geometry on symmetric spaces, Proc. U.S.-Japan Seminar in Differential Geometry, Kyoto, Japan, 1965, pp. 37-56. Nippon Hyronsha, Tokyo 1966.

Princeton University and the Institute for Advanced Study

Multipliers of Banach algebras, Ann. of Math. **64** (1956), 240-254.

The surjectivity of invariant differential operators on symmetric spaces, Ann. of Math. **98** (1973), 451-480.

François Rouvière for the photographs of the Gymnasium in Akureyri (p. xxx) and of Sigurður Helgason and Adam Korányi at the conference in Iceland, 2007 (p. xxxv).

Royal Netherlands Academy of Sciences

The Abel, Fourier and radon transforms on symmetric spaces, Indag. Math. **16** (2005), 531-551.

Boris Rubin for the picture of Sigurður Helgason and Gestur Ólafsson in Tokyo, 2006 (p. xxxiv).

Springer Verlag

Totally geodesic spheres in compact symmetric spaces, Math. Ann. **165** (1966), 309-317. With kind permission of Springer Science and Business Media.

An analogue of the Paley-Wiener theorem for the Fourier transform on certain symmetric spaces, Math. Ann. **165** (1966), 297-308. With kind permission of Springer Science and Business Media.

The American Mathematical Society

Fundamental solutions of invariant differential operators on symmetric spaces, Bull. Amer. Math. Soc. **68** (1963), 778–781.

Radon-Fourier transforms on symmetric spaces and related group representations, Bull. Amer. Math. Soc. **71** (1965), 757-763.

(with A. Korányi) *A Fatou-type theorem for harmonic functions on symmetric spaces*, Bull. Amer. Math. Soc. **74** (1968), 258-263.

Paley-Wiener theorems and surjectivity of invariant differential operators on symmetric spaces and Lie groups, Bull. Amer. Math. Soc. **79** (1973), 129-132.

The totally geodesic Radon transform on constant curvature spaces, Integral geometry and tomography (Arcata, 1989). Contemp. Math. **113** (1990), 141-149.

The Royal Danish Academy of Sciences and Letters

Value-distribution theory for holomorphic almost periodic functions, The Harald Bohr Centenary (Copenhagen 1987). Mat.-Fys. Medd. Danske Vid. Selsk. **42** (1989), 93-102.

Titles in This Series

Volume

22 **Sigurður Helgason:** The Selected Works of Sigurður Helgason (Gestur Ólafsson and Henrik Schlichtkrull, Editors), 2009

21 **Alexandra Bellow:** Selected Papers of Alberto P. Calderón with Commentary, 2008

20 **Yuri I. Manin:** Mathematics as metaphor: Selected essays of Yuri I. Manin, 2007

19.3 **John Milnor:** Collected Papers: Differential topology, Part 3 (edited by the author), 2007

18.4 **Phillip A. Griffiths:** Selected works of Phillip A. Griffiths with Commentary: Differential Systems, Part 4 (Robert Bryand and David R. Morrison, Editors), 2003

18.3 **Phillip A. Griffiths:** Selected works of Phillip A. Griffiths with Commentary: Variations of Hodge Structures, Part 3 (Herbert Clemens and David R. Morrison, Editors), 2003

18.2 **Phillip A. Griffiths:** Selected works of Phillip A. Griffiths with Commentary: Algebraic Geometry, Part 2 (Enrico Arbarello and Joseph Harris, Editors), 2003

18.1 **Phillip A. Griffiths:** Selected works of Phillip A. Griffiths with Commentary: Analytic Geometry, Part 1 (Maurizio Cornalba, Mark Green, and Wilfried Schmid, Editors), 2003

17 **Eberhard Hopf:** Selected works of Eberhard Hopf with Commentary (Cathleen S. Morawetz, James B. Serrin, and Yakov G. Sinai, Editors), 2003

16 **S. A. Amitsur:** Selected Papers of S. A. Amitsur with Commentary, Parts 1 and 2 (Avinoam Mann, Amitai Regev, Louis Rowen, David Saltman, and Lance Small, Editors), 2001

15 **Ilya Piatetski-Shapiro:** Selected Works of Ilya Piatetski-Shapiro (James Cogdell, Simon Gindikin, and Peter Sarnak, Editors), 2000

14 **E. B. Dynkin:** Selected Papers of E. B. Dynkin with Commentary (A. A. Yushkevich, G. M. Seitz, and A. L. Onishchik, Editors), 2000

13 **Frederick J. Almgren, Jr.:** Selected Works of Frederick J. Almgren, Jr. (Jean E. Taylor, Editor), 1999

12 **Ellis Kolchin:** Selected Works of Ellis Kolchin with Commentary (Hyman Bass, Alexandru Buium, and Phyllis J. Cassidy, Editors), 1999

11 **V. S. Varadarajan:** The Selected Works of V. S. Varadarajan, 1999

10 **Maurice Auslander:** Selected Works of Maurice Auslander, Parts 1 and 2 (Idun Reiten, Sverre O. Smalø, and Øyvind Solberg, Editors), 1999

9 **Lipman Bers:** Selected Works of Lipman Bers: Papers on Complex Analysis, Parts 1 and 2 (Irwin Kra and Bernard Maskit, Editors), 1998

8 **Walter E. Thirring:** Selected Papers of Walter E. Thirring with Commentaries, 1998

7 **Robert Steinberg:** Robert Steinberg Collected Papers, 1997

TITLES IN THIS SERIES

6 **Julia Robinson:** The Collected Works of Julia Robinson (Solomon Feferman, Editor), 1996

5 **Freeman Dyson:** Selected Papers of Freeman Dyson with Commentary, 1996

4 **Witold Hurewicz:** Collected Works of Witold Hurewicz (Krystyna Kuperberg, Editor), 1995

3.2 **A. Adrian Albert:** A. Adrian Albert Collected Mathematical Papers: Nonassociative Algebras and Miscellany, Part 2 (Richard E. Block, Nathan Jacobson, J. Marshall Osborn, David J. Saltman, and Daniel Zelinsky, Editors), 1993

3.1 **A. Adrian Albert:** A. Adrian Albert Collected Mathematical Papers: Associative Algebras and Riemann Matrices, Part 1 (Richard E. Block, Nathan Jacobson, J. Marshall Osborn, David J. Saltman, and Daniel Zelinsky, Editors), 1993

2 **Salomon Bochner:** Collected Papers of Salomon Bochner, Parts 1–4 (Robert C. Gunning, Editor), 1992

1 **R. H. Bing:** The Collected Papers of R. H. Bing, Parts 1 and 2 (Sukhjit Singh, Steve Armentrout, and Robert J. Daverman, Editors), 1988

ISBN 978-0-8218-4753-4

9 780821 847534

CWORKS/22